Handbook of

ENVIRONMENTAL CIVIL ENGINEERING

Handbook of

ENVIRONMENTAL CIVIL ENGINEERING

Edited by Robert G. Zilly

VNR Van Nostrand Reinhold Company
New York/ Cincinnati/ Toronto/ London/ Melbourne

Van Nostrand Reinhold Company Regional Offices:
New York Cincinnati Chicago Millbrae Dallas

Van Nostrand Reinhold Company International Offices:
London Toronto Melbourne

Library of Congress Catalog Card Number: 74-26993
ISBN: 0-442-29578-2

Manufactured in the United States of America

Published by Van Nostrand Reinhold Company
450 West 33rd Street, New York, N.Y. 10001

Published simultaneously in Canada by Van Nostrand Reinhold Ltd.

15 14 13 12 11 10 9 8 7 6 5 4 3 2 1

Library of Congress Cataloging in Publication Data

Zilly, Robert G 1920–
Handbook of environmental civil engineering.

Includes bibliographies.
1. Civil engineering–Handbooks, manuals, etc.
2. Environmental engineering–Handbooks, manuals, etc.
I. Title.
TA151.Z49 624 74-26993
ISBN 0-442-29578-2

Contributors

RICHARD W. CHRISTENSEN, Ph.D., P.E., M. ASCE
Associate
Dames & Moore
Park Ridge, Illinois

JAMES L. CLAPP, Ph.D., P.E., M. ASCE
Professor of Civil and Environmental Engineering
Institute for Environmental Studies
The University of Wisconsin
Madison, Wisconsin

MARVIN M. JOHNSON, Professor
Industrial and Management Systems Engineering Department
College of Engineering and Technology
University of Nebraska–Lincoln
Lincoln, Nebraska

MILO S. KETCHUM, P.E., F. ASCE, F. IStructE
Professor of Civil Engineering
University of Connecticut
Storrs, Connecticut

HERMAN A. J. KUHN, Ph.D., P.E., M. ASCE
Associate Professor of Civil and Environmental Engineering
The University of Wisconsin
Madison, Wisconsin

TERENCE J. MC GHEE, Ph.D., P.E., AWWA, WPCF
Associate Professor of Civil Engineering
College of Engineering and Technology
University of Nebraska–Lincoln
Lincoln, Nebraska

B. M. RADCLIFFE, Professor, P.E., M. ASCE
Construction Management Department
College of Engineering and Technology
University of Nebraska–Lincoln
Lincoln, Nebraska

ROBERT G. ZILLY, Professor and Chairman, P.E., M. ASCE
Department of Construction Management
College of Engineering and Technology
University of Nebraska–Lincoln
Lincoln, Nebraska

Preface

The civil engineer has always been at the cutting edge where man and the environment meet. Lately, however, society has shown an increasing reluctance to accept the engineer's decision as to just what is good for mankind and the environment—for the short and the long haul. It is apparent that the civil engineer must now ply his craft in a fishbowl—subject to close scrutiny from all facets of the society to which his profession is dedicated. Now, and hopefully for some time to come, environmental quality is the number one priority on spaceship earth.

This handbook is a beginning step in what must become a massive campaign to convince civil engineers that the words "engineering" and "environment" are indeed inseparable. To the civil engineer this should always have been obvious, and to some it was. Yet, there are many practitioners who have lost sight of broader objectives in the immediate pursuit of specific problems. Let the title of this handbook warn them that they must take a broader approach.

Concentrating on the major fields of civil engineering, this handbook is by necessity limited in scope. Working within tight space limitations, the authors have had to be extremely selective in covering their areas of expertise. The result is a compact and current text. Hopefully, it can be updated as frequently as necessary to keep pace with our rapidly changing technology in the newly emerging world of environmental awareness.

This is a book designed for the engineer in transition—for the young engineer in or just out of college; for the engineer specialist who wishes to broaden his outlook; or for the engineer who has decided to redirect his technical and professional goals. With the inclusion of a major section on engineering management, it is a practical guide to the real-world practice of basic civil engineering, with strong emphasis on the rapidly accelerating environmental aspects of that practice.

ROBERT G. ZILLY

Contents

PREFACE vii

SECTION I **The Civil Engineer and the Human Environment,** *Robert G. Zilly* 1

A Mini-History of Civil Engineering 2

The Civil Engineer and the Environment 5

The Engineer and the Environment 6

SECTION II **Engineering Measurement,** *James L. Clapp* 11

Theory of Measurement 12

Methods of Measurement 12

Definition of Measurement Terms 12

Errors in Measurement 13

Mathematical Errors 13

Physical Errors 14

Systematic and Random Errors 14

Distribution of Accidental Errors 14

Measures of Precision 15

Propagation of Accidental Errors 16

Adjustments of Measurements 18

Distance–Measurement Methods 20

Units of Length 20

Direct Distance Measurement 20

Indirect Distance Measurement 23

Elevation Measuring Systems 34

Curvature and Refraction 35

Differential Leveling 36

Reciprocal Leveling 38

Trigonometric Leveling 39

Barometric Leveling 39

Angle–Measuring Instruments 40

The Transit 43

Theodolites 45

Types of Surveys 46

Profile Leveling 46

Cross-Section Leveling 47

Traverse 47

Triangulation 49

Survey Computations 53
Traverse Computations 53
Area Computation 55
Volume Computations 56
State Plane—Coordinate System 57

Remote Sensing 62
Basic Principles 62
The Remote Sensing Sequence 63
Source Characteristics 64
Interactions at the Object 65
Atmospheric Effects 66
Operational Remote Sensing Systems 66

Photogrammetry 74
Aerial Cameras 74
Vertical Aerial Photographs 75
Radial Triangulation 79
Mosaics 80
Parallax 80
Stereoscopic Viewing and Measuring 81
Stereoscopic Plotting Instruments 83
Flight Planning 85

SECTION III Soil Mechanics and Foundation Engineering,
Richard W. Christensen 87

Soil Composition, Identification, and Classification 87
Soil Components 87
Index Properties 90
Classification 91

Soil Properties 96
Permeability 96
Compressibility 102
Secondary Compression 114
Shear Strength and Deformation Properties 117
Stress at a Point 117
Principle of Effective Stress 118
Stress Systems 120
Drainage Conditions 121
Stress-Strain Properties 123
Shear Strength 130

Site Investigation 140
Scope of Site Investigations 140
Methods of Subsurface Explorations 142

Foundations for Structures 147
General Considerations 147

Shallow Foundations 151
Stress Distribution Beneath Foundations 159
Settlement 164
Deep Foundations 186
Earth Retaining Structures 200
Rigid Structures 200
Flexible Walls 210
Stability of Slopes and Embankments 217
Stability Analysis 218

SECTION IV **Structural Engineering,** *Milo S. Ketchum* 236

Introduction to Structural Engineering 236
The Structural Engineer and the Environment 239
Materials for Structures 242
Types of Structures for Buildings 248
Types of Structures for Bridges 256
The Tools of Structural Engineering 258
Statics 259
Analysis of Trusses 294
Strength of Materials 303
Deflection of Rigid Frames 327
Deflection of Trusses 333
General Method for Indeterminate Structures 342
The moment-Distribution Method 353
Short Cuts and Simplifications for Moment Distribution 366
Buckling of Columns 372
Structural Elements 375
Design of Steel Structural Elements 375
Design of Reinforced Concrete Structural Elements 401
Design of Structures 446
Design of Steel Structures 446
Design of a Single-Span Truss 452
Design of a Continuous Articulated Roof System 457
Steel Office and Apartment Buildings 460
Building-Frame Calculations: The Example of a Steel-Frame Building 463
Steel-Arch Buildings 471
Design of a Small Concrete Structure 477
Design of A Tall Reinforced-Concrete Building 482
Concrete Retaining Walls 493
Bridges 498

SECTION V **Water and Waste Engineering,** *Terence J. McGhee* 516

VA. Engineering Hydraulics 516
Flow in Pipes 516
Minor Pipe Line Losses 519
Open Channel Flow 520
Hydraulic Cross Sections 521
Flow Measurement 524
Pumps 526

VB. Water Supply 526
Sources of Water 528
Estimation of Water Demand 528
Population Estimation 529
Design Values for Water Demand 535
Specific Requirements 535
Water Treatment 536
Screening 537
Hardy Cross Method 567
Circle Analysis 571
Materials and Fittings 571
Corrosion 573
Cleaning and Disinfection 574

VC. Sanitary Sewage and Liquid Wastes 577
Government Regulations 577
Quantity of Sewage 578
Sanitary Sewer Design 580
Sewer Materials 582
Sewage Characteristics 585
Treatment Systems 587
Primary Treatment 587
Secondary Treatment 597
Activated Sludge 604
Oxidation Ponds 617
Sludge Treatment 621
Tertiary Treatment 628
Miscellaneous Waste Treatment Techniques 637
Storm Sewage 641
Treatment and Disposal 645
Industrial Wastes 646
Biological Treatment 648
Joint Treatment 650

VD. Solid Wastes 653
Quantity of Solid Wastes 653
Storage 655
Collection 655

Collection Equipment 656
Volume Reduction and Disposal 656

SECTION VI **Transportation,** *Herman A. J. Kuhn* 660

The Technology-Man Society Interface 660
Social, Economic, and Environmental Effects of Transportation Systems 662
Environmental Capacity (2), (3) 662
Community and Human Needs Related to Transportation 663
Environmental Policy Act of 1969 663
Environmental Effects of Transportation Systems 664
Air Pollution Effects of Automobile Operations 676
Water Quality 680
Displacement of Families and Businesses 681
Miscellaneous Environmental-Social Effects 682

Auto-Truck Transportation 683
Movement Systems (Networks) 683
Flow Characteristics 688
Speed 700
Composition 707

System Planning–Urban 707
The Planning Process 707
Information Analysis 714
Trip Generation Model (15) 727
Trip Generation Modeling Procedures 732
Trip Distribution 735
Trip Assignment (8), (23) 752
Assignment Methods (23) (24) 754
Modal Split (27) (28) 757

Geometric Design 769
Design Controls 770
Elements of Design 800
Intersection Design 824
Details of Design 834

System Operation and Control 846
Signalization 846
Signing and Marking 864

Mass Transportation–Ground, Intracity 876
Present Transit Use 876
Future Transit Use 877
Criteria for Rail Rapid Transit 877
Predicting Demand 878

Improving Transit Usage 878
Mass Transportation Technology, Non-Exclusive Guideway 880
Mass Transportation Technology, Exclusive Guideway (Right-of-Way) Systems 885
Air Transportation 898
Factors Affecting Airport Capacities (5) 898
Airport Facility Requirements 901
Terminal Facilities 905

SECTION VII **Quantitative Engineering Management,** *Marvin M. Johnson, B. M. Radcliffe, R. G. Zilly* 915

Introduction 915
History of the Quantitative Approach to Management 923
Application of Modeling to Decision Making 932
Model Classification 938
Linear Programming 940
Graphic Solution of a Two-Dimensional Simplex Problem 941
Manual Computation Process Involved in Simplex 945
Graphic Representation of the Three-Dimensional Case 950
Waiting Line Problems and Applied Queueing Theory 952
Fundamental Elements of the Queueing Problem 953
Development of Single-Channel Queueing Formulas 955
Engineering Economy 961
Economic Performance Charting 970
Dynamic Programming 980
Critical Path Method 987
Games Against Nature 998
Branch and Bound 1008
Strategy of Competition 1017

INDEX 1023

Handbook of
ENVIRONMENTAL CIVIL ENGINEERING

SECTION I

The Civil Engineer and the Human Environment

Robert G. Zilly, *Professor and Chairman, P.E., M. ASCE*
Department of Construction Management
College of Engineering and Technology
University of Nebraska-Lincoln
Lincoln, Nebraska

"The consulting engineering profession is accustomed to responsibility. It has met well the trust that automatically is placed upon it. For example, consulting engineers were working on effective ways of improving the environment long before some of the more rapt practitioners of things ecological knew what the word meant. We are not only environmentalists, we are also compelled to be part sociologists, part psychologists, part chemists, and part policemen."—Nat P. Turner, President, Consulting Engineers Council of the United States.

Long before the turn of the time scale from B.C. to A.D., he was known in Mesopotamia as "batu." Later, in Greece, he was the "architekton," and to the Romans goes credit for the label "architectus." Who was he? None other than the "master builder;" the ubiquitous architect, the down-to-earth constructor, and, in the context of this Handbook, the visionary environmental-civil engineer. Wherever man had lifted himself from the primal mud, there is evidence that the "master builder" had taught him how to lift himself by his own boot—or sandal—straps.

The early architect-civil engineer-constructor trinity had a longer history than any of its three entities, for it was not until A.D. 200 that the engineer got his first glimmer of recognition. This came by way of the Latin historian Tertullian, who referred to an early military device—the battering ram—as an "ingenium." But, it took another one thousand years before the designer and builder of this military device was given recognition as an "ingeniator." By the 1400s, the Italians were using the terms "engineer" and "architect" with some sense of discrimination, but it was not until the early 1700s that engineers could claim their first book—Belidor's *Science des Ingénieurs.*

A MINI-HISTORY OF CIVIL ENGINEERING

The function of the civil engineer is clearly evident in the Age of the Pyramids, dating back to approximately 3000 B.C. This is the era that nursed into existence a rudimentary knowledge of structural design, surveying, hydraulics, and basic construction management and technology. Later developments in Mesopotamia point to the existence of rather carefully designed public water supply systems, tunnels, and large harbors, in addition to monumental public buildings in the form of religious structures.

It is a long slow crawl from the "engineering mysteries" of Egypt's early temple priests to the free-wheeling developments in Greece that began to surface about 300 B.C. Today, we are enchanted with Greek philosophy and the Greek sense of the aesthetic, but we need to be no less enchanted with Archimedes, the mechanical engineer; Hero, the surveyor; and Cleon, the canal builder. For it is no secret that while Greek philosophers sought truth for truth's sake, Greece was a free-wheeling, free-enterprise democracy where competition reigned supreme. That is perhaps why engineering developed at so rapid a rate under the Greeks that the Romans were never quite able to catch up.

It is easy to fault the Romans for their lack of creativity in the field of science, for which the Greeks had laid the groundwork. However, it is a reasonably safe conclusion that the Romans were far superior to the Greeks as engineers. It was they who put the basic Greek concepts to work in the development of wide-ranging highway systems, lengthy aqueduct lines that fed major water distribution systems in the large cities, and bridges and buildings that carried early Greek concepts to new limits.

Of course, the Romans had the benefit of help from Greek slaves who carried their heritage with them to both military and civilian projects of the emperors. However, it was a Roman, Marcus Vitruvius Pollio, who collected much of the then known knowledge of architecture and engineering in his ten books on architecture. It is perhaps unfortunate that Vitruvius was "hung up" on an idealized version of the architect. But, his writing proved to be extremely durable and was the solid base to which engineering and architecture was anchored through the Middle Ages.

It is a popular notion that when Rome fell, all else fell with it. To the contrary, there are numerous structures that bear mute testimony to the fact that engineering was still alive and well, though moving cautiously, until the time of Gutenberg and the ensuing "Age of Books." Bridging the gap from Rome to Germany, from A.D. zero to A.D. 1450, one need only mention Wilars de Honecourt, a noble example of the engineer–architect–builder triumvirate that kept things rolling through the Middle Ages, the Scholastic Age, and into the Renaissance of Italy and France.

It was a downhill run for engineering, following the invention of the printing press. Galileo Galilei broke the chain of reliance on past authority with his reliance on direct observation and experiment. His contributions to the field of statics are hardly diminished by the fact that much of his work was based on the ideas of Archimedes, Leonardo da Vinci, and others. For the record is clear that his efforts provided the challenge to later scientist/engineers such as Coulomb and Napier.

A mini-history can hardly do justice to the great names of the 15th, 16th, and 17th Centuries who made their mark in the annals of engineering. Names such as Varignon, Descartes, Bernoulli, Hooke, Leibnitz, and Newton come quickly to mind, though the decision to label these men engineers might be hotly contested by the purist in the field of mathematics.

No matter; the age of civil engineering had arrived. Structural engineering flowered in France between 1750 and 1850, and much of its success can be attributed to the fact that engineering was moving from an age of empiricism to a strong underpinning of mathematics and science.

Other branches of engineering had, of course, begun to split off from civil engineering during this era, but civil engineering was closely identified in Great Britain in 1828, when the Institution of Civil Engineers adopted this definition:

"... that species of knowledge which constitutes the profession of a civil engineer; being the art of directing the great sources of power in nature for the use and convenience of man; of the means of production and of traffic in states, both for external and internal trade, as applied in the construction of roads, bridges, aqueducts, canals, river navigation, and docks, for internal intercourse and exchange; and in the construction of ports, harbors, moles, breakwaters, and light-houses; and in the art of navigation by artificial power, for the purposes of commerce; and in the construction and adoption of machinery; and in the drainage of cities and towns."

Also in 1828, in the United States, Noah Webster gave backhanded reference to the civil engineer in his new dictionary. However, the term "engineer" was broadly used in the U.S., and few of its citizens could discriminate between the civil engineer and those who operated new-fangled machinery such as the steam engine.

Meanwhile, in both the U.S. and abroad, the civil engineer and the architect were still inextricably tangled. As late as 1790, the Frenchman Joseph Mangin, was offering his services in the planning of bridges and other improvements, either as an independent consultant or as a contractor. He worked on fortifications in New York City, but later abandoned his engineer's hat for that of the architect.

As in other countries, the civil engineer in America made his mark in the field of public works–highways, canals, and bridges that helped the restless pioneers to satisfy their powerful urge to move westward into the open spaces of the frontier. Though canal building had begun in the late 1700s in America, it was Benjamin Wright's Erie Canal that stirred the imagination of youngsters who were to become the new nation's future engineers. Started in 1817 and completed in 1825, the Erie was probably the longest canal in the world. Connecting the Atlantic and the Greak Lakes via the Hudson River, it was 363 miles long and included 83 locks which made use of a natural cement manufactured from rock adjacent to the canal line.

It was the Erie Canal that created New York City, a modern marvel of engineering triumphs and failures. But the Erie Canal had hardly gone on stream before the railroads rose to challenge its supremacy. The Chesapeake and Ohio Canal was a dream that really began with George Washington–perhaps our greatest civil engineer president after Herbert Hoover–but died aborning when the Baltimore and Ohio Railroad became a reality. Chartered a year before construction on the C & O Canal began, the B & O Railroad soon earned the title "First American Railroad Engineering School." Among its great engineers, the names of Jonathan Knight (chief engineer), Stephen Harriman Long, and Benjamin Henry Latrobe were to be long remembered.

Before, after, and particularly during the age of railroad development, American bridge builders began to write their own pages in the history of civil engineering. Even the layman must recognize names like Howe, Fink, Whipple, Bollman, Pratt, and Warren. These were just a few of the great designers who, as President Lincoln saw it, built fantastic spans with "nothing but bean poles and cornstalks."

With apologies to bridge builders Roebling and Eads, and a passing nod to Arthur Mellen Wellington–the father of American engineering economics–it is now time to turn to the sanitary branch of civil engineering in America. By 1800 there were almost twenty waterworks in the new Republic, only one of which was municipally owned. Ninety-six years later, more than half of the water supply systems serving American cities were still privately owned. But, by this time there were already famous names for modern practitioners to admire or emulate. Birdsill Holly, the pump salesman; Isaac Newton, chief engineer of the "New Croton" system for supplying New York City; and Ellis Sylvester Chesbrough, of the New York City Department of Public Works were all active in the late 1800s. Chesbrough was probably most famous for his sewer plan submitted to the City of Chicago in 1858, for it was inevitable that where there were water works, pollution problems created a demand for sewage treatment.

Meanwhile, civil engineers were moving upward instead of outward in the design of buildings. By 1859, the first elevator building had risen six stories in New York City. The Fifth Avenue Hotel was served by a steam engine driven screw shaft which moved a car from floor to floor, but it was the electric motor that changed vertical transportation after its first use in 1889. The elevator, plus relatively light-framed "skeleton" structures solved the engineer's problem of moving upward, but not before the classic Monadnock Building in Chicago had been erected in the early 1880s. Of brick bearing-wall construction, the Monadnock had walls some five feet thick. It is still in use today, and stands as a symbolic reference point to designers who have revived the masonry bearing-wall concept for modern structures.

Even as it rose, however, the Monadnock was an anachronism. William Le Baron Jenney had already conceived the idea of a structural steel skeleton with light walls that needed only to bear their own weight. So the 14-story Tacoma building had risen to change the Chicago skyline by 1887, and other equally "fantastic" structures followed. The scene soon shifted to New York, and the 925-foot-high Empire State Building completed in 1930 was to stand as a challenge for many years. Today, Chicago and New York City are still vying for the "highest building" title, and will probably do so for some time to come.

High buildings and other massive structures such as the great dams of the American West brought the need for a better understanding of soils and foundations. Thus, it is not surprising that Karl Terzaghi chose to come to the United States from Czechoslovakia in 1925 to pursue his studies of soils and foundations. And so, another branch of civil engineering was born.

The list of engineering disciplines that began with civil engineering is as endless as our modern penchant for using new names to hide old ignorances. Thus, we find much confusion among today's engineers about who they are and what they are supposed to be doing. Vitruvius' record of water and sanitary engineering is the base for modern sanitary engineering, but that became obsolete in 1972 when the Sanitary Engineering Division of the American Society of Civil Engineering renamed itself the "Environmental Engineering Division." Soils engineering or soil mechanics, also a creation of Vitruvius that was given renewed vigor by Terzaghi, is now frequently referred to as "Geotechnical Engineering."

THE CIVIL ENGINEER AND THE ENVIRONMENT

Though the preceding historical section is at best sketchy, it clearly identifies the civil engineer with things environmental. What it does not do is prepare him for the storm of protest that broke about his head in the late 1960s and early 1970s. Trouble was the fate of all engineers in this era, but the civil engineer found his position particularly troublesome. While the demand for engineers from other disciplines fell off significantly, civil engineers were still being requested to do the thing they had always done best–alter the environment–but they were being asked to do it even as their past efforts were being severely criticized.

Reaction from the American Society of Civil Engineers was swift and to the point. In the August 1971 issue of Consulting Engineer Magazine, the editors reported on ASCE policy with regard to the environment. I quote directly: "In relation to his client, the civil engineer . . . must be prepared to relinquish his services in the event the client insists on a course of action which can be demonstrated to have undesirable consequences to the environment, outweighing benefits."

In all, four objectives aimed at attaining the "broad goal of quality living" are urged on civil engineers, said the magazine. These include the obligation to increase knowledge and competence in incorporating ecological considerations into design, and the weighing of social and national considerations and alternatives against apparent lowest cost and technical aspects of a project. Engineers also "must fully utilize mechanisms within the Society [ASCE] which lend support to . . . individual efforts to implement environmental considerations." As citizens they "must . . . take the lead in modifying or supporting governmental programs to insure adequate environmental protection."

Less than a year later, writing in the July 1972 issue of Civil Engineering, C. Maxwell Stanley, F. ASCE, chairman, Stanley Consultants of Muscatine, Iowa, made what will stand for some time as the definitive statement of the engineer's role in relation to the environment. By permission of the author and ASCE, that article is reproduced here in its entirety:

The Engineer and the Environment

"Extending beyond the local and national levels, the crisis of the environment's life-supporting capacity has global dimensions. Airborne and waterborne pollutants pass the boundaries of nations. Accumulated pollution could alter the delicate physiological balances of our biosphere, transform weather patterns and threaten oceanic food production. Uncontrolled pollution has already lowered the quality of human life.

"The quality of human life is the proper criterion for determining the suitability of an environment. It is a more demanding standard than life support of mere existence. Even so, no one knows the finest quality of life that can be maintained indefinitely on this earth. Obviously, it depends upon population growth, standard of living, and future development in science and technology. But there is a finite limit!

"Unless the world's population is stabilized, it will expand beyond a level compatible with an acceptable environment–generating pollution and accelerating the consumption of resources. Conversely, though the aggregate of the world's resources is unknown, there are finite limits to the total population and quality of life that can be maintained. Without conservation or recycling, certain of the world's finite resources will diminish to a level incapable of maintaining an acceptable environment and adequate life support. (The United States with only six percent of the world's population consumes 40 percent of the world's resources.)

Multinational Cooperation "International mechanisms within the United Nations are needed to cope with the threats to global environment and coordinate national environmental programs. Fortunately, the United Nations Conference on Human Environment held in Stockholm in June, 1972 dealt with local and national, as well as international, aspects of the environment. Several nongovernmental conferences have also directed their deliberations to the international aspects of environmental management.

"All effective programs to protect and enhance the environment must, of necessity, be complex and diverse. Since they deal with physical phenomena, they rest on science and technology. But, protection and enhancement of physical environment does require political, social, and economic action. Only through the political process can the will of the public be expressed, decisions taken, and appropriations authorized to implement environmental programs.

Impact on the Civil Engineer "The protection and enhancement of the environment will be a huge, costly, and complicated task, involving numerous engineering functions. There will be much work for engineers, particularly if they are fully qualified to meet the challenges. They will have an important role in the establishment of environmental criteria, emission standards, and the modeling of systems. Environmental protection calls for waste treatment and disposal facilities of all types; the design and manufacture of engines, boilers, and other apparatus meeting stringent emission standards; the development and production of herbicides, insecticides, fuels, chemicals, and consumer products that are more degradable and less harmful; the modification of plants and processes to reduce current pollution; and the implementation of new industry to recycle wastes. Engineers will help manage environmental programs and systems at the local, national, and international level.

"Looking further ahead, other areas demand intensified attention; soil preservation; conservation of resources and development of new ones, including some under the seas; and models and programs for cities, transportation systems, land use, and other problem areas. New scientific and engineering approaches are needed to provide a suitable balance between pollution, population, and resources.

"Environmental concerns add new dimensions to the work practices of the engineers as well. It will be more difficult to obtain permits or licenses from governmental agencies for the construction of new facilities. The owner of a proposed project must now comply with newly established emission standards—not always wise and often tentative—so that the facility will have no detrimental impact on the environment. Though somewhat hesitant and sometimes embarrassed by public protests, licensing agencies have become increasingly tough in their interpretation of standards. The electric utilities have felt this trend with respect to nuclear power plants, fossil-fuel power plants, pumped storage plants, and, on occasion, with transmission and substation facilities. Permits to construct plants can no longer be taken for granted, and alternate concepts must be studied. Applications must be carefully prepared, fully supported, and filed with ample lead time. They must reflect careful attention to environmental impact.

"In fact, all designs for projects, and soon construction processes, will need to show concern for the environment and incorporate measures assuring compliance with environmental standards. Confronted with uncertainties as to future emission standards–and sometimes as to interpretation of existing ones–the designer must often anticipate trends and design so that future standards can be met without unnecessary additional costs.

Protection Plus Enhancement "Designers must increasingly be concerned with the enhancement, as well as the protection, of the environment. Enhancement involves siting, appearance, aesthetics, and other factors contributing to the quality of life. Future projects will not be justifiable solely on need and direct cost/benefit ratios. The broader 'costs' and 'benefits' to the environment will be considered.

"Studies concerned with the environmental impacts of a project or plant are becoming a standard part of preliminary planning and design. Labeled "environmental analysis," "environmental impact study," or "environmental enhancement plan," they examine the multidimensional impact of the proposed facility upon the environment. Often such studies are needed to support applications for permits or licenses.

"Already some engineering colleges are offering such courses in environmental engineering–combining engineering and various social and physical sciences. Civil engineers are well prepared for this field and also for the field of environmental management. Many civil engineers should become the 'generalists' environmental science requires to supplement the various specialists.

Multidisciplinary Approach "Concern for the environment will accelerate the current trend towards a multidisciplinary approach. As projects have grown in size and complexity since the beginning of World War II, engineering organizations have increasingly diversified their talents. Firms originally practicing civil engineering have added electrical, mechanical, and other engineers. The architect-engineer firm has evolved, with both professions under the same roof. Some firms have added urban and regional planning and a few have added economists and social scientists.

"Extensive research on many aspects of the environment requires a multidisciplinary approach. Many areas of study are called for to deal with pollution problems, including meteorology, oceanography, biology, and chemistry. Environmental enhancement involves the social sciences concerned with human behavior and man's reaction to his physical environment.

"Consultants, governmental agencies and industrial concerns need competence in these areas of study, either in-house or through the use of outside specialists. Therefore, whether or not the civil engineer becomes a leader in environmental matters depends in large part upon his willingness to broaden his outlook and his ability to work in a multidisciplinary setting.

From Growth to Conservation "The impacts of pollution, overpopulation, and resource management upon environment may very well lead to a change in our outlook and goals. The idol of exponential growth, so prevalent in the United States and industrialized countries, will be severely challenged. Yet, this growth syndrome has a strong appeal. The efforts of much of the world are directed toward emulating us. Unfortunately, unlimited growth is not consistent with finite resources. Even if pollution is brought under satisfactory control and ecological threats to human life are removed, the resources of the earth will remain finite. The development of new means of exploiting, using, and recycling these resources, cannot alter this fact.

"Greater attention will be given to conservation and to more intelligent use of resources, such as agricultural lands and resources, water resources, fuels, and minerals. The recycling of many types of wastes will recover usable materials. Henceforth, functional solutions must stress quality and minimize material requirements.

"Civil engineers can play an important role in the development and design of facilities for recycling; the conservation of agricultural and oceanic resources; and the development and selection of materials for use in the construction and manufacturing that are consistent with increasing emphasis on intelligent use of resources.

The Ethical Question "Environmental matters demand new ethical principles for civil engineers and other professionals. It is important that we recognize them, not only to better serve our clients and the public, but also to enhance our professional image. Engineers in general, and civil engineers in particular, like to believe they had a favorable impact on the environment over the years. But, from the public's point of view, engineers are the designers of internal combustion engines that create smog, industrial plants that dump wastes into the river, transportation systems that congest our cities, power plants that spew fly ash and gas into the atmosphere, and other polluting mechanisms. It is in our self-interest to dispell these public doubts by a wholehearted acceptance of our environmental responsibilities.

"ASCE adopted an environmental policy statement a few months ago expressing professional awareness and regard for the ecological balance. Its preamble states:

" 'Environmental goals must emphasize the need for assuring a desirable quality of life in the context of the expanding technology necessary to sustain and improve human life. Implicit in these goals is the need to develop resources and facilities to improve the environment of man and the need to abate deleterious effects of technology on the environment.

" 'This environmental policy statement expresses professional awareness and regard for ecological balance. It is clear, therefore, that neither returning the environment to its pristine form, on the one hand, nor allowing technology to develop with total disregard of its long- and short-range environmental effects, on the other hand, is acceptable.'

"The statement goes on to recommend that the civil engineer recognize the effect of his efforts upon the environment by:

- sharpening his awareness and increasing his knowledge and competence regarding ecological considerations.
- informing his client of the environmental consequences of services requested or designs selected and recommending only responsible courses of action.
- ceasing his services in the event the client insists on a course of action which can be demonstrated to have undesirable consequences to the environment that outweigh benefits.
- weighing social and national considerations and alternatives, when appropriate, in addition to the economic and technical aspects of the project.
- utilizing fully the mechanisms within ASCE to support his individual efforts to implement environmental decisions.
- recognizing the urgent need for legislation and enforcement to protect the environment.
- providing leadership in the modification or support of governmental programs to insure adequate environmental protection.

"Civil engineers have a unique opportunity to be leaders, as well as participants, in man's efforts to protect and enhance the environment. In the interests of all humanity, we must earnestly seek an acceptable and tolerable environment not only in our time but for untold generations to follow."

REFERENCES

Calhoun, Daniel H. *The American Civil Engineer.* The Technology Press, MIT, Cambridge, Mass., 1960.

Chalmers, Harvey, II. *The Birth of the Erie Canal.* Bookman Associates, New York, 1960.

Finch, James Kip. *The Story of Engineering.* Anchor Books, Doubleday, Garden City, New York, 1960.

Goodrich, Carter (ed.). *Canals and American Economic Development.* Columbia Univ. Press, New York, 1956.

Kirby, Withington, Darling, and Kilgour. *Engineering in History.* McGraw-Hill Book Company, New York, 1956.

Straub, Hans. *A History of Civil Engineering.* (Translated from the German by E. Rockwell.) Charles T. Branford Company, Newton, Mass., 1960.

SECTION II

Engineering Measurement

James L. Clapp, Ph.D., *P.E., M. ASCE*
Professor of Civil and Environmental Engineering
Institute for Environmental Studies
The University of Wisconsin
Madison, Wisconsin

There is great concern today for the degradation of the earth's physical and biological environment. This degradation has serious social, cultural, economic, and esthetic consequences which could reach catastrophic proportions. The civil engineer has increasingly found his attention directed to this problem. He has joined with other engineers as well as physical, social, and biological scientists to protect our natural resources and to create a better physical and social environment for all people.

The civil engineer is well aware of the complexities of these problems. Of particular significance is the need to develop clear understanding of the quantitative and qualitative relationships of all design projects to their physical and biological surroundings. This is vital in both a fundamental and specific sense to optimum design.

The broad field of engineering measurements must provide the data upon which these relationships are based. In the past, this consisted primarily of measurements directed toward the spatial relationship of points significant to the project. Although this spatial relationship is still very much a requirement, the dimension of the problem must be considerably broadened to include (1) measurements of specific parameters at specific points over a period of time, and (2) measurements of parameters over large areas at an instant in time. It is fortunate that recent advances in the field of engineering measurements provide a variety of methods by which these different types of measurements can be obtained. It is the role of the engineer to determine what parameters should be measured and which method will most economically provide the required data at the required accuracy. Optimum environmental design is not possible without an adequate data base for developing the design.

THEORY OF MEASUREMENT

Any measurement may be broadly defined as a comparison of an unknown quantity with a known standard. Since this can never be done exactly, all measured values must contain some inherent error, that is, the difference between the exact value and the measured value. Because the exact value can never be determined, likewise the inherent error can never be exactly known.

Methods of Measurement

Engineering measurements are made either directly or indirectly. Direct measurements are made by comparing the object being measured with a standard. In many cases direct measurements are used to compute other quantities. Indirect measurements make use of instruments which are sensitive to the quantity being measured and convert the raw data to a form which is detectable by the human senses. This output frequently takes the form of a dial or gauge reading.

Definitions of Measurement Terms

A *single measurement* consists of one or more comparisons made with all conditions constant except time. The exactness of a measurement cannot be refined by increasing the number of readings under identical conditions. In most measurement processes, however, the conditions will rarely be constant with time.

A *multiple measurement* is a measurement of a quantity by a series of independent comparisons. This usually requires a series of different measuring devices. The decision as to whether to rely on a single measurement or to employ a more complex, and usually more expensive, multiple measurement procedure usually rests with the engineer in charge of the project.

Accuracy is defined as the correctness with which a measured value represents the true value. It refers to the quality of the results of a measurement. *Precision* is the reproducibility of a measurement or the consistency of a set of measurements. It refers to nearness of a group of measurements to each other, and is associated with the quality of execution of the measurement procedure. It is possible for a measurement procedure to be accurate without being precise and vice versa.

Reliability is defined as the probability of performing a given measurement in the future to within given accuracy and precision. It is a prediction based on the accuracy and precision of measurements taken previously.

The *sensitivity* of a measuring system refers to the degree to which the system can register discernable differences. The accuracy of any measuring system depends on the sensitivity.

Errors in Measurement

Any measurement, no matter how precise and accurate, must by definition contain an inherent error. This error must be distinguished from a mistake. A mistake is best described as a blunder, usually caused by human or instrument failure. An error is the difference between the measured value and the true value. Since the true value can never be exactly known, likewise the error can never be exactly known. However, error can be analyzed by statistical methods.

The *absolute error* is defined as the difference between the measured value and the true value:

$$E_a = X_t - X_m$$

where

E_a = the absolute error,
X_t = the true value,
X_m = the measured value.

Since the true value can never be known the apparent absolute error is commonly used. The *apparent absolute error* is defined as the difference between the best value of a series of measurements and the measured value.

The *relative error* of a measurement is defined as the absolute error divided by the measured value:

$$E_r = (X_t - X_m)/X_m$$

where E_r is the relative error. Once again, for practical reasons, the true value is replaced by the best value of a series of measurements.

The *percentage error* is simply the relative error expressed as a percentage:

$$E_p = E_r \times 100 = (E_a/X_m) \times 100$$

where E_p is the percentage error.

The *discrepancy* is defined as the difference between two or more measurements in a series of measurements of the same subject. Discrepancy is not an error but can be helpful in a statistical analysis of the measurement.

Mathematical Errors

Discrimination is defined as the least reading or smallest division of the readout scale of a measuring system. Common practice is to estimate with the least reading of the scale. This controls the number of significant figures provided by the measurement and therefore subsequent mathematical calculations.

Physical Errors

Errors may be classified according to their source. *Instrumental errors* are those associated with imperfections in the measuring instrument itself. *Personal errors* are those associated with human limitations; these are not to be confused with mistakes, which can be prevented. *Natural errors* are associated with variations in the environment in which the measurement is being taken; these are frequently caused by variations in temperature, humidity, and pressure.

Systematic and Random Errors

Errors are also classified by their cumulative characteristics. *Systematic errors* are defined as those errors which tend to have the same magnitude and sign as long as the conditions producing the error remain unchanged. Systematic errors follow a physical principle and can usually be eliminated by refining the measurement process or applying a correction. Because they have the same algebraic sign they will tend to accumulate and are frequently referred to as *cumulative errors.*

Accidental errors are defined as those errors for which it is equally probable that the sign of the error is plus or minus. The sign of the error and its magnitude are considered to be matters of chance and governed by the laws of probability. It is not possible to apply a correction to remove the effect of an accidental error. In a series of measurements accidental errors accumulate as the square root of the number of chances for the error to occur. Therefore, there is some tendency for accidental errors to compensate in their total effect. Because of this characteristic accidental errors are frequently referred to as *compensating* or *random errors.*

Distribution of Accidental Errors

Since the effect of accidental errors cannot be eliminated by corrections as the effects of systematic errors can, measurement systems must account for accidental errors through the use of the laws of probability. If a large number of measurements are made of some quantity and the systematic errors are eliminated by procedure or correction, the mean of the measurements yields the *most probable value.* In order to evaluate the errors of the individual measurements it may be assumed that the most probable value represents the true value. The difference between the most probable value and a measured value is termed the *residual.* If the residuals with like sign within a relatively narrow range of values are counted and plotted—with magnitude of the error as abscissa and number of occurrences as ordinate, and with zero error as origin—the distribution of the accidental error results. This distribution has the following characteristics:

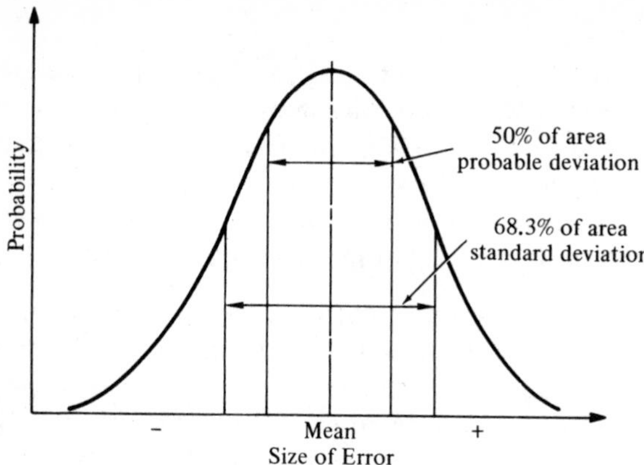

Fig. 1. Probability curve.

1. Positive and negative errors of the same magnitude occur with the same frequency.
2. Small errors occur more frequently than large errors.
3. Very large errors seldom occur.

If it is assumed that an unlimited number of measurements are made, a theoretical probability curve may be derived, as in Fig. 1. The probability curve has the following characteristics:

1. The total area under the curve is unity.
2. The probability of occurrence of an error whose magnitude is given by the abscissa is expressed on the corresponding ordinate as a percentage.
3. The sum of all probabilities is 100%.
4. The relative probability of occurrence of an error within a given range of magnitude is given by the area under the curve within that range.

Measures of Precision

In evaluating any series of measurements, it is desirable to have available a numerical index that is an expression of the precision of the measurement. Two frequently used values are (1) *the standard deviation*, and (2) *the probable deviation*:

$$\sigma_s = \sqrt{\Sigma V^2/(n-1)}$$

$$\sigma_m = \sigma_s/\sqrt{n}$$

$$E_s = 0.6745\ \sigma_s$$

$$E_m = 0.6745\ \sigma_s/\sqrt{n}$$

where

σ_s = the standard deviation of a single measurement
σ_m = the standard deviation of the mean
V = the residual of a single measurement determined from the mean
n = the number of observations
E_s = the probable deviation of a single observation
E_m = the probable deviation of the mean of all the observations

If the true value of the measured quantity is known and the true errors can be substituted for the residuals, then these terms are correctly referred to as (1) the standard error and (2) the probable error. In many cases where the number of observations are sufficiently large to ensure that the mean closely approximates the true value ($n > 30$) these terms are used interchangeably.

Propagation of Accidental Errors

In engineering practice, indirect as well as direct measurements are employed to obtain design and construction data. A direct measurement is made when the quantity itself can be measured. An indirect measurement is made when the quantity itself cannot be measured but can be calculated from quantities which can be measured. It is important, in indirect measurements, to be able to evaluate the error associated with the determined quantity.

The relationship indicated below for the propagation of the standard deviation in indirect measurements is valid, with appropriate substitution for probable error, standard error, or other similar error expressions. *The standard deviation of a sum of independent measurements* may be calculated according to

$$\delta_{\text{sum}} = \sqrt{\delta_1^2 + \delta_2^2 + \cdots + \delta_n^2}$$

where

δ_{sum} = the standard deviation of the indirect measurement quantity
$\delta_1, \delta_2, \cdots, \delta_n$ = the standard deviations of the independent measurements

When a series of similar quantities are measured and their standard deviations are considered to be the same, this equation reduces to

$$\delta_{\text{series}} = \delta\sqrt{N}$$

where

δ = the standard deviation for the measurements
N = the number of measurements.

The standard deviation of a product of two independent measurements such as $A = XY$ may be calculated according to

$$\delta A = \sqrt{X^2(\delta y)^2 + Y^2(\delta x)^2}$$

where

δA = the standard deviation of the indirect measured quantity
δx = the standard deviation of the quantity X
δy = the standard deviation of the quantity Y

The *fundamental expression* which may be used to evaluate the propagation of accidental errors for an indirect measurement is

$$\Delta U = \sqrt{\left(\frac{\delta U}{\delta X}\Delta X\right)^2 + \left(\frac{\delta U}{\delta Y}\Delta Y\right)^2 + \left(\frac{\delta U}{\delta Z}\Delta Z\right)^2 + \cdots + \left(\frac{\delta U}{\delta N}\Delta N\right)^2}$$

where

U = an indirect quantity which is a function of the measured quantities $X, Y, Z, \ldots N$
ΔU = the error in the indirect quantity U
$\Delta X, \Delta Y, \Delta Z, \ldots, \Delta N$ = the errors in the measured quantities X, Y, $Z, \ldots, N$

In the above expression it has been assumed that all the measured quantities were measured under identical conditions with the same procedures; that is, all measurements were of the same quality. Obviously, in practice, some measurements will be made under more favorable conditions, with more accurate procedures or with more accurate instrumentation. Thus, the resulting measurement would be more reliable and should therefore exert more influence on an indirect quantity calculated from the measurements than should a measurement obtained under less favorable conditions. The degree of reliability of a measurement is referred to as the *weight* of the measurement. The weight of a measurement is the relative reliability of the measurement to the other measurements in the computation. It is most commonly expressed as a number, and being relative may be operated on by any factor as long as all other weights in the computation are treated similarly. The general expression for the weighted mean is:

$$\overline{M} = \frac{W_1 M_1 + W_2 M_2 + \cdots + W_n M_n}{W_1 + W_2 + \cdots + W_n}$$

where

$\overline{M}$ = the weighted mean of n measurements,
$M_1, M_2, \ldots, M_n$ = the value of the independent measurements
$W_1, W_2, \ldots, W_n$ = the weights of the independent measurements

The procedures for assigning weights to individual measurements is largely a matter of judgment based on the conditions and instrumentation available at the time of the measurement. However, when calculating the mean value from two or more sets of measurements, it is common practice to use the calculated expression of precision for each of the sets. The weights are considered to be inversely proportional to the square of the standard deviation:

$$W_1/W_2 = \delta_2^2/\delta_1^2$$

where

W_1, W_2 = weights of individual sets of standard deviation
δ_1, δ_2 = standard deviations or probable error, etc., of individual sets

Adjustments of Measurements

Although there are many specialized methods of adjusting measured data, the most widely applicable method is that of *least squares*. The method of least-squares adjustment is derived from the equation of the probability distribution:

$$P(x) = \frac{1}{\delta\sqrt{\pi}} e^{-(x^2/2\delta^2)}$$

The probability of occurrence of an error X_1 in measurement M_1 is

$$P_1 = \frac{1}{\delta\sqrt{2\pi}} e^{-(X_1^2/2\delta^2)}\, dx_1$$

and similarly for $X_2, X_3, \cdots, X_n$. The probability that all the errors, $X_1, X_2, \ldots, X_n$ will be made is

$$P = P_1 \times P_2 \times \cdots \times P_n$$

$$= \frac{1}{\delta\sqrt{2\pi}} e^{[-(1/2\delta^2)(X_1^2 + X_2^2 + \cdots + X_n^2)]}\, dx_1 dx_2 \cdots dx_n$$

When one considers normal distribution, it is apparent that it is more probable for small errors to occur than large errors. Therefore, the set of errors with the greatest probability will be the one which makes the measured values the most likely correct ones. Considering the expression for probability P above, it is apparent that it will have a maximum value when the exponent of e is a minimum, or, since δ is a constant, when $(X_1^2 + X_2^2 + \cdots + X_n^2)$ is a minimum. This is commonly expressed as the least-squares theory: "The most probable value from a set of measurements of equal precision is that value for which the sum of the squares of the errors (residual) is a minimum."

It should be realized that the theory of least squares, based on probability theory, is a very useful tool when dealing with quantities of measured data.

However, it must also be kept in mind that this concept does not apply to systematic errors or blunders, and depends upon repetitions of the measurements, losing much of its validity when the number of repetitions is small.

The least-squares adjustment of measurements frequently falls into one of two classes of solution. One of these is based on *observation equations.* An observation equation may be defined as an equation which relates measured quantities to to their residual observational errors and to independent and unknown parameters. Typically, one such equation is written for each observation. If the number of observation equations is equal to the number of unknowns a unique solution results. However, if the number of observation equations is greater than the number of unknowns, which is most commonly the case, a least-squares solution for the most probable value of the unknown is possible. In matrix notation, the least-squares solution of a set of unweighted observation equations takes the form

$$X = (A^T A)^{-1} \, A^T L$$

where

X = matrix of unknowns
A = coefficient matrix
L = matrix of constants

If weights are used, then the equation takes the form

$$X = (A^T PA)^{-1} \, A^T PL$$

where P is the weight matrix.

The second method of commonly employed least-squares solution is based on the *condition equation*. This approach may be used when one or more conditions in the adjustment of the measurements must be satisfied exactly. The equations which express these conditions are referred to as condition equations. In a least-squares solution by the condition-equation method, the results yield the most probable values for the unknowns and exactly satisfy all imposed conditions. In matrix notation, the least-squares solution of a set of unweighted condition equations takes the form

$$V = B^T (BB^T)^{-1} \, W$$

where

V = matrix of residuals
B = coefficient matrix
W = matrix of constants

If weights are used then the equation takes the form:

$$V = P^{-1} \, B^T (BP^{-1} B)^{-1} \, W$$

where P is the weight matrix. Once the unknown residuals are determined it is necessary to calculate the corrected measurements and the adjusted unknowns.

Both the observation-equation and condition-equation methods lend themselves to computer solution because of the large number of equations involved. The condition-equation method may yield a smaller number of equations, but this is frequently offset by the difficulty in identifying and formulating the independent conditions.

DISTANCE-MEASUREMENT METHODS

Probably the most fundamental measurement associated with engineering practice is length or distance. Distance measurements may be considered in two categories, direct and indirect. Direct measurements are made by determining the number of times a specific unit, such as feet or meters, is physically contained in the length of a line. Indirect measurements are made by accurately measuring a quantity which is a function of the length of the line and calculating the length based upon the measured quantity and its known relationship to the length. In either case, the resultant length determined is the length of the line. In the case of geodetic measurements, the horizontal length projected to the mean sea-level surface is usually desired.

Units of Length

Unfortunately, there are many different units of length in common use in engineering practice. Although it is desirable to adopt the meter as the common unit, at present it is necessary to be familiar with many different units. A summary of the more common length units is presented below:

1 mile = 5,280 ft = 1760 yds = 320 rods = 80 chains
1 chain = 66 ft = 4 rods
1 meter = 39.37 in = 3.2808 ft = 1.0936 yd
1 vara = 33 in. (Calif.) = $33\frac{1}{3}$ in. (Texas)

Direct Distance Measurement

Although many forms of indirect measurement are being used more frequently in the measurement of distance, *direct measurement with a steel tape* is still the most common method. The common surveying tapes are made of steel or steel alloys, have a uniform cross section, and are graduated at regular intervals—with finer graduations provided at the ends. Tapes vary greatly with respect to material, cross section, graduation, length, and other characteristics. The user must take care to be familiar with the characteristics of the particular tape with which he is working. The number of man hours lost by assuming "this tape is like the last one" has never been calculated, but undoubtedly it is significant.

Just as there are a wide variety of tapes, there is also a variety of accessory equipment. The most common include:

Taping pins: metal rods sharpened on one end with a ring on the other; used to mark tape ends or intermediate points on the ground during the taping procedure;
The spring balance: used to apply desired tension to tape when making accurate measurements;
The clamp handle: used to hold steel tape when less than a full tape length is being measured.

Standardization of a tape is the process by which the tape is calibrated. All tapes to be used for accurate work should be standardized. This is usually accomplished by the manufacturer or at the National Bureau of Standards upon request of the owner. Inasmuch as a typical 100-ft tape is 100.00 ft long only under a specific set of conditions, standardization consists of comparing the tape to a known distance under carefully controlled conditions of temperature, tension, and support. For surveys of low accuracy requirements the errors due to differences between field conditions and calibration conditions may be disregarded. However, for more accurate surveys, this must be taken into account. The results of a tape standardization typically include the length of the tape to ±0.001 ft under varied tension and support at a constant temperature.

Taping procedures may vary widely depending upon such factors as the nature of the terrain, project requirements, equipment available, preferences of party chiefs, and standard practice of organizations. However, basic taping procedures can be considered in two categories.

The first category, and most fundamental, is *horizontal taping*, in which the tape itself is maintained in a horizontal position. For moderate precision, on level ground, the tape may be placed directly on the ground and the ends of the tape length marked with taping pins. If the ground is level but vegetation or some other interference prevents the stretching of the tape directly upon the ground, the ends of tape may be equally elevated to a height sufficient to clear the obstruction but preferably not greater than waist height. The graduations are then projected to the ground by means of plumb bobs. When the ground is not level, the tape is held on the ground at the point of greatest elevation while the other end of the tape, or segment being measured, is raised until the tape is level. This may be best accomplished with the use of a hand level. The graduation at the elevated end is then projected to the ground by means of a plumb bob.

When the slope of the ground is great enough to make horizontal taping difficult, or if a high degree of accuracy is required, *slope taping* should be used. The slope taping procedure requires that the measurement be made directly upon the slope and corrections be applied to reduce the slope distance to horizontal. This reduction requires that either the difference in elevation between the end points or the vertical angle be measured in addition to the slope distance (see the discussion of slope correction below).

The required accuracy of the survey will dictate the degree of refinement of the corrections to be applied to the field measurements and what secondary measurements are required to calculate the corrections. The major sources of error associated with taping and their corrections are presented below:

1. *Incorrect length*. Under conditions in the field, the calibrated length of the tape will seldom if ever be its nominal length. Thus, every nominal tape length measured will be in error and the total error will be proportional to the number of tape lengths measured.

2. *Slope*. The slope correction required to reduce the measured slope distance to horizontal may be calculated by means of the following equations:

$$H = s - Ch$$

$$Ch = \frac{h^2}{2s} + \frac{h^4}{(2s)^3} + \cdots$$

where

H = horizontal distance
Ch = correction to be subtracted from inclined distance (this term becomes insignificant on moderate slopes)
s = measured inclined distance
h = difference in elevation of ends of tape or tape segment

or

$$H = s \cos \alpha$$

where α is the vertical angle.

3. *Temperature*. The temperature correction required to account for the difference between calibration temperature and measurement temperature may be calculated by the following equation:

$$C_t = L\alpha (T - Ts)$$

where

C_t = correction to be applied, in feet
L = length of tape used, in feet
α = coefficient of thermal expansion of tape (steel = 0.0000065/°F)
T = measurement temperature, in degrees Fahrenheit
Ts = standardization temperature, in degrees Fahrenheit

4. *Tension*. The tension correction required to account for change in length of the tape due to the difference between calibration tension and measurement

tension may be calculated by

$$Cp = (P_1 - P_0)L/AE$$

where

Cp = the tension correction, in feet
P_1 = the measurement tension, in pounds
P_0 = the standard tension, in pounds
A = the cross-sectional area of the tape, in square inches
E = the modulus of elasticity of the tape material in pounds per square inch (steel = 29,000,000)
L = the length of the tape in feet

5. *Sag.* The sag correction required to account for the difference between tape support during measurement and during standardization may be calculated by:

$$C_s = W^2 L/24P^2$$

where

C_s = sag correction, in feet
W = the weight of tape between supports, in pounds
P = the tension on the tape, in pounds
L = the interval between supports, in feet

6. *Alignment.* The error due to misalignment of the tape is similar to the error due to slope. Alignment errors are most easily eliminated by maintaining alignment during the measurement procedure.

7. *Setting pins.* The errors due to setting pins are accidental and therefore random in nature. No correction can be calculated.

8. *Wind.* A strong wind will tend to bow out a tape which is supported at the ends only. The error introduced is similar to sag. The best technique to reduce this error source is to avoid the conditions which produce the error.

The precision provided by the taping procedures presented above varies depending upon field conditions. In general, the major errors tend to be systematic and their magnitude may be significantly reduced by applying the appropriate corrections. For rough taping, a precision of 1/1000 may generally be achieved with ordinary care in field measurements without regard to corrections. If precisions on the order of 1/5000 are required, the measurements must be taken with extreme care and the appropriate corrections applied.

Indirect Distance Measurement

Distances may be measured indirectly by measuring a quantity which is a function of the unknown length of the line. Having obtained the measured quantity,

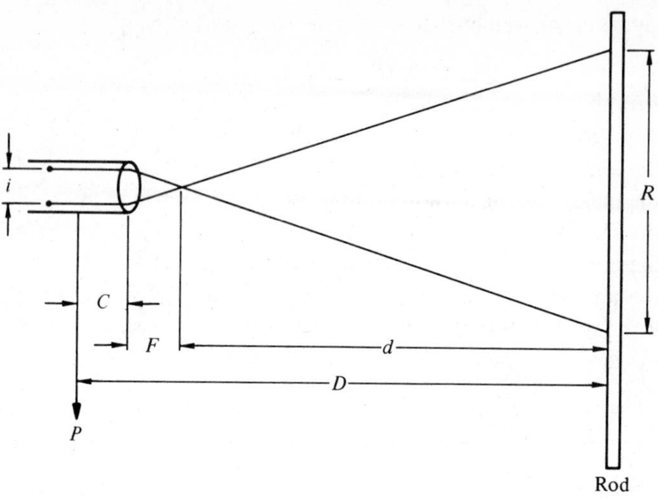

Fig. 2. Horizontal stadia sight.

the unknown length may be readily calculated. There are three common methods of indirect distance measurement: (1) the stadia, (2) subtense, and (3) phase-shift or electromagnetic methods.

The *stadia method* provides a rapid and effective means of obtaining distances to a precision approximately 1/500 under favorable conditions. In combination with measured vertical angles, it provides a means of calculating differences in elevation as well as horizontal distances. Figure 2 illustrates the stadia method for a horizontal line of sight and an external-focus telescope. Two light-ray paths are shown which are projected from the stadia hairs of the telescope parallel to the optical axis. They are refracted by the objective lens, pass through the focal point at a distance F in front of the lens and intersect the rod. The interval between the stadia hairs is represented by i, and the interval between intercepts on the rod is R. The plumb line is represented by P, and C represents the distance between the plumb line and the objective lens. Based upon Fig. 2 the distance can be calculated by

$$D = kR + (F + C)$$

where $k = F/i$ is the stadia interval factor.

When the line of sight is inclined as shown in Fig. 3, the horizontal distance may be calculated by:

$$D = kR \cos^2 \alpha + (F + C) \cos \alpha$$

where α is the measured vertical angle and R is the intercept on the rod. The difference in elevation between ends of the line (V) may be calculated by

$$V = kR \tfrac{1}{2} \sin 2\alpha + (F + C) \sin \alpha$$

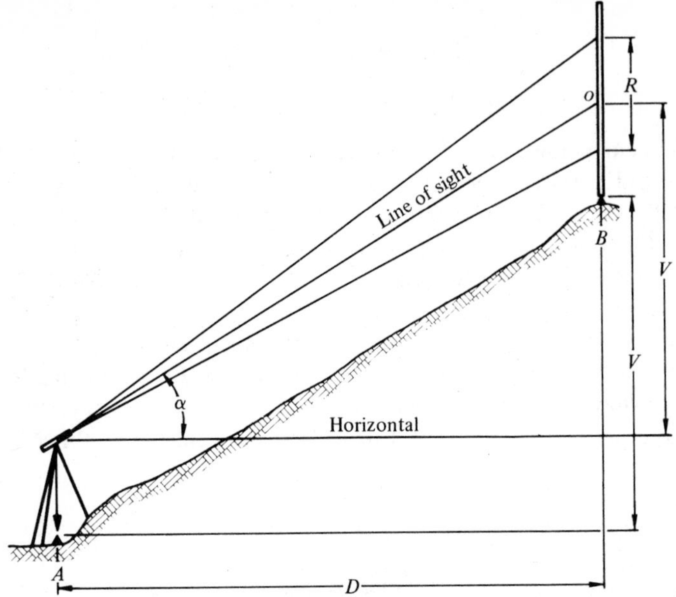

Fig. 3. Inclined stadia sight.

If the instrument is an internal focusing instrument, the term $(F + C)$ may be taken as zero in the above equations. The solution of these equations is greatly facilitated by means of tables and special slide rules.

The *subtense method* of indirect distance measurement consist of measuring the angle subtended by a fixed base and calculating the distance. This is illustrated in Fig. 4. The distance D may be calculated by:

$$D = \tfrac{1}{2} S \cot (\beta/2)$$

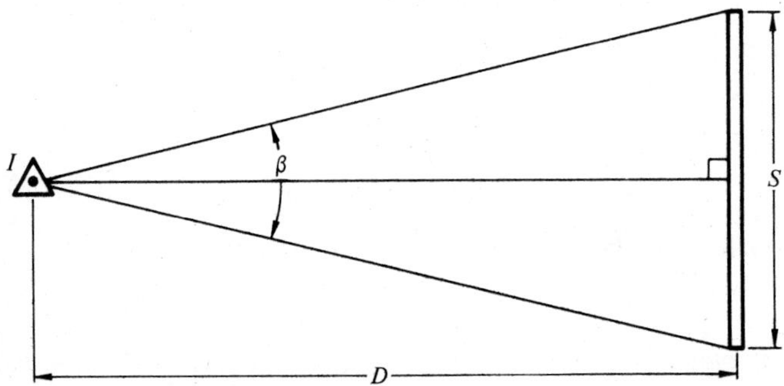

Fig. 4. Subtense method.

where

S = length of fixed base
β = measured subtended angle

The equipment employed in a subtense measurement consists of a transit or theodolite and a *subtense bar.* The subtense bar is commonly 2 m long and constructed of invar so that its length will not be significantly effected by temperature. For the most accurate results, the subtended angle should be measured to ±1 sec.; therefore it is standard practice to employ a high-quality theodolite to measure the subtended angle.

The third method of indirect distance measurement employs electromagnetic energy. Basically, *electromagnetic distance measuring* (EDM) systems determine the time required for electromagnetic energy to travel from one end of the line to the other and return. The distance is then calculated by the relationship

$$D = VT/2$$

where

D = the measured distance
V = velocity of the energy
T = the travel time

In 1957, the International Union of Geodesy and Geophysics (IUGG) adopted 299,792.5 ± 0.4 km/s as the standard in *vacuo* velocity for the propagation of visible light and radio waves. In operational EDM systems, the energy, of course, does not propagate through a vacuum but rather through the atmosphere. Inasmuch as the accuracy of the distance measured is a direct function of the velocity of propagation of the energy, it is necessary to consider the effect of the atmosphere upon the velocity of propagation. The relationship between the velocity of propagation *in vacuo* and in the atmosphere is given by the index of refraction, which is defined as

$$n = V_0/V$$

where

n = index of refraction
V_0 = velocity in vacuo
V = velocity in atmosphere

A typical value for the refractive index n under average conditions would be 1.0003.

The Geodimeter is a typical EDM system employing modulated light as the propagation energy. The system basically consists of a transmitter-receiver unit

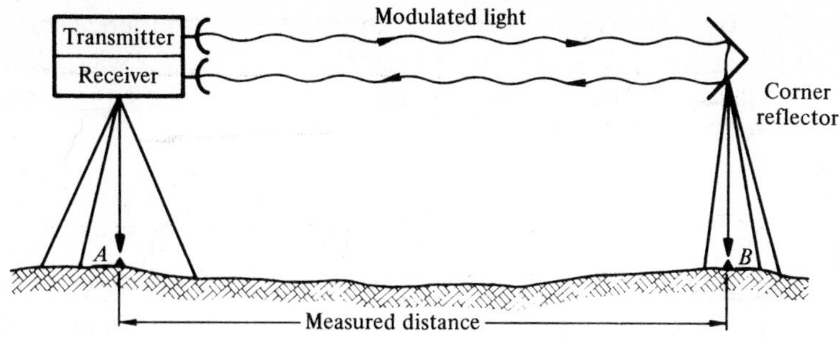

Fig. 5. Basic geodimeter system.

and a passive reflector, as shown in Fig. 5. A modulated light beam is propagated from the transmitter to the reflector, from which the energy is returned to the receiver. The intensity of the signal is modulated at approximately 30 MHz. If we assume that the velocity of the energy is 30×10^8 m/s the length of the modulating waves will be

$$\frac{3 \times 10^8}{3 \times 10^7} \approx 10 \text{ m}$$

The phase of the energy returning to the receiver compared to the phase of the wave at the transmitter at any instant in time must be a function of the distance between the ends of the line. For example, if the returning energy is in phase with the transmitted energy, the double distance (down and back) must be an even number of half wavelengths (5 m). If the returned energy is 180° out of phase with the transmitted energy the double distance must be an odd number of half wavelengths. The system can, therefore, resolve half wavelength changes in the double distance. This is, of course, equivalent to a quarter-wavelength (2.5 m) change in the length of the line. The quarter-wavelength is therefore the basic measuring unit, or "yardstick," of the system.

In the general case, a line to be measured will have a length equal to some integer number of basic units (U) plus a fraction of a unit (L). In order to measure distances to engineering accuracies it is, of course, necessary to evaluate not only the integer number of units but the fractional part as well. The fraction is determined by conducting a portion of the transmitted energy through a variable-inductance delay line. This delay permits the selective change of phase of the portion of the transmitted energy up to 90° which is sufficient to compare the two waves. By converting the delay-line setting to its equivalent length, the fraction L can be determined.

It is also necessary to determine the integer number U of units in the distance. This is accomplished by using three different modulating frequencies (F_1, F_2, and F_3), which will therefore produce three basic measuring units of different lengths (U_1, U_2, and U_3). This will in turn produce three different fractional distances (L_1, L_2, and L_3). The frequencies F_1, F_2, and F_3 are selected so that there is a fundamental relationship between the lengths of the basic measuring units as shown below:

$$400U_1 = 401U_2 = 1000 \text{ m}$$

$$20U_1 = 21U_2 = 50 \text{ m}$$

The method employed is to measure the delay settings L_1, L_2, and L_3 resulting from the three modulating frequencies F_1, F_2 and F_3. The distance may then be expressed by the following set of equations:

$$D = L_1 + n_1 U_1$$

$$D = L_2 + n_2 U_2$$

$$D = L_3 + n_3 U_3$$

Since the relationships between U_1, U_2, and U_3 are known, these equations can be solved for D according to the equation

$$D = L_1 + 401(L_2 - L_1) + 1000P$$

where P is the number of 1000-m intervals in the distance D. It should be noted that three determinations of the distance D are possible from the delay settings on the three modulating frequencies.

The refractive index for modulated light may be calculated by the equation

$$n = 1 + \frac{0.0003036}{1 + (t/273)} \cdot \frac{p}{760} - \frac{0.000000055e}{1 + (t/273)}$$

where

n = refractive index
t = air temperature in degrees Celsius
p = pressure in millimeters Hg
e = water vapor pressure in millimeters Hg

The water-vapor pressure may be calculated by Ferrel's formula:

$$e = e' - 0.000660(1 + 0.00115\, t_w)(t_d - t_w)P$$

where

e = water vapor pressure in millimeters Hg
e' = saturated vapor pressure in millimeters Hg (from Tables)
t_d = psychrometer dry-bulb temperature in degrees Celsius
t_w = psychrometer wet-bulb temperature in degrees Celsius
P = total barometric pressure in millimeters Hg

In many of the modern modulated-light systems, the source of energy is a small laser. Further, since the reductions, as indicated above, are systematic, in most modern systems they are performed internally. Thus, a direct read-out in linear units is provided to the operator. It should be noted, however, that the indicated distance will be a slope distance, and in order to obtain the horizontal distance the difference in elevation between the ends of the line must be taken into account.

The Tellurometer is a typical EDM system employing modulated microwaves as the propagation energy. The model MRA-3 is typical of these systems. The tellurometer system is composed of two active transmitter-receiver units. In the more recent systems the units are identical. During the measurement process each unit is positioned on one end of the line being measured and transmits toward the other unit. The unit at which the readings are obtained is referred to as being in the *master* mode, the other as being in the *remote* mode. The mode of operation may be selected by a switch. The transmissions consist of a microwave carrier frequency modulated by a crystal-controlled "pattern" or measuring frequency. The carrier and modulating frequencies of the master unit differ from the carrier and pattern frequencies of the remote unit by specific amounts. The respective receiver circuits are tuned to respond to the frequency differences produced by mixing the frequencies. Fig. 6 shows a simplified diagram of the Tellurometer system. In both the master and remote units the microwave carrier is produced by the *klystron*. The klystron has the capacity to vary the generated frequency within a range of 10,025–10,450 MHz according to the setting of the cavity. The pattern frequency, which can be selected by a switch, is modulated upon the carrier. The master pattern frequencies are designated A, B, C, D, E, while the remote pattern frequencies are $A+, A-, B, C, D, E$. The remote pattern frequencies are all 1 kHz lower than the corresponding master frequency, with the exception of the $A-$ pattern which is 1 kHz higher. Likewise, the master carrier frequency is always 33 MHz below the remote during operations.

Consider the situation as shown in Fig. 6, with the master pattern selector at A and the remote pattern selector set on $A+$. With this arrangement, the master unit is transmitting a carrier frequency modulated with the A pattern to the remote and the remote is transmitting a carrier frequency modulated with the $A+$ pattern to the master. At the remote unit these two modulated carriers are mixed and the resulting complex wave form contains all the mixed components. This complex wave form is then taken to an intermediate frequency amplifier which is tuned to 33 MHz, the frequency difference between the master and remote carriers. The difference between the pattern frequencies emerges from the mixing as a 1 kHz amplitude modulation on the 33 MHz wave form. After amplification in the IF strip the signal is applied to an AM detector which passes the 1 kHz AM wave but blocks the 33 MHz. The 1 kHz signal is then amplified

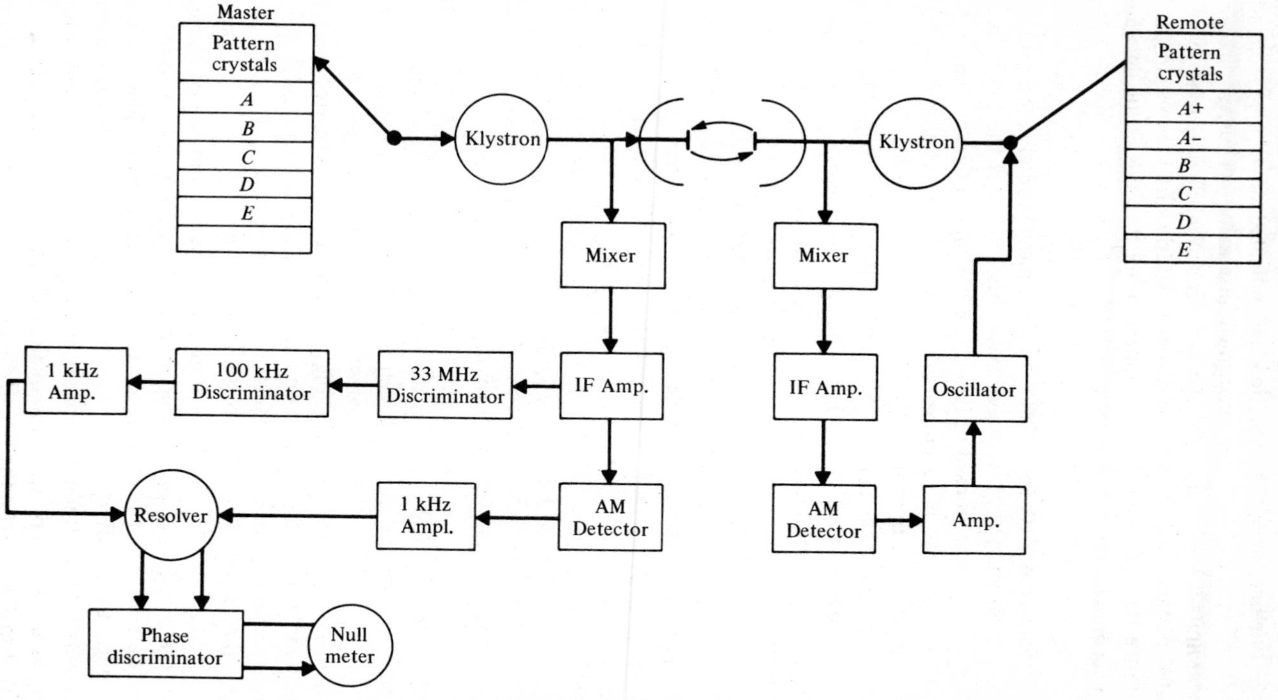

Fig. 6. Tellurometer system.

and used to frequency-modulate a 100 kHz oscillator. The output of this oscillator is then "fed back" to the klystron to produce an additional modulation at the remote carrier. Therefore, at any instant in time, as long as the selector switches are as shown, the master is transmitting a carrier frequency modulated with the A pattern, while the remote is transmitting a carrier frequency modulated with the $A+$ pattern and the 100 kHz + 1 kHz feedback signal.

At the master unit the frequencies are mixed just as they were in the remote. Likewise, the master IF amplifier is tuned to 33 MHz and the 1 kHz resulting from the pattern frequencies difference is detected by the AM detector, amplified, and applied to the resolver. The feedback signal, frequency-modulated on the remote carrier as 100 kHz + 1 kHz, is detected from the IF amplifier by a discriminator which passes the 100 kHz + 1 kHz but rejects the 33 MHz. The 100 kHz is in turn rejected by a discriminator, leaving the 1 kHz feedback signal. This signal is amplified and applied to the resolver.

From the above it can be seen that the master receiver has two functions. First, to generate the 1 kHz signal resulting from the difference between the A master pattern and the $A+$ remote pattern and to process this signal by AM stages to the resolver. Second, to detect the 1 kHz feedback signal generated at the remote unit and process this signal by FM stages to the resolver. At the resolver, the phases of the two signals are compared in order to obtain the distance information.

The function of the carrier frequencies is to provide the vehicle upon which the pattern frequencies are propagated. The carriers contain no distance information and can, therefore, be disregarded in determining the distance. The master pattern, in traveling from the master to the remote, experiences a phase change of some whole number of cycles plus a fractional part of a cycle. Similarly, the remote pattern experiences a phase change. The problem is to resolve the fraction and count the whole number of cycles. The phase relationship at the resolver, between the 1 kHz signal generated at the master and the 1 kHz feedback signal generated at the remote, is illustrated in Fig. 7. This figure is a vector

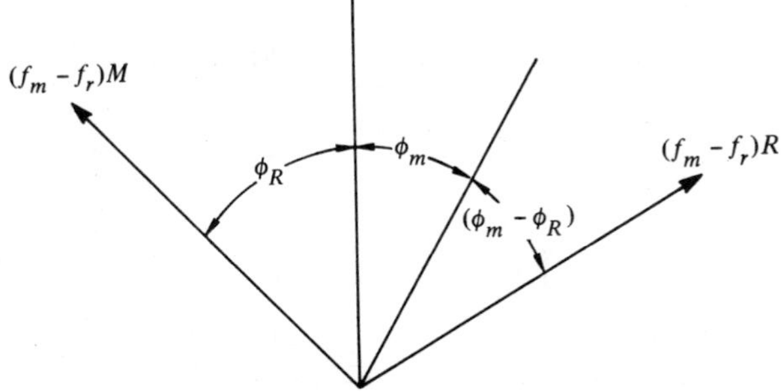

Fig. 7. Phase comparison at resolver.

diagram representing the phase differences at the master resolver. The vector $(f_m - f_r)M$ represents the 1 kHz signal generated by mixing the master and remote pattern frequencies at the master. The vector $(f_m - f_r)R$ represents the signal generated at the remote and returned to the master. The angle ϕ_R represents the phase shift of the remote pattern traveling to the master. The angle ϕ_M represents the phase shift of the master pattern traveling to the remote. The angle $(\phi_M - \phi_R)$ therefore represents the phase of the feedback signal when it arrives at the master resolver. The total phase difference between the two 1 kHz signals is seen to be

$$\phi_R + \phi_M + (\phi_M - \phi_R) = 2\phi_M$$

This is twice the phase shift of the master pattern in the distance. This angle is accurately measured by the resolver, and since the master pattern wavelength is accurately known, the fractional distance can be determined.

It is still necessary to determine the whole number of cycles contained in the measured distance. Theoretically this could be accomplished by choosing the pattern frequencies so as to have longer and longer wave lengths until a pattern wave length would be reached which was longer than the distance being measured. This is not practical, however, because of electronic and economic limitations. The system in essence accomplishes this same result through the use of the subtractive properties of the pattern wave lengths. The principle involved is that if the phase shift at one frequency is subtracted from the phase shift at a different frequency the resulting phase shift will be of a frequency equal to the difference between the original frequencies. By using this principle, the resolvable distance can be significantly increased without having to cover wide bands of frequencies. This is illustrated in Table 1 which shows the half wavelengths resulting from the subtractions indicated. The half wavelengths are used because they relate to the single rather than the double distance. It should be noted that the subtraction $A - (A-)$ produces essentially a double reading upon the A pattern, which is used to minimize internal errors.

The measurement procedure is to measure the $2\phi_M$ values on each of the

TABLE 1. HALF WAVELENGTHS

Subtraction	Value
$A - B$	28---.--
$A - C$	73--.--
$A - D$	36-.--
$A - E$	42.--
$A - (A-)$	1.97
Distance:	27,341.97m

pattern settings (A, B, C, D, E). The subtractions are made internally and readout consists of the phase difference of the half wavelengths in order to give single-distance values. When the five subtractive values are combined, they yield all digits of the distance from between 100,000 and 0.01 m. For example, assume the subtractive values are as indicated in Table 1. Note that the system produces an overlap in the readout. The last digit in each value should be the same as the first digit in the row below, the fineness increasing by a factor of 10 in each row.

The distance resulting from the measurement must be considered a raw measurement. If the line has a significant slope it should be reduced to horizontal. Further, if a high degree of accuracy is required, or if the line is long, it is necessary to apply a correction for the difference between the assumed refractive index and the refractive index at the time of measurement. The most commonly used formula for microwave refractive index is that of Essen and Froome:

$$n = 1 + \left[\frac{103.49\,(P - e)}{t + 273.15} + \frac{86.26}{t + 273.15}\left(1 + \frac{5748}{t + 273.15}\right)e\right] \times 10^{-6}$$

where

n = refractive index
t = temperature in, degrees Celsius
e = partial water vapor pressure in millimeters Hg

Another factor which should be taken into account when high accuracies are desired from microwave measurements is *ground swing*. Ground swing results when some of the transmitted energy is reflected from the earth's surface and interferes with the energy propagating directly between the units. This interference produces a variation in the fine readings which is a function of the strength of the reflected component, the excess length of the reflected path, and the carrier frequency. Inasmuch as the klystron permits a range of carrier frequencies to be generated, the ground swing can be eliminated by acquiring fine readings over a range of carrier frequencies. These are commonly plotted vs. klystron setting to produce a ground-swing curve. If a whole cycle or more is present, the zero axis of the curve will approach the true value. If less than a full cycle is presented, consideration should be given to the selection of an alternate site.

When extreme accuracy is desired on long lines, further refinements may be employed. These would account for the fact that the energy travels in a curved path due to atmospheric refraction. According to the work of Saastamoiner this correction, which takes into account both ray-path curvature and atmospheric density variations, is

$$C = (1 - k)^2\, K^3/24R_e^{\,2}$$

where

C = combined ray-path and atmospheric-variation correction
k = ratio of radius of the earth to ray path radius
K = chord distance
R_e = radius of the earth

When this equation is employed with a microwave system k is usually taken as 0.25 and the equation becomes

$$C = K^3/43R_e^2$$

When the equation is employed with an optical system k is usually taken as 0.2–0.15, with the larger value associated with night observations.

Inasmuch as many EDM-measured distances are relatively long, it is frequently desirable to reduce them to sea-level rather than horizontal distances. One formula commonly used to accomplish this reduction is

$$s = \left\{ \frac{(D + h)(D - h)}{[1 + (h_1/R)]\ [1 + (h_2/R)]} \right\}^{1/2} + \frac{K^3}{24R^2}$$

where

s = the sea-level distance
D = the measured slope distance
h_1 = elevation of one station
h_2 = elevation of second station
$h = (h_1 - h_2)$
R = radius of curvature of line on reference spheroid (usually taken as radius of earth–R_e)
K = sea-level chord distance

ELEVATION MEASURING SYSTEMS

Measurements of elevation are associated with almost every engineering project. The elevation of a point is its vertical distance above or below a datum or surface of reference. Some of the basic principles and definitions which apply to elevations are illustrated in Fig. 8. Mean sea level is universally considered as the elevation reference surface although local surveys may use an arbitrary local reference. The mean sea-level surface is the average height of the sea for all stages of tide over a long period of years. This surface is closely approximated by the spheroid which is the mathematical surface upon which geodetic computations are made. A level surface, for example the sea-level surface, is an equipotential surface, being everywhere perpendicular to the direction of gravity. Note that the direction of gravity and a normal to the spheroid do not usually

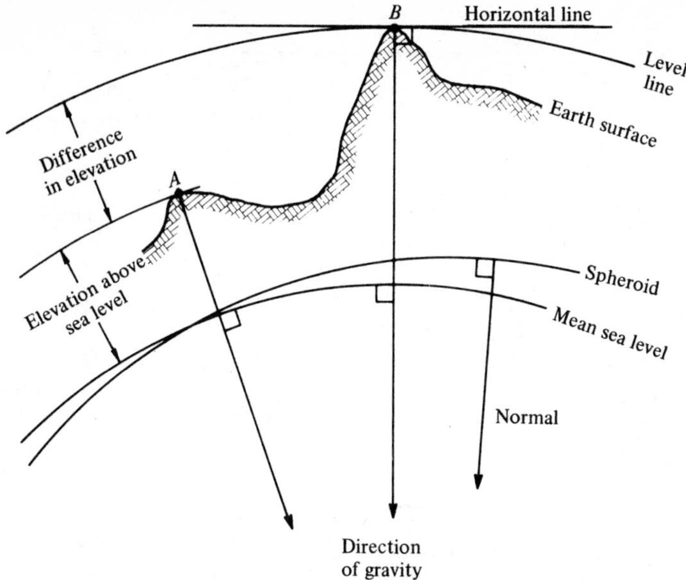

Fig. 8. Basic principles of elevation.

exactly coincide at any point. The *elevation* of station *A* is the vertical distance along the direction of gravity between the sea-level datum and the point. The *difference in elevation* between stations *A* and *B* is the vertical distance between the imaginary curved surfaces through *A* and *B* everywhere parallel to the mean sea-level surface. A horizontal line is a straight line tangent to a level surface at any given point.

Curvature and Refraction

The relationship between a level line, a horizontal line, and a line of sight through an instrument is illustrated in Fig. 9. At station *A*, the horizontal line, the level surface, and the line of sight are all perpendicular to the direction of gravity. Because of atmospheric refraction, the line of sight will not coincide with a horizontal line but will follow a concave curve. The curvature of the line of sight will, of course, vary with atmospheric conditions. For operations to determine elevations, however, it is normally considered a constant. For the distance *AB* in Fig. 9 the effect of earth curvature may be seen as the distance *BD*, while the effect of atmospheric refraction is the distance *BC*. The combined effect of curvature and refraction may be calculated by:

$$h = 0.574M^2 = 0.0206S^2$$

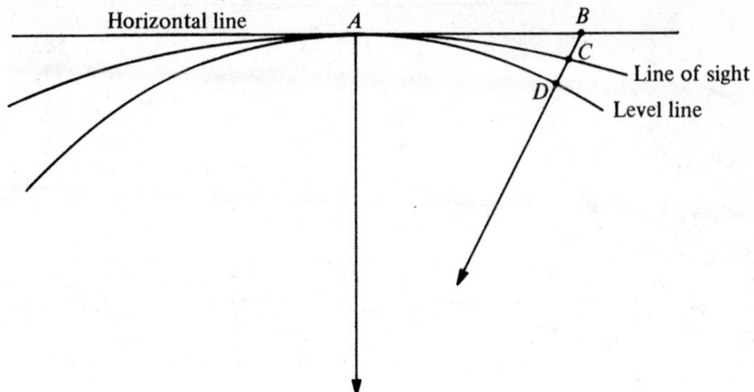

Fig. 9. Curvature and refraction.

where

h = the combined effect of curvature and refraction, in feet
M = length of sight in miles
S = length of sight in thousands of feet

Differential Leveling

Differential leveling is the most commonly used method for determining differences in elevations. Figure 10 illustrates the basic procedures of differential leveling. A level is set up midway between stations *A* and *B*. A rod is read at *A* and *B*, producing the readings *a* and *c*. A level line tangent to the line of sight at the instrument would intersect the rods at *b* and *c*. It is evident that if the instrument is midway between *A* and *B* then *ab* = *cd*. This would also hold true for most instrument errors and for refraction effects if the assumption is made that the atmosphere is uniform over the distance *AB*. Therefore, the difference

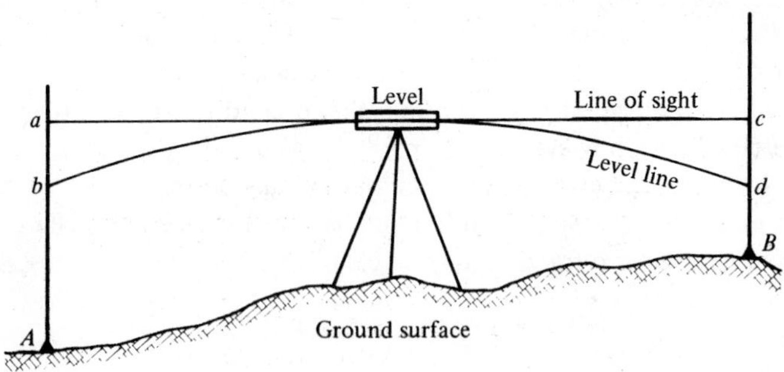

Fig. 10. Basic leveling.

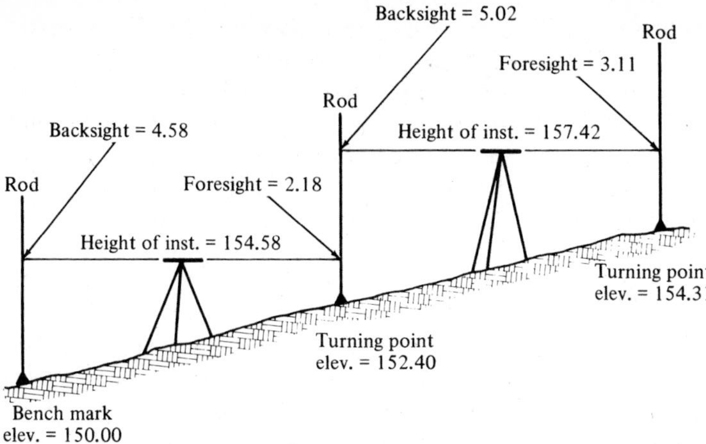

Fig. 11. Differential leveling.

in elevation between *A* and *B* may be calculated as the rod reading *a* less the rod reading *c*. In order to determine elevation differences over longer intervals, the procedure illustrated is simply repeated from *B* to *C*, etc. Terminology associated with differential leveling is illustrated in Fig. 11. A bench mark is a point of known elevation. A backsight is a reading upon a point of known elevation. The height of instrument is the elevation of the line of sight. A foresight is a reading taken upon a point whose elevation is to be determined.

In standard differential procedure level loops are run. A level loop consists of starting from a bench mark, either known or assumed, carrying the elevations through the points to be determined, and returning either to the original bench mark or another bench mark in the area. If the differential leveling is done perfectly, the final or closing elevation of the bench mark will agree exactly with the original elevation or, in the case of closing upon a different bench mark, with the elevation of that bench mark. Any difference between the original elevation and the closing elevation is termed the error of closure.

Relative accuracy is commonly used as a standard for defining the quality of differential leveling. Table 2 gives some generally accepted values. First and

TABLE 2. RELATIVE ACCURACY FOR DIFFERENTIAL LEVELING

Class of levels	Relative accuracy
1st-order, class I	0.5 mm $\sqrt{\text{Kilometers}}$
1st-order, class II	0.7 mm $\sqrt{\text{Kilometers}}$
2nd-order, class I	1.0 mm $\sqrt{\text{Kilometers}}$
2nd-order, class II	1.3 mm $\sqrt{\text{Kilometers}}$
3rd-order	2.0 mm $\sqrt{\text{Kilometers}}$
4th-order	varies

second orders are usually employed in geodetic work. Third and fourth orders are generally associated with engineering construction.

If the error of closure falls within the maximum allowable error of closure, the level loop is adjusted. In the case of single loops, this is usually accomplished by distributing the error of closure equally over the number of turning points. In the case of interconnected loops, this is most easily accomplished by a least-squares adjustment employing the observation-equation method.

Reciprocal Leveling

If, when running a line of levels, a topographic feature such as a river is encountered which makes it impossible to employ satisfactory differential leveling methods, reciprocal leveling may be used to carry elevations over the obstacle. Figure 12 illustrates the reciprocal leveling method. The level is set up at posi-

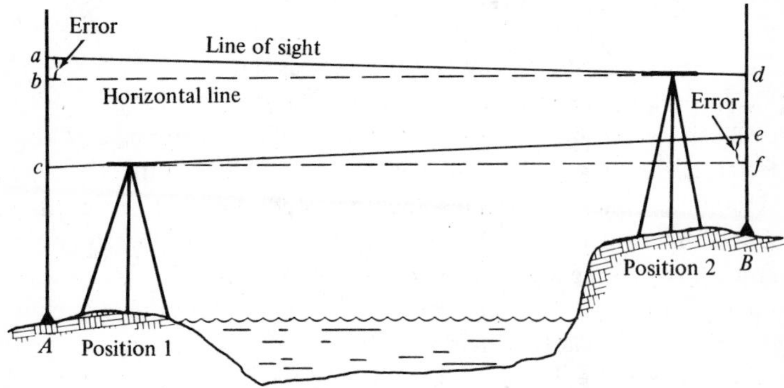

Fig. 12. Reciprocal leveling.

tion 1 a short distance from A and readings c and e are taken. The reading c should be relatively free of error since the sight is short, while that at e would contain significant error due to nonadjustment of the instrument, refraction, and curvature. The same instrument is then set up at position 2 near B and readings d and a are obtained. The errors would be similar to those entailed at position 1. The difference in elevation with the error removed between A and B would be given by

$$\Delta h = [(c - e) + (a - d)]/2$$

where

$$\Delta h = \text{difference in elevation,}$$
$$a, c, d, e = \text{rod readings.}$$

It should be noted that when reciprocal leveling is used it is common practice to use a target on the rod and to obtain multiple sightings.

Trigonometric Leveling

The determination of differences in elevation using measured distances and vertical angles is termed trigonometric leveling and is illustrated in Fig. 13. The vertical angle α, the horizontal distance D, and the height of the instrument and

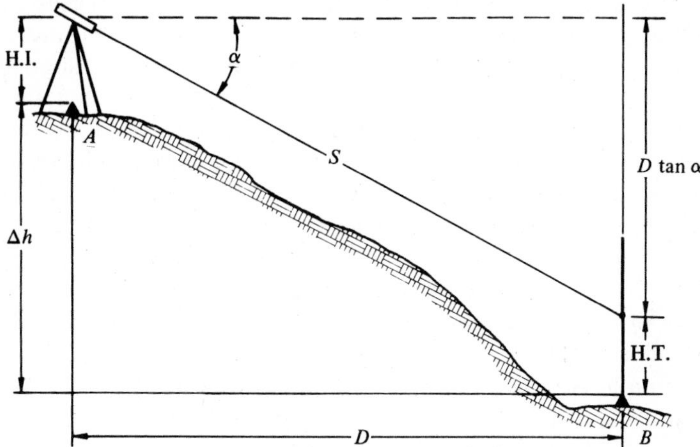

Fig. 13. Trigonometric leveling.

target are determined. Then the difference in elevation between A and B may be calculated by

$$\Delta h = D \tan \alpha + (HT) - (HI)$$

On long lines the correction for curvature and refraction should be employed and the equation becomes

$$\Delta h = D \tan \alpha - 0.0206S^2 + (HT) - (HI)$$

Note that the sign of the refraction and curvature correction will be positive for an inclined line of sight.

Barometric Leveling

Since the atmospheric pressure varies inversely with elevation, surveying altimeters may be used to measure differences in elevation. The survey altimeter is simply an improved version of the aneroid barometer. Although there are many techniques by which altimeter surveys can be run, only the single-base method will be described.

The *single-base method*, as the name implies, employs one station of known or assumed elevation and determines the elevations of the unknown stations relative to this base. The method requires at least two altimeters and psychrometers. Field procedure is as follows. Both altimeters are read simultaneously at the base station. One altimeter remains at the base and continues to take readings periodically or at designated times. The second altimeter, termed the rover, proceeds to the unknown stations where readings are taken. In addition to the altimeter readings, time and psychrometer readings are obtained. Upon completion of the readings at the unknown stations, a final simultaneous reading is obtained at the base. In the reduction of the readings, instrument drift is calculated from the simultaneous readings at the base at the beginning and end of the survey and appropriate corrections are applied to the rover readings with respect to time. The difference in readings between the base at the time of the rover reading and the corrected rover reading is calculated. Based upon the average psychrometer readings, a correction is calculated for the difference in the atmosphere during observations and the calibration atmosphere. When this correction is applied to the calculated difference in reading the results should be the difference in elevation between the base and the unknown station. When the difference in elevation is added to the base elevation, the elevation of the unknown station results.

Barometric leveling techniques lend themselves to exploratory or reconnaissance work since they can cover large areas in relatively short periods of time. They are not, however, well suited to situations which require high degrees of accuracy, since they will not provide reliable results within several feet. In recent years barometric leveling has been successfully employed to provide differences in elevation for the reduction to horizontal of slope distances measured electronically.

ANGLE-MEASURING INSTRUMENTS

The basic purpose of field surveys is to determine the relative locations of points on or near the earth's surface. The measurements required to accomplish this ordinarily include distance and angle. The angular measurements may be horizontal or vertical or both and are made with a transit or theodolite.

The unit of angular measurement in the United States is the degree (°). There are 360° in the circumference of a circle. The degree is further divided into 60 minutes (′) and the minute is divided into 60 seconds (″).

Many different horizontal angles can be used to express changes in direction. For a closed figure the interior angles are most commonly used. For an open traverse, angles to the right or deflection angles, as shown in Fig. 14, are most often employed.

It is frequently desirable to base the directions of all lines in a network upon a single reference line. Such a reference line is termed a *meridian.* A magnetic meridian is defined by the direction in a horizontal plane taken by a compass needle in the earth's magnetic field. An astronomic meridian is a meridian

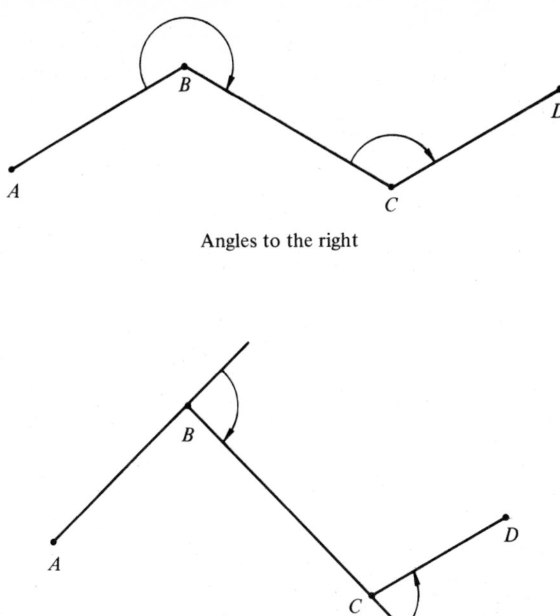

Fig. 14. Open traverse angles.

through a given point joining the North and South Poles of the earth's axis determined by observations on an astronomic body. A grid meridian is a line parallel to the central meridian or North-South axis of a system of plane-rectangular coordinates. An assumed meridian is a direction chosen for convenience for a particular survey or locality.

The acute angle which any line makes with a meridian is called its *bearing.* In Fig. 15 the bearing of *AB* is N45°E, *AC* is S45°E, *AD* is S45°W, and *AE* is N45°W. If the meridian of reference is magnetic, then the bearings would be magnetic bearings. Likewise for astronomic bearings, grid bearings, and assumed bearings. In order to convert from magnetic bearings to "true" bearings it is necessary to know the magnetic declination. The declination is defined as the angle between the compass needle and the "true" meridian. If the compass needle points east of the "true" meridian, it is termed east declination; if it points west, it is west declination.

The azimuth of a line is the clockwise angle measured from the meridian. Azimuths may be measured from either the north or south end of the meridian, although in the United States it is most common to measure azimuths from the north. In Fig. 16 the azimuth of *AB* is 45° while the azimuth of *AC* is 225°.

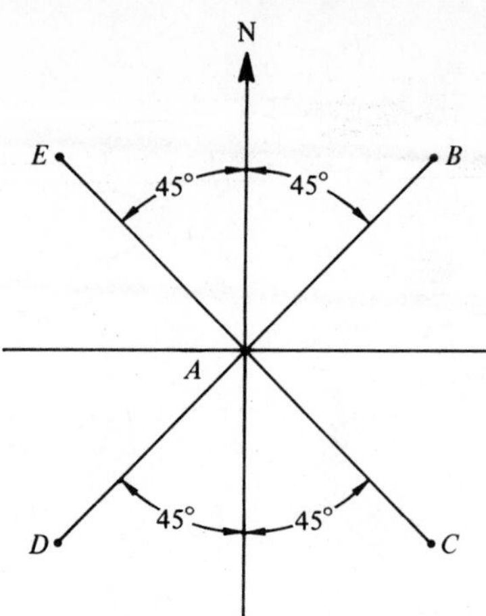

Fig. 15. Bearings.

Azimuths, like bearings, may be magnetic, astronomic, grid, or assumed, depending upon the meridian to which they are referenced.

A vertical angle is an angle in a vertical plane between two intersecting lines. Normally one of the lines is a horizontal line and the other the line of sight to a point. The vertical angle is generally considered positive when the line of sight is inclined above the horizontal and negative when depressed below the horizontal. The zenith distance is the angle measured in a vertical plane from the zenith, defined by the extension of the plumb line, to the line of sight to the point. The zenith is taken as 0°; thus a horizontal line would have a zenith angle of 90°.

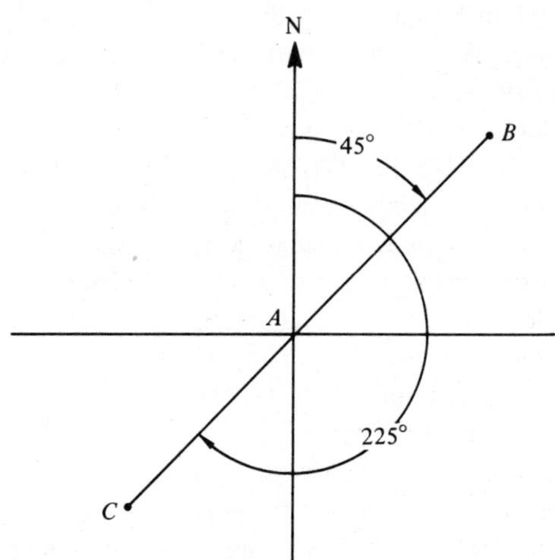

Fig. 16. Azimuths.

The Transit

The transit, shown in Fig. 17, because of its versatility, has been called the universal surveying instrument. It is principally used to measure horizontal angles but also can be used to prolong straight lines, determine differences in elevation, and obtain vertical angles and stadia readings.

The transit consists of three main assemblies: (1) the upper plate, (2) the lower plate, and (3) the leveling head. The upper plate is a horizontal circle combined with a vertical spindle upon which the circle revolves. Attached to the plate are two level vials, one parallel to the telescope, the other perpendicular to it, and two verniers separated by 180°. Two vertical standards are included which support the telescope. The telescope revolves on a vertical plane about an axis through the center of the standards termed the horizontal axis. The telescope contains a reticle containing one vertical and three horizontal cross hairs. A sensitive vial is attached to the telescope to permit the instrument's use as a level.

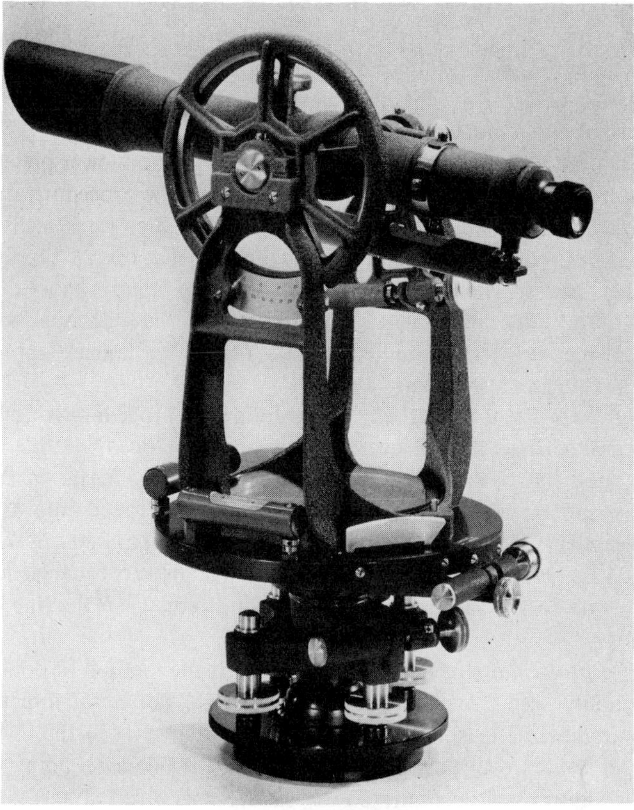

Fig. 17. The transit.

The vertical circle and vernier are mounted so as to provide measurement of the rotation of the telescope about the horizontal axis. A compass box containing a magnetic compass is usually included in the upper-plate assembly. A clamp and tangent screw are provided to permit rotation of the upper plate relative to the lower.

The lower plate consists of a graduated horizontal circle which is mounted upon a hollow spindle that is fitted to a ball socket in the leveling head. A clamp and tangent screw are provided to permit rotation of the lower plate relative to the leveling head.

The leveling head supports the instrument upon the tripod and provides the means for leveling the instrument. Typically, four leveling screws are provided. The bottom of the leveling head is a collar threaded to fit the tripod head.

In order to *measure a horizontal angle* with a transit, the following procedure is used: (1) the instrument is centered and leveled over the station to be occupied; (2) the lower clamp is loosened, and the *A* vernier set to zero using the upper clamp and tangent screw; (3) using the lower clamp and tangent screw the vertical cross hair is set on the backsight station; (4) using the upper clamp and tangent screw the vertical cross hair is set on the foresight station; (5) the horizontal angle is read from the *A* vernier.

In order to obtain a more refined reading of the horizontal angle, the angle can be repeated. This is accomplished by leaving the upper and lower plates clamped together, turning back to the backsight and repeating the procedure above. This can be done any number of times, but standard practice rarely calls for more than doubling the angle. The value of the angle, in that case, is simply one half the *A* vernier reading–including 360° if the doubling requires more than a complete revolution of the instrument. For best results, the telescope should be inverted when measuring the doubled angle. This will eliminate the effects of nonadjustment of the instrument.

In order to *measure a vertical angle*, the following procedure is used: (1) the transit is centered and leveled over the station to be occupied; (2) the center horizontal cross hair is set on the point being sighted by means of the vertical circle clamp and tangent screw; (3) the vertical angle is read from the vertical circle and vernier; (4) the angle is read again with the instrument in the inverted position; (5) the average of the two readings yields the vertical angle free of any index error due to the vertical circle not reading zero when the line of sight is horizontal.

In order to *prolong a straight line* with a transit the method of double centering is commonly used. Assume that points *A* and *B* define the direction of the line to be extended. The following procedure is employed: (1) the instrument is centered and leveled over point *B*; (2) a backsight is taken on point *A* with the telescope in the normal position; (3) the telescope is inverted and a temporary point C_1 is set in the direction of the line; (4) the instrument is rotated about the vertical axis and a second backsight is made to point *A*, this time with the instrument inverted; (5) the instrument is then rotated about the horizontal axis to the normal position and a second temporary point C_2 set next to point C_1; (6) the position of point C on the extension of the line AB is located midway between temporary points C_1 and C_2. If the instrument is in perfect adjustment the two temporary points will coincide. This is rarely the case.

Theodolites

A modern theodolite, an example of which is shown in Fig. 18, might be defined as a high-precision transit. There are two basic types of theodolites

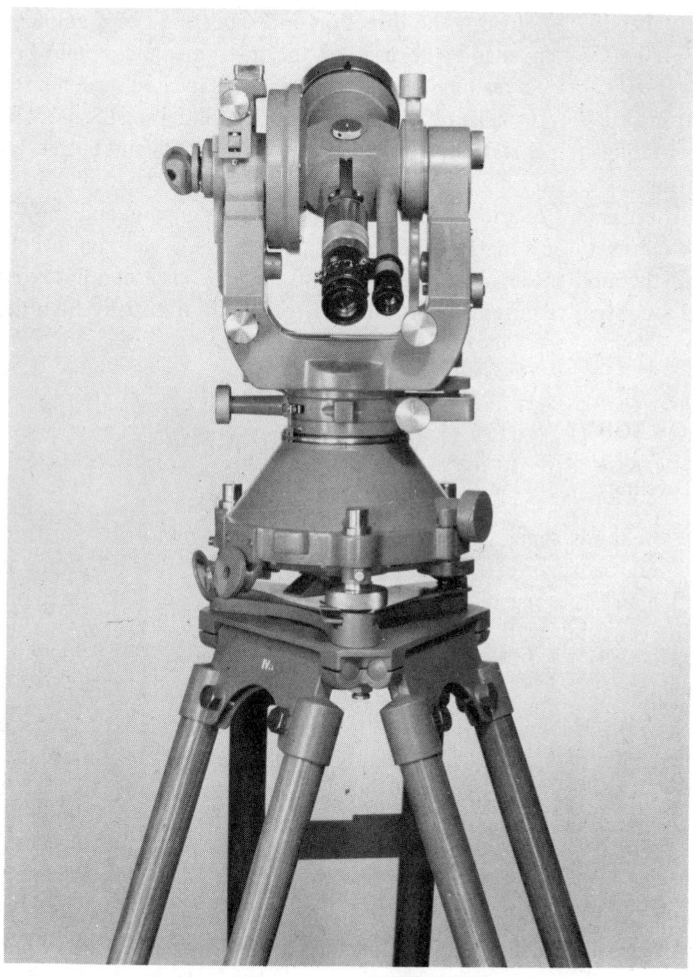

Fig. 18. The theodolite.

available. First, the *direction theodolites*, which are designed so that the horizontal circle remains fixed during a series of observations. The telescope is pointed at the desired stations in succession, and the horizontal circle is read for each pointing. Angles are calculated by the difference between adjacent pointings. The reference direction is arbitrary, although most direction theodolites are provided with a rough lower motion so that sets of observations may be made using different portions of the horizontal circle.

The second basic type of theodolite is a *repeating theodolite*, which is designed like a transit in that it has a lower motion which permits the successive accumulation of repeated angle measurements on the circle. The operating procedures are similar to repeating angles with a transit.

Modern theodolites of both the direction and repeating types employ optical reading systems which enable the user of the instrument to obtain both horizontal and vertical circle readings through an eyepiece located near the telescope. In addition, they are usually provided with optical plummet to allow more accurate centering over a station, particularly in high winds, and battery-powered illumination systems for night work.

The highest grades of theodolites are direction instruments and they are used almost exclusively for higher-order geodetic surveys. Neither direction nor repeating theodolites are particularly well suited for engineering layout work unless the control requirements of the job require an unusually high degree of accuracy.

TYPES OF SURVEYS

Profile Leveling

Profile leveling is generally performed prior to the design and construction of highways, railroads, sewers, and similar linear projects. Elevations are obtained at regular intervals along the proposed alignment. Project stakes are commonly set at intervals of 100 ft. These 100-ft intervals are termed stations and designated as 23+00, 24+00, etc. The notation 23+00 indicates that the stake is 2300 ft from the beginning of the project (0+00). However, in practice 0+00 is seldom used, in order to prevent negative stations when a change order extends the job back beyond the original beginning. If it becomes necessary in the eyes of the field-crew chief to mark a point not a full station, say 35.45 ft beyond 24+00, this intermediate point would be designated 24+35.45.

Assuming that the centerline has been located on the ground, profile levels are generally run by carrying the height of the instrument through turning points by differential levels. Foresights, which are not used as turning points, are taken upon the ground surface at full stations and all intermediate points where there is a significant change in slope or other feature which would influence design and/or construction. These foresights on the centerline are typically referred to as sideshots, since they will not be used for a backsight. In practice, the height of instrument is normally carried to one-order-of-magnitude-greater accuracy than the sideshots. For example, for most route location work, the ground profile is read to ±0.1′, while the height of instrument is carried to ±0.01′. Profile level loops should be closed and the error of closure distributed to the turning points prior to calculation of side shot elevations.

Cross-Section Leveling

Cross sections are ground profiles taken perpendicular to the centerline of route projects. They provide the basic terrain data upon which earthwork computations are based. Typical cross-section procedure carries the height of instrument by differential levels. Cross sections are taken at each station along the centerline and at all significant intermediate locations between stations. Elevations at changes in slope along the cross sections are taken by side shots usually to ±0.1′. The distance left or right of the centerline to the points where elevations are obtained is measured with a cloth tape to ±0.5′. The cross sections are typically extended to the construction limits or to include any feature which would influence the design or construction. Similar cross sections may be taken on borrow pits.

Traverse

In any large engineering construction project it is necessary to have a framework of points whose horizontal and vertical positions are known. The network of points whose horizontal positions are known is termed the *horizontal control*. A traverse is a commonly used method for establishing horizontal control for engineering works. A traverse is basically a series of lines between points whose positions are to be determined. Once the length and directions of the lines have been measured, the positions of the points can be calculated. Traverses are classified in many different ways, including the accuracy of the survey, the method employed, the purpose served, and the geometric form of the net.

Traverses may be closed or open. A *closed traverse* is one which begins and ends at the same point or at points whose horizontal positions are known. This provides the advantage of allowing a check of the accuracy of the traverse by comparison of the calculated position of the end station with its known position. Any difference is termed the horizontal closure. In the adjustment of the traverse the known stations are held fixed.

An *open traverse* is a traverse which begins at a point whose horizontal position is known and terminates at a point whose position is unknown. Although the open traverse minimizes the work required to extend horizontal control into an area, this is more than offset by the fact that no means is provided for calculating the horizontal closure. Thus, it is impossible to evaluate the accuracy of the traverse.

The route followed by a traverse upon the ground will depend upon whether control is to be established for some particular project or is intended to determine the location of previously existing points. All traverses should be closed upon points whose positions are established with a greater accuracy than the traverse. Careful planning is required to ensure that the traverse data provided will meet the requirements of the project with maximum economy of men and equipment. Although most control problems arise from inadequate surveys, it is sometimes the case that the surveys are too precise for the project. This is a waste of men and materials. The ideal traverse will provide a sufficient number of adequately monumented, properly described, well positioned control points located with an accuracy consistent with the project needs.

Each point at which an angular change in direction is measured is termed a traverse station. The requirements of the project will dictate which of the traverse stations are "permanently" monumented. Those stations which are to be monumented should be referenced to sufficient points in order that the station can be reset should it be destroyed. If the monument is inside the limits of construction the reference points should be taken outside the limits. Description of the stations should be included in the notes in sufficient detail to provide for the future use of the monument. In some cases it is desirable to provide a line of known azimuth. In such cases the direction from a station to a clearly visible feature such as a church spire should be measured and recorded, as well as a description of the mark.

The measurements required for traversing are basically lengths and angles. The lengths may be measured by taping or EDM. The angles may be measured with a transit or theodolite depending upon the accuracy required. The measured angles may be interior, angles to the right, deflection angles, or any other angles which meet the requirements. It is desirable that there be a logical relationship between the accuracy of the length measurements and angle measurements in order to obtain the most economical results. In general, if the relative error in distance equals the angular error in radians, a consistent relationship between the linear and angular measurements will result.

In any traverse operation it is desirable to obtain checks upon the accuracy of the measured lengths and angles to prevent the accumulation of errors. For a traverse closing upon the initial point the computed closure will indicate the precision of the work but will not reveal systematic distance errors. Therefore, on important traverses of this type, it is common practice to double the angles and measure the distances in both directions. On traverses connecting different horizontal control points, the closure is an excellent means of assessing the accuracy of the traverse. Nevertheless, it is common practice to include astronomic azimuth checks at regular intervals on traverses of this type. It should be noted that if the interval between azimuth checks is large, there will be a difference between the traverse direction and the astronomic direction of a line even if the traverse angle is measured perfectly. This is due to the convergence of the meridians to the north geographic pole. The convergence may be calculated by

$$\theta = 52.13L \tan \phi$$

where

θ = the convergence in seconds
L = the east-west distance in miles
ϕ = the mean latitude

There are many grades or qualities of traverse work. It is important to establish traverse specifications which are adequate for the project but not too restrictive. Table 3 gives the specifications which are used by the federal surveying and mapping agencies and many other similar organizations. Most engineering work would fall in the third-order category.

TABLE 3. SPECIFICATIONS FOR TRAVERSE

	First order	Second order Class I	Third order Class I
Number of lines between azimuth checks	5–6	10–12	20–25
Standard error of astronomic azimuth	0.45″	0.45″	3.0″
Maximum azimuth error per station at check points	1.0″	1.5″	3.0″
Length measurement standard error	1/600,000	1/300,000	1/60,000
After azimuth adjustment position closing error not to exceed	1/100,000	1/50,000	1/10,000

Triangulation

Triangulation, like traverse, is a method of determining the horizontal position of points on the surface of the earth. It is well suited to large-area coverage because it eliminates the need to measure the lengths of all lines. Basically, a triangulation net consists of a series of triangles having common vertices. All, or at least most of the angles, are measured, along with a few sides, termed base lines. Based upon the known angles and sides, the unknown sides and therefore the location of the vertices can be calculated if known coordinates and azimuths exist in the net.

Triangulation as well as other surveys can be classified as geodetic or plane depending upon whether or not the effect of the earth's curvature can be ignored. Plane triangulation is satisfactory only over a limited area; however, most engineering triangulation falls in this category.

A basic principle of any control extension, be it horizontal or vertical, is that the basic control upon which other control is to be based should be established with a higher degree of accuracy than the subsequent systems. The reverse

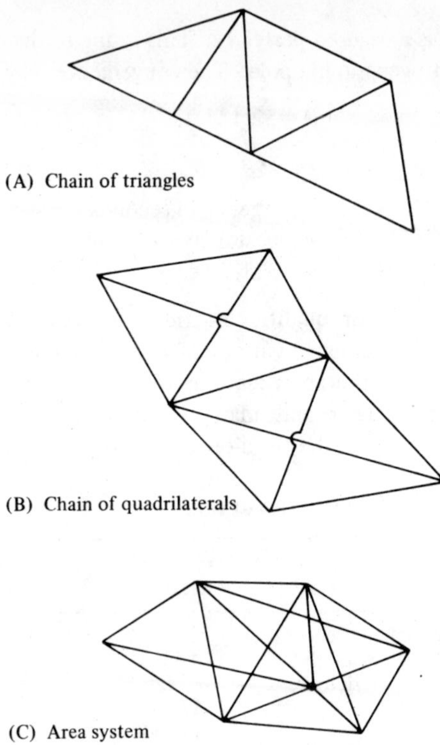

Fig. 19. Triangulation configurations.

would produce large and costly errors. Triangulation, and to some extent traverse, is used to provide this basic control in the United States.

Many different configurations, as indicated in Fig. 19, may be used in a triangulation net depending upon the nature of the requirements and the terrain. A *chain of triangles* is a simple and economic means of covering a narrow strip. A *chain of quadrilaterals* provides a very strong network, since checks are provided by the redundancy introduced. An area system might be employed where a relatively dense distribution of control is desired over a broad area. Combinations of the above systems may also be employed when the situation warrants.

The *distances between triangulation stations* may vary widely. On the primary control net for the United States, lengths of 100 miles are not uncommon. On the other hand, when triangulation is used to provide horizontal control for bridge construction, lengths of a few thousand feet are frequently used. The significant point is that the network must be designed to meet the requirements of the project.

The fundamental consideration for the *accuracy classification* of any method of control extension is the accuracy of the measured angles and distances. Triangulation classifications cover a wide range of values due to the variety of uses to which triangulation can be applied. The most commonly accepted classification for accuracy of triangulation is first, second, and third order. The specifications associated with this classification are shown in Table 4. The primary horizontal control network for the United States is first-order, class II triangulation. Second-order, class I triangulation subdivides the regions between first-order nets. Third-order triangulation is extended from higher-order stations and can be used for a wide variety of engineering projects, including mapping and construction.

In any triangulation survey, a *planning and reconnaissance phase* should precede the actual triangulation. This should be done in order to select the most satisfactory sites, considering both visibility and strength of figure. Inasmuch as the measurement of angles depends upon the *intervisibility between stations*, this must be established before the field observations begin. This can be accomplished by analyzing the effects of the intervening terrain based upon topographic maps, barometric leveling, or trigonometric leveling. The influence of earth's curvature and atmospheric refraction should be taken into account. It is frequently necessary to employ towers to provide clear lines of sight. Typically, the towers are constructed to a height of 10 feet above the minimum required height in order to avoid extreme refraction effects and to provide for uncertainties in the terrain analyses.

In the extension of triangulation the accuracy of the computed sides deteriorates progressively as the computation moves away from the base. The

TABLE 4. SPECIFICATIONS FOR TRIANGULATION

Classification	First Order	Second Order Class I	Third Order Class I
Strength of Figure R between bases, maximum	25	80	130
Base Measurement standard error	1/1,000,000	1/900,000	1/500,000
Triangle Closure			
Average not to exceed	1.0″	1.2″	3.0″
Maximum seldom to exceed	3.0″	3.0″	5.0″
Azimuth Check Spacing-figures	6–8	6–10	10–12
Closure in Length after angle and side conditions satisfied should not exceed	1/100,000	1/50,000	1/10,000

geometric configuration bears directly upon the rate of deterioration or loss of "strength." The strength of a triangle can be evaluated by the *strength of figure*. This should also be done prior to selecting the triangulation stations. The factor for comparing the strengths of alternative chains of triangles can be computed by

$$R = \frac{D - C}{D} \sum (\delta_a^2 + \delta_a \delta_b + \delta_b^2)$$

where

R = the strength of figure
D = the number of directions observed in the figure not including the known line
C = the number of conditions to be satisfied in the figure
δ_a, δ_b = respective logarithmic differences of the sines of the distance angles

The number of conditions to be satisfied in the figure may be calculated by

$$C = (n' - s' + 1) + (n - 2s + 3)$$

where

n = total number of lines
n' = number of lines observed in both directions
s = total number of stations
s' = total number of stations occupied

Tables may be found in most surveying texts which tabulate $(\delta_a^2 + \delta_a \delta_b + \delta_b^2)$ for combinations of distance angles, and list $(D - C)/D$ for typical triangulation configurations.

The lower the value of R, the stronger is the triangulation net. Table 4 indicates maximum allowable strengths of figure for the various orders of triangulation.

Triangulation angles are commonly measured with direction theodolites. A variety of such instruments are available. Following is a typical angle-observing procedure: (1) the theodolite is centered and leveled over the station; (2) one line is selected as the initial or reference direction; (3) the horizontal circle is set approximately for the first "position;" (4) the observer makes a pointing on the reference station, records the direction, and proceeds in a clockwise direction, taking pointings and readings upon all stations to be sighted from the occupied stations; (5) the telescope is inverted and the sequence is repeated. The procedure constitutes one "position." Several "positions" are normally taken at each station, each beginning with an initial reading on the reference line at a different setting upon the horizontal circle. This ensures that all parts of the circle will be used in the determination of the angles. The number of positions turned will vary with the accuracy requirements of the triangulation.

Triangulation bases may be measured either with tapes or by electronic instruments. In either case, it is essential that the methods employed be consistent with the accuracy requirements of the triangulation. Typically this means that field procedures must be designed to minimize accidental error and corrections must be applied to remove all effects of systematic errors.

If electronic distance-measuring systems are available, *trilateration* may be employed to extend horizontal control. Trilateration is similar to triangulation except that rather than measuring angles, all the distances are measured. With the known distances, the angles can be calculated based upon the law of cosines. Trilateration, like triangulation, requires a reference direction and beginning coordinates if the positions of unknown stations are to be determined.

In addition to triangulation and trilateration, it is possible to extend horizontal control by a combination of the two. This could take the form of measuring all the angles and distances or any desired combination. This combined procedure would provide a very strong network, although it would typically be more time-consuming.

SURVEY COMPUTATIONS

Survey computations are the necessary mathematical manipulations required to convert the raw field data into such form as may be required. The digital computer is ideally suited for most survey computations because it can handle the large quantities of data typically generated by a field survey.

Traverse Computations

Plane-survey computations are generally based upon a *rectangular coordinate system.* The Y axis is assigned the North-South direction, and the X axis assigned East-West. The origin is typically placed to the South and West in order to avoid negative values. The Y coordinates of a point are termed North coordinates and the X coordinates are termed East coordinates.

The *forward and inverse problems* are basic to survey computations. These are illustrated in Fig. 20. In the forward problem the coordinates of point A, the bearing of AB, and the length of AB are known and it is desired to calculate the coordinates of point B. These may be calculated from

$$N_b = N_a - s \cos \alpha$$

$$E_b = E_a + s \sin \alpha$$

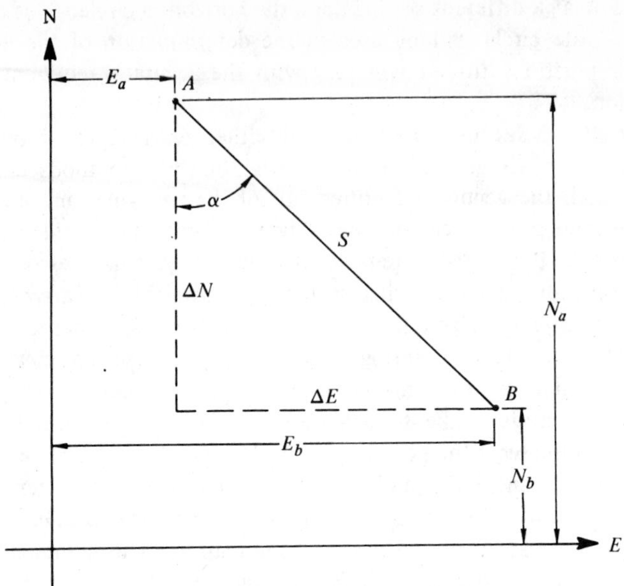

Fig. 20. The forward and inverse problem.

The inverse problem is to determine the length and bearing of *AB* when the coordinates are known. The solution is

$$\tan \alpha = \Delta E/\Delta N$$

$$s = \Delta E/\sin \alpha$$

$$s = \Delta N/\cos \alpha$$

The *latitude of a line* is defined as the projection of the line on the reference meridian. The *departure of a line* is defined as the projection of the line on an East-West line perpendicular to the reference meridian. Latitudes and departures of a line are calculated by

$$\text{Latitude} = \text{length} \times \text{cosine of bearing angle}$$

$$\text{Departure} = \text{length} \times \text{sine of bearing angle}$$

Latitudes are taken as positive when the bearing is northerly and negative when the bearing is southerly. Departures are positive when the bearing is easterly and negative when the bearing is westerly.

In any traverse which is closed upon the initial station, the algebraic sum of the latitudes and that of the departures should be zero. In the case of a traverse which is run between two stations whose positions are known, the algebraic sum of the latitudes and that of the departures should equal the latitude and departure of the line connecting the known stations. However, because of errors in measurement, this will rarely, if ever, occur. There will generally be a small residual in both latitude and departure. The linear error of closure may be calculated as follows:

$$\text{error of closure} = \sqrt{(\Sigma \text{ latitudes})^2 + (\Sigma \text{ departures})^2}$$

The *precision of the traverse* may then be calculated:

$$\text{precision of traverse} = \frac{\text{error of closure}}{\text{total length of traverse}}$$

If the precision of the traverse is not within acceptable limits, the measurements must be checked in the field.

If the precision is within acceptable limits the traverse is then adjusted. *Traverse adjustment* refers to the distribution of the error of closure over the latitudes and departures to bring their sums to zero. One method of accomplishing this is *adjustment by the compass rule.* The compass rule states that the correction to the latitude (or departure) of a line is to the total error in the latitudes (or departures) as the length of the line is to the total length of the traverse. Other methods, such as the transit rule and least squares, also may be used to adjust a traverse. Once the traverse is adjusted the coordinates of the unknown points may be calculated by

$$N_b = N_a + \text{adjusted latitude of } AB$$

$$E_b = E_a + \text{adjusted departure of } AB$$

where

N_b, E_b = coordinates of an unknown station
N_a, E_a = coordinates of a known station

Area Computation

It is frequently desirable to calculate the *area enclosed by a traverse.* The procedure for any closed figure bounded by straight lines can be expressed as follows:

> The area is equal to one-half the algebraic sum of the products of each ordinate multiplied by the difference between the two adjacent abscissas, always subtracting the preceeding from the following abscissa.

The area of a five-sided closed traverse may therefore be calculated by

$$A = \tfrac{1}{2}\,[N_1(E_2 - E_5) + N_2(E_3 - E_1) + N_3(E_4 - E_2) + N_4(E_5 - E_3) + N_5(E_1 - E_4)]$$

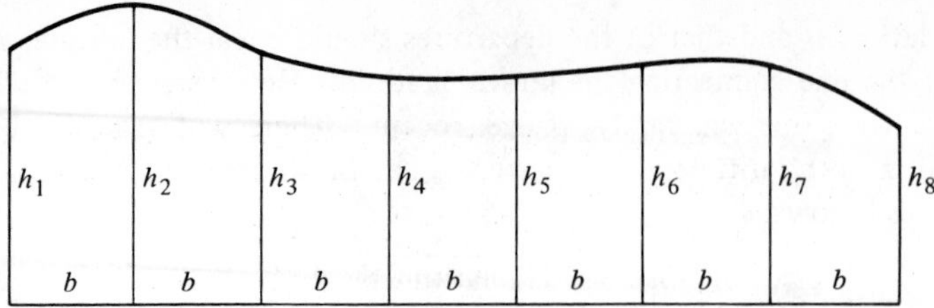

Fig. 21. Area of irregular parcel.

Areas which include an *irregular or curved boundary* are usually determined by offsets at regular intervals from a convenient base line; as illustrated in Fig. 21. The figure is approximated by a series of trapezoids which yields the equation:

$$A = b\,[(h_1 + h_n)/2 + (h_2 + h_3 + \cdots + h_{n-1})]$$

where

$$A = \text{area of irregular parcel}$$
$$b = \text{interval between offsets}$$
$$h_1, \ldots, h_n = \text{length of offsets}$$

Obviously, the approximation becomes more accurate when the interval between offsets is reduced.

Volume Computations

In many engineering projects, particularly those involving earthwork, volume computations are required. The most commonly used procedure is the *average-end-area method.* This is typically used in conjunction with a cross-section survey. Volumes are calculated by the equation:

$$V = \frac{L}{27}\,\frac{(A_1 + A_2)}{2}$$

where

$$V = \text{the volume, in cubic yards}$$
$$L = \text{the distance between end sections, in feet}$$
$$A_1 \text{ and } A_2 = \text{the area of the end sections, in square feet}$$

The average-end-area method is not exact, but if the sections are taken at the correct location by the field party, the volumes provided are typically accurate to ±2%. In cases where higher accuracy is required, say in rock excavation, the *prismoidal formula* is often employed. The prismoidal equation takes the form

$$V = \frac{L}{27}\,\frac{(A_1 + 4A_m + A_2)}{6}$$

where

V = the volume, in cubic yards
L = the distance between end sections, in feet
A_1 and A_2 = the areas of the end sections, in square feet
A_m = the area of the midsection based upon the average dimensions of A_1 and A_2

State Plane-Coordinate System

For most engineering surveys, the area covered is small enough that the effect of earth's curvature can be neglected. However, for large area projects this assumption is no longer justified. A State Plane-Coordinate System, developed by the United States Coast and Geodetic Survey, provides a common datum for horizontal control for all surveys in a large area. It provides a system for handling geodetic problems with plane rectangular coordinates.

The mean sea-level surface is closely approximated by a mathematical surface of revolution, the spheroid. In order to convert geodetic coordinates in terms of latitude and longitude to plane rectangular coordinates, it is necessary to mathematically project points from the spheroid to some plane surface. In order that the surface be plane, it must be "developable," that is, it must be a solid surface which can be "unrolled" on a plane surface without distortion. If the surface is developed, a rectangular grid can be superimposed and the positions of points on the plane located with respect to rectangular axes.

Two basic projections are used in the State Plane Coordinate System: the Lambert conformal conic projection and the transverse Mercator projection. The former is used for states having their greatest dimension in the East–West direction, while the latter is used for states having their greatest dimension in the North–South direction. The Lambert projection employs a cone and the Mercator projection employs a cylinder as their developable surfaces, as shown in Fig. 22. When the surfaces shown are developed, they have the characteristics illustrated in Fig. 23. Note that the scale of the projections varies over the surface of the projection. When computing the projection of points they are projected mathematically from a point near the center of the earth along radial lines to the developable surface. In areas where the surface falls inside the spheroid, the scale will be less than the true spheroid scale; if the surface falls outside the spheroid the scale will be greater than the true spheroid scale. It is impossible to project points from a spheroid to a developable surface without introducing some distortion. However, if the system is designed correctly, these distortions can be held to a negligible minimum.

The *Lambert conformal conic projection* is used in 31 of the 50 States. It is conformal in that true angular relationships are retained about all points. Inasmuch as its scale varies from North to South, it is suited for states having

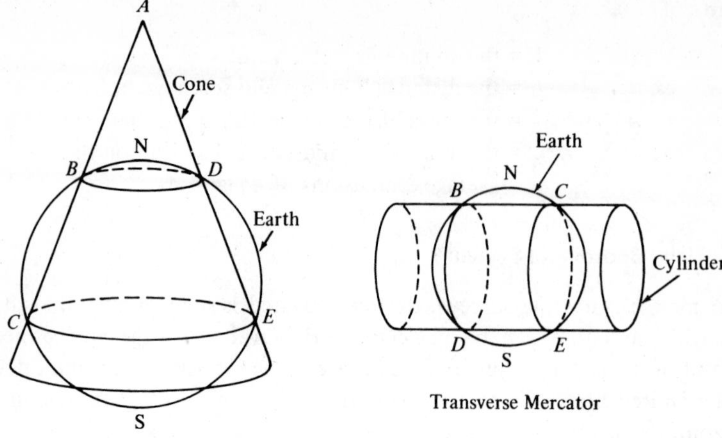

Fig. 22. State plane coordinate projections.

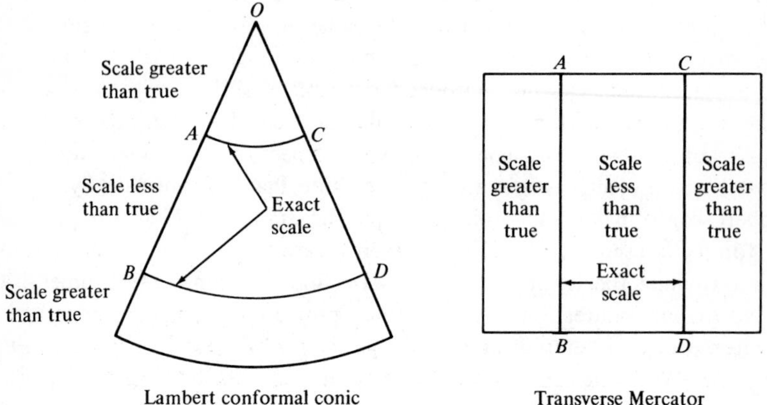

Fig. 23. Developed State plane surfaces

their maximum dimension in the East-West direction. The Lambert projection is illustrated in Fig. 24. The cone is selected so that it intersects the spheroid along two parallels of latitude at one sixth of the North-South zone width. These parallels are termed "standard parallels." On the developed surface, all meridians are straight lines converging at A, the apex of the cone. All parallels of latitude are segments of concentric circles whose centers are at the apex of the cone. The projection is centered in the East-West direction by selecting a *central meridian* whose longitude is close to the middle of the zone. Grid North

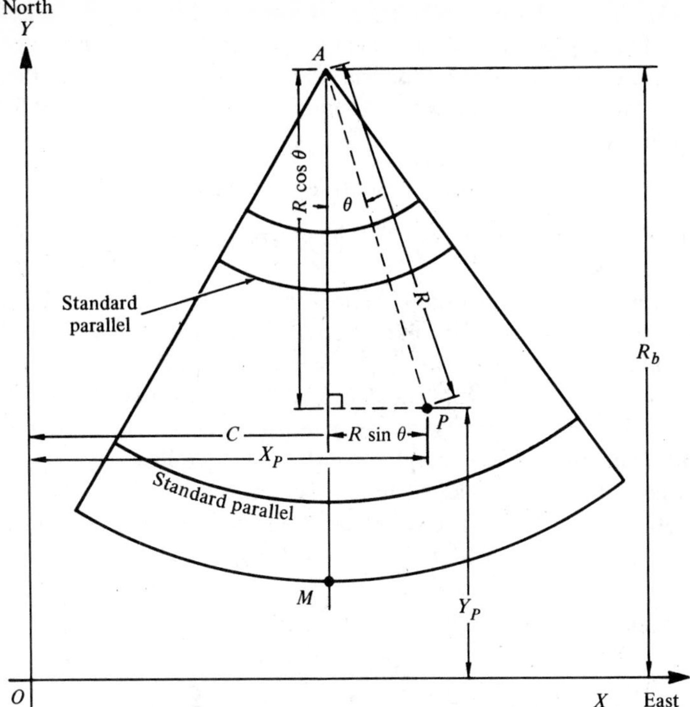

Fig. 24. The Lambert conformal conic projection.

is established by the direction of the central meridian. Any line parallel to the central meridian is considered a grid North–South line. Thus, with the exception of the central meridian, grid North and true North do not coincide.

If the latitude and longitude of any point P are known, the state plane coordinates on the Lambert projection may be calculated. With reference to Fig. 24, the coordinate equations are

$$X_p = R \sin \theta + C$$

$$Y_p = R_b - R \cos \theta$$

where

X_p, Y_p = the state plane coordinates of P
R = the distance on the meridian through P from the apex of the cone to P
θ = the angle between the central meridian and the meridian through P, positive when P is east of central meridian
C = the constant distance between the Y axis and the central meridian
R_b = the constant distance between the apex of the cone and the X axis

The USCGS has published tables by states to be used in the solution of these equations. For any point P whose geodetic position is known, values of R, θ, C and R_b are given. For lines less than five miles in length, the grid azimuth may be calculated from the geodetic azimuth with sufficient accuracy for most surveys by the equation

$$\text{Grid Azimuth} = \text{Geodetic Azimuth} - \theta$$

Note that θ is positive when P is east of the central meridian, and negative when P is west of central meridian.

The *transverse Mercator projection* is used in 22 States. It is a conformal projection and since its scale varies from East to West, it is used in states having their maximum dimension in the North-South direction. The Mercator projection is illustrated in Fig. 25. The axis of the cylinder lies in the plane of the equator. The cylinder intersects the spheroid along two circles (*BD* and *CE*) equidistant from the central meridian (*AM*). All meridians and parallels are curves, as indicated by the dashed lines. The central meridian establishes the direction of grid North. The Y axis is established parallel to the central meridian and the X axis perpendicular to it.

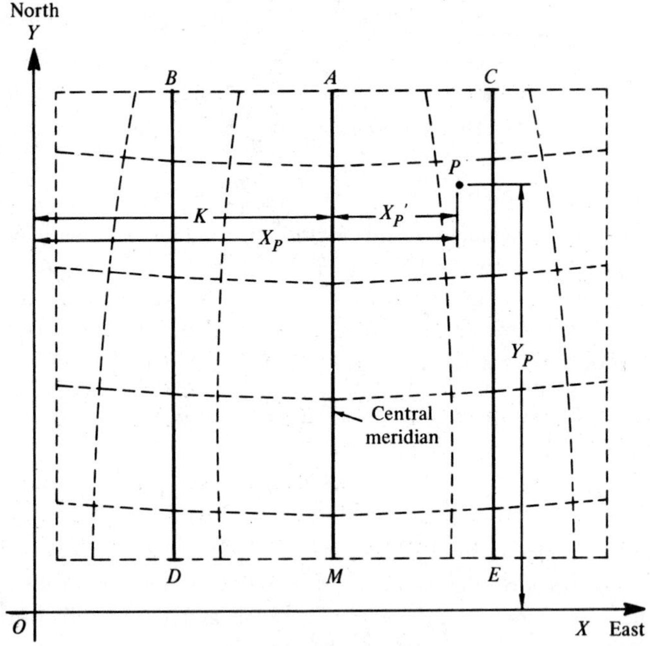

Fig. 25. The transverse Mercator projection.

If the geodetic position of any point P is known, the state plane coordinates on the transverse Mercator projection may be calculated. With reference to Fig. 25, the equations are

$$X_p = H \times \Delta\lambda'' \pm ab + K$$

$$Y_p = Y_0 + V\left(\frac{\Delta\lambda''}{100}\right)^2 \pm c$$

where

X_p, Y_p = the State plane coordinates of P

$\Delta\lambda''$ = the difference in longitude between the central meridian and point P in seconds, positive when P is east of the central meridian

K = the constant distance between the central meridian and the Y axis

H, a, Y_0, V are tabulated in USCGS tables vs. latitude of P, and b, c are tabulated in USCGS tables vs. $\Delta\lambda''$

For lines less than five miles in length, the grid azimuth may be calculated from the geodetic azimuth with sufficient accuracy for most surveys by the equation

$$\text{Grid Azimuth} = \text{Geodetic Azimuth} - \Delta\alpha''$$

where

$$\Delta\alpha'' = \Delta\lambda'' \sin(\text{Lat.}P) + g$$

where $\Delta\lambda''$ is as defined above, Lat. P is the latitude of P, and g is tabulated in USCGS tables vs. $\Delta\lambda''$.

When placing any survey on the State Plane Coordinate System, normal procedure calls for beginning and ending on stations whose state plane coordinates are known or can be computed from the known geodetic position. If this is done, all resulting azimuths will be grid azimuths, and corrections for convergence of meridians will not be required.

Measured lengths of lines must be reduced to lengths at their respective positions on the State coordinate grid. This is most easily accomplished by reducing the measured length to sea level and then reducing the sea-level length to grid. The following equation may be used to reduce the measured length to sea level:

$$L_s = L_m\left(\frac{R_e}{R_e + h}\right)$$

where

L_s = the sea-level length
L_m = the measured length
R_e = the mean radius of the earth (20,906,000 ft)
h = the average elevation of the line above mean sea level

The sea-level length is then multiplied by a *scale factor* which is obtained from the projection tables for the area in which the line falls.

REMOTE SENSING

Remote sensing has been defined as the process by which the characteristics of an object may be studied without contact with the object. Of particular value to the engineer are remote sensing systems which provide environmental data from an aircraft or space vantage point. In the past this was accomplished by the systematic interpretation and analysis of photographic emulsions sensitive only to visible light. At the present time emulsions are available which extend the photographic capability beyond the visible into the near infrared portion of the electromagnetic spectrum. In addition, systems are available which detect and record electromagnetic radiation well beyond the range covered by photography. It is fortunate that this technology has developed during the time when man has become increasingly aware of the need to preserve and protect the environment. Remote sensing provides a practical means of frequent and accurate monitoring of large areas in relatively short periods of time. Thus, remote sensing can provide a data base which has not been available in the past.

Basic Principles

Almost all remote sensing systems operate on the principal of measuring and recording the *electromagnetic energy* emitted or reflected from the object under study. All electromagnetic energy travels at the speed of light in an approximate sinusoidal pattern. Visible light is electromagnetic energy in a narrow range of wavelengths to which the human eye is sensitive. The following equation shows the relationship between frequency, wavelength, and velocity of electromagnetic energy:

$$c = f\lambda$$

where

c = velocity of electromagnetic energy in a vacuum
f = the frequency of the energy
λ = the wavelength of the energy

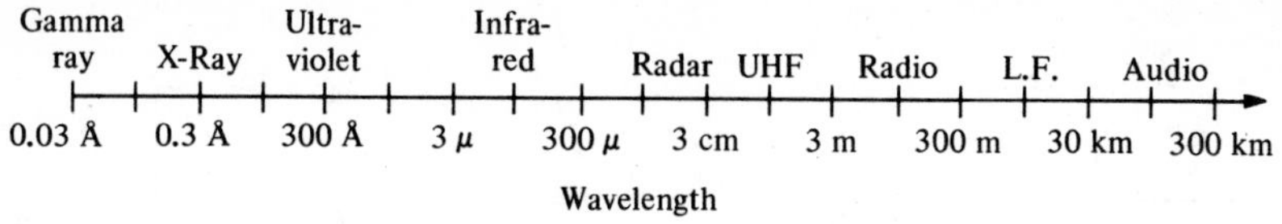

Fig. 26. Wavelength classification of electromagnetic energy.

Electromagnetic energy is potentially available at all wavelengths. The entire range of electromagnetic energy according to wavelength is termed the *electromagnetic spectrum* and is illustrated in Fig. 26.

The basic property of electromagnetic energy which makes it suitable for remote sensing is that the nature of its interaction with matter is a function of (1) the physical-chemical structure of the matter, and (2) the wavelength of the energy. Further, the energy does not interact with itself or other wavelengths of electromagnetic energy. Fundamentally then, if the wavelength of the energy is known, the characteristics of the object can be determined by the nature of the interaction.

When radiant energy strikes an object it can (1) be reflected, (2) be absorbed, (3) be transmitted, or (4) undergo some combination of the above. The amount of energy in each reaction is a function only of the wavelength of the energy and the nature of the object. Because of the wavelength dependence on the interaction, remote sensing systems have been developed which respond to a wide range of wavelengths. Thus, if an object has a distinguishing characteristic interaction at a particular wavelength, the remote sensing system which operates at that wavelength would be used to detect and monitor the object.

The Remote Sensing Sequence

In order to fully understand an operational remote sensing system, it is best to consider an idealized or "perfect" system. Figure 27 illustrates a "perfect" remote sensing system and the sequence of events necessary to obtain useful data. It should be emphasized that such a system does not exist. The sequence can be outlined as follows: (1) a uniform source produces electromagnetic energy of known uniform intensity at all wavelengths; (2) the energy propagates through the atmosphere to the object without loss; (3) the energy selectively interacts with a homogeneous object depending upon wavelength and the physical-chemical nature of the object; (4) the returned signal, consisting of reflected and emitted energy, propagates through the atmosphere without loss; (5) the sensor detects and records the returned signal over all wavelengths without distortion; (6) the data is processed in real time into a desirable format; (7) the processed data is provided to a knowledgeable user who can correctly apply it to the solution of a problem.

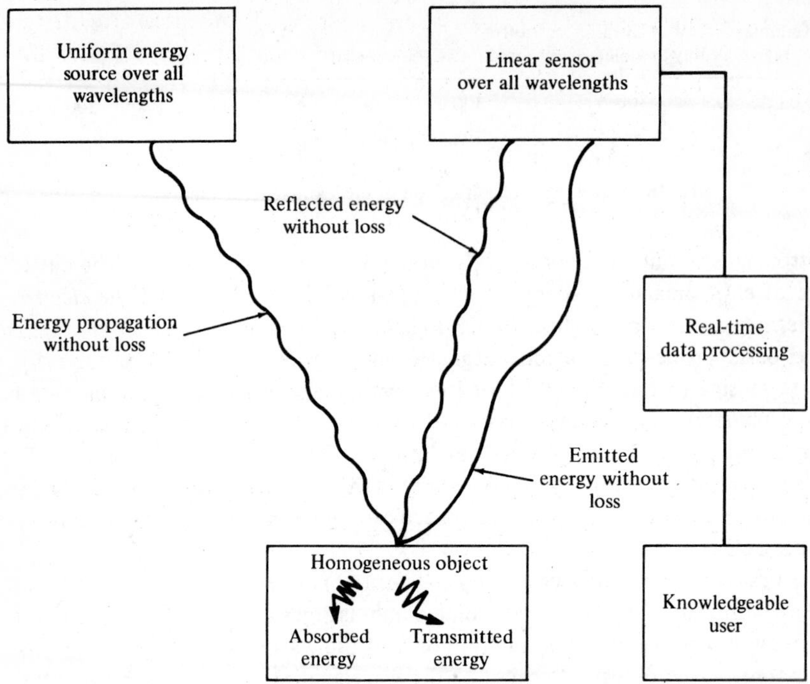

Fig. 27. Ideal remote sensing sequence.

The successful application of remote sensing technology depends upon an understanding of the differences between each of the steps in an ideal sequence and the corresponding steps in an operational system. The steps in an operational system are, of course, modified by the constraints of nature and the limitations of technology.

Source Characteristics

All matter at a temperature above absolute zero is a source of electromagnetic radiation. A theoretical *black body* is frequently used as a model source. A black body may be simply defined as an object which will absorb and reemit all energy incident upon it. A black body radiates energy according to the *Stefan–Boltzmann law*;

$$W_B = \sigma T^4$$

where

W_B = the black-body spectral radiant emittance per unit
σ = proportionality constant
T = the temperature, in degrees Kelvin

The energy radiated by a black body covers a wide range of wavelengths. However, the wavelength at which the maximum energy is radiated may be calculated by the *Wein displacement law* as follows:

$$\lambda_m = k/T$$

where

λ_m = the wavelength at which maximum energy is radiated
k = constant
T = temperature, in degrees Kelvin

A black body at a temperature of 6000°K is a close approximation of the sun as a source of energy. The maximum energy is radiated in the visible portion of the spectrum. The earth is closely approximated by a black body at 300°K, which peaks at a wavelength of approximately 10 μ (microns), which is far beyond the visible portion of the spectrum.

Since no source is a perfect radiator or black body, it is important to understand the relationship between a real-body and a black-body radiator. The radiant-energy emittance for a real body is given by

$$W = \epsilon \sigma T^4$$

where

W = real-body radiant emittance
ϵ = the emissivity factor of the source
σ = the proportionality constant
T = the temperature, in degrees Kelvin

The emissivity is defined as the ratio of the emittance of a real body to the emittance of a black body at the same temperature. Typical emissivity factors are: water, 0.99; sand, 0.96; dry soil, 0.90; and polished silver, 0.02.

Interactions at the Object

As shown in Fig. 27, in an ideal system electromagnetic energy can interact with the object in four basic ways. These are: (1) reflection, in which the energy returns to the atmosphere fundamentally unchanged; (2) absorption, in which the energy is converted to some other form within the object, typically heat; (3) transmission, in which the energy propagates through the object; and (4) emission, in which the absorbed energy is reemitted, usually at a longer wavelength. The proportion of the energy in each of these categories is a function of the wavelength (or lengths) of the incident energy and the atomic, molecular and crystalline structure of the object. The intensity of the energy returned from the object to the atmosphere at various wavelengths defines a theoretically unique *spectral signature*. The differences among the spectral signatures of objects is the primary means by which remote sensing systems distinguish different objects.

Atmospheric Effects

Inasmuch as the electromagnetic energy must typically pass through a portion of the atmosphere twice in a remote sensing system, the effect of the atmosphere upon the energy must be considered. Since the atmosphere contains matter in the form of particles and gases, it will interact with the electromagnetic energy in a manner similar to objects on the surface of the earth. If it is desired to study the atmosphere, it is these interreactions that are the primary interest. However, since most engineering projects are located at or near the surface of the earth, engineers generally desire to look through the atmosphere. The extent to which the atmosphere blocks electromagnetic energy can be seen in Fig. 28, which is a plot of atmospheric transmission vs. wavelength. Atmospheric "windows," through which the energy passes with relatively little loss, can be seen to exist in various wavelength bands, e.g., the visible portion of the spectrum. It is necessary, therefore, in designing a remote sensing system intended to monitor the earth's surface, to select a source and sensor which operate in a wavelength band compatible with an atmospheric window.

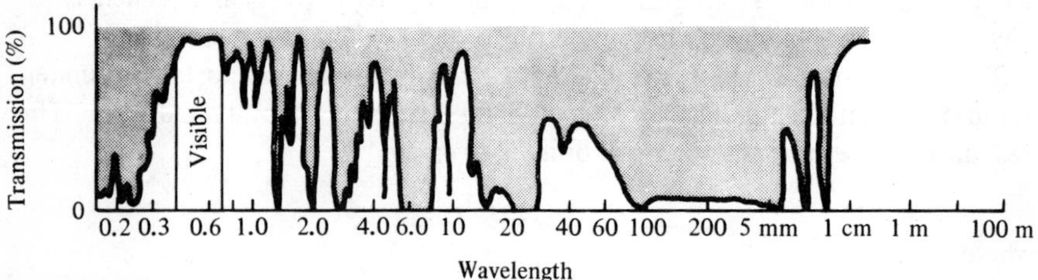

Fig. 28. Atmospheric transmission.

Operational Remote Sensing Systems

In the design of an operational remote sensing system, the nature of electromagnetic radiation, the source characteristics, the atmospheric effects, and the interactions at the object must all be taken into account. Therefore, it should be apparent that a variety of systems are required to meet the many combinations of parameters. The following discussion will present the basic characteristics of the currently available remote sensing systems. The systems may be considered in two basic categories: active systems and passive systems. An *active system* is a system which carries its own source of energy, e.g., radar. A *passive system* is one which depends upon some outside source of energy, e.g., photography. Figure 29 shows the basic systems currently available and indicates the spectral ranges in which they operate.

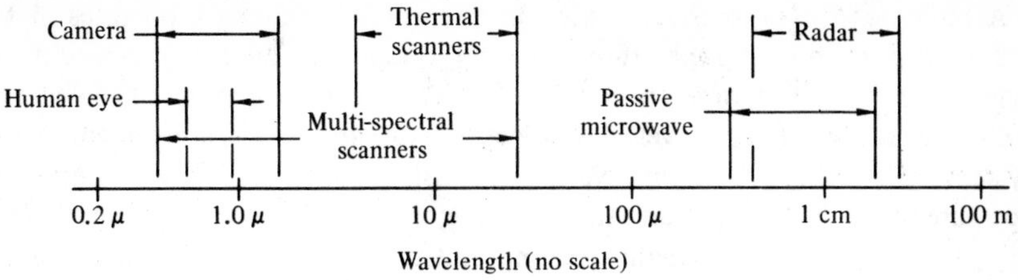

Fig. 29. Current remote sensing systems.

In the use of any of the remote sensing systems, the validity of the data obtained is a function of the strength of the associated "ground truth." Ground truth refers to any surface parameters that are used to calibrate, interpret, and evaluate remote sensed data.

Photography is the most commonly used remote sensing system. It is generally a passive system employing the sun as a source of energy. As shown in Fig. 29, the photographic portion of the spectrum extends from about 0.3 to 1.2 μ. Multispectral photographic systems employ from two to nine cameras mounted in a bank. The system is designed to trigger all the cameras in the bank simultaneously, thus producing multiple exposures of the same scene. Each camera in the bank is provided with a different film–filter combination. These are selected in such a manner that each camera records a specific wavelength band or color. The response differences obtained from simultaneous multispectral photography must be due to variations in the interactions at the object, since the atmospheric and target conditions would be identical for each of the exposures. Multispectral photography has many practical applications. For example, it has been used by agriculturalists for crop surveillance, by foresters for species identification, and by engineers for urban growth studies and wastewater plume identification.

Quantitative analysis of multispectral photography may be obtained by measuring the emulsion density variations with a densitometer. The resulting quantified variations can be used as an index to significant ground parameters such as vegetation type and vigor, soil type and moisture content, and effluent type and concentration. When photographic densities are used for quantitative spectral analysis, the effects of all parameters which influence the photographic density variations must be taken into account.

Photography is a relatively economical, simple sensing system and is currently the most frequently employed. It provides excellent resolution and geometric fidelity. However, such systems have the following limitations, which render them useless for some projects: (1) the spectral range is limited to the photographic portion of the electromagnetic spectrum; (2) field operations are limited to daylight hours during favorable weather; (3) the recording format consists of images which do not lend themselves to electronic processing.

A *radiometer* is a passive, nonimaging sensor. It generally operates in the 0.3–14 μ portion of the spectrum, although the specific range may vary with the type of project. Radiometers consist of three basic components: (1) an energy-collecting system; (2) a wavelength-selection system; and (3) a signal-converting system. The energy-collecting system establishes the field of view and effective aperture of the radiometer, and transmits the collected energy to the wavelength-selecting unit. The wavelength-selecting unit breaks down the incoming energy into discrete spectral bands by means of diffraction gratings and/or interference techniques, depending upon the wavelengths to be selected. Energy from the selection unit is used to activate a sensor in the signal-converting unit. The sensor, phototubes, or heat-sensitive transducers convert the energy to an electrical signal, which is typically amplified and recorded.

Of particular concern to engineers is the airborne thermal radiometer, which produces a profile of thermal infrared emittance in the 8–14 μ portion of the spectrum. Figure 30 illustrates the basic characteristics of a thermal radiometer. An aircraft carrying a thermal radiometer flies over an area as shown in Fig. 30a. The record of the signal generated by the radiometer as it passes over the area is shown in Fig. 30b. This signal is a profile of the apparent temperature, which can be converted to temperature if the emissivities of the objects are known and atmospheric effects are correctly taken into account. If the data provided by

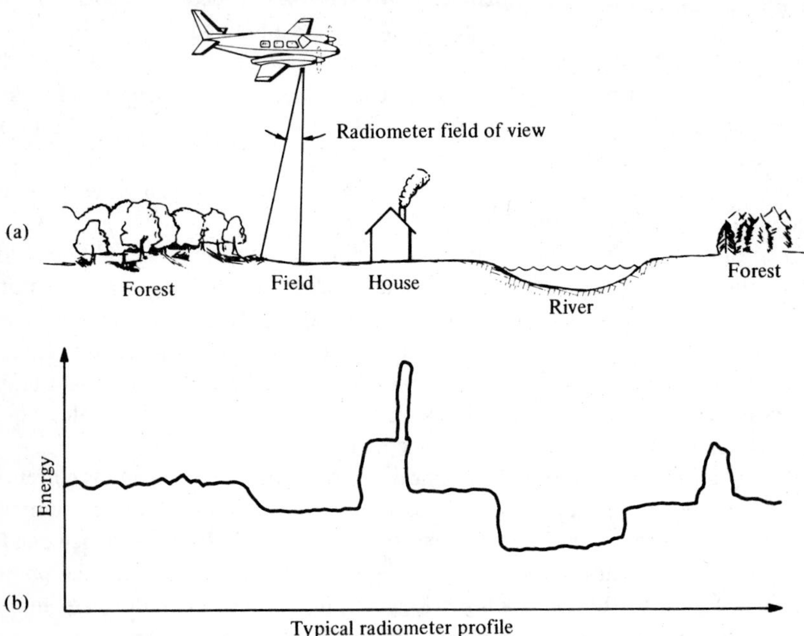

Fig. 30. Radiometer. (a) Field of view. (b) Typical profile.

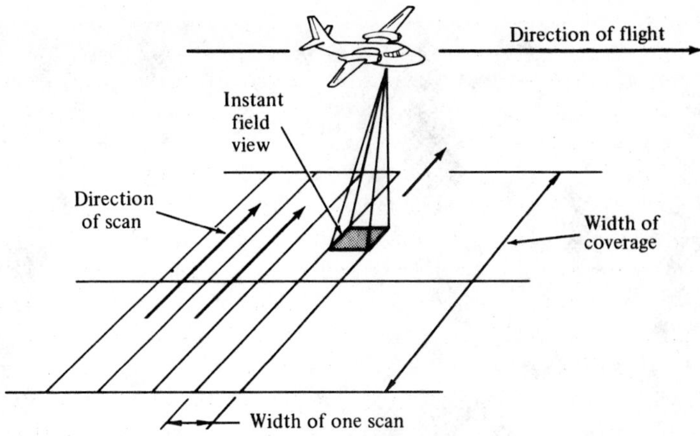

Fig. 31. Basic scanner operation.

thermal radiometers is properly reduced and related to ground-truth temperatures, accuracies to within 1°C are possible.

A *scanner* is a passive, imaging sensor. Conceptually, it is a radiometer which has been provided with a scanning mechanism. The scans, as illustrated in Fig. 31, are taken perpendicular to the direction of the flight at a rate proportional to the velocity of the aircraft in order to produce a series of contiguous electrical response profiles.

A simplified collection and recording sequence typical of most single-channel scanners is shown in Fig. 32. In-flight monitoring can be accomplished by displaying the amplitude of the signal vs. time on a cathode-ray oscilloscope. An image can be produced in flight by intensity-modulating a glow tube with the signal and recording the modulation photographically with a strip camera. The

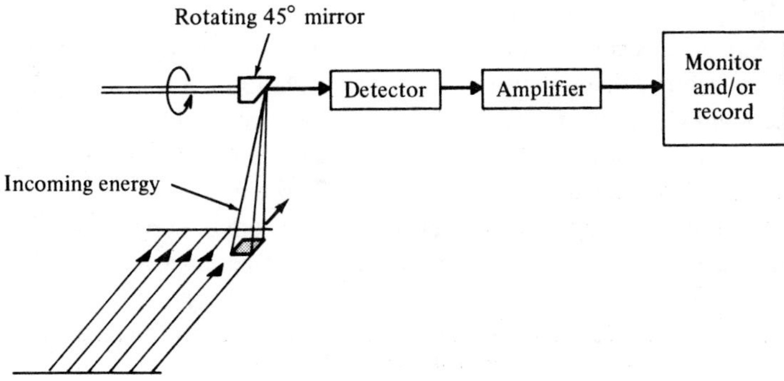

Fig. 32. Scanner collecting and recording.

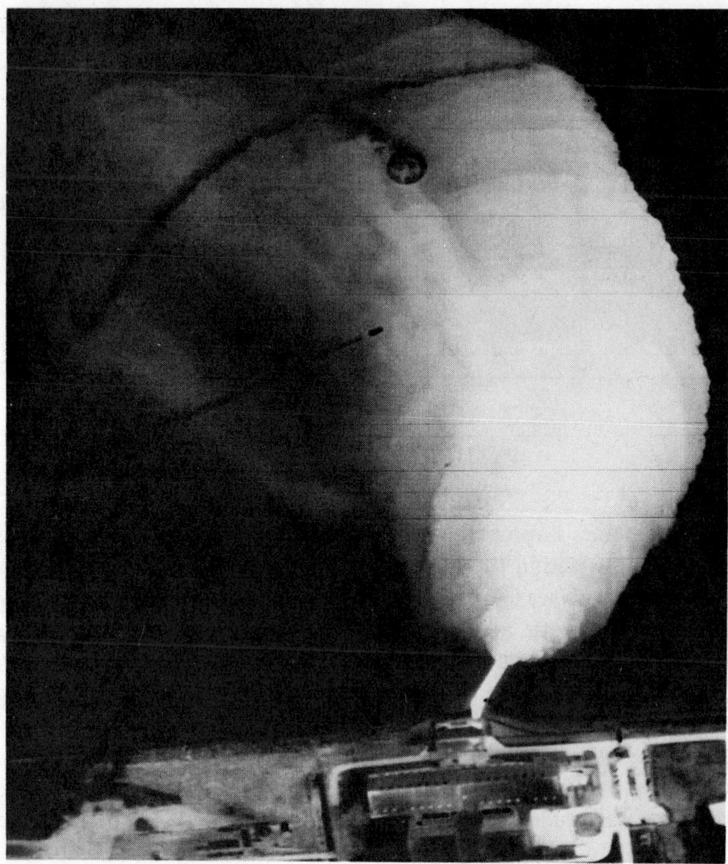

Fig. 33. Thermal image of cooling water plume.

resulting imagery will have tonal variations which are a function of the energy received. Since the signal is in an electrical format, it is frequently recorded on tape for later ground manipulation. Such manipulation may include geometric rectification and image enhancement using electronic computers.

The most common single-channel scanners are the thermal scanners, which operate in either the 3-5 or 8-14 μ windows. The output of a thermal scanner depicts the variation of the emitted thermal energy from the area covered by the scan. Figure 33 is a thermal image of a power-plant cooling-water plume being discharged into a lake. This imagery can be analyzed in conjunction with ground data to study the structure of the plume and its effect upon the receiving water body.

A multispectral scanner is basically a combination of single-channel scanners. Figure 34 illustrates the basic operation of a multispectral scanner. The multi-

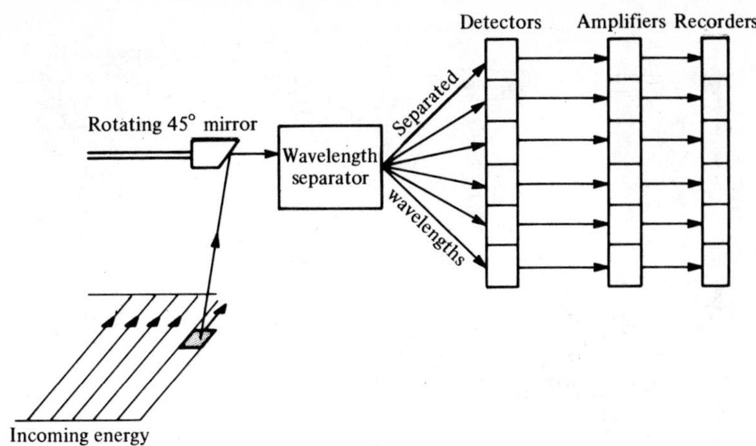

Fig. 34. Multispectral scanner.

spectral scanner employs multiple detectors and senses energy in multiple wavelength bands. The imagery or recorded signals at each band show the identical scene viewed at different wavelengths. Atmospheric effects, other than those which are wavelength-dependent, and variations of object conditions with time, are eliminated from the response differences. Therefore, the response differences are only, in theory, a function of the nature of the object. By the application of computer algorithms, multispectral scanner data can be used to rapidly discriminate between various object types. The accuracy of crop identification by such techniques has been shown to be comparable to the accuracies obtained by trained photo interpreters.

Scanners offer the following advantages relative to photography in the acquisition of environmental data: (1) they may be operated day or night at many wavelengths; (2) they provide an output which is compatible with computer analysis; and (3) they provide the capability of imaging energy beyond the photographic portion of the section. On the other hand, scanners, at the present time, have some distinct limitations: (1) the resolution is significantly less than that of photography, and (2) the geometric distortions are significantly more complex and subject to more variation than those of photography.

There are many types of *radar systems*. The side-looking airborne radar (SLAR) system is of most interest and utility in remote sensing operations. SLAR is an active imaging sensor typically operating in the 1mm to 1m wavelength range. As with all radar systems, SLAR transmits a microwave pulse of energy and monitors the travel time for the pulse to reach the object, be reflected, and return to the receiver. In SLAR systems the pulse is transmitted in a narrow beam focused by a directional antenna system. Since the propagation velocity of the energy is known, travel times can be related to the sensor–object distance.

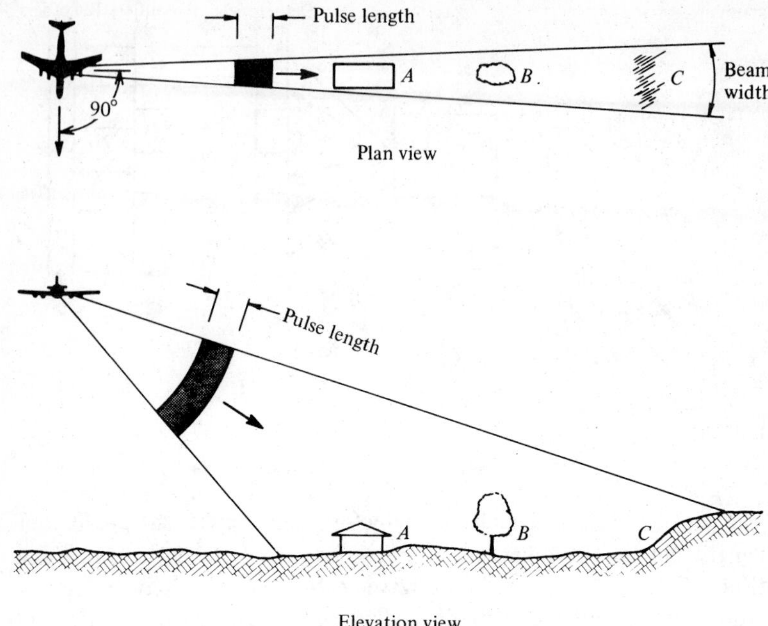

Fig. 35. Side looking airborne radar.

The basic operation of a SLAR system is illustrated in Fig. 35. The system transmits and receives over a beam centered at right angles to the direction of flight. The pulse transmitted from the aircraft initiates a sweep on a cathode ray oscilloscope (CRT). The pulse propagates to the object area, where it is reflected from the surface. The strength of the reflected component is a function of the physical characteristics of the reflecting surface. The "echoes" are detected by the SLAR and are intensity-modulated on the CRT. The echo from *A* will return before the echo from *B*, and likewise for the echo from *C*. Therefore, the CRT scan will indicate varying intensities as a function of distance and the physical nature of the objects. Each individual trace may be recorded on film by means of a strip camera. Since each traçe is associated with a pulse transmission, a continuous image of the microwave reflections from the surface is produced as the aircraft advances. The flight height and velocity of the aircraft must be controlled so that gaps or undesired overlap in the scans do not occur.

In operational systems, the returned echoes are frequently recorded on tape and later processed by computer to produce images. This permits greater resolution and geometric control of the quality of the imagery.

The imagery produced by SLAR systems has excellent clarity and good spatial orientation, as illustrated by Fig. 36. The areal coverage of SLAR strips can be

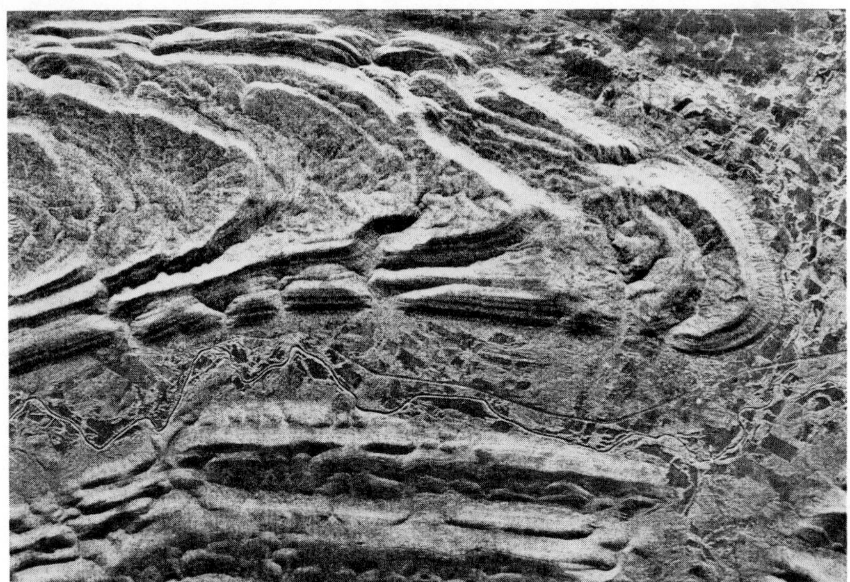

Fig. 36. Side-looking airborne radar image.

extensive. Since SLAR is an active system, it can operate day and night. Further, because of the relatively long wavelengths employed, SLAR has the ability to penetrate light cloud cover. It has been used to map areas that are perpetually overcast. The principal disadvantage of SLAR as an operational remote sensor is its relatively high cost and complex data reduction. Further, the highest-quality systems are classified and not available to the general public.

There are many other types of remote sensing systems. *Passive microwave* systems are designed to detect and record the emitted radiation from the earth in the spectral region from 1 mm to 1 m. Inasmuch as these are passive systems and the emitted radiation at these wavelengths is of very low intensity, the resolution of these systems is poor. However, because of the longer wavelengths employed, the detected signal is associated with characteristics to some depth beneath the surface. Therefore, passive microwave systems have been used to monitor soil temperature and moisture conditions.

Atmosphere gas monitors are designed to monitor the distribution and concentration of specific gases in the atmosphere. Since gases exhibit characteristic absorption spectra it is possible to compare the spectrum of a specific gas with the spectrum of the energy coming from the atmosphere beneath the aircraft. If the spectra match, the gas is present. The concentration of the gas can be determined by analyzing the relative strength of the energy absorbed by the atmosphere.

Another class of remote sensing systems monitors variations in force fields of the earth. The *magnetometer* measures variations in the strength of the earth's magnetic field. These variations can be related to subsurface structure. *Gravitymeters*, similarly, measure the variations in the earth's gravitational field. These variations likewise can be related to subsurface structure.

PHOTOGRAMMETRY

Photogrammetry is defined as the art, science, and technology of obtaining reliable measurements and qualitative data from photographs. Photogrammetry is a branch of the broad field of remote sensing. It may be subdivided into *geometric photogrammetry*, which is concerned with measurements, and *photo interpretation*, which is concerned with the recognition and identification of objects appearing on photographic images.

The photographs used in photogrammetric procedures may be classified in a variety of ways. The term *terrestrial photograph* refers to photographs taken with the camera located on the ground, frequently with a *phototheodolite* which is a combination precision camera and theodolite. The term *aerial photograph* refers to photographs taken with the camera located above the surface in an aircraft or similar vehicle. *Vertical photographs* are aerial photographs taken with the camera axis vertical or as nearly vertical as possible. *Oblique photographs* are aerial photographs taken with the axis of the camera intentionally inclined at some angle from the vertical. A *high oblique* refers to an oblique photograph in which the horizon appears. The horizon does not appear in a *low oblique.*

Aerial Cameras

The basic tool of aerial photogrammetry is the camera. An aerial camera, commonly referred to as a "mapping camera," is a precision instrument. It is designed to expose a large number of photographs in a short period of time while moving in an aircraft at a high velocity. Geometric distortions are held to a minimum. Aerial cameras are classified by type, angular field of view, focal length, and frame size. Types include: (1) single-lens frame, (2) multi-lens frame, (3) strip, and (4) panoramic. Angular fields of view include: (1) normal angle, for fields up to 75°; (2) wide-angle, for fields from 75° to 100°; and (3) superwide-angle, for fields greater than 100°. Common focal lengths include: (1) $3\frac{1}{2}''$, (2) 6″, (3) $8\frac{1}{4}''$, and (4) 12″, with the 6″ being most common. The common frame sizes are: (1) 7″ × 7″ and (2) 9″ × 9″, with the 9″ × 9″ being most common.

A frame camera is the type most generally used for aerial photogrammetry. The frame camera exposes the entire format simultaneously through a lens at a fixed distance from the focal plane. The principal components of the frame camera are: (1) the *lens system*, which collects the incident light rays and brings them to a focus on the focal plane; (2) the *shutter*, which controls the exposure interval; (3) the *diaphragm*, which controls the size of the lens opening; (4) a *filter* to control the wavelength of the light entering the system and to distribute it uniformly over the format; (5) the *cone*, which supports the lens-shutter–diaphragm assembly and prevents external light from reaching the film; (6) *fiducial marks*, which provide for the location of the photographic principal point; (7) the *focal plane*, which is formed by the upper surface of the fiducial marks and the focal-plane frame; (8) the *drive mechanism*, which operates the shutter, advances the film prior to exposure, and flattens the film during exposure; (9) the *camera body*, which houses the drive mechanism and supports the magazine; and (10) the *magazine*, which holds the supply of exposed and unexposed film and the film-flattening mechanism.

In order to obtain reliable measurements, aerial cameras are precisely calibrated under laboratory conditions. Calibrated values are obtained for the focal length, lens distortions, format dimensions, flatness of the focal plane, and the position of the principal point in relation to the fiducial marks.

Vertical Aerial Photographs

A vertical aerial photograph, as illustrated in Fig. 37, is a perspective projection. The figure represents the *geometry of a vertical photograph* taken at exposure station L. The positive, or contact print, is a reversal of the negative and is generally used in the development of photogrammetric equations. The length $o'L = oL$ is the focal length of the lens system. The photo coordinate reference system is defined by establishing the positive X axis as the straight line joining opposite fiducial marks most nearly coincident with the direction of flight. The positive Y axis is defined as a line 90° counterclockwise from the positive X axis.

The scale of an aerial photo is defined as the ratio of photo distance to ground distance. Since a photograph is a perspective projection the photo scale varies from point to point within the photo because of differences in elevation of the terrain. The photo scale at a point may be calculated by:

$$S_p = f/(H - h_p)$$

where

S_p = the photo scale at any point P
f = the focal length of the camera
H = the height of the exposure station above datum termed flight height
h_p = the elevation of point P above datum

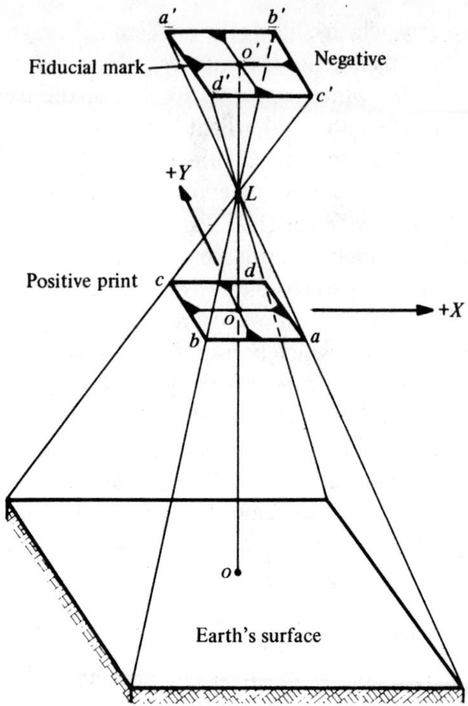

Fig. 37. A vertical aerial photograph.

Frequently it is desirable to develop an average photo scale for an aerial photograph. In this case the equation becomes

$$S_{\text{ave}} = f/(H - h_{\text{av}})$$

where h_{av} is the average elevation of the terrain covered by the photograph. This equation must be used with caution, particularly in areas where there is significant relief. At best, the equation provides only an approximate scale.

Vertical aerial photographs can be used to obtain relative *coordinates of ground points* which appear in the photographs. Figure 38 illustrates a vertical photograph taken at exposure station L, a height H above datum. The ground axes are established at the intersection of the vertical planes containing the photo axes and the datum plane. The origin of the system, O, is thus the intersection of the vertical through the exposure station with the datum plane. The images of ground points A and B appear on the photo at a and b, respectively. From similar triangles, the ground coordinates may be calculated by

$$X_A = \frac{(H - h_a)}{f} x_a$$

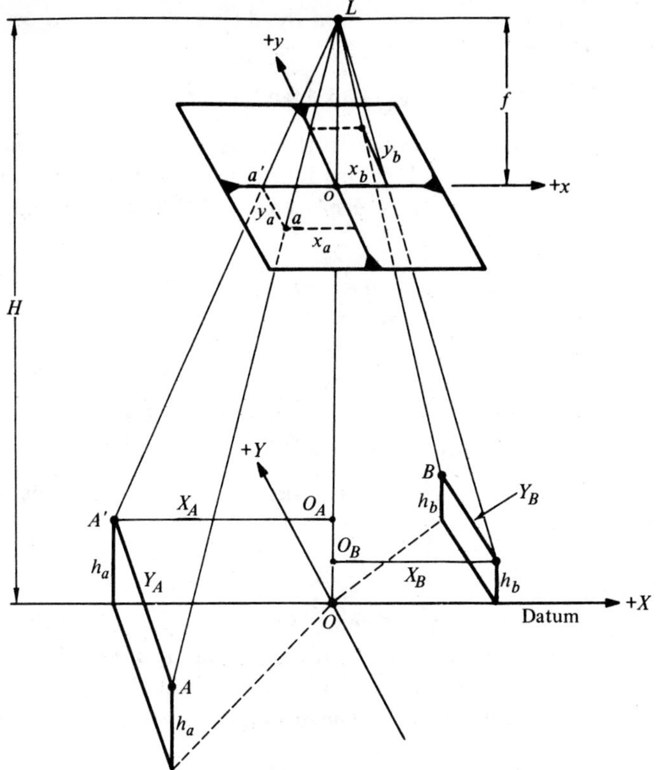

Fig. 38. Ground coordinates from a vertical photograph.

$$Y_A = \frac{(H - h_a)}{f} y_a$$

$$X_B = \frac{(H - h_b)}{f} x_b$$

$$Y_B = \frac{(H - h_b)}{f} y_b$$

where

X_A, Y_A, X_B, Y_B = the ground coordinates of points A and B
H = the flight height of the exposure station above datum
h_a, h_b = the elevation of points A and B above datum
f = focal length
x_a, y_a, x_b, y_b = the photo coordinates of the images of points A and B

The distance AB may be calculated from the difference in ground coordinates, and the angle AOB may be calculated by

$$\angle AOB = \tan^{-1} (X_A/Y_A) + \tan^{-1} (Y_B/X_B) + 90°$$

The difference between the location of the actual image of a point on an aerial photograph and the theoretical datum location of the point is termed *relief displacement*. The effect of relief displacement is illustrated in Fig. 39. The relief displacement is defined as

$$d = r - r'$$

where

d = the relief displacement
r = the radial distance from O to the image of A
r' = the radial distance from O to the datum position of A

Inasmuch as the datum position of A is generally unknown, the relief displacement is typically calculated by

$$d = rh/H$$

where all terms are previously defined. It should be noted that the relief displacement on a vertical photograph acts radially from the principal point.

The *flight height H* is an important parameter in most photogrammetric computations. It is typically an unknown for which a value must be obtained.

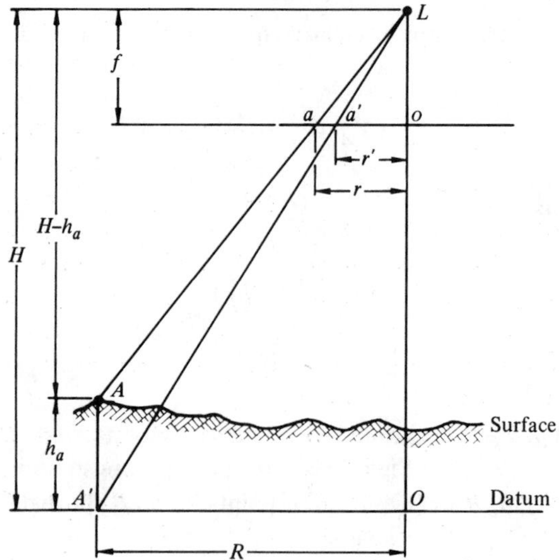

Fig. 39. Relief displacement.

There are three basic methods of determining H: (1) aircraft altimeter readings can be used to obtain an approximate value; (2) the average photo scale equation can be used with a known length of line to obtain an approximate value; or (3) the images of two ground control stations may be used to compute a precise value according to the equation

$$L^2 = \left[\frac{(H - h_b)x_b - (H - h_a)x_a}{f}\right]^2 + \left[\frac{(H - h_b)y_b - (H - h_a)y_a}{f}\right]^2$$

where

L = the known ground distance between A and B
H = the flying height above datum
h_a, h_b = the datum elevations of A and B
x_a, y_a, x_b, y_b = photo coordinates of the images of A and B
f = the focal length

If a computer is not available this equation is most easily solved by trial and error.

Radial Triangulation

Radial triangulation is based upon the photogrammetric principle that on a vertical photograph relief displacement acts radially from the principal point. Therefore, angles measured with the principal point as a center will be true horizontal angles regardless of the relative elevations of the points. Radial triangulation refers to the use of this principle to the extension of horizontal control. In the past, radial triangulation was performed by graphic methods. Since the advent of the digital computer, however, radial triangulation has been done numerically employing the principle of least squares.

Basically, radial triangulation consists of successive resection and intersection. In order to be effective it requires a strip of aerial photography with at least 50% overlap. The sequence of radial triangulation is as follows: (1) On the vertical photographs, measure the angles about the principal points to known ground stations, unknown stations to be located, and the principal points of adjacent photographs. (2) Based upon the known ground control points and the angles to these points about the principal points, calculate the position of principal points. (3) From the known location of the principal points and the angles to unknown points from the principal points, calculate the positions of the unknown points. (4) These unknown points now become known points and can be used to establish the position of principal points further in the strip; these principal points can in turn be used to establish the positions of other unknown points. (5) The network should be tied into additional known stations at the far end of the strip and at sufficient intermediate points to provide the required accuracy.

Mosaics

An aerial mosaic is a composite of multiple overlapping photographs. It provides a continuous image of the area covered by the photography. Inasmuch as it is made up from a series of perspective projections it is not a map, but it can be used as a map substitute when the true horizontal positions of features are not critical.

Mosaics are classified according to the method of preparation as (1) controlled, (2) semicontrolled, or (3) uncontrolled. A controlled mosaic is constructed from rectified (corrected for tilt and brought to a common scale by a rectifier) photographs and laid to an accurate horizontal scale. A semicontrolled mosaic is constructed like a controlled mosaic but one or more of the controlling features is neglected, e.g., the rectification of the photographs. An uncontrolled mosaic is prepared from unrectified photographs by simply matching terrain features on adjacent photographs. No effort is made to adjust the resulting assembly to a horizontal scale.

Mosaics may be further classified according to the purpose for which they were constructed. A *strip mosaic* is a compilation of a single strip of overlapping photographs prepared for qualitative study of various route locations. An *index mosaic* is a compilation of all the photographs of a particular set, laid so that the identification number of each photograph is visible. The mosaic is then, typically, photographed and the resulting print used as an index to aid in the selection of desired prints from the set.

Parallax

Parallax is defined as the apparent displacement of an object due to a change in the position of the point of observation. If the focal plane of an aerial camera is taken as a frame of reference, parallax will exist for all images on successive photographs due to the motion of the aircraft between exposures. This parallax effect is illustrated in Fig. 40. The displacement due to parallax occurs in the direction of the flight line or the photographic X axis. The parallax of point A in Fig. 40 may be calculated by the relationship

$$p = x_2 - x_1$$

where

p = the parallax of point A
x_2 = the x photographic coordinate of the image of A on photograph 2
x_1 = the x photographic coordinate of the image of A on photograph 1

Parallax measurements can also be used to calculate coordinates according to the equations

$$X = (B/p)x_1$$

$$Y = (B/p)y_1$$

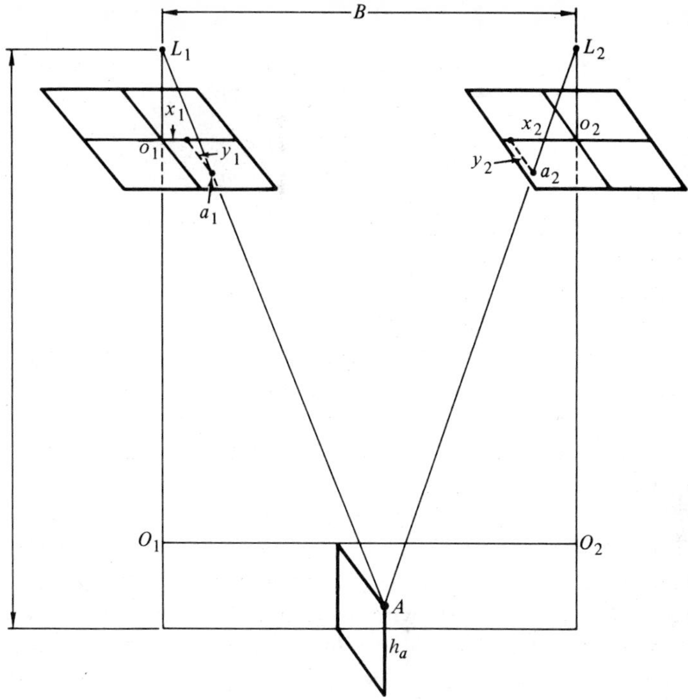

Fig. 40. Parallax.

where

X, Y = the relative ground coordinates of point A
B = the distance between exposure stations (the air base)
p = the parallax of point A
x_1, y_1 = the photo coordinates of the image of point A on photograph 1

Stereoscopic Viewing and Measuring

True depth perception is achieved only by binocular vision. In Fig. 41 two eyes are separated by a distance b, termed the *eye base*. When focused upon point A, the eyes converge at the *parallactic angle* ϕ_a. The mind converts this to the distance to A. Likewise, when the eyes are focused on point B, the parallactic angle ϕ_b is converted to the distance to B. Thus, the depth or distance AB is seen by the viewer.

When two overlapping photographs are viewed in such a manner that the left eye of the observer focuses on the left photograph and the right eye the right photograph, a mental three-dimensional model is produced. The observer, in effect, has placed each eye at a different exposure station. This process is

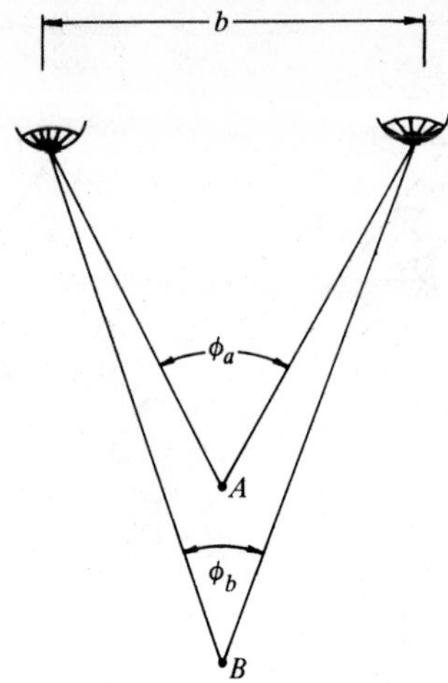

Fig. 41. Stereoscopic viewing.

greatly facilitated by the use of a *stereoscope*, which permits the eyes to be focused at a distance compatible with the parallactic angles being viewed.

The parallax of points on photographs being viewed stereoscopically can be quickly and accurately measured. Two identical marks are etched on two small plates of clear glass which are placed over the photographs. The images of the marks will merge and the viewer will see a single mark which will appear to be at a particular elevation with respect to the terrain. As shown in Fig. 42, the

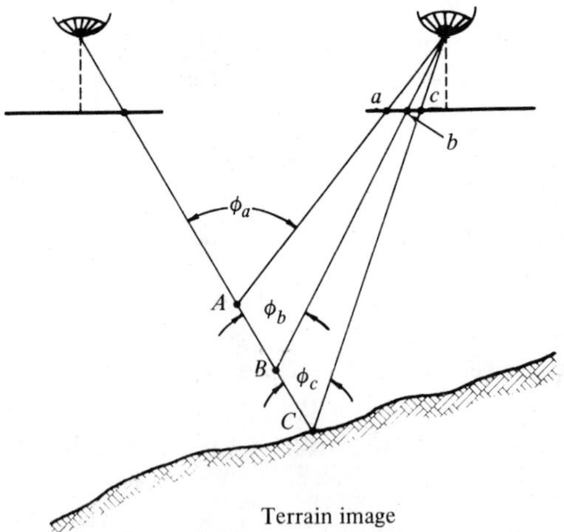

Fig. 42. The floating mark.

mark can be made to "float" up and down by varying the distance between the glass plates. As the mark on the right photo is moved from a to b to c the associated parallactic angles ϕ_a, ϕ_b and ϕ_c are viewed and the image of the mark appears to "float" from A to B to C, where it coincides with the terrain surface. A parallax bar can be used to facilitate these measurements. It is simply a bar upon which the glass plates are mounted in such a way that the right mark can be moved with respect to the left by turning a micrometer screw. When the apparent position of the mark is in coincidence with the terrain, the resultant readings are related to parallax by the equation

$$C = p - r$$

where

C = parallax-bar set up constant
p = parallax of the point
r = the parallax-bar reading

Stereoscopic Plotting Instruments

A stereoscopic plotting instrument is a precise optical-mechanical instrument used primarily for the compilation of topographic maps from overlapping aerial photographs. Basically a plotting instrument provides an optical-mechanical solution to the parallax equations. A stylized stereoscopic plotter is illustrated in Fig. 43. The photographs are typically developed from negatives onto glass termed diapositives. Light is projected through the diapositives of a pair of

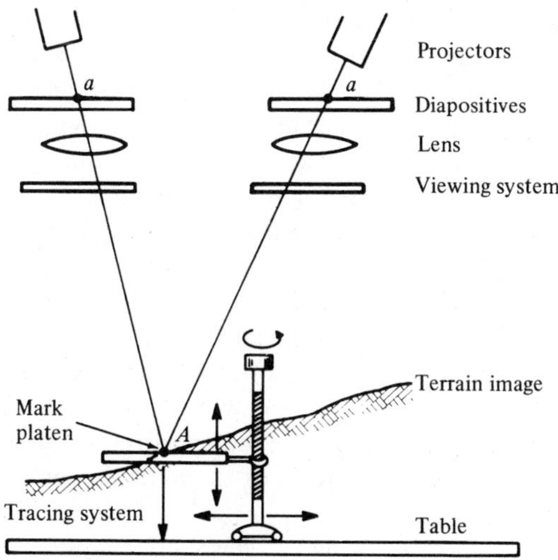

Fig. 43. Stylized stereoscopic plotter.

overlapping aerial photographs in such a manner that the light rays intersect to form an accurate stereoscopic model of the overlap area in the measuring space of the instrument. The operator employs a movable measuring mark and tracing system to measure and record the features of the model.

There is great variety in types and models of stereoscopic plotting instruments. However, they all include the same basic components: (1) a projection system, (2) a viewing system, (3) a measuring system, and (4) a recording system.

Stereoscopic plotting projectors are precise optical projectors. Each of the projectors is provided with six motions, three translational and three rotational. By combinations of these motions, it is possible to position the diapositives in the same relative spatial attitude as the aerial camera at the instants the photographs were exposed. This is termed *relative orientation.* When this has been accomplished, corresponding rays intersect to form a perfect three-dimensional optical model of the overlapping portions of the photographs in the measuring space. This model is brought to the desired scale and leveled with respect to the measuring system by adjusting the motions of the projectors until the model fits known ground control at the desired scale. This process is termed *absolute orientation* and requires a minimum of two horizontal control points and three vertical control points.

In order that the model be seen stereoscopically, the viewing system must be designed so that the operator views one image with each eye. Two methods of accomplishing this are (1) the anaglyphic method, and (2) the stereo image alternator (SIA). The *anaglyphic method* filters one image through a blue filter, the other through a red filter. The operator views the model through glasses on an optic train which places a red and blue filter in front of the corresponding eye. Thus, the left eye sees only the image from the left projector and the right eye sees only the image from the right projector, and therefore the operator sees the image at the correct parallactic angle. The SIA method accomplishes the same thing by placing rapidly rotating shutters on the projectors and the viewing optics. The shutters are synchronized so that the left and right eye can see only the corresponding images. The SIA system has the advantages of being able to project images both in color, and with no light intensity lost due to filters.

The basic function of the measuring system is to measure the model. Many different mechanical techniques have been devised to accomplish this. The simplest system incorporates a platen, which is a small screen supported by the tracing table. A pin-hole in the center of the platen emits a small light which is the measuring mark. When the platen is moved to a position where the light rays from the same image point on the two diapositives intersect, the measuring mark will appear to rest on the surface of the model according to the principle of the floating mark.

Recording systems also vary widely. The simplest system employs a tracing device directly beneath the measuring mark which traces the horizontal position of the mark on a map sheet placed upon the table. The vertical position of the platen is read from a counter geared to the vertical motion of the platen. In more complex systems, the horizontal and vertical position of the measuring mark can be recorded by an *XYZ* digitizer for direct linkage to a computer. This is particularly advantageous for such things as earthwork volumes or area computations.

Stereoscopic plotting instruments which are designed for higher orders of accuracy usually employ a somewhat different system than the one described above. Their projection system is based upon two metal rods, termed space rods. These simulate the light rays from the respective diapositives. The floating mark is derived from half marks imposed in the direct binocular viewing chain. Figure 44 illustrates a first order stereoscopic plotter.

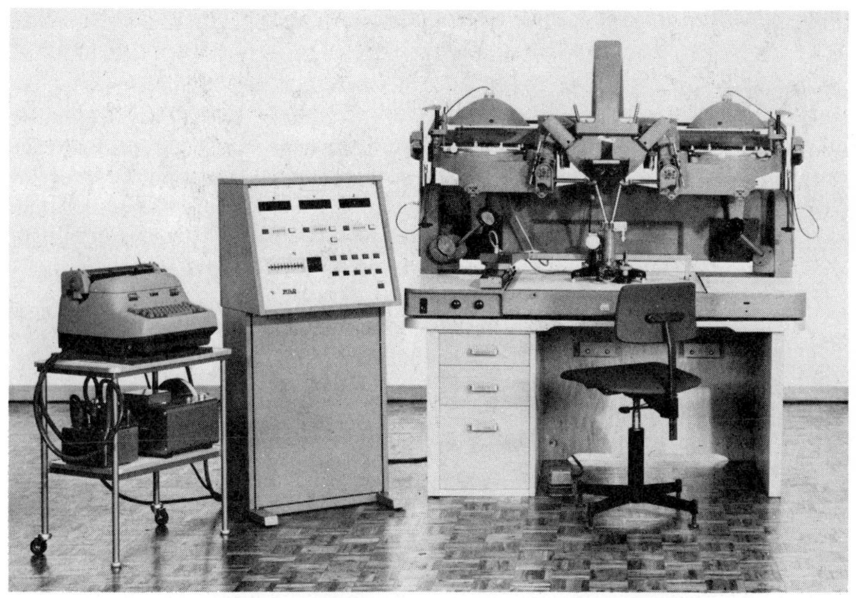

Fig. 44. Wild B8 aviograph with Wild EK8 coordinate printer.

Since the development of high-speed computers much of the control work formerly done on stereoscopic plotting instruments can be done economically with mathematical models. This procedure, termed *analytical photogrammetry*, requires as basic data precise measurements of photographic coordinates.

Flight Planning

In order that the photography obtained from an aerial photographic mission be satisfactory for the intended use, it is necessary to carefully design the mission. Based upon the purpose for which the photography is being obtained, the following basic factors must be determined: (1) the boundaries of the area; (2) the camera focal length and format size; (3) the required scale of photography; (4) the required overlap of adjacent photographs in the same flight line (endlap); (5) the required overlap of photographs in adjacent flight lines (sidelap); and (6) the aircraft speed in operation. Many other factors, such as magazine capacity and duration of the aircraft's stay on site, are significant but not basic. Once the above factors have been determined with respect to the purpose of the photography, the flight plan may be drawn up. A typical flight plan includes (1) the camera to be used; (2) the flight height above sea level; (3) the time interval between exposures; (4) the number of flight lines; (5) the number of photographs per flight line; (6) the total number of photographs; and (7) the flight map, which indicates the location of the flight lines with respect to topographic features and associated ground control points.

ACKNOWLEDGMENTS

In a field as broad as engineering measurements, a vast body of reference material is available. In order to condense this material into handbook format only the most significant portions of the available material have been included. The divisions of this section were adopted with permission from appropriate sections of the following more complete and detailed references.

Engineering Fundamentals in Measurements, Probability, Statistics and Dimensions, by Crandall, K. C. and Seabloom, R. W., McGraw-Hill Book Co., 1970.

Engineering Measurements, by B. Austin Barry, John Wiley & Sons, Inc., 1964.

Fundamentals of Surveying, by William H. Rayner and Milton O. Schmidt, Van Nostrand Reinhold Co., 1969.

Surveying, by Harry Bouchard and Francis H. Moffitt, Intext Educational Pubs., 1965. 5th ed.

Surveyor's Guide to Electromagnetic Distance Measurement, edited by J.J. Saastamoinen, University of Toronto Press, 1967.

An Introduction to Remote Sensing for Environmental Monitoring, by Scherz and Stevens, University of Wisconsin.

Elements of Photogrammetry, by Paul R. Wolf, McGraw-Hill Book Co.., 1974.

Elementary Surveying, 5th Ed., by Russell C. Brinker, Intext Educational Pubs., 1969.

SECTION III

Soil Mechanics and Foundation Engineering

Richard W. Christensen Ph.D., *P.E., M. ASCE*
Associate
Dames & Moore
Park Ridge, Illinois

SOIL COMPOSITION, IDENTIFICATION, AND CLASSIFICATION

The environmental aspects of soil mechanics and foundation engineering have become increasingly important in recent years because of our concern with the social aspects of land use. In addition to the fact that good building sites are increasingly difficult to find, the emphasis on planned urban and rural development often dictates the location of buildings with little regard to the adequacy of the site in terms of proper soil conditions. A good example is the growing development in the "Jersey Meadows," where the need for structures of various types has forced engineers to the limit of modern technology in overcoming unfavorable foundation conditions. As the cost of land continues its upward trend in our major urban centers, there is growing pressure for taller and taller buildings, with accompanying foundation problems. Thus, the soil mechanics and foundation engineer must perform at an extremely high level of technical competence, without losing sight of the social, economic, and esthetic impact of his work on the environment.

Soil Components

General Description In an engineering sense, the term "soil" applies to any unconsolidated inorganic or organic materials found in the ground, whereas "rock" refers to a material which does not disintegrate upon exposure. Soils are a mixture of solid particles and pore fluids, usually water and air. The relative proportions of solids and pore fluids are important factors in the engineering properties of soil masses.

In terms of origin, soils are further described as sedimentary (transported), residual (accumulated in place), or fill (man-made deposits). The way in which a particular soil deposit was created frequently has an important bearing upon the engineering properties of that deposit (see, e.g., U.S.B.R., 1968 pp. 99–111).

Grain Size Soil types may be distinguished according to grain size. According to the unified soil classification system, the fraction of soil particles larger than the No. 200 U.S. standard sieve size (0.074 mm) is considered *coarse-grained*, and that smaller than the No. 200 sieve is considered *fine-grained.* In the coarse-grained category, particles larger than $\frac{3}{4}$ in. are termed *gravel* and particles between $\frac{3}{4}$ in. and the No. 200 sieve are termed *sands.* The fine-grained particles, silts and clays, are not distinguished according to grain size but rather by their behavior, namely, their plasticity characteristics.

Gradation The grain-size, i.e., gradation, of the coarse-grained fraction is determined in the laboratory by sieve analysis, and that of the fine-grained fraction by hydrometer analysis. The results of the grain-size analysis are presented in the form of grain-size distribution curves, as shown in Fig. 1. Depending on the shape of the grain-size curve, the soil may be termed *well graded* (good representation of all particle sizes), or *poorly graded* (uniform particle size or gap-graded, i.e., absence of one or more particle sizes).

The range of particle sizes is represented by the coefficient of uniformity, $C_u = D_{60}/D_{10}$, and the shape of the grain-size curve is represented by the coefficient of curvature, $C_c = D_{30}{}^2/(D_{60})\,(D_{10})$. In addition, the D_{10} size (see Fig. 1) is often used as an indicator of the permeability and capillarity characteristics of coarse-grained soils. The percentage of fines (percent finer

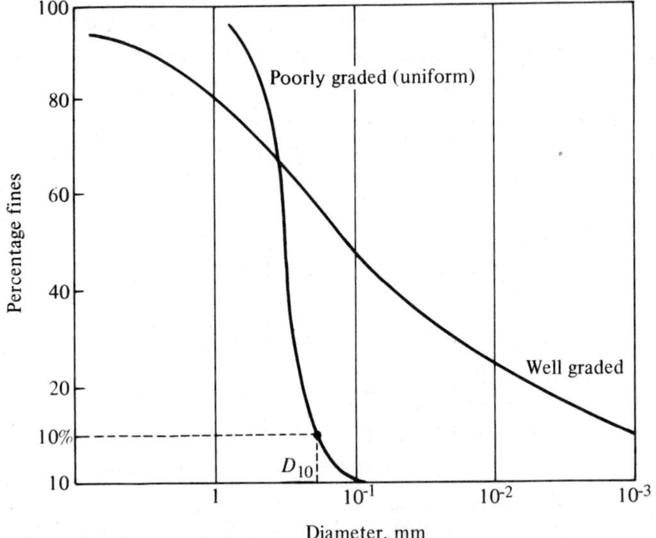

Fig. 1. Grain-size distribution curves.

than the No. 200 sieve) in otherwise coarse-grained soils is another significant feature of the grain-size distribution, since the presence of fines, even in relatively small amounts, can have an important effect on the engineering properties of the soil.

Gradation characteristics can provide useful information regarding the engineering properties of coarse-grained soils, e.g., permeability, compactibility, and frost susceptibility. However, in the case of fine-grained soils, i.e., silts and clays, grain size is a relatively insignificant property. Similarly, the effects of organic materials which may be present in the soil are not related to grain size.

Grain Shape The shapes of the individual particles in a soil mass may be identified as *bulky*, *platelike*, or *elongated.* Bulky grains are more or less equidimensional in shape and are characteristic of coarse-grained soils. They may be further differentiated as *rounded*, *subrounded*, or *angular.* The engineering properties of coarse-grained soils, such as shear strength and compressibility, are, to a certain extent, affected by the shapes of the grains.

Platelike particles are characteristic of the fine-grained fraction. This characteristic grain shape, along with the influence of electrochemical forces, allows for the formation of a great variety of soil structure types (flocculated, dispersed, etc.). In turn, the different types of soil structure may exhibit vastly different behavior.

Elongated particles are found principally in the clay mineral halloysite and in the fibers of organic soils. These types of particles, if found in sufficient quantity, may impart distinctive characteristics to the soil mass as a whole.

Weight-Volume Relationships In general, a soil mass is composed of solids, water, and air. The relationship between these soil components may be conveniently represented in the form of the diagram in Fig. 2.

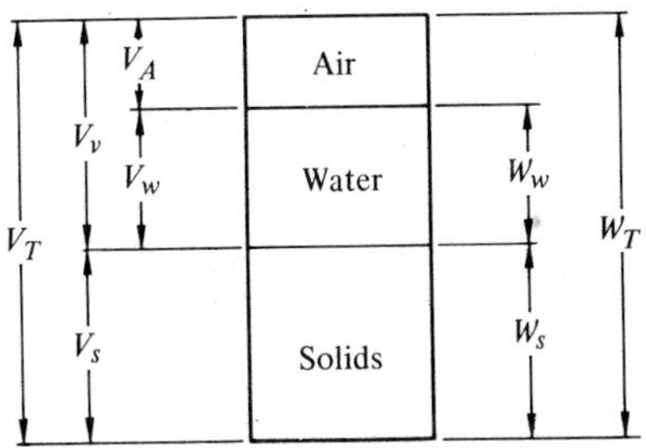

Fig. 2. Weight-volume diagram.

From the quantities shown in the diagram in Fig. 2, certain useful definitions have been established in soil-engineering terminology. A partial list of these definitions follows:

unit weight:

$$\gamma = W_T/V_T$$

saturated unit weight:

$$\gamma_{sat} = (W_s + V_v\gamma_w)/V_T$$

dry unit weight:

$$\gamma_{dry} = W_s/V_T$$

buoyant (effective) unit weight:

$$\gamma' = \gamma_{sat} - \gamma_w = (W_s - V_s\gamma_w)/V_T$$

specific gravity of solids:

$$G_s = W_s/V_s\gamma_w$$

water content (%):

$$\omega = (W_w/W_s) \times 100$$

void ratio:

$$e = V_v/V_s$$

porosity (%):

$$n = (V_v/V_T) \times 100$$

degree of saturation (%):

$$S = (V_w/V_v) \times 100$$

The relationships given in the above list, and others that may be derived from those given, are used extensively in soil-engineering computations.

Index Properties

Soil Consistency Depending upon the nature of the mineral grains and the amount of moisture present, a given soil mass may be found in any of three states of consistency: liquid, plastic, or solid. In the *liquid state*, the soil exhibits little or no shear strength; in the *plastic state*, the soil deforms without crumbling; and in the *solid state*, the soil either crumbles when deformed or exhibits elastic rebound. The range of moisture content over which soils exhibit the various states of consistency has been found to be a useful indicator of engineering properties.

Atterberg Limits The Atterberg limit tests are a series of simple laboratory tests (see, e.g., Lambe, 1951 for test procedures) designed to delineate the transition points between the various states of consistency. Since these transitions are actually gradual, the transition points defined by the Atterberg limit tests are somewhat arbitrary.

The Atterberg limits consist of: (a) the liquid limit (LL)–the water content at which the soil passes from the liquid to the plastic state; (b) the plastic limit (PL)–the water content at which the soil passes from the plastic to the semisolid state; and (c) the shrinkage limit (SL)–the water content at which the soil passes from the semisolid to the solid state. In addition, the plasticity index (PI = LL - PL) is defined as the range in water content over which the soil remains in the plastic state, and the liquidity index [LI = (ω_n - PL)/PI] indicates the nearness of a natural soil to the liquid limit. In practice, the Atterberg limit tests are performed on the soil fraction finer than the No. 40 sieve size and the results are applicable only to the fine-grained fraction.

It has been found from experience that the Atterberg limits correlate quite well with certain engineering properties and are very useful for classifying fine-grained soils. For example: the liquid limit is directly related to compressibility; for a given liquid limit the plasticity index is directly related to the strength at the plastic limit; the shrinkage limit is related to swelling characteristics; and the liquidity index may be correlated with the effective consolidation pressures of a natural soil. It should be noted, however, that the Atterberg limit tests are performed on remolded soil specimens and, therefore, do not reflect the properties of the natural soil structure, a very important factor in fine-grained soils.

Classification

It is often useful to classify soils according to the properties which are most important for a specific purpose. There are many such classification systems available for different purposes, e.g., agricultural uses, highways, and airfields. The Unified Soil Classification System (Wagner, 1957), presented in Table 1 was developed by the U.S. Bureau of Reclamation and the U.S. Army Corps of Engineers for use in all types of soil engineering applications. The classification of soils according to the unified system is based on grain size, gradation, plasticity, and compressibility and it divides soils into three major categories: coarse-grained, fine-grained, and highly organic soils. Coarse-grained soils are further subdivided according to gradation and presence of fines. Fine-grained soils are categorized as predominantly silt, clay, or organic, and are subdivided according to their plasticity characteristics. The classification may be done in the laboratory or in the field by the procedures described in Table 1. As a further aid to the engineer, an engineering use chart (Table 2) has been prepared for use with the Unified Soil Classification System. This chart relates, in a general way, the engineering properties to the classification of the soil.

TABLE 1. UNIFIED SOIL CLASSIFICATION SYSTEM (AFTER WAGNER, 1957)

Field Identification Procedures (Excluding particles larger than 3 in. and basing fractions on estimated weights)						Group Symbols a	Typical Names	Information Required for Describing Soils	Laboratory Classification Criteria	
Coarse-grained soils More than half of material is *larger* than No. 200 sieve size[b] (The No. 200 sieve size is about the smallest particle visible to naked eye)	Gravels More than half of coarse fraction is larger than No. 7 sieve size (For visual classification, the $\frac{1}{4}$ in. size may be used as equivalent to the No. 7 sieve size)	Clean gravels (little or no fines)	Wide range in grain size and substantial amounts of all intermediate particle sizes			*GW*	Well graded gravels, gravel-sand mixtures, little or no fines	Give typical name; indicate approximate percentages of sand and gravel; maximum size; angularity, surface condition, and hardness of the coarse grains; local or geologic name and other pertinent descriptive information; and symbols in parentheses For undisturbed soils add information on stratification, degree of compactness, cementation, moisture conditions and drainage characteristics Examples: *Silty sand*, gravelly; about 20% hard, angular gravel particles $\frac{1}{2}$-in. maximum size; rounded and subangular sand grains coarse to fine, about 15% nonplastic fines with low dry strength; well compacted and moist in place; alluvial sand; (*SM*)	Use grain size curve in identifying the fractions as given under field identification Determine percentages of gravel and sand from grain size curve Depending on percentage of fines (fraction smaller than No. 200 sieve size) coarse grained soils are classified as follows: Less than 5% *GW, GP, SW, SP* More than 12% *GM, GC, SM, SC* 5% to 12% *Borderline* cases requiring use of dual symbols	$C_U = \frac{D_{60}}{D_{10}}$ Greater than 4 $C_C = \frac{(D_{30})^2}{D_{10} \times D_{60}}$ Between 1 and 3
			Predominantly one size or a range of sizes with some intermediate sizes missing			*GP*	Poorly graded gravels, gravel-sand mixtures, little or no fines			Not meeting all gradation requirements for *GW*
		Gravels with fines (appreciable amount of fines)	Nonplastic fines (for identification procedures, see *ML* below)			*GM*	Silty gravels, poorly graded gravel-sand-silt mixtures			Atterberg limits below "A" line, or *PI* less than 4 / Above "A" line with *PI* between 4 and 7 are *borderline* cases requiring use of dual symbols
			Plastic fines (for identification procedures, see *CL* below)			*GC*	Clayey gravels, poorly graded gravel-sand-clay mixtures			Atterberg limits above "A" line, with *PI* greater than 7
	Sands More than half of coarse fraction is smaller than No. 7 sieve size	Clean sands (little or no fines)	Wide range in grain sizes and substantial amounts of all intermediate particle sizes			*SW*	Well graded sands, gravelly sands, little or no fines			$C_U = \frac{D_{60}}{D_{10}}$ Greater than 6 $C_C = \frac{(D_{30})^2}{D_{10} \times D_{60}}$ Between 1 and 3
			Predominantly one size or a range of sizes with some intermediate sizes missing			*SP*	Poorly graded sands, gravelly sands, little or no fines			Not meeting all gradation requirements for *SW*
		Sands with fines (appreciable amount of fines)	Nonplastic fines (for identification procedures, see *ML* below)			*SM*	Silty sands, poorly graded sand-silt mixtures			Atterberg limits below "A" line or *PI* less than 5 / Above "A" line with *PI* between 4 and 7 are *borderline* cases requiring use of dual symbols
			Plastic fines (for identification procedures, see *CL* below)			*SC*	Clayey sands, poorly graded sand-clay mixtures			Atterberg limits below "A" line with *PI* greater than 7
	Identification Procedures on Fraction Smaller than No. 40 Sieve Size		Dry Strength (crushing characteristics)	Dilatancy (reaction to shaking)	Toughness (consistency near plastic limit)					
Fine-grained soils More than half of material is *smaller* than No. 200 sieve size	Silts and clays liquid limit less than 50		None to slight	Quick to slow	None	*ML*	Inorganic silts and very fine sands, rock flour, silty or clayey fine sands with slight plasticity	Give typical name; indicate degree and character of plasticity, amount and maximum size of coarse grains; colour in wet condition, odour if any, local or geologic name, and other pertinent descriptive information, and symbol in parentheses For undisturbed soils add information on structure, stratification, consistency in undisturbed and remoulded states, moisture and drainage conditions Example: *Clayey Silt*, brown; slightly plastic; small percentage of fine sand; numerous vertical root holes; firm and dry in place; loess; (*ML*)	Plasticity chart (see below)	
			Medium to high	None to very slow	Medium	*CL*	Inorganic clays of low to medium plasticity, gravelly clays, sandy clays, silty clays, lean clays			
			Slight to medium	Slow	Slight	*OL*	Organic silts and organic silt-clays of low plasticity			
	Silts and clays liquid limit greater than 50		Slight to medium	Slow to none	Slight to medium	*MH*	Inorganic silts, micaceous or diatomaceous fine sandy or silty soils, elastic silts			
			High to very high	None	High	*CH*	Inorganic clays of high plasticity, fat clays			
			Medium to high	None to very slow	Slight to medium	*OH*	Organic clays of medium to high plasticity			
Highly Organic Soils			Readily identified by colour, odour, spongy feel and frequently by fibrous texture			*Pt*	Peat and other highly organic soils			

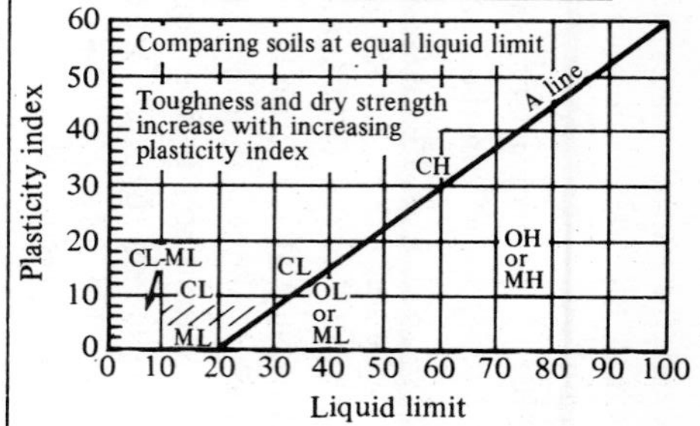

Plasticity chart for laboratory classification of fine-grained soils

[a]*Boundary classifications*. Soils possessing characteristics of two groups are designated by combinations of group symbols. For example *GW-GC*, well graded gravel-sand mixture with clay binder.
[b]All sieve sizes on this chart are U.S. standard.

Field Identification Procedure for Fine Grained Soils or Fractions

These procedures are to be performed on the minus No. 40 sieve size particles, approximately $^1/_{64}$ in. For field classification purposes, screening is not intended, simply remove by hand the coarse particles that interfere with the tests.

Dilatancy (Reaction to shaking):

After removing particles larger than No. 40 sieve size, prepare a pat of moist soil with a volume of about one-half cubic inch. Add enough water if necessary to make the soil soft by not sticky.

Place the pat in the open palm of one hand and shake horizontally, striking vigorously against the other hand several times. A positive reaction consists of the appearance of water on the surface of the pat which changes a livery consistency and becomes glossy. When the sample is squeezed between the fingers, the water and gloss disappear from the surface, the pat stiffens and finally it cracks or crumbles. The rapidity of appearance of water during shaking and of its disappearance during squeezing assist in identifying the character of the fines in a soil.

Very fine clean sands give the quickest and most distinct reaction whereas a plastic clay has no reaction. Inorganic silts, such as a typical rock flour, show a moderately quick reaction.

Dry Strength (Crushing characteristics):

After removing particles larger than No. 40 sieve size, mould a pat of soil to the consistency of putty, adding water if necessary. Allow the pat to dry completely by oven, sun or air drying, and then test its strength by breaking and crumbling between the fingers. This strength is a measure of the character and quantity of the colloidal fraction contained in the soil. The dry strength increases with increasing plasticity.

High dry strength is characteristic for clays of the CH group. A typical inorganic silt possesses only very slight dry strength. Silty fine sands and silts have about the same slight dry strength, but can be distinguished by the feel when powdering the dried specimen. Fine sand feels gritty whereas a typical silt has the smooth feel of flour.

Toughness (Consistency near plastic limit):

After removing particles larger than the No. 40 sieve size, a specimen of soil about one-half inch cube in size, is moulded to the consistency of putty. If too dry, water must be added and if sticky, the specimen should be spread out in a thin layer and allowed to lose some moisture by evaporation. Then the specimen is rolled out by hand on a smooth surface or between the palms into a thread about one-eight inch in diameter. The thread is then folded and re-rolled repeatedly. During this manipulation the moisture content is gradually reduced and the specimen stiffens, finally loses its plasticity, and crumbles when the plastic limit is reached.

After the thread crumbles, the pieces should be lumped together and a slight kneading action continued until the lump crumbles.

The tougher the thread near the plastic limit and the stiffer the lump when it finally crumbles, the more potent is the colloidal clay fraction in the soil. Weakness of the thread at the plastic limit and quick loss of coherence of the lump below the plastic limit indicate either inorganic clay of low plasticity, or materials such as kaolin-type clays and organic clays which occur below the A-line.

Highly organic clays have a very weak and spongy feel at the plastic limit.

TABLE 2. ENGINEERING USE CHART (AFTER WAGNER, 1957)

Typical Names of Soil Groups	Group Symbols	Important Properties: Permeability when Compacted	Shearing Strength when Compacted and Saturated	Compressibility when Compacted and Saturated	Workability as a Construction Material
Well-graded gravels, gravel-sand mixtures, little or no fines	*GW*	pervious	excellent	negligible	excellent
Poorly graded gravels, gravel-sand mixtures, little or no fines	*GP*	very pervious	good	negligible	good
Silty gravels, poorly graded gravel-sand-silt mixtures	*GM*	semipervious to impervious	good	negligible	good
Clayey gravels, poorly graded gravel-sand-clay mixtures	*GC*	impervious	good to fair	very low	good
Well-graded sands, gravelly sands, little or no fines	*SW*	pervious	excellent	negligible	excellent
Poorly graded sands, gravelly sands, little or no fines	*SP*	pervious	good	very low	fair
Silty sands, poorly graded sand-silt mixtures	*SM*	semipervious to impervious	good	low	fair
Clayey sands, poorly graded sand-clay mixtures	*SC*	impervious	good to fair	low	good
Inorganic silts and very fine sands, rock flour, silty or clayey fine sands with slight plasticity	*ML*	semipervious to impervious	fair	medium	fair
Inorganic clays of low to medium plasticity, gravelly clays, sandy clays, silty clays, lean clays	*CL*	impervious	fair	medium	good to fair
Organic silts and organic silt-clays of low plasticity	*OL*	semipervious to impervious	poor	medium	fair
Inorganic silts, micaceous or diatomaceous fine sandy or silty soils, elastic silts	*MH*	semipervious to impervious	fair to poor	high	poor
Inorganic clays of high plasticity, fat clays	*CH*	impervious	poor	high	poor
Organic clays of medium to high plasticity	*OH*	impervious	poor	high	poor
Peat and other highly organic soils	*Pt*	–	–	–	–

Relative Desirability for Various Uses									
Rolled Earth Dams			Canal Sections		Foundations		Roadways		
							Fills		
Homogeneous Embankment	Core	Shell	Erosion Resistance	Compacted Earth Lining	Seepage Important	Seepage not Important	Frost Heave not Possible	Frost Heave Possible	Surfacing
–	–	1	1	–	–	1	1	1	3
–	–	2	2	–	–	3	3	3	–
2	4	–	4	4	1	4	4	9	5
1	1	–	3	1	2	6	5	5	1
–	–	3 if gravelly	6	–	–	2	2	2	4
–	–	4 if gravelly	7 if gravelly	–	–	5	6	4	–
4	5	–	8 if gravelly	5 erosion critical	3	7	8	10	6
3	2	–	5	2	4	8	7	6	2
6	6	–	–	6 erosion critical	6	9	10	11	–
5	3	–	9	3	5	10	9	7	7
8	8	–	–	7 erosion critical	7	11	11	12	–
9	9	–	–	–	8	12	12	13	–
7	7	–	10	8 volume change critical	9	13	13	8	–
10	10	–	–	–	10	14	14	14	–
–	–	–	–	–	–	–	–	–	–

A more detailed description of the Unified Soil Classification System may be found in the U.S. Bureau of Reclamation's *Earth Manual* (U.S.B.R., 1968).

SOIL PROPERTIES

In many respects, soil is a unique engineering material. Because of the particulate, multiphase nature of soils, their behavior is governed by interaction between discrete particles and by the physical and chemical interaction between the mineral skeleton and the pore fluid. In this section, the most significant engineering aspects of soil behavior are examined.

Permeability

Flow of Water Through Soils The void spaces in soils are interconnected with neighboring voids, which makes it possible for water to move through the mass in the presence of a hydraulic gradient. Since the void spaces in a soil mass are irregularly shaped, the water actually travels a somewhat tortuous path from point to point, but experience has shown that, for practical purposes, the flow can be assumed to be linear.

Under most conditions water flow in soils can be described by Darcy's law, stated as follows:

$$Q = k \frac{h}{L} A = kiA$$

where

Q = the rate of flow
k = a constant, known as the coefficient of permeability
h = the head loss (the difference in total head between two reference points)
L = the distance between the reference points through which the head is lost
A = the gross cross-sectional area through which the flow occurs
i = the hydraulic gradient, i.e., the ratio of the head lost to the distance in which it is lost

Darcy's law, which assumes laminar flow, has been experimentally validated for medium sands, silts and clays. For more pervious soils, such as coarse sands or gravels, the actual flow relationship can be established experimentally for the particular soil of interest.

Darcy's law can also be expressed in terms of the "discharge velocity" v, where

$$v = Q/A = ki$$

or the "effective seepage velocity" v_s, where

$$v_s = \frac{Q}{A_v} = \frac{v}{n} = \frac{ki}{n}$$

in which

A_v = the average cross-sectional area of voids
n = the porosity of the soil

Although the discharge velocity and seepage velocity are fictitious quantities, since the flow does not actually follow a straight line, both are useful means for expressing the time required for water to move through a given distance in a soil mass.

Measurement of Permeability Soil permeability can be measured either in the field or in the laboratory. Laboratory permeability tests, although easier and cheaper to perform, are less reliable than field tests because soil permeability is very dependent upon the natural structure of the soil and minor geologic details that may not be reflected in small laboratory samples. Nevertheless, laboratory permeability tests are commonly performed and can be quite useful in evaluating flow characteristics in many cases.

The most common types of laboratory permeability tests are:

1. Falling-head permeameter tests
2. Constant-head permeameter tests
3. Direct or indirect measurement during oedometer tests

The range of applicability of the various methods of determining permeability is shown in Table 3.

The test setups for falling-head and constant-head permeability tests are shown schematically in Fig. 3. Detailed descriptions of the testing apparatus

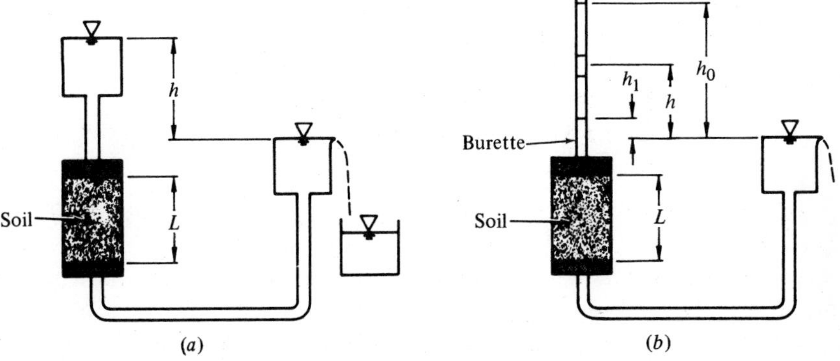

Fig. 3. Permeability tests. (a) Constant head. (b) Falling head.

TABLE 3. PERMEABILITY AND DRAINAGE CHARACTERISTICS OF SOILS (AFTER CASAGRANDE AND FADUM, 1940).

<table>
<tr><th></th><th colspan="11">Coefficient of Permeability k in cm per sec (log scale)</th></tr>
<tr><th></th><th>10^2 – 10^1</th><th>10^1 – 1.0</th><th>1.0 – 10^-1</th><th>10^-1 – 10^-2</th><th>10^-2 – 10^-3</th><th>10^-3 – 10^-4</th><th>10^-4 – 10^-5</th><th>10^-5 – 10^-6</th><th>10^-6 – 10^-7</th><th>10^-7 – 10^-8</th><th>10^-8 – 10^-9</th></tr>
<tr><td>Drainage</td><td colspan="6">Good</td><td colspan="2">Poor</td><td colspan="3">Practically Impervious</td></tr>
<tr><td rowspan="2">Soil types</td><td colspan="2" rowspan="2">Clean gravel</td><td colspan="3">Clean sands, clean sand and gravel mixtures</td><td colspan="4">Very fine sands, organic and inorganic silts, mixtures of sand silt and clay, glacial till, stratified clay deposits, etc.</td><td colspan="2" rowspan="2">“Impervious” soils, e.g., homogeneous clays below zone of weathering</td></tr>
<tr><td colspan="2"></td><td colspan="5">“Impervious” soils modified by effects of vegetation and weathering</td></tr>
<tr><td rowspan="3">Direct determination of k</td><td colspan="6">Direct testing of soil in its original position–pumping tests. Reliable if properly conducted. Considerable experience required</td><td colspan="5"></td></tr>
<tr><td colspan="5">Constant-head permeameter. Little experience required</td><td colspan="6"></td></tr>
<tr><td colspan="2"></td><td colspan="3">Falling-head permeameter. Reliable. Little experience required</td><td colspan="3">Falling-head permeameter. Unreliable. Much experience required</td><td colspan="3">Falling-head permeameter. Fairly reliable. Considerable experience necessary</td></tr>
<tr><td>Indirect determination of k</td><td colspan="5">Computation from grain-size distribution. Applicable only to clean cohesionless sands and gravels</td><td colspan="4"></td><td colspan="2">Computation based on results of consolidation tests. Reliable. Considerable experience required</td></tr>
</table>

and experimental procedures for the different types of laboratory permeability tests are given by Lambe (1951).

A number of field procedures have been developed to measure soil permeability *in-situ.* An excellent discussion of field permeability testing has been presented by Hvorslev (1962).

Ranges of Permeability The coefficient of permeability of soils covers an extremely broad range. Table 4 presents representative values of the permeability coefficient for several natural formations. It can be seen from this list that the variation in permeability for different types of formations covers at least eight orders of magnitude.

Figure 4 gives laboratory permeability test data on a variety of soils. These plots show that the log of permeability varies linearly with void ratio for most soils. Alternatively, permeability is often related to some "representative particle diameter", usually D_{10}. This type of correlation generally works better for sands and silts than for clays because the natural structure of clay soils exerts a dominating influence on the flow characteristics. Figure 5 shows the relationship between the permeability of clean, coarse-grained soils and the D_{10} size.

In addition to the wide range of permeability between different soil types, natural soil formations frequently exhibit vastly different permeabilities in the

TABLE 4. COEFFICIENT OF PERMEABILITY OF COMMON NATURAL SOIL FORMATIONS (AFTER TERZAGHI AND PECK, 1967).

Formation	Value of k (cm/sec)
River deposits	
Rhone at Genissiat	up to 0.40
Small streams, eastern Alps	0.02 to 0.16
Missouri	0.02 to 0.20
Mississippi	0.02 to 0.12
Glacial deposits	
Outwash plains	0.05 to 2.00
Esker, Westfield, Mass.	0.01 to 0.13
Delta, Chicopee, Mass.	0.0001 to 0.015
Till	less than 0.0001
Wind Deposits	
Dune sand	0.1 to 0.3
Loess	0.001 ±
Loess loam	0.0001 ±
Lacustrine and marine offshore deposits	
Very fine uniform sand, U = 5 to 2	0.0001 to 0.0064
Bull's liver, Sixth Ave., N.Y., U = 5 to 2	0.0001 to 0.0050
Bull's liver, Brooklyn, U = 5	0.00001 to 0.0001
Clay	less than 0.0000001

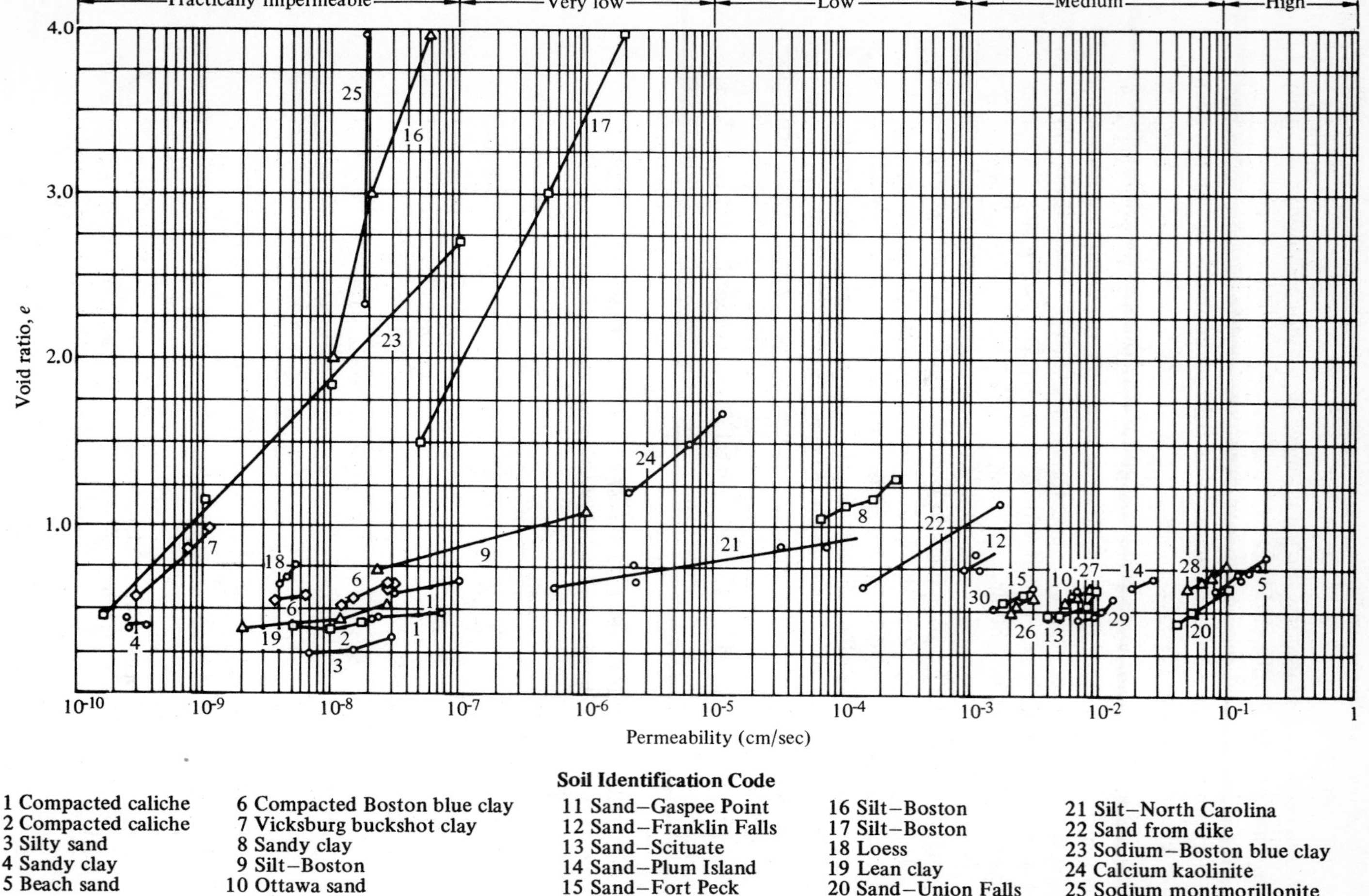

Fig. 4 Permeability test data on several soils (after Lambe and Whitman, 1969)

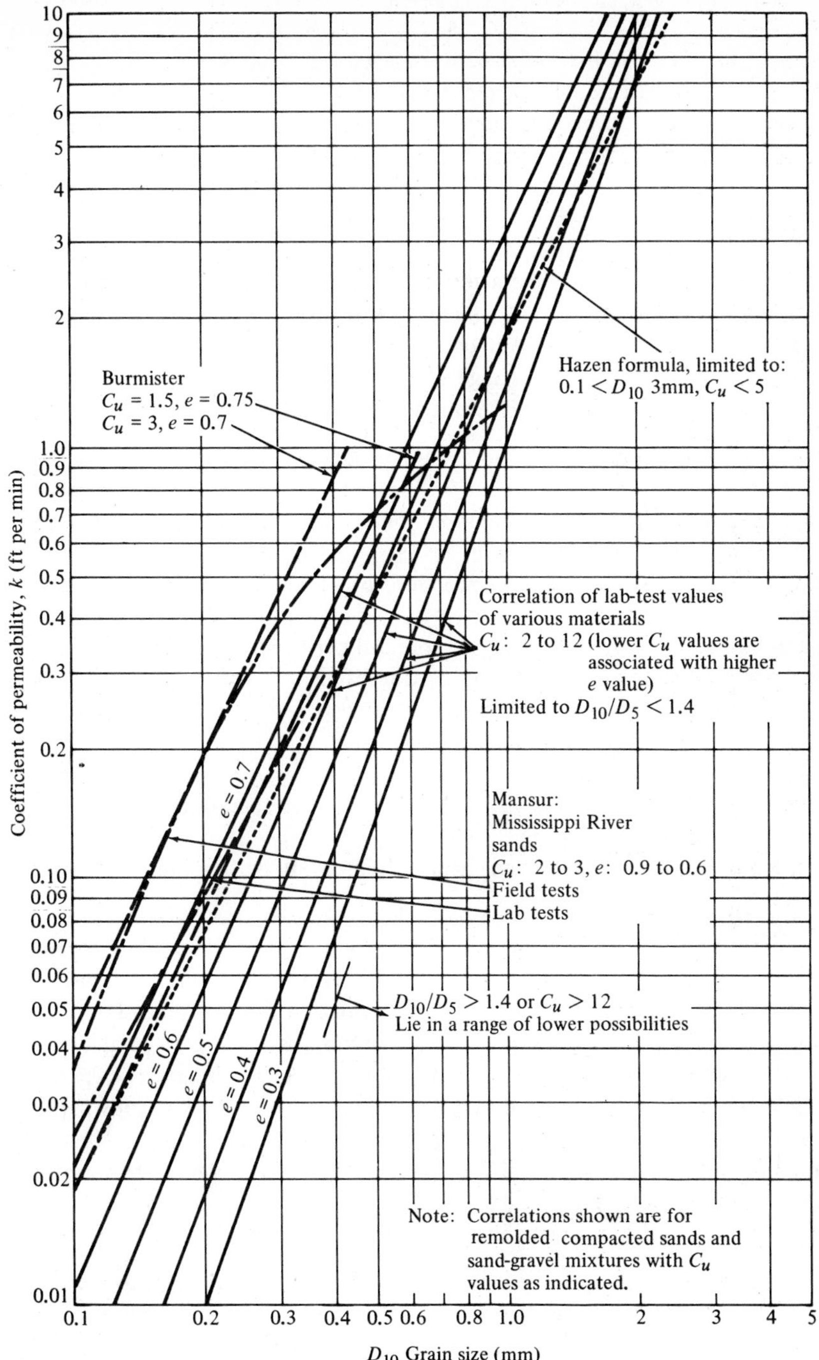

Fig. 5. Permeability of granular soils (after Navdocks DM-7, 1962).

horizontal and vertical directions; occasionally as much as two or three orders of magnitude. In most cases the horizontal permeability is greater than the vertical because of horizontal stratification. However, in exceptional cases such as loess deposits, the vertical permeability exceeds that in the horizontal direction.

Compressibility

Definitions Volume-change characteristics represent one of the most important aspects of soil behavior. These volume changes are, in general, a function of applied load, time, density, water content, and soil types. A list of terms describing the different volume-change phenomena that occur in soils has been compiled by the U.S. Bureau of Reclamation (1968) as follows:

Compression defines the volume change produced by application of a static external load.
Consolidation defines volume change that is achieved with the passage of time.
Shrinkage is the volume change produced by capillary stresses during drying of a soil.
Compaction is the volume change produced artificially by momentary load application, such as rolling, tamping, or vibration.
Rebound as opposed to compression.
Expansion as opposed to consolidation.
Swell as opposed to shrinkage.
Loosening, *scarifying*, and similar terms describe the operation used in opposition to compaction.
Heave is used to describe volume change produced by frost action or expansive soils.

Soil compression is accompanied by the expulsion of water and/or air from the voids. If the voids are filled primarily with air, compression occurs quickly, as in the case of compaction, whereas in a saturated clay, compression occurs slowly as a consolidation process that may take months or years to complete.

Volume changes in soils may result from particle rearrangement, particle breakdown, and absorption or desorption of moisture. In most soils particle rearrangement is the most important mechanism of volume change. However, particle breakdown may be significant in granular soils subjected to very high stresses or in residual soils, and moisture absorption or desorption is an important factor in expansive clays such as montmorillonite.

The One-Dimensional Compression Test The one-dimensional or confined compression behavior of soils is important because it represents a situation that is common in natural soil formations and is also relatively easy to measure in the laboratory. The one-dimensional compression test is performed in the

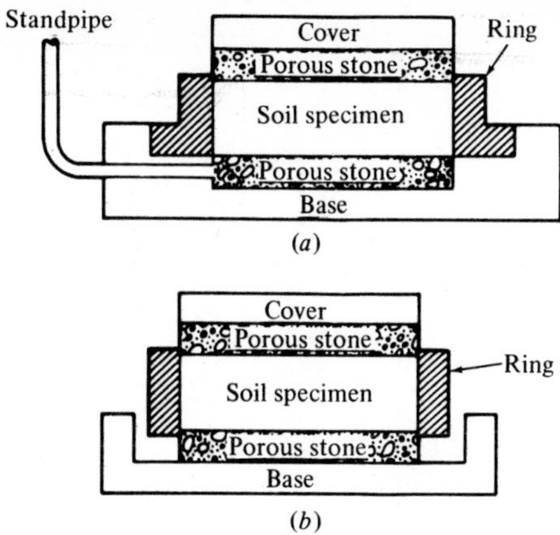

Fig. 6. Common forms of oedometer. (a) Fixed-ring. (b) Floating ring. (After Lambe, 1951).

laboratory with an oedometer, shown schematically in Fig. 6, in which a cylindrical soil sample is loaded vertically while lateral deformations are prevented by a rigid confining ring. In order to minimize side friction between the soil sample and the confining ring, the thickness to diameter ratio of the sample is kept as small as possible, typically 1 : 2.5 to 1 : 4. Porous disks are placed on the top or bottom of the sample, or both, to permit drainage of water from the sample, and the load is applied through a rigid disk on the top of the sample. The compression is measured with a dial indicator.

The results of the one-dimensional compression test are presented graphically in the form of void ratio or vertical strain vs. applied pressure, plotted to either a natural or logarithmic scale. In this test the ratio of lateral stress to vertical stress is K_0, the coefficient of earth pressure at rest, which is representative of natural stress conditions when there has been no lateral strain in the ground subsequent to deposition. Normally, the lateral stress is not measured in the confined compression test; however, such measurements are possible and have been made occasionally for research purposes (see, e.g., Hendron, 1963).

One-Dimensional Compression Characteristics of Granular Soils Compression behavior in granular soils can be explained in terms of certain basic mechanisms which occur at various levels of stress applied to the material, as illustrated in Fig. 7.

1. At very low stress levels (1 psi or less) the stress–strain curve is concave downward and the constrained modulus decreases with increasing stress. This phenomenon, known as yielding, is attributed to elastic distortion of the individual grains.
2. At higher stress levels (up to 500–5000 psi), a sliding and reorientation of

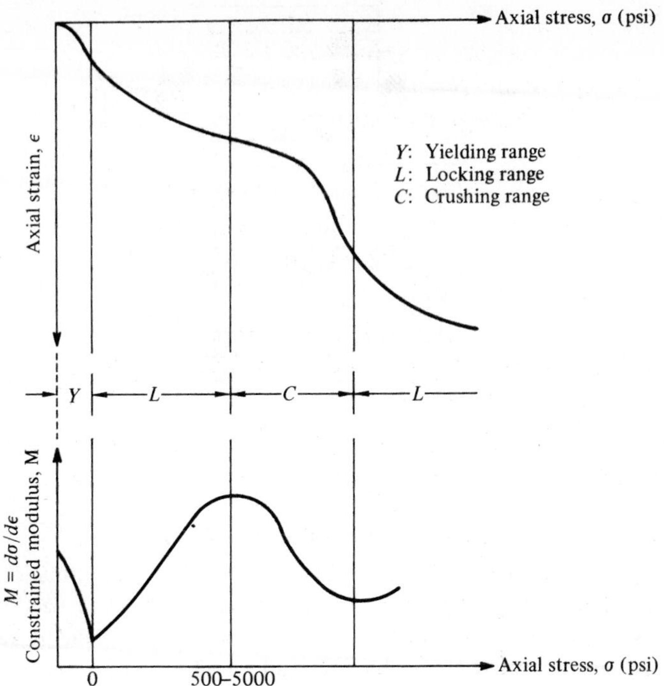

Fig. 7. Compression behavior of sands at various stress levels (after Gardner, 1966).

grains occurs, resulting in a more stable packing arrangement. In this stress range, the stress–strain curve is concave upward and the constrained modulus increases with increasing stress. This type of behavior is called *locking*.

3. Once the densest possible state has been achieved in the locking range, further stress increase causes *crushing* of the individual grains, which produces a stress–strain curve that is concave downward, and the constrained modulus again decreases. The phenomenon of grain crushing permits further sliding and reorientation of grains and, as stresses are further increased, the locking trend is reestablished.

On the basis of both theoretical and experimental studies (Wilson and Sutton, 1948; Hendron, 1963) exponential relationships between stress and strain and between constrained modulus and stress have been postulated. These relationships can be expressed as follows:

$$\epsilon = \epsilon_1 \sigma^n$$

where

ϵ = vertical strain,
ϵ_1 = vertical strain at a stress level of unity,
σ = vertical stress,
n = an exponent, largely dependent upon soil type, density and stress history,

and

$$M = M_1 \sigma^m$$

where

M = constrained modulus,
M_1 = constrained modulus at a stress level of unity,
m = an exponent = $1 - n$.

The proposed relationships indicate that log–log plots of stress versus strain and constrained modulus versus stress should appear as straight lines, as shown in Fig. 8. Experimental studies by a number of investigators (Schultze and Moussa, 1961; Hendron, 1963; Gardner, 1966; Janbu, 1967) substantiate this for a variety of granular soils.

The following variables have been found to exert some degree of influence on the compression behavior of granular soils (references to selected papers treating these topics are given in parentheses):

1. Stress level (Roberts and DeSouza, 1958; Hendron, 1963).
2. Stress history [precompression] (Hendron, 1963; Seaman, et al., 1963).

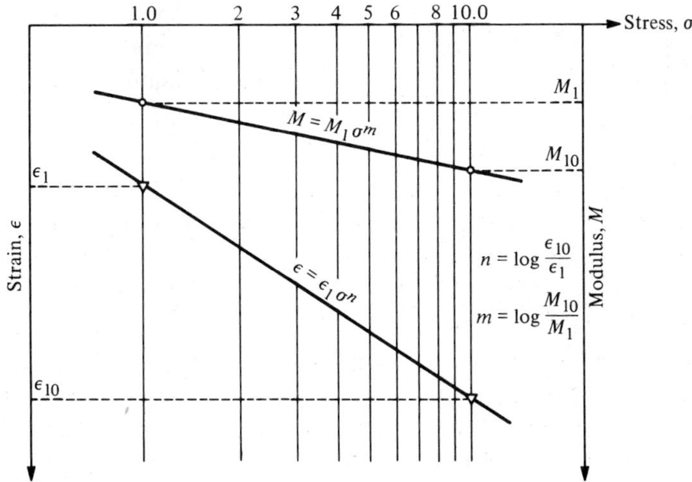

Fig. 8. Stress–strain and modulus–stress relationships for sand.

3. Initial relative density (Schultze and Moussa, 1961).
4. Gradation and grain shape (Schultze and Moussa, 1961); (Burmister, 1956).

An extensive series of tests performed by Schultze and Moussa (1961) involving 25 different sands has provided valuable information as to the variation in numerical values of the parameters ϵ_1, n, M_1, and m that might be expected for a broad range of granular soils. It was found that, for practical purposes, these parameters could be correlated with a single soil variable, (initial relative density) for all 25 sands tested. The results of their tests are reproduced in Table 5.

TABLE 5. COMPRESSION PARAMETERS FOR 25 SANDS (AFTER SCHULTZE & MOUSSA, 1961).

	ϵ_1		n		M_1		m	
R_D	from	to	from	to	from	to	from	to
0.0	0.0106	0.0480	0.206	0.432	82.5	309	0.568	0.794
0.1	0.0086	0.0370	0.232	0.450	98	336	0.550	0.768
0.2	0.0070	0.0290	0.264	0.470	118	391	0.530	0.736
0.3	0.0058	0.0220	0.292	0.486	142	430	0.514	0.708
0.4	0.0046	0.0170	0.320	0.508	168	489	0.492	0.680
0.5	0.0038	0.0132	0.346	0.528	204	578	0.472	0.654
0.6	0.0030	0.0102	0.372	0.552	252	669	0.448	0.628
0.7	0.0024	0.0078	0.394	0.578	304	784	0.422	0.606
0.8	0.0020	0.0060	0.418	0.606	379	897	0.394	0.582
0.9	0.0018	0.0047	0.440	0.640	464	1 042	0.360	0.560
1.0	0.0013	0.0037	0.460	0.680	571	1 255	0.320	0.540

One-Dimensional Compression Behavior of Saturated, Cohesive Soils When loads are applied to saturated, cohesive soil strata the resulting settlements are time-dependent, as shown in Fig. 9. For convenience, these settlements are generally divided into three categories: immediate, primary, and secondary. Immediate settlement is considered to be that which occurs under undrained conditions (constant volume). This type of settlement is absent in one-dimensional loading. Primary compression, or consolidation settlement, is due to volume changes accompanying the dissipation of excess pore water pressure. Secondary compression is the settlement which occurs after the excess pore water pressure is essentially dissipated. These divisions are somewhat arbitrary, however, since in reality there is usually no clear distinction between the different types of settlement. The degree to which the different types of settlement will develop in any given situation will depend upon a number of factors, including clay type, stress history of the deposit, and magnitude and rate of load increase.

The oedometer test is commonly used in soil mechanics to investigate the compression behavior of saturated clays. It is generally believed that the con-

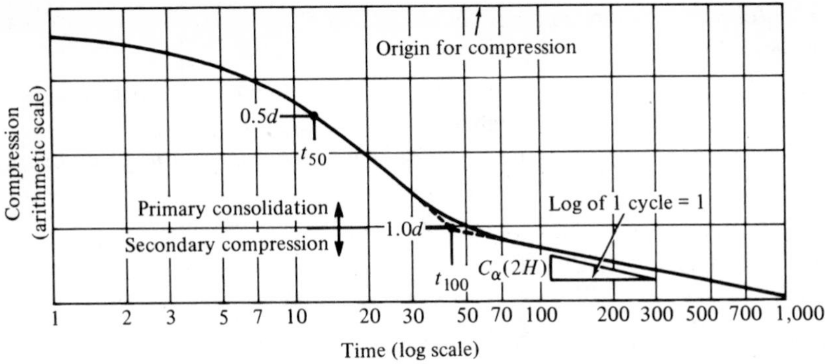

Fig. 9. Time rate of compression for saturated, cohesive soils.

ditions of vertical drainage and no lateral strain imposed in this test result in a suitable approximation of field conditions in most cases.

The results of oedometer tests on clays may be presented in a variety of ways. Figure 10 shows three alternative methods of presentation: void ratio vs. log of effective consolidation stress, vertical strain vs. log stress, and stress vs. strain.

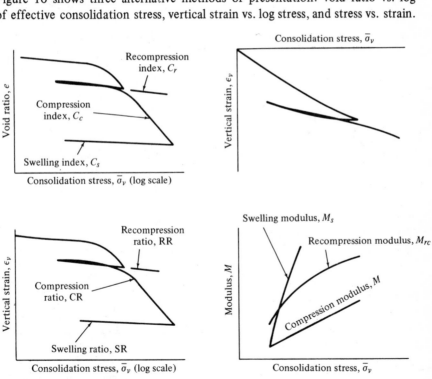

Fig. 10. Compression curves for saturated clay.

The following parameters are used to describe the confined compression behavior:

e vs. $\log \bar{\sigma}_v$ plot: C_c = compression index (slope of compression curve in virgin compression region),
C_r = recompression index (average slope of unloading-reloading cycle),
C_s = swelling index (slope of rebound curve);

ϵ_v vs. $\log \bar{\sigma}_v$ plot: $\text{CR} = C_c/(1 + e_0)$ = compression ratio,
$\text{RR} = C_r/(1 + e_0)$ = recompression ratio,
$\text{SR} = C_s/(1 + e_0)$ = swelling ratio;

ϵ_v vs. $\bar{\sigma}_v$ plot: M = compression modulus,
M_{rc} = recompression modulus,
M_s = swelling modulus.

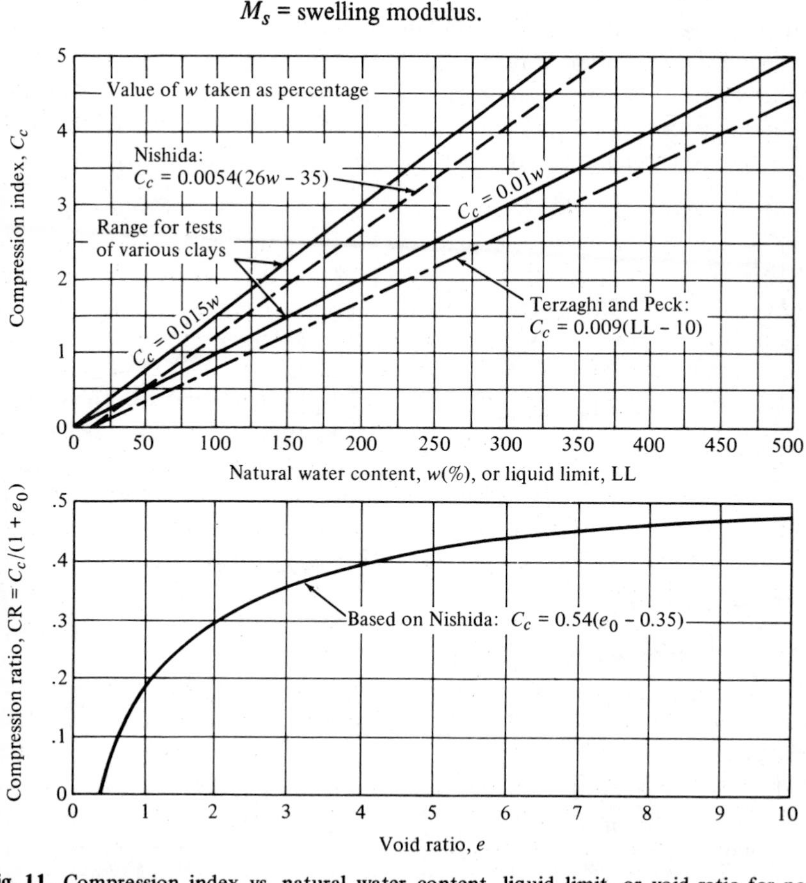

Fig. 11. Compression index vs. natural water content, liquid limit, or void ratio for normally consolidated silts and clays (after Navdocks DM-7, 1962).

The numerical values of these parameters may vary greatly depending upon soil type. However, as shown in Fig. 11, under certain conditions, preliminary estimates may be made from empirical correlations developed from compression tests on samples from many different parts of the world.

One of the most important aspects of the compression behavior of clay is the determination of maximum past pressure or preconsolidation pressure. The preconsolidation pressure represents the stress level at which the compressibility undergoes a sharp increase. Since the compressibility increases by a factor of at least five to ten times at stresses above the preconsolidation pressure, it is extremely important that this value be determined as accurately as possible.

The existence of a maximum past pressure may be due to a variety of factors; e.g., previous loads that have subsequently been removed, changes in water-table elevation, desiccation, secondary compression, and chemical alterations. Although the laboratory procedures for determining maximum past pressure do not differentiate between these various mechanisms, it is often possible to infer the cause from a knowledge of the age, depositional conditions, and stress history of the deposit and from other data. This information is often useful in interpreting the results of oedometer tests.

Several methods have been suggested for determining the maximum past pressure from laboratory tests. Three of these methods are illustrated in Fig. 12. The first and most widely used procedure was proposed by Casagrande (1936); the second method, presented by Schmertmann (1955), also includes a procedure for reconstructing the *in-situ* compression curve; and the third method, suggested by Janbu (1969), is based on the behavior of the compression modulus as a function of the effective vertical stress. All of these methods require good-quality samples, i.e., samples that are reasonably free of sampling disturbance. In addition, the compression curve used to determine maximum past pressure should correspond to the end of primary compression because, if secondary compression is included, the resulting value of maximum past pressure may be significantly reduced. Further, more accurate results may also be obtained by using reduced load increments until the break in the curve occurs.

The standard oedometer test is performed by doubling the load with each increment and keeping the load constant for a convenient length of time, typically 24 or 48 hours. However, this procedure has several disadvantages: the maximum past pressure may be obscured by such a large load-increment ratio, some secondary compression is usually included, and the tests are time-consuming and expensive. Recently, new test procedures have been introduced which utilize continuously varying load- and pore-pressure measurements (Lowe, et al., 1969; Smith and Wahls, 1969; Wissa and Heilberg, 1969). Although the equipment required to perform such tests is more complex and expensive to operate, the disadvantages of the standard procedure mentioned above are largely overcome.

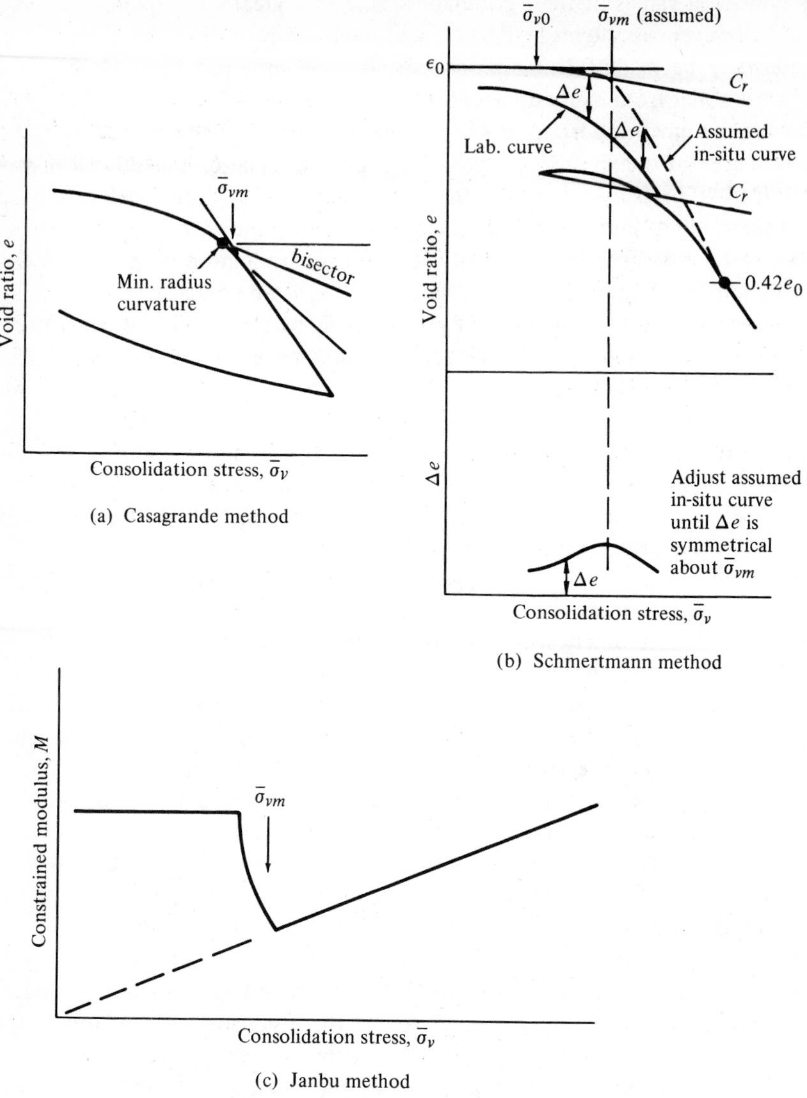

Fig. 12. Methods for determining maximum past pressure.

Time Rate of Compression in Saturated Clays The compression of saturated clay proceeds very slowly. This is due, in part, to the gradual readjustment of the soil structure to the increased load (Mitchell, 1964). However, the major part of the time delay is caused by the low permeability of the material. Thus, volume changes in saturated clay require a long time to complete because the excess water drains out very slowly. This gradual expulsion of water under a constant load is called consolidation. The consolidation process can be described as follows.

When loads are applied to a saturated clay mass, the excess stresses are initially carried by the pore fluid, since no volume change can occur immediately. If one or more drainage surfaces are present, a pore pressure gradient is established and the pore fluid begins to flow toward the drainage surface(s). As the pore fluid gradually drains out of the soil mass, pore pressures are reduced and the excess stresses are transferred to the soil skeleton. This process continues until the pore pressures are completely dissipated and all of the excess stresses are carried by the soil skeleton. The transfer of stress from the pore fluid to the soil skeleton causes compression of the mass and, if the clay is initially saturated, the amount of compression at any time during the consolidation process (volume change) is equal to the amount of water that has drained from the soil.

The consolidation process just described can be modeled mathematically. If the soil skeleton is assumed to be linearly elastic (an approximation) and Darcy's law is assumed to hold true for the flow process, the following differential equation results for one-dimensional consolidation:

$$c_v \frac{\partial^2 u}{\partial z^2} = \frac{\partial u}{\partial t}$$

where

u = excess pore water pressure
z = vertical distance from nearest drainage boundary
t = time

and c_v is the coefficient of consolidation which is equal to:

$$c_v = \frac{k(1+e)}{a_v \gamma_w} = \frac{k}{m_v \gamma_w}$$

where

k = coefficient of permeability
a_v = coefficient of compressibility = $-(\Delta e/\Delta\bar{\sigma}_v)$
m_v = coefficient of volume change = $a_v/(1+e) = \Delta\epsilon_v/\Delta\bar{\sigma}_v$

In order to derive the differential equation for one-dimensional consolidation, it is necessary to assume that m_v is a constant and that k is independent of z. In general, neither of these assumptions holds true for real soil masses and, to the extent that the real soil behavior deviates from the mathematical model, solutions of the equation will only give an approximation of the consolidation process.

Solutions of the differential equation for one-dimensional consolidation with the appropriate boundary conditions are usually presented in terms of the average degree of consolidation $U(\%)$ as a function of a dimensionless time factor, i.e.,

$$U(\%) = f(T_v)$$

where

$$T_v = (c_v/H^2)\,t$$

and H is the maximum length of the drainage path. Solutions for some typical drainage and loading conditions are shown graphically in Fig. 13.

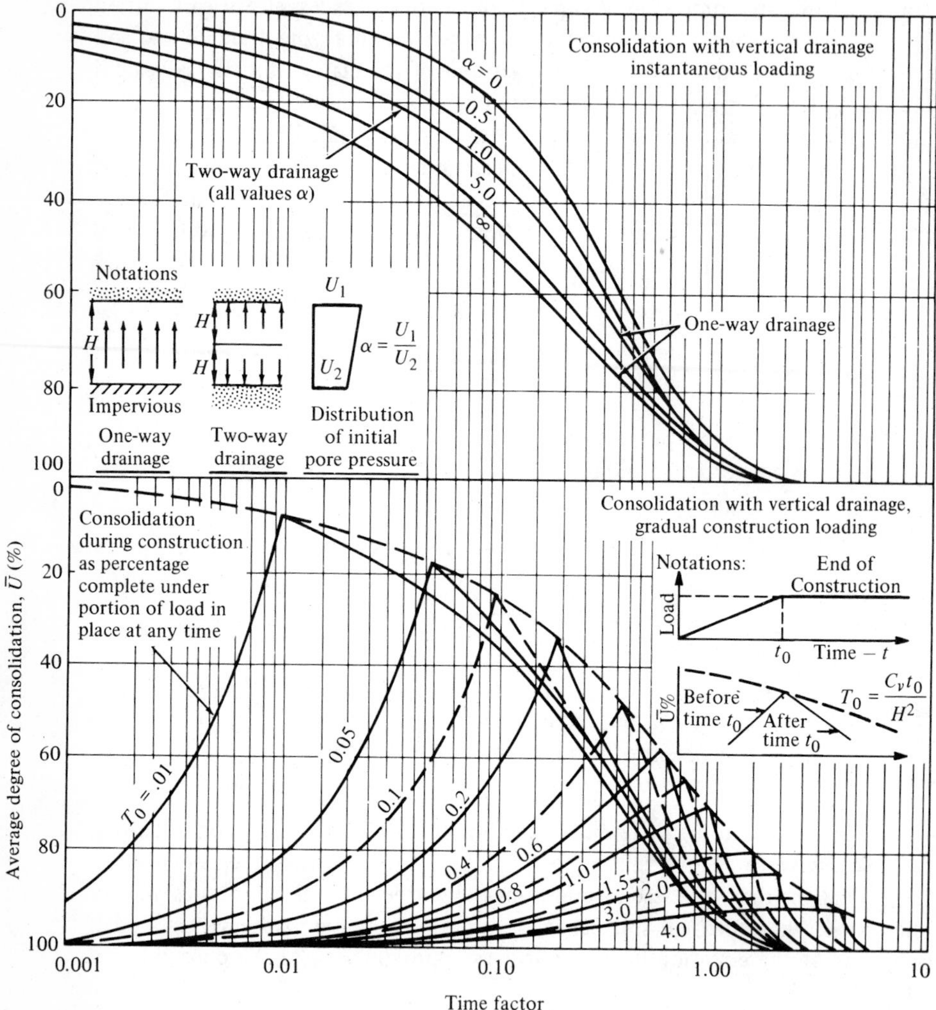

Fig. 13. Degree of consolidation vs. time factor for various conditions of drainage and loading (after Navdocks DM-7, 1962).

The material property that governs the time rate of consolidation is c_v. It is determined experimentally with curve-fitting techniques, by matching the characteristics of the laboratory consolidation curves with those of the theoretical curves. The most commonly used curve-fitting methods, the square root of time and log time methods, are illustrated in Fig. 14. Typically c_v values from the square root of time method are about twice those obtained by the log time method; an average of those two values is probably more nearly correct. In the newer types of consolidation tests mentioned previously, the applied load is not constant and special techniques are necessary for the evaluation of c_v.

The coefficient of consolidation is substantially affected by stress history. It is generally found that c_v is of the order of five to ten times greater for recompression than for virgin compression. Similarly, c_v for swelling is very large initially and then decreases rapidly with further rebound.

Sample disturbance also has an important effect on c_v. In the normally consolidated region, sample disturbance will cause some reduction in the value of

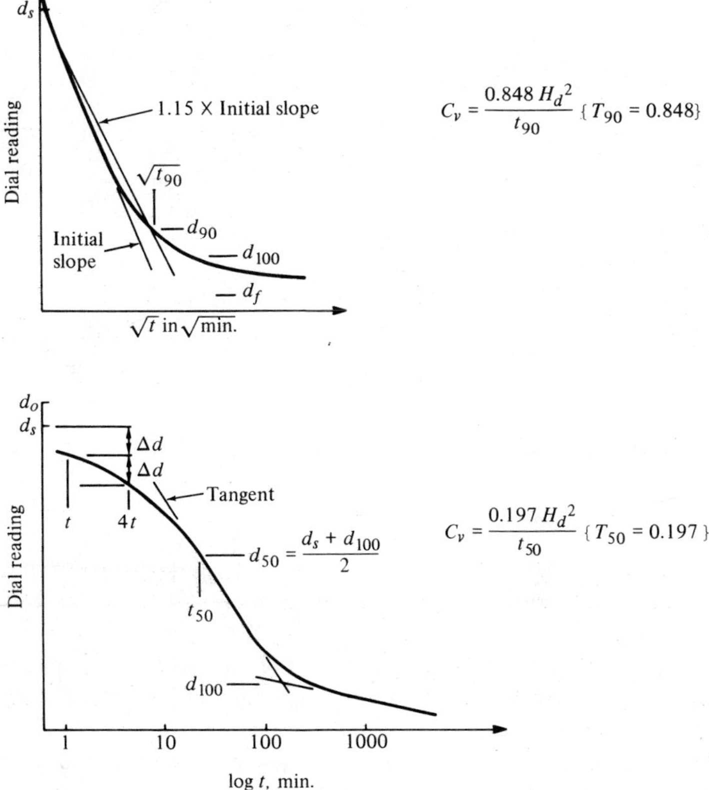

Fig. 14. Methods of calculating coefficient of consolidation.

c_v; however, the effect in this region is usually of minor importance. On the other hand, sample disturbance has a large effect on c_v during initial loading and the laboratory value of c_v in this range is usually unreliable. The coefficient of consolidation for initial loading *in situ* can be more reliably predicted from an unload-reload cycle beyond the maximum past pressure.

For normally consolidated clays a fairly good estimate of the coefficient of consolidation can be made from emperical correlations between c_v and the liquid limit (Navdocks DM-7, 1962) as shown in Fig. 15.

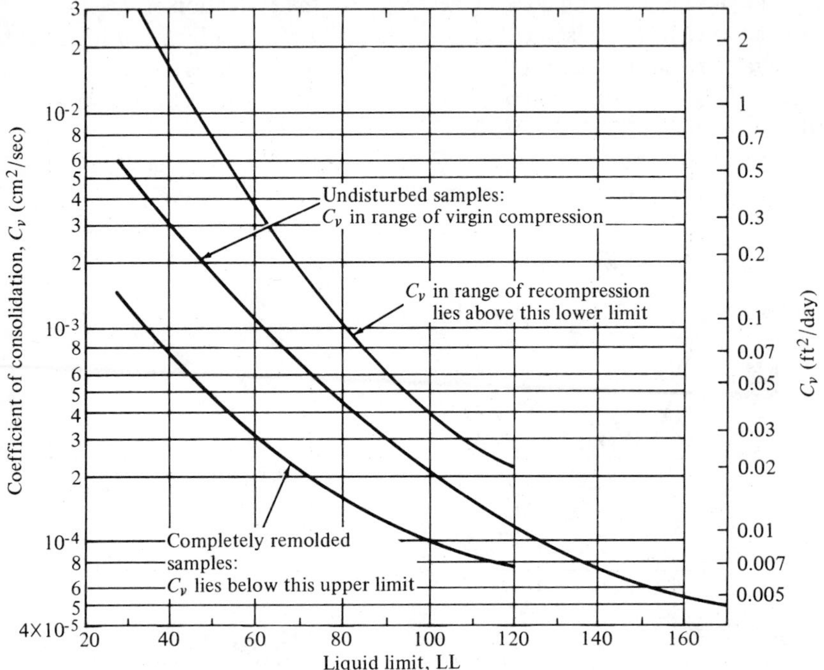

Fig. 15. Coefficient of consolidation vs. liquid limit (after Navdocks DM-7, 1962).

Secondary Compression

The volume reduction that occurs after the excess pore water pressure has essentially dissipated is called secondary compression. Although the basic mechanisms of deformation probably do not change, the distinction between primary and secondary compression is made because the rate of primary compression is governed by hydrodynamic effects while the rate of secondary compression depends upon other factors. The several hypotheses that have been advanced to explain secondary compression have been reviewed by Barden (1969); however, a detailed discussion of this subject is beyond the scope of this book. Nevertheless, it is quite apparent that at the end of primary compression, the soil skeleton has not achieved a state of equilibrium under the applied loads, thus causing deformations to continue with the passage of time.

It is generally observed that the rate of secondary compression follows a relationship of the form

$$\Delta\epsilon_v = C_\alpha \log \frac{t}{t_p}$$

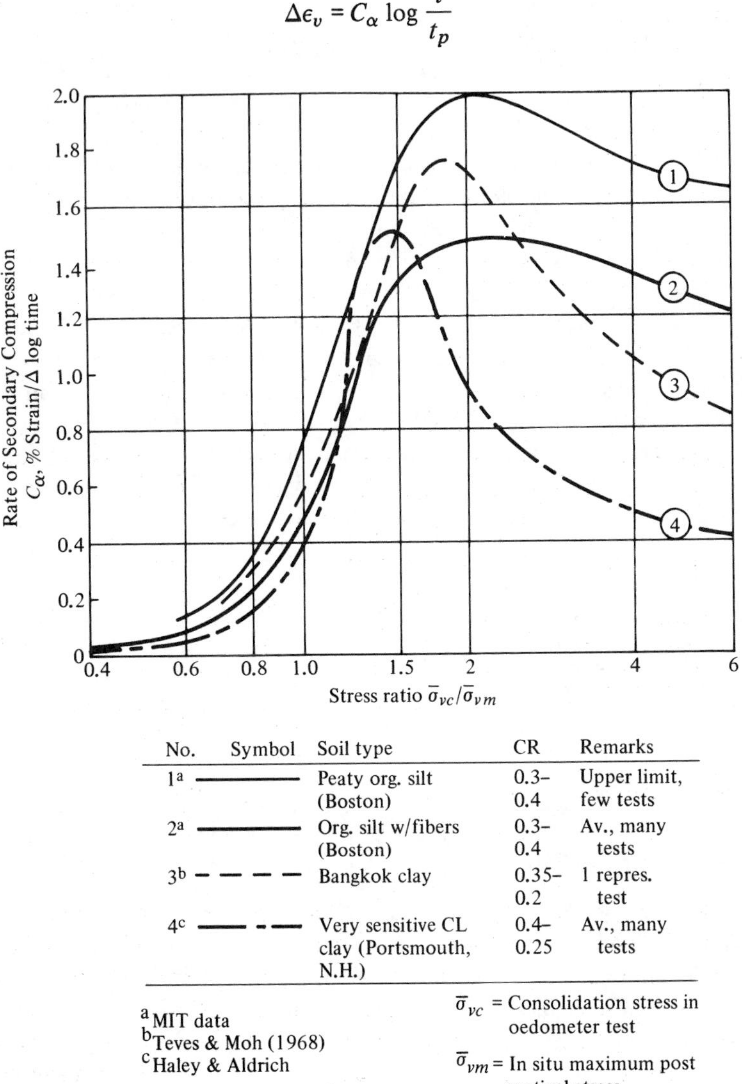

Fig. 16. Effect of stress history on rate of secondary compression (after Ladd, 1971a).

where

$\Delta\epsilon_v$ = vertical strain in the secondary compression range up to time t
t_p = time required for the completion of primary compression
C_α = slope of the ϵ_v vs. log t curve in the secondary range (see Fig. 9)

Furthermore, C_α for most soils is independent of time, thickness of the compressible layer and the consolidation stress (in the normally consolidated range).

The major factors which influence the rate of secondary compression for a given soil appear to be: stress history, level of shear stress (for non-K_0 conditions), and temperature. The effect of stress history on C_α is shown in Fig. 16 for several undisturbed cohesive soils. The general trend is for C_α to increase with decreasing overconsolidation ratio in the preconsolidated range and to remain fairly constant in the normally consolidated range if the compression index remains constant. If the compression index decreases in the normally consolidated range, C_α also decreases. For non-K_0 conditions, C_α tends to increase with increasing levels of shear stress (Bishop, 1966; Bishop and Lovenbury, 1969). C_α also tends to increase somewhat with increasing temperature.

For one-dimensional loading conditions, C_α can be determined from the results of oedometer tests (see Fig. 9). However, high-quality undisturbed samples and careful testing procedures are required, since the value of C_α is influenced by sample disturbance in much the same way as the compression index. Approximate values of C_α can be obtained from the empirical correlations given in Fig. 17.

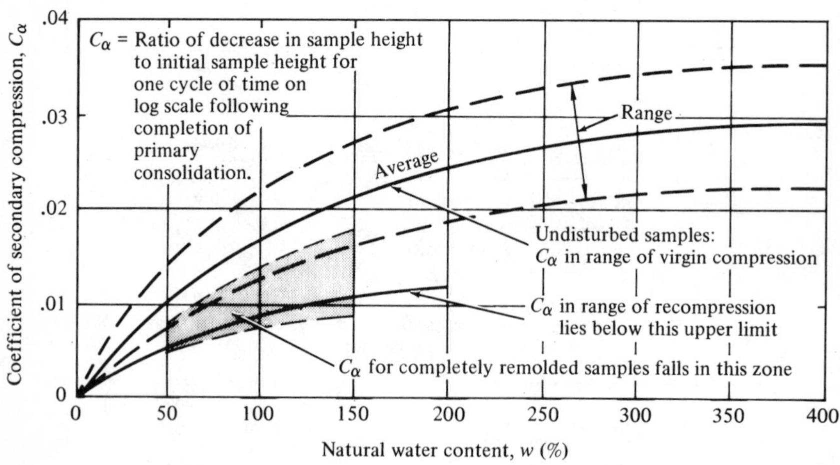

Fig. 17. Rate of secondary compression for fine-grained soils (after Navdocks DM-7, 1962).

Shear Strength and Deformation Properties

On most construction projects the ground movements associated with the construction are an important consideration. To avoid an outright failure the soil mass must be stable under the applied loads and, even if the safety factor is adequate to prevent a stability failure, the possibility exists that the ground movements may be large enough to endanger the project or adjacent structures. An analysis of the stability of a soil mass is a problem that requires a knowledge of the shear strength of the soil. On the other hand, an analysis of ground movements short of outright failure requires a knowledge of the deformation (stress-strain) properties of the soil. In this section the stress-strain and strength properties of soils subjected to various types of loadings will be examined.

Stress at a Point

The state of stress at a point can be represented in terms of the magnitude and orientation of the principal stresses σ_1, σ_2, and σ_3. For two-dimensional cases,

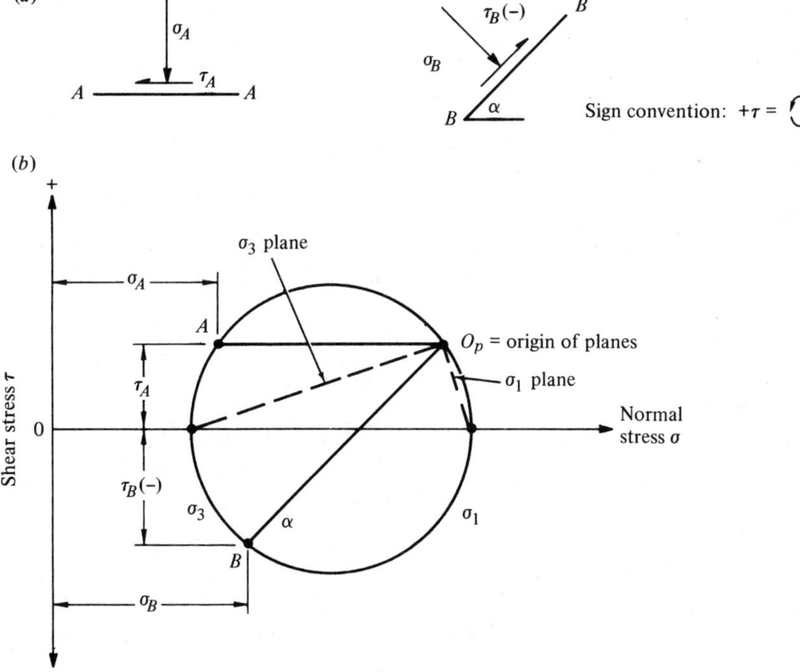

Fig. 18. Stress at a point. (a) Shear and normal stresses of planes A-A and B-B. (b) Mohr's circle of stress.

a common situation in soil engineering problems, Mohr's circle, as shown in Fig. 18 is a convenient method of graphically illustrating the stress system. Points on Mohr's circle represent the shear and normal stresses on different planes whose orientation is given by the lines drawn between the point and the point at the "origin of planes." When changes in stress state are being considered, Mohr's circles of stress can be replaced by a single point having the coordinates

$$q = (\sigma_1 - \sigma_3)/2; \quad p = (\sigma_1 + \sigma_3)/2$$

as shown in Fig. 19. Lines connecting the *q-p* values for each stress state are known as *stress paths*.

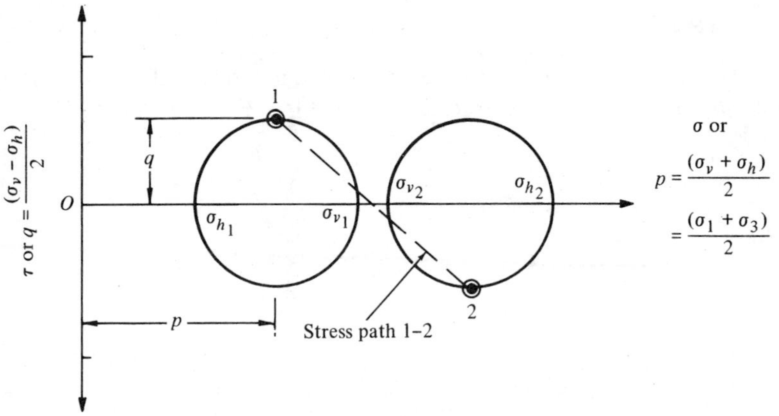

Fig. 19. Stress paths–*q-p* diagram.

Principle of Effective Stress

In a multiphase material like soil, the applied stress is shared by the different phases. This concept may be expressed by the equation

$$\sigma = \bar{\sigma} + u$$

where

σ = total stress
$\bar{\sigma}$ = effective (intergranular) stress
u = neutral stress, or pore pressure

Total, neutral, and effective stresses can be represented in the Mohr diagram or the *q-p* diagram without difficulty. Since the neutral stress has no shear component, the presence of a neutral stress simply causes the Mohr circles (or *q-p* values) for total and effective stresses to be displaced horizontally from one another by an amount equal to the neutral stress, as in Fig. 20.

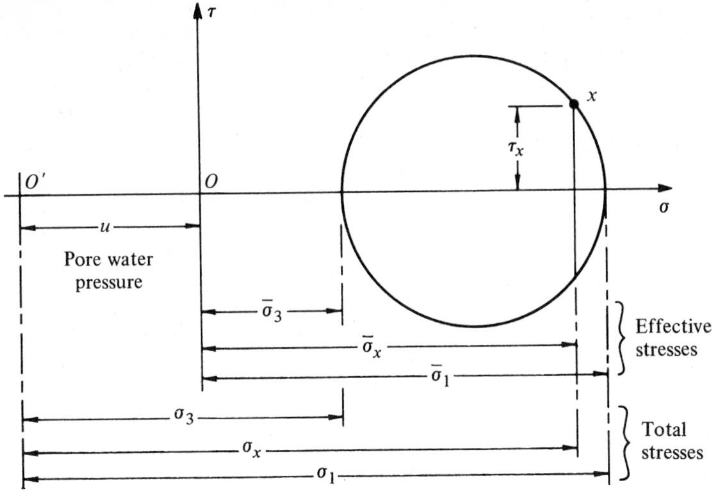

Fig. 20. Relationships between total, neutral and effective stresses in Mohr diagram.

Pore-Pressure Parameters When a soil is subjected to undrained loading, the resulting excess pore pressure is given by the expression (Skempton, 1954)

$$\Delta u = B[\Delta\sigma_3 + A(\Delta\sigma_1 - \Delta\sigma_3)]$$

where

Δu = excess pore pressure resulting from total stress changes $\Delta\sigma_1$ and $\Delta\sigma_3$ in undrained loading,
A, B = pore-pressure parameters.

The parameter B relates to the relative compressibility of the pore fluid and the soil skeleton, i.e.,

$$B = \frac{1}{1 + n(C_v/C_s)}$$

where

n = soil porosity
C_v = compressibility of soil voids
C_s = compressibility of soil skeleton

Thus, for saturated soils, $C_s >> C_v$ and $B = 1$; whereas, for very low degrees of saturation, $C_v >> C_s$ and B approaches zero.

The parameter A is largely a function of the stress history of the soil and the proportion of the failure stress applied. A at failure is approximately unity for normally consolidated soils of average sensitivity, and decreases with increasing overconsolidation ratio. For normally consolidated soils, A increases with increasing levels of shear stress, reflecting the tendency of the soil to consolidate. However, for heavily overconsolidated soils, A decreases with increasing shear stress, reflecting, in this case, the tendency of the soil to expand. Typical behavior patterns of the A and B parameters are shown in Fig. 21.

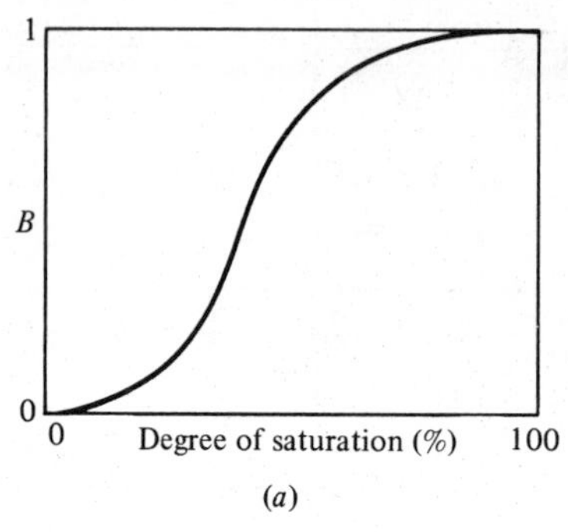

(a)

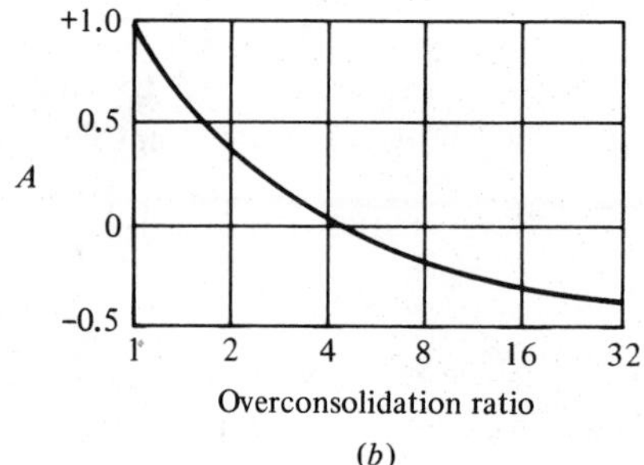

(b)

Fig. 21. Pore pressure parameters. (a) B parameter as a function of degree of saturation. (b) A parameter (at failure) as a function of overconsolidation ratio.

Stress Systems

Geostatic Stresses The stresses in a horizontal soil deposit due to its own weight are called geostatic stresses. In this condition there is no lateral strain, so that the principal stresses act in the horizontal and vertical directions. The ratio of horizontal to vertical (effective) stresses for the geostatic condition is defined as

$$K_0 = \frac{\bar{\sigma}_{hc}}{\bar{\sigma}_{vc}}$$

where K_0 is the coefficient of earth pressure at rest. For normally consolidated clays, K_0 may be estimated from the empirical correlation

$$K_0 = 1 - \sin \bar{\phi}$$

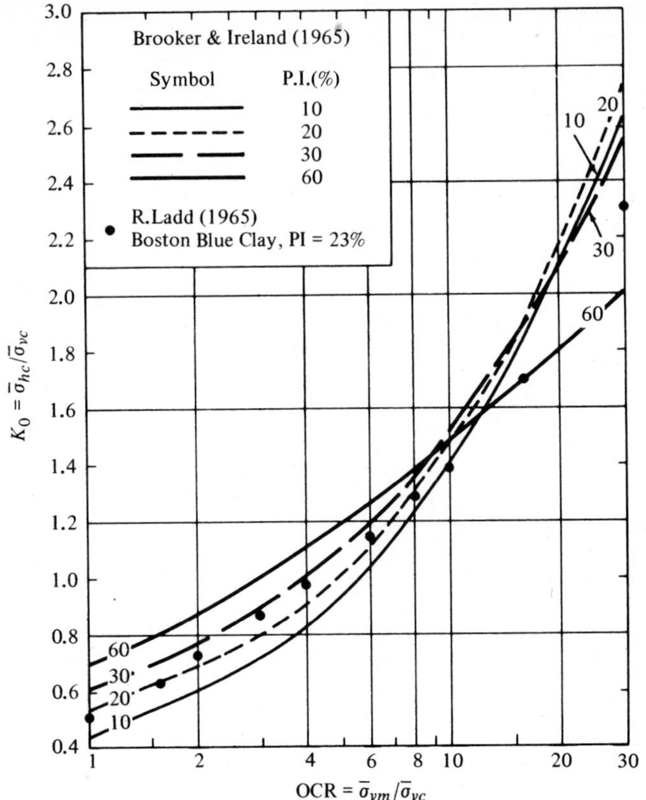

Fig. 22. Variation of K_0 with overconsolidation ratio (after Ladd, 1971a).

K_0 increases with increasing overconsolidation ratio. Ladd (1971a) reports data, shown in Fig. 22, on the variation of K_0 with overconsolidation ratio for soils of varying plasticity.

Typical Stress Systems in the Field A great variety of stress systems may be encountered in the field, a situation that results in variations in magnitude and orientation of principal stresses and in pore pressure. Some cases are too complicated to be duplicated in existing laboratory testing apparatus; however, many typical cases fall into categories for which laboratory tests are available. These categories are triaxial compression and extension, plane strain (active and passive cases), and simple shear. A few typical examples are shown in Fig. 23.

Drainage Conditions

In a field situation the rate of dissipation of excess pore pressure during and after construction depends upon many factors, e.g., type of loading, rate of

(*a*) Under Centerline of a Circular Footing

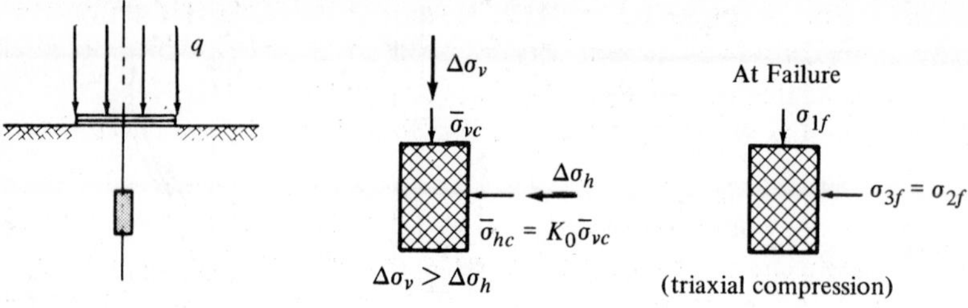

(*b*) Under Centerline of a Circular Excavation

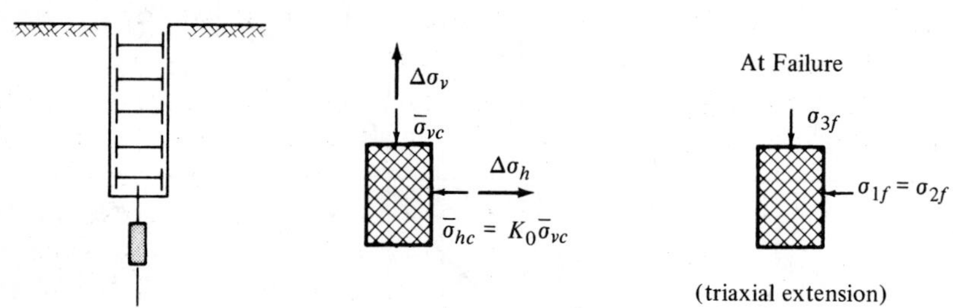

(*c*) Under Centerline of a Strip Footing

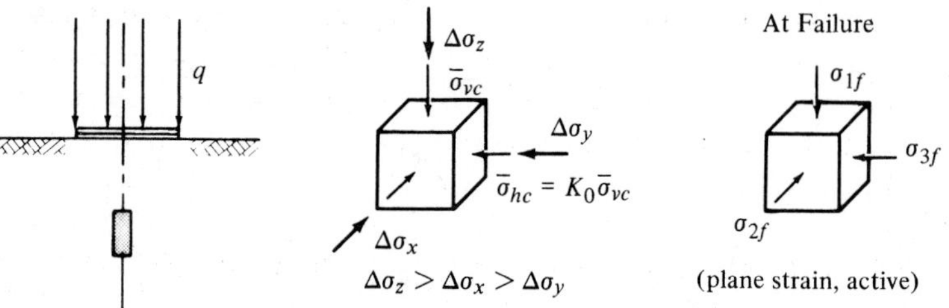

(*d*) Retaining Wall with a Passive Pressure

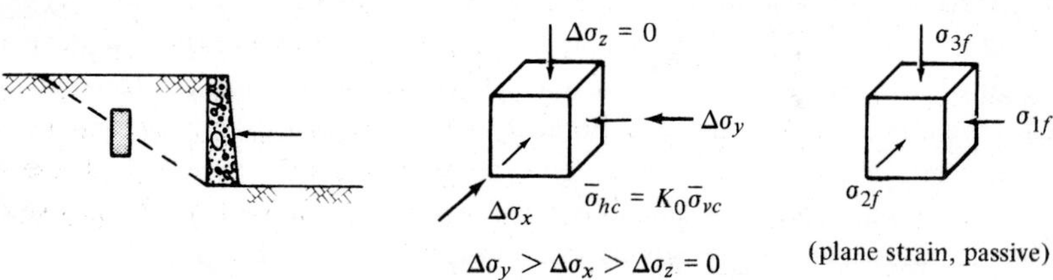

Fig. 23. Typical stress systems in the field for a normally consolidated clay (after Ladd, 1965). Note: Footings and walls are assumed to have frictionless surfaces.

loading, permeability of the soil, and boundary conditions with respect to drainage. For purposes of analysis, however, the drainage conditions are usually approximated as one of the following cases: unconsolidated-undrained (UU), consolidated-undrained (CU), or consolidated-drained (CD). These represent limiting conditions that can be investigated with conventional laboratory testing devices.

Unconsolidated-Undrained In this case, construction occurs so rapidly that no drainage can occur during the construction period. Problems of this type are appropriately treated by *total stress analysis* since the stress–strain and strength properties that apply under these conditions are those existing prior to construction and the excess pore pressure that develops is a function (dependent variable) of the applied loads. An example of UU conditions is rapid construction of an embankment over a soft clay foundation.

Consolidated-Undrained This condition corresponds to a situation where additional loads are applied rapidly after the soil has first been allowed to consolidate under a previously applied load. The strength and deformation properties that apply in this case are those existing after consolidation under the initial loading and prior to the application of the additional loads; the excess pore pressure that develops is, again, a function of the applied (additional) loads. These problems are treated by either total or *effective stress analysis*, depending upon the preference of the engineer. Total stress analysis is easier to apply but, since the pore pressures are not evaluated separately, they cannot be checked against field measurements during construction. On the other hand, effective stress analysis permits utilization of field measurements of pore pressure dissipation, but it also requires somewhat more elaborate testing procedures to measure pore pressure behavior during undrained shear. An example of CU conditions is stage construction of an embankment over a soft clay foundation where the foundation soil is permitted to consolidate (partially or completely) under one lift before the next lift is placed.

Consolidated-Drained This condition applies when there are no excess pore pressures as a result of applied loads. The *in-situ* effective-stress soil parameters apply in this case and the pore pressure is an independent variable (dependent on ground water or steady seepage conditions). An effective stress analysis is required because the existing pore pressures are unrelated to the applied loads. Examples of CD conditions are cases where the loads are applied so slowly that all excess pore pressures completely dissipate during construction and long-term stability problems where any excess pore pressure that may have developed during construction has fully dissipated.

Stress–Strain Properties

The stress–strain behavior of soils is affected by many factors, such as the type of soil, stress history prior to loading, type of loading, rate of loading, and drainage conditions. Because of the large number of important variables involved, generalizations about soil stress–strain behavior are difficult. Therefore, for a complete investigation of any particular case, laboratory and/or field tests are usually required with appropriate consideration given to the factors just mentioned. The following is a brief discussion of typical stress–strain behavior of various types of soils and the effects of some of the variables that influence this behavior.

Typical Stress-Strain Behavior in Triaxial Compression The behavior of soils in triaxial compression following isotropic consolidation will form the basis for a description of typical stress–strain behavior. This particular loading condition is selected because it is a common laboratory testing procedure and serves well to illustrate the basic stress–strain properties of various soil types under different drainage conditions (for a comprehensive treatment of triaxial testing procedures, the reader is referred to Bishop and Henkel, 1962).

For purposes of discussion, soils may be categorized as *nondilatant*, i.e., those which tend to undergo volume decrease during shear; and *dilatant*, i.e., those which tend to expand during shear. Examples of nondilatant soils are normally consolidated clays and loose sands; examples of dilatant soils are stiff, overconsolidated clays and dense sands. Typical stress–strain behavior for soils in both of these categories is shown in Fig. 24 for drained and undrained triaxial compression following isotropic consolidation.

The nondilatant soils are characterized by a strain-hardening type of stress–strain curve over the entire range of strain that would normally be of practical interest. Typically, failure (in terms of maximum principal stress difference) occurs at strains of 10% or more. As a general rule the strain at failure is greater for drained tests than for undrained.

By definition, the nondilatant soils undergo volume decrease in the drained test; in the undrained condition the tendency of these soils to compress during shear gives rise to positive excess pore pressure ($A_f > 0$). As a result, the effective normal stresses increase in the drained case and decrease in the undrained case. Therefore, for a given soil at a given consolidation pressure, the maximum principal stress difference is greater for drained loading than for undrained loading.

Dilatant soils are usually strain-hardening in undrained loading but may exhibit pronounced strain-softening characteristics in the drained condition; i.e., a peak stress is reached at relatively low strain followed by a reduction in strength with increasing strain. This feature of stress–strain behavior in dilatant soils has important implications for the long-term stability of slopes in stiff overconsolidated clays.

Typically, dilatant soils undergo a slight volume decrease (or pore pressure increase) at small strains, followed by a volume increase (or pore pressure decrease) at large strains. For these soils, A_f is less than zero.

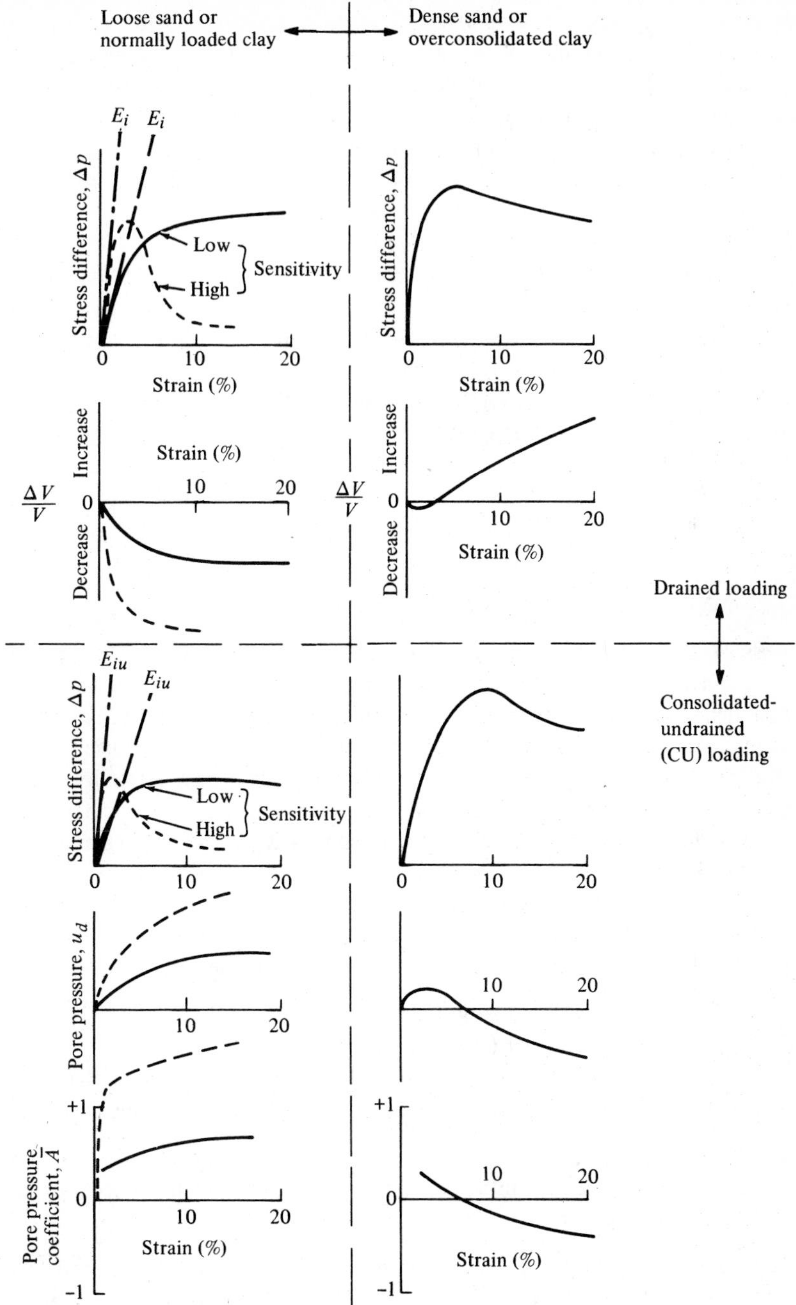

Fig. 24. Typical stress–strain behavior of soils in triaxial compression (after Terzaghi and Peck, 1967).

Effects of Anisotropic Consolidation The stress system for soils *in-situ* during consolidation is generally anisotropic; i.e., for horizontal soil deposits, the K_0 condition applies and for sloping deposits the ratio of principal stresses is somewhere between K_0 and K_f, where, K_f is the principal stress ratio at failure. Moreover, anisotropic consolidation can have a significant effect on the subsequent stress–strain behavior of the soil when additional loads are applied. This is particularly true for normally consolidated clays because K_0 is very different from unity ($K = 1$ for isotropic consolidation).

As an example of the effect of anisotropic consolidation, Fig. 25 shows a comparison of stress–strain behavior in CU triaxial compression for two normally consolidated clays. In this case, anisotropic compression results in much smaller strain at failure and a more pronounced strain-softening effect.

To account for this effect of anisotropic consolidation, Bjerrum and Lo (1963) suggested that stress–strain behavior in CU triaxial compression tests for isotropic and anisotropic consolidation could be correlated with the equation

$$\frac{\sigma_1 - \sigma_3}{\sigma_{1_c}} = \left(\frac{\sigma_1}{\sigma_3} - 1\right)\left[1 - \left(\frac{\Delta u}{\sigma_{1_c}} + 1 - K\right)\right]$$

provided $\Delta\sigma_3 = 0$ and $A = 1$ (Ladd and Varallyay, 1965).

Effects of Stress System During Shear Variations in the stress system during shear from the conventional triaxial compression test may involve (1) changes in the magnitude of the intermediate principal stress (σ_2) and (2) rotation of principal planes. Some of the common loading conditions that cause variations in the magnitude of σ_2 are triaxial extension, plane–strain active loading and plane–strain passive loading. Rotation of principal planes occurs in triaxial extension and plane–strain passive loading (assuming $K_0 \leqq 1$), and in direct simple shear.

In a very general sense, changes in magnitude of σ_2 and rotation of principal planes during undrained loading following K_0 consolidation tend to produce larger strains to failure and increased pore pressures for the following reasons:

1. Variation of σ_2 will cause greater variation of the octahedral shear stress ($\Delta\sigma_{oct} = [(\Delta\sigma_1 - \Delta\sigma_2) + (\Delta\sigma_2 - \Delta\sigma_3) + (\Delta\sigma_1 - \Delta\sigma_3)]$).
2. Rotation of principal stresses requires greater readjustment of the internal structure of the soil to accommodate the changing directions of the principal stresses.

The more K_0 differs from unity, the more pronounced are the effects of variations in the stress system.

Effects of Loading Rate Excluding very high rates of loading, i.e., dynamic loading, strain rate effects on the stress–strain behavior of soils are generally

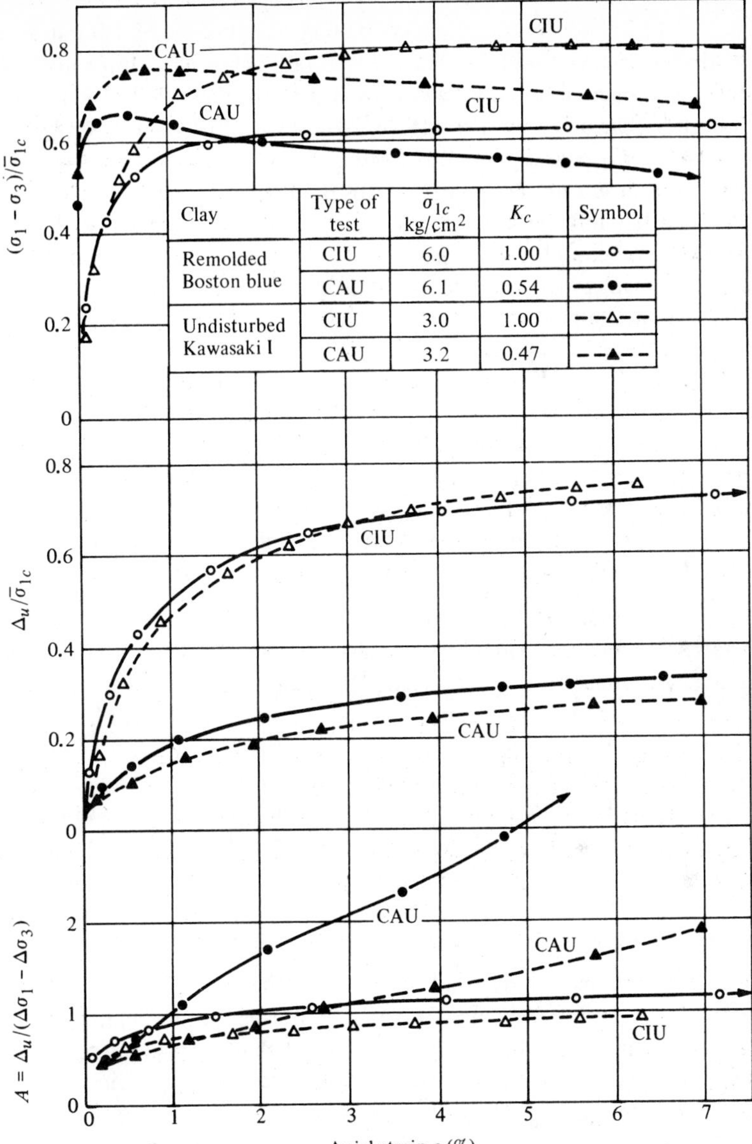

Fig. 25. Effect of anisotropic consolidation on the stress–strain behavior of two normally consolidated clays (after Ladd, 1965).

minor. Most of the effects of time of loading observed in the laboratory (e.g., membrane leakage, incomplete pore pressure redistribution or incomplete drainage and insufficient secondary compression prior to shear; all of which lead to increased pore pressures and, therefore, reduced strength) appear to be related to the testing procedures themselves rather than the true behavior of the soil as it would occur *in situ*. Field data on this point is, so far, inconclusive.

In a somewhat different sense, however, creep under constant load or stress relaxation under constant deformation may cause important time effects. For the case of creep, Singh and Mitchell (1969) have proposed a three-parameter model based on rate-process theory as follows:

$$\dot{\epsilon} = Ae^{\alpha D}\left(\frac{t_1}{t}\right)m$$

where

$\dot{\epsilon}$ = creep strain rate
D = "deviator stress" = $\sigma_1 - \sigma_3$
t = time after application of deviator stress
A, α, m = model parameters

This equation describes a creep relationship that is linear on a log-log plot of strain rate vs. time. The model parameters are empirical quantities that can be determined from laboratory creep tests. Based on available data, the Singh and Mitchell model appears to offer the most promise for describing soil-creep phenomena to date.

Singh and Mitchell (1969) found that if $m \geqslant 1$, creep failure would not occur. However, if $m < 1$, failure will eventually take place and the time to failure can be found by plotting log $\dot{\epsilon}t$ vs. t, where, from the preceding equation,

$$\dot{\epsilon}t = At_1{}^m e^{\alpha D} t^{1-m}$$

Failure is defined as the point where log $\dot{\epsilon}t$ undergoes a marked increase.

Elastic Constants In a number of applications the equations of the theory of elasticity may be used to estimate stresses and deformations in soil masses. To do this, it is necessary to express the stress-strain behavior in terms of the elastic modulus and Poisson's ratio. For uniaxial compression (in the z direction),

$$E = \sigma_z/\epsilon_z = \text{Young's modulus}$$

$$\mu = -(\epsilon_x/\epsilon_z) = -(\epsilon_y/\epsilon_z) = \text{Poisson's ratio}$$

and, for shear (in the xz plane),

$$G = \tau_{xz}/\gamma_{xz} = \tau_{zx}/\gamma_{zx} = \text{shear modulus}$$

Actually G and E are interrelated, i.e.,

$$G = E/2(1 + \mu)$$

so that only two of the three elastic constants is needed for a complete description of the elastic properties of the soil.

For soils these quantities are not, in fact, constants. As was pointed out previously in this section, there are numerous factors which influence stress–strain behavior of soils. Thus, treating these quantities as constants is merely a necessary expedient in order to be able to apply the equations of theory of elasticity.

The application of the theory of elasticity works best at small strains where the assumption of linear stress–strain behavior is reasonably correct. Hardin and Black (1968) have found that, in the small-strain range, the shear modulus is a function only of the void ratio and effective consolidation stress, i.e.,

$$G = \frac{2630\,(2.17 - e)^2}{1 + e}(\bar{\sigma}_0) \qquad \text{(round-grained sands, } e < 0.80\text{)}$$

$$G = \frac{1230\,(2.97 - e)^2}{1 + e}(\bar{\sigma}_0) \qquad \text{(angular-grained materials)}$$

These values may be considered as upper limits, since the modulus decreases with increasing strain. If Poisson's ratio is known or can be estimated with reasonable accuracy, the values of the shear modulus can be converted to values for Young's modulus using the equations relating G and E. The shear modulus is often a more convenient parameter for experimental determination than Young's modulus because it is essentially independent of moisture conditions (shear stresses are transmitted entirely through the soil skeleton, whereas compressive stresses are carried by both the soil skeleton and the pore fluid).

Poisson's ratio for sand typically varies from 0.1–0.2 in the early stages of loading to 0.5 or more near failure; a fairly representative average value for sand would be approximately $\frac{1}{3}$. Fortunately, the value of Poisson's ratio used in the calculations does not, as a rule, have much influence on the result. For normally consolidated saturated clays, Poisson's ratio is 0.5 (incompressible); for undrained loading, decreasing to $\mu \approx \frac{1}{3}$ for drained conditions.

For larger strains, the modulus is commonly taken as the slope of the secant line from zero stress to the stress level appropriate for the problem under consideration. A glance at any of the stress–strain curves presented previously will reveal that the secant modulus decreases with increasing stress level (or strain). Konder (1963) and Konder and Żelasko (1963) report that the stress–strain curves for many soils, both granular and cohesive, have a hyperbolic form up to the failure stress, which can be described as

$$\sigma_1 - \sigma_3 = \epsilon_1/(a + b\epsilon_1)$$

where

$\sigma_1 - \sigma_3$ = "deviator" stress
ϵ_1 = axial strain
a, b = constants determined by curve fitting

This equation, which has now been verified for a wide variety of soil types, provides a simple mathematical expression for the strain dependence of the secant modulus. Rewriting it:

$$(\sigma_1 - \sigma_3)/\epsilon_1 = E = 1/(a + b\epsilon_1)$$

it can be seen that the initial tangent modulus is $E_i = 1/a$, and for larger strains the parameter b describes the secant-modulus variation with strain. An expression of this type is extremely useful in connection with finite-element computer programs in which nonlinear elastic properties can be utilized.

Stress history affects the modulus in several ways. Laboratory results indicate that the modulus increases with successive load cycles. Most of this increase is due to the gradual elimination of seating imperfections, sample defects, and the effects of sample disturbance. These are all factors that do not exist *in situ*. Therefore, it is recommended that several cycles of loading be employed in the laboratory determination of E.

A second stress-history effect is that of overconsolidation. As a general rule, an increase in the overconsolidation ratio causes some increase in the modulus, particularly at low stress, levels. However, the data on this point are somewhat inconclusive and, in any case, the effect is quite small.

Anisotropic consolidation also has a definite effect on the modulus, as can be seen from the stress–strain curves presented in Fig. 25. However, just what that effect will be depends primarily on the type of stress system employed in subsequent loading. For example, anisotropic consolidation produces opposite effects on the modulus for triaxial extension (E decreases) and compression (E increases).

From the data and discussion presented in this section, it should be apparent that modulus values for soils are affected by many factors and can vary over wide limits. Therefore, if accurate values are important, and particularly where complicated stress systems are concerned, the preferred approach is to determine the modulus (or strains) directly in laboratory tests on undisturbed samples, duplicating as closely as possible the *in-situ* conditions.

Shear Strength

Shear strength of soils is important in those cases where the stability of a soil mass is to be analyzed. Examples of this class of problems include: ultimate bearing capacity of foundations, stability of slopes and embankments, and earth-pressure problems.

Drainage conditions influence the shear-strength characteristics of soils more than any other single factor. In turn, the drainage conditions in a field situation depend upon the rate of loading and the permeability of the soil. Thus, the evaluation of field drainage conditions and choice of the appropriate counterpart in the laboratory testing program is extremely important.

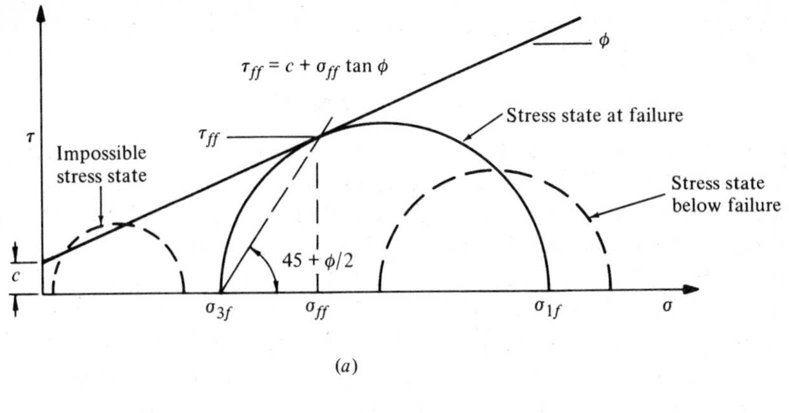

(a)

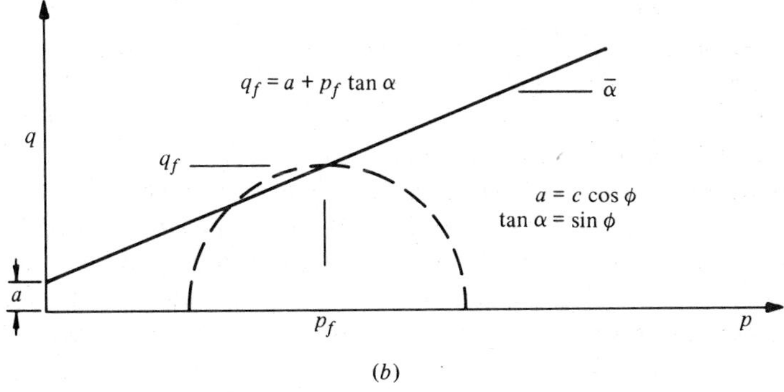

(b)

Fig. 26. Mohr–Coulomb failure criterion. (a) Mohr diagram. (b) q-p diagram.

Mohr-Coulomb Failure Criterion For practical purposes, the Mohr–Coulomb failure criterion adequately expresses the shear-strength characteristics of soils. This failure criterion, shown graphically in Fig. 26, may be expressed mathematically as follows:

$$\tau_{ff} = c + \sigma_{ff} \tan \phi$$

where

τ_{ff} = shear stress on the failure plane at failure
c = "cohesion" intercept on the τ axis
σ_{ff} = normal stress on the failure plane at failure
ϕ = angle of "internal friction" (slope of failure envelope)

or

$$(\sigma_1 - \sigma_3)/2 = [(\sigma_1 + \sigma_3)/2] \sin \phi + c \cos \phi$$

or

$$q = a + p \tan \alpha$$

where

$q = (\sigma_1 - \sigma_3)/2$
$p = (\sigma_1 + \sigma_3)/2$
a = intercept on the q axis ($a = c \cos \phi$)
α = slope of the failure envelope on the $q - p$ plot ($\tan \alpha = \sin \phi$)

The failure criterion may be expressed in terms of total or effective stresses depending upon the method of analysis being used. It should be noted, however, that the shear strength is governed by effective stresses, not total stresses, and the corresponding shear strength parameters $\bar{c}$ and $\bar{\phi}$ (or, alternatively, $\bar{a}$ and $\bar{\alpha}$) represent fundamental soil properties for the conditions under which they were determined. On the other hand, the total stress parameters are subject to change if the pore pressures are changed. Nevertheless, in certain situations the total stress representation may be more convenient.

Granular Soils The failure envelope for granular soils is characterized by a zero cohesion intercept (except when moisture causes "apparent cohesion" by capillary action) and a friction angle that varies between the approximate limits of $28°$ to $50°$ depending upon the composition of the soil and the conditions of the test.

The major factors that influence the friction angle of granular soils are described below:

1. Effect of Composition. Composition refers to mineral type, grain size, and grain shape. Mineral type has very little influence on friction angle except for mica sands, which may exhibit low friction angles due to high void ratio. The main effect of composition on friction angle is due to grain shape and gradation, as demonstrated by the values compiled in Table 6.

2. Effect of Initial Void Ratio. In granular soils, a substantial component of

TABLE 6. ANGLE OF INTERNAL FRICTION OF COHESIONLESS SOILS COMPOSED LARGELY OF QUARTZ (AFTER SOWERS AND SOWERS, 1961).

Description	Angle of Internal Friction	
	D_d less than 20	D_d over 70
Rounded, uniform	29	35
Rounded, well-graded	32	38
Angular, uniform	35	43
Angular, well-graded	37	45

the shear strength is due to interlocking of grains, which increases as the soil is made more dense. Therefore, the initial void ratio exerts a large influence on the friction angle. Typically, the variation in ϕ for a granular soil from the loosest to the densest possible states is on the order of eight degrees. This type of variation is approximated by the empirical equation

$$\phi = \phi_r + 8R_D$$

where ϕ_r is the angle of repose (roughly equivalent to friction angle for loosest condition).

Saturated Cohesive Soils The shear strength of saturated, cohesive soils depends primarily upon (1) the drainage conditions and (2) the prior stress history of the soils. Typical shear-strength characteristics for these soils are illustrated in Fig. 27.

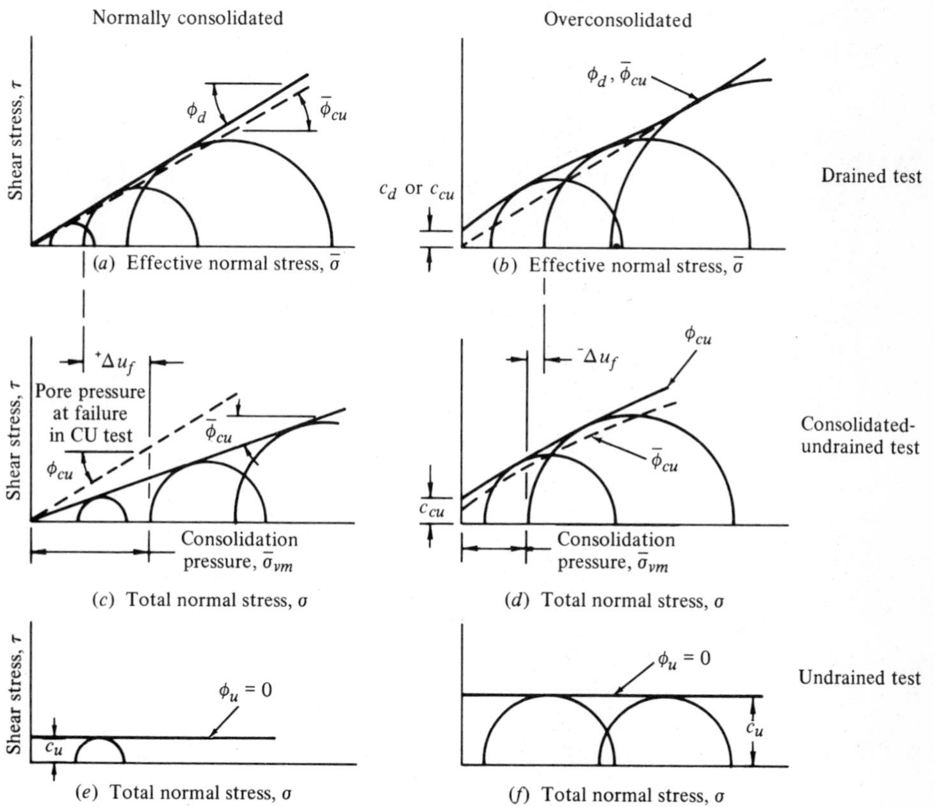

Fig. 27. Shear strength characteristics of saturated, cohesive soils.

Depending upon the drainage conditions of the test, the shear strength may be expressed as:

1. drained test: $\tau_{ff} = c_d + \bar{\sigma}_{ff} \tan \phi_d$
2. consolidated-undrained test:
   a. effective stress parameters:

$$\tau_{ff} = \bar{c} + \bar{\sigma}_{ff} \tan \bar{\phi}$$

   b. total stress parameters:

$$\tau_{ff} = c_{cu} + \sigma_{ff} \tan \phi_{cu}$$

3. undrained test: $\tau_{ff} = c_u = s_u$

In a drained test, all stresses are effective stresses and c_d and ϕ_d are effective stress parameters. In the case of uncemented normally consolidated clays, c_d is equal to zero. The value of ϕ_d can be expected to vary with the plasticity index approximately as shown in Fig. 28. However, the statistical relation illustrated in Fig. 28 should not be used in lieu of laboratory tests because exceptions are common.

For overconsolidated clays with a preconsolidation pressure of less than 5-10 kg/cm^2, c_d probably should not exceed 0.05-0.10 kg/cm^2 at low stresses (Ladd, 1971b). However, the value of c_d can be extremely important in stability problems under low confining stresses.

In performing drained shear tests in the laboratory, the most important consid-

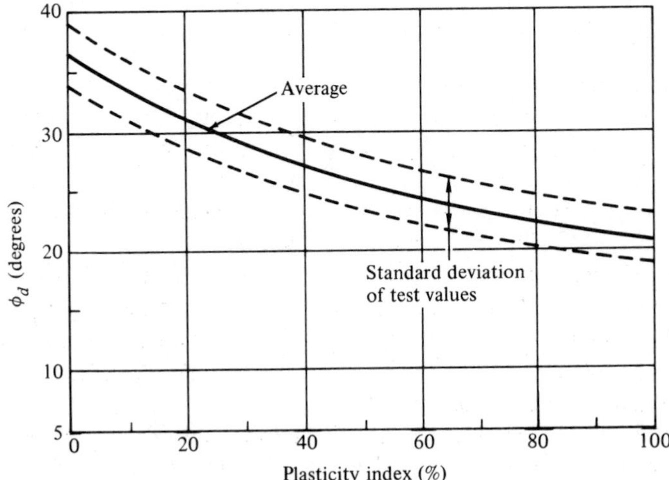

Fig. 28. Approximate relationship between ϕ_d and plasticity index for normally consolidated clays.

eration is to ensure complete dissipation of excess pore water pressures. Failure to do so can lead to very misleading results.

Consolidated-undrained tests can be used to determine either total or effective stress parameters. When CU tests are used to determine the effective stress parameters, care must be taken to avoid possible errors in pore-pressure measurements due to membrane leakage, incomplete saturation of the soil and the measuring system, or nonequalization of pore pressures in the sample. If the tests are properly conducted, the effective stress parameters determined by CU tests agree fairly well with the results of drained tests. However, CU tests are more susceptible to varying interpretations of failure criteria and the effects of different stress systems during shear. In certain circumstances, particularly with overconsolidated clays, these differences can be quite significant.

The total stress parameters c_{cu} and ϕ_{cu} obtained from conventional interpretation of CU tests can be very misleading because of sample disturbance and differences in the stress system *in situ* and in the laboratory tests. However, these effects can be largely overcome by applying more appropriate stress systems in the laboratory (see Fig. 23) and by using parameters which are "normalized" with respect to the consolidation stress. The basic elements of this approach, which has been proposed by Ladd (1971b), are as follows:

1. The properties are normalized, i.e., divided by the consolidation stress.
2. The samples are subjected to stresses sufficiently greater than the *in-situ* stresses to overcome the effects of sample disturbance.
3. The samples are consolidated under the same stress ratio (e.g., K_0 stresses) and degree of aging (i.e., the same number of cycles of secondary compression).

The application of this approach is illustrated in Fig. 29. Figure 29a shows typical undrained shear-strength data which might be obtained from a series of CU tests. The open circles represent samples tested in the overconsolidated range. Samples tested in the normally consolidated range define a constant ratio of undrained shear strength to consolidation stress; whereas, in the overconsolidated range, the relationship is nonlinear. These data are compiled and plotted in Fig. 29b, showing the normalized undrained shear strength $(S_u/\bar{\sigma}_{vc})$ as a function of the overconsolidation ratio $(\bar{\sigma}_{vm}/\bar{\sigma}_{vc})$. For practical purposes, these data should be unique for a given deposit. With a knowledge of the preconsolidation pressure and present overburden pressure *in-situ*, the undrained shear strength can be obtained directly from Fig. 29b.

The undrained shear strength can also be determined from unconsolidated-undrained tests, such as unconfined-compression, triaxial UU, or field-vane tests. However, the results are usually less satisfactory than those obtained by procedures outlined above. Field-vane tests frequently produce strengths that are too high, particularly if the clay is heterogeneous, whereas triaxial UU and uncon-

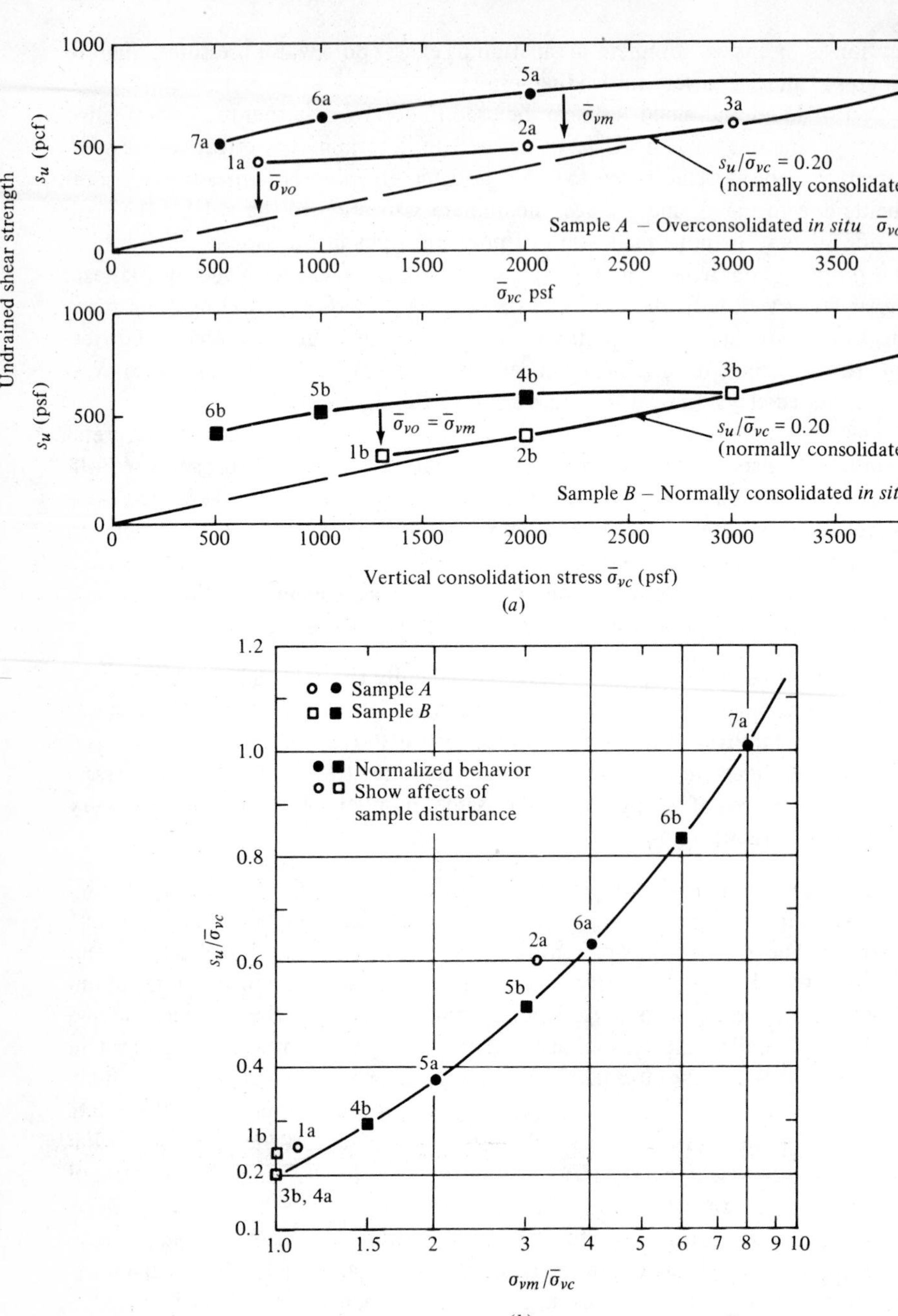

Fig. 29. Normalized undrained shear strength parameters (after Ladd, 1971b).

fined compression tests often produce strengths that are too low because of sample disturbance.

A rough approximation of the undrained shear strength of normally consolidated clays can be obtained from the statistical relation between the ratio of undrained shear strength to the effective overburden pressure and the plasticity index compiled by Skempton (1957). This relation, shown in Fig. 30, has been verified for a variety of clays. However, as with any statistical relation, it should be used as a guideline, not as a substitute for laboratory tests.

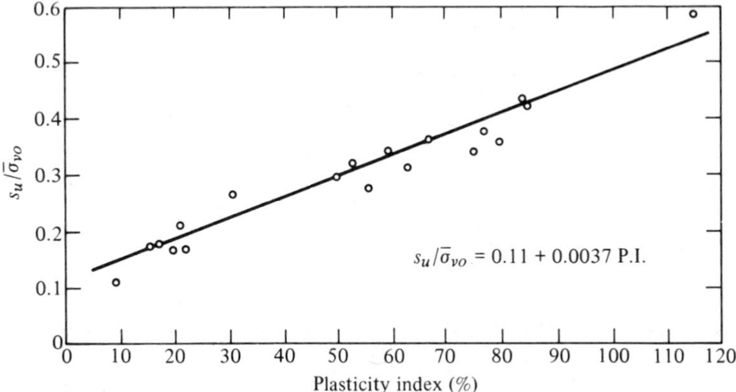

Fig. 30. Statistical relation between $S_u/\bar{\sigma}_{vo}$ ratio and plasticity index (after Skempton, 1957).

Stiff, Fissured Clays Heavily overconsolidated clays often contain networks of minute cracks. If the pressure on such clays is reduced (e.g., by excavation or natural erosion) the strength may be reduced to a fraction of the original value. Furthermore, laboratory tests conducted on small samples may overestimate the strength of fissured clays because the smaller the sample, the less chance there is that it will contain fissures.

Some heavily overconsolidated clays are susceptible to a process known as "progressive failure" (Bjerrum, 1967). Progressive failure, which is the result of a gradual reduction in shear strength, has produced numerous slope failures in overconsolidated clays; often these failures occur years after the cut is made. The gradual reduction in strength in heavily overconsolidated clays is the result of the typical stress-strain-strength behavior shown in Fig. 31. It can be seen from Fig. 31 that the shear strength, after reaching a peak strength at relatively low strain, progressively decreases with increasing strain. At large strains, the strength approaches a constant value known as the residual strength. Similarly, failure envelopes plotted for both peak strength and residual strength show a marked reduction in the values of the shear-strength parameters at large strain.

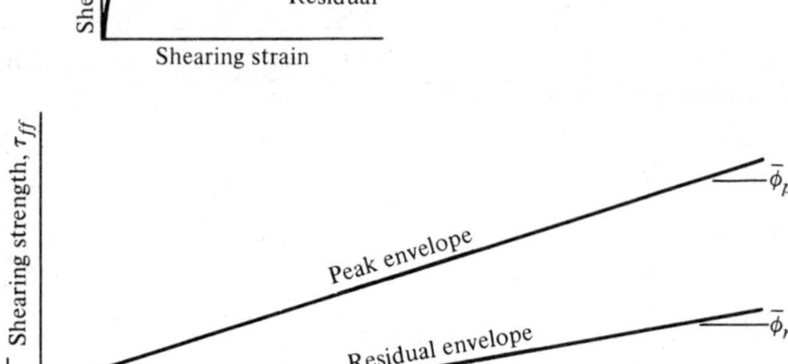

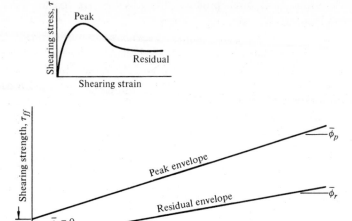

Fig. 31. Shear characteristics of typical heavily overconsolidated clay.

Typically, the residual cohesion is zero and the residual angle of internal friction is reduced to one-half or less of the peak value.

Experience has shown that some heavily overconsolidated clays are susceptible to progressive failure, while some are not. Bjerrum (1967) suggests that the potential for progressive failure of slopes depends upon three factors:

1. ratio of horizontal stresses to peak shear strength (p_H/s_{peak}),
2. ratio of horizontal strain to strain at peak strength ($\epsilon_H/\epsilon_{\text{peak}}$), and
3. ratio of peak strength to residual strength ($s_{\text{peak}}/s_{\text{res}}$).

Bjerrum's evaluation of the influence of these factors on the potential for progressive failure is given in Table 7. At the present state of knowledge, the choice of shear strength must be based upon prior experience and geologic evidence.

Partially Saturated Cohesive Soils The voids of partially saturated soils contain both air and water, and the pressures in these two fluid phases are different. Therefore, the Mohr–Coulomb failure criterion in terms of effective stresses must be revised to account for the pore pressures in both the air and the water. According to Bishop et. al. (1960),

$$\tau_{ff} = \bar{c} + \sigma_{ff} - u_g - \chi(u_w - u_g) \tan \bar{\phi}$$

where

u_g = pressure in the gaseous phase
u_w = pressure in the water
χ = a parameter which depends upon the degree of saturation for a given soil

TABLE 7. POTENTIAL FOR PROGRESSIVE FAILURE OF VARIOUS TYPES OF CLAY (AFTER BJERRUM, 1967).

	Overconsolidated plastic clay with weak bonds		Overconsolidated plastic clay with strong bonds		Overconsolidated clay with low plasticity
	Unweathered	Weathered	Unweathered	Weathered	
p_H/s_{peak}	2	3	0–1	3	1
$\epsilon_H/\epsilon_{peak}$	2	2	1	3	0–1
s_{peak}/s_{res}	2	1	3	2	0–1
Relative danger of progressive failure	high	high	low	very high	very low

Potential for progressive failure of various types of clay, based on an evaluation of the degree to which the three significant ratios are fulfilled. Notations used:
0 Fulfillment not pronounced
1 Fulfillment less pronounced
2 Fulfillment pronounced
3 Fulfillment very pronounced

However, this equation has not come into general use for practical problems because of the difficulty of determining u_g, u_w, and χ. Therefore, a majority of analyses involving partially saturated soils are based on total stress parameters.

Typical shear-strength characteristics of partially saturated soils in undrained shear are shown in Fig. 32. It may be noted that the failure envelopes are distinctly curved and eventually become horizontal at high values of normal stress. The reason for this behavior is that increasing the normal stress causes the pore pressures to increase and air to be driven into solution. When the normal stress becomes large enough, all the air is dissolved and the sample is saturated. After all the air is disolved, the sample behaves as a saturated clay under undrained loading, i.e., $\phi = 0$. The samples with lower initial degrees of saturation achieve higher strengths because of lowei pore pressures and, therefore, higher effective stresses at a given total normal stress.

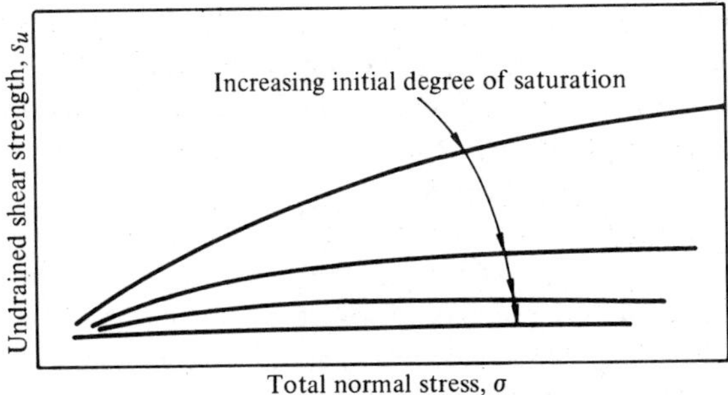

Fig. 32. Undrained shear strength of partially saturated clay in terms of total stress.

SITE INVESTIGATION

Scope of Site Investigations

The design of any structure requires some form of site investigation. The extent of the investigation depends on the size and importance of the project, the complexity of the soil conditions, and past experience with similar structures in the same area. Thus, the scope of the work may range from visual examination of open trenches or excavations, possibly with a few auger borings (as in the case of small, lightweight structures in familiar soil conditions), to deep borings with extensive laboratory testing (for tall, heavy structures or deep excavations).

The nature of the soil conditions also influences the type of site investigation that is likely to provide the most useful information. Homogeneous soil conditions lend themselves to high-quality sampling techniques and laboratory testing, particularly where important structures are involved. On the other hand, where heterogeneous soil conditions are encountered, highly refined sampling and testing techniques may be both wasteful and inconclusive, since results obtained from a few select samples are not likely to be representative of the soil deposit as a whole. In these cases, large numbers of soundings and simple laboratory tests may prove to be more informative as well as cheaper to obtain.

A detailed site investigation is comprised of the following elements.

a. Reconnaissance. This involves an examination of the topography of the site, general geology of the area, performance of existing structures, location of utility lines, and availability of construction materials. This information can usually be obtained from a site visit, study of available soil and geologic maps and aerial photographs, and discussion with residents and local authorities.

b. Exploratory borings. These usually consist of auger or wash borings with soundings and representative (disturbed) samples taken periodically for identification of the individual strata and determination of index properties. On many projects, reconnaissance and a few exploratory borings is all that is necessary. However, if additional study of the soil conditions is required the exploratory borings will serve as a guide for further investigation. The further investigation may take the form of additional soundings and index-property tests if the soil conditions are erratic; if the soil conditions are fairly uniform, undisturbed samples and more elaborate laboratory tests may be indicated.

c. Undisturbed samples. If the project calls for the accurate determination of the physical properties of the various soil strata, laboratory tests on high-quality undisturbed samples are usually required. Such samples are obtained with thin-walled tube samplers or are hand-cut from test pits. Extreme care must be exercised in securing, handling, and transporting these samples, since even a minor amount of disturbance may adversely affect subsequent test results. Certain physical properties, such as permeability and relative density of sand strata, are extremely difficult to measure in the laboratory; in such cases, *in-situ* testing may produce better results.

TABLE 8. REQUIREMENT FOR BORING LAYOUT (AFTER NAVDOCKS DM-7, 1962).

1. On large sites where subsurface conditions are relatively uniform, preliminary borings at 100 to 500 ft. spacing may be adequate. Spacing is decreased in detailed exploration by intermediate borings as required to define variations in subsoil profile. Final spacing of 25 ft. usually suffices for even erratic conditions.

2. Where factors such as cavities in limestone or fractures and joint zones in bedrock are being investigated, wash boring or rotary borings without sample recovery, or soundings and probings are spaced as close as 10 ft.

3. Where detailed settlement, stability or seepage analyses are required, include a minimum of one boring to obtain undisturbed samples of critical strata. Provide sufficient preliminary dry sample borings to determine most respresentative location for undisturbed sample borings.

4. Inclined borings are required in special cases when surface obstructions prevent use of vertical holes or subsurface irregularities such as buried channels, cavities or fault zones are to be investigated.

Investigation for:	Boring layout
New site of wide extent	Space preliminary borings so that area between any four borings includes approximately 10% of total area. In detailed exploration, add borings to establish geological sections at the most useful orientations.
Development of site on soft compressible strata.	Space borings 100 to 200 ft. at possible building locations. Add intermediate borings when building sites are determined.
Large structure with separate closely spaced footings.	Space boring approximately 50 ft. in both directions, including borings at possible exterior foundation walls, at machinery or elevator pits and to establish geologic sections at the most useful orientations.
Low load warehouse building of large area.	Minimum of four borings at corners plus intermediate borings at interior foundations sufficient to define subsoil profile.
Isolated rigid foundation, 2,500 to 10,000 sq. ft. in area.	Minimum of three borings around perimeter. Add interior borings depending on initial results.
Isolated rigid foundation, less than 2,500 sq. ft. in area.	Minimum of two dry sample borings at opposite corners. Add more for erratic conditions.
Major waterfront structures, such as dry docks.	If definite site is established, space borings generally not farther than 100 ft. adding intermediate borings at critical locations, such as deep pumpwell, gate seat, tunnel or culverts.
Long bulkhead or wharf wall	Preliminary borings on line of wall at 400 ft. spacing. Add intermediate borings to decrease spacing to 100 ft. or 50 ft. Place certain intermediate borings inboard and outboard of wall line to determine materials in scour zone at toe and in active wedge behind wall.
Slope stability, deep cuts, high embankments.	Provide three to five borings on line in the critical direction to establish geological section for analysis. Number of geological sections depends on extent of stability problem. For an active slide, place at least one boring upslope of sliding area.
Dams and water retention structures	Space preliminary borings approximately 200 ft. over foundation area. Decrease spacing on centerline to 100 ft. by intermediate borings. Include borings at location of cut-off and critical spots in abutment.
Highways and airfields	See DM-5 and DM-21 for general requirements of highways and airfields. For slope stability, deep cuts and high embankments see layout recommended above.

A conscious effort should be made to get the most out of every soil exploration program, consistent with economy. This means planning the layout of boreholes and depth of borings so that sufficient information is obtained for a rational foundation design and, at the same time, unnecessary expense is avoided.

If the structural layout has been predetermined, the boreholes should be placed as close as possible to the proposed foundations. If, on the other hand, the structural layout is not known in advance, an evenly spaced grid of boreholes may be used. The spacing of the grid will, in general, depend on the size of the area involved, the variability of the soil conditions, and the importance of the project. Suggested borehole layouts for various situations are given in Table 8.

The required depth of the borings is determined by the size and spacing of the loads, the type of foundation to be used, and, most important, the nature of the subsurface strata. The basic considerations are that the borings extend through all unsuitable foundation strata to reach a level of suitable bearing capacity and that borings extend to such a depth that the stresses from foundation loads will not cause significant settlement of the structure. Table 9 lists requirements for boring depths for several common situations.

If the foundations are to bear on rock, the rock should be cored to a sufficient depth that boulders or compact or cemented soil layers are not mistaken for bedrock. The required depth of rock coring may range from about five to twenty feet depending on local geology and the importance of the structure.

Methods of Subsurface Explorations

A summary of the most common methods of subsurface exploration is given in Table 10. These methods are categorized according to group (indirect, semi-direct, or direct methods) and type (geophysical, probing or sounding, borings or accessible explorations).

The geophysical methods, including gravitational, magnetic, electrical, seismic, and continuous vibration methods, are used principally in reconnaissance surveys and in determination of dynamic properties. The chief advantage of the geophysical methods lies in the ability to cover large areas in short periods of time. The main disadvantage is that the subsurface materials and their properties cannot be positively identified; therefore, supplementary borings are usually required.

Sounding methods are used to differentiate between dissimilar soil strata and to estimate some of the physical properties of soils. Soundings are faster and cheaper than borings, allowing large areas to be explored rapidly and economically. In some cases, where sufficient correlation data are available, soundings can be used to approximate bearing capacity of soil strata and settlement of foundations on cohesionless soils. However, soundings alone are insufficient for

TABLE 9. REQUIREMENTS FOR BORING DEPTHS (AFTER NAVDOCKS DM-7, 1962).

1. Extend all borings through unsuitable foundation strata, such as unconsolidated fill, peat, highly organic materials, soft fine grained soils and loose coarse grained soils to reach hard or compact materials of suitable bearing capacity.

2. Borings in potentially compressible fine grained strata of great thickness should extend to a depth where stress from superposed load is so small that corresponding consolidation will not significantly influence surface settlements.

3. Where stiff or compact soils are encountered at shallow depths, extend one or more boring through this material to a depth where the presence of an underlying weaker strata cannot effect stability or settlement.

4. If bedrock surface is to be determined but character and general location of rock are known, extend borings 5 ft. into sound, unweathered rock. Where character of rock is not known or where boulders or irregularly weathered material overlie bedrock, core 10 ft. into sound rock and include 20 ft. of coring in one or two selected borings. In cavitated limestone, extend borings through strata suspected of containing solution channels.

Investigation for	Boring extend to
Large structure with separate closely-spaced footings.	Depth where increase in vertical stress for combined foundations is less than 10% of effective overburden stress. Generally all borings should extend no less than 30 ft. below lowest part of foundation unless rock is encountered at shallower depth.
Isolated rigid foundations	Depth where vertical stress decreases to 10% of bearing pressure. Generally all borings should extend no less than 30 ft. below lowest part of foundation unless rock is encountered at shallower depth.
Long bulkhead or wharf wall	Depth below dredge line between 3/4 and 1 1/2 times unbalanced height of wall. Where stratification indicates possible deep stability problem, selected borings should reach top of hard stratum.
Slope stability	An elevation below active or potential failure surface and into hard stratum, or to a depth for which failure is unlikely because of geometry of cross-section.
Deep cuts	Depth between 3/4 and one times base width of narrow cuts. Where cut is above ground water in stable materials, depth of 4 to 8 ft. below base may suffice. Where base is below ground water, determine extent of pervious strata below base.
High embankments	Depth between 1/2 and 1 1/4 times horizontal length of side slope in relatively homogeneous foundation. Where deep or irregular soft strata are encountered, borings should reach hard materials.
Dams and water retention structures	Depth of 1/2 base width of earth dams or 1 to 1 1/2 times height of small concrete dams in relatively homogeneous foundations. Borings may terminate after penetration of 10 to 20 ft. in hard and impervious stratum if continuity of this stratum is known from reconnaissance.
Highways and airfields	Auger holes to extend 6 ft. below top of pavement in cuts, 6 ft. below existing ground in shallow fills. For high embankments or deep cuts, follow criteria given above.
Airfields	Auger borings to extend 10 ft. below top of pavement in cuts, 10 ft. below existing ground in shallow fills, or to a depth at which CBR for proposed loading is 1, whichever is greater.

TABLE 10. METHODS OF SUBSURFACE EXPLORATION (AFTER HVORSLEV, 1961).

Group	Type	Method		Measurements or Methods of Advance	Indication of Change in Material	Type of Formation	Use in Civil Engineering
Indirect Methods	Geophysical Methods (1)	Gravitational	Gravimeter, Pend.	Intensity of gravitational field	Anomalies grav. field No depth control	Rock ridges, domes, intrusions, faults, steeply inclined strata	Not used in Civil Engineering
			Torsion Balance	Curvature of gravitational field			
		Magnetic Methods		Intensity of magnetic field supplemented by inclination, declination	Anomalies magn. field Limited depth control	Ore bodies, faults, ridges and intrusions igneous, mag. rocks	Reconnais. rock ridges, faults; rapid, econ., application limited
		Electrical (Galvanic)	Resistivity	Current and potential drop	Variat. in resistivity	Rock, soils, and ground water; horizontal and inclined strata at shallow to medium depths	Reconn. general stratigraphy; detection of irregularities. Rapid, fairly reliable with correlation borings, incl. represent. samples
			Pot. Drop Ratio	Ratio pot. drop betw. three points	Variat. pot. drop ratio		
		Seismic	Refraction	Travel times of refracted waves	Veloc. compres. waves		
			Reflection	Travel times of reflected waves	Veloc. compres. waves	Deposits at depths over 2000 ft	Not used in Civil Engineering
		Continuous Vibration		Contin. waves, variable frequency phase, amplitude, power, settlement	Variat. velocity, amplitude etc. of shear waves	Soil and rock, shallow depths, horizontal and inclined strata	Reconn. general stratigraphy, dynamic properties; in development
	Probing or Sounding	Rod alone	Simple Point	Driving by drop hammer	Blows : penetration	All soils without large stones	Reconn. rock and rough soil profiles; rapid not always reliable
			Screw Point	Static pressure and rotation	Rev's : penetration	Medium to hard cohesive soils	
			Cone or Disk	Static pressure, constant speed	Force : penetration	Soft to stiff and dense soils	Shallow reconn. and control tests
		Rod with a Sleeve Pipe	Wash Point	Alternating jetting and jacking	Variations in point resistance alone	Primarily cohesionless soils	Compactness profiles; sand, silt
			Large Cone Point	Alternating jacking and driving of rod and sleeve pipe. In some cases concurrent jacking of rod and sleeve		Soft to hard and dense soils without stones and boulders	Reconnaissance, detection irregularities, detail stratigraphy but without positive identification; correlate with borings. Fast inexpensive, indication of compactness, strength, bearing capacity
			Flush Cone Point		Variat. point resistance and skin friction		
			Cone and Collar				
		Kjellman "Insitu" Method		Insertion and withdrawal of resistor	Withdrawal resistance	Primarily soft and loose soils	

Semi-Direct Methods	Borings	Displacement Boring	Slit, Cup Sampler Piston Samplers	Driving closed sampler into the soil, rotation, release of piston, sampling	Blows or static force versus penetration	Loose to medium cohesionless soils; soft to stiff cohesive soils	Reconn. and detailed exploration Rapid under favorable conditions
		Wash Boring (3)		Light chopping, strong jetting; removal of cuttings by circulating water	Cuttings in water, rate of progress (2)	Soft to stiff cohesive and fine to coarse cohesionless soils	Reconn. to special explorations, ground water; inexpensive equip.
		Percussion Drilling–also called Cable-Tool Drilling		Power chopping; periodic removal of slurry with bailers or sandpumps	Cuttings in slurry, rate of progress (2)	Soil and rock but difficult in soft sticky clay or loose sand	Penetrat. gravel, boulders, rock; supplementing wash, auger borings
		Rotary Drilling		Power rotation of bit; cuttings removed by circulating drilling fluid	Cuttings in fluid, rate of progress (2)	Soil and rock, except stony or very porous soil, fissured rock	Detailed and special exploration; fast; water observations difficult
Direct Methods		Auger Boring		Hand or power operat. with periodic withdrawal or use of contin. auger	Soil removed constitutes representative sample	Medium to stiff cohesive soils Part. saturated sand and silt	Shallow reconn. or detail explor. Power operat., fast; special expl.
		Continuous Sampling		Alternating sampling and cleaning with drive samplers or core barrels	Samples obtained are represent. or undisturb.	All soils and rock–cohesionless soils may require freezing	Best method for detail soil expl. Majority of explorations in rock
	Accessible Explorations	Test Pits and Trenches, Caissons, Drifts, Tunnels		Excavation by hand and power tools, use of explosives; sheeting of walls	Inspection, mapping, sampling, and testing material in situ	Soil and rock; unstable soils require ground water control, compressed air, or freezing	Detailed and special exploration. Expensive but best of all methods except when load reduction causes soil displacement and disturbance
		Accessible Borings		Power operated disk or bucket augers; single tube core barrels; mucking			

(1) Only principal methods listed. (2) Samples of cuttings, settled from wash water, slurry, or drilling fluid, are called "Wet Samples." They are non-representative and inadequate for positive identification of soil strata; however, the borings make separate sampling operations possible. (3) Wash borings with representative samples taken each stratum often called "Dry Sample Borings."

TABLE 11. TYPES OF TEST BORINGS (AFTER NAVDOCKS DM-7, 1962).

Boring method	Procedure utilized	Applicability
Displacement type	Repeatedly driving or pushing tube or spoon sampler into soil and withdrawing recovered materials. Changes indicated by examination of materials and resistance to driving or static force for penetration. No casing required.	Used in loose to medium compact sands above water table and soft to stiff cohesive soils. Economical where excessive caving does not occur. Limited to holes $<3''$ in diameter.
Auger boring	Hand or power operated augering with periodic removal of material. In some cases continuous auger may be used requiring only one withdrawal. Changes indicated by examination of material removed. Casing generally not used.	Ordinarily used for shallow explorations above water table in partly saturated sands and silts, and soft to stiff cohesive soils. May be used to clean out hole between drive samples. Very fast when power-driven. Large diameter bucket auger permits examination of hole.
Wash type boring for undisturbed or dry samples	Chopping, twisting and jetting action of a light bit as circulating drilling fluid removes cuttings from hole. Changes indicated by rate of progress, action of rods and examination of cuttings in drilling fluid. Casing used as required to prevent caving.	Used in sands, sand and gravel without boulders and soft to hard cohesive soils. Most common method of subsoil exploration. Usually can be adapted for inaccessible locations, such as over water, in swamps, on slopes or within buildings.
Rotary drilling	Power rotation of drilling bit as circulating fluid removes cuttings from hole. Changes indicated by rate of progress, action of drilling tools and examination of cuttings in drilling fluid. Casing usually not required except near surface.	Applicable to all soils except those containing much large gravel, cobbles and boulders. Difficult to determine changes accurately in some soils. Not practical in inaccessible locations because of heavy truck mounted equipment, but applications are increasing since it is usually most rapid method of advancing bore hold.
Percussion drilling (Churn drilling)	Power chopping with limited amount of water at bottom of hole. Water becomes a slurry which is periodically removed with bailer or sand pump. Changes indicated by rate of progress, action of drilling tools and composition of slurry removed. Casing required except in stable rock.	Not preferred for ordinary exploration or where undisturbed samples are required because of difficulty in determining strata changes, disturbance caused below chopping bit, difficulty of access, and usual higher cost. Sometimes used in combination with auger or wash borings for penetration of coarse gravel, boulders and rock formations.
Rock core drilling	Power rotation of a core barrel as circulating water removes ground-up material from hole. Water also acts as coolant for core barrel bit. Generally hole is cased to rock.	Used alone and in combination with boring types to drill weathered rocks, bedrock and boulder formations.

the final design of important structures or in any case where consolidation or pore water pressures must be considered. Soundings can also produce very misleading results if they are unproperly performed or if correlation with physical properties is attempted outside the region of their applicability.

The most common types of borings are summarized in Table 11. The choice of boring method in a given situation depends upon (1) the relative efficiency of the boring procedure in the materials encountered, (2) the ability to detect changes in soil and groundwater conditions, and (3) possible disturbance of material to be sampled.

Accessible explorations provide the most complete information on soil and rock conditions *in-situ*. They also provide valuable information on possible construction difficulties and costs of excavation. However, this method of exploration is generally more expensive than the other methods and should not be undertaken until the results of reconnaissance and detailed explorations have clearly established the need and the proper procedure for carrying out accessible explorations.

The various methods of obtaining samples of subsurface materials are summarized in Table 12, which also indicates the materials in which the various types of sampling can be used and the probable condition of the recovered samples.

This section has dealt only briefly with the subject of subsurface exploration. For a detailed treatment of this subject, the reader is referred to Hvorslev (1962).

FOUNDATIONS FOR STRUCTURES

General Considerations

Definitions For purposes of analysis and design, foundations for structures are usually divided into two main categories: shallow and deep. Shallow foundations are those which transfer the structural loads to the underlying soil or rock at relatively shallow depths; they include *spread footings*, *strip footings*, *combined footings*, and *mat foundations*. Deep foundations, including *piers*, *caissons*, and *pile foundations*, transfer the structural loads to the underlying soil or rock at considerable depth.

Requirements for a Satisfactory Foundation Every foundation, if it is to function properly must satisfy the following basic requirements:

1. The foundation must be properly located so that it will not be adversely affected by environmental factors such as frost, seasonal volume change, and ground-water levels.
2. The foundation must be adequately safe against bearing failure.

TABLE 12. METHODS OF SAMPLING SUBSURFACE MATERIALS (AFTER HVORSLEV, 1962).

Group	Type or Purpose	Sampler or Method		ID Boring OD Sampler in.	Sample Diameter in.	Materials in Which Used	Condition of Samples
Exploration	Clean-out Tools	Bailers		2.9–up	–	Very soft soils, loose cohesionless soils, and slurry of all materials	Often non-representative with soil constituents mixed and segregated
		Sandpumps		2.0–up			
	Slit Samplers	Longitudinal Slit		1.3–4.0	–	Soft soils, silt, and loose sand	Representative of average conditions but adjacent strata are often mixed
		Circumferential Slit or Cup		2.3– –	–		
	Augers	Helical or Worm Type Augers		1.5–16	–	Medium soft to stiff cohesive soils Partially saturated sand and silt	Seriously disturbed and often partially mixed but generally representative of the average condition
		Iwan or Post Hole Augers		4–9	–		
		Barrel Augers, Helical or Iwan.		2.5–5.6	–	All soils including gravelly soils	
		Disk and Bucket Power Augers		12–40	–		
Drive Sampling	Open Drive Samplers	Thick-Wall, Solid-Barrel		1.4–8.0	1.0–7.0	All soils except coarse gravel. Retainers req'd in soft or loose soils	Top often non-representative; rest part. disturbed but representative
		Thick-Wall, Split-Barrel		2.0–5.6	1.4–5.0		
		Thin-Wall Samplers		1.0–8.0	0.94–7.8	Thick-wall sampler as above. Thin-wall: Soft to stiff and loose to medium dense soils. Special methods to prevent loss required in some soils	Representative to undisturbed, depending on type of soil and design and method of operation of sampler
		Composite Samplers–Liners		1.3–8.0	0.94–7.0		
	Piston Samplers	Thin-Wall or Composite	Retracted Piston	$^7/_8$–6.0	$^3/_4$–4.9		
			Free Piston	$^3/_4$–6.0	$^5/_8$–5.9		As above; less disturbance and danger of loss; better recovery data
			Stationary Piston**	$^3/_4$–6.0	$^5/_8$–5.9	As above but incl. very soft soils	

Core Boring, Rotary Core Barrels: Soil Core Barrel	Hard-Metal Teeth. Flush or protruding inner tube with liner	3.8–8.9	2.8–7.4	Stiff to hard clays, brittle soils, dense sand, partial cemented soils	Probably less disturbance than by drive sampling. Method in develop.
Core Boring, Rotary Core Barrels: Shot Core Boring	Chilled steel shot in soft steel bit. Single tube with calyx	2.8*–4.8 36 –up*	1.5*–3.4 34 –up*	All except very soft, fissured, or cavernous rock; slow in hard rock	Cores of sound rock generally undisturbed but cores of non-uniform rock are often broken and soft and erodible sections ground up and removed by the circulating fluid unless special precautions are taken
Core Boring, Rotary Core Barrels: Diamond Bit	Single tube or double tube barrel with retracted inner tube	1.5*–1.9 3.0 –up*	7/8*–1.1 2.1 –up*	All sound rock but best suited for small cores of medium to hard rock	
Core Boring, Rotary Core Barrels: Hard-Metal Teeth	Tungsten Carbide Surface or Inserts: Single Tube	2.5–38	2.0–36	Soft to medium rock and frozen soil occasionally hard and dense soils	
	Tungsten Carbide Surface or Inserts: Double Tube	2.5–8.9	1.8–7.4		
Core Boring, Rotary Core Barrels: Hard-Metal Bit Oilfield Core Barrels	Standard, stationary inner tube	3.9–12	1.2–5.5	Bladed Bit: Hard soils or soft rock Roller or Cone bit: Hard formations	
	Wire-Line, retractable inner tube	5.4–8.0	1.0–2.5		
	Pressure, inner tube with valves	6.25	1.5–1.7	Developed for sampling of oil sands	Fluid and gas pressure maintained
Core Boring: Percussion Core Boring Cable-Tool Core Barrel	Double Tube. Sliding outer barrel with hard-surfaced steel bit	3.8–7.3	1.6–3.8	Medium soft to medium hard rock	Fair recovery, but the cores are often broken into small sections
Side Wall Samplers	Open Drive Samplers. Operation hydraulic, wire-line, or shooting	Bore-Hole 4.8–8.5	7/16–1 1/4	Stiff and compact soils to soft or medium rock; side walls of borings	Partially disturbed to undisturbed depending on formation and design
Surface Sampling in Accessible Explorations, Earth Structures	Bag Sample, Field Density Tests	4–10	–	Primarily sandy and gravelly soils	Representative but natural density
	Short Open Drive Samplers	2–6	1.9–5.9	Soft to medium stiff clayey soils, loose sand and silt. Control tests	As below but occasionally some disturbance and especially compaction
	Short Piston Samplers	3/4–6	5/8–5.9		
	Sampling by Advance Trimming	–	4–8	Stiff, brittle or dense soils. Compacted or partially saturated soils	Undisturbed excepting influence of stress changes and soil movements before or exposure during sampling
	Auger Core Barrels	6.3–6.5	3.7–4.0		
	Block or Box Sampling	–	8–12	Coarse, dense, brittle to hard soil	
Submarine Bottom Explorations	Scrapers and Clamshell Buckets	–	–	Only materials from bottom surface	Disturbed except the larger chunks
	Restricted Gravity / Free-Fall Gravity / Driven by Shooting — Composite Open Drive Samplers	1.8–3.3	1.0–2.4	All soils to soft rock. Samples of hard and gravelly soils short but of very soft soils up to 18 ft long	Often partially disturbed, depending on the character of the material and the design of the sampler
	Free-Fall Piston Sampler	3.5–3.7	1.8	As above but sample length to 50 ft	Some disturbance, in development

The dimensions shown are approximate and represent the commonly used range. Lower and upper limits marked * are rarely used in explorations for civil engineering purposes. **A drive sampler with liner and steel foils attached to a stationary piston, recently developed by the Swedish Geotechnical Institute, permits taking 20 m long samples of soft soils in a single operation.

3. The foundation must not experience excessive settlements or other movements (e.g., heave due to swelling soils).
4. The foundation must be both technically feasible and economically practical to build.

Factors Affecting Foundation Selection The selection of a suitable foundation system requires careful consideration of the many factors involved. Aldrich (1968) has provided a checklist of factors which should be considered in the selection. Table 13 contains this checklist in abbreviated form. However, since

TABLE 13. CHECKLIST OF FACTORS AFFECTING FOUNDATION SELECTION (AFTER ALDRICH, 1968).

Structural requirements
Sensitivity of structure to differential settlement
Useful life of structure
Foundation and floor elevations
Miscellaneous–freezing, drying, etc.
Foundation loads
Magnitude of dead and design live load
Negative skin friction (deep foundations)
Load compensation by excavation
Distribution of loads (differential settlement)
Type of loads (eccentric and unbalanced lateral loads, wind loads, vibratory loads, seismic loads)
Soil and ground-water conditions
Transfer of load through structure and underlying soil
Strength and compressibility of soil and rock layers
Variability of subsurface conditions
Swelling soils
Existing fills (settlement, corrosion, deterioration of organic material, etc.)
Ground-water level and effect of changes (drainage, waterproofing, settlement, deterioration of untreated wood piles, etc.)
Corrosion and chemical effects
False driving resistance (piles driven through dense layers and underlying soft soils)
Negative skin friction
Underground defects (solution cavities, abandoned mines and pipe lines, steep rock surfaces)
Site and environmental conditions
Topography (stability of natural slopes, fill requirements)
Flood levels and site drainage (grade requirements, erosion, scour)
Proximity of adjacent structures and property lines (underpinning, effects of dewatering and vibrations during construction, working room)
Frost action and thermal effects
Earthquakes and seismic effects
Construction requirements
Time available for construction
Space available for construction
Feasibility (excavation and lateral support, dewatering, effects of pile driving, blasting, obstructions, etc.)
Economic considerations and miscellaneous
Construction costs and estimating
Reliability of foundation system
Inspection and field control

the list is intended to be as comprehensive as possible, many of these factors may not apply in any given situation.

Shallow Foundations

Teng (1964) has presented a systematic procedure for the design of shallow foundations as follows:

1. Calculate the loads applied at top of footings.
2. Sketch a soil profile or soil profiles showing the soil stratification at the site. On this profile superimpose an outline of the proposed foundation scheme.
3. Establish the maximum water level.
4. Determine the minimum depth of footings.
5. Determine the bearing capacity of the supporting stratum.
6. Proportion the footing sizes.
7. Check for danger of overstressing the soil strata at greater depths.
8. Predict the total and differential settlement.
9. Check stability against horizontal forces.
10. Check uplift on individual footings and basement slab.
11. Design the footings.
12. Check the need for foundation drains, waterproofing or dampproofing.

Depth and Position of Foundations In general, shallow foundations must be placed so that their performance will not be adversely affected by environmental factors, construction operations, or proximity to adjacent footings or existing structures. A listing of factors to be considered in positioning spread footings is given in Table 14.

Bearing Capacity The theoretical bearing capacity for a shallow foundation resting on homogeneous soil and subjected to vertical, concentric loads may be expressed as

$$q_{\text{ult}} = \gamma B N_\gamma S_\gamma R_\gamma + \gamma D_f N_q R_q + c N_c S_c$$

where

q_{ult} = ultimate bearing capacity,
γ = unit weight of soil,
B = width or diameter of footing,
D_f = depth of footing below grade,
$N_\gamma, N_q, N_c = f(\phi)$ = dimensionless bearing-capacity factors,
ϕ = angle of internal friction of soil,
c = cohesion of soil.
S_γ, S_c = empirical shape factors (see Table 15),
R_γ, R_q = approximate correction factors for position of water table (see Table 16).

TABLE 14. GENERAL REQUIREMENTS FOR SPREAD FOUNDATIONS (MODIFIED AFTER NAVDOCKS, 1962).

1. The base of spread foundations shall be located below the depth to which the soil is subject to seasonal volume changes caused by alternate wetting and drying or within which frost may cause a preciptible heave.

2. Where foundations are to be placed in an excavation below the water table, provide for drawdown of water levels so that the work can be done in the dry and no piping, boiling or heaving occurs in soils which will support the foundations. Ordinarily this requires lowering the ground water within the foundation area to an elevation no higher than 2 ft below subgrade.

3. For foundations supported in fine-grained soil which would be disturbed and softened by construction activities upon it, provide a working mat at subgrade of lean concrete or cohesionless, coarse-grained materials.

4. For clays or shales which will expand and soften with release of overburden, place working mat at subgrade immediately after completion of excavation. Provide surface drainage facilities to prevent collection of water in excavation as it nears the subgrade.

5. Where adjacent footings in the same structure bottom on materials of substantially different bearing quality, such as medium compact soil and rock, provide a cushion of yielding material beneath the footing on the harder foundation. For the footings on rock, place an 18-in.-thick layer of uncompacted sand above the bearing surface in rock. Where practical, consider separating the portions of the structure on dissimilar materials by expansion joints.

6. For ordinary warehouse floor slabs on grade, compact the subgrade in natural soils to a depth of 8 in. To avoid dampness on the warehouse floor, provide a vapor barrier of plastic sheeting at base of slab. In addition, where ground-water table is within about 8 ft of base of slab provide a base course 8 in. thick of clean, coarse-grained material beneath the floor slab, if underlying soils are fine grained or are silty fine sands with considerable capillary potential.

7. For warehouse floors supporting wheel loads, provide floor slab and base course plus subgrade compaction.

8. Adjacent footings at different bearing levels shall be separated as follows:

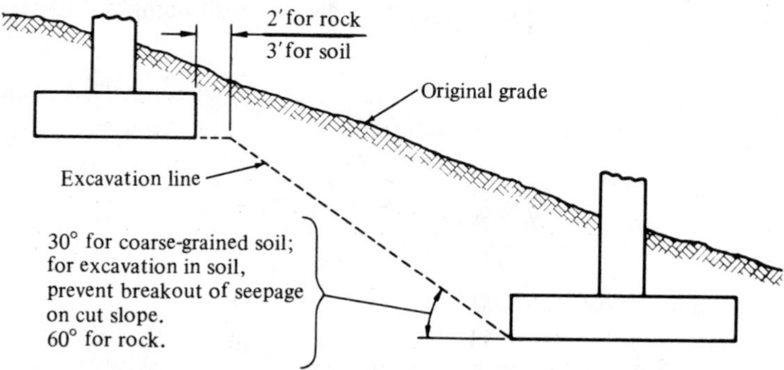

This requirement shall not apply where adequate lateral support by bracing is provided for the material beneath the higher footing. Where the higher footing is supported on cohesive soils, evaluate its bearing capacity.

The dimensionless bearing-capacity factors N_γ, N_q, and N_c depend only on the angle of internal friction of the soil. Values given for these factors by Terzaghi and Peck (1967) are shown in Fig. 33.

Modifications of the general bearing equation are necessary to account for the effect of the shape of the foundation and the position of the water table. Empirical shape factors, as recommended by Terzaghi and Peck, and approximate correction factors for the position of the water table are given in Tables 15 and 16, respectively.

In purely cohesive soils, the bearing capacity can be conveniently expressed in the form (Skempton, 1951)

$$q_{net} = 5c[1 + 0.2\,(D_f/B)]\;[1 + 0.2\,(B/L)]$$

where $D_f/B \leqslant 2.5$.

For determination of ultimate bearing capacity of eccentrically loaded footings, the concept of *useful width* may be used. In the case of cohesive soils, the portion of the footing which is symmetrical about the load is considered useful for the purpose of computing bearing capacity. In granular soils the reduction in useful width with the eccentricity of the load is somewhat greater; i.e., the reduction is parabolic rather than linear. Accordingly, the useful widths for eccentric loading may be expressed as follows (Meyerhof, 1953):

$$\left.\begin{aligned} B' &= B - 2e_b \\ L' &= L - 2e_l \end{aligned}\right\} \text{ useful widths for cohesive soils}$$

$$\left.\begin{aligned} B' &= [1 - (2e_b/B)]^2 \\ L' &= [1 - (2e_l/L)]^2 \end{aligned}\right\} \text{ useful widths for granular soils}$$

TABLE 15. SHAPE FACTORS FOR ULTIMATE BEARING CAPACITY.

Shape factor	Shape of footing		
	Strip	Square or rectangular	Circular
S_γ	0.5	0.4	0.3
S_c	1.0	$1 + 0.3(B/L)$	1.3

TABLE 16. CORRECTION FACTORS FOR POSITION OF WATER TABLE.

Water table Correction factor	Water table at d_a above base of footing	Water table at d_b below base of footing
R_γ	0.5	$1 - 0.5(d_b/B)$
R_q	$1 - 0.5(d_a/D_f)$	1.0

(a)

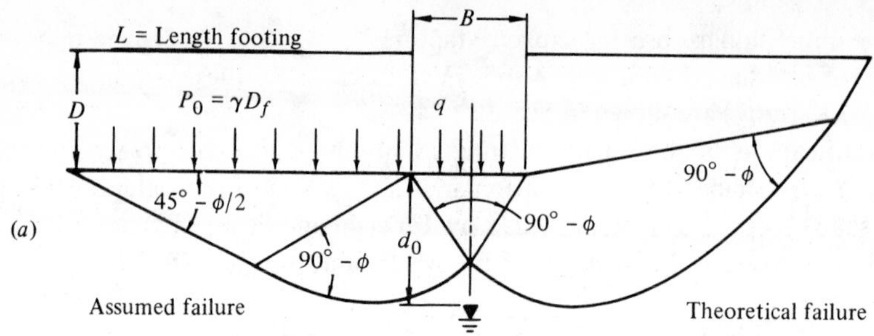

Assumed conditions:

1. $D_f \leqslant B$.
2. Soil is uniform to depth $d_0 > B$.
3. Water level lower than d_0 below base of footing.
4. Vertical load concentric.
5. Friction and adhesion on vertical sides of footing are neglected.
6. Foundation soil with properties c, ϕ, γ.

(b)

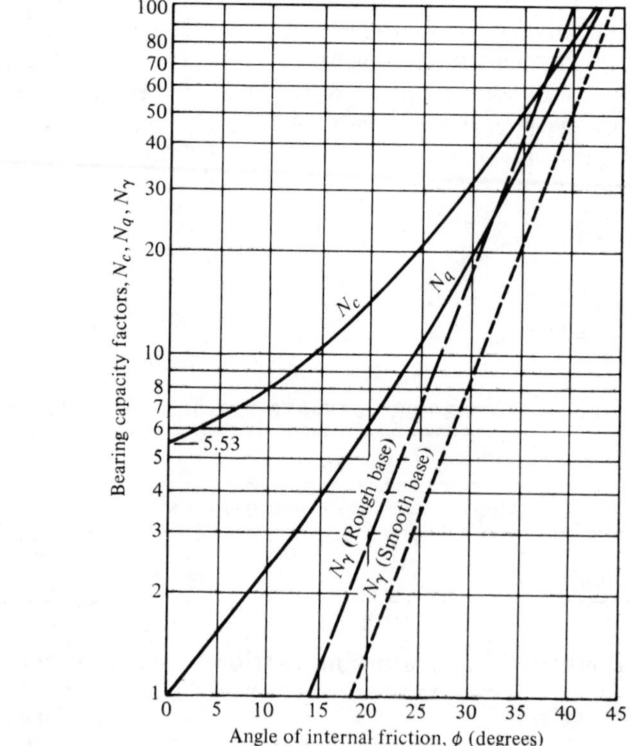

Fig. 33. Ultimate bearing capacity of shallow foundations. (a) Mode of failure. (b) Bearing-capacity factors.

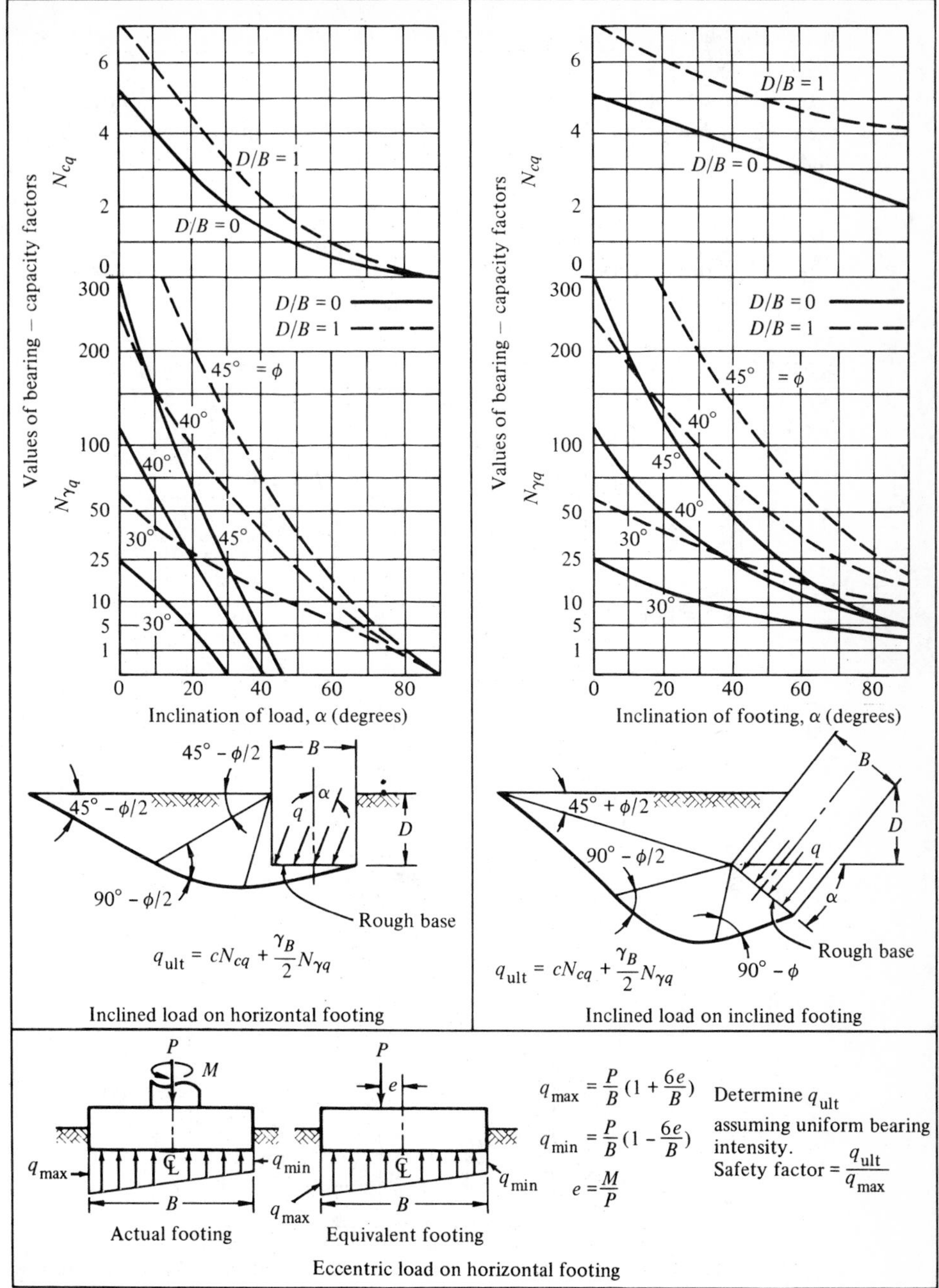

Fig. 34. Bearing capacity of footings with eccentric or inclined loads (after Meyerhof, 1953).

where e_b and e_l are the eccentricities of the load in the B and L directions, respectively.

For inclined loading, the ultimate bearing capacity may be computed by the methods of Meyerhof (1953) or Janbu (1957). According to Meyerhof, the ultimate bearing capacity is expressed as

$$q_{\text{ult}} = \tfrac{1}{2}\,\gamma B N_{\gamma q} + C N_{cq}$$

where $N_{\gamma q}$ and N_{cq} are the dimensionless bearing capacity factors which depend on the angle of internal friction, the inclination of the load, and the embedment depth as shown in Fig. 34. Janbu's method is based on the Terzaghi bearing-capacity theory and utilizes the conventional Terzaghi bearing-capacity factors N_γ, N_q, and N_c and an additional factor N_h as given in Fig. 35. Janbu's equation for ultimate bearing capacity for inclined loads is given by

$$(Q + N_h Q_h)/A = \tfrac{1}{2}\,\gamma B N_\gamma + D_f N_q + cNc$$

where

Q = total inclined load
Q_h = horizontal component of load (not to exceed $Q_v \tan\phi$)
A = area of footing

Meyerhof (1957) obtained a solution for the ultimate bearing capacity of footings on slopes. The bearing capacity equation has the form given previously, and the bearing capacity factors $N_{\gamma q}$ and N_{cq} depend upon the angle of internal

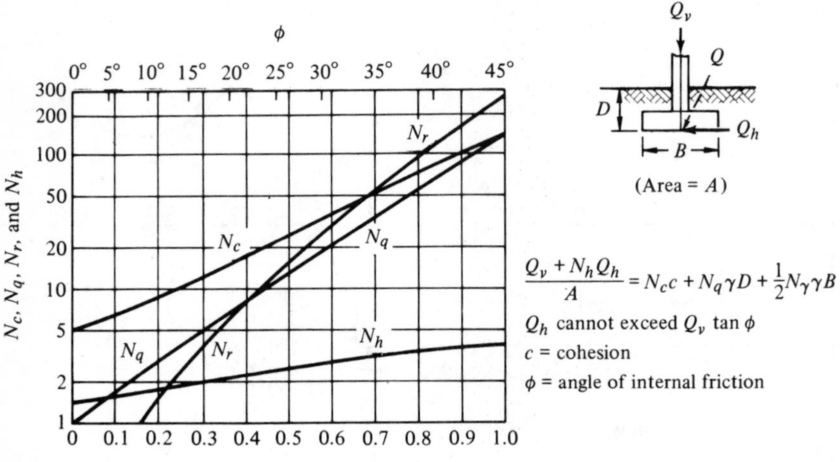

Fig. 35. Bearing capacity of continuous footing subjected to inclined load (after Janbu, 1957).

friction, the slope angle, the embedment depth, and the distance of the footing from the edge of the slope, as shown in Fig. 36. Footings should not be constructed on slopes which are, themselves, on the verge of instability nor on slopes which exhibit a tendency for creep.

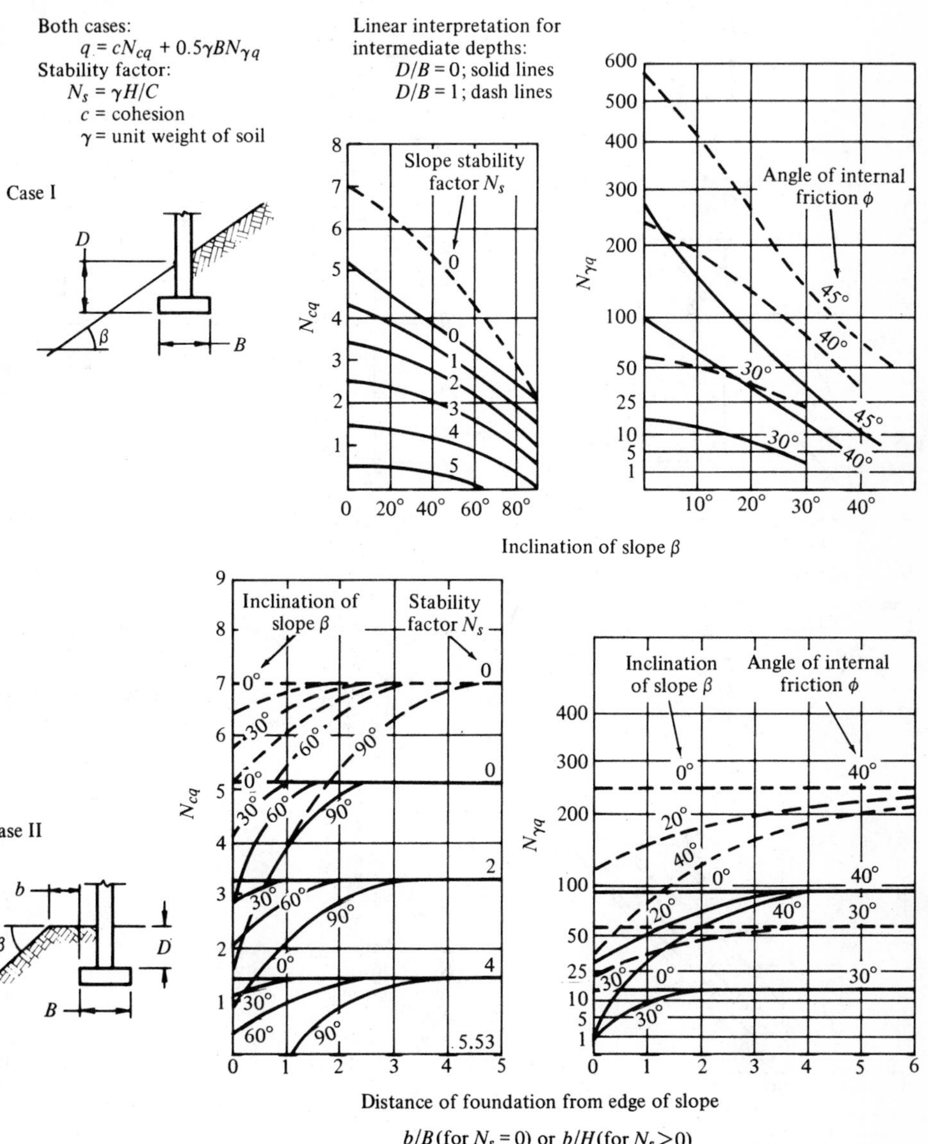

Fig. 36. Ultimate bearing capacity of continuous footings on slopes (after Meyerhof, 1957).

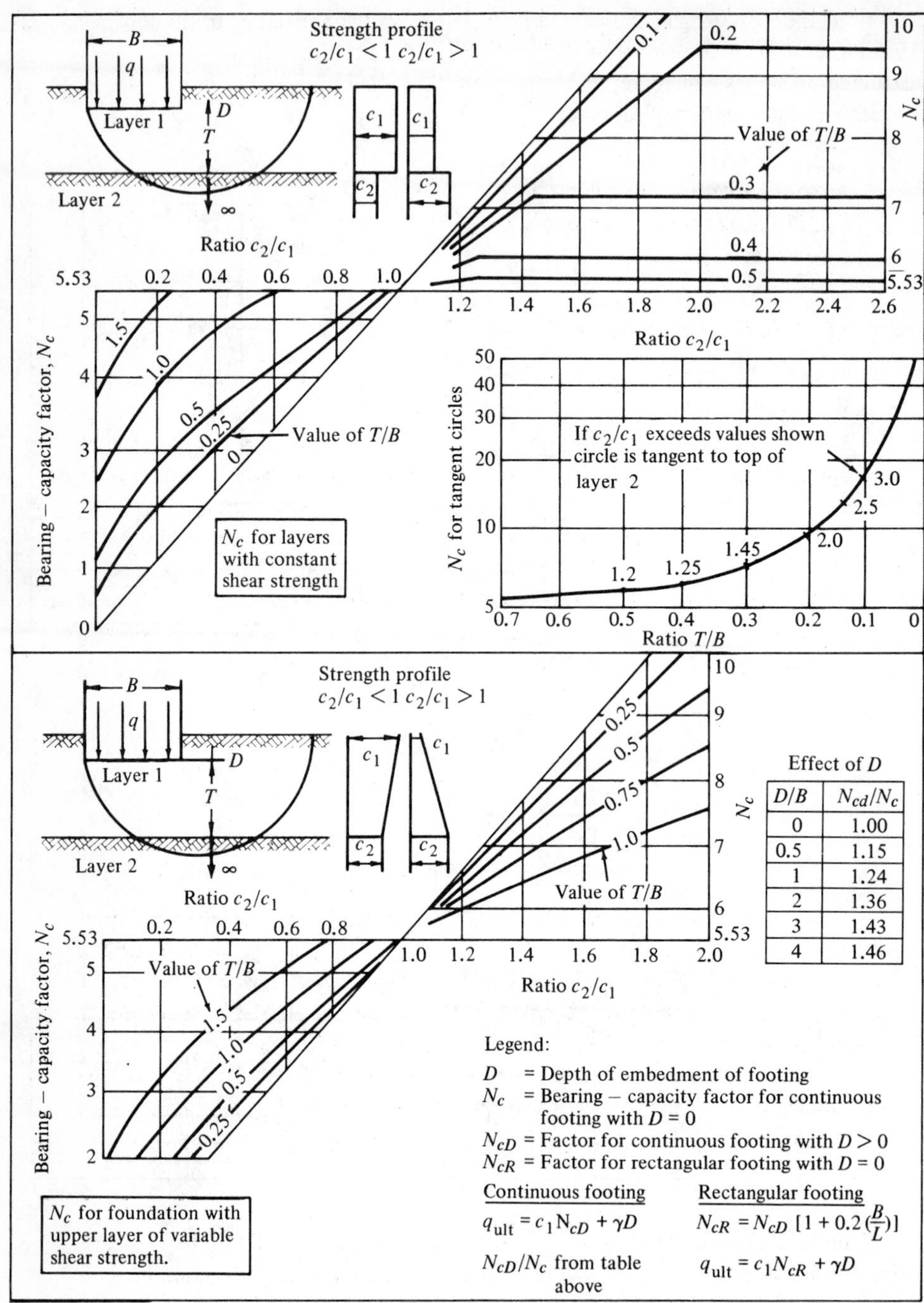

D/B	N_{cd}/N_c
0	1.00
0.5	1.15
1	1.24
2	1.36
3	1.43
4	1.46

Fig. 37. Ultimate bearing capacity of two-layer cohesive soil (after Button, 1953).

The ultimate bearing capacity of shallow foundations on layered cohesive soils can be computed from the equation (Button, 1953)

$$q_{\text{net}} = c_1 N'_{cD}$$

where

c_1 = cohesion of upper layer of soil
N'_{cD} = bearing capacity factor of two-layer system for footing at depth D_f (Fig. 37)

Stress Distribution Beneath Foundations

Analysis of foundation settlements usually requires the determination of stresses on compressible strata below the base of the foundation. For simplicity, computation of stresses produced at depth by foundation loads is generally based on the assumption that the subsoil behaves as a homogeneous, isotropic, elastic half-space. Closed-form solutions for various types of surface loadings are available from the theory of elasticity and can be conveniently expressed in the form of influence diagrams. Figures 38, 39, and 40 give the influence values for vertical stresses produced by uniformly loaded rectangular and circular areas and embankment-type loading, respectively (Fadum, 1948, Foster and Ahlvin, 1954; Osterberg, 1957). Stresses are computed from the equation

$$\Delta\sigma = I_\sigma p$$

where

$\Delta\sigma$ = stress increase due to applied loads
I_σ = influence value from the chart
p = intensity of applied stresses for the particular loading condition

A different type of influence chart has been devised by Newmark (1951), as shown in Fig. 41. In using the Newmark chart, the plan dimensions of the loaded area are drawn to a scale such that the depth at which the stress is to be computed is equal to the scale length indicated on the chart. The drawing is then placed over the chart with the point below which the stress is desired located at the center of the chart. The stress increase is determined by counting the number of influence areas covered by the footing and multiplying by the influence value given on the chart by the applied footing pressure; i.e.,

$$\Delta\sigma_v = N I_\sigma p$$

If more than one load is applied (e.g., in a foundation plan where several footing loads are involved), the total stress increase due to all applied loads is

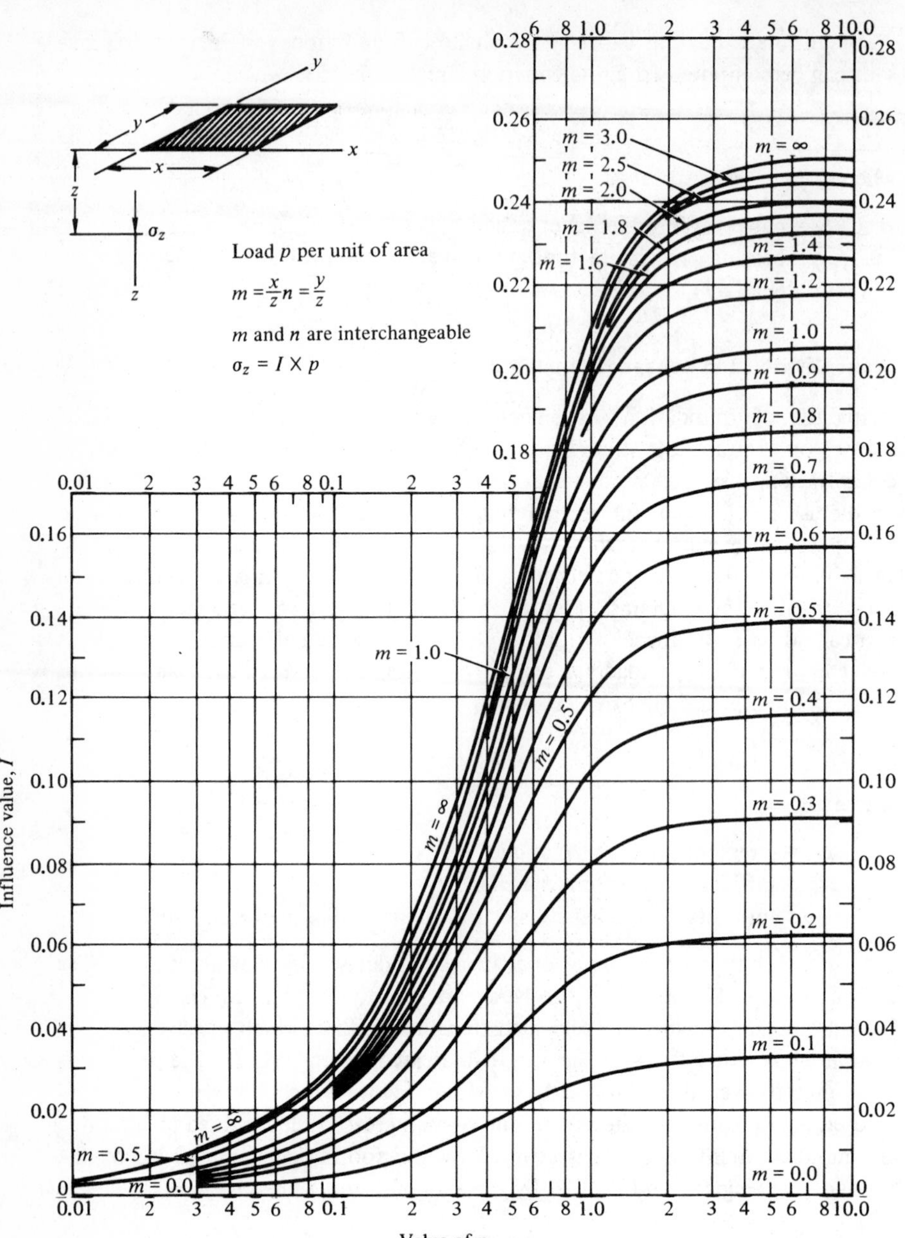

Fig. 38. Influence values for vertical stress under corner of uniformly loaded area (after Fadum, 1948).

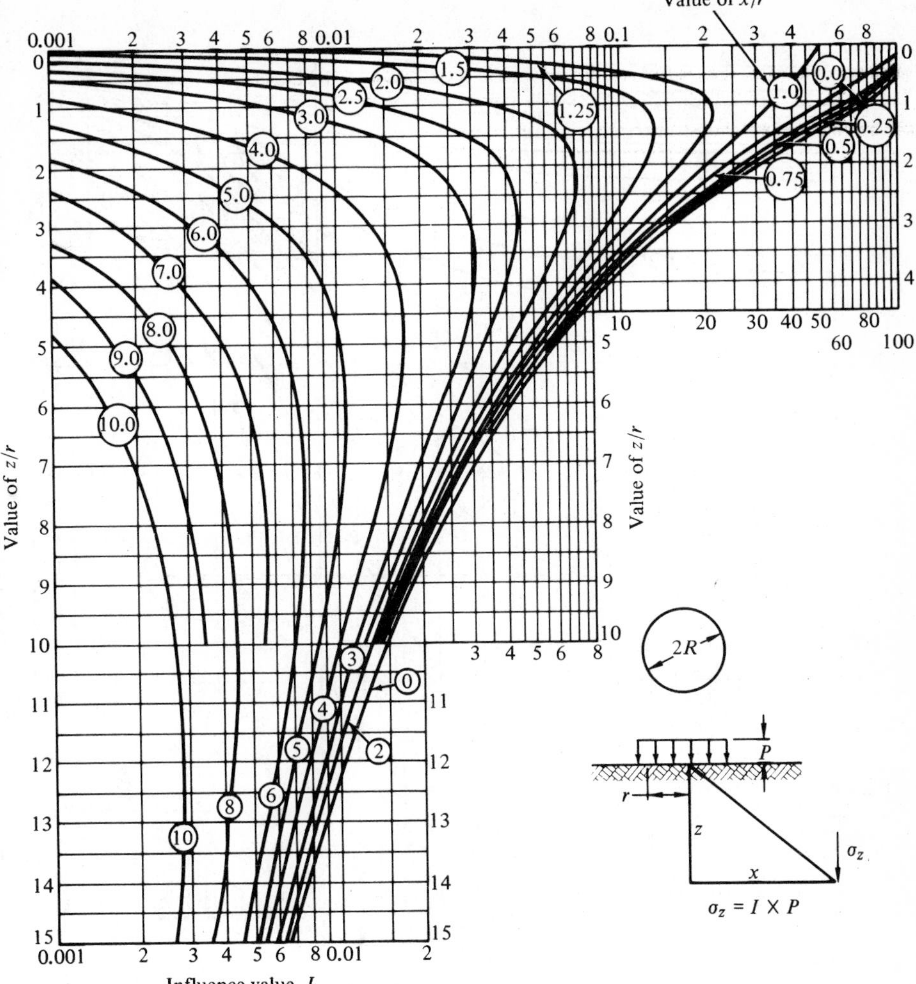

Fig. 39. Influence values for vertical stress under uniformly loaded circular area (after Foster and Ahlvin, 1954).

obtained by superposition; i.e., the influence values for all loads are summed up to obtain the total influence values for stress at a given point. The Newmark chart is particularly well suited to such cases, since the entire foundation plan can be drawn on the same sheet and the influence value for stresses at a given depth determined in one step.

When the loads are to be applied at some distance below the ground surface, e.g., at the bottom of an excavation, the usual practice is to reduce the applied

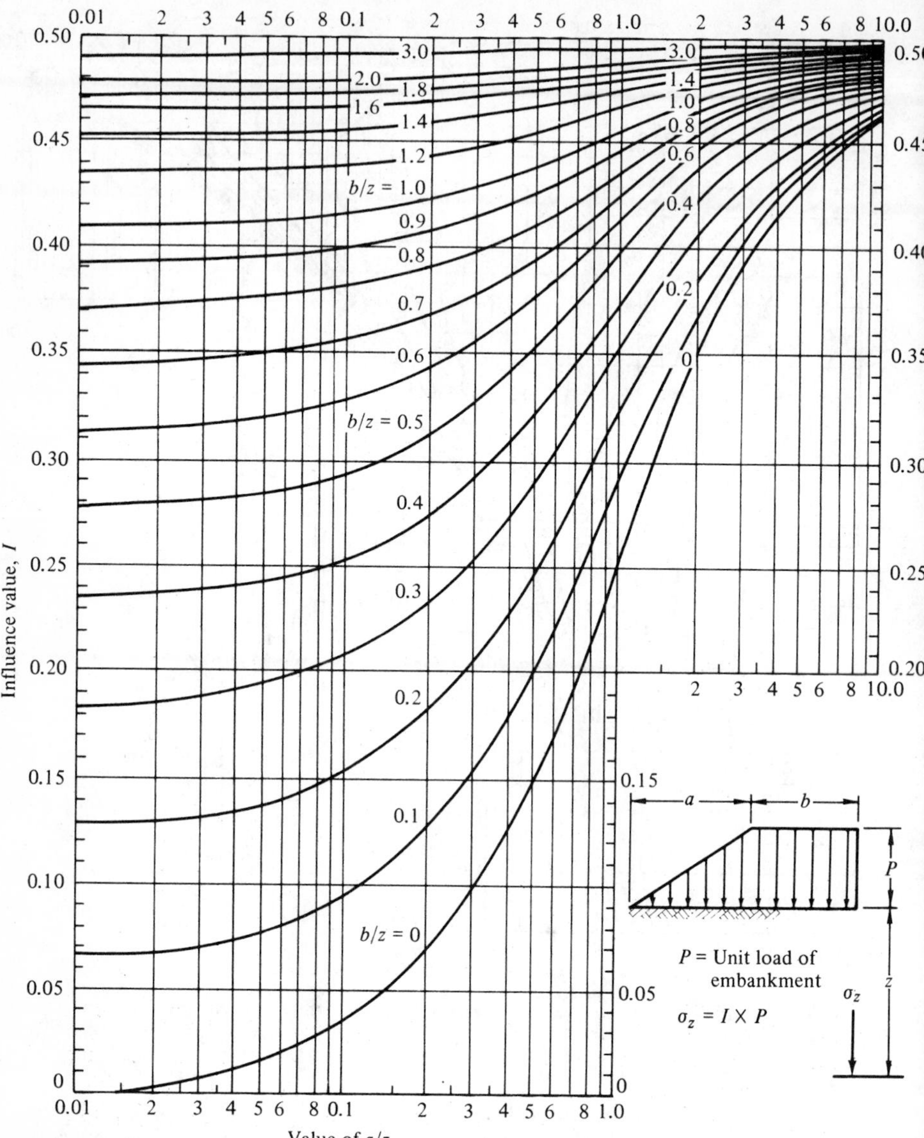

Fig. 40. Influence values for vertical stress under embankment load of infinite length (after Osterberg, 1957).

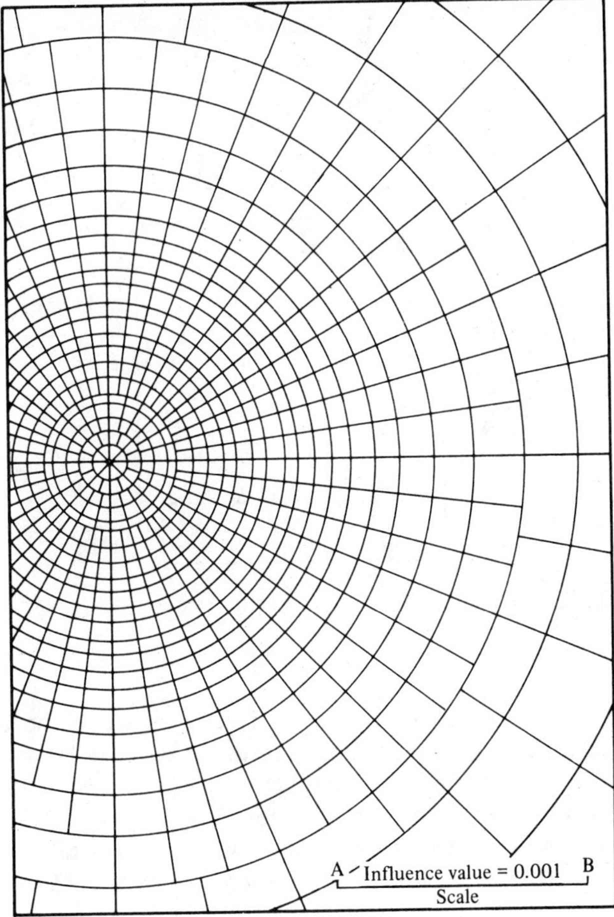

Fig. 41. Newmark influence chart for vertical stress.

load intensity by the weight of soil to be permanently removed, and then to calculate the stresses from elastic theory as though the ground surface coincided with the level at which the loads are actually applied. Although this method is not strictly correct, it is a necessary practical expedient, since correct solutions for this type of problem are not readily available.

For approximate calculation of vertical stresses at depth, the pressure applied by the footing may be assumed to spread out on a slope of 2 (vertical) : 1 (horizontal), as shown in Fig. 42. This method is often used to determine whether deeper strata may be overstressed by the application of surface loads. However, for settlement calculations, particularly where many loads are involved, this method may be too inaccurate.

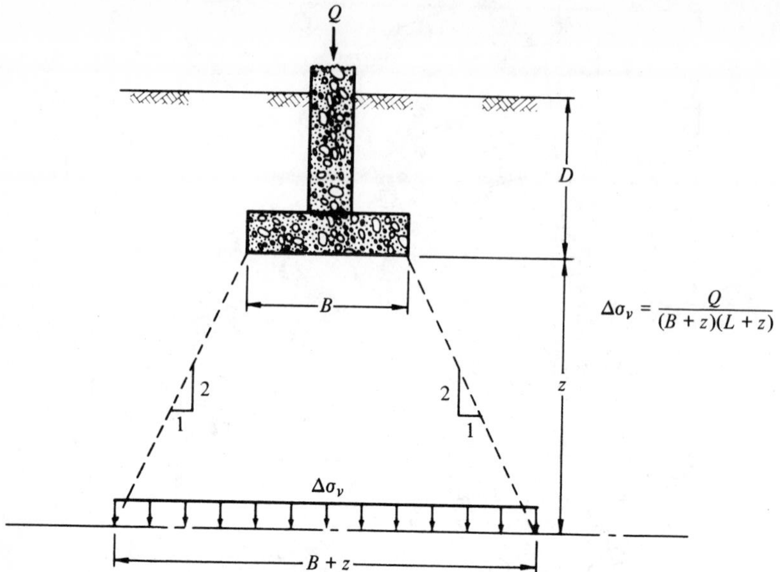

Fig. 42. Approximate distribution of vertical stress under footing.

Settlement

Settlement of foundations is the result of compression of the underlying strata. Fundamentally, the compression process in soils can be viewed as the vertical deformation of soil strata caused by stresses created in the soil by the foundation loads. The application of foundation loads (including excavation, if any) produces a unique distribution of stresses within the compressible soil mass. This stress distribution can be estimated with sufficient accuracy using elastic theory even though, strictly speaking, soil is not an elastic material. The elastic stress distribution in a homogeneous soil mass for a typical isotropic surface loading is shown in Fig. 43. It may be noted from Fig. 43 that the resulting stress distribution involves different principal stress ratios and different orientation of principal stresses throughout the soil mass. Unfortunately, soils respond differently to different principal stress ratios and orientations of principal stress; i.e., soils are not generally linearly elastic, isotropic, and homogeneous. This fact is one of the main sources of difficulty in predicting foundation settlements. Moreover, the compression of soil strata is, in general, a complicated, time-dependent process, since soils are multiphase materials and have a tendency to creep under constant load.

In current practice, several methods of settlement calculation are employed. These methods vary considerably depending on the type of soil encountered; i.e., granular or cohesive soils, the degree of accuracy warranted by the project,

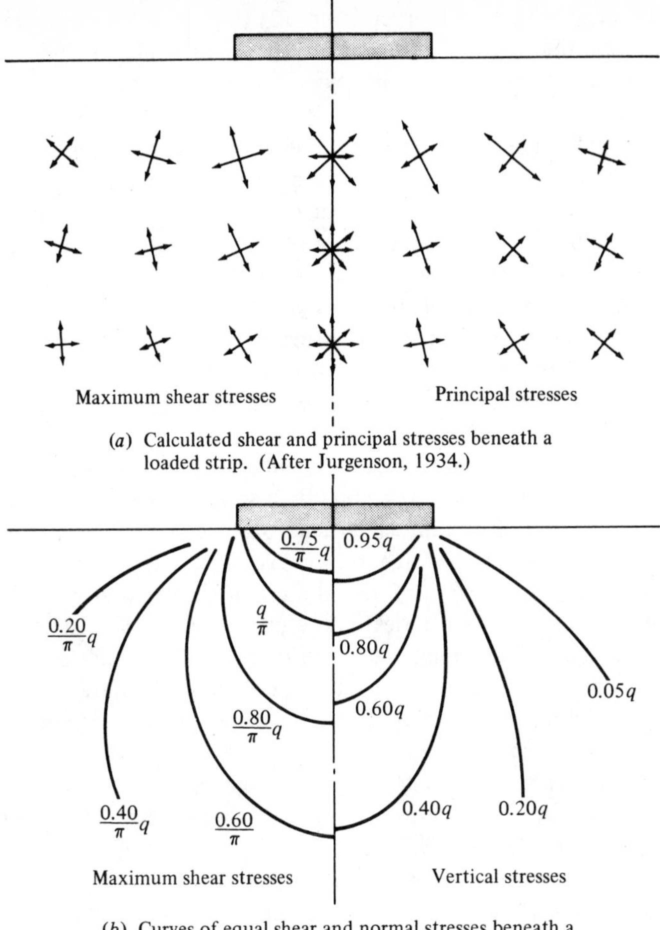

(a) Calculated shear and principal stresses beneath a loaded strip. (After Jurgenson, 1934.)

(b) Curves of equal shear and normal stresses beneath a loaded strip. (After Jurgenson, 1934.)

Fig. 43. Elastic stress distribution in a soil mass due to surface loading.

and the past experience of the engineer. A brief review of these methods is presented in the following discussion.

Stress-Path Method Of all the methods of settlement analysis currently available, the stress-path method (Lambe, 1964) appears to be the most fundamentally sound. In this method, the stress changes (including excess pore pressures) to be experienced by the soil strata in the field are determined, and then undisturbed soil samples are subjected to the anticipated stress changes in the laboratory. The resulting laboratory behavior is used to estimate field performance. The accuracy of the method depends on (1) the accuracy of the prediction of stress changes in the field, (2) the amount of disturbance of the laboratory samples, and (3) the degree to which field stress changes can be reproduced in the laboratory. In a typical application, the stress-path method might be employed as follows.

1. Estimate the field stress changes to be produced by the foundation loading. This is generally done for a representative point (e.g., a point at mid-depth in a compressible stratum along the centerline of the foundation) or, in some cases, for several points if greater accuracy is required. These stress changes may be approximated by means of the closed-form elastic solutions discussed in the previous section or by finite-element analysis if the problem warrants.

2. Obtain high-quality undisturbed samples for laboratory testing. This step can be the weakest link in the entire process since, without samples which are truely representative of the soil *in-situ*, the laboratory test results may be very misleading.

3. Run sufficient laboratory tests to obtain reliable data on probable performance of the soil *in-situ*. The applicability of the laboratory test results to field performance will be limited by such factors as sample disturbance, any differences between field and laboratory loading conditions (e.g., plane strain vs. axi-symmetric loading), and orientation of principle stresses (if soil has anisotropic stress–strain properties). The effects of sample disturbance can be minimized by using proper laboratory testing techniques and interpretation of results as outlined by Ladd (1971b), but the effects of different loading conditions or orientation of principal stresses may be largely indeterminate if these effects are not specifically investigated.

4. Total settlement is estimated by integrating the vertical strain observed in the laboratory tests (for the range of stress changes anticipated in the field) over the thickness of the compressible strata.

The stress-path method may be carried out in various ways and with any desired degree of sophistication. Two examples will serve to illustrate this point.

EXAMPLE 1 CIRCULAR LOADED AREA ON HOMOGENEOUS SAND (FIG. 44).

In this example only immediate settlement is considered, since the compressible stratum is a sand. The stress changes due to surface loading (see Fig. 44a) are computed from elastic theory and, for simplicity, the *in-situ* principal stresses, $\bar{\sigma}_{1_0}$ and $\bar{\sigma}_{3_0}$ are assumed to be equal to the present overburden pressure ($K_0 = 1$). The representative point is assumed to be at mid-depth in the sand layer along the centerline of the loading.

A conventional triaxial test is run on a representative sample of the sand (with confining pressure equal to the present overburden pressure on the representative point in the field). The stress–strain behavior of the sand is shown in Fig. 44b and d. Figure 44b is a plot of the "deviator stress" $(\sigma_1 - \sigma_3)/2$ vs.

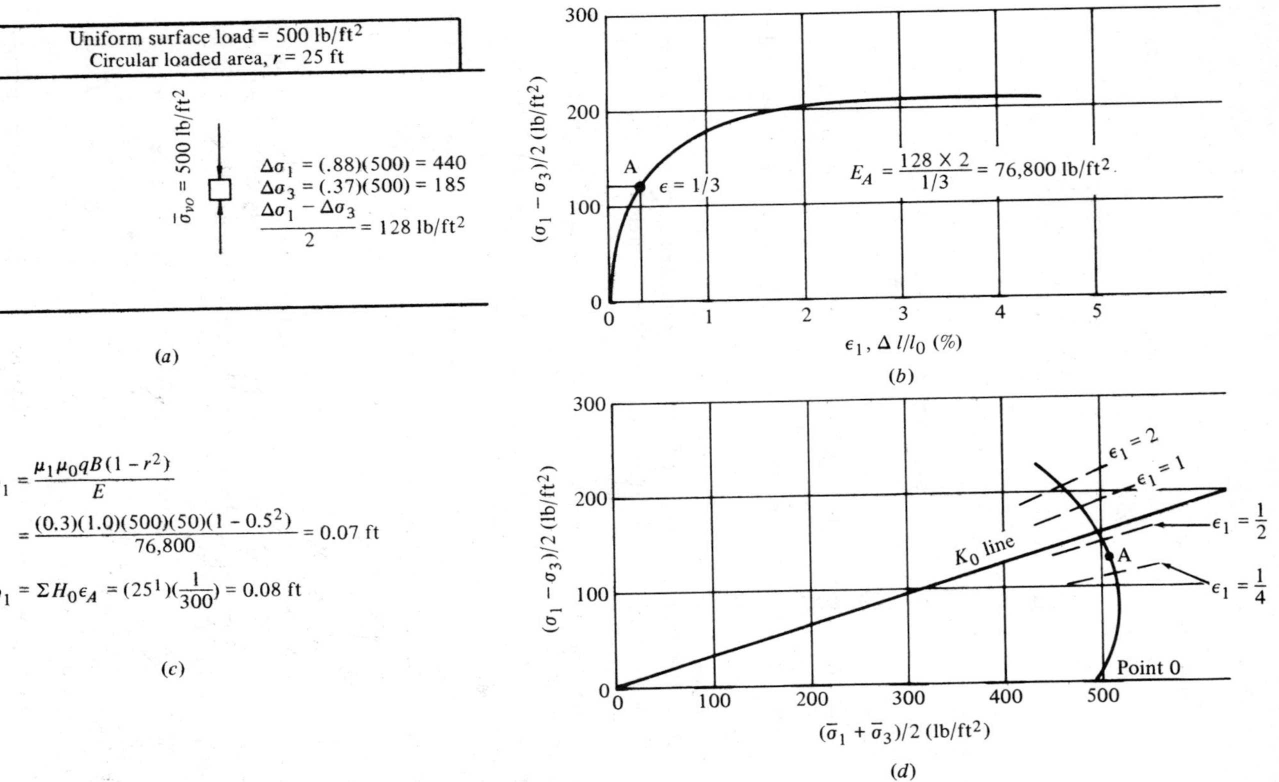

Fig. 44. Example: Settlement of circular loaded area on homogeneous sand by stress-path method (after Lambe, 1964). (a) Stress increments at average point. (b) Compression test results on average element. (c) Computation of immediate settlement. (d) Stress path for average element.

axial strain and Fig. 44d is a $q = (\sigma_1 - \sigma_3)/2$ vs. $p = (\sigma_1 + \sigma_3)/2$ plot with the axial strains for various q/p ratios shown as dashed lines. In both plots, point A corresponds to the field stress conditions after application of the surface loading.

The settlement due to surface loading can be computed in either of two ways (Fig. 44c). The first method utilizes the elastic solution for settlement (with correction factors for the geometry of the problem) with the modulus of the soil obtained from Fig. 44b. In the second method, the strain measured in the triaxial test corresponding to the field stress changes is integrated over the thickness of the sand stratum. The two methods give essentially the same result.

EXAMPLE 2 CIRCULAR LOADED AREA ON COHESIVE SOIL (FIG. 45)

The loading conditions for this example consist of rapid (undrained) loading followed by consolidation under constant load, followed by rapid unloading and expansion under constant load. Since the compressible stratum is a cohesive soil, both the immediate settlement (or heave) and consolidation settlement (or expansion) will be estimated by the stress-path method.

The movements at the various stages of loading and unloading are computed as follows:

1. The average point is selected for the compressible stratum.
2. The initial stresses and pore pressures are determined for the average point.
3. The total stress changes at the average point are estimated by elastic theory for the various stages of loading and unloading (see Fig. 45b.) (In this example, the total stress changes occurring during consolidation due to the change in Poisson's ratio are neglected for simplicity.)
4. Laboratory tests are run on undisturbed samples, duplicating the initial stresses and stress changes computed for the average point. The stress–strain data for the various stages of loading and unloading are given in Fig. 45c and d. The effective stress paths for the average point are shown in Fig. 45b.
5. The vertical movement of the center of the loaded area is computed by summing the vertical strains measured in the laboratory over the entire thickness (72 m) of the compressible stratum. The results of the laboratory measurements and the computed movements are summarized in Fig. 45d.

The two preceding examples have been selected to illustrate the application of the stress-path method to field settlement problems. This method can be a powerful tool for analyzing settlements, particularly in cases where the stress paths differ greatly from the commonly assumed one-dimensional compression. However, the method also has certain shortcomings. For instance, field pore-pressure changes may have to be estimated, stress paths close to failure are difficult to execute in the laboratory, and the selection of appropriate "average points" in the compressible strata is not always a straightforward matter.

Methods of Computing Settlements in Granular Soils Settlements in granular soils are the results of shear and volumetric distortions, both of which occur under essentially drained conditions because of the relatively high permeability

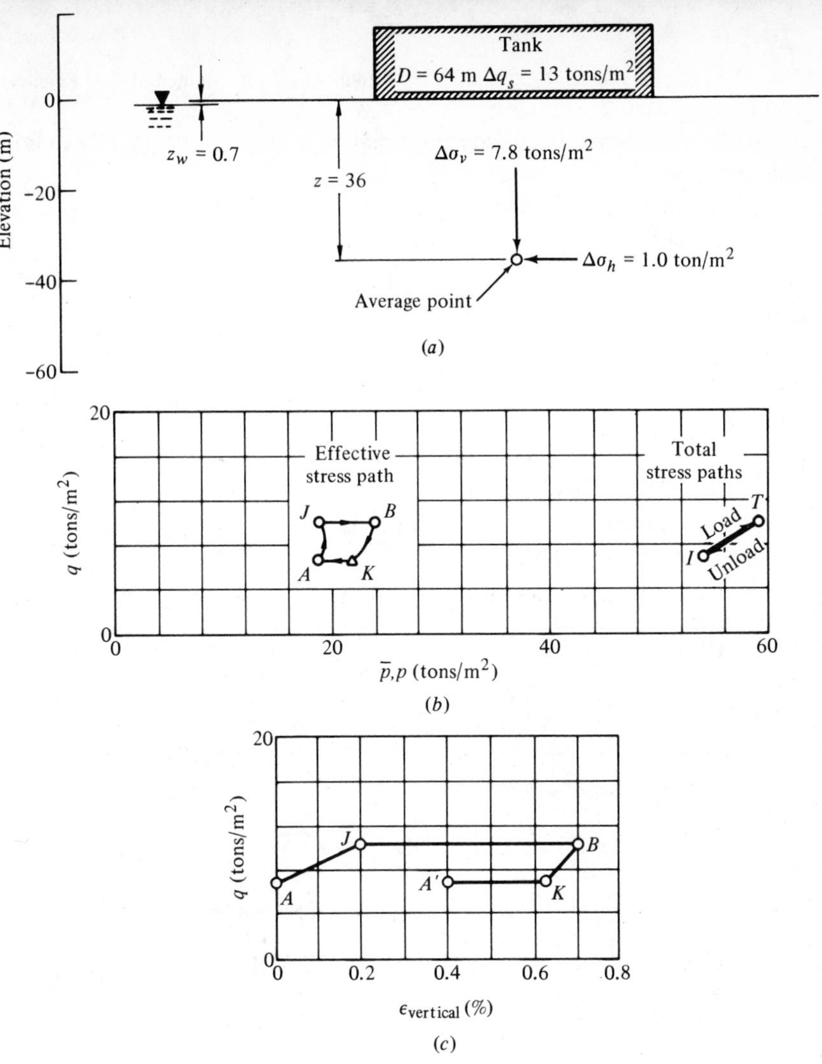

Conditions at average point					Vertical movement of tank center
Stage	Type of stress	Change in Δu	Type of strain	Vertical strain	
1	*AJ* (undrained loading)	from 0 to +4.4 tons/m^2		0.2%↓	72 m × 0.002 = 0.144 m↓
2	*JB* (consolidation at constant load)	from +4.4 to 0		0.5%↓	72 m × 0.005 = 0.360 m↓
3	*BK* (undrained unloading)	from 0 to −2.9		0.07%↑	72 m × 0.0007 = 0.050 m↑
4	*KA* (expansion at constant load)	from −2.9 to 0		0.23%↑	72 m × 0.0023 = 0.166 m↑

(*d*)

Fig. 45. Example: Settlement of circular loaded area on cohesive soil by stress-path method (after Lambe and Whitman, 1969). (a) Stress increments for average point. (b) Stress paths for average point. (c) Strains at average point. (d) Vertical movement of tank center.

of granular soils. Several methods of estimating settlements in granular soils have been proposed. These methods are discussed below (in no special order).

1. Elastic theory. According to elastic theory, settlements may be computed from the equation (Terzaghi, 1943; Bjerrum, 1958)

$$\rho_i = \mu_1 \mu_0 (qB/E)(1 - \nu^2)$$

where

ρ_i = elastic settlement
μ_1, μ_0 = dimensionless factors depending on the geometry of the problem (see Fig. 46)
q = intensity of surface load
B = width of loaded area
E = stress–strain modulus of the soil
ν = Poisson's ratio

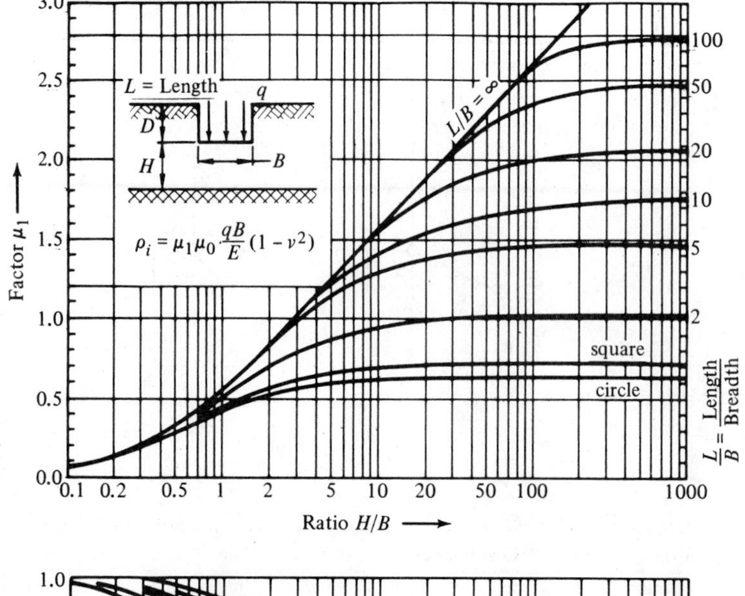

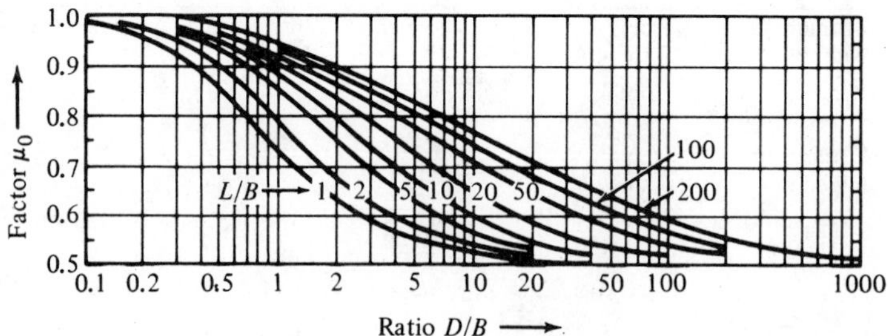

Fig. 46. Dimensionless factors for elastic settlement (after Bjerrum, 1958).

The most important aspect of this method of settlement analysis is the laboratory determination of the "elastic" constants E and ν, since the stress–strain modulus of granular soils is particularly sensitive to the relative density of the soil and the stress path followed during loading. The proper use of this method has already been illustrated in Example 1 in the discussion of the stress-path method.

2. Compression modulus. This method is based on the concept that if the compression modulus of the soil is determined under appropriate loading conditions, the settlement can be estimated by integrating the vertical strains over the thickness of the compressible stratum. Hough (1959) recommends the use of the compression index measured in a one-dimensional compression test. In this method, the compression index is determined from the e vs. log $\bar{\sigma}$ curve and the settlement computed from the equation

$$\rho = \int_0^H \frac{C_c}{1 + e_0} \log_{10} \left(\frac{\bar{\sigma}_{vo} + \Delta\sigma_v}{\sigma_{vo}} \right) dz$$

where

ρ = total settlement
H = thickness of compressible stratum
C_c = compression index
e_0 = initial void ratio
$\bar{\sigma}_{vo}$ = present overburden pressure
$\Delta\sigma_v$ = vertical stress increase due to applied surface load

Approximate values of $(1 + e_0)/C_c$ as a function of soil type and standard penetration resistance have been provided by Hough (1959) (Fig. 47). The main shortcoming in Hough's method is that only one-dimensional loading conditions are considered, with the result that settlements could be seriously underestimated where the field loading conditions are three-dimensional, particularly if the safety factor is low.

Janbu (1967) has presented a method of settlement analysis based on the tangent modulus of the soil. Janbu's definition of the modulus is expressed as

$$M = \frac{d\sigma}{d\epsilon}$$

where

M = deformation modulus
σ = stress in a given direction
ϵ = strain in the same direction

The modulus definition is illustrated in Fig. 48. The tangent molulus is determined from laboratory tests which best represent the field loading conditions

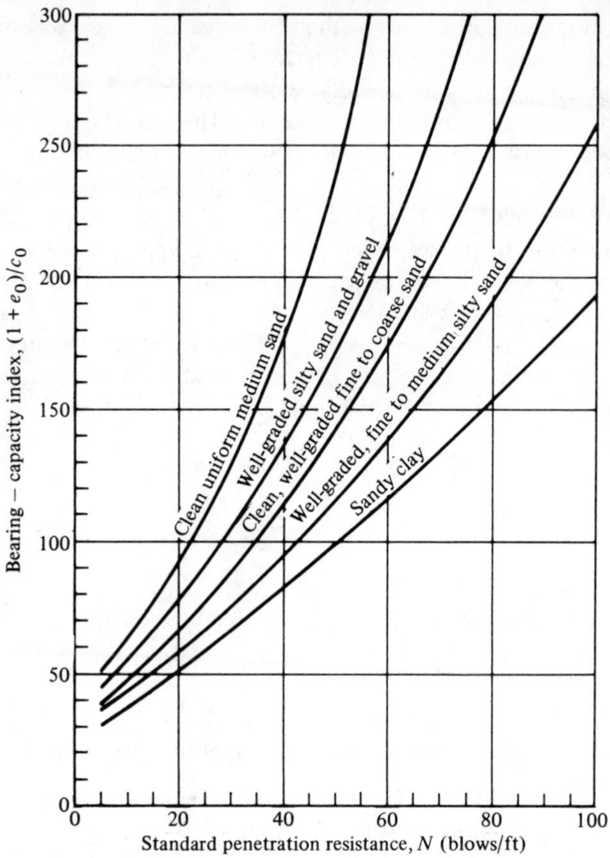

Fig. 47. Relation between N value and bearing-capacity index $(1 + e_0)/C_c$ (after Hough, 1959).

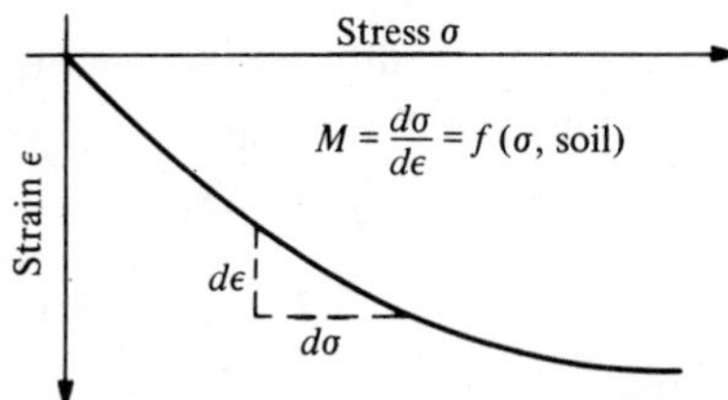

Fig. 48. Definition of deformation modulus (after Janbu, 1967).

(e.g., one-dimensional compression, triaxial compression, etc.). Two examples of field loading conditions and their counterparts in laboratory tests are shown in Fig. 49. For cases like Case B in Fig. 49, the modulus can be obtained from triaxial tests in which the ratio of lateral stress to vertical stress is governed by the factor of safety with respect to shear failure (e.g., for $F_s = 1, \bar{\sigma}_h/\bar{\sigma}_v = K_A$).

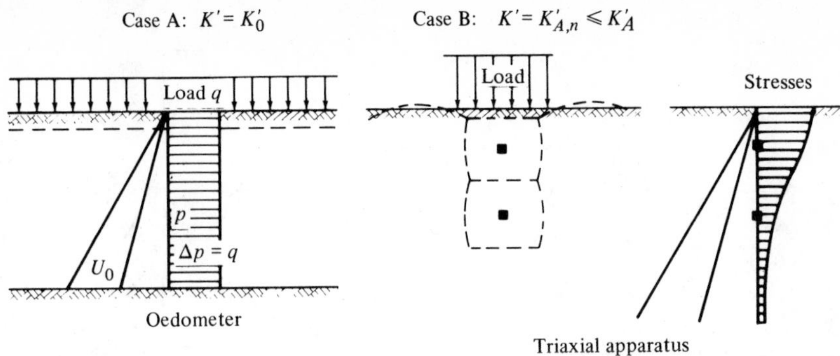

Fig. 49. Laboratory simulation of field loading conditions (after Janbu, 1967).

Typical stress-strain and modulus-strain curves for a sand, as measured in the oedometer test, are shown in Fig. 50. Janbu reports that the modulus variation for granular soils can be adequately expressed by the equation

$$M = m\sqrt{\bar{\sigma}_v \sigma_a}$$

where

m = modulus number
$\bar{\sigma}_v$ = applied vertical stress (effective)
σ_a = reference pressure ≈ 1 atmosphere

This form of expression for the compressibility modulus for granular soils is basically in agreement with that reported by Schultze and Moussa (1961) (see

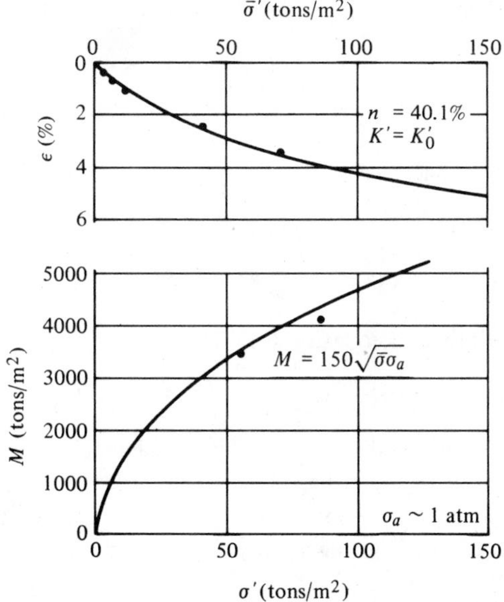

Fig. 50. Results of oedometer test on Valgrinda sand (after Janbu, 1967).

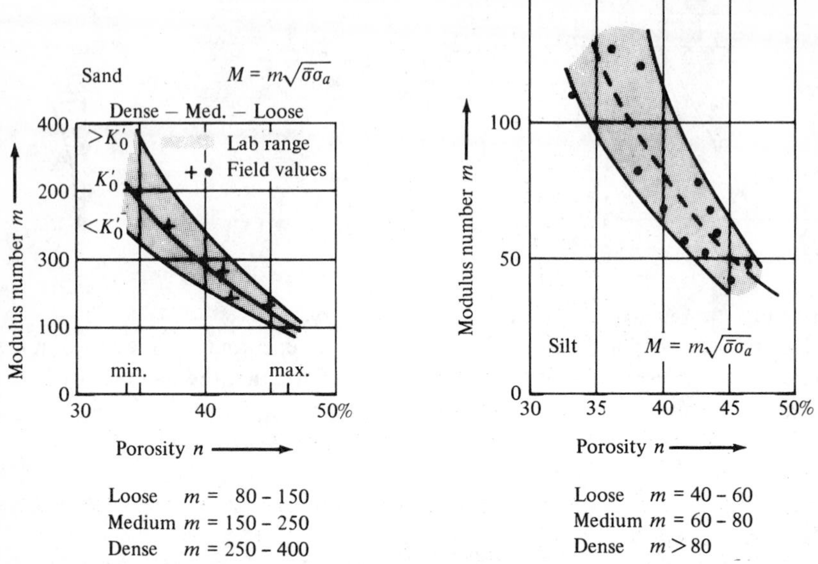

Fig. 51. Modulus numbers for inorganic sands and silts (after Janbu, 1967). (a) Valgrinda sand. (b) Trondelag silt.

Table 5). Figure 51 shows the variation of modulus number with porosity and principal stress ratios for typical inorganic sands and silts in Norway.

Using Janbu's method, settlements are calculated from the equation

$$\rho = \int_0^H \epsilon \, dz$$

where, from Janbu's first equation for M,

$$\epsilon = \int_{\bar{\sigma}_{vo}}^{\bar{\sigma}_{vo} + \Delta\sigma_v} \frac{d\bar{\sigma}_v}{M}$$

Substituting Janbu's second equation for M,

$$\epsilon = (2/m)[\sqrt{(\bar{\sigma}_{vo} + \Delta\sigma_v)/\sigma_a} = \sqrt{(\bar{\sigma}_{vo}/\sigma_a)}]$$

Figure 52 illustrates the method of settlement calculation. The procedure consists of (a) estimating the profile of initial effective vertical stress and stress increase (elastic theory), (b) computing the vertical strain ϵ for several depths z beneath the base of the footing from the last equation to obtain the strain

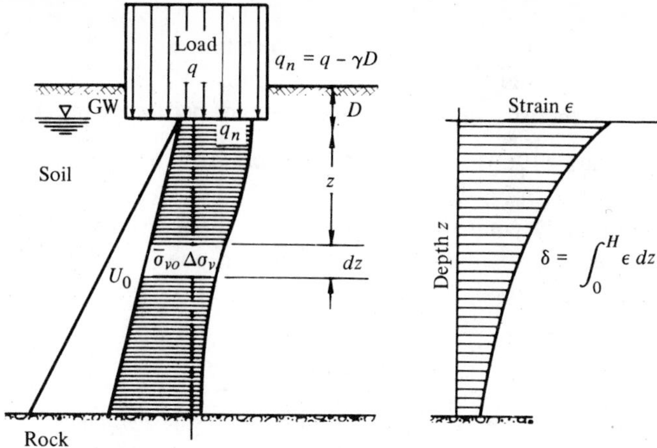

Fig. 52. Settlement calculation using tangent modulus (after Janbu, 1967).

profile, and (c) computing the total settlement from Janbu's equation for ρ (area under the strain profile curve).

3. Empirical Methods. Empirical methods of estimating settlements in granular soils are based on the following types of *in-situ* tests:

a. Standard penetration test.
b. Static-cone penetration tests.
c. Plate-loading tests.

Meyerhof (1964) has summarized the *standard penetration test* method in the form of the equations

$$q_a = NS_a/12 \qquad (B \leqslant 4 \text{ ft})$$

$$q_a = (NS_a/8)\,[B/(B+1)]^2 \qquad (B > 4 \text{ ft})$$

where

q_a = allowable bearing pressure (tons per square foot)
N = blows per foot from standard penetration test
S_a = allowable settlement (inches)
B = width of footing (feet) = $\sqrt{A}$ for rectangular footings

In granular soils the N value measured in the standard penetration test is influenced by the confining pressure at the depth at which the test is performed. Gibbs and Holtz (1957) have recommended the following correction factor for the N value determined under low confining pressure:

$$N = N'\,[50/(\bar{p} + 10)], \qquad \bar{p} \leqslant 40 \text{ psi}$$

where

$$N = \text{corrected standard penetration resistance}$$
$$N' = \text{measured standard penetration resistance}$$
$$\bar{p} = \text{effective confining pressure (pounds per square inch)}$$

Peck and Bazara (1969) have recently suggested more conservative N-value corrections, as shown in Fig. 53.

In very fine or silty sands of low permeability the measured N value may be too high because of the possibility of negative pore pressures temporarily strengthening the soil. For this case Terzaghi and Peck (1967) recommend the following correction:

$$N = 15 + \tfrac{1}{2}(N' - 15); \quad N' > 15$$

where N and N' are the corrected and measured values, respectively.

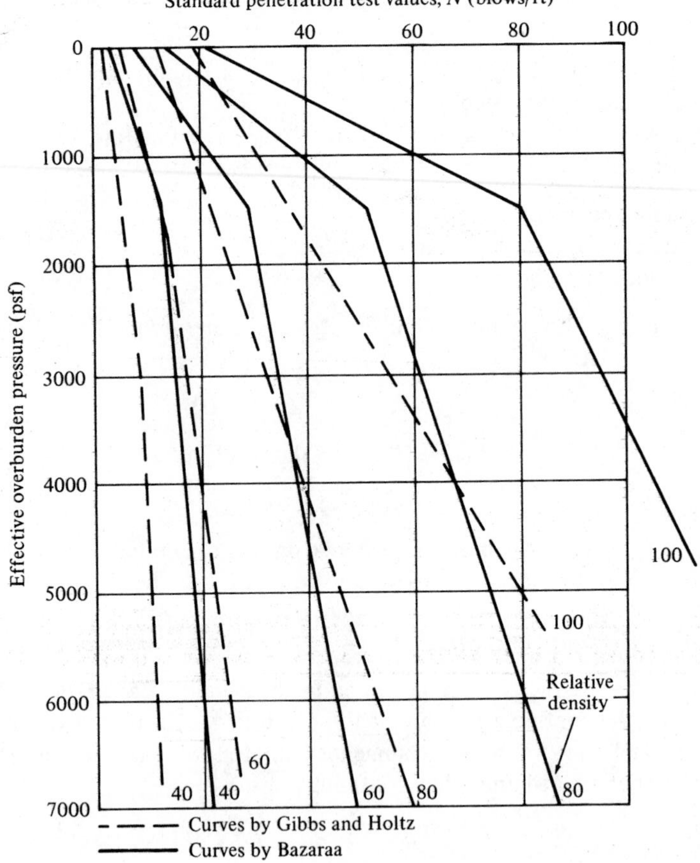

Fig. 53. Effect of overburden pressure on standard penetration resistance in granular soils (after Peck and Bazara, 1967).

Terzaghi and Peck (1967) further recommend a 50% reduction in the allowable bearing pressure when the water table is less than a depth B below the base of the footing. Other investigators, however, have indicated that the presence of the water table will be reflected in reduced N values, so that a water-table correction is unnecessary (Meyerhof, 1964; D'Appolonia, et. al., 1968).

If the footing is placed below ground level, Meyerhof (1964) recommends a maximum increase of 33% in the allowable bearing pressure; i.e.,

$$q_a = q_a' \, [1 + 0.33(D/B)]\,; \qquad D \leqslant B$$

where

q_a = allowable bearing pressure at depth D,
q_a' = allowable bearing pressure at ground surface (see Meyerhof's equation).

In a recent comprehensive field investigation of settlements of footings on sand, D'Appolonia et al. (1968) found that the best agreement between predicted and measured settlements using the standard penetration resistance were obtained by (a) using Gibbs and Holtz's correction for confining pressure, (b) increasing the allowable bearing pressures given by the Meyerhof equation by 50%, and (c) eliminating the water-table correction. However, this procedure allows no margin of safety, so that, in some cases, actual settlements will exceed predicted values.

The *static-cone penetration test*, although not commonly used in the United States, has a long history in other parts of the world, particularly in Europe (DeBeer and Martins, 1957). This method utilizes the static-cone resistance as a measure of soil compressibility expressed as follows:

$$E_s = \tfrac{3}{2}\,(q_c/\bar{\sigma}_0)$$

where

E_s = compression modulus
q_c = static-cone resistance (kilograms per square centimeter)
$\bar{\sigma}_0$ = effective overburden pressure at point of measurement

The compression modulus is used in the standard formula for calculating consolidation settlement; i.e.,

$$\rho = \int_0^H \left\{ \frac{1}{E_s} \log_{10} \left(\frac{\bar{\sigma}_{vo} + \Delta\sigma_v}{\bar{\sigma}_{vo}} \right) \right\} dz$$

where

ρ = total settlement
$\bar{\sigma}_{vo}$ = effective overburden pressure
$\Delta\sigma_v$ = vertical stress increase due to footing load (from elastic theory)

In practice the stressed zone is subdivided into layers having approximately equal values of cone resistance, the settlement is calculated separately for each layer, and the calculated settlements are summed to obtain the final settlement. The stressed zone is considered to extend to the depth at which the vertical stress increase equals 10% of the effective overburden pressure. Experience has shown that settlements calculated by this procedure are generally conservative by a factor of about 2.

Schmertmann (1970) recently proposed a new method of settlement prediction for sands using the static-cone test. Schmertmann's method is theoretically sounder, and published results indicate that it is more accurate than other penetrometer methods.

In Schmertmann's method, settlements are computed from the equation

$$\rho = C_1 C_2 \sum_{0}^{2B} (I_z/E_s)\,\Delta z$$

where

ρ = total settlement

$C_1 = 1 - 0.5\,(p_0/\Delta p)$ = correction factor for foundation embedment (p_0 is the overburden pressure at foundation level; $\Delta p = p - p_0$ is the net foundation pressure increase)

$C_2 = 1 + 0.2 \log (t \text{ yrs}/0.1)$ = correction factor for creep

$$I_z = \text{strain influence factor} = \begin{cases} 1.2\,(z/B) & \text{for } z/B \leqslant 0.5 \\ [0.8 - 0.4\,(z/B)] & \text{for } 0.5 \leqslant z/B \leqslant 2, \end{cases}$$

E_s = compression modulus = $2q_c$

The strain influence factor in this equation was developed as an approximation of the vertical strain distribution obtained from theoretical (Boussinesq), model-study, experimental, and computer-simulation (finite-element) results. The correlation between static-cone resistance and compression modulus was obtained by test loading a screw plate *in-situ* and backfiguring the modulus with this equation, and then comparing with static-cone resistance.

Schmertmann's method is simple to use and appears to give accurate predictions of settlement in sands. Therefore, it should become an important tool in foundation engineering in the future.

Plate-loading tests can be used to estimate settlement provided the soil conditions are uniform to a depth at least equal to the width of the largest footing. If the plate size and footing size are roughly the same, the results can be used directly. However, if, as is usually the case, the footings are considerably larger than the test plate, the load-test results must be extrapolated, taking into account the effect of size.

The nature of the size effect governing settlements on granular soils can be seen, for example, in the Meyerhof equation. If the settlements for two different-sized footings subjected to the same bearing pressure on the same soil are computed from that equation, the settlement ratio will be

$$S_2/S_1 = [B_1(B_2 + 1)/B_2(B_1 + 1)]^2$$

where S_2/S_1 is the settlement ratio of two footings of widths B_2 and B_1, respectively. However, Bjerrum and Eggestad (1963) and D'Appolonia et al. (1968) report settlement ratios which deviate considerably from this ratio. Moreover, in terms of estimating full-sized footing settlements from plate-load tests, this ratio equation usually errs on the unsafe side. Figure 54 shows the observations reported by Bjerrum and Eggestad and D'Appolonia et al., compared to the ratio equation. Judging from the scatter in the observed settlement ratios, as shown in Fig. 54, estimating foundation settlements from plate-load tests would seem to be at best unreliable and at worst dangerous.

Methods of Computing Settlements in Cohesive Soils Settlements in cohesive soil can be divided into three categories:

1. Immediate settlement. Settlement due to shear distortions during undrained loading, occurring immediately upon application of the load.

2. Consolidation settlement. Settlement due to expulsion of fluid from the pore spaces. This type of settlement is time-dependent, depending upon the compressibility of the soil skeleton and the permeability of the soil.

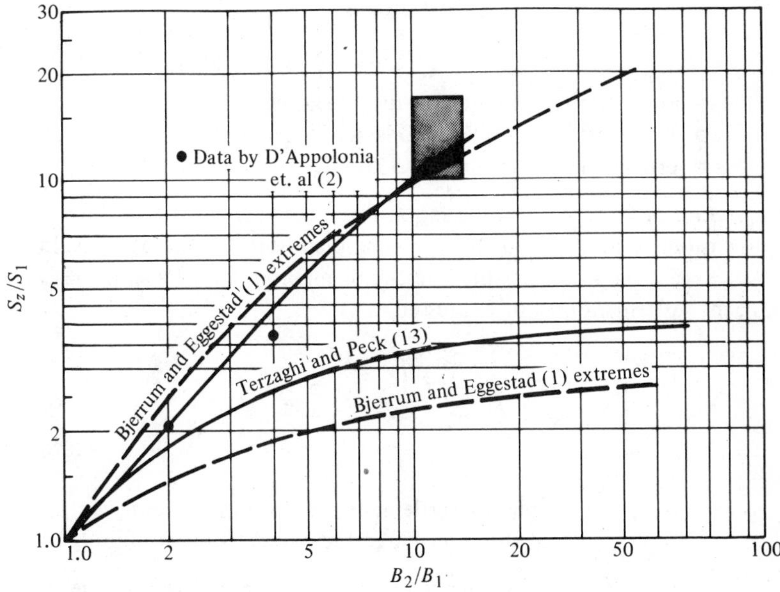

Fig. 54. Settlement ratios for different size footings on granular soils (after D'Appolonia et al., 1971).

3. Secondary compression. Settlement due to gradual adjustment of the soil skeleton to the applied loads after excess pore pressures are essentially dissipated. The rate of secondary compression depends upon soil type and how close the soil is to failure.

Immediate Settlement Immediate settlement is estimated by using elastic theory as described in a previous section, except that the undrained modulus E_u is used and Poisson's ratio is equal to $\frac{1}{2}$ for undrained loading.

Unfortunately, the undrained modulus is very difficult to measure experimentally because of the effects of sample disturbance and the many test variables that must be considered (Ladd, 1964). Because of these difficulties, empirical correlations are recommended (e.g., the ratio of the undrained modulus to the undrained shear strength, E_u/S_u). D'Appolonia et al. (1971) reported the values of this ratio from several case studies. As a general rule of thumb, E_u/S_u = 1000–1500 for normally consolidated or slightly overconsolidated clays of low plasticity. However, for clays of higher plasticity or organic clays, the value may be much lower.

In soft clays, considerable local yielding may occur during undrained loading even though the overall factor of safety is greater than unity. Based on finite-element solutions, D'Appolonia et. al. (1971) presented charts for computing immediate settlements when local yielding occurs (Fig. 55). From these charts, immediate settlement is computed from the equation

$$\rho_i = \rho_e/S_R$$

where

ρ_i = immediate settlement with local yielding
ρ_e = elastic settlement from the elastic-theory settlement equation
S_R = settlement ratio (from Fig. 55)

Consolidation Settlement Consolidation settlements are usually estimated from one-dimensional consolidation theory, using the results of laboratory oedometer tests (an exception would be the stress-path method discussed previously). In general, the settlement may be expressed as

$$\rho_{\text{oed}} = \int_0^H \epsilon_v \, dz$$

where

ρ_{oed} = consolidation settlement from oedometer tests
ϵ_v = vertical strain profile in the compressible stratum

The computation of vertical strains may be done in several ways, depending on how the oedometer test results are presented. The various methods are sum-

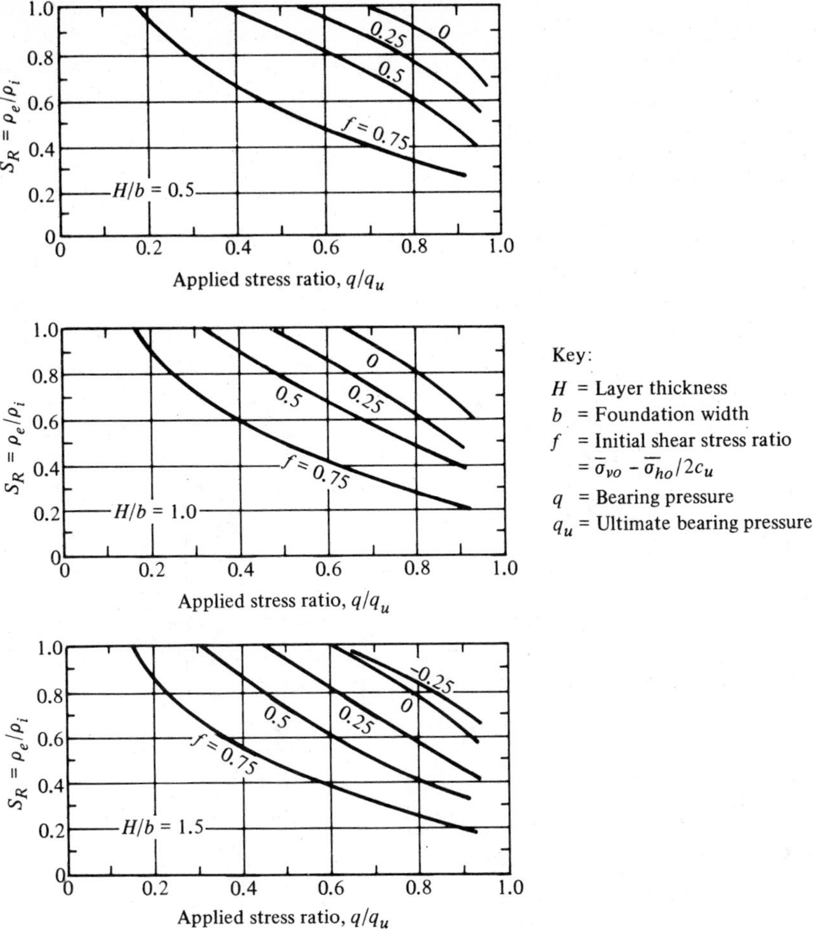

Fig. 55. Relationship between settlement ratio and applied stress ratio for strip foundations on homogeneous isotropic layer (after D'Appolonia et al., 1971).

marized in Table 17. The choice of method is largely a matter of personal preference. The procedure for settlement calculation may be described as follows:

1. Sketch the soil profile, indicating soil type, soil properties and layer thickness.
2. Determine the vertical stress profile, including present (effective) overburden pressure and (net) vertical stress increase due to applied loads as a function of depth.
3. Compute the vertical strain profile from the vertical stresses and the measured oedometer stress-strain behavior.

TABLE 17. DETERMINATION OF VERTICAL STRAINS FROM OEDOMETER TESTS.

Presentation of oedometer test results	Compression parameters: Virgin compression	Compression parameters: Recompression	Vertical strain: Case 1: normally consolidated	Vertical strain: Case 2: preconsolidated, $(\bar\sigma_{vo} + \Delta\sigma_r < \bar\sigma_{vm})$	Vertical strain: Case 3: preconsolidated, $(\bar\sigma_{vo} + \Delta\sigma_v > \bar\sigma_{vm})$
e vs. $\log_{10} \bar\sigma_v$	$C_c = \dfrac{\Delta e}{\Delta \log_{10} \bar\sigma_v}$	$C_r = \dfrac{\Delta e}{\Delta \log_{10} \bar\sigma_v}$	$\dfrac{C_c}{1+e_0} \log_{10} \dfrac{\bar\sigma_{vo} + \Delta\sigma}{\bar\sigma_{vo}}$	$\dfrac{C_r}{1+e_0} \log_{10} \dfrac{\bar\sigma_{vo} + \Delta\sigma_v}{\bar\sigma_{vo}}$	$\dfrac{C_r}{1+e_0} \log_{10} \dfrac{\bar\sigma_{vm}}{\bar\sigma_{vo}} + \dfrac{C_c}{1+e_0} \log_{10} \dfrac{\bar\sigma_{vo} + \Delta\sigma_v}{\bar\sigma_{vm}}$
ϵ_v vs. $\log_{10} \bar\sigma_v$	$CR = \dfrac{\Delta\epsilon_v}{\Delta \log_{10} \bar\sigma_v}$	$RR = \dfrac{\Delta\epsilon_v}{\Delta \log_{10} \bar\sigma_v}$	$CR \log_{10} \dfrac{\bar\sigma_{vo} + \Delta\sigma_v}{\bar\sigma_{vo}}$	$RR \log_{10} \dfrac{\bar\sigma_{vo} + \Delta\sigma_v}{\bar\sigma_{vo}}$	$RR \log_{10} \dfrac{\bar\sigma_{vm}}{\bar\sigma_{vo}} + CR \log_{10} \dfrac{\bar\sigma_{vo} + \Delta\sigma_v}{\bar\sigma_{vo}}$
e vs. $\bar\sigma_v$	a_{vn}	a_{vp}	$\dfrac{a_{vn}}{1+e_0} \Delta\sigma_c$	$\dfrac{a_{vp}}{1+e_0} \Delta\sigma_v$	$\dfrac{a_{vp}}{1+e_0} (\bar\sigma_{vm} - \bar\sigma_{vo}) + \dfrac{a_{vn}}{1+e_0} (\bar\sigma_{vo} + \Delta\sigma_v - \bar\sigma_{vm})$
	$a_v = \Delta e/\Delta\bar\sigma_v$				
ϵ_v vs. $\bar\sigma_v$	m_{vn}	m_{vp}	$m_{vn} \Delta\sigma_v$	$m_{vp} \Delta\sigma_v$	$m_{vp}(\bar\sigma_{vm} - \bar\sigma_{vo}) + m_{vn}(\bar\sigma_{vo} + \Delta\sigma_v - \bar\sigma_{vm})$
	$m_v = \Delta\epsilon_v/\Delta\bar\sigma_v$				
ϵ_v vs. $\bar\sigma_v$	M	M_{rc}	$\displaystyle\int_{\bar\sigma_{vo}}^{\bar\sigma_{vo}+\Delta\sigma_v} \frac{d\bar\sigma_v}{M}$	$\displaystyle\int_{\bar\sigma_{vo}}^{\bar\sigma_{vo}+\Delta\sigma_v} \frac{d\bar\sigma_v}{M_{rc}}$	$\displaystyle\int_{\bar\sigma_{vo}}^{\bar\sigma_{vm}} \frac{d\bar\sigma_v}{M_{rc}} + \int_{\bar\sigma_{vm}}^{\bar\sigma_{vo}+\Delta\sigma_v} \frac{d\bar\sigma_v}{M}$
	$M = \dfrac{d\bar\sigma_v}{d\epsilon_v} = m\sigma_a \left(\dfrac{\bar\sigma_v}{\sigma_a}\right)^{1-n}$		$\left\{\epsilon_v = \displaystyle\int_{\sigma_1}^{\sigma_2} \frac{d\sigma_v}{M} = \frac{1}{mn}\left[\left(\frac{\sigma_2}{\sigma_a}\right)^n - \left(\frac{\sigma_1}{\sigma_a}\right)^n\right] \quad (n \neq 0) \qquad \epsilon_v = \frac{1}{m} \ln \frac{\sigma_2}{\sigma_1} \quad (n = 0)\right\}$		

4. Sum vertical strains (area under the vertical strain vs. depth curve) to obtain total settlement.

Skempton and Bjerrum (1957) recognized that actual settlement, where the loading is three-dimensional, may differ from that computed from oedometer tests (one-dimensional loading). The Skempton–Bjerrum method may be stated as

$$\rho_c = C_\rho \rho_{\text{oed}}$$

where

ρ_c = total (field) consolidation settlement,
ρ_{oed} = consolidation settlements computed from oedometer tests (any of the methods given in Table 17),
C_ρ = correction factor to account for the fact that the actual $\Delta\bar{\sigma}_v$ developed in the field may be different than the $\Delta\sigma_v$ applied in the oedometer test; C_ρ depends upon the pore-pressure parameter A and the geometry of loading, as shown in Fig. 56.

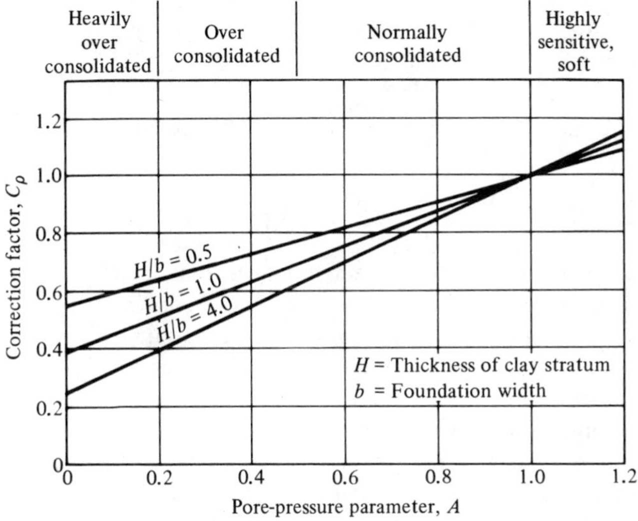

Fig. 56. Correction factors for consolidation settlement (after Skempton and Bjerrum, 1957).

The procedure for calculating consolidation settlements is illustrated by an example in Fig. 57.

The time rate of consolidation settlement is usually computed from one-dimensional consolidation theory; it may be expressed in the form

$$U\,(\%) = f(T)$$

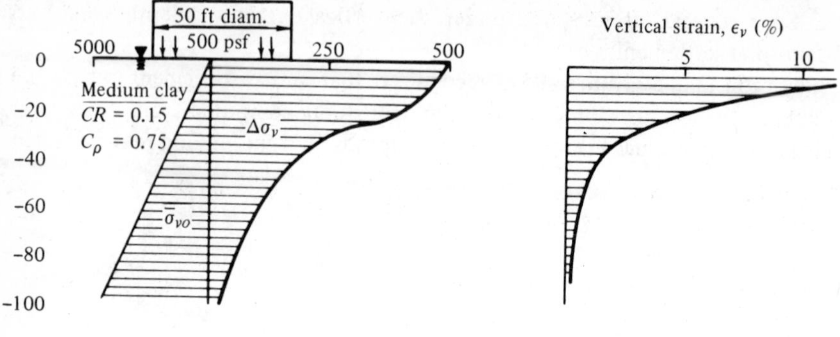

Layer no.	Depth (ft)	Thickness (ft)	Original effective overburden pressure, $\bar{\sigma}_{vo}$ (psf)	Vertical stress increase, $\Delta\sigma_v$ (psf)	Vertical strain, ϵ_v (%)	Settlement (in.)
1	0–20	20	450	470	4.6	11.0
2	20–40	20	1350	273	1.2	2.9
3	40–60	20	2250	143	0.4	1.0
4	60–80	20	3150	82	0.2	0.5
5	80–100	20	4050	55	0.1	0.2

Total oedometer settlement, ρ_{oed} = 15.6 in.
Estimated settlement *in situ* = 0.75 × 15.6 ≈ 12 in.

Fig. 57. Example: Calculation of consolidation settlements.

where

$$T = (c_v/H^2)^t$$

The coefficient of consolidation c_v is determined from oedometer tests (Fig. 14) or, alternatively, it may be estimated from empirical correlations (Fig. 15).

Solutions of the preceding equations have been obtained for a variety of drainage conditions, including horizontal as well as vertical drainage (Fig. 13). In addition, one-dimensional consolidation theory is readily adaptable to numerical computation, permitting special cases to be solved as needed with relative ease (see, for example, Wu, 1966).

When vertical and horizontal drainage occur simultaneously, the average degree of consolidation is given by

$$U_T = 1 - (1 - U_x)(1 - U_y)(1 - U_z)$$

or

$$U_T = 1 - (1 - U_r)(1 - U_z)$$

where

U_T = average degree of consolidation for combined vertical and horizontal drainage,

U_x, U_y, U_z, U_r = average degree of consolidation in each coordinate direction.

The rates of consolidation in each coordinate direction are computed independently and then combined to obtain the effect of combined drainage, as indicated in the two preceding equations.

These equations are only valid for one-dimensional consolidation in which the total stresses are constant throughout the consolidation process. A true three-dimensional consolidation theory is complicated by the fact that the total stresses vary with time. Figure 58 shows the effect of three-dimensional

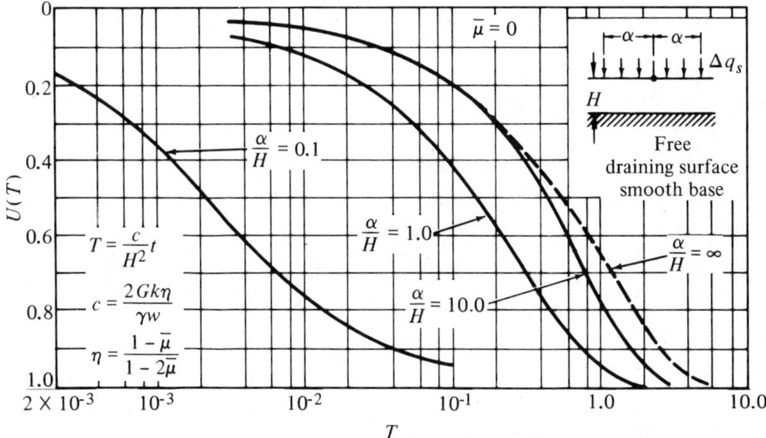

Fig. 58. Consolidation of stratum under circular load (after Gibson et al., 1967).

consolidation on the rate of consolidation for a circular loaded area. It is apparent that the one-dimensional consolidation theory can give very conservative results depending on the geometry and boundary conditions of the problem.

Secondary Compression Secondary compression is the time-dependent settlement which takes place following the completion of primary consolidation. It is computed as follows:

$$\rho_{sec} = C_\alpha H \log_{10} (t_{sec}/t_p)$$

where

ρ_{sec} = secondary compression settlement
C_α = coefficient of secondary compression
H = thickness of compressible stratum
t_{sec} = time for which secondary compression is being computed, measured from the start of loading
t_p = time required for completion of primary consolidation

The coefficient of secondary compression C_α is determined from the slope of the vertical strain vs. $\log_{10}$ time curve from oedometer tests or, as an approximation, from empirical correlations (Fig. 17).

Deep Foundations

When adequate foundation support cannot be obtained at shallow depths, deep foundations are required. This situation may arise, for example, when the soils near the surface are weak and compressible, when the loads are extremely heavy, producing unusually high bearing pressures, or when large lateral loads must be resisted.

The load-carrying capacity of deep foundations may be achieved in the following ways:

1. By transferring the load from the surface to deeper, firmer, strata by means of end-bearing.
2. By distributing the load gradually by means of friction along the shaft through strata which would not be capable of supporting the loads if concentrated near the surface.
3. By compacting loose granular soils to higher densities, thereby increasing resistance to settlement.

These are the basis mechanisms by which deep foundations function to achieve the desired foundation support. Depending upon soil conditions and the type of foundation used, more than one of these mechanisms may contribute to the overall load-carrying capacity of the foundation. For example, most driven piles function in both end-bearing and skin friction.

Types of Deep Foundations There is a considerable variety of deep foundation types available commercially. However, for purposes of analysis, the different types may be generally categorized as (1) driven piles and (2) drilled piers or caissons.

Driven Piles Driven piles are available in many forms. Wooden piles, concrete piles, steel H piles, and steel shells filled with concrete are commonly used. Normal design loads range from approximately 25 tons for wooden piles to 100+ tons for steel H piles or cylinder piles, although higher capacities are not uncommon. Practical lengths range from approximately 60 ft for wooden piles to 150+ ft for concrete-filled pipe piles.

Driving equipment may consist of (1) drop hammers, (2) single-acting steam hammers, (3) double-acting steam hammers, (4) differential-action steam hammers, or (5) diesel hammers. For a detailed description of the various types of driving equipment, the reader is referred to Chellis (1961). Chellis also provides extensive data on the characteristics of different types of pile-driving hammers.

Driven piles are very often the least expensive and simplest type of deep foundation to install. The technology of pile driving is well developed and piles and pile-driving equipment are readily available. On the other hand, surface heave caused by displacement of soil and ground vibrations during pile driving operations may be objectionable. Installation of piles by driving also does not permit inspection of the bearing stratum.

Drilled Piers or Caissons Deep foundations may also be installed by placing concrete in a bored hole. These foundations are referred to as drilled piles, piers, or caissons. Some of the most common forms of this type of foundation are (1) cast-in-place piles, (2) drilled or drilled-and-belled caissons, and (3) auger-cast* piles. Caissons, both straight-shafted and drilled-and-belled, are coming into increasing use in this country. The use of caissons offers several advantages under the appropriate conditions. For example, the installation of caissons produces very little ground vibration and less noise than pile driving; one caisson can carry the load of several piles, thereby eliminating the need for pile caps; and the bearing stratum can be inspected first-hand before the caisson is installed.

On the other hand, caissons have certain disadvantages, such as the difficulty of penetrating strata of water-bearing granular soils and dewatering problems which may occur when attempting to establish caissons in water-bearing soil and rock formations.

The selection of the most suitable type of deep foundation for a particular set of conditions is often a complex matter. It depends upon soil conditions at the site, the magnitude of loads to be supported, and the economics of various foundation types in a particular locality. Therefore, it is usually desirable to evaluate a number of foundation types on the basis of both technical considerations and economy before arriving at a final decision.

Load-Carrying Capacity of a Single Pile The load-carrying capacity of a single pile or caisson may be estimated by static analysis of the skin friction resistance between the shaft and the surrounding soil and the end-bearing capacity at the tip. In the case of driven piles, dynamic pile-driving formulas are often used to estimate load-carrying capacity from the dynamic response of the pile during driving. Load tests are also used to determine pile capacity both in compression and uplift.

Static Analysis The load-carrying capacity of a pile or caisson is the sum of point resistance and shaft resistance. From a knowledge of the strength characteristics of the soil surrounding the shaft and beneath the tip, the load-carrying capacity may be estimated by the methods of soil mechanics.

The total capacity of a single deep foundation may be expressed as

$$Q = Q_p + Q_s$$

*This is a patented process.

where

$$Q_p = \text{point (end-bearing) resistance}$$
$$Q_s = \text{shaft resistance.}$$

The point resistance is computed in a manner analogous to bearing capacity of shallow foundations; i.e.,

$$Q_p = A_p\,(cN_c + (\gamma B/2)N_\gamma + \gamma d N_q)$$

where

$$A_p = \text{area of the tip}$$
$$c = \text{cohesion of soil beneath the tip}$$
$$\gamma = \text{density of soil beneath the tip}$$
$$B = \text{width (or diameter) of the tip}$$
$$d = \text{depth of penetration}$$
$$N_c, N_\gamma, N_q = \text{bearing-capacity factors for deep foundations}$$

Several theories have been developed for computing the point resistance of a deep foundation. These theories differ primarily in the failure patterns that are assumed to develop around the tip of the foundation, as shown in Fig. 59. Bearing capacities computed from these different failure patterns may differ considerably, as reflected in the bearing-capacity factors N_q plotted in Fig. 60. In particular it may be noted that, for most cases shown, N_q is considerably higher for deep foundations than for shallow foundations.

The bearing-capacity factor N_c also increases with the foundation depth. As shown in Fig. 61, $N_c = 9$ for $d/B > 10$ for circular or square foundations.

The contribution of the N_γ term becomes very small for deep foundations and is usually neglected. Therefore, the equation for Q_p simplifies as follows:

$$Q_p = A_p\,(cN_c + \gamma d N_q)$$

Furthermore, for cohesionless soils, $c = 0$ so that

$$Q_p = A_p(\gamma d N_q)$$

and, for cohesive soils, $\phi = 0$ (undrained analysis) so that

$$Q_p = A_p(cN_c + \gamma d)$$

Shaft resistance of deep foundations is the result of friction and adhesion between the shaft and the surrounding soil. The magnitude of skin friction depends upon several factors; e.g., the shear-strength properties of the soil, the roughness of the shaft, the method of placement, and the effect of disturbance on the strength of the soil.

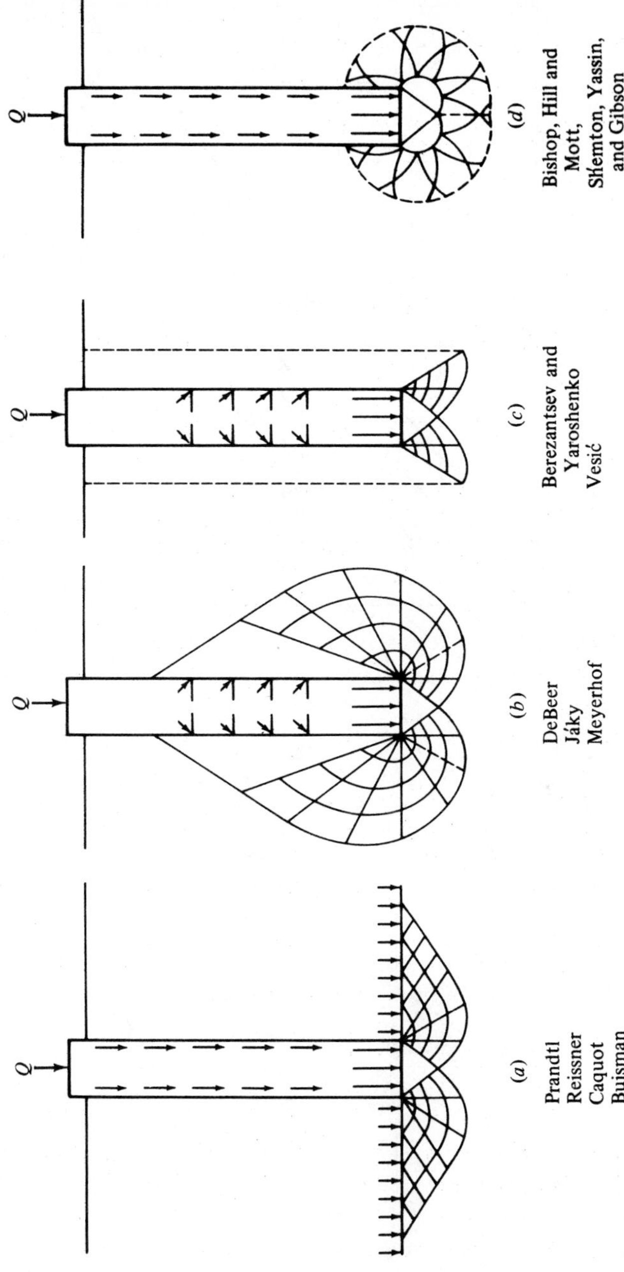

Fig. 59. Assumed failure patterns under deep foundations (after Vesic, 1967).

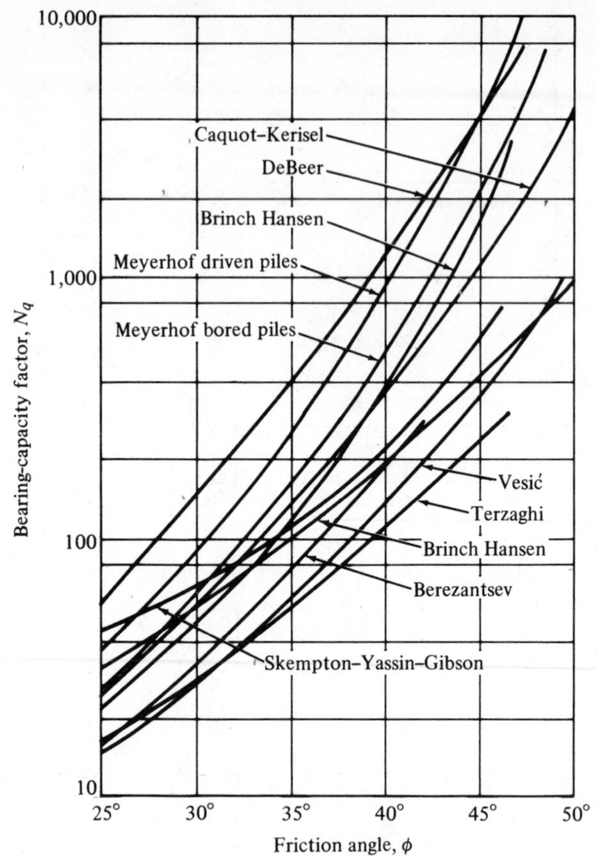

Fig. 60. Bearing-capacity factors N_q for deep circular foundation (after Vesic, 1967).

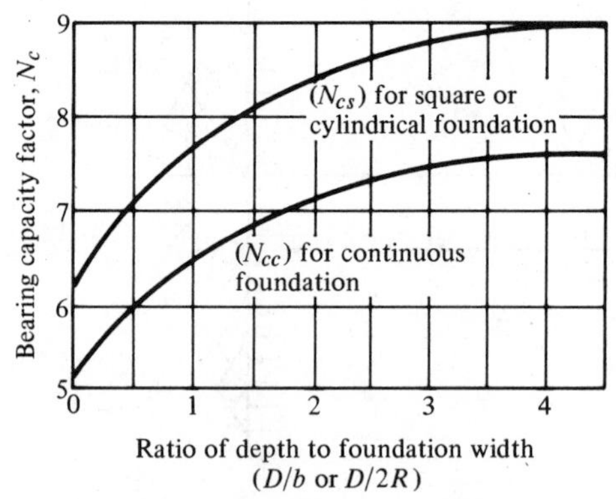

Fig. 61. Bearing-capacity factor N_c.

The shaft resistance may be expressed as

$$Q_s = \int_0^L (2\pi r f) dl$$

where

L = embedded length of shaft
$2\pi r$ = perimeter of shaft
f = unit shaft resistance

In general, the unit shaft resistance equals

$$f = c_a + \sigma \tan \delta$$

where

c_a = unit adhesion between shaft and surrounding soil
σ = normal stress on shaft
$\tan \delta$ = coefficient of skin friction between shaft and surrounding soil

For a cohesionless soil, $c_a = 0$ and $\sigma = \bar{\sigma}_h = K_h \bar{\sigma}_v$,

$$f = K_h \bar{\sigma}_v \tan \delta$$

where

σ_v = effective vertical stress
$K_h = \bar{\sigma}_h / \bar{\sigma}_v$ = coefficient of horizontal earth pressure at soil–shaft interface.

The value of K_h can be extremely variable depending upon the amount of densification produced during driving. Table 18 contains values recommended by several investigators.

The skin-friction coefficient $\tan \delta$ depends on the frictional resistance of the

TABLE 18. HORIZONTAL STRESS ON PILES DRIVEN IN SAND. (AFTER LAMBE AND WHITMAN, 1969; REFERENCE: HORN, 1966).

Reference	Relationship	Basis of relationship
Brinch Hansen and Lundgren (1960)	(a) $\bar{\sigma}_h = \cos^2 \bar{\phi} \cdot \bar{\sigma}_v = 0.43 \bar{\sigma}_v$ if $\bar{\phi} = 30°$ (b) $\bar{\sigma}_h = 0.8 \sigma_v$	(a) theory (b) pile test
Henry (1956)	$\bar{\sigma}_h = K_p \cdot \bar{\sigma}_v = 3 \bar{\sigma}_v$	theory
Ireland (1957)	$\bar{\sigma}_h = K \cdot \bar{\sigma}_v = (1.75 \text{ to } 3) \cdot \bar{\sigma}_v$	pulling tests
Meyerhof (1951)	$\bar{\sigma}_h = 0.5 \bar{\sigma}_v$; loose sand $\bar{\sigma}_h = 1.0 \bar{\sigma}_v$; dense sand	analysis of field data
Mansur and Kaufman (1958)	$\bar{\sigma}_h = K \bar{\sigma}_v$; $K = 0.3$ (compression) $K = 0.6$ (tension)	analysis of field data

TABLE 19. VALUES OF δ/ϕ FOR VARIOUS FOUNDATION MATERIALS IN CONTACT WITH DENSE SAND (AFTER TOMLINSON, 1969).

Foundation material	Surface finish	Values of δ/ϕ for Dry sand	Values of δ/ϕ for Saturated sand
Steel	smooth (polished)	0.54	0.64
	rough (rusted)	0.76	0.80
Wood	parallel to grain	0.76	0.85
	right angles to grain	0.88	0.89
Concrete	smooth (made in metal formwork)	0.76	0.80
	grained (made in timber formwork)	0.88	0.88
	rough (cast on ground)	0.98	0.90

soil and the roughness of the shaft. Representative values of δ/ϕ for various soil types and pile materials are given in Table 19.

In the case of cohesive soils, the unit adhesion between the shaft and the surrounding soil depends on the cohesion of the soil, the effect of disturbance on the soil, the method of placement, and the type of pile. Soft, normally consolidated soils reconsolidate after the foundation is in place and develop adhesion which is as great as or greater than the original strength of the soil. On the other hand, stiff, overconsolidated soils tend to lose strength when disturbed and are less likely to heal and form a strong bond along the shaft after having been disturbed. Similarly, bored piles generally develop less adhesion with the surrounding soil than driven piles, particularly in stiff clays. These effects are illustrated in Fig. 62.

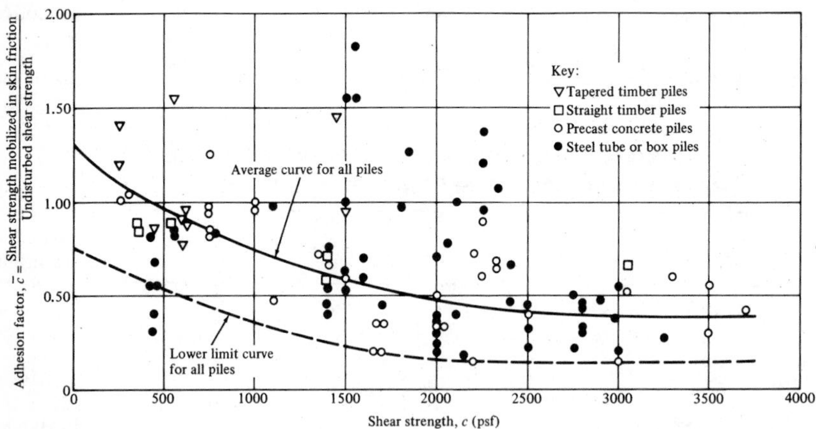

Fig. 62. Relationship between the skin friction on the shaft of piles driven into clay soils and the shear strength of the clays (after Tomlinson, 1969).

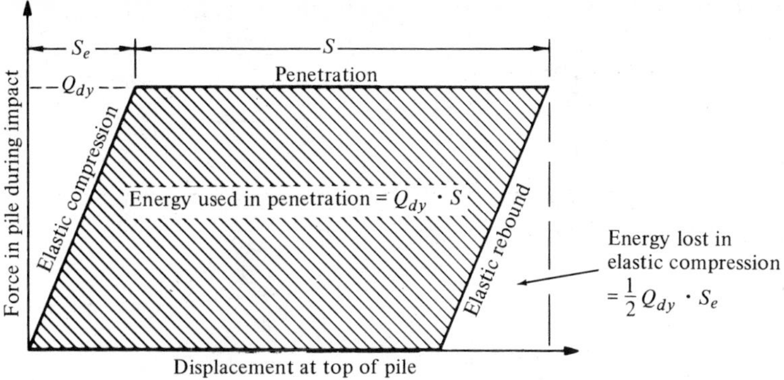

Fig. 63. Simplified pile response during driving.

Pile-Driving Formulas Pile-driving formulas are frequently used to determine the capacity of a pile under static loads. These formulas are based on the concept that the energy input is equal to the energy used in driving the pile plus losses. For example, one of the simpler pile-driving formulas, the Danish formula, is derived as follows (see Fig. 63):

Energy input (drop hammer) = Weight of ram × height of fall = $W_H H$
Energy used = Resistance of pile × permanent set = $Q_{dy} s$
Energy lost (in elastic compression of pile) = $\frac{1}{2} Q_{dy} s_e$

from which

$$Q_{dy} = W_H H / (S + \tfrac{1}{2} S_e)$$

where

$$S_e = W_H H / AE$$

In the Danish formula, the only losses considered are those due to elastic compression of the pile. In spite of its simplicity, studies have shown that it is one of the more reliable of the many pile-driving formulas available (Agerschou, 1962).

The most commonly used pile-driving formula, known as the *Engineering News* formula, is similar to the previous formula except that the elastic compression term is replaced by a constant. Thus,

$$Q_{dy} = \frac{2 W_H H}{S + c}, \qquad c = \begin{cases} 1 \text{ in. for drop hammers} \\ 0.1 \text{ in. for steam hammers} \end{cases}$$

where H is expressed in feet and S in inches. The *Engineering News* formula contains a built-in safety factor of 6, but studies (Agerschou, 1962; Flaate, 1964)

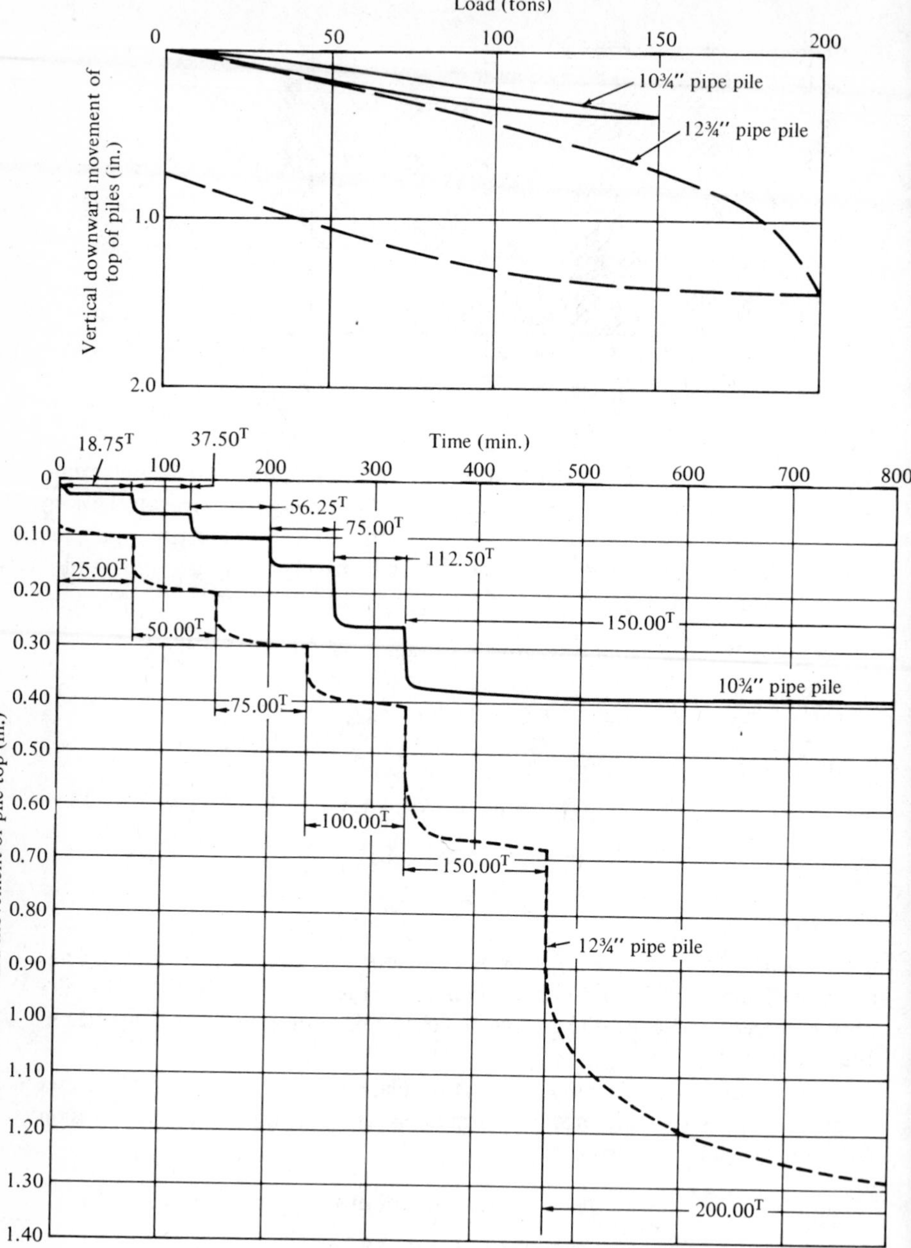

Fig. 64. Static pile load test (courtesy Dames & Moore).

have shown that the actual safety factor ranges from approximately 1.0 to more than 30.

The major difficulties associated with most dynamic pile-driving formulas are (1) the difficulty of accurately evaluating energy losses, (2) the difference in the dynamic behavior of the pile during driving and the static behavior after driving, and (3) the failure to account for wave-propogation phenomena in the pile–soil system. However, in spite of these difficulties, pile-driving formulas provide a useful tool for the soil engineer in controlling pile-driving operations, particularly when correlated with pile load tests.

Pile Load Tests Pile load tests are conducted by applying increments of static load to a pile and measuring the resulting deflection. The loads may be applied either by means of dead loads, or by driving reaction piles, connecting them with a beam, and applying the loads with a hydraulic jack. In the latter method, piles may be tested either in compression or in tension. A detailed procedure for conducting static pile load tests is described in ASTM Specification No. D1143-69. The results of a pile load test performed in accordance with these specifications is shown in Fig. 64.

The reasons for conducting pile load tests are (1) to develop driving criteria for subsequent production piles and (2) to provide evidence that the piles are capable of supporting the design load. However, because piles are always driven in clusters, the results of pile load tests cannot be extrapolated to pile groups without exercising extreme caution.

Behavior of Pile Groups The behavior of piles acting as a group is frequently quite different from that of the piles acting individually. Consequently, pile foundations should be analyzed as a unit.

Pile-Group Capacity Under certain conditions, pile groups may fail as units before the load per pile becomes equal to the design load. Therefore, the bearing capacity of the group must be determined in addition to that of the individual piles.

In order for a pile group to fail as a unit, the shearing resistance of the soil around the perimeter of the group and the end bearing resistance at the tips of the piles must be overcome, as shown in Fig. 65.

The ultimate bearing capacity of a pile group may be expressed as

$$Q_g = q_{\mathrm{ult}} BL + (2B + 2L) \sum_{0}^{D_f} (s \times \Delta z)$$

where

B = width of pile group
L = length of pile group

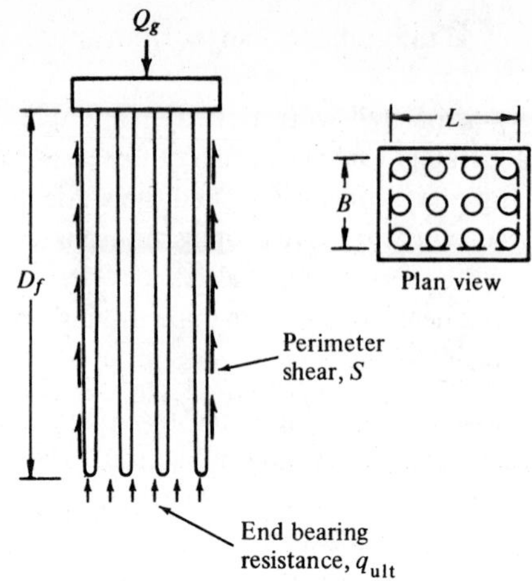

Fig. 65. Ultimete bearing capacity of pile groups.

q_{ult} = ultimate bearing capacity of a rectangular loaded area of width B, length L, and depth D_f (see equation for q_{ult} for a shallow foundation resting on a homogeneous soil, given earlier)

s = unit shearing resistance of soil surrounding the pile group

Δz = thickness of each soil layer through which the piles are driven

Theoretically, the most efficient use of piles would occur when the group capacity Q_g and the sum of the capacities of the individual piles are equal. However, as a practical matter, in most soil conditions Q_g is much greater because of the large end-bearing resistance developed by a pile group. Exceptions to this rule may occur, for example, with friction piles in soft clay or end-bearing piles embedded in a thin bearing station underlain by soft soils.

Negative Skin Friction Under normal circumstances, the direction of movement of the pile or caisson is downward relative to the surrounding soil. This produces upward forces along the shaft which tend to add support to the foundation. However, under certain conditions, the soil may move downward relative to the shaft, producing, in effect, an additional load on the foundation. This phenomenon is known as *negative skin friction* and the additional load must be accounted for in the design of the foundation.

The two most common situations in which negative skin friction may be expected to develop are (1) consolidation due to placement of new fill over compressible soils, and (2) consolidation due to disturbance of soft soils, as illustrated in Fig. 66. As indicated in Fig. 66, when negative skin friction is produced by consolidation under the weight of a fill, the maximum downdrag force is equal to the total adhesion force that can be mobilized on the shaft acting in the downward direction, whereas, when the negative skin friction

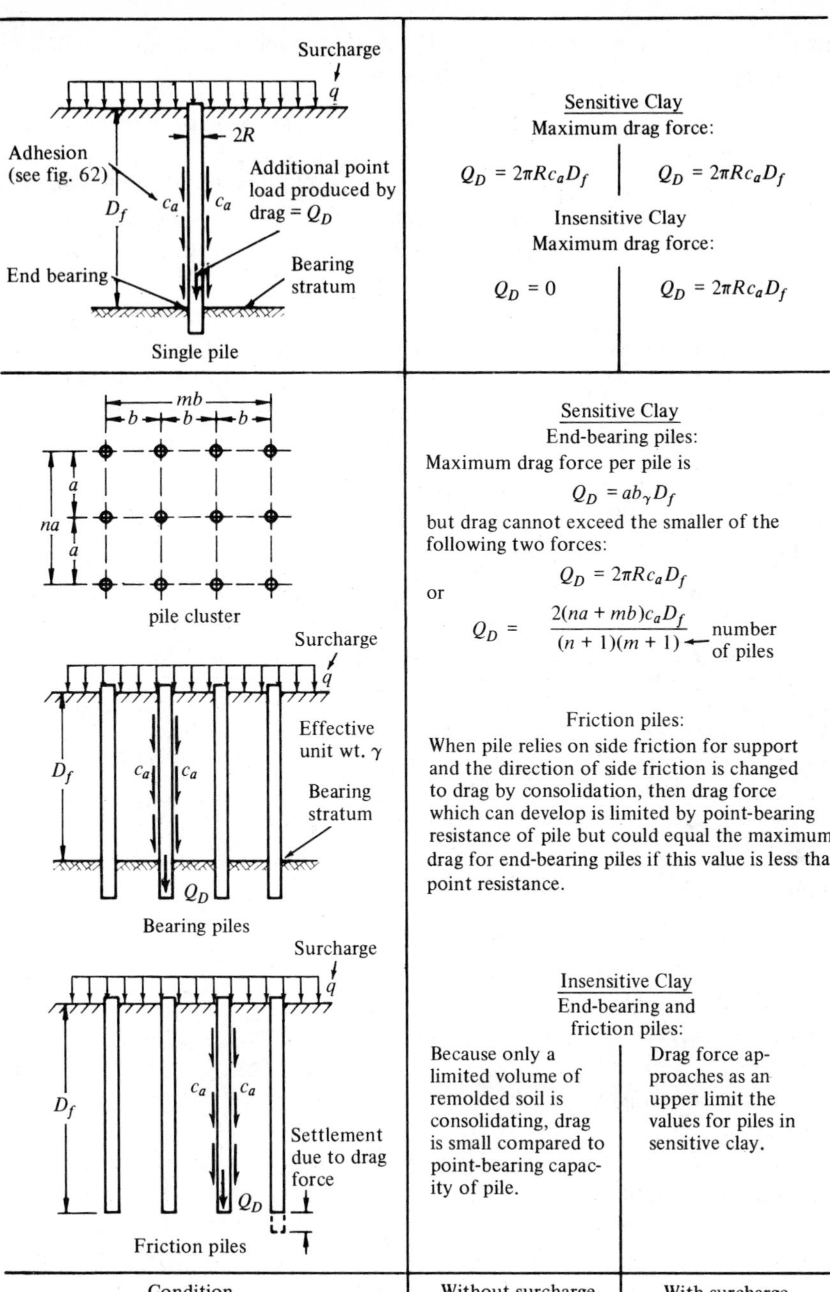

Fig. 66. Analysis of negative skin friction on piles in clay (after Navdocks DM-7, 1962).

results from disturbance of the soil, the maximum downdrag force cannot exceed the weight of the soil volume enclosed with the pile group.

Settlement of Pile Groups The load-settlement behavior of a single isolated pile usually bears little resemblance to that of the same pile in a pile group. The main reason for the discrepancy is that the "zone of influence," i.e., the volume of soil affected, is much larger for a pile group than for a single isolated pile. Therefore, in analyzing settlements of pile foundations, the computations must consider the pile group as a unit.

The settlement of pile groups is determined by the methods previously described for shallow foundations. However, in the case of pile groups, the problem is made more difficult by the necessity of estimating the load-transfer characteristics of the pile-soil system. For example, in the case of friction piles, the loads are gradually transferred from the piles to the surrounding soil along

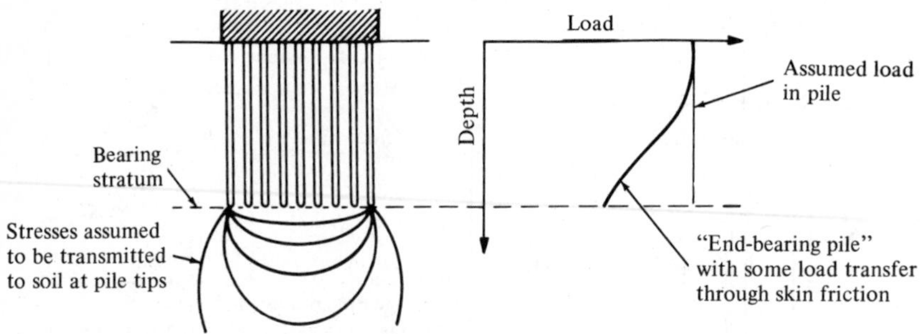

(a) End-bearing pile groups

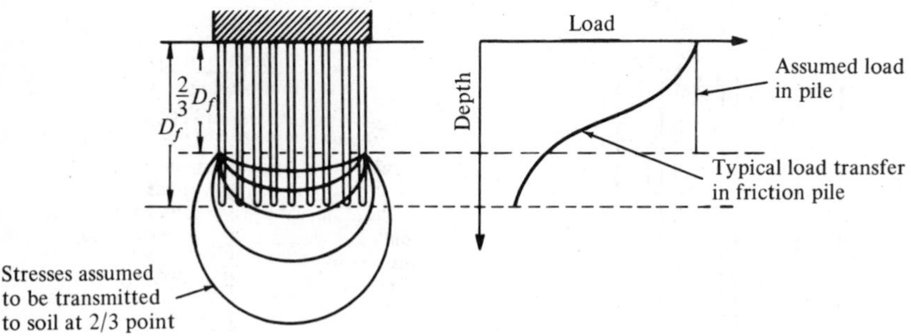

(b) Friction pile groups

Fig. 67. Load transfer concepts for settlement analysis of pile groups. (a) End-bearing pile groups. (b) Friction pile groups.

the length of the piles. Thus, for purposes of estimating settlement of friction-pile groups, it is generally assumed that the load is transferred to the soil at a depth of approximately two-thirds of the embedded length of the piles. On the other hand, for groups of end-bearing piles, the entire load is assumed to be transferred to the bearing stratum at the pile tips. These general concepts are illustrated in Fig. 67.

In the case of pile groups in sand, approximate empirical relationships have been developed relating the settlement of a single isolated test pile to the expected settlement of pile groups of various widths with the same load per pile (Skempton, 1953). These relationships are shown in Fig. 68.

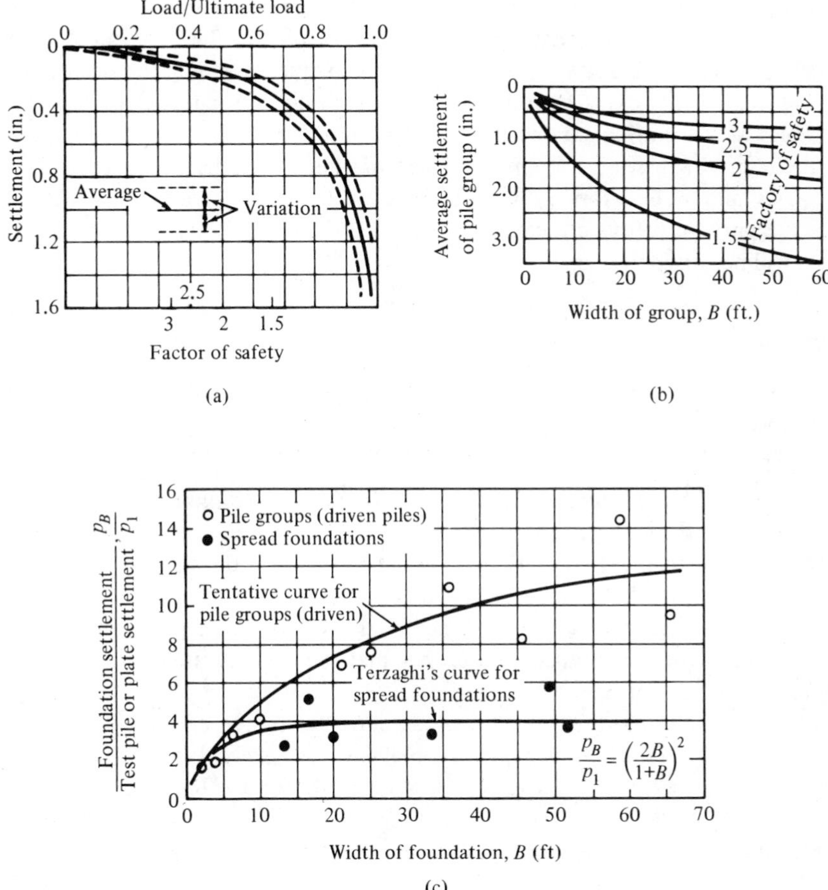

Fig. 68. Settlement of pile groups in sand. (a) Load settlement curves for piles driven in sand. (b) Approximate relationship between the settlement of pile groups and the factor of safety of a single pile. (c) Design curve for settlement of foundations in sand.

EARTH-RETAINING STRUCTURES

The safe and economical design of earth-retaining structures requires a knowledge of the lateral earth pressures that will be exerted against the structure. The determination of lateral earth pressures is, from a theoretical point of view, one of the most difficult problems in soil mechanics, yet it is one which soil engineers must deal with on a day-to-day basis. The problem of computing lateral earth pressures is complicated by the interdependence of the earth pressures and the deformation of the structure and by sometimes unpredictable construction details.

Ideally, it would be desirable to be able to predict the complete construction sequence in the analysis of the load–deformation response of a structure subjected to lateral earth pressures. However, because of the large number of variables involved, this is a difficult, if not impossible, task in a practical sense. Therefore, it is necessary to consider certain limiting cases with respect to such variables as wall movement and the stress–strain properties of the soil. Solutions of these limiting cases can then be used to bracket the probable range of earth pressures and wall deformations that may be expected in a given situation. For the purpose of this discussion, flexible structures and relatively rigid structures will be considered separately.

Rigid Structures

A *rigid structure* is defined as one which experiences very small flexural deformations as a result of lateral earth pressures. Translation or rotation of the wall may occur, but flexure of the wall is primarily in the elastic range.

Limiting-Equilibrium Methods of Analysis The limiting-equilibrium methods of analysis of earth pressures are based on the ultimate shear strength of the soil, without regard to the deformations required to mobilize the strength. In many practical problems, knowledge of the deformation is not specifically required. However, recognition of the importance of the deformations in modifying the mobilized shearing resistance of the soil may be incorporated in design by appropriate choice of safety factors.

All of the limiting-equilibrium methods currently in use employ the Mohr–Coulomb failure criterion as described previously. Ideally, laboratory tests to determine the shear-strength parameters c and ϕ for earth-pressure analysis should incorporate loading conditions which duplicate, as closely as possible, the loading conditions in the field. For example, the typical active case would involve plane strain loading conditions and failure by *decreasing* the lateral stress; the typical passive condition would involve plane strain loading with failure by *increasing* the lateral stress. In practice, however, this type of testing procedure is rarely available. More frequently, the available shear strength data are likely to consist of conventional triaxial or unconfined compression test results. The designer should at least be aware of the possible differences in shear strength that may be obtained by different loading conditions, as discussed in the section on soil properties.

The limiting-equilibrium methods incorrectly assume that the ultimate shear strength of the soil is mobilized simultaneously along the entire failure surface. In the case of strain-softening materials, this assumption may lead to an overestimation of the safety factor because, along portions of the failure surface where large strains may have developed, the strength will decrease after reaching the peak value. Thus, the average shear strength along the failure surface at failure may be significantly less than the peak strength. This problem can be minimized by choosing a safety factor large enough to ensure that deformations remain small.

The two most common methods of earth-pressure analysis applied to rigid walls are (1) the Rankine method and (2) the wedge (Coulomb) method. Other methods have been developed (e.g., Hansen, 1953; Sokolovski, 1960) but they are not used extensively in the U.S., and will not be considered further here. The reader may consult Morgenstern and Eisenstein (1970) for an interesting comparison of the results obtained by the various methods.

The Rankine Method The Rankine method treats the case of a smooth, vertical wall with a plane backfill surface. The effect of a uniform surcharge loading can also be included in the analysis. The Rankine method is illustrated in Fig. 69 for the case of a level backfill with uniform surcharge loading. The method may be described briefly as follows. The vertical effective stress at a point behind the wall is given by

$$\sigma_v = q + \gamma z$$

where

q = intensity of surcharge loading
γ = effective unit weight of soil (total weight above the water table, buoyant unit weight below the water table)
z = depth below backfill surface

Using the Mohr-Coulomb failure criterion in the form

$$\sigma_1 = \sigma_3 \tan^2 [45° + (\bar{\phi}/2)] + 2\bar{c} \tan [45° + (\bar{\phi}/2)]$$

and noting that

active case:

$$\bar{\sigma}_1 = \bar{\sigma}_v$$

$$\bar{\sigma}_3 = \bar{\sigma}_h = \bar{\sigma}_a$$

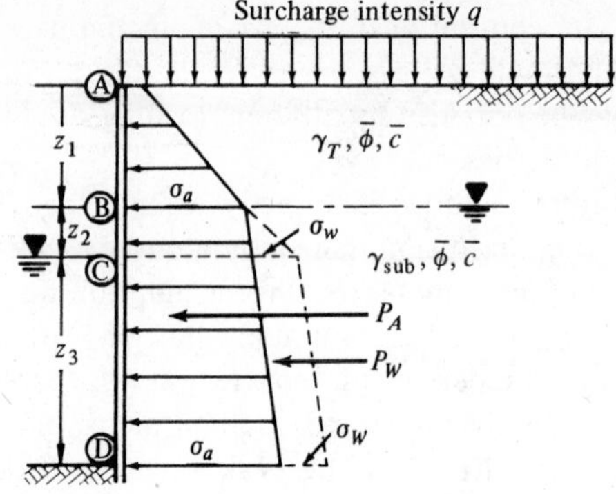

(a) Active earth pressure:

at A: $\sigma_a = q \tan^2 (45° - \bar{\phi}/2) - 2\bar{c} \tan (45° - \bar{\phi}/2)$

at B: $\sigma_a = (q + \gamma_T z_1) \tan^2 (45° - \bar{\phi}/2) - 2\bar{c} \tan (45° - \bar{\phi}/2)$

at C: $\sigma_a = (q + \gamma_T z_1 + \gamma_{sub} z_2) \tan^2 (45° - \phi/2) - 2\bar{c} \tan (45° - \bar{\phi}/2)$

$\sigma_w = \gamma_w z_2$

at D: $\sigma_a = [q + \gamma_T z_1 + \gamma_{sub} (z_2 + z_3)] \tan^2 (45° - \bar{\phi}/2) - 2\bar{c} \tan (45° - \bar{\phi}/2)$

$\sigma_w = \gamma_w z_2$

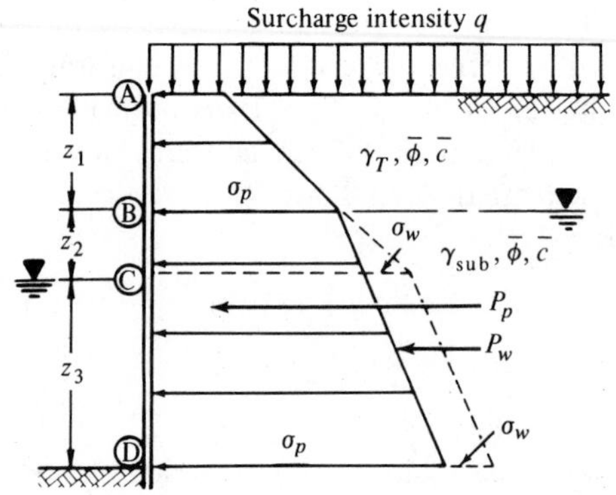

(b) Passive earth pressure

at A: $\sigma_p = q \tan^2 (45° + \bar{\phi}/2) + 2\bar{c} \tan (45° + \bar{\phi}/2)$

at B: $\sigma_p = (q + \gamma_T z_1) \tan^2 (45° + \bar{\phi}/2) + 2\bar{c} \tan (45° + \bar{\phi}/2)$

at C: $\sigma_p = (q + \gamma_T z_1 + \gamma_{sub} z_2) \tan^2 (45° + \bar{\phi}/2) + 2\bar{c} \tan (45° + \bar{\phi}/2)$

$\sigma_w = \gamma_w z_2$

at D: $\sigma_p = [q + \gamma_T z_1 + \gamma_{sub} (z_2 + z_3)] \tan^2 (45° + \bar{\phi}/2) + 2\bar{c} \tan (45° + \bar{\phi}/2)$

$\sigma_w = \gamma_w z_2$

Fig. 69. Rankine method of earth pressure analysis.

passive case:

$$\bar{\sigma}_1 = \bar{\sigma}_h = \bar{\sigma}_p$$

$$\bar{\sigma}_3 = \bar{\sigma}_v$$

the equations for active and passive pressures become

$$\bar{\sigma}_a = (q + \gamma z) \tan^2 [45° - (\bar{\phi}/2)] - 2\bar{c} \tan [45° - (\bar{\phi}/2)]$$

$$\bar{\sigma}_p = (q + \gamma z) \tan^2 [45° + (\bar{\phi}/2)] + 2\bar{c} \tan [45° + (\bar{\phi}/2)]$$

Where the backfill contains soil layers with different properties, the appropriate values of γ, $\bar{c}$, and $\bar{\phi}$ must be used for each layer. Furthermore, if the water table is located above the base of the wall, hydrostatic pressures must be added to the effective earth pressures to obtain the total pressure on the wall.

In the case of a horizontal backfill surface, the failure planes are inclined at an angle of

$$\alpha = 45° + (\bar{\phi}/2)$$

with respect to the direction of the major principal plane. Thus, in the active case, α is measured from the horizontal and, in the passive case, α is measured from the vertical.

The Rankine formulas can also be derived for the case of an inclined backfill, but the algebraic expressions become considerably more complex.

Wedge Method Earth-pressure problems are frequently encountered in which the assumptions of the Rankine method are not valid. In these cases the wedge method provides a more flexible method of analysis. Some examples of factors which violate the assumptions of the Rankine method but which can be treated by the wedge method include (1) friction between the wall and the backfill, (2) nonvertical walls, (3) nonplanar backfill surfaces, (4) surcharge loadings of nonuniform intensity or limited extent, (5) layered soil conditions, and (6) sloping water level.

In the wedge method, a trial failure wedge is selected and then the requirements of static equilibrium applied to compute the magnitude and direction of the resultant thrust on the wall. The procedure is repeated with several trial wedges until the critical wedge is found. Planar failure surfaces are generally used, but the method can also be adapted to nonplanar surfaces such as circles or logarithmic spirals.

Calculation of active and passive earth pressures by the wedge method are illustrated in Figs. 70 and 71, respectively. The example shown for the active case in Fig. 70 includes wall friction, a nonplanar backfill, nonuniform surcharge loading, layered soil conditions, and a sloping water level.

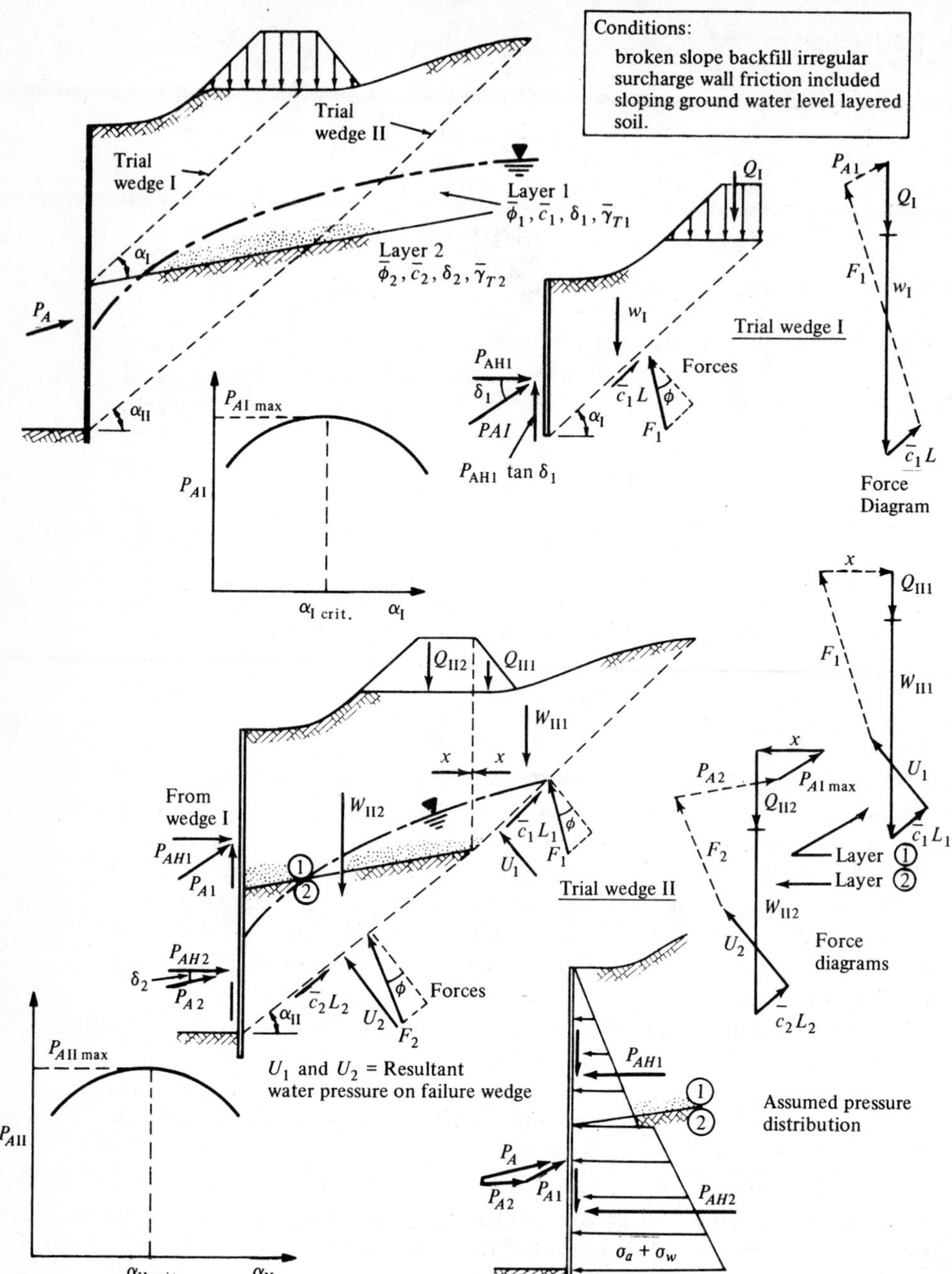

Fig. 70. Calculation of active pressures by the wedge method.

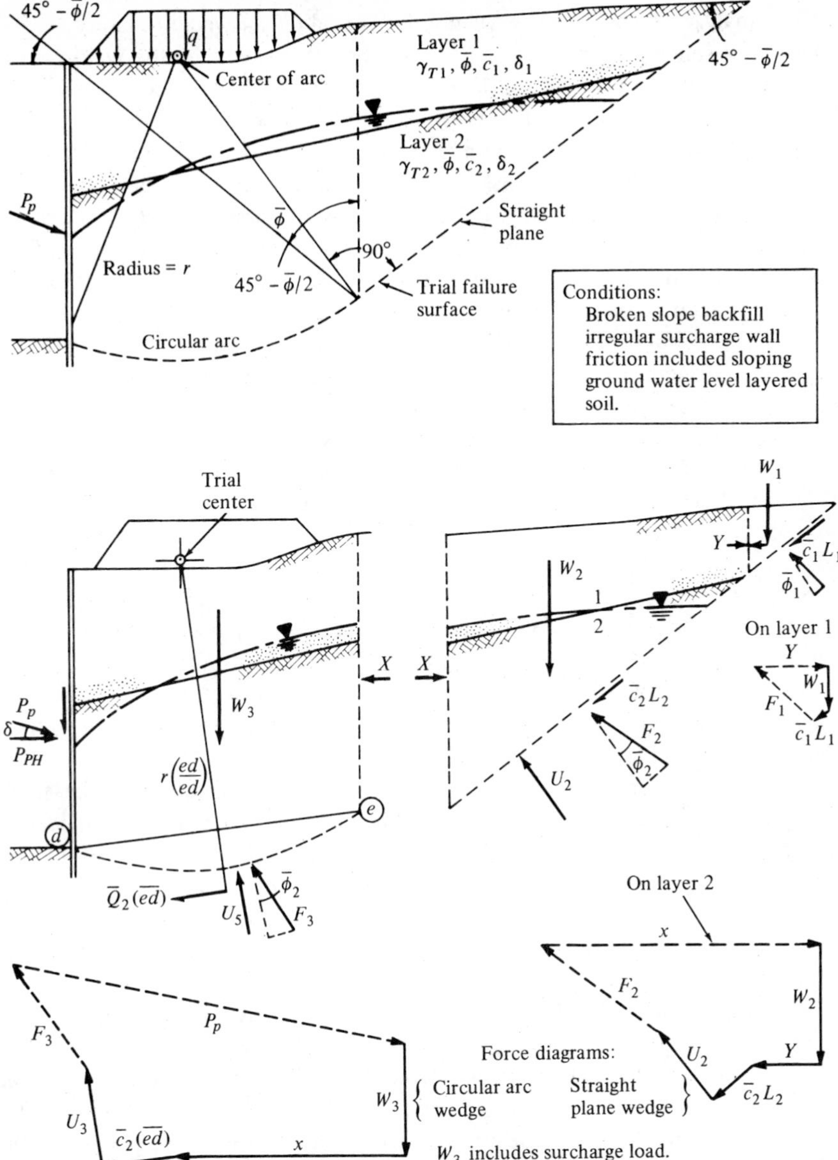

Fig. 71. Calculation of passive pressures by the wedge method. Note: Minimum value of P_p is obtained by trial and error using several different trial center locations.

The analysis begins with the selection of a trial failure wedge (trial wedge I) in the upper soil layer which terminates at the intersection of the wall and the boundary between the two soil layers. The analysis of the forces acting on this trial wedge as shown in the free-body diagram of the wedge and the corresponding force diagram establishes the magnitude and direction of the resultant force P_{A1} imposed by the upper soil layer on the wall for the trial wedge shown. A second trial wedge (trial wedge II) is selected which passes through the base of the wall. Two force diagrams are necessary for this wedge because of the different soil parameters in layer 1 and layer 2. The force diagrams include resultant water forces acting on the assumed failure surface, the magnitudes of which can be obtained from a flow net for the appropriate seepage conditions. The force diagram for layer 1 is used to obtain the magnitude of the horizontal resultant force X acting on the imaginary vertical boundary dividing trial wedge II into two segments. This resultant then becomes a part of the force diagram for layer 2. The force P_{A1} obtained from the previous analysis of trial wedge I is also included in the force diagram for layer 2, from which the resultant P_{A2} is obtained. The total resultant on the wall, P_A, is the vector sum of the resultants P_{A1} and P_{A2}.

Since the critical failure surface is not known in advance, a trial-and-error procedure is required to determine the surface which gives the maximum resultant thrust on the wall. In a case like the one shown in the example, where more than one trial wedge is involved, the trial-and-error process is carried out one wedge at a time, starting with the upper wedge and progressing downward until all the resultants have been determined.

The method just described gives the magnitude and direction of the earth-pressure resultant. However, it does not determine the point of application of the resultant. For the example shown in Fig. 70, the approximate point of application of the resultant may be determined on the assumption that the pressure distribution on the wall is linear, as shown. Other approximate methods for determining the location of the resultant are discussed by Terzaghi and Peck (1967).

Figure 71 illustrates the calculation of passive pressure by the wedge method. In the case of passive pressure, it is well known that the use of planar failure surfaces introduces considerable error (on the unsafe side) if there is appreciable wall friction (see, e.g., Morgenstern and Eisenstein, 1970). Most of this error can be eliminated by choosing a curvilinear failure surface for the portion of the failure mass influenced by wall friction, i.e., the zone bounded by a line through the top of the wall sloping downward at an angle of $45° - (\bar{\phi}/2)$. In the example shown, the curvilinear portion is approximated by a circular arc; however, a logarithmic spiral could be used with equal validity.

After the trial failure surface has been selected, the failure mass is subdivided into the necessary segments to account for the different soil properties in the backfill, the respective force diagrams are constructed, and the resultant passive thrust P_p obtained. Several trial failure surfaces are analyzed until the critical surface (the surface which gives the minimum passive thrust) is found. The use of separate trial surfaces in the different soil layers to aid in determining the direction and location of the resultant thrust is not convenient in the passive case because of the curvilinear failure surface. As an approximation, the passive resultant is assumed to act at the lower-third point of the wall in the case of cohesionless soils or at the midpoint of the wall for cohesive ($\phi = 0$) soils. For intermediate cases the position of the resultant may be adjusted accordingly. Similarly, a weighted average may be taken for the angle of wall friction to determine the direction of the resultant.

Effects of Wall Restraint and Construction Procedures In foundation engineering practice, the engineer encounters an almost endless variety of methods of installing and supporting rigid walls. However, for the purpose of this discussion, two limiting cases (restrained walls and translating or rotating walls) will serve to illustrate the influence that varying degrees of wall restraint and different construction procedures have on the lateral earth pressures.

Restrained Walls A restrained wall is one that is, for all practical purposes, prevented from translating or rotating relative to the soil behind it. A very common example of this type of wall is a backfilled foundation wall.

If the condition of complete restraint is satisfied, the lateral earth pressures are governed by the at-rest conditions for the soil, which are expressed in terms of the ratio of the horizontal effective stress to the vertical effective overburden pressure, i.e.,

$$K_0 = \bar{\sigma}_h / \bar{\sigma}_v$$

In the case of normally consolidated clays, or cohesionless soils, the at-rest coefficient is approximated by the expression

$$K_0 = 1 - \sin \bar{\phi}$$

On the basis of the first equation, the ratio of horizontal to vertical effective stresses behind a nonyielding wall could be expected to range from a minimum of approximately 0.35 for a dense sand and gravel with $\bar{\phi} = 40°$ to a maximum of approximately 0.7 for a highly plastic clay with $\bar{\phi} = 17°$ (Gould, 1970).

Several factors, including overconsolidation, heavy compaction, and swelling pressures in expansive clays may tend to produce at-rest lateral pressures greater than would be predicted by this equation. The effect of overconsolidation on the K_0 value is shown in Fig. 72. For the case shown, K_0 varies from approximately 0.7 for the normally consolidated condition to approximately 2.0 for the heavily overconsolidated condition. Similar effects due to preloading, although of reduced magnitude, have been reported for granular soils (Obrcian, 1969).

Heavy compaction, carried out in a confined space adjacent to a nonyielding

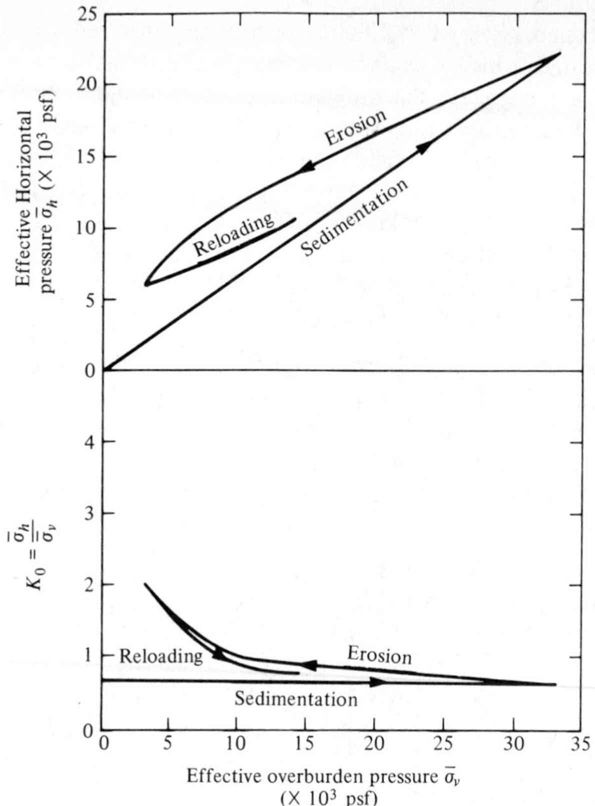

Fig. 72. Typical relation between vertical effective overburden pressure and horizontal effective stress (after Morgenstern and Eisenstein, 1970).

wall, can produce effects similar to overconsolidation. A specific case where lateral pressures in excess of normal at-rest values were observed has been reported by Sowers et. al. (1957). On the other hand, if compaction is carried out within a very narrow zone between stiff natural soil and a nonyielding wall, friction forces along the boundaries of the compacted zone may reduce the lateral pressures to less than the at-rest value (Grant, 1961).

Clays and clay-shales with swelling tendencies can produce pressures on rigid walls considerably in excess of nominal at-rest values. Furthermore, in contrast to normal earth pressure due to gravity loading, lateral pressures due to swelling tend to persist in spite of wall movement. In the case of swelling clays, an increase in moisture content leads to the development of swelling pressures, frequently of considerable magnitude. Similar effects may occur with heavily overconsolidated clay or compacted clay, apart from the high lateral pressures generated by the compaction process itself.

Excavations in heavily overconsolidated clays and clay shales may involve the release of considerable locked-in strain energy (Bjerrum, 1967). If a wall is placed against a cut in these materials, high lateral stresses may result unless a substantial "buffer zone" of granular backfill is placed between the wall and the cut face.

Translating or Rotating Walls When rigid walls are not completely restrained, some type of movement results. The magnitude and direction of the wall movement, in turn, affects the lateral earth pressures. Terzaghi (1934) demonstrated, experimentally, the effects of wall movement on lateral earth pressures in sand. The results of Terzaghi's experiments are shown in Fig. 73. The experiments were conducted for wall rotations in the active and passive directions; however, the results are generally applicable for translation as well. With the at-rest condition as a starting point, the earth pressure approaches the active value with comparatively little wall movement, whereas much greater wall movement is required to fully mobilize the passive resistance because the stress change is much greater.

Since the active and passive states are limiting conditions governed by the ultimate shear strength of the soil, they represent the extreme values of lateral earth pressure that can be sustained by a given soil regardless of the amount of

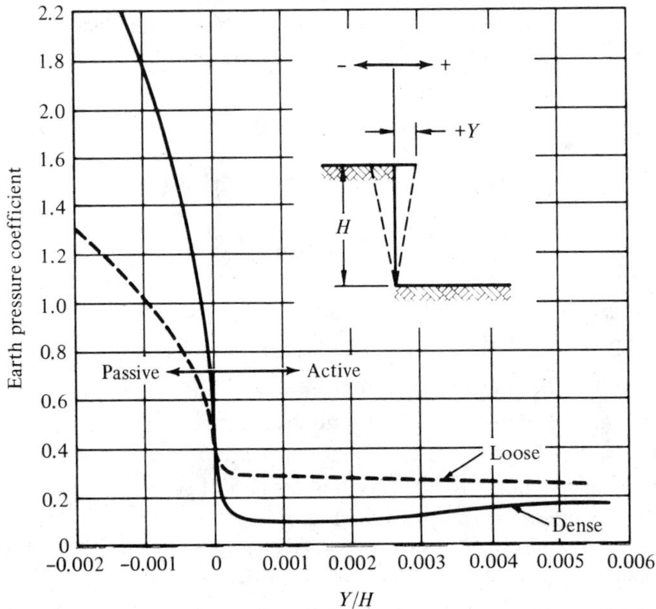

Fig. 73. Typical relationship between movement of wall and earth pressure for different densities (after Terzaghi, 1954).

wall movement. However, wall pressures lower than the active value may occur in certain instances, e.g., when part of the load can be shifted downward to an unyielding base. Active earth pressures are commonly used for the design of cantilever retaining walls and other types of walls where there is reasonable certainty that sufficient wall movements will occur. On the other hand, most designers use a reduced "working value" for passive pressure because of the large movements required to mobilize full passive resistance.

In certain types of construction, the wall movement consists essentially of rotation about the top of the wall. Although there is no completely satisfactory theoretical solution for the distribution of lateral earth pressure created by this type of wall movement, the overall effect is to raise the location of the earth pressure resultant from the lower third point on the wall to somewhere near mid-height. The shape of the earth-pressure distribution in this case is generally thought to be approximately parabolic if the movements are fairly small. If the rotation becomes large and is accompanied by some outward translation at the top of the wall, the pressures return to the active state.

In cases where wall movement is partially restricted by construction procedures, the earth-pressure distribution is extremely difficult to predict accurately. For these cases, the idealized modes of wall movement described above can provide useful guidelines in the form of limiting earth-pressure values that can be expected for varying degrees of wall movement. An excellent discussion of lateral earth pressures on various types of rigid walls has recently been presented by Gould (1970).

Flexible Walls

There are two important types of earth-supporting structures which may serve as illustrative examples for lateral earth pressures on flexible walls, i.e., anchored bulkheads and braced sheet-pile walls. In the case of anchored bulkheads, the wall movement is generally sufficient to develop active pressures behind the wall; however, in the case of braced sheet-pile walls, the movement may be sufficiently restricted to prevent full development of the active condition.

Anchored Bulkheads Anchored bulkheads are constructed of flexible sheet piles, restrained by tieback anchors and by passive soil resistance below the dredge line. Depending on the depth of penetration of the sheet piles, the soil in the embedded portion provides varying degrees of support. The two extreme types of earth support, as shown in Fig. 74 are designated as (a) free earth support or (b) fixed earth support. The classical design assumptions concerning earth-pressure distributions for the two extremes are shown in Fig. 75. These earth-pressure distributions are based on the tacit assumption that any amount of wall movement, no matter how small, is sufficient to develop the active or

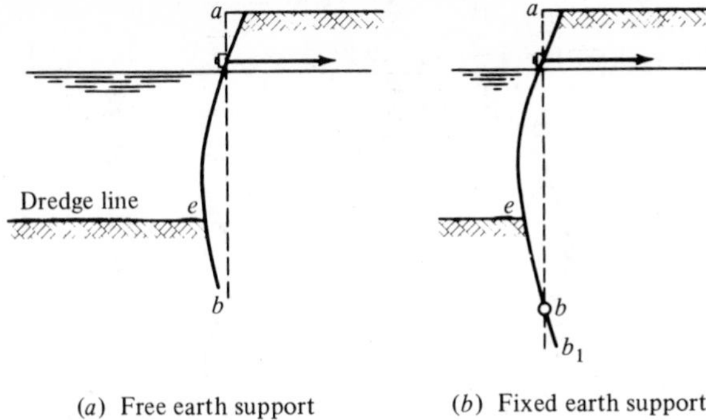

Fig. 74. Two extreme types of earth support for anchored bulkheads (after Terzaghi, 1954).

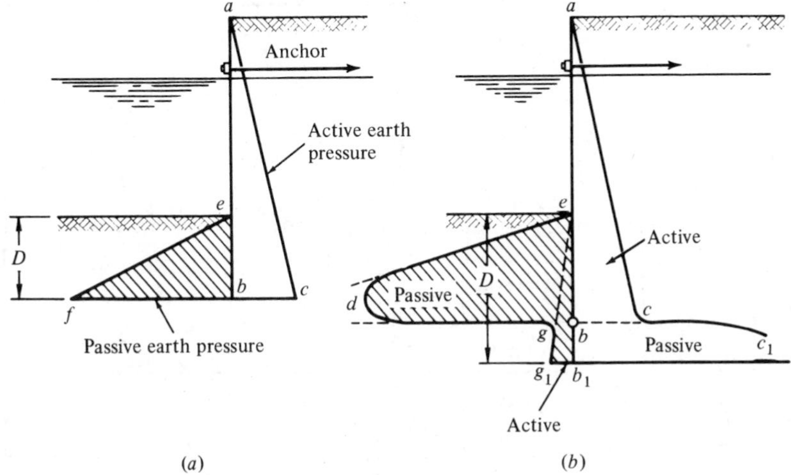

Fig. 75. Classical design assumptions concerning earth pressure distributions on anchored bulkheads with (a) free earth and (b) fixed earth support (after Terzaghi, 1954).

passive condition. Depending on the flexibility of the sheet piles, the amount of yielding of the anchorage, and the degree of fixity provided by the soil at the lower end of the sheet piles, the actual pressure distributions may vary considerably from those shown in Fig. 75. However, for design purposes, it is safe to assume that the earth-pressure distribution agrees with that assumed for free-earth-support conditions (Terzaghi, 1954). The following discussion describes the design of anchored bulkheads based on the free-earth-support concept.

The first step in the design involves determining all forces acting on the wall (Fig. 76). The forces that may be acting on the back of the wall include

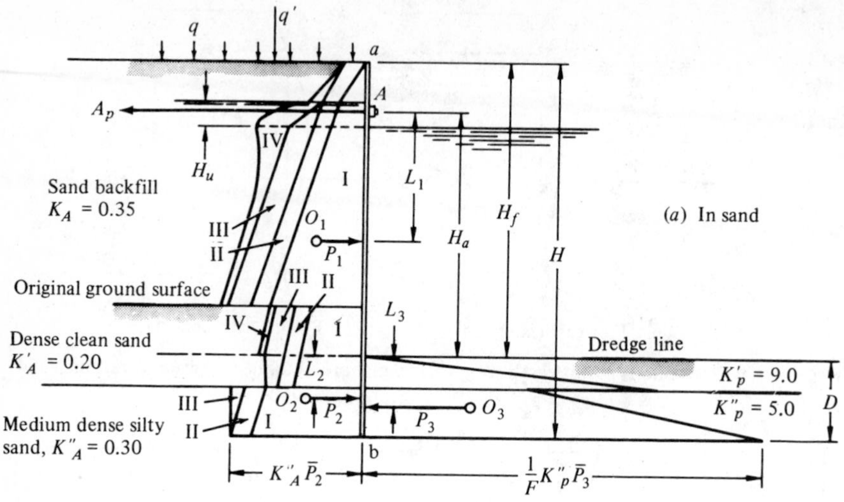

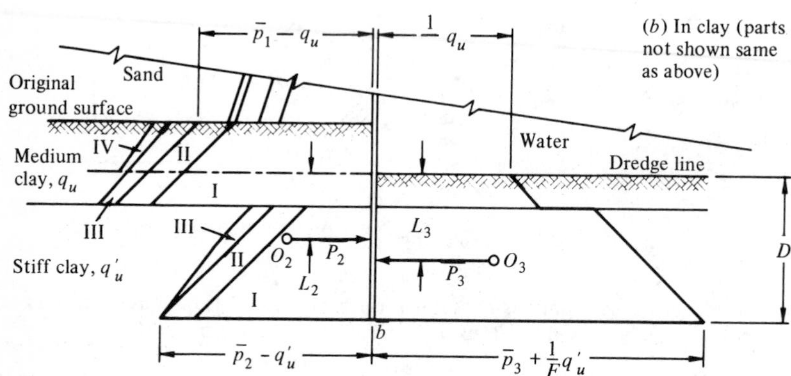

Key:

I. Active earth pressure produced by the weight of the backfill.
II. Active earth pressure produced by the uniformly distributed surcharge, q.
III. Unbalanced water pressure.
IV. Lateral pressure caused by line load q'.

Fig. 76. Forces acting on anchored bulkheads by free earth support method (after Terzaghi, 1954).

(1) active earth pressures, (2) pressures due to surcharge loads, and (3) unbalanced water pressures (the net difference between the water pressures acting on the back and the front of the wall). The restraining force on the front of the wall is the passive soil resistance developed below the dredge line. In order to

provide a margin of safety against a toe failure, the passive pressures should be reduced by a suitable safety factor (usually approximately 2.0).

The minimum required depth of penetration may be computed by examining moments about the anchorage point; i.e.,

$$\sum M_A = \sum M_p$$

where

$M_A = f(P_a, \Delta U_w, q, 0)$ = movements produced by earth pressure P_a plus unbalanced water pressure ΔU_w and surcharge loads q behind the wall,

$M_p = f[(P_p/F, D]$ = moments produced by passive resistance P_p/F in front of the wall, below the dredge line.

This equation results in a cubic equation in D, the solution of which results in the required minimum depth of penetration. The computed penetration D is normally increased by about 20% to account for the possibility of over-dredging, scour, etc.

After the minimum depth of penetration has been determined, the anchor pull can be computed by summing horizontal forces acting on the sheeting; i.e.,

$$A_p = \sum(P_A + \Delta U_w + R_Q) = \sum(P_p/F)$$

where

A = anchor pull
P_A = active earth-pressure resultants
ΔU_w = unbalanced water-pressure resultant
R_Q = resultant forces on wall due to surcharge loading not accounted for in the earth-pressure diagram

The maximum moment in the sheeting is computed at the point of zero shear, from which a suitable sheet-pile section can be selected. If the sheet piles are driven into moderately dense to dense homogeneous sand, a condition of partial or complete fixity is likely to develop, causing a reduction in the maximum bending moment in the sheeting. Rowe (1952) has shown that the moment reduction that occurs under such circumstances is a function of the amount of embedment and the flexibility of the sheeting. On the basis of model experiments, Rowe developed the moment reduction curves shown in Fig. 77.

The procedure for selecting the required sheet-pile section using moment reduction involves a trial-and-error process of selecting a trial section, determining the design moment from the moment-reduction curves, and then checking the maximum fiber stress in the trial section against the maximum allowable stress in the steel. If necessary, different sections are tried until the computed maximum stress in the steel agrees with the allowable stress.

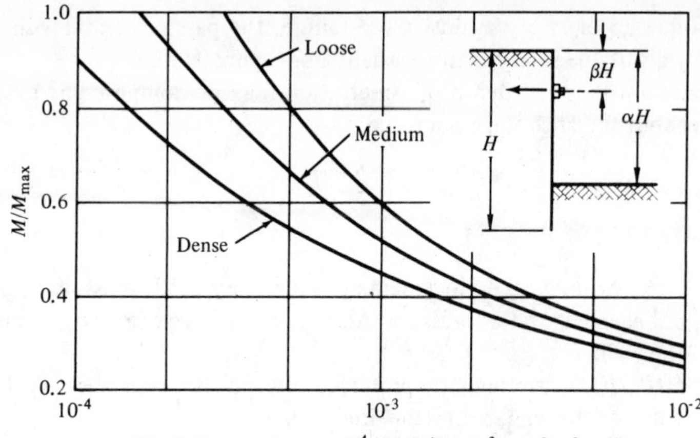

Fig. 77. Moment reduction curves for sheet piles in sand (after Rowe, 1952).

Moment reduction should not be applied to loose, silty sands because the compressibility of such soils precludes the development of the fixed-end condition. Similarly, in the case of clay soils, the initial fixity that may occur immediately upon driving the piles may disappear with time because of long-term consolidation.

Braced Sheet-Pile Walls Braced sheet-pile walls are temporary structures used to support the sides of excavations during construction. In conjunction with the construction of temporary bracing systems, it is usually desirable to restrict lateral movements of the wall as much as possible because large lateral movements may lead to encroachment on the space required for permanent construction and detrimental settlements of adjacent structures or roadways.

A common method of construction involves driving vertical sheeting and placing horizontal bracing or tiebacks as the excavation progresses. If the first row of bracing is placed before the excavation proceeds too far, lateral movements near the top of the excavation can usually be effectively minimized. However, as the excavation is deepened, lateral movements of the sheeting may occur below the depth of the excavation before bracing can be installed. As a result, the typical profile of wall movement resembles a rotation about the top of the wall.

The resulting earth-pressure distribution may bear little resemblance to the typical triangular active-pressure diagram. The pressures near the top of the wall are likely to exceed the active pressure because of restricted wall movement. On the other hand, near the base of the excavation, the wall pressures may be less than active because of stress transfer to underlying strata.

Because of the uncertainties involved in the construction process, it is usually impossible to predict the lateral earth pressures in advance of construction.

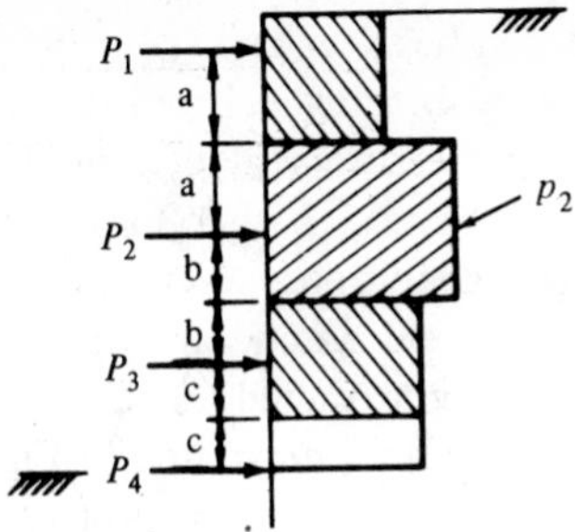

(a) Development of actual pressure diagram:

P_1, P_2, P_3 are measured loads in the three struts of a vertical profile divided by the strut horizontal spacing; e.g., in kips per foot.

P_4 is the assumed load at the bottom of the excavation: $P_4 = P_3c/(b + c)$.

The actual pressure diagram is obtained by distributing P_1 to P_4 as shown; e.g., $p_2 = P_2/(a + b)$.

The total load $P = \sum_1^4 P_i$ is often between P_A and 1.2 P_A (P_A is the Rankine active force per unit length of excavation).

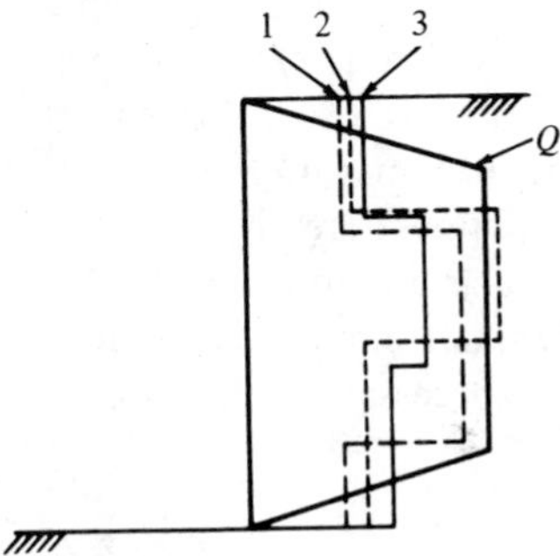

(b) Development of apparent pressure diagram:

1, 2, and 3 are actual pressure diagrams for different vertical profiles.

The apparent total load Q is equal to the area of the apparent pressure diagram which envelopes the actual pressure diagrams.

Q is often 1.5P.

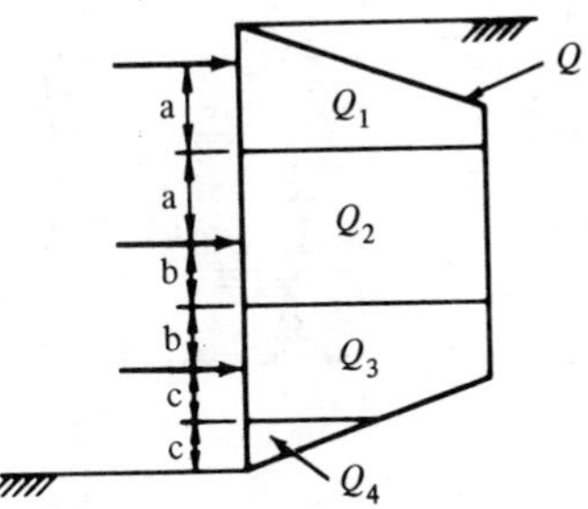

(c) Maximum probable strut load

The maximum probable strut loads Q_1, Q_2, and Q_3 are calculated from the apparent pressure diagram the same way (but in reverse) the actual pressure diagram was obtained from the loads P_1, P_2, and P_3.

Fig. 78. Actual and apparent pressure diagrams for temporary bracing (Lacroix and Jackson, 1972).

Therefore, the procedure that has developed over the years for estimating lateral earth pressures on temporary bracing is based largely on field measurements. This procedure involves the measurement of strut loads in an excavation and back-calculating the earth pressures. If such measurements are made along several profiles in the same soil conditions, several pressure diagrams are obtained, from which an apparent-pressure diagram, representing an envelope of the actual pressure distributions, can be constructed. The development of apparent-pressure diagrams for design purposes is illustrated in Fig. 78. Once the apparent-pressure diagram is available, the maximum probable strut loads may be calculated. Since the apparent-pressure diagram envelopes the actual-pressure diagrams, its resultant is often 50% greater than the sum of the strut loads in any given vertical section. Such conservatism is justified in design, however, because even with uniform construction procedure, the actual loads in sets of struts at various locations in the same (or very similar) soil conditions may deviate as much as 30% to 60% from the average (Terzaghi and Peck, 1967).

A possible alternative to the conservatism implicit in the use of pressure envelopes for the design of temporary bracing systems would be to design for lower strut loads (perhaps based on the average-pressure diagram rather than an envelope of pressures) and monitor all strut loads so that corrective action could be taken if any of the struts becomes overloaded. A recent case where this procedure was followed has been described by Peck (1969).

Apparent-pressure diagrams are now available for certain broadly defined soil types Fig. 79. These diagrams should be considered only as guidelines; considerable modifications may be necessary for specific soil conditions.

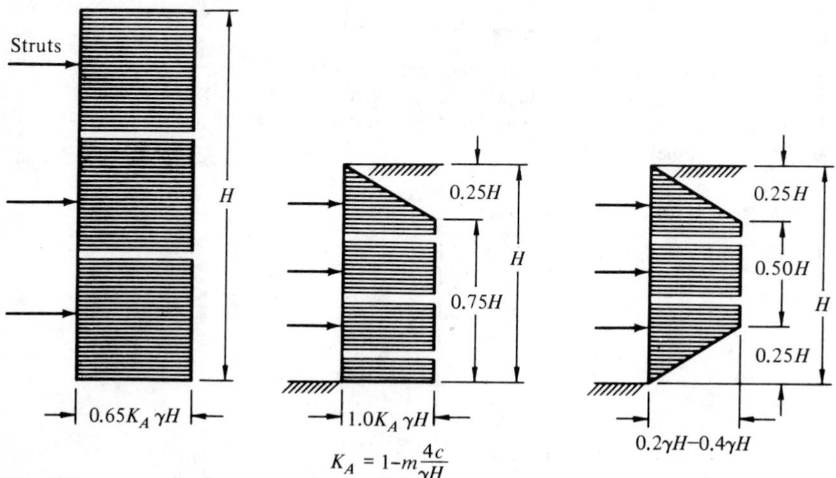

Fig. 79. Suggested apparent pressure diagrams for design of struts in braced cuts (after Terzaghi and Peck, 1967).

The apparent-pressure diagrams presented in Fig. 79 do not include the effects of surcharge loading or water pressure. If these effects are present, resulting pressures may be computed by the appropriate methods previously discussed and added to the apparent-earth-pressure diagram.

Some settlement of the ground surface adjacent to the excavation cannot be avoided. However, the following construction procedures can help to reduce movements to a minimum:

1. Prestress struts or tiebacks (50% of design load is commonly used).
2. Keep vertical strut spacing to a minimum.
3. Place struts as soon as possible after excavation.

Peck (1969) has presented field data from many sites on settlements adjacent to braced excavations Fig. 80. Depending on the depth of the excavation and the soil conditions, this plot provides a rough indication of the probable settlements as a function of the distance from the edge of the excavation.

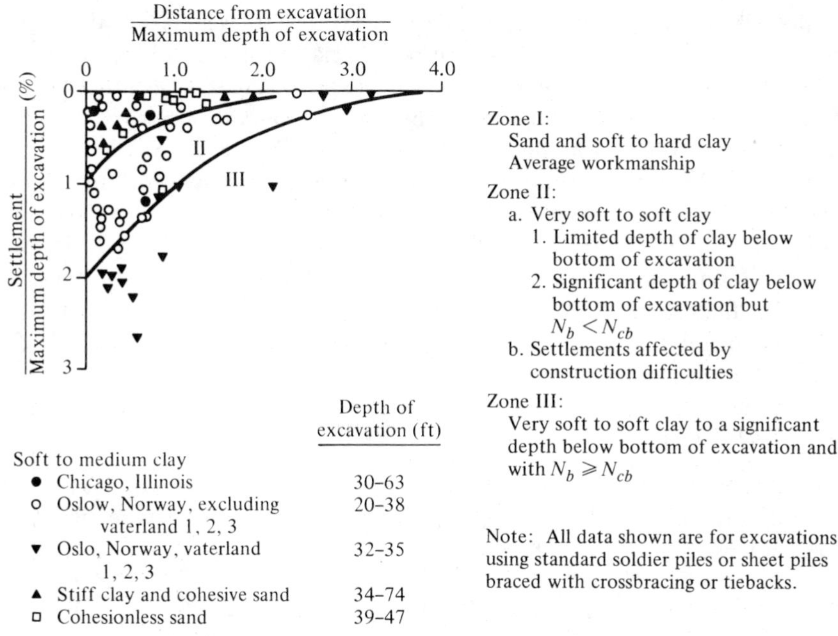

Fig. 80. Settlements outside of braced excavations (after Peck, 1969).

STABILITY OF SLOPES AND EMBANKMENTS

The stability of a soil mass is governed by the relative magnitudes of the shear stresses and the shear strength along a potential surface of sliding. Thus, instability in a soil mass may develop as a result of either (1) an increase in shear stress or (2) a decrease in shear strength along the potential sliding surface. Some of the factors which may contribute to instability of a soil mass are steepening or undercutting of slopes, additional loads placed on the surface, seepage pressures, earthquakes, and weathering processes.

Stability Analysis

Stability analysis of slopes and embankments is generally based on the concept of limiting equilibrium, wherein a plausible mode of failure is assumed, the shear stresses along the potential sliding surface are computed from static equilibrium requirements, and the safety factor is determined by comparing the magnitude of the shear strength of the soil with the shear stresses necessary for static equilibrium. Different modes of failures are tried until the one with the minimum safety factor is found. This method offers the advantages of (relative) ease of computation and allowing the engineer to exercise his experience and judgment in the selection of possible failure modes. The disadvantages are that the actual stress–strain properties of the soil are not considered (the limiting-equilibrium method assumes that the soil behaves as a rigid plastic material) and that the absolute minimum safety factor may not be obtained, since the choice of trial failure surfaces is, of necessity, limited.

Alternatively, the stresses and strains in a slope or embankment can be analyzed with the use of electronic computers. This approach is highly desirable, especially where transient loads such as those generated by earthquakes are concerned. In such cases, the shear strength of the soil is only temporarily exceeded and the "failure" may involve fairly large displacements in the soil mass but stop short of complete and catastrophic failure. However, such analyses are extremely costly and require sophisticated testing methods to completely define the stress–strain–time characteristics of the soil. As a result, at the present time most stability analyses are based on limiting equilibrium.

Factor of Safety In stability analysis, the factor of safety may be defined as the ratio of the available shear strength of the soil to the shear stress required for equilibrium; i.e.,

$$F = \frac{S}{S_d} = \frac{c + \sigma_n \tan \phi}{(c/F) + [\sigma_n (\tan \phi)/F]} = \frac{c + \sigma_n \tan \phi}{c_d + \sigma_n \tan \phi_d}$$

where c and ϕ are total or effective shear-strength parameters, depending on the method of analysis being used.

Acceptable factors of safety depend on the degree of uncertainty with regard to the strength of the soil and the mode of failure, the potential danger involved if a failure occurs, the likelihood that certain loading conditions will develop, and the expected duration of loading. For example, if failure would involve small risk to lives and property, or if the duration of the loading will be short (as in the case of earthquakes), relatively low factors of safety are usually acceptable.

Analysis of Forces Several methods have been devised for analyzing the forces acting on a potential sliding mass. Among these are (1) the ϕ-circle method, (2) the sliding-block method, (3) the logarithmic-spiral method, and (4) the method of slices. Because it offers the greatest degree of flexibility, only the method of slices will be described herein.

The general procedure of the method of slices may be described as follows:

1. A potential failure surface is selected (any plausible shape of failure surface may be used).
2. The potential failure mass is subdivided into a series of slices separated by imaginary vertical lines (the number of slices chosen will depend on the geometry of the failure mass and the degree of accuracy desired).
3. The forces on the individual slices are computed, satisfying static equilibrium. In general, this step involves trial and error because there are more unknowns than equations available for each individual slice. (The problem is actually statically indeterminate; however, the necessary additional conditions are obtained by considering the equilibrium of the failure mass as a whole and making reasonable assumptions regarding the side forces on the slices.)
4. The factor of safety is computed from the previously given equation.
5. The process is repeated for different trial failure surfaces until the minimum factor of safety is found.

Stability analysis by the method of slices is depicted in Fig. 81. Figure 81a shows a potential sliding mass which has been subdivided into several slices. Of the forces shown, the weight W is known, as are the water forces U_R, U_L, and U_B (from analysis of the pore water pressures). The side forces E_R and E_L are unknown and an assumption must be made to complete the analysis. For example, the direction of the side forces may be assumed (midway between the slopes of the upper and lower boundaries of the slices is one reasonable choice). A safety factor is also assumed in order to establish the direction of the frictional component $(N \tan \phi)/F$ of the developed shear strength. The two remaining unknowns, i.e., the magnitudes of the side forces and $(N \tan \phi)/F$ are determined by equilibrium requirements for the individual slice (summation of horizontal and vertical forces must be equal to zero). This is shown in the force polygon in Fig. 81b.

The analysis of the individual slices is continued, proceeding from the top to the bottom of the slope. Since the side force E_L on one slice must be the same as E_R on the adjacent slice on its immediate left, the force polygons may be joined as a "chain," as shown in Fig. 81c. When the chain of force polygons for

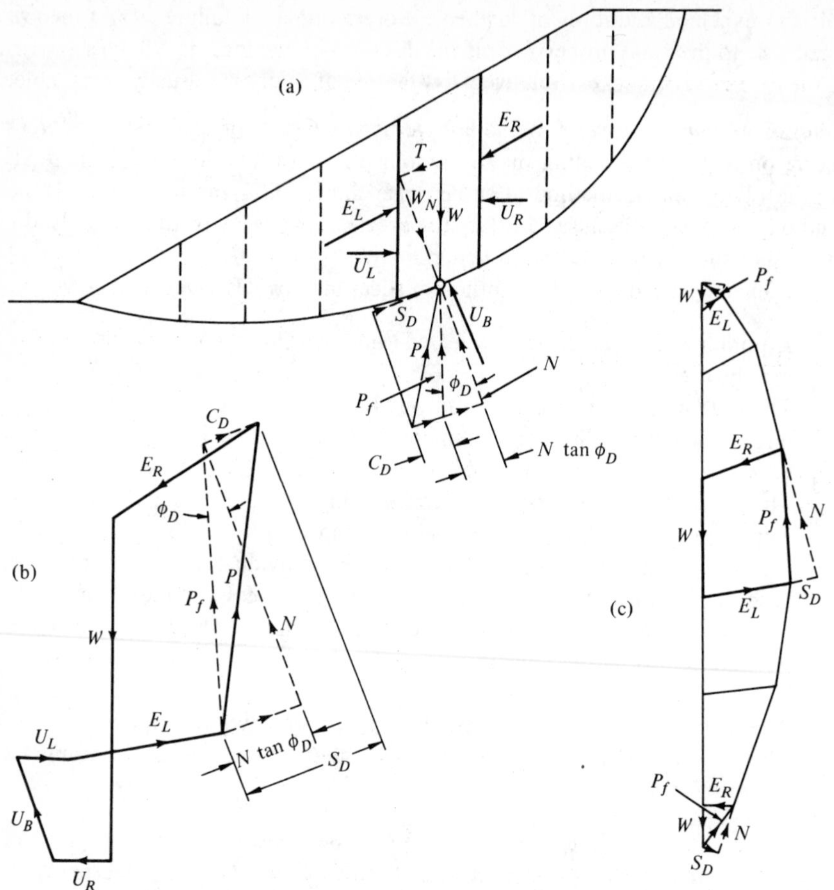

Fig. 81. Stability analysis by the method of slices.

all slices has been completed, it will close if the assumed safety factor was the correct value for that sliding surface. If it does not close, the assumed safety factor must be revised and the computations repeated until it does close. Of course, the closure of the force polygon is also affected by the assumptions that were made regarding the side forces. However, experience has shown that the safety factor does not change drastically, provided the assumptions made regarding the side forces are realistic.

If the side forces are omitted from the analysis, trial and error is not required and the safety factor can be computed directly from the expression

$$F = (cL + \Sigma N \tan \phi)/\Sigma T$$

where

L = length of sliding surface
N = normal component of weight W
T = tangential component of weight W

However, if this method is used, the magnitudes of the normal stresses on the failure surface are seriously in error and, if laboratory specimens are to be consolidated under the normal stress acting on the failure surface, the resulting strengths will also be erroneous. The fact that this method sometimes gives approximately correct results is largely due to a fortunate cancellation of errors in determining the normal stress along the entire failure surface.

Stability analysis using the method of slices can also be made analytically (Bishop, 1955). The analytical methods are adaptable to electronic computation (Little and Price, 1958; Morgenstern and Price, 1965). An excellent discussion on the use of computers for stability analysis has been presented by Whitman and Bailey (1967).

Stability analysis by the wedge method, shown in Fig. 82 involves basically the

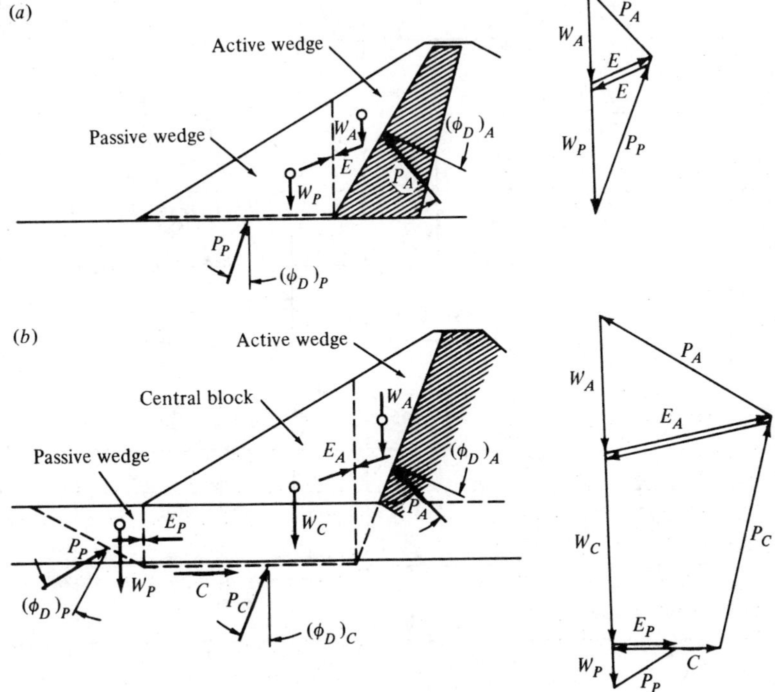

Fig. 82. Stability analysis by the wedge method.

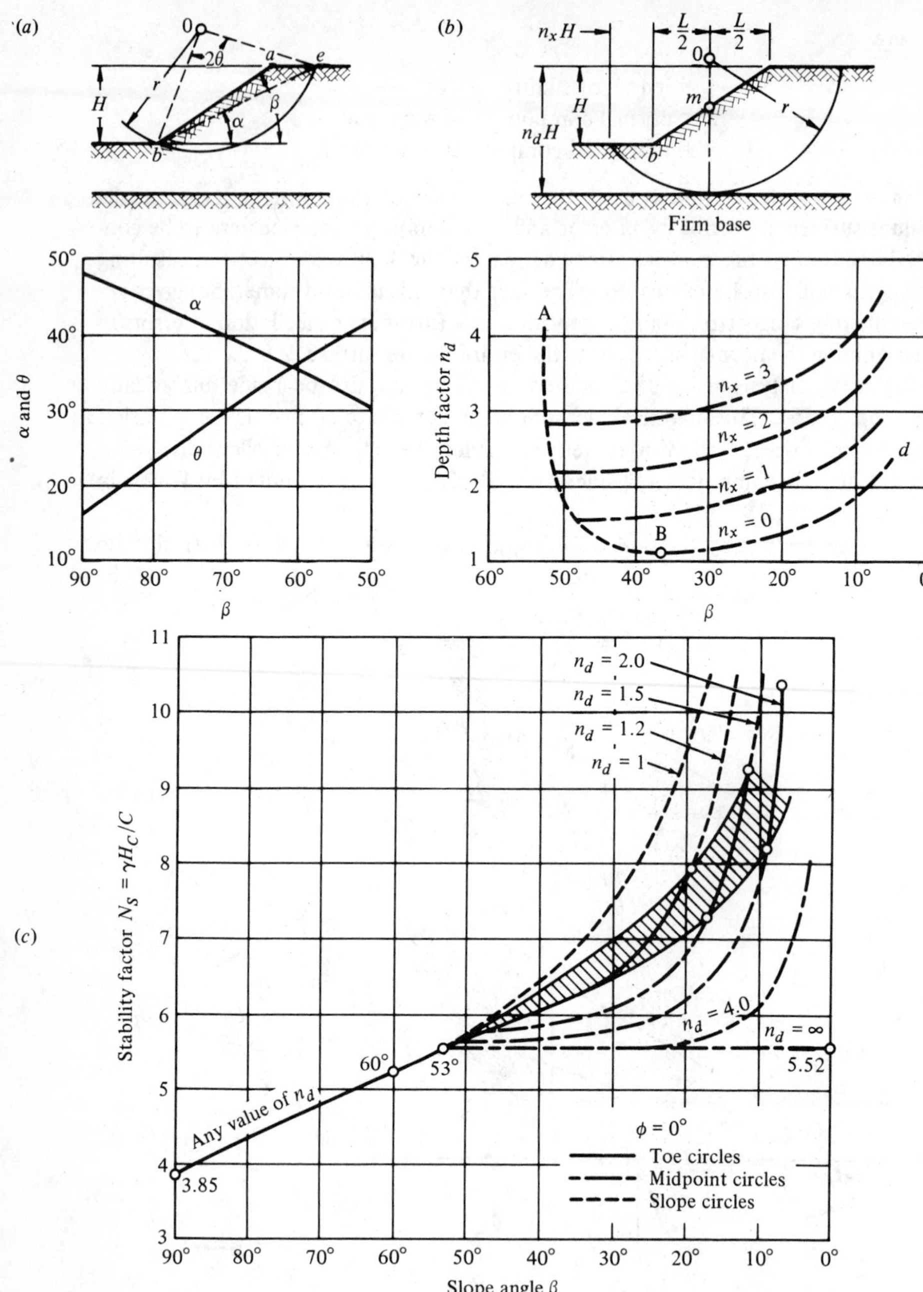

Fig. 83. Chart solutions for simple, homogeneous slopes (after Terzaghi and Peck, 1967).

(*d*)

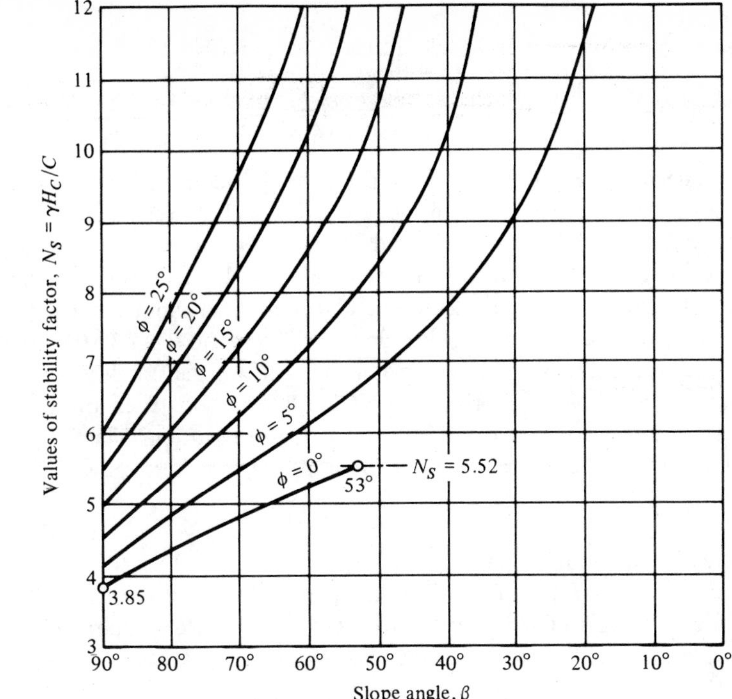

Fig. 83 *(Continued)*

same procedure as described previously. Sultan and Seed (1967) and Seed and Sultan (1967), who recently investigated the stability of sloping-core earth dams, report that the accuracy of the wedge method may be improved by considering nonvertical boundaries between the sliding blocks. If this procedure is followed, an additional trial-and-error process is required to determine the critical slope of the boundary between the sliding blocks.

Chart Solutions Chart solutions for stability analysis of slopes were first introduced by Taylor (1937). Taylor's solutions were for simple, homogeneous slopes and were obtained by the ϕ-circle method, using total stress analysis. Charts for Taylor's solution are presented in Fig. 83.

The geometry of the slope is expressed in terms of the slope angle β and the depth factor n_d as shown in Fig. 83a and b. The stability factor for the slope is expressed as

$$N_s = \gamma H_c / c$$

where

$$\gamma = \text{unit weight of soil}$$
$$H_c = \text{critical height of slope}$$
$$c = \text{cohesion of soil}$$

For cohesive soils ($\phi = 0$ condition), the stability factor is a function of the slope angle and the depth factor (see Fig. 83c). For soils possessing both friction and cohesion, the stability factor depends upon the slope angle and the angle of internal friction (Fig. 83d). In the latter case, the depth factor is immaterial because all failures are toe circles where ϕ is more than about 3°. The location of the critical circles are given in Fig. 83c (toe failure) and Fig. 83b (base failure).

If the slope angle and soil parameters are known, the stability number can be obtained from the appropriate chart (Fig. 83c or d). Then the critical slope height can be computed from the stability-factor equation. The factor of safety with respect to the critical height of the slope may be expressed as

$$F_H = H_c/H$$

where H is the actual slope height.

To obtain the factor of safety as defined previously using the chart solutions requires a process of trial and error (unless $\phi = 0$), since F appears on both sides of the equation. The procedure may be described as follows:

1. Estimate the factor of safety.
2. Compute $c_d = c/F$ and $\tan \phi_d = (\tan \phi)/F$ and use c_d and ϕ_d in the charts and subsequent calculations.
3. Determine the stability factor from Fig. 83d, using ϕ_d.
4. Compute the factor of safety and compare the computed value against the assumed value.
5. Repeat steps 1-4 until the assumed and computed factors of safety agree.

The factor of safety F_H, from the preceding equation, is always greater than F, except when both are equal to unity. If $\phi = 0$, F and F_H are identical.

In addition to Taylor's original work, numerous other chart solutions are now available for a variety of slope conditions. Hunter and Schuster (1971) have summarized many of the recent developments in the use of chart solutions for stability analysis.

Stability Problems* In this section the analysis of slope stability for different conditions of loading will be illustrated by means of several examples. The examples will serve to demonstrate the effects of loading rates, loading vs. unloading, and pore-pressure conditions.

*Examples used in this section are taken from Bishop and Bjerrum (1960).

Embankment Constructed Over Soft Clay Foundation This problem is illustrated in Fig. 84. The loading conditions consist of the rapid construction of the fill with the loads remaining constant thereafter. The average shear stress on a potential failure surface increases as the fill is placed and then remains constant after the fill has reached full height.

The pore water pressure also increases as the fill is placed. If the load is applied rapidly enough that no dissipation of pore water pressure can occur, the excess pore water pressure increases in accordance with the equation

$$\Delta u = B\Delta\sigma_3 + A(\Delta\sigma_1 - \Delta\sigma_3)$$

If the soil is saturated, $B = 1$, and, as shown in Fig. 84, the magnitude of the pore pressure increase depends upon the value of the A parameter. After completion of construction, the excess pore water pressure gradually dissipates by consolida-

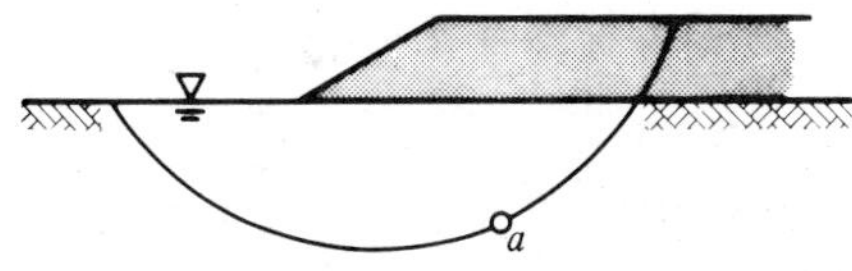

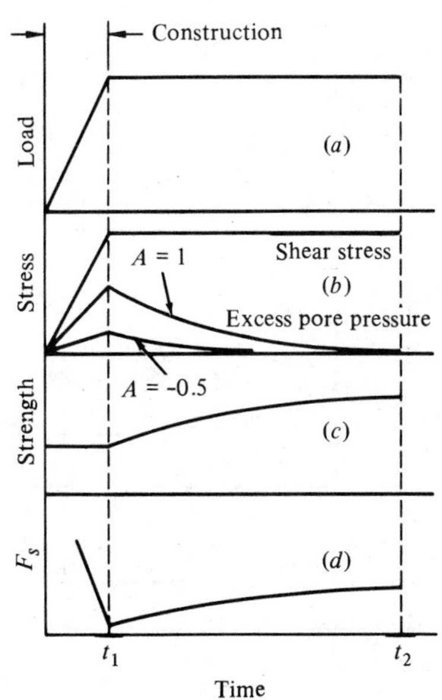

Fig. 84. Stability conditions of an embankment load (after Bishop and Bjerrum, 1960).

tion and eventually becomes equal to zero. The consolidation process also causes an increase in the shear strength of the foundation soil.

The shear strength of the soil can be determined for any stage of the loading from the effective stress parameters $\bar{c}$ and $\bar{\phi}$ if the excess pore pressure is measured in the field or can be calculated. However, as shown in Fig. 84, the lowest factor of safety occurs at the end of construction because at that stage the shear stresses and pore water pressures reach maximum values and the shear strength remains unchanged. Thus, the end of construction represents the most critical stage and the factor of safety may be computed from total stress ($\phi = 0$) analysis.

Excavations in Clay This case is illustrated in Fig. 85. Excavation results in an unloading condition along the potential failure surface, which also causes a reduction in excess pore water pressures. As in the previous case, the magnitude of the excess pore pressure depends on the value of A. However, the shear stress in-

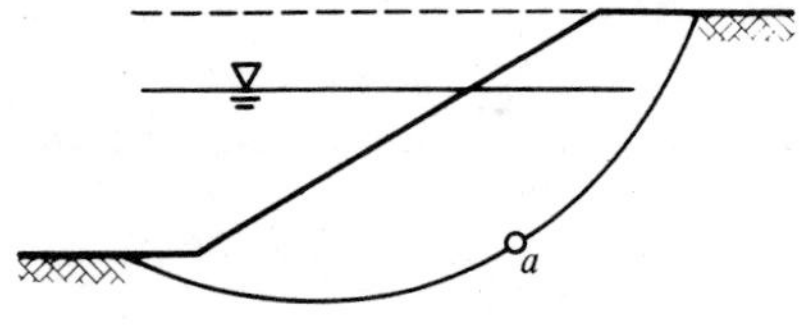

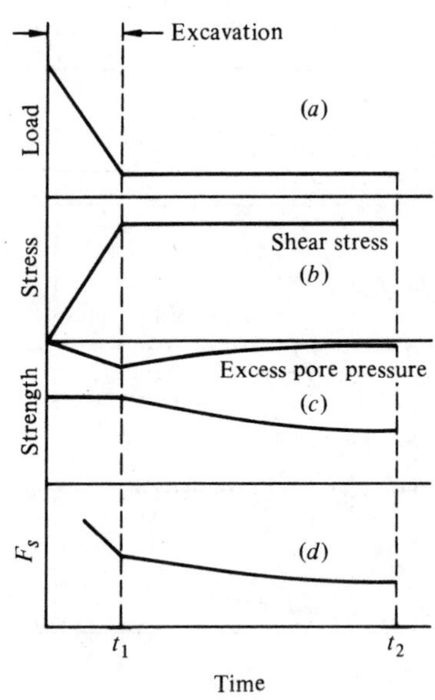

Fig. 85. Stability conditions of an excavation (after Bishop and Bjerrum, 1960).

creases and the net effect is a decrease in the factor of safety during the construction stage. As in the previous example, the factor of safety at the end of construction may be determined by total stress analysis.

Because of the reduction in pore water pressure caused by unloading, the soil has a tendency to swell, causing a reduction in the shear strength with time. Hence, the critical stage for an excavation occurs some time after construction when the pore pressures have achieved equilibrium. Since, at this stage, the pore pressures are independent of the applied loads, the effective stress analysis must be used, with the pore water pressures determined from the specific groundwater conditions.

Earth Dams The stability conditions for a typical earth dam are shown in Fig. 86. Three distinct stages must be considered in the case of an earth dam: (1) end of construction, (2) full reservoir (steady seepage), and (3) rapid drawdown.

1. End of construction. During construction the shear stresses and pore water pressures increase and the shear strength remains relatively constant. Various approaches have been suggested for stability analysis of the end-of-construction stage.

Lowe (1967) recommends a total stress analysis with the shear strength at the end of construction determined from unconsolidated-undrained triaxial tests conducted under a confining pressure equivalent to that in the embankment. This method automatically accounts for the excess pore water pressure that develops under the weight of the fill.

Hilf (1948) and Bishop and Bjerrum (1960) recommend effective stress analysis with estimated pore water pressures. Hilf relates the construction pore water pressure to the compression of the embankment material and determines the compression characteristics in a consolidation test. This method involves some degree of approximation, since the stress conditions along the potential failure surface differ considerably from those in one-dimensional compression. Nevertheless, this method has been used successfully by the U.S. Bureau of Reclamation on a number of large earth-dam projects.

Bishop and Bjerrum use the pore-pressure parameters A and B determined in laboratory tests, to estimate construction pore water pressures. The principal drawback of this approach is the difficulty of measuring pore air and pore water pressures in partially saturated soils.

Because of the relatively long construction period required for most earth dams, some dissipation of pore water pressures occurs during construction. In some cases, particularly when the embankment material is compacted on the wet side of optimum water content and construction pore water pressures build up rapidly, pore pressure dissipation is essential in order to be able to complete the embankment without a failure. Unfortunately, at the present time, rates of pore pressure dissipation in partially saturated soils cannot be reliably predicted, so

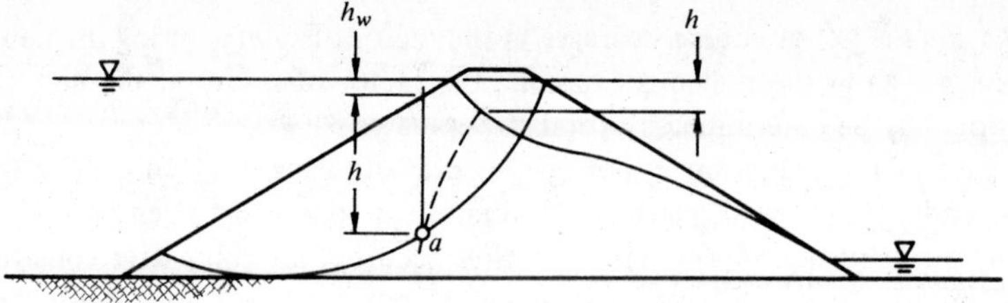

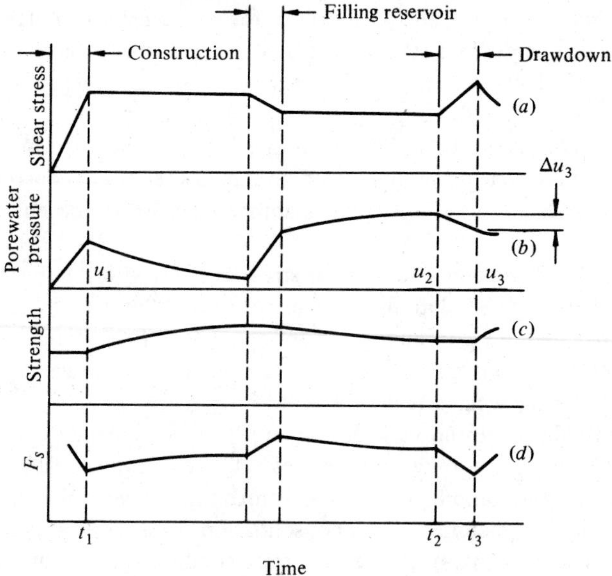

Max. value of u, (zero dissipation) is $u_1 = u_0 + \bar{B}(h \cdot \gamma)$

$$u_2 = \gamma_w (h + h_w - h')$$

$$u_3 = \gamma_w (h - h')$$

Fig. 86. Stability conditions of an earth dam (after Bishop and Bjerrum, 1960).

that it becomes necessary to make field measurements when dissipation rates are required.

2. Steady seepage. After the reservoir has remained full for some time, a pattern of steady seepage develops through the embankment. For this condition, pore water pressures can be determined from the flow net.

The factor of safety of the upstream slope increases for the steady-seepage condition. However, this condition is critical for the downstream slope. Since the pore water pressure for the steady-seepage condition is independent of the loads, the appropriate type of analysis is the effective stress method. The effective stress parameters $\bar{c}$ and $\bar{\phi}$ for this condition should be determined from laboratory tests on saturated samples consolidated under the stresses present in the embankment.

3. Rapid drawdown. Rapid lowering of the reservoir level causes a sudden, large increase in the shear stresses in the upstream slope and reduction in the factor of safety. This condition is often critical for the upstream slope. Either effective or total stress analysis may be used for this case. However, if total stress analysis is used, it is important to recognize that the undrained shear strength must be determined for the conditions existing along the potential failure surface *prior* to drawdown.

REFERENCES

Agerschou, H. A., Analysis of Engineering News Pile Formula, *Soil Mechanics and Foundations Division,* ASCE, Vol. 88, No. SM5, Oct., 1962, pp. 1–34.

Aldrich, H. P., Selection of Foundation Systems, *Proc. Soil Mechanics Lecture Series: Foundation Engineering*, Soil Mechanics and Foundations Division, Illinois Section ASCE and Department of Civil Engineering Northwestern Univ., Evanston, Illinois, Jan., 1968, pp. 19–38.

Barden, L., Time Dependent Deformation of Normally consolidated Clays and Peats, *Soil Mechanics and Foundations Division*, ASCE, Vol. 95, No. SM1, Proc. Paper 6337, Jan., 1969, pp. 1–31.

Bishop, A. W., The Use of the Slip Circle in the Stability Analysis of Slopes, *Geotechnique*, Vol. 5, No. 1, 1955.

Bishop, A. W., The Strength of Soils as Engineering Materials, 6th Rankine Lecture, *Geotechnique*, Vol. 16, No. 2, 1966, pp. 91–128.

Bishop, A. W., and Bjerrum, L., The Relevance of the Triaxial Test to the Solution of Stability Problems, *Proc. ASCE Research Conf. Shear Strength of Cohesive Soils*, Boulder, Colorado, 1960.

Bishop, A. W., and Henkel, D. J., The Measurement of Soil Properties, *The Triaxial Test*, Edward Arnold Ltd., London, 2nd ed., 1962.

Bishop, A. W., et. al., Factors Controlling the Strength of Partly Saturated Cohesive Soils, *Proc. ASCE Research Conf. Shear Strength of Cohesive Soils*, Boulder, Colorado, 1960, pp. 503–532.

Bishop, A. W., and Lovenbury, H. T., Creep Characteristics of Two Undisturbed Clays, *Proc. 7th International Conf. Soil Mechanics and Foundation Engineering*, Mexico, Vol. 1, 1969.

Bjerrum, L., Progressive Failure in Slopes of Overconsolidated Plastic Clay and Clay Shales, *Soil Mechanics and Foundations Division*, ASCE, Vol. 93, No. SM5 (Part 1), Proc. Paper 5456, 1967, pp. 1–49.

Bjerrum, L., and Eggestad, A., Interpretation of Loading Tests on Sand, *Proc. European Conf. Soil Mechanics and Foundation Engineering,* Wiesbaden, Vol. 1, 1963.

Bjerrum, L., and Lo, K. Y., Effect of Aging on the Shear-Strength Properties of a Normally Consolidated Clay, *Geotechnique*, Vol. 13, No. 2, 1963, pp. 147–157.

Burmister, D. M., Stress and Displacement Characteristics of a Two-Layer Ridge Base Soil System: Influence Diagrams and Practical Applications, *Proc. Highway Research Board*, Vol. 35, 1956.

Button, S. J., The Bearing Capacity of Footings on a Two-Layer Cohesive Subsoil, *Proc. 3rd International Conf. Soil Mechanics and Foundation Engineering*, Zurich, Vol. 1, 1953.

Casagrande, A., The Determination of the Preconsolidation Load and its Practical Significance, *Proc. 1st International Conf. Soil Mechanics and Foundation Engineering*, Cambridge, Vol. 3, 1936.

Casagrande, A., and Fadum, R. E., Notes on Soil Testing for Engineering Purposes, Harvard Univ. Soil Mechanics Series, No. 8, Bulletin 268, Cambridge, Mass., 1940.

Chellis, R. D., *Pile Foundations*, McGraw-Hill Book Company, Inc., New York, 1961.

D'Appolonia, D. J., et. al., Settlement of Spread Footings on Sand, *Soil Mechanics and Foundations Division*, ASCE, Vol. 94, No. SM3, Proc. Paper 5959, May, 1968, pp. 735-760.

D'Appolonia, D. J., et. al., Initial Settlement of Structures on Clay, *Soil Mechanics and Foundations Division*, ASCE, Vol. 97, No. SM10, Proc. Paper 8438, Oct., 1971, pp. 1359–1378.

DeBeer, E., and Martins, A., Method of Computation of an Upper Limit for the Influence of the Heterogeneity of Sand Layers in the Settlement of Bridges, *Proc. 4th International Conf. Soil Mechanics and Foundation Engineering*, London, Vol. 1, 1957.

Fadum, R. E., Influence Values for Estimating Stresses in Elastic Foundations, *Proc. 2nd International Conf. Soil Mechanics and Foundation Engineering*, Rotterdam, 1948.

Flaate, K. S., An Investigation of the Validity of Three Pile-Driving Formulae in Cohesionless Materials, *Norwegian Geotechnical Institute Publ. 56*, 1964.

Foster, C. R., and Ahlvin, P. G., Stresses and Deflections Induced by Uniform Circular Load, *Highway Research Board Proc.*, Highway Research Board, Washington, D.C., 1954.

Gardner, W. S., Stress Strain Behavior of Granular Soils in One-Dimensional Compression, *ASCE Annual Structural Engineering Conf.*, Miami, Feb., 1966.

Gibbs, H. J., and Holtz, W. G., Research on Determining the Density of Sands by Spoon Penetration Testing, *Proc. 4th International Conf. Soil Mechanics and Foundation Engineering*, London, Vol. 1, 1957.

Gibson, R. E., et. al., Plain Strain and Axially Symmetric Consolidation of a Clay Layer of Limited Thickness Univ. Illinois, Chicago Circle, MATE Report 67–4, 1967.

Gould, J. P., Lateral Pressures on Ridge Permanent Structures, *Lateral Stresses in the Ground and Design of Earth-Retaining Structures*, ASCE, June, 1970.

Grant, D. W., An Experimental Investigation of the Effect of Thickness of Backfill on Lateral Earth Pressures against a Vertical Wall, M. S. Thesis, University of Washington, Seattle, 1961.

Hansen, B. J., *Earth Pressure Calculation*, Danish Technical Press, Copenhagen, 1953.

Hough, B. K., Compressibility as the Basis for Soil Bearing Value, *Soil Mechanics and Foundations Division,* ASCE, Vol. 85, No. SM4, Aug., 1959.

Hendron, A. J., The Behavior of Sand in One-Dimensional Compression, Ph.D. thesis, Univ. Illinois, Urbana, 1963.

Hilf, J. W., Estimating Construction Port Pressures in Rolled Earth Dams, *Proc. 2nd International Conf. Soil Mechanics and Foundation Engineering*, Rotterdam, Vol. 3, 1948.

Horn, H. M., Influence of Pile Driving and Pile Characteristics on Pile Foundation Performance, notes for lectures to N.Y. Metropolitan Section ASCE, Soil Mechanics and Foundations Group, 1966.

Hunter, J. H., and Schuster, R. L., Chart Solutions for Analysis of Earth Slopes, *Highway Research Record*, No. 345, 1971.

Hvorslev, M. J., *Subsurface Exploration and Sampling of Soils for Civil Engineering Purposes,* Waterways Experiment Station, Vicksburg, Mississippi, 1962.

Janbu, N., Earth Pressures and Bearing Capacity Calculations by Generalized Procedure of Slices, *Proc. 4th International Conf. Soil Mechanics and Foundation Engineering*, London, Vol. 2, 1957.

Janbu, N., Settlement Calculations Based on the Tangent Modulus Concept, *Soil Mechanics and Foundation Engineering,* The Technical Univ. Norway, Trondheim, Bulletin 2, 1967.

Janbu, N., The Resistance Concept Applied to Deformation of Soils, *Proc. 7th International Conf. Soil Mechanics and Foundation Engineering*, Mexico, Vol. 1, 1969.

Janbu, N., et. al., Soil Mechanics Applied to Some Engineering Problems, *Publ. No. 16,* Norwegian Geothechnical Institute, Oslo, Norway, 1956.

Konder, R. L., Hyperbolic Stress-Strain Response: Cohesive Soils, *Soil Mechanics and Foundations Division,* ASCE, Vol. 89, No. SM1, Proc. Paper 3429, Feb., 1963, pp. 115–143.

Konder, R. L., and Zelasko, J. S., A Hyperbolic Stress-Strain Formulation for Sands, *Proc. 2nd Pan-Am. Conf. Soil Mechanics and Foundation Engineering,* Brasil, Vol. 1, 1963.

Lacroix, Y., and Jackson, W., Supported Temporary Excavation in Urban Areas, Paper presented at seminar on construction excavation, Univ. Wisconsin, Madison, Wisconsin, Apr., 1972.

Ladd, C. C., Stress-Strain Modulus of Clay from Undrained Triaxial Tests, *Soil Mechanics and Foundations Division,* ASCE, Vol. 9, No. SM5, Proc. Paper 4039, Sept., 1964, pp. 103–132.

Ladd, C. C., Stress-Strain Behavior of Anisotropically Consolidated Clays During Undrained Shear, *Proc. 6th International Conf. Soil Mechanics and Foundation Engineering,* Montreal, Vol. 1, 1965.

Ladd, C. C., Settlement Analysis for Cohesive Soils, *Soil Mechanics Division, Dept. Civil Engineering Research Report R 71-2*, Soils Publ. 272, 1971a.

Ladd, C. C., Strength Parameters and Stress-Strain Behavior of Saturated Clays, *Soil Mechanics Division, Dept. Civil Engineering Research Report R 71-23*, Soils Publ. 278, 197lb.

Ladd, C. C., and Varallyay, J., The Influence of Stress System on the Behavior of Saturated Clays during Undrained Shear, *Dept. Civil Engineering, Research Report R 65-11*, MIT, 1965.

Lambe, T. W., *Soil Testing for Engineers*, John Wiley & Sons, Inc., New York, 1951.

Lambe, T. W., Methods of Estimating Settlement, *ASCE Settlement Conf.* (Preprint), Northwestern Univ., June, 1964.

Lambe, T. W., and Whitman, R. V., *Soil Mechanics*, John Wiley & Sons, Inc., New York, 1969.

Little, A. L., and Price, V. E., The Use of an Electronic Computer for Slope Stability Analysis, *Geotechnique*, Vol. 8, 1958.

Lowe, J. III, Stability Analysis of Embankments, *Soil Mechanics and Foundations Division*, ASCE, Vol. 93, No. SM4, Proc. Paper 5305, July, 1967, pp. 1–33.

Lowe, J. III, et. al., Controlled Gradient Consolidation Test, *Soil Mechanics and Foundations Division*, ASCE, Vol. 95, No. SM1, Proc. Paper 6327, Jan., 1969, pp. 77–97.

Meyerhof, G. G., The Bearing Capacity of Foundations under Eccentric and Inclined Loads, *Proc. 3rd International Conf. Soil Mechanics and Foundation Engineering*, Zurich, Vol. 1, 1953.

Meyerhof, G. G., The Ultimate Bearing Capacity of Foundations on Slopes, *Proc. 4th International Conf. Soil Mechanics and Foundation Engineering*, London, Vol. 1, 1957.

Meyerhof, G. G., Shallow Foundations, *Design of Foundations for Control of Settlement*, ASCE, Proc. Paper 4271, Mar., 1965.

Mitchell, J. K., Shearing Resistance of Soils as a Rate Process, *Soil Mechanics and Foundations Division*, ASCE, Vol. 90, No. SM1, Feb., 1962, pp. 29–61.

Morgenstern, N. R., and Eisenstein, Z., Methods of Estimating Lateral Loads and Deformations, *Lateral Stresses in the Ground and Design of Earth-Retaining Structures*, ASCE, June, 1970.

Navdocks DM-7, Soil Mechanics, Foundations, and Earth Structures, *Design Manual*, Department of the Navy, Bur. of Yards and Docks, Washington, D.C., 1962.

Newmark, N. M., Influence Charts for Computation of Stresses in Elastic Foundations, *Engineering Experiment Sta., Bulletin*, No. 338, Univ. Illinois, Urbana, Illinois, 1951.

Obrcian, V. F., Determination of Lateral Pressures Associated with Consolidation of Granular Soils, *Highway Research Record*, No. 284, 1969.

Osterberg, J. O., Influence Values for Vertical Stresses in a Semi-Infinite Mass Due to an Embankment Loading, *Proc. 4th International Conf. Soil Mechanics and Foundation Engineering*, London, 1957.

Peck, R. B., Deep Excavation and Tunneling in Soft Ground, *Proc. 7th International Conf. on Soil Mechanics and Foundation Engineering*, Mexico, 1969.

Peck, R. B., and Bazara, A. R. S., discussion of Settlement of Spread Footings on Sand, by D'Appolonia, D. J., et. al., *Soil Mechanics and Foundations Division*, ASCE, Vol. 94, No. SM3, Proc. Paper 5959, May, 1968, pp. 735–760.

Roberts, J. E., and De Sousa, J. M., The Compressibility of Sands, Paper presented to the 61st annual meeting of the A.S.T.M., 1958.

Rowe, P. W., Anchored Sheet-Pile Walls, *Proc. Institute of Civil Engineers*, London, Vol. 1, 1952.

Seaman, L., et. al., Stress Propagation in Soils—Part III, Reported by Stanford Research Institute to Defense Atomic Support Agency, DASA-1266-3, 1963.

Schmertmann, J. H., The Undisturbed Consolidation of Clay, *Transactions ASCE*, Vol. 120, 1955.

Schmertmann, J. H., Static Cone to Compute Static Settlement over Sand, *Soil Mechanics and Foundations Division*, ASCE, Vol. 96, No. SM3, Proc. Paper 7302, May, 1970, pp. 1011–1043.

Schultze, E., and Moussa, A., Factors Affecting the Compressibility of Sands, *Proc. 5th International Conf. Soil Mechanics and Foundation Engineering*, Zurich, 1961.

Seed, H. B., and Sultan, H. A., Stability Analyses for a Sloping Core Embankment, *Soil Mechanics and Foundations Division*, ASCE, Vol. 93, No. SM4, Proc. Paper 5308, July, 1967, pp. 69–83.

Singh, A., and Mitchell, J. K., Creep Potential and Creep Rupture of Soils, *Proc. 7th International Conf. Soil Mechanics and Foundation Engineering*, Vol. 1, 1969.

Skempton, A. W., The Bearing Capacity of Clays, Bldg. Research Congress, England, 1951.

Skempton, A. W., The Pore-Pressure Coefficient *A* and *B*, *Geotechnique*, Vol. 4, 1954, pp. 143–147.

Skempton, A. W., Discussion: The Planning and Design of the New Hong Kong Airport, *Proc. Institute of Civil Engineers*, London, Vol. 7, 1957.

Skempton, A. W., and Bjerrum, L., A Contribution to the Settlement Analysis of Foundations on Clay, *Geotechnique*, Vol. 7, No. 4, 1957.

Skempton, A. W., et. al., Théorie de la Force Portante des Pieux dans le Sable, *Annales Institut Technique Batim.*, Vol. 6, Mar.–Apr., 1953.

Smith, R. E., and Wahls, H. E., Consolidation under Constant Rates of Strain, *Soil Mechanics and Foundations Division*, ASCE, Vol. 95, No. SM2, Proc. Paper 6452, Mar., 1969, pp. 519–539.

Sokolovski, V. V., *Statics of Soil Media*, Butterworth, London, 1960.

Sowers, G. B., and Sowers, G. F., *Introductory Soil Mechanics and Foundations*, The Macmillan Company, New York, 2nd ed., 1971.

Sowers, G. F., et. al., The Residual Lateral Pressures Produced by Compacting Soils, *Proc. 4th International Conf. Soil Mechanics and Foundation Engineering*, London, Vol. 2, 1957.

Sultan, H. A., and Seed, H. B., Stability of Sloping Core Earth Dams, *Soil Mechanics and Foundations Division*, ASCE, Vol. 93, No. SM4, Proc. Paper 5307, July, 1967, pp. 45–67.

Taylor, D. W., Stability of Earth Slopes, *Boston Society of Civil Engineers*, Vol. 24, 1937.

Teng, W. C., *Foundation Design*, Prentice-Hall, Inc., Englewood Cliffs, New Jersey, 1964.

Terzaghi, K., Large Retaining-Wall Tests, *Engineering News-Record*, Vol. 112, 1934.

Terzaghi, K., *Theoretical Soil Mechanics*, John Wiley & Sons, Inc., New York, 1943.

Terzaghi, K., Anchord Bulkheads, *Transactions ASCE*, Vol. 119, 1954.

Terzaghi, K., and Peck, R. B., *Soil Mechanics in Engineering Practice*, John Wiley & Sons, Inc., New York, 2nd ed., 1967.

Tomlinson, M. J., *Foundation Design and Construction*, Wiley-Interscience, New York, 2nd ed., 1969.

United States Bureau of Reclamation, *Earth Manual*, U.S. Department of the Interior, Bureau of Reclamation, U.S. Government Printing Office, Washington, D.C., 1968, pp. 99–111.

Vesic, A. S., Ultimate Loads and Settlements of Deep Foundations in Sand, *Proc. Symposium on Bearing Capacity and Settlements of Foundations*, Duke Univ., Durham, North Carolina, 1967.

Wagner, A. A., The Use of the Unified Soil Classification System by the Bureau of Reclamation, *Proc. 4th International Conf. Soil Mechanics and Foundation Engineering*, London, Vol. 1, 1957.

Whitman, R. V., and Bailey, W. A., Use of Computers for Slope Stability Analysis, *Soil Mechanics and Foundation Division*, ASCE, Vol. 93, No. SM4, Proc. Paper 5327, July, 1967, pp. 475–528.

Wilson, G., and Sutton, J. L. E., A Contribution to the Study of the Elastic Properties of Sand, *Proc. 2nd International Conf. Soil Mechanics and Foundation Engineering*, Rotterdam, Vol. 1, 1948.

Wissa, A. E. Z., and Heiberg, S., A New One-Dimensional Consolidation Test, *Dept. Civil Engineering Research Report 69-9*, Soils Publ. 229, MIT, 1969.

Wu, T. H., *Soil Mechanics*, Allyn and Bacon, Inc., Boston, 1966.

ADDITIONAL REFERENCES*

Barkan, D. D., *Dynamics of Bases & Foundations* (translated from Russian), McGraw-Hill Book Co., N.Y., 1962.

Foundations Subject to Vibratory Loads, U.S. Army Office Chief of Engineers, EM 1110-345-310.

Handbook of Drainage and Construction Products, Armco Drainage and Metal Products, Inc., Middletown, Ohio.

Hetenyi, M., *Beams on Elastic Foundation*, University of Michigan Press, Ann Arbor, Michigan.

Hvorslev, M. J., *Subsurface Exploration and Sampling of Soils for Civil Engineering Purposes*, Department of the Army, Corps of Engineers, Waterways Experiment Station, Vicksburg, Miss.

Jaeger, J. C. and Cook, N.G.W., *Fundamentals of Rock Mechanics*, Methuen & Co. Ltd., London, 1969.

Ketchum, M. S., *The Design of Walls, Bins and Grain Elevators*, Engineering News Publishing Co., New York, N.Y.

*From *Design Manual: Soil Mechanics, Foundations, and Earth Structures*; NAVFAC DM-7, March 1971; revised January 1971

Mehta, M. R., and Veletsos, A. S., *Stresses and Displacements in Layered Systems*, University of Illinois, Structural Research Series No. 178.

Obert, L., and Duvall, W. I., and Merrill, R. H., *Design of Underground Openings in Competent Rock*, Bulletin 587, U.S. Bureau of Mines.

Soil Mechanics Design, Seepage Control, Chapter 1, Part CXIX, Engineering Manual, Civil Works Construction, Department of the Army, Corps of Engineers.

Stagg, K. G. and Zienkiewicz, O. C., *Rock Mechanics in Engineering Practice*, John Wiley & Sons Ltd., London, 1968.

Taylor, D. W., *Pressure Distribution Studies on Soils*, Soil Mechanics Fact Finding Study Progress Report, Department of the Army, Corps of Engineers, Waterways Experiment Station, Vicksburg, Miss.

Terzaghi, K., *Rock Defects and Loads on Tunnel Supports, Section 1 of Rock Tunnelling with Steel Supports*, The Commercial Shearing and Stamping Co., Youngstown, Ohio.

Van der Veen, C., and Boersma, L., *The Bearing Capacity of a Pile Pre-Determined by a Cone Penetration Test*, Proceedings, Fourth International Conference on Soil Mechanics and Foundation Engineering, Butterworths Scientific Publications, London.

Woods, K. B., *Highway Engineering Handbook*, McGraw-Hill Book Company, Inc., New York, N.Y.

Zanger, C. N., *Theory and Problems of Water Percolation*, Engineering Monograph No. 8, U.S. Bureau of Reclamation.

SECTION IV

Structural Engineering

Milo S. Ketchum, *P. E., F. ASCE, F. IStructE*
Professor of Civil Engineering
University of Connecticut
Storrs, Connecticut
and the staff of
Ketchum, Konkel, Barrett, Nickel, Austin
Consulting Engineers
Denver, Colorado

INTRODUCTION TO STRUCTURAL ENGINEERING

Structural engineering is usually considered as a part of civil engineering, which in turn may be defined as "engineering for the permanent physical plant of the nation." Therefore structural engineering is "engineering of the structures that sustain the buildings and bridges and other structures which are a part of the permanent physical plant of the nation." In the case of buildings, the design of the facilities is in the hands of the profession called architecture. There are some exceptions as, in the case of buildings which serve to house engineering equipment, and some industrial buildings which are designed completely by the engineer. Bridges are designed by structural engineers normally working for highway departments or consulting engineering firms specializing in highway structures.

Most structural engineers, therefore, prepare their designs for buildings under the direction of or at the request of the architect, so that all of the decisions on structure cannot be made entirely by the engineer. This calls for close cooperation of architect and engineer if the result is to be successful. The aims of the architect and the structural engineer may be divergent and compromises must be made by both parties. It is interesting and instructive to prepare flow charts of the interaction of the engineer and architect, and such a diagram is shown in Fig. 1. Events are indicated by a square symbol with a number in the middle for the description on the right side of the page. Logical decisions are indicated by circles with branches showing "yes" or "no." This chart shows that many of the decisions are made by the architect before the engineer is called into the design team. This may be a source of weakness for the solution finally achieved.

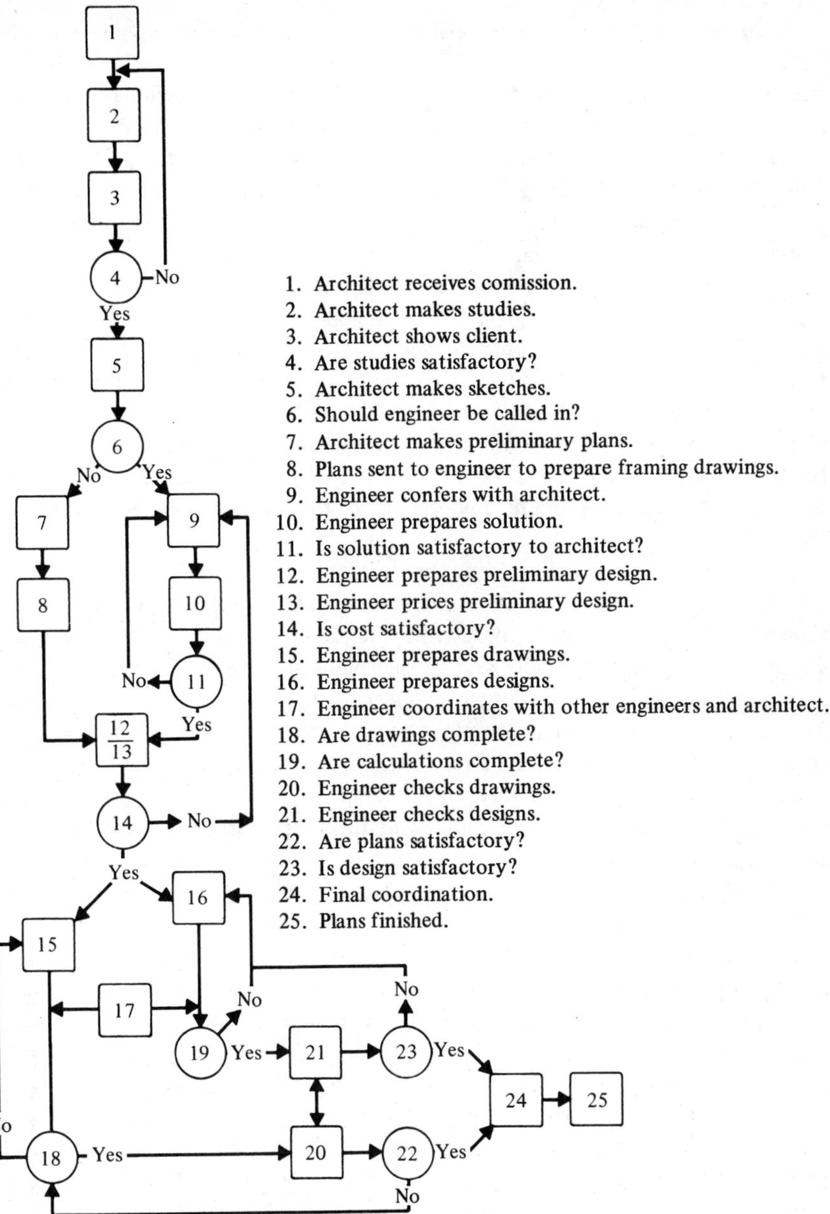

1. Architect receives comission.
2. Architect makes studies.
3. Architect shows client.
4. Are studies satisfactory?
5. Architect makes sketches.
6. Should engineer be called in?
7. Architect makes preliminary plans.
8. Plans sent to engineer to prepare framing drawings.
9. Engineer confers with architect.
10. Engineer prepares solution.
11. Is solution satisfactory to architect?
12. Engineer prepares preliminary design.
13. Engineer prices preliminary design.
14. Is cost satisfactory?
15. Engineer prepares drawings.
16. Engineer prepares designs.
17. Engineer coordinates with other engineers and architect.
18. Are drawings complete?
19. Are calculations complete?
20. Engineer checks drawings.
21. Engineer checks designs.
22. Are plans satisfactory?
23. Is design satisfactory?
24. Final coordination.
25. Plans finished.

Fig. 1. The structural design process.

A structural engineer must, therefore, not only be capable of handling mathematical and scientific problems, but also have the ability to work with other individuals to obtain solutions to his problems. A total introvert is not usually a good structural engineer. Furthermore, a structural engineer must be creative, because the problems are varied and seldom routine. One of the principal attributes of a good structural engineer is the ability to visualize the structure and the loads on the structure in space.

Structural engineers normally receive their education in schools of engineering as civil engineers, although there are some departments of architectural engineering. The engineer receives a grounding in general physics, chemistry, and mathematics, as well as mechanics, which includes the fields of statics (forces at rest) and dynamics (forces in motion). His engineering studies of structures start with mechanics of materials, which describes the strength of simple structural systems. The remaining courses cover structural theory, simple and indeterminate structures, and design, particularly of steel and concrete structures. After the bachelor's degree the student may go into practice or may go for further graduate work.

On graduation, the engineer should work for a practicing engineer so that he can learn many details of practice that the engineering school did not have the time to supply, and learn to sharpen his judgment in making correct decisions. To practice engineering legally and to sign plans and specifications, one should have registration as a professional engineer in the state where he will practice. This may be obtained after an examination at the end of four years of experience. It is generally in two parts, the first of which may be taken at graduation, and which confers the title of Engineer in Training.

Much of the structural engineering of buildings is performed by relatively small consulting firms consisting of one or two principals and employee engineers. It is common practice for all engineers to work in one large room at drafting tables with only the principals having private offices. While this might seem to be demeaning to the employee, it provides for considerable interaction between all members of the group and promotes teamwork. Engineers who have worked under this system find it difficult to work alone in private offices on a group project.

The first drawings of a design project for a building will usually come from the architect as a floor plan of the building with perhaps several small elevations to indicate heights. The engineer must devise framing systems for this plan, and indicate to the architect the depths of floor construction. The selection of the actual framing system may depend on studies of cost made by the engineer and perhaps prices by a contractor. Drawings may be made by a draftsman, but in many offices, the engineer does his own drawing. The advantage is that the engineer may do the design and drawing at the same time, and can study the structure in a more thorough way, and so be more efficient. Structural engineers, therefore, must have the ability to make presentable drawings. One of the problems of the structural engineer is that many of the decisions must be made by the architect, so that it is not always possible to proceed at full speed on the design, and the engineer must put design problems aside. It is not good planning to get too far ahead of the architect in the design, or the engineer may have to do it over.

Structural calculations should be made in a neat and workmanlike manner; sloppy calculations are an indication of lack of understanding of the process of design. In many cases it is not possible to have calculations checked in detail. The checker usually confines his efforts to the detection of gross errors in design or in the preparation of the drawings, particularly the dimensions and the completeness of the drawings. If items are omitted it may in some cases be necessary for the consulting engineer or the architect to pay the cost of the item omitted.

Most structural engineers perform some, but not necessarily complete, inspection of the construction. This is an important phase of the design process, because it is here that gross errors may be detected which otherwise might not be seen.

The Structural Engineer and the Environment

The design and construction of a suitable environment for man is becoming more and more a concern of the engineering profession because, although it does not bear all of the responsibility, it is often blamed for the final result. A similar position is occupied by the scientific profession and their work on the fearful instruments of destruction.

The structural engineer has been blamed for many of the ugly structures that are all too evident in this country, particularly those associated with factories and industrial installations. As long as the industrial complex insists that first cost is the only criterion for selection of facilities, then these ugly structures will continue to be built.

It is incumbent on the structural engineer to know the principles of aesthetics so that he can design structures that will be beautiful in appearance and that will enhance the environment rather than degrade it.

Some of the factors that will enhance the appearance and consequently improve the environment may be listed. They are not necessarily in the order of their importance.

1. Not only is the structure itself important, but its relationship to other structures near it is also important. Open spaces around and through the project should be provided.
2. The materials used to finish the structure should have a life equal to the economic life of the project so that they will not become an eyesore before the structure is removed.

3. The cost of maintenance for the materials provided should not be so great that the appearance will be allowed to deteriorate.
4. Forms must be simple. Masses must be arranged in pleasing shapes. The eye must perceive some order and logical arrangement to the composition.
5. Colors must be selected for compatibility and harmony with the colors of other structures surrounding.
6. Decoration should be used with restraint and then only if there is some logical reason for using it.
7. Provision should be made in the design for growing things such as grass, trees, and shrubs to take away the stark, utilitarian look.
8. The access to and from the project must be logical, simple, and functional. Provision must be made for parking that will be convenient.
9. The inside of a building is as important as the outside, if not more so. The surfaces and the structural elements that show should be simple, light, and clean. There should be adequate artifical and also natural light and proper heating and ventilation. Colors should be pleasing and as light as possible, consistent with the maintenance of the interior.
10. To be beautiful, a structure must function properly. If it does not, it is an eyesore no matter how carefully it has been designed for aesthetic considerations.

The design of elevated water tanks and towers is an area where a forward-looking and ingenious structural engineer can use his knowledge of structures to effect interesting and different shapes that may improve the environment. In the past, the cheapest design for elevated water tanks was that shown in Fig. 2a, which is a cylindrical steel tank set on four steel columns, braced with a multiplicity of diagonal steel rods and struts between the columns. This design with variations is to be found in almost every small midwestern town. New and cleaner designs are now available from the leading companies that produce steel tanks and several are suggested by the sketches in Fig. 2b and c. These new tanks may

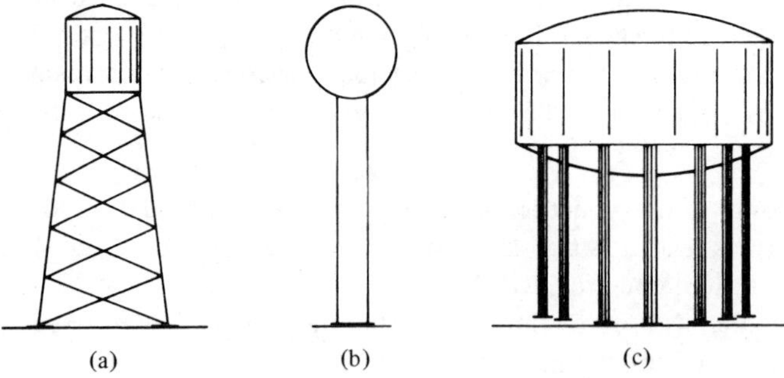

Fig. 2.

be more expensive for the first cost but accurate engineering analysis of long-time costs may show that the newer design will cost the least in the long run. These decisions should be made only on the basis of sound principles of engineering economy with proper consideration being given to the monetary value of aesthetics.

Sketches of shapes of some recent reinforced concrete tanks are shown in Fig. 3. They are formed from geometrical shapes arranged so that the required

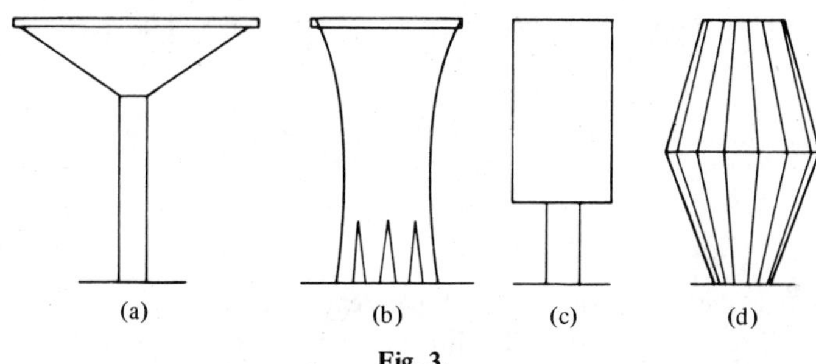

Fig. 3.

structural properties are obtained. The key to the proper design lies in the forming of the tank for placing the steel and concrete. Another interesting possibility is to combine the tank with some other function such as a building to house the office of the waterworks officials or other city offices, or to provide a space for a restaurant on the top. It is most important that the site for the tank be properly considered and the land around it be landscaped to provide an interesting setting.

The automobile has become a dominant factor in our environment and has created some of the most horrendous problems that man has had to face, such as the smog condition in urban areas. One area where the structural engineer has had an impact has been the appearance of bridges, especially those that cross our interstate freeways at every mile. The approach to the aesthetic design has changed considerably since the first parkways were built near urban areas 30 to 40 years ago. Considerable time and talent was spent to make them as attractive as possible and architectural assistance was enlisted. Expensive materials such as cut stone and brick were often used to finish them. The results were about as shown in Fig. 4, with thick, massive piers and heavy retaining walls. Elaborate,

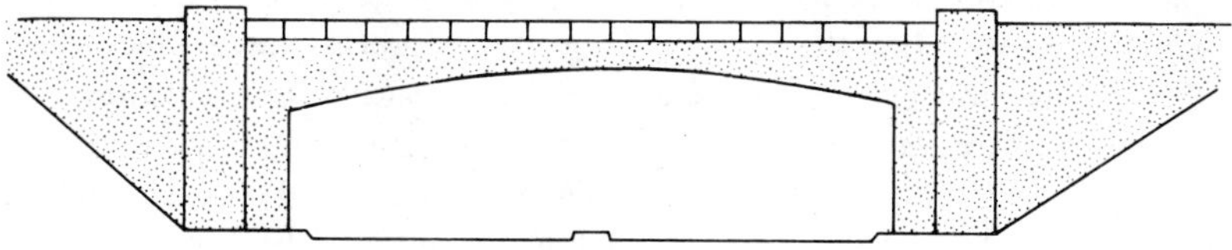

Fig. 4.

decorated hand rails were designed. Now these bridges are white elephants because they can not be widened to fit modern widths of roadway, median strip, and shoulders. When you drive up to them at high speed, you feel that you are going through a tunnel, and if they are on a curve, it is like driving into a stone wall.

The modern approach is to make the bridge as unobtrusive as possible, as sketched in Fig. 5. A thin center pier is used and small side piers are placed as far

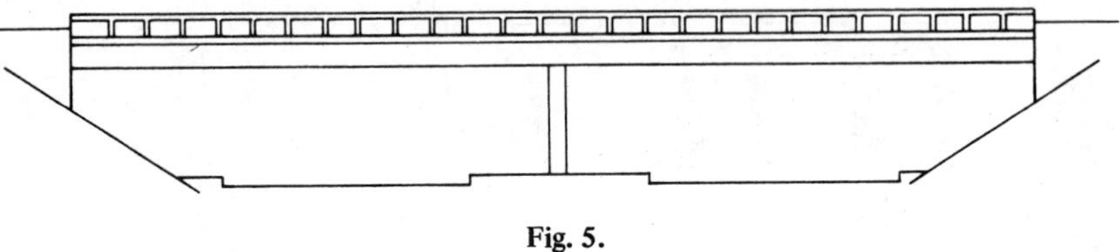

Fig. 5.

from the roadway as possible. The slopes near to the bridge are paved with concrete or concrete blocks. The depth of the girders is made to appear as shallow as possible and the handrail, often of steel or aluminum, is unobtrusive and functional. To properly perform its mission, this bridge must do as little to distract the driver of the automobile passing under it as possible. Even colors should be neutral and blend with local environment such as green grass, trees, or rocks.

Materials for Structures

The materials most used for bridge and building structures are steel, reinforced concrete, prestressed concrete, and timber. Aluminum and plastics are used to a lesser extent because of the higher cost. The choice of one of these materials will depend on many factors, such as size, use, and particularly cost.

Steel structures may be either (1) structural steel having members which are rolled at the steel mill to various shapes such as I, H, or circular sections; (2) steel plates in thin flat rectangular section with thicknesses up to one inch and width of up to six feet; (3) thin steel sheet with thicknesses up to $\frac{1}{8}$ in. and bent in the form of trough shaped corrugations; or (4) cables of high-strength wire. Many steel structures will use all of these in combination to achieve a satisfactory structural system. Several rolled steel sections are sketched in Fig. 6, and a section made from steel plate is shown in Fig. 7. The figure on the left is called a wide flange section; the official designation for this depth and weight is W12 X 27, as shown below the sketch. The first letter designates the type of section, in this case a "wide flange" section. The first number, the approximate depth; note that the precise depth is given at 11.96 in.; and the last number, the weight, is pounds per foot. The X is used just to separate the numbers. It is useful to remember that steel weighs 3.4 lb for each square inch a foot long. The

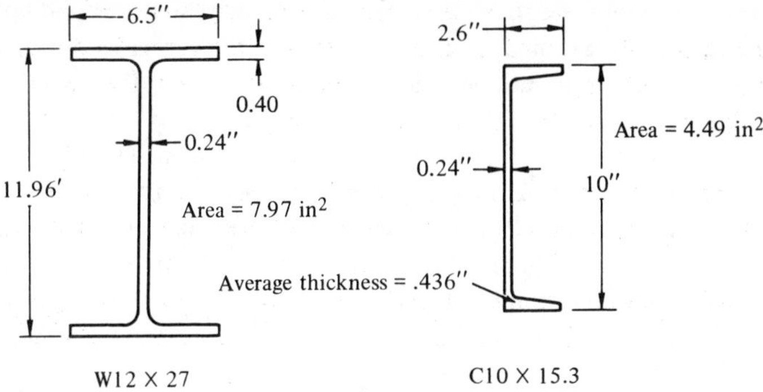

Fig. 6.

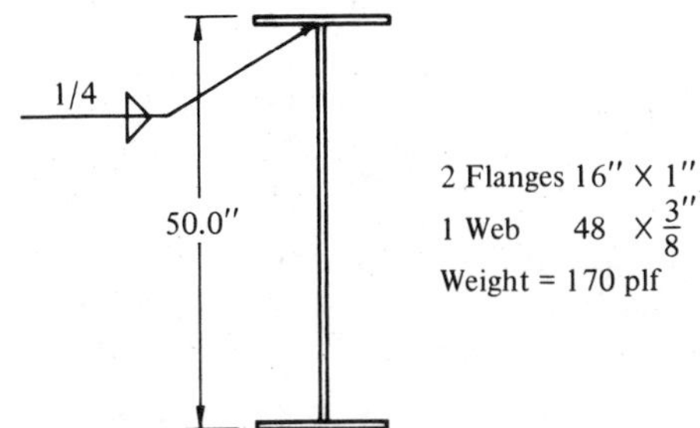

Fig. 7. Welded plate girder.

official designation only recently has been changed so that if you see this section on old plans it might be called 12WF27 instead of W12 X 27.

The other section shown is a 10-in. channel and has the designation C10 X 15.3. You will note that for both sections, the various rectangular sections are joined by small fillets, with additional curved areas to make the sections stronger at their intersections. The properties (area, dimensions, etc.) of steel sections are listed in tables furnished by the steel mills. Since the sections are now standardized, the best source of information is the *Manual of Steel Construction*, 7th ed., published by the American Institute of Steel Construction, Inc., 101 Park Avenue, New York, N.Y. 10017. This book also includes the rules for design embodied in: *Specification for the Design, Fabrication and Erection of Structural Steel for Buildings*, formulated by the American Institute of Steel Construction (AISC). The *Steel Manual* also includes many design tables and is a required reference for designing steel structures.

The joining of steel sections into elements that will create steel structures is performed by *steel fabricators* who have the plant, equipment, and knowledge to perform this work. The fabricator usually will furnish to the building contractor who will build the structure at a price for the fabrication and erection of the structural steel based on drawings and specifications prepared by an architect or engineer. This is a competitive field and often several fabricators will be asked for bids.

Structural steel sections come in stock lengths of 40 and 60 ft, but for larger projects, the fabricator will order the lengths direct from the steel mill. Steel sections are joined by rivets, bolts, or welds. Rivets are used less frequently than in the past years. A round bar of steel, usually $\frac{3}{4}$ in. in diameter is heated to a red glow and placed in holes punched or drilled between two steel plates, as shown in Fig. 8. A power hammer with a die to form the rounded head and a similar

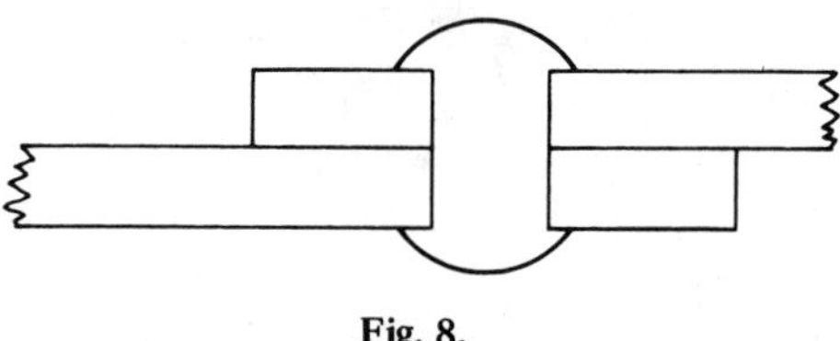

Fig. 8.

bucking tool on the opposite side are used to form the head. As the rivet cools it grips the two steel plates and the friction between the plates furnishes additional strength and rigidity.

Bolts for steel may be either machine bolts of mild steel or may be high-strength bolts with a much higher capacity obtained by tightening the bolts so that they grip the plates and the bolt is under a very high tension stress.

Several types of welds are sketched in Fig. 9. The fillet weld joins the end of one plate and the side of another. The butt weld joins the two ends, and for

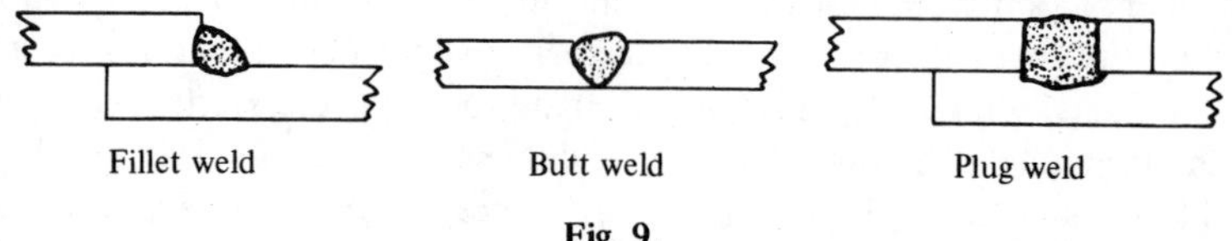

Fig. 9.

plates over $\frac{1}{4}$ in. in thickness the ends must be prepared to form a V or an X where the plates join. A plug weld is used to join two plates by welding through a previously prepared hole. Of these three welds, the butt is obviously the strongest, but is also the most expensive because of the time required for preparation.

Structural steel is available in several strengths. The type of steel most used has a yield strength of 36,000 psi and has the designation A36.

Sheet metal is used mostly in buildings and bridges for roof or floor surfaces in the form of corrugated sections, as sketched in Fig. 10. On floors it is covered

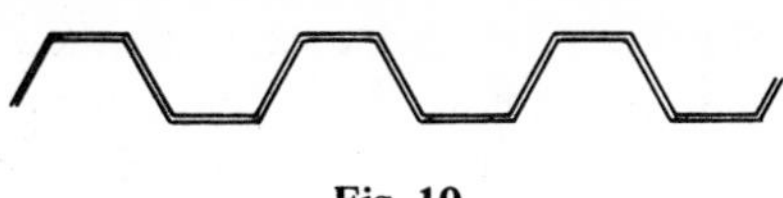

Fig. 10.

with concrete to provide the floor surface. A sheet-metal roof may span from 6 ft to 16 ft for the deeper sections, and floors may span up to 12 ft.

A *reinforced concrete* structure is essentially a concrete beam with steel bars used to take the tension forces which may be present. In this discussion the term "concrete" will be assumed to mean *portland cement concrete*. The term "concrete" may apply to many different types of materials which are mixtures of ingredients (for example there are asphaltic concretes) so it is important to properly define the term here. Portland cement is made by grinding limestone and other natural materials such as clay together and heating nearly to fusion. This material is then finely ground, and when mixed with water, sand (fine aggregate), and rock or gravel (coarse aggregate) in the proper proportions and allowed to cure for several days forms a very hard, rock-like substance called concrete.

Concrete is strong in compression (particles pushed together) but weak in tension, (particles pulled apart). The steel bars, therefore, supply the tension elements and must be placed where tension occurs. A portion of a reinforced concrete beam and column is sketched in Fig. 11. The steel bars appear as solid

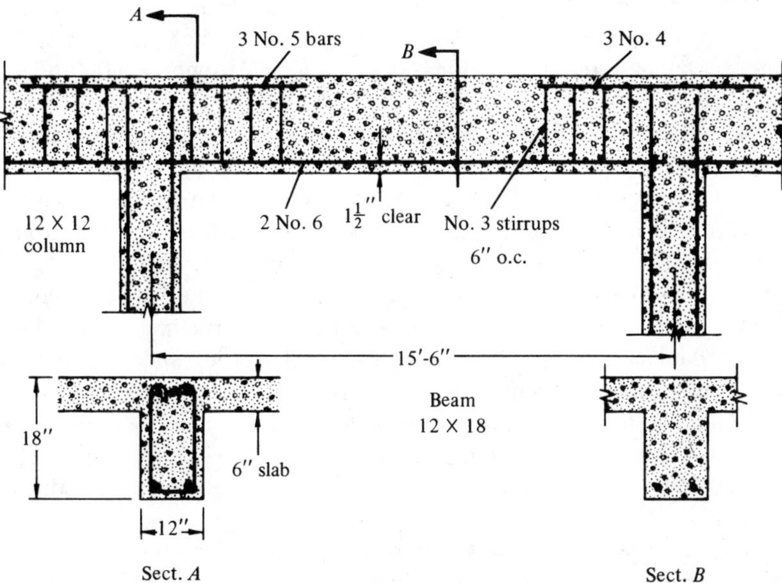

Fig. 11.

black lines and the width of the line is indicative of the relative size of the bar. The tension forces in the beam are usually at the top over the supports and at the bottom in the center of the beam. There are also tension forces requiring vertical bars near the supports. These bars are called stirrups and often come in a U shape. The size of the bars is indicated by a number representing the diameter in eighths of an inch. For example a No. 5 bar would have a diameter of approximately $\frac{5}{8}$ in. These are special steel bars used only for reinforcing of concrete and have regular bumps and ridges rolled into them to increase the bond of the steel to the concrete.

To build a concrete building it is necessary to construct wood or metal formwork which functions as a mould for obtaining the proper shape of the structure. The reinforcing is placed in these forms, and often has to be supported in mid-air on metal devices called chairs. The reinforcing must be securely fastened to keep it in the proper position when the concrete is placed. The concrete (sand, cement, and aggregate) is usually mixed at a separate location and hauled through the streets to the job site in special trucks with large drums which mix the ingredients together as they move toward the building site. On the job, the concrete is dumped into buckets and lifted by cranes to the forms or is placed into wheel barrows or mechanical buggies and moved over wood plank runways to the required location.

The concrete must be allowed to cure for several weeks, or until it has gained a specified strength, before the forms may be removed. It is important to keep the concrete wet during this period to improve its strength. The strength of concrete is measured by compression tests on 6-in.-diameter cylinders, 12 in. long, and the actual strength of the material is determined on this basis. For example, concrete may be specified as requiring a strength of 3,000 psi at 28 days.

An advantage of reinforced concrete as compared with structural steel is that it can be shaped in many different forms such as beams of variable depth; slabs (flat thin sections), walls (straight as well as curved), and shells (curved surfaces).

Sometimes it is expedient to cast small short concrete members away from the construction site in moulds that can be reused many times. This is called *precast concrete.* A disadvantage is that these members must be joined together by special methods so that they will form a satisfactory structure. In many cases, however, the end of the precast member can be placed on a wall for support and small buildings are often constructed in this way.

Rules for the design of reinforced concrete buildings are embodied in *Building Code Requirements for Reinforced Concrete, ACI 318–71,* formulated by the American Concrete Institute, P. O. Box 4754, Redford Station, Detroit, Michigan, 48219.

The second type of concrete structural element is called *prestressed concrete*, which has properties that are considerably different from those of the reinforced concrete elements just described. Prestressed concrete members are constructed

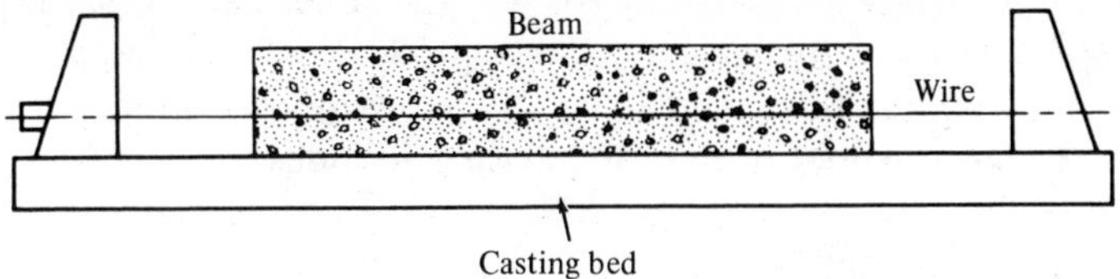

Fig. 12.

by two methods. In the first method, called *pretensioning*, steel wires or cables are stretched between concrete or steel uprights and concrete is then placed between, as shown in Fig. 12. Then the wires are cut and the elements of the member are pressed together. In this way tension stresses are eliminated when the member is placed in bending. In the second method, called *post-tensioning*, wires or cables are placed in a tube or sheath and placed in the formwork for the building. The concrete is then placed in the usual way. After the concrete has hardened, the wires are tightened by hydraulic jacks at the ends. This procedure is illustrated in Fig. 13. A demonstration of the principles of prestressed con-

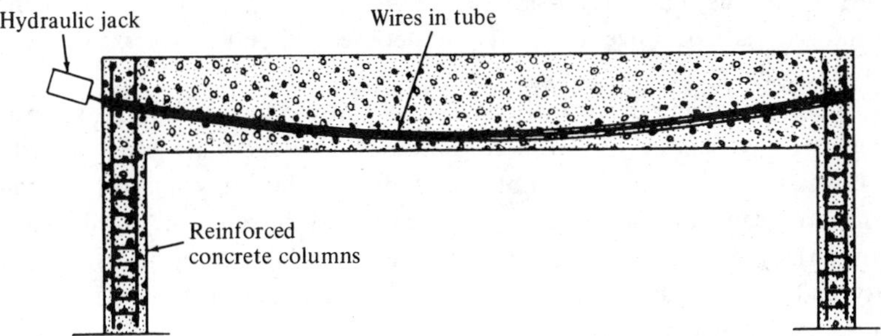

Fig. 13.

crete may be made by picking several books by the first and last books only, as shown in Fig. 14. As long as a force is exerted against the ends, the middle books will not fall out of the pile.

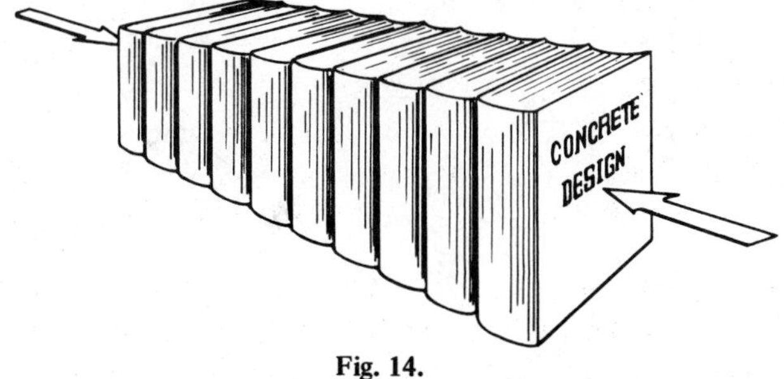

Fig. 14.

The advantage of precast, pretensioned, prestressed concrete is that it can be constructed in a manufacturing plant under controlled conditions with minimum materials. Post-tensioned members may be necessary in reinforced concrete structures for especially long spans or to reduce deflections.

Types of Structures for Buildings

To design a building, one must first select a structural system that will serve as the skeleton on which the walls, floors, and roofs are placed. There are many possibilities and only a few typical ones can be mentioned here. Structures most suitable for steel buildings will be described first, and to do this we start with the basic elements of a steel structure and then turn to the systems made from a combination of these elements.

Beams are members, usually horizontal, carrying vertical loads between end supports. Some typical beam sections were described in the previous section.

Columns are usually vertical members that carry loads that come to them mostly from beams. Column sections in steel are usually H-shaped, but round pipe and rectangular box columns are also used.

Girders are beams that carry loads from other beams, and so are heavier than beams and generally of longer span. It is sometimes difficult to say exactly when a beam is a girder.

Plate girders are made up of sections of steel plate and are usually deeper than the deepest rolled section (36 in.). A section of a welded plate girder was illustrated in the preceding section. In buildings, plate girders may support columns.

Trusses are beams or girders made from light steel members arranged in triangular patterns. A truss will usually weigh less than a steel beam of the same capacity and span because it can be deeper. Line diagrams of typical trusses are shown in Fig. 15. The pattern of the diagonals determines the name of the truss, usually derived from the name of the inventor or someone who first used it extensively.

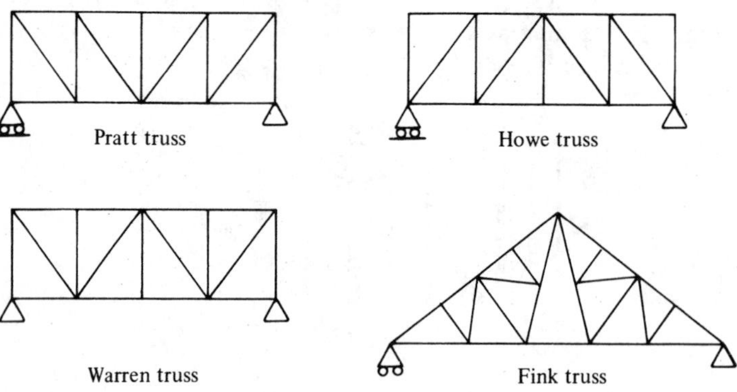

Fig. 15.

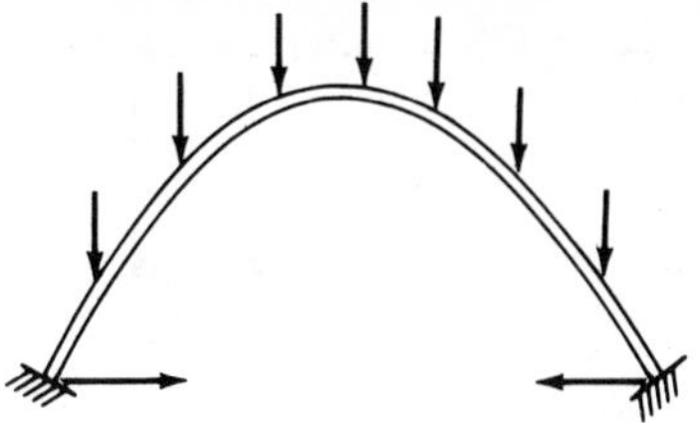

Fig. 16.

Arches are curved structural members that carry loads mostly in compression and are characterized by horizontal forces at the supports, as shown in Fig. 16. An arch, of course, must be supported in the other direction and this is usually achieved by placing several arches side by side with bracing between.

Cable-supported structures (Fig. 17) are composed mostly of cables with additional beams to make them stiffer or to spread the loads over the cable. The cable can take only direct tension forces (pull).

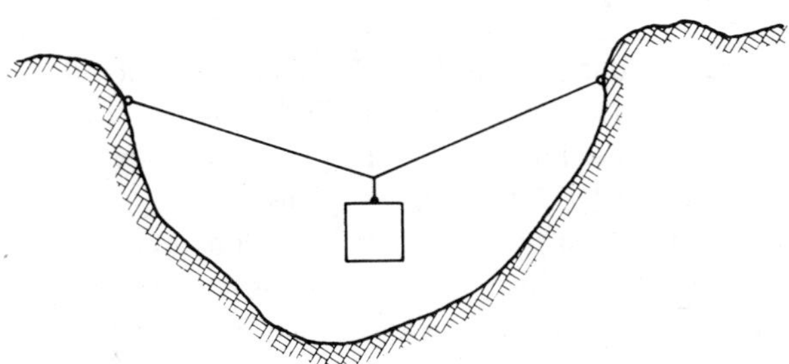

Fig. 17.

Tanks are cylindrical or spherical sheet or plate structures, generally used to hold liquids. The ends may be flat, conical or spherical.

The elements just described are used in combinations to achieve structures that will best serve the particular function required in the building. Several examples will be given to illustrate some of the combinations. The first is the section of the steel office building shown in Fig. 18. Trusses span the entire width of the building at the top. There are three rows of columns but the middle row rests on a plate girder above the first floor. Note that the columns are continuous from the first floor to the top. However, there are field splices just above every other floor. The framing in the opposite direction (perpendicular to the paper) is shown in section. A concrete slab placed on a sheet-metal deck, not shown in these sketches, would be used between the girders. A cable carries the canopy over the entrance. This structure rests on concrete walls on the outside of the building and on concrete pad footings for the inside columns.

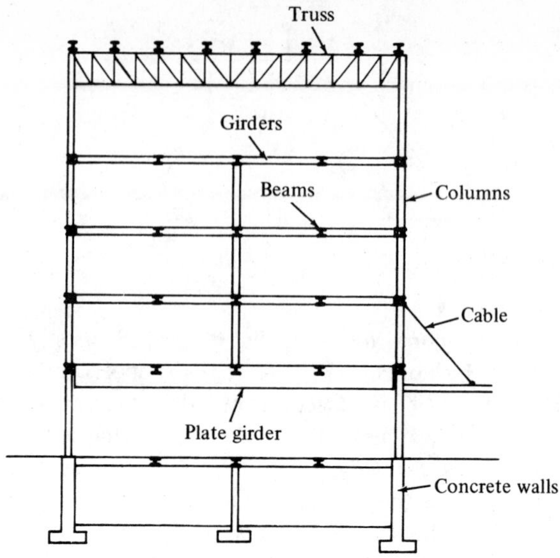

Fig. 18.

The second structure illustrated is the rigid frame sketched in Fig. 19. It is like an arch except the members are not curved. The columns are wider at the top to carry larger stresses at the knee where the joints are made rigid to carry both sidewise wind and vertical forces. The beams used to support the roof surface in a roof structure are called *purlins.* The roof material in this case is a wood-plank deck spanning the purlins. The walls are sheet-metal siding carried from frame to frame by channel-section beams called *girts*. The structure rests on concrete pads called *footings*. The frame exerts a horizontal force at the base which in this case is resisted by a steel tie bar buried in the concrete floor.

Reinforced concrete structures also use beams and girders but trusses are not as practicable. Unique to concrete is the *slab*, which is a wide, thin, flat member

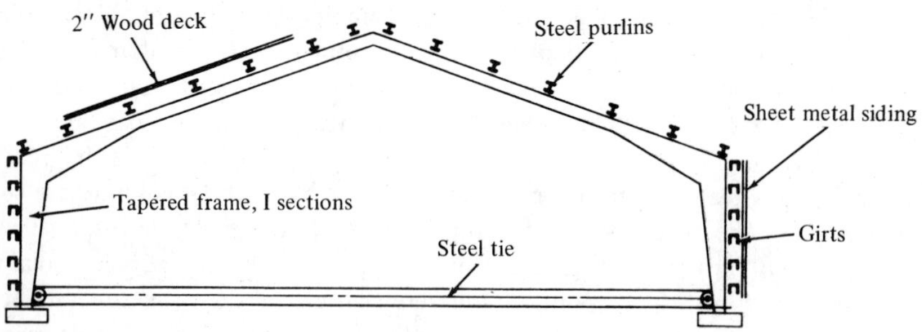

Fig. 19.

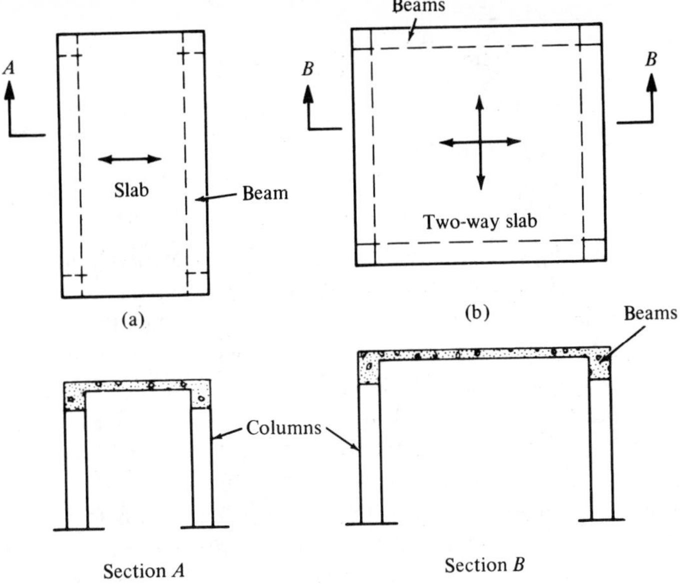

Fig. 20.

that spans from beam to beam. A *one-way* slab is a long rectangle and spans in one direction, as indicated by the arrows sketched in Fig. 20 (a). A *two-way* slab is more nearly square and is supported on four sides, Fig. 20 (b). The direction of the arrows indicates the span of the slabs. These structures, both beams and slabs, are usually continuous over several supports and the beams or slabs may be carried by walls or columns.

Another slab system is the *flat slab* shown in Fig. 21. In this case the slab is supported by columns only, with beams sometimes used around the edges. For additional strength, the slab may be thicker at the junction of the slab and the column (this is called a *drop*) and the column may be flared at the top to reduce the stresses and increase the strength (this is called a *column cap*). These

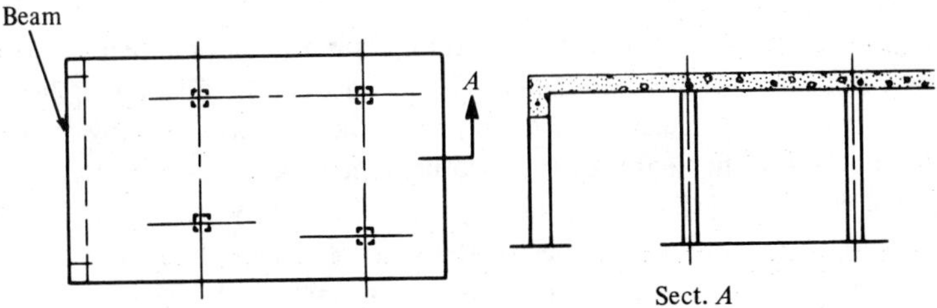

Fig. 21.

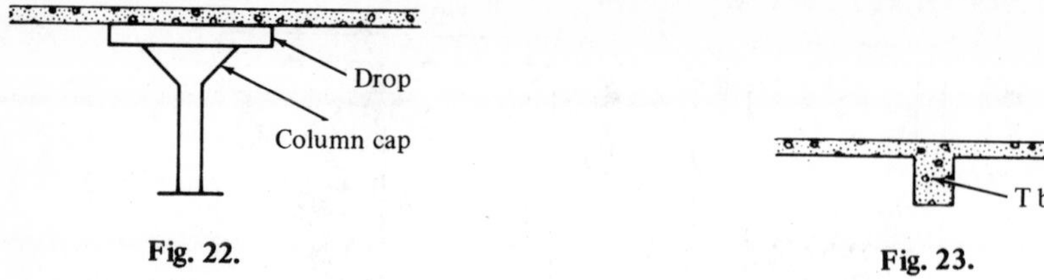

Fig. 22.

Fig. 23.

are illustrated in Fig. 22. If the drop and cap are not used, the structure is sometimes called a *flat plate*.

If concrete beams are used in combination with slabs, then the concrete in the slab and in the beam is placed at the same time and the slab contributes to the area of the top of the beam and makes it stronger and stiffer. This combination is called a T beam and is sketched in Fig. 23. When T sections are used close together, say 2 to 3 ft, the system is called *concrete joists* and is usually formed with removable sheet-metal pans that may be reused many times. These are placed on wood forms or posts, and the resulting structure is shown for typical dimensions in Fig. 24. Note that the sides of the pans are slanted so that they

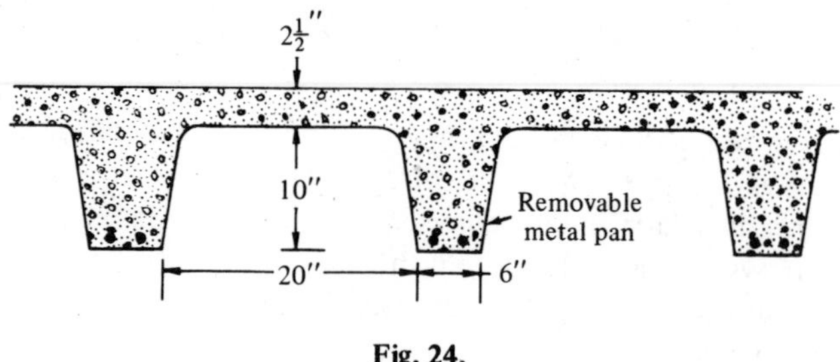

Fig. 24.

may be more easily removed. They must be coated with a special oil so the form does not stick to the concrete. If the pans are square, the structure is called a *waffle slab*, for obvious reasons, and is used in structural systems similar to two-way and flat slabs. However, instead of the drops used for an ordinary flat slab, pans are omitted near the columns, so that the entire ceiling of the waffle slab is level.

Concrete walls are an important structural element in both steel and concrete buildings, not only for the basement and foundation but for carrying horizontal wind or earthquake forces. Used in this way, they become shear walls and may be erected either in the center of the building (the core) or as outside walls.

A curved slab is called a shell and may be either singly curved (a cylinder) or doubly curved; e.g., a dome similar to the shape of an eggshell. There are many different shapes and forms of shell structures, each with different visual and structural characteristics, but all are characterized by thin sections (2–5 in.) and

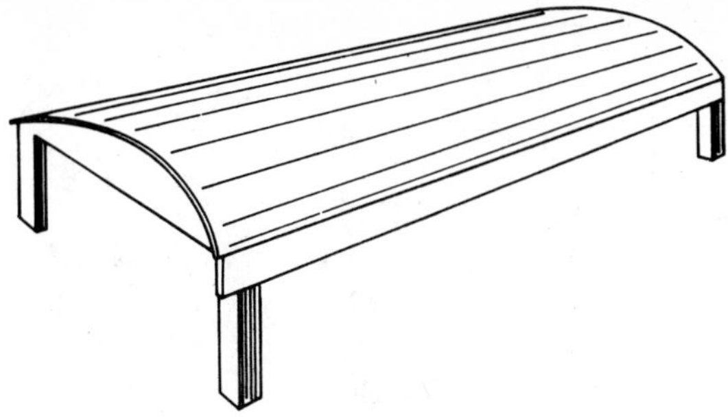

Fig. 25.

a supporting system that will make the structure work satisfactorily with these thin sections. A typical barrel shell is sketched in Fig. 25.

An intermediate type of shell using flat sections is called a *folded plate*, and a single element is sketched in Fig. 26. Loads are carried by slab action from fold to fold and by plate action to the end supports.

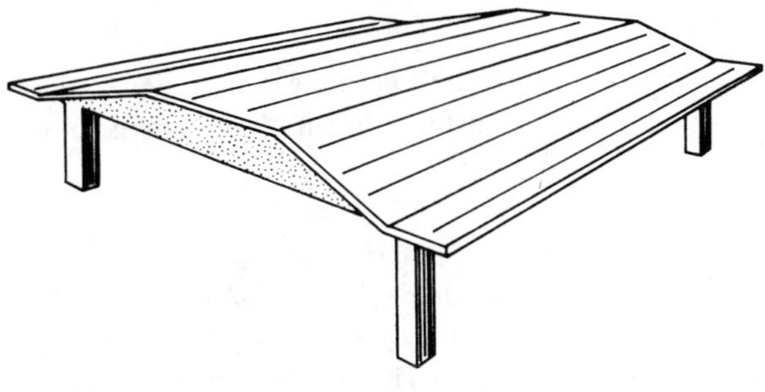

Fig. 26.

We come now to structural systems for buildings that could be made up of some of the concrete structural elements just described. A concrete building similar to the steel building already discussed is shown in Fig. 27. The top floor is clear without columns across the building and is covered with a barrel shell, shown in section to the right. This substitutes for the steel trusses and would be about the same depth but would not require an additional ceiling as would the trusses. Four types of floor systems are shown. The top floor is a one-way-slab-and-beam system with additional girders at the column lines to pick up the beams. The next down is a two-way slab with beams on four sides, so the heavier girders are not required. The waffle slab (next floor down) has square pans, so it acts like a two-way slab, but the beams to support it are made the same depth as the waffle slab. The second floor has a one-way concrete-joist system. The central column at this floor is supported by a concrete rigid frame

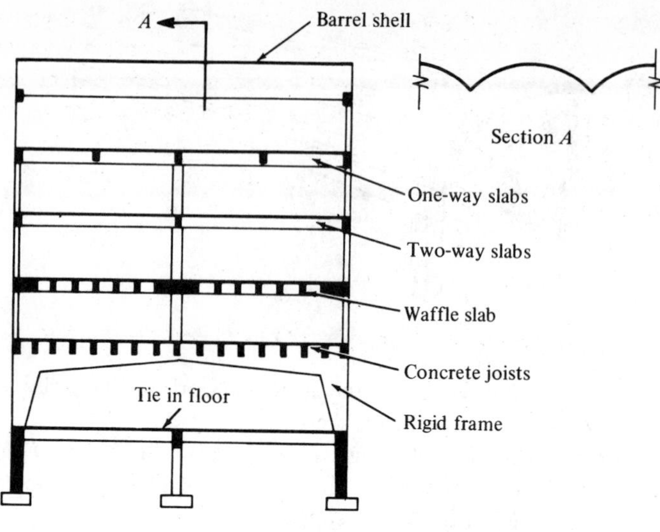

Fig. 27.

with tapered members, instead of the steel plate girders used in the steel building. This frame will cause a horizontal force in the floor system below, so a steel tie will be required.

The second structure to illustrate the types of structures that can be designed with concrete uses precast arches, as shown in Fig. 28. The roof is a series of precast slabs and beams, so the entire roof does not need to be formed in place. Therefore the weight of these elements must be small enough that they can be picked up and erected with a crane. The arch has hinges at the supports (called the *springing*) and at the top (called the *crown*). These hinges could be made of welded structural steel fastened to the concrete before it is cast. The arches must be supported by concrete abutments, and in this case the thrusts are carried by the earth rather than by a steel tie as in the case of the steel arch.

The basic structural elements for timber structures are boards, planks and timbers. These elements may be made by gluing wood together; for example,

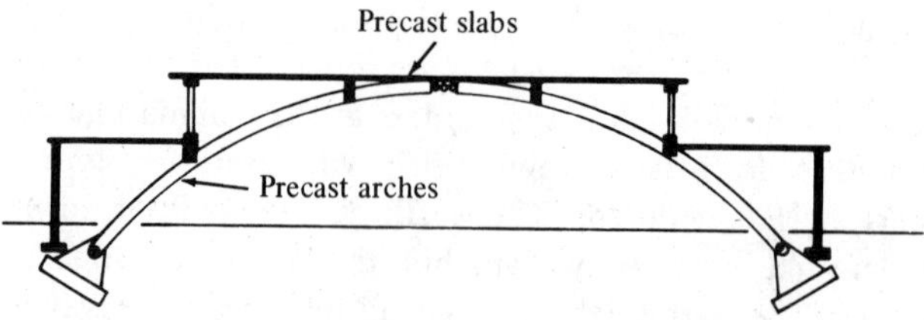

Fig. 28.

sheets of plywood or glued and laminated beams. Plywood is made from thin layers of wood shaved from large logs and glued under pressure with waterproof glue. The standard size is a sheet 4 ft by 8 ft, and in thicknesses from $\frac{1}{4}$ to $1\frac{1}{8}$ in. Timber planks are nominally 1 in. and 2 in. thick but actual thicknesses of timber are $\frac{1}{2}$ in. less; so a 2 X 4 is actually more like $1\frac{1}{2}$ by $3\frac{1}{2}$. Planks are used for floors and the most common type of floor is the wood-joist floor (as illustrated in Fig. 29) used in most houses. The flooring is $\frac{5}{8}$-in. plywood resting on 2 X 10

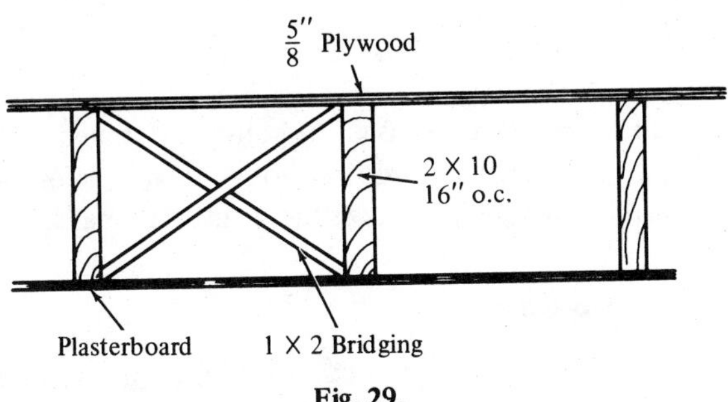

Fig. 29.

joists at 16-in. centers. The ceiling is plasterboard. To prevent the joists from moving with respect to each other, wood strips 1 in. by 4 in. are placed diagonally at the center of the spans.

A floor with heavier planks and wood beams is shown in Fig. 30. The floor is made with 2-in. nominal (actual $1\frac{1}{2}$ in.) tongue-and-groove planks spanning

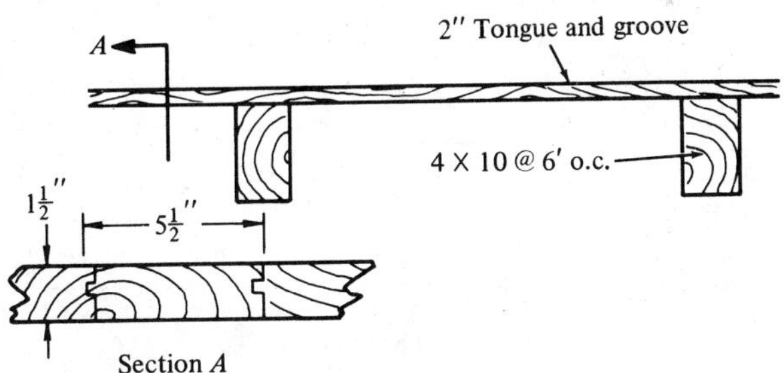

Fig. 30.

6 ft and supported by 4 in. by 10 in. ($3\frac{1}{2}$ X $9\frac{1}{2}$) timber beams. Large timber beams of almost any length can be made by gluing planks together under pressure with waterproof glue, as sketched in Fig. 31. Special shapes can be curved for arches or rigid frames. Timber trusses are made from planks or beams and are joined by bolts, nails, or special timber connectors in the shape of rings

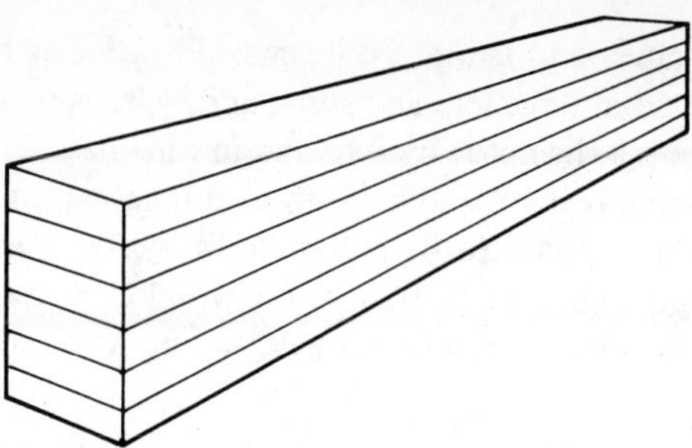

Fig. 31.

set between the planks. Small trusses used for houses are joined with plywood or metal plates called *gussets*. The best and time-honored way to join wood is to lay one piece on top of the other, if this is possible, and to use nails.

Types of Structures for Bridges

Bridges use the same structural elements as do buildings, including the beam, the girder, the arch, and the cable-supported structure, but in different combinations. The spans are generally greater, so the structural members may be very large. A steel-girder bridge is illustrated in Fig. 32. The typical section shows a

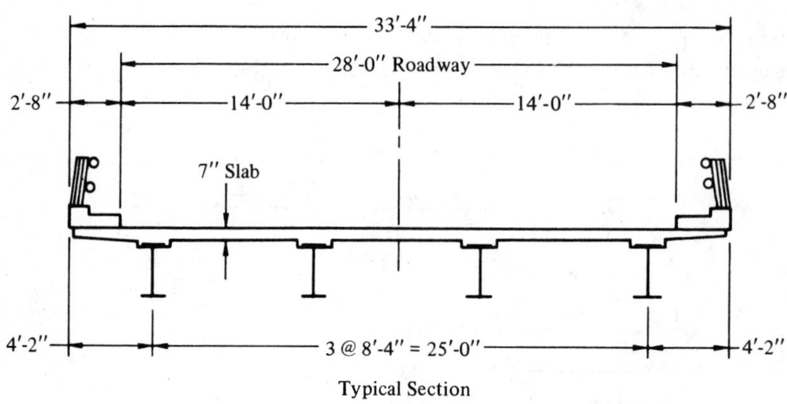

Typical Section

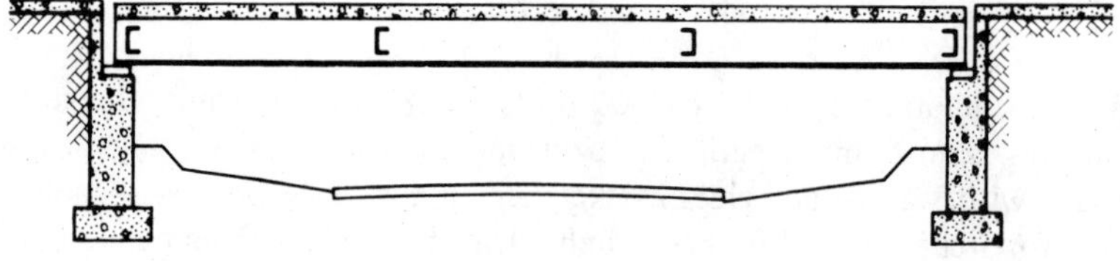

Fig. 32.

roadway width of 28 ft with a concrete slab for the deck. The ends of the slab support the concrete curbs and hand railings. Rolled-steel girders support the slabs and a longitudinal section indicates that these girders are tied together by channel sections at four points on the span. The girders rest on concrete walls that retain the earth. The slab is constructed to participate with the girders in bending to provide composite action. Special connectors called studs are welded to the top of the girders to make the connections to the slab. A concrete bridge would be built in a similar way, but prestressed concrete beams would be substituted for the steel girders.

Bridges are sometimes built with box girders of steel or concrete rather than the I sections described above. A concrete bridge of this type is drawn in Fig. 33.

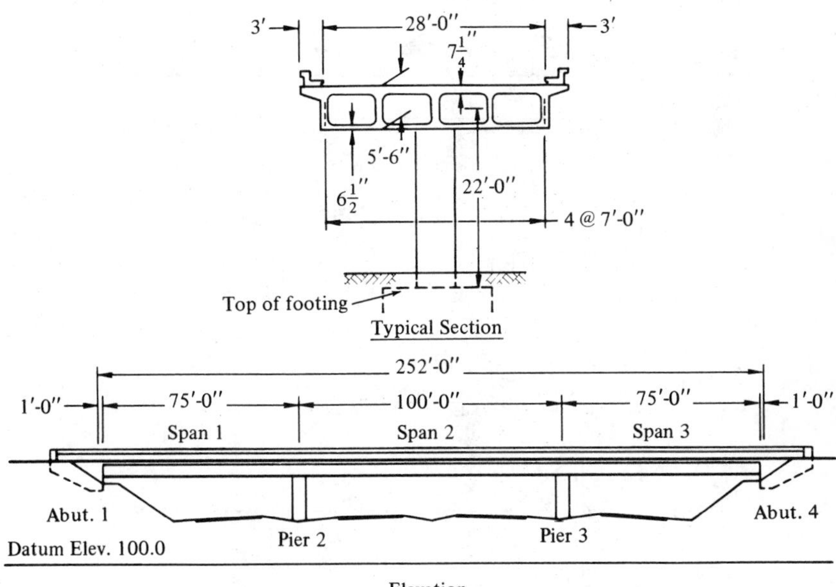

Fig. 33.

The middle piers are round concrete posts, providing minimum obstruction to the sight of the motorists passing underneath. The spans of these girders may be increased if the structures are made deeper at the interior supports or if prestressing is used.

A number of possible profiles of long-span bridges are shown in Fig. 34, and a name has been given to each type. Both *deck* and *through* bridges are shown. These terms refer to the location of the roadway with respect to the remainder of the structure. The *deck arch*, for example, has a roadway above the arch, but in the *through truss* the roadway passes between the trusses. Each type of bridge

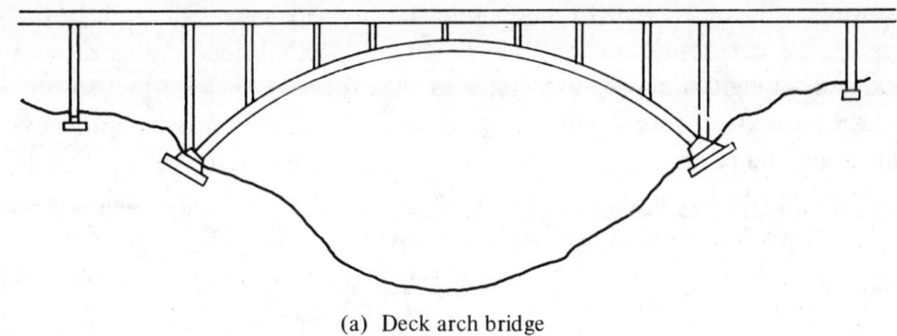

(a) Deck arch bridge

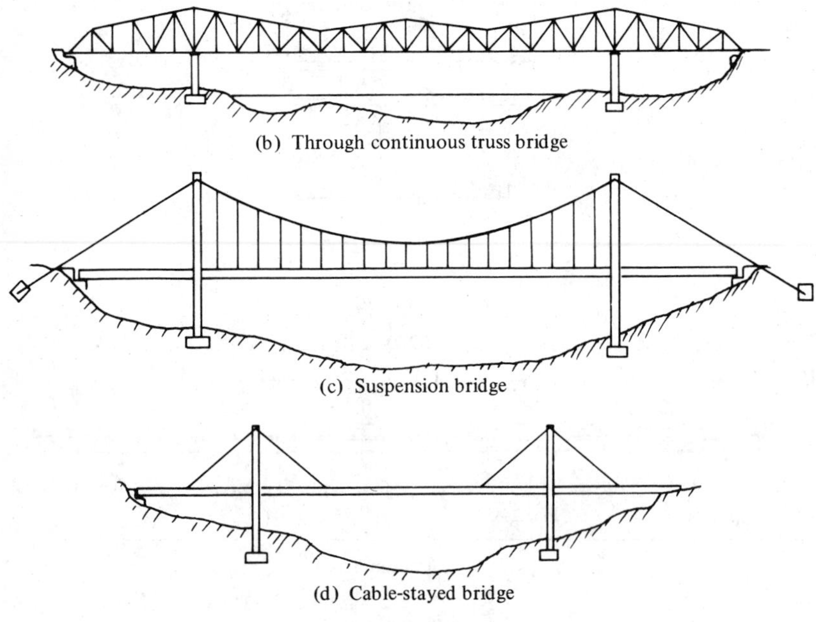

(b) Through continuous truss bridge

(c) Suspension bridge

(d) Cable-stayed bridge

Fig. 34.

is particularly suitable for a special foundation condition, length of span, and horizontal and vertical clearance, as well as the erection requirements of the bridge.

THE TOOLS OF STRUCTURAL ENGINEERING

In this section, methods of evaluating the strength and capacity of structural elements (individual members of a structure) and structural systems (the entire structure) will be presented. The general term used for these studies is called *analysis*. The methods of choosing the size of members and systems is called *design*, and includes analysis as a part of its discipline.

Before we can determine the strength of structures we must be able to work with forces either at rest (statics) or in motion (dynamics). The determination of internal forces and internal deformations is called either *mechanics of materials* or *strength of materials.* A mastery of these forces puts us in a position to study the forces in actual structural systems; first, those where the internal forces can be determined from the principles of statics only (statically determinate structures), or second, systems that depend on the elastic properties of parts of the systems (statically indeterminate structures). Another important class of problems is concerned with elements such as columns, where the deflections of the member will increase the internal forces so that, if this increase in internal forces is allowed to continue, the member will buckle out of the way of the forces. The final part of this section will discuss some special problems in the strength of structural systems and methods for their solution.

The importance of these fundamentals cannot be overemphasized. When problems in analysis or design occur, the designer must fall back on the basic understanding of fundamentals to solve the problem.

Statics

Forces result from gravity, the effect of the earth's pull on objects; or from inertia, the tendency for an object to remain in motion if already in motion or to remain at rest if already at rest.

A force has three independent properties: (1) *direction*, (2) *magnitude*, and (3) *point of application.* For example, in Fig. 35, the world is represented by a two-dimensional coordinate system x, y. The force A, represented by an upward arrow has a direction to the right and is inclined at 60° to the x axis. The magnitude of A is 30 lb and the point of application is at $x = 10$ ft from the origin or $y = -10 \tan 60° = -10\,(1.732) = -17.32$ ft. Note that the force, with the properties described, could act anywhere along its line of action. The magnitude of a force may be represented graphically by the length of the arrow.

The *moment* of a force about a point is the magnitude times the perpendicular distance to the point. For example, the moment of the force shown in Fig. 35 about point B is the distance $a = (10 + 17.32) \sin 30° = (27.32)\,(0.5) = 13.66$ times the magnitude of the force; now $A = 30$ lb, so that the moment $M_B = Aa = 30(13.66) = 409.8$ ft-lb.

Forces may be added to each other either by (1) vector addition (graphically) or (2) by addition of x and y components. (Components of forces will be explained below.) By the first method, the force A shown in Fig. 36 is 60° to the x axis and force B lies parallel to the x axis. The vector sum of these forces is the diagonal force C resulting from the completion of the *force parallelogram*

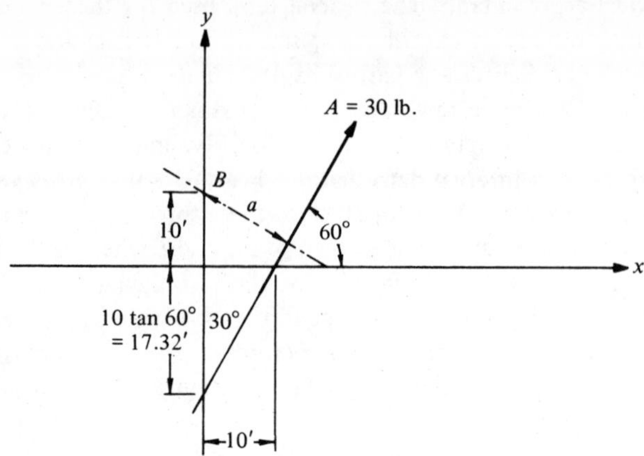

Fig. 35.

with side $\overline{13}$ parallel to side B and side $\overline{23}$ parallel to side A. In most cases, vector addition is easiest to accomplish by graphical procedures, but in this case, since one side is parallel to the x axis, the magnitude and direction can be calculated without too much effort by trigonometry:

$$C = \sqrt{(B + A \cos 60°)^2 + (A \sin 60°)^2} = \sqrt{[50 + 30(0.5)]^2 + [30(0.866)]^2}$$

$$= 70.0 \text{ lb}$$

A force may be *divided* or *resolved into components* by the reverse procedure from addition. For example, the force C in Fig. 36 may be resolved into components A and B. If C is resolved into the x and y directions the components are called x and y components. In the case of the force C shown in Fig. 36, these components would be

$$C_x = 50 + 30 \cos 60° = 50 + 30(0.5) = 65 \text{ lb}$$

$$C_y = 30 \sin 60° = 30(0.866) = 25.98 \text{ lb}$$

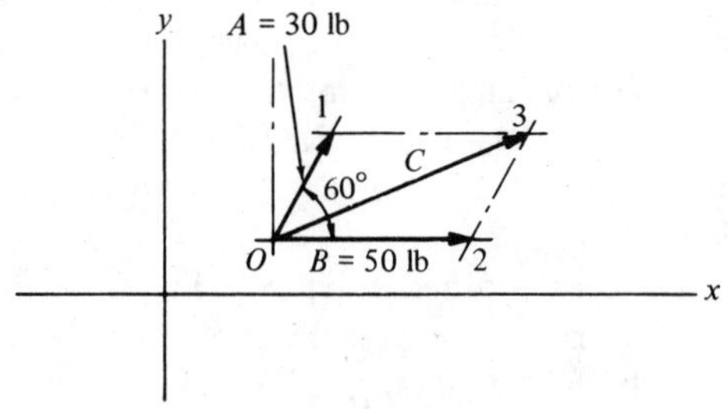

Fig. 36.

The second method of adding forces is by adding their x and y components and then finding the vector sum of the components. In this case the total force is easy to calculate because it is the square root of the sum of the squares of the x and y components.

For forces in space the procedure for finding the components is a little more complicated. The *direction cosines*, the projection of a unit length of the force on the x, y, and z axes, must be determined, and the component is the force times the direction cosine for the axis.

Equilibrium A single force, if applied to an object, would cause that object to move. So, if an object is to be considered at rest, there must be another force acting on the object equal and opposite to the first force to prevent the object from moving. The study of such forces is called *equilibrium.* If several forces are applied to the object in Fig. 37, the vector sum of the forces, $A + B + C$, must be resisted by a force D equal and opposite to $A + B + C$.

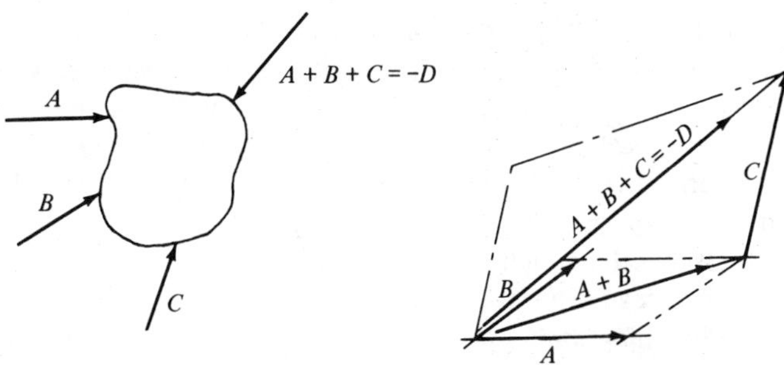

Fig. 37.

If x and y components of forces on an object are used instead of those in a random direction, then components can be added directly. In Fig. 38, if the forces $A + B + C$ are equal and opposite to force D, then $A_x + B_x + C_x = D_x$ and $A_y + B_y + C_y = D_y$, assuming that these forces do not turn the object. To

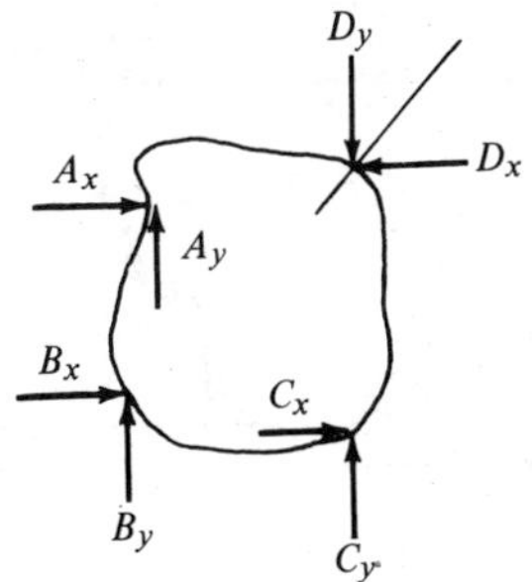

Fig. 38.

satisfy this condition, the moments of the components of all of the forces on the object must be zero.

These relations between forces and moments may be summarized by the *equations of equilibrium:*

$$\sum F_x = 0, \quad \sum F_y = 0, \quad \sum M = 0$$

The symbol $\sum$ (sigma) indicates a summation. F_x and F_y indicate forces in the x and y directions, and M indicates the moments about any point, on or off the object.

To show how these equations may be applied to solve for the equilibrium of an object, the bar shown in Fig. 39a will be analyzed. It is supported at both ends and the manner of support is shown to be knife-edged. At the right the support is fixed, allowing forces in either the horizontal or vertical direction but no moment. At the left the knife edge is on rollers, so there can be no horizontal force or moment. The forces on the structure are a vertical load of 10 lb, 15 in. from the left end, and a horizontal force at the left end of 5 lb. The forces at A and B are called *reactions. What are the magnitudes of the reactions at A and B?*

The first step is to draw a *free-body diagram* of all the forces on the structure. This device will be used constantly throughout this chapter, and is shown in Fig. 39b. Note that there are three unknown forces: V_A, V_B, and H_B. There are three equations of equilibrium, so the problem can be solved by the equations. We will assume the directions of the unknown forces shown in the sketch. If the signs of the forces are negative, then the wrong direction has been assumed. This method of handling directions of unknown forces will be used throughout the text. Positive directions are to be *down* and to the *right.* Apply the equations of equilibrium:

$$\sum F_x = 0; \quad 5 - H_B = 0$$

$$\sum F_y = 0; \quad 10 - V_A - V_B = 0$$

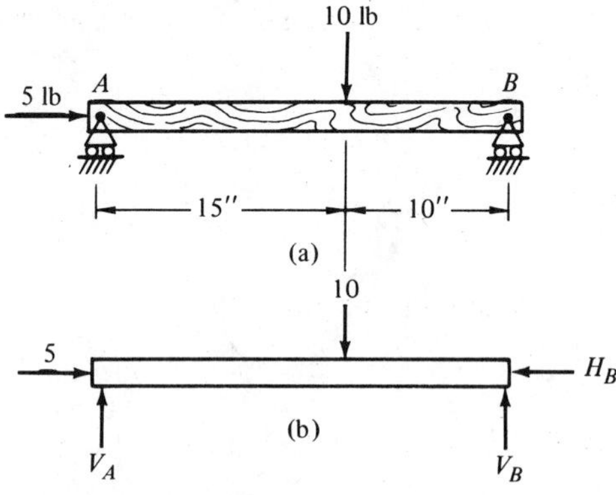

Fig. 39.

For the third equation we may take moments about any point. However, we can locate this point so that one or more of the components goes through the point and its moment is zero. If we choose point A, then V_A and H_B will be eliminated from the moment equation. Assume that clockwise moments are positive:

$$\sum M_A = 0; \qquad 10(15) - 25V_B = 0$$

The solution of the three equations proceeds as follows:

$$\sum F_x = 0; \qquad H_B = 5 \text{ lb}$$

(the correct direction has been assumed) and

$$\sum M_A = 0; \qquad V_B = \frac{10(15)}{25} = 6 \text{ lb}$$

Use this value in the last equation:

$$\sum F_y = 0; \qquad 10 - V_A - 6 = 0; \qquad V_A = 4 \text{ lb}$$

The last value also has a positive sign, so the correct direction has been assumed.

Note that there are three reactions and they correspond to the three equations of equilibrium. If there are more than three reactions, then the structure may be either statically indeterminate, or some other condition may be present which either gives a relationship between reactions, such as a support on rollers on a sloping plane, or stipulates a condition such as a hinge in the middle of a beam.

In the problem just studied, the direction of the reaction at A was vertical, as specified by the rollers. In Fig. 40a the direction of the right reaction (point B) has been specified as slanting upward, to the left so that there are two components, H_B and V_B, as indicated by the free-body diagram in Fig. 40b. This picture of the forces would seem to indicate that there are four forces with only three equations of equilibrium. However, there is a relationship between H_B and V_B because they are components of the force B. This relationship is: $3H_B = 4V_B$ and $B = 5V_B/3 = 5H_B/4$.

The equations of equilibrium applied to the beam in Fig. 40 are

$$\sum F_x = 0; \qquad H_A + 7 - H_B = 0$$

$$\sum F_y = 0; \qquad -V_A + 24 - V_B = 0$$

and, using the point A as the center of moments,

$$\sum M_A = 0; \qquad 24(12) - V_B(30) = 0$$

Solve the last equation first:

$$V_B = 24(12)/30 = 9.6 \text{ lb}$$

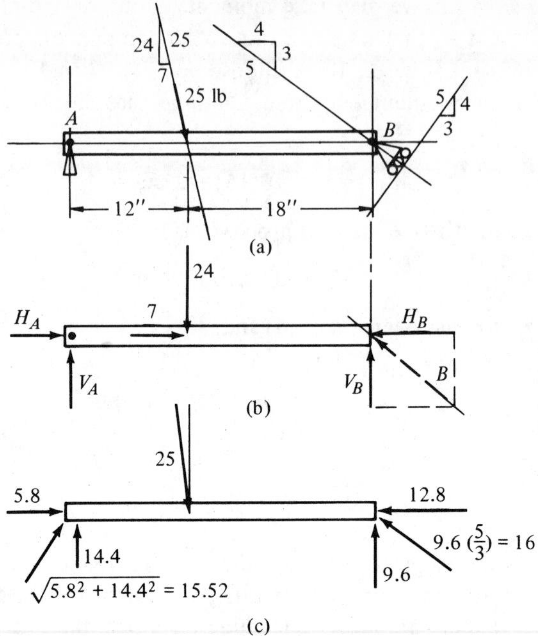

Fig. 40.

The value is positive, so the correct direction has been assumed for V_B. Put V_B in the second equation:

$$-V_A + 24 - 9.6 = 0, \qquad V_A = 14.4 \text{ lb}$$

Finally, for the first equation,

$$H_B = 4V_B/3 = 9.6(4)/3 = 12.8 \text{ lb}$$

$$H_A + 7 - 12.8 = 0, \qquad H_A = 5.8 \text{ lb}$$

The positive sign indicates that we have guessed the right direction for H_A and that it is acting to the right.

The forces on the structure are shown in Fig. 40c. The components H_B and V_B are combined into the single force B, and those at A are combined into a single force A by taking the square root of the sum of the squares of the components.

There is a hinge in the structure shown in Fig. 41a. The condition that the moments to either side of the hinge must be zero provides an additional equation, so that the structure can be solved using the equations of equilibrium, even though there are four reactions as shown in the free-body diagram Fig. 41b. The beam is built into the wall at the left so that one of the reactions is a moment,

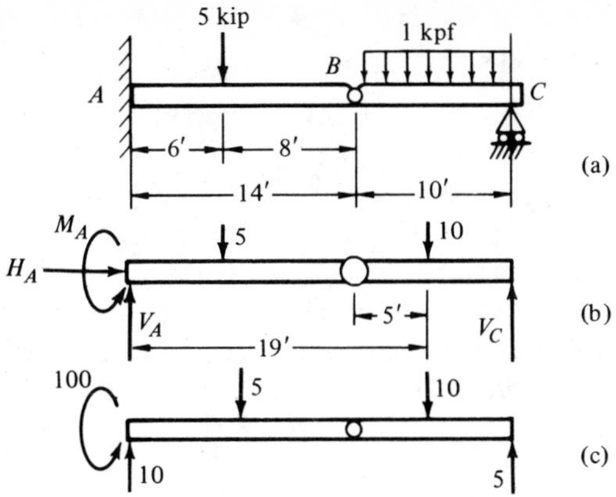

Fig. 41.

M_A. The load on the right side of the beam is now a uniform load of 1 kip per foot, so the total load is 10 kip (often written 10 k) applied at the center of the uniform load, and is shown on the free-body diagram as a concentrated load (1 kip = 1000 lb).

Apply the equations of equilibrium starting with $\sum F_x = 0$. It is immediately evident that $H_A = 0$ because there are no other forces on the beam except the horizontal reaction at A. There are three more equations to be applied, and the ease with which the problem is solved depends on the order in which these are selected. The best procedure is to select a direction to take forces or a point to take moments that will create an equation having only one unknown. In this case it seems expedient to take the condition that the moments at the hinge, point B, must be zero to the right of the point:

$$\sum M_B = 0; \quad 5(10) - 10V_C = 0; \quad V_C = 5 \text{ kip}$$

This value is positive, so the correct direction has been assumed. Next, use the relation that the moments about A must be zero:

$$\sum M_A = 0; \quad -M_A + 5(6) + 10(19) - 24V_C = 0;$$

$$-M_A + 30 + 190 - 5(24) = 0; \quad M_A = 100 \text{ ft-kip}$$

Again this value is positive, so the arrow on the free-body diagram is correct. The left reaction can be determined from the equation $\sum F_y = 0$:

$$\sum F_y = 0; \quad -V_A + 5 + 10 - V_C = 0; \quad V_C = 5 \text{ kip}$$

$$V_A = 5 + 10 - 5 = 10 \text{ kip}$$

These calculations can be checked by taking moments about some other point than A or B; for example, point C:

$$\sum M_C = 0; \quad -M_A + 24V_A - 5(18) - 10(5) = 0;$$

$$-100 + 10(24) - 90 - 50 = 0 \qquad \text{Check}$$

This check calculation should be made as often as possible. The reactions on the structure are shown in a free-body diagram, Fig. 41c.

It is not always possible to solve for reactions without using simultaneous equations, and the following example illustrates such a case. The structure shown in Fig. 42 is a three-hinged arch. There are four reactions, as shown by the sketch.

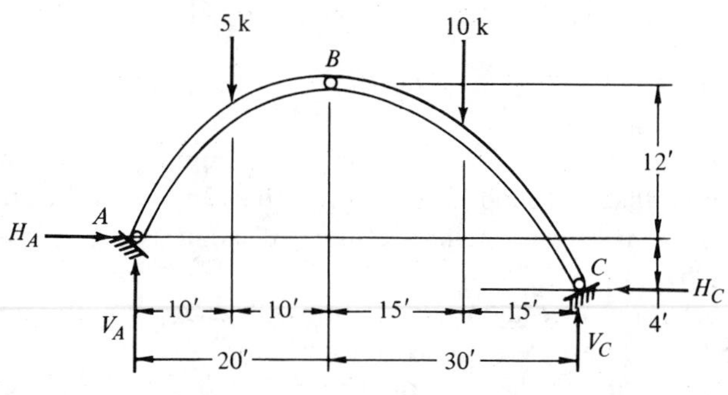

Fig. 42.

An additional condition is that at the crown, the top of the arch, there is a hinge, and the moments to either side of this point must be zero. The supports are at different levels, so it is not convenient to eliminate some of the forces to reduce the moment equation to one unknown as we did in previous examples. There are single loads at the centers of each half of the arch. The four equations that will be used are $\sum F_x = 0$, $\sum F_y = 0$, $\sum M_A = 0$, and $\sum M_B = 0$, for loads to the right of point B:

$$\sum F_x = 0; \quad -V_A + 5 + 10 - V_C = 0 \tag{1}$$

$$\sum F_y = 0; \quad H_A - H_C = 0, \quad H_A = H_C \tag{2}$$

$$\sum M_A = 0; \quad 5(10) + 10(35) + 4H_C - 50V_C = 0 \tag{3}$$

$$\sum M_B = 0; \quad 10(15) + 16H_C - 30V_C = 0 \tag{4}$$

These four equations must be solved simultaneously, and the procedure will be shown here in detail.

Rewrite equations (3) and (4):

$$-50V_C + 4H_C = -50 - 350 = -400 \qquad (3)$$

$$-30V_C + 16H_C = -150 \qquad (4)$$

Multiply (4) by $-50/30$:

$$50V_C - 26.67H_C = 250$$

Add to (3):

$$4H_C - 26.67H_C = -150; \qquad H_C = 6.62$$

Use this value of H_C in Equation 4:

$$-30V_C + 16(6.62) = -150; \qquad V_C = 8.53$$

Check these values by equation (3):

$$-50(8.53) + 4(6.62) = -400: \text{ Check}$$

Finally, from equations (1) and (2):

$$V_A = 15 - 8.53 = 6.47; \qquad H_A = H_C = 6.62$$

All values are positive, so the directions of forces are those assumed in Fig. 42.

Space Structures The two-dimensional structures just described used three equations of equilibrium. For structures in space with three dimensions x, y, and z there are six equations of equilibrium:

Force equations $\quad \sum F_x = 0, \quad \sum F_y = 0, \quad \sum F_z = 0$

Moment equations $\quad \sum M_x = 0, \quad \sum M_y = 0, \quad \sum M_z = 0$

The moments now are taken about *axes* parallel to x, y, or z, rather than about points.

For an example of a space structure let us study the structure shown in Fig. 43, which is a box supported on hinged struts which define the direction of forces at the reaction points a, b, c, d, e, f. Call the forces at these points A, B, C, D, E, F. There is a 20-kip load at the center of the box acting downward. Apply each of the above equations in a sequence that will eliminate as many unknowns from the equations as possible. Positive directions and moments are indicated by the arrows at the coordinate axes (Fig. 43b).

$\sum M_x = 0.$ Use the x axis (A, C, D, E, and F have no moment about this axis):

$$10B - 20(5) = 0; \qquad B = 10, \qquad \text{upward as assumed}$$

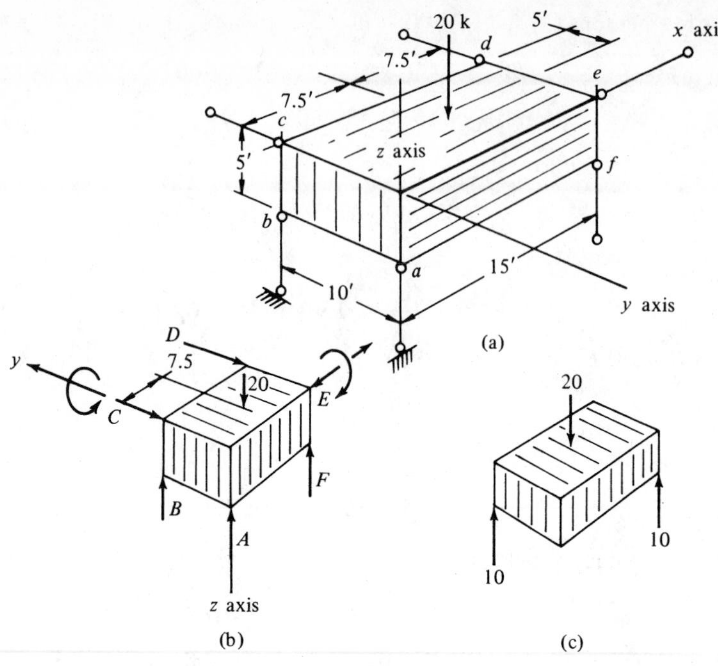

Fig. 43.

$\sum M_y = 0$. Use the y axis (A, B, C, D, and E have no moment about the y axis):

$$15F - 20(7.5) = 0; \quad F = 10, \quad \text{upward}$$

$\sum M_z = 0$. Use the z axis (A, B, C, E, and F have no moment):

$$D(15) = 0, \quad D = 0$$

Now apply the force equations

$$\sum F_x = 0, \quad E = 0:$$

$$\sum F_y = 0; \quad -C - D = 0; \quad C = 0$$

$$\sum F_z = 0; \quad A + B + F = 20; \quad A + 10 + 10 = 20; \quad A = 0$$

The forces acting on the structure are shown in Fig. 43c. Evidently the structure balances on the supports b and f. This result could have been anticipated if the structure had been studied carefully before the calculations were started.

Determinate and Indeterminate Structures As described in the previous pages, a problem in equilibrium is solved by finding several unknown forces using given known forces. For any rigid body with only horizontally and vertically applied forces, a solution is possible if there are not more than three unknown quantities. A structure for which the reactions can be obtained by employing the equilibrium equations of statics is called statically determinate. If a solution cannot be obtained by the equations of statics alone, the structure is called indeterminate. If the body is not in equilibrium it is called unstable.

In Fig. 44 are presented several structures which are to be analyzed for their

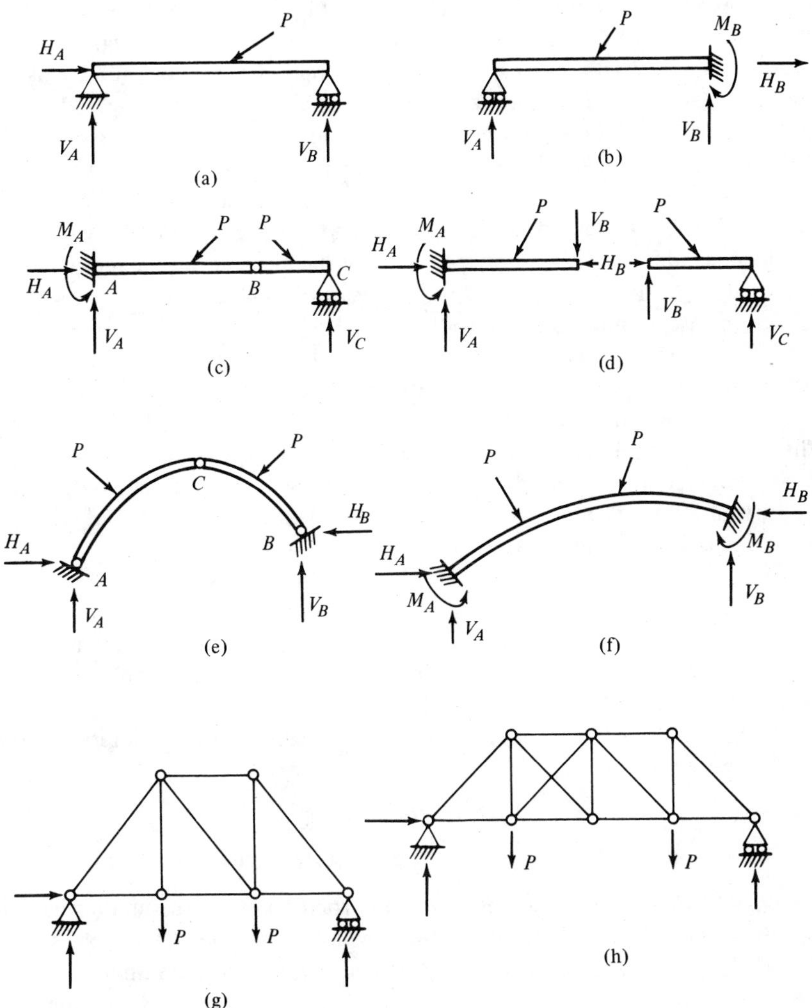

Fig. 44.

statical determination. They are discussed below in the order indicated by the subdivisions a through h of the figure.

a. This structure is determinate. There are only three reactions; they may be solved by the three equations of equilibrium. Note that it is necessary to place one end of the beam on rollers.

b. There are four reactions on this structure and only three equations of equilibrium. The structure is indeterminate.

c and d. A first examination of the structure in Fig. 44c will reveal four reactions. Actually, however, this is not one rigid body but is separated by a hinge; therefore, it may be considered as two structures, as sketched in Fig. 44d. Solve for the reactions on the member *BC* first, then apply the reactions at *B* as loads on member *AB*. Note that they are opposite in direction. A hinge added to an indeterminate structure gives another equation of equilibrium, thus permitting another reaction, provided the hinge is in the proper place.

e. The structure shown is a typical three-hinged arch and is statically determinate. There are four external reactions. To the three equations of equilibrium is added an additional condition, provided by the hinge at the top (crown).

f. There are six reactions on this hingeless arch, and it is three times statically indeterminate, or indeterminate to the third degree.

g. The previous structures considered have been beam structures. This example is a truss structure. There are three reactions, so that it is statically determinate, externally. It is also determinate internally, because forces in all the bars can be obtained. A truss structure is usually made up of a series of triangles. To the basic triangle of three bars there will be added two bars for each additional joint. Therefore, if b is number of bars and j is number of joints, the structure is determinate internally if

$$b - 3 = 2(j - 3) \qquad \text{or} \qquad b = 2j - 3$$

If $b < 2j - 3$, the structure is unstable; if $b > 2j - 3$, the structure is indeterminate. There are special cases where the above test does not apply, so that the test should not be considered as final.

h. This structure is determinate externally but indeterminate internally, since by the test equation:

$$b = 14, \qquad j = 8, \qquad 14 > 16 - 3$$

There are more bars than necessary. The structure is indeterminate internally.

Graphic Statics Graphic statics is an important tool in structural engineering. As the title implies, graphical methods will be described in this subsection. Forces are represented by arrows. Such an arrow, or force, has three properties: *magnitude*, the length of the arrow; *direction*, given by the angle of the arrow to a coordinate system; *location*, or point of application, given by the location of the arrow, with reference to the coordinate system. These three properties are

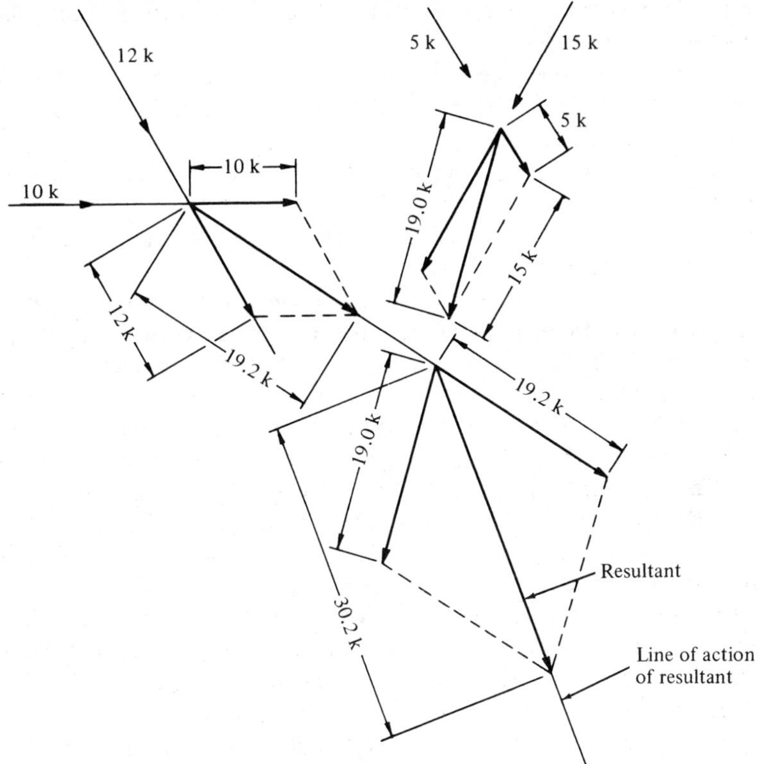

Fig. 45. A series of four forces has been combined by means of the force parallelogram. Forces may be combined in any order and the resultant will be the same.

analogous to the three equations of equilibrium used in the previous analytical method.

The first exercise in graphic statics will be to combine a series of forces together by means of force parallelograms, as shown in Fig. 45. The forces have been combined in pairs first. Note that the parallelograms have been laid off at the intersections of the lines of action of the pairs of forces. The resultants of these pairs of forces are then combined into a single force. Note that the forces in this problem have been arranged so that no difficulty was experienced in keeping the forces on the paper. If some of the forces or resultants had been parallel to each other, they could not have been combined by this method.

The Force Triangle The method just presented gets rather cumbersome for complicated problems because so many lines must be drawn which often lie on top of each other. Therefore, a simplification will be made. The location of the forces will be shown on one diagram, the space diagram, and the solution for

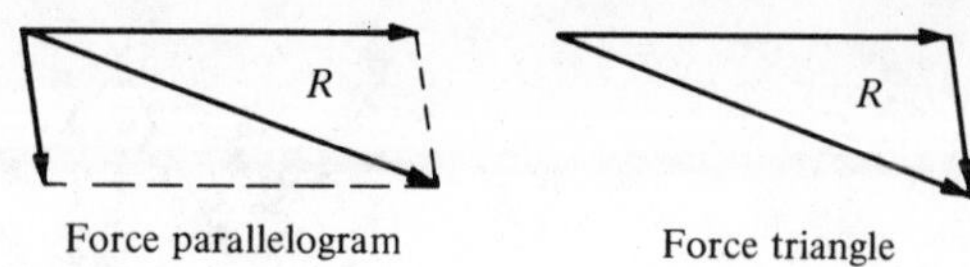

Fig. 46. In a force triangle, the forces are arranged in order. The magnitude and direction of the force are given but not the location.

magnitude will be made on another diagram, the force diagram. The force triangle will be used for the latter diagram.

A comparison of a force triangle and force parallelogram is presented in Fig. 46. Note that the force triangle is merely two sides of the parallelogram. The force triangle gives the magnitude and direction of the resultant but not the correct point of application.

Now all our graphical problems will be solved using two separate diagrams:

1. The space diagram, which shows the location of all forces; and
2. The force diagram, which solves for the magnitude and direction of forces.

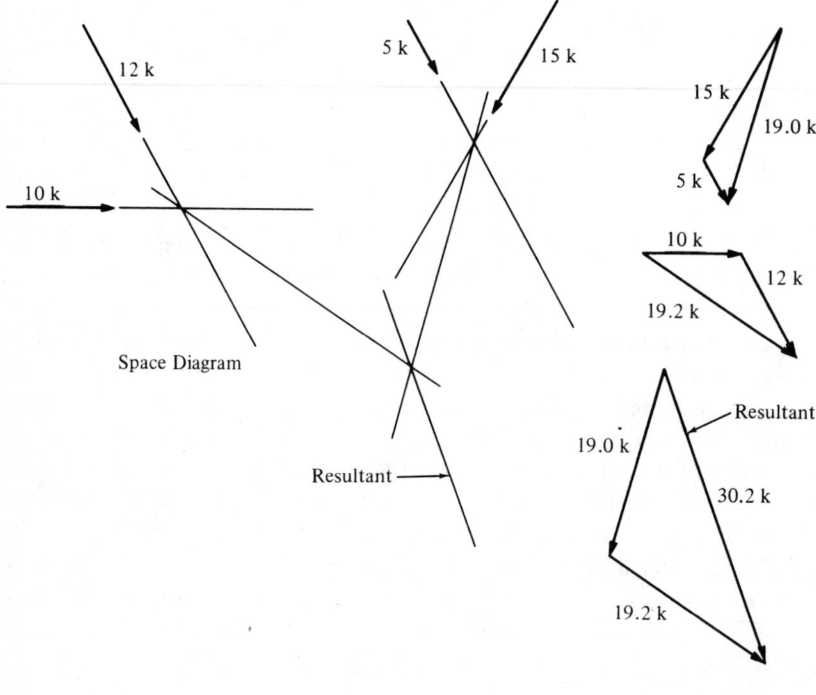

Fig. 47. This is the same problem as in Fig. 45, but is now solved by separate space and force diagrams. Directions are transferred from the force to the space diagrams.

To illustrate the construction of space and force diagrams the same force system as solved in Fig. 45 has been shown in Fig. 47. Force triangles are shown on the right and the space diagram on the left. The two diagrams are constructed at the same time and directions must be transferred from the force diagram to the space diagram or vice versa.

An advantage of force triangles is that a number of forces may be combined into a single diagram called a force polygon. The force triangles of the problem of Fig. 47 are shown as force polygon in Fig. 48. Note that the triangles are simply added to each other, and it has not been necessary to draw the third force triangle as in Fig. 47.

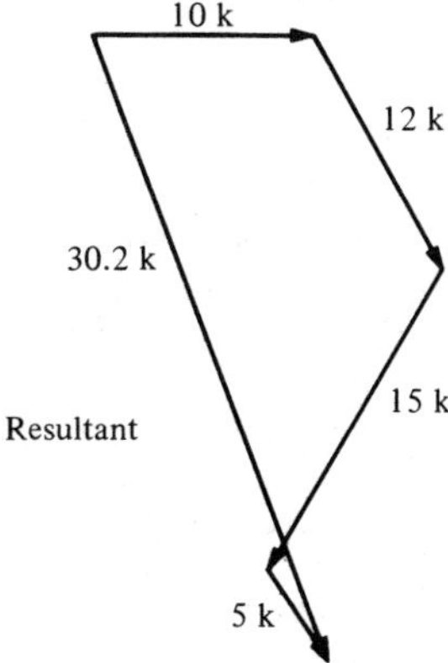

Fig. 48. These are the same forces as in Fig. 47, now drawn as a force polygon.

Substitute Forces In many cases the forces may be parallel or nearly parallel. A special trick is used to solve these problems. One of the forces is resolved into two components and one of these components is disregarded at first and the other component is combined with the next forces so that a solution becomes possible with all construction lines on the paper.

The arrangement of forces for this method is shown in Fig. 49, and the complete solution in Fig. 50. Note that the 10 kip force has been resolved into two components, 1 and 2. Then force 2 is combined with the next 8 kip force to give force 3, which in turn is combined with the 12 kip force to give force 4. The final resultant is the combination of force 4 and force 1. The final position and magnitude of the resultant is independent of the way the forces 1 and 2 were picked.

Reactions The examples to follow will involve problems in the equilibrium of bodies. The graphical condition for equilibrium is that the resultant of all forces on a body shall be zero. Usually several forces are unknown, so the problem is to

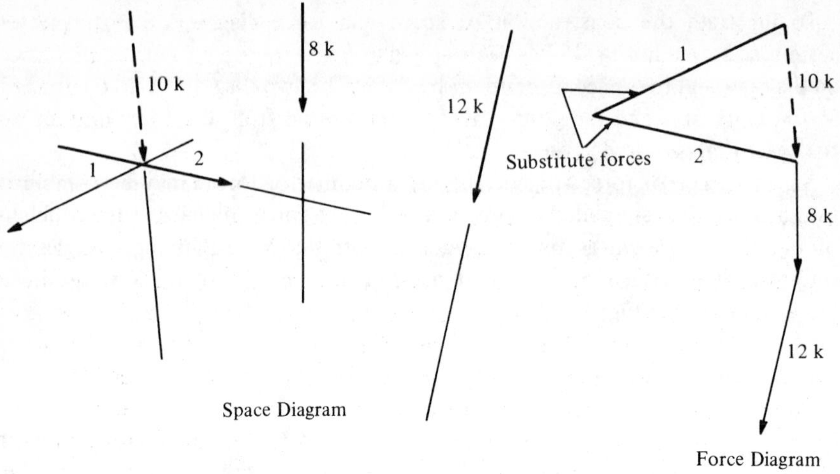

Fig. 49. The basis of the method of combining nearly parallel forces is illustrated here. Resolve the 10-kip force into the two components 1 and 2; then combine with the 8-kip force. The complete solution is shown in Fig. 50.

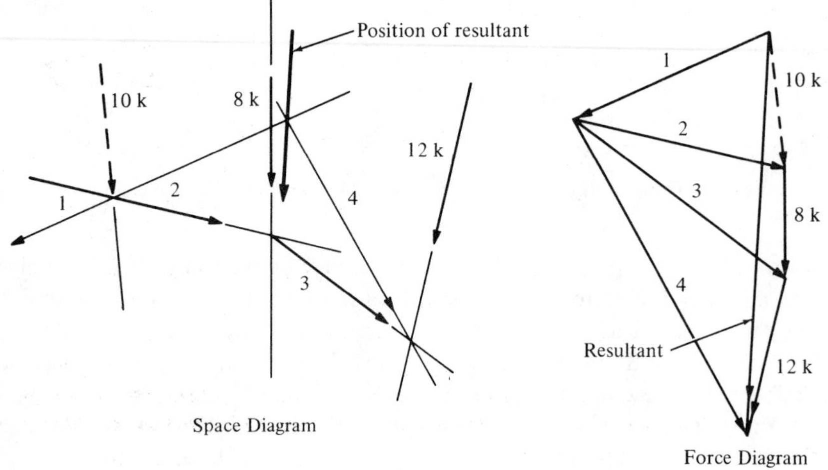

Fig. 50. The complete solution of the problem of Fig. 49 is shown here. Force 2 is combined with the 8-kip force to give Force 3. Then Force 3 is combined with the 12-kip force to give Force 4. The final force is the resultant of Forces 1 and 4.

find the unknown forces in direction, magnitude, and point of application. The equations of equilibrium ($\sum H = 0$, $\sum V = 0$, and $\sum M = 0$) will make it possible to determine only three components of reactions. Correspondingly, in graphic analysis only three characteristics of reactions may be obtained. For example, only one force can be completely determined in direction, magnitude, and point of application. If there are two unknown forces, the direction and point of application of one of these forces must be known, and the point of application of the other must also be known before a solution is possible. Three unknown forces must each be already known in, say, direction and point of application before a solution can be obtained.

The first step in the process of finding the unknown forces, which hold in equilibrium a given set of known forces, is to reduce all the individual known forces to a single resultant. A single unknown force which holds this force in equilibrium is a force equal and opposite to the known force. This condition is illustrated in Fig. 51. Note that in this problem the forces P_1 and P_2 have been combined by the method of substitute forces. The (graphic) sum of the known forces is called the resultant, and the unknown force is called the reaction.

Another type of force diagram may now be illustrated. Up to this point resultant force diagrams have always been drawn. Figure 52 shows an equilibrium diagram. In this case the reaction force is included instead of the resultant. The

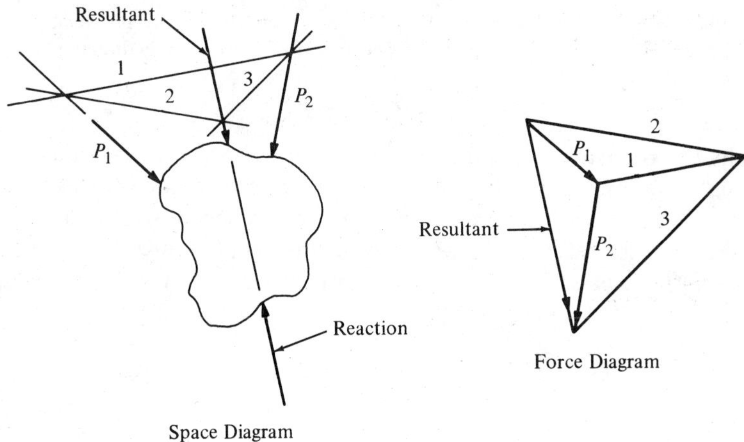

Fig. 51. The solution of a problem involving a single reaction is shown. First combine the known forces P_1 and P_2 into a single resultant. The reaction is equal and opposite to this resultant.

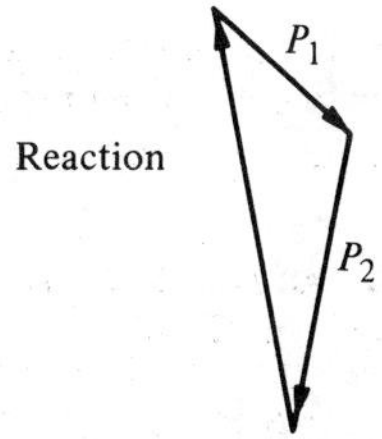

Fig. 52. An equilibrium diagram for the force system of Fig. 51 is shown here. The resultant of all forces must be zero.

graphic sum of all forces in an equilibrium diagram must be zero. In reaction problems it is necessary to distinguish between resultant and equilibrium diagrams, so the difference should be kept in mind at all times.

Reactions on a Beam The usual beam problem has two unknown reactions, as sketched in Fig. 53. One of the reactions is known in point of application only and the other is known in direction and point of application. This beam is

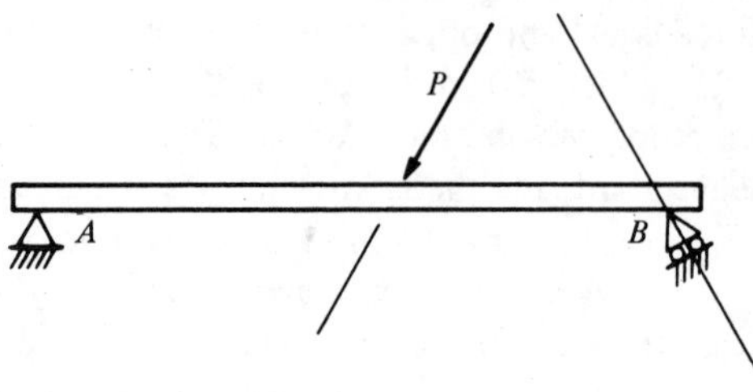

Fig. 53. This sketch illustrates the conditions found in the usual beam problem. Reaction at A is known only in point of application. Reaction at B is known in direction and point of application.

statically determinate because there are only three unknown characteristics. The principle used for the solution of this problem is that three forces to be in equilibrium must meet at a common point; this is shown in Fig. 54. In the space diagram, the reaction R_B was extended to meet the force P. The reaction R_A was made to pass through the common point and point A. Then the equilibrium diagram was drawn using the force P as the base of a triangle and the directions of R_A and R_B.

Graphic problems may be solved by this method if it is possible to obtain a

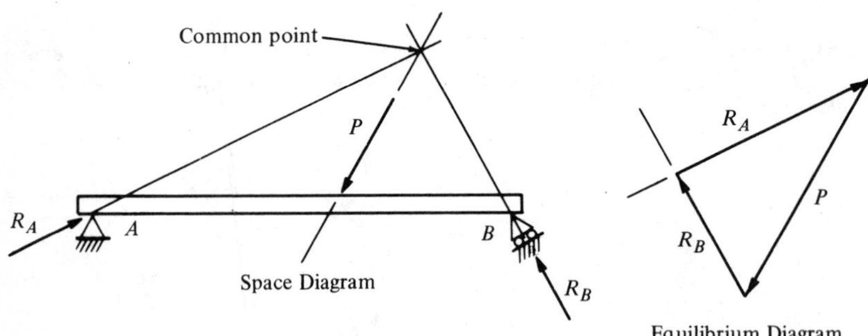

Fig. 54. Three forces to be in equilibrium must meet at a common point. The direction of force R_A is obtained from the space diagram and the magnitudes of R_A and R_B are obtained from the equilibrium diagram.

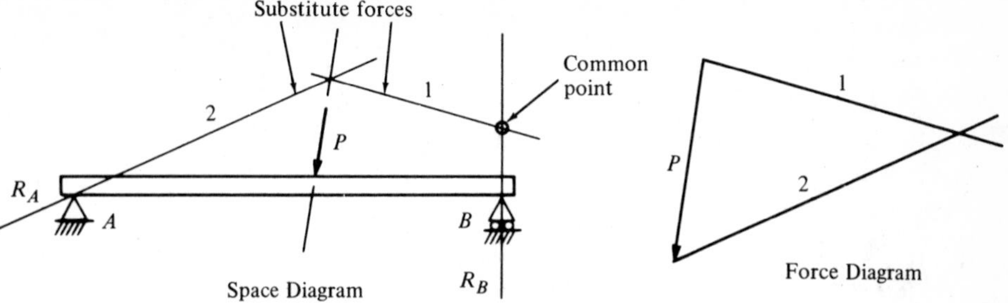

Fig. 55. The first step in the solution of a beam problem by the use of substitute forces is illustrated. Break the force P into two forces 1 and 2, such that force 2 passes through point A.

common point on the paper that is used. To solve a problem involving nearly parallel forces, the method of substitute forces may be used, as illustrated in Fig. 55. Only the first step is shown in this figure. Break up force P into two components, force 1 and force 2. The basis of the solution is that the system may be considered as three forces—a force through point A unknown in direction and magnitude; a force R_B known in direction and point of application; and force 1 known in direction, magnitude, and point of application. The common point for three forces is the intersection of R_B and force 1.

The complete solution of this problem is given in Fig. 56. Draw on the space diagram force C, which is called the closing line, through the common point and A. Then solve the equilibrium diagram involving R_B, force C, and force 1. The final step in the solution is to solve for R_A, which with force 2 is one of the components of force C.

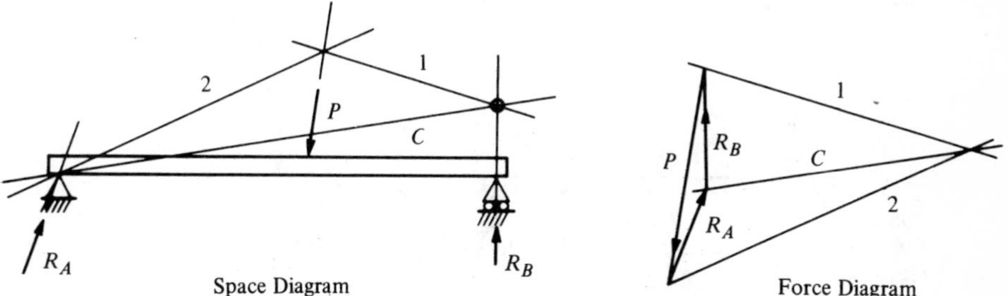

Fig. 56. The final step in the solution is to draw force C through A and the common point. The equilibrium diagram will involve R_B, force C, and force 1. To complete the force diagram, draw R_A, which added to force 2 will give force C.

In Fig. 57a and b is sketched just this part of the force diagram. Force C and force 2 are known in direction, magnitude and point of application. It is required to find R_A. The solution is shown in Fig. 57b.

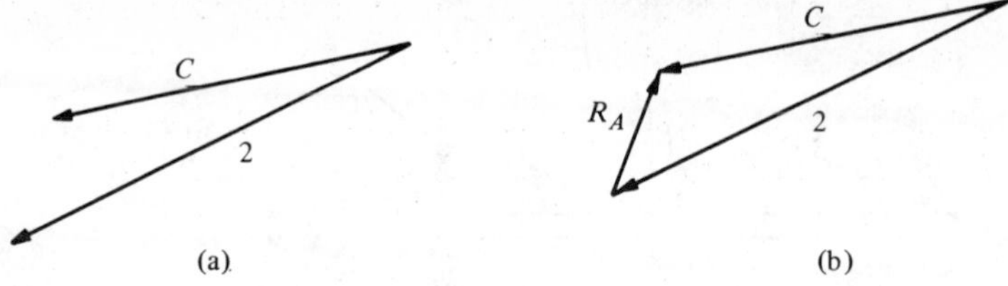

Fig. 57. In this sketch the final step in the solution of the beam of Fig. 56 is shown. The force C is the resultant of R_A and force 2.

A Series of Forces The usual reaction problem will be a beam loaded with a series of forces. A solution for this type of problem is shown in Fig. 58. Note the following points in the solution:

1. The task of finding the resultant of the forces on the structure and solving for reactions is combined into one operation. Divide the 5 kip force into components 1 and 2. Then pass force 1 through point A. Then continue combining the substitute forces until force 4 intersects the line of action of R_B.
2. Forces 1 and 4 will now be the components of the resultant of forces on the structure.
3. The common-point solution for three forces will be the intersection of R_B and force 4. Draw, on the space diagram, line C from A to the common point. Transfer force C from the space diagram to the force diagram and draw through the pole.
4. The remainder of the solution is as in the previous problem. Solve the equilibrium diagram for force C, force 4, and R_B.
5. Solve for R_B, which is a component of force C.

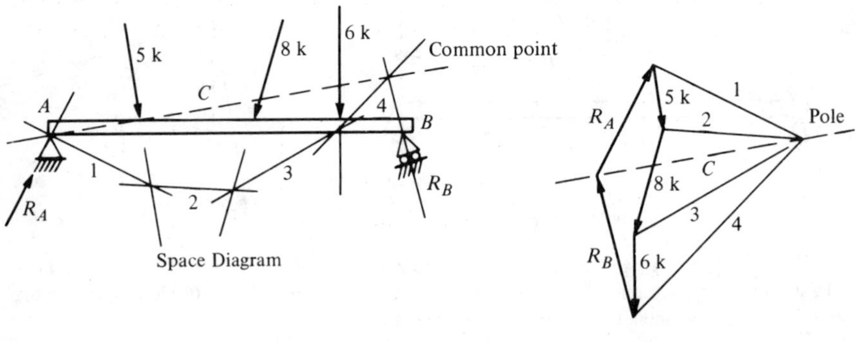

Fig. 58. The solution for the reactions for several loads on a beam is shown. Start with force 1 through A.

Note that it has not been necessary to find the actual resultant of the three forces either on the space diagram or the force diagram.

Center of Gravity A very useful concept which derives from equilibrium is the *center of gravity*. Consider the parallel loads shown in Fig. 59. The magnitude

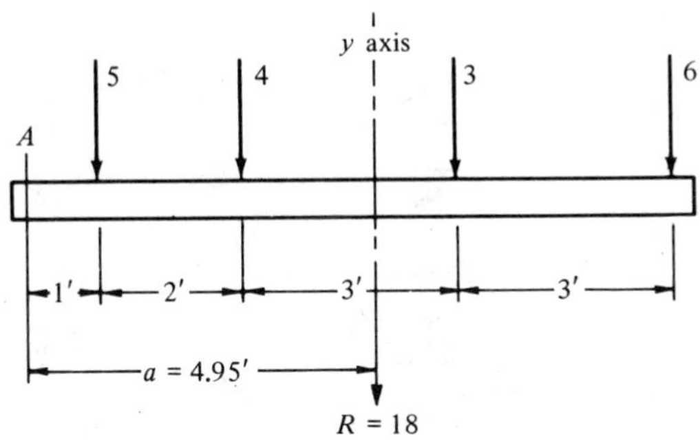

Fig. 59.

and point of application of the resultant of these loads can be obtained from the equations of equilibrium as follows:

$$F_y = 0; \quad R = 5 + 4 + 3 + 6 = 18$$

The location of this resultant is obtained from the moment about some convenient point, so we will use point A which is 1 ft to the left of the first load:

$$M_A = 0; \quad 5(1) + 4(3) + 3(6) + 6(9) = 89$$

Divide the moment by the force to obtain the distance from point A to the resultant $a = 89/18 = 4.95$ ft.

This point may be called the *center of gravity* of the loads. The same term is used for areas. In Fig. 60 are arranged some rectangular figures which have areas

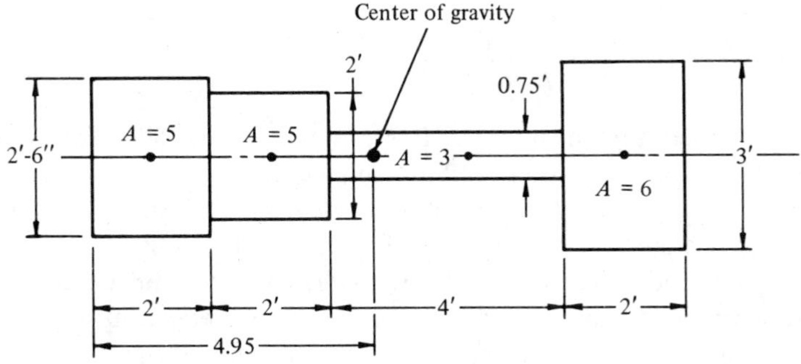

Fig. 60.

TABLE 1. AREA AND CENTER OF GRAVITY.

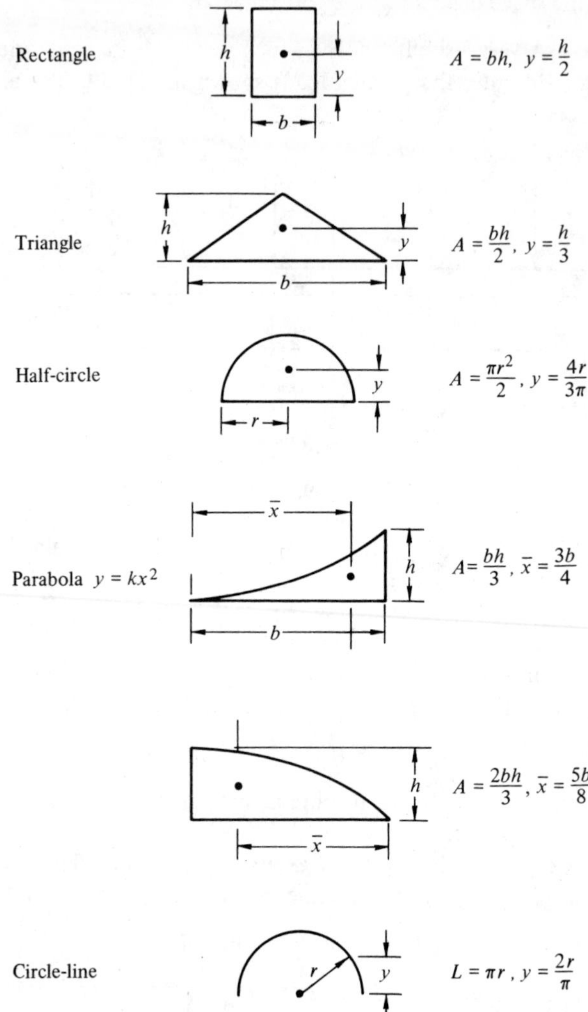

which are numerically equal to the forces shown in Fig. 59. Therefore, the center of gravity of these areas is 4.95 ft from the left edge.

The position of the center of gravity of some common figures is summarized in Table 1. The areas of most irregular figures can be broken down into these elements and the area and center of gravity obtained. An example follows.

Find the center of gravity of the section of the concrete beam shown in Fig. 61a. It is 30 in. wide and 20 in. deep and is composed of rectangles and triangles. It is symmetrical about the vertical axis, so the center of gravity about that axis

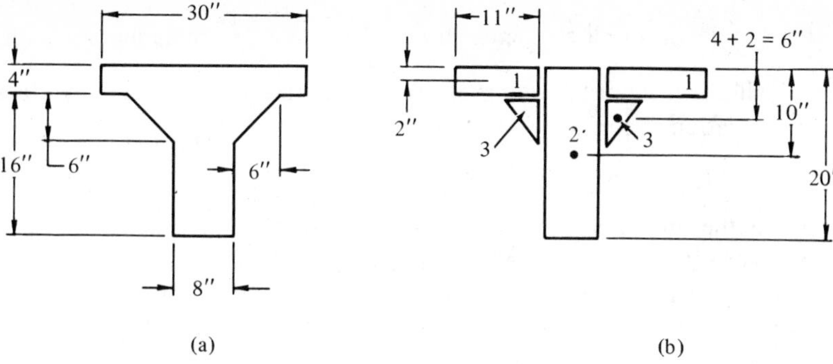

Fig. 61.

lies on the center line and calculations are not required. The section has been rearranged in Fig. 61b to show the components, and each is numbered for convenience. Take moments about the top of the section and summarize all calculations in the following table:

Component number	Dimensions	Number of elements	Arm from top, a	Area A	Aa
1	4(11) = 44	2	2	88	176
2	8(20) = 160	1	10	160	1600
3	$6(6)(\frac{1}{2}) = 18$	2	6	36	216
			Sum:	284	1992

The distance of the center of gravity from the top of the section is the summation of the individual moments Aa divided by the area. In many center-of-gravity calculations this is called $\overline{a}$ (a-bar). For this case $\overline{a} = 1992/284 = 7.01$ in.

Moment of Inertia Another property of an area which will be used extensively in structures is called the *moment of inertia.* Consider a rectangular area (Fig. 62) and an axis BB parallel to one side of the rectangle. The moment of inertia of a small element dA about this axis is $y^2\ dA$, and the total moment of inertia

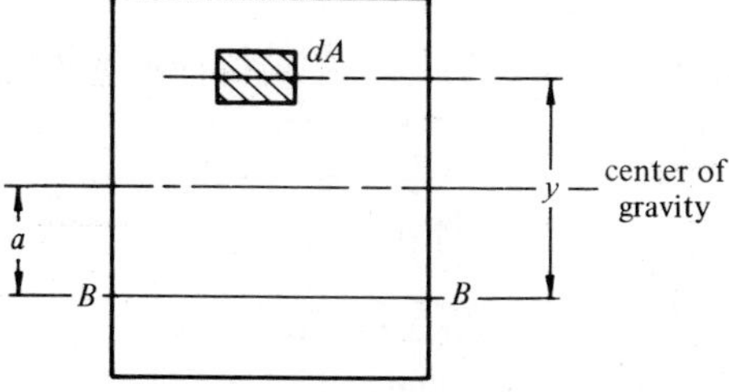

Fig. 62.

I is the summation of all elements, so $I_B = \int y^2\, dA$. The minimum moment of inertia I will occur about the center-of-gravity axis, and the value can be obtained by the translation formula

$$I = I_B - Aa^2$$

where A is the total area of the figure and a is the distance from axis BB to the center-of-gravity axis, as indicated in Fig. 62.

Methods for calculating moments of inertia for sections are described in the section on the design of beams. Values of I for some typical shapes are shown in Table 2.

TABLE 2.

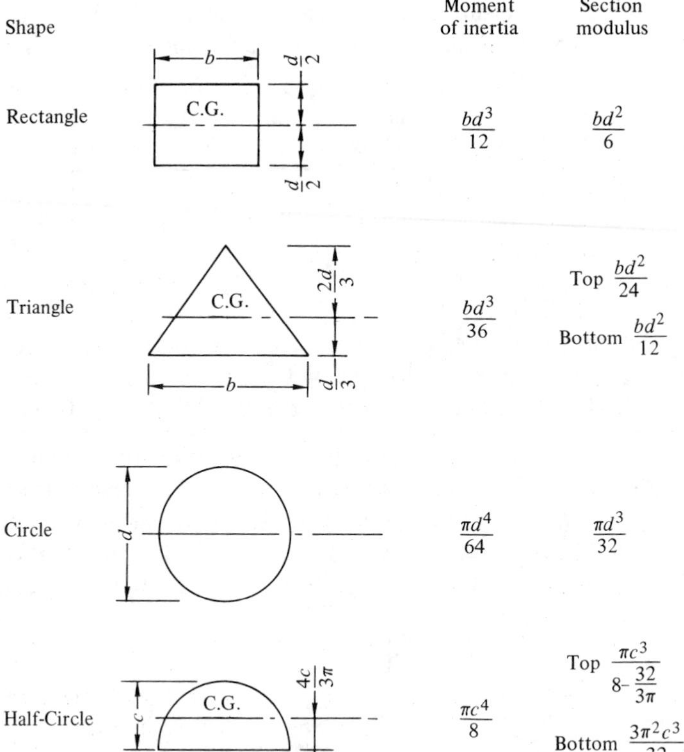

Reactions, Shears, and Moments for Statically Determinate Beams and Frames

In the preceding subsections, the equations of equilibrium were described and their use with several structures was explained. In this section, the forces in statically determinate beam and frame structures will be studied with more

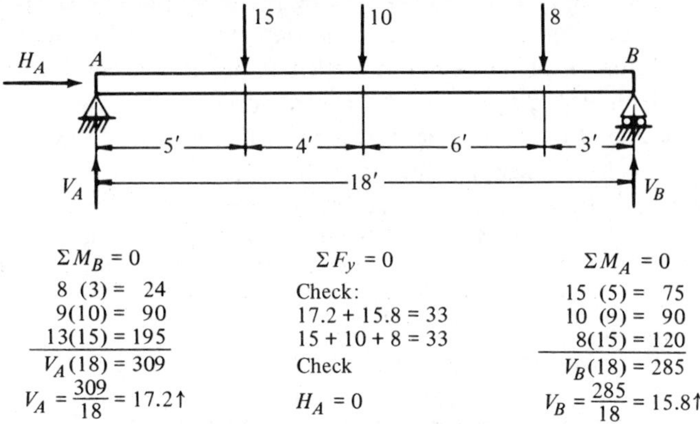

Fig. 63. Typical calculations for reactions on a simple beam.

complicated examples. We will start with the beam shown in Fig. 63 and determine the reactions. The calculations are written under the sketch and are arranged for ease in making and checking the numbers.

Note that all of the possible forces on the structure are shown. Such a diagram is called a *free-body diagram*, and will be used extensively in the determination of reactions, shears, and moments. The first equation to be applied is $\sum M = 0$ about the right end, point B. Note that equations are placed adjacent to the force for which the computations are being made, and they are written vertically. The first equation gives the value of the left reaction. The second equation is a solution of $\sum M_A = 0$ about the left reaction and gives the value of the right reaction. A check on the computations is to use $\sum F_y = 0$, which should give the same value as obtained by the first two steps. In these computations, a vertical reaction upward and a horizontal reaction to the left will be assumed to have a positive value. Note that by the equation $\sum F_x = 0$, the horizontal reaction $H_A = 0$.

The usual manner of writing equations is to proceed horizontally, as follows:

$$\sum M_B = 0; \quad V_A(18) - 8(3) - 10(9) - 15(13) = 0; \quad V_A = 17.2$$

To solve this equation it was necessary to use a piece of scratch paper, record the individual multiplications, and total them. It is much easier to follow the practice shown in the above example by writing each term in a column with extensions in the next column. These calculations can easily be checked by someone who has not made the original calculations.

The calculation of reactions in a beam structure with a hinge is illustrated in Fig. 64. The procedure is first to solve the simple structure AB on the left, and use the reactions obtained as loads on the structure BCD. In many cases it requires some study before the proper starting place becomes apparent. In this problem the concept and the symbol for a *uniform load* is introduced. The resultant of this load is at the center of the length of the load. When a question exists about a structure, a free-body diagram of forces on the part of the structure under consideration should be drawn as in Fig. 64. After solving several problems it will not be necessary always to draw the free-body diagrams, but in all cases the forces should be visualized. The horizontal reaction must be taken by the extreme right reaction, because this is the only reaction not on rollers.

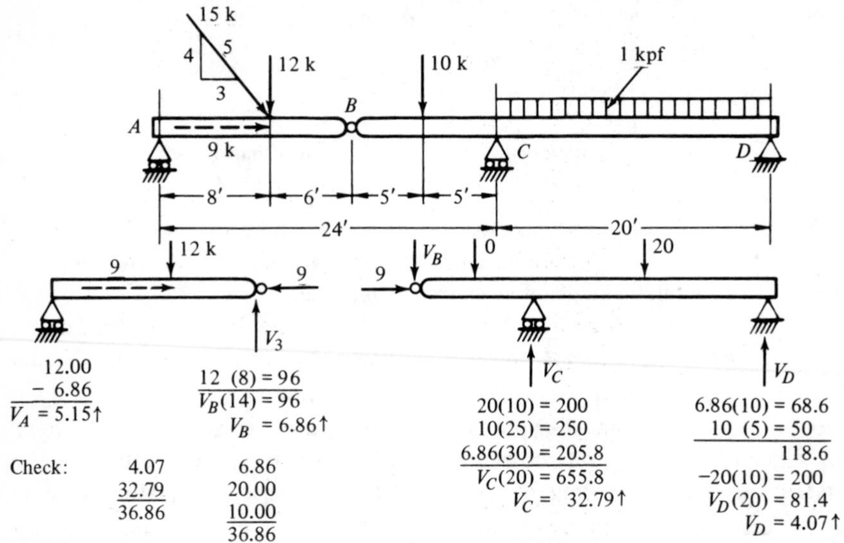

Fig. 64. Calculation of reactions in a beam structure with a hinge.

Reactions on Two Levels In some of the more complicated cases to be described later it is necessary to use simultaneous equations to solve for reactions. In many problems they can be avoided by a judicious choice of moment centers and force equations and by the trick of resolving forces at strategic points. The object is to obtain equations with only one unknown value. An example follows.

Refer to Fig. 65. The right reaction is inclined as shown. Instead of writing two simultaneous equations for reactions at A and C, extend the inclined reaction R to point B and resolve it at the point. Now the reaction H_B goes through point A and is eliminated from the moment equation $\sum M_A = 0$. The remainder of the problem follows the examples in Figs. 65 and 66.

A link as sketched in Fig. 66 is equivalent to the roller on an inclined plane, and the method of analysis is the same. The reactions may be resolved at any point provided there are no forces on the link except at the ends.

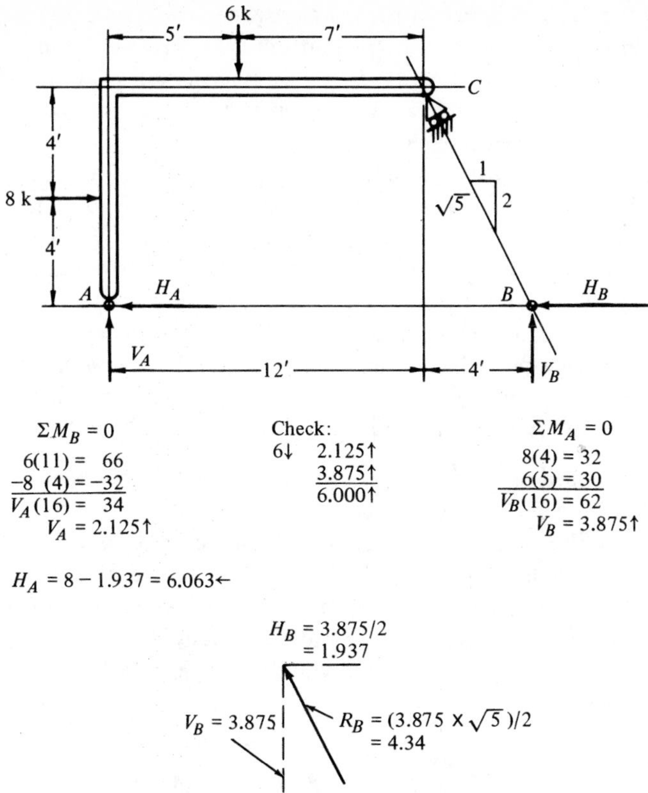

Fig. 65. The right reaction in this problem is resolved at the level of point A. An equation for moments about A will contain only one unknown reaction.

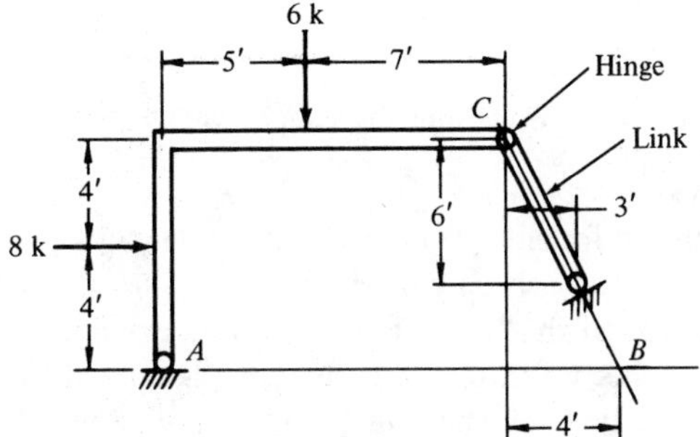

Fig. 66. This structure has the same dimensions and loads as that in Fig. 65, because the action of the sloping member on the right is the same as the roller of Fig. 65. For a member having no lateral forces on it, the line of reaction must pass through the two pins. This type of member will be called a link.

Stress After having computed the reactions, the next step in the analysis is to find the internal forces. For the present only the forces in beams will be considered, and trusses will be reserved for discussion later. A beam is a structure which resists forces by bending. The members of a truss resist external force by longitudinal forces in each of the members.

The internal forces just mentioned are called *stresses* and may be either direct stress or bending stress. The word "stress" is used both to indicate the total force on a member and the force on a unit area of a body. The latter is called a *unit stress*. The dimensions of stress are pounds for direct stress and foot-pounds for bending stress. Unit stresses are in pounds per square inch.

Another classification of stresses is by their tendency to deform an element of the structure. A stress which tends to elongate is called *tension*. A stress which shortens is called *compression*. A third type of deformation, *shear*, is the tendency of two elements to slide past each other.

Shear Two types of internal stresses in beams will be considered here: shear and bending moment.

The shear at a section cut through a beam is the summation of all forces (acting parallel to the section) on one side of the section. Since the beam is in equilibrium the sum of forces on one side of the section must equal the sum of the forces on the other. This is shown by the free-body diagrams in Fig. 67. The beam is the same structure as that for which reactions were calculated in Fig. 63.

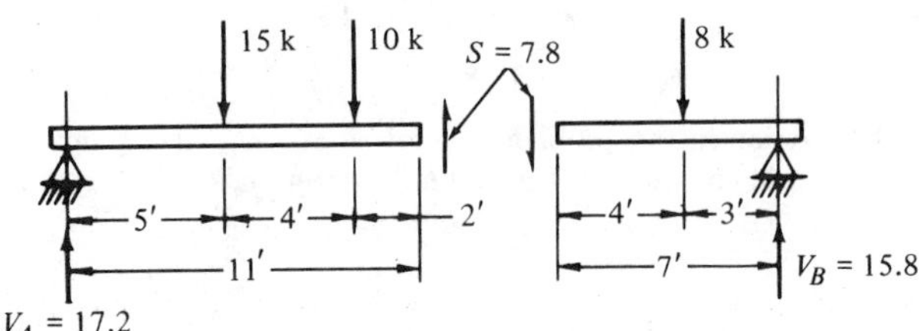

Fig. 67. Free-body diagrams, showing the shear on either side of a section in a beam.

The sign convention for shear is based on the tendency of the beam to deform. A shear which tends to push part of the beam to the left of the section up and the part of the beam to the right of the section down is positive. The reverse is negative. These sign conventions are shown diagrammatically in Fig. 68. According to this definition the shear in the section chosen for Fig. 67 is negative Note that this rule states the direction of the external forces and not the internal forces. To apply this convention to vertical beams, assume that the beam is rotated clockwise until it is horizontal.

Fig. 68. Diagrammatic representation of the sign convention for shear.

Shear Diagrams To show the magnitude of the shear at all points in a beam it is usual to construct a shear diagram. A typical diagram is sketched in Fig. 69 and is the same beam and loads as in Figs. 63 and 67. Positive values of shear are plotted above the line and negative below. At concentrated loads the shear diagram will have a vertical drop equal to the value of the load. At uniform loads the shear diagram will have a constant slope. The inclination of this line in pounds per foot is the magnitude of the uniform load. An example is shown

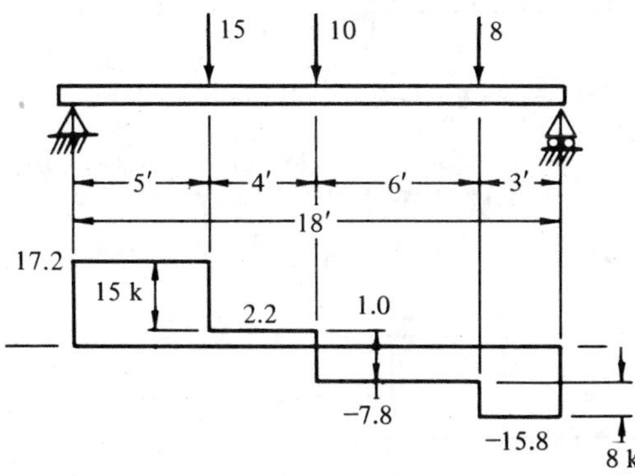

Fig. 69. A typical shear diagram for concentrated loads.

in Fig. 70 of the shear diagram for a uniform load and a single concentrated load. Note that to dimension this diagram completely, it is necessary to locate the points of zero shear. The simplest method is to divide the shear at one end of the sloping diagram by the uniform load per foot.

Bending Moments The second type of stress to be computed in beams is called bending moment and is defined as the sum of the moments of all forces acting to one side of a section. The center of gravity of the cross section is used as the center of moments because unit stresses are computed from this point. The bending moment on one side of a section will equal the moments on the other side, so that either side may be used for computing the moment.

As in the case of shear, draw a free-body diagram of the section under consideration and show all the forces. Then tabulate the forces and distances and

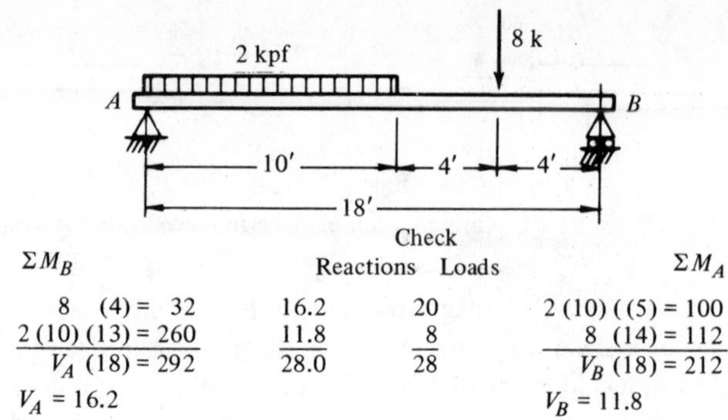

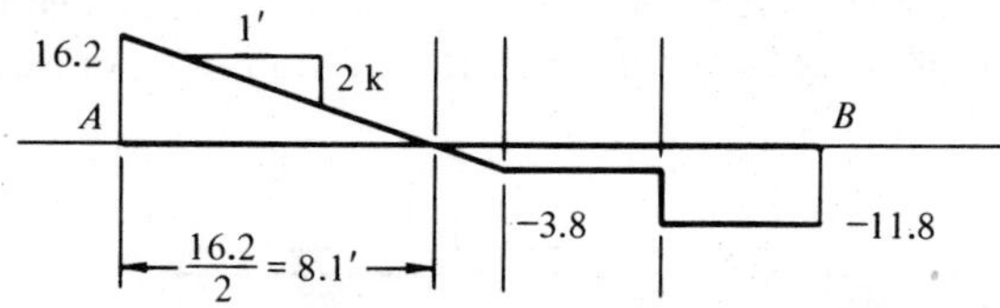

Fig. 70. A shear diagram for a uniform load. The slope of the shear diagram is equal to the load per foot. As an exercise, draw several free-body diagrams for this beam.

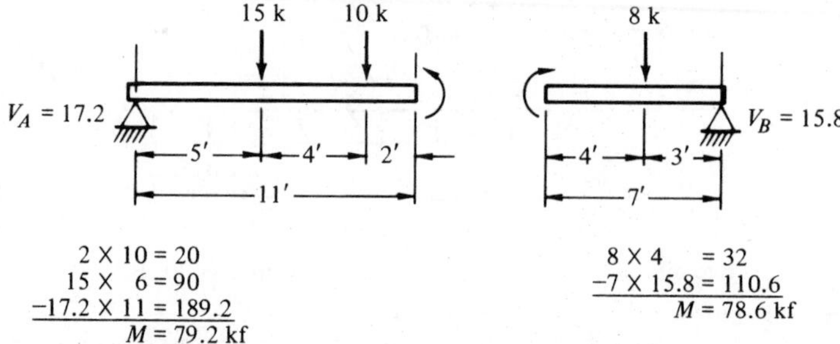

Fig. 71. Typical computations for bending moments. The moments to either side of the section should be equal.

compute the bending moment. Figure 71 is an example using the same beam and loads as in Fig. 63. The discrepancy in the two values computed is due to the number of significant places to which the computations for reactions were carried in Fig. 63.

The sign convention for moment is based on the type of stress in the beam. Bending moments which produce compression in the upper fibers of a beam and tension in the lower fibers are positive. Conversely, bending moments which produce tension in the upper fibers and compression in the lower fibers are negative.

The easiest way to remember these conventions is by the visualization of the forces on an element, as sketched in Fig. 72. In case of doubt, always draw a

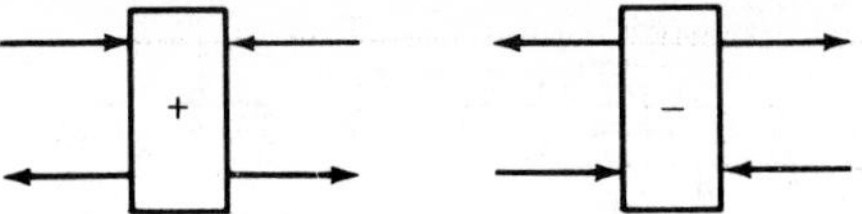

Fig. 72. A method of visualizing the sign conventions for bending moments. In case of doubt, draw a free-body diagram.

free-body diagram showing the forces acting on the beam and the direction of the internal forces.

Bending-Moment Diagrams The values of the bending moment for all points in the beam may be recorded in a diagram as for shear. The diagram in Fig. 73 is for the beam of Fig. 69. Computations are not shown but resemble those of Fig. 71. Values have been computed only for the points under the load because for a series of concentrated loads the diagram will be a series of straight lines.

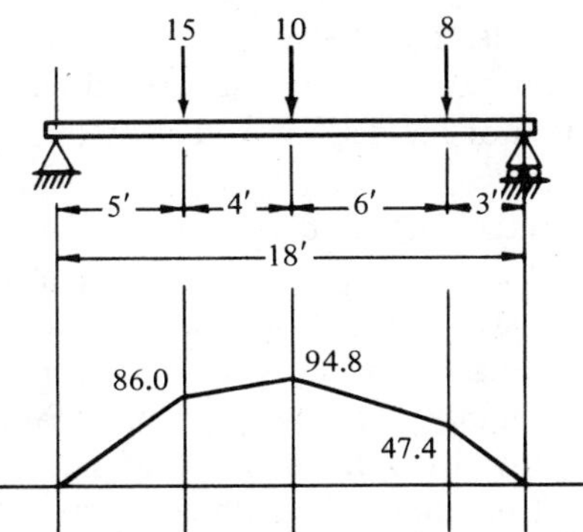

Fig. 73. A typical bending-moment diagram for a series of concentrated loads. Values are computed only at the concentrations. The diagram is a straight line between loads.

For uniform loads, the bending-moment diagram is a curved line. In Fig. 74 a diagram is shown for the beam of Fig. 70. The equation of the curved portion is $M = 16.2x - (wx^2/2)$. Substitutions in this equation are: for $x = 0; M = 0$; for $x = 10, M = 62.0$. To find the maximum value of the moment set the derivative equal to zero. Then $dM/dx = 16.2 - (wx/2) = 0, x = 8.1$ for $w = 2$. Therefore, the bending moment will be a maximum at a distance from the support of 8.1 ft. Note from Fig. 70 that this is also the point of zero shear.

Relationships Between Load, Shear, and Bending Moment Certain relationships between load, shear and bending moment are very useful in drawing shear and moment diagrams. The shear diagram results from the progressive summation of all loads to one side of a section. Mathematically, the shear diagram is the integral of the load diagram. The reverse of the process of integration is

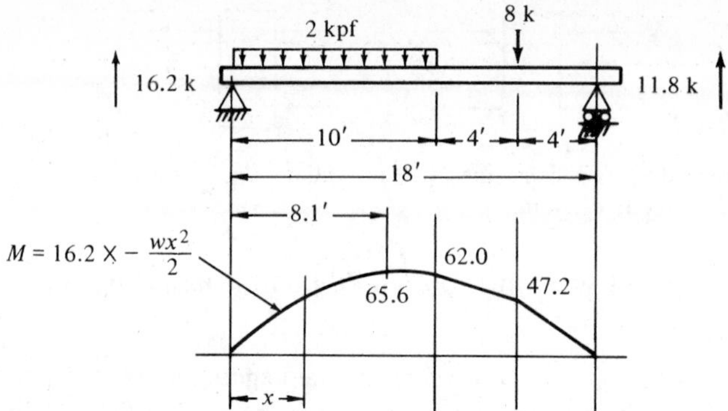

Fig. 74. The moment diagram for a uniform load. The maximum moment is obtained by differentiating the equation of the moment diagram.

differentiation. The load diagram is the derivative of the shear diagram, and the rate of change of the shear diagram is equal to the load. Thus in Fig. 70 the rate of change of the shear under the uniform load is equal in magnitude to the uniform load. At the concentrated load the shear diagram is a discontinuous function, but an inspection of the diagram will show the change of the shear is equal to the load.

In the previous examples the moment diagram was obtained by summing up all the forces times their respective distances from a given section. These same forces and distances are also given by the shear diagram. The progressive summation of the shear diagram will give the moment diagram or, mathematically, the moment diagram is the integral of the shear diagram. Conversely the shear diagram is the derivative of the moment diagram.

The most useful results from these relationships are the following propositions:

1. The area under the shear diagram is equal to the change in bending moment.
2. The point of zero shear is the point of maximum moment.

To compute bending-moment diagrams, then, it is necessary only to consider the shear diagram. Start from a point of known bending moment, say at a hinge, compute the areas under the shear curve, and progressively add these values up to give the bending moment at any point. Fig. 75 is an illustration of the method used. The computation of areas for each section of the shear diagram is placed near the areas to which they apply and the final value is placed in the area. Points of zero shear are computed as previously mentioned. The check on these computations is that the progressive summation of the areas should finally give zero at another point of zero moment. It is unnecessary to compute the values of the moment to establish the shape of the curved lines if the shape of the shear diagram is first studied. For example, in Fig. 75 the line must curve as shown because the shear diagram is larger at the left end and the slope of the moment diagram must be greater.

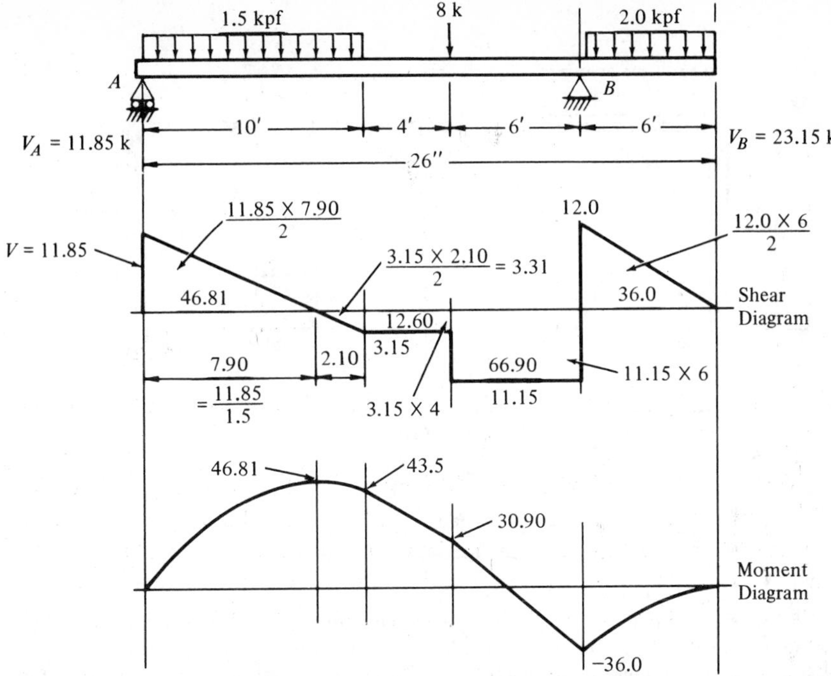

Fig. 75. Bending moments in this example are computed by the area of the shear diagram. Place the multiplications near the area to which they apply. The algebraic sum of the shear areas should finally return to zero.

Bending Moments in Framed Beams Structures in which beams are continuous with columns, as in Figs. 65 and 76 will be called framed beams or *frames*. When the structure is for the most part indeterminate with stiff joints the structure will be called a *rigid frame*. The convention for obtaining the sign of bending moments for the vertical members of rigid frames is to assume that the frame is rotated clockwise through an angle of 90°. A vertical member will then be horizontal and the sign convention for horizontal members applies. In all cases for rigid frames, free-body diagrams should be drawn to show the actual forces inside the structure.

In Fig. 76 the same structure is presented for which reactions were calculated in Fig. 65. Shear and moment diagrams have been led off to one side. Sometimes, however, it is expedient to draw a complete sketch of the frame and show the shears and bending moments on this sketch. The latter method of presentation is shown for the beam in Fig. 77. In both of these problems the check on the computations for both reactions and for the diagrams is that the sum of the areas starting from a point of zero moment should again return to zero at another point of zero moment.

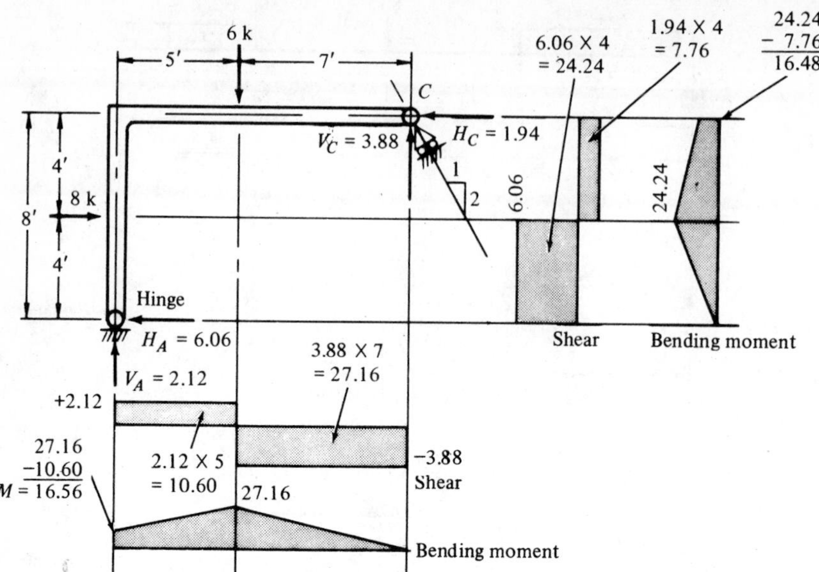

Fig. 76. Shear and moment diagrams for the structure of Fig. 65. The area under the shear diagram is equal to the change in bending moment. The discrepancy between the moment at the top of the column and the left end of the beam is caused by the rounding off of certain values during the calculation. These two moments are actually equal.

Shear and Moment Diagrams for Continuous Structures The equations of equilibrium presented here do not provide sufficient information to make it possible to determine reactions, shears, and bending moments for indeterminate structures. However, given the magnitude and location of bending moments, reactions can be calculated. A typical example is presented in Fig. 77. This is a three-span beam on knife edges at three supports and fixed at the left end. The bending moments at the supports have been figured by methods which will be described later in this text and are, respectively, -32 ft-kip, (abbreviated as -32 kf), -24 ft-kip, and -52 ft-kip, as indicated in the sketch.

On this structure there are five possible reactions, two at *A* and one each at *B*, *C*, and *D*. Therefore, only two additional conditions are actually needed to find the reactions and one of the bending moments is redundant (not required). Cut the structure into five free-body diagrams, as shown in the figure. Indicate on these diagrams the known bending moments. Then take a summation of the moments about a center such that in the resulting equation only one unknown reaction is involved. The calculations are very similar to the previous reaction calculations and are shown in the figure. The next step after computing reactions is to draw shear and moment diagrams as for simple beams.

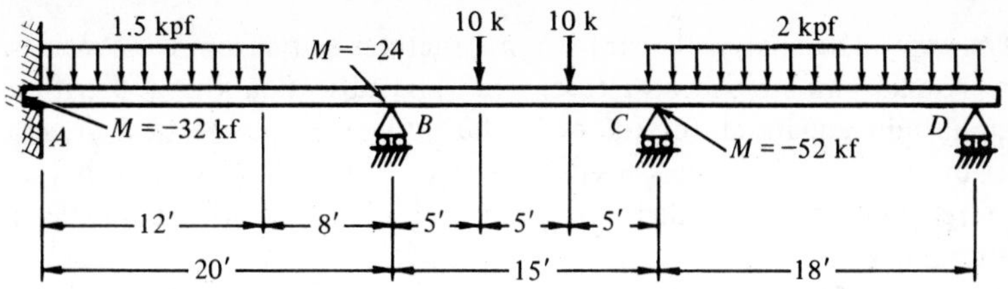

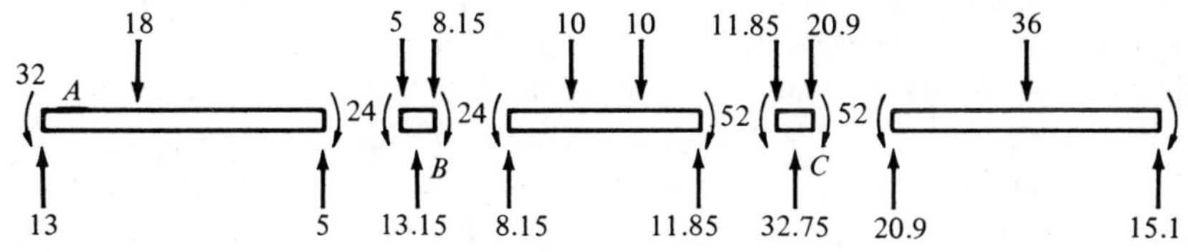

18(6) = 108
  24
132
− 32
100 ÷ 20 = 5

10(10) = 100
10 (5) = 50
  24
174
− 52
122 ÷ 15 = 8.15

36(9) = 324
  52
376 ÷ 18 = 20.9

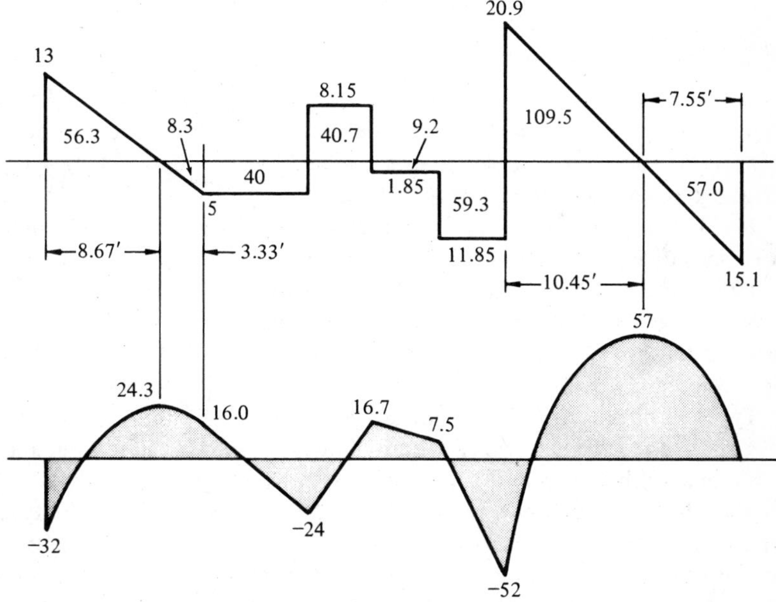

Fig. 77.

Summary The analysis of stresses in structures starts from the consideration of the equilibrium of the structure. In this subsection the application of the equilibrium equations for forces in the horizontal and vertical directions and the moment of forces has been presented. These equations enable us to find the reactions from applied forces on a structure if the structure is statically determinate.

Reactions for structures with supports on more than one level can be determined without using simultaneous equations if the moment centers are properly chosen to eliminate all but one unknown in each equation.

The internal forces in beams are shear and bending moment. These forces may be determined from free-body diagrams showing all forces to one side of the sections. They are most conveniently shown for beams by shear and moment diagrams. An important relation for constructing these diagrams is that the area under the shear diagram between two points is equal to the change in bending moment between the same two points, and the point of zero shear is the point of maximum (or minimum) moment. The shear and moment diagrams for continuous and indeterminate structures may be determined by statics if the moments are given at a sufficient number of locations on the beams.

Analysis of Trusses

The idealized truss is a framework of straight bars connected at the ends by hinges and pins. Loads are applied only at the joints, so the members are in tension or compression. The shapes outlined by the bars are usually triangles, although other shapes are permissible in some cases.

The joints of most trusses used in structural design are actually bolted, or welded, rather than being joined by pins. It can be shown by methods for the solution of indeterminate structures that this affects the tension and compression stresses in trusses very little. In the design of heavy bridges with thick members the bending stresses arising from this condition are investigated, but such problems will not be treated here.

Types of trusses and the names assigned to them by tradition are shown in Fig. 78. The types associated with buildings are shown in Fig. 78a, and those associated with bridges in Fig. 78b. The essential characteristic of the Pratt truss is that the diagonals are in tension for dead loads. Note that in the Pratt roof truss and Pratt bridge truss, the diagonals slope in opposite directions. In the Howe truss the diagonals are in compression. The earliest timber structures were usually of this type because the diagonal could be made much stiffer if of wood.

Parts of a Truss The names shown in Fig. 78c are given to various parts of a truss to describe their position and function. The members which lie along the top of the truss are called the *top chord* or the *upper chord*, and those along the

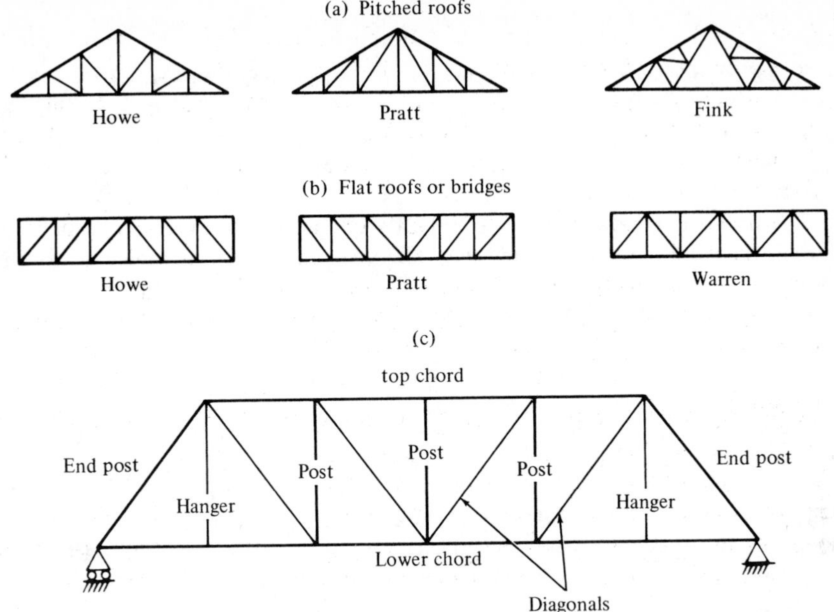

Fig. 78. The most common types of trusses (a and b) and the names of various members of a bridge (c).

bottom, the *bottom chord* or the *lower chord*. Vertical members are called *posts*, except for the member marked *hanger*, which serves a different function. Inclined members are called *diagonals*. The end diagonal, however, is usually called an end post. Members which connect two trusses are called bracing. Slabs, beams, and girders which serve to carry loads to the truss are called the *floor system* in the case of highway or railway bridges.

Analysis The method to be used in the determination of stresses is the same general method as for reactions. Cut the structure and replace the cut ends with forces which hold the structure in equilibrium. The problem is then identical with the calculation of reactions. Problems in the analysis of trusses are chiefly concerned with the proper way to cut the structure and the most expeditious way to solve for the stresses.

EXAMPLE

Consider the structure shown in Fig. 79a, which is an 80-ft-span truss. Loads are applied at both the top and bottom chords. First calculate the reactions. Note that since the loads are all vertical and at panel points, the *length of one panel* may be used as the unit of measure. The next preliminary step is to calculate the lengths of all members and record them on the sketch of the structure.

The first member for which stress is to be calculated will be the top chord member U_1U_2. Draw the free-body diagram shown in Fig. 79b and show the forces acting in the cut members. Note that in all cases the forces lie along the members and are not at an angle to the members. Any other force system will cause shear and bending stresses in the bars.

The problem as now set up is simply a problem in reactions. There are three forces for which only the magnitude is unknown. There are three equations of equilibrium, and therefore the unknown stresses may be obtained. The same procedures used for reactions are applicable here: select an equation of equilibrium and a resolution of forces which will enable the stresses to be obtained by one equation and one unknown. Either the force equation or the moment equation may be used.

For the problem under consideration, take moments about the intersection of the diagonal U_1L_2 and the lower chord L_1L_2. Resolve the force in the upper chord in two components at a point directly above the intersection of the diagonal and the lower chord. (Before the truss was cut and the free-body diagram was drawn, this point would have been U_2. However, this part of the truss has been thrown away and does not exist as far as the free-body diagram is concerned.) One of these components is eliminated in the equation $\sum M = 0$ about the intersection of U_1L_2 and L_1L_2. This equation is now

$$\sum M = 0, \qquad \begin{array}{r} 10(20) = 200 \\ 13.75(40) = 550 \\ \hline H(30) = 350 \\ H = \ 11.67 \end{array}$$

The direction of this force should be established in the same way that the direction of reactions was established. Look at the free-body diagram and observe that the moments are greater counterclockwise, about the center of moments, than they are clockwise; therefore, the reaction acts to the left.

To determine the magnitude of the vertical reaction, draw the force diagram for the resolved forces as shown in Fig. 79c. These forces are in proportion to the lengths of the sides. The length of one side is given and the general proportions are established by the slope triangle shown on the force U_1U_2; therefore,

$$V = 11.67\ (10/20) = 5.83$$

and

$$U_1U_2 = 11.67\ (22.3/20) = 13.0$$

From a consideration of the above example, a formula may be established for the stress in a member given the horizontal component:

$$\text{stress in member} = \text{horizontal component}\ \frac{\text{length of member}}{\text{horizontal projection of member}}$$

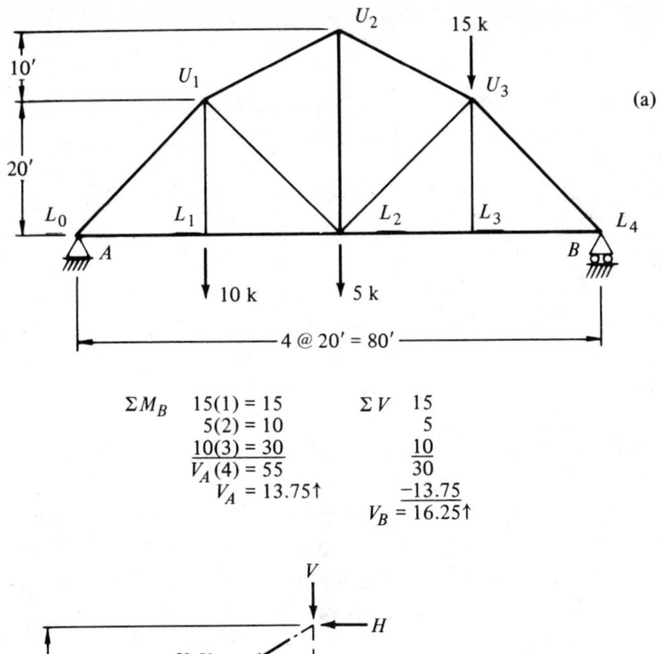

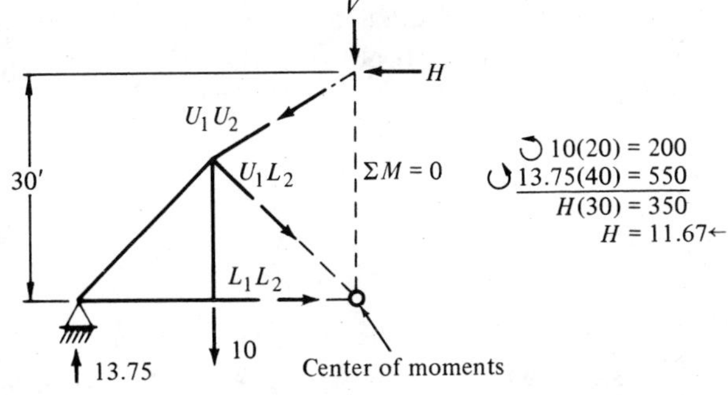

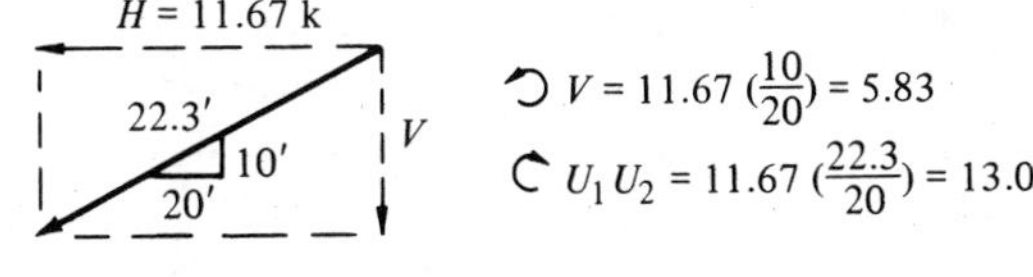

Fig. 79.

The formula for stress, given the vertical component is

$$\text{stress in member} = \text{vertical component}\ \frac{\text{length of member}}{\text{vertical projection of member}}$$

To obtain a horizontal component from a vertical component:

$$\text{horizontal component} = \text{vertical component}\ \frac{\text{horizontal projection}}{\text{vertical projection}}$$

These formulas make it very convenient to obtain stresses and components in trussed structures.

Signs The convention used for the sign of direct stress in truss members is that tension is positive (+) and compression is negative (−). The free-body diagram should always be consulted when establishing the type of stress in a bar. *A compression stress will push on the free-body diagram of the structure*, and the fibers of the member will be pushed together. *A tension stress will always pull on the free-body diagram.*

In the case of the free-body diagram shown in Fig. 79b, the force in the member obviously pushes on the structure and, therefore, member U_1U_2 is in compression. On trusses with vertical loads the upper chord is usually in compression and the lower chord is in tension. The stress in the diagonals is less obvious.

The next member for which stress is to be computed is member U_1L_2 of the truss in Fig. 79a. Draw the free-body diagram shown in Fig. 80. The position of the center of moments is the intersection of the lower chord and the upper chord and lies at a point outside the truss. Resolve the force in U_1L_2 at a point on the lower chord. The horizontal component will then go through the center of moments, and only the vertical force will remain in the equation of moments, which becomes

$$\begin{array}{r} 13.75(20) = 275 \\ 10(40) = 400 \\ \hline V(60) = 125 \\ V = \quad 2.1\uparrow \end{array}$$

The direction of this force is based on the following reasoning:
The moment of the 13.75-kip force is counterclockwise, and the moment of the 10-kip force is clockwise; therefore the vertical component is up. The horizontal component and the stress in the member are computed as for the previous

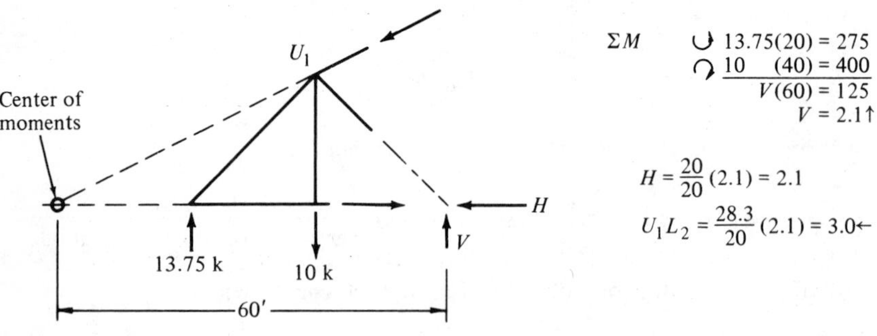

Fig. 80.

example from the relationships between the length of the member and the horizontal and vertical projections. It should be remembered that these distances also give the ratio between the sides of the force diagram for the resolution of forces in the member.

The type of stress (tension or compression) again is determined from the manner in which the force in the member acts on the joint. In this case it is compression (−), negative.

Stresses may be calculated in the end posts, member L_0U_1 by the solution of the equations $\sum H = 0$ and $\sum V = 0$. This is often called the *method of joints,* while the method already described is called the *method of sections*. The free-body diagram is shown in Fig. 81 and computations are shown alongside. Note that the stress in L_0U_1 is determined before the stress in L_0L_1 is calculated. The type of stress again is obtained by the way the forces act on the *joint*.

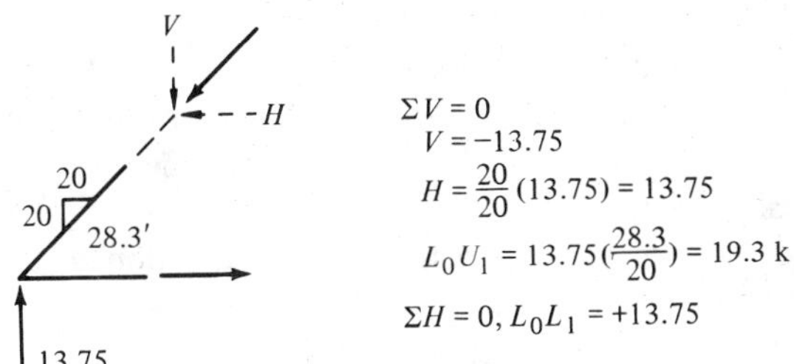

Fig. 81.

Stresses in other members of the structure have been computed and are recorded directly on the members as shown in Fig. 82.

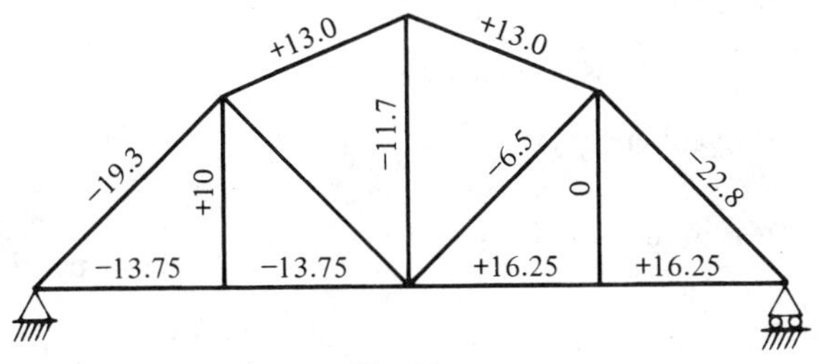

Fig. 82.

Summary of Procedure The steps followed in the previous examples of the calculation of stresses in trusses are as follows:

1. Draw an accurate sketch of the structure and show all forces acting.

2. Compute the reactions due to the applied loads.
3. Select a member for which the stress is to be calculated. Draw a free-body diagram for a portion of the structure cut through the member. Be sure that for the cut chosen, there are not more than three unknown forces. In some cases it may be necessary to calculate the stresses in one of the other members from another free-body daigram before proceeding to the given member.
4. Study the free-body diagram and the equations of equilibrium so that an equation can be written with only one unknown force. There are two possible methods:
   a. Take a summation for forces in a horizontal or vertical direction.
   b. Take moments about the intersection of two of the three unknown stresses. In this case make the additional simplification of resolving the force in the member into horizontal and vertical components at a point such that either the horizontal or the vertical component goes through the center of moments.
5. Compute the stress in the member from the horizontal or vertical stress by multiplying by the ratios of the lengths of the member, and its horizontal and vertical projections.
6. Determine the sign of the stress by a consideration of the action on the free-body diagram.
7. Record the stresses on a sketch of the truss.

Stress Coefficients For parallel-chord trusses, a method is available for the rapid calculation of stresses in trusses which will be called the *method of stress coefficients*. The stress in each member is expressed in terms of the *vertical forces* acting at any joint. The relation between the actual stress in the member and the stress coefficient is:

For vertical members:

$$\text{stress in member} = \text{stress coefficient}$$

For horizontal members:

$$\text{stress in member} = \text{stress coefficient}\ \frac{\text{length of member}}{\text{height of panel}}$$

For diagonals:

$$\text{stress in member} = \text{stress coefficient}\ \frac{\text{length of member}}{\text{vertical projection of length}}$$

EXAMPLE

To illustrate the calculation of stresses by this method, the truss shown in Fig. 83a will be solved. Note that this is a parallel-chord truss and is unsym-

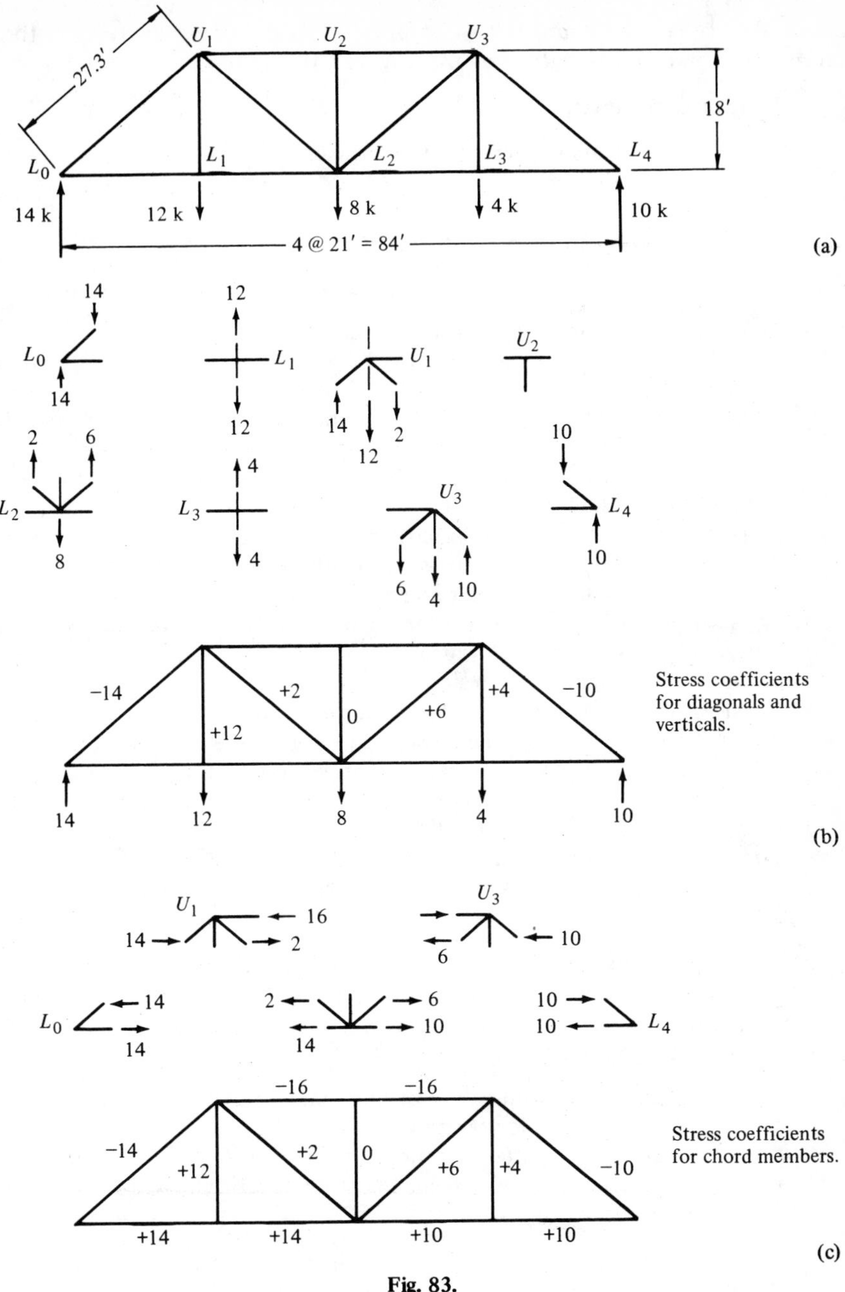
U1
U2
U3
27.3′
18′
L0
L1
L2
L3
L4
14 k
12 k
8 k
4 k
10 k
4 @ 21′ = 84′
(a)
Stress coefficients for diagonals and verticals.
−14
+2
0
+6
+4
−10
+12
(b)
−16
−16
Stress coefficients for chord members.
+14
+14
+10
+10
(c)

Fig. 83.

metrically loaded. As a preliminary step, calculate the ratios described in the previous paragraph (the length of all the diagonals is 27.3 ft):

For horizontal members:

$$\text{stress in member} = \text{stress coefficient} \times (21/18)$$

For diagonals:

$$\text{stress in member} = \text{stress coefficient} \times (27.3/18)$$

The calculation of the stress coefficients is accomplished in the following steps:

1. Calculate the reactions by the usual methods.
2. Consider joint L_0 and determine the *vertical component* of stress in the end post, member L_0U_1 = 14 kip, which is the stress coefficient in both member L_0U_1 and L_0L_1. Write these values on a diagram of the truss as shown in Fig. 83b. The sign of the stress coefficients should be determined by making a free-body diagram of the joint and observing the direction of the forces for equilibrium.
3. Consider joint L_1. The stress coefficient in L_1U_1 is the stress in the member of 12 kip. Write this value on the member with a tension (+) sign. Now it is possible to determine the stresses at joint U_1.
4. Consider joint U_1. A summation of vertical forces gives the vertical component in the diagonal, which is also the stress coefficient. The value is 2 kip and the sign is (+), tension.
5. Consider joint U_2. The stress coefficient in U_2L_2 is zero because there is no load at the joint. It is now possible to solve for the stresses at joint L_2.
6. Consider joint L_2. Take a summation of vertical forces. The stress coefficient in L_2U_3 is +6 kip.
7. Consider joint L_3. The stress coefficient in L_3U_3 is +4 kip.
8. Consider joint U_3. A summation of vertical forces gives a stress coefficient in U_3L_4 of −10 kip. This value now agrees with the solution for stresses at joint L_4.

The stress coefficients for all the vertical and diagonal members have now been determined. The next step is to compute the coefficients for horizontal members (Fig. 83c). The rule is to *add the coefficients for diagonal members at each joint* with proper respect to the direction of each force:

9. Consider joint L_0. A summation of horizontal forces gives a stress coefficient of +14 kip in the lower chord member.
10. Consider joint U_1. The stress coefficient in member U_1U_2 is −16 kip.
11. Consider joint L_2. The only remaining stress coefficient is that in L_2L_3 and is +10 kip.

The final step in the solution of this problem is to compute the real stresses. Multiply all horizontal members by $2^{1}/_{18}$ and all diagonals by $27.^{3}/_{18}$. The index stress in the vertical members will be the real stress.

Strength of Materials

This section discusses the methods of evaluating the strength and elastic properties of materials that are used for structures and how the size and shape of structural elements may be selected for the loads they are to support.

Stress One measure of strength is *stress*, the force on a small element of a body. The term is sometimes used as the total force but for the present we will work with *unit stress*, with dimensions in force per unit area. An example is pounds per square inch (abbreviated psi). In the case of earth materials where the forces are low we use pounds per square foot (psf).

Stress may be either *tension*, the tendency to pull the element apart, or *compression*, the tendency to push the element together. In this discussion tension will be taken as positive (+) and compression as negative (−). Elements with stresses acting on them are shown in Fig. 84.

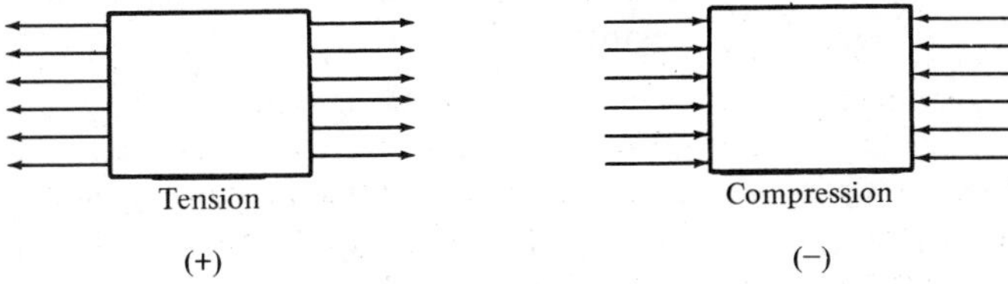

Fig. 84

The stress f due to a concentric tension force P at the center of a long bar having a cross-sectional area A, as sketched in Fig. 85, is

$$f = P/A$$

This equation assumes that the stress is uniformly distributed over the section. It is one of the most useful equations of the strength of materials. At the end of the bar, where the force must be concentrated to a smaller dimension, this

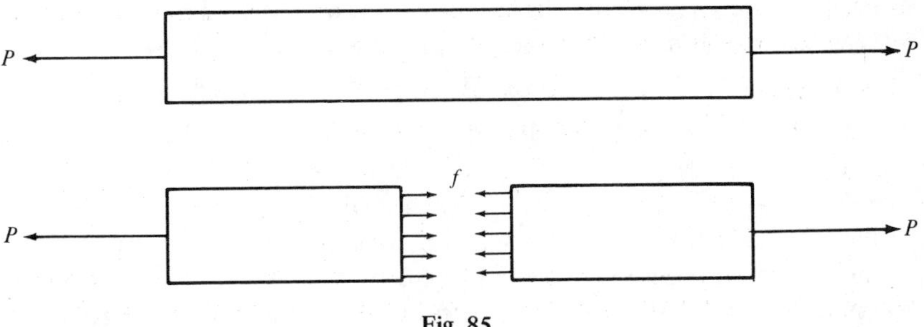

Fig. 85.

uniform stress may not hold. Also the force must pass through the center of gravity of the area of the bar, otherwise the stress f and the force P will not be in equilibrium.

Strain Associated with the uniform stress f is a *strain e*, a unit change in length of the member, with dimensions of inches per inch or feet per foot, (a dimensionless quantity). The term strain is sometimes used for the total change in length, but here it will be used as a unit change in length. In an elastic material, there is a constant relationship between stress f and strain e, called the *modulus of elasticity E*, the ratio of stress to strain:

$$E = f/e$$

The units of E are the same as those of stress because strain is dimensionless. Each elastic material has a value of E characteristic of its action in the elastic range. For example, for steel the value used for many calculations is $E = 29{,}000{,}000$ psi. The total change in length d of a bar can be obtained by multiplying the strain by the length L:

$$d = eL$$

For example, find the stress f, the strain e, and the total change in length d, of a steel bar with a cross-sectional area A of 2 sq. in., due to a force P, of 40,000 lb, the length L of the bar being 116 in.

Stress: $f = P/A = 40{,}000/2 = 20{,}000$ psi
Strain: $e = f/E = 20{,}000/29{,}000{,}000 = 0.00069$
Total change: $d = eL = 116(0.00069) = 0.08$ in.

There is another characteristic of elastic materials. A bar which is pulled lengthwise will also contract crosswise. The ratio of crosswise strain to lengthwise strain is called *Poisson's ratio*, and the symbol μ (mu) is used for this ratio. For example, if the value of Poisson's ratio for steel is $\mu = 0.25$ in the previous problem, the crosswise strain is $0.25(0.0069) = 0.0017$ in. per inch. This characteristic becomes of importance if we wish to measure the stress by means of gauges which will measure the strain. If there are stresses in two directions, then the direct measurement of strain will not correspond to the stress.

Stress-Strain Diagram for Elastic Materials Elastic materials are elastic only in a given range of stresses, after which they begin to yield. This point is called the *elastic limit* and in many materials the stress at which the material begins to fail is called the *yield stress*. For steel, the plot of stress to strain for a typical specimen tested in tension is shown in Fig. 86. The slope of the curve remains almost constant until the yield stress, after which the specimen elongates with little additional load. Eventually, however, a phenomenon called *strain hardening* occurs and it begins to pick up more load until it fails at a stress

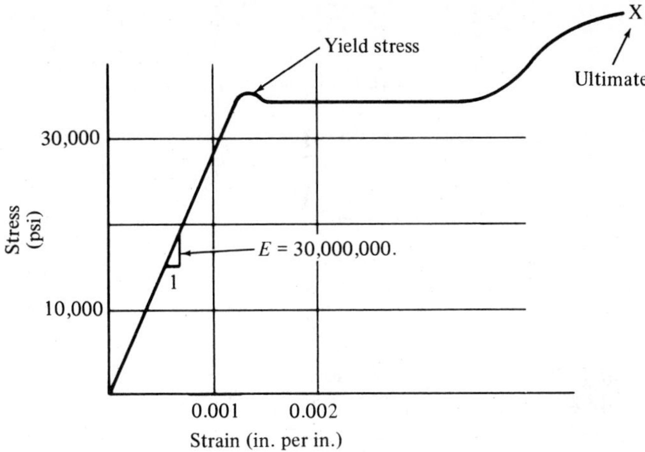

Fig. 86.

called the *ultimate*. This property of continued yielding, called *ductility*, is very useful in a material used for structures, because it makes possible simple connections where high concentrations of stress may occur.

Shear Stress and Strain In addition to direct stress where elements of a material are pushed together (compressed) or pulled apart (stretched), there is another type of stress called shear stress. An element distorted by shear forces is shown in Fig. 87. The shear stress is measured by the force per unit area along (parallel to) the sides of the element. An example of a structural element subject to shear is shown in Fig. 88. Across the section *AB*, there is a force of 3000 lb tending to shear the section. The average shearing stress is

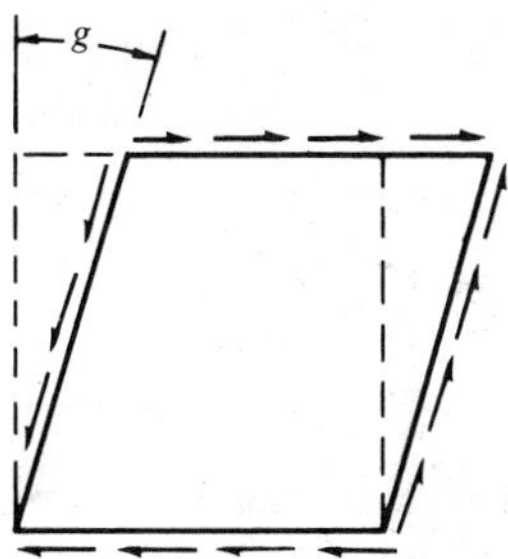

Fig. 87.

$$s = P/A = 3000/(2)(3) = 500 \text{ psi}$$

For this example the actual shearing stress across the section would probably not be uniform but the average as calculated above will give useful information to the designer of the structural element.

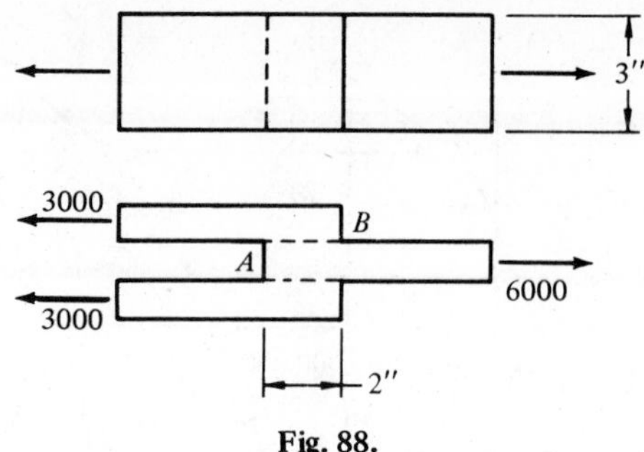

Fig. 88.

The relation between shear stress and shear strain is similar to that of direct stress and strain except that a new constant must be defined:

$$G = f/g$$

where G is the shearing modulus of elasticity and g is the shear strain. In Fig. 87, the shear strain is the angle g between the original position of the element and the final position and is measured in radians (a dimensionless quantity). There is a relationship between G and E, which also includes Poisson's ratio μ:

$$G = E/2(1 + \mu)$$

so that the shearing modulus is less than half of the modulus of elasticity.

Relationship Between Stresses at a Point A plate subjected to a uniform tensile stress in the x direction and a uniform compressive stress in the y direction is sketched in Fig. 89a. Consider a small rectangular element at the center of the plate, as shown in Fig. 89b. The stresses at any angle may be obtained by the equations of equilibrium. For example take a section at any angle of 30° to the x axis (which is 60° to the y axis), as sketched in Fig. 89c. Assume that the element is 1 in. long on the diagonal, so that the length of its projection on the x axis is 0.866 in. and on the y axis is 0.5 in. What are the stresses on the inclined face? Figure 89c shows the total forces on the x and y faces of the element. Figure 89d shows the components of f the direct stress, and g the shear stress. Writing the equilibrium equations for the x and y directions:

$$\sin 30^\circ = 0.5, \qquad \cos 30^\circ = 0.866$$

$$\sum F_x = 0, \qquad -1000 - f \sin 30^\circ + s \cos 30^\circ = 0$$
$$-1000 - 0.5f + 0.866s = 0$$

$$\sum F_y = 0, \qquad -866 + f \cos 30^\circ + s \sin 30^\circ = 0$$
$$-866 + 0.866f + 0.5s = 0$$

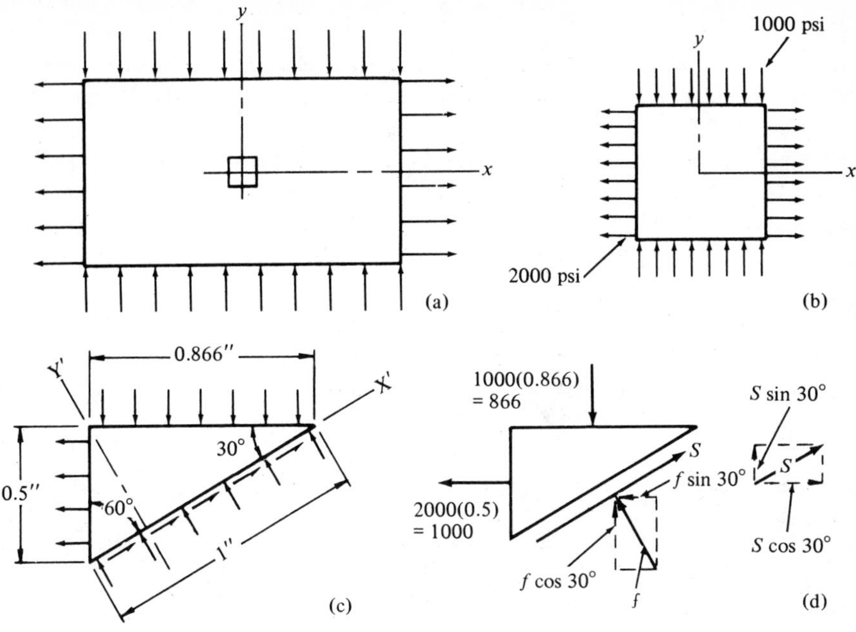

Fig. 89.

Solving these equations, $f = 250$ psi and $s = 1299$ psi. Both values are positive, so the directions assumed for the stresses are correct. The sign for the direct stress on the inclined face is negative. The shearing stress is positive if it produces a counterclockwise rotation about the element. Therefore the shear on this element is negative.

Mohr's Circle of Stress The stresses on a small element can be determined from a graphical construction using circles developed by Otto Mohr (1835–1918). Mohr's circle, for the above example, will now be discussed. The rules for construction are as follows:

1. Plot direct stresses f on the x (horizontal) axis with tension plotted as positive and compression plotted as negative. In Fig. 90a the values of +2000 and −1000 are shown plotted.

2. Plot shear stresses s on the y (vertical) axis. In the example we are discussing, the shear on the faces of the element is zero where the direct stresses exist.

3. Values on a perimeter of the circle will represent all values of direct stress and shear stress. The center of the circle is on the x axis at the average value of the stresses, −1000 and +2000, so in our example the center of the circle is at

+500 psi and the x axis. The radius of the circle is 1500 psi and is drawn in Fig. 90a.

4. To find the value of the shear at other angles, say 30° counterclockwise, by Mohr's Circle, draw a diagonal through the center of the circle. The angle of this diagonal to the x axis must be *twice* the angle of new face to the old face. In our example the angle becomes $2(30°) = 60°$. The diagonal is shown plotted in Fig. 90b. The y distances represent shear values, so the value of the shear stress s from the figure is

$$s = 1500 \sin 60° = 1500(0.866) = 1299 \text{ psi}$$

which agrees with the previous example.

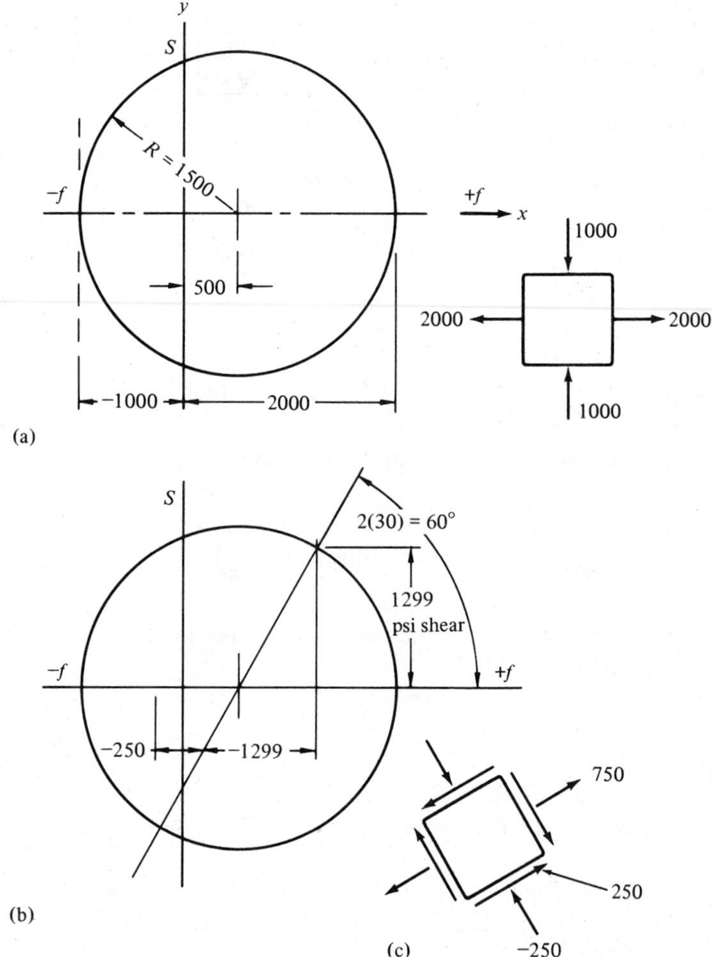

Fig. 90.

The value of the direct tensile stress corresponding to this angle is the x distance to the intersection of the circle with the diagonal:

$$f = 500 + 1500 \sin 30° = 500 + 750 = 1250$$

and the compressive stress is

$$f = 500 - 1500 \sin 30° = 500 - 750 = 250 \text{ psi}$$

A free-body diagram with the stresses on the inclined element is shown in Fig. 90c. Note that the diagonal in the construction has been rotated counterclockwise. The sign of the shear is clockwise about the center of the element if on the positive (tension) face of the element.

In this example, the largest possible value of the direct stress must be +2000 psi and the smallest −1000 psi, and with this combination the shear stress must be zero. These stresses are defined as *principal stresses*, and to designate them in calculations we will sometimes use the notation f_1 for the maximum principal stress and f_2 for the minimum principal stress. Two other useful facts also are evident: (1) The shear must be zero on the element having principal stresses. (2) The shear is a maximum when the axis is at an angle of 45° to the principal stresses and is equal to one-half the algebraic difference between the principal stresses.

Often stresses are given on sections other than those having principal stresses, and the principal stresses and their angle must be obtained.

As an alternate to Mohr's circle of stress, algebraic formulas may be used. Given f_x, f_y, and shear stress s_{xy} on the x and y axes, the principal stresses are

$$f_1 = (f_x + f_y)/2 + \sqrt{[(f_x - f_y)/2]^2 + s_{xy}^2}$$

$$f_2 = (f_x + f_y)/2 - \sqrt{[(f_x - f_y)/2]^2 + s_{xy}^2}$$

The shear stress is one-half the sum of the above stresses:

$$s = (f_x + f_y)/2$$

The angle θ of the plane of principal stresses is given by

$$\tan 2\theta = (f_x - f_y)/2s_{xy}$$

Torsion in Circular Bars A bar, circular in cross section, subjected to moments on each end, has shear stresses on sections perpendicular and parallel to the axis of the bar. Such a structure is shown in Fig. 91a, and a small element in Fig. 91b. Experimental evidence and mathematical studies indicate that for elastic materials, the magnitude of the shear stress and strain varies directly as the distance from the central axis of the bar. An element on the circular cross section of the bar is sketched in Fig. 91c.

To satisfy equilibrium, the total moment M on the cross section must be the summation of the products of the small area dA times the stress s_r on this

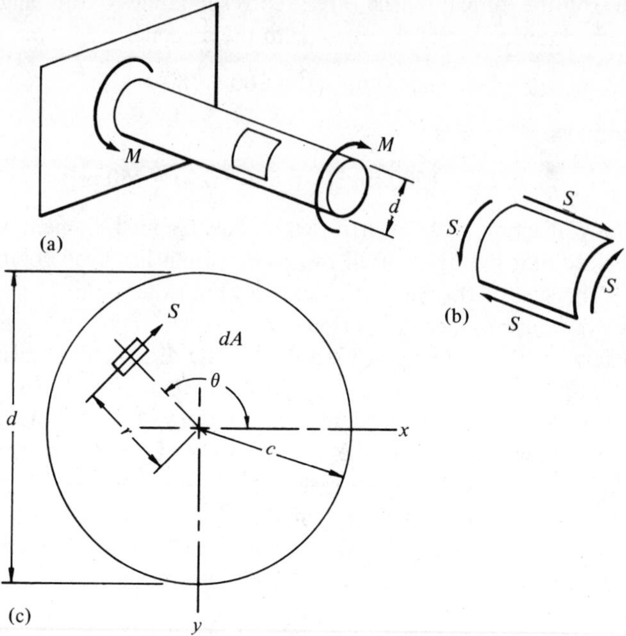

Fig. 91.

element (which varies directly as the radius r):

$$M = \int r s_r \, dA$$

Let s be the stress on the outer edge of the section; then

$$M = (s/c) \int r^2 \, dA$$

where c is the radius of the circle. The quantity $\int r^2 \, dA$ is called the polar moment of inertia J, and, solving for s,

$$s = Mc/J$$

For a circular cross section,

$$J = \int r^2 \, dA = \int_0^c 2\pi r^3 \, dr = \pi c^4/2 = \pi d^4/32$$

It is useful to remember that the polar moment of inertia is equal to the sum of the moments of inertia about the x and y axes. For a thin-walled tube the polar moment of inertia is approximately equal to $J = 2\pi c^3 t$, where t is the thickness of the tube and c is the average radius of the section.

Note that the stresses on each cross section are the same from one end of the member to the other. However, near the point of application of the forces, the method of attachment of the loading device will determine the stress pattern. Also note, from Fig. 91b, that the shear stresses on a plane parallel to the axis of the bar are the same as those perpendicular to the axis. This must be true from the equations of equilibrium. Other stresses may be determined by Mohr's circle of stress or from the equations.

An example will now be studied. The bar shown in Fig. 92a is 1.75 in. in diameter and is subjected to a torque moment of M = 1000 in.-lb. What is the shear stress on the surface of the bar? A length of 120 in. is also shown but will not be needed until we calculate the strain and twist of the bar.

The value of the polar moment of inertia J for a bar 1.75 in. in diameter is

$$J = \pi d^4/32 = \pi(1.75^4)/32 = 0.921, \qquad c = d/2 = 0.875$$

and the shear stress is

$$s = Mc/J = 1000(0.875)/0.921 = 950 \text{ psi}$$

The maximum direct stress is at 45° and may be obtained from Mohr's circle of stress (Fig. 92b). It is numerically equal to the shear stress, 950 psi, and may be either tension or compression. A free-body diagram is shown in Fig. 92c.

The deflection of one end of the shaft with respect to the other, which we will call the angle of twist, may be obtained from the relationship between the shear stress and the shear strain previously discussed. For this example, the shear strain g is an angle in radians per unit length at the surface of the bar for a shear stress s = 950 psi, assuming a modulus of elasticity of E = 10,000,000 (a value often used for aluminum), and a Poisson's ratio of $\mu = 0.25$:

$$G = E/2(1 + \mu) = 10(10^6)/2(1 + 0.25) = 4(10^6)$$

$$g = s/G = 950/4(10^6) = 2.37(10^{-4}) \text{ radians per inch}$$

The total angle of twist from A to B is

$$2.37(10^{-4})(120) = 0.0284 \text{ radians} = 0.0284(180)/\pi = 1.63°$$

Direct Stresses in Beams The determination of shear and moment forces has been presented in a previous subsection. We will now study the direct and shear stresses resulting from these shears and moments.

The basic assumptions for determining the direct stress in beams have been confirmed by experiment and by theoretical studies.

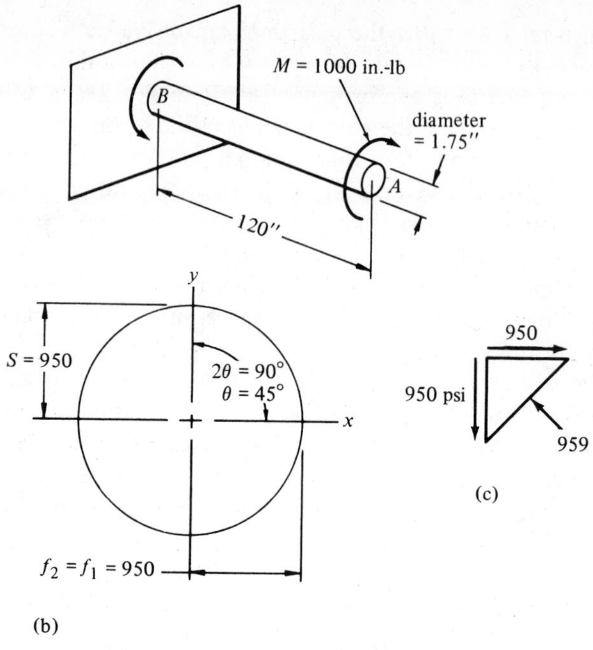

Fig. 92.

1. The longitudinal strain on an element in a cross section of a beam is proportional to the distance of that element from the neutral axis (that point where the strains are zero).

2. The direct longitudinal stress is proportional to the strain. Therefore the stresses are also proportional to their distance from the neutral axis. The formulas for stress result from the application of the equations of equilibrium using these assumptions. A section of a beam is shown in Fig. 93. On a small element dA at a distance y from the neutral axis the stress is f_y, and the stress at the top will be represented as f, so that $f_y = fy/c$.

The first equation of equilibrium to be applied is $\sum F_x = 0$. The force on a

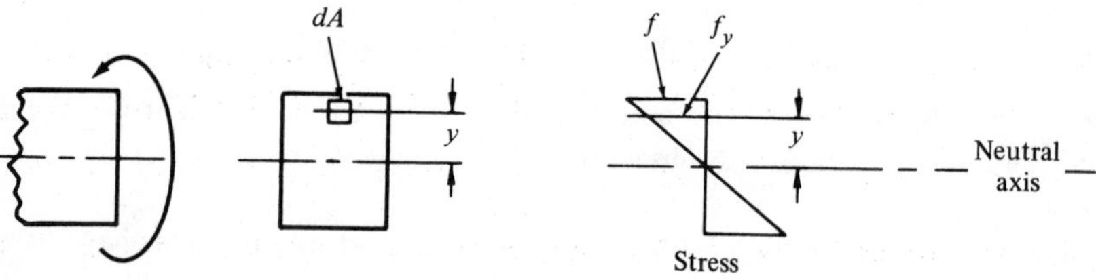

Fig. 93.

small element is $f_y\, dA$ so the total over the area is

$$\sum F_x = 0, \quad \int f_y\, dA = 0$$

Replace the stress f_y with fy/c;

$$\int_c^{fy} dA = 0 = \frac{f}{c}\int y\, dA$$

The term f/c will not be zero, so the quantity $\int y\, dA$ must be zero. This equation is the same one used to find the center of gravity of a section. Therefore, we can state that the neutral axis must lie on the center of gravity of the section.

The second equation of equilibrium to be considered is $\sum M = 0$. The sum of the external moment M and the moment of the internal forces is

$$M + \int (f_y\, dA)y = 0, \quad M + \frac{f}{c}\int y^2\, dA = 0$$

The term $\int y^2\, dA$ is the moment of inertia I of the section, so that the equation for stress reduces to

$$f = -Mc/I$$

This equation is called the *flexure formula* and is one of the most useful equations in structural engineering.

The concept of the moment of inertia has been discussed previously. For a rectangular section, $I = bd^3/12$. The term I/c is also useful, and is called the section modulus. For a rectangle the value of $I/c = Z = bd^2/6$.

Find the stress on a rectangular section 8 in. wide by 15 in. high due to a bending moment of 20,000 ft-lb:

$$I = bd^3/12 = 8(15^3)/12 = 2250 \text{ in.}^4$$

The value of c is half the depth, $c = 7.5$ in. In determining the stress, remember to multiply the moment in foot-pounds to convert it to inch-pounds.

$$f = 20000(12)(7.5)/2250 = 800 \text{ psi}$$

The sign convention will be the same as that used for beam analysis. The stress will be compression (−) on the top fiber and tension (+) on the bottom fiber of the beam if the bending moment is positive.

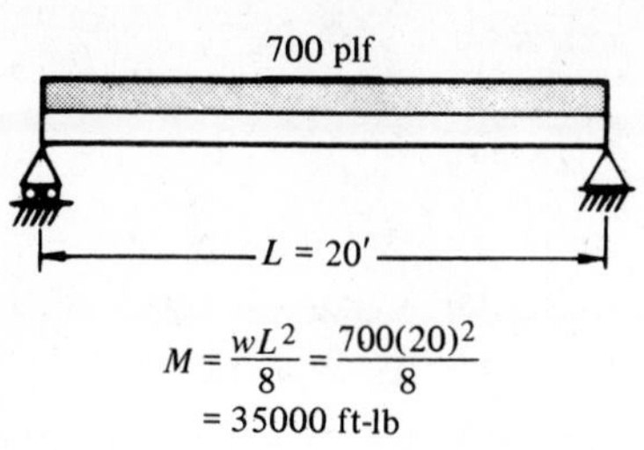

$$M = \frac{wL^2}{8} = \frac{700(20)^2}{8}$$
$$= 35000 \text{ ft-lb}$$

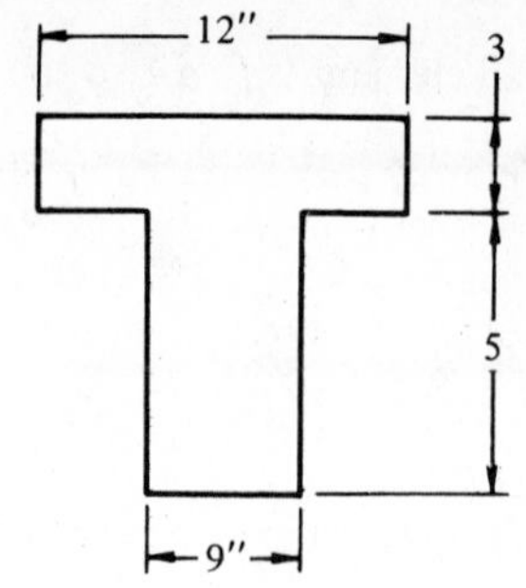

Moment of Inertia of Section

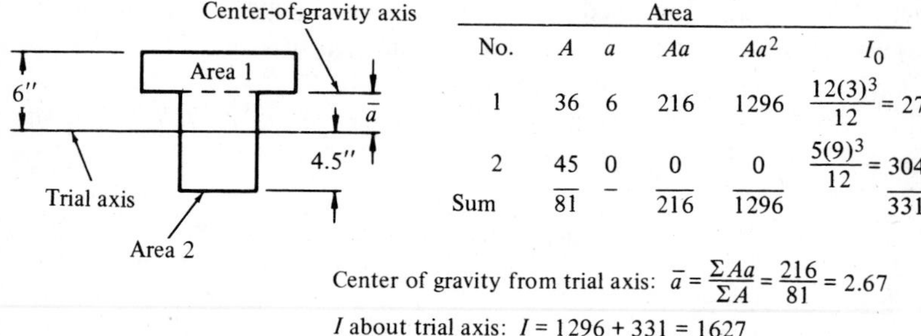

No.	A	a	Aa	Aa^2	I_0
1	36	6	216	1296	$\frac{12(3)^3}{12} = 27$
2	45	0	0	0	$\frac{5(9)^3}{12} = 304$
Sum	81	–	216	1296	331

Center of gravity from trial axis: $\bar{a} = \frac{\Sigma Aa}{\Sigma A} = \frac{216}{81} = 2.67$

I about trial axis: $I = 1296 + 331 = 1627$

I about center of gravity: $I = 1627 - 81(2.67)^2 = 1049$

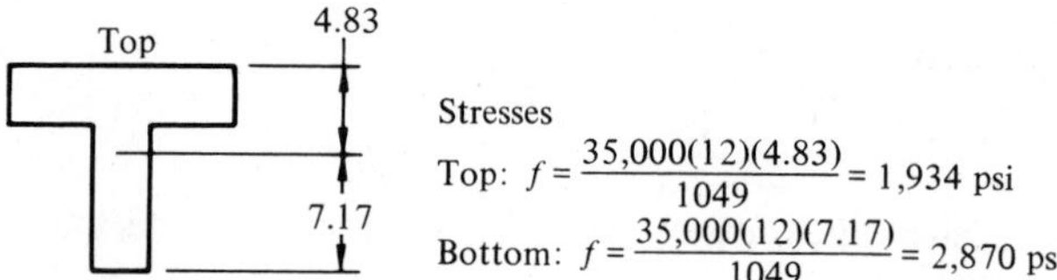

Stresses

Top: $f = \frac{35,000(12)(4.83)}{1049} = 1,934 \text{ psi}$

Bottom: $f = \frac{35,000(12)(7.17)}{1049} = 2,870 \text{ psi}$

Fig. 94.

Another example will be presented for a beam which is unsymmetrical about a horizontal axis. In Fig. 94, a beam is shown which has a span of 20 ft and a load of 700 lb per lineal foot (700 plf). A T section is used as shown in Fig. 94. Determine the stresses at the top and the bottom of the beam. All calculations are shown on the figure. The bending moment at the center is determined from the equation: $M = wL^2/8$ and is 35,000 ft-lb.

The following are the steps for calculation of the section properties:

1. Divide the area of the cross section into convenient rectangular areas. There are several possibilities but each should give the same final result.

2. Select any convenient trial axis. If it goes through the center of one of the rectangular areas it will make some of the figures zero in the calculations, so you will have an advantage. In our case, the center of area no. 2 was selected.

3. In a table, calculate for each element, the area A, the arm of the area about the trial axis a, the first moment Aa, the second moment Aa^2, and the moment of inertia of the area about its own axis, $I_0 = bd^3/12$.

4. Sum the values of A, Aa, Aa^2, and I_0.

5. The moment of inertia of the section about the trial axis is $I_t = \sum Aa^2 + \sum I_0$.

6. The distance of the center of gravity axis from the trial axis is $\bar{a} = \sum Aa/\sum A$.

7. The moment of inertia about the center of gravity axis is

$$I = I_t - \sum A(a^2) = 1049 \text{ in.}^4.$$

8. The last step is to determine the stresses top and bottom by the formula $f = Mc/I$. In this case the c value will be 4.83 in. for the top and 7.17 in. for the bottom.

Shear Stresses In Beams If the direct stresses on a section are known, then the shearing stresses may be determined by the equations of equilibrium. Consider an element of a beam of length dy, as sketched in Fig. 95a. It is subjected to moments M, to the left and an increase in moment $M + dM$, to the right. The direct stresses at the top of the beam are, by the flexure formula, $f = Mc/I$, where the c corresponds to the distance from the neutral axis to the top. On the right the corresponding stress is $f = (M + dM)\,c/I$. In Fig. 95c is shown a partial element with a horizontal shear s on the section at a short distance above

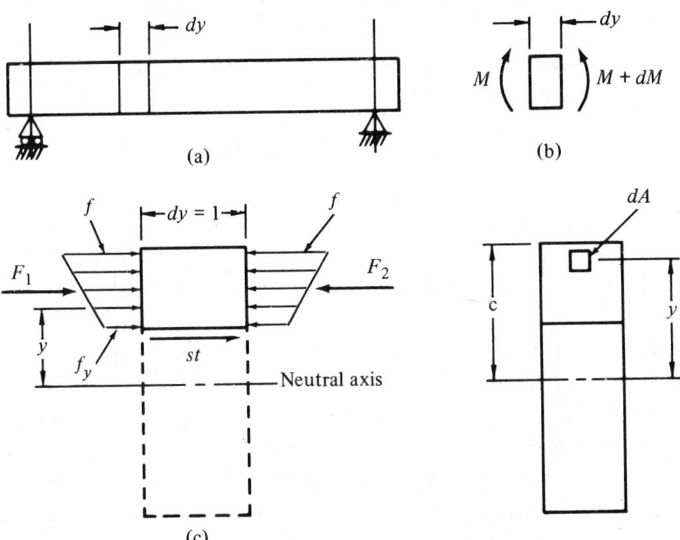

Fig. 95.

the neutral axis. The shear on this section may be obtained from the difference between forces F_1 and F_2 (Fig. 95b) divided by the horizontal area of the element. Let f_y represent the stress at a distance y from the neutral axis. Then $F_1 = \int f_y \, dA$. Substitute in the expression for the flexure formula and then $F_1 = (M/I) \int y \, dA$. The expression for the difference in stresses, $F_1 - F_2$ becomes $(dM/I) \int y \, dA$.

The term $\int y \, dA$ is *the first moment of the area about the neutral axis*, for which the symbol Q will be used. The expression for the shearing stress now becomes $s = dMQ/It$, but the value of dM may be replaced by the external shear V at the section because the rate of change of moment is equal to the shear. Finally the equation for the shearing stress becomes

$$s = VQ/It$$

From the previous example (Fig. 94), the shear at several points on the cross section of the unsymmetrical section will be calculated:

1. Determine the external shear at a point 2 ft from the left end of the beam: $V = (10 - 2)(700) = 5{,}600$ lb.

2. Find the shear at the neutral axis of the section. Either the area above or that below the neutral axis may be used, but the area below will be more convenient. This area is shown shaded in Fig. 96a. The value of Q, the first moment of the area below the neutral axis, is the distance from the center of the area to the neutral axis:

$$Q = 7.17(5)(7.17)/2 = 128.5$$

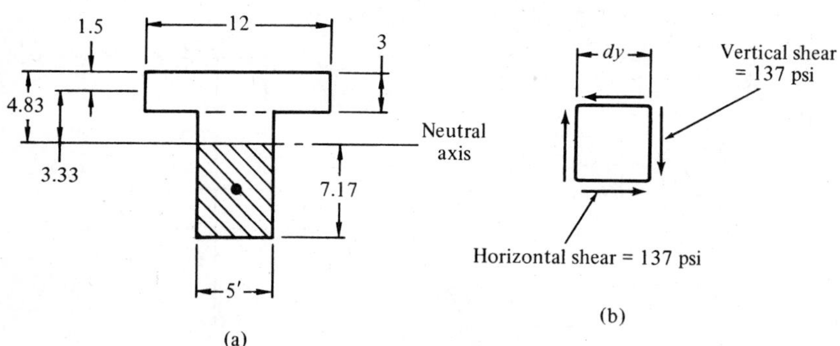

Fig. 96.

The horizontal shearing stress is

$$s = VQ/It = 5600(128.5)/1049(5) = 137 \text{ psi}$$

The vertical shearing stress, by equilibrium of the small element, has the same numerical value as the horizontal stress, as sketched in Fig. 96b.

3. Find the shear at the junction of the stem (5 in. wide) and the flange (12 in. wide). The values of Q and s are

$$Q = 12(3)(3.33) = 120$$

$$s = 5600(120)/1049(5) = 128 \text{ psi}$$

Note that the shear in the stem was determined by using a value of $t = 5$ in. The shear in the flange would be based on a 12-in. width, but practically speaking, at this point the shearing stress would have very little meaning because it would have to spread out to the 12-in. width in a very short distance.

Deflections of Beams The calculation of the deflection of beams due to bending stresses is important in structural engineering because the design must be made so that the deflection is not excessive and the beam will function properly. In the following presentation the effect of shear stresses have not been considered.

In Fig. 97a is shown a beam for which a small element of length dl is subjected to a bending moment M. The deformation and stresses on this element are sketched in Fig. 97b. The bending moment is positive, so the lower fiber

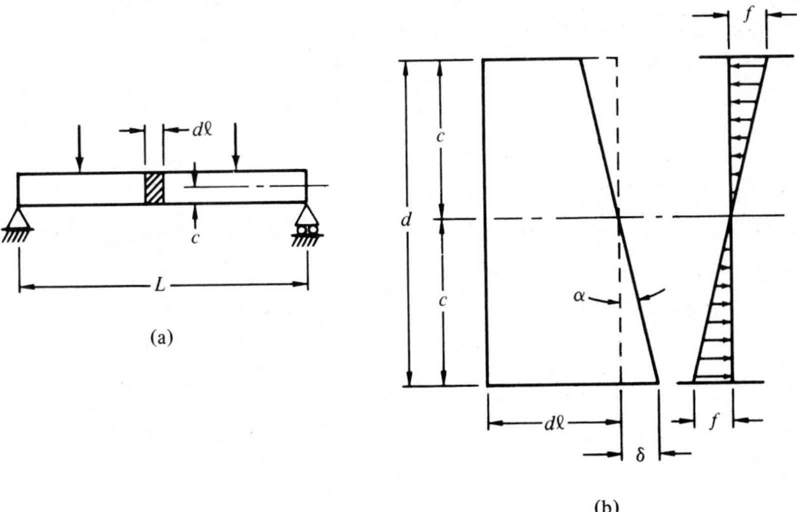

Fig. 97.

elongates and the upper fiber contracts, both a distance δ, which may be computed from the known unit stress ($f = Mc/I$) on the fiber. This change in length is $\delta = dl\, f/E$, where E is the modulus of elasticity. The angle α for such small angles is

$$\alpha = \delta / c \text{ (radians)} = dl\, f/Ec = M\, dl/EI$$

This is a small angular distortion for any point in a beam, and from this value the deflections of beams may be determined.

Deflections By Geometry The most simple and basic method of determining the deflection of a beam with a single angular change is by geometry. The first example will be the cantilever beam shown in Fig. 98, which has received an

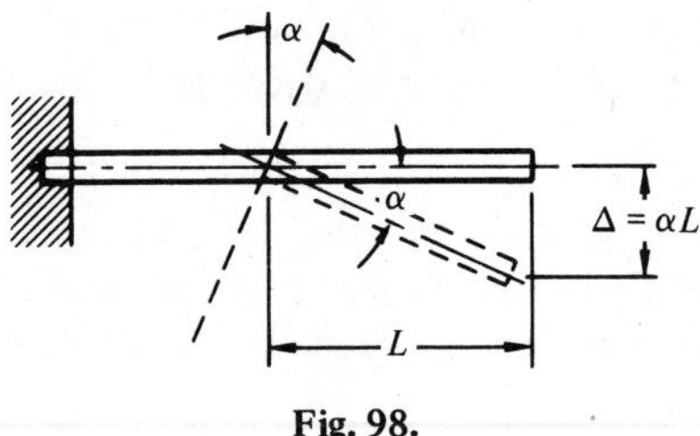

Fig. 98.

angle change α in the middle of the span, a distance L from the end. Note that the angle between the faces of the element is also the angle between the center lines of the member. The angles are assumed to be very small, so that the sine of the angle and the tangent of the angle are nearly equal, and the deflection is the angle α times the distance L from the end; therefore

$$\Delta = \alpha L$$

The next case is a little more complicated, and is the simple beam shown in Fig. 99, with an angle change α at a distance a from the left end. As in the previous case, the angle between the faces of the element is the same as between the center lines of the beam. This problem is solved by finding first Δ_b, the

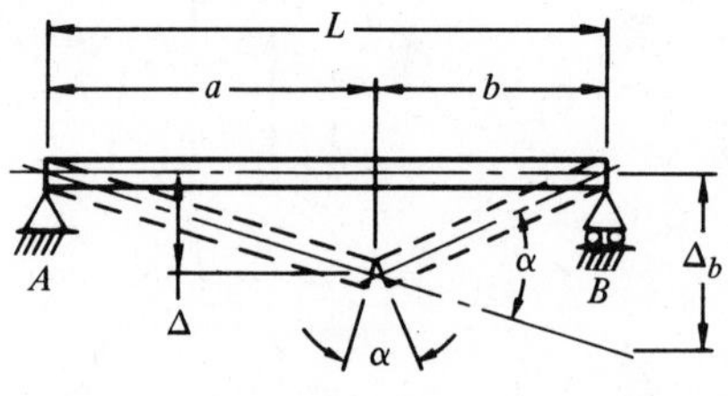

Fig. 99.

deflection at point B for the axis of the part of the beam on the left. By geometry, $\Delta_b = \alpha b$. Now the deflection Δ may be computed by the ratio of distances a to L:

$$\Delta : a = \Delta_b : L; \qquad \Delta = \Delta_b a/L = \alpha ba/L$$

Thus the deflection has been obtained by a completely geometrical method.

It is possible to solve any deflection problem by geometry, but it would take considerable study to become accustomed enough to this method to solve working problems. Therefore, the *elastic-weight analogy* will be used. This method is based on the condition that the equations used to find the deflection due to a small angle change α are the same as those used to find the bending moments due to a load α, called an "elastic weight."

The Elastic-Weight Analogy The simplest case to demonstrate the analogy is the cantilever beam of Fig. 98. The deflection at the end of the beam, determined by geometry, was $\Delta = \alpha L$. In Fig. 100, the same beam is shown with

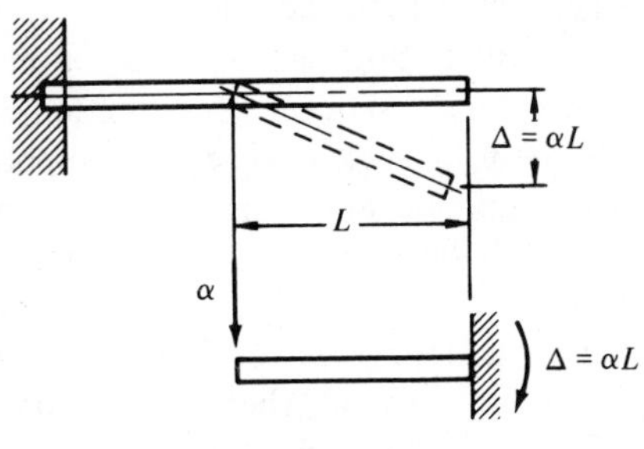

Fig. 100.

another beam sketched below it, called the conjugate beam.* Note that this beam is supported rigidly at the right end, corresponding to the free end of the real beam.

Because of an elastic load α on the conjugate beam, the bending moment in the conjugate beam at the right end is $\Delta = \alpha L$. The bending moments in the conjugate beam correspond to the deflections in the real beam. Thus it is possible to calculate deflections by methods already developed for computing reactions, shears, and moments. The calculations required will be simpler than those for computing deflections by geometrical methods. Two new concepts have been introduced: (1) the elastic weight and (2) the conjugate beam. It will take a bit of study before these concepts become completely clear.

The simple beam in Fig. 99, with a single angle change, has already been

*"Deflection of Beams by the Conjugate Beam Method" by H. M. Westergard, *J. Western Society of Engineers*, November 1921.

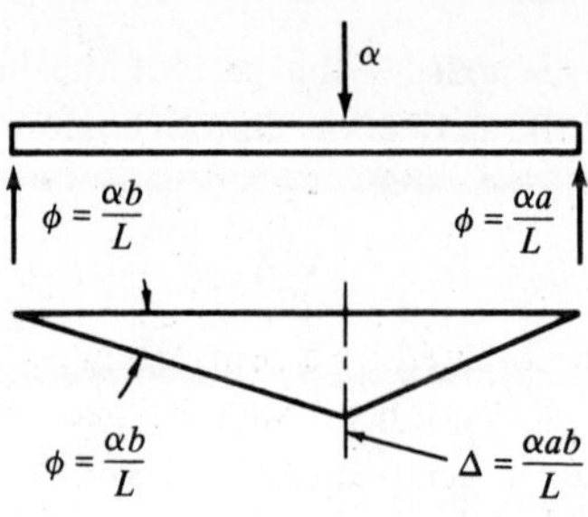

Fig. 101.

analyzed by the methods of geometry. The conjugate beam in this case is also a simple beam, as sketched in Fig. 101. The bending moment, Δ, under the elastic load, α, is $\Delta = \alpha ab/L$. To the engineer acquainted with the calculation of shears and moments, it is much more convenient to use the elastic-weight analogy than it is to use geometrical calculations.

An additional property of the analogy is that the shear on the conjugate beam corresponds to the slope ϕ, of the real beam. The reaction of the elastic weights on the conjugate beam is the total angle change at the support of the real beam. Examine the angles in the simple beam on Fig. 101 to verify this fact. A table summarizing the relationship between the elastic weight analogy and geometry is given in Table 3.

The elastic-weight analogy is applicable to all types of problems where the angular changes are known (e.g., a truss). When this method is applied to beams, it is sometimes called the *moment-area method.* The elastic weight has been

TABLE 3 NOMENCLATURE–ELASTIC WEIGHT ANALOGY*

Alpha, α	=	*Geometry:* A small angular distortion in a beam or other structural framework. *Elastic-weight analogy:* An elastic weight.
Theta, θ	=	*Geometry:* The slope of the beam at any point. *Elastic-weight analogy:* The shear at any point on the conjugate beam, due to the elastic loads.
Phi, ϕ	=	*Geometry:* A large angular change at a support or at a hinge in the real beam. *Elastic-weight analogy:* A reaction due to the elastic loads on the conjugate beam.
Delta, Δ	=	*Geometry:* A deflection at any point. *Elastic-weight analogy:* The bending moment due to the elastic loads.

*The terminology is given here with two meanings (1) geometrical and (2) according to the elastic-weight analogy. In this discussion, these terms without a subscript refer to an angle change, slope, or deflection at any point. A subscript refers to a specific location; e.g., ϕ_A refers to an angle change at point A.

defined as $\alpha = M\,dl/EI$. The quantity $M\,dl$ is a small element of area of a moment diagram. Hence, the name "moment area."

Moment areas are used as distributed loads on the conjugate beam. The technique of computing the bending moments due to uniform or distributed loads is well known to most designers. The corresponding geometrical problem, however, is quite difficult without developing special techniques. Examples of the calculation of deflections follow.

EXAMPLE: CANTILEVER BEAM

As an example of the application of the moment-area method, the deflection of a cantilever beam, loaded with a single concentrated load P at the end, will be described. The beam, the moment diagram, and the conjugate beam and elastic weights are sketched in Fig. 102. The I value for the beam is the same at all points.

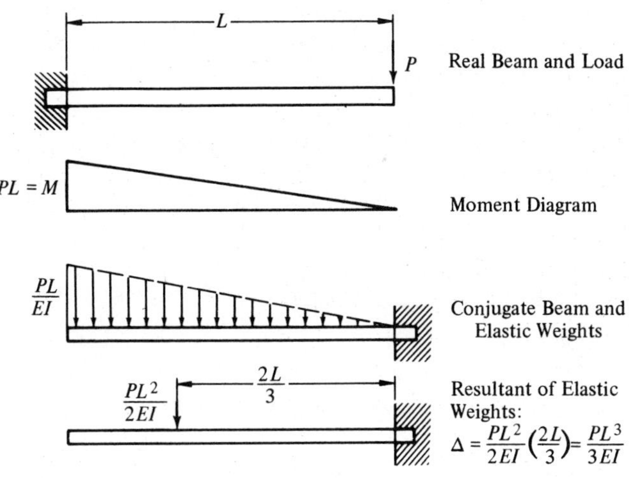

Fig. 102.

The bending moment at the support of the real beam is PL; therefore, the elastic weight per unit length at that point is PL/EI. The deflection in the real beam is the bending moment due to the elastic weights acting on the conjugate beam. The resultant of all elastic weights is the dimension of the elastic weight at the left end, PL/EI, times the span L, divided by two. The position of the resultant of elastic weights on the beam is at a point two-thirds of the distance from the right end. The bending moment due to this elastic load is

$$\Delta = \frac{PL}{EI}\left(\frac{L}{2}\right)\frac{2L}{3} = \frac{PL^3}{3EI}$$

The moment-area method may be used to solve problems algebraically, as in the above example, or the deflection may be obtained for a definite set of loads, spans and moments of inertia, as in the next example.

EXAMPLE: SIMPLE BEAM

This problem, shown in Fig. 103, is to determine the deflection of a simple beam of 20-ft span due to a single load of 12 kip at the center. The moment of

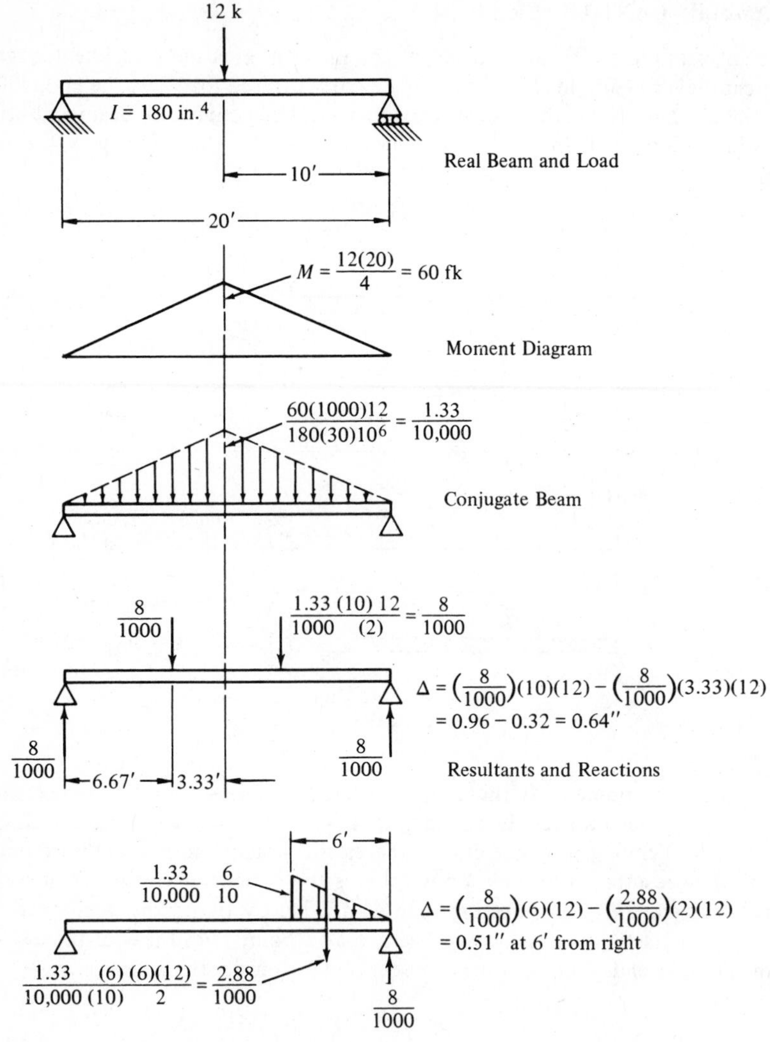

Fig. 103.

inertia of the section is 180 in.^{4}, and the modulus of elasticity is $E = 30 \times 10^6$ psi. Determine the deflection under the load, and at a point 6 ft from the end.

The calculations for this problem are shown in the figure. The following steps are required for the solution:

1. Compute the bending moment at the center of the span. $M = PL/4 =$ 60 ft-kip. Draw the bending moment diagram.
2. Compute the size of the elastic weight at the center and draw the conjugate beam.
3. Determine the resultant of the elastic weights to each side of the center.
4. The reaction at each end is by observation the same as the elastic weights.
5. The deflection at the center, Δ, will be the moment of the reaction less the moment of the loads between.

The Conjugate Beam The supports for two typical conjugate beams have been described: that for a simple beam and that for a cantilever beam. The conjugate beam for any other type of real beam, determinate or indeterminate, may be set up from the following rules:

a. A free end becomes fixed.
b. A fixed end becomes free.
c. An exterior knife-edged support remains a knife-edged support.
d. An interior knife-edged support becomes a hinge.
e. An interior hinge becomes a knife-edged support.

All these rules can be proven by consideration of the elastic-weight analogy and the cantilever beam used to demonstrate the analogy. Take, for example, rule *d*. An interior knife-edged support becomes a hinge. At a knife-edged support the deflection is zero. Therefore, there must be no bending moment at this point in the conjugate beam. The only condition by which this requirement can be fulfilled is that there be a hinge in the conjugate beam.

In Fig. 104 are shown a number of examples of real beams and corresponding conjugate beams. In Fig. 104c and d there are continuous beams which are statically indeterminate. Note that apparently the conjugate beams for these structures are unstable, but the elastic weights must keep the beam in equilibrium. This provides a key to the solution for continuous beams. Also note that none of these conjugate beams are indeterminate.

Signs for Elastic Weights and Deflections The sign convention follows the sign convention for shear and moment. For vertical deflections (horizontal deflections will be discussed later) a positive real bending moment corresponds to a downward elastic weight. A positive bending moment due to elastic weights produces a downward deflection. The positive direction is downward for both elastic weights and deflections.

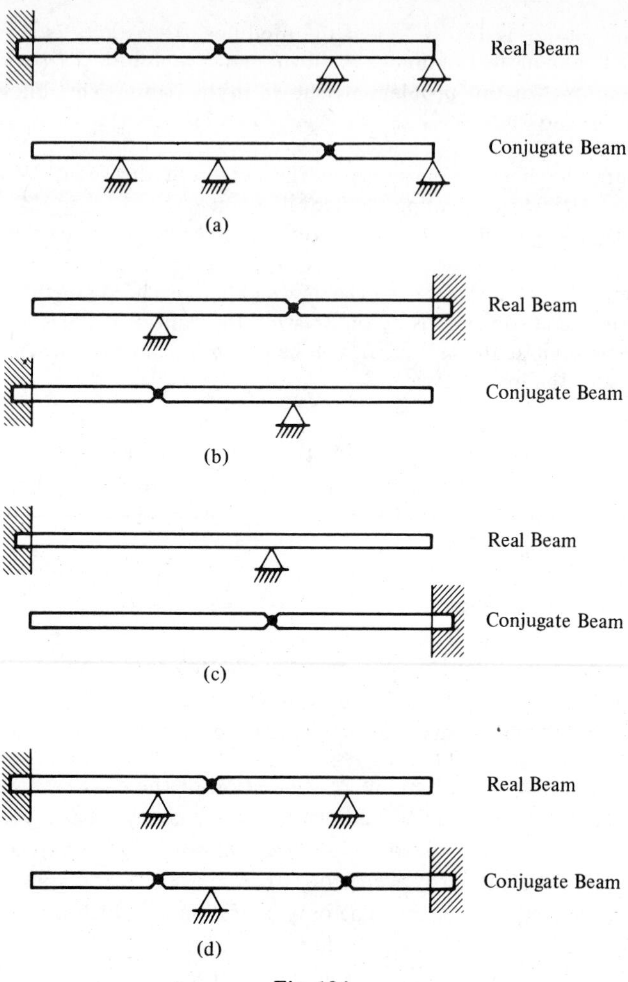

Fig. 104.

With respect to rotation and slope, a positive shear due to elastic weights on the conjugate beam corresponds to a clockwise rotation from the original undeflected position.

Uniform Loads Uniform loads need special consideration here, because the moment diagrams consist of curved figures, usually parabolas. The best way to treat these cases is to break the moment diagram into squares, triangles and parabolas. The areas and centers of gravity of figures bounded by common (second-degree) and third-degree parabolas are shown in Fig. 105.

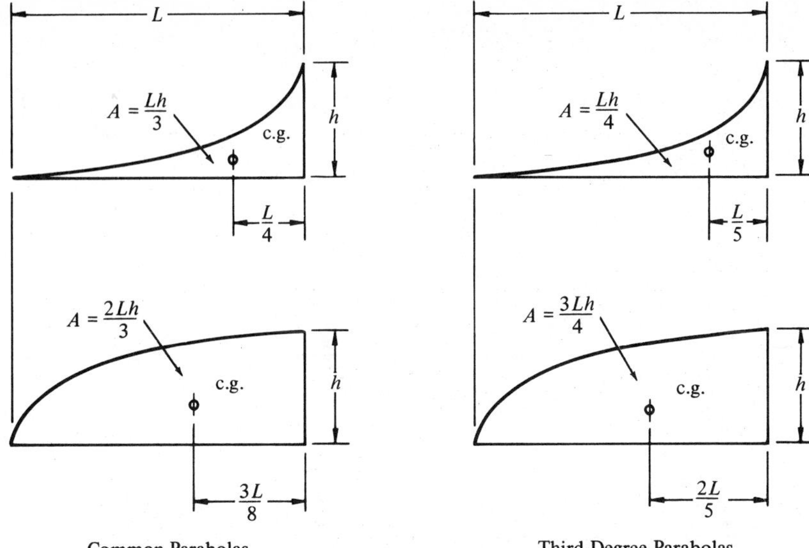

Fig. 105.

EXAMPLE:

A typical example is the simple beam loaded with a uniform load over a portion of the span, as shown in Fig. 106. The problem is to find the deflection at the center of the beam. The shear and moment diagrams are computed by the usual methods. The moment diagram has an irregular shape. Rather than use the diagram as shown, we have broken it up into two equivalent diagrams, a parabola with a negative value and a triangle with a positive value. The equations of these diagrams are very simple to write. For example, the equation of the parabola is $M = 1.75X^2/2$, where X is measured from the point B, 6 ft from the right end. To find the moment at the center, simply substitute $X = 4$, which gives $M = 14$ ft-kip.

For convenience, all computations are made in units of inch-kips. The term EI has been left as an algebraic quantity, and the numerical value has been used only at the final computation of deflection.

The next step in the solution of this problem after the moment diagrams have been computed is to determine the resultants of the elastic weights for the entire parabolic area and the triangular area. These elastic weights are then applied as concentrated loads on the conjugate beam, which in this case is a simple beam. The reactions due to these resultant weights are then computed. These figures represent the slope of the deflected beam at the ends.

To obtain the bending moment at the center of the beam, determine the sum of elastic weights to one side of the center line. The weights to the right are

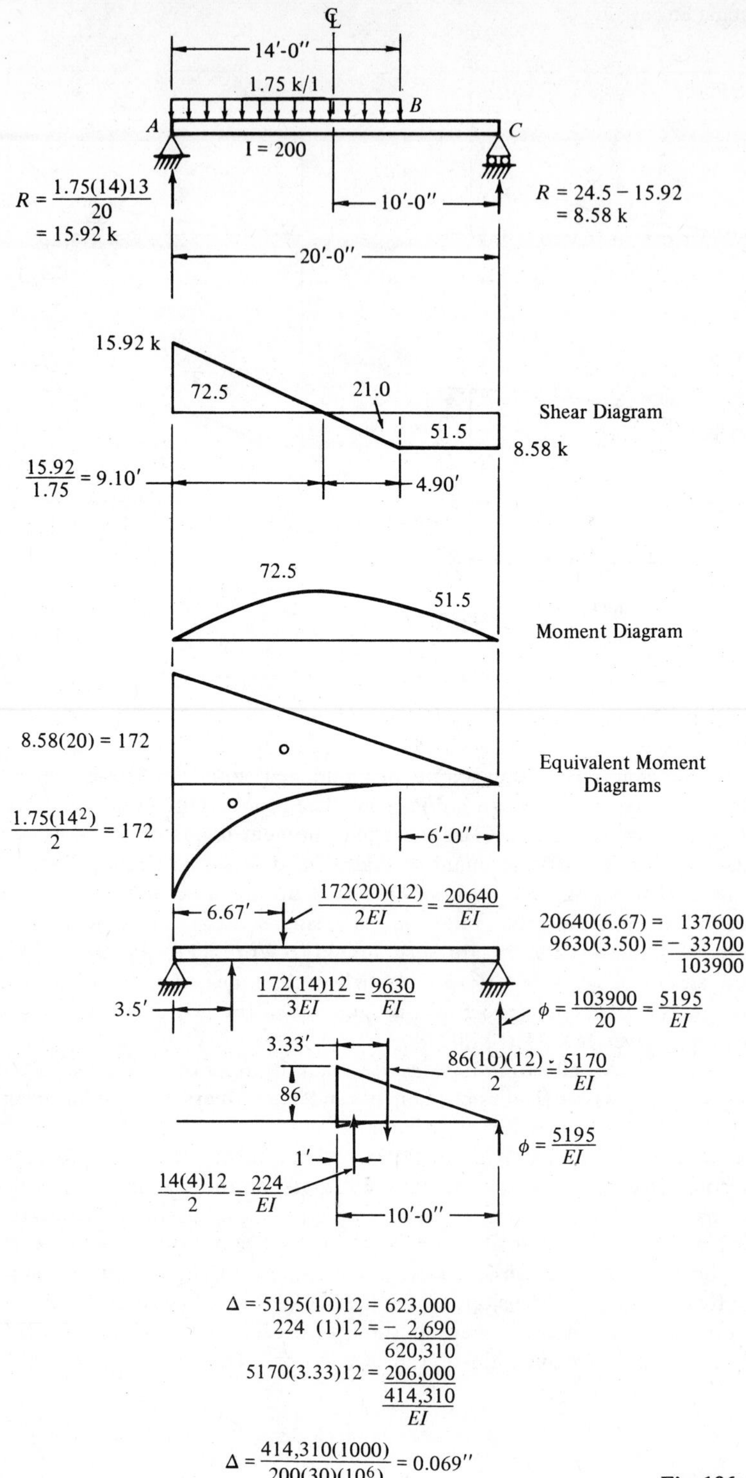

℄
14′-0″
1.75 k/1
B
A
C
I = 200
R = 1.75(14)13 / 20
= 15.92 k
10′-0″
R = 24.5 − 15.92
= 8.58 k
20′-0″
15.92 k
72.5
21.0
Shear Diagram
51.5
8.58 k
15.92 / 1.75 = 9.10′
4.90′
72.5
51.5
Moment Diagram
8.58(20) = 172
Equivalent Moment
Diagrams
1.75(14²) / 2 = 172
6′-0″
6.67′
172(20)(12) / 2EI = 20640 / EI
20640(6.67) = 137600
9630(3.50) = − 33700
103900
172(14)12 / 3EI = 9630 / EI
3.5′
φ = 103900 / 20 = 5195 / EI
3.33′
86(10)(12) / 2 = 5170 / EI
86
φ = 5195 / EI
1′
14(4)12 / 2 = 224 / EI
10′-0″
Δ = 5195(10)12 = 623,000
224 (1)12 = − 2,690
620,310
5170(3.33)12 = 206,000
414,310 / EI
Δ = 414,310(1000) / 200(30)(10⁶) = 0.069″

Fig. 106.

the simplest to use, because there is a single triangle and a small parabola, for which the area and center of gravity are known. The deflection at the center is the moment of these weights and the right reaction about the center.

Summary The deflection of beams requires a consideration of the elastic properties of the beams. The small angular change in a length dl of a beam may be expressed by the equation

$$\alpha = M\,dl/EI$$

The deflection of a beam due to this small angular change may be computed by geometry, but a less complicated method is to use the elastic-weight analogy. In this method, the small angle changes are assumed to act as loads called elastic weights, on a beam called the conjugate beam. The bending moments due to these elastic weights on the conjugate beam are the deflections on the real beam. When used in connection with beams, the elastic-weight method is sometimes called the moment-area method.

Deflection of Rigid Frames

In the preceding subsection, methods for computing the deflections of horizontal beams were presented. In this subsection deflections will be computed for beam structures for which members are at an angle to each other and have supports on several levels. Such structures will be called *rigid frames*, and it will be necessary to consider both horizontal and vertical deflections.

Deflections By Geometry The presentation of the deflections of rigid frames will start with the simple cantilever beam shown in Fig. 107, which may be solved by geometry. This beam is on a slope and has received an angular deformation at a point C in the middle. Determine the deflections Δ_x and Δ_y at the end due to the angle change α. The horizontal distance from the end of the member will be called x and the vertical distance y.

The algebraic expression for the solution of this problem by geometry is

$$\Delta_x = y\alpha; \qquad \Delta_y = x\alpha.$$

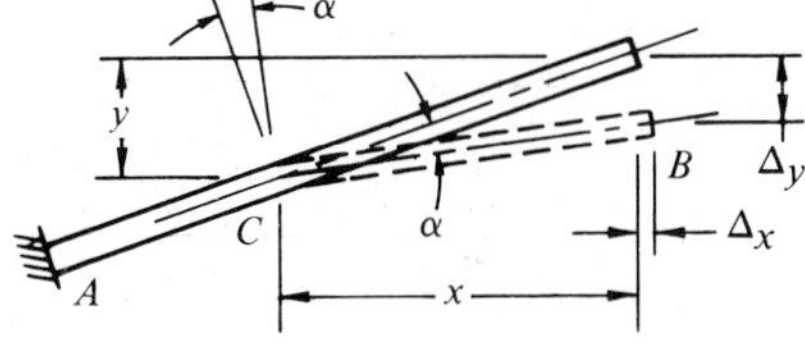

Fig. 107.

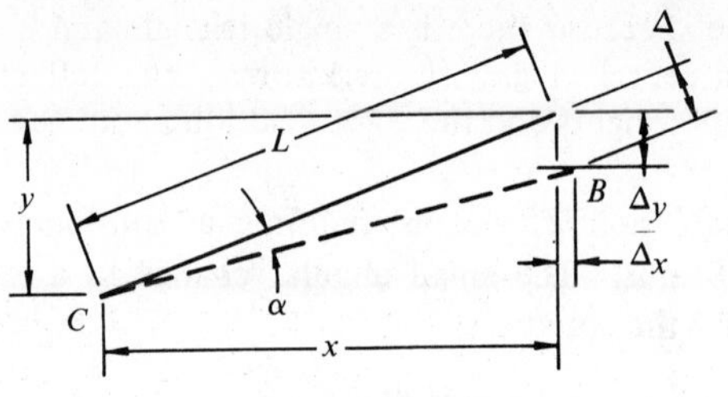

Fig. 108.

The proof of this statement is given by the construction in Fig. 108. Let L be the distance from the deformation to the end of the beam and Δ be the total deflection of the end. Then

$$\Delta = L\alpha$$

But by geometry

$$\Delta : \Delta_x = L : y \qquad \text{and} \qquad \Delta : \Delta_y = L : x$$

Therefore

$$\Delta_x = y\alpha \qquad \text{and} \qquad \Delta_y = x\alpha$$

The horizontal and vertical deflections for any frame can be obtained by geometrical methods, but the method will be too cumbersome for practical use. Again the elastic-weight analogy will enable complicated problems to be solved simply and expeditiously.

The Elastic-Weight Analogy To apply the elastic-weight analogy to vertical and horizontal deflections the following rules for the direction of the elastic weights must be observed:

a. For vertical deflections the elastic weight must act, vertically.
b. For horizontal deflections the elastic weight must act horizontally.

The elastic weights for the beam just discussed are shown in Fig. 109. The conjugate beam is fixed at point B (the free end of the real beam). The bending

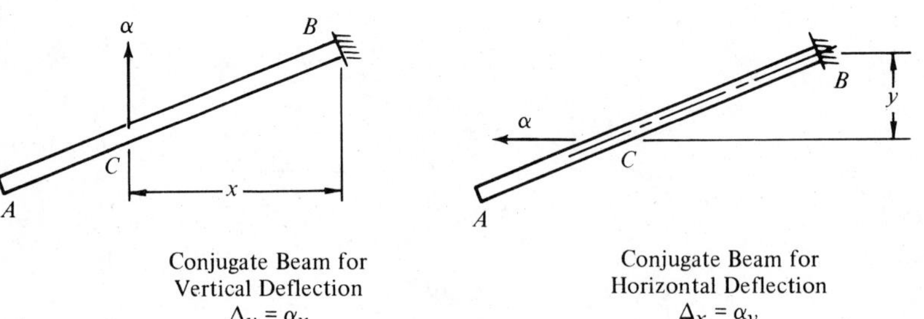

Conjugate Beam for Vertical Deflection $\Delta_y = \alpha_x$

Conjugate Beam for Horizontal Deflection $\Delta_x = \alpha_y$

Fig. 109.

moment at the support due to the elastic weight α is the deflection at that end. Therefore

$$\Delta_y = \alpha x \qquad \text{and} \qquad \Delta_x = \alpha y$$

A typical example to illustrate is the frame shown in Fig. 110a. The deflection conditions at the supports are that there is a horizontal deflection at point A and a rotation at points A and B. The two conjugate beams are shown in Fig. 110b and c. On the conjugate beam for vertical deflection, the only reactions are the end rotations ϕ_A and ϕ_B. Note that all the elastic weights, including the reactions, are vertical.

On the conjugate beam for horizontal deflection there are three reactions: ϕ_A; ϕ_B, which represent rotations; and the moment Δ_A, which represents deflection. All the forces are horizontal. The magnitude of the elastic weights on the top member has been indicated by dotted lines. The elastic weights are applied along the member.

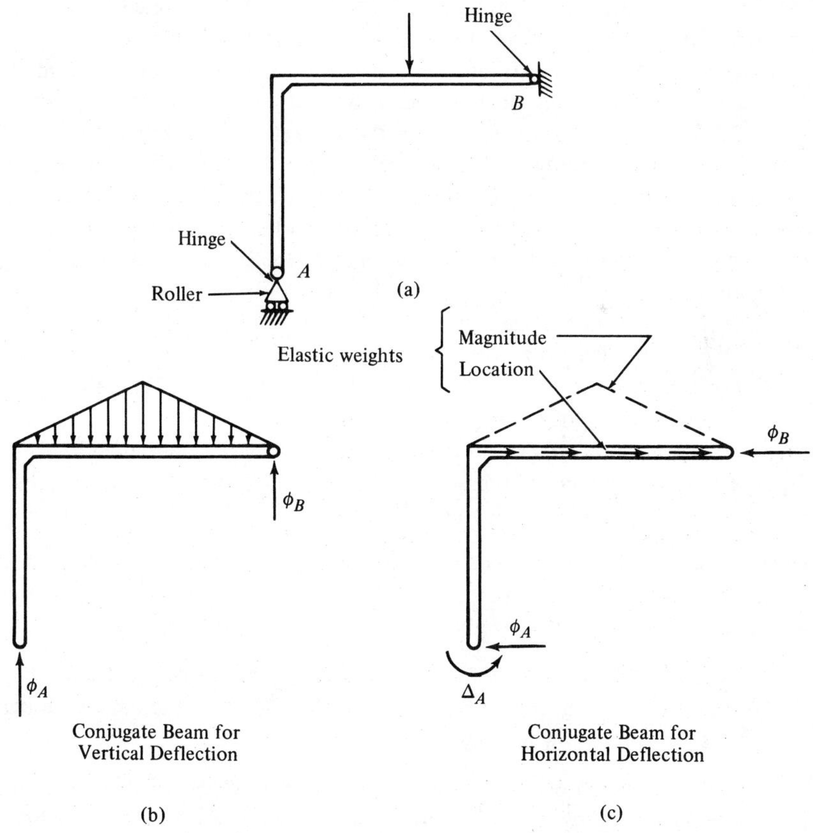

Fig. 110.

To solve this problem completely it is necessary first to solve the conjugate beam for vertical deflection and obtain ϕ_A and ϕ_B. Then these forces are placed on the conjugate beam for horizontal deflection. It is then possible to solve for the deflection Δ_A.

In Fig. 111, the frame just described has been given loads and dimensions so that the procedure for determining deflections may be illustrated. A description of each step follows:

a. Draw the bending-moment diagram for the structure. In this example the units used are inch-kips.
b. Draw the conjugate beam for vertical deflections and show the magnitude and directions of the elastic weights. The bending moment is positive; therefore, the applied elastic loads are downward. In order to hold these forces in equilibrium the reactions ϕ_A and ϕ_B must act upward. These reactions are obtained by taking moments about either point A or point B. Note that in this example the modulus of elasticity, $E = 30 \times 10^3$ in.-kip, has been omitted until the final deflection has been calculated.
c. Draw the conjugate beam for horizontal deflections and show the elastic loads. The values of ϕ_A and ϕ_B are already known, so that they may be indicated immediately on the sketch, but this time their direction is to the left. The applied elastic loads have the same magnitude as before, but they act to the right through the member. The unknown deflection Δ_A is obtained by a summation of the moments of all forces about point A. The sign of Δ_A, the bending moment of elastic loads, is negative because there is compression on the inside fiber. Therefore the direction of the deflection is to the left.
d. The rotation of the hinge at point A should now be examined. The rotation of the ground is zero and the elastic weight ϕ_A is negative, to the left. The hinge is in a counterclockwise sequence from the ground. Therefore the rotation is counterclockwise.
e. The deflection at other points on the structure may be obtained by finding the bending moment at the point due to the elastic weights acting in the direction of the desired deflection.

Signs It is now possible to establish the sign conventions for deflections and elastic weights. The signs for vertical deflections are the same as previously presented and will be summarized here again:

a. A positive real bending moment corresponds to a downward elastic weight.
b. A positive bending moment due to elastic weights on the conjugate beam corresponds to a downward deflection.

It should be understood that the converse of the above definitions is also true. A negative real bending moment corresponds to an upward elastic weight and negative bending moment due to elastic weights produces an upward deflection.

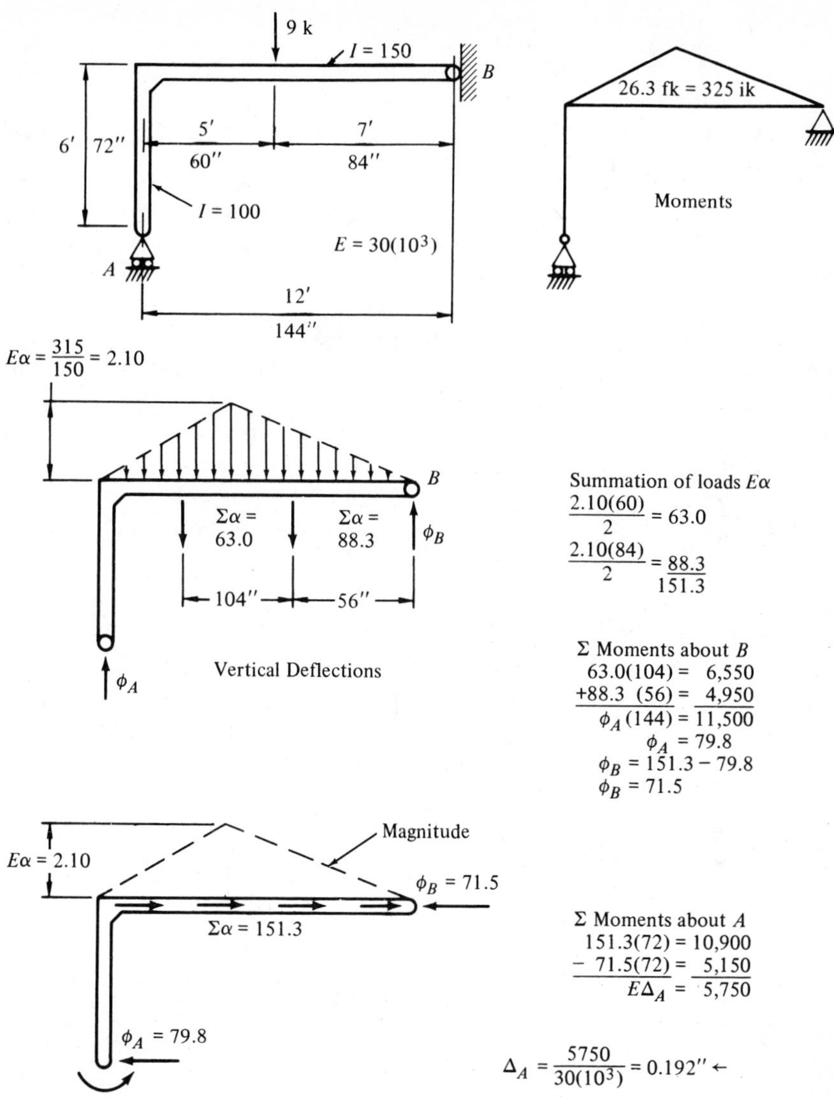

Horizontal Deflections:
The sign of the elastic bending moment is negative; therefore the deflection is to the left.

Fig. 111.

The sign convention for horizontal deflections may be set up as follows:

a. A positive real bending moment corresponds to an elastic weight acting to the right.
b. A positive bending moment due to elastic weights on the conjugate beam corresponds to a deflection to the right.

The key to the above rules for signs is that the positive directions are down and to the right, and the whole system is consistent; i.e., positive real bending moments give positive elastic weights and positive moments due to elastic weights on the conjugate beam give positive deflections.

It is now necessary to present a general definition for real bending moments. Previously, in this discussion positive moments have been defined for horizontal beams as moments which produce tension on the lower fiber. Now, for vertical and horizontal deflections, positive bending moment is defined as moment which tends to produce tension in the fiber on the inside of the frame.

For horizontal beams, the inside of the frames is usually arbitrarily chosen to be the lower side. A typical example of a rigid frame is shown in Fig. 112; the inside and outside have been designated on this sketch. Note that the earth con-

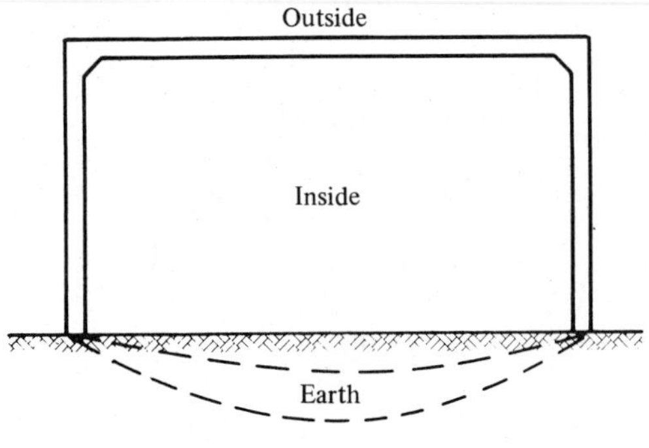

Fig. 112.

sidered with the frame constitutes a closed ring. The conjugate-beam method is limited to structures with a single closed ring. Note also that the definition given above for deflections must be revised to read "to the center of the ring" instead of "to the right."

The final point to clarify is the sign of the rotation. The rule is that positive moments produce counterclockwise rotations if the rotations are considered with respect to points which are in a counterclockwise sequence. Usually signs of rotations can be determined from inspection of the particular problem.

Summary The deflections of rigid frames may be calculated by the elastic-weight method. For vertical deflections, the elastic weights act vertically; for horizontal deflections, the elastic weights act horizontally. It is necessary to draw separate and different conjugate beams for each direction, sometimes solving for the elastic weights in one direction before the other can be completely solved. These separate conjugate beams have different supports depending on the deflection conditions for the real beam.

Deflections of Trusses

In this subsection on truss structures, the principal methods for computing deflections, which will be presented, are the virtual-work analogy and the Williot diagram. The first method lends itself to algebraic calculation and the second to graphical construction.

The Virtual-Work Analogy Deflections result from the internal distortions of structures. The simplest method for computing deflections is by application of the principles of geometry. For example, consider the truss sketched in Fig. 113. It has a small distortion δ at point A in the top chord, as shown. This

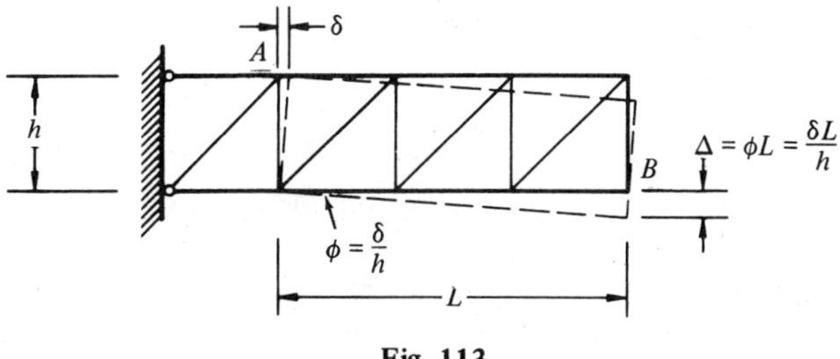

Fig. 113.

causes the truss to rotate through an angle $\theta = \delta/h$ and the point B to deflect a distance Δ. By the geometry of this problem,

$$\Delta = \delta L/h$$

The geometrical method becomes rather complicated, however, when applied to many problems. Again it is expedient to use an analogy which is a much more powerful tool. This analogy is called *virtual work*, and by it the calculation of deflections is transferred to a procedure in the calculation of stresses in trusses.

Figure 114 shows the same cantilever truss of Fig. 113. At point B apply a unit load (say, 1 kip). In the member that has been distorted an amount δ, the

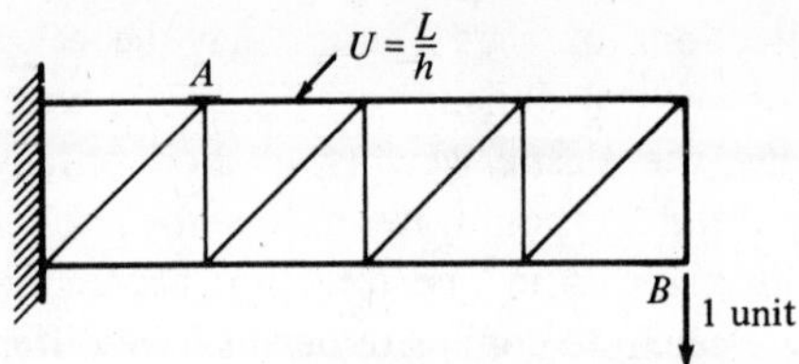

Fig. 114.

stress* will be:

$$u = 1 \times L/h$$

The work due to the movement of the unit load through a distance Δ must equal the internal work due to the force u moving through a distance δ if the fundamental theorem of the conservation of energy is to be preserved. Therefore

$$\Delta \times 1 = u \times \delta \qquad \text{so that} \qquad \Delta = u\delta$$

Substituting, we get $u = L/h$ and $\Delta = \delta L/h$, which is the same result as obtained by the geometrical solution.

The equation of the virtual-work method ($\Delta = u\delta$) applies to any type of truss (or beam) structure. To find the deflection due to a number of elements, the equation is written

$$\Delta = \sum u\delta$$

where the summation sign indicates that all the members of the structure must be included.

To obtain the distortions δ resulting from stresses due to loads on the structure, call the stresses due to this cause S, the length of each member L, and the area A. Then

$$\delta = SL/AE$$

where E is the modulus of elasticity. Finally we have the equation

$$\Delta = \sum (uSL/AE)$$

EXAMPLE: DEFLECTIONS OF A TRUSS

To illustrate the application of this theory, the deflections of the truss sketched in Fig. 115 will be calculated for two cases. The first involves arbitrary changes

*The total stress in kips, not the unit stress

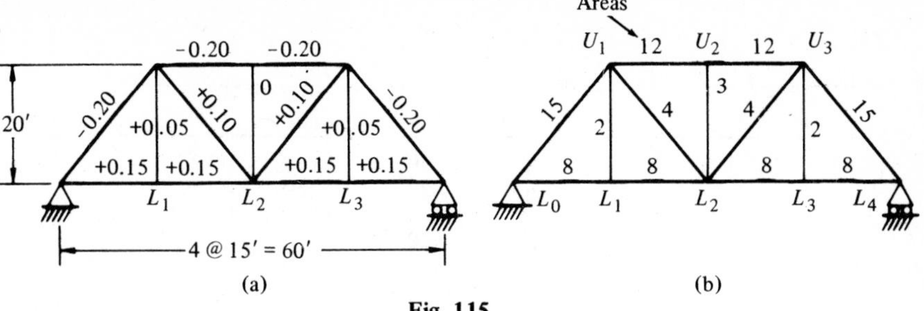

Fig. 115.

in length of all members due, say, to errors in fabrication or to an allowance for camber. The changes will be:

- −0.20 in. in the top chord
- +0.15 in. in the bottom chord
- +0.05 in. in the hangers
- +0.10 in. in the diagonals

These values are indicated on the sketch of the structure, Fig. 115a.

In the second part of the problem the deflection at the center will be determined for the loads shown in Fig. 116b. The areas of the individual members in square inches are indicated in Fig. 115b.

The calculations for the first part of this problem are summarized in Table 4. The members have been numbered to correspond to the panel points.

The stresses u due to a unit load at the center of the truss, are calculated by the stress-coefficient method, and the results are given in Fig. 116a. The stress coefficients are written above the line, and the real stresses are shown below the

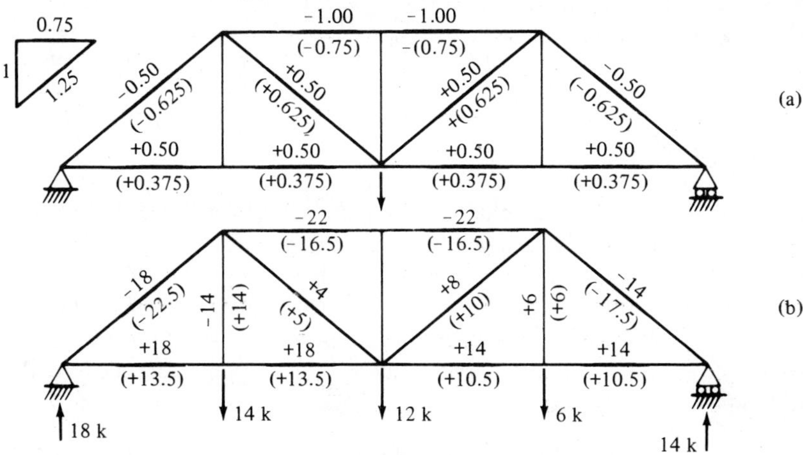

Fig. 116.

TABLE 4

	Member	δ	u	δu
Top chord	U_1L_0	−0.20	−0.625	0.125
	U_1U_2	−0.20	−0.750	0.150
	U_2U_3	−0.20	−0.750	0.150
	U_3L_4	−0.20	−0.625	0.125
Web	U_1L_1	+0.05	0	0
	U_1L_2	+0.10	+0.625	0.062
	U_2L_2	0	0	0
	U_3L_2	+0.10	+0.625	0.062
	U_3L_3	+0.05	0	0
Bottom chord	L_0L_1	+0.15	+0.375	0.056
	L_1L_2	+0.15	+0.375	0.056
	L_2L_3	+0.15	+0.375	0.056
	L_3L_1	+0.15	+0.375	0.056
			$\sum \delta u$ = 0.898 in.	

line in brackets. The multipliers are shown to the left. All stresses are listed in Table 4. Note that the stress in each vertical member is zero. The values of δ and u are multiplied together and the column is added to give $\sum \delta u = 0.898$ in., the deflection of the truss at the center.

EXAMPLE: DEFLECTIONS DUE TO STRESSES

The summary of computations for the deflections due to the loads of Fig. 116b is shown in Table 5. The index stresses and real stresses for S are shown in Fig. 116b. In Table 5 the values of S, L, A, and u are recorded. The value of the deformation in each member, SL/AE, is recorded here to show the change in length of each member, although this intermediate step is not really necessary. The deflection at the center is equal to the summation of the quantities SLu/AE, and is given in inches.

The Williot Diagram The graphical method for determining the deflection of trusses is called the *Williot diagram*. The procedure may be explained by considering the deflection of the truss sketched in Fig. 117. Each member of this truss has changed in length by the amount written beside the member. A negative sign indicates a compression strain, and a positive sign a tension strain.

The following steps describe the basic theory for the construction of a deflection diagram:

1. Lay off the deformation in the members as shown in Fig. 118a. The scale of the members is small in comparison with the scale of deformations.

TABLE 5

$E = 30 \times 10^3$ kip/in.^{2}

	Member	S (kip)	L (in.)	A (sq. in.)	u	SL/AE (in.)	SLu/AE (in.)
Top chord	U_1L_0	−22.5	275	15	0.625	−0.0137	0.0086
	U_1U_2	−16.5	180	12	0.750	−0.0082	0.0061
	U_2U_3	−16.5	180	12	0.750	−0.0082	0.0061
	U_3L_4	−17.5	275	15	0.625	−0.0107	0.0067
Web	U_1L_1	+14.0	240	2	0	+0.0560	0
	U_1L_2	+ 5.0	275	4	+0.625	+0.0114	0.0071
	U_3L_2	+10.0	275	4	+0.625	+0.0228	0.0142
	U_2L_2	0	240	3	0	0	0
	U_3L_3	+ 6.0	240	2	0	+0.0240	0
Bottom chord	L_0L_1	+13.5	180	8	+0.375	+0.0102	0.0038
	L_1L_2	+13.5	180	8	+0.375	+0.0102	0.0038
	L_2L_3	+10.5	180	8	+0.375	+0.0079	0.0030
	L_3L_4	+10.5	180	8	+0.375	+0.0079	0.0030
							$\sum (SLu/AE) = 0.0624$

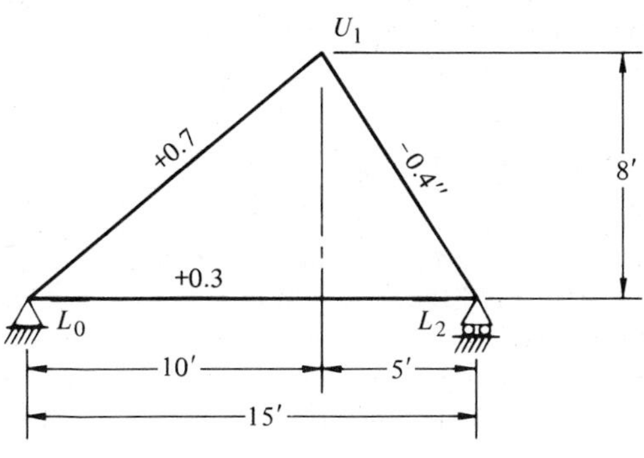

Fig. 117.

2. With a compass use a radius equal to the new length of U_1L_2 and swing an arc with the center in the new position of L_2. This construction is sketched in Fig. 118b.
3. With a radius the new length of L_0U_1 swing an arc with a center at L_0. The intersection of these areas will be the new position of point U_1.

It is not necessary to draw the deflections on a sketch of the truss, and lines erected perpendicular to the deformations may be used instead of arcs of circles. The construction shown in Fig. 118c is the same as the previous construction of

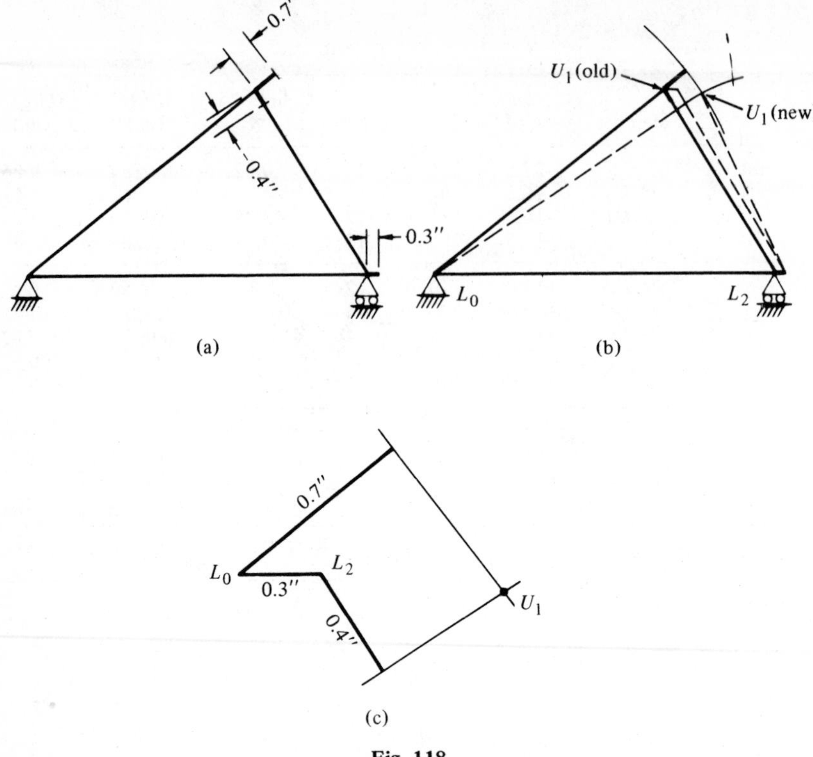

Fig. 118.

Fig. 118b. The perpendicular lines are almost the same as an arc of a circle drawn to a very large radius compared to the length of the lines.

In the Williot diagram, only the relative positions of points are plotted. The slopes of the members are obtained from a small-scale sketch of the truss. Plotting is started from a base which is known (or sometimes assumed) to be fixed against rotation. Each succeeding point is obtained by triangulation from the points previously found. An example of the use of the Williot diagram on a cantilever truss follows.

EXAMPLE

The truss shown in Fig. 119 is fixed in a wall at points U_0 and L_0. The magnitude of the deformations in the individual members is indicated on the sketch. The Williot diagram is sketched in Fig. 119b. A description of the procedure follows:

1. There are no relative deformations or deflections between U_0 and L_0; plot this point first.
2. Now, find the new position of point L_1 from the point first plotted: Lay off the deformation 0.10 in. in U_0L_1 down and to the right of the first

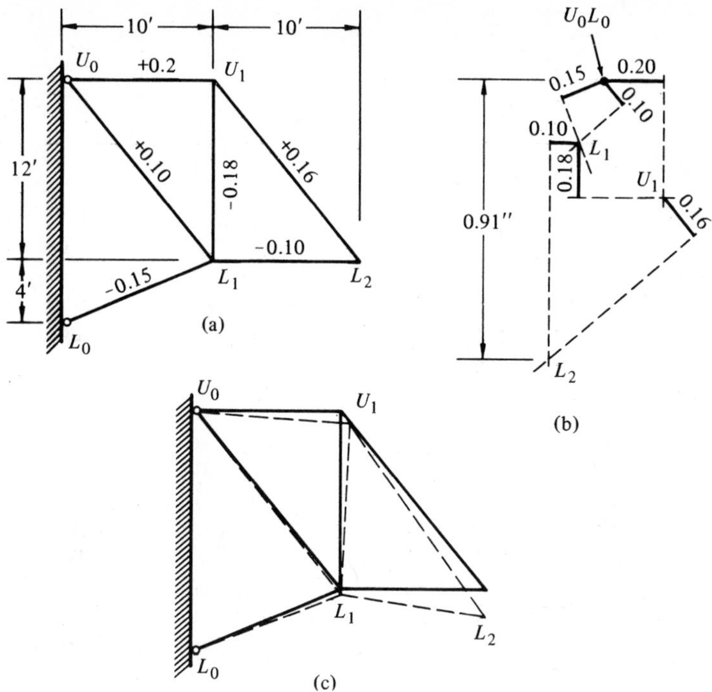

Fig. 119.

point (U_0, L_0). Also plot the deformation 0.15 in. in L_0L_1 down and to the left of the first point (U_0, L_0). Erect perpendiculars from the ends of these deformations. The intersection will be the new position of L_1.

3. With point U_0 and L_1 now known, find the new position of point U_1: From L_1 lay off 0.18 in. downward and erect a perpendicular. From U_0L_0 plot 0.2 in. to the right and erect a perpendicular. The intersection of these lines is the new position of U_1.
4. Now, find the new position of L_2: The triangulation is made from U_1 and L_1. Lay off from L_1 the deformation 0.10 in. to the left. From U_1 plot L_2 downward and to the right. The intersection of the perpendiculars from these lines gives the new position of point L_2. The vertical deflection of L_2, scaled from this diagram, is 0.91 in. The deflections for this structure have been sketched to a larger scale in Fig. 119c.

EXAMPLE: SIMPLE TRUSS

The Williot diagram must be started from a member which has no rotation. If the member actually rotates, the deflections obtained from the diagram will not represent the true deflection of the structure. In the truss shown in Fig. 120, all of the members deflect. However, if the structure is symmetrical and is loaded symmetrically, the center member U_2L_2 will not rotate. The deflection diagram may be drawn with this member as a base.

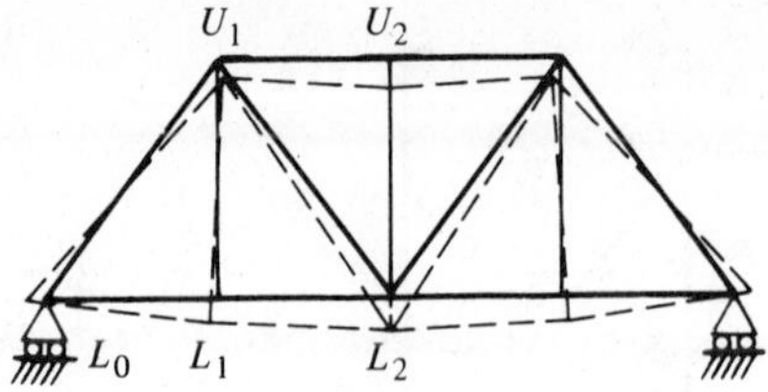

Fig. 120.

In Fig. 121 is shown the deformation in each of the members and the deflection diagram started from U_2L_2 as a fixed base. The construction will be described in the following steps:

1. Plot point U_2L_2, the fixed base.
2. Lay off deformations to find point U_1: Plot the value 0.20 in. on U_1U_2 to the right and the value 0.10 in. on U_1L_2 up and to the left. The intersection of the perpendiculars will give the deflection of point U_1.
3. Determine the location of point L_1: From U_1 lay off 0.05 in., the deformation of U_1L_1 downward. From L_2 plot 0.15 in. to the left. The new position of L_1 will be at the intersection of the perpendicular lines drawn from the ends of these deformations.
4. Find point L_0: plot from L_1 0.15 in. to the left and from U_1 plot 0.20 in. upward and to the right. Draw perpendiculars to find point L_0.

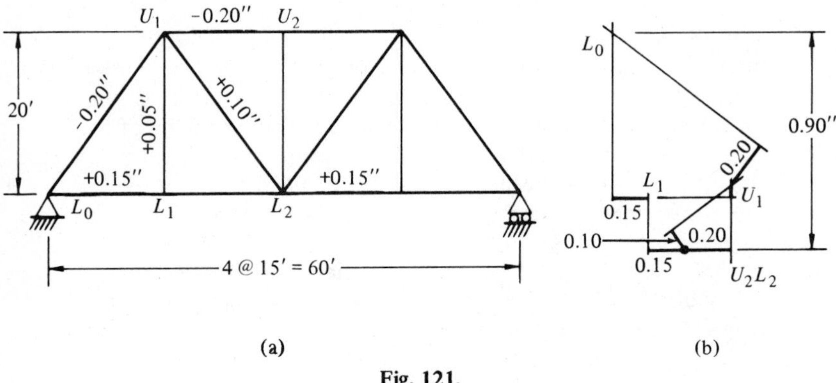

Fig. 121.

The deflections obtained from this diagram are sketched in Fig. 122. Instead of being fixed in position, point L_0 has risen. The correction to make point L_0 fixed is very simple. All that is necessary to do is to shift the diagram down and to the right and to consider L_0 as the fixed point. The deflections for this condition are sketched in Fig. 123.

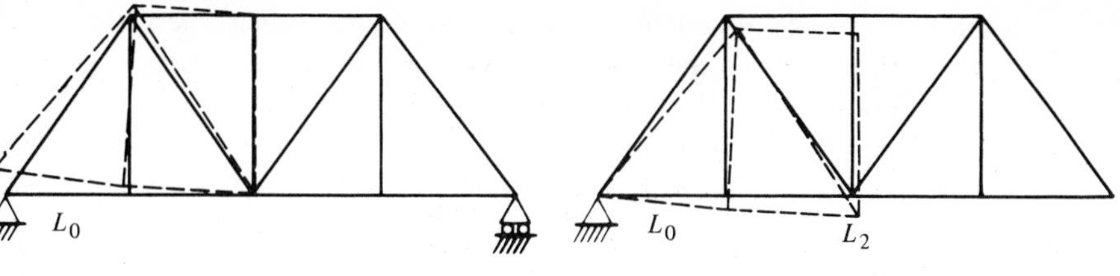

Fig. 122. Fig. 123.

The Rotation Diagram It is necessary in some cases to correct Williot diagrams for rotation. Therefore, the deflection diagram for a truss rotated through a small angle will be explained. This type of diagram will be added to other deflection diagrams in order to give the deflections from the required points.

In Fig. 124a is shown the truss of Fig. 121. It will be assumed that point L_4 will move upward a distance of 3 in. The angle of rotation will be $\theta = 3/(60 \times 12) = 0.00417$ radians. The problem is to draw the Williot diagram for this structure. There are no deformations in the members. A description of each of the steps follows:

1. Start the diagram with point L_0 as shown in Fig. 124b. The deflection of point L_1 will be

$$\Delta = 15(12)0.00417 = 0.75 \text{ in.}$$

This point will be directly above L_0 because the angle of rotation is quite small.

2. Now find the new position of point U_1 by triangulation from L_0 and L_1. If there had been, say, a small positive deformation in L_0U_1 it would have

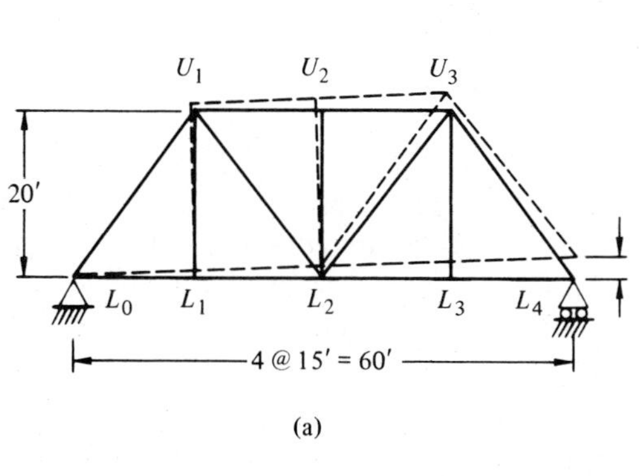

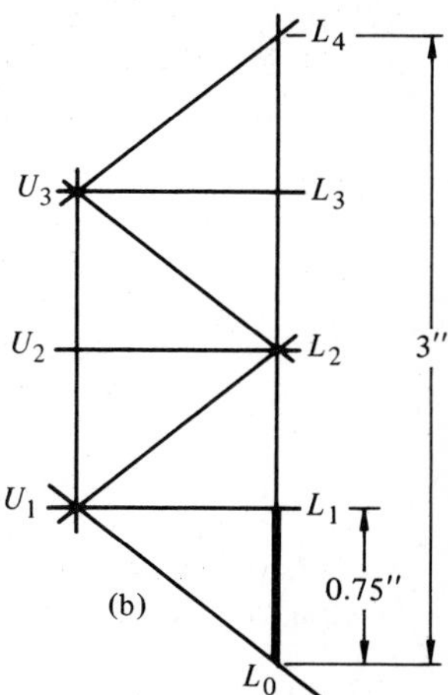

Fig. 124.

been plotted upward and to the right. The perpendicular line would then have been drawn. If the deformation is greater, the perpendicular line is drawn through L_0. The same procedure is carried through for member L_1U_1. The intersection of the perpendiculars gives the new position of U_1.

3. Repeat this procedure for the other points, and a Williot diagram will result as sketched in Fig. 124b. Note that the final diagram resembles the truss, but is turned through an angle of 90°. The final deflection of L_4 is 3 in. above point L_0.

This device was presented by Mohr and is therefore often called the *Mohr correction diagram*. In Fig. 125, it has been applied to the correction of the deflection diagram for the truss of Fig. 121, starting with L_0L_1 assumed fixed in

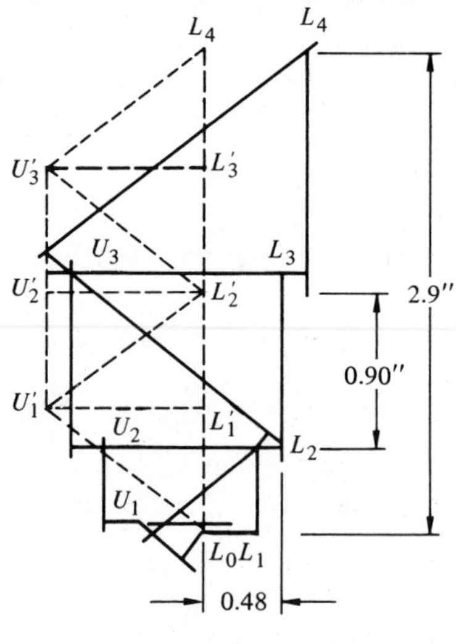

Fig. 125.

position. The point L_4 is thus caused to rise a distance 2.9 in. above its true position. One end of the Mohr correction diagram is placed on L_0. The top of the diagram is placed level with L_4. *The true deflections of the truss are the difference between the Williot diagram and the rotation diagram.* For example, the deflection of point L_2 has been indicated as 0.9 in. vertical and 0.48 in. horizontal.

General Method for Indeterminate Structures

The previous subsections of this discussion presented methods of determining deflections in beams, frames, and trusses which were statically determinate.

With these tools available, it is now possible to determine the reactions, shears, and moments in indeterminate structures.

The first structure to be considered will be statically indeterminate to the first degree. (See the section on equilibrium for a discussion of the degree of indeterminancy.) Such a structure has one more reaction than can be solved from the equations of equilibrium. The two-span beam in Fig. 126 is an example.

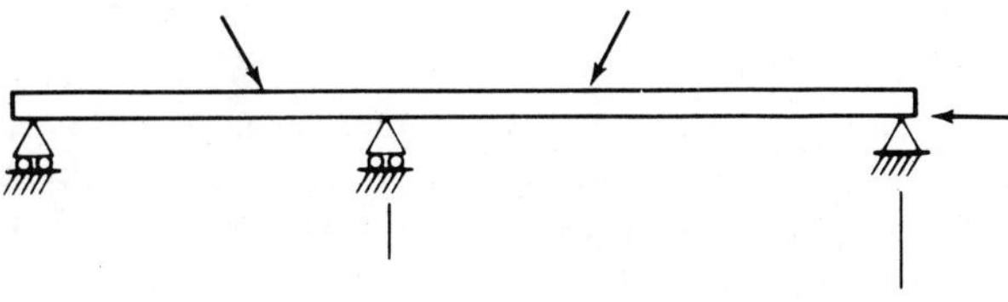

Fig. 126.

There are four possible reactions and only three equations of equilibrium are available. (In this structure there are no additional hinges to furnish another equation.)

The method of solution for this structure is as follows:

1. Remove one of the reactions and calculate the vertical deflection at the point where the reaction has been removed. This step is illustrated in Fig. 127a. The deflection will be noted as Δ.

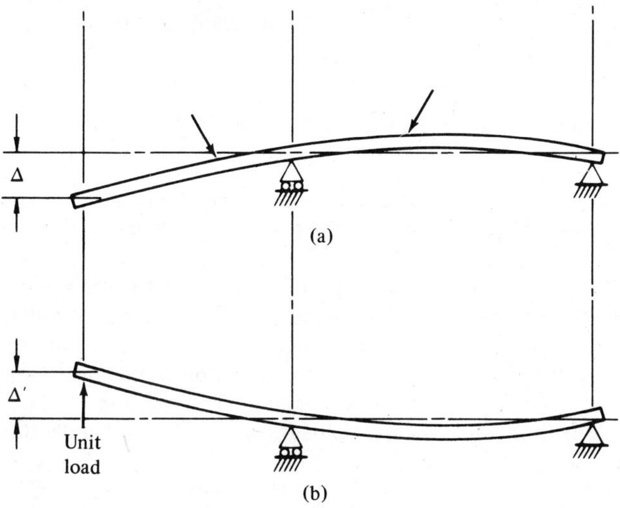

Fig. 127.

2. The next step is to place a unit load in the direction of the reaction and find the deflection due to this load. Call this Δ' (Fig. 127b).
3. Then the true reaction will be the force which brings the structure just back into place or $R = \Delta/\Delta'$.

If one of the reactions on the beam has been determined, the other reactions may be solved from the equations of equilibrium. An example follows.

EXAMPLE

The beam under investigation is shown in Fig. 128a. A single load of 8 kip has been placed on the left span. The problem is to determine the reactions at A, B, and C which will be called R_A, R_B, and R_C, respectively. Note that the depths of the beams vary for each of the spans.

The first step is to remove the reaction at A. The structure and load are shown in the second diagram of Fig. 128a. The moment diagram is sketched in Fig. 128b with the conjugate beam and the elastic weights. Calculation of the deflection Δ follows the previous example given for calculation of beam deflections. Note that the value of E is omitted until the final step. Calculations have been made in inch-kips.

In Fig. 128c the deflection Δ' due to a unit load of 1 kip acting upward at point A is calculated. The results of these deflection calculations are that Δ = 1.58 in. and Δ' = 0.38 in. Therefore, the reaction at A is $\Delta/\Delta' = 1.58/0.38 =$ 4.16 kip. The final step in the solution of this problem is to draw the moment diagram for the two-span continuous beam.

Another example will demonstrate how this method may be applied to rigid frame problems. The structure must be statically indeterminate to the first degree only.

EXAMPLE: RIGID-FRAME STRUCTURE

The frame to be investigated is shown in Fig. 129 and is the same structure that was used as an example in the previous section on the deflection of rigid frames. However, the reaction at A is no longer on rollers and is instead on a fixed hinge.

The first step in the analysis is to cut the structure at some point to make it statically determinate. In this case the cut has been made at point A, and the hinge put back on rollers. Then the deflection at the cut end will be calculated. To bring the structure back into its original position, a unit load will be placed on the structure and its deflection determined. The reaction at the cut member will be the deflection due to applied loads, divided by the deflection due to a unit load.

The deflection due to applied loads has already been determined in the previous section and is equal to 0.192 in. to the left. This deflection is sketched in Fig. 129b. The unit load for calculating Δ' is assumed to act to the right, as

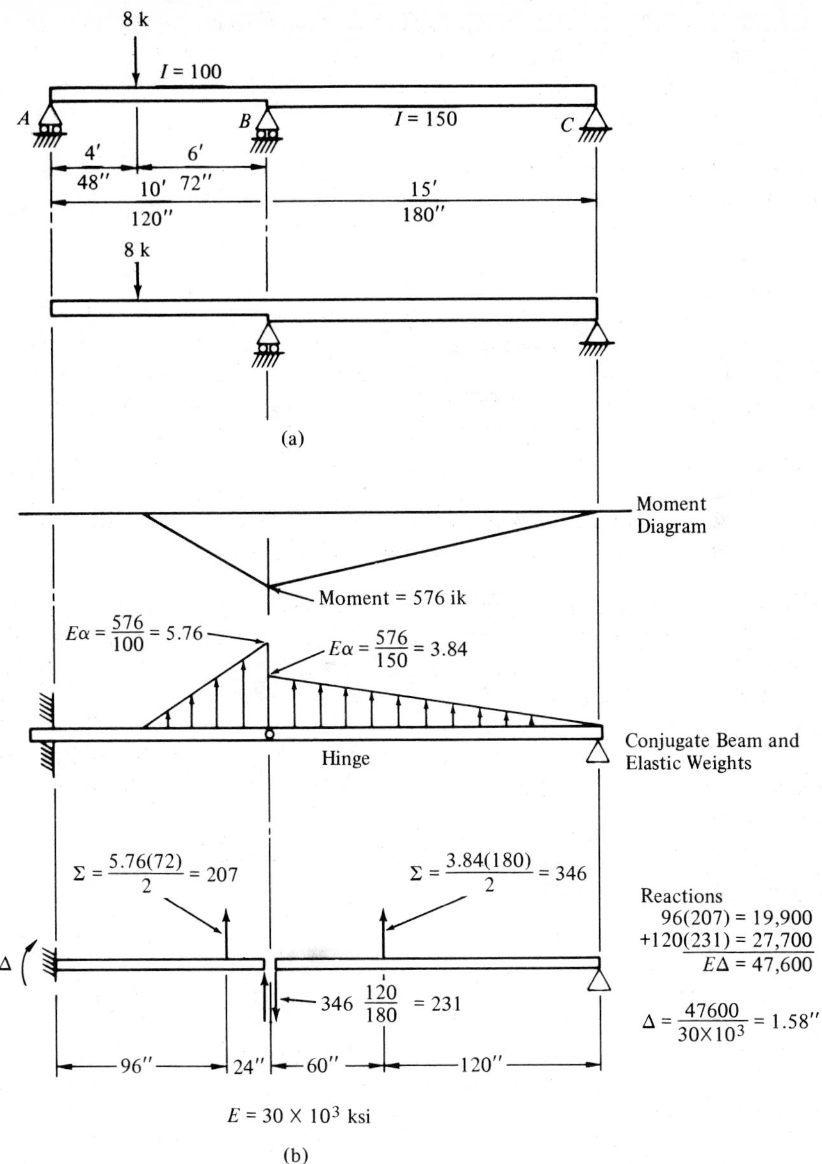

8 k
I = 100
A
B
I = 150
C
4′
6′
48″
10′ 72″
120″
15′
180″
8 k
(a)
Moment Diagram
Moment = 576 ik
Eα = 576/100 = 5.76
Eα = 576/150 = 3.84
Hinge
Conjugate Beam and Elastic Weights
Σ = 5.76(72)/2 = 207
Σ = 3.84(180)/2 = 346
Reactions
96(207) = 19,900
+120(231) = 27,700
EΔ = 47,600
Δ
346 120/180 = 231
Δ = 47600/30×10³ = 1.58″
96″
24″
60″
120″
E = 30 × 10³ ksi
(b)

Fig. 128.

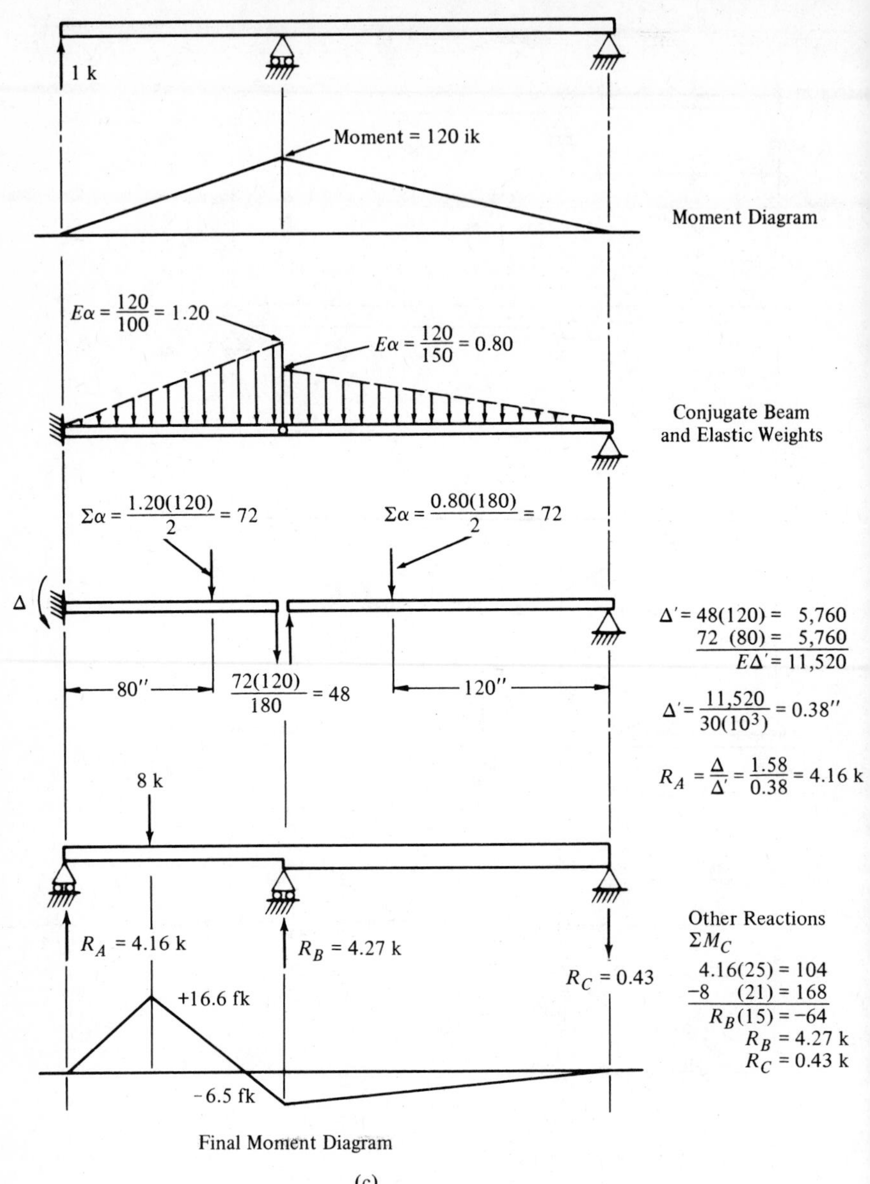

(c)

Fig. 128. (*Continued*)

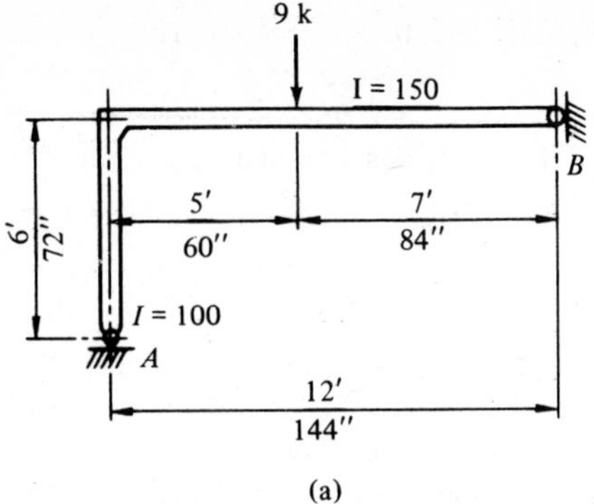

(a)

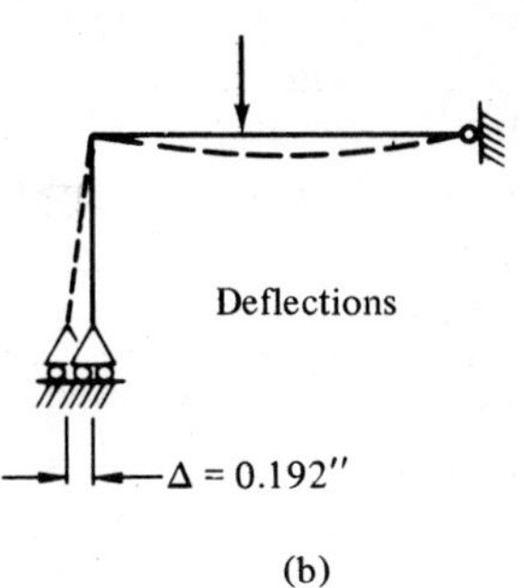

(b)

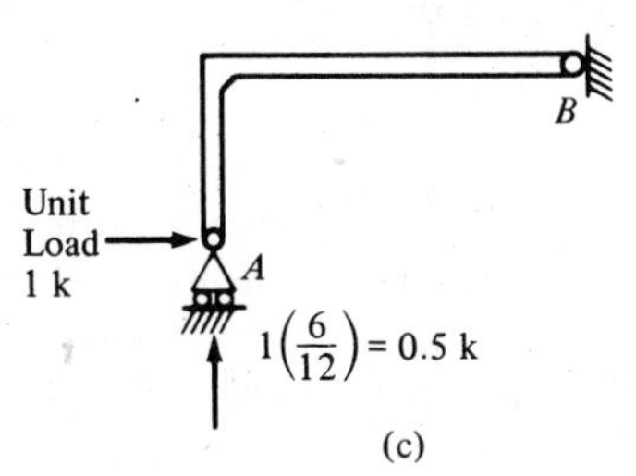

(c)

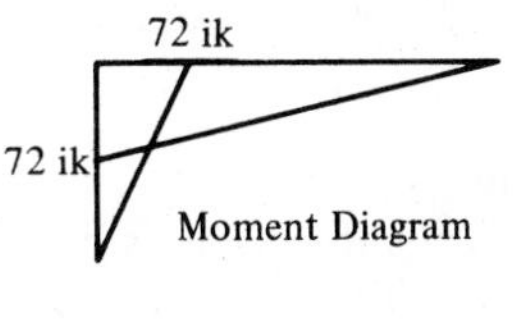

(d)

$E\alpha = \frac{72}{150} = 0.48$

$\Sigma E\alpha = \frac{0.48(144)}{2} = 34.6$

ϕ_B

48″

$\Sigma E\alpha = \frac{0.72(72)}{2} = 26.0$

$E\alpha = \frac{72}{100} = 0.72$

ϕ_A

Summation of Moments about A

$34.6(48) = 1660$

$\phi_B(144) = 1660$

$\phi_A = 34.6 + 26.0 - 11.5 = 49.1$

(e)

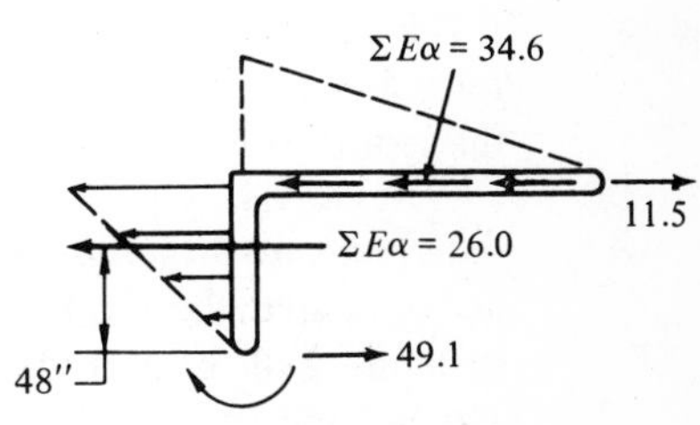

Summation of Moments about A

$26.0(48) = 1250$

$34.6(72) = 2490$

$-11.5(72) = -823$

2917

$\Delta' = \frac{2917}{30(10^3)} = 0.098''$

(f)

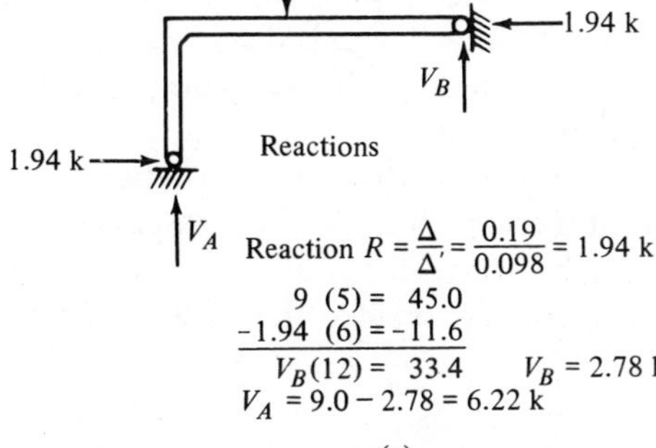

Reaction $R = \frac{\Delta}{\Delta'} = \frac{0.19}{0.098} = 1.94$ k

$9\ (5) = 45.0$

$-1.94\ (6) = -11.6$

$V_B(12) = 33.4 \qquad V_B = 2.78$ k

$V_A = 9.0 - 2.78 = 6.22$ k

(g)

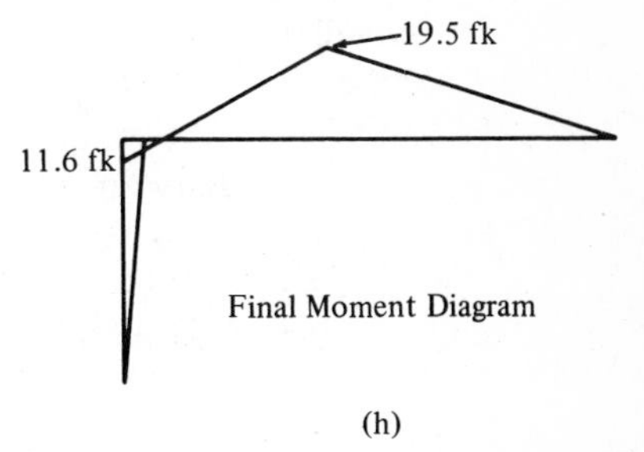

(h)

Fig. 129.

sketched in Fig. 129c. The moment diagram due to this load is sketched in Fig. 129d. The conjugate beams and deflection computations are given in Fig. 129e and f. The deflection due to a unit load is $\Delta' = 0.098$ in. Therefore, the reaction is $R = 0.192/0.098 = 1.94$ kip. The reactions and bending moments have been calculated in Fig. 129g and h.

EXAMPLE: MOMENTS AS UNKNOWNS

Either reactions or bending moments may be used as unknowns for the solution of indeterminate structures by the general method. In Fig. 130, the same two-span continuous beam as solved in Fig. 128 is shown. The method is to find the angular rotation ϕ at the support B when the beam is cut through at this point but still held up by reactions. Then a unit moment is placed on the cut ends and the angular rotation ϕ' determined. The final moment at B will be

$$M_B = \phi/\phi'$$

Computations for this problem are shown in Fig. 130d and e. Note that the deflections and rotations have been computed in units which are not consistent. The value of I is in inches to the fourth power and the lengths of members are in feet. The value of E has been neglected entirely. All that is required is that the deflections ϕ and ϕ' be in the correct ratio. The exact value in inches or feet is not necessary.

The final result of the computations gives $\phi = 0.442$ and $\phi' = 0.066$. Therefore, $M = \phi/\phi' = 0.442/0.066 = 6.6$ ft-kip. This result agrees substantially with the results of Fig. 128.

Indeterminate Structures with Two or More Unknowns The preceding problems have required the solution of only one unknown reaction or moment. Next, structures will be considered which are indeterminate to two or more degrees. The method is fundamentally the same except that now many more deflections must be considered, so that it becomes necessary to set up a formal procedure. Consider the continuous beam shown in Fig. 131a. There are six possible reactions, five vertical and one horizontal. There are three more reactions than can be solved by the three equations of equilibrium. Therefore, remove three of the reactions, say at A, B, and C. The deflections at these points will be called Δ_A, Δ_B, and Δ_C. These are sketched in Fig. 131b.

In order to get the beam back into position, place a unit load at point A. The corresponding deflections at the three reactions will be called Δ_{AA}, Δ_{AB}, Δ_{AC}, as sketched in Fig. 131c. Note that the first subscript indicates the position of the unit load and the second the location of the deflection.

It is also necessary to place a unit load at B to obtain the deflections Δ_{BB}, Δ_{BA}, and Δ_{BC} and to place a unit load at C to obtain Δ_{CC}, Δ_{CA}, and Δ_{CB}.

Let R_A, R_B, and R_C represent the unknown reactions at A, B, and C. The condition that must be fulfilled for this structure, so that the reactions do not

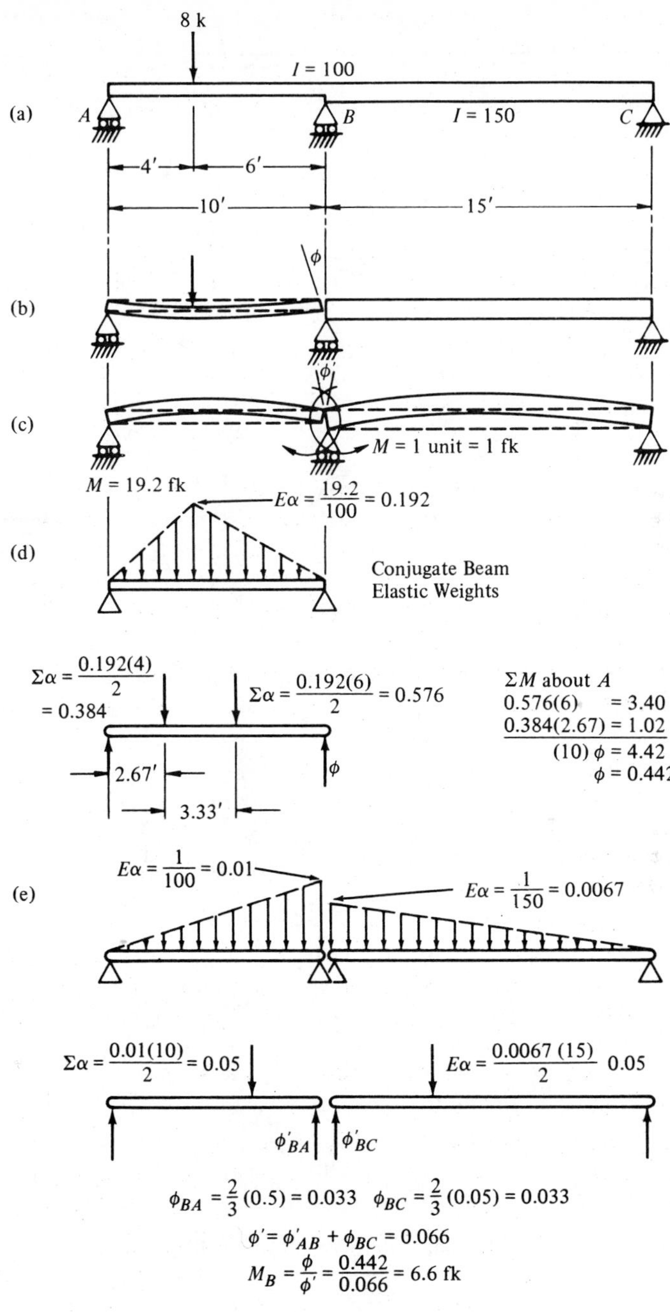

8 k
I = 100
(a)
A
B
I = 150
C
4′
6′
10′
15′
(b)
φ
(c)
φ′
M = 1 unit = 1 fk
M = 19.2 fk
Eα = 19.2/100 = 0.192
(d)
Conjugate Beam
Elastic Weights
Σα = 0.192(4)/2 = 0.384
Σα = 0.192(6)/2 = 0.576
ΣM about A
0.576(6) = 3.40
0.384(2.67) = 1.02
(10) φ = 4.42
φ = 0.442
2.67′
3.33′
φ
Eα = 1/100 = 0.01
(e)
Eα = 1/150 = 0.0067
Σα = 0.01(10)/2 = 0.05
Eα = 0.0067 (15)/2 0.05
φ′BA
φ′BC
φBA = 2/3 (0.5) = 0.033 φBC = 2/3 (0.05) = 0.033
φ′ = φ′AB + φBC = 0.066
MB = φ/φ′ = 0.442/0.066 = 6.6 fk

Fig. 130.

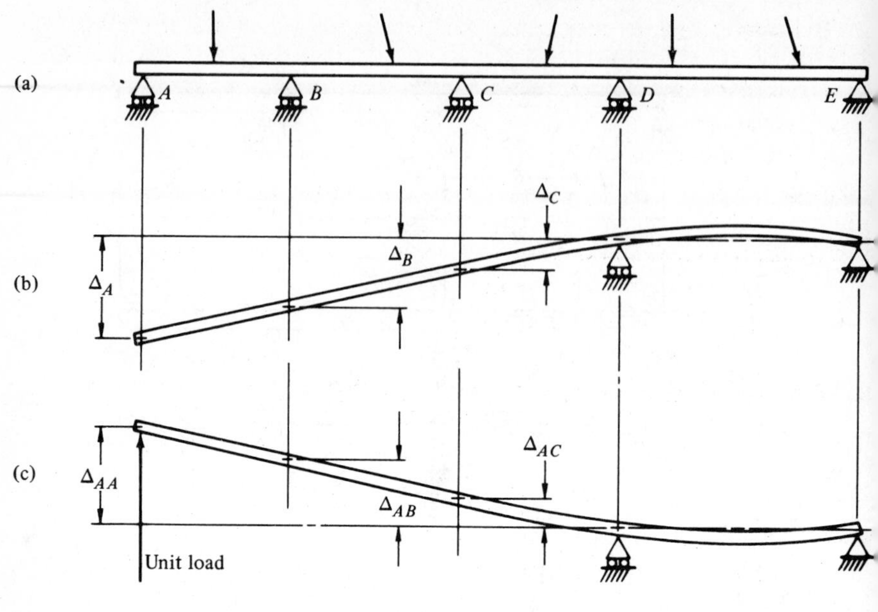

Fig. 131.

move, is that the sum of the deflections from applied loads and reactions must be zero at A, B, and C.

This condition will be satisfied by the following equation involving the reactions and deflection:

$$R_A \Delta_{AA} + R_B \Delta_{BA} + R_C \Delta_{CA} + \Delta_A = 0$$

Two similar equations at B and C may also be written:

$$R_B \Delta_{BB} + R_A \Delta_{AB} + R_C \Delta_{CB} + \Delta_B = 0$$

$$R_C \Delta_{CC} + R_A \Delta_{AC} + R_B \Delta_{BC} + \Delta_C = 0$$

The simultaneous solution of these three equations will give values of R_A, R_B, and R_C.

It would appear from the above equations that it is necessary to solve for nine deflections in order to obtain the three reactions. However, the deflections having the same subscripts are equal; i.e., $\Delta_{AB} = \Delta_{BA}$, $\Delta_{CA} = \Delta_{AC}$, and $\Delta_{BC} = \Delta_{CB}$. This may be proved by Maxwell's theorem of reciprocal displacements. Therefore, it will be necessary to digress momentarily to present this theorem.

Maxwell's Theorem The formal statement of Maxwell's theorem is as follows:

The deflection at A due to a load at B is equal to the deflection at B due to the same load at A.

There are several methods of proving this theorem, the usual one being a consideration of the energy required to deflect the structure. In this discussion, however, the proof will be made by simple geometrical considerations similar to the procedure by which the elastic-weight analogy was derived.

Consider the cantilever beam shown in Fig. 132. Assume that the only elastic portion of this beam is a small section at point C near the support. The rotation of this element is, say, K radians per foot-pound. Next place a force of P pounds at point A, which is at a distance a from point C. The rotation at C is KPa radians. The deflection at B is $\Delta_{AB} = PaKb$.

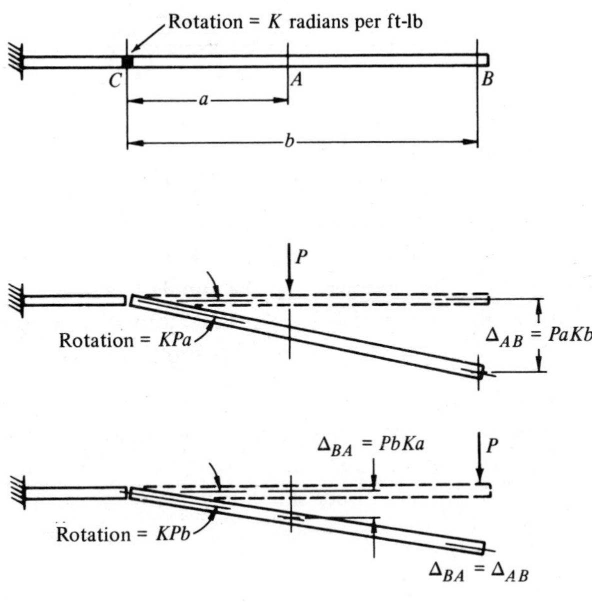

Fig. 132.

Now place the force, P, at B; the rotation at C is KPb and deflection at A is $\Delta_{BA} = PbKa$. Therefore, $\Delta_{BA} = \Delta_{AB}$.

This simple calculation by geometry demonstrates that the quantities Δ_{AB} and Δ_{BA} are identical and the deflection at B due to a load P at A is equal to the deflection at A due to a load P at B. The theorem proved here for a cantilever with a single elastic element may be extended to other beam structures by the following considerations:

1. If the entire beam is considered elastic, the identity of these deflections will be unaltered. The equations for deflections become $\Delta_{AB} = \sum PaKb$ and $\Delta_{BA} = \sum PbKa$.

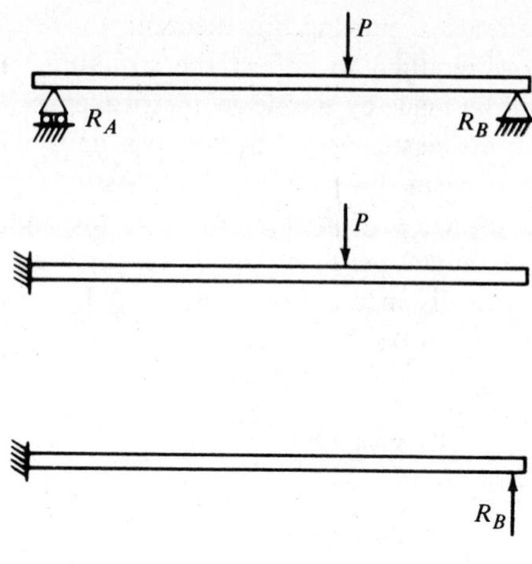

Fig. 133.

2. Any other beam structure may be considered as the summation of a series of cantilevers, as was done in the proof of the elastic-weight analogy. For example, the shears and bending moments on the simple beam shown in Fig. 133 may be considered as the sum of the two cantilever structures shown below the simple beam. Any other beam structure may be broken down into component cantilevers.

The use of Maxwell's theorem either reduces the calculations or provides a check on the calculations. In the case of the analysis of arch structures, it makes it possible to solve for reactions without simultaneous equations.

EXAMPLE: THREE-SPAN BEAM

An example of a structure with more than one unknown is the three-span beam sketched in Fig. 134. Use the method of redundant moments, rather than redundant reactions. Cut the structure at points A and B and determine the rotations ϕ_A and ϕ_B due to the applied loads. Then place unit moments (in this case, 1000 in.-kip) alternately at the cut ends. The equations resulting from the sum of the rotations, being equal to zero, gives the solution for the moments at points A and B.

The complete calculations for this problem are given in Fig. 134, including the solution of the simultaneous equations. Note that $\phi_{AB} = \phi_{BA}$, so a check is provided.

The use of moments instead of reactions as unknowns is a little simpler because the bending moments are obtained directly.

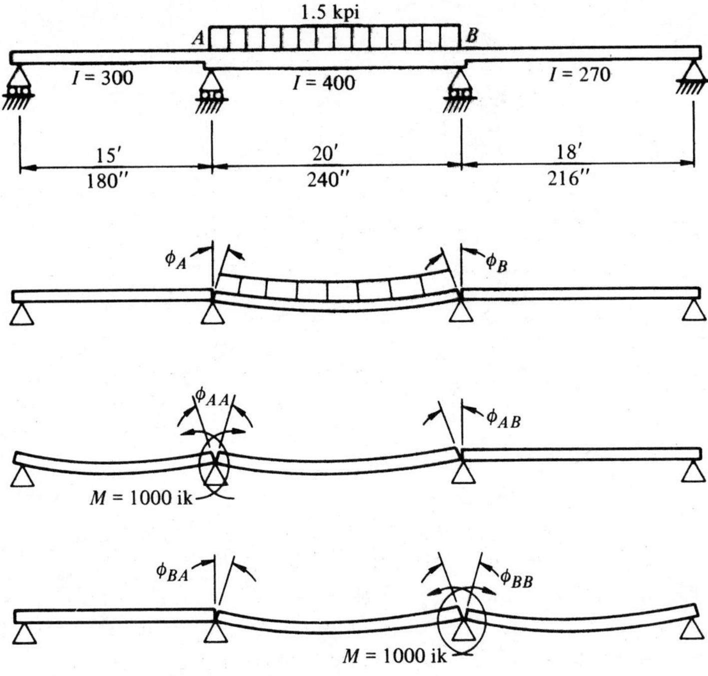

Fig. 134.

Summary The general method is: (1) Cut the indeterminate structure back to a determinate condition. (2) Find the deflections, or rotations, on the cut ends. (3) Apply the required forces to bring the cut ends together and find the deflections, or rotations, for these forces. (4) Write simultaneous equations in terms of forces and deflections which state that the deflections at the cut ends are zero. The solution of the equations gives the value of forces which bring the cut ends together. Maxwell's theorem of reciprocal deflections will assist in checking the calculations. The unknown values may be either forces or moments.

The Moment-Distribution Method

The moment-distribution method is the development of Professor Hardy Cross and was set forth in a now classic paper, "Analysis of Continuous Frames by Distributing Fixed End Moments" in the *Transactions of the American Society of Civil Engineers*, Vol. 96 (1932) pages 1-158.

The method has captured the imagination of designers of continuous structures and rigid frames because it reduces analysis to a simple numerical process.

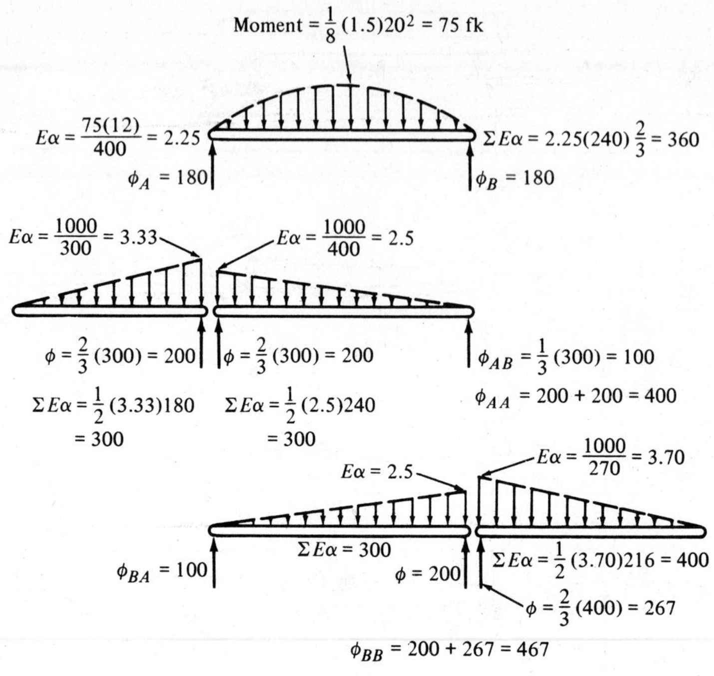

Simultaneous Equations

$$(1)\quad M_A\phi_{AA} + M_B\phi_{BA} + \phi_A = 0; + 400M_A + 100M_B + 180 = 0$$

$$(2)\quad M_A\phi_{AB} + M_B\phi_{BB} + \phi_B = 0; + 100M_A + 467M_B + 180 = 0$$

$$\text{Equation (1)} \times 400: \quad M_A + 0.25M_B + 0.45 = 0$$

$$\text{Equation (2)} \times (-100): \quad -M_A - 4.67M_B - 1.80 = 0$$

$$-4.42M_B - 1.35 = 0$$

$$M_B = -0.304$$

$$\text{Equation (1)}: \quad +M_A - 0.076 + 0.45 = 0; \; M_A = -0.374$$

$$\text{Check Equation (2)}: \quad -100(0.0374) - 467(0.0304) + 18 = 0$$

$$-3.74 - 14.26 + 18 = 0: \text{ Check}$$

Moments $M_A = -374$ in.-kip = 31.2 ft.-kip

$M_B = -304$ in.-kip = 25.3 ft.-kip

Fig. 134. *(Continued)*

Each step in the process has a definite physical significance which is readily understood and easily remembered. Many complex problems can be solved by the moment-distribution method.

The general outline to be followed in this discussion is: (1) the definition of certain constants; (2) derivation of formulas for these constants for members with uniform section; (3) a general statement of the method; (4) a step-by-step outline of the procedure for a continuous beam without sidesway movements at a joint; (5) illustrative problems for several cases.

Moment-Distribution Constants In the moment-distribution method, the elastic properties of each of the members of a continuous beam or frame are defined by three constants: (1) stiffness; (2) carry-over factor; and (3) fixed-end moment.

1. *Stiffness* is defined as the moment at the end of a member necessary to produce a unit rotation of that end when the other end is fixed. The stiffness factor at A in Fig. 135 is $K = M_{AB}/\phi_A$. At a joint where two or more members intersect, any change in moment will be distributed to the members in proportion to the stiffness factors of the members. For example, in Fig. 136 joint A is subjected to a moment M. The portion of this moment which will go to member AC will be

$$M_{AC} = M \frac{K_{AC}}{K_{AC} + K_{AB} + K_{AD} + K_{AE}}$$

For a given structure, the stiffness factor need only be proportional to the constant $K = M_{AB}/\phi_A$. In the above equation the numerator and denominator may be multiplied by the same constant without affecting the value of M_{AC}. For members of constant moment of inertia, the stiffness is proportional to I/L, where I is the moment of inertia of the cross section and L is the span length.
2. The *carry-over factor* is defined as the ratio of the moment at point A (Fig. 135) to the moment at point B. For members with the same moment of inertia I, the carry-over factor is $+\frac{1}{2}$. Note that the sign of the quantity is positive. The sign convention will be explained later.
3. The *fixed-end moment* is the moment at the end of a member due to applied loads if both ends of the member are fixed against rotation. For a member having a constant moment of inertia I and span L, the fixed-end

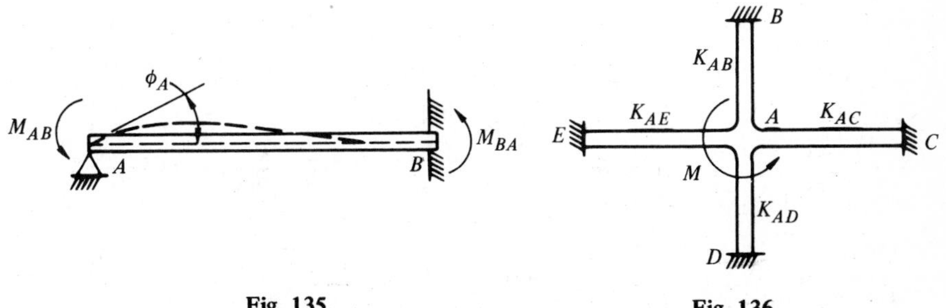

Fig. 135. Fig. 136.

moment is equal to

$$M^F = wL^2/12$$

for a uniform load of w pounds per lineal foot. For a concentrated load P at a distance a from the left and b from the right end, the fixed-end moments are

$$M_B{}^F = Pa^2 b/L^2$$

at the right end and

$$M_A{}^F = Pab^2/L^2$$

at the left end.

Derivation of Constants for Members of Uniform Section The above definitions and values of the constants for members of uniform section are all that is required at the beginning of this discussion of moment distribution. The following derivations may be passed over temporarily, to be studied later.

For the derivation of these constants, the conjugate-beam method, as described previously, will be used. The equilibrium of a *conjugate* beam, similar to the *real* beam, will be determined. The moment diagram divided by EI (the M/EI diagram) on the real beam becomes the load on the conjugate beam. The bending moment on the conjugate beam is the deflection on the real beam; also the reaction on the conjugate beam becomes the slope of the real beam.

Stiffness The beam shown in Fig. 135 and again in Fig. 137 is statically indeterminate to the first degree and both the magnitude of the bending moments and the rotation at the hinged end must be obtained. The conjugate beam for

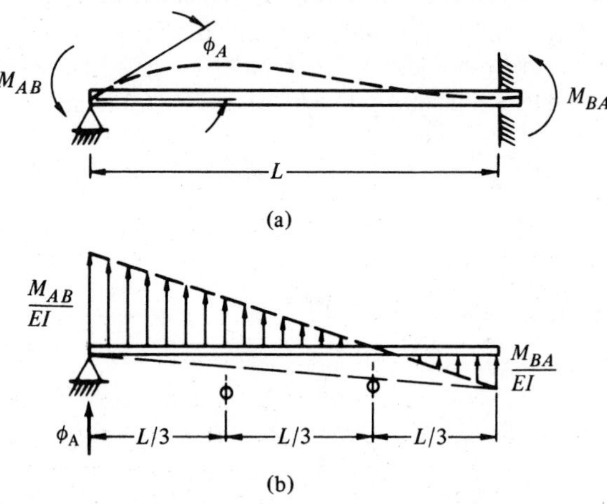

Fig. 137.

this problem is shown in Fig. 137b. The exterior hinge remains a hinge and the fixed end of the real beam becomes a free end in the conjugate beam. The moment areas acting as elastic weights must hold the beam in equilibrium. The exterior moments acting are a moment M_{AB}, of known size, at the left end, and a moment M_{BA} at the right, which is to be determined. Assume, as shown in the figure, that M_{BA} is of opposite sign to M_{AB}. For convenience in calculation, use as elastic weights the two triangles whose heights are M_{AB} and M_{BA}, respectively, and whose bases are the dotted line. Take moments about the center of gravity of the M_{BA} area at a distance $L/3$ from B. Then M_{BA} will be eliminated in the equation of equilibrium.

$$\phi_A \times \frac{2L}{3} = \frac{M_{AB}}{EI}\left(\frac{L}{2}\right)\frac{L}{3}, \qquad \phi_A = \frac{M_{AB}L}{4EI}$$

The stiffness constant has previously been defined as $K = M_{AB}/\phi_A$. Solve for M/ϕ in the above equation:

$$K = M_{AB}/\phi_A = 4EI/L$$

Since only proportional values of K are required, it is customary, if the structure is composed only of members of uniform section, to use a factor $K = I/L$.

Carry-over Factor In Fig. 137b take moments about A, thus eliminating ϕ.

$$\frac{M_{AB}}{EI}\left(\frac{L}{2}\right)\frac{L}{3} = \frac{M_{BA}}{EI}\left(\frac{L}{2}\right)\frac{2}{3}L$$

$$M_{AB} = 2M_{BA}$$

Note that M_{BA} appears with positive sign, indicating that its sign has been chosen correctly to produce equilibrium of the conjugate beam.

The moment at B is half the moment at the free end, and therefore, by our definition, the carry-over factor is $\frac{1}{2}$. According to Fig. 137b, M_{BA} is of opposite sign to M_{AB}. A new sign convention will be developed which will be different from the beam convention used for this derivation.

Fixed-End Moments The fixed-end moment M^F is derived here for a single concentrated load, but the procedure for a uniform load or any combination of uniform and concentrated loads is essentially the same.

Consider Fig. 138, showing the deflection and the conjugate beam for the problem to be solved. Call the fixed-end moments M_{AB} and M_{BA}, respectively. The bending moments due to P, neglecting end restraint, are shown by the triangular area whose base is the lower dotted line. At the highest point of this diagram the moment is Pab/L. The moment areas representing the elastic weights due to restraint are made up of triangular areas, as in the previous example. Take moments about the center of gravity of the area corresponding

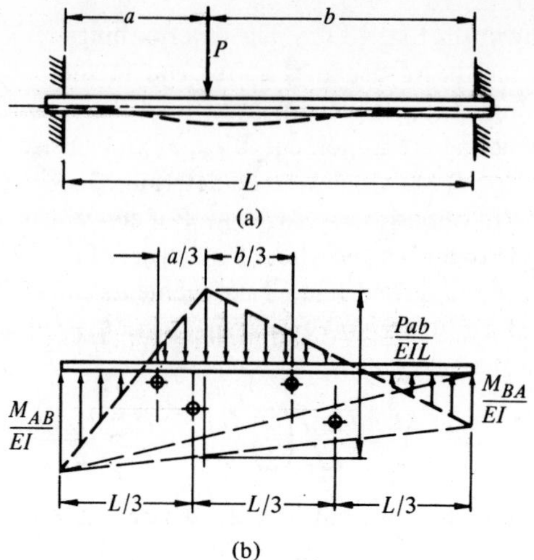

Fig. 138.

to M_{BA}, thus eliminating this term in the equation of equilibrium for the conjugate beam:

$$\frac{M_{AB}}{EI}\left(\frac{L}{2}\right)\frac{L}{3} = \frac{Pab}{EIL}\left[\frac{a}{2}\left(b + \frac{a}{3} - \frac{L}{3}\right)\right] + \frac{Pab}{EIL}\left[\frac{b}{2}\left(\frac{2b}{3} - \frac{L}{3}\right)\right]$$

$$M_{AB} = Pab^2/L^2$$

It is not necessary to solve for M_{BA} because a and b may be interchanged so that

$$M_{BA} = Pa^2 b/L^2$$

The fixed-end moment for a uniform load may be obtained by a procedure similar to that above. The moments are the same on each end, and

$$M = wL^2/12$$

where w is the uniform load per foot and L is the span.

This completes the study of the constants in the moment-distribution method. The same type of analysis may be used to obtain the constants for members of nonuniform section.

Side-sway The moment-distribution method considers the effect of bending moments on the rotation and translation of joints in a continuous beam or rigid frame. A broad definition of a joint is: a junction between two portions of a structure which must be considered separately in analyzing the structure. In Fig. 139a is shown a continuous beam having joints at A, B, and C. Consideration of these points, only, is necessary in the analysis. The same can also be said of the lettered points for the structures shown in b and c.

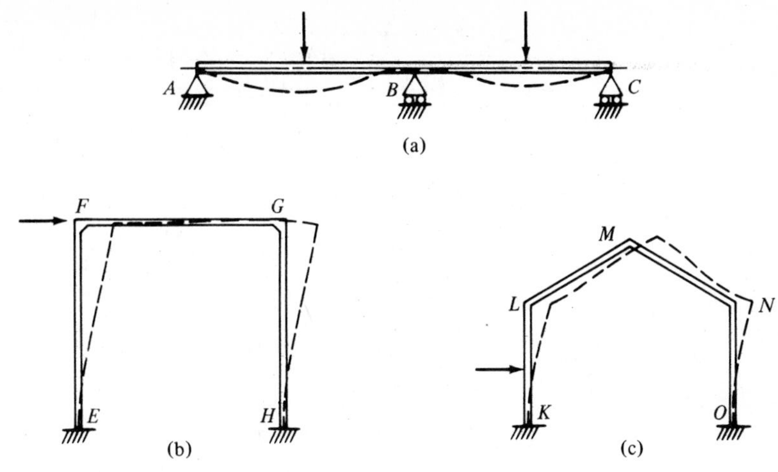

Fig. 139.

In the structure shown in Fig. 139b, the joints F and G have moved to the right. In Fig. 139c, points L and N have moved to the right and point M has not only moved to the right, but up as well.

This type of action in a structure is called side-sway. Analysis by the moment-distribution method for structures with side-sway is more complex than for structures having only rotation at joints, and will not be included in this presentation.

Moment Distribution The moment-distribution procedure operates with numerical quantities representing the actual bending moments in the structure. Therefore the numerical example shown in Fig. 140 will be used to explain procedures and prove the theory. The structure is a three-span continuous beam with fixed ends and carries a uniform load on the center span and concentrated load on the left span. A preliminary step is to compute the constants for each member. The stiffness factors I/L are in the ratio 2 : 3 : 2, as shown in the figure. All members are assumed to be of uniform moment of inertia. The carry-over factor is $+\frac{1}{2}$. The fixed-end moments are $M_{AB}{}^F = M_{BA}{}^F = 10$ ft-kip, and $M_{BC}{}^F = M_{CB}{}^F = 18.7$ ft-kip as computed in the figure.

The sign convention to be used is that moment tending to turn a joint clockwise is positive (+). Therefore, the fixed-end moments are positive on the left end of a beam and negative on the right. Note that this convention refers to joints and not to members, and for convenience it will be referred to as the *joint convention.*

Fundamental Theory The moments in continuous structures are correct if (1) the structure and every part of it is in equilibrium, and (2) if the structure is continuous, i.e., the distortions caused by the bending moments and forces agree

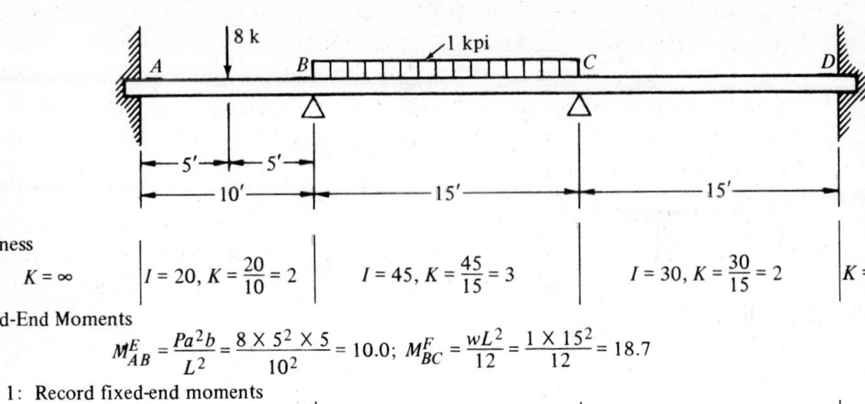

Stiffness

$K = \infty$	$I = 20,\ K = \frac{20}{10} = 2$	$I = 45,\ K = \frac{45}{15} = 3$	$I = 30,\ K = \frac{30}{15} = 2$	$K =$

Fixed-End Moments

$$M_{AB}^{F} = \frac{Pa^2b}{L^2} = \frac{8 \times 5^2 \times 5}{10^2} = 10.0; \quad M_{BC}^{F} = \frac{wL^2}{12} = \frac{1 \times 15^2}{12} = 18.7$$

Step 1: Record fixed-end moments

A	AB	BA	BC	CB	CD	DC
0	+10.0	−10.0	+18.7	−18.7	0	0

Step 2: Determine unbalanced moment and distribute in proportion to stiffness of each beam at the joint. Change the sign.

A	AB	BA	BC	CB	CD	DC
+10.0		+8.7		−18.7		0
		$\frac{2}{5}$	$\frac{3}{5}$	$\frac{3}{5}$	$\frac{2}{5}$	
−10.0	0	−3.5	−5.2	+11.2	+7.5	0

Step 3: Carry over half the distributed moment to the far end of the beam (same sign).

A	AB	BA	BC	CB	CD	DC
0	−1.7	0	+5.6	−2.6	0	+3.7

Step 2(a): Distribute unbalanced moments.

A	AB	BA	BC	CB	CD	DC
+1.7	0	−2.3	−3.3	+1.5	+1.1	0

Step 3(a): Carry over.

A	AB	BA	BC	CB	CD	DC
0	−1.1	0	+0.7	−1.6	0	+0.5

Step 2(b): Distribute.

A	AB	BA	BC	CB	CD	DC
+1.1	0	−0.3	−0.4	+1.0	+0.6	0

Final Moments: Sum of all previous moments.

A	AB	BA	BC	CB	CD	DC
0	+10.0	−10.0	+18.7	−18.7	0	0
−10.0	0	− 3.5	− 5.2	+11.2	+7.5	0
0	− 1.7	0	+ 5.6	− 2.6	0	+3.7
+ 1.7	0	− 2.3	− 3.3	+ 1.5	+1.1	0
0	− 1.1	0	+ 0.7	− 1.6	0	+0.5
+ 1.1	0	− 0.3	− 0.4	+ 1.0	+0.6	0
− 7.2	+ 7.2	−16.1	+16.1	− 9.2	+9.2	+4.2

Fig. 140.

with the distortions permitted by the supports of the structure. These requirements are the fundamental bases of all methods of analysis of indeterminate structures.

Procedure The procedure may be described in the following steps; all references are to Fig. 140.

Step 1. Record the fixed-end moments using some scheme which will indicate the location of the end of the member corresponding to the recorded figure. In this example the figures are placed below the sketch of the beam.

Discussion: This set of figures represents the condition of the structure with all joints locked against rotation. The structure may be said to be continuous, but the joints are not in equilibrium.

Step 2. Compute the unbalanced fixed-end moment at each joint. Distribute this unbalanced moment to each member at the joint in proportion to the relative stiffness of that member. Record this value with a sign opposite to the sign of the unbalanced moment.

Example: At the joint B the unbalanced moment is +8.7 ft-kip. The portion that will be distributed to BA is

$$\frac{8.7K_{BA}}{(K_{BA}+K_{BC})}=\frac{8.7(2)}{(2+3)}=3.5 \text{ ft-kip}$$

Discussion: This step puts each individual joint in equilibrium, but has destroyed the continuity of each member. To restore the continuity, the next step will be to add to the end of the member away from the joint a portion of the distributed moment as defined by the carry-over factor. The continuity will then be restored.

Step 3. Divide the moment by two and record at the far end of the member with the same sign.

Example: At B, in member BC, the distributed moment is −5.2. Record −2.6 at C, in BC.

Discussion: Each member is continuous (the moment agrees with the deflection) but joints are now not in equilibrium.

Step 2a. Repeat step 2.

Discussion: The joints are in equilibrium but the structure is not continuous. Therefore step 3 must be repeated.

Procedure. Repeat steps 2 and 3 alternately until the moments distributed and carried over become very small. In the example shown, this procedure has been carried out for one more cycle (to step 2b). Note that for the joints to be in equilibrium the procedure must stop with a distribution.

Discussion: The structure is now practically in equilibrium and has also satisfied the requirements of continuity.

Final Moments in Structure. Add the fixed-end moments, distributed moments and carry-over moments (Note: do not add the total unbalanced fixed-end moment shown for convenience in the calculations.)

Example: In member BC, at B, the final moment is

$$+18.7-5.2+5.6-3.3+0.7-0.4=16.1$$

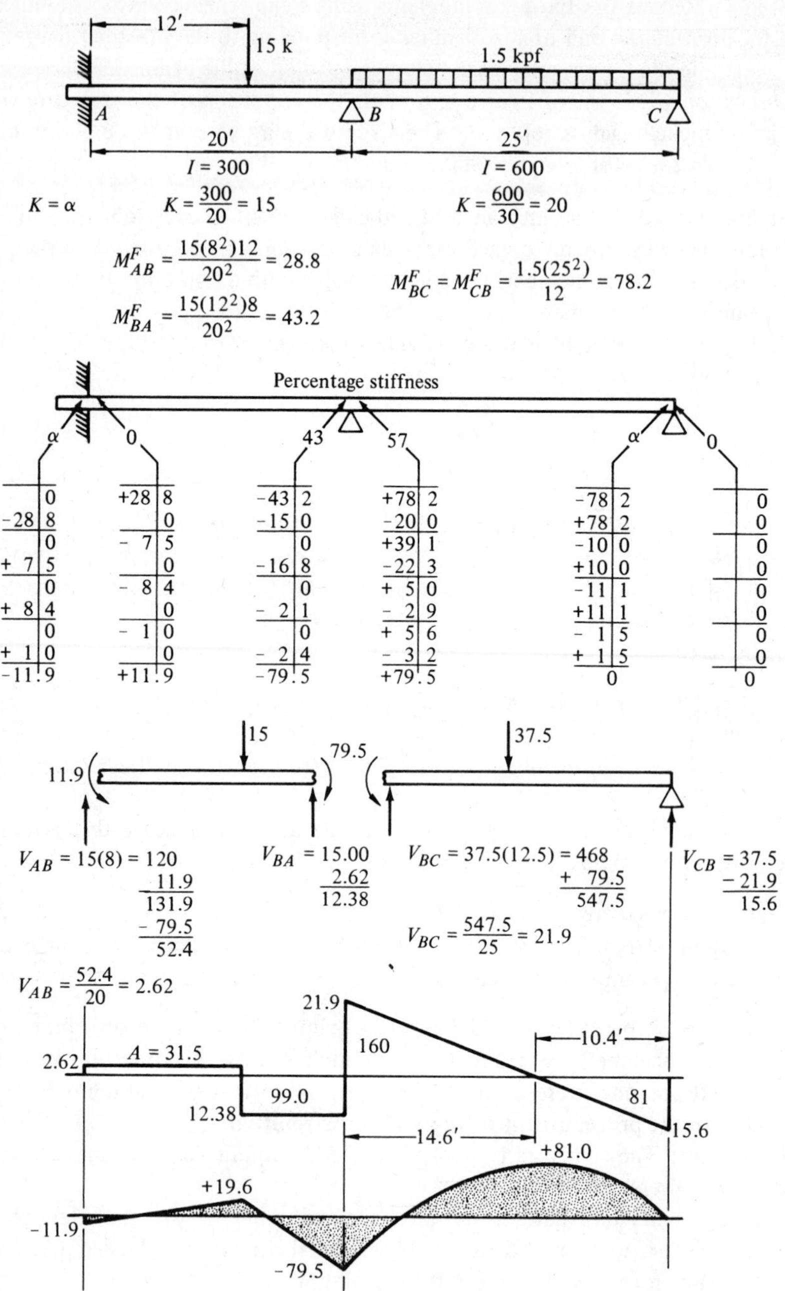

12′
15 k
1.5 kpf
A
B
C
20′
25′
I = 300
I = 600
K = α
K = 300/20 = 15
K = 600/30 = 20
M^F_AB = 15(8²)12/20² = 28.8
M^F_BA = 15(12²)8/20² = 43.2
M^F_BC = M^F_CB = 1.5(25²)/12 = 78.2
Percentage stiffness
α
0
43
57
α
0
11.9
15
79.5
37.5
V_AB = 15(8) = 120
11.9
131.9
– 79.5
52.4
V_AB = 52.4/20 = 2.62
V_BA = 15.00
2.62
12.38
V_BC = 37.5(12.5) = 468
+ 79.5
547.5
V_BC = 547.5/25 = 21.9
V_CB = 37.5
– 21.9
15.6
21.9
160
10.4′
2.62
A = 31.5
99.0
81
12.38
14.6′
15.6
+81.0
+19.6
-11.9
-79.5

Fig. 141.

Discussion: Note that this method is a converging process and the answer is obtained by successive approximations. The final answer may be as accurate as required, and this method is not an approximate method.

The treatment of the fixed ends should be noted in this problem. The walls into which the beams are embedded are, in the distribution process, assumed to be members of infinite moment of inertia. Therefore, the entire moment is distributed to the wall.

Two examples will now be presented illustrating various problems which will be encountered, and the methods of treating these problems.

EXAMPLE: CONTINUOUS BEAM WITH ONE HINGED END

In Fig. 141 is shown a continuous beam with a hinged end on the right. This problem is treated simply as if there were a dummy member to the right of the hinge with zero stiffness. The stiffness and fixed-end moments for this problem are calculated as in the previous problem. Also, the writer's style of recording calculations is shown. A light guide line is drawn, with an arrowhead to the point where the moment is acting so that the location of the particular moment is clearly indicated. This guide line is used as a decimal point for the figures to be recorded. After each distribution, a horizontal line is drawn so that a distribution will not be confused with a carry-over, and vice versa.

In this example, the stiffness factors are recorded in percentage of stiffness at each joint. The sum of the recorded figures is added for each column. The check on the computations is that on each side of the joint the moments are equal but of opposite sign. That is, the sum of the moments at each joint must be zero.

Shear and moment diagrams have been drawn for the beams in this problem. First draw free-body diagrams for each beam and show all the forces on the beam. Then take the sum of moments about one of the ends and divide by the span length to find the shear. The sum of shears at a joint gives the reaction. The shear diagram is drawn and the areas of each of the figures are recorded inside the appropriate figures. The moment diagram is drawn by adding areas progressively from a point of known moment. A check on the computations is that the moments should add up to their previously assigned values.

EXAMPLE: BEAM CONTINUOUS WITH COLUMNS

The solution of problems involving columns as well as beams does not present any difficulty. In Fig. 142 is a structure, one joint of which is continuous with a column which has a lateral load acting on it. Note also that the right end has a cantilever with a load on the end.

The procedure for calculating stiffness factors and fixed-end moments is the same as in the previous example. The moment M_{CF}, for the cantilever, is also calculated and recorded as if it were a fixed-end moment.

An innovation is introduced here which eliminates several steps. In a fixed end, for example joints A and D, no moment is carried over from the adjacent

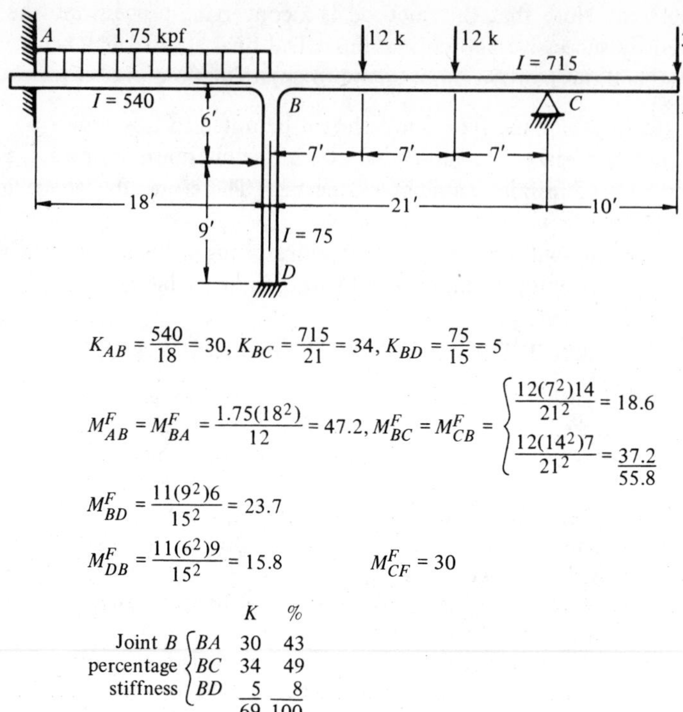

$$K_{AB} = \frac{540}{18} = 30,\ K_{BC} = \frac{715}{21} = 34,\ K_{BD} = \frac{75}{15} = 5$$

$$M^F_{AB} = M^F_{BA} = \frac{1.75(18^2)}{12} = 47.2,\ M^F_{BC} = M^F_{CB} = \begin{cases} \dfrac{12(7^2)14}{21^2} = 18.6 \\ \dfrac{12(14^2)7}{21^2} = \underline{37.2} \\ \qquad\qquad 55.8 \end{cases}$$

$$M^F_{BD} = \frac{11(9^2)6}{15^2} = 23.7$$

$$M^F_{DB} = \frac{11(6^2)9}{15^2} = 15.8 \qquad\qquad M^F_{CF} = 30$$

		K	%
Joint B percentage stiffness	BA	30	43
	BC	34	49
	BD	5	8
		69	100

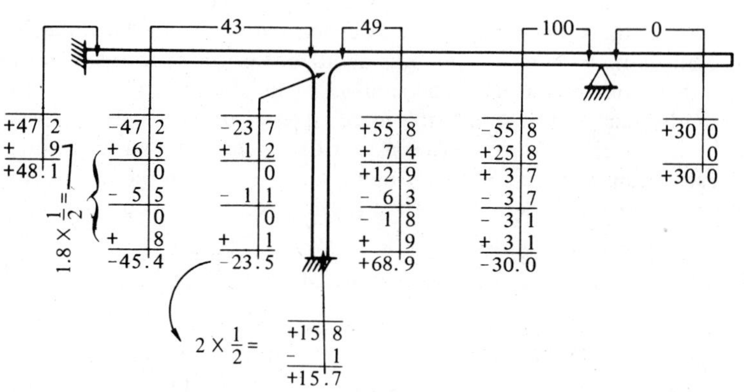

Fig. 142.

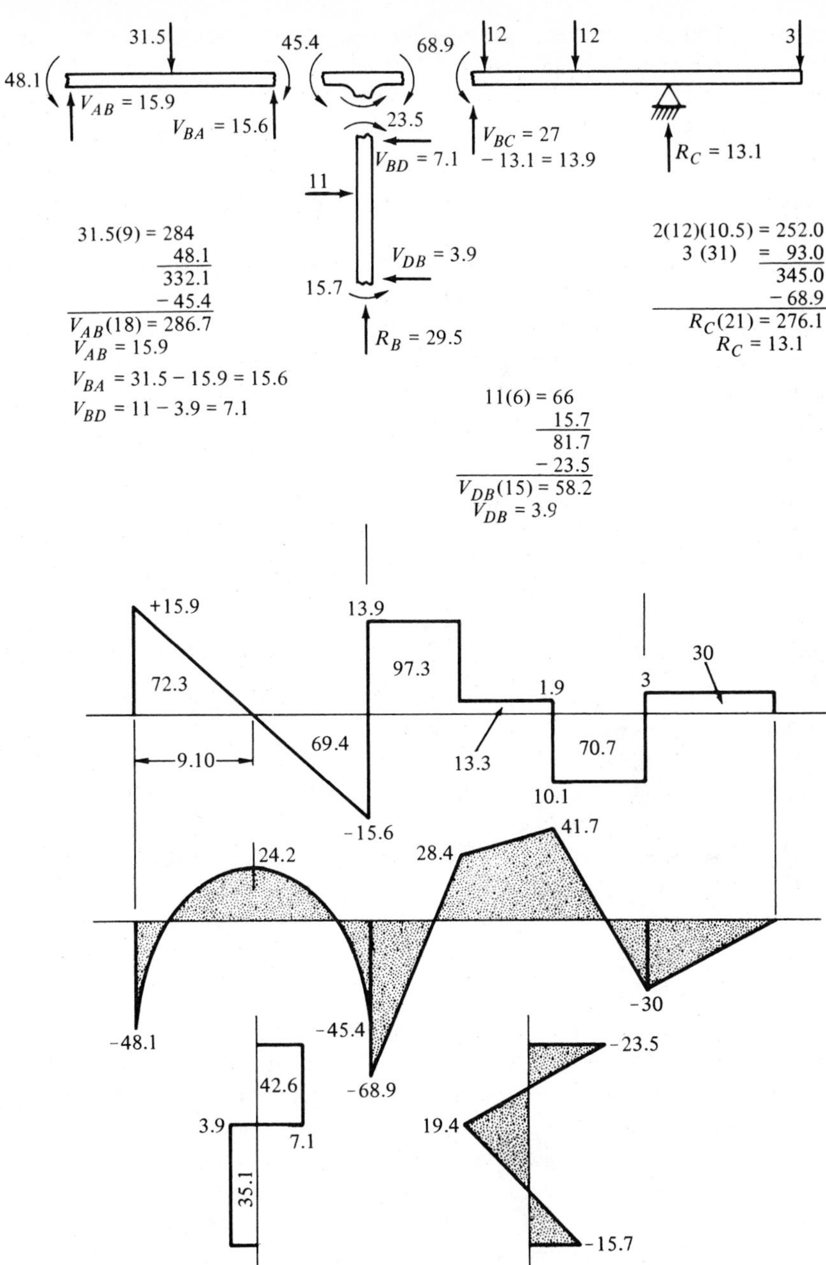
31.5
48.1
V_{AB} = 15.9
V_{BA} = 15.6
45.4
68.9
23.5
V_{BD} = 7.1
11
V_{DB} = 3.9
15.7
R_B = 29.5
12
12
3
V_{BC} = 27
− 13.1 = 13.9
R_C = 13.1
31.5(9) = 284
48.1
332.1
− 45.4
V_{AB}(18) = 286.7
V_{AB} = 15.9
V_{BA} = 31.5 − 15.9 = 15.6
V_{BD} = 11 − 3.9 = 7.1
2(12)(10.5) = 252.0
3 (31) = 93.0
345.0
− 68.9
R_C(21) = 276.1
R_C = 13.1
11(6) = 66
15.7
81.7
− 23.5
V_{DB}(15) = 58.2
V_{DB} = 3.9
+15.9
72.3
9.10
69.4
−15.6
13.9
97.3
13.3
1.9
70.7
10.1
3
30
24.2
−48.1
−45.4
−68.9
28.4
41.7
−30
42.6
3.9
7.1
35.1
−23.5
19.4
−15.7

Fig. 142. (*Continued*)

joint until all the moments have been distributed. Then the entire sum of distributions (not including the original fixed-end moment) is added up and divided by two and carried over. Reactions have been calculated and shear and moment diagrams have been drawn as in the previous example. The construction of free-body diagrams greatly aids the interpretation of the forces on the structure.

Summary In the moment-distribution method, fixed-end moments are computed for all members with joints locked against rotation. Each joint is then unlocked and the unbalanced moment is distributed in proportion to the stiffness of each member. Then this moment is carried over to the far end of each member to preserve continuity. These steps are repeated until there is no longer appreciable moment to distribute. Each step of the procedure represents a physical action of the beam. Construction of shear and moment diagrams will give a picture of the stresses in the structure.

Short Cuts and Simplifications for Moment Distribution

It requires considerable practice to become proficient in solving moment-distribution problems expeditiously. Each person will gradually evolve his own style of computation. The previous section has introduced the writer's methods of showing computations, which are characterized by the following points:

1. All computations possible are shown on the sheet, including calculations for fixed-end moments.
2. A clear sketch of the structure is first made. Then a vertical line is drawn with an arrow directed at each point where a distribution will occur, so that there is no question about just where the figures apply.
3. The stiffness of each joint is given as a percentage of the total stiffness.
4. As far as possible, all the fixed-end moments are placed on the same horizontal line.
5. A horizontal line is drawn after each distribution to indicate the completion of that cycle.

Often the most tedious part of this method is the adding up of columns of figures after the distribution has been made. The writer's custom is to turn himself into an automatic adding machine as completely as possible. Take a piece of paper with horizontal ruling (not cross-section paper), and turn the paper so the rules are vertical. Now figures may be recorded with numbers far apart and all lined up vertically, as shown in the example sketched in Fig. 143. Write down all the positive (+) values first and add. Do not try to carry sums in your head or the process ceases to become atuomatic. This method will enable the adding to be carried out with the speed and precision of an adding machine with little fatigue to the computer.

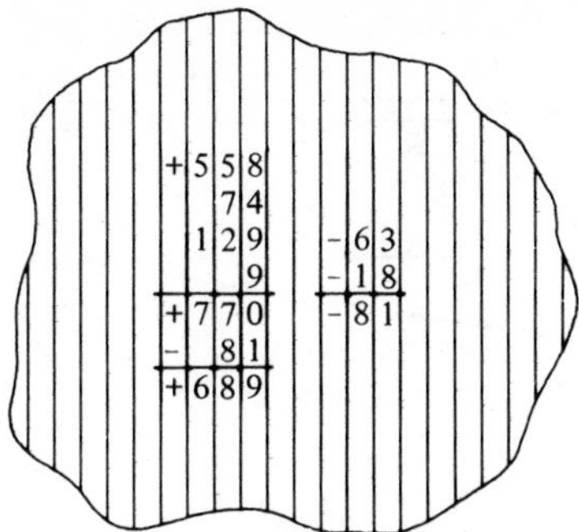

Fig. 143. The easiest procedure for adding columns of figures is to use paper with ruling turned vertical.

Modified Stiffness Factors In the previous subsection, which introduced the moment-distribution method, the stiffness constant was defined as the moment at the end of a member necessary to produce a unit rotation at that end when the other end is fixed. The formula for the stiffness K, given the rotation ϕ, and the moment M, is

$$K = M/\phi$$

The stiffness may be determined for other cases than for the far end of the member fixed. The stiffness in any of these cases will be called *modified stiffness* and will be designated by the symbol K^M. Three cases of modified stiffness will be demonstrated: (1) hinged end, (2) symmetrical loading on symmetrical structure, and (3) antisymmetrical loading on symmetrical structure. Derivation of these constants and an illustration of their use by simple examples will now be presented.

Hinged Ends Consider the simple beam with a uniform moment of inertia and with hinged ends shown in Fig. 144. A moment M is applied to the left end. The moment diagram is a triangle and the values of K are

$$K = M/\phi_A = 3EI/L$$

The value of K for the fixed end is $K = 4EI/L$. Therefore the relation between K and K^M for hinged ends is

$$K^M = \tfrac{3}{4}K$$

This value applies only to members of uniform moment of inertia.

As an example of the use of this modified stiffness constant, the beam shown in Fig. 145 will be solved. This is the same structure and loads as in Fig. 141 of the previous section, and the results should check.

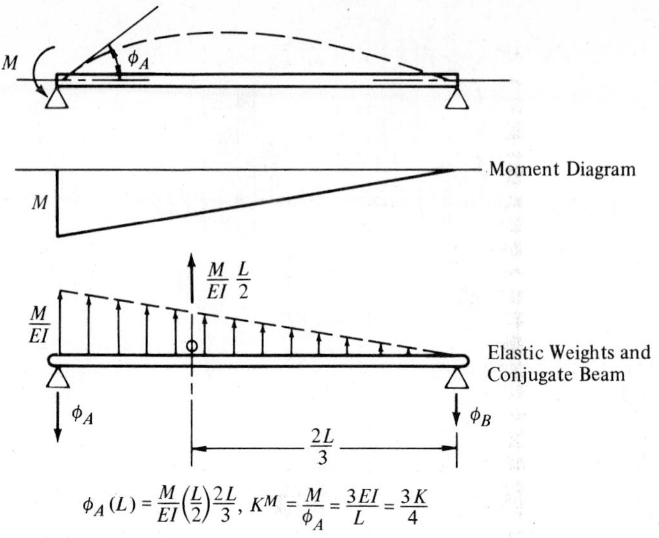

Fig. 144.

The stiffness and fixed-end moments for span AB are the same as for the previous example.

The modified stiffness for span BC is three-fourths of the original value, or $K^M = \frac{3}{4}(20) = 15$. The fixed-end moment for span BC must be changed for the case with end C hinged. This has been carried out by the moment-distribution method as a separate operation. The moment at point C is released and half of it is carried to point B. However, it may be combined with the final distribution. Only one distribution is required and a carry-over to end A. Since no moments are carried back to B, the distribution is complete. It may be observed from this example that modified stiffness factor shortened the distribution process considerably.

Symmetrical Structure and Loading If the entire structure is symmetrical and the loading is symmetrical about the centerline, it is obvious that the calculations will be the same for each side of the structure. At a beam in the center, a change of moment on one side will be accompanied by a similar change on the other. A modified stiffness factor may be used for the central beam of this structure.

Consider the beam shown in Fig. 146, which is symmetrically loaded by bending moments M on each end. The bending-moment diagram and the elastic-weight diagram are uniform across the beam. The angle ϕ at each end of the beam is the same and

$$\phi = \frac{ML}{2EI}, \qquad \text{so that} \qquad K^M = \frac{M}{\phi} = \frac{2EI}{L}$$

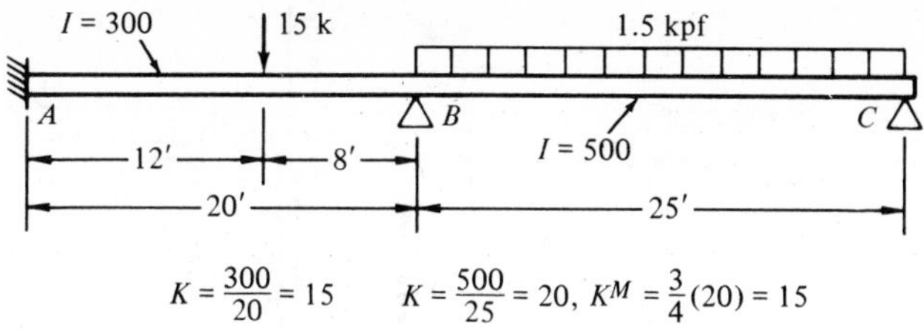

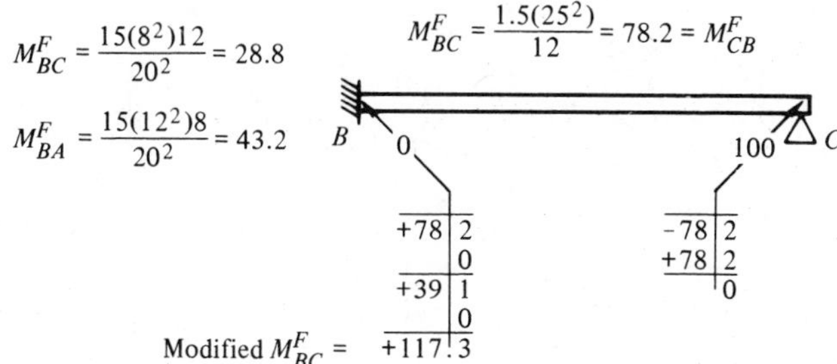

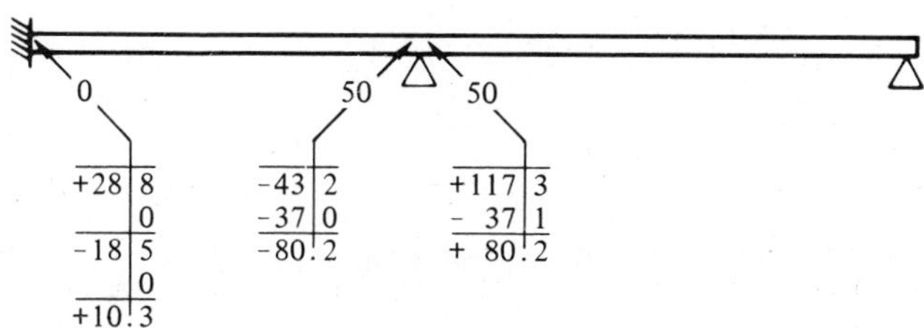

Fig. 145.

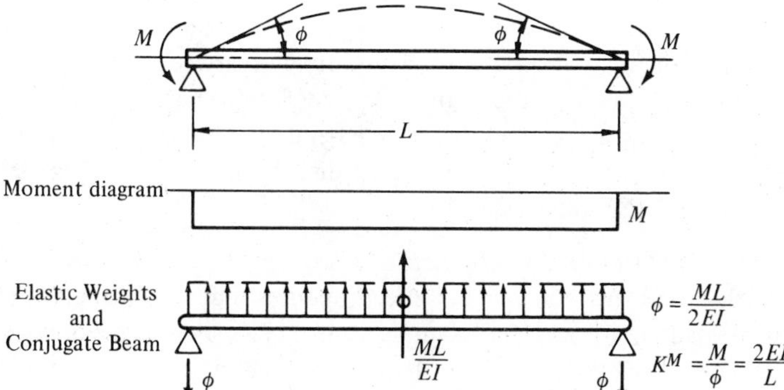

Fig. 146.

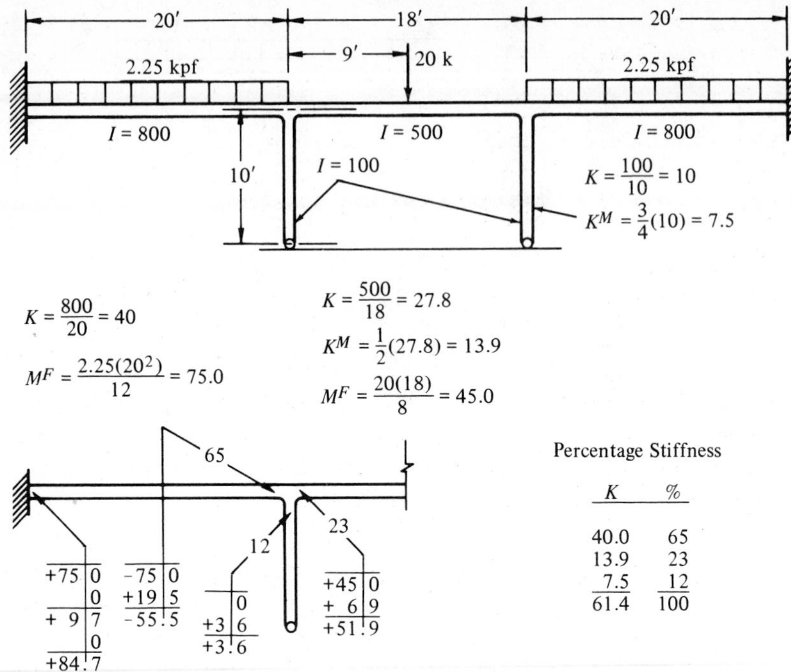

Fig. 147.

In terms of the original stiffness factor the modified factor is

$$\frac{K^M}{K} = \frac{2EI}{L}\left(\frac{L}{4EI}\right); \qquad K^M = \frac{K}{2}$$

The frame shown in Fig. 147 will be solved as an example of the symmetrical case. Note, also, that the columns are hinged at the base so that the hinged stiffness coefficient may also be used. It is not necessary to make any change in the fixed-end moments with the symmetrical coefficient because the ends of the members are fixed as far as the original fixed-end moments are concerned. This example converges after the first cycle of distribution, there being only one joint where distribution is necessary.

Symmetrical Structure, Antisymmetrical Loading In some cases a structure may be symmetrical with respect to its elastic properties, but is subjected to an antisymmetrical loading. A typical example is the bent shown in Fig. 148. The bending moments on the right side of the frame will be equal and opposite to those on the other side of the frame. The stiffness of the horizontal center member may be modified so that only one cycle of distribution is necessary.

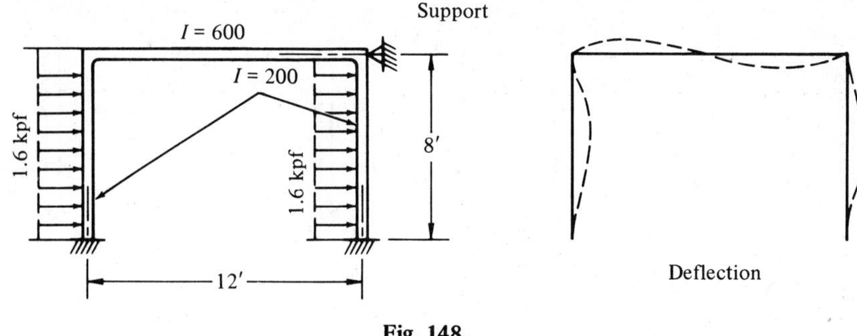

Fig. 148.

Note that a support has been provided for this frame at the top in order to prevent side-sway.

The modified coefficient may be derived from the structure, moment diagram, and conjugate beam sketched in Fig. 149; from this figure:

$$K^M = 6EI/L; \quad \frac{K^M}{K} = (6EI/L)\,(L/4EI)$$

and

$$K^M = 3K/2$$

In Fig. 150 this modified factor has been used to solve the structure shown in Fig. 148. It is necessary to distribute moments on only one side of the frame. The distribution closes in one cycle.

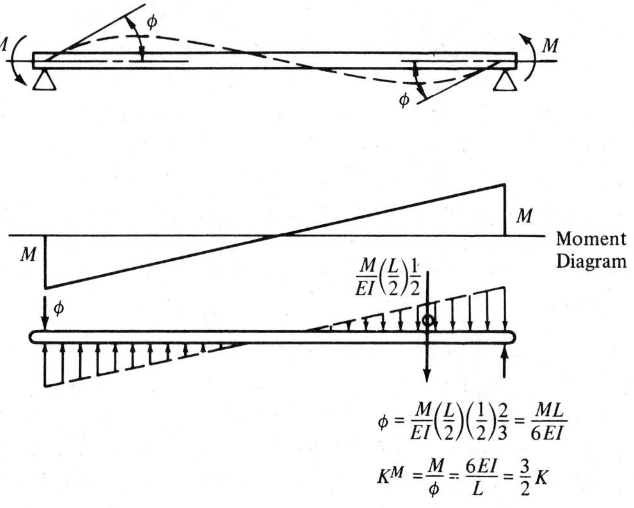

Fig. 149.

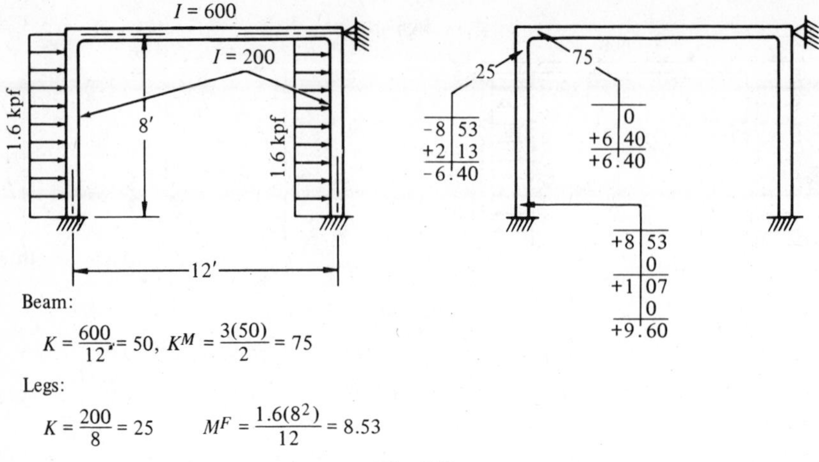

Fig. 150.

Buckling of Columns

It is recognized that a long structural column element, subjected to both bending and compression forces, will have stresses greater than the sum of the stresses due to the individual bending and compression. Consider the column shown in Fig. 151. It is subjected to a vertical load *P* and a lateral uniform load *w*. The length is *L* and the ends are hinged. The moment due to the lateral load is M_L and the deflection is Δ_L. For this condition there is also a moment due to the eccentricity of the load *P* which is initially equal to $\Delta_L P$. This moment, however, increases the deflection and consequently increases the additional moment. Thus the deflection will continue to increase and will finally come to rest or will become so great that the column will buckle out of shape and cease to carry load. This action can be demonstrated by using a long flexible stick and

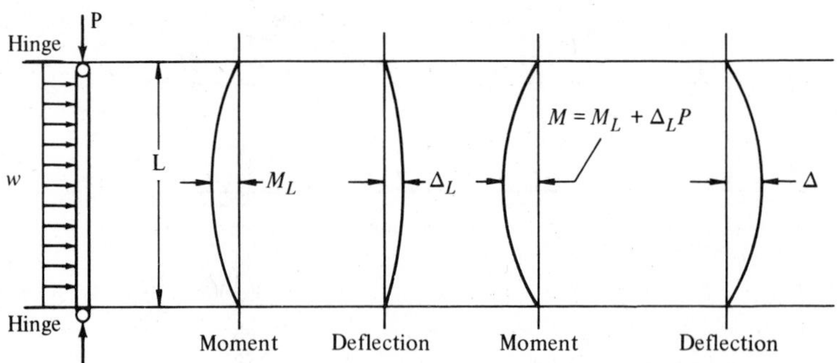

Fig. 151.

pushing on the ends. It is not necessary to have an actual lateral load if the column has a small initial eccentricity.

We can study the action of a column under lateral and compressive load by the following procedure: (1) determine the deflection Δ_L, and the moment M_L; (2) determine the additional moment and deflection, and add to the original deflection; (3) repeat until the additional deflections are no longer significant or the column has buckled.

EXAMPLE: COLUMN WITH LATERAL LOAD

Determine the deflection in a steel column, 1 in. square and 60 in. long, subjected to a lateral load of 5 lb/in. and a vertical compression force of 1000 lb. Assume that $E = 29(10^6)$ psi. The moment of inertia of a 1-in. square bar is $I = bd^3/12 = 0.0833$, and the section modulus is $S = I/c = 0.1667$.

The initial deflection due to the lateral load is

$$\Delta_L = \frac{5wL^4}{384EI} = \frac{5(5)(60^4)}{384(29)(10^6)(0.0833)} = 0.349 \text{ in.}$$

The moment due to a load $P = 1000$ lb with an eccentricity of 0.349 in. is

$$M = 1000(0.349) = 349 \text{ in.-lb}$$

at the center of the column.

Now we must determine the deflection due to this moment. If we assume that the moment curve is a parabola then

$$\Delta = \frac{5ML^2}{48EI} = \frac{5(349)(60^2)}{48(29)(10^6)(0.0833)} = 0.0541 \text{ in.}$$

The combined deflection will be:

$$0.0541 + 0.3490 = 0.4031 \text{ in.}$$

Now the moment to the vertical load must be revised due to the greater deflection, by proportion:

$$\Delta = 0.0541(0.4031)/(0.3490) = 0.0624 \text{ in.}$$

and the total is

$$\Delta_T = 0.0624 + 0.3490 = 0.4114 \text{ in.}$$

If this same procedure is repeated for two more times the total deflection becomes $\Delta_T = 0.4128$ for the first time and 0.4129 for the second, showing that the deflection has nearly reached its final value.

This deflection can be determined approximately without the successive steps by making the assumption that the shape of the moment diagram is similar to the shape of the deflection curve:

$$\Delta_T = \Delta_L + (\Delta_L/M_L)\,\Delta_T P$$

Solve for Δ_T:

$$\Delta_T = \frac{\Delta_L}{1 - (\Delta_L P/M_L)}$$

For our example,

$$M_L = 5(60^2)/8 = 2250 \text{ ft-lb}$$

$$\Delta_T = \frac{0.349}{1 - [0.349(1000)/2250]} = 0.4131 \text{ in.}$$

which agrees with the result obtained from the successive deflection calculations.

The stresses in the square bar are:

Due to lateral load:

$$f = M/S = 2250/0.1667 = 13{,}497 \text{ psi}$$

Due to P:

$$f = P/A = 1000/1.0 = 1{,}000 \text{ psi}$$

Due to deflection:

$$f = P\,\Delta_T/S = 1000(0.4131)/0.1667 = 2{,}478 \text{ psi}$$

Total: 16,975 psi

The Euler Buckling Load If we assume that Δ_T in the preceding equation is very large, then the denominator for the right side approaches zero and

$$1 - (\Delta_L P/M_L) = 0, \quad \text{so that} \quad P = M_L/\Delta_L$$

For the above example, $P = 2250/0.3490 = 6446$ lb. This is the load that would cause the column to buckle. The size of the lateral load will not matter since only the ratio of load to deflection is used.

We can write an equation using

$$M_L = wL^2/8 \quad \text{and} \quad \Delta_L = 5wL^4/384EI$$

so that

$$P = M_L/\Delta_L = 384EI/40L^2 = 9.6EI/L^2$$

The well known *Euler formula* for columns can be derived by assuming that the load and moment diagrams are sine curves for which

$$M_L = wL^2/\pi^2 \quad \text{and} \quad \Delta_L = \frac{wL^4}{\pi^4 EI}$$

so that

$$P = \pi^2 EI/L^2 = 9.87EI/L^2$$

This value is 3% higher than the method using the uniform load.

Columns With Fixed Ends If a column is fixed at one or more of the ends, the assumption can be made that the effective length of the column is a fraction of the length of a pin-ended column. This coefficient is called K. Values for various configurations of support are given in the subsection on steel-column design.

STRUCTURAL ELEMENTS

Design of Steel Structural Elements

The selection of structural steel members and connections will in general be based on the *Specification for the Design, Fabrication and Erection of Structural Steel for Buildings*, issued by the American Institute of Steel Construction, 101 Park Avenue, New York City. This code, together with tables of members and connections and design examples, is included in the *Manual of Steel Construction* of the AISC. A copy of the *Manual* is indispensable for the design of steel members and connections. For the first subsections here it will not be absolutely necessary to possess a copy, but for later subsections it will be assumed that copies are available to the reader. A summary of some of the regulations will be given here, but the requirements are in many cases so complex that only those that directly affect the design of the simplest members will be mentioned.

Allowable Stresses in Steel Beams Allowable stresses are generally based on the designated yield strength in tension in accordance with ratings of the American Society for Testing and Materials. The type of steel used for most buildings has the ASTM designation A36 and has a yield strength in tension of $F_y = 36{,}000$ psi. In the design problems that follow, this type of steel will be used unless some other is mentioned at the start of the problem.

Following is a summary of allowable stresses from Section 1.5 of the *Manual* for several conditions. The stresses are for $F_y = 36{,}000$ psi. Note that, in some cases, stresses have been rounded to the nearest 1000 psi.

1. Tension on the net section of tension members, except at pin holes — $F_t = 0.60F_y = 22{,}000$ psi
2. Tension on net section at pin holes — $F_t = 0.45F_y = 16{,}000$ psi
3. Shear on gross section — $F_v = 0.40F_y = 14{,}400$ psi
4. Bending in laterally supported rolled compact sections. A compact section must satisfy certain requirements for thickness of projecting elements and depth to thickness ratios for the web. Most, A36 steel, wide flange sections satisfy this requirement. — $F_b = 0.66F_y = 24{,}000$ psi
5. Bending is noncompact sections — $F_b = 0.60F_y = 22{,}000$ psi

Design of laterally unsupported members, members in compression and members with both compression and bending require rather complicated expressions for allowable stresses, and their design will be discussed later.

TABLE 6. W SHAPES.

			Flange		Web	Elastic Properties					
						Axis X-X			Axis Y-Y		
	Area A	Depth d	Width b_f	Thickness t_f	Thickness t_w	I	S	r	I	S	r
Designation	In.^{2}	In.	In.	In.	In.	In.^{4}	In.^{3}	In.	In.^{4}	In.^{3}	In.
W36×160	47.1	36.00	12.000	1.020	0.653	9760	542	14.4	295	49.1	2.50
×150	44.2	35.84	11.972	0.940	0.625	9030	504	14.3	270	45.0	2.47
×135	39.8	35.55	11.945	0.794	0.598	7820	440	14.0	226	37.9	2.39
W33×130	38.3	33.10	11.510	0.855	0.580	6710	406	13.2	218	37.9	2.38
×118	34.8	32.86	11.484	0.738	0.554	5900	359	13.0	187	32.5	2.32
W30×108	31.8	29.82	10.484	0.760	0.548	4470	300	11.9	146	27.9	2.15
× 99	29.1	29.64	10.458	0.670	0.522	4000	270	11.7	128	24.5	2.10
W27× 94	27.7	26.91	9.990	0.747	0.490	3270	243	10.9	124	24.9	2.12
× 84	24.8	26.69	9.963	0.636	0.463	2830	212	10.7	105	21.1	2.06
W24× 76	22.4	23.91	8.985	0.682	0.440	2100	176	9.69	82.6	18.4	1.92
× 68	20.0	23.71	8.961	0.582	0.416	1820	153	9.53	70.0	15.6	1.87
W24× 61	18.0	23.72	7.023	0.591	0.419	1540	130	9.25	34.3	9.76	1.38
× 55	16.2	23.55	7.000	0.503	0.396	1340	114	9.10	28.9	8.25	1.34
W21× 73	21.5	21.24	8.295	0.740	0.455	1600	151	8.64	70.6	17.0	1.81
× 68	20.0	21.23	8.270	0.685	0.430	1480	140	8.60	64.7	15.7	1.80
× 62	18.3	20.99	8.240	0.615	0.400	1330	127	8.54	57.5	13.9	1.77
× 55	16.2	20.80	8.215	0.522	0.375	1140	110	8.40	48.3	11.8	1.73
W21× 49	14.4	20.82	6.520	0.532	0.368	971	93.3	8.21	24.7	7.57	1.31
× 44	13.0	20.66	6.500	0.451	0.348	843	81.6	8.07	20.7	6.38	1.27
W18× 70	20.6	18.00	8.750	0.751	0.438	1160	129	7.50	84.0	19.2	2.02
× 64	18.9	17.87	8.715	0.686	0.403	1050	118	7.46	75.8	17.4	2.00
W18× 60	17.7	18.25	7.558	0.695	0.416	986	108	7.47	50.1	13.3	1.68
× 55	16.2	18.12	7.532	0.630	0.390	891	98.4	7.42	45.0	11.9	1.67
× 50	14.7	18.00	7.500	0.570	0.358	802	89.1	7.38	40.2	10.7	1.65
×45	13.2	17.86	7.477	0.499	0.335	706	79.0	7.30	34.8	9.32	1.62
W18× 40	11.8	17.90	6.018	0.524	0.316	612	68.4	7.21	19.1	6.34	1.27
× 35	10.3	17.71	6.000	0.429	0.298	513	57.9	7.05	15.5	5.16	1.23
W16× 58	17.1	15.86	8.464	0.645	0.407	748	94.4	6.62	65.3	15.4	1.96
W16× 50	14.7	16.25	7.703	0.628	0.380	657	80.8	6.68	37.1	10.5	1.59
× 45	13.3	16.12	7.039	0.563	0.346	584	72.5	6.64	32.8	9.32	1.57
× 40	11.8	16.00	7.000	0.503	0.307	517	64.6	6.62	28.8	8.23	1.56
× 36	10.6	15.85	6.992	0.428	0.299	447	56.5	6.50	24.4	6.99	1.52
W16× 31	9.13	15.84	5.525	0.442	0.275	374	47.2	6.40	12.5	4.51	1.17
× 26	7.67	15.65	5.500	0.345	0.250	300	38.3	6.25	9.59	3.49	1.12
W14× 38	11.2	14.12	6.776	0.513	0.313	386	54.7	5.88	26.6	7.86	1.54
× 34	10.0	14.00	6.750	0.453	0.287	340	48.6	5.83	23.3	6.89	1.52
× 30	8.83	13.86	6.733	0.383	0.270	290	41.9	5.74	19.5	5.80	1.49
W14× 26	7.67	13.89	5.025	0.418	0.255	244	35.1	5.64	8.86	3.53	1.08
× 22	6.49	13.72	5.000	0.335	0.230	198	28.9	5.53	7.00	2.80	1.04
W12× 36	10.6	12.24	6.565	0.540	0.305	281	46.0	5.15	25.5	7.77	1.55
× 31	9.13	12.09	6.525	0.465	0.265	239	39.5	5.12	21.6	6.61	1.54
× 27	7.95	11.96	6.497	0.400	0.237	204	34.2	5.07	18.3	5.63	1.52
W12× 22	6.47	12.31	4.030	0.424	0.260	156	25.3	4.91	4.64	2.31	0.847
× 19	5.59	12.16	4.007	0.349	0.237	130	21.3	4.82	3.76	1.88	0.820
W10× 33	9.71	9.75	7.964	0.433	0.292	171	35.0	4.20	36.5	9.16	1.94

TABLE 6. *(Continued)*

			Flange		Web	Elastic Properties					
						Axis X-X			Axis Y-Y		
	Area A	Depth d	Width b_f	Thickness t_f	Thickness t_w	I	S	r	I	S	r
Designation	In.^{2}	In.	In.	In.	In.	In.^{4}	In.^{3}	In.	In.^{4}	In.^{2}	In.
W10× 29	8.54	10.22	5.799	0.500	0.289	158	30.8	4.30	16.3	5.61	1.38
× 25	7.36	10.08	5.762	0.430	0.252	133	26.5	4.26	13.7	4.76	1.37
× 21	6.20	9.90	5.750	0.340	0.240	107	21.5	4.15	10.8	3.75	1.32
W10× 19	5.61	10.25	4.020	0.394	0.250	96.3	18.8	4.14	4.28	2.13	0.874
× 17	4.99	10.12	4.010	0.329	0.240	81.9	16.2	4.05	3.55	1.77	0.844
W8× 28	8.23	8.06	6.540	0.463	0.285	97.8	24.3	3.45	21.6	6.61	1.62
× 24	7.06	7.93	6.500	0.398	0.245	82.5	20.8	3.42	18.2	5.61	1.61
W8× 20	5.89	8.14	5.268	0.378	0.248	69.4	17.0	3.43	9.22	3.50	1.25
× 17	5.01	8.00	5.250	0.308	0.230	56.6	14.1	3.36	7.44	2.83	1.22
W8× 15	4.43	8.12	4.015	0.314	0.245	48.1	11.8	3.29	3.40	1.69	0.876
× 13	3.83	8.00	4.000	0.254	0.230	39.6	9.90	3.21	2.72	1.36	0.842

Properties of Steel Sections Because the AISC *Manual* may not be immediately available, the section modulus $S = I/c$ of a number of steel beams is reproduced in Table 6. These members are generally the lightest for the depths shown, so should be the most economical.

An approximate value for $S = I/c$ for I sections may be obtained if the depth and weight are known. The value of S is approximately

$$S = \text{(weight) (depth)}/10$$

For example, Table 6 gives the value of S for a W16×36 as 56.5 in.^{3}; then, by the approximate formula,

$$S = (36)(16)/10 = 57.6 \text{ in.}^3$$

Formulas for Moments in Simple Beams Most floor systems have regularly spaced beams and girders, so it is much easier to determine the bending moments from formulas if the beams or girders are pin-ended (simple connections), rather than going through the procedure of the construction of shear and moment diagrams. Some typical examples follow.

For a uniform load w on a span L, the moment at the center is

$$M = wL^2/8$$

A beam with two equally spaced, concentrated loads P, has a moment $M = PL/3$. Compare this with the equivalent uniform load for which $P = wL/3$, $M = (wL/3)(L/3) = wL^2/9$. The moment due to the concentrated loads is $\frac{8}{9}$ or about 89% of the moment due to the uniform load. Therefore it is more accurate to design girders with two loads.

Now consider a beam with three equal loads P, equally spaced. Then $M = PL/2$. Again compare this with the uniform load where $P = wL/4, M = (wL/4)(L/2) = wL^2/8$. The moment at the center with concentrated loads will be the same as for uniform loads.

In general, if there are more than two equally spaced, concentrated loads, it is more convenient to use the equivalent uniform load. For an even number of spaces the results will be exact. For odd numbers there will be a small error. For example, the error for four loads (five spaces) is about 4%, on the conservative side. For calculation of stresses in continuous beams the use of uniform loads is much quicker.

Simplified Formulas for Deflection of Simple Beams The deflection of a simple beam of span L, due to a uniform load w with a modulus of elasticity E and a moment of inertia I is

$$D = 5wL^4/384EI$$

This formula is not convenient for routine calculations, so a simplified formula can be devised that is easy to remember and to apply.

If L is in feet; if d, the beam depth, is equal to $2c$ and is in inches; if the value of E is 30,000 ksi (kips per square inch); and if the stress on the beam is given in kips per square inch; then, using the relations $M = wL^2/8$ and $f = Mc/I$, the following formula results:

$$D = L^2 f/1000d$$

For example, if a beam has a stress of 21.6 ksi and a span of 30 ft, and is 15 in. deep, then

$$D = 30^2(21.6)/1000(15) = 1.3 \text{ in.}$$

The value of $E = 30{,}000{,}000$ psi is slightly higher than the 29,000,000 normally used for structural steel, and it makes the formula easier to use, but remember it will slightly underestimate the deflection.

Requirements for Deflection of Steel Beams The AISC code, Section 1.13 requires that the deflection for live loads for beams or girders supporting plaster ceilings should not exceed 1/360 of the span of the beam or girder. The commentary to the code suggests that the depth of fully stressed beams or girders in floors should be not less than $F_y/800$, and for roof members $F_y/1000$, times the span. These ratios for F_y = 36 ksi are 1/22.2 and 1/27.8, respectively. If the members are not fully loaded, then the ratios can be exceeded.

Types of Steel Beams Section 1.2 of the AISC code recognizes three basic types of framing systems for steel structures as follows:

Type 1. Continuous rigid frames with rigid joints. This will include continuous beams on knife-edged supports.

Type 2. Simple framing with unrestrained ends free to rotate under gravity loads. Actually, joints may have some rigidity if wind loads are carried by frame action, but this effect is neglected in the design of beams or girders for gravity loads.

Type 3. Semirigid framing with partial fixity at the joints.

EXAMPLE: SIMPLE BEAM WITH UNIFORM LOAD

A typical steel beam in a floor system has a span of 20 ft and a spacing between adjacent beams of 8 ft. The dead load is 25 psf including the weight of the steel beam. The live load is 100 psf. Select a beam size based on bending moment and check for deflection based on 1/360 of the span. Use A36 steel, $F_y = 36$ ksi and $F_b = 24$ ksi.

The total load per square foot will be 25 + 100 = 125 psf, and the load on the beam will be the load per square foot multiplied by the 8-ft spacing, so that $w = 8(125) = 1000$ plf (pounds per lineal foot). It is more convenient to use forces in kips per lineal foot: $w = 1$ klf.

The maximum moment on the beam at the center is

$$M = wL^2/8 = 1(20^2)/8 = 50 \text{ ft-kip}$$

The section modulus required is

$$S = M/F_b = 50(12)/24 = 25.0 \text{ in.}^3$$

The minimum depth-to-span ratio for a floor system is 1/22.2 so the minimum depth of the beam should be 20(12)/22.2 = 10.8 in.

Scan Table 6. A W14×22 steel beam will provide a section modulus of $S = 28.8 \text{ in.}^3$, and therefore should be satisfactory for moment. The actual stress on the section is $f = M/S = 50(12)/28.8 = 20.8$ ksi.

The deflection by the simplified formula is

$$D = \frac{L^2 f}{1000d} = \frac{(20^2)(20.8)}{1000(14)} = 0.60 \text{ in.}$$

The ratio of deflection to span is $D/L = 0.60/(20)(12) = 1/400$. However, this is for the full dead and live load. For live load only, by ratio of loads, $D = 0.60(100)/125 = 0.48$ and the deflection–span ratio is 1/500. The deflection of the beam is satisfactory.

EXAMPLE: GIRDER IN FLOOR SYSTEM

Select the size of the girder *G1* for the floor system shown in Fig. 152. The dead load of the floor including the weights of steel beams *B1* and *G2* is 45 psf. The weight of member *G1* must be estimated and added to the dead load. The live load is 80 psf. The total load is then 45 + 80 = 125 psf. The uniform load on *G1* for 16-ft section to the left is 125(6) = 750 plf = 0.75 klf, and the load on 10-ft section to the right is 125(3) = 375 plf = 0.375 klf. The concentrated

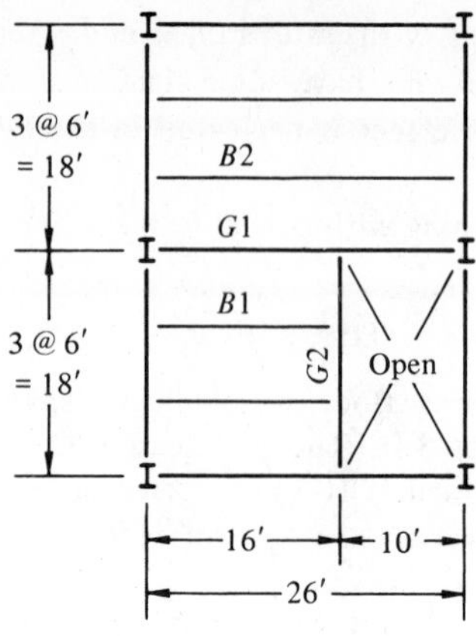

Fig. 152.

load from girder *G2* is equal to the reaction of *B1* on *G2:*

$$125(6)(8) = 6{,}000 \text{ lb} = 6.0 \text{ kip}$$

The weight of the girder must be estimated and then checked later to see if the correct guess has been made. This can usually be done on scratch paper. If we assume the load to be uniform and equal to slightly more than the 0.75 klf above, say 0.8 klf, the moment will be $M = 0.8(26)^2/8 = 68$ ft-kip. The section modulus required would be $S = 68(12)/24 = 34$ in.^{3}. Use the approximate relationship S = (weight) (depth)/10, so that weight = 34(10)/12 = 28. Use 30 plf.

The loads on the girder and the calculations for shear and moment are shown in Fig. 153. Note that reactions have been checked by taking moments about each end and summing the vertical forces.

The moments have been determined by a summation of areas under the shear diagram. The maximum moment occurs at a point 15.05 ft from the left end and is 88.3 ft-kip.

The required section modulus for an allowable stress F_b = 24 ksi is M = 89(12)/24 = 44.5. A W14X34 was selected from Table 6 and this section has a value of $S = 48.6$ in.^{3}; so the stress is, by proportion,

$$f = 44.5(24)/48.6 = 22.0 \text{ ksi}$$

The determination of the deflection was not included as a part of this problem. The precise calculation will require considerable detail. An approximate value can be obtained by assuming the load to be uniform. For the total load

$$D = L^2 f/1000d = 26^2(22.0)/1000(14) = 1.06 \text{ in.}$$

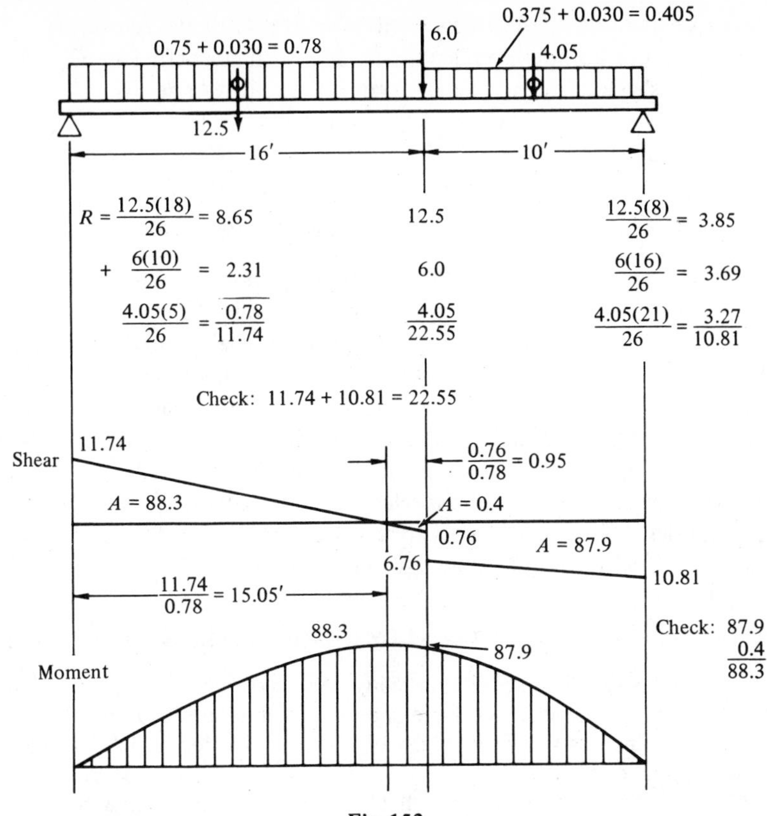

Fig. 153.

Design of Laterally Unsupported Beams If the compressive flange of a steel beam is not supported laterally, then it acts as a column and the stress must be reduced depending on the laterally unsupported length. The following presentation follows, in general, the requirements of the AISC code, but neglects certain aspects of the problem which are of minor importance.

A beam is considered laterally supported if the laterally unsupported length L does not exceed the following formulas:

$$L = 76.0 b_f / \sqrt{F_y} \quad \text{or} \quad L = 20{,}000/(d/A_f)F_y$$

In these equations, b_f is the width of the flange, A_f is the area of the flange, and d is the depth of the section. For F_y = 36 ksi these become:

$$L = 12.7 b_f \quad \text{or} \quad L = 556 A_f/d$$

If the length of lateral support is greater than these figures, then the allowable stress must be reduced in accordance with the area A_f of the flange, with the depth d, with the unsupported length L, and with a coefficient C_b which depends on the shape of the moment diagram. The reduced stress is

$$F_b = 12{,}000 A_f C_b / Ld$$

where

$$C_b = 1.75' + 1.05(M_1/M_2) + 0.3(M_1/M_2)^2$$

The maximum value of C_b is 2.3. For the bending moments shown, M_1 is the smaller and M_2 is the larger bending moment at the ends of the unbraced length under consideration. The factor M_1/M_2 is positive if the length is in reverse curvature and is negative if in single curvature. Simple beams will always be in single curvature, but portions of continuous beams may have the curvature reversed. An additional requirement is that if the moment is larger at a point between the bracing points, then the value of C_b shall be taken as unity.

In the following example, a beam is selected and tested against these formulas. For practical design it is much faster to use the charts in the AISC *Manual*, but this example will serve to explain the basic procedures.

EXAMPLE: SIMPLE BEAM, LATERALLY UNSUPPORTED AT TWO POINTS

A beam of 30-ft span is supported laterally only at the ends and at two points, one-third (10 ft) of the distance from the ends. There are loads of 9 kip and 6 kip at these points, as shown in Fig. 154. Select a section which will satisfy the lateral support conditions.

The shear and moment diagrams are shown in Fig. 154; the maximum moment is 80 ft-kip at the first-third point from the left. If the section had been

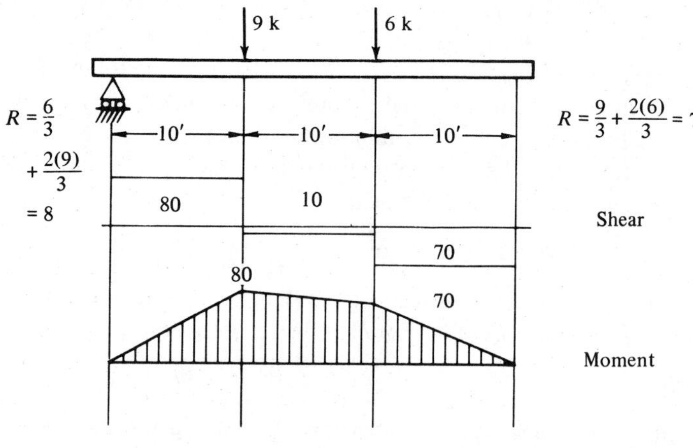

Fig. 154.

laterally supported, the section modulus based on an allowable unit stress of $F_b = 24$ ksi would be $S = 80(12)/24 = 40$ in.^{3}.

A W14X30 for which $S = 41.9$ in.^{3} will satisfy this requirement. This section, from Table 6, has the following properties: $b_f = 6.733$ in. and thickness of flange = 0.383 in., so the area of the flange is $A_f = 0.383(6.733) = 2.58$ in.^{2}. The actual depth is 13.86 in.

The maximum unbraced length is the smallest of the following:

$$L = 12.7b_f = 12.7(6.733)/12 = 7.12 \text{ ft}$$

or

$$L = 556A_f/d = 556(2.58)/(12)(13.86) = 8.62 \text{ ft}$$

The unbraced length is exceeded, so the section chosen is not satisfactory. A wider, heavier, or deeper section is required, so we will try a W16X36, which, from Table 6, has the following properties:

$$d = 15.85 \text{ in.,} \qquad b_f = 6.992 \text{ in.,} \qquad \text{thickness} = 0.428 \text{ in.}$$

so the area of the flange is $A_f = 2.99$ in.^{2}, and $S = 56.5$ in.^{3}.

Now the limiting values of the unbraced length L become

$$L = 12.7(6.992)/12 = 7.4 \text{ ft} \qquad \text{or} \qquad L = 556(2.99)/12(15.85) = 8.7 \text{ ft}$$

The unbraced length is still too large, so we shall see if the stress in the beam is satisfactory. The allowable stress becomes

$$F_b = 12{,}000A_fC_b/Ld$$

where all terms are known except C_b. From Fig. 154, $M_1 = 70$ and $M_2 = 80$ and the member is in single curvature, so the ratio M_1/M_2 is negative and $C_b = 1.75 - 1.05(70)/80 + 0.3(70/80)^2 = 1.07$.

The allowable stress becomes

$$F_b = 12{,}000(2.99)(1.07)/10(12)(15.85) = 20.18 \text{ ksi}$$

Now we must check this stress against the stress on the W16X36.

$$f = M/S = 80(12)/56.5 = 16.99 \text{ ksi}$$

This is less than the 20.18, so perhaps a lighter section could be selected, but there is not much difference between the weight of the W16X36 and the original W14X30.

Requirements for Plate Girders Plate girders are beams with box or I sections made with steel-plate web members and either plate flanges, if welded, or with angles and plates, if riveted or bolted. Channels or T sections are sometimes used for the flanges. Plate girders are required when more capacity or greater depth is needed than that of the largest (36-in.-deep) rolled sections. The most common method of fabrication is by welding, with steel-plate flanges welded directly to the web. Stiffeners, consisting of steel bars or angles, may be required if the web is too thin to resist the buckling of diagonal compressive forces. Bearing stiffeners are used at the ends to transfer loads to supports, or the plate girder's loads are transferred to columns through connection angles.

The important elements in the design are the selection of the flange to resist bending moments, the size of the web for shear, and the selection of the spacing and size of the stiffeners. In addition, the welding of all components must be selected. The regulations for the design of plate girders are contained in Section 1.10 of the AISC code. The principal requirements are summarized here, but references to hybrid girders (those having different types of steel incorporated in one section) have been omitted for clarity.

1. The size of both the tension and compression flange of a plate girder may be selected, using the moment of inertia of the gross section without any allowance in the tension flange for possible holes, provided that the area of the holes does not exceed 15% of the gross flange area.
2. The clear distance h between flanges must be less than $14{,}000t/\sqrt{F_y(F_y + 16.5)}$, so for $F_y = 36$ ksi, h = $320t$, where t is the web thickness.
3. If transverse stiffeners are used and are spaced at a distance not less than $1.5d$, where d is the total depth, the clear distance between flanges need not be less than $h = 2000t/\sqrt{F_y}$, so for $F_y = 36$ ksi, $h = 333t$
4. The allowable design stress in the tension or compression flange shall be $F_b = 0.6F_y$ (for $F_y = 36$ ksi use 22 ksi) unless the ratio h/t exceeds $760/\sqrt{F_y}$ (for $F_y = 36$ ksi, h/t must not exceed 162).
5. For the condition that the h/t value is exceeded, the maximum stress in the compression flange, F_b', shall be

$$F_b' = F_b \left[1.0 - 0.0005 \frac{A_w}{A_f} \frac{h}{t} - \frac{760}{\sqrt{F_b}}\right]$$

6. The largest average web shear F_v (in kips per square inch) for any loading condition shall not exceed the stress:

$$F_1 = \frac{F_y}{2.89} C_v \leqslant 0.4F_y$$

where

$$C_v \begin{cases} = \dfrac{45{,}000k}{F_y(h/t)^2} & \text{when } C_v \text{ is less than 0.8} \\ = \dfrac{190}{h/t} \sqrt{\dfrac{k}{F_y}} & \text{when } C_v \text{ is more than 0.8} \end{cases}$$

$$k = \begin{cases} = 4.00 + \dfrac{5.34}{(a/h)^2} & \text{when } a/h \text{ is less than 1.0} \\ = 5.34 + \dfrac{4.00}{(a/h)^2} & \text{when } a/h \text{ is more than 1.0} \end{cases}$$

t = thickness of web, in inches
a = clear distance between transverse stiffeners, in inches
h = clear distance between flanges, in inches

7. Intermediate stiffeners are not required if h/t is less than 260 and the maximum web shear stress F_v is less than given by the above formula.
8. If stiffeners are required, the ratio a/h shall not exceed the value $(260/(h/t))^2$ nor 3.0.
9. The moment of inertia of a single stiffener or a pair of stiffeners shall be not less than $(h/50)^4$.
10. Intermediate stiffeners may be stopped short of the tension flange a distance four times the web thickness unless needed for bearing. It is not considered good practice to make welds on the tension flange, because of the possibility of the weld cutting into the flange material and reducing its thickness.

Design of a Plate Girder Without Stiffeners We will design a plate girder of 50-ft span to carry a uniform load of 2 klf dead load and 2 klf live load. The weight of the girder itself is not included. The girder will be laterally supported by the floor system and will be carried by steel columns connected by bolted double angles. A thickness of web will be selected so that stiffeners will not be required and no reduction in the design stress will be necessary.

The first step is to select an economical depth. The greater the depth, the smaller the flange, but the thicker and heavier the web. The girder must not be so shallow that deflection will be excessive. We shall select three suitable depths and find the approximate comparative weights for each, say web depths of 40, 50, and 60 in. It will be easier for this study to select the size of the flange on the basis of the equivalent flange area required. This is equal to the area of the flange plus one-sixth of the area of the web. The limiting thickness of the flange, from requirement 7 above, will require an h/t ratio of 162.

For the dead load of the girder we will allow a conservative value of 0.3 klf, so the total load on the girder is 2 + 2 + 0.3 = 4.3 klf. The moment at the center of the 50-ft span is $M = wL^2/8 = 4.3(50^2)/8 = 1344$ ft-kip.

Following is a table of design for the 40-, 50-, and 60-in. depths to give the weight per foot of girder for each depth. The area of flange based on a unit stress of 22 ksi is selected first. Then the web thickness based on $t = h/162$ is calculated and a web thickness to the next largest $\frac{1}{16}$ in. is selected. The equivalent area of web which will serve as flange is determined and subtracted from the gross area of the flange to give the actual area required. Then a flange section is selected. All the weights are then calculated based on a weight of steel equal to 3.4 lb for 1 in.^{2} of steel 1 ft long.

Comparative Weights of Plate Girders

Depth of girder	40 in.	50 in.	60 in.
Area of flange, $A = (M/dF_G) = 1344(12)/d(22) = 733/d$	18.3	14.7	12.2
Minimum web thickness $t = h/162$	0.247	0.309	0.370
Web thickness used to next $\frac{1}{16}$ in.	0.25	0.312	0.375
Equivalent area of web as flange = $\frac{1}{6}$ area of web	1.7	2.6	3.8
Net area required for flange	16.6	12.1	8.4
Size of flange plate	$22 \times \frac{3}{4}$	$20 \times \frac{5}{8}$	$18 \times \frac{1}{2}$
Area furnished	16.5	12.5	9.0
Weight of flange per foot = $A(3.4)$	56.1	42.5	30.6
Weight of web = $ht(3.4)$	34.0	53.0	76.5
Total weight of two flanges and a web, pounds per foot	146.2	138.0	137.7

The greatest depth has the least weight by a small amount, but the curve of economy has a very flat shape. The 50-in. girder will be easier to handle, so we will use that depth. For determining the deflection requirements, the ratio of live to total load is 2.0/4.3 = 0.46, so the proportion of the design stress for F_b = 22 ksi will be 0.46(22) = 10.2 and the deflection by the simplified formula is $D = L^2f/1000d = 50^2(10.2)/1000(50) = 0.51$ in. The deflection to span ratio is 0.51/50(12) = 1/1176, which is satisfactory.

Design of Columns The allowable stresses in columns designed by the AISC code are obtained by formulas giving the relationship between the allowable axial stress F_a; the length L; the least radius of gyration of the column, r; and the equivalent pin-ended-column length factor K. If the column is subjected also to bending, it is known as a beam-column. Typical beam-columns in building frames with rigid joints are subjected to both vertical and lateral loads.

The actual column formulas in the AISC code are long and complicated and will not be reproduced here. The reader is referred to the AISC *Manual*. We will use tables, also in the code, for obtaining the allowable stresses. In Table 7 is shown the allowable stress for A36 steel having a basic stress of F_y = 36 ksi, and this table will be used for problems in this text.

There are many possible column sections; in fact, almost any steel member may be used as a column. Here we will show, as examples, only single or double angles for truss members and H columns for members of building frames. In Fig. 155 is shown a table of approximate radii of gyration of single and double

TABLE 7. ALLOWABLE STRESS (KSI) FOR COMPRESSION MEMBERS OF 36 KSI SPECIFIED YIELD STRESS STEEL.

Main and Secondary Members Kl/r not over 120						Main Members Kl/r 121 to 200				Secondary Members* l/r 121 to 200			
$\frac{Kl}{r}$	F_a (ksi)	$\frac{Kl}{r}$	F_a (ksi)	$\frac{Kl}{r}$	F_a (ksi)	$\frac{Kl}{r}$	F_a (ksi)	$\frac{Kl}{r}$	F_a (ksi)	$\frac{l}{r}$	F_{as} (ksi)	$\frac{l}{r}$	F_{as} (ksi)
1	21.56	41	19.11	81	15.24	121	10.14	161	5.76	121	10.19	161	7.25
2	21.52	42	19.03	82	15.13	122	9.99	162	5.69	122	10.09	162	7.20
3	21.48	43	18.95	83	15.02	123	9.85	163	5.62	123	10.00	163	7.16
4	21.44	44	18.86	84	14.90	124	9.70	164	5.55	124	9.90	164	7.12
5	21.39	45	18.78	85	14.79	125	9.55	165	5.49	125	9.80	165	7.08
6	21.35	46	18.70	86	14.67	126	9.41	166	5.42	126	9.70	166	7.04
7	21.30	47	18.61	87	14.56	127	9.26	167	5.35	127	9.59	167	7.00
8	21.25	48	18.53	88	14.44	128	9.11	168	5.29	128	9.49	168	6.96
9	21.21	49	18.44	89	14.32	129	8.97	169	5.23	129	9.40	169	6.93
10	21.16	50	18.35	90	14.20	130	8.84	170	5.17	130	9.30	170	6.89
11	21.10	51	18.26	91	14.09	131	8.70	171	5.11	131	9.21	171	6.85
12	21.05	52	18.17	92	13.97	132	8.57	172	5.05	132	9.12	172	6.82
13	21.00	53	18.08	93	13.84	133	8.44	173	4.99	133	9.03	173	6.79
14	20.95	54	17.99	94	13.72	134	8.32	174	4.93	134	8.94	174	6.76
15	20.89	55	17.90	95	13.60	135	8.19	175	4.88	135	8.86	175	6.73
16	20.83	56	17.81	96	13.48	136	8.07	176	4.82	136	8.78	176	6.70
17	20.78	57	17.71	97	13.35	137	7.96	177	4.77	137	8.70	177	6.67
18	20.72	58	17.62	98	13.23	138	7.84	178	4.71	138	8.62	178	6.64
19	20.66	59	17.53	99	13.10	139	7.73	179	4.66	139	8.54	179	6.61
20	20.60	60	17.43	100	12.98	140	7.62	180	4.61	140	8.47	180	6.58
21	20.54	61	17.33	101	12.85	141	7.51	181	4.56	141	8.39	181	6.56
22	20.48	62	17.24	102	12.72	142	7.41	182	4.51	142	8.32	182	6.53
23	20.41	63	17.14	103	12.59	143	7.30	183	4.46	143	8.25	183	6.51
24	20.35	64	17.04	104	12.47	144	7.20	184	4.41	144	8.18	184	6.49
25	20.28	65	16.94	105	12.33	145	7.10	185	4.36	145	8.12	185	6.46
26	20.22	66	16.84	106	12.20	146	7.01	186	4.32	146	8.05	186	6.44
27	20.15	67	16.74	107	12.07	147	6.91	187	4.27	147	7.99	187	6.42
28	20.08	68	16.64	108	11.94	148	6.82	188	4.23	148	7.93	188	6.40
29	20.01	69	16.53	109	11.81	149	6.73	189	4.18	149	7.87	189	6.38
30	19.94	70	16.43	110	11.67	150	6.64	190	4.14	150	7.81	190	6.36
31	19.87	71	16.33	111	11.54	151	6.55	191	4.09	151	7.75	191	6.35
32	19.80	72	16.22	112	11.40	152	6.46	192	4.05	152	7.69	192	6.33
33	19.73	73	16.12	113	11.26	153	6.38	193	4.01	153	7.64	193	6.31
34	19.65	74	16.01	114	11.13	154	6.30	194	3.97	154	7.59	194	6.30
35	19.58	75	15.90	115	10.99	155	6.22	195	3.93	155	7.53	195	6.28
36	19.50	76	15.79	116	10.85	156	6.14	196	3.89	156	7.48	196	6.27
37	19.42	77	15.69	117	10.71	157	6.06	197	3.85	157	7.43	197	6.26
38	19.35	78	15.58	118	10.57	158	5.98	198	3.81	158	7.39	198	6.24
39	19.27	79	15.47	119	10.43	159	5.91	199	3.77	159	7.34	199	6.23
40	19.19	80	15.36	120	10.28	160	5.83	200	3.73	160	7.29	200	6.22

*K taken as 1.0 for secondary members.

angle members and these will be used with Table 8, which gives the sizes of equal-leg and unequal-leg angles available.

Also in this table is shown, for some of the thinner angles, the stress reduction and the maximum L/r for elements of a column made up of several angles

TABLE 8. ANGLE SIZES AVAILABLE.

Numbers in boxes indicate: above, stress reduction for F_y = 36 ksi and below, maximum L/r for a braced element

	Thickness													
Size	1⅛	1	⅞	¾	⅝	9/16	½	7/16	⅜	5/16	¼	3/16	5/32	⅛
8 × 8	X	X	X	X	X	X	.995 126							
6 × 6		X	X	X	X	X	X	X	0.995 126	0.912 132				
5 × 5			X	X	X		X	X	X	0.995 126				
4 × 4			X	X	X		X	X	X	X	0.995 126			
3½ × 3½							X	X	X	X	X			
3 × 3							X	X	X	X	X	0.995 126		
2½ × 2½							X		X	X	X	X		
2 × 2									X	X	X	X		0.995 126
9 × 4		X	X	X	.954 129	.911 132	.857 136							
8 × 6		X	X	X	.997 126	.959 129	.911 132	.850 137						
8 × 4		X	X	X	.997 126	.959 129	.911 132	.850 137						
7 × 4			X	X	X	X	.965 128	.911 132	.839 138					

6 × 4			X	X	X	X	X	.972 128	.911 132	.825 139	.696 151			
6 × 3½							X		.911 132	.825 139	.696 151			
5 × 3½				X	X	X	X	X	.982 127	.911 132	.804 144			
5 × 3							X	X	.982 127	.911 132	.804 141			
4 × 3½					X		X	X	X	.997 126	.911 132			
4 × 3					X		X	X	X	.997 126	.911 132			
3½ × 3							X	X	X	X	.965 128			
3½ × 2½							X	X	X	X	.965 128			
3 × 2½							X	X	X	X	X	.911 132		
3 × 2							X	X	X	X	X	.911 132		
2½ × 2									X	X	X	.982 127		
2½ × 1½											X	.982 127		
2 × 1½											X	X	.911 132	
2 × 1¼											X	X	.911 132	
1¾ × 1¼											X	X	.965 128	

Numbers in boxes indicate; above, stress reduction for F_y = 36 ksi; below, maximum L/r for a braced element.

Radius of Gyration

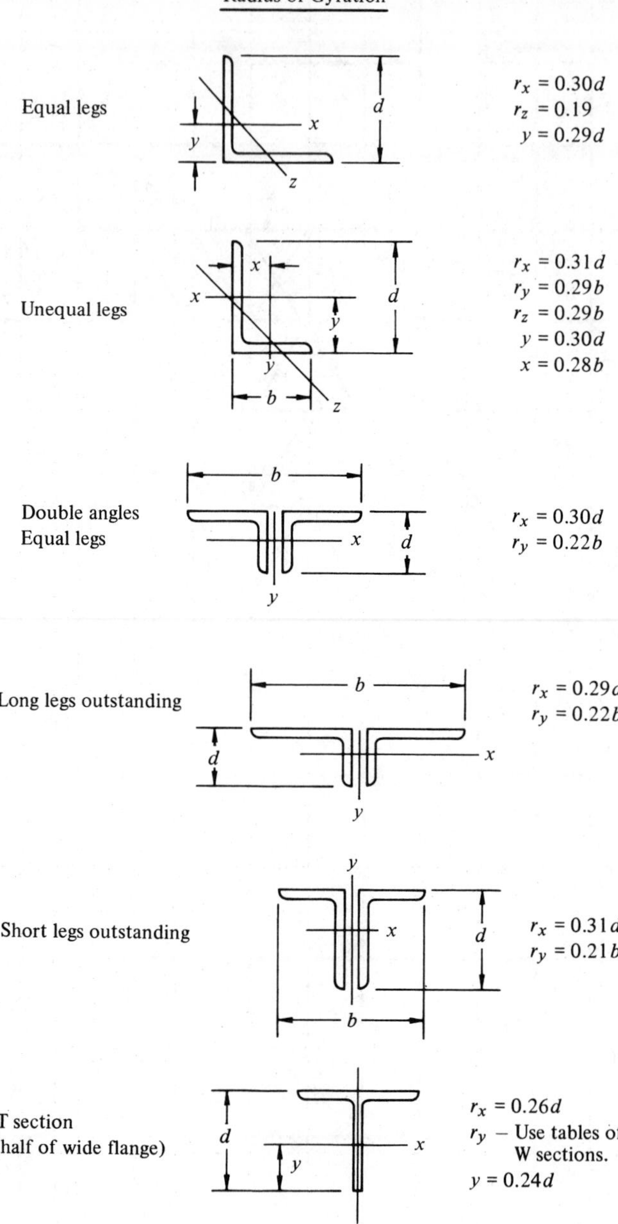

Equal legs
x
y
z
d
$r_x = 0.30d$
$r_z = 0.19$
$y = 0.29d$
Unequal legs
x
y
b
z
d
$r_x = 0.31d$
$r_y = 0.29b$
$r_z = 0.29b$
$y = 0.30d$
$x = 0.28b$
Double angles
Equal legs
b
x
d
y
$r_x = 0.30d$
$r_y = 0.22b$
Long legs outstanding
b
d
x
y
$r_x = 0.29d$
$r_y = 0.22b$
Short legs outstanding
y
x
d
b
$r_x = 0.31d$
$r_y = 0.21b$
T section
(half of wide flange)
d
y
x
$r_x = 0.26d$
r_y – Use tables of W sections.
$y = 0.24d$

Fig. 155.

TABLE 9.

			Flange		Web	Elastic Properties					
						Axis X-X			Axis Y-Y		
	Area A	Depth d	Width b_f	Thickness t_f	Thickness t_w	I	S	r	I	S	r
Designation	In.^{2}	In.	In.	In.	In.	In.^{4}	In.^{3}	In.	In.^{4}	In.^{3}	In.
W14×136	40.0	14.75	14.740	1.063	0.660	1590	216	6.31	568	77.0	3.77
×127	37.3	14.62	14.690	0.998	0.610	1480	202	6.29	528	71.8	3.76
×119	35.0	14.50	14.650	0.938	0.570	1370	189	6.26	492	67.1	3.75
×111	32.7	14.37	14.620	0.873	0.540	1270	176	6.23	455	62.2	3.73
×103	30.3	14.25	14.575	0.813	0.495	1170	164	6.21	420	57.6	3.72
× 95	27.9	14.12	14.545	0.748	0.465	1060	151	6.17	384	52.8	3.71
× 87	25.6	14.00	14.500	0.688	0.420	967	138	6.15	350	48.2	3.70
W14× 84	24.7	14.18	12.023	0.778	0.451	928	131	6.13	225	37.5	3.02
× 78	22.9	14.06	12.000	0.718	0.428	851	121	6.09	207	34.5	3.00
W14× 74	21.8	14.19	10.072	0.783	0.450	797	112	6.05	133	26.5	2.48
× 68	20.0	14.06	10.040	0.718	0.418	724	103	6.02	121	24.1	2.46
× 61	17.9	13.91	10.000	0.643	0.378	641	92.2	5.98	107	21.5	2.45
W14× 53	15.6	13.94	8.062	0.658	0.370	542	77.8	5.90	57.5	14.3	1.92
× 48	14.1	13.81	8.031	0.593	0.339	485	70.2	5.86	51.3	12.8	1.91
× 43	12.6	13.68	8.000	0.528	0.308	429	62.7	5.82	45.1	11.3	1.89
W12×106	31.2	12.88	12.230	0.986	0.620	931	145	5.46	301	49.2	3.11
× 99	29.1	12.75	12.192	0.921	0.582	859	135	5.43	278	45.7	3.09
× 92	27.1	12.62	12.155	0.856	0.545	789	125	5.40	256	42.2	3.08
× 85	25.0	12.50	12.105	0.796	0.495	723	116	5.38	235	38.9	3.07
× 79	23.2	12.38	12.080	0.736	0.470	663	107	5.34	216	35.8	3.05
× 72	21.2	12.25	12.040	0.671	0.430	597	97.5	5.31	195	32.4	3.04
× 65	19.1	12.12	12.000	0.606	0.390	533	88.0	5.28	175	29.1	3.02
W12× 58	17.1	12.19	10.014	0.641	0.359	476	78.1	5.28	107	21.4	2.51
× 53	15.6	12.06	10.000	0.576	0.345	426	70.7	5.23	96.1	19.2	2.48
W12× 50	14.7	12.19	8.077	0.641	0.371	395	64.7	5.18	56.4	14.0	1.96
× 45	13.2	12.06	8.042	0.576	0.336	351	58.2	5.15	50.0	12.4	1.94
× 40	11.8	11.94	8.000	0.516	0.294	310	51.9	5.13	44.1	11.0	1.94
W10× 89	26.2	10.88	10.275	0.998	0.615	542	99.7	4.55	181	35.2	2.63
× 77	22.7	10.62	10.195	0.868	0.535	457	86.1	4.49	153	30.1	2.60
× 72	21.2	10.50	10.170	0.808	0.510	421	80.1	4.46	142	27.9	2.59
× 66	19.4	10.38	10.117	0.748	0.457	382	73.7	4.44	129	25.5	2.58
× 60	17.7	10.25	10.075	0.683	0.415	344	67.1	4.41	116	23.1	2.57
× 54	15.9	10.12	10.028	0.618	0.368	306	60.4	4.39	104	20.7	2.56
× 49	14.4	10.00	10.000	0.558	0.340	273	54.6	4.35	93.0	18.6	2.54
W8× 67	19.7	9.00	8.287	0.933	0.575	272	60.4	3.71	88.6	21.4	2.12
× 58	17.1	8.75	8.222	0.808	0.510	227	52.0	3.65	74.9	18.2	2.10
× 48	14.1	8.50	8.117	0.683	0.405	184	43.2	3.61	60.9	15.0	2.08
× 40	11.8	8.25	8.077	0.558	0.365	146	35.5	3.53	49.0	12.1	2.04
× 35	10.3	8.12	8.027	0.493	0.315	126	31.1	3.50	42.5	10.6	2.03
× 31	9.12	8.00	8.000	0.433	0.288	110	27.4	3.47	37.0	9.24	2.01
W6× 25	7.35	6.37	6.080	0.456	0.320	53.3	16.7	2.69	17.1	5.62	1.53
× 20	5.88	6.20	6.018	0.367	0.258	41.5	13.4	2.66	13.3	4.43	1.51
× 15.5	4.56	6.00	5.995	0.269	0.235	30.1	10.0	2.57	9.67	3.23	1.46

fastened together at intermediate points. For regular H column sections, we will use the properties shown in Table 9 showing just a few of the sections available. For others see the AISC *Manual*.

It is easier to calculate the area of angles than to furnish a table of values. The area of a steel angle having legs equal to d and b and a thickness equal to t is $A = (d + b - t)t$. For example, a $L3\frac{1}{2} \times 3\frac{1}{2} \times \frac{5}{16}$ has an area of $A = (3.5 + 3.5 - 0.3125)0.3125 = 2.09 \text{ in.}^2$, which agrees with the tables in the AISC *Manual*.

The equivalent column length K will vary with the end conditions of the columns. For most trusses it is satisfactory to use the value $K = 1$. In Fig. 156 is reproduced a table from the AISC *Manual*, showing the values of K for a number of different end conditions. For column members continuous with beams, the rotation may either be fixed or free or the value of K may lie between the values shown in Fig. 156 or, in some cases, outside these values. The value of the least radius of gyration r is the lowest value for the particular cross section. The axis for which r is smallest will be called the weak axis. To avoid very flexible columns, the value of KL/r is limited to less than $KL/r = 200$ for main compression members. There is also a limit on the flexibility of tension members of $KL/r = 240$ for main members and 300 for secondary bracing members.

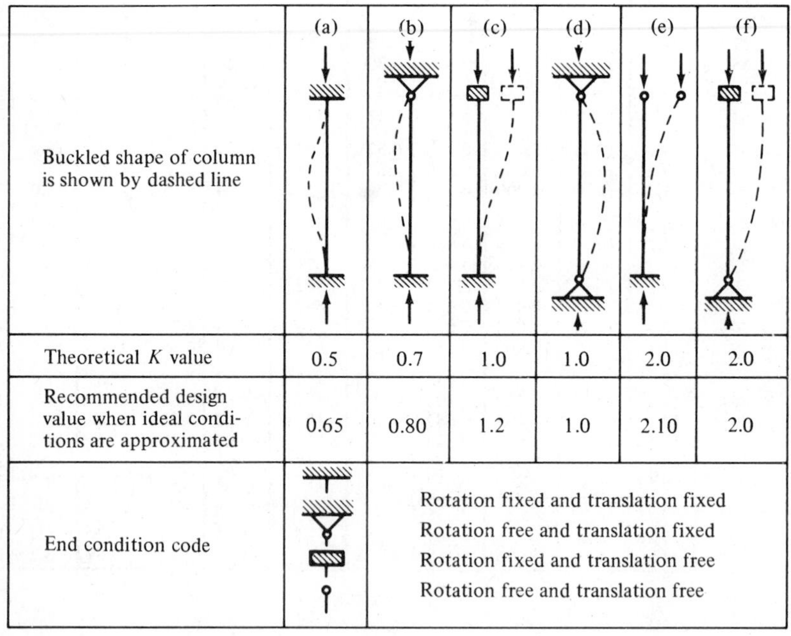

Fig. 156.

Some examples of the design of steel columns follow.

EXAMPLE–SINGLE-ANGLE TRUSS MEMBER

A compression member of a steel roof truss is 6 ft long and carries a compressive force of P = 20 kip. Select a single-angle section for this member. Use A36 steel with $F_y = 36$ ksi. Assume that the value of K is 1.

The solution of this design problem cannot be made directly. We must get there by steps. The easiest approach is to make a wild but educated guess of the allowable stress for which the angle might be designed. A single angle is quite flexible, so we will assume that the stress is quite low, say $F_a = 10$ ksi. Therefore the area required will be $A = P/F = 20/10 = 2.0$ in.^{2}. If the angle is $\frac{1}{4}$ in. thick, the combined length of both legs will have to be $1 = A/t = 2.0/0.25 = 8$ in. We shall try a L$3\frac{1}{2}$X$3\frac{1}{2}$X$\frac{5}{16}$ which has an area of 2.09 in.^{2}; from Fig. 155 the minimum value of the radius of gyration about the diagonal axis is $r_z = 0.19d = 0.19(3.5) = 0.67$.

The value of KL/r for $K = 1$ and $L = 6(12) = 72$ in. is $KL/r = 1(72)/0.67 = 107$. From Table 7, select the allowable stress for $KL/r = 107$, which is $F_a = 12.07$ ksi. This will require an area of $A = P/F_a = 20/12.07 = 1.65$ in.^{2}. The angle we selected initially has a greater area than this. We can reduce the thickness so the area will be at least equal to 1.65 in.^{2}. Try an L$3\frac{1}{2}$X$3\frac{1}{2}$X$\frac{1}{4}$, which will have an area of $A = (3.5 + 3.5 - 0.25)0.25 = 1.69$ in.^{2}.

Use an L$3\frac{1}{2}$X$3\frac{1}{2}$X$\frac{1}{4}$.

EXAMPLE: DOUBLE-ANGLE TRUSS MEMBER

Design a truss member using two angles separated by a $\frac{3}{8}$-in. gusset plate. The member carries a compressive force of 70 kip. The length between ends is 10 ft, but it is supported in one direction at the center, so the unsupported length in this direction is 5 ft.

The ideal section for this condition would be a member with a radius of gyration in one direction that is twice that in the other direction. This can best be accomplished by using double angles with unequal legs, long legs outstanding, similar to the fourth sketch in Fig. 155. The short unsupported length must correspond to the weak axis.

We will start off this design problem by again making an educated guess of the allowable stress. We know that it should be larger than the stress developed from the previous example, so we will try $F_a = 14$ ksi. The required area of the two angles would be $A = P/F_a = 70/14 = 5$ in.^{2}. For these larger stresses we should avoid getting the members too thin, otherwise the stresses must be reduced in accordance with the schedule in Table 8. Therefore we will assume a thickness of $\frac{3}{8}$ in. The total length of the sides of the angles, neglecting the effect of the thickness of this length, is $1 = 5/0.375 = 13.3$ in. An L4X$3\frac{1}{2}$X$\frac{3}{8}$ will have an area of $A = 2(4 + 3.5 - 0.375)0.375 = 5.34$ in.^{2}.

The dimensions used to determine the radius of gyration from Fig. 155 are

$$b = 4 + 4 + 0.375 = 8.375 \text{ in.,} \qquad d = 3.5 \text{ in.}$$

The radii of gyration are

$$r_x = 0.29d = 0.29(3.5) = 1.02 \text{ in.,} \qquad r_y = 0.22b = 0.22(8.37) = 1.84 \text{ in.}$$

The latter value should have been twice that of r_x, so r_y will control the maximum allowable stress and the unsupported length will be 10 ft. The value of KL/r is $1(120)/1.84 = 65$, and from Table 7, $F_a = 16.94$ ksi. This is larger than our first guess of 14 ksi, so our angle is too large. We now require an area of $A = 70/16.94 = 4.13$ in.^{2}. Let us try two L4X$3\frac{1}{2}$X$\frac{5}{16}$, which have an area of $A = 2(4 + 3\frac{1}{2} - 0.3125)0.3125 = 4.49$ in.^{2}. This is close to the required area of 4.13 in.^{2} and will be considered satisfactory. There is one more consideration. Note that in Table 8 an angle of this size and thickness requires a stress reduction of 0.997, but this will not affect the result just obtained.

It might be possible, by making more trials, to obtain a section that will have a smaller area than that chosen. There are many possibilities.

EXAMPLE: BUILDING COLUMN

Select a column section for a building frame. The design load is $P = 200$ kip and the unsupported length for both the weak and the strong axes is 12 ft. The building frame is not restrained by bracing, so the tops of the columns are free to move sideways in accordance with the third diagram in Fig. 156, so that $K = 1.2$. Use A36 steel.

Again we will make a wild guess of the allowable stress and refine this guess as we get closer to the answer. Building columns with heavy loads are relatively stiff, so we will try a stress of $F_a = 18$ ksi, which will require an area of $A = P/F_a = 200/18 = 11.1$ in.^{2}.

From Table 9, a W8X40 will have an area $A = 11.8$ in.^{2} with radii of gyration $r_x = 3.53$ in. and $r_y = 2.04$ in. The unsupported length of the column is 12 ft in both directions, so the y axis will control and the value of KL/r is $KL/r = 1.2(12)(12)/2.04 = 85$. From Table 7, the allowable stress for this value of KL/r is 14.79 ksi. Our wild guess of 18 ksi is too high. We have the choice of using a heavier 8-in. section, or going to a 10-in. section. Wider sections are usually more efficient, so we will try a W10X49, which, from Table 9, has an area of $A = 14.4$ in.^{2} and a value of $r_y = 2.54$ in.^{2}; $KL/r = 1.2(12)(12)/2.54 = 68$, and from Table 7 the design stress is $F_a = 16.64$ ksi. The required area is now $A = 200/16.64 = 12.02$. The W10X49 furnishes more than this required area, so the section is satisfactory. It is possible that a W8X48 will also work, and it is suggested that the reader try this section.

EXAMPLE: BUILDING COLUMN SUPPORTED AT AN INTERMEDIATE POINT ON THE WEAK AXIS

Columns may have lateral support for the weak axis at intermediate points, so the value of r_x controls the design. Use the same data as above, but assume that there is a lateral support at the middle of the weak axis. This column will be stiffer than the previous column, so we will try the W8X40, for which we have already recorded the properties.

Use the same value of $K = 1.2$. Then $KL/r = 1.2(12)(6)/(2.04) = 42$ for the y axis, and $KL/r = 1.2(12)(12)/3.53 = 49$ for the strong axis, so the strong axis controls and, from Table 7, $F_a = 18.44$ ksi. The required area of the column is $A = 200/18.33 = 10.8$ in.^{2}. Therefore, the W8X40, with an area of 11.8 in.^{2}, will be satisfactory.

Design of Beam–Columns The combination of direct compression load and bending on a beam-column will increase the buckling effect, so that stresses cannot be added directly to determine if the section is satisfactory. The regulations for beam-columns are given in the AISC code in Section 1.6. The formulas shown include the effects of bending in two planes, which will be disregarded in this discussion.

The capacity of a steel section in combined bending and direct stress is determined by interaction formulas. If the axial stress is low, the following formula can be applied:

$$(f_a/F_a) + (f_b/F_b) \leqslant 1.0$$

where f_a is the calculated axial stress, F_a is the allowable axial stress if only direct stress were present, f_b is the calculated bending stress, and F_b is the allowable bending stress. This formula can be used if $f_a/F_a \leqslant 0.15$. For larger values of f_a/F_a, a similar but more complicated interaction formula is required, which will be presented after an example.

EXAMPLE: BEAM-COLUMN FOR $f_a/F_a < 0.15$

Design a beam-column for a direct load of 40 kip and a bending moment at the center of the section of 160 ft-kip. The column length is 15 ft, but is supported laterally at the midpoint of the height on the weak axis, so the laterally unsupported length for bending and compression about the weak axis is 7.5 ft. Use A36 steel with $F_b = 24$ ksi. Assume that $K = 1.0$.

We will select a section on the basis that all the stresses are caused by bending, and then guess the increase in size to account for the direct compression. The section modulus S required for bending is

$$S = Mf_b = 160(12)/24 = 80 \text{ in.}^3$$

From Table 6, a W18X45 would just satisfy this requirement. Let us assume that a section 20% larger in area would be required to take into account the direct stress. This will be a W18X55, which has the following properties needed for this problem: $A = 16.2$ in.^{2}, $S = 98.4$ in.^{3}, $b_f = 7.53$ in., $t_f = 0.63$ in., $d = 18.12$ in., $r_x = 7.42$ in., and $r_y = 1.67$ in.

For this section, the bending stress is $f_b = M/S = 160(12)/98.4 = 19.51$ ksi. The allowable bending must take into account that the beam is laterally unsupported for 7.5 ft = 90 in., and the formulas used for the design of laterally unsupported beams are applicable. The stress must be reduced if the laterally unsupported length is greater than either of the following:

$$L = 12.7b_f = 12.7(7.53) = 95.6 \text{ in.,} \quad \text{or} \quad 8.0 \text{ ft}$$

$$L = 556A_f/d = 556(7.53)(0.63)/18.12 = 145.6 \text{ in.,} \quad \text{or} \quad 12.1 \text{ ft}$$

Both criteria are satisfied because $L = 90$ in. If they had not been satisfied, then it would be well to try a wider section rather than use a reduced stress.

Now we must determine the calculated and allowable stresses for compression. For a length of 7.5 ft for the weak axis with $r_y = 1.67$ in., the value of KL/r is $KL/r = 1.0(7.5)(12)/1.67 = 54$. Table 7 gives a value of $F_a = 17.99$ ksi. The calculated axial stress is $f_a = P/A = 40/16.2 = 2.47$ ksi. We must test the provision that f_a/F_a be less than 0.15. Here $f_a/F_a = 2.47/17.99 = 0.137$, which is under the requirement of 0.15, so the interaction formula can be used:

$$\frac{f_a}{F_a} + \frac{f_b}{F_b} = \frac{2.47}{17.99} + \frac{19.51}{24} = 0.137 + 0.813 = 0.95 \leqslant 1.0$$

and the section is satisfactory. The value above is so close to 1.0 that further tries seem unnecessary.

Beam–Columns for Large Column Loads If the ratio f_a/F_a is greater than 0.15 for a beam-column, then the interaction formula becomes

$$\frac{f_a}{F_a} + \frac{C_m}{1 - (f_a/F_e')} \frac{f_b}{F_b} \leqslant 1$$

The modifying factor for the second term is called the *amplification factor*, and C_m is a coefficient to be defined later. Also

$$F_e' = \frac{12\pi^2 E}{23(kL_b/r_b)^2}$$

where L_b and r_b are, respectively, the length and radius of gyration in the plane of bending.

In addition to the interaction formula modified above, another formula must be satisfied which is similar to the first formula:

$$(f_a/0.60F_y) + (f_b/F_b) \leqslant 1$$

The term C_m is a coefficient whose value is taken as follows:

1. For compression members in frames subject to joint translation (side-sway), $C_m = 0.85$.
2. For restrained compression members in frames braced against joint translation and not subject to transverse loading between their supports in the plane of bending,

$$C_m = 0.6 - 0.4\,(M_1/M_2), \qquad \text{but not less than } 0.4$$

where M_1/M_2 is the ratio of the smaller to larger moments at the ends of that portion of the member unbraced in the plane of bending under consideration. M_1/M_2 is positive when the member is bent in reverse curvature and negative when it is bent in single curvature.

3. For compression members in frames braced against joint translation in the plane of loading and subjected to transverse loading between their supports, the value of C_m may be determined by rational analysis. However, in lieu of such analysis, the following values may be used: (a) for members whose ends are restrained, $C_m = 0.85$; (b) for members whose ends are unrestrained, $C_m = 1.0$.

EXAMPLE: BEAM-COLUMN WITH LARGE COLUMN LOAD

Design a beam-column for the loads and bending moments shown in Fig. 157. The length is 18 ft and is laterally unsupported for this distance for both the strong and the weak axes. The frame is fixed against joint translation so that $K = 1$. The maximum moment comes at the base of the column, so it will not be necessary to consider the beam element to be laterally unsupported as in the previous problem. Note that the beam is bent in reverse curvature; this will affect the value of C_m.

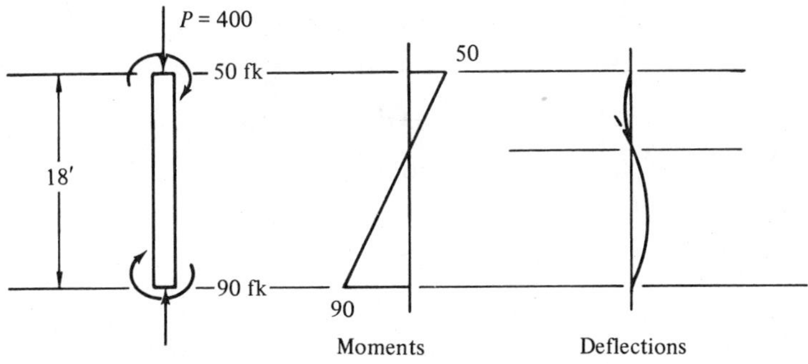

Fig. 157.

To arrive at a preliminary size, we could guess at a column stress as in the previous example, but there is a better method, which is to express the section modulus used in the bending equation in terms of the radius of gyration and the area and then solve for the required area as follows:

$$F = M/S = Mc/I, \quad \text{but} \quad I = Ar^2, \quad \text{so} \quad F = Mc/Ar^2 \quad \text{and} \quad A = Mc/Fr^2$$

Let us select a column dimension, say 14 in. The value of r will vary little for similar columns and, for a 14-in.-wide flange, r_x is about 6.2 in. and r_y is about 3.7 in.

With these values of r_x and r_y we can pick an allowable compression stress from Table 7 as follows:

$$KL/r = 1(18)(12)/3.7 = 58, \qquad F_a = 17.62.$$

Neglect for the present the amplification factor. The required area will be, including the factors for direct stress and bending,

$$A = (P/F_a) + (Mc/F_b r_x^2) = 400/17.62 + 90(12)(7)/24(6.2^2) = 30.9 \text{ in.}^2$$

We can select a column section with this area, but we do not know at this point how the amplification factor will affect the design. Let us take another wild guess and say that the area will be increased 10%, requiring an area of 35 in.^{2}. From Table 9 a W14X119 has the following properties:

$$A = 35.0 \text{ in.}^2, \qquad S_x = 189 \text{ in.}^3, \qquad r_x = 6.26 \text{ in.}, \qquad r_y = 3.75 \text{ in.}$$

We are now in a position to check the stresses in this section against the modified interaction formula. First we must determine the value of C_m. For this problem, the joints are braced against translation and not subject to transverse loading between supports, so method 2 is applicable.

The ratio M_1/M_2 is 50/90 = 0.55, and the member is in reverse curvature as sketched in Fig. 157, so the ratio M_1/M_2 is positive and

$$C_m = 0.6 + 0.4(M_1/M_2) = 0.6 - (0.4)(.55) = 0.38$$

with a minimum value of 0.4, which controls.

The value of F_e' for $r_b = 6.26$ and $L = 18(12) = 216$ in.

$$F_e' = 12\pi^2 E/23(KL_b/r_b)^2, \qquad KL/r = 1(216)/6.2 = 34.83$$

$$= 12\pi^2(29{,}000)/23(34.83)^2 = 123$$

The stress due to the compression force is $f_a = P/A = 400/35.0 = 11.42$ and the amplification factor becomes

$$\frac{C_m}{1 - (f_a/F_c')} = \frac{0.4}{1 - (11.42/123)} = 0.44$$

The allowable stress F_a for the column for $KL/r = 1(18)12/3.75 = 57.6$ is, from Table 7, $F_a = 17.7$. The stress due to bending is $f_b = M/S_x = 90(12)/189 = 5.71$.

The equation for the interaction formula becomes

$$(f_a/F_a) + 0.44(f_b/F_b) = (11.42/17.7) + [0.44(5.71)/24] = 0.749$$

This value is considerably under the maximum value of 1.0. However, we must check the alternative formula:

$$(f_a/0.6F_y) + (f_b/F_b) = [11.42/(0.6)(36)] + (5.71/24) = 0.53 + 0.24 = 0.77$$

It is evident that we have picked too large a section, so must try again. With the ratios of the interaction formula as a guide, the next section should have an area 0.749 times the area of the W14X119. This will be left for the reader to follow.

Connections The principal methods of connecting structural steel elements are rivets, bolts, and welds.

Rivets Rivets are round bars from $\frac{1}{2}$ to $1\frac{1}{4}$ in. in diameter, heated to a red heat and driven into holes in adjacent plates. A special hammer is used to form a rivet head over both ends, as shown in Fig. 158. As the rivet cools, it shortens and grips the plates together. Figure 158 shows two plates connected so they will take tension forces in each plate, and Fig. 159 shows a single plate connected to two plates. The latter arrangement is stronger because the forces on the rivet do not tend to turn it and more bearing and shear area is provided.

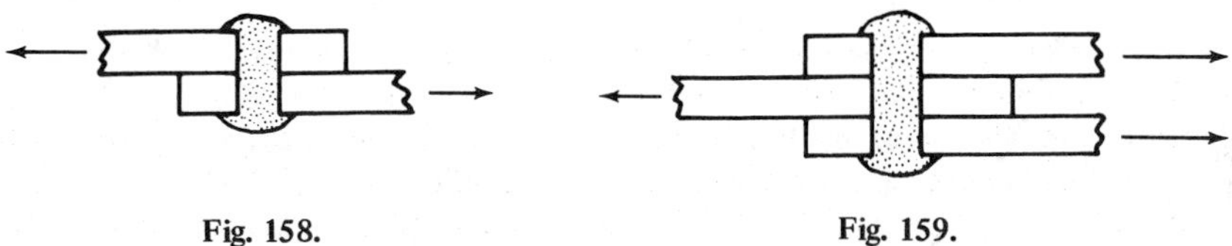

Fig. 158. **Fig. 159.**

Rivets are designed on the basis of the stress on (1) the bearing area between the rivet and the plate and (2) the cross-sectional area of the rivet in shear. The rivet in Fig. 158 is in single shear and will deform as illustrated in Fig. 160a, while the rivet of Fig. 159 will deform as in Fig. 160b.

The allowable bearing stress given by the AISC code is $F_b = 1.35F_y$, where F_y is the yield stress of the plate (not the rivet). For A36 steel $F_b = 48.6$ ksi.

Assume that the outer plates shown in Fig. 159 are $\frac{1}{4}$ in. thick, the inner plate is $\frac{3}{8}$ in. thick, and the rivet is $\frac{3}{4}$ in. in diameter. The bearing area on both the outer plates is $0.75(2)(0.25) = 0.375$ in.^{2} and the inner plate is $(0.75)(0.375) = 0.281$ in.^{2}. The allowable force on the rivet in bearing is the smaller area times the allowable stress, $R = 0.281(48.6) = 13.7$ kips. No distinction is made between the bearing capacity of rivets in single shear and double shear.

The allowable load on a rivet must also be checked for its capacity in shear. For AISC Grade 1 hot-driven rivets, the allowable stress is $F_v = 15$ ksi. For the

Fig. 160.

example above, as shown in Fig. 159, there are two shear areas per bolt, $A = 2\pi D^2/4 = 2\pi(0.75^2)/4 = 0.883$ in.^{2}, and the allowable load is: $R = 0.883(15) = 13.2$ ksi, which is slightly less than the capacity in bearing. For a connection in single shear, Fig. 158 and Fig. 160a, there is only one-half the shear area, so the capacity would only be half of that shown above.

Bolts Bolts are like rivets except that the head is preformed and the bolt is tightened by a nut working on threads and is placed cold. There are several grades of bolts for structural purposes, including ordinary machine bolts and special high-strength bolts which are tightened to a high tension in the bolt. They obtain much of their capacity from the friction between the plates that the bolts are connecting. The reader is referred to the AISC *Manual* for tables of allowable loads.

Machine bolts have an allowable shear stress of 10 ksi, and the lowest strength of the high-strength bolts has a shear stress of 15 ksi, the same as for rivets.

The minimum distance between centers of rivet or bolt holes by the AISC Code is $2\frac{2}{3}$ times the nominal diameter, but preferably not less than 3 diameters. The minimum edge distance varies with the diameter and type of edge. For $\frac{3}{4}$-in. rivets or bolts, the distance for a sheared edge is $1\frac{1}{4}$ in., and for a rolled edge or gas-cut edge, it is 1 in.

Welds Electric-fusion welding is used to connect steel plates directly, and the most common types of weld are the fillet weld (Fig. 161) and the butt weld (Fig. 162). The fillet weld is placed along the edges of overlapping plates, and

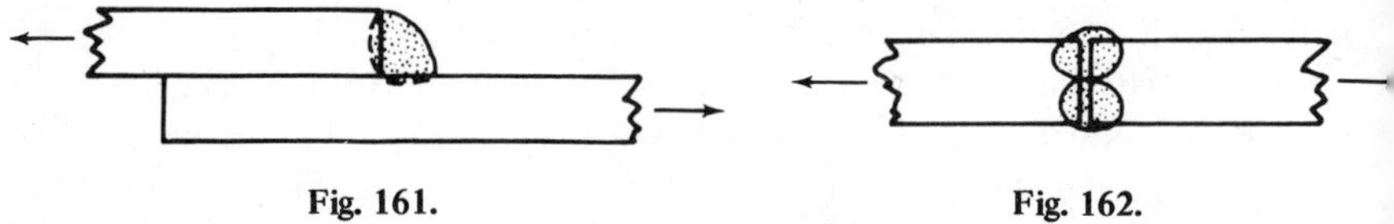

Fig. 161. Fig. 162.

the welder deposits a triangular area of metal and at the same time melts some of the steel in the plates. The butt weld connects two plates having edges butted together. Thin plates or thicker plates that do not require the full strength of the material can be welded by making a pass on each side of the butted plates, as shown in Fig. 162. Wide plates require special preparation to form a V groove at the joint before welding is started, as sketched in Fig. 163. Several passes may be required to build up the weld metal, and after each pass the weld must be cleaned so that inclusions of slag will not be left in the weld.

The strength of butt welds is usually considered to be equal to that of the plates they connect, provided a complete penetration weld is achieved; otherwise the weld is rated in percentage of the strength of the plates it connects. Fillet welds may be either side welds or end welds, as shown in Fig. 164, and the capacity is assumed to be the same per linear inch of weld. However, it is not

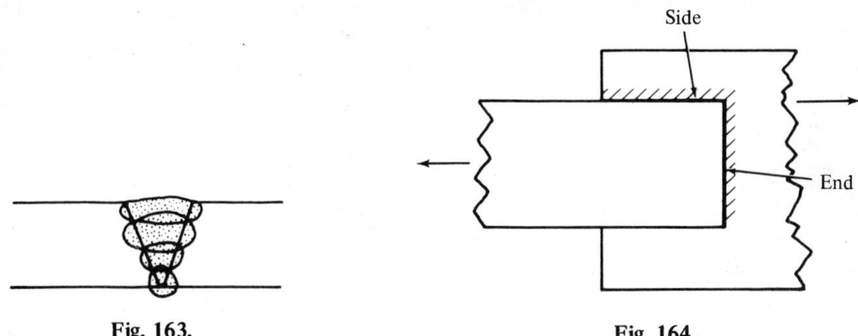

Fig. 163.

Fig. 164.

considered desirable to use both side and end welds on the same plates, because the side welds will pick up the load first. The strength of a fillet weld is measured by the allowable stress times the length times the minimum distance across the triangular fillet. For a 45° weld, the distance will be the weld size times 0.707. The allowable stress given by the AISC code for electrodes suitable for A36 steel is 21 ksi, so the allowable load per inch on a $\frac{1}{4}$-in. weld would be: $R = 0.25(0.707)21 = 3.7$ kip per inch.

The maximum nominal size of fillet welds must be $\frac{1}{16}$ in. less than the thickness of the plate except for plates less than $\frac{1}{4}$ in. The minimum effective length must be not less than 4 times the nominal size. The minimum amount of lap is 5 times the thickness of the thinner part but not less than 1 in. Side or end fillet welds should, wherever practicable, be returned around the corners for a distance not less than twice the nominal size of the weld.

Design of Reinforced Concrete Structural Elements

The design of reinforced concrete building structures is governed by the *Building Code Requirements for Reinforced Concrete*, prepared by the American Concrete Institute, Box 4754, Redford Station, Detroit, Michigan 48219. The design of highway bridge structures is governed by the *Standard Specifications for Highway Bridges* prepared by the American Association of State Highway Officials, 341 National Press Building, Washington, D.C. 20004. These regulations differ in considerable detail, and in this section we will prepare all designs in accordance with the ACI *Code* and will discuss the AASHO *Specifications* in the subsection on bridge design. The ACI *Code* is a long and complicated document, and the space available here will not allow complete coverage of all the requirements. Examples will deal with simple structures, using a conservative design approach.

Reinforcing Steel and Concrete Cover Reinforcing bars for concrete are special steel bars rolled for this purpose with deformations on the surface to increase the bond between concrete and steel. Bars are designated by numbers indicating

TABLE 10 PROPERTIES OF REINFORCING BARS.

	Bar number											
	2	3	4	5	6	7	8	9	10	11	14S	18S
Weight (lb.)	0.17	0.38	0.67	1.04	1.50	2.04	2.67	3.40	4.30	5.31	7.65	13.60
Diameter (in.)	0.25	0.375	0.50	0.625	0.75	0.875	1.00	1.128	1.27	1.41	1.693	2.257
Area (in.^{2})	0.05	0.11	0.20	0.31	0.44	0.60	0.79	1.00	1.27	1.56	2.25	4.00
Perimeter (in.)	0.79	1.18	1.57	1.96	2.36	2.75	3.14	3.54	3.99	4.43	5.32	7.09

their approximate diameter in eights of an inch. For example, a No. 6 bar will have an approximate diameter of $\frac{3}{4}$ in. The properties of available reinforcing bars are shown in Table 10.

The strength of reinforcing bars is measured by the yield strength f_y as measured by standard tests. Currently, most reinforcing bars furnished have a yield strength of $f_y = 60{,}000$ psi, but for purposes of illustration, lesser strengths may be used in examples.

Bars come in stock lengths of 40–60 ft. If splices are required, they may be made by butt or lap welds, or if made in regions of compression, they may be lapped with space between for the concrete to bond with the lapped steel bars. It is permissible to use several bars tied together to form larger steel areas if considerations of bond are fulfilled. These are called bundled bars.

There must be adequate spaces between bars and there must be cover over bars at the surface of reinforced concrete to provide protection against the weather and possible fire. The ACI *Code* requires that the minimum clear distance between parallel bars be one bar diameter, or a minimum of one inch. The maximum clear distance for walls or slabs is three times the slab, or wall thickness, or 18 in.

For cast-in-place concrete, the following minimum cover must be provided:

Cast against and permanently exposed to earth	3 in.
Exposed to earth or weather:	
No. 6 through No. 18 bars	2 in.
No. 5 bars, $\frac{5}{8}$ in. wire, and smaller	$1\frac{1}{2}$ in.
Not exposed to weather or in contact with the ground:	
Slabs, walls, joists:	
No. 14 and No. 18 bars	$1\frac{1}{2}$ in.
No. 11 and smaller	$\frac{3}{4}$ in.
Beams, girders, columns:	
Principal reinforcement, ties, stirrups or spirals	$1\frac{1}{2}$ in.
Shells and folded plate members:	
No. 6 bars and larger	$\frac{3}{4}$ in.
No. 5 bars, $\frac{5}{8}$ in. wire, and smaller	$\frac{1}{2}$ in.

Working-Stress or Ultimate-Strength Design The ACI *Code* permits the proportioning for strength of reinforced concrete, either by the working-stress method, using the ordinary principles of structural mechanics, or by ultimate-strength concepts, basing the strength or the capacity of the concrete at failure loads. These considerations apply to the cross sections only and not to the determination of bending moments and shears. The ACI *Code* is written around the ultimate-strength provisions, and those for working stress are contained in Section 8.1.2. Design requirements for compression, for compression combined with bending, for shear, and for other considerations, are given in working-stress design as a proportion of those used for ultimate-strength design.

In this discussion, the working-stress method will be described for the selection of dimensions and area of reinforcing bars only.

Ultimate-strength design requires that the design loads be multiplied by load factors which differ for dead load and live load. Then the dimensions and reinforcing are selected for strengths approaching the yield strength of the materials. In working-stress design, the loads are used directly and the yield strength is reduced by a factor of safety to give a working stress to be used for design. Working stress is used if a conservative approach to design is sufficient, if local building codes do not permit ultimate-strength design, or, as in the case of lightly loaded slab structures of short span, where multiplication of design loads by load factors seems complicated and unnecessary.

Concrete Strengths The strength of concrete is based on f'_c, the crushing strength of 6-in. by 12-in. cylinders at the age of 28 days, cured and tested in accordance with standard methods. These strengths vary from 2500 psi to 5000 psi or larger. A value of $f'_c = 3000$ psi is often used for design, and most of the examples in this discussion will use this value.

The modulus of elasticity E_c of concrete will vary with strength and weight. For normal-weight concrete (150 pcf) the ACI *Code* gives the formula: $E_c = 57{,}000\sqrt{f'_c}$, so for $f'_c = 3{,}000$ psi, $E_c = 3{,}123{,}000$ psi. This gives a ratio of the modulus of elasticity of steel to that of concrete of $n = E_s/E_c = 29{,}000/3{,}123 = 9.28$. However, the nearest whole number should be used, so n becomes 9.

Working-Stress Deisgn The following stresses in reinforcing steel are permitted for working-stress design:

$$f_y = 40 \quad \text{or} \quad 50 \text{ ksi}, \qquad f_s = 20 \text{ ksi}$$

$$f_y = 60 \text{ ksi}, \qquad f_s = 24 \text{ ksi}$$

For concrete the maximum compressive stress in flexure shall be $f_c = 0.45 f'_c$, so for $f'_c = 3{,}000$ psi, $f_c = 1350$ psi.

For working-stress design, the usual assumptions for stresses in beams are made:

1. A section which is plane before bending remains plane after bending.

2. Strains vary as the distance from the neutral axis.
3. Stress is proportional to strain.

In addition, for reinforced concrete, two further assumptions are made:

4. Steel takes all the tension.
5. The ratio between the transformed area of steel and that of concrete is given by the modular ratio, $n = E_s/E_c$.

These assumptions result in a beam with the cross section shown in Fig. 165. The concrete acts to resist bending only above the neutral axis, and the steel bars resist the tension forces. The next task will be to assign names and symbols to the areas of steel and concrete to develop formulas for design. The dimensions of the beam are shown in Fig. 166. The width is b and the distance from the top of the beam to the center of the steel area is d. That portion of the depth of the section in compression is called kd.

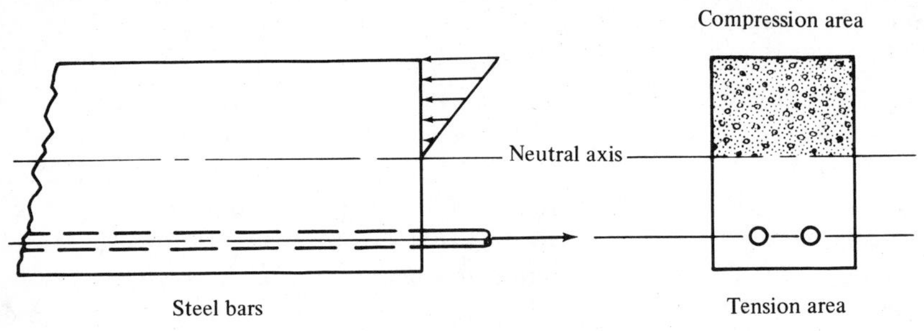

Fig. 165.

The distribution of stresses is shown on the right of Fig. 166, and they vary directly with the distance from the neutral axis. Therefore, the center of the

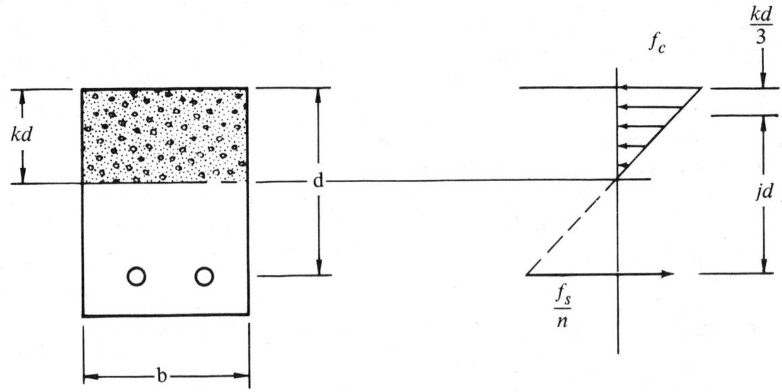

Fig. 166.

compressive forces for a rectangular beam is $kd/3$ from the top of the beam. The distance from the center of compressive forces to the center of the tensile forces is called jd. The relationship between j and k is, therefore,

$$jd = d - kd/3$$

$$j = 1 - k/3$$

The compressive unit stress is called f_c and the tensile unit stress, f_s. The steel and the concrete are assumed to have the same unit strain. The ratio of the modulus of elasticity in the steel (E_s) to the modulus of elasticity in the concrete (E_c) is $E_s/E_c = n$. The apparent tensile stress in the beam is, therefore, f_s/n.

A relationship can now be set up between the value of k and the tensile and compressive forces:

$$kd : d = f_c : [f_c + (f_s/n)]$$

or

$$k = \frac{f_c}{f_c + (f_s/n)}$$

Usually the values of the allowable stress in the concrete and steel are established by the specifications. For example, assume that $f_c = 1000$, $f_s = 20{,}000$, and $n = 12$. The value of k can immediately be established for this set of stresses:

$$k = \frac{1000}{1000 + (20{,}000/12)} = 0.375$$

Now a formula for the dimensions of the beam may be set up in terms of the bending moment M on the beam. The force on the compression side of the beam is the bending moment M divided by the lever arm jd, i.e., $C = M/jd$. But, in terms of the compressive stress f_c, the total compression is $C = f_c bkd/2$. Therefore

$$M/jd = f_c bkd/2$$

$$bd^2 = 2M/f_c kj$$

It is customary to simplify this formula by making the substitution

$$K = f_c kj/2$$

so that

$$bd^2 = M/K \quad \text{or} \quad d = \sqrt{M/Kb}$$

The area of steel required in a beam may be obtained by dividing the moment M by the lever arm jd to get the total tensile force $T = M/jd$. The area of steel

A_s required for this tensile force is

$$A_s = T/f_s$$
$$= M/f_s jd$$

The following example is given to illustrate the design of a beam for moment.

EXAMPLE

Find the size of beam and the area of steel required for a simple beam with a span of 20 ft carrying a load of 1.5 klf. Design for the following set of stresses:

$$f_c = 1000 \text{ psi}, \quad f_s = 20{,}000 \text{ psi}, \quad n = 12$$

The bending moment for the load and span is

$$M = \tfrac{1}{8}\, wL^2 = 1.5 \times 20^2/8 = 75 \text{ ft-kip}$$

The values of k, j, and K are

$$k = \frac{f_c}{f_c + (f_s/n)} = \frac{1000}{1000 + (20{,}000/12)} = 0.375$$
$$j = 1 - (k/3) = 0.875$$
$$K = f_c kj/2 = 1000(0.375)0.875/2 = 164$$

Assume that the width of the beam is 12 in. The depth required for this width is

$$d = \sqrt{M/Kb} = \sqrt{75(1000)12/164(12)} = 21.4 \text{ in.}$$

The required depth of beam is, therefore, 21.4 in. The beam may be made shallower by increasing the width; e.g., if $b = 15$ in., $d = 19.1$ in. The area of steel required for a depth of 21.4 in. is

$$A_s = M/f_s jd = 75(12)/20(.875)21.4 = 2.40$$

The required area is 2.4 in.^{2}, or, say, three No. 8 bars.

The General Method for the Analysis of Stresses by the Working-Stress Method For T-shaped, L-shaped, or other irregular sections, or those containing compressive reinforcement, the general method for the analysis of elastic beams may be used. The unknown factor is the size of the area in tension. It is fairly easy to write a quadratic equation having the neutral axis as an unknown, but in many cases it is simpler to guess the position, even though the guess will allow some tensile stress in the concrete. Following is an example of the procedure for finding the stresses given the external bending moment, the size and shape of the beam and the amount of reinforcing.

EXAMPLE OF THE GENERAL METHOD

A sketch of the cross section is shown in Fig. 167. Determine the stresses in the steel and in the concrete for an external bending moment of 150 ft-kip. Use a modular ratio of 9. Allow twice the area of compressive steel as is permissible by the ACI *Code*. The modular ratio then must be decreased by 1 to take into account the area displaced by the bar.

We will guess the neutral axis to be 10 in. below the top of the section. The transformed section is shown in Fig. 168 with the appropriate areas determined. In Fig. 169 the properties of the cross section are calculated in accordance with the methods shown in previous chapters. In Fig. 170 the stresses are calculated.

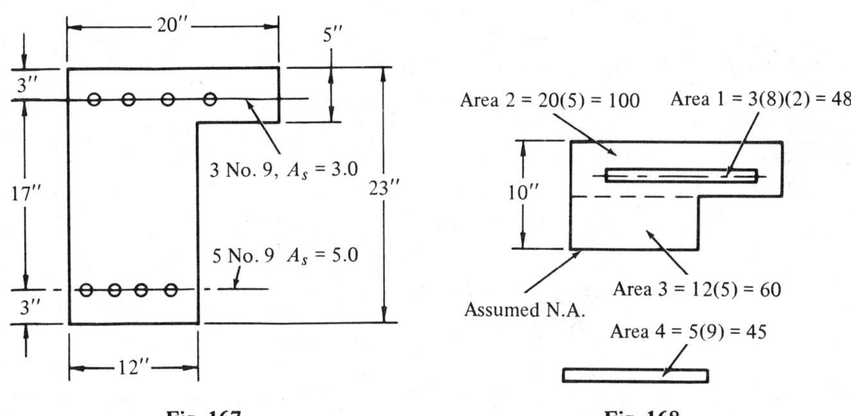

Fig. 167. **Fig. 168.**

Properties of Section
a about Top

Area	A	a	aA	a^2A	I_0
1	48	3.0	144	432	–
2	100	2.5	250	625	$\frac{20(5)^3}{12} = 208$
3	60	7.5	450	3,375	$\frac{12(5)^3}{12} = 125$
4	45	20.0	900	18,000	–
	253		1744	22,432	333

$\bar{a} = \frac{1744}{253} = 6.89$ $\qquad I_T = 22{,}432 + 333 = 22{,}765$

$I_{CG} = I_T - A\bar{a}^2 = 22{,}765 - 253(6.89)^2 = 10{,}755$

$I/C\ (\text{top}) = \frac{10{,}755}{6.89} = 1561; I/C\ (\text{bottom}) = \frac{10755}{(20-6.89)} = 820$

$I/C\ (\text{comp. steel}) = \frac{10{,}755}{6.89-3} = 2764$

Fig. 169. Calculation of the properties of the cross section of Fig. 168.

$$\text{Top: } f_c = \frac{Mc}{I} = \frac{150{,}000(12)}{1561} = 1{,}153 \text{ psi}$$

$$\text{Bottom: } f_c = \frac{150{,}000(12)}{820} = 2{,}195$$

so $f_s = f_c n = 2{,}195(9) = 19{,}755$

Concrete stress at compression steel:

$$f_c = \frac{150{,}000(12)}{2764} = 651$$

Steel stress:

$f_s = 651(9) = 5859$

Fig. 170. Calculation of the stresses on the cross section of Fig. 168.

Note that the stress on the bottom at the tensile steel must be multiplied by the modular ratio, $n = 9$, to obtain the steel stress.

The calculated stresses are indicated on the section in Fig. 171. Our guess of 10 in. from the top for the neutral axis was too large, and the tensile stress in the concrete is 520 psi. The stress at the top is 1153 psi, and in the steel it is 19,755 psi. The compressive steel stress is 5859 psi, which is satisfactory because

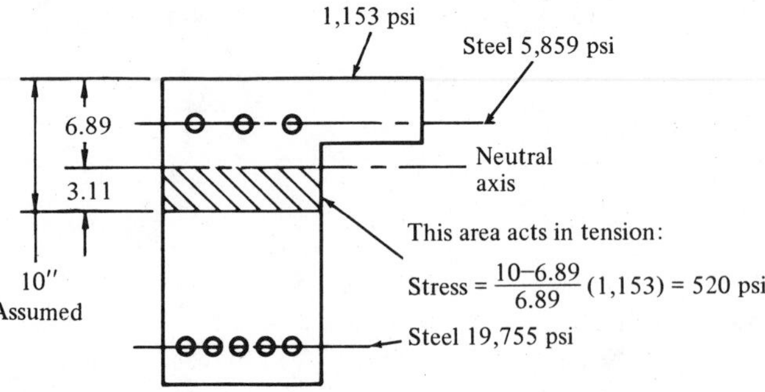

Fig. 171.

it must be less than the design tensile stress. We assumed that the neutral axis was 10 in. from the top, but the calculations show that it is 6.89 in. There is an area 3.11 in. wide that is in tension. The affect of this area is small on the stresses. It is left up to the reader to revise the design for a smaller value of the distance from the top to the neutral axis.

Distribution of Shearing Stresses, Elastic Beams Until the revision of the ACI *Building Code* in 1963, the design of stirrups to resist the shearing forces was based on the elastic equation of the distribution of shearing stresses using the transformed area with no tension in the concrete. The distribution of shear in a rectangular section is shown in Fig. 172. The values of shear can be obtained from the formula $v = VQ/Ib$, where V is the external shear, Q is the first moment

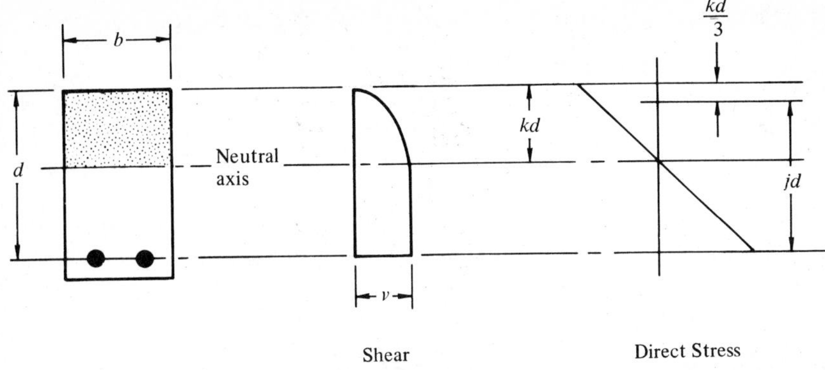

Fig. 172.

(area times distance) of the area away from the neutral axis about the point where shear is obtained, I is the moment of inertia of the section, and b is the width of the section. However, it is easier to use some of the coefficients already developed for concrete, kd the distance to the neutral axis and jd, the distance from the center of compression to the center of the steel area.

Refer to Fig. 172. The upper half of the shear diagram is similar to that of an elastic beam of rectangular cross section and has a parabolic distribution. The shear for the lower half of the section is constant. The area under this diagram times the width must equal the total external shear: $V = \frac{2}{3}\, vkdb + vdb(1 - k)$. Solve for the stress, v:

$$v = \frac{V}{bd[1 - (k/3)]}$$

But the term $1 - (k/3)$ is equal to the value of j. So $v = V/bjd$. We will refer to this equation in our discussion of the design of stirrups by the strength method.

Ultimate-Strength Design–Load Factors (For the remainder of this discussion, the term "ultimate-strength design" will be replaced by "strength design," in accordance with the usage in the ACI *Code*.)

With this method, the design loads and stresses will be multiplied by different factors for dead loads and live loads. Terms are defined as follows:

U = total load or force used for strength design
D = design dead loads or forces as calculated from weights of materials
L = design live loads as prescribed by applicable building codes
T = loads or forces resulting from the cumulative effects of temperature, creep, shrinkage, or differential settlement
W = wind loads or forces

There are several combinations of these loads which must be satisfied.

1. If D and L only are present, $U = 1.4D + 1.7L$.
2. If W is included with D and L, $U = 0.75(1.4D + 1.7L + 1.7W)$ and cases where L is present or absent must be checked; if L is absent, $U = 0.9D + 1.3W$.
3. If earthquake loads or forces are present a value of $1.1E$ shall be substituted for W.
4. For earth pressure H, $U = 1.4D + 1.7L + 1.7H$, but where D reduces the effect of H, the corresponding load factors shall be taken as 0.9 for D and zero for L.
5. For liquids substitute $1.4F$ for $1.7H$ above.

The application of these load factors will be demonstrated in subsequent examples.

In most design it is expedient to form the ultimate loads $1.4D$ and $1.7L$ first and use these to determine moments, thrusts, and shears rather than to make the conversion later.

Ultimate-Strength Design–Capacity-Reduction Factors Another provision of the ACI *Code* is that the strength of a member must be reduced by a *capacity reduction factor* ϕ which depends on the type of stress as follows:

1. Bending or axial tension, $\phi = 0.9$.
2. Axial compression or axial compression with bending with
   a. spirally reinforced members, $\phi = 0.75$,
   b. other members, $\phi = 0.70$.
3. Where there is a very small compression compared to the bending, there is a transitional formula which increases the value of ϕ from 0.70 for compression members to 0.90 for members having no compression. See the ACI *Code* for this formula.
4. Shear or torsion, $\phi = 0.85$.
5. Bearing, $\phi = 0.70$.

In the following discussion of design by the strength method, if it is an analysis problem, the capacity-reduction factors will be used to reduce the moments or forces for the member sizes assumed, or conversely, if it is a design problem, the moments will be increased by the reciprocal of the capacity-reduction factor and then the member will be designed. In this way the capacity-reduction factor will not appear in the formulas for design. The increased moment or force will be given a prime symbol; e.g., $M_u' = M_u/0.9 = 1.11M_u$. Thus if $M_D = 100$ ft-kip, $M_L = 50$ ft-kip, then $M_u = 1.4(100) + 1.7(50) = 225$ ft-kip and $M_u' = 225/0.9 = 250$ ft-kip.

Basic Assumptions of Strength Design The strength design of concrete beams and compression members is based on the structural action at failure as determined by many laboratory and field tests. Following are the assumptions made for strength analysis:

1. The concrete takes no tensile forces.
2. The unit strain on a cross section, for both the steel and the concrete, varies directly as the distance from the neutral axis of the section, as shown in Fig. 173b.
3. The maximum strain in the concrete is 0.003.
4. The strain in the steel is related linearly to the yield stress of the steel. For example, at yield, if $f_y = 60{,}000$ psi and if $E_s = 29{,}000{,}000$ psi, the strain in the steel reinforcing is $e = 60{,}000/29{,}000{,}000 = 0.0021$.
5. The distribution of stress in the concrete at ultimate conditions may be represented by any convenient stress block. In Fig. 173c is shown a parabolic-shaped block which most nearly represents the actual distribution. This distribution may be used, but it is more convenient to use a rectangular stress block, as shown in Fig. 173d.
6. For the rectangular block the stress at ultimate load shall be $0.85 f'_c$.
7. The maximum depth of the stress block, $a = k_1 c$, as shown in Fig. 173d will vary with the concrete strength f'_c. For f'_c up to 4000 psi, $k_1 = 0.85$, and for strengths over this value, reduce the value of k_1 at a rate of 0.05 for each 1000 psi over 4000 psi. For example, if $f'_c = 6000$ psi, the maximum value of a shall be $a = [0.85 - 0.05(2)]\, c = 0.75c$.

For bending, the area of steel shall not be greater than 0.75 times the area for balanced reinforcing, i.e., for the moment producing maximum strains in the concrete and steel simultaneously.

The above assumptions are applicable to both bending and bending plus compression, as will be demonstrated later.

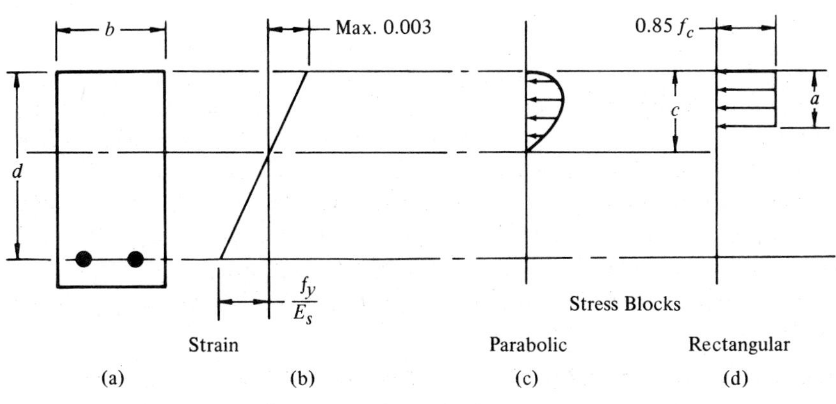

Fig. 173.

EXAMPLE: ANALYSIS OF A RECTANGULAR BEAM

The above assumptions will be applied to the determination of the capacity of a rectangular concrete beam which is 10 in. wide and 18 in. deep from the top to the center of the steel, as sketched in Fig. 174. It has an area of steel of 3.0 in.^{2}, (three No. 9 bars). Assume that the cylinder strength of the concrete f_c' is 3000 psi and that the yield strength of the steel f_y is 40,000 psi. Determine the allowable bending moment M_u on the section.

The size and reinforcing for this beam have already been selected so that we know the force that the steel can furnish. With $A_s = 3.0$ in.^{2} and $f_y = 40{,}000$ psi, the force in the steel is $T = A_s f_y = 3.0(40{,}000) = 120{,}000$ lb. In a beam the compression force C must equal the tension force T. Equate the reduced concrete stress $0.85 f_c'$ times the area ab (with the width $b = 10$ in.) to the tension force T, and solve for a, the effective depth of the stress block:

$$C = 0.85 f_c' ab = T, \qquad 0.85(3000)(10)a = 120{,}000, \qquad a = 4.70 \text{ in.}$$

The distance jd, the arm of the center of the stress block about the reinforcing steel, is $jd = 18.0 - 4.70/2 = 15.65$ in. The moment M_u' resulting from the couple made by T and C is

$$M_u' = Tjd = 120{,}000(15.65) = 1{,}878{,}000 \text{ in.-lb} = 156.5 \text{ ft-kip}$$

Now apply the capacity-reduction factor to determine the allowable bending moment: $M_u = 156.5(0.9) = 140.9$ ft-kip.

We have two final checks to make: (1) Be sure that a is less than $0.85c$ (point 7 above), and (2) check to see that the area of steel is less than 0.75 times the area for balanced reinforcement (point 8 above).

The dimension c, the distance to the neutral axis from the top of the beam for steel and concrete at their full stress (the balanced condition), can be determined from the plot of maximum allowable strain as sketched in Fig. 174c, the

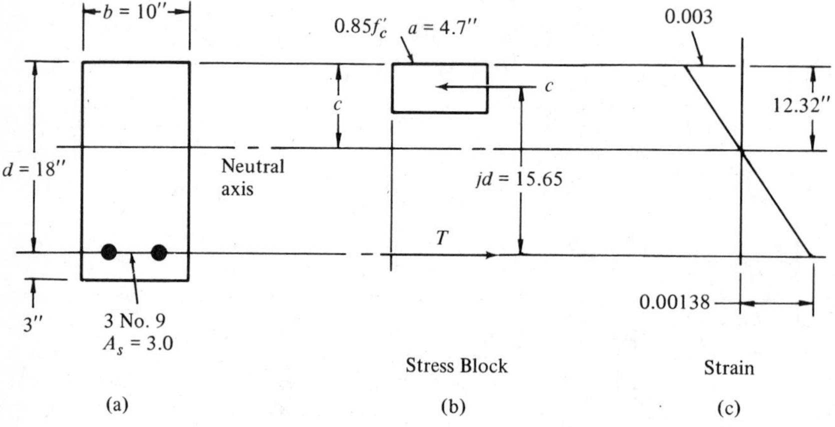

Fig. 174.

strain at the compression edge being, for the concrete, $e = 0.003$, and for the steel, $e_s = f_y/E_s = 40{,}000/29{,}000{,}000 = 0.00138$.

The value of c based on the triangles for the strain diagram shown in Fig. 174 is $c = 18(0.003)/(0.003 + 0.00138) = 12.32$ in.

Apply the requirement that the distance a shall not be greater than $0.75(0.85)c = 0.75(0.85)(12.32) = 7.85$ in. The calculated value, $a = 4.70$ in., is less than the maximum value of 7.85, so the depth of the stress is satisfactory.

With reference to the second requirement, it is evident that the moment is not even close to that possible from balanced reinforcement, so the analysis is satisfactory. See the next subsection, on the design of rectangular beams, for a further discussion of this problem.

EXAMPLE: AREA OF STEEL REQUIRED IN A RECTANGULAR BEAM

In the preceding problem, the size of the beam and the area of steel were given. In this problem we will again give the size of the beam, but ask for the area of steel required if the bending moment is given.

Assume that $b = 12$ in., $d = 20$ in., and $M_u = 250$ ft-kip. Use $f'_c = 5000$ psi and $f_y = 60{,}000$ psi. The section is sketched in Fig. 175.

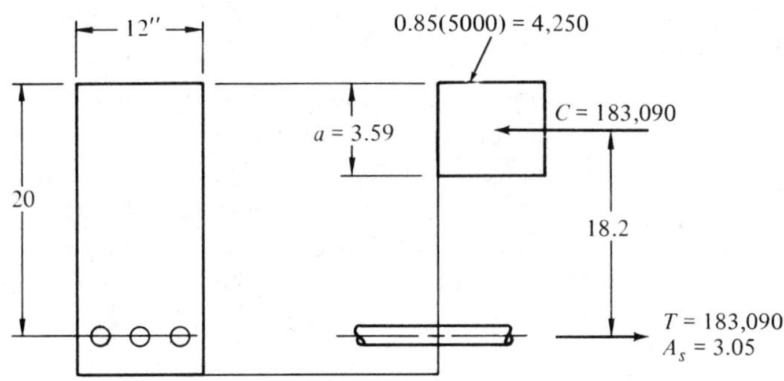

Fig. 175.

The moment shown above must be increased for the capacity-reduction factor so we will use $M'_u = 250/0.9 = 277.8$ ft-kip. We do not know the value of the tension force T as in the previous problem, so we cannot directly determine the value of a. We can either guess the value of a and then refine that guess or write a quadratic equation and solve it for the value of a. We shall use the latter method.

The allowable reduced stress on the stress block is $0.85f'_c = 0.85(5000) = 4250$ psi. The available force in compression can be obtained in terms of the distance a: $C = a(12)(4250) = 51{,}000a$. The arm of the center of the stress block from the steel is $d - (a/2) = 20 - (a/2)$. The moment of the force in the stress block is the arm times the force, and must equal the external moment. The moment must be converted into inch pounds:

$$[20 - (a/2)]\ 51{,}000a = 277.8(12)(1000)$$

$$-25{,}500a^2 + 1{,}020{,}000a - 3{,}333{,}000 = 0$$

To solve this use the binomial theorem, which has the general form

$$x = \frac{-B \pm \sqrt{B^2 - 4AC}}{2A}$$

where A, B, and C are the coefficients of each term above. Dividing by $-10{,}000$, $A = +2.55$, $B = -102.0$, and $C = +333.3$.

From this equation, the value of a is 3.64 in. We should check this value by seeing if the proper amount of moment is furnished for this size of stress block. The force in the stress block is $C = 3.64(12)(4250) = 185{,}500$. The moment arm is $20 - (a/2) = 18.2$ and the moment is

$$M'_u = 18.2(185{,}500) = 3{,}380{,}000 = 282 \text{ ft-kip}$$

so the analysis is satisfactory.

Now we must determine the area of steel required, which is equal to the tension force (which, in turn, is equal to the compression force) divided by the yield stress in the steel. $A_s = C/f_y = 185{,}500/60{,}000 = 3.10$ in.^{2}. Therefore we will need the same three No. 9 bars used in the previous problem. It must be obvious, also, that the criteria for the size of the stress block are satisfied.

Design of Rectangular Beams If the bending moment is given, and we are required to select dimensions and area of steel, then the problem is more complicated than previous examples. It is possible and sometimes desirable simply to guess the size of the beam and to refine this guess until the size and area of steel are satisfactory. Instead of guessing, however, we will develop equations that are quicker and more accurate. A concrete beam, showing the stress block and the symbols used for dimensions, is sketched in Fig. 176. The following relations

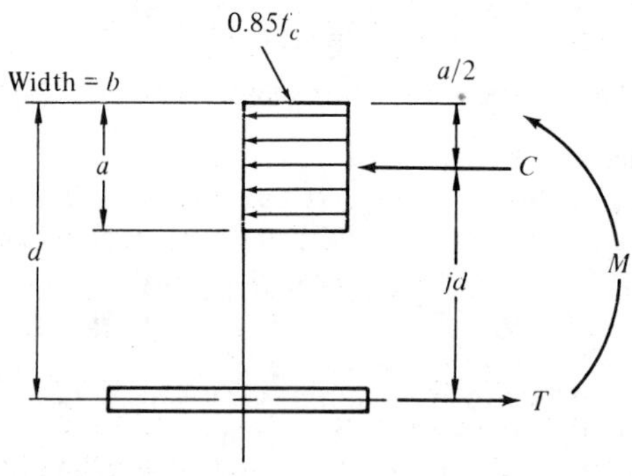

Fig. 176.

are demonstrated:

$$jd = d - (a/2), \quad \text{so that} \quad a = 2d(1 - j)$$

$$M = Cjd, \quad C = 0.85f'_c ab, \quad \text{so that } M = (0.85f'_c)[2d(1 - j)bjd]$$

Our method of design is based on using the coefficient $K = M/bd^2$ as developed in the subsection on working-stress design. Insert M in the equation for K:

$$K = M/bd^2 = (0.85f'_c)[2d(1 - j)bjd]/bd^2 = 1.7f'_c(1 - j)j$$

Solve for j in terms of K and f'_c using the binomial theorem:

$$j^2 - j + (K/1.7f'_c) = 0, \quad \text{so that}$$

$$j = 0.5 \pm \sqrt{0.25 - (K/1.7f'_c)}$$

With this equation, the effective depth j can be obtained, given the value of K and f'_c. Note that f_y does not enter into this equation. However, it may govern the limits of the use of this equation.

We are now in a position to check the computations in the previous problem to see if they correspond to this formula. For that example:

$$M'_u = 277.8 \text{ ft-kip}, \quad f'_c = 5000, \quad b = 12, \quad \text{and} \quad d = 20 \text{ in.}$$

$$K = M/bd^2 = 277.8(1000)(12)/(12)(20^2) = 694$$

The value of j is

$$j = 0.5 + \sqrt{0.25 - 694/(1.7)(5000)} = 0.5 + \sqrt{0.1684} = 0.91$$

The value of j for the previous example was the moment arm of the stress block about the steel divided by the depth, $j = 18.2/20 = 0.91$, so the two methods agree.

To use the equation for j, it may be solved directly for each problem or a set of tables for j, K, and f'_c with limiting values for f_y may be calculated and used to shorten the calculations. We will use the formula. Before we can design beams, however, it will be necessary to investigate the limiting depths of beams for deflection.

Limiting Depths of Concrete Beams The calculation of the deflections of concrete beams becomes very complicated under the procedures described by the ACI *Code* and will not be discussed here. It is not necessary to determine deflections if certain proportions of span to depth are selected. The following table is taken from the ACI *Code* and shows the maximum ratio of span to depth permissible unless deflections are computed and they are satisfactory. This table applies only to members not supporting or attached to other construction or likely to be damaged by large deflections.

Maximum Ratio of Span to Thickness

	Method of support			
	Simply supported	One end continuous	Both ends continuous	Cantilever
Solid one-way slabs	20	24	28	10
Beams or ribbed one-way slabs	16	18.5	21	8

This table applies only to normal-weight concrete ($w = 145$ pcf) and for $f_y = 60$ ksi. For light-weight concrete with $w = 90–120$ pcf, multiply all values by $1.65 - 0.005w$ but not less than 1.09.

For f_y other than 60 ksi, divide the values in the table by a factor $0.4 + 0.01f_y$, where f_y is in kips per square inch.

In the design of most reinforced concrete it is better to keep the depth as large as feasible to reduce the possibility of excessive deflection for dead loads. We will use the factors given above for design of beams and slabs unless strength controls.

Limiting Values of the Coefficient K The ACI *Code* provides that if the stress block is deeper than that which would result in the same member with steel equal to $0.75p$, where p is the percentage of steel, then compressive reinforcement shall be used. The largest size of the stress block is based on the limiting strains in steel and concrete, f_y/E_s for the steel and 0.003 for the concrete. For example, for $f_y = 40{,}000$ psi and $E_s = 29{,}000{,}000$ psi, the maximum size of the stress block is calculated in Fig. 177 with values of $c = 0.683d$, $a = 0.5835d$, and $j = 0.7082$. Note that if f_c' is greater than 4,000 psi, the ratio of a to c is k_1, less than 0.85 and equal to $0.85 - (f_c' - 4{,}000)(0.05)/1000$.

If the area of steel is reduced to 0.75 times that required for balanced reinforcing, then, since the compressive force equals the tension force, the value of a must be 0.75 times that required for balanced reinforcing. For the example in Fig. 177, $a = 0.5835d$ for balanced reinforcing, so the maximum value is now $0.75(0.5835)d = 0.437d$ and the corresponding value of j is $1 - 0.437/2 = 0.782$.

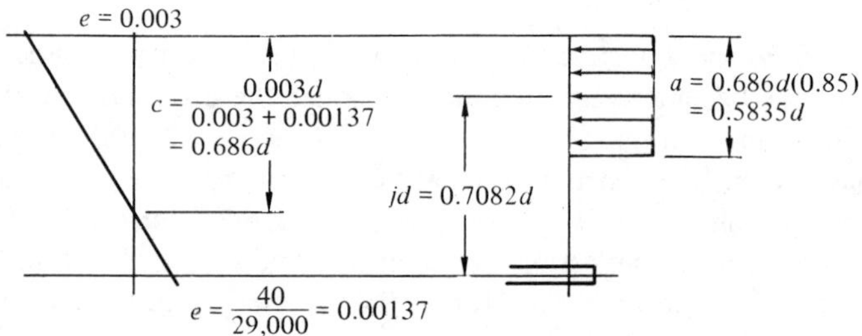

Fig. 177.

TABLE 11 LIMITING VALUES OF a/d, j, K FOR $p = 0.75p_b$.

$$a/d = \frac{87(0.85)(0.75)}{87 + f_y}, \quad j = 1 - (a/2d), \quad K = 1.7f_c'(1 - j)j$$

f_y (ksi)	a/d	j	K for f_c' = (ksi) 3.0	3.5	4.0	4.5	5.0
40	0.437	0.782	870	1015	1161	1306	1451
50	0.405	0.798	823	961	1098	1235	1372
60	0.377	0.811	781	911	1041	1171	1301

The value for K, if $f_c' = 3{,}000$ psi, is

$$K = 1.7f_c'(1 - j)j = 1.7(3{,}000)(1 - 0.782)0.782 = 869$$

There will be limiting values of a, j, and K for each combination of f_y and f_c'. These are recorded in Table 11 for $f_y = 40$, 50, and 60 ksi and for $f_c' = 3.0$, 3.5, 4.0, 4.5, and 5.0 ksi.

Minimum Steel Areas for Beams The ACI *Code* also provides for a minimum area of steel in beams of a percentage $p = 200/f_y$, so for $f_y = 40$ ksi, 50 ksi, and 60 ksi, the percentages are 0.005, 0.004, and 0.003, respectively. If you provide an area of steel at least one third greater than the area required for analysis, then less can be used.

Many structural members are governed by architectural requirements, particularly beams at ground level which carry walls, often called grade beams. An example follows.

EXAMPLE

Design a continuous 20-ft-span grade beam which is 8 by 36 in. total depth to carry a load not including its own weight of 2.15 kpf. Assume all of this load is dead load. Use a moment design coefficient of $M = wL^2/11$, $f_c' = 3{,}000$ psi, and $f_y = 40$ ksi.

The weight of the beam, using a weight of concrete of 150 pcf, is 0.30 klf, so the dead load is 2.45 klf and the ultimate load is $U = 2.45(1.4) = 3.43$ klf. The moment is $M_u = 3.43(20^2)/11 = 124.7$ ft-kip, and $M_u' = 124.7/0.9 = 138.6$ ft-kip. Assume that the effective depth is 4 in. less than the total depth so $d = 32$ in. Solve for K:

$$K = M/bd^2 = 138.6(12)(1000)/8(32^2) = 203$$

The corresponding value of j is

$$j = 0.5 + \sqrt{0.25 - K/1.7f_c'}$$

$$= 0.5 + \sqrt{0.25 - 203/1.7(3000)} = 0.96$$

The required area of steel is

$$A_s = M/djf_y = 138.6(12)/32(0.96)(40) = 1.35 \text{ in.}^2$$

The percentage of steel is 1.35/(32)(8) = 0.0053, which is just above the minimum requirement of 0.005 for 40-ksi steel.

Values of K to be Used in Design In the method of design proposed here, the coefficient K is used as a parameter for the selection of the width and depth of the section. If K exceeds the values shown in Table 11, then the beam does not conform to the ACI *Code*. However, it has been the writer's experience that a conservative approach should be used for the selection of depth, and it is not generally advisable to use values of K near the values of Table 11. Beams designed by the working-stress method have generally been satisfactory. For the working-stress method with $f'_c = 3{,}000$ psi with a corresponding value of $f_c = 1350$, and with $f_y = 40$ ksi with a corresponding value of $f_s = 20$ ksi, the value of K is 236. If we multiply this by the average-load factor, say 1.55, the value of K is 366. It is the writer's practice to start with values of K from 500 to 600, and these values are used in the following design problems. If the beam is short and heavily loaded, or if the depth is restricted, then the maximum value of K may be necessary.

EXAMPLE: DESIGN OF A RECTANGULAR BEAM

Design a beam with a simple span of 20 ft to carry a dead load of 1 kpf and a live load of 1.5 klf. Use $f'_c = 3500$ psi and $f_y = 50{,}000$ psi. Disregard shear.

First we must obtain the total ultimate load, $U = 1.4D + 1.7L = 1.4(1.0) + 1.7(1.5) = 3.95$ klf. The moment due to this load at the center of a simple beam is $M_u = wL^2/8 = 3.95(20^2)/8 = 197.5$, and with the capacity-reduction factor $M'_u = M_u/0.9 = 219.4$ ft-kip.

Now we must select dimensions for this beam. Often the width is given by architectural considerations. Otherwise, a ratio of width to depth of 1:2 is considered good practice. We will use $b = 10$ in.

The depth will be determined by using a value of $K = 600$, and then we will check to see if the length-to-depth ratio is under the requirement for deflection calculations. We have already made sure that we are under the requirements of Table 11 because K is less than the value in that table. Now solve for d, given K:

$$K = 600 = M/bd^2 = 219.4(12)(1000)/10(d^2)$$

$$d = \sqrt{439} = 21.0 \text{ in.}$$

The limiting value for length over depth, from the table in the subsection on the limiting depth of concrete beams (above), disregarding the note about reduced stresses, is $L/d = 16$. For this example $L/d = 20(12)/21 = 11$, so the depth is satisfactory.

The area of steel required is obtained by dividing the moment by the moment arm jd and the stress f_y. The value of j is

$$j = 0.5 + \sqrt{0.25 - 600/(1.7)(3500)} = 0.887$$

so the area of steel required is

$$A_s = M/jdf_y = 219.4(12)/21\,(0.887)(50) = 2.82 \text{ in.}^2$$

Three No. 9 bars will furnish 3.0 in.^{2}, which is well above the required area. If we use two No. 9 and one No. 8, we will be closer to the required area. The actual selection will depend somewhat on the type of reinforcing details used.

In the selection of the depth for this beam we did not consider the requirements for shear, which may affect the depth. This will be discussed in a later subsection.

Compression Reinforcing The use of strength design usually relieves the need for compression reinforcing except for members subjected to compressive and bending forces. The discussion of design for combined bending and direct stress (below) will show the principles involved, so this subject will not be discussed here.

T ***Beams*** If slabs or joists are used in concrete structures, the top of the beam at the center will be quite wide and should be considered as part of the section of the beam. The ACI *Code* provides rules for the width of the top of the T as follows: The width of the T shall not be greater than:

1. one-fourth the span of the beam
2. eight times the thickness of the slab on either side plus the width of the web of the beam
3. one-half the clear distance to the next beam

There are similar requirements for inverted L beams.

In making an analysis of the strength of a T beam, if the value of a, the depth of the stress block, is less than the depth of the slab, the total width of the T may be used as the beam width. If a is greater than the depth of the slab, then some of the area below the slab must be included in the stress block. The procedure for determining the strength of a T beam follows the methods already discussed. No example will be presented here.

Shear and Diagonal Tension Reinforcing Reinforced concrete beams will develop diagonal cracks, as sketched in Fig. 178, and fail in areas near the supports at loads below those predicted by the strength-design method for longitudinal reinforcement. To provide reinforcing for these cracks, U-shaped bars, called stirrups, are placed perpendicular, or inclined, to the longitudinal reinforcing steel. The design of stirrups is empirical (based on tests); the following observations were made after a large number of tests:

1. There is little or no stress in the stirrups until the beam cracks as shown in Fig. 178.

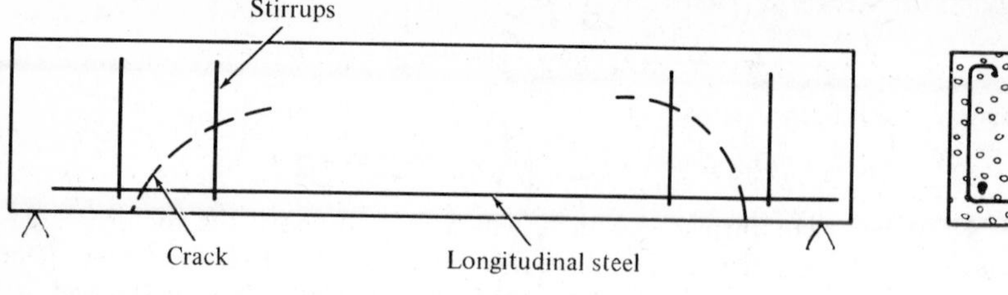

Fig. 178.

2. The stress in the reinforcing is then proportional to the load added *after* the crack has occurred.
3. Thick webs will crack at higher loads than thin webs, in proportion to the thickness of the web.
4. There is an increase in the capacity of beams to resist cracking near the ends of short beams where the ratio of V to M is high. (V is the shear and M is the bending moment at the section.)

Extensive studies have resulted in the following rules for the design of vertical stirrups given by the ACI *Code*. (For inclined stirrups, refer to the *Code*.) The rules as stated here assume that the moments and shears have been increased to take into account the capacity-reduction factor, which for shear is $\phi = 0.85$. The increased shear will be termed V_{ui}.

1. A minimum area of reinforcement shall be provided in concrete members except: (a) for slabs and footings, (b) in joist construction, (c) in beams where the total depth does not exceed 10 in., 2.5 times the thickness of the flange, or one-half the width of the web, or (*d*) v_u is less than half of v_c (see description to follow).

2. The maximum spacing shall be $0.5d$ but not more than 24 in.

3. The total design shearing stress in the concrete is $v_u = V/b_w d$, where b_w is the width of the beam web.

4. The shear stress v_c carried by the concrete may be determined by either of the following formulas:

$$v_c = 2\sqrt{f'_c}, \quad \text{or} \quad v_c = 1.9\sqrt{f'_c} + (2500 p_w V_u d/M_u)$$

but is not to exceed $3.5\sqrt{f'_c}$. In the above equation, M_u is the bending moment occurring simultaneously with the shear V_u taken at the section considered, but $V_u d/M_u$ shall not be greater than 1. The term p_w is the percentage of steel in the web, $p_w = A_s/b_w d$. For most design, it is easiest to use the first formula, but there may be cases where the second will allow a better design if stirrups are crowded.

5. When the reaction introduces vertical compressive forces into the end regions of the beam, then sections located less than a distance d from the face of

the support may be designed for the same v_u that has been computed at a distance d from the face of the support.

6. The area of vertical stirrups may be determined from the formula

$$A_v = (v_u - v_c)\, b_w s/f_y$$

7. If stirrups are required, then they shall be designed for a stress of not less than 50 psi.

8. The value of $(v_u - v_c)$ shall not exceed $8\sqrt{f'_c}$. If $(v_u - v_c)$ exceeds $4\sqrt{f'_c}$, then the maximum spacings shall be $0.25d$ and 12 in. Otherwise, the maximum spacing shall be $0.5d$.

EXAMPLE: SHEAR REINFORCING

We will use the same size beam (and load) as in the previous example of the rectangular beam. The simple span was 20 ft, the total load was $U = 3.95$ klf, $b_w = 10$ in., $d = 21.0$ in. We also need to know the width of the support of the beam which we will assume to be 6 in. from the center of the support to the edge. The concrete stress is $f'_c = 3500$ psi and the steel is $f_y = 50{,}000$ psi. The size of stirrups we will assume to be No. 3 U stirrups, having a total area of $A_v = 2(0.11) = 0.22$ in.^{2}

First we will increase the load for the capacity reduction factor for shear, which is $\phi = 0.85$, so $w = 3.95/0.85 = 4.64$ klf. Now draw the shear diagram for the left half of the beam, as sketched in Fig. 179. The shear V_{ui} at a distance $d = 21$ in. from the face of the support is 35.96 kip. The unit shearing stress v_u is $v_u = V_{ui}/b_w d = 35{,}960/10(21.0) = 171$ psi.

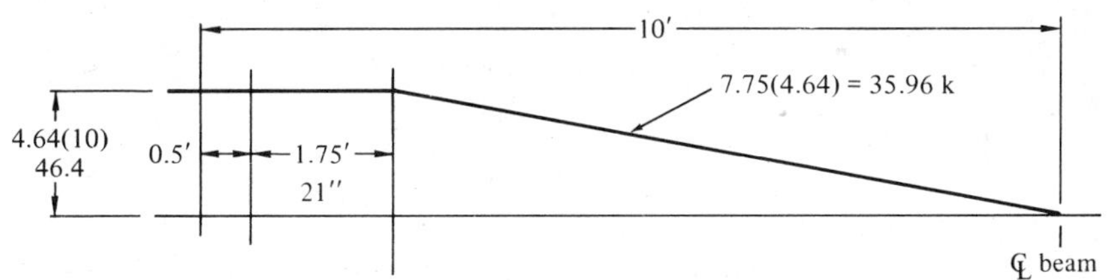

Fig. 179.

The shears to be used in design are shown in Fig. 180. Note that v_u is a constant 171 psi to the support.

The shear v_c that can be taken by the concrete is

$$v_c = 2\sqrt{f'_c} = 2\sqrt{3{,}500} = 118.3 \text{ psi}$$

so the stirrups must take $v_u - v_c = 171 - 118.3 = 52.7$ psi. The required spacing can be determined from the formula $A_v = (v_u - v_c)b_w s/f_y$, but since A_v is given as 0.22 in.^{2}, we will rewrite this equation to solve for s:

$$s = A_v f_y/(v_u - v_c)b_w = 0.22(50{,}000)/52.7(10) = 20.9 \text{ in.}$$

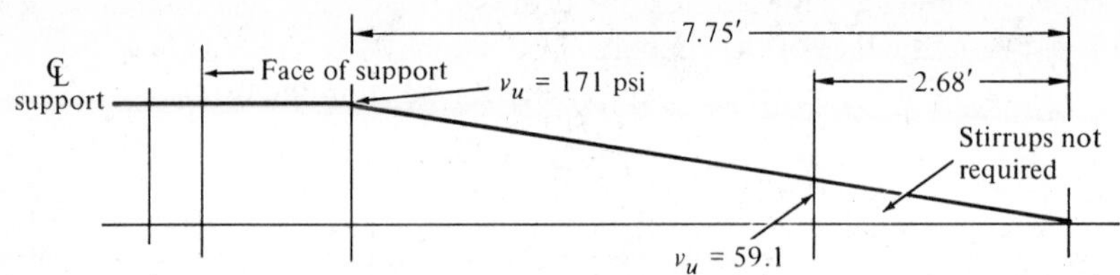

Fig. 180.

The maximum spacing is limited to either $0.5d$ or 24 in., so we must use a maximum of 10.5 in. Stirrups will be required for those areas where v_u is more than one-half v_c, that is, $v_u = 118.3/2 = 59.1$ psi. We can find this point by the following ratio: If the stress $v_u = 171$ psi at a point 7.75 ft from the center of the beam, then when $v_u = 59.1$ the distance from the center must be $L = 59.1(7.75)/171 = 2.68$ ft.

The net shear stress $v_u - v_c = 52.7$ psi is very close to the minimum stress (50 psi), so we will use a spacing of 10 in. at all points where stirrups are required. The reinforcing pattern selected is shown in Fig. 181. Note that two No. 4 bars have been used in the top of the beam to hold the stirrups. Also, the spacing starts with a half space at the support.

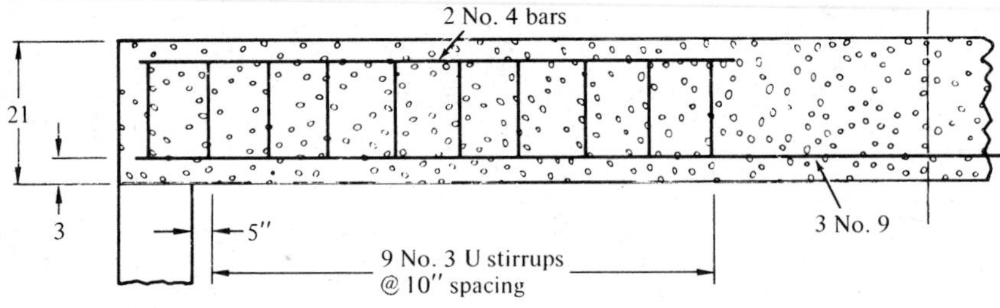

Fig. 181.

This beam proved to have about the minimum number of stirrups and did not illustrate the problem of variable spacing of stirrups, but it did illustrate the general procedure. The next problem will be more heavily loaded to produce higher shears.

EXAMPLE: STIRRUPS REQUIRED FOR CONTINUOUS BEAM

Diagrams for maximum moments and shears for a continuous beam of 24-ft span are shown in Fig. 182. The ultimate dead load is 2.1 klf and live load is 3.4 klf. The width of support is 1.33 ft, so the half-width is 0.67 ft. Use $f'_c = 4000$ psi and $f_y = 60{,}000$ psi.

We will select the required depth for bending moment and then investigate the shear stresses. Try a web width of $b_w = 16$ in. and determine the depth d re-

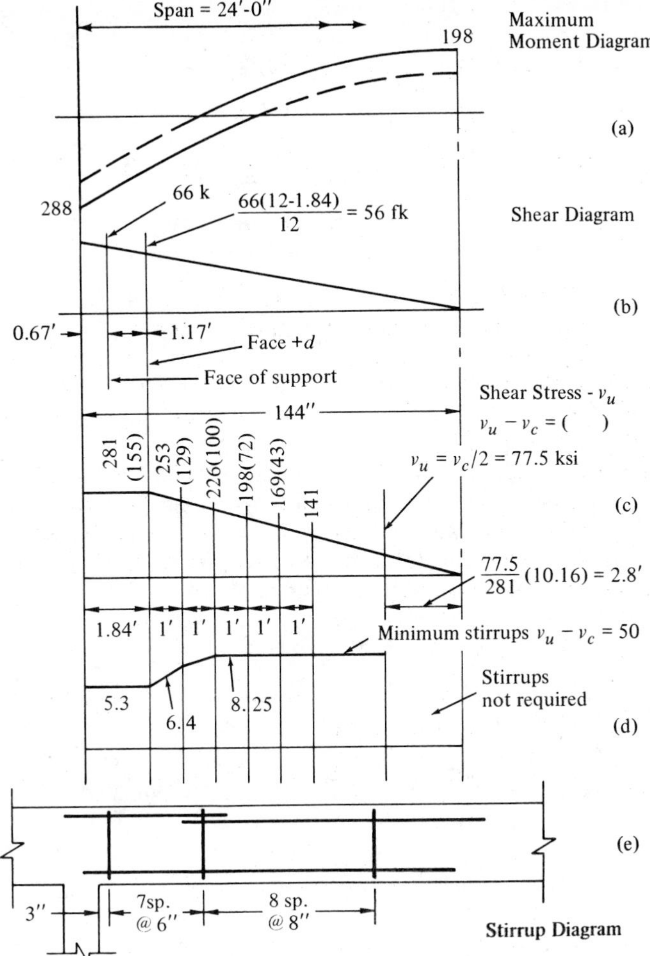

Fig. 182.

quired for the moment at the face of the support. For this value we will use an approximate formula: $M_f = M_c - Vt/3$, where M_c is the moment at the center of the support, V is the shear at the center of the support, and t is the width of the support. The moment at the face of the support is $M_f = 288 - 66(1.33)/3 =$ 259 ft-kip.

Now we must select an appropriate value of K. We will choose a value near to the maximum shown in Table 11 for $f_c' = 4000$ psi, $f_y = 60{,}000$ psi, say $K =$ 1000. We must correct M by the capacity reduction factor $M/0.9$, so $M =$ $259/0.9 = 288$ ft-kip. The required depth will be

$$d = \sqrt{M/b_w K} = \sqrt{288(12)(1000)/16(1000)} = 14.7 \text{ in.}$$

Now find the shearing stress at a distance d from the face of the support. The shear V is $V_u = 66(12 - 1.84)/12 = 56$ kip. Increase V by the capacity reduction factor, $\phi = 0.85$, for shear. $V_{ui} = V_u/0.85 = 64.7$, so that

$$v_u = V_{ui}/bd = 64.7(1000)/16.0(14.7) = 281 \text{ psi}$$

The shear that can be carried by the concrete is $v_c = 2\sqrt{f'_c} = 2\sqrt{4000} =$ 126 psi. The net stress is $v_u - v_c = 281 - 126 = 155$ psi. The value is well below the maximum $3.5\sqrt{f'_c}$, so our design to this point is satisfactory.

As in the previous problem, we will use No. 3 U stirrups with $A_v = 0.22$ in.^{2}, so the required spacing for $v = 155$ psi is

$$s = A_v f_y/(v_u - v_c)b_w = 0.22(60)(1000)/155(16) = 5.3 \text{ in.}$$

Another point we need to know is where stirrups are no longer needed and $v_u = v_c/2 = 155/2 = 77.5$ psi. By the ratio of this stress to the value of 281 calculated above, the distance from the centerline is 2.8 ft.

Our next task is to plot values of v, at one-foot intervals, so we can determine the stirrup spacing. These are shown in Fig. 182c. The calculations are not reproduced here. When we get to a value of $v_u - v_c = 50$ psi we must use minimum stirrups, which in this case require a spacing of one-half the effective depth d, or 8 in. The plot of required spacings is shown in Fig. 182d, and the spacing actually used is shown in Fig. 182e. If the depth had been greater, the spacings for the minimum could have been larger.

Moment and Shear Coefficients To avoid the laborious calculations necessary for determining the maximum bending moments in continuous beams for reinforced concrete design, coefficients have been developed for equal or nearly equal spans. The exterior support of the string of beams may be a knife-edged support, a column, a spandrel beam, or a girder. Coefficients are shown in Fig. 183 for both shear and moment. The value L_n is the clear span for positive moment or for shear, and the average of adjacent clear spans for negative moment. The larger of adjacent spans must not exceed the shorter by more than 20% and the unit live load must not exceed three times the dead load. All loads must be uniform.

Concrete-Column Design All reinforced concrete columns are required to resist a minimum bending moment due to an eccentric force with an eccentricity $e = 0.1t$, where t is the width of the column in the direction for which the principal moments might be expected. In addition, the effect of the length-to-width ratio must be evaluated, and in a long column, the slenderness ratio may require a reduction in the allowable capacity.

Disregarding the effects of slenderness, the relationship between the capacity P of a column in compression and the capacity M in moment can be defined by a characteristic interaction diagram, as shown in Fig. 184. The allowable values of P are plotted on the vertical axis and the corresponding values of M on the horizontal axis. For $M = 0$, then $P = P_0$ as indicated. As the moment increases,

Moments

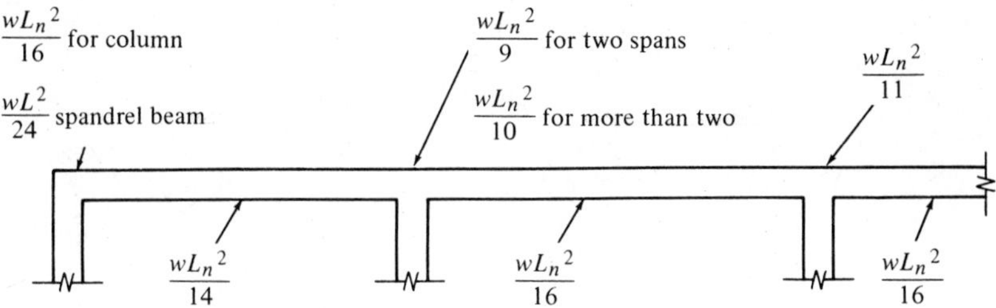

Other values are the same.

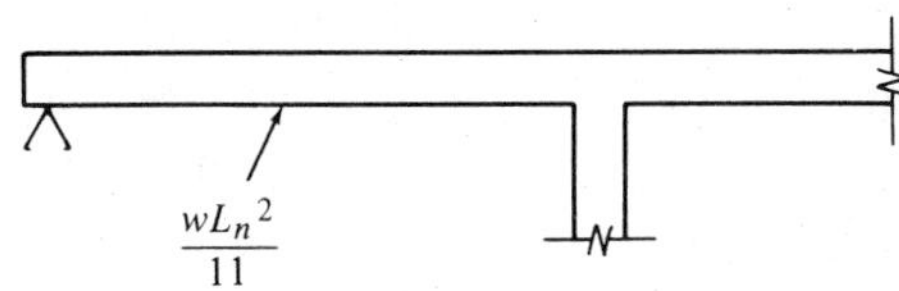

Shears

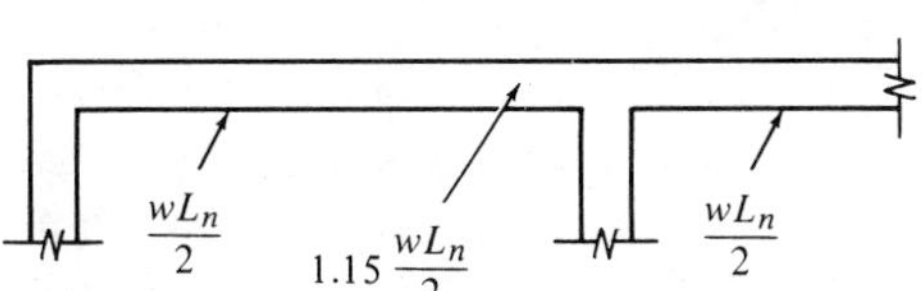

Fig. 183.

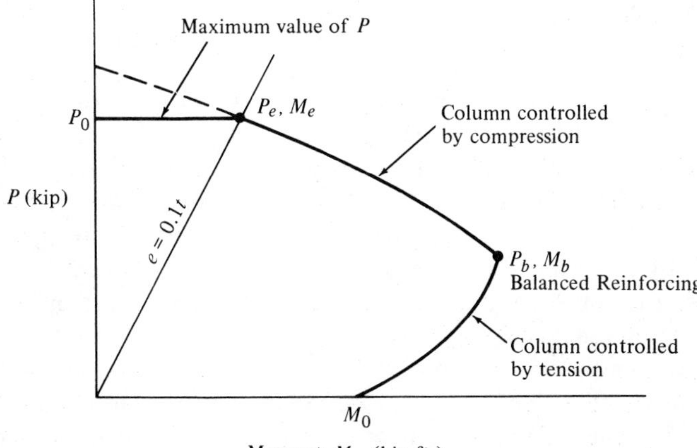

Fig. 184.

the value of P decreases almost, but not quite, linearly until the balanced condition (P_b, M_b) is reached. Again, as for beams, the balanced condition indicates that the strain in the concrete is 0.003 and the strain in the steel is f_y/E_s. Finally, as the compression force P is reduced, the moment capacity falls off until the capacity M_0 of the section with moment only is reached.

From P_0 to P_b the section is said to be controlled by compression, and below P_b it is said to be controlled by tension. These areas are indicated in Fig. 184. In this diagram a slanting line for $e = 0.1t$ is shown which intersects the interaction diagram at P_e, M_e. All columns must be reinforced for a minimum moment M_e.

In the calculations to follow, it is assumed that an adjustment has been made for the capacity-reduction factor, which for tied columns is $\phi = 0.70$.

There are minimum and maximum limits on the area of reinforcing in columns. The minimum is 1% of the gross area and the maximum is 8%. If the section is larger than required for considerations of loading, a reduced effective gross area, not less than one-half the total area, may be used to determine minimum steel area and load capacity. A minimum of four bars is required.

Ultimate Capacity, No Moment The value of P_0 in Fig. 184 is obtained from the ultimate capacity of the section for the concrete at a stress 0.85 f_c', and for a steel force of $A_s f_y$. Consider a column 14 by 14 in. with four No. 7 bars, A_s = 2.4 in.^{2}; $f_c' = 4$ ksi and $f_y = 60$ ksi. The capacity of the bars will be $A_s f_y$ = 2.4(60) = 144 kip. The capacity of the concrete will be the effective stress $0.85 f_c'$ times the net area deducting the area of steel: [14(14) - 2.4] 0.85 (4) = 658 kip. The total capacity is 144 + 658 = 802 kip. Later, when we involve moments of areas in the calculations, it will be more convenient to subtract the effective stress in the concrete, $0.85 f_c'$, from the stress in the steel, so that P_0 = 14(14)(4)(0.85) + [60 - (4)(0.85)] 2.4 = 802 kip. This may seem awkward here, but will save calculations later.

Balanced Reinforcing After the determination of the force P_0, the next important point on the interaction diagram, Fig. 184, is the force and moment capacity at balanced strains, P_b and M_b, also indicated. To do this, we establish the capacity of the steel and the capacity of the concrete, based on the allowable strains, and then find the resulting eccentricity for the forces thus determined. We will investigate the section for which P_0 has just been determined, which is a 14 by 14 in. column with a No. 7 bar in each corner. We will now need the location of these bars, which we will assume to be 3 in. from each edge.

Calculations are shown in Fig. 185. The first step is to draw the strain diagram and to determine the distance to the neutral axis, c = 6.51 in., and the value of the depth of the stress block, $a = 0.85c = 5.53$ in. (Remember that if f_c' is larger than 4 ksi, the value 0.85 must be reduced at the rate of 0.05 for each increment of 1 ksi). We are no longer limited by the requirement for members in bending

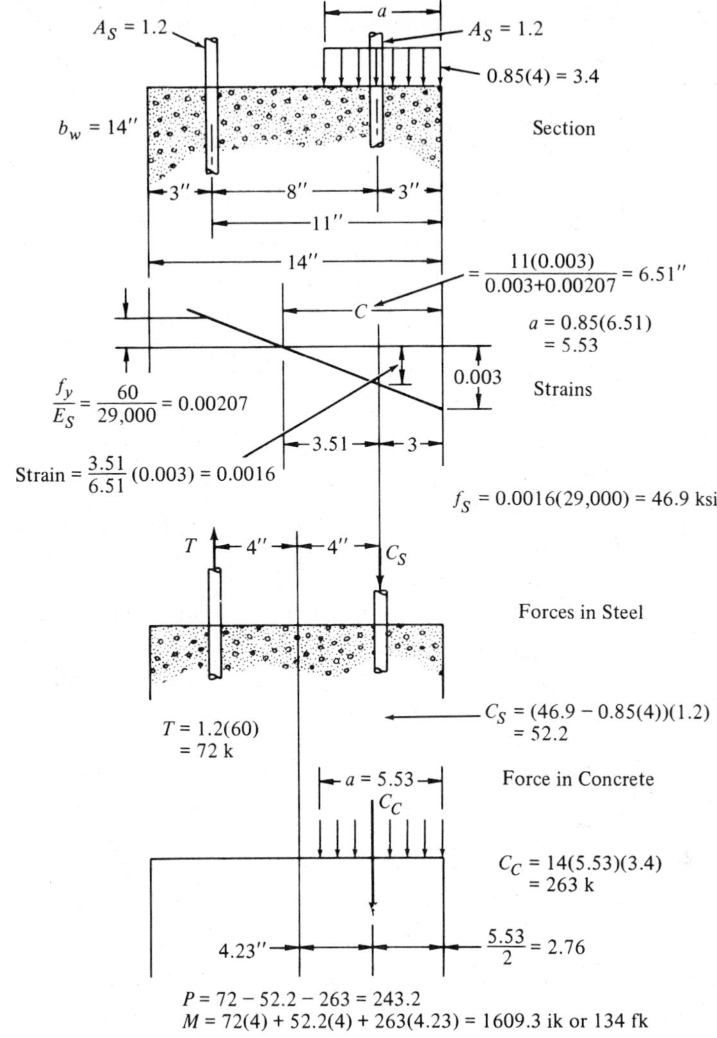

Fig. 185.

that a be smaller than 0.75 times the value for balanced reinforcement because we have furnished compressive reinforcement. Also shown are the strain in the compressive reinforcement (= 0.0016) and the resulting stress (= 46.9 ksi). If this latter value had been larger than 60 ksi, the lower value would have been used.

Now the forces in the steel and in the concrete are determined as shown in the figure. Note that the stress in the concrete must be deducted from the stress in the compressive steel to take account of the area of concrete deducted by the hole occupied by the steel. The sum of tension and compression forces is (taking tension to be positive)

$$P = 72 - 52.2 - 263 = -243.2 \text{ kip}$$

The moment of these forces about the centerline of the section is

$$M = 72(4) + 52.2(4) + 263(4.23) = 1609.3 \text{ in-kip} = 134 \text{ ft-kip}$$

Note that all of the moments are in the same direction. The final available capacity of the section must be reduced by the capacity-reduction factor which, for tied columns, is $\phi = 0.70$, so for design purposes, $P_b = 0.70(243.2) = 170.0$ kip and $M_u = 0.70(134) = 93.8$ ft-kip. Another useful piece of information is the eccentricity, the location of the resultant about the center of the section: $e = M/P = 93.8/170.0 = 0.552$ ft = 6.62 in.

Capacity of Section Controlled by Concrete We will now demonstrate how points on the interaction diagram (Fig. 184) in the "compression controls" area may be calculated. Again we will use the same 14 by 14 in. column. If compression controls, the strain in the tension steel must be below the 0.00207 used for the balanced condition. We will arbitrarily select a value of 0.75×0.00207 and find the resulting capacity in direct stress and moment.

Calculations are shown in Fig. 186, and the procedure is practically the same as before except for the value of the steel strains in both compression and tension. The compression force (not corrected for ϕ) is 296.4 kip, an increase of 53.2 kip, and the moment is reduced to 132.7 ft-kip from 134 ft-kip. The eccentricity is 5.37 in.

Another way to find a point on the interaction curve is to assume a value of the eccentricity and then determine the strain. However, this results in a quadratic equation which is inconvenient to solve, so the above method is superior unless the value of the eccentricity is actually given.

Capacity Controlled by Tension In the region of the interaction diagram where tension controls, strain in the steel will be larger than that for the balanced condition, but the size of the stress block will be smaller.

For this example, we will set the depth of the stress block a to be 0.75 times the depth of the block for balanced reinforcing, so that $a = 0.75(5.53) = 4.15$ in. The resulting calculations are shown in Fig. 187. The force P is 162.46 kip, i.e., 66 kip less than for the balanced condition, and M is 117.4 ft-kip, down from 134 ft-kip.

Design of Columns The above demonstrations of the capacity of columns are hardly suitable for routine calculation of concrete columns. This is often accomplished by tables of values or by charts. If these are not available, there is a

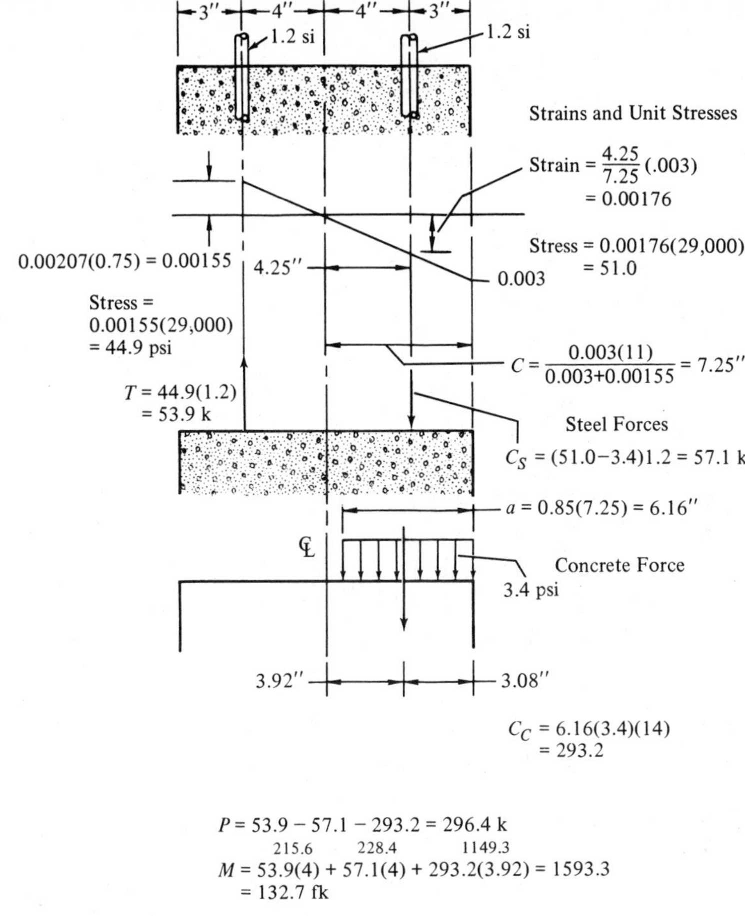

$P = 53.9 - 57.1 - 293.2 = 296.4$ k

215.6 228.4 1149.3

$M = 53.9(4) + 57.1(4) + 293.2(3.92) = 1593.3$

$= 132.7$ fk

$$e = \frac{1593.3}{296.4} = 5.37''$$

Fig. 186.

short cut permitted by the ACI *Code*. In the "compression controls" region of the interaction curve, Fig. 184, values may be assumed to vary as a straight line from P_0 to the point with the coordinates P_b, M_b. We need only calculate these two points; then intermediate values may be determined either graphically (by plotting), or by ratios of values.

In Fig. 188, the values of the 14 by 14 in. column are plotted. We wish to determine the value P_e for the minimum eccentricity $e = 0.1t$. First express M_e in terms of P_e. $M_e = 0.1tP_e = 0.1(14)P_e/12 = 0.1167P_e$.

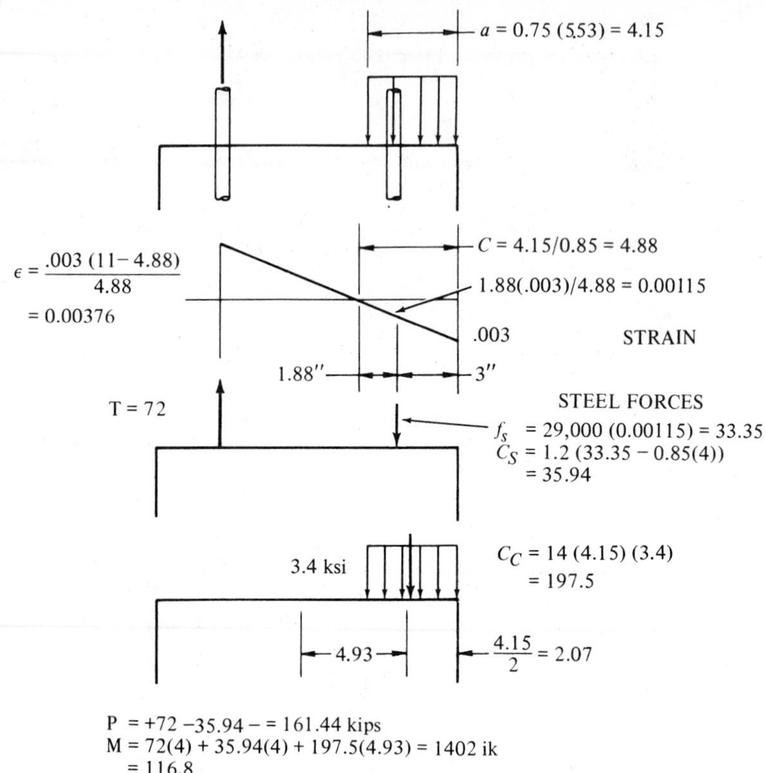

Fig. 187.

By similar triangles,

$$(P_0 - P_b)/(P_e - P_b) = M_b/(M_b - M_e)$$

$$(P_0 - P_b)(M_b - M_e) = M_b(P_e - P_b) = (802 - 243.2)(134 - 0.1167P_e)$$

$$= 134(P_e - 243.2)$$

$$P_e = 539.4 \text{ kip} \qquad \text{and} \qquad M_e = 539.4(0.1167) = 62.95 \text{ ft-kip}$$

These values agree with those plotted in Fig. 188. This diagram may be used for design if you can use a 14 by 14 column with four No. 7 bars. Otherwise, it is necessary to have a number of these charts for several sizes of columns and sets of reinforcing bars.

Tables for Concrete-Column Design The method of determining force and moment capacity in concrete columns described in this text is indirect, so that

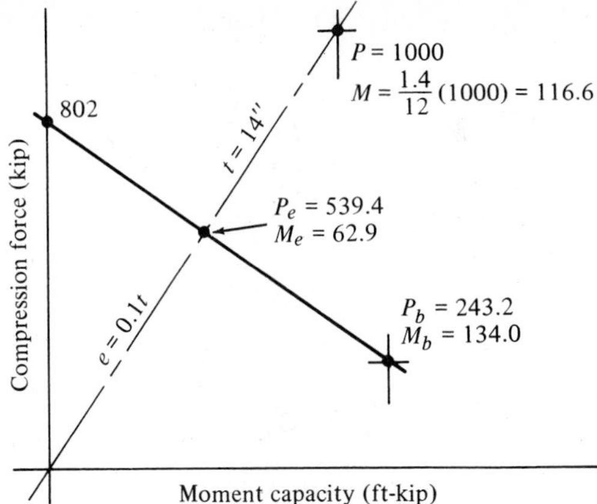

Fig. 188.

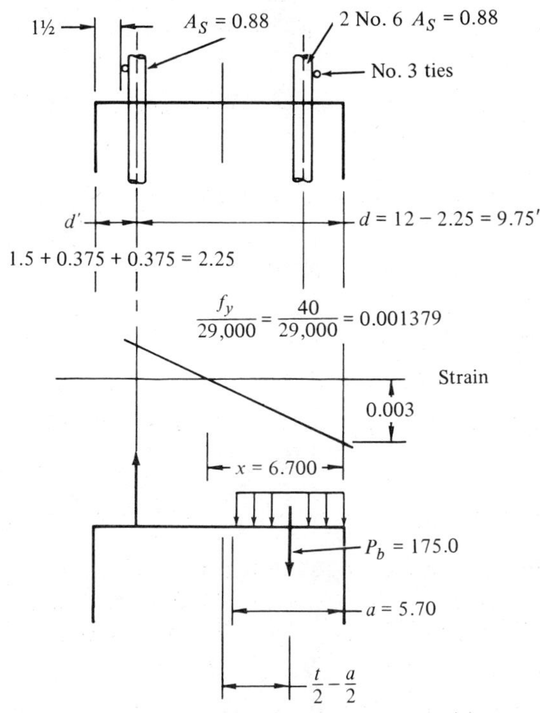

Fig. 189.

TABLE 12. CAPACITY OF CONCRETE COLUMNS.

CAPACITY OF CONCRETE COLUMNS

* VALUES DO NOT INCLUDE THE CAPACITY REDUCTION FACTOR
* RESULTS ARE IN FEET AND KIPS

CONCRETE STRESS = 3.000 STEEL STRESS = 40.

CONCRETE COVER = 1.5 INCHES OVER NO. 3 TIES

COLUMN SIZE, WIDTH =12.00 DEPTH = 12.00

BARS NO.	SIZE	AREA	P ZERO	P BAL	M BAL	E BAL	M ZERO
4	5	1.24	414.	178.	62.	0.350	20.
4	6	1.76	433.	177.	69.	0.389	27.
6	5	1.86	437.	178.	71.	0.397	29.
4	7	2.40	457.	176.	77.	0.437	36.
6	6	2.64	466.	177.	80.	0.455	40.
4	8	3.16	486.	175.	86.	0.492	46.
6	7	3.60	502.	176.	92.	0.525	52.
4	9	4.00	517.	174.	96.	0.551	56.
6	8	4.74	545.	175.	106.	0.606	65.

CONCRETE STRESS = 3.000 STEEL STRESS = 60.

CONCRETE COVER = 1.5 INCHES OVER NO. 3 TIES

COLUMN SIZE, WIDTH =12.00 DEPTH = 12.00

BARS NO.	SIZE	AREA	P ZERO	P BAL	M BAL	E BAL	M ZERO
4	5	1.24	438.	154.	70.	0.451	29.
4	6	1.76	468.	153.	79.	0.518	40.
6	5	1.86	474.	154.	82.	0.532	42.
4	7	2.40	505.	152.	91.	0.599	52.
6	6	2.64	519.	153.	97.	0.631	57.
4	8	3.16	549.	151.	105.	0.693	65.
6	7	3.60	574.	152.	114.	0.752	73.
4	9	4.00	597.	150.	119.	0.795	78.
6	8	4.74	640.	151.	135.	0.893	89.

TABLE 12. *(Continued)*

CAPACITY OF CONCRETE COLUMNS

* VALUES DO NOT INCLUDE THE CAPACITY REDUCTION FACTOR
* RESULTS ARE IN FEET AND KIPS

CONCRETE STRESS = 4.000 STEEL STRESS = 40.

CONCRETE COVER = 1.5 INCHES OVER NO. 3 TIES

COLUMN SIZE, WIDTH =12.00 DEPTH = 12.00

BARS NO.	SIZE	AREA	P ZERO	P BAL	M BAL	E BAL	M ZERO
4	5	1.24	535.	238.	78.	0.327	20.
4	6	1.76	554.	236.	84.	0.357	28.
6	5	1.86	558.	238.	86.	0.362	30.
4	7	2.40	577.	235.	92.	0.393	37.
6	6	2.64	586.	236.	96.	0.406	41.
4	8	3.16	605.	233.	101.	0.434	48.
6	7	3.60	621.	235.	108.	0.459	54.
4	9	4.00	636.	232.	111.	0.479	58.
6	8	4.74	663.	233.	121.	0.520	68.

CONCRETE STRESS = 4.000 STEEL STRESS = 60.

CONCRETE COVER = 1.5 INCHES OVER NO. 3 TIES

COLUMN SIZE, WIDTH =12.00 DEPTH = 12.00

BARS NO.	SIZE	AREA	P ZERO	P BAL	M BAL	E BAL	M ZERO
4	5	1.24	560.	205.	84.	0.411	30.
4	6	1.76	589.	204.	94.	0.462	41.
6	5	1.86	595.	205.	97.	0.472	43.
4	7	2.40	625.	203.	106.	0.522	54.
6	6	2.64	639.	204.	111.	0.547	59.
4	8	3.16	668.	201.	120.	0.593	68.
6	7	3.60	693.	203.	129.	0.637	77.
4	9	4.00	716.	200.	134.	0.670	83.
6	8	4.74	758.	201.	150.	0.743	96.

TABLE 12. *(Continued)*

CAPACITY OF CONCRETE COLUMNS

* VALUES DO NOT INCLUDE THE CAPACITY REDUCTION FACTOR
* RESULTS ARE IN FEET AND KIPS

CONCRETE STRESS = 3.000 STEEL STRESS = 40.

CONCRETE COVER = 1.5 INCHES OVER NO. 3 TIES

COLUMN SIZE, WIDTH =14.00 DEPTH = 14.00

BARS NO.	SIZE	AREA	P ZERO	P BAL	M BAL	E BAL	M ZERO
4	6	1.76	566.	248.	102.	0.410	34.
6	5	1.86	569.	249.	104.	0.416	36.
4	7	2.40	590.	247.	112.	0.453	45.
6	6	2.64	599.	248.	116.	0.469	49.
4	8	3.16	618.	246.	124.	0.503	58.
6	7	3.60	635.	247.	131.	0.532	65.
4	9	4.00	650.	244.	136.	0.558	71.
6	8	4.74	677.	246.	149.	0.606	83.
4	10	5.08	690.	243.	152.	0.626	87.
6	9	6.00	724.	244.	168.	0.687	101.

CONCRETE STRESS = 3.000 STEEL STRESS = 60.

CONCRETE COVER = 1.5 INCHES OVER NO. 3 TIES

COLUMN SIZE, WIDTH =14.00 DEPTH = 14.00

BARS NO.	SIZE	AREA	P ZERO	P BAL	M BAL	E BAL	M ZERO
4	6	1.76	601.	214.	115.	0.536	49.
6	5	1.86	607.	216.	118.	0.548	52.
4	7	2.40	638.	213.	130.	0.609	65.
6	6	2.64	651.	214.	137.	0.637	71.
4	8	3.16	681.	212.	147.	0.694	83.
6	7	3.60	707.	213.	159.	0.746	93.
4	9	4.00	730.	211.	166.	0.787	101.
6	8	4.74	772.	212.	185.	0.873	116.
4	10	5.08	792.	210.	190.	0.906	121.
6	9	6.00	844.	211.	214.	1.012	138.

TABLE 12. *(Continued)*

CAPACITY OF CONCRETE COLUMNS

* VALUES DO NOT INCLUDE THE CAPACITY REDUCTION FACTOR
* RESULTS ARE IN FEET AND KIPS

CONCRETE STRESS = 4.000 STEEL STRESS = 40.

CONCRETE COVER = 1.5 INCHES OVER NO. 3 TIES

COLUMN SIZE, WIDTH =14.00 DEPTH = 14.00

BARS NO.	SIZE	AREA	P ZERO	P BAL	M BAL	E BAL	M ZERO
4	6	1.76	731.	331.	126.	0.381	34.
6	5	1.86	734.	333.	128.	0.385	36.
4	7	2.40	754.	329.	136.	0.414	45.
6	6	2.64	763.	331.	141.	0.425	50.
4	8	3.16	782.	327.	148.	0.452	59.
6	7	3.60	798.	329.	156.	0.473	67.
4	9	4.00	813.	326.	160.	0.493	73.
6	8	4.74	840.	327.	173.	0.529	85.
4	10	5.08	852.	324.	176.	0.545	90.
6	9	6.00	886.	326.	192.	0.590	105.

CONCRETE STRESS = 4.000 STEEL STRESS = 60.

CONCRETE COVER = 1.5 INCHES OVER NO. 3 TIES

COLUMN SIZE, WIDTH =14.00 DEPTH = 14.00

BARS NO.	SIZE	AREA	P ZERO	P BAL	M BAL	E BAL	M ZERO
4	6	1.76	766.	286.	139.	0.485	50.
6	5	1.86	772.	287.	142.	0.494	53.
4	7	2.40	802.	284.	154.	0.540	67.
6	6	2.64	816.	286.	160.	0.561	73.
4	8	3.16	845.	283.	171.	0.605	85.
6	7	3.60	870.	284.	183.	0.643	97.
4	9	4.00	893.	281.	190.	0.675	105.
6	8	4.74	935.	283.	209.	0.739	122.
4	10	5.08	954.	280.	214.	0.764	128.
6	9	6.00	1006.	281.	237.	0.844	148.

TABLE 12. *(Continued)*

CAPACITY OF CONCRETE COLUMNS

* VALUES DO NOT INCLUDE THE CAPACITY REDUCTION FACTOR
* RESULTS ARE IN FEET AND KIPS

CONCRETE STRESS = 3.000 STEEL STRESS = 40.

CONCRETE COVER = 1.5 INCHES OVER NO. 3 TIES

COLUMN SIZE, WIDTH =16.00 DEPTH = 16.00

BARS NO.	SIZE	AREA	P ZERO	P BAL	M BAL	E BAL	M ZERO
4	7	2.40	743.	330.	156.	0.473	53.
6	6	2.64	752.	331.	161.	0.486	58.
4	8	3.16	771.	328.	170.	0.518	69.
6	7	3.60	788.	330.	179.	0.544	78.
4	9	4.00	803.	327.	185.	0.568	85.
6	8	4.74	830.	328.	201.	0.611	100.
4	10	5.08	843.	325.	205.	0.631	105.
6	9	6.00	877.	327.	224.	0.685	123.
8	9	8.00	952.	327.	262.	0.802	157.
6	10	7.62	938.	325.	253.	0.779	150.
6	11	9.36	1003.	324.	284.	0.878	177.

CONCRETE STRESS = 3.000 STEEL STRESS = 60.

COLUMN SIZE, WIDTH =16.00 DEPTH = 16.00

BARS NO.	SIZE	AREA	P ZERO	P BAL	M BAL	E BAL	M ZERO
4	7	2.40	791.	285.	178.	0.623	78.
6	6	2.64	804.	286.	186.	0.648	86.
4	8	3.16	834.	284.	199.	0.701	100.
6	7	3.60	860.	285.	213.	0.747	113.
4	9	4.00	883.	282.	222.	0.786	123.
6	8	4.74	925.	284.	245.	0.863	143.
4	10	5.08	945.	281.	251.	0.894	150.
6	9	6.00	997.	282.	279.	0.990	173.
8	9	8.00	1112.	282.	337.	1.193	216.
6	10	7.62	1091.	281.	323.	1.151	207.
6	11	9.36	1191.	280.	370.	1.322	238.

TABLE 12. *(Continued)*

CAPACITY OF CONCRETE COLUMNS

* VALUES DO NOT INCLUDE THE CAPACITY REDUCTION FACTOR
* RESULTS ARE IN FEET AND KIPS

CONCRETE STRESS = 4.000 STEEL STRESS = 40.

CONCRETE COVER = 1.5 INCHES OVER NO. 3 TIES

COLUMN SIZE, WIDTH =16.00 DEPTH = 16.00

BARS NO.	SIZE	AREA	P ZERO	P BAL	M BAL	E BAL	M ZERO
4	7	2.40	958.	440.	192.	0.437	54.
6	6	2.64	967.	441.	197.	0.447	59.
4	8	3.16	986.	438.	206.	0.471	70.
6	7	3.60	1002.	440.	216.	0.490	79.
4	9	4.00	1017.	436.	222.	0.509	87.
6	8	4.74	1044.	438.	237.	0.541	102.
4	10	5.08	1056.	434.	241.	0.557	108.
6	9	6.00	1090.	436.	260.	0.597	126.
8	9	8.00	1163.	436.	298.	0.685	164.
6	10	7.62	1149.	434.	290.	0.668	156.
6	11	9.36	1213.	432.	321.	0.743	186.

CONCRETE STRESS = 4.000 STEEL STRESS = 60.

COLUMN SIZE, WIDTH =16.00 DEPTH = 16.00

BARS NO.	SIZE	AREA	P ZERO	P BAL	M BAL	E BAL	M ZERO
4	7	2.40	1006.	380.	213.	0.562	79.
6	6	2.64	1020.	381.	221.	0.580	87.
4	8	3.16	1049.	378.	234.	0.620	102.
6	7	3.60	1074.	380.	248.	0.654	116.
4	9	4.00	1097.	376.	257.	0.684	126.
6	8	4.74	1139.	378.	280.	0.742	148.
4	10	5.08	1158.	375.	287.	0.765	156.
6	9	6.00	1210.	376.	315.	0.837	181.
8	9	8.00	1323.	376.	372.	0.990	231.
6	10	7.62	1302.	375.	359.	0.958	221.
6	11	9.36	1400.	373.	405.	1.087	258.

TABLE 12. *(Continued)*

CAPACITY OF CONCRETE COLUMNS

* VALUES DO NOT INCLUDE THE CAPACITY REDUCTION FACTOR
* RESULTS ARE IN FEET AND KIPS

CONCRETE STRESS = 3.000 STEEL STRESS = 40.

CONCRETE COVER = 1.5 INCHES OVER NO. 3 TIES

COLUMN SIZE, WIDTH =18.00 DEPTH = 18.00

BARS NO.	SIZE	AREA	P ZERO	P BAL	M BAL	E BAL	M ZERO
4	8	3.16	945.	423.	227.	0.536	80.
6	7	3.60	961.	424.	237.	0.559	91.
4	9	4.00	976.	421.	245.	0.582	99.
6	8	4.74	1004.	423.	262.	0.621	117.
4	10	5.08	1016.	419.	268.	0.639	123.
6	9	6.00	1051.	421.	290.	0.689	144.
8	9	8.00	1126.	421.	335.	0.795	187.
6	10	7.62	1112.	419.	325.	0.775	178.
6	11	9.36	1177.	418.	362.	0.866	212.
8	10	10.16	1207.	419.	381.	0.910	228.
8	11	12.48	1294.	418.	430.	1.031	268.

CONCRETE STRESS = 3.000 STEEL STRESS = 60.

COLUMN SIZE, WIDTH =18.00 DEPTH = 18.00

BARS NO.	SIZE	AREA	P ZERO	P BAL	M BAL	E BAL	M ZERO
4	8	3.16	1008.	365.	260.	0.713	117.
6	7	3.60	1033.	367.	277.	0.755	132.
4	9	4.00	1056.	364.	288.	0.791	144.
6	8	4.74	1099.	365.	314.	0.861	169.
4	10	5.08	1118.	362.	322.	0.890	178.
6	9	6.00	1171.	364.	355.	0.977	207.
8	9	8.00	1286.	364.	423.	1.162	263.
6	10	7.62	1264.	362.	407.	1.125	251.
6	11	9.36	1364.	361.	462.	1.282	294.
8	10	10.16	1410.	362.	492.	1.359	314.
8	11	12.48	1543.	361.	566.	1.568	360.

TABLE 12. *(Continued)*

CAPACITY OF CONCRETE COLUMNS

* VALUES DO NOT INCLUDE THE CAPACITY REDUCTION FACTOR
* RESULTS ARE IN FEET AND KIPS

CONCRETE STRESS = 4.000 STEEL STRESS = 40.

CONCRETE COVER = 1.5 INCHES OVER NO. 3 TIES

COLUMN SIZE, WIDTH =18.00 DEPTH = 18.00

BARS NO.	SIZE	AREA	P ZERO	P BAL	M BAL	E BAL	M ZERO
4	8	3.16	1217.	563.	278.	0.494	81.
6	7	3.60	1233.	566.	289.	0.511	92.
4	9	4.00	1248.	561.	296.	0.528	101.
6	8	4.74	1275.	563.	314.	0.557	119.
4	10	5.08	1288.	559.	320.	0.572	126.
6	9	6.00	1321.	561.	341.	0.608	148.
8	9	8.00	1394.	561.	386.	0.689	193.
6	10	7.62	1380.	559.	376.	0.673	183.
6	11	9.36	1444.	557.	413.	0.742	220.
8	10	10.16	1473.	559.	433.	0.775	238.
8	11	12.48	1558.	557.	482.	0.866	283.

CONCRETE STRESS = 4.000 STEEL STRESS = 60.

COLUMN SIZE, WIDTH =18.00 DEPTH = 18.00

BARS NO.	SIZE	AREA	P ZERO	P BAL	M BAL	E BAL	M ZERO
4	8	3.16	1280.	487.	311.	0.640	119.
6	7	3.60	1305.	489.	328.	0.670	135.
4	9	4.00	1328.	485.	339.	0.698	148.
6	8	4.74	1370.	487.	365.	0.750	174.
4	10	5.08	1389.	483.	373.	0.773	183.
6	9	6.00	1441.	485.	406.	0.837	214.
8	9	8.00	1554.	485.	474.	0.977	276.
6	10	7.62	1533.	483.	458.	0.949	263.
6	11	9.36	1631.	481.	513.	1.067	312.
8	10	10.16	1677.	483.	543.	1.125	335.
8	11	12.48	1808.	481.	617.	1.282	392.

Values do not include the capacity reduction factor. Results are in feet and kips.

tables or charts for determining the capacity of columns become necessary. We will show a simplified method and present a set of tables that can be checked with very little extra effort. In Table 12 are shown a series of column sections with the number and size of bars and corresponding values of P_0, P_b, M_b, e_b, and the moment M_0 for the condition of no external force P. Properties are given for several combinations of f_c' and f_y. From these values, charts are constructed for design as shown in Fig. 190. The computer program used to prepare these charts is given in Fig. 191.

It is assumed that all reinforcement is symmetrical, in single layers parallel to the axis of bending. We will neglect the small part of the area of concrete displaced by the steel in the computation of moment values.

We will check one of the entries in Table 12, say a 12 by 12 column, with four No. 6 bars, using $f_c' = 3$ ksi and $f_y = 40$ ksi. The concrete cover is 1.5 in. over No. 3 ties, so d' is 1.5 + 0.375 + 0.375 = 2.25 in. A section through the column showing the strain diagram and the stress block is shown in Fig. 189.

The value of P_0 is the total capacity of the section:

$$P_0 = 0.85(bt - 2A_s)f_c' + 2A_sf_y = 0.85(3)(12(12) - 2(0.88)) + 2(0.88)(40) = 433 \text{ kip}$$

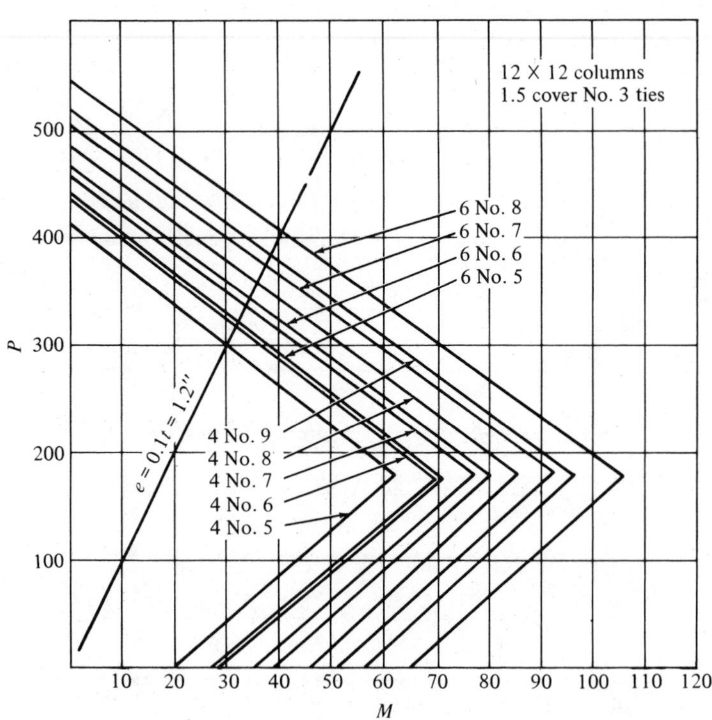

Fig. 190.

```
C           PROGRAM FOR CONCRETE COLUMN TABLES
C           M. S. KETCHUM, AUGUST 1971
C
        WRITE(6,101)
  101 FORMAT(1H1,///)
        WRITE(6,102)
  102 FORMAT(4X,'CAPACITY OF CONCRETE COLUMNS',/,1H0, '* VALUES DO NOT I
     1NCLUDE THE CAPACITY REDUCTION FACTOR',/,' * RESULTS ARE IN FEET AN
     2D KIPS',/)
        READ(5,104) FCU, FY
  104 FORMAT(2F10.0)
        WRITE(6,108) FCU, FY
  108 FORMAT(1H ,' CONCRETE STRESS =', F6.3,'    STEEL STRESS = ', F4.0,
     1//,'  CONCRETE COVER = 1.5 INCHES OVER NO. 3 TIES',/ )
        COV = 1.5
        READ(5,104) B,T
        WRITE(6,109) B,T
  109 FORMAT('    COLUMN SIZE, WIDTH ='F5.2,'    DEPTH = ' F5.2 )
        WRITE(6,103)
  103 FORMAT(1H0 4X 4HBARS 8X 4HAREA 2X 6HP ZERO 2X 5HP BAL   2X   5HM BAL
     2 2X   5HE BAL 1X   6HM ZERO,/, 4X   3HNO. 1X   4HSIZE)
    1 READ(5,105,END=40) NN, NS, AS
  105 FORMAT(2I4, F8.3)
        PO  = 0.85*FCU*(B*T-AS) + AS*FY
        RS  = NS
        RS  = RS/16.
        DP  = COV + RS + 0.1875
        D = T - DP
        A = 0.85*0.003*D/(0.003 + FY/29000.)
        PB  = A*B*0.85*FCU
        AMB = AS*FY*(T/2. - DP) + PB*(T-A)/2.
        EB  = AMB/(12.*PB)
        TEN = AS*FY/2.
        A = TEN/(B*0.85*FCU)
        AMO = TEN*(D-A/2.)/12.
        AMB = AMB/12.
        WRITE(6,106) NN, NS, AS, PO, PB, AMB, EB, AMO
  106 FORMAT(1H0, 2I5, 3X, F7.2, 3F7.0, F7.3, F7.0)
        GO TO 1
   40 WRITE(6,107)
  107 FORMAT(1H1)
        CALL EXIT
        END
```

Fig. 191.

The distance to the neutral axis is based on the strain diagram:

$$x = \frac{0.003d}{(f_y/29{,}000) + 0.003} = \frac{0.003(9.75)}{(40/29{,}000) + 0.003} = 6.700$$

The depth of the stress block (provided f_c' is not greater than 4 ksi) is $a = 0.85x = 0.85(6.700) = 5.70$ in.

In determining P_b, we can neglect the area of steel because the tension and compression will cancel:

$$P_b = 0.85(3)(5.70)(12) = 175.0 \text{ kip}$$

The moment M_b for the balanced condition is

$$M_b = P_b\,[(t/2) - (a/2)] + 2A_s f_y\,[(t/2) - d'] = 175.0\,[6 - (5.70/2)]$$
$$+\, 2(0.88)(40)(6 - 2.25) = 816.0 \text{ in. kip} = 68.0 \text{ ft-kip}$$

The eccentricity is M_b/P_b:

$$e_b = 68.0/175.0 = 0.388 \text{ ft}$$

To calculate the moment for zero direct force, we must go back to the section on design for bending only. The tensile force must equal the compression force (we will neglect the force of the compression steel):

$$T = A_s f_y = 0.88(40) = 35.2 \text{ kip}$$

$$C = T = ab(0.85)f'_c = a(12)(0.85)(3.0) = 35.2, \qquad \text{so that } a = 1.15 \text{ in.}$$

The moment that this system will furnish is

$$M_0 = C[d - (a/2)] = 35.2[9.75 - (1.15/2)] = 323.0 \text{ in-kip} = 27.0 \text{ ft-kip}$$

A chart plotted for 12-by-12-in. columns for f'_c = 3 ksi and f_y = 40 ksi is shown in Fig. 190. All values are shown in Table 12, and the computer program in Fig. 191. On the chart, the capacities are plotted as straight lines between P_0, P_b, and M_b and between P_b, M_b, and M_0. The minimum eccentricity is shown by the slanting line, $e = 0.1t = 1.2$ in.

A 12 by 12 column with a load P = 300 kip and a moment M = 50 ft-kip will require four No. 8 bars. A column with P = 350 and M = 30 must actually be designed for $M = P(0.1)t = 350(1.2)/12 = 35$ ft-kip, so will need the same four No. 8 bars.

Design of Two-Way Rectangular Concrete Footings A typical two-way concrete footing is shown in Fig. 192. The column may be either concrete or steel. The base is usually square or rectangular as necessary to fit space requirements. Large footings may have a step to increase the thickness under the column. The reinforcement is placed in the bottom on chairs or blocks to maintain the required cover of 3 in. Straight bars are used at a uniform spacing. The base is connected to the column with L-shaped bars called dowels which must develop the strength of the column bars. The footing must be strong enough to resist both bending moment and shear across critical sections.

The requirements of the ACI *Code*, with respect to design for bending moments, are as follows:

1. The critical section for moment is obtained by passing a plane completely through the footing in accordance with the following rules:
   a. at the face of the column for concrete columns (or wall for one-way footings);
   b. halfway between the middle and edge of the wall for masonry walls;
   c. halfway between the face of the column and edge of the steel base for footings under steel base plates.
2. In one-way footings and in square footings, the reinforcement shall be distributed evenly across the section for the full width of the footing.

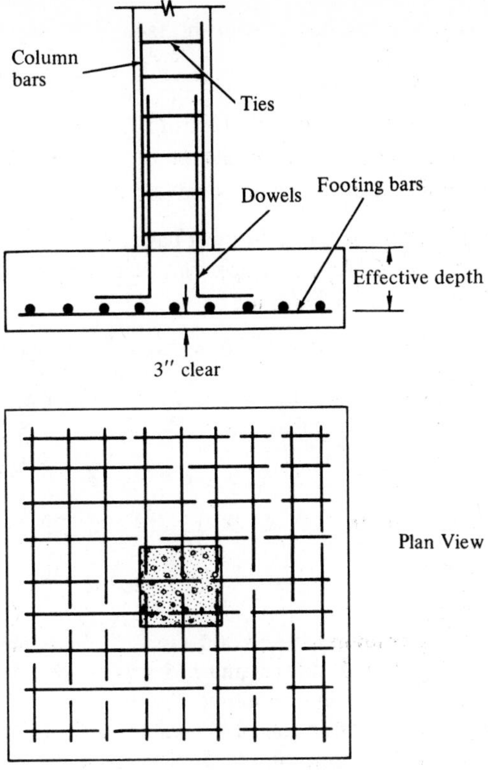

Fig. 192.

3. In two-way rectangular footings, the reinforcement in the long direction shall be distributed uniformly across the full width. In the short direction it shall be centered in a band of width B equal to the length of the short side of the footing (see Fig. 193). The reinforcement in this band shall be

$$\text{Reinforcing in band } B = \frac{2(\text{Total reinforcement})}{S + 1}$$

where S is the ratio of the long side to the short side. For example, if the ratio of long side to short side is 1.5, the reinforcement in band B will be $2/(1.5 + 1) = 0.8$ of the total. The remainder is placed in the outer bands.

4. The design for reinforcement and thickness of concrete follows that for beam design. The requirements of the ACI *Code* with respect to shear are:

a. Shear reinforcement is not mandatory for footings.

b. The footing is to be considered as follows:

1. the footing is a wide rectangular beam, and the shearing stress is

checked at a distance d from the face of the support as for regular beams; or

2. the shearing stress is computed on the planes of a rectangular box which is at a distance of $d/2$ from the edge of the column: the maximum shear in this case must not be greater than $v_c = 4\sqrt{f'_c}$, unless reinforcing is provided.

An additional requirement is that bars must have a proper development length in accordance with the ACI *Code*, which specifies that, for No. 11 bars or smaller, the development length L_d in inches must be

$$L_d = 0.04 a_s f_y / \sqrt{f'_c} \qquad \text{where} \qquad a_s = \text{area on an individual bar}$$

For the set of stresses $f_y = 40{,}000$ psi and $f'_c = 3{,}000$ psi, the development length of bars from No. 3 to No. 11 is:

Size	3	4	5	6	7	8	9	10	11
Development length	3	6	9	13	18	23	29	37	46

EXAMPLE

Design a footing for a column load of 300 kip. The maximum allowable soil pressure of 5.0 ksf. Use $f'_c = 3{,}000$ psi and $f_y = 40{,}000$ psi. The column size is 16 by 16 in. The maximum dimension in one direction is 7 ft.

The required area of the footing for an allowable soil pressure of $p = 5$ ksf is $A = P/p = 300/5 = 60$ ft.^{2} If one of the dimensions must be 7 ft, the other dimension must be $60/7 = 8.57$ ft. Use a footing 7 ft by 9 ft. The pressure is $p = P/A = 300/7(9) = 4.76$ ksf. A sketch of the footing is shown in Fig. 193.

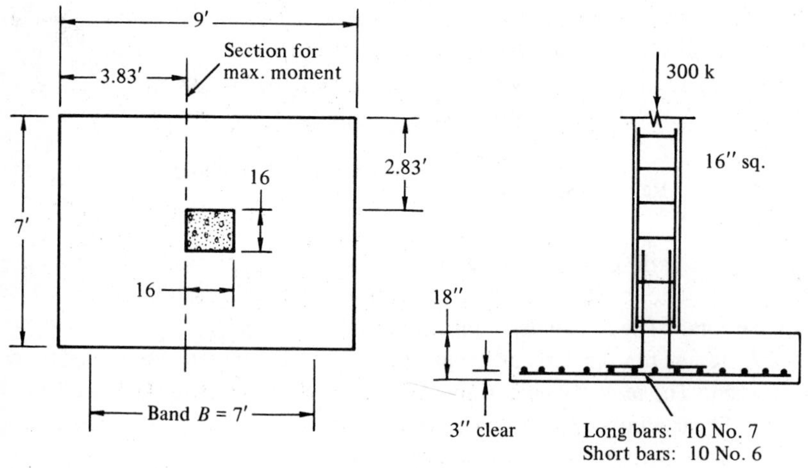

Fig. 193.

There are several ways to approach this problem. It is possible to write equations to find the depth directly, but we will use a trial-and-error approach and start with a depth required for moment and then check for shear.

The moment at the face of the column for the longest span is

$$M = wL^2/2 = 7(4.76)(3.83^2)/2 = 244.4 \text{ ft-kip}$$

With a capacity-reduction factor for moment of $\phi = 0.9$, $M = 244.4/0.9 = 271.5$ ft-kip. We will choose a depth based on a conservative value of the thickness coefficient, $K = 400$; then the effective depth will be $d = \sqrt{M/bK}$, and with $b = 7$ ft, $d = \sqrt{271.5(12)(1000)/7(12)(400)} = 9.8$ in. Use $d = 10$ in.

The distance from the edge of the base to the face of the column is 3.83 ft and the shear at a distance $d/2$ from the face of the column is

$$V = (3.83 - 5/12)(7)(4.76) = 113.7 \text{ kip}$$

With a capacity-reduction factor of $\phi = 0.85$, V is equal to $113.7/0.85 = 133.8$, and the shearing stress is $v_u = V/bd = 133{,}800/7(12)(10) = 159$ psi. The maximum is $f_c = 2\sqrt{f'_c} = 2\sqrt{3000} = 109.5$ psi. The depth must be greater to satisfy the shear requirement. If the location of the section did not change, the value of d would be, by proportion, $d = 10(159)/109.5 = 14.5$ in.; but the shear will be reduced by the change in location of the section, so we will try a depth of 14 in. The shear at $d/2$ from the face of the column is

$$V = (3.83 - 7/23)7(4.76)/0.85 = 127.4 \text{ kip}$$

$$v_u = 127{,}4000/7(12)(14) = 108.3$$

so the shearing stress is satisfactory.

The shear around the column must now be checked. The shear on the face of the box is equal to the total force on the footing less the pressure on the horizontal area within the box. The plan dimension of the square box is 16 + 14 = 30 in. = 2.5 ft, so that

$$V = 300 - 4.76(2.5^2) = 270.3 \text{ kip}$$

and with the capacity-reduction factor, $V = 270.3/0.85 = 317.9$ kip. The perimeter of the box is 30(4) = 120 in. and the shearing stress v_u is

$$v_u = V/bd = 317.9(1000)/120(14) = 189.3 \text{ psi}$$

The allowable is $4\sqrt{f'_c} = 4\sqrt{3000} = 219$ psi, so the section is satisfactory.

We must now determine the reinforcing. For the long (3.83-ft) span, we have already determined that $M = 271.5$ ft-kip. We now redetermine K based on a new depth, $d = 14$ in., and from this we can find the value of j.

$$K = M/bd^2 = 271{,}500(12)/7.0(12)(14^2) = 197.9$$

$$j = 0.5 + \sqrt{0.25 - (K/1.7f'_c)} = 0.5 + \sqrt{0.25 - [197.9/1.7(3000)]} = 0.95$$

The area of steel required is

$$A_s = M/jdf_y = 271.5(12)/0.95(14)(40) = 6.12 \text{ in.}^2$$

Use 10 No. 7 bars. The development length, $L_d = 18$ in., is much less than the clear span, $L = 3.83(12) = 46$ in.

For the short span (the 9-ft-wide section), the clear span is 2.83 ft and the total moment is

$$M = 4.76(2.83^2)(9)/2 = 171.5 \text{ ft-kip}$$

and the design moment is

$$M = 171.5/0.9 = 190.6 \text{ ft-kip}$$

The portion that is distributed to a band of width $B = 7$ ft is

$$2/(S + 1) = 2/(9/7 + 1) = 0.87$$

We will use the same value of the effective depth jd, even though j will be larger with the lower moment. We then can find the area of steel by proportion with the previous area:

$$A_s = 6.12(190.6)/271.5 = 4.29 \text{ in.}^2$$

This will require 10 No. 6 bars, of which 8 should be evenly spaced in the 7-ft middle band. The development length for a No. 6 bar is 13 in.; therefore, it is satisfactory. A sketch of the footing is shown in Fig. 193. The dowels will depend on the column design. The total depth should be 14 + 3 + 0.875/2 = 17.4 in., but a depth of 18 in. is shown for simplicity.

DESIGN OF STRUCTURES

Design of Steel Structures

This subsection will describe some typical steel structural systems and will illustrate the design problems and the resulting calculations for several of these structures. We will use the methods of analysis of stresses and forces in structures and the methods of design of structural elements that were developed in the previous chapters and, in addition, devise special methods for each structure.

The types of structures described will be industrial buildings, steel office and apartment buildings, and arches for long-span construction. Complete calculations for each structure would require much more space than is available in this text, so only the most significant and controlling structural elements will be investigated. The results will be summarized in small-scale preliminary drawings which show the sizes of members and the general dimensions selected.

In these structures, no attempt has been made to obtain the best design from the point of view of cost and, in some cases, additional study may be suggested to solve unresolved problems.

Types of Structural Systems–Industrial Buildings There are many possible structural systems and only a few can be illustrated here. We will start with the most common and discuss their relative advantages.

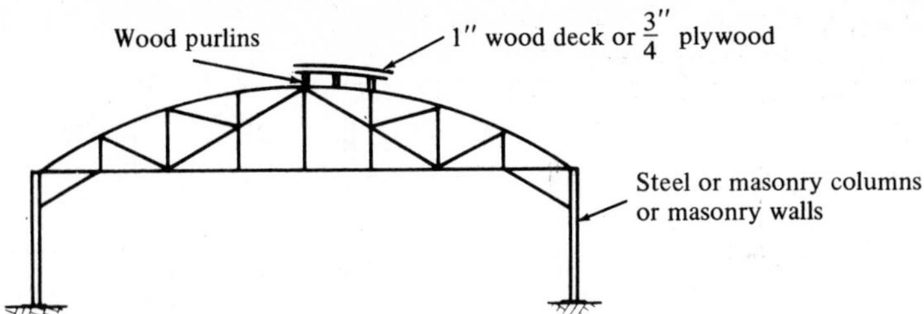

Fig. 194. Bow-string trusses.

Figure 194–Bow-String Trusses The bow string is probably the least expensive industrial building type available and these trusses are produced by almost every steel fabricator. The ratio of height to span is usually 1 to 8. The knee braces shown may be eliminated if the columns are fixed at their bases or the lateral loads are transferred to the ends of the building. Spans may be from 40 to 80 ft, and spacing from 12 to 20 ft. If 2-in. wood deck is used, the purlin spacing will be from 5 to 8 ft. The 2-in. deck may be hard to bend for short truss spans with considerable curvature of the top chord of the truss.

Figure 195–Typical Beam-and-Column Framing In this sketch, the roof beams frame into the columns. The $1\frac{1}{2}$-in. metal deck spans from 5 to 8 ft, and is virtually a substitute for 2-in. wood deck. This type of framing can easily

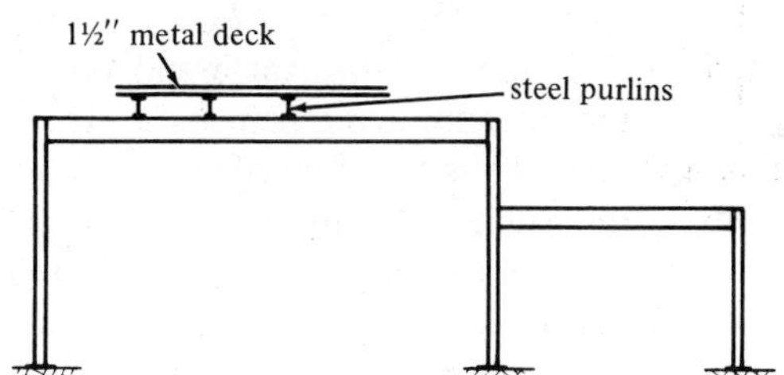

Fig. 195. Beam and column framing.

accommodate changes in level of the roof. For elastic design, the beams are normally considered as simple spans and the lateral loads are carried by the special connections, which are assumed to carry the lateral forces only. Bent spacing is usually from 16 to 25 ft. The purlins may either rest on top of the girders, as shown, or be framed into the web to reduce the height of the structure.

Figure 196–Continuous Articulated Framing Considerable structural efficiency and lower cost may be obtained if the beams are made to run over the tops of columns and hinges are placed at locations which will make the moments equal over the supports and at the centers of spans. A further refinement is to use additional plates, top and bottom, over the columns, but these have the disad-

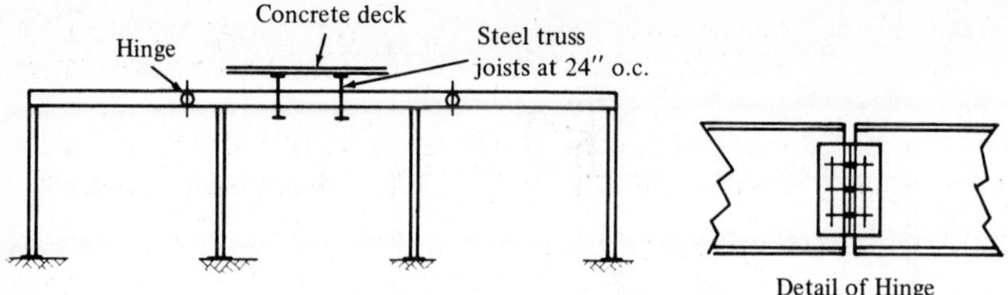

Fig. 196. Continuous articulated framing.

vantage that they raise the level of the roof at the column and cause a bump. The hinge detail shown is made of two standard connections. If only web plates are used, there may be some slip between the faces of the joint. The advantage of this system is that the structural efficiency is good and moment splices are not required at the joints as in continuous construction, either elastic or plastic. A disadvantage is that deflections may be much greater than for continuous beams.

It is quite a task to select the location of the hinges to obtain maximum efficiency and yet keep the structure statically determinate. It is possible to have a system in which, although the system is stable, the failure of one element would cause failure of a long string of other members. For example, this might happen if a hinge is placed at the same place in each beam.

Normally, the columns are not made continuous with the frame. Spans for this type are from 20 to 50 ft, and bay spacings are 20 to 30 ft.

Figure 197–Continuous Framing (Plastic or Elastic) Continuous framing is similar to the articulated framing except that the beams are spliced for moment. Unless plastic design is used, or reinforcing plates are used over columns, the efficiency of continuous beams is less than for articulated beams. The deflection characteristics, however, are better. The most severe condition for deflection is for the possibility of ponding on a flat roof due to rain. This may be especially severe if the roof is designed for a low live load (e.g., 12 psf in California).

It may be necessary to change sizes of beams if spans vary. This should be done at a point off the column at the point of contraflexure. If a hinge is used

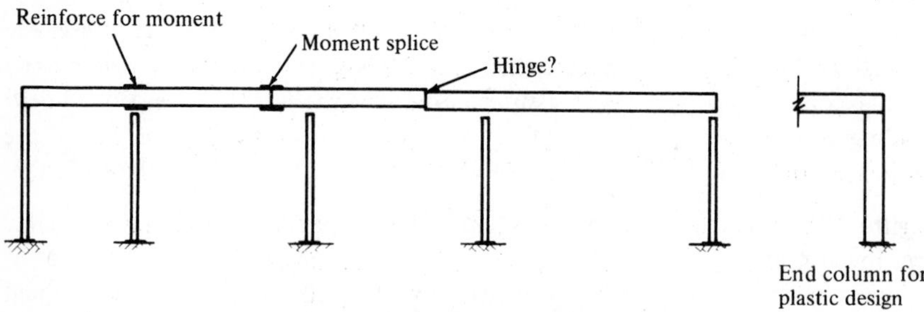

Fig. 197. Continuous framing.

then this should be taken into account in the analysis by appropriate stiffness and carry-over factors, and by fixed-end moments if the moment-distribution method is used.

Plastic design will generally require smaller members. Outside columns tend to become quite wide, since the same maximum moment occurs in all members.

Figure 198–Single-Span Trusses This system is used for the better-class long-span industrial building, particularly for high bays. The trusses usually have some slope to outside gutters. The depth-to-span ratio is on the order of 1 to 10, and any number of roofing systems may be used. Because the slope is small, a built-up roofing material is usually necessary. The columns are framed into the trusses so that the lateral load resistance on this type of building is excellent and loads do not need to be carried to the end walls.

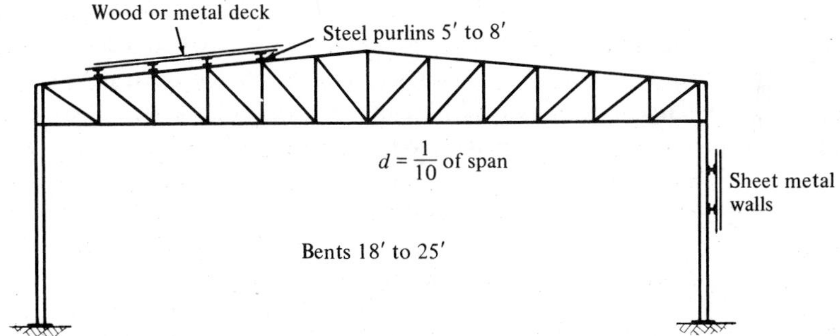

Fig. 198. Single-span trusses.

Figure 199–Standardized Rigid-Frame Buildings There are many competing types of rigid-frame building now on the market. They are usually designed and built to use a minimum of materials to lower the cost. Also, they are often available with substandard (for the region in which they are sold) live loadings. They must be designed so that they can be shipped in straight lengths, so ingenuity is required in the selection and design of the splice at the knee. Also, the frame

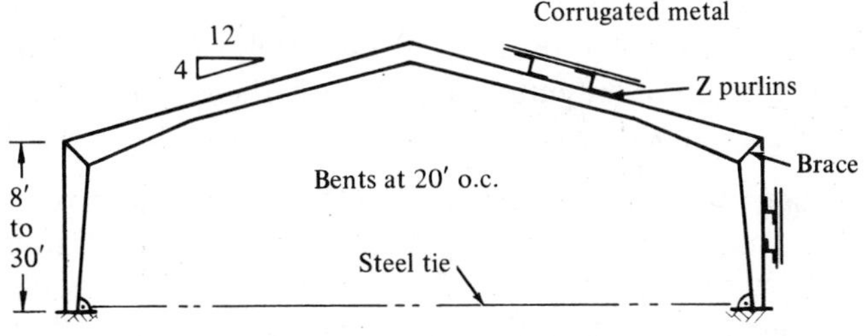

Fig. 199. Rigid frames.

must be braced at the knee, and this brace may not be adequate. The legs are usually tapered from the knee, but the rafters are tapered only to the point of inflection at about a third of the distance to the crown.

Usually standard rigid frames are designed by elastic methods, there being no appreciable savings by plastic design unless tapered members are used.

Corrugated metal is used for roofing and the purlins are Z shapes for ease of fabrication. Often colors are used for roof and walls. The standard slope of the rafters is 4 in 12, but "low-profile" buildings are available.

Steel-rod ties are required in the floor to carry the horizontal thrusts of the frames. In cut-rate buildings, the floor slab may be utilized as a tie or the tie may be omitted.

There have been a number of failures of this type of building, principally as a result of modification for special use (e.g., by raising the height), or when they are used as grain-storage buildings and grain is placed against the walls.

Figure 200–Long-Span Truss Systems For very large industrial projects requiring long spans in both directions, it is customary to use simple-span trusses in both directions. The large trusses used to pick up the smaller ones are called "jack" trusses to distinguish them. The trusses are not normally made continuous. They

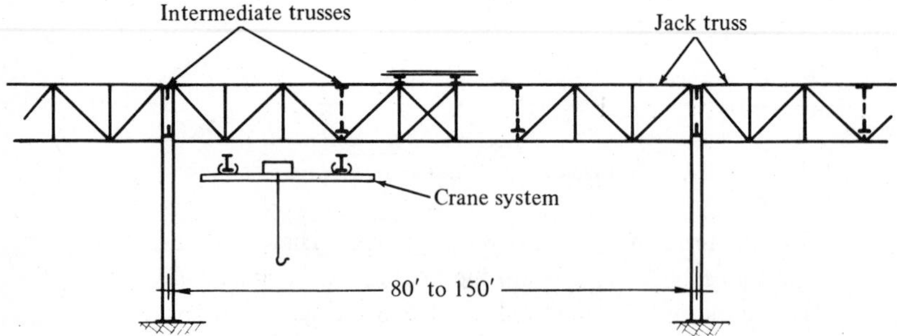

Fig. 200. Long-span truss system.

often support extensive monorail crane systems or other materials-handling equipment.

Long spans can be easily justified on the basis of the ratio of cost of the structural frame to the cost of the entire building. An expenditure of about 5% extra will provide spans double the usual spans and give an excellent flexibility of use.

Figure 201–Long-Span Joists Long-span joists may either be mass-produced as industrial building components or may be made up by local fabricators. They may be used for spans from 50 to 90 ft. They are not necessarily the most economical system, but do provide a minimum-depth structure. Beams framing in the other direction are usually of much shorter span. To provide lateral resis-

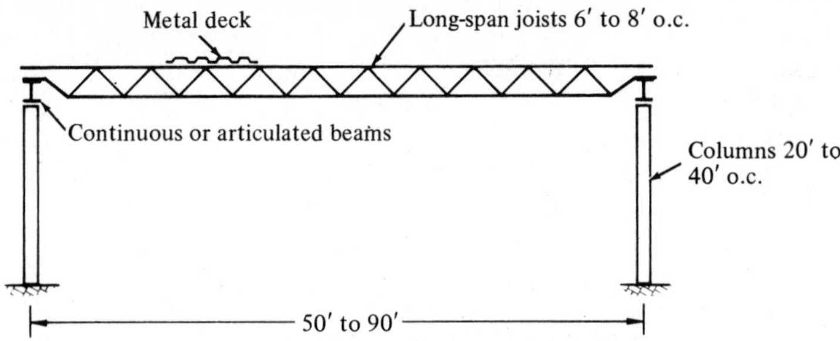

Fig. 201. Long-span joists.

tance, the joists should be detailed so that there are joists at each column and the lower chord is extended to the columns to be used for the lateral bracing. An alternative is to provide a rigid connection of the column to the base.

Figure 202–Steel-Mill Buildings Industrial buildings for heavy mill work, such as steel mills, are in a class by themselves and have their own peculiar problems. A typical cross section of a mill building is shown in Fig. 202. The design is characterized by the following:

1. Roofs are steel trusses of long span. For a steel mill the pitch of the roof is great enough that corrugated metal can be used. Heating is usually not a problem so insulation is not provided.
2. Rigid frames are not generally used, possibly because of the lack of faith in their stiffness and extra cost.
3. The truss and columns form a rigid bent. There may be adjacent bays which participate in the lateral loads. Some possible combinations are shown in Figure 203. There also may be cranes in the side bays.
4. The heavy columns support the cranes which run on trucks resting on steel

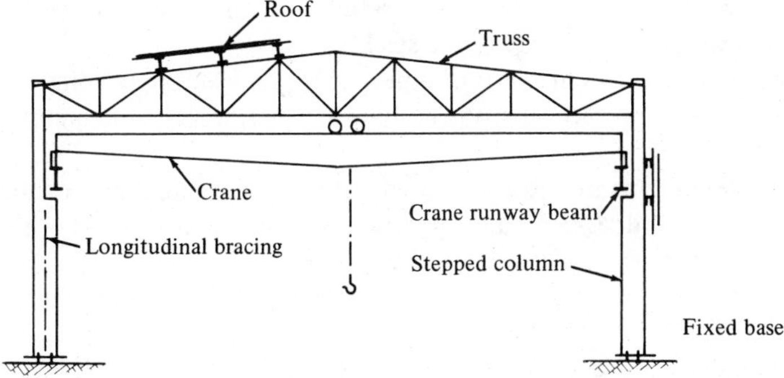

Fig. 202. Typical steel mill building.

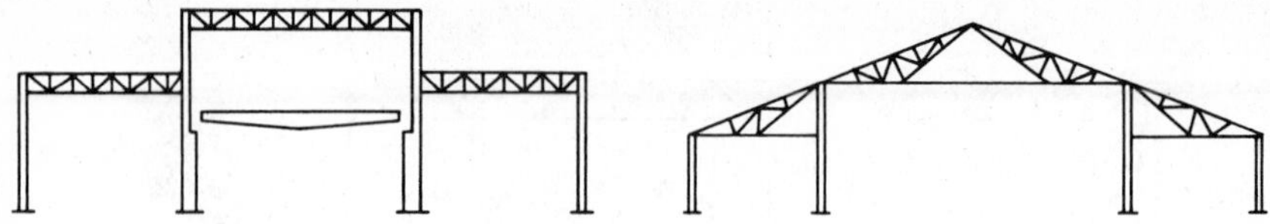

Fig. 203. Multiple-span buildings.

rails. The rails are carried by the crane runway beams between columns. These beams must be designed for lateral loads as well as vertical live load and impact. Since these members are subject to repeated loads and to vibration, the proper design becomes very important.

5. If the crane loads are small, ordinary columns without the stepped offset may be used with the crane resting on a bracket attached to the column. This induces bending moments in the columns.
6. Because of the great height and exposure to wind, very large bases and footings may be required. This brings up some sticky and theoretical problems in the detail design and proportioning of these elements.
7. Longitudinal X bracing at selected locations must be provided both for lateral loads and for the moving loads due to the crane. This involves the solution of problems of change in length of longitudinal members due to temperature. Because these industrial buildings are normally unheated, the range of temperature may be quite large. It is better to place a single braced bent at the middle of the building so that expansion can occur each way.
8. Provision must be made for expansion in very long buildings. Expansion joints are not quite as much of a problem in this type of building as for office buildings because leakage of joints will normally not cause much trouble before they can be fixed.
9. Roofing and wall material are normally of the corrugated-metal type, supported by purlins and girts. There are a great many proprietary materials on the market for this purpose. Natural light is often provided by corrugated plastic used instead of the steel sheets.
10. Provisions must be made for framing of the ends of the building as well as for doors. The vertical spans become very large, and so columns may be quite heavy.
11. Double columns are sometimes used instead of the single large column, but such a detail does not materially increase the lateral stiffness of the bent.

Design of a Single-Span Truss

Design a truss for a typical interior bay of an industrial building with a 60-ft span supported by columns. The truss will have a slope of $\frac{1}{4}$ in./ft for the top chord from the center of the truss for drainage to the columns. Spacing of the truss will be 20 ft and the roofing system will be metal deck with a span not to exceed 8 ft. Design for a live load of 30 psf plus a possible load for mechanical equipment of 10 psf. Use A36 steel.

The layout of the truss is shown in Fig. 204. The depth at the center is one-tenth of the span, i.e., 6 ft, and 8 panels are used, so the purlin spacing is 7.5 ft. The slope of the top chord indicates a depth at the support of 4.75 ft. A Warren truss system is used for the arrangement of members, because of appearance and economy. The length of each member is written directly on the sketch.

The first task will be to select suitable purlins, so we must estimate the dead loads to be used as follows: roofing, 6 psf; insulation, 3 psf; steel deck, 2 psf; steel purlins, 2 psf; for a total of 13 psf. The total load will be 13 + 30 + 10 = 53 psf.

The required bending moment and section for a typical purlin is

$$M = wL^2/8 = 53(7.5)(20^2)/8 = 19{,}875 \text{ ft-lb}$$

$$S = M/f_s = 19{,}875(12)/24{,}000 = 9.9 \text{ in.}^3$$

The unit stress $f_s = 24{,}000$ psi assumes a compact section. Table 6 indicates a W8X13 furnishing $S = 9.90$ in.^{3}. Check the deflection for live load. The ratio of live load to total load is 40/53 = 0.75. The stress for the total load is 24 ksi, so the deflection is $\Delta = L^2 f/1000d = 20^2(24)(0.75)/1000(8) = 0.9$ in. The span-to-deflection ratio is $L/\Delta = 20(12)/0.9 = 267$, which is satisfactory for roof structures. Note that the guess made for the weight of the purlins is satisfactory.

Forces in Truss To the previous load of 53 psf we must add the weight of the truss which we will guess to be 3 psf so the total load is 56 psf. The panel point load for the truss is 56(7.5)(20)/1000 = 8.4 kip each.

We will use the same section for all members of the top chord because it is usually uneconomical to splice chord members. The same is true of the bottom chord. The maximum chord force will be at the center of the truss. We will design only three diagonals, numbers 1, 2, and 3, as indicated in Fig. 204.

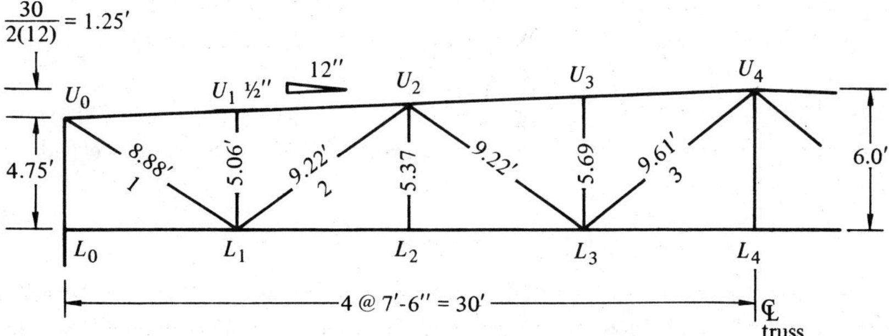

Fig. 204. Layout of a sample single-span truss.

The force in top-chord member U_2U_4 will now be determined:

1. The reaction at the support is half of the load on the truss, $R = 8.4(3.5) = 29.4$ kip. Note that we have neglected the load directly over the reaction.
2. Cut the truss to the left of the center member and draw a free-body diagram as shown in Fig. 205.

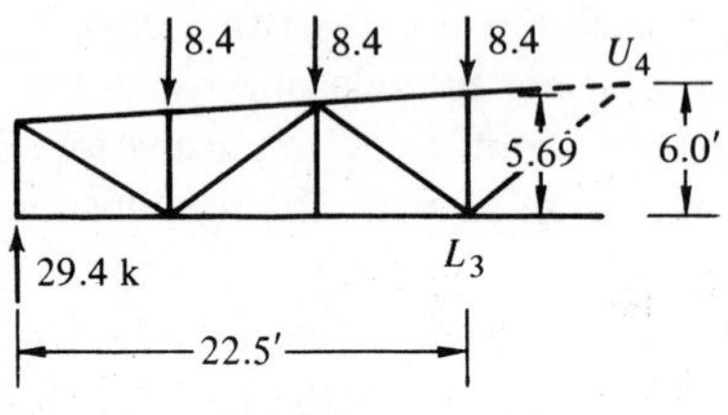

Fig. 205.

3. Take moments about point L_3 of all forces shown on the free-body diagram:

$$M = 29.4(22.5) - 8.4(7.5)(1 + 2) = 472.5 \text{ ft-kip}$$

4. To find the horizontal component of force in the top chord, divide the moment by the vertical distance (5.69 ft) to the top chord from point L_3:

$$H = 472.5/5.69 = 83.0 \text{ kip}$$

5. The force in the member is the horizontal component multiplied by the length of the member divided by the horizontal length. The inclined length is 7.51 ft, so the force in the member is $P = 83.0(7.51)/7.5 = 83.1$ kip compression.

The force in the bottom chord is determined with the same free-body diagram, but the center of moments is at point U_4 on the top chord. The calculations are:

$$M = 29.4(7.5)(4) - 8.4(7.5)(1 + 2 + 3) = 504 \text{ ft-kip}$$

$$T = 504/6.0 = 84 \text{ kip tension}$$

To determine the force in member 1 we will use the free-body diagram shown in Fig. 206. To exclude the effect of the top and bottom chords we must find the intersection of the chords and use this as the center of moments. This point will be 114 ft from the support; this number is obtained from the following ratio:

$$0.5/12 = 4.75/x; \quad \text{solving for } x: \quad x = 114 \text{ ft}$$

The only force in the free-body diagram is the reaction $R = 29.4$ kip. Take moments about the intersection of the chords. Call the vertical component of the

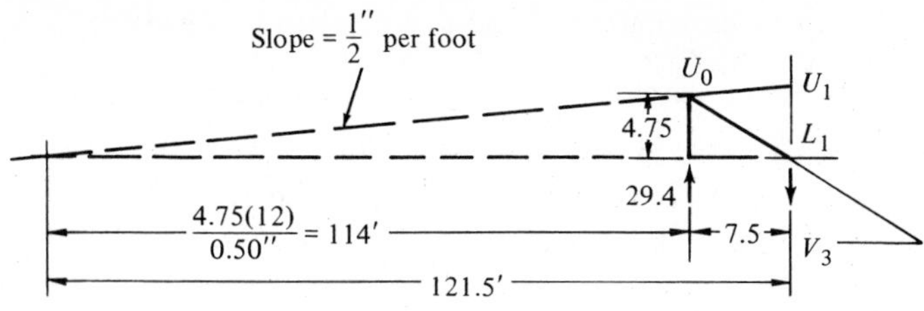

Fig. 206.

force in member 1 V:

$$29.4(114) = V(121.5), \qquad V = 27.6 \text{ kip}$$

The force in the member is equal to the vertical component times the length of the member divided by the vertical projection of the member.

$$P = 27.6(8.88)/4.75 = 51.5 \text{ kip}$$

This member is in tension.

The force in members 2 and 3 will involve the same center of moments, but there will be additional forces at the panel points U_1, U_2, and U_3. The calculated force is 31.0 kip compression in member 2 and 1.00 kip compression in member 3.

Selection of Member Sizes For the top chord we will use a section made by cutting a wide-flange beam in half; this will be called a T section. We do not have tables of properties of T sections in this text, so we will have to use one-half the area of the corresponding wide-flange section and the value of r_y, which is the same. For the value of r_x we will use an approximate value $r_x = 0.26d$, where d is the depth of the section. The value of r_y often controls.

The top chord is supported by purlins at spacings of 7.5 ft in the horizontal plane and at the same distance in the vertical plane. As recommended by commentary to the AISC *Specification*, use $K = 1$ for trusses. We start the selection of the top-chord member by making a guess for the allowable stress. This column will be slender, so we will start with a low stress, say $F_a = 12$ ksi, which gives a required area $A = P/F = 83.1/12 = 7.0$ in.^{2}. Table 6 indicates one-half of a W18X50. The value of r_y from these tables is 1.65, and, using the approximate formula, $r_x = 0.26d = 9(0.26) = 2.3$ in. The value of r_y controls and the value of KL/r is $1.0(7.5)(12)/1.65 = 55$. Table 7 gives $F_b = 17.9$ ksi.

Evidently our original guess of 12 ksi was rather poor, so we must guess again, say 16 ksi. Then $A = 83.1/16 = 5.2$ in.^{2}, and now one-half of a W16X36 is indicated, for which $r_y = 1.52$ in., $KL/r = 1(7.5)(12)/1.52 = 59$, and $F_a = 17.5$ ksi. This is closer, but not satisfactory. By studying the tables we find that one-half of a W14X30 with an area of 4.42 in.^{2} will have about the same value of

r_y, so the stress will be about the same and the area required for $F_a = 17.5$ is $A = 83.1/17.5 = 4.74$ in.^{2}. The W14X30 is slightly low, but the next larger size, a W14X34, with $A/2 = 5.0$ in.^{2}, will be satisfactory. We should make a final check: $r_x = 7(0.26) = 1.82$ in., which is greater than r_y, so r_y controls.

The bottom chord has a force of 84 kip in tension, so column formulas are not required. The allowable stress is $0.6F_y = 22$ ksi, so that $A = 84/22 = 3.8$ in.^{2}. We should allow some area for possible holes required for hanging equipment from the trusses. Assume that there may be two holes, each one inch in diameter, and the thickness of the flange might be 5/16 in. (This number was selected by studying the tables of W sections.) We must add $A = 2(5/16)(1) = 0.6$ in.^{2} to the required area in tension, so the total area required is $3.8 + 0.6 = 4.4$ in.^{2}. Using Table 6, we select one-half of a W14X30 with an area of 4.41 in.^{2}.

Now we must face the problem of the lateral support of the bottom chord for which, by the AISC code, KL/r preferably should not exceed 240. The value of r_y for the W14X30 is 1.49 in., so, solving for the maximum length for $KL/r = 240$, $L = 240r_y = 240(1.49) = 358$ in. = 30 ft. It works out very neatly that there should be a horizontal brace in the plane of the bottom chord and perpendicular to the truss at the center. The selection of this member is left up to the reader.

The selection of member sizes for the diagonals is very similar to the problems given in the section on the design of steel structural elements. Double-angle members should be used for both members 1 and 2. The selection of these members is left to the reader.

The center member is evidently very lightly stressed, with a force of 1.00 kip, compression. However, this is probably not the highest force, considering partial live load on one-half the truss. Diagonals at the center of the truss should be designed as minimum-compression members even though the stress as indicated is positive.

We will first try a single-angle member. The maximum KL/r for main members by the AISC code is $KL/r = 200$. With $L = 9.61$ ft, solve for $r = KL/200 = 9.61(12)/200 = 0.57$ in. Use the approximate formula from Fig. 155 for equal leg angles, $r_z = 0.19d$, so that $d = r_z/0.19 = 0.57/0.19 = 3.0$ in. This would indicate a $3 \times 3 \times \frac{1}{4}$ angle with an area of $A = (3.0 + 3.0 - \frac{1}{4})\frac{1}{4} = 1.43$ in.^{2}. The capacity will be for $KL/r = 200$ in Table 7, $F_a = 3.73$ ksi and $P = 3.73(1.43) = 5.33$ kip. It is possible that a double-angle member might be less in weight and furnish more reserve capacity.

Connections This truss will have welded connections with the diagonals welded directly to the web of the T section of both the top and bottom chords. The connection of the truss to the steel column will be a plate welded to the T, as shown in Fig. 207, and bolted to a cap plate on the column with four bolts.

Connections are usually drawn to scale to see if sufficient length of weld can be developed. The end diagonals, which are $2L3X3X\frac{1}{4}$, have a thickness of $\frac{1}{4}$ in. The net thickness that can be used for welding is $\frac{1}{16}$ in. less, or $\frac{1}{4} - \frac{1}{16} = \frac{3}{16}$ in. We will use a basic stress for a fillet weld of 21 ksi, so the allowable stress on a

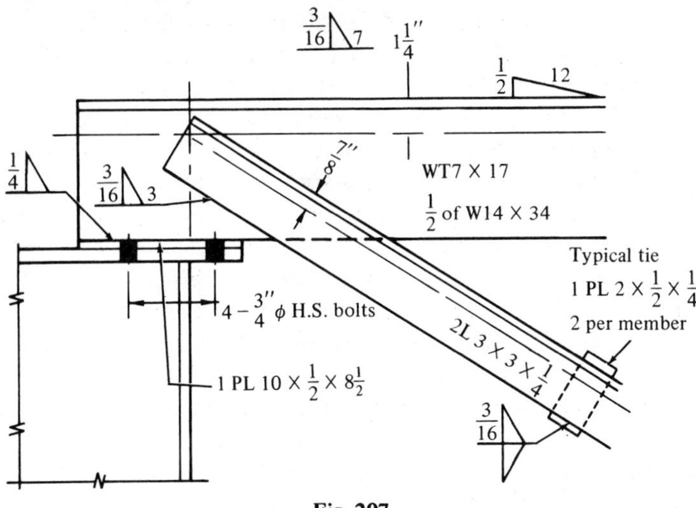

Fig. 207.

weld $\frac{3}{16}$ in. thick and 1 in. long is $21(\frac{3}{16})(0.707) = 2.8$ kip per inch. In the above we used the net diagonal distance across the fillet. The first diagonal member U_0L_1 has a force of 51.5 kip, so the total number of inches required is $L = 51.5/2.8 = 18.4$ in. We are now ready to lay out the connection.

The working lines of the truss should lie at the center of gravity of the members. The distance from the back of the member to the center of gravity may be determined either from the tables in the AISC *Manual* or from the approximate formulas in Fig. 155. The Y value for the T is 1.29 in. For the angles, the approximate formula for $d = 3.0$ in. is $0.29(3.0) = 0.87$ in. Use $\frac{7}{8}$ in.

The intersection of the centerline of the top chord and the diagonal is placed directly over the flange of the column and the slope of the diagonal must be slightly changed to take this into account. A plate is welded directly to the T and is bolted to a similar cap plate on the column. The welding of the diagonal to the top chord is shown by the conventional weld symbols. The total weld furnished is $2(7 + 3) = 20$ in., which is somewhat more than the 18.4 in. required by the calculations. Welds should be distributed to the angle so that the moment of the weld group is approximately zero at the centroid of the angle.

The design of the connections for the other members will follow a similar procedure. The double-angle members will require welded spacers at the third-points of the length. Typical spacers are shown in Fig. 207.

Design of a Continuous Articulated Roof System

Design a flat roof for an industrial building having five equal spans of 30 ft in one direction and six equal spans of 25 ft in the other. Use articulated steel beams with splices arranged to achieve the maximum economy. The roofing is steel deck; weight of the roofing and deck is 10 psf; live load due to snow is 30 psf. Use A36 steel.

A plan of column centerlines is shown in Fig. 208. The first task is to select the direction of spans for purlins and for girders. Should the purlins be 25 ft or 30 ft long? To accurately determine the framing with the least weight and so presumably the least cost, it will be necessary to make alternate designs. We will arbitrarily select the 30-ft spans for the purlins and 25-ft spans for the girders. The purlin spacing for three spaces in 25 ft is 8 ft 4 in.

The location of hinges for the purlins must now be determined. The most logical arrangement is with alternate suspended spans as sketched in either Fig. 209a or 209b. We will study only alternative B, although A might give a better design.

The economy of an articulated beam system is achieved by choosing a cantilever length such that the positive and negative moments in the span with cantilevers are equal, thus using the material more efficiently. Hence, in Fig. 209b we will solve for b_1 and b_2 such that $M_B = M_C = \frac{1}{16} wL_{BC}^2$ (one-half the simple beam

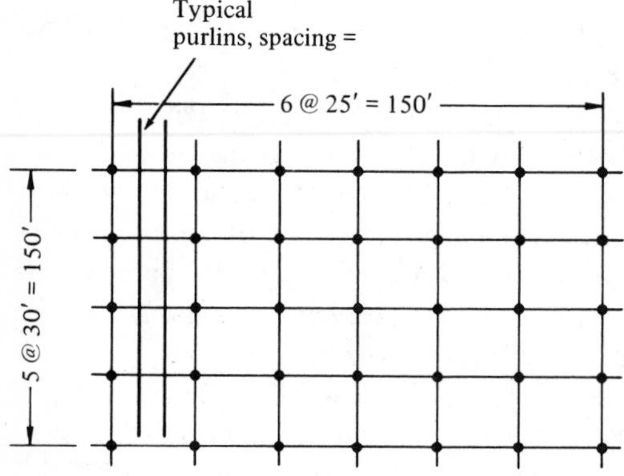

Fig. 208.

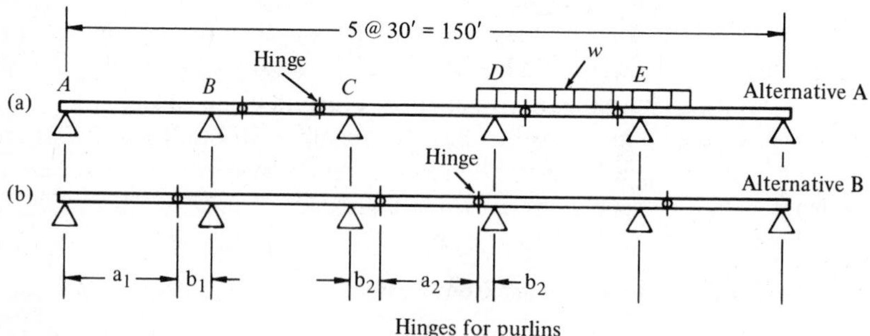

Fig. 209.

moment in span *BC*) for a condition of uniform load in all spans:

$$(30 - b)\frac{w}{2}b_1 + \frac{1}{2}wb_1{}^2 = \frac{1}{16}w(30)^2 = 56.25w$$

$$b_1 = 3.75 \text{ ft}$$

and

$$(30 - 2b_2)\frac{w}{2}b_2 + \frac{1}{2}wb_2{}^2 = 56.25w$$

$$b_2 = 4.39 \text{ ft}$$

The design loads are: roofing and insulation, 10 psf; steel purlins (guess) 3 psf; snow live load 30 psf; mechanical-equipment live load 10 psf. The dead load is 13 psf and the total load is 53 psf. The load on a purlin is 13(8.33)/1000 = 0.108 klf for the dead load and 53(8.33)/1000 = 0.441 klf for the total load.

We now determine the design moments in the various spans as follows:

End span:

$$L' = 30 - b_1 = 26.25 \text{ ft}$$

$$M = \tfrac{1}{8} \times 0.441(26.25)^2 = 38.0 \text{ ft-kip}$$

Second span: Normal design procedures do not consider skip live loadings for a roof. However, since a portion of our live load is for mechanical equipment, which will probably not exist on all spans, we will place this portion of the live load to produce a maximum design moment at the center of the second span:

$$M = 56.25 \times 0.441(2 - \tfrac{43}{53}) = 29.5 \text{ ft-kip}$$

Center span:

$$L' = 30 - 2b_2 = 21.11 \text{ ft}$$

$$M = \tfrac{1}{8} \times 0.44(21.22)^2 = 24.8 \text{ ft-kip}$$

Assuming F_b = 24 ksi for compact sections, adequately braced, we now select member sizes as follows:

End span:

$$S = 38.0 \times 12/24 = 19.0 \text{ in.}^3 \text{ req'd.}$$

Use W10X19, S = 18.8 in.^{3} (1% overstressed).

Second span:

$$S = 29.5 \times 12/24 = 14.8 \text{ in.}^3 \text{ req'd.}$$

Use W10X17, S = 16.2 in.^{3}.

Center span:

$$S = 24.8 \times 12/24 = 12.4 \text{ in.}^3 \text{ req'd.}$$

Use W10X17, $S = 16.2$ in.^{3}.

Note that the above member selection was based on the assumption that the compression flange is supported such that no reduction in the allowable bending stress is necessary. This may not be true in all cases for the negative moments in the cantilever spans. Verification of the above for this example is left to the reader.

Because of the weight savings achieved by the use of articulated systems, they are generally quite flexible and are not recommended for floor construction. Careful consideration should be given to problems that may arise due to excessive deflections of the members. In particular, if the roof system is flat, the possibility of the occurrence of ponding should be investigated.

It is obvious that there are other solutions to this problem that may give less material, and the reader may want to investigate some of these.

The design of the girders follows a similar pattern but loads are concentrated. The location of the hinges may be determined by trial and error.

The connections suggested for this structure are a pair of double angles welded to each purlin or girder with bolts connecting beams together. The connection to the column can be made by welding a cap plate to the column and bolting the girders to it.

Steel Office and Apartment Buildings

This sub-section will describe the structure and the design of steel building frames for office buildings and apartment buildings. The structural system must be chosen to solve the aesthetic, functional, structural and economic problems, and there are a number of possible systems available from which this choice may be made.

Types of Structures The AISC *Specification* permits the use of three general types of construction as described in the AISC *Manual*, page 5-12. These are as follows:

Type 1 uses rigid frames with fully moment-resisting connections. Plastically designed frames are included under this type.

Type 2 allows the use of simple beam design with no allowance for joint restraint and with the joints free to rotate under load.

Type 3 designates "semirigid" framing having connections that are partially restrained.

The AISC *Specification* allows type 2 construction for use in frames that resist moments at their connections, provided that special wind connections are used for resisting the wind only. Presumably, these connections do not participate in the moment resistance against vertical loads. This assumption has been used to design tall steel building frames in this country for many years. Its justification is based on the following considerations:*

1. The beams are overdesigned considering the moment resistance of the joints and, therefore, have a large reserve capacity.
2. Connections are ductile and permit adjustment of stresses if the wind-moment connection is not large enough to carry the moment from gravity loading.

The same considerations apply to semirigid connections.

Lateral Resistance Aside from strength for vertical loads, the principal design problem in tall building frames is to develop adequate resistance to lateral loadings. The common types of structures are, (a) moment-resisting frame, (b) braced frame, with either X-braced panels distributed throughout the frame or a core structure of either steel X-bracing or concrete walls, usually at the center of the structure, (c) exterior bracing to form a vertical tube, or (d) exterior walls with closely spaced columns which act as a moment-resisting frame of considerable stiffness. This type of design is essentially a vertical steel bearing wall.

In very tall buildings (over 200 ft high), it is desirable to carry the lateral loads in the outside walls. The stresses in an interior core structure may be very high and the foundation conditions may preclude development of sufficient resistance to lateral loads. The cost of moment-resisting frames may be double that of a properly designed structure where lateral loads are carried by the outside walls. A description of several types of frames follows.

Type 1 with Moment-Resisting Connections, Fig. 210 Beams and girders are designed as a fully continuous structure including resistance to lateral loads. Each bent carries essentially its own vertical load and lateral load in proportion to the relative stiffness of the bent. Columns must be designed for bending moments caused by their continuity. This system is suitable for structures of from low to medium height, although for medium height (up to 200 ft) the members and connections may get rather heavy. However, this system may be necessary if a braced frame or a core system is not feasible for architectural reasons.

Type 2 or 3 with Simple Beams and Wind Connections, Fig. 211 This system is much easier to design because determination of bending moments from continuity is not necessary and column calculations are less involved. This system is suitable for low to low-medium heights (100′).

Type 1 and 2 with Braced Bents, Fig. 212 Lateral loads are taken by special bents containing X bracing or, as in this sketch, K bracing. There must be

*Beedle, L. S., et al., *Structural Steel Design*, p. 699.

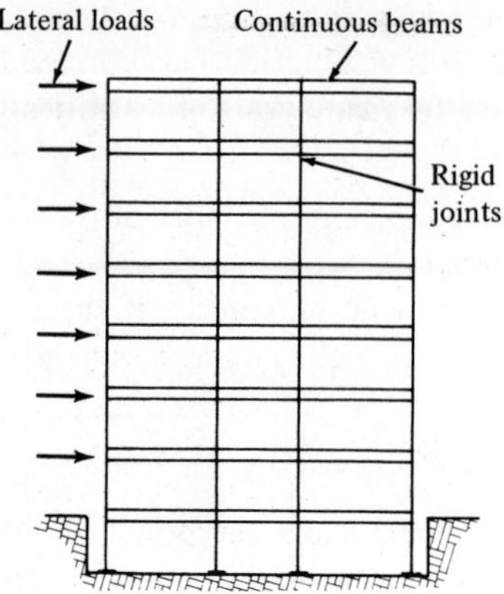

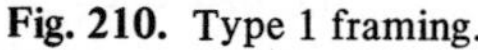

Fig. 210. Type 1 framing.

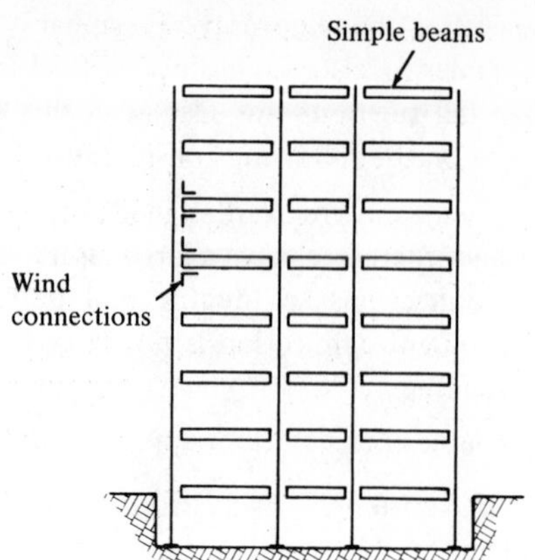

Fig. 211. Types 2 and 3 framing.

enough of these bents to carry the lateral loads in each direction. This, of course, poses a special problem in the utilization of the space blocked by the bracing and is more satisfactory for buildings with a fixed floor plan, such as apartments. The foundation must be designed for overturning moments at the base. Concrete walls may be substituted for the steel X or K bracing.

Core-Type Buildings, Fig. 213 This framing system contains a central tower of steel or concrete to carry the lateral loads. With concrete the tower may be erected first with slip forming. For buildings with fairly small plan areas, it is desirable to exclude all interior columns outside the core area so there is uninterrupted floor space with more flexibility for partitions. Spans up to 70 ft are sometimes used and heavy beams with holes to pass utilities may be required

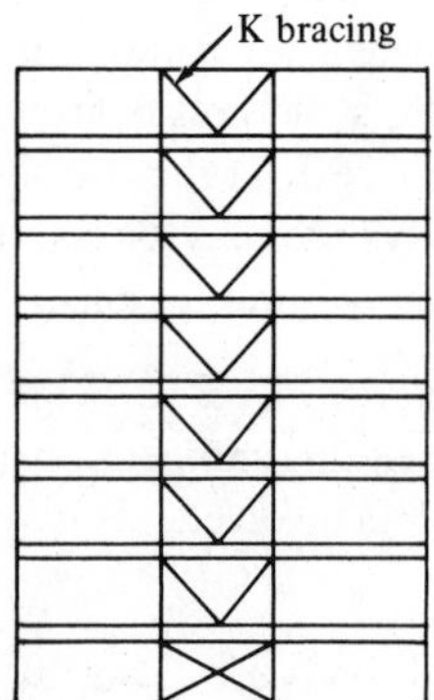

Fig. 212. Braced bent.

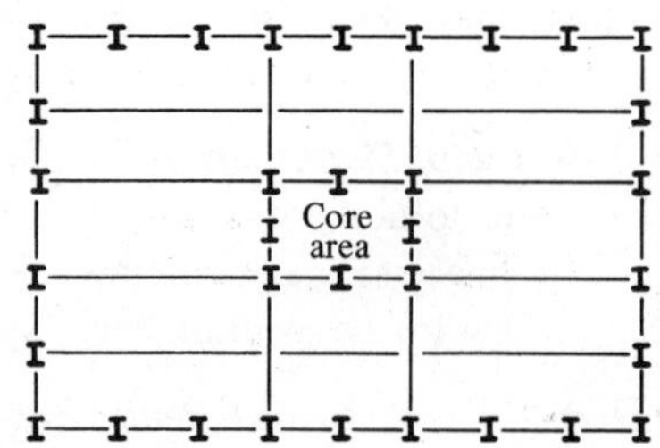

Fig. 213. Framing plan, building with braced co1

adjacent to the corners of the core. This system lends itself to fairly closely spaced outside columns.

Tube-type Buildings, Fig. 214 The stresses in tall buildings from lateral loads are considerably reduced if the outer walls are used for bracing so that the structure acts virtually as a tube. Cross bracing may be used, as sketched in Fig. 214a (braced tube), or closely spaced columns may be used, as in Fig. 214b (framed tube). In the latter case, the calculations must be for a moment-resisting frame which, due to the close spacing of members, should be much stiffer than the interior framing.

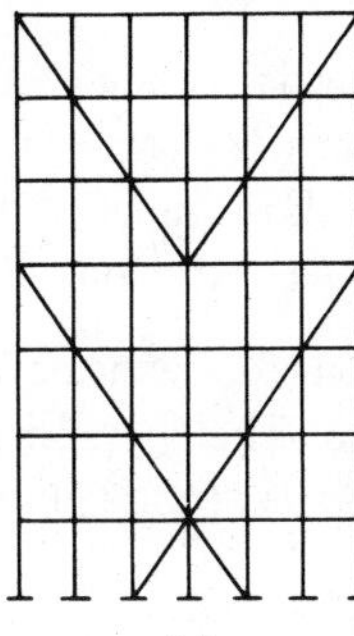

(a)

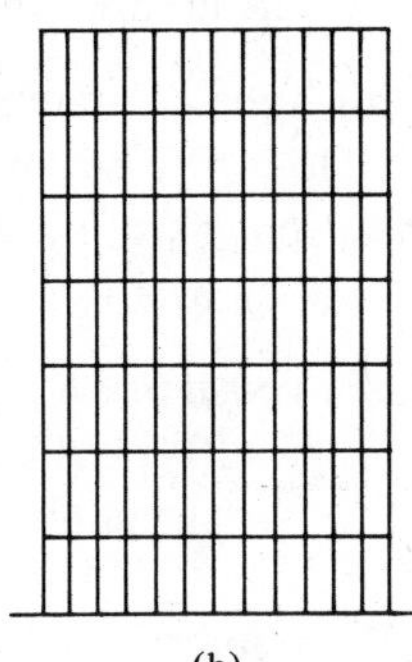

(b)

Fig. 214.

The beams and columns of these structures may be of either type 1 or type 2 construction. If composite design having steel floor members integral with a concrete floor slab is used, then type 2 (simple) connections should be used because of the inherent difficulty of developing continuity with composite design. The interior core of such buildings may also be used to develop part of the resistance to lateral loads.

Building-Frame Calculations: The Example of a Steel-Frame Building

The design of tall building frames will be illustrated with an example. Some of the most interesting problems in structural design are concerned with the selection of a structural system that best satisfies the requirements of aesthetics, function, and economy. In the following example, unfortunately, it will not be possible to have the active participation of an architect, so that the engineer will have to serve as his own aesthetic consultant, which means that function may take precedence over aesthetics.

We will design a few typical members for a steel-frame building having a basement, four floors, and a roof. The area of a typical floor is to be about 8000 ft^2. It is to have a stair at each end with a bank of two elevators 15 ft by 20 ft at one end of the building, and is to be designed to accommodate two additional floors at some later date.

The following will be accomplished:

1. Preliminary drawings will be prepared including framing plans showing the location of beams, girders, and columns on a typical floor and typical building elevations.
2. Appropriate loads and stresses will be selected for the design.
3. Calculations for a typical beam and girder will be presented.
4. Wind loads will be resisted by moment connections (type 2) and calculations for a typical connection will be shown.

A building with type 2 framing requiring special moment connections is sketched in Fig. 215. The spacing of floor beams is 7 ft 4 in. Columns are oriented with the strong axis of the column in the short direction to resist lateral moments. There will be over twice as many columns in the long direction.

Data for Design of Typical Structural Elements The following notes and calculations show the design of typical beams and girders. The lateral moments will be determined by the portal method. Members on the second floor only will be designed.

The following loads and stresses are assumed for this design:

Loads Floor dead load:

$1\frac{1}{2}$-in. metal deck	2 psf
4-in. concrete slab	50 psf
ceiling	8 psf
dead load	60 psf

Preliminary value for additional weight of steel of floor beams:

(Weight of steel applied as a general load)	3 psf
Exterior wall load	60 psf

Roof: Use the same design as for the floor since provision is to be made for vertical expansion at a later date.

Live load: 50 psf plus a general partition load of 20 psf for a total of 70 psf.

For live-load reduction use "Minimum Design Loads in Buildings and Other Structures," AISC *Manual*, page 5-163. The percentage reduction of load must be the least of the following:

1. $R = 0.08$ times floor area carried for members supporting more than 150 ft^2 of area.
2. $R = 100(D + L)/4.33L$ where D is the dead load and L is the live load per square foot.
3. $R = 60\%$ maximum.

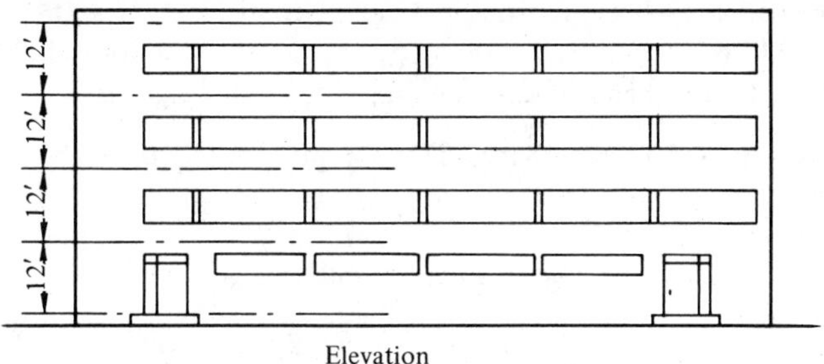

Elevation

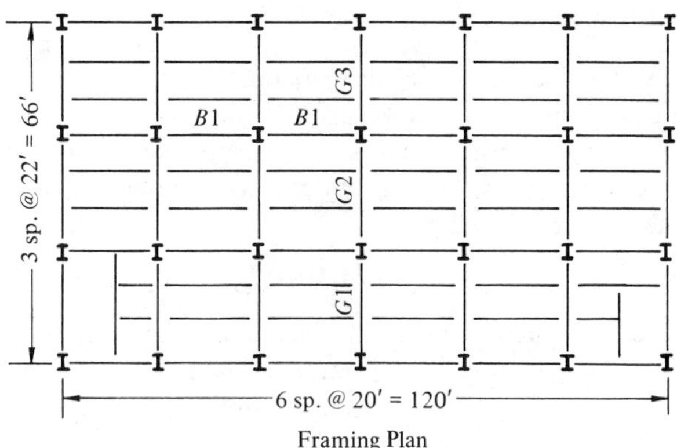

Framing Plan

Fig. 215.

Lateral loads: Wind load = 20 psf
Design for two additional floors.

Stresses Use AISC *Specification* with A-36 steel. For beams, the allowable stresses are:

Noncompact sections	22 ksi
Compact sections	24 ksi

Beam B1 (see Fig. 215) This is a typical interior floor beam with a span of 20 ft and a spacing of 7 ft 4 in. It will be designed for a dead load of 60 psf and a live load of 70 psf, for a total of 130 psf, and a section will be used which has a satisfactory section modulus based on the maximum bending moment.

The live load may be reduced for areas of more than 150 ft^2. The area is 20(7.33) = 147 ft^2, which is less than 150 ft^2, so no reduction is indicated. The total load per foot of beam adding the 3 psf for beam weight is

$$w = (70 + 60 + 3)(7.33) = 975 \text{ plf} \quad \text{or} \quad 0.98 \text{ klf}$$

The bending moment is

$$M = 0.98(20)^2/8 = 49.0 \text{ ft-kip}$$

The required section modulus S for a stress $F = 24$ (compact section) is

$$S = M/F = 49.0(12)/24 = 24.5 \text{ in.}^3$$

The table of beam properties gives the following choices:

$$\begin{array}{ll} \text{W10X25} & S = 26.5 \text{ in.}^3 \\ \text{W12X22} & S = 25.3 \text{ in.}^3 \\ \text{W14X22} & S = 28.9 \text{ in.}^3 \end{array}$$

Unless there is a restriction on headroom, the obvious choice is the W14X22, which, although no lighter than the W12X22, will be stiffer.

Occasionally, a beam must be checked for deflection when it is particularly long or is reduced in depth by headroom or clearance requirements. The easiest formula available is the following:

$$\Delta = L^2 f/1000d$$

where L is in feet, $E = 30 \times 10^6$, d is in inches, and the actual stress f is in kips per square inch. This may be derived from the usual formula for deflection at the center of a simple beam with a uniform load. In this case, the stress in the beam is

$$f = M/S = 49(12)/28.9 = 20.3 \text{ ksi}$$

$$\Delta = (20^2)(20.3)/1000(14) = 0.58 \text{ in.}$$

The L/Δ ratio for full load is

$$20(12)/0.58 = 414$$

Based on the usual requirement (AISC *Specification*, Section 1.13) of 1/360 of span for live load, this beam is satisfactory. This completes the essential design of this member. It will be interesting to see if our guess for the weight of steel beams (3 psf) is accurate. The actual value is exactly

$$w = 22/7.33 = 3.0$$

Girder G1 A typical girder has a span of 22 ft and carries two concentrated loads. To the dead loads must be added the weight of the girder, which we will guess to be 3 psf, again making a total dead load of 60 + 6 = 66 psf.

The live load may be reduced for area as follows:

$$R = 0.08\% \text{ per square foot, or}$$
$$R = 100(D + L)/4.33L, \text{ or}$$
$$R = 60\%, \text{ whichever is smallest}$$

where R is the percentage reduction, D is the dead load (pounds per square foot), L is the live load. For this girder, the loaded area is $A = 22(20) = 440\ \text{ft}^2$. Then

$$R = 0.08(440) = 35\%$$
$$R = 100(66 + 70)/4.33(70) = 44.9\%$$
$$R = 60\%$$

The first formula controls, so that the live load is $70(1 - 0.35) = 46$ psf. The concentrated load from a beam on the girder is

$$P = (66 + 46)(7.33)(20) = 16{,}419 \text{ lb} = 16.4 \text{ kip}$$

The moment under the load is

$$M = PL/3 = 16.4(22)/3 = 120.3 \text{ ft-kip}$$

Assuming a compact section with $F_b = 24$ ksi,

$$S = M/F = 120.3(12)/24 = 60.2 \text{ in.}^3$$

The following choices are indicated:

W16X40	$S = 64.6$ in.^{3}
W14X43	$S = 62.7$ in.^{3}

There being no headroom restrictions, the W16X40 can be used.

Lateral Load Stresses For type 2 construction, the girder is usually sized for vertical load, but possibly for lateral load. The connections are designed for moments due to lateral loads. For this structure, the lateral moments will be determined by a simplified "portal method," which assumes that the outer columns take half as much lateral shear as the interior columns and points of contraflexure are at the centers of beams and columns.

A sketch of the structure is shown in Fig. 216. The frame is designed for two extra stories. The girder we are designing is at the second floor so that the shear above that floor is $4.0 + 4(4.8) = 23.2$ kip and below the floor the shear is $23.2 + 4.8 = 28.0$ kip. A free-body diagram of an element of the frame is shown in Fig. 217. The shear at each level is divided so that half the shear is in the outside columns. The remainder of the solution is merely a matter of statics, as follows:

Above A	$M = 6.0(3.87) = 23.2$ ft-kip
Below A	$M = 6.0(4.67) = 28.0$ ft-kip
Above B	$M = 6.0(7.73) = 46.4$ ft-kip
Below B	$M = 6.0(9.33) = 56.0$ ft-kip

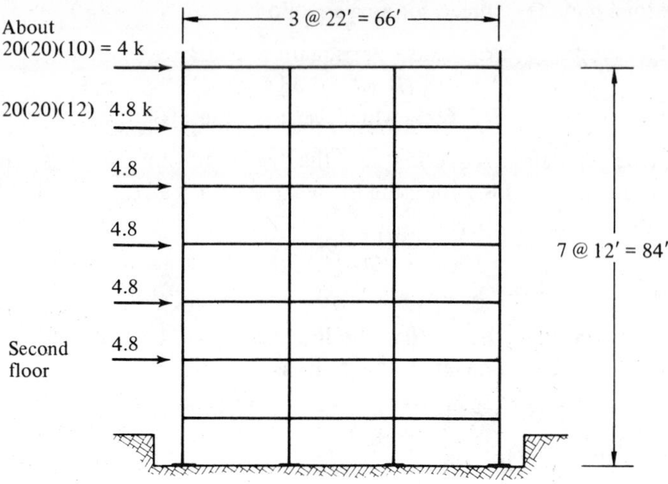

Fig. 216.

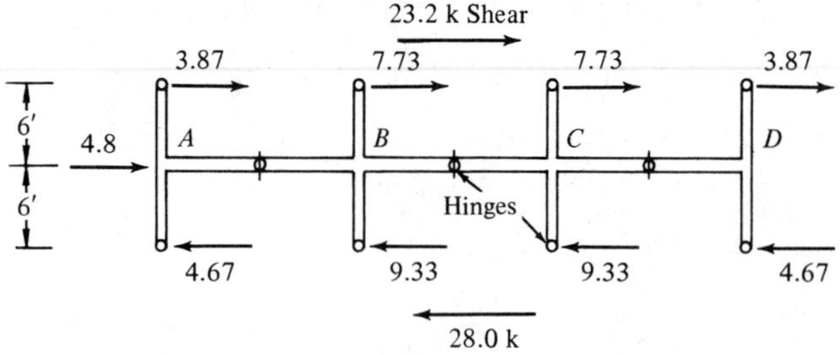

Fig. 217.

The moment at A in the girder is the sum of moments in the column:

$$M = 23.2 + 28.0 = 51.2 \text{ ft-kip}$$

There is a point of contraflexure at the center of girder AB, so the moment at B in AB is the same as the other end, $M = 51.2$ ft-kip. The moment in BC at B may be obtained from equilibrium of joint B as sketched in Fig. 218b:

$$M = 46.4 + 56.0 - 51.2 = 51.2 \text{ ft-kip}$$

Note that the moment in BC is the same as in AB with this assumption of shearing force distribution.

Selection of Lateral Load Connections The bending moment for lateral load by these calculations is 51.2 ft-kip. Allowable stresses may be increased a third if

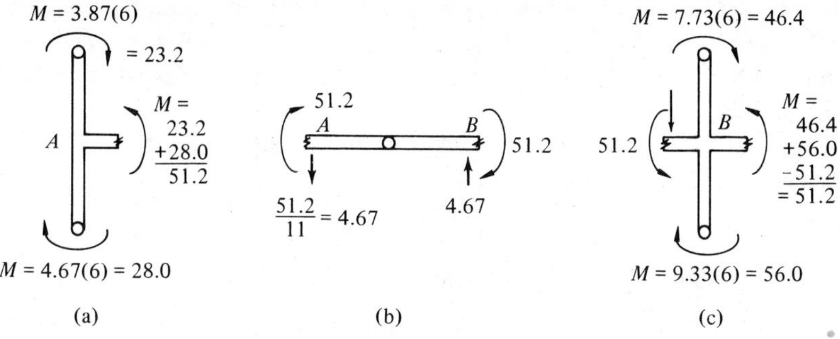

Fig. 218.

wind or seismic forces are considered (see AISC *Specification*, Section 1.5.6). Instead of increasing stresses, the design moment may be reduced to three-quarters of the calculated moment and normal stresses used; then $M = 0.75(51.2) = 38.4$ ft-kip.

There are many different types of lateral load connections. The first step is to try a standard connection because the ratio of lateral load moment to the strength of a beam is quite low. The beam strength for a W16X40, $S = 64.6$ in.^{3} is

$$M = 64.6(24)/12 = 129.2 \text{ ft-kip}$$

The ratio of required capacity to maximum is $38.4/129.2 = 0.30$.

We will try a four-row framed beam connection bolted to the column and to the beam, as shown in Fig. 219. The connection is analyzed as an analagous beam,

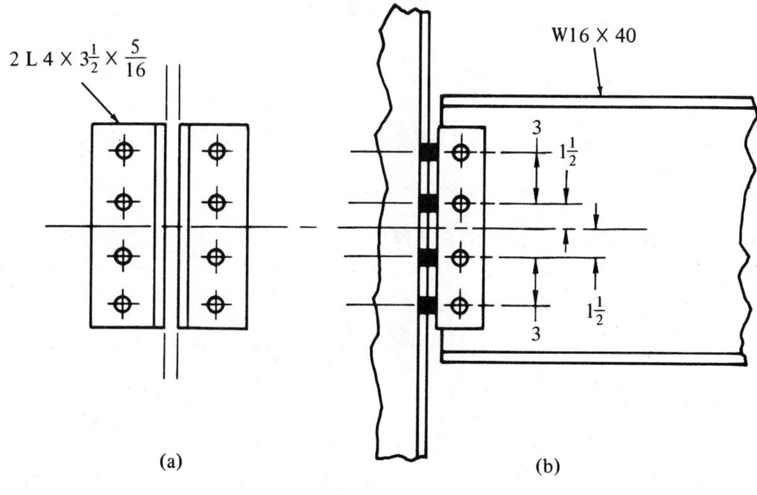

Fig. 219.

the moment of inertia of the equivalent beam being that of the unit areas of each bolt as sketched in Fig. 219b. The bolts in the legs of the angles attached to the beam will be analyzed. Calculations for the other set of bolts from angles to column are similar.

The equivalent area of the beam is $4A$, where A is the area of one bolt. The equivalent moment of inertia is

$$I = 2A(4.5^2 + 1.5^2) = 45.0A$$

The section modulus for the bolts farthest from the centroid is

$$S = I/C = 45A/4.5 = 10A$$

The wind moment on the connection is M = 38.4 ft-kip, so the bolt stress due to the moment is

$$R = M/S = 38.4(12)/10A = 46.1/A$$

For A = 1, then, R = 46.1 kip. To this must be added (vectorially) the vertical shear stress on the connection:

$$V = 16.4 \times 0.75 = 12.3 \text{ kip}, \qquad R = 12.3/4 = 3.08 \text{ kip}$$

and the vector sum is

$$R = \sqrt{3.08^2 + 46.1^2} = 46.2 \text{ kip}$$

The allowable value for a $\frac{3}{4}$-in. diameter rivet or high-strength bolt in double shear is 13.2 kip. This capacity is considerably below the indicated force, so it is evident that the connection is inadequate.

An alternate design uses a plate welded to the top and the bottom of the column as sketched in Fig. 220. The force on the plate is

$$P = M/d = 38.4(12)/16.25 = 28.4 \text{ kip}$$

The capacity of a $\frac{3}{4}$-in.-diameter high-strength bolt is $R = \pi(\frac{3}{4})^2(15)/4 = 6.63$ kip, so that N = 28.4/6.63 = 4.3 rivets, nearly 4 rivets, as shown in Fig. 220.

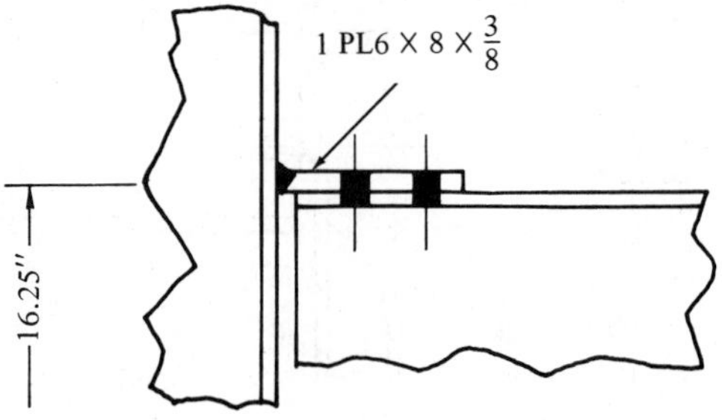

Fig. 220.

The bottom connection might utilize a stiffened beam seat rather than the plate.

We have now designed a connection to resist the lateral force moment. In addition, it is necessary to design a connection to resist the shear force from vertical loads. This can be a standard framed connection, or the beam seat mentioned above may be utilized.

Steel-Arch Buildings

Long-span steel arches are a structural system that offers considerable economy for spans over 100 ft where the function of the building will fit the curved shape of the roof. There are a number of configurations and the arrangement of structural elements will depend on the clearance required underneath the arch. Two hinged arches are used most often because the problem of restraining the arch at the abutment is usually rather difficult unless the abutments are placed on bedrock. There are a number of abutment configurations depending on the foundations.

Basically, for economy, the effective span of the arch should be as short as the function of the building will permit, so that bending moments are reduced. The circle is most used for the profile because of the simple geometry and ease of fabrication. Rolled beams require considerable camber for a smooth curve; alternatively, the arch may be constructed in segments with the purlins placed at the junction of segments. It may be less expensive to use welded built-up members, either in the form of an I section or a box. Trusses are used for longer spans but they are not as clean looking as rolled or built-up beams.

The advantages of the various types and forms of arches cannot be adequately discussed without reference to the thrust line, which shows the position of the resultant of the internal forces relative to the centerline of the arch.

In Fig. 221 is sketched a typical thrust line. Loads are unbalanced, with the greater loads on the left. Graphical solution of the force system by means of the space diagram (left) and the force diagram (right) show the position of the loads on the space diagram and their magnitude on the force diagram. A section through the arch at A shows the force T with relation to the section. This force is at a distance a from the centerline. It can be resolved into a moment $M = aT$ and a direct stress T at the center of the section.

The thrust line shows a number of things about the stresses in the arch. (1) The maximum moment occurs at the 10-kip load and is positive (tension in the low fiber of the beam). The maximum negative moment is on the right and is probably less than that under the 10-kip load.

The thrust line for a fixed arch is sketched in Fig. 222. Note that the thrust line hugs the arch, and at the abutment it is *above* the centerline. This principle will be used later in describing the detail for the precise location of hinges in three-hinged arches.

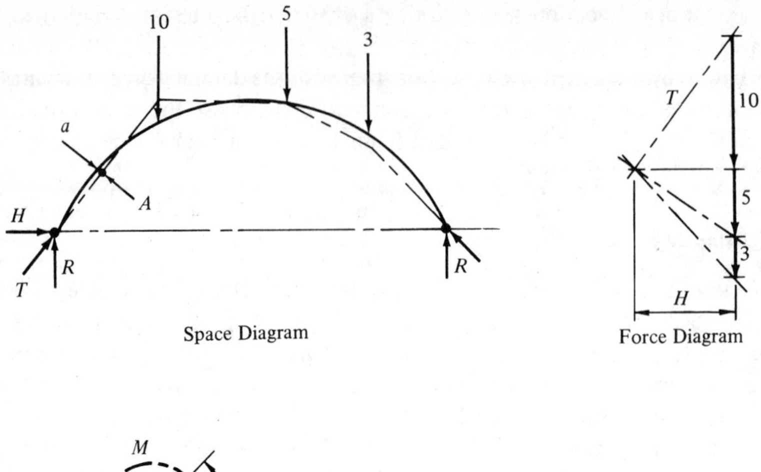

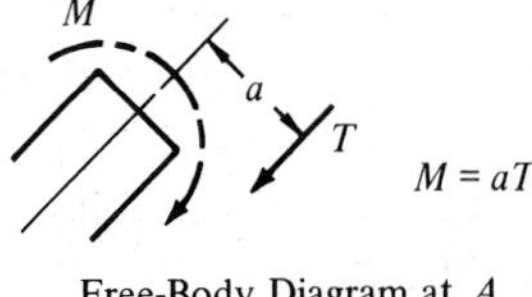

Fig. 221. Thrust line of a two-hinged arch.

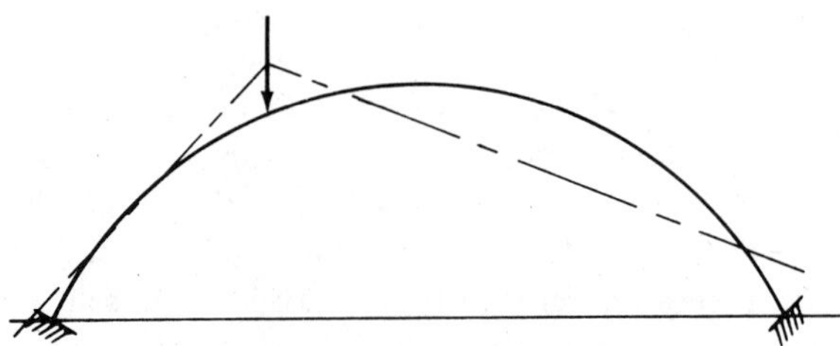

Fig. 222. Thrust line of a fixed arch.

There is a stress condition for arches that is not present in beam structures (except for columns where it is the prime factor in the selection of member sizes). This is sometimes called *amplification*. Amplification may be defined as the nonlinear increase in stress due to movement of the arch away from the loads. In Fig. 221, for example, the arch at the 10-kip load will move downward away from the thrust line, which remains effectively fixed. The moment arm A will be increased, and consequently the moment M will increase. This, in turn, increases the deflection, which again increases the moment, and so on. If the arch is stable, the increases will be progressively smaller until finally the position of the arch stabilizes. The total increase of stress from the first position is called the amplification. Fortunately, there is an approximate, simple method of calculating these stresses without repeated calculation of deflections. On arches of satisfactory stiffness, the amplification should be no more than 5% to 10% of the stresses due to vertical loads.

Types of Arch Structures

Typical Building Structure Figure 223 shows an arch used for a gymnasium or similar occupancy. The rise of the arch is determined by means of a clearance diagram representing, for example, the requirements for basketball. The arch is

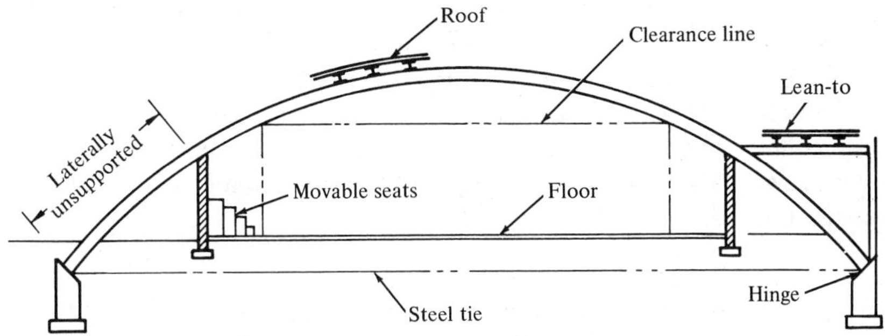

Fig. 223. Arch for a gymnasium.

extended to ground level *outside* the building, as shown on the left, or it may be enclosed by a lean-to, as sketched on the right. There are a number of structural disadvantages to either of these arrangements: (a) the span of the arch is considerably greater than the effective span of the building; (b) the ribs are laterally unsupported from the edge of the main roof to the abutments; (c) there may be differential movements between the walls and roof. The architectural disadvantages are: (a) the spectators have low headroom; (b) the arches are exposed to the weather or cut up the space under the lean-to so that it cannot be used for anything but small offices, classrooms or lockers.

One solution to these problems is to cut down the arch span and raise the springing line to gain more headroom.

Triangular Frame The span of the arch can be shortened considerably and the springing can be raised if a triangular frame is used to pick up the thrust of the arch (Fig. 224). If the slope of the outside member is greater than the slope of the arch, then the vertical member will be in tension and will require a tie-down with a footing sunk below the floor to pick up the weight of the arch to counteract the tension. The shorter span of this structure will result in a considerable reduction in maximum thrusts, moments and amplification stresses in the arch.

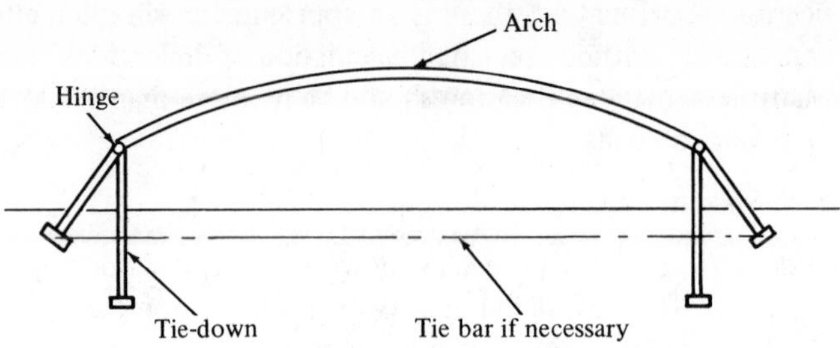

Fig. 224. Triangular frame for an abutment.

Bleachers Used as Abutments The triangular frame can be reversed and the members of the frame used as supports for bleacher seats, as in Fig. 225. Now, the vertical member is in compression and the slanting member is in tension and may require a tie-down if the weight of the bleachers is not sufficient to hold the frame down. The span under the bleachers may be used for locker rooms, classrooms or offices.

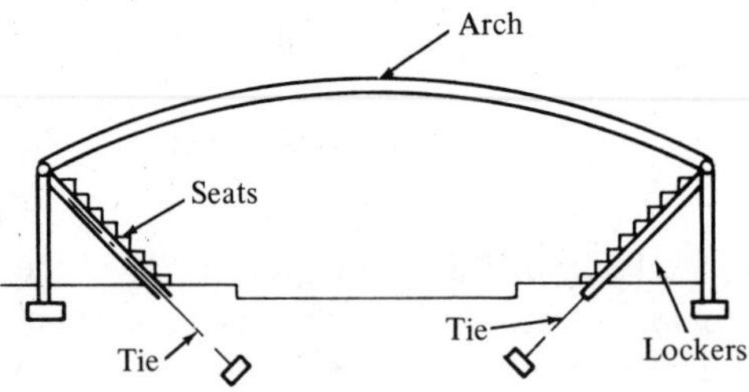

Fig. 225. Bleachers used as abutments.

Inverted L Frame The clear space under the arch may be increased by using an inverted L frame, as in Fig. 226. The foundation of the frame should be arranged so that the thrust line of the arch intersects the center of the base to avoid moments in the base. The frames may become very heavy, and reinforced concrete might be a better material for the legs because of the extra weight available for counteracting the thrust of the arch.

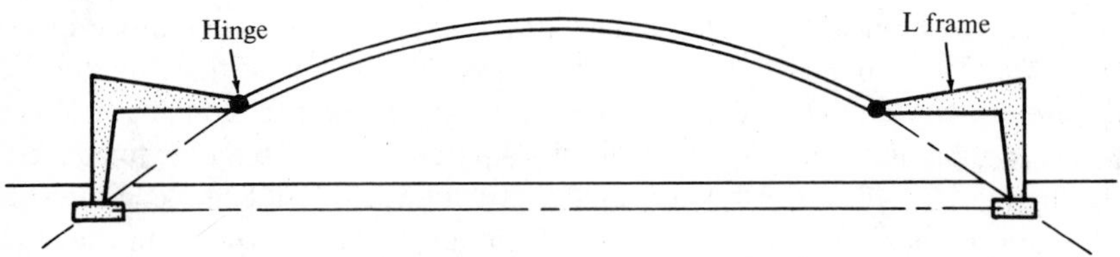

Fig. 226. Arch carried by L frame.

Preliminary Analysis of Two-Hinged Arches The two-hinged arch can be analyzed on the basis of some very simple principles as follows:

1. The influence line for horizontal thrust due to vertical loads may be generated by unit horizontal movement of the abutment, as sketched in Fig. 227.

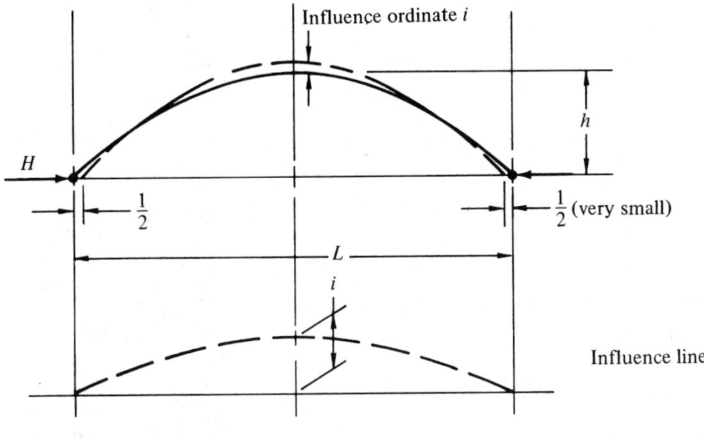

Fig. 227.

The upward deflection of the arch is the influence line for horizontal thrust. In other words, if a vertical load of unity is placed at any point, the horizontal thrust at the abutment is i.

2. Assume that the influence line is a parabola and that the arch is very close to a parabola in shape.

3. For a parabolic arch we know that for uniform vertical load, the thrust line fits the arch exactly and that the horizontal thrust H is the bending moment M divided by the rise h:

$$M = WL^2/8, \qquad H = WL^2/8h$$

However, the horizontal thrust on the basis of the influence line is the area under the influence line: $A = 2iL/3$, so that

$$H = AW = (2iL/3)W$$

Equating values of H:

$$H = WL^2/8h = 2iLW/3, \qquad i = 3L/16h$$

To find the maximum bending moment at a point on the arch, we add the influence line for bending in a simple beam due to a unit load to the influence line for bending due to the horizontal thrust. For example, say the span is 100 ft and the rise at center is 20 ft; then the arch is a parabola, so the rise at the quarter-point is 15 ft, as sketched in Fig. 228.

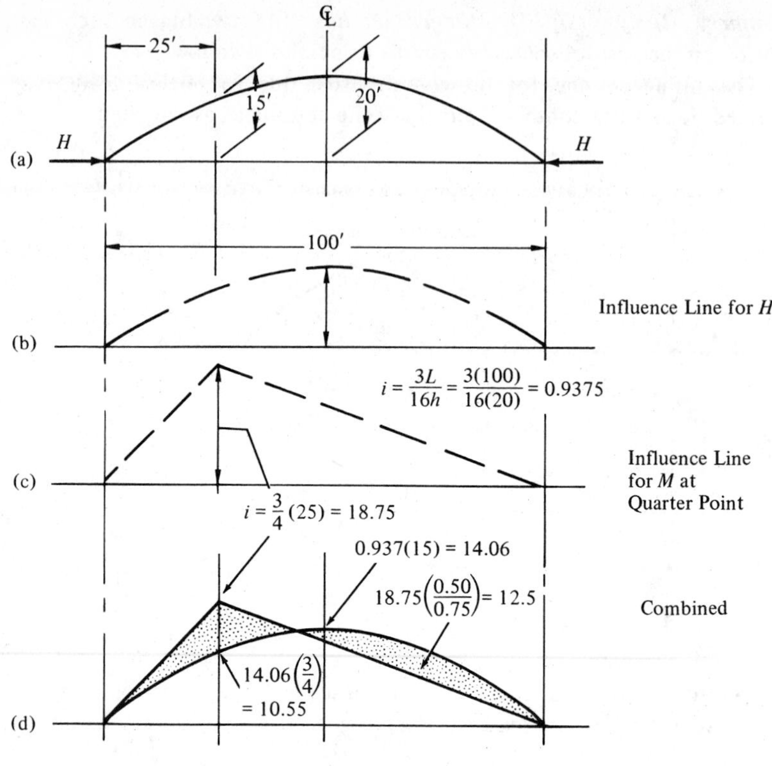

Fig. 228.

The maximum ordinate for the influence line on a simple beam with a span of 100 ft at the quarter-point, is $1(\frac{3}{4})(25) = 18.75$, which is the amount due to a unit load. The maximum ordinate on the parabola for the influence line for bending due to the horizontal thrust is

$$i = \frac{3}{16}(L/h)y$$

where y is the height from the base line to the point on the arch

$$i = \frac{3}{16}(100/20)(15) = 14.06$$

In Fig. 228d, the two diagrams are shown laid on top of each other to form the influence line for moment in the arch rib. The principal ordinates have been calculated.

This influence line can be drawn to scale and the areas determined graphically for design purposes. A check on the calculations is that the area under the line above the curve must be equal to the area under the line.

Design of a Small Concrete Structure

In this and the next subsections we will design several reinforced concrete structures using the methods of design and analysis developed in previous subsections.

The initial project is to design a small park shelter in reinforced concrete. A sketch of the proposed structure is shown in Fig. 229, a building 80 ft long and 48 ft wide, column center to column center. The roof is framed with slabs of 12-ft span carried by beams of 20-ft span. The beams in turn are carried by girders of 24-ft span with the exception of the span on one end, which is 48 ft, column center to column center. The building has a height of 10 ft to the under-

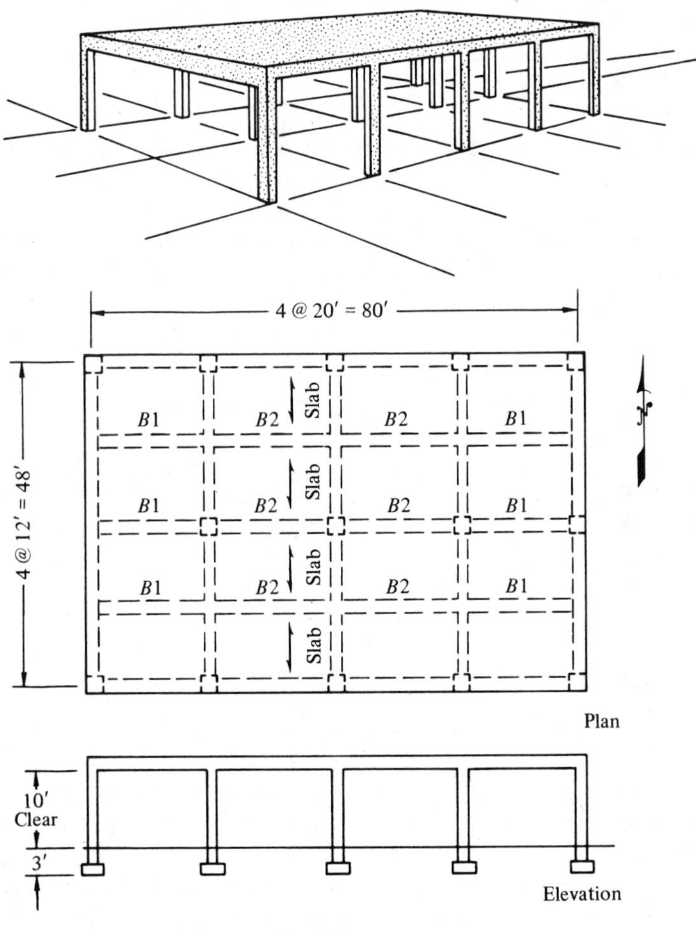

Fig. 229.

side of the deepest beams. The footings are shown as 3 ft deep to the bottom of concrete.

Select the loads and the working stresses to be used. These would normally be based on the building code that governs this particular location plus the judgment of the engineer. Special loads imposed by mechanical equipment would also have to be considered. After some study, we decide on the following requirements:

Roof live load: 100 psf. This is based on the possibility of people swarming on top.
Additional roof load for roofing material: 10 psf.
Concrete cylinder strength for design: $f_c' = 3000$ psi.
Reinforcing-steel yield point: $f_y = 40{,}000$ psi.
Weight of concrete: 150 psf.

Other requirements will, of course, be governed by the ACI *Code*.

Design of Slab We will initially base the choice of the depth of the slab on the requirements of the ACI *Code* for length-to-depth ratio, which have already been tabulated. Then we will check to see if it is within the requirements for maximum value of $K = M/bd^2$. For the code requirements we will assume that the slab is half-way between conditions with one end continuous and with both ends continuous, so the ratio L/d is assumed to be $(24 + 28)/2 = 26$. This figure can be increased because we are using a value of $f_y = 40$ ksi in accordance with the equation $L/d = 26(1)/(0.4 + 0.1f_y) = 26/0.8 = 32.5$. We will use the clear span, so we must assume the width of the supporting beams, say 1.0 ft. The required depth becomes: $t = (12 - 1)(12)/32.5 = 4.0$ in.

The ultimate design loads are calculated next. It is convenient to remember that a 1-in. thickness of concrete represents 12.5 psf of dead load, and the load factors are 1.4 for dead load and 1.7 for live load:

Slab	4(12.5) (1.4)=	70 psf	
Roofing	10(1.4)	= 14 psf	
Live load	100(1.7)	= 170 psf	
	Total:	254 psf;	use 0.25 klf

To obtain the effective depth for the slab we must make some assumption for the diameter of the bar, so we will assume that No. 4 bars are used with a cover of $\frac{3}{4}$ in., and that the effective depth is $d = 3.0$ in. to the center of the bar.

For the bending moments in the slab, we will use the design coefficients in the ACI *Code*. These are shown in Fig. 230 for each significant point of design and apply to the clear span. The coefficient of $\frac{1}{10}$ over the first interior support will govern the design and the moment will be: $M = -wL^2/10 = 0.25(11^2)/10 = -3.02$ ft-kip/ft. We must increase this moment for the capacity-reduction factor, so we will use a value of $M = 3.02/0.9 = 3.36$ for the moment calculations.

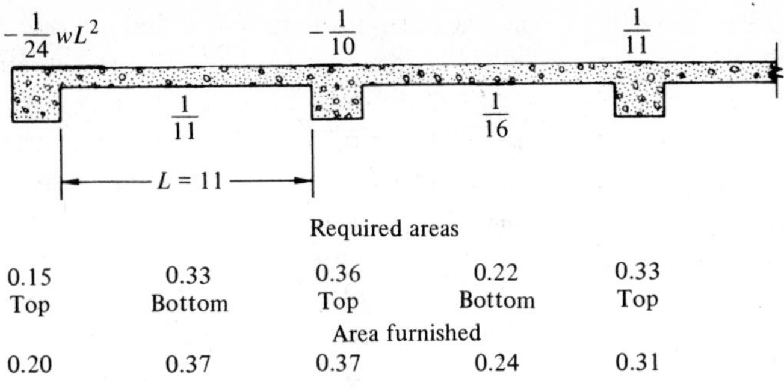

Required areas

0.15	0.33	0.36	0.22	0.33
Top	Bottom	Top	Bottom	Top
		Area furnished		
0.20	0.37	0.37	0.24	0.31

Fig. 230.

We will now check to see if the required depth is satisfactory for bending moment. Solve for

$$K = M/bd^2 = 3360(12)/12(3.0)^2 = 373$$

This value is quite low and should be satisfactory. The maximum for $f_c' = 3000$ and $f_y = 40$ ksi is 869.

To determine the area of steel required for this moment we must calculate a value of the coefficient j, where jd is the effective depth. Use the formula:

$$j = 0.5 + \sqrt{0.25 - K/(1.7f_c')} = 0.5 + \sqrt{0.25 - 373/(1.7)(3000)} = 0.92$$

The required area of reinforcement is the moment divided by the moment arm and the allowable stress:

$$A_s = M/jdf_y = 3.36(12)/3.0(0.92)(40) = 0.37 \text{ in.}^2/\text{ft}$$

Reinforcing required for other points on the beam may be obtained by the ratio of moment coefficients. Actually, the value of j will change slightly but the minimum value will be on the safe side.

The required areas are shown in Fig. 230 on the basis of the ratios of coefficients. The reinforcing in the slab with the areas furnished is shown in Fig. 231. In the sketch it is convenient to exaggerate the depth of the slab, so there

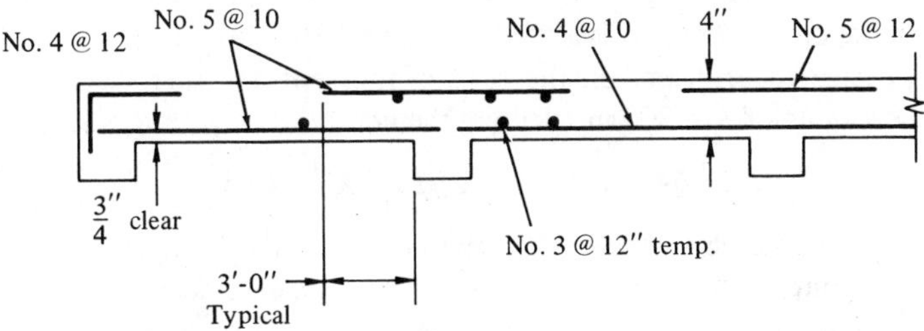

Fig. 231.

will be space to show the bars. Bars also are required in the opposite direction even though there are no calculated stresses. This is called temperature reinforcing and is specified by the ACI *Code* to be not less than 0.2% of the gross cross-sectional area, $A = 4(12)(0.2)/100 = 0.096$ in^2/ft. We have used No. 3 bars at $12(0.11)/(0.096) = 13.75$ in. (say 12 in.). The maximum spacing must not be greater than 5 times the slab thickness, or 18 in.

The shear stresses in slabs are normally quite low. The coefficient for maximum shear on end spans by the ACI *Code* is $V = 1.15wL/2 = 1.14(0.25)(11)/2 =$ 1.58 kip/ft width. The capacity-reduction factor is 0.85, so $V = 1.58/0.85 =$ 1.86 kip, and for $d = 3.0$, $v_u = 1860/3.0(12) = 51.7$ psi, which is well below the value of $v_c = 2\sqrt{f_c'} = 109$ psi.

Design of Interior Beams In the framing plan of Fig. 229, beams *B*1 and *B*2 are shown as typical interior beams. All of the interior beams are not precisely alike because some will be continuous with the columns, but this factor will be neglected and beams will be assumed to have knife-edged supports. The same moment coefficients will be used as were used for the design of the slab.

The load on a typical beam is the dead and live load on the slab, plus the weight of the beam. An exterior slab carries more than half its load to an interior beam because of the continuity of the slab. The ACI coefficient for maximum shear is 1.15. Also, we must guess the depth of the beam. As a first guess we will assume an L/t ratio based on the ACI minimum requirements, say a ratio of 21, which gives a depth of $(20 - 1)(12)/21 = 10.9$. Use 12 in. as a preliminary figure. The ultimate loads are:

Dead load of slab	$4(12.5)(11)(1.15 + 1)(1.4)/2$	= 828 plf
Roofing	$10(11)(1.15 + 1)(1.4)/2$	= 166 plf
Beam	$12(12)(150)(1.4)/144$	= 210 plf
Live load	$100(1 + 11)(1.15 + 1)(1.7)/2$	= 2193 plf
	Total:	3397 plf

The same moment coefficients will be used for the beam as the slab. The maximum moment will occur at the face of the first interior support with a clear span of 19 ft assuming a 1-ft width of beam or column. The moment is $M = wL^2/10 = 3.40(19^2)/10 = 122.7$ ft-kip. The capacity-reduction factor is $\phi = 0.9$, so the design moment is $M = 122.7/0.9 = 136.4$ ft-kip.

We will design these beams on the basis of moment and then check for shear stress. Use a value of $K = 500$ in selecting the depth:

$$d = M/bK = 135{,}500(12)/12(500) = 16.5 \text{ in.}$$

This value is considerably greater than the depth assumed for finding the loads on the beam, but we will not change our previous loads, since the error is small. We will use a total depth of 20 in. and assume that a 2-in. cover is required plus one-half the thickness of a 1-in. bar, giving a net depth of $d = 20 - 2.0 - 0.5 =$ 17.5 in.

We will pause now to check the shear stresses. For an end span using a coefficient of 1.15 for maximum shear, the shear at the support is $V =$ 3.40(1.15)(19.0)/2 = 37.1 kip. At a distance d from the support, the shear will be

$$V = 37.1 - 3.40(17.5)/12 = 32.1 \text{ kip}$$

The capacity-reduction factor is $\phi = 0.85$, so the design shear is $V = 32.1/0.85 =$ 37.8 kip. The unit shearing stress is $v_u = V/bd = 37{,}800/(12)(17.5) = 180$ psi. Stirrups will be required, but the stresses will not be too high. Therefore, the value of $d = 17.5$ in. is satisfactory and will be used for design.

We are ready, now, to calculate the area of steel. At the first interior support, $M = 136{,}400$ ft-lb, $d = 17.5$ in., and the revised value of the constant K for determining the effective depth jd is

$$K = M/bd^2 = 136{,}400(12)/12(17.5^2) = 445$$

The values of j and A_s for design of the beam are

$$j = 0.5 + \sqrt{0.25 - K/(1.7f_c')} = 0.5 + \sqrt{0.25 - 445/(1.7)(3000)} = 0.90$$

$$A_s = M/jdf_y = 136.4(12)/0.094(17.5)(40) = 2.6 \text{ in.}^2$$

Corresponding areas of steel for other design points, assuming that j is the same, are

Center exterior, $M = 1/11$ $\quad A_s = 2.4$ in.^{2}
Center, interior, $M = 1/16$ $\quad A_s = 1.6$ in.^{2}
Interior support, $M = 1/11$ $\quad A_s = 2.4$ in.^{2}
Exterior support, $M = 1/24$ $\quad A_s = 1.1$ in.^{2}

A sketch of the beam and the longitudinal reinforcing is shown in Fig. 232.

The design of the stirrups will follow closely the examples given in the introductory material on reinforced-concrete design (Fig. 182).

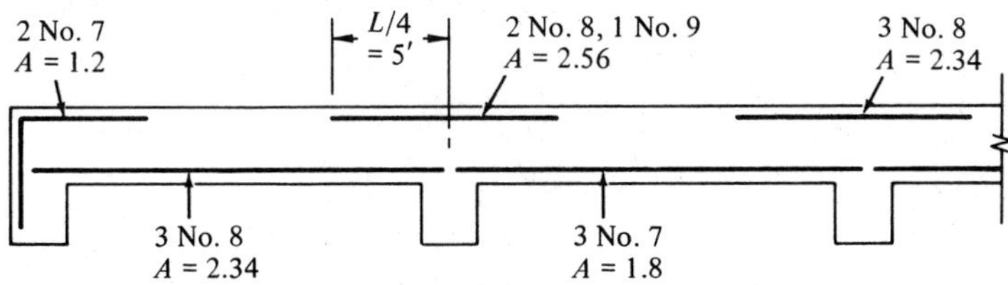

Fig. 232.

Design of a Tall Reinforced-Concrete Building

The drawings in Fig. 233 show the floor plan of a reinforced-concrete building with a floor system using concrete joists and wide, shallow beams. The beams are flush with the bottom of the joists, as shown in the sketch of the cross section. It is assumed that there are six stories.

Lateral loads on the structure are carried by the concrete walls at the corners,

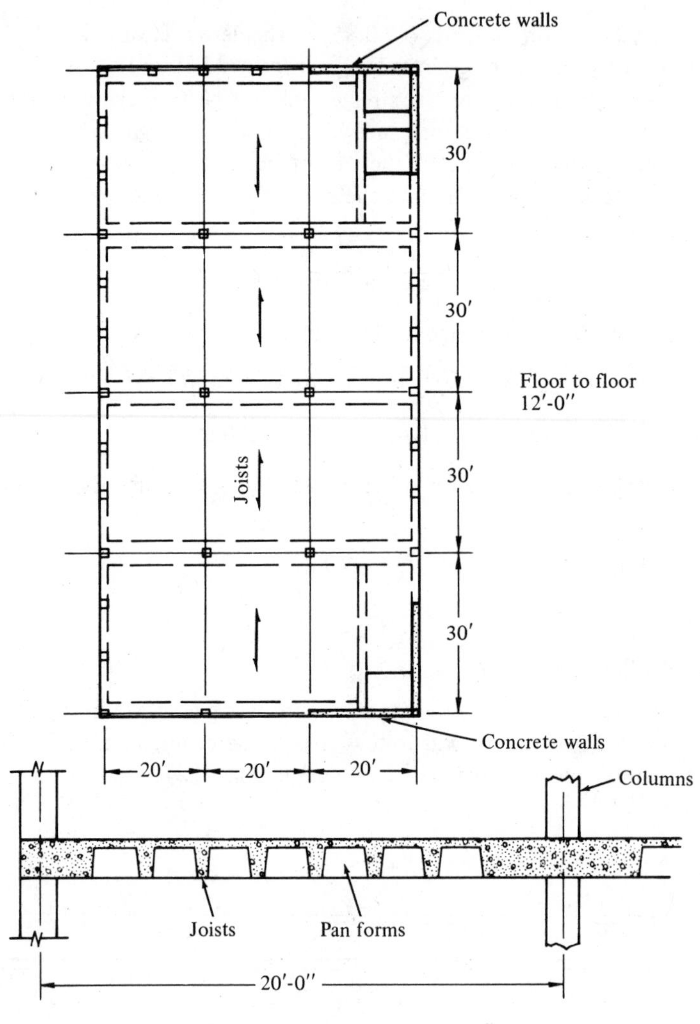

Typical Section

Fig. 233.

so it will not be necessary to design the beams and columns for bending due to lateral loads. The new problems introduced in the design of this structure are the joists, the design of beams using moment distribution for the analysis (rather than coefficients as in the last example), and the design of columns.

The joists span 30 ft center to center. The width of the pans used to form the joists will be 20 in. The choice of depth of joists will also affect the design of the beams and, therefore, it will be better to use fairly deep joists. The flush ceiling of this design will greatly assist the planning of mechanical equipment for heating and air-conditioning. The spandrel (outside) beams have only a 10-ft span, so they can be the same depth as the interior beams.

Concrete-Joist Sizes The concrete joists are formed using prefabricated, reusable steel pans. They are available in depths (of pans) from 6 in. to 20 in., and the standard width of pans is 20 in., although 30-in. pans are also available. Cross sections of standard pans are shown in Fig. 234 (from the catalog of the

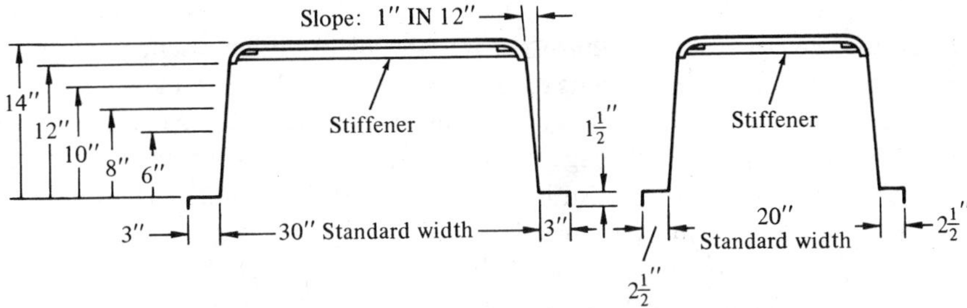

Fig. 234.

Ceco Corporation, Chicago, Illinois). Special pan widths of 10 and 15 in. are available for special cases where the standard widths are not suitable. Special tapered units are used to provide an extra width of stem at the supports if required for shear capacity, and end caps are available for creating bridging beams between joists.

Width of the joists may be varied by the spacing of the pans. The usual widths vary from 4 to 6 in. for 6 in. deep pans, to 7 to 9 in. for 20 in. deep pans. The ACI *Code* requires joists at least 4 in. wide and a depth not more than $3\frac{1}{2}$ times their minimum width. Concrete quantities for 20-in.- and 30-in.-width pans are shown in Table 13, also from the Ceco Corporation catalog.

The depth of the slab over the pans normally varies from $2\frac{1}{2}$ to $4\frac{1}{2}$ in., but may be larger if necessary. The ACI *Code* requires a minimum slab of 2 in. for removable fillers.

The following concrete and steel stresses will be used in this design: $f_c' = 4000$ psi, $f_y = 60{,}000$ psi.

TABLE 13.

CONCRETE QUANTITIES/20″ WIDTHS

Depth of Steelform	Width of Joist	Cubic feet of concrete per square foot for various slab thicknesses			Additional concrete for Tapered Endforms cu. ft. per lin. foot of beaming wall or beam (One side only)
		2½″	3″	4½″	
6″	4″	.303	.345	.470	.13
	5″	.319	.361	.486	.12
	6″	.334	.376	.501	.12
8″	4″	.339	.381	.506	.17
	5″	.361	.402	.527	.16
	6″	.380	.422	.547	.16
10″	4″	.377	.419	.544	.21
	5″	.404	.445	.570	.20
	6″	.428	.470	.595	.19
12″	4″	.418	.459	.584	.25
	5″	.449	.491	.616	.24
	6″	.479	.520	.645	.23
14″	5″	.497	.538	.664	.28
	6″	.531	.573	.698	.27
	7″	.562	.604	.729	.26
16″	6″	.585	.627	.752	.31
	7″	.621	.663	.788	.30
	8″	.654	.695	.820	.29
20″	7″	.744	.786	.911	.37
	8″	.785	.826	.951	.36
	9″	.822	.864	.989	.35

CONCRETE QUANTITIES/30″ WIDTHS

Depth of Steelform	Width of Joist	Cubic feet of concrete per square foot for various slab thicknesses			Additional concrete for Tapered Endforms cu. ft. per lin. foot of beaming wall or beam (One side only)
		2½″	3″	4½″	
6″	5″	.288	.329	.454	.11
	6″	.299	.341	.466	.10
	7″	.310	.352	.477	.10
8″	5″	.317	.359	.484	.14
	6″	.333	.374	.499	.14
	7″	.347	.389	.514	.14
10″	5″	.348	.390	.515	.18
	6″	.367	.409	.534	.17
	7″	.386	.427	.552	.17
12″	5″	.381	.422	.547	.21
	6″	.404	.445	.570	.21
	7″	.425	.464	.592	.20
14″	5″	.415	.456	.581	.25
	6″	.441	.483	.608	.24
	7″	.467	.508	.633	.24
16″	6″	.481	.522	.647	.28
	7″	.509	.551	.676	.27
	8″	.537	.578	.703	.26
20″	7″	.599	.641	.766	.34
	8″	.633	.675	.800	.33
	9″	.665	.707	.832	.32

Dead Loads The depth of the thin slab over the pans will be assumed to be 2.5 in. Sometimes electrical conduits or raceways must be placed in the slab and the thickness may be up to 4.5 in., but we will assume that these facilities can be placed below the joists or through the joists. Also, some architects prefer to place a concrete topping for the finished surface of the floor, and this may add another $1\frac{1}{2}$ in. to the thickness of the slab.

For a preliminary estimate of weights, we will assume that the 20-in.-wide pans are 14 in. deep. A dropped ceiling will be required which we will guess to weigh 10 psf. No additional load will be added for mechanical equipment hung from the ceiling. Assume a 6-in. width for the joists. Table 13 gives an equivalent solid slab thickness of 0.531 feet. Following are the dead loads in pounds per square foot for the joists:

Ceiling load	10 psf
14 inch joists 0.531(150)	80 psf
Total:	90 psf

We need a preliminary estimate of the loads on the beam which is not yet designed. They must be quite wide, so a first guess is 60 in. The loads on the beam then become

Joists	$(90)(30 - 5)$	= 2250 plf
Beam	$[(14 + 2.5)/12]\,(5)(150)$	= 1031 plf
Ceiling	$5(10)$	= 50 plf
	Total:	3331 plf

Live Loads The live load used in the design of the floor will be 100 psf. Building codes may require less load than this value, but with heavy files and computing equipment it is sometimes wise to have more capacity than is required. Corridor loads are also usually 100 psf. Also, we will disregard the general partition load, usually on the order of 20 psf.

Most building codes allow a reduction of live load on beams or girders, depending on the area carried by the floor. We will use the document "Minimum Design Loads," ANSI A 58.1-1955, sponsored by the National Bureau of Standards and reprinted in the AISC *Manual*. Use the minimum reduction according to the following alternatives:

1. For areas of 150 ft^2 or more, reduce the load at the rate of 0.08% per square foot of area supported by the member.
2. Reduce the load by the formula $R = 100(D + L)/4.33L$, in which D is the live load.
3. Maximum reduction 60%.

For our building, the area supported by a beam of 20-ft. span is $A = 20(30) = 600\ ft^2$, and for rule 1 above, $R = 600(0.08) = 48\%$. For rule 2, $D = 3.33$, and $L = 100(30)/1000 = 3.0$, so that

$$R = 100\ \frac{3.33 + 3.0}{4.33(3.0)} = 49\%$$

The two rules give nearly the same result, and a live load of $100(1.00 - 0.48) = 52$ psf will be used for the beams.

Design of Joists The joists will have four spans of 30 ft column center to column center. They may be haunched near the beams to provide increased shear capacity, but for the present we will assume prismatic sections in computing the stiffness of the joists. The joist calculations will be made for the 20-ft strip, and column stiffness will be used, assuming the far ends of the column are fixed against rotation as suggested by the ACI *Code*.

For the moment of inertia of the T elements we will multiply the moment of inertia I of an individual joist by a factor of 2 to take into account the effect of the flange rather than make an accurate calculation. For a 6-in. web widtn and a 16.5-in. total depth, the value of I will be

$$I = 2(6)(16.5^3)/12 = 4492 \text{ in.}^4 \text{ per } 26 \text{ in.}$$

For the full panel width of 20 ft and for the span of 30 ft, the value of the relative stiffness K will be

$$K = 4I/L = 4(4492)(20)(12)/(26)(30)(12) = 460.7$$

We do not know the size of the columns, so we must make a guess. The column size will vary, increasing for the lower stories. The heavier columns will probably decrease the moments at the center of the beams, so it would appear better to choose a relatively small column for our analysis. We will pick a 14 in. by 14 in. column. The exterior columns will probably all be the same for reasons of architecture, and we will assume the same size (14 in. square). There will be two exterior columns per panel. The values of I and K for an interior column are

$$I = bd^3/12 = 14^4/12 = 3201 \text{ in.}^4$$

$$K = 4I/L = 4(3201)/12(12) = 88.9$$

There are two exterior columns per 20-ft panel, so $K = 2(88.9) = 177.8$. The next step is to determine the distribution factors around each joint. We will lump both columns above and below in one distribution to shorten the calculations and then divide the moments by two for the columns. The distribution factors are:

				K	D.F. = $K_i/\Sigma K$
Exterior joint	columns	2(177.8)	=	355.6	0.436
	joist			460.7	0.564
		Totals:		816.3	1.000
Interior joint	columns	2(88.9)	=	177.8	0.162
	joist, left			460.7	0.419
	right			460.7	0.419
		Totals:		1099.2	1.000

These values are recorded on the moment-distribution calculations shown in Fig. 235.

The ultimate loads on a 20-ft width of joists are:

Dead load	90(20) (1.4)/1000 =	2.52 klf
Live load	100(20) (1.7)/1000 =	3.40 klf
	Total	5.92 klf

The fixed-end moments for the 30-ft span are:

Dead-load moment	$2.52(30^2)/12$ =	189.0 ft-kip
Live-load moment	$3.40(30^2)/12$ =	255.0 ft-kip
	Total	444.0 ft-kip

We now must devise loading patterns that will represent probable maximum loadings on the beams for the most significant design points. We can put the loadings in the required patterns or we can load each span separately. We will use the latter method.

The moment-distribution calculations for the joists are shown in Fig. 235. The first distribution is for dead loads on all spans, so there is very little distribu-

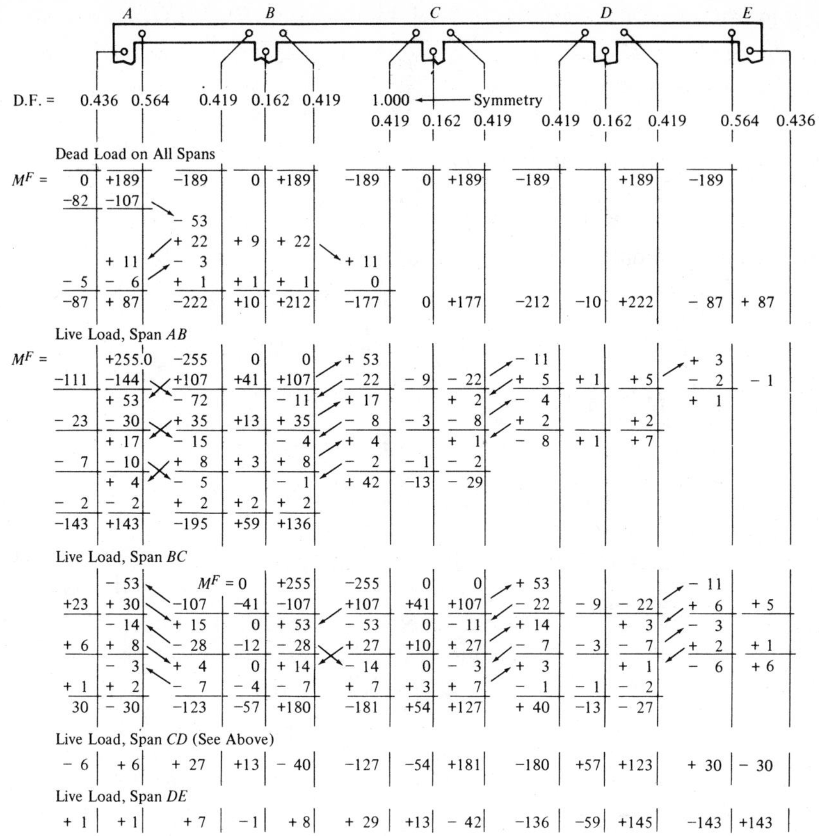

Fig. 235.

tion and the structure is symmetrical about point C. The live load is placed first on span AB and then on BC. The moments for CD and DE are obtained from these by symmetry. To reduce the number of lines of distribution and to obtain faster convergence, a carry-over may be moved upward, as indicated by the arrows.

To obtain the design moments at each significant point on the string of beams, the results of these three distributions are added as required to produce maximum moments at supports, and at the center of the span. The dead loads are, of course, always on the beam.

An examination of the distributions will indicate the following combinations give the maximum moments:

For maximum moment at A: live load on AB and CD,

$$M_A = 87 + 143 + 6 = 236 \text{ ft-kip}$$

For maximum moment at center of *AB*: live load on *AB* and *CD*,

$$M_A = 87 + 143 + 6 = 236 \text{ ft-kip}$$

$$M_B = -222 - 195 + 27 = -390 \text{ ft-kip}$$

For maximum moment at *B*: live load on *AB* and *BC*,

$$M_B = -222 - 195 - 123 = -540 \text{ ft-kip}$$

For maximum moment at center of *BC*: live load on *BC* and *DE*,

$$M_B = 212 + 180 + 8 = 400 \text{ ft-kip}$$

$$M_C = -177 - 181 + 29 = -329 \text{ ft-kip}$$

For maximum moment at *C*: live load on *BC* and *CD*,

$$M_C = -177 - 181 - 127 = -485 \text{ ft-kip}$$

The signs of these moments must be converted to the designers' convention (tension on bottom of beam is positive).

The most accurate method of finding the moments at the center of the spans is to construct shear and bending-moment diagrams. However, if the end moments are very nearly equal, it is satisfactory to subtract the average of the end moments from the simple-beam moment. This procedure will be used here. The simple-beam moment for each span for full load is:

$$M_s = 5.97(30^2)/8 = 666 \text{ ft-kip}$$

The moment at the center of span *AB* is

$$M = -(236 + 390)/2 + 666 = 353 \text{ ft-kip}$$

At the center of span *BC* the moment is

$$M = -(400 + 329)/2 + 666 = 302 \text{ ft-kip}$$

The maximum negative moment for design of the joists will occur at point *B*. We must check both the moment at the middle of the beam and at the face of the beam. We have made a preliminary guess for the beam size of 60 in. wide, so we will use this to check the moment at the face of the beam. A shear and partial moment diagram for span *AB* for the loading which gives the maximum moment at point *B* is shown in Fig. 236. The moment at the face of the beam is 308 ft-kip; and we will use this to check the joist size.

For the 26-in. width (6-in. joists plus 20-in. pans), the moment is

$$M = -308(26)/20(12) = 33.4 \text{ ft-kip}$$

and the shear at 30-in. from the beam centerline is, from Fig. 236,

$$V = 85.3(26)/20(12) = 9.24 \text{ kip per 26-in. width}$$

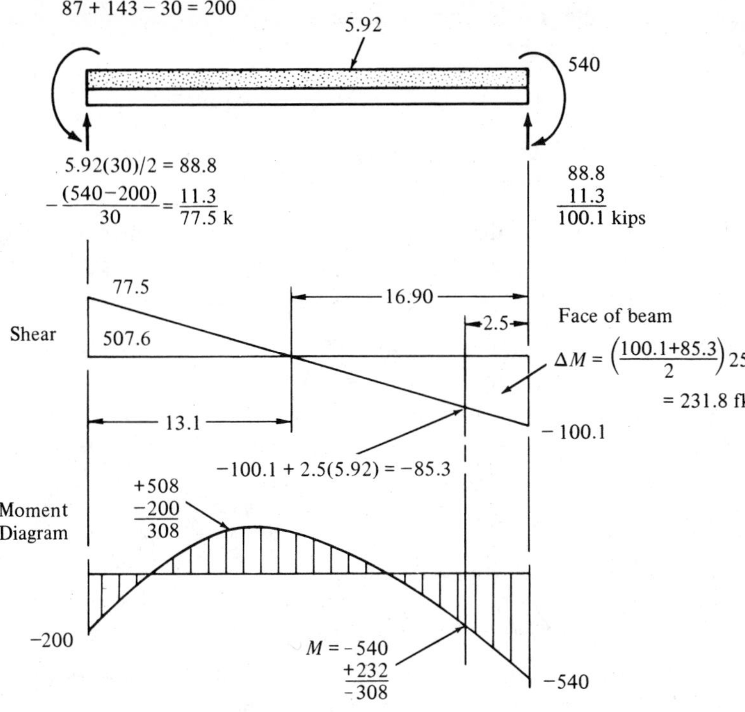

Fig. 236.

At the center of the beam the moment is

$$M = 540(26)/20(12) = 58.5 \text{ ft-kip}$$

These values must be divided by the appropriate capacity-reduction factors for use in design, so that $M = 33.4/0.9 = 37.1$ ft-kip and $V = 9.24/0.85 = 10.87$ kip. At the center of the beam, $M = 58.5/0.9 = 65.0$ ft-kip.

The effective depth d for a total depth $t = 16.5$, assuming that we might use No. 8 1-in.-diameter bars and $\frac{3}{4}$-in. clearance, is $d = 16.5 - 0.5 - 0.75 = 15.25$ in. As a measure of the capacity of the section, we will use the value K for the moment at the face of the beam:

$$K = M/bd^2 = 37{,}100(12)/6(15.25^2) = 319$$

The values of K are low, indicating that a shallower beam section could have been used as far as moment is concerned. Now we must check the shear stress. The shear at a distance d from the face of the support is

$$V = 10.87 - (5.92)(26)(15.25)/(20)(12)(0.85)(12) = 9.91 \text{ kip}$$

$$v_u = V/bd = 9910/6(15.25) = 108 \text{ psi}$$

The maximum value of the shear that can be carried by the concrete is 10% greater for joists than for beams:

$$v_c = 1.1(2)\sqrt{f_c'} = 1.1(2)\sqrt{4000} = 139 \text{ psi}$$

The value of v_u is less than v_c, so the joist is satisfactory. It is evident that the joist is oversize and could be reduced.

Beam Design We will use the same width and depth of beam as originally assumed, even though it is evident from the previous calculations that the depth of joists could be reduced. The ultimate loads on the beam are:

Dead load of joists	90(30 − 5)(1.4)/1000	= 3.15 klf
Beam	60(16.5)(150)(1.4)/144(1000)	= 1.44 klf
	Total dead load	4.59 klf
Reduced live load	52(30)(1.7)/1000	= 2.65 klf
	Total dead plus live	7.24 klf

The values of the moment of inertia I and the relative stiffness K of the beam are

$$I = bd^3/12 = 60(16.5^3)/12 = 22460 \text{ in.}^4$$

$$K = 4I/L = 4(22{,}460)/20(12) = 374$$

The column stiffness for the columns above and below from the joist calculations is $K = 177.8$. The distribution factors for the end are

	K	D.F. = $K_i/\Sigma K$
Column	177.8	0.322
Beam	374.0	0.678
Total	551.8	1.000

For the interior supports, not assuming symmetry, the distribution factors are

Column	177.8	0.192
Beam, left	374.0	0.404
right	374.0	0.404
Total:	925.8	1.000

For the interior supports, assuming symmetry and a value of K for the center span of half of its normal value; the distribution factors are

Column	177.8	0.241
Beam, left	374.0	0.506
right	187.0	0.253
Total:	738.8	1.000

The fixed-end moments are, for both spans,

Dead load	$4.59(20^2)/12 =$	153.0 ft-kip
Live load	$2.65(20^2)/12 =$	88.3 ft-kip
	Total:	241.3 ft-kip

Distributions are shown in Fig. 237. Following are the combinations that give maximum design moments:

For maximum moment at A: live load on AB and CD,

$$M_A = 54 + 36 + 1 = 91 \text{ ft-kip}$$

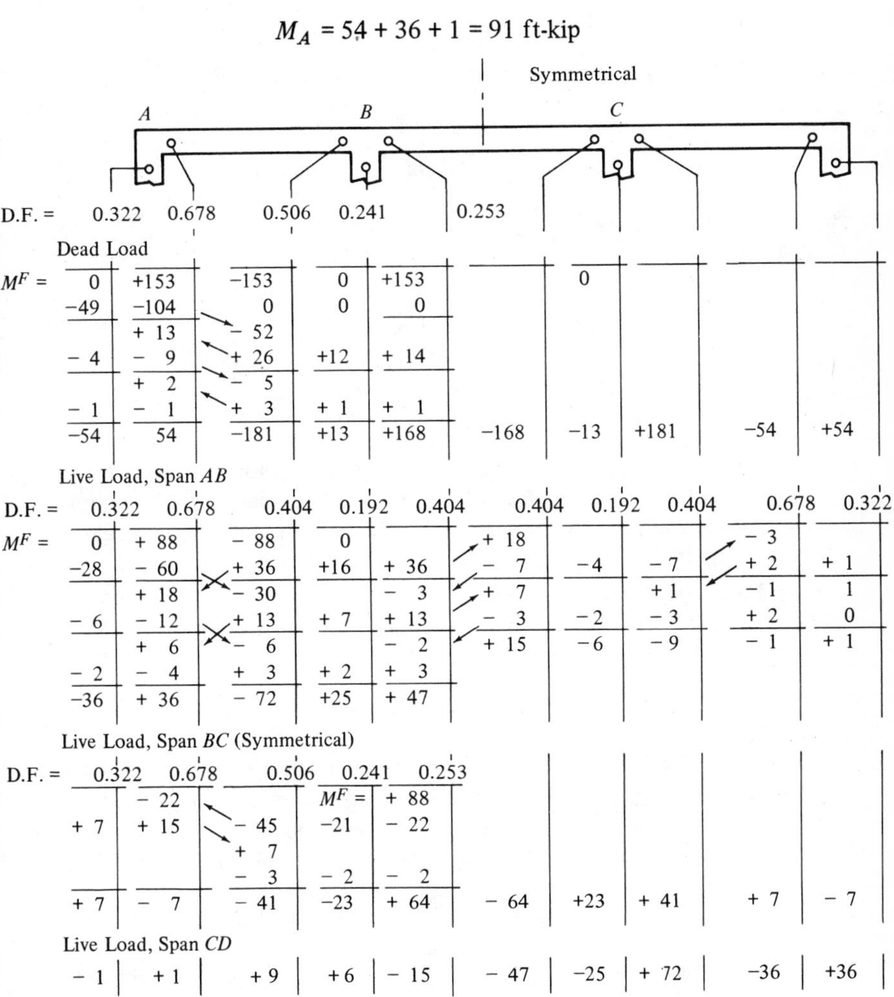

Fig. 237.

For maximum moment at center of AB: live load on AB and CD,

$$M_A = 91 \text{ ft-kip}$$

$$M_B = -181 - 72 + 9 = -244 \text{ ft-kip}$$

For maximum moment at B: live load on AB and BC,

$$M_B = -181 - 72 - 41 = -294 \text{ ft-kip}$$

For maximum moment at center of BC: live load on BC,

$$M_B = -168 - 64 = 232 \text{ ft-kip}$$

$$M_C = -168 - 64 = -232 \text{ ft-kip}$$

The simple-beam moment for the total load is

$$M_s = 7.24(20^2)/8 = 362 \text{ ft-kip}$$

The maximum moments at the center are

span AB,

$$M = -(91 + 244)/2 + 362 = 195 \text{ ft-kip}$$

span BC,

$$M = -232 + 362 = 130 \text{ ft-kip}$$

We are now ready to check the size of the beam for moment. We are allowed to reduce the moment to the face of the support, but because the columns are narrow with respect to the total width, we will use the moment at the centerline, $M = 294$ ft-kip. For the design use $M = 294/0.9 = 327$ ft-kip.

The effective depth will be reduced over that required for the joists because of the larger cover required (1.5 in.). Assume No. 8 bars, $D = 16.5 - 0.5 - 1.5 = 14.5$ in. For $b = 60$ in.,

$$K = M/bd^2 = 327{,}000(12)/60(14.5^2) = 311$$

The beams are larger than necessary for moment, but we also must check the shear stress. The calculations for maximum shear in span AB at a distance d from the face of the column are shown in Fig. 238. The shear is $V = 74.9$ kip, and for design we use $V = 74.9/0.85 = 88.1$ kip. The shearing stress is:

$$v_u = V/bd = 88{,}100/60(14.5) = 101 \text{ psi}$$

which must be less than the value of the shear that can be taken by the concrete,

$$v_c = 2\sqrt{f_c'} = 2\sqrt{4000} = 126 \text{ psi}$$

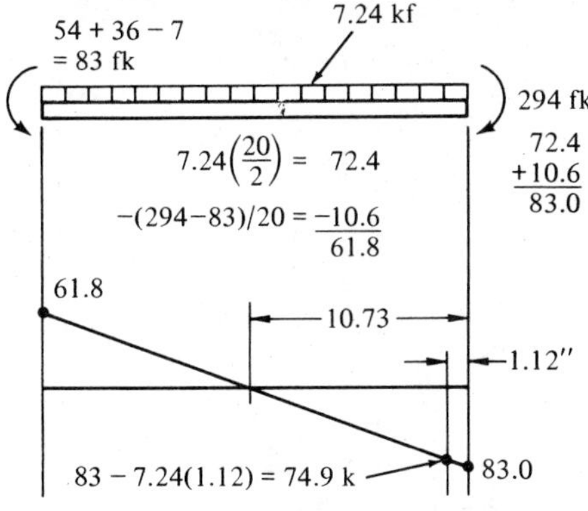

Fig. 238.

The width of the beam is such that stirrups are not required for shearing stress; furthermore, the total depth is less than one-half the width, so that minimum stirrups are not required.

The design that has resulted from our initial guesses for depth of joist (16.5 in.) and the width of the beam (60 in.) is not satisfactory because the beam is wider than necessary. A smaller width will raise the shear stress in the joists and will increase the stresses in the beam, both for shear and for moment. There appears to be no particular reason to decrease the depth of the joists.

The selection of reinforcing for the joists and the girders will be similar to the previous problem, and it will be left to the reader to complete the design calculations.

Concrete Retaining Walls

Most low to medium-height concrete retaining walls are of the cantilever type shown in Fig. 239, but for high walls the counterfort wall, as sketched in Fig. 240, may be necessary or economical. We will confine our discussion to cantilever walls and the problems that arise from their design. The discussion will be in the form of an example.

The theory of the determination of the lateral loads due to earth pressure is outside of the scope of this section. We will use for design an equivalent fluid pressure of 30 psf per foot of depth. If heavy loads are placed next to a wall, these will be taken into account by an additional height of earth fill above the top of the wall, called the *surcharge*. The edge of the base in front of the wall is called the *toe* and the edge of the base behind the wall is called the *heel*.

If properly designed, the base of the wall will be large enough that the soil pressure at the toe will not be too great; it must have an adequate factor of safety

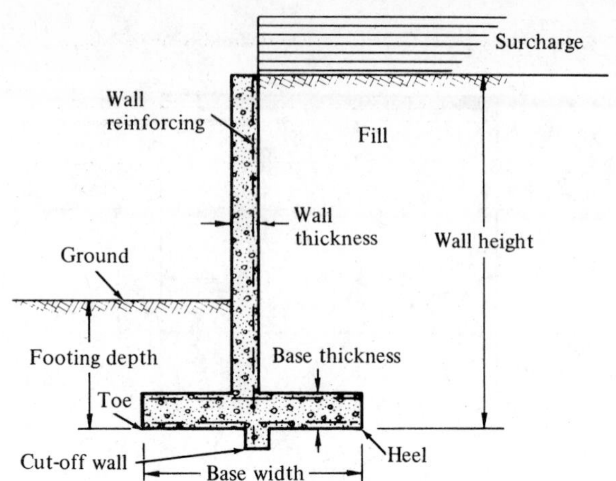

Fig. 239.

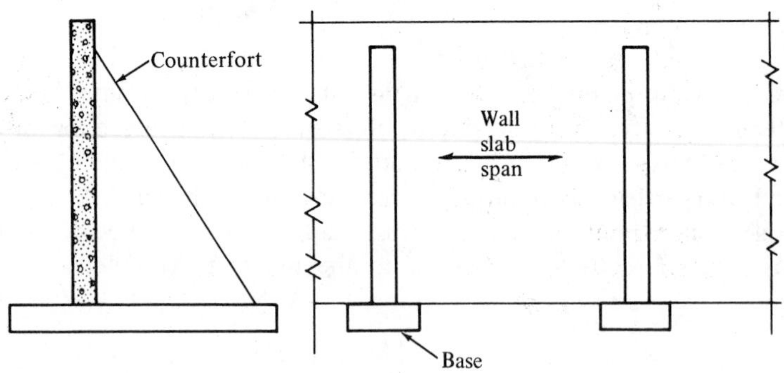

Fig. 240.

against overturning, and must not slide on its base. Also, the wall must be strong enough as a cantilever from its base to resist the lateral earth pressure.

In this presentation we will investigate these problems in the following order:

1. Determine the width of the cantilever wall just above the base and then guess the thickness of the base so that the weight of the wall may be calculated.
2. Guess the width of the base and make a stability analysis of the wall to determine the soil pressures on the base. The position of the resultant of all forces should lie within the middle third of the base and the maximum allowable soil pressure should not be exceeded.
3. Revise the base width if necessary.
4. Check stability against overturning.
5. Check the resistance to sliding and provide a cut-off wall under the base if necessary.

6. Select the reinforcing in the stem and in the base.

Calculations for cantilever walls are usually made on the basis of a length of wall of 1 ft, both for the stability and for the reinforcing in the wall.

A sketch of the proposed retaining wall is shown in Fig. 241. The wall is 16 ft high from the top of the base to the top of the wall. The bottom of the base is 4 ft below the ground level to get below the frost line. The lateral pressure of the fill and the surcharge will be assumed as an equivalent fluid pressure of 30 psf per foot height, and the surcharge height is 3 ft. The weight of the earth fill is assumed as 100 pcf. The maximum allowable soil pressure on the toe of the wall is 4000 psf. Use $f_c' = 3000$ psi and $f_y = 40{,}000$ psi. The design will be made on the basis of a strip of wall one foot wide.

The lateral-soil-pressure diagram is also shown in Fig. 241. The lateral pressure at the base may be assumed to be the result of a triangle with a base of 480 psf with a total lateral load of $P_1 = 480(16)/2 = 3840$ lb, and a rectangle width of 90 psf with a total load $P_2 = 90(16) = 1440$ lb. The moment of these forces about the base after applying the appropriate load factors is

$$M_u = 3840(5.33)(1.4) + 1440(8)(1.7) = 48{,}238 \text{ ft-lb}$$

We must divide by the capacity-reduction factor $\phi = 0.9$, so for design

$$M = 48{,}238/0.9 = 53{,}598 \text{ ft-lb}$$

There is no particular reason to use a thin member for the cantilever wall. The additional concrete will help the stability; the formwork will be no more expensive; the reinforcing will be reduced; and the concrete can be placed more easily. Therefore, we will use a conservative value for the thickness factor $K = M/bd^2$,

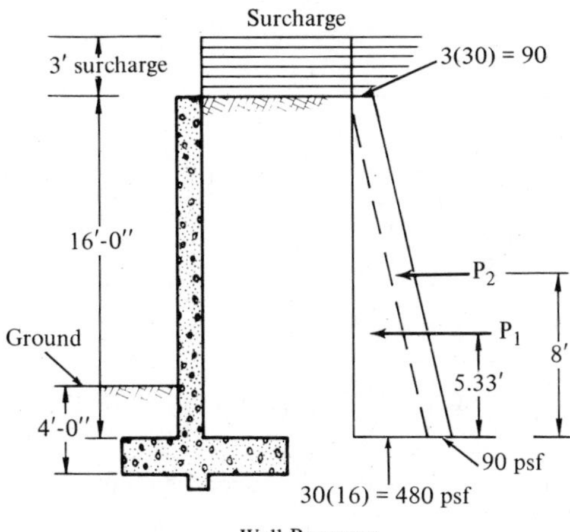

Fig. 241.

say $K = 350$. The effective depth d will be

$$d = \sqrt{M/bK} = \sqrt{53{,}598(12)/12(350)} = 12.4 \text{ in.}$$

A cover over the reinforcing steel of 2 in. will be required, so the total depth of the section assuming No. 6($\frac{3}{4}$-in.) bars would be $t = 12.4 + 0.38 + 2.0 = 14.78$ in. We will use $t = 14$ in., so that the effective depth is $d = 12 - 2.38 = 11.62$ in.

It is not probable that the shearing stress will govern, but we will check. The shear, including the load factors and the capacity reduction factor, is

$$V = (3840 \times 1.4 + 1440 \times 1.7)/0.85 = 9205 \text{ lb}$$

$$v_u = 9205/11.62(12) = 66.0 \text{ psi}$$

The maximum is $2\sqrt{f_c'} = 109$ psi. We could proceed to design the reinforcing in the wall, but first we will investigate the stability of the wall. We will assume that the base is one-half the total height of the wall, or about 9 ft. The vertical and horizontal loads on the wall are shown in Fig. 242. The thickness of the

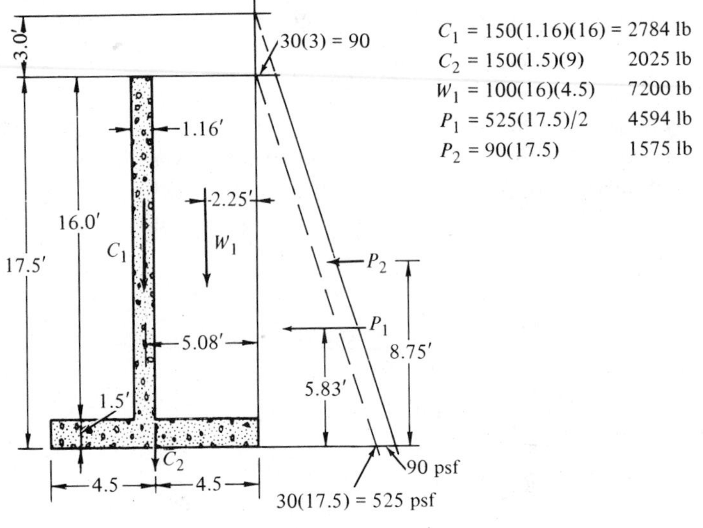

Fig. 242.

base will be assumed to be 18 in., and the inside face of the wall is assumed to lie on the centerline of the base.

The vertical and horizontal forces and the moments of all forces about the heel are

Force	Vertical force (lb)	Horizontal force (lb)	Arm (ft)	Moment (ft-lb)
C_1	2,784		5.08	14,143
C_2	2,025		4.50	9,112
W_1	7,200		2.25	16,200
P_1		4594	5.83	26,783
P_2		1575	8.75	13,781
Totals	12,009	6169		80,019

The horizontal distance of all forces from the heel will be the moment divided by the vertical force:

$$a = 80{,}019/12{,}009 = 6.66 \text{ ft}$$

If the resultant is to be within the middle third, then the width of the base must be 1.5 times the above value of 6.66 ft, so $b = 6.66(1.5) = 9.99$ ft. The width of 9 ft that we have chosen is not satisfactory and so should be increased. A 10 ft width should be satisfactory, so we must revise our calculations.

Place the center of the base at the inside edge of the wall again. The revised values of the loads and moments are:

Force	Vertical loads (lb)	Horizontal force (lb)	Arm (ft)	Moment (ft-lb)
C_1	(no change)	2,784	5.38	15, 535
C_2	150(1.5)(10)	2,250	5.00	11,250
W_1	100(5)(16)	8,000	2.50	20,000
P_1	(no change)		(no change)	26,783
P_2	(no change)		(no change)	13,781
Totals		13,034		87,349

The location of the resultant is now at $a = 87{,}349/13{,}034 = 6.70$ ft from the heel, and the width of the base for the resultant at the middle third is $1.5(6.70) = 10.05$ ft, so the 10-ft width is satisfactory.

To determine the base pressures due to the vertical loads, use the formulas for stress due to a concentric load and due to bending:

$$f = P/A \pm Mc/I$$

For a rectangular area which is b wide and d deep, this equation becomes, using the symbol p for pressure,

$$p = f = P[1 \pm (6e/d)]\ bd$$

where e is the eccentricity of the force about the center of the rectangle. Applying this formula to the above base with $e = 6.70 - 5.0 = 1.70$ ft, we obtain

$$p = 13{,}034\,[-1 \pm 6(1.7/10)/1(10)] = -2633 \quad \text{or} \quad +26 \text{ psf}$$

A slight tension force is indicated. This value of the footing pressure is well within the maximum allowable pressure of $p = 4000$ psf.

The customary value of the factor of safety for overturning of the wall about the toe of the base is 2. The distance a of the center of the vertical loads from the toe is, using the moments and forces in the table above,

$$a = (10.0 - 46{,}785/13{,}034) = 6.41 \text{ ft}$$

The ratio R of the moment of the vertical forces to the amount of the lateral forces about the toe is

$$R = 13{,}034(6.41)/(26{,}783 + 13{,}781) = 2.06$$

so the wall is satisfactory for resistance to overturning.

The criterion we shall use for testing the tendency of the wall to slide sidewise on its base is that if the ratio R of lateral force to vertical force is greater than 0.4, a cut-off wall should be used below the base, as shown in Fig. 239. For this problem,

$$R = 6{,}169/13{,}034 = 0.47$$

so a cut-off wall will be required.

We will make the cut-off the same width and thickness as the base width and place it so that the vertical reinforcing from the wall above extends into the cut-off wall so that it will be effective as reinforcing. This has the additional advantage that it develops anchorage for the vertical-wall steel in a much better manner than if the wall steel is hooked or bent at the level of the bottom of the base.

The next task would be to determine the final base thickness and the moments and shears for determining the reinforcing in the base. This will follow the methods used for designing footings, described under the heading "Design of Concrete Structural Elements."

Bridges

Bridges are one of the most important features of our environment planned by the civil engineer. A bridge can either be a delight or an eyesore. The shape, the alignment, and the elevation of the bridge and how it fits into other elements in the landscape are factors to be considered. If the materials are not long-lasting, or if the bridge is poorly maintained, it will be ugly. No matter how beautiful a bridge may be in the eyes of the designer or the casual viewer, if it does not fulfill its function as a roadway for people, automobiles, or for railroad trains, it cannot be truly beautiful. It is important that both the engineer and the public understand the factors involved in the aesthetics of bridge design.

There are many types of bridges and materials for construction, so several types will be described along with the arrangement of structural elements. With this background, it is possible to discuss some of the problems and rules for aesthetics of bridge design and to indicate where some of the types discussed may be successful in solving environmental problems. Next, the engineering problems of bridge design are presented, including the basic rules of the American Association of State Highway Officials. Finally, an example for the design of a small beam bridge is given.

Types of Bridges A general classification system for bridges, while rather dull reading, serves to reveal the many types of bridges that are possible. We will be generally concerned with highway bridges or special bridges over highways. This list is not intended to be exhaustive, but to serve as an indicator of some of the possible types. There are many combinations not shown here. Sketches, where necessary, to show the configuration are indicated by figure numbers.

Classified by Materials in the Main Structural Members

Steel with concrete deck either laid on the steel or made composite with steel by means of connectors.

Steel with steel-plate deck, the so-called orthotropic plate. This greatly reduces the dead load and may be economical for long spans.

Reinforced concrete, cast in place at the site. The high cost of formwork has made this type uneconomical for many applications in this country.

Prestressed concrete—either *pretensioned*, cast at a manufacturing yard and trucked to the site; or *post-tensioned*, cast at the site with the prestressing tendons tightened after the concrete has hardened.

Timber, for small spans on local roads, particularly in the Pacific northwest where timber is plentiful.

Aluminum has been used for experimental bridges.

Classified by Principal Structural Elements

Slab (*Fig. 243*). Concrete slabs with simple spans are suitable for short spans over small streams, or continuous spans with intermediate piers may be used for wide, shallow streams where the foundations are suitable.

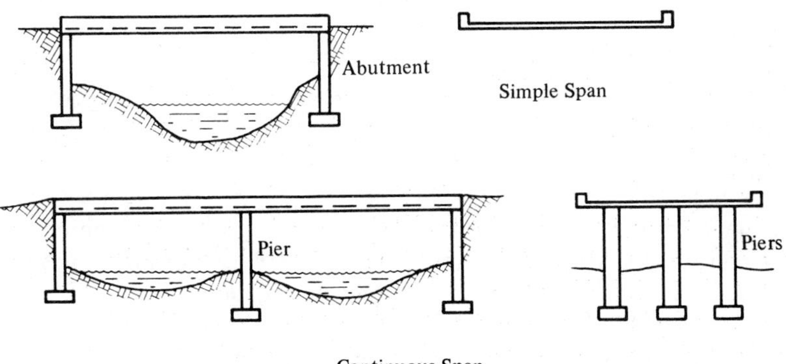

Fig. 243. Bridges constructed using concrete slabs.

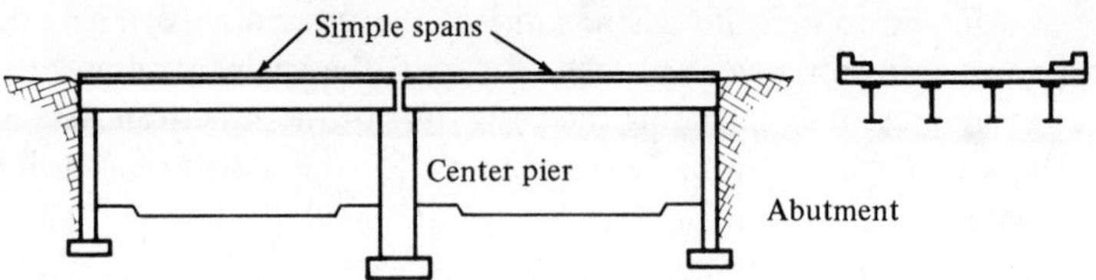

Fig. 244. Bridge constructed using beams.

Beam (*Fig. 244*). Steel or concrete beams with: (1) simple spans for short spans with intermediate piers; or (2) continuous, cantilever spans with intermediate hinges, or (3) continuous spans, where settlement of foundations will not affect the stresses.

Pile. For marsh or swamp areas, piles are used with a timber or concrete deck for short spans.

Plate Girder. Beams are fabricated from steel plate with large or variable depths, so are suitable for long spans. A deck-plate girder has the roadway at the top of the girder, and a through girder has the roadway running between the girders to improve the clearance under the bridge.

Rigid Frame (*Fig. 245*). Steel or concrete. The beam is continuous with the abutment walls, so the clearance under the roadway at the center may be reduced. Often used at grade separations for multilane highways.

Fig. 245. Rigid-frame bridge.

Truss (*Fig. 246*). Usually steel. Suitable for longer spans and may be simple, cantilever-continuous, or continuous and may have deck or through trusses. There are many configurations of trusses.

Box Girders (*Fig. 247*). Steel or concrete, simple or continuous spans. Box girders are particularly suitable where torsion stresses are high, as in curved girders.

Suspension (*Fig. 248*). Steel, suitable for very long spans. A disadvantage is the amount of deflection under live load for a heavy concentration of load, as for a railroad bridge. Stresses are reduced by making the stiffening girder or truss more flexible at the expense of additional deflection.

Cable-Stayed Girders (*Fig. 249*). Steel-plate girders with a few stays to reduce the bending. Suitable for intermediate spans. Has the virtue of simplicity over the suspension bridge.

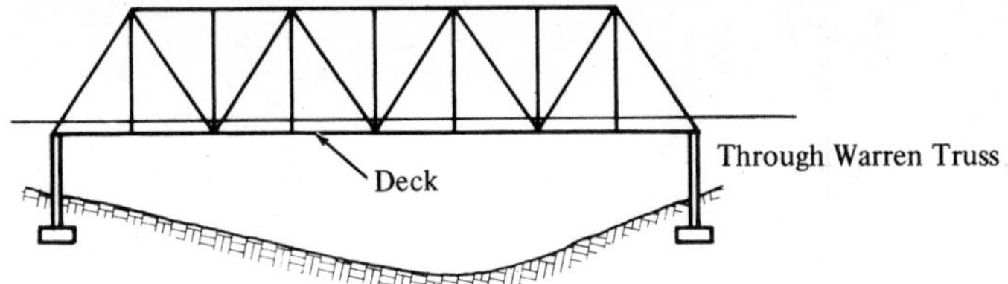

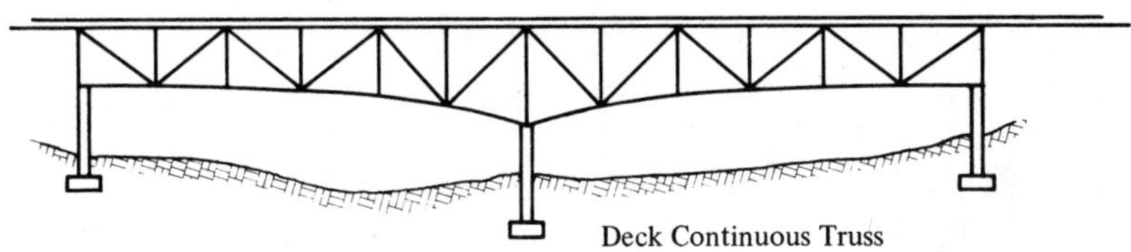

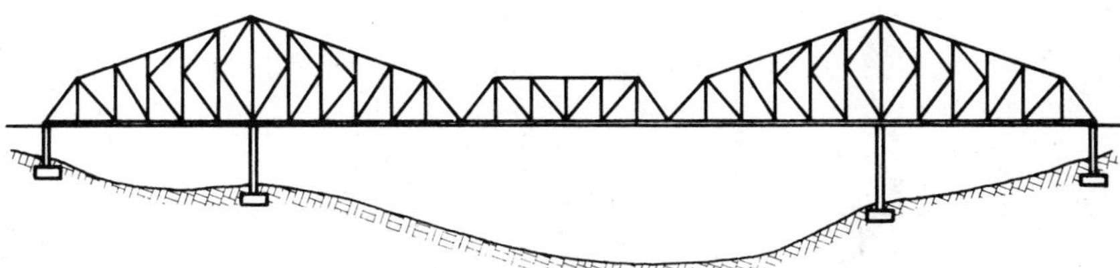

Fig. 246. Three types of truss bridges.

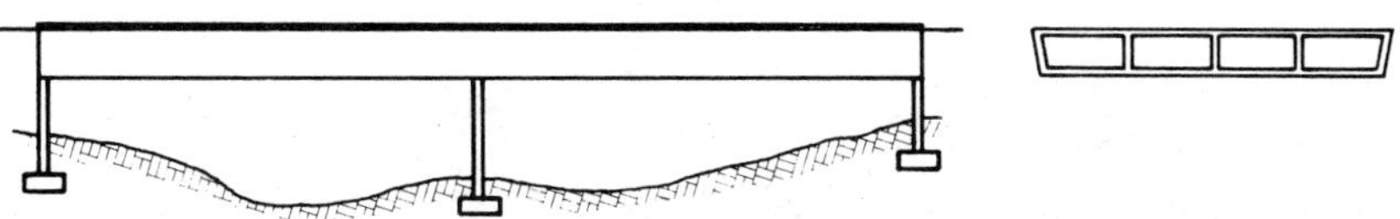

Fig. 247. Continuous box-girder bridge.

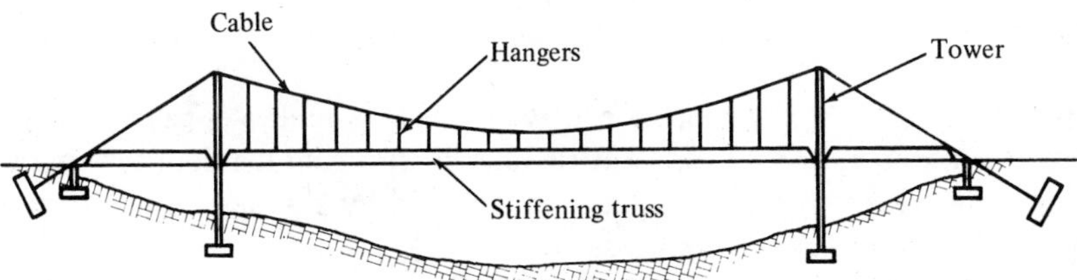

Fig. 248. Suspension bridge.

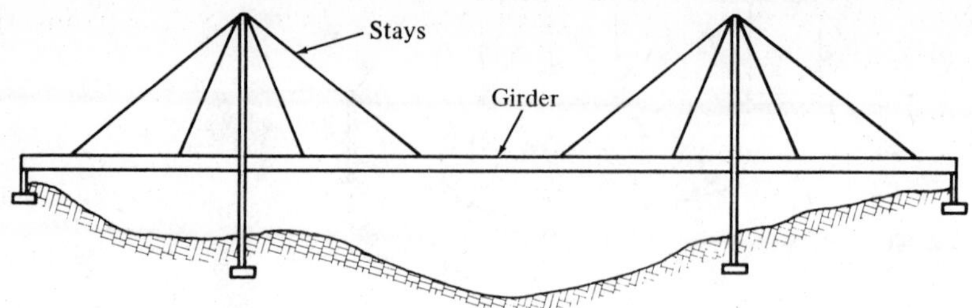

Fig. 249. Cable-stayed bridge.

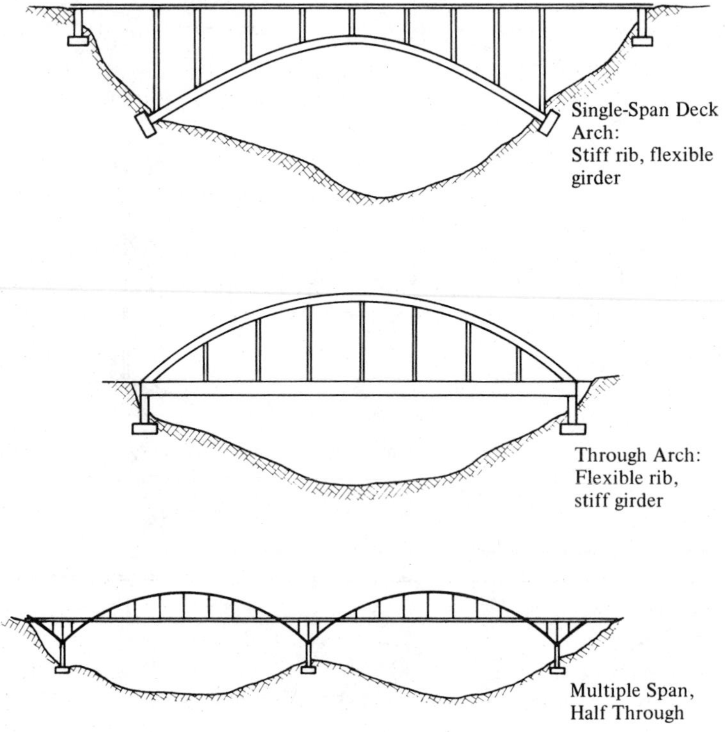

Fig. 250. Three types of arch bridge.

Arch (*Fig. 250*). Steel or concrete; may be single- or multiple-span and may be deck, half-through, or through. The rib may be thick and the floor system thin, or the rib may be thin and flexible and stiffened by a girder. Arches may be made with solid ribs or may be trussed.

Classified by Function and Type of Crossing

Types of crossings are small streams, rivers, estuaries, canyons, freeway overpasses, railroad yards, or railroad overpasses. Bridges may carry automobiles, railways, pedestrians, and even canals. Movable bridges may be swing, lift, bascule, or floating.

Classified by Plan Considerations

Bridges may be straight, skewed (at an angle to the stream or roadway it is crossing), or curved. The bridge may be level, on a grade, or curved in elevation.

Aesthetics Everyone agrees that it is important for our natural environment that bridges be beautiful, but all will not agree on what constitutes beauty. The cliche "Beauty is in the eye of the beholder" still has a great deal of meaning to the bridge engineer, both with respect to himself and to the layman. Probably some of the ugly bridges that we see were beautiful to their designers because they saw how cheap the bridge was to build. If a bridge user is irked by slow traffic on a bridge, it is not beautiful. If a yachtsman cannot get under a bridge and it has ruined his sailing, then the bridge is ugly. What is beautiful to an engineer who can see the inner workings of a bridge may not be understood by a layman.

Therefore, one of the primary requirements for aesthetics is that the bridge must serve its function properly. Most bridges are seen more by the people that cross them than from the side, so it is first important that the roadway be clean and free from obstructions. One very frustrating aspect of modern bridges is that often the hand rails are so high you cannot see over the bridge. If the scenery is interesting, this can annoy the passengers in an automobile. Of course the rail must serve its function of holding cars and people on the roadway, but if the designer starts with this requirement of good vision for the passengers, then the problem can be solved during the process of design.

In most small bridges, there is little scenery to be observed and the most important consideration is maintenance of the traffic speed and volume. Therefore, the bridge should continue the same shoulders and guard rail used on the approaches, and the roadway surface should be the same.

Many bridges will never be seen from the side, so it is of little importance how they look as long as they function properly for traffic, and are economical to build and to maintain. Bridges over freeways, however, are another matter. They should be as inconspicuous as possible so the driver of the automobile going underneath is not bothered by the bulk of the bridge flying toward him at 70 miles an hour. The spans should be as long as is economically possible and the depth should be small. The bridge should extend over the shoulders so the normal roadway is not contracted. A center pier, if necessary, should be as thin and inconspicuous as possible. The alignment of the road below the bridge should be such that you do not drive directly toward a wall or a pier, so there is not a mental hazard which tends to slow traffic, or a potential safety hazard if a car gets out of control.

If a bridge can be seen from the side, and especially if it is large and monumental, then the shapes and arrangement of the individual structural elements become important. Systems must be as simple as possible, and the method of carrying the forces must be evident to the layman as well as to the engineer. Truss bridges, while economical, are not easy to understand and so may appear ugly to some. Eventually, however, if there are enough small elements in a truss, they will blend into one whole and the bridge will appear as a single structural element. A stiffening truss on a suspension bridge is a good example.

Ornamentation on bridges now serves no useful function, as no one has time to see or study it. On bridges subject to a high volume of pedestrian traffic, as in London or Paris, ornament and statuary are a joy because there is time for observation.

Architects and artists have rules of order and proportion that have been developed over the ages. These should be studied and used where applicable. Architects have much to contribute to bridge design provided they, and the engineer, understand that their function is not to add ornament to the engineer's design but to contribute to the basic design in terms of function and proportion.

Finally, the materials and methods of construction used must be good and long-lasting. Otherwise, the bridge will be ugly in a few years. It seems improbable that we will ever arrive at the age of the disposable bridge that lasts only five years and is torn down to make room for the latest model from Washington.

AASHO Live Loads The standard load for highway bridges is a single truck or a tractor-trailer combination, as shown in Fig. 251. Smaller loadings (H15, HS15 or H10) are in proportion to the H20 or HS20 loads. The number 44 indicates

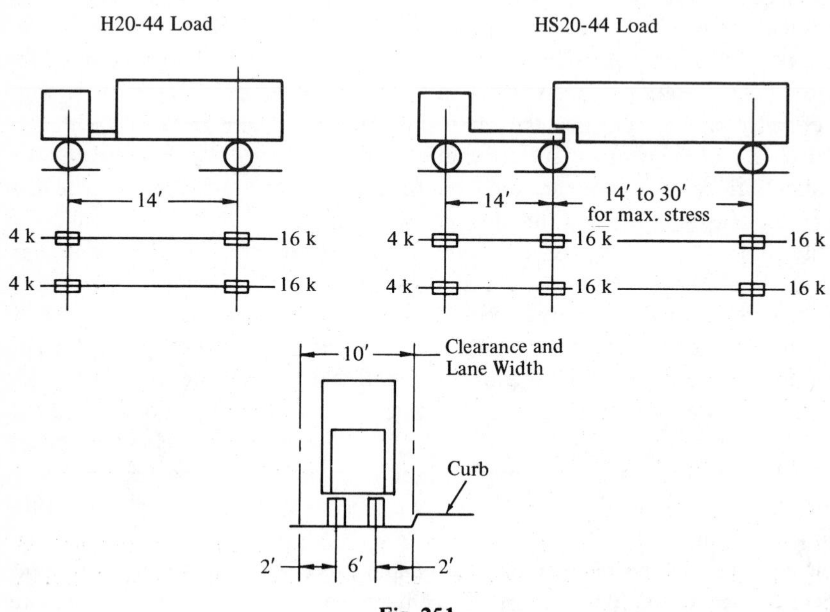

Fig. 251.

the year this loading was established. For the HS loading, the rear wheel is placed so that it produces the maximum stress in the structural element of the bridge. Usually the 14 ft will be used. A standard lane is 10 ft wide and the center of a set of wheels can get within 2 ft of the curb, as shown in the sketch. If a bridge is from 20 to 30 ft wide, then only two lanes need be placed on the bridge; if it is from 30 to 42 ft wide, only three.

It is not necessary to design for a string of trucks. Only a single truck is placed on the bridge at one time. To take into account extended loads, a "lane load" is provided which is a uniform load of 640 lb per lineal foot of lane with a concentrated load to be placed for maximum stress. If the stress involves moment, this load is 18 kip, and if it involves shear, the concentrated load is 26 kip. The lane load is the same for both H loading and HS loading. The H15 and HS15 and the H10 are in proportion. Either the single-truck or the lane load should be used, depending on which gives the greatest stress. For example, for single-span bridges, the truck load gives a greater stress than the lane load up to a span of 130 ft, as may be observed from the tables for maximum moment in the AASHO *Specifications*, page 414 of the 1973 edition. If lane loads are used for continuous spans, then two concentrated loads should be used.

In addition to the live loads as described, an additional impact load must be included which is inversely in proportion to the span. The percentage of impact is obtained from the following formula:

$$I = 50/(L + 125)$$

where L is the span length in feet. The maximum percentage of impact required is 30%. For continuous spans use only the span immediately under the truck for positive moment, and for negative moment, usually over the interior support, use the average of the two adjacent spans. For shear use the length of the loaded position of the span from the point under consideration to the far reaction.

For more than two lanes on a bridge, the loads may be reduced in accordance with the number of lanes as follows:

One or two lanes	100% of basic load
Three lanes	90% of basic load
Four or more lanes	75% of basic load

AASHO Code Provisions for Design The *AASHO Standard Specifications for Highway Bridges* now allow alternate design methods, for both steel and concrete. The working stress method uses the yield stress with an appropriate factor of safety. The load factor uses increased loads and higher stresses similar to the strength method of the ACI Code. We will discuss only the working stress method.

Structural Steel The provisions for the design of structural steel are long and detailed, so that only those provisions that directly affect the design of girders and composite design will be discussed.

Types of Steel The code makes provision for three types of steel: A36, A441, and a group including A242, A440, and again A441. The ultimate tensile strength for A36 is 58 ksi and the yield point is 36 ksi. For A441 steel with thicknesses over 4 in. and up to 8 in., the ultimate tensile strength is 60 ksi and yield point is 40 ksi. The steels in the third group have ultimate tensile strengths of 63 ksi for thicknesses of $1\frac{1}{2}$ to 4 in., with a yield point of 42 ksi; for thicknesses of from $\frac{3}{4}$ in. to $1\frac{1}{2}$ in., an ultimate tensile strength of 67 ksi and yield point of 46 ksi; and finally, for thicknesses of $\frac{3}{4}$ in. and under, an ultimate tensile strength of 70 ksi and a yield point of 50 ksi. Therefore, it is incumbent on the designer to visualize the thickness of plates before selecting the area of material required.

Allowable Stress For axial tension and tension in beams and girders, the allowable stress is $0.55F_y$, where F_y is the yield strength, so that A36 steel has a usable stress of 20 ksi. Listing the formulas for other types of stress with A36 steel:

Shear in beams and girder webs gross section, $F_v = 0.33F_y$	12 ksi
Bearing on milled stiffeners, $0.80F_y$	29 ksi
Stress in pins, $0.80F_y$	29 ksi
Shear in pins, $F_v = 0.40F_y$	14 ksi
Bearing on pins not subject to rotation, $0.80F_y$	29 ksi
Bearing on pins subject to rotation, $0.40F_y$	14 ksi
Bearing on power-driven rivets and high-strength bolts (parent metal), $1.22F_y$	44 ksi

Ordinary bolts are allowed 13.5 ksi in tension at the root of the thread, 20 ksi in bearing, and 11 ksi in shear.

There are two grades of rivets: A-502 grade 1 with a bearing stress of 40 ksi and a shear stress of 13.5 ksi, and A-502 grade 2 with a bearing stress of 40 ksi and a shear stress of 20 ksi.

High-strength bolts are allowed a stress of 36 ksi in tension and 40 ksi in bearing. For shear there are two types of connections: friction-type connections are assigned 13.5 ksi in shear, and bearing-type 20 ksi in shear.

Depth Ratios The following limiting depth ratios are given:

Beams and girders	1/25
Composite girders, overall	1/25
steel only	1/30
Trusses	1/10

For continuous spans, the length shall be the distance between dead load points of contraflexure.

Deflections The ratio of deflection to span for live load plus impact is limited to 1/800 in general and 1/1000 for urban areas used in part by pedestrians.

Cantilevers are limited to 1/300 and 1/375. The gross cross-sectional area may be used in computations and for composite members the live load may be considered to be acting on the composite section.

Minimum Thicknesses Thicknesses must be at least $\frac{5}{16}$ in. except for web thicknesses of rolled beams, which shall not be less than 0.23 in.

Welding Butt welds may be designed for the same strength as the parent metal. For fillet welds on A36 steel $F_v = 13{,}400$, and for the other steels $F_v = 14{,}700$ psi. There are provisions for fatigue stresses in welds depending upon the number of cycles of maximum stress.

Rolled Beams Bearing stiffeners should be provided if the shear is larger than 75% of the allowable shear for the web. Cover plates should be increased in length by $2D + 3$ in., where D is the depth of the beam. Cover plates are limited to one per flange and the maximum thickness not greater than $1\frac{1}{2}$ times the thickness of the flange.

Composite Girders The value of n for composite design varies as follows:

$$f'_c = \begin{cases} \text{3–3.9 ksi}, & n = 10 \\ \text{4–4.9 ksi}, & n = 8 \\ \text{5 ksi or more}, & n = 6 \end{cases}$$

For dead-load deflection, these values should be multiplied by three.

Shear connectors are designed on the basis of flexure theory and ultimate load. Appropriate formulas are used for the capacity of the connectors. For example, studs are allowed a value of $Z = 27\,HDf'_c$, where H is the height and D is the diameter.

Code Provisions for Concrete Design Concrete-cylinder strengths are set with a minimum (basic) strength of 3000 psi with a maximum of 4400 psi. The working strength in compression is $0.4 f'_c$ instead of the $0.45\, f'_c$ used in building design. The allowable shear on beams without reinforcement is $0.02 f'_c$ with a maximum of 75 psi if longitudinal bars are not anchored, or $0.03\, f'_c$ with a maximum of 90 psi if they are anchored. (Anchorage requirements will be described later). The maximum shear with web reinforcement is $0.075\, f'_c$, with $0.03\, f'_c$ taken by the concrete.

For reinforcing there are essentially two grades, structural grade (33,000 psi yield) with an allowable stress of 18,000 psi in tension, and intermediate and hard grades (40,000 and 50,000 psi yield) with an allowable stress of 20,000 psi in tension.

Bond stresses for either straight or hooked bars (but not including top bars) are $0.10\, f'_c$ in beams, slabs, and one-way footings with a maximum of 350 psi, and $0.08\, f'_c$ in two-way footings with a maximum of 280 psi. Bond for top bars in beams or slabs (having less than 12 in. cover) is $0.06\, f'_c$, with a maximum of 210 psi. For bars larger than 11S, the bond values are reduced by $\frac{1}{3}$.

Working-Stress Theory The value of n for calculation of strength is 10, 8, and 6 for cylinder strengths of, respectively, 3–3.9 ksi, 4–4.9 ksi, and 5 ksi or more. For deflections, the value of n is 8. If plastic flow is to be included, the value of n is 30.

For the determination of the elastic properties of sections for beam or frame calculations, the gross concrete section may be used, neglecting the effect of steel reinforcement. An exception is columns or other compressive members where the transformed area shall be used. For the bases of columns a hinge may be assumed unless some actual restraint is provided.

T Beams The effective width of a T-beam flange follows the usual requirements: either one-fourth the span length of the beam, the distance center to center of beams, or twelve times the least thickness of the slab plus the width of the girder stem, whichever is least. For an L beam the requirements are one-twelfth the span length, one-half the clear distance to the next beam, or six times the thickness of the slab. For T beams over 40 ft in length, diaphrams are required at either the middle or the third points.

Reinforcing Requirements The clear distance between parallel bars shall not be less than the diameter of the bar, $1\frac{1}{3}$ times the maximum size of the coarse aggregate, or 1 in. The distance between layers shall not be less than 1 in. and the top layers shall be directly above the bottom layers.

The *maximum* spacing of bars shall be not greater than $1\frac{1}{2}$ times the thickness of the wall or the slab. The maximum spacing of temperature bars shall be 18 in.

The minimum cover is 2 in., except for slabs, in which case the requirements are: top of slab $1\frac{1}{2}$ in.; bottom, 1 in. Stirrups or ties in T beams may be spaced $1\frac{1}{2}$ in.

Bar splices for bars in tension shall be lapped not less than 24 bar diameters for 40-ksi-yield-strength steel, and 30 diameters for 50-ksi-yield-strength steel, nor less than 12 in.

For compression splices for cylinder strengths of 3 ksi or over, the splice length may be 20 diameters.

Reinforcing bars shall be extended at least 15 diameters but not less than $\frac{1}{20}$ of the span length beyond the point at which computations indicate it is no longer needed to resist stress. One-third of the positive reinforcement shall extend beyond the face of support for simple beams or the freely supported ends of continuous beams, and the distance shall be sufficient to develop one-half the allowable stress in the bars. If the beam is restrained, the ratio shall be 1:4.

Webs of beams shall have temperature reinforcement on each face of not less than $\frac{1}{8}$ in.^{2} per foot of height with a spacing of not more than 18 in.

Compression Reinforcement Ties and stirrups shall be used for compression reinforcement spaced not more than 16 bar diameters apart. Twice the calculated stress may be used, but not greater than 16,000 psi.

Shear Reinforcement Shear reinforcement follows the usual method for working-stress design: the unit shearing stress is $v = Vbjd$ and bond $u = V/jd\Sigma_0$. The value of j may be assumed to be $\frac{7}{8}$. For bent-up bars the spacing shall be measured at the neutral axis and shall not exceed the depth of the beam. The angle may be from 20° to 45°.

Columns The minimum column reinforcement is 1%, and not more than 8% for spiral reinforced columns, and 4% for tied columns.

For short columns (length-to-width ratio less than 10), the allowable load for spiral columns is for A_g = gross area, A_s = area of steel, f_s = allowable stress in concrete,

$$P_s = 0.225 f_c' A_g + A_s f_s$$

and 0.8 times this value shall be used for tied columns.

For long columns (ratio of length to depth is 10 to 20) the load shall be

$$P_1 = P[1.3 - (0.03\, L/d)]$$

where P is the load on the short column. For a ratio over 20 the column shall be investigated for stability.

For combined bending and axial stress, if the ratio e/t is less than 0.5, the section can be designed on the basis of the uncracked section.

Load Distribution to Beams and Slabs Trucks and cars cannot be made to run directly over beams or girders and, in addition, the beams will deflect under load and pass some of the load to adjacent beams, so generally a truck is carried by several beams or girders. The precise analysis of the distribution is a very complex problem, which has been resolved by the adoption of simple formulas and rules for each of the several types of bridge deck structures. The rules are based on a long series of theoretical and experimental studies by F. E. Richart, N. M. Newmark and C. P. Siess, which is published in the *Transactions of the American Society of Civil Engineers*, 1949, pages 980 to 1023. Also studied was the distribution of bending moments in concrete slabs.

For steel I-beam stringers or prestressed-concrete girders, the portion of a set of wheel loads that goes to an interior girder is $S/7.0$, where S is the stringer spacing. This applies to bridges designed for one lane of traffic (e.g., a single-lane ramp). For normal two-lane bridges the portion is $S/5.5$. For example, if the stringer spacing is 6.0 ft, a girder for a two-lane bridge would have placed on it 6.0/5.5 or 1.09 wheel loads. If the value of S exceeds 10 ft for one lane or 14 ft for the two lanes, then each stringer is designed for the reaction of the wheel loads, as on a simple beam.

Other materials and structural systems have different ratios and limiting values as follows:

System	One lane	Maximum	Two lanes	Maximum
Concrete T beams	$S/6.5$	6 feet	$S/6.0$	10 feet
Concrete box girders	$S/8.0$	12 feet	$S/7.0$	16 feet

For outside stringers, the loads are found by assuming that the floor slab acts as a simple beam between stringers; if there are four or more steel stringers, the fraction of the wheel load shall not be less than $S/5.5$ if S is 6 ft or less, and $S/(4 + 0.25S)$ if S is more than 6 ft and less than 14 ft. For outside stringers of box girders, the wheel distribution is $W_e/7.0$, where W_e is the top width as measured from the outside edge to the middle of the first interior slab span. Also, the additional dead load carried by the exterior stringer, e.g., curbs, railings, and wearing surface, can be considered to be carried evenly by all the roadway stringers or beams. This greatly simplifies the design of exterior stringers, so that usually all stringers will be the same size, resulting in a simpler and more economical structure.

The distribution of moments in concrete slabs is obtained from similar formulas involving the slab spans. For slabs with the main reinforcement parallel to traffic, for H20 or HS20 loading, the moment in foot pounds per foot width is

$$M = (S + 2)P/32$$

where S is the slab span and P = kip, the wheel load for the H20 loading. If slab spans are continuous these moments are multiplied by 0.8 both for the positive and negative moments for live load.

Design of a Single-Span Steel-Beam Bridge Select the slab thickness and reinforcing and the size of the longitudinal steel beams for the small single-span bridge shown in Fig. 252. Design for HS20-44 loading. Use the following design stresses: concrete, f_c = 1200 psi, f_s = 20 ksi; steel, f_s = 20 ksi. For the ratio of the modulus of elasticity of steel to that of concrete use n = 10. Design for an additional wearing surface of 24 psf. Use a cover of reinforcing steel of 1.5 in. for the top of the slab and 1 in. for the bottom.

The layout of the slab, rail, curbs, and beams is shown in Fig. 252. The roadway width is 28 ft. Beams are spaced at 8 ft 4 in. on centers with a 4-ft-2-in. overhang of the slab at the outside girders.

The Floor Slab The first task is to design the floor slab. We will guess that the total depth is 7 in. and revise this depth to obtain the required capacity. The effective depth with 1.5 in. cover over the steel for maximum moment over the support, assuming that we may have a 1-in.-diameter bar, is $d = 7 - 1.5 - 0.5 = 5.0$ in.

Dead loads, using 12.5 lb per inch of slab thickness, are:

Slab	7(12.5) =	88 psf
Additional wearing surface		24 psf
	Total:	112 psf

For the bending moment in the slab due to this dead load we will use a coefficient $M = wL^2/10 = 112(8.33^2)/10 = 777$ ft-lb.

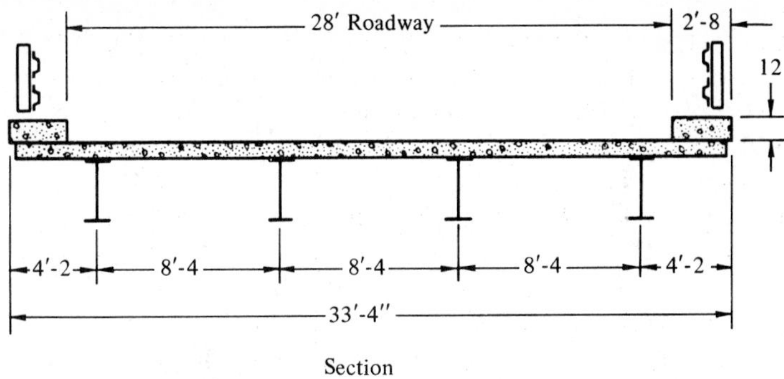

Section

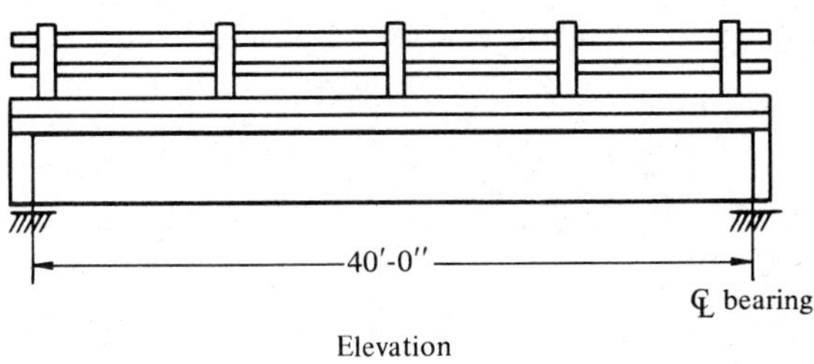

Elevation

Fig. 252.

The AASHO formula for live load moment in the slab has been described previously:

$$M = (S + 2)(P)/32 = (8.33 + 2)(16{,}000)/32 = 5160 \text{ ft-lb}$$

The slab spans are continuous so we can use 0.8 times this figure. In addition, we must add an impact factor in accordance with the formula

$$I = 50/(L + 125) = 50/(8.33 + 125) = 0.375$$

but the maximum is 0.30, which we shall use.

The moments in the slab over a support are:

Dead load		777 ft-lb
Live load	5160(0.8) =	4128 ft-lb
Impact	4128(0.30) =	1238 ft-lb
	Total:	6143 ft-lb

We are now ready to check the total depth of 7 in. we guessed and to select the reinforcing. We will assume that the stresses in the slab are at the design values f_c = 1200 psi and f_y = 20 ksi, and then we will find the capacity. If it is not sufficient we must select a greater depth. The stress diagram of the trans-

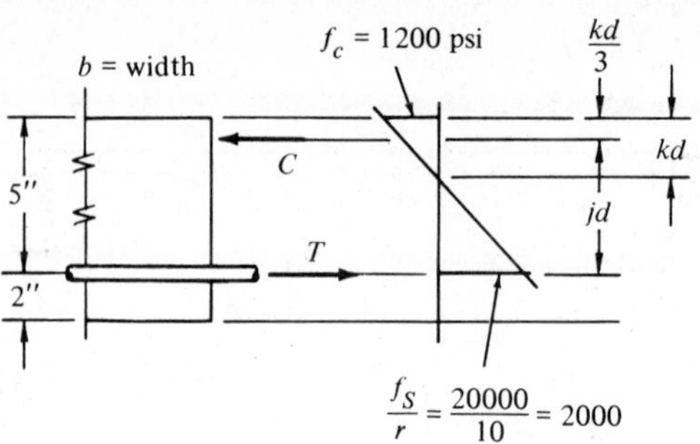

Fig. 253.

formed section is shown in Fig. 253. Steel is shown on the bottom and compression on the top because that is the most familiar way in which reinforcing is arranged, but actually for our example the steel may also be on the top of the slab.

From the previous description of working-stress theory for reinforced concrete, and from the relationships shown in Fig. 253,

$$k = f_c/[f_c + (f_s/n)] = 1200/[1200 + (20{,}000/10)] = 0.375$$
$$j = 1 - (k/3) = 0.875$$

The value of the total concrete force for a section 12 in. wide will be the total force in the triangular wedge:

$$C = bdkf_c/2 = 12(5)(0.375)(1200)/2 = 13{,}500 \text{ lb}$$

The moment of this force about the steel is

$$M = Cjd = 13{,}500(0.875)(5) = 59{,}062 \text{ in.-lb} \quad \text{or} \quad 4{,}922 \text{ ft-lb}$$

It appears that we have made a poor guess. Our capacity is less than the M = 6143 ft-lb required, so we will have to increase the depth. We will now try t = 8 in., so that d = 6.0 in. For the present we will neglect the additional dead load for the extra 1-in. thickness. We have

$$C = 12(6.0)(0.375)(1200)/2 = 16{,}200 \text{ lb}$$
$$M = 16{,}200(0.875)(6.0)/12 = 7{,}087 \text{ ft-lb}$$

We are now well over the requirement. A scratch-pad computation will indicate that 7.5 in. is not sufficient so we will use the 8-in. depth.

The dead load must be increased by the additional depth of concrete. The additional weight is 12.5 psf, so the additional dead load is

$$M = 12.5(8.33)^2/10 = 87 \text{ ft-lb}$$

and the total moment is

$$M = 6143 + 87 = 6230 \text{ ft-lb}$$

The slab is satisfactory. For determining the area of steel required, we will use the above moment with a value of $j = 0.875$ rather than slightly more accurate figure based on the exact stress distribution.

$$A_s = M/jdf_s = 6230(12)/0.875(6.0)(20{,}000) = 0.712 \text{ in.}^2/\text{ft}$$

To furnish this area of steel, No. 7 bars spaced at 10-in. centers will be required: $A = 0.6(12)/10 = 0.72$ in.^{2}. The positive reinforcement will have a greater effective depth and the moment will be less because of the dead load. Therefore, less area will be required; however, we will use the same area as above for positive moment.

In the opposite direction, parallel to the steel floor beams, in the bottom of the slab, distribution reinforcement will be required. The AASHO *Specifications* require the following percentage of the main positive reinforcement:

$$\text{Percentage} = 220/\sqrt{S}$$

with a maximum of 67%. For this case $P = 220/\sqrt{8.33} = 76$, so we will use 67% and $A_s = 0.71(0.67) = 0.47$. Use No. 6 bars at 12-in. centers.

A sketch of the reinforcing with two possible arrangements of bars is shown in Fig. 254. The second arrangement with the bent bars has the virtue that the bends hold up the top bars over the support and special high chairs are not required for support.

Longitudinal Beams All beams will be of the same size, and the design will be based on an interior girder. The weight of curbs and rails will be divided between all of the beams.

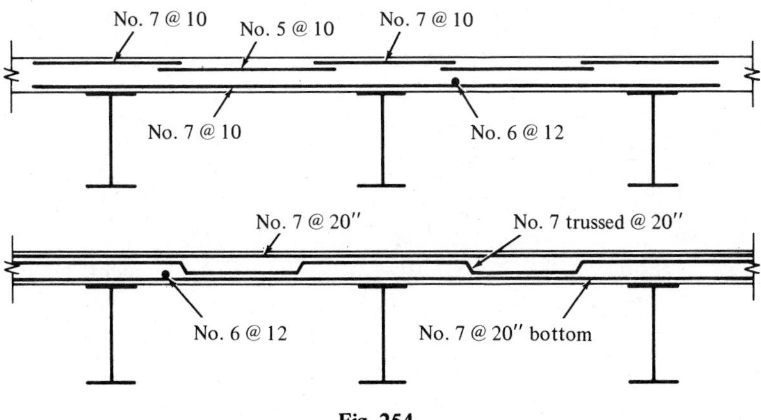

Fig. 254.

Curb	150(1)(2.67) = 400
Rail	30

The load on one beam for two curbs is 430(2)/4 = 215 plf. The total dead load, composed of the weights of the 8-in. slab and the wearing surface and an estimated weight of the beam is:

Curbs		215 plf
Slab	8(12.5)(8.33) =	833 plf
Wearing surface	24(8.33) =	200 plf
Assumed weight of beam		150 plf
	Total:	1398 plf

The bending moment due to this load for a 40-ft span is

$$M = 1.40(40^2)/8 = 280 \text{ ft-kip}$$

The live load will be the HS20-44 loading with wheel loads of 4 kip, 16 kip, and 16 kip, with 14 ft between loads. To obtain the loads that act on one beam, we must use the AASHO load-distribution formula $S/5.5$, which is the ratio for one set of wheel loads:

$$\text{Ratio} = S/5.5 = 8.33/5.5 = 1.51 \text{ wheels}$$

The largest loads are now 1.51(16) = 24.2 kip each and the small 6.1 kip.

The loads are placed on the girder in accordance with the following criterion: "The maximum moment under a given load occurs when that load and the center of gravity of all loads are equidistant from the centerline." We will now determine the center of gravity of loads, preferably from the middle load because it seems from inspection that this load will cause the maximum moment. To find the center of gravity we need not use the exact loads, only the ratio 1 to 4. Take moments about the center load:

$$\begin{aligned} -1(14) &= -14 \\ 4(14) &= 56 \\ \text{Total:}\quad & 42 \end{aligned}$$

Divide by the total load (1 + 4 + 4 = 9) and the arm (a = 42/9 = 4.67 ft). The loads must be arranged as in Fig. 255, with the centerline of the beam 2.33 ft from both the center load and the center of gravity of all loads.

The left reaction is 54.5(17.67)/40 = 24.08 kip. Take moments about the center load:

$$\begin{aligned} M = 24.08(17.67) &= 425 \text{ ft-kip} \\ -6.1(14) &= -85 \text{ ft-kip} \\ \text{Total:}\quad & 340 \text{ ft-kip} \end{aligned}$$

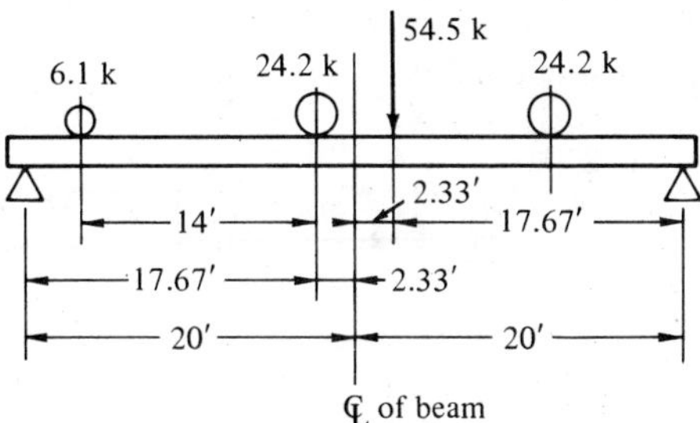

Fig. 255.

The impact ratio will be $i = 50/(L + 125) = 50/(40 + 125) = 0.303$; therefore, use 0.30 (maximum). The impact moment is 0.3(340) = 102 ft-kip.

The sum of the moments is:

Dead load	280 ft-kip
Live load	340 ft-kip
Impact	102 ft-kip
Total:	722 ft-kip

The required section modulus for a steel stress, $f_s = 20$ ksi, is

$$S = M/f_s = 722(12)/20 = 433 \text{ in.}^3$$

Table 6 indicates a W36 X 135 section which has a section modulus, $S = 440$ in.^{3}. We guessed the weight to be 150 plf for the beam, which agrees substantially with the weight finally selected.

The calculation of the deflection of this beam under the applied loads is left to the reader. The deflection-to-span ratio should be greater than 1 to 800 by the AASHO *Specifications.*

SECTION V

Water and Waste Engineering

Terence J. McGhee Ph.D., *P.E., AWWA, WPCF*
Associate Professor of Civil Engineering
College of Engineering and Technology
University of Nebraska-Lincoln
Lincoln, Nebraska

VA. ENGINEERING HYDRAULICS

Flow in Pipes

The development of fluid mechanics has led to many theoretical formulations describing the generalized behavior of fluids. Engineers have, however, continued to rely primarily upon empirical formulas for the design of water distribution systems. The most commonly used formula is that of Hazen and Williams, which is normally expressed as

$$V = 1.318\, CR^{0.63}S^{0.54} \tag{1}$$

in which V is the average velocity in feet per second, R the hydraulic radius in feet, S the head loss per unit length in feet per feet, and C the Hazen-Williams roughness coefficient, which is dependent upon the nature of the pipe, as shown in Table A1.

The Hazen-Williams formula may be used directly in the solution of engineering problems; however, its fractional exponents make it unwieldy and have led to the widespread use of tables and nomographs. A nomograph which may be used in the solution of the Hazen-Williams formula for $C = 100$ is presented in Fig. A1. For values of C other than 100 the results yielded by the nomograph may be modified as shown in Table A2.

EXAMPLE.

Determine the head loss in a 3000-ft-long 3-in.-diameter pipeline at a flow rate of 90 gal/min. Assume $C = 140$.

TABLE A1. HAZEN-WILLIAMS COEFFICIENT FOR VARIOUS MATERIALS.

Material	C
Plastic	150–160
Cement lining	130–150
Bitumen lining	135–155
Concrete	130–150
Cast iron (new)	125–135
Cast iron (old, stable water)	110–120
Cast iron (old, corrosive water)	60–110

$$V = Q/A = 4.08 \text{ ft/sec}$$

$$R = D/4 = 0.0625 \text{ ft}$$

$$\therefore 4.08 = 1.318\ (140)\ 0.0625^{0.63} S^{0.54}$$

$$\therefore S = 0.0218 = h_L/L$$

$$\therefore h_L = 0.0218 \times 3000 = 65.3 \text{ ft}$$

Alternately, using the nomograph, place a straight edge on the graph intersecting the flow at 90 gal/min and the diameter at 3 in. Then directly read from the second two lines the loss of head in feet per 1000 feet and velocity in feet per second. From this,

$$V = 4.08 \text{ ft/sec}$$

and

$$h_L = 40 \text{ ft/1000 ft} \quad \text{or} \quad 120 \text{ ft in 3000 ft}$$

The head loss obtained from the nomograph is that which would be obtained for $C = 100$. This must be appropriately modified as shown in Table A2 for $C = 140$:

$$h_L = 120 \times 0.546 = 65.5 \text{ ft}$$

Slight differences in the results obtained are due to a small loss of accuracy in reading the nomograph.

TABLE A2. NOMOGRAPH RESULT MODIFICATIONS.

$h_L = (h_L @ C = 100) \times f$

C	60	70	80	90	100	110	120	130	140	150
f	2.58	1.34	1.51	1.22	1.00	0.840	0.716	0.615	0.546	0.472

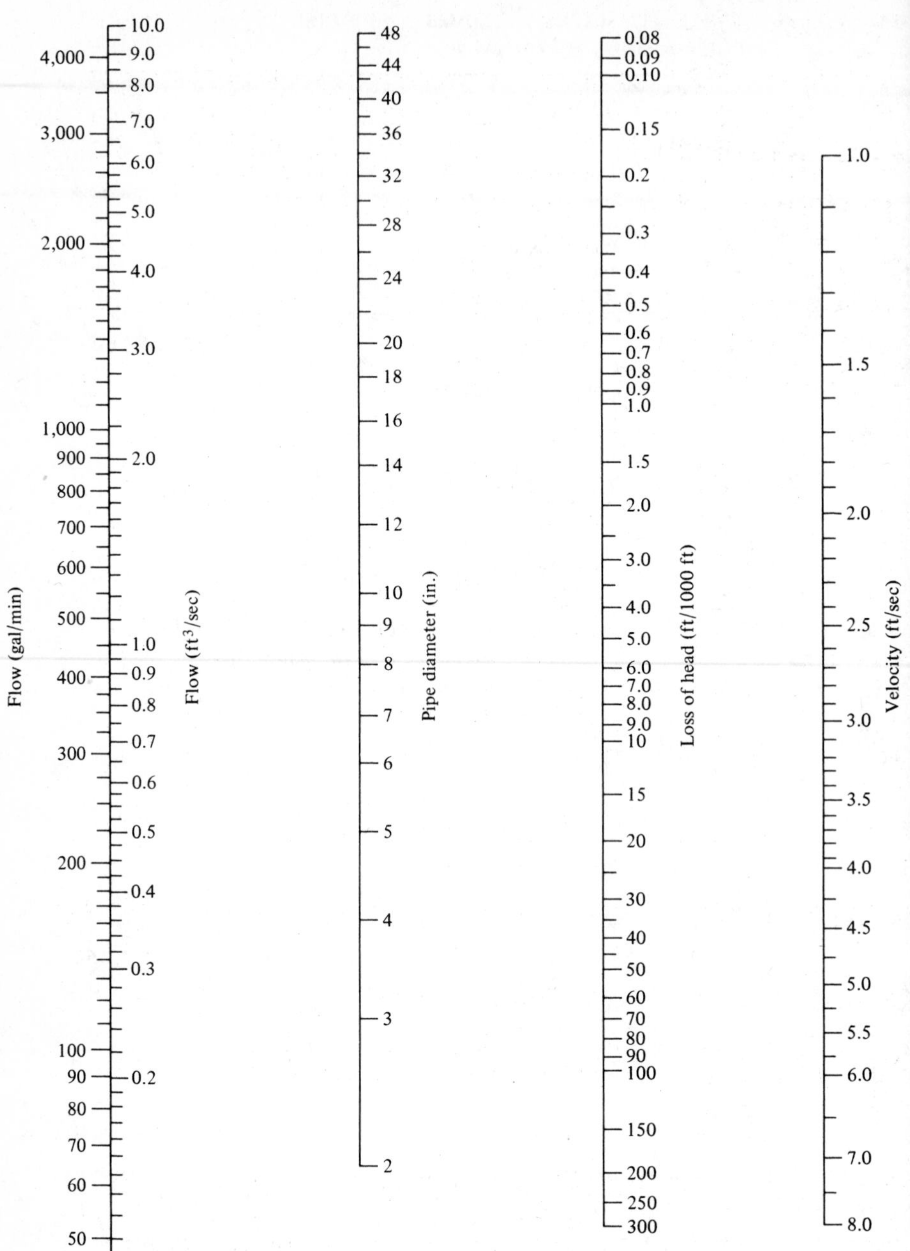

Fig. A1. Flow in pipes by Hazen–Williams formula ($C = 100$).

Minor Pipe Line Losses

Minor pipe line head losses include those produced by changes in cross section, bends, valves and other fittings. In long lines such as those found in municipal distribution systems these losses are negligible. In shorter lines with higher velocities they may become significant. In general, minor losses in fittings may be approximated with sufficient accuracy by the equation

$$h_L = K_L\ V^2/2g \qquad (2)$$

in which h_L is the head loss in feet, K_L the loss coefficient for the particular fitting, V the average velocity in feet per second, and g the gravitational constant in feet per second squared. The value of K_L for typical fittings is given in Table A3.

TABLE A3. K_L FOR TYPICAL PIPE FITTINGS.

Fitting	K_L
90° elbow	0.9
45° elbow	0.4
Tee	1.8
Return bend	2.2
Globe valve (open)	10.0
Angle valve (open)	5.0
Gate valve, open	0.2
$\frac{3}{4}$ open	1.2
$\frac{1}{2}$ open	5.6
$\frac{1}{4}$ open	24.0

The head loss at enlargements may be determined in most cases from the relation

$$h_L = (\Delta V)^2/2g \qquad (3)$$

in which ΔV is the change in average velocity across the enlargement. The head loss at gradual constrictions may be determined from

$$h_L = 0.04\ V_2^2/2g \qquad (4)$$

in which V_2 is the average velocity downstream of the contraction. Sudden constrictions, on the other hand, produce venae contractae with subsequent high head losses. For these, K_L may be as high as 0.50. Such constructions and the even more inefficient reentrant structures (K_L = 0.8) are normally avoided in major water distribution lines.

Open Channel Flow

Open channel flow refers to that special class of problems in which the pressure at the surface of the fluid is a constant, usually equal to the atmospheric pressure. Examples of open channel flow include rivers, canals, aqueducts, and storm and sanitary sewers which are not surcharged. The basic equation describing open channel flow is that of Chezy:

$$V = C(RS)^{1/2} \tag{5}$$

or

$$Q = CA(RS)^{1/2} \tag{6}$$

Evaluation of Chezy's C by Manning led to the development of the Chezy-Manning equation:

$$Q = \frac{1.49}{n} AR^{2/3} S^{1/2} \tag{7}$$

in which Q is the flow in cubic feet per second, A the channel area in square feet, R the hydraulic radius in feet, S the channel slope, and n the roughness coefficient, for which typical values are presented in Table A4.

TABLE A4. TYPICAL VALUES OF MANNING'S COEFFICIENT.

Type of Channel	n
Sewers	0.012–0.015
Smooth cement	0.010
Gravel	0.020
Rubble	0.017
Vitrified clay	0.015

Kutter's formula has also been proposed for the description of open-channel flow. The results it yields are for practical purposes equivalent to those given by the Chezy-Manning equation but the form of the expression makes it far more difficult to manipulate.

EXAMPLE.

A rectangular channel is 10 ft wide and laid on a slope of 0.0001. The total flow is 200 ft^3/sec and $n = 0.015$. Calculate the depth of flow.

Let

$$d = \text{depth of flow}$$

Then

$$A = 10d, \quad R = \frac{10d}{10 + 2d}$$

$$200 = \frac{1.49}{0.015}(20d)\left(\frac{10d}{10 + 2d}\right)^{2/3}(0.0001)^{1/2}$$

$$\left(\frac{10d}{10 + 2d}\right)^{2/3}(d) = 10.07$$

Solving by trial,

$$d = 5.35 \text{ ft.}$$

The solution to problems involving the Chezy-Manning formula may be more readily obtained through the use of nomographs similar to that shown in Fig. A2. The figure shown is for circular sections flowing full since these are of particular significance in sewer design.

Hydraulic Cross Sections

From an examination of the Chezy-Manning formula [Eq. (7)] it is apparent that for a given roughness, area, and slope the flow will be a maximum when R is a maximum. Simple applications of differential calculus will yield solutions for the optimum proportions of various cross sections. For rectangular channels, for example, the best proportions exist when the depth is one-half the width.

Of particular interest to the engineer are the properties of circular sections under conditions other than full flow. Both R and A in the Chezy-Manning formula are functions of the depth, as shown in Fig. A3 and hence the flow and velocity of flow are unusually affected by variations in depth. From Fig. A4 maximum flow is seen to occur when the pipe is 0.94 full and maximum velocity when the pipe is 0.80 full. Similar curves may be drawn for noncircular closed cross sections.

EXAMPLE.

An 18-in. sewer line carries a flow of 1.8 ft^3/sec. The sewer is on a slope of 0.001 and has roughness coefficient of 0.015. Calculate the depth and velocity of flow.

From Eq. (7), if the line were full,

$$Q_F = \frac{1.49}{0.015}\left(\frac{\pi}{4}\right)(1.5)^2\left(\frac{1.5}{4}\right)^{2/3}(0.001)^{1/2}$$

$$= 2.90 \text{ cfs}$$

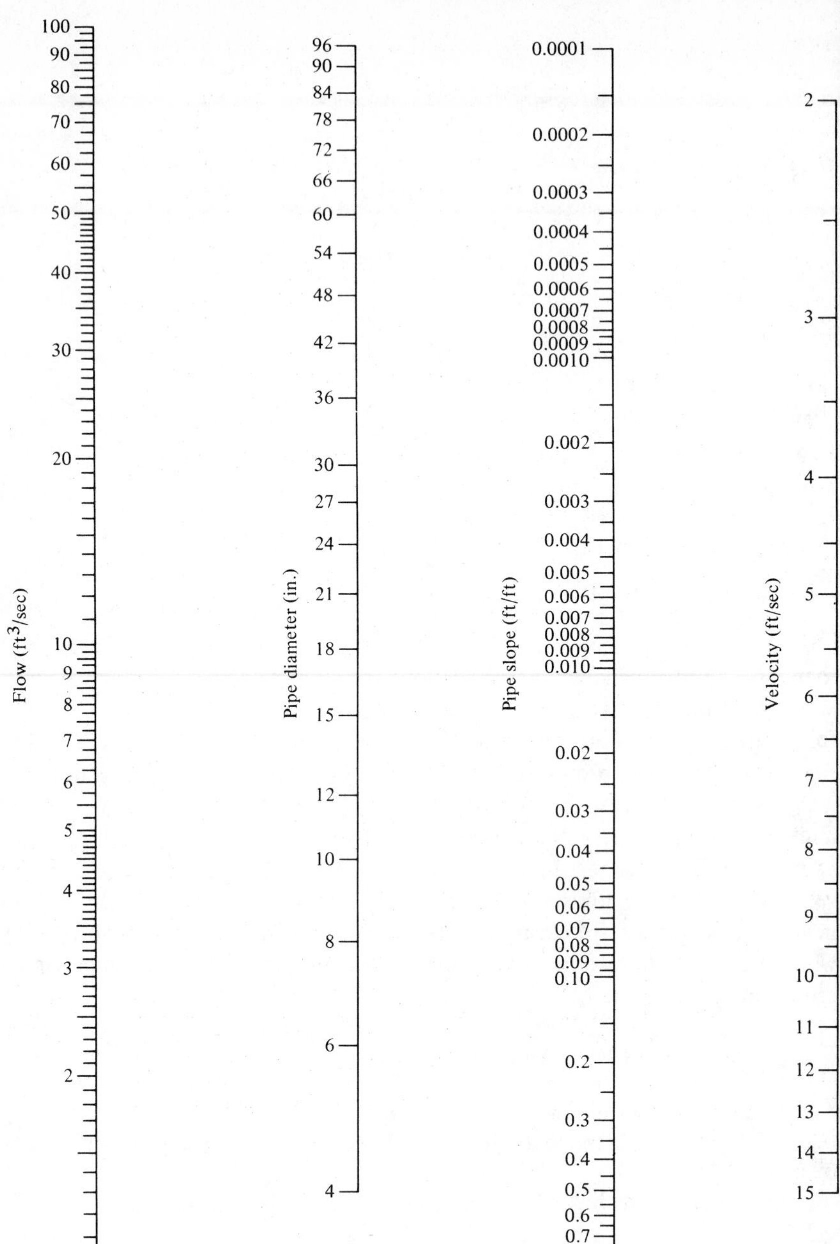

Fig. A2. Flow in pipes by Chezy-Manning formula ($n = 0.013$).

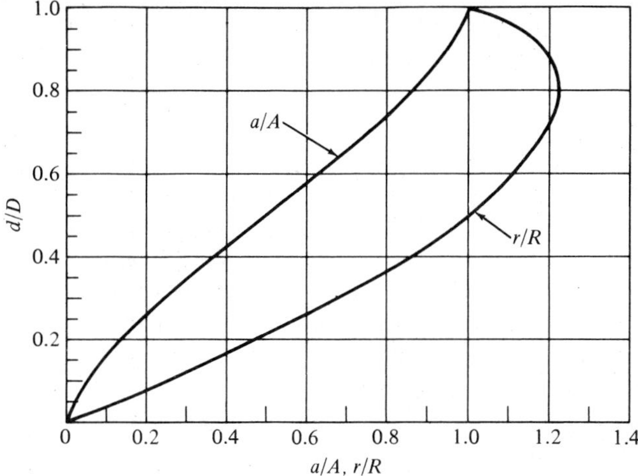

Fig. A3. Variation of area and hydraulic radius for circular sections partially full.

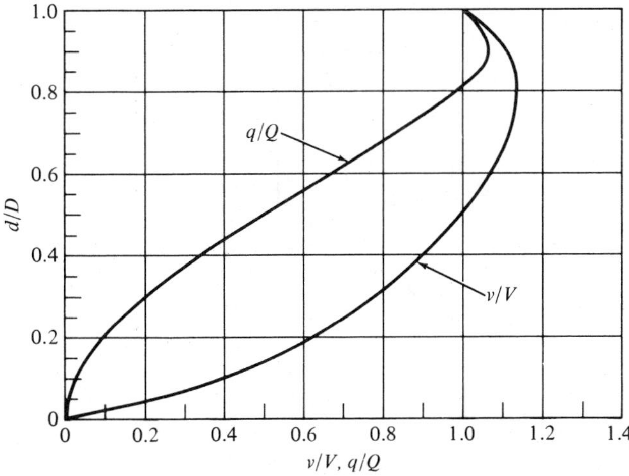

Fig. A4. Variation of velocity and flow for circular sections partially full.

Since the actual flow is 1.8 ft^3/sec,

$$Q/Q_F = 1.8/2.9 = 0.621$$

From Fig. A4 for $Q/Q_F = 0.621$, $d/D = 0.58$. Therefore the depth is $0.58 \times 18 =$ 10.45 in. When the pipe is flowing full the velocity is $Q/A = 2.90/(\pi/4)(1.5)^2 =$ 1.63 ft/sec.

For $d/D = 0.58$, from Fig. A4, $V/V_F = 1.05$. Therefore, $V = 1.05 \times 1.63 = 1.71$ ft/sec. The nomograph (Fig. A2) may also be used for the initial solution of the Chezy–Manning Formula.

Flow Measurement

The measurement of flow is important in the control of water and waste-water treatment plants as well as in water distribution systems. Flow measurements in pipe lines are normally made with meters calibrated to measure the average flow in standard pipe lines, while open-channel flows are determined by construction of special weirs and flumes in which the depth of the water bears a known relation to the discharge. Empirical formulas have been established for many typical weir shapes. Among those of interest are rectangular weirs (Fig. A5); triangular weirs (Fig. A6); Cipolletti (trapezoidal) weirs (Fig. A7); and proportional weirs (Fig. A8).

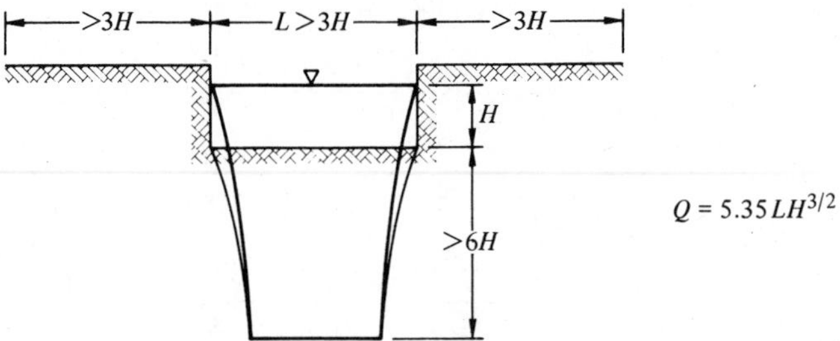

Fig. A5. Rectangular weir.

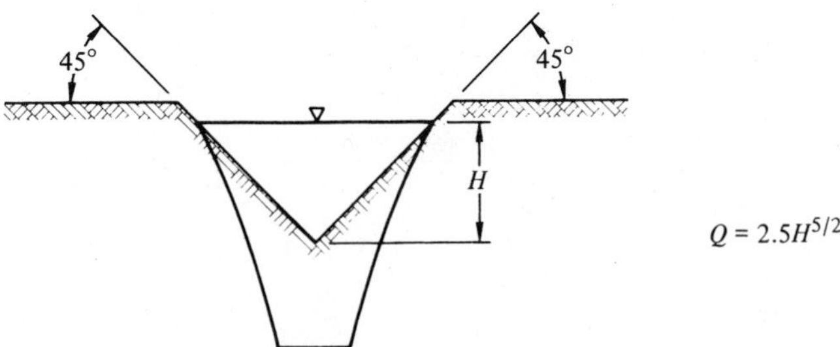

Fig. A6. Triangular weir.

The Parshall flume (Fig. A9) has wide application in the measurement of controlled open channel flow. The advantages of the flume include a lower head loss than that encountered with most weirs and the maintenance of velocities sufficient to prevent deposition of solids. In general, the flow may be determined either from the upstream head, or, if the flow is submerged, from the difference in head above and below the throat of the flume. Rating curves are available for Parshall flumes of standard dimensions.

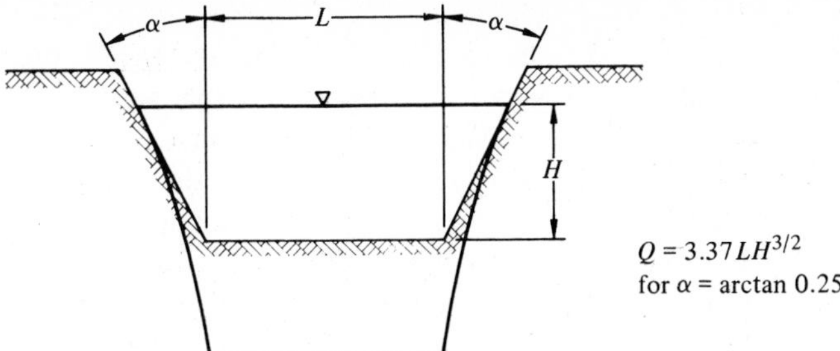

Fig. A7. Cipolletti (trapezoidal) weir.

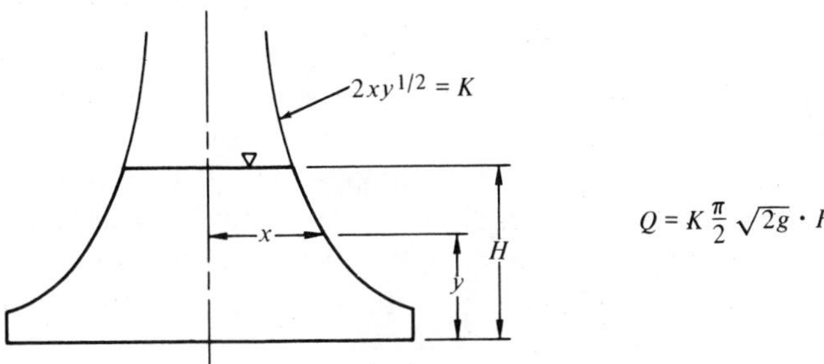

Fig. A8. Proportional weir.

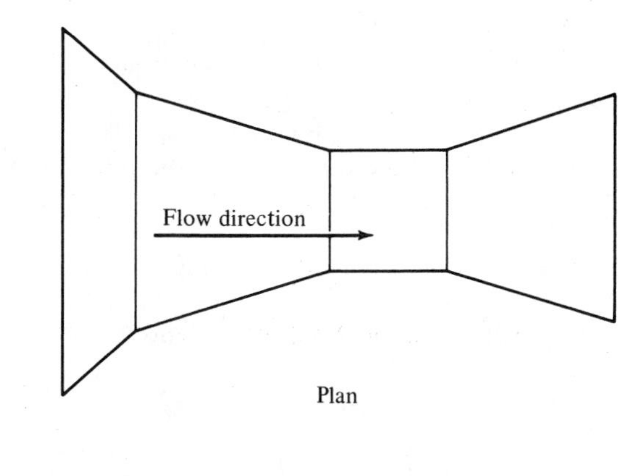

Plan

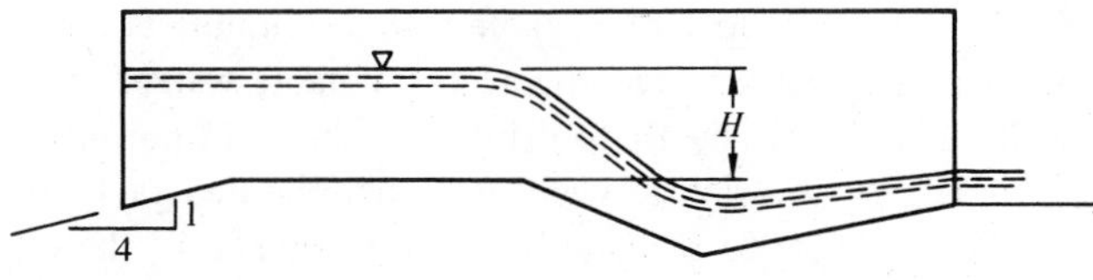

Elevation

Fig. A9. Typical Parshall flume.

Pumps

Centrifugal pumps are generally to be preferred for the transmission of water provided it contains no significant quantity of abrasive particles. In general pumps are best selected for specific applications from data furnished by the manufacturers. Such data will normally include characteristic curves which relate horsepower, head, discharge, efficiency and speed. The normal maximum head of such pumps is about 250 ft; however, they may be placed in series to attain virtually any desired lift. A major consideration in the use of centrifugal pumps is their limited suction lift. The maximum theoretical suction lift is approximately 26 ft, but this is rapidly reduced with increasing velocity of flow. Excessive velocities at high suction lifts will produce cavitation with resultant damage to the pump.

Sewage may also be pumped more advantageously with centrifugal pumps than with pumps of other types. Centrifugal pumps specially designed for pumping sewage are variously called "nonclogging" or "trash" pumps. Sewage solids (normally 2-10% solids) are frequently pumped with positive-displacement "sludge" pumps. Such units normally are used only in low-head and low-flow situations in which the concentration of both solids and abrasive particles is high. Such pumps cannot compare with centrifugal pumps in efficiency but may still have limited applications.

REFERENCES

Engineering Hydraulics

American Society of Civil Engineers, Design and Construction of Sanitary and Storm Sewers, *ASCE Manual of Engineering Practice*, **37** (1960).

Clark, J. W., W. Viessman Jr., and M. J. Hammer, *Water Supply And Pollution Control*, 2nd ed., International Textbook Co., Scranton, Pa. (1971).

Crane Co., Flow of Fluids Through Valves, Fittings, and Pipes, Crane Co. Tech. Paper 410, Chicago, Ill. (1957).

Olson, R. M., *Essentials of Engineering Fluid Mechanics*, International Textbook Co., Scranton, Pa. (1961).

Parshall, R. L., The Parshall Measuring Flume, Colorado Exp. Sta. Bull. No. 423 (1936).

VB. WATER SUPPLY

The development of a supply of potable water is doubtless one of the most important tasks undertaken by civil engineers. Potable water is water which is bacteriologically and chemically fit to drink, but the term normally connotes a water which is pleasant to drink as well. Although the Federal Government has no constitutional authority to regulate local water standards, the establishment of standards for waters used in interstate commerce has led most states and municipalities to adopt regulations paralleling those of the U.S. Public Health Service. These regulations establish limits not only for microorganisms and toxic chemicals but also for other contaminants which are undesirable mainly from an esthetic standpoint. Typical limits for contaminants are presented in Table B1. The first column in the table lists levels of contamination which are not to be exceeded when an alternative source of water is available. The second column indicates concentrations which are unacceptable under any circumstances.

TABLE B1. TYPICAL DRINKING WATER STANDARDS.

	Standard	
Quality Parameter	Limit not to be Exceeded	Cause for Rejection
Biological		
Membrane filter technique	1 coliform colony/100 ml	
Physical		
color	15 units	
taste	unobjectionable	
threshold odor No.	3	
turbidity	5 units	
Chemical		
ABS	0.5 mg/l	
arsenic	0.01 mg/l	0.05 mg/l
barium		1.0 mg/l
cadmium		0.01 mg/l
chloride	250 mg/l	
chromium		0.05 mg/l
copper	1.0 mg/l	
CCE	0.2 mg/l	
cyanide	0.01 mg/l	0.20 mg/l
fluoride	0.7-1.2 mg/l*	1.4-3.4 mg/l*
iron	0.3 mg/l	
lead		0.05 mg/l
manganese	0.05 mg/l	
nitrate	45 mg/l	
phenol	0.001 mg/l	
selenium		0.01 mg/l
silver		0.05 mg/l
sulfate	250 mg/l	
total dissolved solids	500 mg/l	
zinc	5 mg/l	
Radiological		
strontium-90	10 pc/l	
radium-226	3 pc/l	

*Dependent on temperature.

Sources of Water

Sources for public water supply systems include surface waters such as streams and impoundments, which may be either natural or man-made, and ground waters from either pumped or artesian wells. In many respects ground waters are preferable to surface waters as sources but in some areas the ground water supply may not be sufficient; additionally, in some sections of the country, such as New England, the surface water is of exceptionally high quality. Table B2 provides a basis for comparison between average ground and surface waters in general terms.

TABLE B2. COMPARISON OF GROUND AND SURFACE WATERS.

Quality Parameter	Ground water	Surface water
Coliform bacteria	low	moderate–high
Total bacteria	low	high
Color	low	variable
Taste	pleasant	variable
Threshold odor No.	low	variable
Turbidity	low	moderate–high
Temperature	low	variable
Dissolved solids	high	low–moderate
Radioactives	low	variable
Dissolved oxygen	low	variable

Ground waters are normally quite uniform in quality throughout the year, which is a considerable advantage in the operation of water treatment plants. The major disadvantage of such water is the high dissolved-solids concentration. This may produce excessive concentrations of sulfate and chloride, both of which require sophisticated and expensive treatment for their removal. Additionally, such waters will normally require removal of calcium, magnesium, iron and manganese.

Surface waters exhibit wide variation in quality with time. All of the water quality parameters listed will fluctuate both in streams and impoundments. This variability demands greater flexibility in treatment and more careful control of operation than would be required for ground waters. It should be remembered that water taken from shallow wells driven in river flood plains may have characteristics more typical of surface than of ground water.

In selecting a source of water it is necessary to be sure not only that the quality is satisfactory but also that a sufficient quantity will be available to meet the forseeable demand. The prediction of future water requirements is an intrinsic part of any water supply system design.

Estimation of Water Demand

Water use in a community may normally be related to the population. On the average, water demand will approximate 150 gal per person per day, but individ-

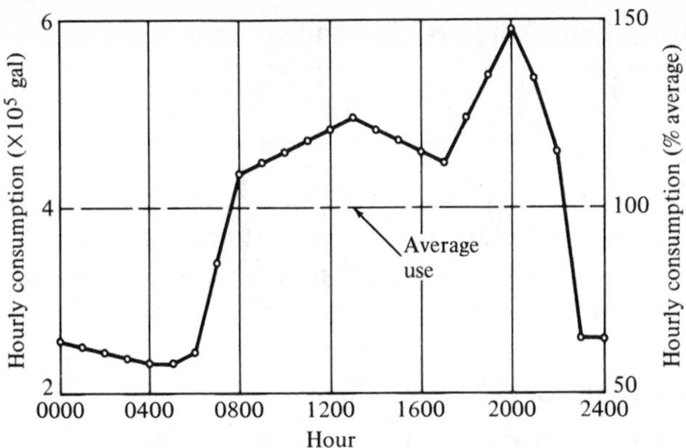

Fig. B1. Typical hourly consumption and percentage average consumption vs. time.

TABLE B3. FACTORS INFLUENCING WATER CONSUMPTION.

Variable	Effect
Climate	Warm climate increases consumption.
Quality	Improved quality increases consumption.
Economic level of consumers	Consumption increases with income.
Cost	Small: very high cost may decrease use.
Metering of households	Metering decreases use.
Pressure in system	Low pressure decreases use.
Industry	Dependent on industry: may cause large increase in average consumption.

ual communities may vary widely from this norm. Usage will also vary during the day as shown in Fig. B1 and from day to day during the year. These fluctuations will have a considerable effect upon treatment plant and distribution system design. Average per capita use will be affected by the factors enumerated in Table B3.

Engineers normally strive to produce a water of high quality, delivered at a useful pressure, at a minimum cost. It should be noted, however, that in areas with limited water supplies the demand may be strongly influenced by the design and operation of the treatment and distribution systems. The average per capita water consumption for a particular community may be best estimated from the pumping records of the existing plant or from the records of similar communities.

Population Estimation

A number of methods of population forecasting have been applied in the design of water supply systems. These are (1) graphical projection, (2) arithmetic

TABLE B4. POPULATION GROWTH OF A MIDWESTERN CITY.

Year	1880	1890	1900	1910	1920	1930	1940	1945	1950	1960
Population	8,951	10,724	12,586	16,082	17,476	18,587	19,863	19,412	25,715	34,360

increase, (3) geometric increase, and (4) graphical comparison. The methods will be applied to a selected midwestern city as illustrative examples. Since all are dependent to some degree on the past growth of the community this must first be determined from census data as shown in Table B4.

EXAMPLE–GRAPHICAL PROJECTION.

The past population of the city may be plotted as shown in Fig. B2. Projecting the trend of recent years (but neglecting variations directly attributable to external events such as the decrease from 1940 to 1945) one obtains a projected population of 68,500 in 1990.

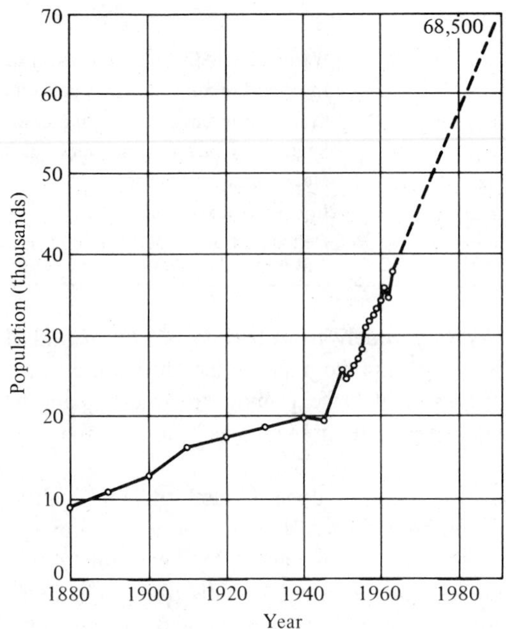

Fig. B2. Population vs. time, 1880–1963, projected to 1990.

EXAMPLE–ARITHMETIC INCREASE.

The method of arithmetic increase assumes that the rate of growth is a constant, independent of population. That is, $dP/dt = K$. The assumption may be tested for a given community by an examination of the pattern of past growth. For the community cited above one obtains the results indicated

TABLE B5. AVERAGE POPULATION GROWTH PER YEAR.

Period	1880–1890	1890–1900	1900–1910	1910–1920	1920–1930
$\Delta P/\Delta t$	177	186	350	139	111
Period	1930–1940	1940–1945	1945–1950	1950–1960	
$\Delta P/\Delta t$	128	-90	1261	865	

TABLE B6. POPULATION AND ANNUAL POPULATION GROWTH.

Year	Population	$\Delta P/\Delta t$	Year	Population	$\Delta P/\Delta t$
1950	25,715	–967	1957	31,690	755
1951	24,748	677	1958	32,445	758
1952	25,425	896	1959	33,203	1157
1953	26,321	645	1960	34,360	1578
1954	26,966	1367	1961	35,938	–1297
1955	28,333	2692	1962	34,641	3261
1956	31,025	665	1963	37,902	
1957	31,690				

in Table B5. The data appear to fail the test. However, further examination may lead to a different conclusion. In this case local population records were available on an annual basis for the years 1950 to 1963, as shown in Table B6.

The data still seem to fail the test although the average annual growth during this period (937 persons/year) is nearly equal to many individual annual increases. If the method were to be applied, the estimated population for any year after 1963 would be

$$P = 37{,}902 + 937\Delta t.$$

For the year 1990 this would yield an estimated population of approximately 63,200.

EXAMPLE–GEOMETRIC INCREASE.

The method of geometric increase assumes that the rate of growth is proportional to the population, that is,

$$dP/dt = KP \tag{1}$$

Integration of eq. (1) between the appropriate limits yields

$$\log P_n = \log P_2 + (\log P_2 - \log P_1)(t_n - t_2)/(t_2 - t_1) \tag{2}$$

in which

P_n is the estimated population in year n

P_2 is the recorded population in year t_2

P_1 is the recorded population in year t_1

t_1, t_2, and t_n are the years corresponding to P_1, P_2, and P_n.

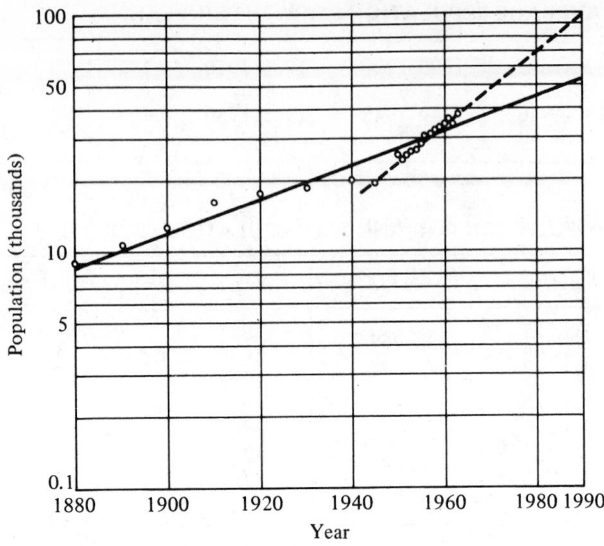

Fig. B3. Population vs. time, example community, 1880–1963. Geometric population growth (line shown is least-squares line of best fit).

The calculation may be made directly if the population in any two years is known, but the hypothesis of geometric growth is best tested by plotting the population data on semilog paper as shown in Fig. B3. If the data plot as a straight line the hypothesis is confirmed. In the case shown the plot is far from linear, although a line may be fitted as shown. It should be noted that in this instance the data for the shorter period 1945–1963 demonstrate an excellent linearity. From this it might be concluded that the community had entered upon a period of geometric growth at the conclusion of the Second World War. Projection of this segment of the data or calculation from Eq. (2) using 1945–1963 as the base period yields an estimated population in 1990 of approximately 100,000.

EXAMPLE–GRAPHICAL COMPARISON

Graphical comparison is based on the assumption that similar cities will exhibit similar patterns of growth. It is assumed that the more points of similarity which exist, the better the comparison will be. The cities chosen are shown in Fig. B4, with the selection based upon geographical proximity and likenesses in economic base. The recorded growth of these communities from the time at which they reached the current population of the example community is superimposed upon the growth record as shown in Fig. B5. The future population is then projected by comparison with the recorded growth of the other cities. By this method the estimated population in 1990 is 72,500.

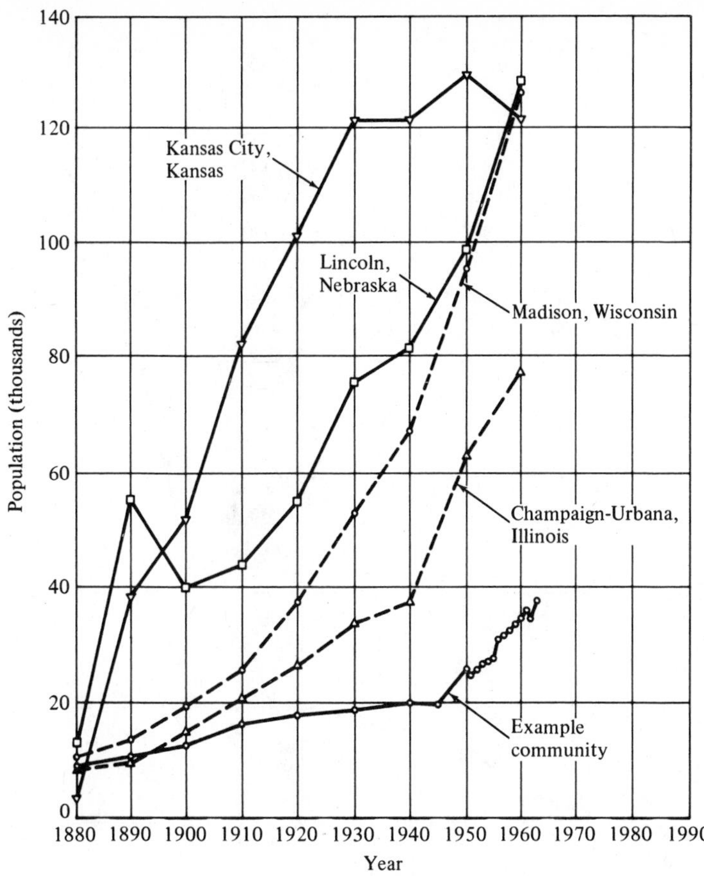

Fig. B4. Population vs. time for five midwestern cities.

The selection of the method of population estimation will be dependent upon the nature of each community. Many engineers favor graphical comparison since in most instances suitable cities may be selected with little difficulty. Arithmetic and geometric projection will be favored in those cases where the past growth of the community fits one of the mathematical hypotheses. In applying the method of geometric growth care must be exercised not to extend the projection beyond a reasonable period of time, since no population can increase logarithmically forever. Each community will have a saturation population governed by available land area, natural resources, transportation, etc. The estimated population for the community evaluated here varied as shown in Table B7. For purposes of design the 1990 population was selected as 70,000 since in this instance arithmetic and geometric increase appeared to be less applicable than the other methods.

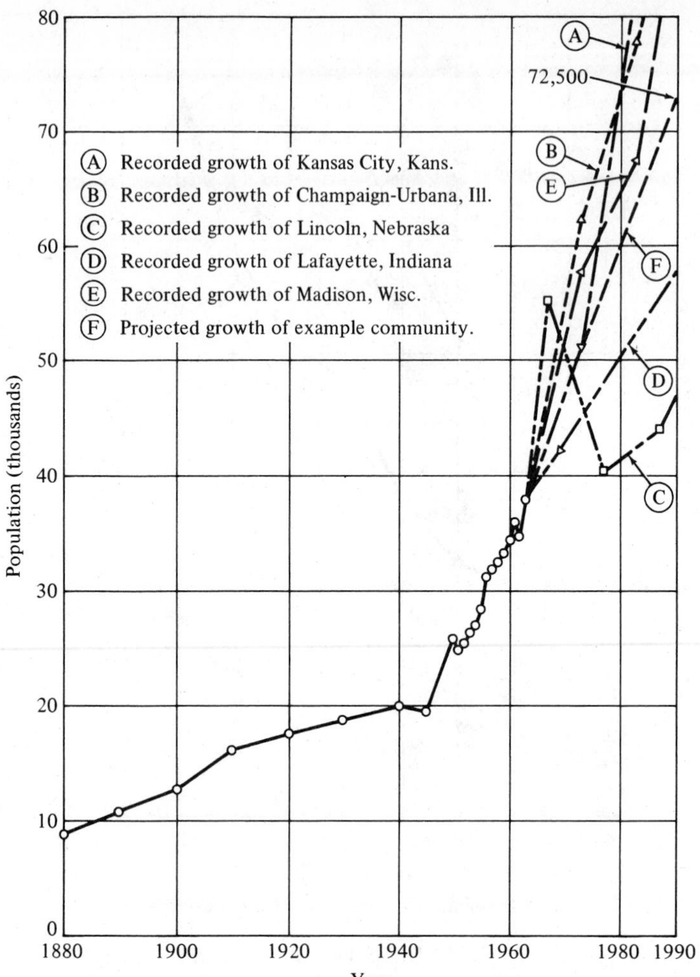

Fig. B5. Comparison of growth rates of similar cities.

TABLE B7. ESTIMATION POPULATION OF A MIDWESTERN CITY IN 1990.

Method	Estimated population
Graphical projection	68,500
Arithmetic increase	63,200
Geometric increase	100,000
Graphical comparison	72,500

TABLE B8. AVERAGE WATER DEMAND RATE (GAL/PERSON/DAY).

Period	Normal range	Average
Average day	50–250	150
Maximum day	1.2 × avg.–2.0 × avg.	1.5 × avg.
Maximum hour	2.0 × avg.–3.0 × avg.	2.5 × avg.

Design Values for Water Demand

Based upon the recorded water demand in many communities the average values given in Table B8 have gained common acceptance, but local records are always to be preferred to national averages. Variations may also be expected to be far greater in small communities.

The Goodrich formula has also been used to estimate the ratio of short term flow rates to the average rate. The formula is

$$P = 180t^{-0.10} \tag{3}$$

In which P is the percentage of average annual consumption in time t, and t is the length of the period in days. Application of this relation gives the ratio of maximum day, maximum week, and maximum month to the average as 1.80 : 1, 1.48 : 1, and 1.28 : 1, respectively.

Specific Requirements

In addition to providing sufficient capacity to meet the expected domestic requirements, provision must also be made for other specific requirements. Local industries with a high water use are an example of such demand. The most important additional water demand which must always be considered is fire demand. The required fire flow may be determined from the formula of the Insurance Services Office Guide,

$$F = 18C(A)^{0.5} \tag{4}$$

in which F = flow in gpm
A = total building floor area, excluding basements
C = a coefficient ranging from 0.6 for fire resistant construction to 1.5 for wood frame buildings

A maximum value of 8000 gpm and a minimum of 500 gpm are also specified. Additional modifications are based upon the fire hazard potential of the particular tenant. The fire flow must be maintained for a period of 2 to 10 hours, depending on population. This fire demand must be supplied in addition to all other requirements. Since it is unlikely that a major fire will coincide with the rare maximum hourly rate it is customary to provide for meeting the maximum daily rate plus the fire demand.

EXAMPLE.

Determine the design requirements for the various constituents of a water supply system to serve a population of 100,000.

Structure	Required capacity	Capacity ($\times 10^3$ gal/day)
Source	maximum day	22.5
Raw water main	maximum day	22.5
Treated water main to storage	maximum day	22.5
Treated water main to distribution system	maximum day plus fire demand	35.7
Distribution system	maximum day plus fire demand	35.7
High-service pumps	maximum day plus excess capacity	45.0
Low-service pumps	maximum day plus excess capacity	33.8

Water Treatment

While water for human consumption must meet standards similar to those listed above, few water supplies will be found to be completely suitable in their natural state. For this reason water treatment is directed toward improving the quality of water through both physical and chemical means. The major techniques of water treatment and their goals are as listed in Table B9.

TABLE B9. MAJOR WATER TREATMENT TECHNIQUES.

Process	Purpose
Screening	Removal of coarse suspended material
Sedimentation	Removal of finely divided suspended material
Coagulation	Removal of colloidal material
Flocculation	Agglomeration of finely divided particles
Softening	Precipitation of hardness
Oxidation of iron and manganese	Removal of iron and manganese
Disinfection	Removal of pathogenic microorganisms
Fluoridation-defluoridation	Adjustment of fluoride content
Filtration	Removal of turbidity, microorganisms, and chemical precipitates

Screening

Coarse screens for removal of large suspended material are normally constructed of steel mesh with openings of approximately $\frac{1}{8}''$-$\frac{1}{4}''$ in the minor dimension. These screens serve to prevent floating trash and fish from entering the treatment system. Typical designs such as that shown in Fig. B6 employ a moving mesh which carries the screenings to a collection point and replaces the clogged section with a clear one either at a predetermined head loss or on a timed basis. Such screens provide little or no improvement in water quality but serve rather to protect the pumps from materials which might cause damage.

In northern climates ice formation may cause rapid clogging of screens. Reversal of flow, shaking of screens, and other mechanical means may be used to dislodge the ice, but the simpler expedient of heating the screens seems generally to be more practical.

Fig. B6. Traveling water intake screen (courtesy Rex Chainbelt, Inc., Process Equipment Division).

Sedimentation Sedimentation will usually play a part in any water treatment plant. It is employed in the reduction of turbidity in raw water and in the removal of chemical precipitates generated in other treatment processes such as coagulation and softening. The settling of a discrete particle may be described by Stokes' law:

$$V = g/18 \, \frac{(\rho_s - \rho_w)}{\mu} \, d^2 \tag{5}$$

In which

V is the settling velocity
g is the gravitational constant
μ is the absolute viscosity of the liquid
ρ_s is the mass density of the particle
ρ_w is the mass density of the liquid
d is the diameter of the particle.

The equation is applicable to particles of a diameter of 1 mm or less, which from Table B10 may be seen to include most suspended solids of interest.

Sedimentation theory is based upon the concept of the "ideal settling basin" in which all particles are considered to move on a path defined by the horizontal velocity of the water and the settling velocity of the particle given by Eq. (5). Considering such a basin as shown in Fig. B7, with length L, depth D, and width W, the following observations may be made:

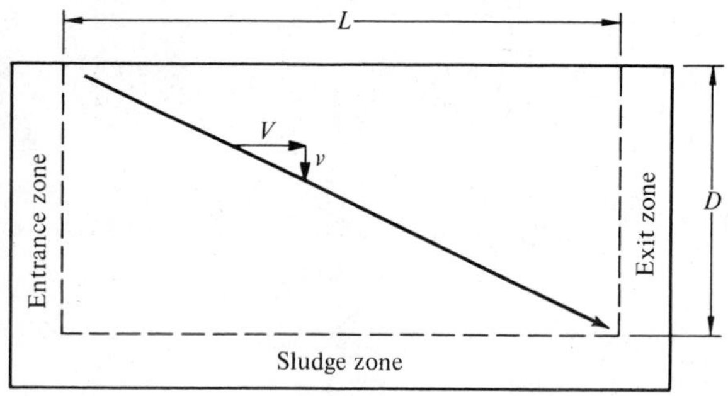

Fig. B7. Ideal sedimentation basin.

TABLE B10. SIZE DISTRIBUTION OF SUSPENDED MATERIALS.

Material	Diameter (mm)
Gravel	2–1
Sand	1–0.1
Fine Sand	0.1–0.05
Silt	0.05–0.005
Clay	0.01–0.0001
Bacteria	0.001
Colloids	< 0.0001

The horizontal velocity V is a function only of the flow Q and the tank geometry, thus:

$$V = Q/WD \tag{6}$$

Also, for removal of a particle entering at the top of the basin,

$$v/V = D/L \quad \text{or} \quad v = DV/L \tag{7}$$

Combining (6) and (7),

$$v = Q/WL \tag{8}$$

where v is the settling velocity of those particles which will be 100% removed and is seen to be mathematically equal to the hydraulic flow divided by the plan area of the tank. This quantity, called the surface overflow rate, is the basic design parameter for sedimentation basins. Theoretically, but not practically, the depth has no effect on the efficiency of sedimentation. Although a rectangular design was used in the development above, the tank geometry does not affect the theoretical result. The required plan area will be the same for a tank of any shape. Typical design values for standard sedimentation basins in water treatment are as shown in Table B11.

TABLE B11. SEDIMENTATION BASIN PARAMETERS.

Process	Detention time (hr)	Surface overflow rate (gal/ft^2/day)
Plain sedimentation	6–24	<500
Coagulation and flocculation	2–8	500–800
Lime–Soda softening	4–8	500–1000

Certain modifications of the standard sedimentation basin have been made in order to decrease the required size. These include the placing of trays in the basin which effectively increase the area, the use of upflow clarifiers (Fig. B8) in which the flow is filtered, in a sense, by a layer of settled material, and tube settlers (Fig. B9) which presumably combine the features of the first two modifications.

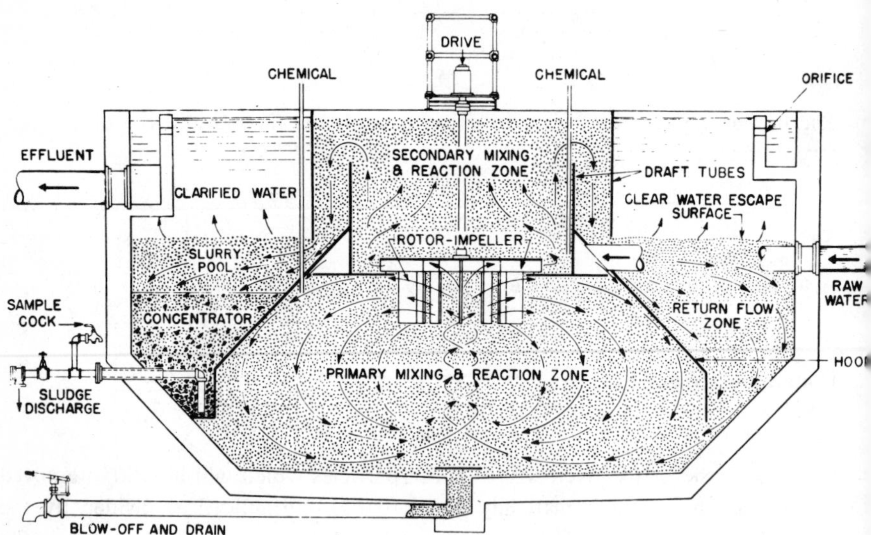

Fig. B8. Infilco Accelator (courtesy Westinghouse Electric Corp., Infilco Division).

All sedimentation basins should incorporate sludge removal equipment. A typical design for a circular basin is shown in Figs. B10 and B11. The settled material is normally scraped to a pit from which it may be pumped as necessary.

The flow into and from a sedimentation basin must be carefully controlled in order to prevent short circuiting. Baffles at the entrance and careful design of the effluent structure are required. Effluent weirs are normally designed for a maximum overflow rate of 50,000 gal/lineal ft./day. The total weir length is considered to be effective although in most designs the flow is through V-notch sections located 4–12 in. apart on center. Peripheral and radial weirs are normally used for circular basins. In rectangular designs the weirs may be distributed over as much as one-quarter to one-third of the plan area.

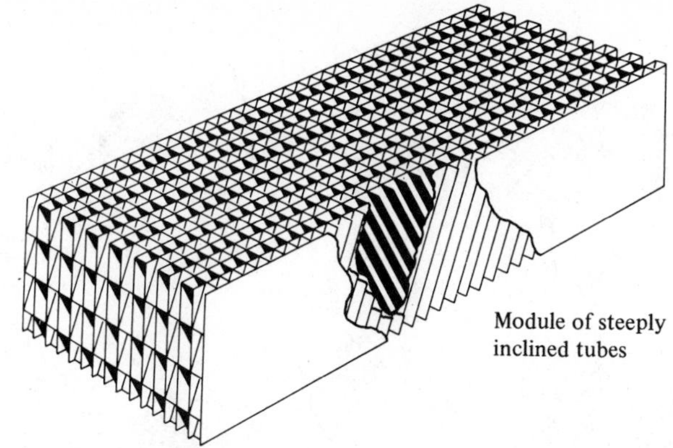

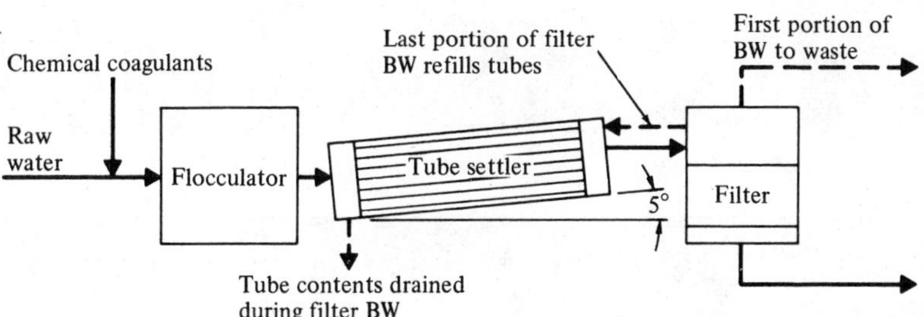

Essentially Horizontal Tube Settler

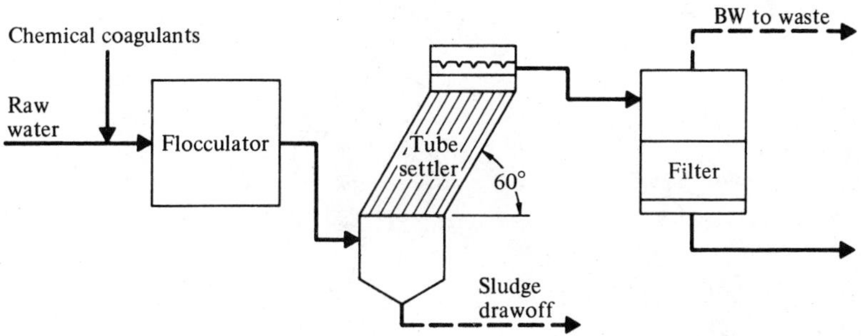

Steeply Inclined Tube Settler

Fig. B9. Tube settlers (courtesy Neptune Microfloc, Inc.).

Fig. B10. Circular clarifier (courtesy Walker Process Equipment).

Fig. B11. Circular clarifier showing sludge removal equipment (courtesy Walker Process Equipment).

Coagulation and Flocculation Coagulation may be defined as a chemical destabilization technique in which colloidal material in stable suspension is neutralized in order to permit agglomeration. Flocculation is a physical process in which the destabilized colloidal particles are brought into contact in order to speed agglomeration.

Colloidal particles are of such a size (less than 0.0001 mm) that their rate of sedimentation is slower than the rate of diffusion produced by their repulsion by the similar charges on their surfaces. These surface charges are complex in nature (Fig. B12) but may be effectively reduced by addition of specific chemicals. Following the neutralization of the surface charge the colloidal particles, if brought into contact, will adhere due to van der Waals forces with an effective increase in particle size and rate of sedimentation.

The overall chemical reactions which occur in coagulation may be represented as shown in Eqs. (9)–(11).

Aluminum sulfate (alum):

$$Al_2(SO_4)_3 \cdot 18H_2O + 3Ca(HCO_3)_2 \longrightarrow 2Al(OH)_3 + 3CaSO_4 + 6CO_2 + 18H_2O \tag{9}$$

Ferric chloride:

$$2FeCl_3 + 3Ca(HCO_3)_2 \longrightarrow 2Fe(OH)_3 + 3CaCl_2 + 6CO_2 \tag{10}$$

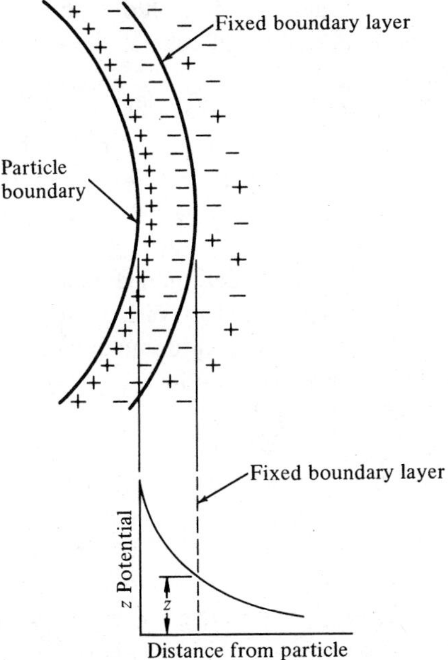

Fig. B12. Colloidal surface charge.

TABLE B12. ALKALINITY REQUIRED FOR COAGULATION (PARTS/PART OF COAGULANT)

Coagulant	Alkalinity required
Alum	0.46 HCO_3^- as $CaCO_3$ or 0.26 lime as CaO or 0.34 lime as $Ca(OH)_2$ or 0.49 soda ash as Na_2CO_3
Ferric chloride	1.39 HCO_3^- as $CaCO_3$ or 0.78 lime as CaO or 1.03 lime as $Ca(OH)_2$ or 1.47 soda ash as Na_2CO_3
Ferric sulfate	0.75 HCO_3^- as $CaCO_3$ or 0.43 lime as CaO or 0.56 lime as $Ca(OH)_2$ or 0.81 soda ash as Na_2CO_3

Ferric sulfate:

$$Fe_2(SO_4)_3 + 3Ca(HCO_3)_2 \longrightarrow 2Fe(OH)_3 + 3CaSO_4 + 6CO_2 \qquad (11)$$

The aluminum and ferric hydroxides which are formed are the active agents of coagulation. These are actually more complex molecules than is indicated by the stoichiometric equations (positively charged hydrous oxides). Ferrous salts may also be used with lime alone to form ferrous hydroxide or with lime and chlorine to form ferric hydroxide. In general the trivalent ferric salts are the more effective coagulants.

From Eqs. (9)–(11) it is evident that the water must contain sufficient alkalinity or the reactions will not occur. In the absence of the required alkalinity lime or soda ash is normally added to supply the difference. Table B12 lists the required alkalinity from various sources for the aluminum and ferric salts.

The dosage of coagulant must be determined experimentally for each water. The usual procedure involves the trial of various doses, normally in the range of 5–85 mg/l, on a jar test basis. The optimum dosage may then be determined by observation of the floc formation and the residual turbidity. The several coagulants have distinct optimum pH ranges, as shown in Table B13. Alum and the ferric salts are normally suitable for most natural waters. Various commercial devices are available for the feeding of chemical coagulants.

Coagulation occurs quite rapidly following the addition of the necessary chemicals. A rapid mix period of about 15 sec is normally adequate. The mixing may be effected by mechanical means or by turbulent open channel flow.

Flocculation is provided by establishing velocity gradients in the coagulated water which increase the probability of contact among the destabilized particles.

TABLE B13. OPTIMUM pH FOR COAGULATION.

Coagulant	pH
Alum	4.0–7.0
Ferric chloride	3.5–6.5 and > 8.5
Ferric sulfate	3.5–7.0 and > 9.0
Ferrous sulfate	>8.5

Fig. B13. Paddle-wheel flocculator (courtesy Rex Chainbelt, Inc., Process Equipment Division).

The usual method of flocculation involves mechanical agitation with slowly rotating paddle wheels. Typical flocculation units are shown in Figs. B13 and B14. The maximum tangential velocity of the paddles is usually 1.0–2.5 ft/sec. Higher velocities will tend to shear the agglomerated floc particles. The retention time in a flocculation chamber is normally 30–60 min. In general the efficiency will increase with retention time.

Lime-Soda Ash Softening The term hardness refers to the presence in water of those ionic species which interfere with the lathering of soap. In general, any divalent cation will contribute to hardness but the term normally denotes

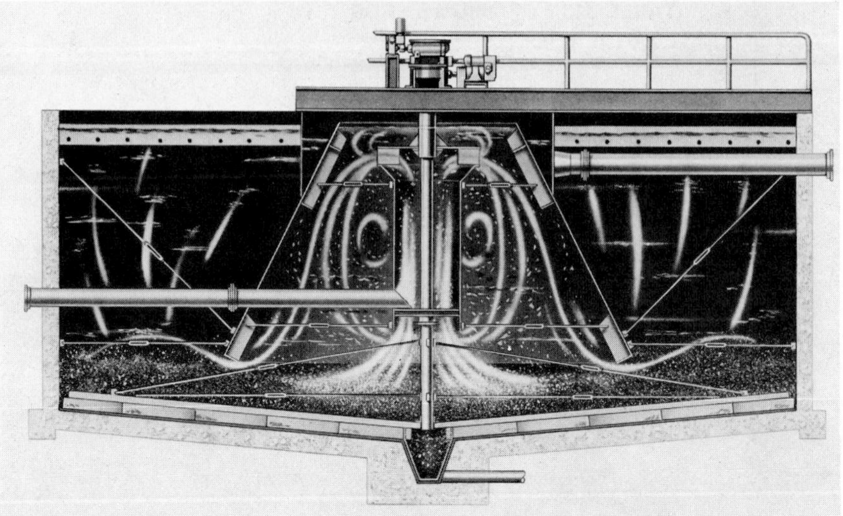

Fig. B14. Solids contact unit (courtesy Rex Chainbelt, Inc., Process Equipment Division).

the presence of calcium and magnesium. Hardness is frequently divided into carbonate and noncarbonate or temporary and permanent fractions. These terms refer to that portion of the divalent cations which may be considered to be associated with the bicarbonate alkalinity and the remainder which is associated with any other anion or anions. Such a division is purely arbitrary and does not reflect the actual ionic condition.

Both calcium and magnesium may be readily precipitated by simple chemical reactions. Calcium carbonate and magnesium hydroxide are quite insoluble, hence the formation of either compound will reduce the hardness of the water. The precipitation of calcium may be described by the equation

$$Ca^{2+} + CO_3{}^{2-} \longrightarrow CaCO_3 \tag{12}$$

The carbonate ion required may be added in the form of soda ash (Na_2CO_3) or may be produced from the bicarbonate alkalinity (if present) by reaction with the hydroxyl ion:

$$HCO_3{}^{-} + OH^{-} \longrightarrow H_2O + CO_3{}^{2-} \tag{13}$$

Magnesium may be precipitated by addition of hydroxyl ion alone:

$$Mg^{2+} + 2OH^{-} \longrightarrow Mg(OH)_2 \tag{14}$$

The amount of calcium or magnesium removed will be a function of the amount of carbonate ion or hydroxyl ion added, but is not a function of any hypothetical combination of anions and cations in the raw water.

The usual source of excess alkalinity in water softening is soda ash, while the hydroxyl ion is normally supplied by addition of lime. Lime has been selected because of cost considerations but in itself is a contributor to hardness. Lime will precipitate magnesium as shown in Eq. (14), but the calcium will simply replace the magnesium in solution and must be removed in turn as shown in Eq. (12).

Both calcium carbonate and magnesium are slightly soluble, with the amount left in solution primarily a function of the carbonate ion concentration and hydroxyl ion concentration, respectively. Since both of these variables are a function of pH, careful control is necessary for proper softening. Calcium removal is normally most effective at a pH of 9 to 10. Magnesium removal is optimum at a pH of 10 to 11. Higher pH values will increase the removal of magnesium but the increase is neither required nor economically practical. Effective removal of calcium and magnesium cannot be carried out simultaneously because of the different pH values required. Removal of both ions requires either sequential treatment or split treatment.

A water of zero hardness is not desirable for domestic purposes. Extremely soft waters are flat tasting, corrosive, and poor rinsing agents. Waters with a total hardness of 1.7-2.0 meq/1 (85-100 mg/l as $CaCO_3$) are considered most suitable for domestic purposes. The desired hardness is usually obtained by leaving the magnesium hardness unremoved or by treating only a portion of either the calcium or magnesium hardness. Typical situations are detailed below.

Calcium Removal When the pH of the water is raised above 9.0 virtually all the bicarbonate alkalinity present will be converted to the carbonate form; thus, an amount of calcium equivalent to twice the bicarbonate ion concentration may be precipitated. If lime is used to raise the pH the amount of calcium added in the lime will be equivalent to the bicarbonate concentration. The net reduction in hardness will then be equivalent to the bicarbonate alkalinity. If this reduction is insufficient an amount of soda ash equivalent to the desired additional calcium removal will precipitate the remainder.

Magnesium Removal At a pH above 10 the hydroxyl ion concentration is sufficient to insure a very low residual magnesium concentration. Elevation of the pH is sufficient to assure the precipitation of magnesium, thus all the water so treated will contain only negligible quantities irrespective of the other ionic species normally present. A partial removal of magnesium can thus be obtained by the expedient of treating only a portion of the flow.

Recarbonation The effluent from a softening basin will have a high pH and may be supersaturated with magnesium hydroxide, calcium carbonate, or both. In order to prevent later precipitation of these salts the pH must be adjusted to convert them to a soluble form. This adjustment may be effected with any acid but ease of control and handling normally dictates the use of carbon dioxide. Upon

Calcium Removal

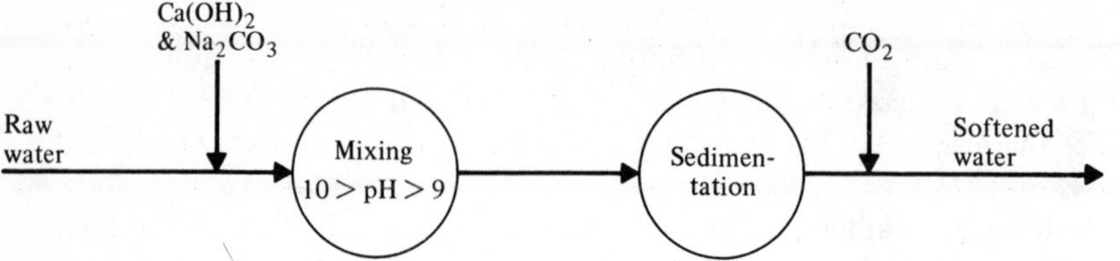

Magnesium Removal

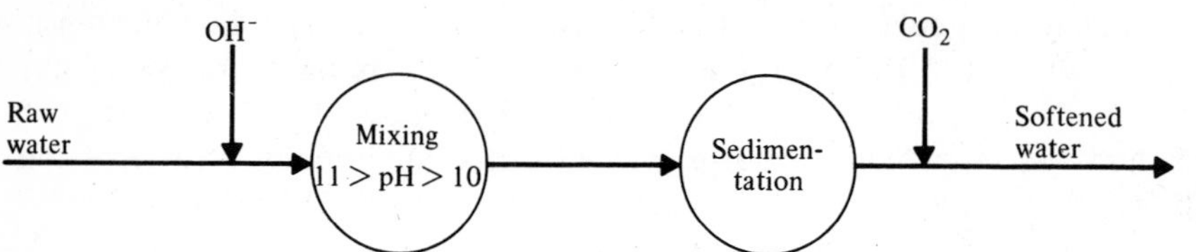

Sequential Removal

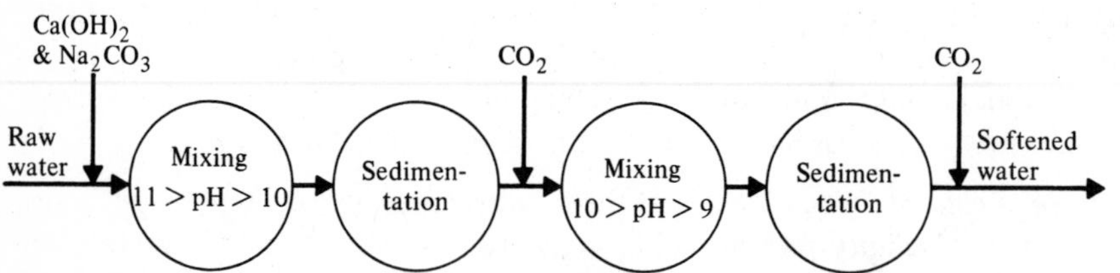

Split Treatment

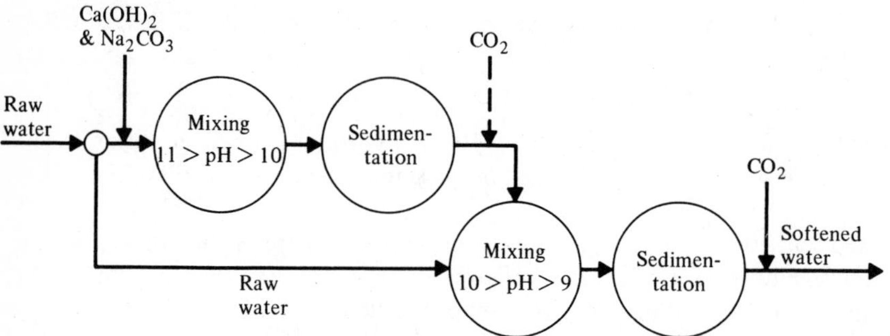

Fig. B15. Typical softening systems.

entering water, CO_2 forms carbonic acid as shown in Equation (15):

$$CO_2 + H_2O \longrightarrow H_2CO_3 \tag{15}$$

The carbonic acid then reacts with the magnesium hydroxide and calcium carbonate as indicated below.

$$Mg(OH)_2 + H_2CO_3 \longrightarrow MgCO_3 + 2H_2O \tag{16}$$

$$CaCO_3 + H_2CO_3 \longrightarrow Ca(HCO_3)_2 \tag{17}$$

Recarbonation is also used to adjust the pH of the flow between chambers for sequential removal of magnesium and calcium.

Typical Softening Systems Line diagrams of typical softening systems are presented in Fig. B15. The diagrams are largely self explanatory with the possible exception of the split treatment system. In split treatment the desired hardness is obtained by removing only a portion of the magnesium hardness and as much of the calcium hardness as possible.

EXAMPLE

A water has a magnesium concentration of 30 mg/l as Mg^{2+} and a calcium hardness of 75 mg/l as Ca^{2+}. Determine the fraction of the water to be treated for magnesium removal in order to obtain a water with a final hardness of 100 mg/l as $CaCO_3$.

The total hardness is:

$$\begin{aligned} Ca^{2+} &= 75 \text{ mg/l as } Ca^{2+} = 187.5 \text{ mg/l as } CaCO_3 \\ Mg^{2+} &= 30 \text{ mg/l as } Mg^{2+} = \underline{125.0 \text{ mg/l as } CaCO_3} \\ & \qquad\qquad\qquad \text{Total} = 312.5 \text{ mg/l as } CaCO_3 \end{aligned}$$

The minimum calcium hardness attainable is equal to the solubility of $CaCO_3$, which is approximately 35 mg/l. The required magnesium removal is thus:

$$\begin{aligned} Mg^{2+} \text{ to be removed} &= 125 - (100 - 35) \\ &= 60 \text{ mg/l as } CaCO_3 \end{aligned}$$

This will yield a final hardness of

$$\begin{aligned} Ca^{2+} &= 35 \\ Mg^{2+} &= \underline{65} \\ \text{Total} &= 100 \text{ mg/l as } CaCO_3 \end{aligned}$$

Assuming Mg^{2+} to have a solubility of 10 mg/l, the effluent from the magnesium removal will have a magnesium hardness of 10 mg/l. Thus for a fraction X being treated,

$$X(10) + (1 - X)(125) = 65$$

$$\therefore X = 0.546$$

Thus approximately 55% of the total flow will be treated for magnesium removal, with the remainder being bypassed to the calcium softening basins.

Iron and Manganese Removal Ground water frequently contains significant concentrations of divalent iron and manganese. Since the allowable concentrations of these species are 0.3 and 0.05 mg/l, respectively, treatment for their removal is necessary. Both iron and manganese will be precipitated in the lime-soda ash softening process, but their conversion to the trivalent and tetravalent forms, respectively, will greatly speed their removal since Fe^{3+} and Mn^{4+} are quite insoluble at pH values above neutrality. The oxidation may be effected in a variety of ways. Aeration, chlorination, and application of other chemical oxidants such as permanganate and ozone are all used in current practice. Typical reactions are

$$\left.\begin{aligned} 4Fe^{2+} + O_2{}^0 &\longrightarrow 4Fe^{3+} + 2O^{2-} \\ 2Mn^{2+} + O_2{}^0 &\longrightarrow 2Mn^{4+} + 2O^{2-} \end{aligned}\right\} \tag{18}$$

$$\left.\begin{aligned} 2Fe^{2+} + Cl_2{}^0 &\longrightarrow 2Fe^{3+} + 2Cl^- \\ Mn^{2+} + Cl_2{}^0 &\longrightarrow Mn^{4+} + 2Cl^- \end{aligned}\right\} \tag{19}$$

$$3Mn^{2+} + 2Mn^{7+} \longrightarrow 3Mn^{4+} + 2Mn^{4+} \tag{20}$$

Aeration may consist of simply spraying the water into the air or of allowing the water to cascade over a series of trays which may or may not contain a catalyst to aid the oxidation. The presence of dissolved oxygen in the water is sufficient to oxidize both iron and manganese, although the reaction may be speeded by the use of a suitable catalyst. The precipitation of both iron and manganese is pH dependent, with manganese being far slower and far more sensitive to pH than iron.

In order to obtain a satisfactory rate of reaction the pH should be above 7 for oxidation of iron and above 9 for oxidation of manganese. The reaction will be more rapid at higher pH values for either ion.

The precipitation of both iron and manganese is aided by the presence of the oxidized end products of the reaction. The oxidation is thus autocatalytic in a sense, although either end product will increase the rate of either reaction. This autocatalytic nature has led to the use of systems which retain the oxidized end products as a means of speeding the reaction. The efficiency of such systems is dependent on the presence of the end products and a "ripening" is thus necessary before they are put into service. A layer of MnO_2 or $Fe(OH)_3$ will build up on the trays of a cascade aerator or will coat the grains of the filter which normally follows chemical oxidation. This accumulation is necessary for proper operation of the system (Table B14).

TABLE B14. IRON AND MANGANESE OXIDATION PARAMETERS.

Oxidation parameter	Iron	Manganese
Minimum pH	6.5	8.5
Optimum pH	>7.2	>9.5
Catalysts	Cu^{2+}	Mn^{4+}
	Co^{2+}	
Sorbants	MnO_2	MnO_2
	$Fe(OH)_3$	$Fe(OH)_3$

Disinfection Disinfection is the removal of pathogenic microorganisms and does not imply that a water will be free of all biological life. Pathogenic microorganisms are, in general, more sensitive to variations in their environment than other species and hence are more readily removed from water supplies. Water is seldom tested for the presence of specific pathogens since such determinations, based on form, biochemical reactions, and stain reactions, are both difficult and time consuming. Rather, the presence of pathogens is inferred from the presence of coliform bacteria, particularly *Escherichia coli.* The test for the presence of coliforms most frequently performed is the MPN determination. This test is at best a statistical evaluation and does not possess the accuracy frequently attributed to it. A superior test which is now gaining acceptance involves the use of membrane filters and yields absolute values for the number of coliform bacteria rather than the most probable number by statistical techniques.

Significant removal of all bacteria occurs in many water treatment processes. Storage alone may produce a reduction in coliform count of as much as 90%. Coagulation and flocculation in addition to lime–soda ash softening may also produce a reduction of more than 90%, and filtration (discussed later) may remove in excess of 99% of all microorganisms. All these processes are, in a sense, disinfection techniques, but they fail to provide residual protection against later recontamination. The same observation may be made with respect to such techniques as pasteurization, ultraviolet irradiation, etc.

The nearly universal technique of disinfection of water involves the addition of chlorine either as a gas or as the salt of hypochlorous acid. In either case the predominant form will be largely dependent on pH, as shown below:

for chlorine:

$$Cl_2 + H_2O \longrightarrow HOCl + HCl \tag{21}$$

$$HOCl \longrightarrow H^+ + OCl^- \tag{22}$$

for hypochlorite:

$$NaOCl \longrightarrow Na^+ + OCl^- \tag{23}$$

$$HOCl \longrightarrow H^+ + OCl^- \qquad (22)$$

The equilibrium between the hypochlorous acid and hypochlorite ion is defined by the equation

$$\frac{[OCl^-]}{[HOCl]} = \frac{2.85 \times 10^{-8}}{[H]} \tag{24}$$

or

$$\log [HOCl] - \log [OCl^-] = 7.55 - pH \tag{25}$$

Since the hypochlorous acid is the more effective disinfectant and since the relative concentration of HOCl increases with decreasing pH as shown in Eq. (24)-(25) and Fig. B16, the pH will have a great effect on the efficiency of disinfection with chlorine.

The mechanism of disinfection by chlorination involves chemical oxidation of the extracellular bacterial enzymes. This reaction, like most chemical reactions, is dependent upon the temperature, the concentration of chlorine, the concentration of oxidizable materials, and the pH as noted above. In general the rate of the reaction will double with each 10°C increase in temperature.

Since disinfection with chlorine will require a significant time of contact, waters which are heavily contaminated should be chlorinated as early as possible in the treatment process in order to provide a maximum contact period. Prechlorination normally consists of the addition of sufficient chlorine to maintain a residual concentration throughout the treatment plant. The dosage required will be a function primarily of the concentration of oxidizable materials present and must be determined empirically for each water. Ordinarily the range of dosage necessary to maintain a residual of 0.2 mg/l will be 1-5 mg/l.

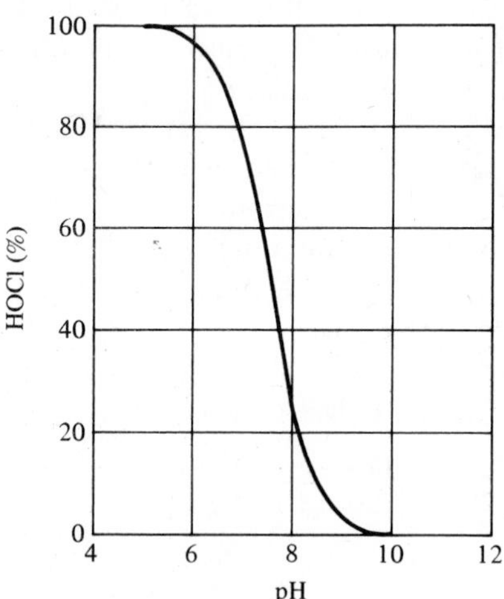

Fig. B16. Percentage HOCl vs. pH.

Free available chlorine (HOCl and OCl^-) tends to oxidize materials other than the bacterial enzymes for which it is intended. Thus, while it is a good disinfectant, it will provide little residual protection against subsequent contamination after the water leaves the treatment plant. For this reason chlorine is frequently combined with ammonia to form chloramines or combined available chlorine. The reactions involved may be represented as follows:

$$NH_3 + Cl_2 \longrightarrow HCl + NH_2Cl \tag{26}$$

$$NH_2Cl + Cl_2 \longrightarrow HCl + NHCl_2 \tag{27}$$

$$NHCl_2 + Cl_2 \longrightarrow HCl + NCl_3 \tag{28}$$

The monochloramine (NH_2Cl) and dichloramine ($NHCl_2$) are effective disinfectants. Nitrogen trichloride, the product of the third reaction, is not. The production of the mono- and dichloramines is governed by the molar ratio of the two reactants and by the pH of the solution. At equimolar concentrations the dichloramine will predominate at a pH below 5.5–6 while the monochloramine will be the major product at pH values above 6.

The chloramines are less efficient disinfectants than free available chlorine in that the time required for disinfection will be greater. Their advantage lies in their lack of reaction with inorganic materials which permits them to remain in solution until required for bactericidal action. Normally free chlorine will be used for prechlorination and combined chlorine for maintenance of a residual chlorine concentration.

Fluoridation Fluorides have been found to be effective in the reduction of dental caries when present in drinking water. The desired effect may be obtained by other applications of fluorides in toothpastes, tablets, or by direct painting, but public policy has generally favored the administration of the desired dosage through the water supply.

The concentration which has been found effective is approximately 1 mg/l as F^-, while concentrations in excess of about 2 mg/l as F^- have reportedly produced adverse reactions in some persons. The fluoride concentration must therefore be carefully controlled and must fall within the limits tabulated in Table B15. The variation in allowable fluoride concentration with temperature is based

TABLE B15. FLUORIDE CONCENTRATION VS. TEMPERATURE.

	Recommended			Mandatory
Temperature (°F)	Lower limit (mg/l)	Optimum (mg/l)	Upper limit (mg/l)	Upper limit (mg/l)
79.3–90.5	0.6	0.7	0.8	1.6
63.9–70.6	0.7	0.9	1.2	2.4
50.0–55.7	0.9	1.2	1.7	3.4

TABLE B16. COMMON FLUORIDE COMPOUNDS.

Chemical	Formula	% F	Solubility	Feeding mechanism
Sodium fluoride	NaF	45.2	4%	liquid
Sodium silicofluoride	Na_2SiF_6	60.6	very low	dry
Fluorosilicic acid	H_2SiF_6	79.0	30%	liquid

upon a presumed increase in water drunk with increasing air temperature. Fluorides may be added to water supplies as any of a large number of salts. The sodium salts are most commonly used since they add no other undesirable ions to the water. Sodium fluoride, sodium silicofluoride, and fluorosilicic acid are frequently used. The properties of these compounds are presented in Table B16. Fluorosilicic acid is particularly corrosive and, if used, will require the use of lined pipes and considerable care in handling.

Many waters contain natural fluoride, hence any additions must consider the original concentration in order that the therapeutic dosage not be exceeded. In some instances the water may contain quantities of fluoride in excess of the prescribed limits. In such a case it may be desirable to remove a portion of the natural fluoride.

Defluoridation may be effected by deionization (discussed later), but fluorides may be specifically removed by ion exchange using crushed, degreased, protein-free bone or tricalcium phosphate, consisting of granules of $Ca_3P_2O_8 \cdot Ca(OH)_2$ and $Ca_3P_2O_8 \cdot H_2O$. The exchange material may be regenerated as shown in the discussion of ion-exchange. A partial reduction in fluorides may also be obtained in conjunction with precipitation of $Mg(OH)_2$. The mechanism is uncertain but may be due to adsorption of fluoride on the $Mg(OH)_2$ floc. The percentage fluoride removal is related to the amount of magnesium precipitated and may thus be enhanced in relatively soft waters by addition of magnesium prior to softening.

Taste and Odor Control Tastes and odors in water are produced by decaying organic material; algae and actinomycetes; iron, manganese, and metallic oxides; industrial wastes such as phenols; and disinfectants such as chlorine. In general water should possess no noticeable taste or odor and treatment should be directed toward this end.

Aeration will remove tastes and odors due to gases or other volatile substances but will have little effect upon those contributed by algae or industrial phenols. Devices used include injection aerators similar to those employed in waste treatment, cascade aerators as used in iron and manganese removal, and jet or fountain aerators.

Chemical oxidation may be effected with chlorine, chlorine dioxide, or ozone. The technique depends upon the oxidation of the odor or taste producing compounds to simpler, innocuous substances. In some cases chlorination may be counterproductive. Phenols, for example, may be partially oxidized to chlorophenols which are even more offensive than the original compound (threshold odor level <1 μg/l).

Activated carbon has proven to be valuable in the removal of taste and odor produced by organic materials of both natural and industrial origin. The removal is effected by adsorption on the surface of the activated carbon; hence a finely divided material is most effective. Granular activated carbon consists of particles less than a millimeter in diameter, while powdered activated carbon is ground to pass a 200-mesh sieve.

Adsorption on activated carbon may be represented by the Langmuir isotherm:

$$\log \frac{C_o - C}{m} = \log k + \frac{1}{n} \log C \tag{29}$$

in which C_o is the taste or odor concentration in the water, C is the residual concentration produced by a dosage m of activated carbon, and k and n are constants to be experimentally determined.

EXAMPLE

A raw water has an odor concentration of 20 units. A dose of 4 mg/l of activated carbon reduces the odor concentration to 10 units. A dose of 12 mg/l reduces the odor concentration to 2 units. Determine the dosage required to obtain a residual odor concentration of 5 units.

$C_o = 20$, $C = 10$ for $m = 4$, $C = 2$ for $m = 12$,

$$\therefore \log \frac{20 - 10}{4} = \log k + \frac{1}{n} \log 10$$

and

$$\log \frac{20 - 2}{12} = \log k + \frac{1}{n} \log 2$$

Solving,

$$k = 1.205, \qquad 1/n = 0.317$$

Therefore, for $C = 5$,

$$\log \frac{20 - 5}{m} = \log 1.205 + 0.317 \log 5$$

$$m = 7.5 \text{ mg/l}$$

Alternatively the experimental data may be plotted on log–log paper, in which case the y intercept will be equal to k and the slope of the line equal to $1/n$. Required dosages for desired odor concentrations may then be read directly from the graph.

TABLE B17. ION EXCHANGE MEDIA.

Exchanger	Approximate capacity (grains/ft^3 as $CaCO_3$)	Regenerant
Cationic		
natural zeolite	5,000	NaCl
artificial zeolite	10,000	NaCl
artificial resin (Na)	20,000	NaCl
artificial resin (H)	20,000	H_2SO_4
Anionic		
artificial resin	20,000	NaOH

Ion Exchange The technique of ion exchange has been developed from the observed capacity of certain natural greensands to exchange monovalent for divalent ions and thus to remove the latter from solution. Both the natural glauconite greensands and artificial exchange media have come to be referred to as zeolites (Table B17). The term is inappropriate but its use is widespread.

Ion exchange relies upon the varying affinity of the different ions for the zeolite medium. In general, ammonium will replace sodium, divalent cations will replace monovalent cations, and trivalent cations will replace mono- or divalent species. The reaction can, however, be reversed under the proper circumstances. Using the letter Z to represent the portion of the exchange medium which is not affected, the ion exchange reaction may be represented as:

$$Na_2Z + Ca^{2+} \rightleftharpoons CaZ + 2Na^+ \tag{30}$$

The reaction may be reversed by application of a strong brine solution. Such a process will in no way affect the ion balance of the water but may either decrease or increase the total dissolved-solids concentration depending upon the species exchanged.

Synthetic resins have been developed which are effective in the removal of both cations and anions. The cationic resins are similar in action to the zeolite while the action of the anionic resins may be represented by

$$A + SO_4{}^{2-} \rightleftharpoons A \cdot SO_4$$

$$A \cdot SO_4 + Na_2CO_3 \rightleftharpoons A + Na_2SO_4 + CO_3{}^{2-} \tag{31}$$

The regenerant may be either sodium carbonate or sodium hydroxide.

Hydrogen exchange resins are also available and react as follows:

$$H_2 \cdot R + Ca^{2+} \rightleftharpoons Ca \cdot R + 2H^+ \tag{32}$$

Regeneration is effected with a strong acid.

Deionization Cation and anion exchange resins may be used in series to yield ion-free water as follows:

$$H_2 \cdot R + Ca^{2+} + SO_4{}^{2-} \rightleftharpoons Ca \cdot R + 2H^+ + SO_4{}^{2-} \quad (33)$$

$$CO_3 \cdot A + 2H^+ + SO_4{}^{2-} \rightleftharpoons SO_4 \cdot A + 2H^+ + CO_3{}^{2-} \quad (34)$$

$$2H^+ + CO_3{}^{2-} \rightleftharpoons H_2CO_3 \rightleftharpoons H_2O + CO_2 \quad (35)$$

A line diagram of a typical process is shown in Fig. B17.

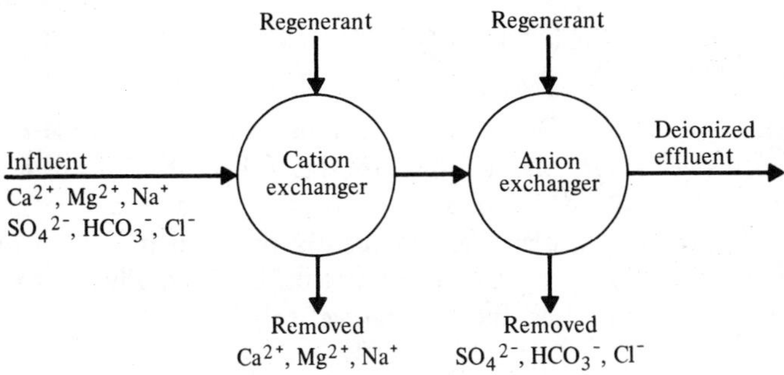

Fig. B17. Deionization schematic.

Exchange capacity of zeolites and resins is normally expressed in kilograins per cubic foot as $CaCO_3$ since zeolites were initially used for water softening and still have applications in that area. For other ions the conversion may be made by converting the value in grains (7000 grains = 1 lb) to equivalents and multiplying by the equivalent weight of the ion in question.

EXAMPLE

A synthetic ion exchange resin has a capacity of 10,000 grains/ft^3 as $CaCO_3$. Determine its capacity in grams of magnesium per cubic foot and what volume of water containing 40 mg/l of Mg^{2+} may be treated before regeneration of the bed.

$$10{,}000 \text{ grains} = 1.43 \text{ lb} = 648 \text{ g}$$

$$648 \times \frac{1}{50} = 12.96 \text{ eq}$$

$$\therefore \text{Capacity} = 12.96 \text{ eq/ft}^3$$

$$= 12.96 \times 24.3/2$$

$$= 157.5 \text{ g/ft}^3 \text{ as } Mg^{2+}$$

A total of 157.5 g of Mg^{2+} may be removed; thus

$$V = 157{,}500/40 = 3940 \text{ l} \quad \text{or} \quad 1300 \text{ gal}$$

may be treated per cubic foot of media before regeneration.

Stabilization Waters following softening, coagulation, or ion exchange may be either supersaturated with or deficient in certain ionic species. Such waters are said to be either corrosive or depositing and must be adjusted to chemical stability. Stabilization normally is effected by the addition of CO_2 if the water is depositing, or by the addition of Na_2CO_3 if it is corrosive. The amounts to be added must be determined experimentally. This is normally determined by laboratory jar tests.

Other Desalting Processes Desalinization has historically been achieved by distillation and many modern plants still utilize this technique. The power requirements even for sophisticated systems such as triple-effect and vapor compression stills are high, hence such plants are frequently operated in conjunction with nuclear power generation. Distillation is unselective and totally salt-free waters are undesirable for many uses, thus alternative techniques are currently gaining wider acceptance.

Dialysis employs semipermeable membranes which act in a sense as ion exchangers. The driving force in the reaction is electrical, with cations being drawn to the cathode and anions to the anode through membranes permeable to only anions or cations. Alternate cells are either demineralized or concentrated to a brine (Fig. B18).

Reverse osmosis relies upon the exertion of mechanical force to oppose the tendency of water molecules to pass through a semipermeable membrane to a region of higher salt concentration in order to equalize the ionic strength. If sufficient pressure is applied to the vessel containing the salt solution the flow of

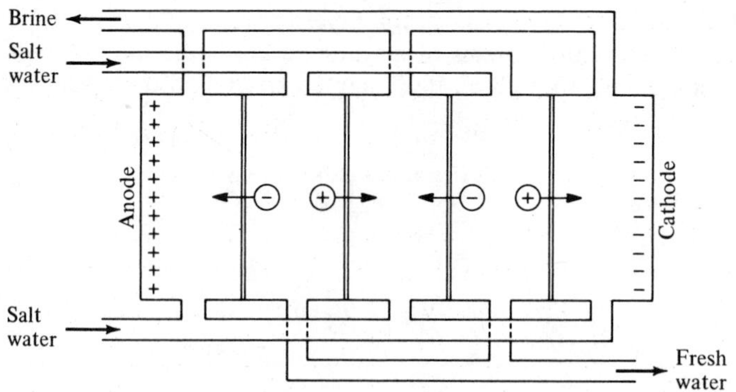

Fig. B18. Dialysis schematic.

water will be reversed with a resultant concentration of the salt and production of salt-free water. The method of reverse osmosis is particularly valuable in that the membrane may be selected for exclusion of particular ionic species. Thus it may be possible to reduce the concentration of chloride and sulfate, for example, without demineralizing the water. The pressure required will depend upon the membrane used, the rate of flow desired, and the concentration of the salt. Working pressures may be as high as 25 atmospheres although pressures of 2 to 3 atmospheres will suffice for nominal flow at low salt concentrations. Reverse osmosis is now being used in the treatment of municipal waters in areas where the raw water would be otherwise unacceptable for domestic use.

Filtration Although it may be only marginally effective in improving the quality of a water which has already been chlorinated, coagulated, and perhaps softened, filtration is considered by many engineers to be the most important process in the production of potable water. Filtration alone will remove suspended solids, much colloidal material, and in excess of 99% of bacterial contamination. Hence the filters serve not only to polish the water but also to guard against any failure of other systems.

Rapid sand filters serve as far more than the mechanical straining devices implied by their name. In addition to straining, coagulation, flocculation, and adsorption also occur within the filter bed. This complex action results in the removal of material which is significantly smaller than the pore size of the sand.

The sand used in filters should be quartz or quartzite, free of dirt and fines, and relatively insoluble in acid. The depth of the sand is normally 2-$2\frac{1}{2}$ ft. Sand size and uniformity are specified by effective size and uniformity coefficient, as shown in Table B18.

Medium sands are most widely used at present, although coarse sands are more suitable for waters which are not highly polluted. Fine sands may be used where an increase in grain size due to incrustation with calcium carbonate is expected to occur.

Filtration rates are specified in gallons per square foot per minute. Original rates were about 2 gal/ft^2/min but current design favors 4 gal/ft^2/min. Experiments have indicated that rates in excess of 4 gal/ft^2/min will yield a satisfactory water. It may therefore be reasonable to design plants to meet the lower rate in

TABLE B18. FILTER SAND SIZE.

	Coarse		Medium		Fine	
	Min.	Max.	Min.	Max.	Min.	Max.
Effective size	0.55	0.65	0.45	0.55	0.35	0.45
Uniformity coefficient	1.5	1.7	1.5	1.7	1.5	1.7

the immediate future with a tacit understanding that the filters will be nominally overloaded before the end of the design period.

When sand alone is used as a filter media it will settle following washing with the finer grains on top. This gradation has the effect of placing the most effective filtering material (i.e., that with the smallest pore size) at the top of the filter. The bulk of the filter material would thus appear to be ineffective and this is largely the case. The depth does serve to remove those particles which pass the upper layer and aids in assuring uniform flow across the plan area, but it cannot be denied that a reversal in stratification would aid in a more effective use

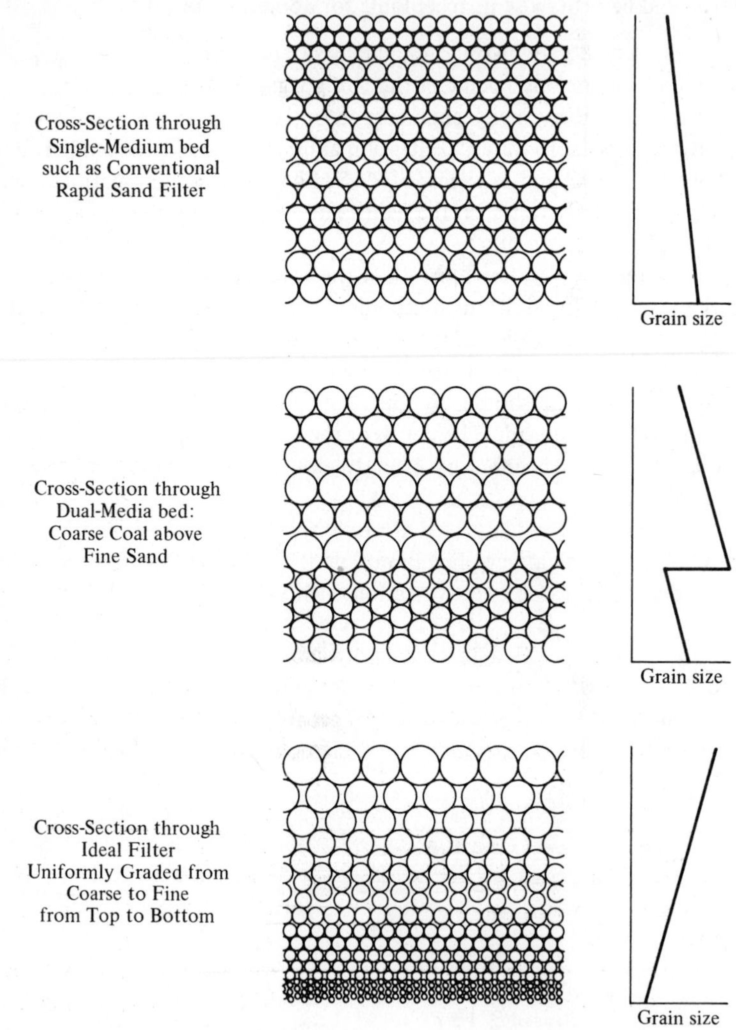

Fig. B19. Filter gradation (courtesy Neptune Microfloc, Inc.).

of the total filter volume. Upflow filters have been used in eastern Europe and the Soviet Union but such a flow pattern constitutes a crossconnection and is thus not acceptable in the United States.

Mixed media filters afford a partial solution to the unfavorable gradation of single medium filters. Anthracite coal, crushed and graded, has proven to be a useful filter media when used alone and has a density only slightly more than half that of sand. Relatively coarse particles of coal will thus tend to settle less rapidly than finer sand particles. Garnet sand is particularly dense and will settle more rapidly than either coal or quartz sand even when its particle size is smaller. Combinations of these materials will thus yield a bed in which the normal gradation is partially reversed with the coarse anthracite on top, the medium quartz in the middle, and the fine garnet on the bottom. The usual gradation is, of course, maintained within the individual layers. Such filters have longer runs between washing cycles and thus are more economical in operation than single medium filters (Fig. B19).

The filter is underlain with gravel to a depth of 15–24 in. in graded layers with the larger sizes on the bottom. The gravel should be rounded and free of foreign material. A typical gravel layer is shown in Fig. B20. The gravel serves to support the filter media and aids in the uniform distribution of backwash water. When specialized proprietary filter bottoms are used the gravel layer may be greatly reduced.

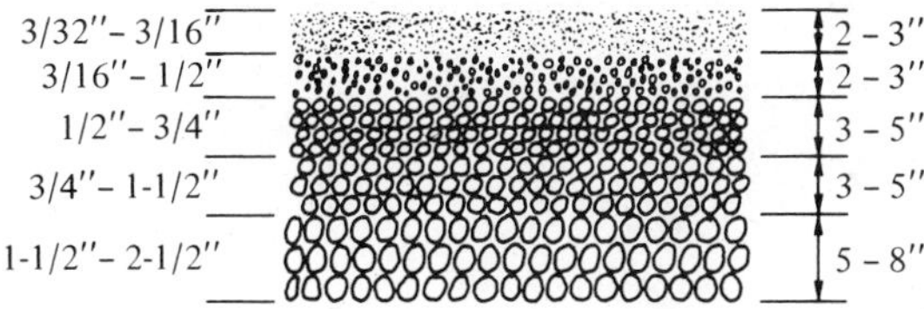

Fig. B20. Filter gravel gradation.

Backwashing consists of reversing the flow and passing filtered water upward through the bed at a velocity sufficient to cause a 30–50% expansion. This expansion and the turbulent flow cause the grains to abrade each other with subsequent removal of filtered solids. Backwashing is normally begun before the head loss through the filter becomes sufficiently high to produce negative head within the bed. Negative head, if produced, may cause short-circuiting of the filter. A typical backwash rate for quartz sand is 15 gal/ft^2/min. Agitation of the bed surface before and during the first few moments of backwash either with compressed air or water jets aids the washing procedure by breaking up mud accumulations which might otherwise form balls which would remain in the filter. Backwash rates for other media will depend upon the grain size and specific gravity of the material. The rate must be such that little or no medium is carried out of the bed. For anthracite the rate is about 10 gal/ft^2/min. Typical details for a rapid sand filter are shown in Figs. B21 and B22, with filter underdrains and air diffusers in Fig. B23.

Pressure filters are sand filters enclosed in a container and operated under pressures higher than those obtainable by the static head in open filters.

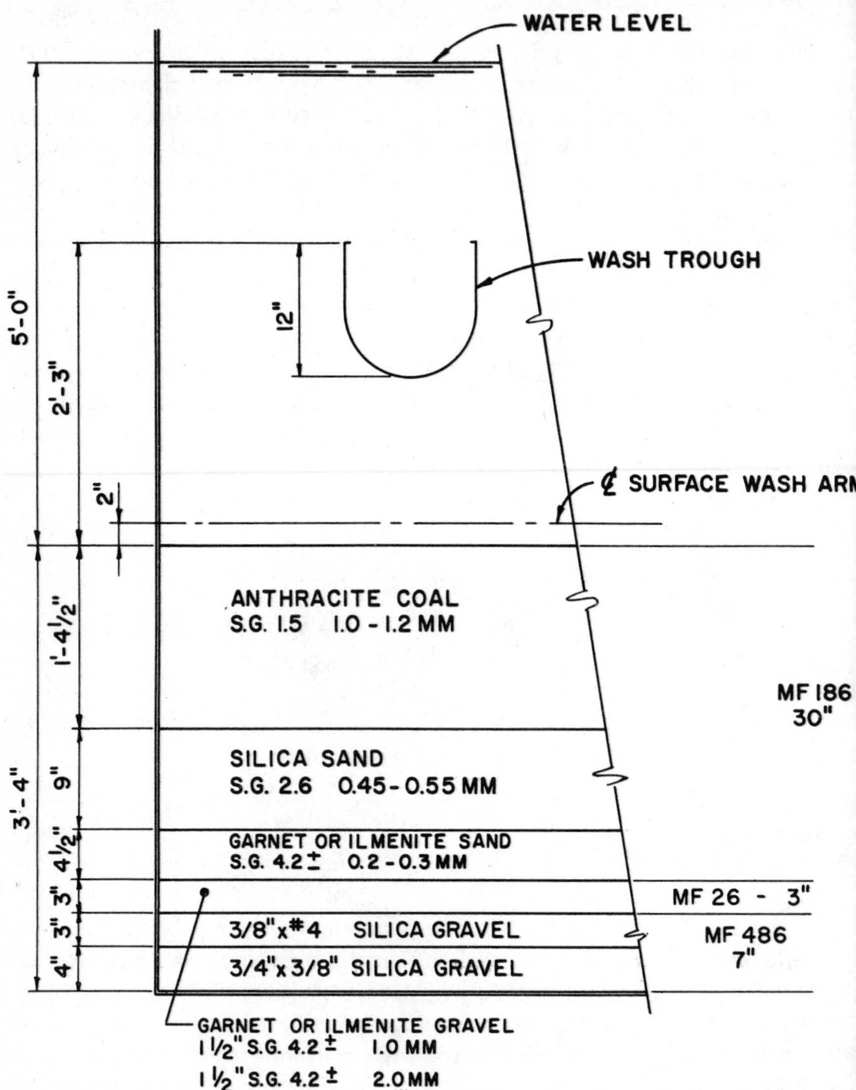

Fig. B21. Typical cross section of rapid sand filter (courtesy Neptune Microfloc, Inc.).

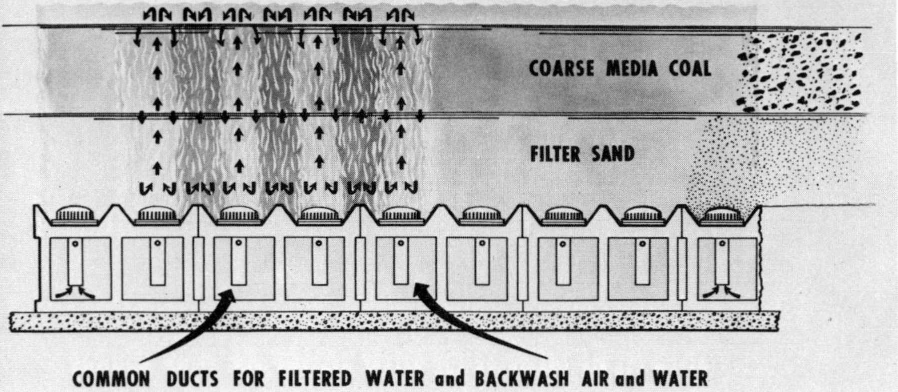

Fig. B22. Dual-medium filter with air backwash (courtesy Walker Process Equipment.)

(Fig. B24). The sand used is normally the same as that in ordinary filters although the depth used is somewhat less (18–24 in.). Coagulants may be added prior to introduction of the water to the filter. Pressure filters are seldom used in treatment of public water supplies but are applicable to industrial and swimming pool waters. Filtering rates of up to 5 gal/ft^2/min are commonly used. As in the standard sand filter, backwash is dictated by head loss through the filter. The advantage of pressurized systems lies in their capacity to permit higher pressure losses without producing negative head in the filter bed.

Diatomaceous earth filters are pressure units which consist of rigid elements of wire, fabric, porous stone, or alumina. A precoat of diatomaceous earth (essentially consisting of the irregularly shaped shells of microscopic crustaceans) is placed on the filter area by placing it in the first water passed through the unit. The precoat is normally about 1oz per square foot of filter area. The precoat carriage water is wasted.

A "body coat" of diatomaceous earth is added continuously with the water to be treated. The body coat dosage is dependent on the turbidity of the raw water and is normally about 2 mg/l per mg/l of turbidity. This continuous coating renews the prime filtering surface and maintains the medium in a porous condition. The filter is backwashed by reversing the flow when the head loss approaches 10 psi. Diatomaceous earth filters are not uniformly approved by state regulatory agencies and may not be permissible in some localities. They are frequently used for swimming pool installations.

Water Distribution A water distribution system includes the mains, branches, submains, pumps, substations, and storage facilities. The system must be so

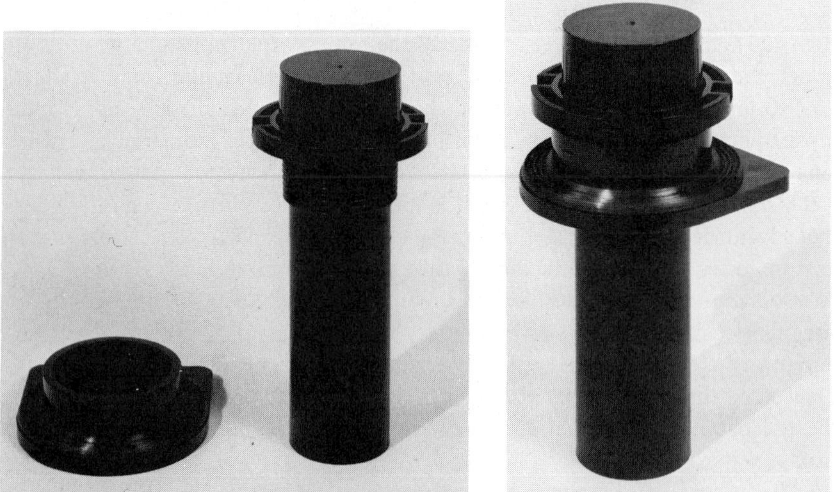

Fig. B23. Camp filter underdrains with air diffusers (courtesy Walker Process Equipment).

designed that it will meet the maximum demand rate and deliver water at a useful pressure, but should not incorporate undue excess capacity.

In the analysis of pipe networks a number of simplifications can be made without introducing a significant error. Of these, the concept of the equivalent pipe is particularly useful. Equivalent pipe analysis applies the premise that a section of a network may be replaced with a theoretically equivalent section which carries the same flow at the same head loss.

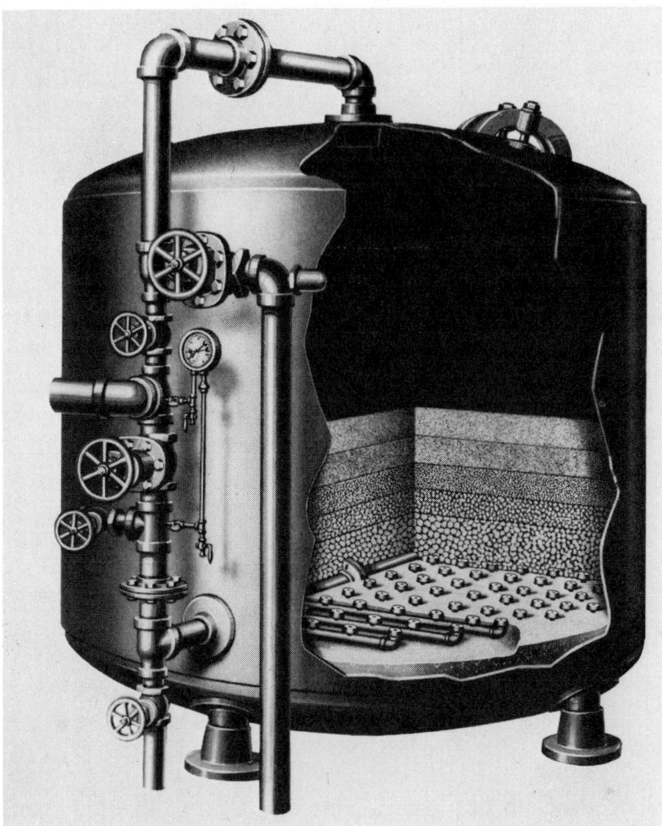

Fig. B24. Pressure filter (courtesy Westinghouse Electric Corp., Infilco Division).

EXAMPLE

Determine the diameter and length of the equivalent pipe which may replace the system shown in Fig. B25 in later computations.

First determine a pipe equivalent to loop *BCED*. To do this pipes equivalent to leg *BDE* and leg *BCE* must first be found.

Assume a flow through leg *BDE* of 500 gpm (gal/min; note that any assumption may be made). From the Hazen-Williams formula or the nomograph (see opening section on engineering hydraulics),

$$\begin{aligned}\text{Head loss in } BD &= 2.8 \times 1.0 = 2.80 \text{ ft}\\ \text{Head loss in } DE &= 7.8 \times 0.75 = \underline{5.85 \text{ ft}}\\ \text{Head loss in } BDE &= 8.65 \text{ ft}\end{aligned}$$

The average loss of head is 8.65/1.75 = 4.94 ft/1000 ft

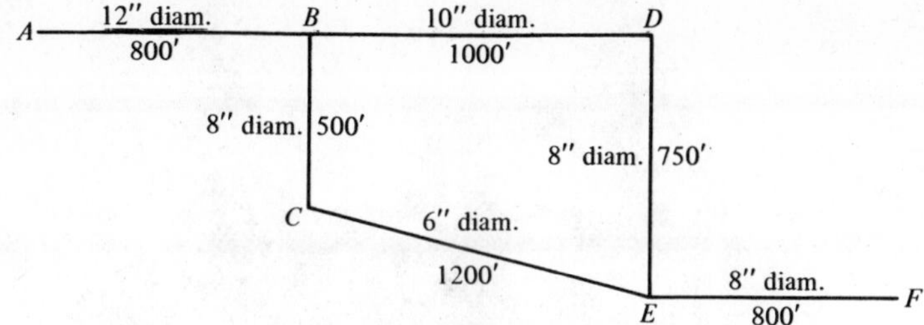

Fig. B25. A pipe network section.

Referring again to the nomograph, for a head loss of 4.94 ft/1000 ft at a flow of 500 gpm the equivalent pipe is 1750 ft long and has a diameter of 8.9″. The noncommercial diameter is of no significance. Using the same method leg *BCE* is found to be equivalent to a line 1700 ft long with a diameter of 6.4″. The section may now be considered to be as shown in Fig. B26.

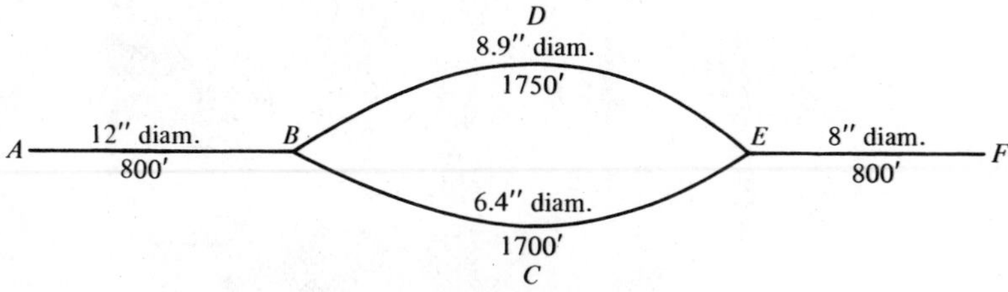

Fig. B26. Equivalent pipes.

It is now necessary to find a single pipe equivalent to the two branches *BDE* and *BCE*. Since the loss of head must be the same in either branch it may be taken to be any value. Choosing 10′ as the head loss, the head loss in *BDE* is

$$10/1.75 = 5.72 \text{ ft}/1000 \text{ ft}$$

and in line *BCE* it is

$$10/1.7 = 5.88 \text{ ft}/1000 \text{ ft}$$

From the nomograph the flows are

$$\begin{aligned} BDE &= 550 \text{ gpm} \\ BCE &= \underline{230 \text{ gpm}} \\ BE &= 780 \text{ gpm} \end{aligned}$$

The equivalent pipe carrying a flow of 780 gpm at a head loss of 10′ may be found from the nomograph to be 1700′ long with a diameter of 10.2″. The section may now be represented as in Fig. B27. Lines *AB*, *BE*, and *EF* may now

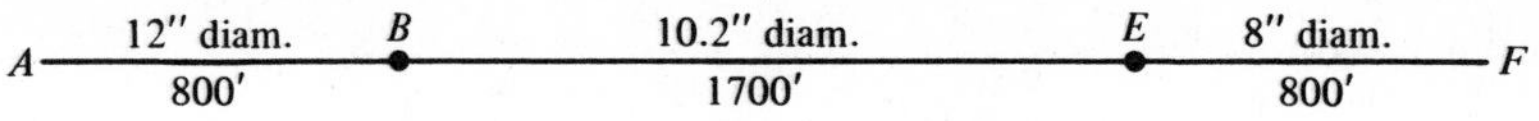

A — 12" diam. 800'	B — 10.2" diam. 1700'	E — 8" diam. 800' — F

Fig. B27. Equivalent pipes.

be combined into a single equivalent pipe. Assuming a flow of 1000 gpm,

$$\begin{aligned}\text{Head loss in } AB &= 4.1 \times 0.8 = 3.28 \text{ ft}\\ \text{Head loss in } BE &= 9.2 \times 1.7 = 15.64 \text{ ft}\\ \text{Head loss in } EF &= 28 \times 0.8 = \underline{22.40 \text{ ft}}\\ \text{Total head loss} &= 41.32 \text{ ft}\end{aligned}$$

This is equivalent to a 3300-ft pipe with a diameter of 9.8". The equivalent pipe method is useful in the analysis of many systems, particularly those which are long and narrow. Smaller cross pipes may be eliminated to simplify the calculations since these carry relatively little water.

Hardy Cross Method

The Hardy Cross method of pipe network analysis is an iterative method in which successive approximations are made for the actual pipeline flows. The Hazen–Williams formula is normally used to calculate the head loss in each reach of pipe. The formula, from above, is

$$v = 1.318\, Cr^{0.63} s^{0.54}$$

Since $Q = vA$ this may be rewritten as

$$Q = 1.318\, CAr^{0.63} s^{0.54} \tag{36}$$

A and r are both functions of only the diameter for circular pipes flowing full, hence this may be rearranged as:

$$Q = 1.318\, C \cdot \pi d^2/4\, (d/4)^{0.63} s^{0.54} \tag{37}$$

Since C and d will be constant for any given pipe this may be reduced to:

$$Q = ks^{0.54} \tag{38}$$

From which the head loss in feet per 1000 ft is:

$$h = kQ^{1.85} \tag{39}$$

The actual flow in any line will be Q, which is equal to the assumed flow, Q_1, plus the error in the assumption. Thus,

$$Q = Q_1 + \delta \tag{40}$$

$$\therefore h = K(Q_1 + \delta)^{1.85} \tag{41}$$

Equating (39) and (41),

$$KQ^{1.85} = K(Q_1 + \delta)^{1.85}$$
$$= K(Q_1{}^{1.85} + 1.85Q_1{}^{0.85}\delta + \cdots) \tag{42}$$

Neglecting the smaller terms in the expansion,

$$KQ^{1.85} = KQ_1{}^{1.85} + 1.85KQ_1{}^{0.85}\delta \tag{43}$$

For any closed pipe loop the sum of the head loss must be equal to zero, thus,

$$\sum KQ^{1.85} = \sum KQ_1{}^{1.85} + \sum 1.85KQ_1{}^{0.85}\delta = 0 \tag{44}$$
$$\therefore \delta = - \sum KQ_1{}^{1.85}/1.85 \sum KQ_1{}^{0.85} \tag{45}$$

The correction to the assumed flow in any loop may thus be determined from the assumed flow and the properties of the individual pipes.

The application of the Hardy Cross method is simple but care must be taken to assure internal consistency. The procedure may be outlined as follows:

1. Assume any flow distribution with respect to magnitude and direction. The net flow into any junction must be zero and the net flow from the loop must equal the flow into the loop. Any assumption may be made but the number of iterations will be decreased if the initial assumption is reasonable.
2. Determine the head loss in each line from the Hazen–Williams formula by calculation or from the nomograph. This yields the $K_iQ_{1_i}{}^{1.85}$
3. Divide the head loss obtained in each line by the flow assumed for that line. This yields the $K_iQ_{1_i}{}^{0.85}$.
4. Sum with attention to sign the $K_iQ_{1_i}{}^{1.85}$. Yielding the term $\sum KQ_1{}^{1.85}$.
5. Sum the $K_iQ_{1_i}{}^{0.85}$. Note that $K_iQ_{1_i}{}^{0.85}$ will always be a positive quantity. This yields the term $\sum KQ_1{}^{0.85}$.
6. Calculate the loop correction from Eq. (45).
7. Apply the correction δ to the assumed flow in each line of the loop and repeat steps 2–6. This procedure is continued until the calculated correction is 50 gpm or less. Normally three iterations will suffice. *Note:* In applying corrections to lines common to two loops the corrections for both loops are made to the common line. If this is not done the net flow at the junctions will not be zero and the method will fail.

EXAMPLE

The method of Hardy Cross will be applied to a simple network. Larger systems are to be handled in identical fashion. Figure B28 is a pipe network in which lines less than 8″ in diameter have been eliminated. The flows from the loop are assumed to be concentrated at the junctions of major lines and are determined from the estimated water demand of the area served.

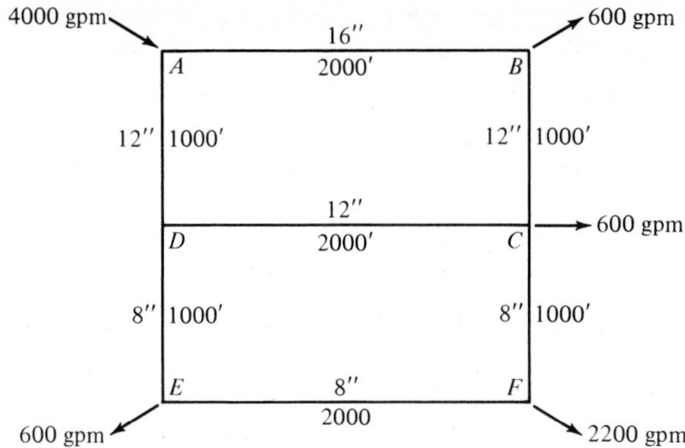

Fig. B28. Pipe network for Hardy Cross analysis.

The initial flows may be assumed to have any magnitude and direction consistent with the input and output from the system. The method is more rapid however if the initial assumptions are approximately correct. Experience will aid greatly in the selection of a good first assumption but the investigator may be partially guided by the relative flows which will be carried at the same head loss in different sizes of pipe as presented in Table B19.

TABLE B19. RELATIVE FLOW AT IDENTICAL HEAD LOSS.

Pipe diameter	8"	12"	16"	20"	24"	36"
Relative flow	1	3	6	12	20	60

The assumed flows for this example are as shown in Fig. B29 and Table B20. The calculations are tabulated in Tables B20 and B21. Figure B30 shows the balanced network.

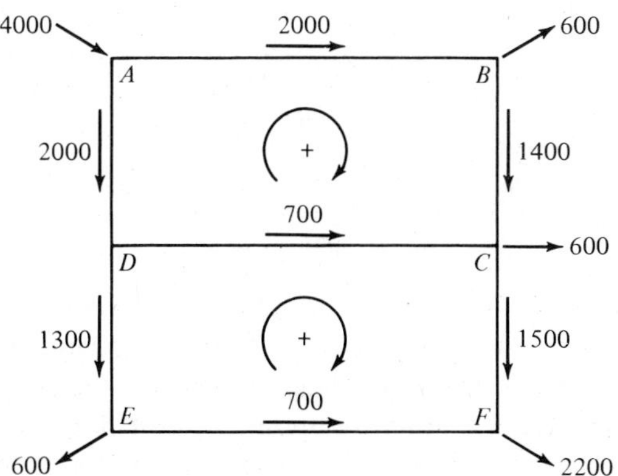

Fig. B29. Assumed flow distribution.

TABLE B20. FIRST INVESTIGATION FOR EXAMPLE PROBLEM.

Line	Diam. (in.)	Length (×1000 ft)	Assumed Flow (gpm)	Head loss (ft/1000 ft)	Head loss (ft)	Head loss/Q
AB	16	2	2000	3.6	7.2	0.0036
BC	12	1	1400	8.0	8.0	0.0057
CD	12	2	−700	−2.1	−4.2	0.0060
DE	12	1	−2000	−15.0	−15.0	0.0075
DC	12	2	700	2.1	4.2	0.0060
CF	8	1	1500	60.0	60.0	0.0400
FE	8	2	−700	−15.0	−30.0	0.0429
ED	8	1	−1300	−48.0	−48.0	0.0369

$$\delta_{\text{loop}ABCD} = -\frac{\Sigma \text{H.L.}}{1.85\Sigma \text{H.L.}/Q} = -\frac{-4.0}{1.85(.0228)} = +95 \approx +100$$

$$\delta_{\text{loop}DCFE} = -\frac{\Sigma \text{H.L.}}{1.85\Sigma \text{H.L.}/Q} = -\frac{-13.8}{1.85(0.1258)} = +59.2 \approx +60$$

The assumed flows must now be corrected with due attention to sign. The new flow in the common line will be: with respect to loop *ABCD*, −700 + 100 − 60 = −660; with respect to loop *DCFE*, 700 − 100 + 60 = 660.

TABLE B21. SECOND INVESTIGATION FOR EXAMPLE PROBLEM.

Line	Diam. (in.)	Length (×1000 ft)	Assumed Flow (gpm)	Head loss (ft/1000 ft)	Head loss (ft)	Head loss/Q
AB	16	2	2100	4.0	8.0	0.0038
BC	12	1	1500	9.0	9.0	0.0060
CD	12	2	−660	−2.0	−4.0	0.0061
DE	12	1	−1900	−14.0	−14.0	0.0074
DC	12	2	660	2.0	4.0	0.0061
DF	8	1	1560	62.0	62.0	0.0397
FE	8	2	−640	−12.5	−25.0	0.0391
ED	8	1	1240	−43.0	−43.0	0.0347

$$\delta_{\text{loop}ABCE} = -\frac{-1.0}{1.85(.0233)} = +23$$

$$\delta_{\text{loop}DCFE} = -\frac{-2.0}{1.85(.1196)} = +9$$

Both of which are negligible. The balanced flow will be as shown in Fig. B30. The head loss in each line calculated from the balanced flow is also shown.

The Hardy Cross method may be applied as shown; however the time required has led to the common use of digital computers to perform the calculations. The application to computer calculation is direct. A subprogram is required for the calculation of the head loss from the Hazen-Williams formula. The iteration is performed, as above, until the corrections obtained are less than 50 gpm.

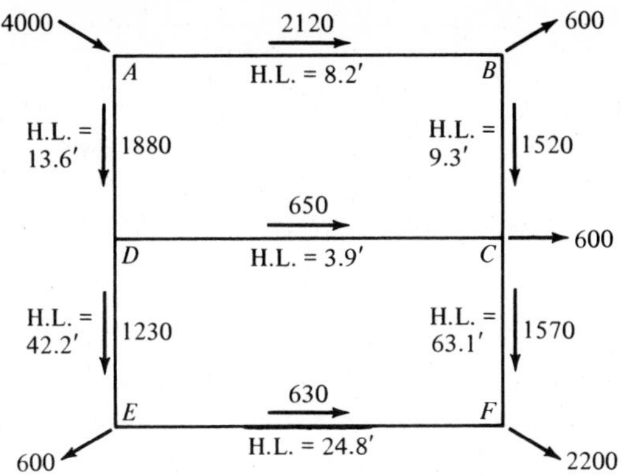

Fig. B30. Balanced network.

An analog computer specifically designed for the analysis of pipe networks is useful in distribution system design. This device, called the McIlroy Analyzer, employs special vacuum tubes called fluistors which respond to electrical current in a manner analogous to a pipe's response to flow. In this system the flow is proportional to the current and the head is proportional to the voltage. Since ordinary resistors do not vary with current, the fluistors must be used in such a circuit. The data output is normally in the form of voltage drops which are proportional to the head loss. The circuit is automatically balanced for any applied voltage, thus the major time requirement lies in setting up the analog. If excessive head losses are encountered the circuit may be quickly modified by replacing the fluistors with others analagous to larger pipes.

Circle Analysis

The water available at a given point within a distribution network to meet fire demand may be determined by the method of circle analysis. A circle of radius equal to the maximum hose length centered at the point in question is constructed on a grid of the distribution system. Each line intersected by the circle is assumed to be available to provide water to the central point. From the pipe diameters the amount of water available may be determined. Any line tangent to the circle is considered to be cut twice and thus counts as two lines.

Materials and Fittings

Pipe lines for water distribution may be of steel, cast iron, concrete, asbestos cement or plastic. Copper is normally used only for service connections and internal plumbing, while wood and lead are seldom used in modern practice.

Lead should not be used since it may dissolve slowly in soft or acid waters. Other metals may also dissolve, but of those commonly used in plumbing only lead is known to dissolve to an extent sufficient to cause poisoning. Plumbism may result from habitual consumption of waters containing over 0.3 mg/l of lead.

Pipe lines must be designed to resist both internal and external loads. The internal pressure will produce tensile stresses as given by

$$\sigma = rp/t \tag{46}$$

in which σ is the stress in pounds per square inch, r is the radius of the pipe in inches, p is the internal pressure in pounds per square inch, and t is the thickness of the pipe wall in inches. The pressure to be used for design must include that due to water hammer produced when lines are closed quickly. Such surge pressures can be extremely high, hence slow-closing valves should be used wherever possible. As an upper limit the maximum pressure producible by water hammer in pounds per square inch will be 50 to 60 times the velocity in the pipe in feet per second. Where the direction of flow changes (as at tees, wyes, and bends) an external buttress should be used to react against the locally unbalanced internal pressure.

External loads are of particular importance in the selection of brittle pipe materials. Manufacturers will provide pipe of various wall thicknesses for differing cover and loading conditions.

Cast iron pipe exhibits good resistance to corrosion and has a useful life of as much as 100 years. Cast iron is coated with tar at about 300°F both inside and out as normally produced, but may be obtained with a cement mortar inner lining or an external and internal coating of bituminous enamel. Standard lengths are 12 and 16 ft for pit cast pipe, and 16 and 18 ft for centrifugally cast pipe. Cast iron pipe is designed for working pressures of 50–300 psi in increments of 50 psi. Pipe thickness is based upon the working pressure, the depth of cover, and the laying method, all of which must be specified when ordering. Standard diameters commonly used in water distribution are 4–20 in. in 2-in. increments and 24, 30, 36, 42, and 48 in. Smaller sizes are also available.

The most common jointing used in cast iron pipe is the bell and spigot type, although flanged and mechanical joints are also used, particularly for larger lines. Changes of direction and diameter are made by means of standard fittings. Those available include 90°, 45°, and $22\frac{1}{2}°$ bends, wyes, tees, crosses, reducers, etc. Connections are usually made with lead and oakum, although lead substitutes and cement have been used. Plastic couplers are also available which greatly speed assembly.

Steel pipe, because of its higher strength, will be far lighter than cast iron designed to carry the same working pressure. Because of its thinner wall, steel pipe is more susceptible to damage from corrosion, external loads, and negative heads which may occasionally be drawn. Steel pipe may be expected to have a maximum useful life of 50 years. It is available in sizes up to 96 in. in diameter. Steel pipe may be coated with tar, bituminous enamel, graphite paint, asphalt, cement mortar, or plastic.

Concrete pipe is available in sizes from 16 to 108 in. for static heads of 100-600 ft (43-260 psi). Joints consist of steel rings with rubber gaskets. The joints, as with cast iron pipe, may be deflected slightly to form curves of large radius. Standard fittings for bends, branches, etc. are available. Concrete pipe may be expected to have a useful life of as much as 75 years.

Asbestos cement pipe is made of portland cement and asbestos fiber and is available in 4–16 in. diameters for working pressures of 100, 150, and 200 psi. Joints consist of external sleeves with rubber rings compressed between sleeve and pipe. The joints are flexible, permitting deflections of 10° to 12°. Standard cast iron pipe fittings may be used. Asbestos cement pipe has been in use for over forty years without significant problems. A useful life equivalent to that of concrete pipe appears to be a reasonable projection.

Plastic pipe, while not yet accepted by all regulatory agencies, has been found to be useful in many applications. Its use in service connections is largely restricted by local building codes and union work rules. Plastic pipe (particularly PVC) has been used in small distribution systems and appears to be sufficiently durable and corrosion resistant to have a long service life.

Corrosion

Corrosion of metals (Table B22) is an electrochemical phenomenon which may be represented as shown below:

$$4Fe \rightleftharpoons 4Fe^{3+} + 12e^- \tag{47}$$

$$12H^+ \rightleftharpoons 6H_2 - 12e^- \tag{48}$$

$$6H_2O \rightleftharpoons 12H^+ + 60^{2-} \tag{49}$$

$$4Fe^{3+} + 60^{2-} \rightleftharpoons 2Fe_2O_3 \tag{50}$$

TABLE B22. ORP OF COMMON METALS.

Metal	Reaction	E^0 (Volts)
Aluminum	$Al \rightarrow Al^{3+} + 3e^-$	+1.66
Zinc	$Zn \rightarrow Zn^{2+} + 2e^-$	+0.76
Iron	$Fe \rightarrow Fe^{2+} + 2e^-$	+0.44
Lead	$Pb \rightarrow Pb^{2+} + 2e^-$	+0.13
Iron	$Fe \rightarrow Fe^{3+} + 3e^-$	+0.04
Hydrogen	$H_2 \rightarrow 2H^+ + 2e^-$	0
Copper	$Cu \rightarrow Cu^{2+} + 2e^-$	−0.34

The overall reaction is then:

$$4Fe + 6H_2O \rightleftharpoons 2Fe_2O_3 + 6H_2 \tag{51}$$

Corrosion may be prevented by any mechanism which interferes with any of the general reactions. Reaction (47) may be reversed by driving electrons into solution and plating out iron. Coatings will prevent contact between water and pipe thus preventing reaction (50). Protective coatings include asphaltic or bituminous paints over red lead, tar, cement mortar, and galvanizing. A sacrificial anode with a higher oxidation-reduction potential than iron will be oxidized in preference to the pipe.

Galvanic corrosion occurs when two dissimilar metals are placed in contact in water solution. The metal with the higher oxidation-reduction potential will act as the anode and hence will corrode while the other metal will become cathodic. Direct connections between copper service lines and an iron distribution system will thus result in corrosion of the main. Such galvanic action can be prevented by insertion of an inert material between the two lines.

Cleaning and Disinfection

In placing new mains and in repairing old lines it is practically impossible to avoid contamination. Such lines should be flushed at a velocity of over 2.5 fps to remove any mud, silt, or other material which may have entered. Lines should then be filled with a strong chlorine solution (25–100 mg/l as chlorine) which is then retained for 24–72 h until a residual of at least 10 mg/l is maintained. The lines are then flushed with treated water before being placed in service.

Accumulations of calcium carbonate, rust, and sediment can greatly reduce the diameter of water mains and may adversely affect the friction factor. Scrapers may be used to restore such lines; however, the improvement will usually be temporary if severe corrosion has begun.

Leakage of new water mains is specified by the American Water Works Associa-

TABLE B23. ALLOWABLE LEAKAGE IN NEW MAINS (GAL/MILE/H/IN. DIAM.).

Pipe length (ft)	Test pressure (psi) 100	150	200
12	2.38	2.91	3.36
14	2.04	2.50	2.88
16	1.78	2.19	2.52
18	1.58	1.94	2.24

tion to be less than

$$L = \frac{NDP}{1850} \tag{52}$$

in which L is the allowable leakage in gallons per hour, N is the number of joints in the length of line tested, D is the diameter of the pipe in inches, and P is the pressure during the test in pounds per square inch. Table B23 presents the allowable leakage for pipe of various lengths.

REFERENCES

Water Quality

American Public Health Association, American Water Works Association, and Water Pollution Control Federation, *Standard Methods For The Examination of Water and Wastewater*, 13th ed. (1970).

Hopkins, O. C. and O. Gullans, New U.S. Public Health Service Standards, *J. AWWA*, **52** (1960).

Task Group Report, Physiologic and Health Aspects of Water Quality, *J. AWWA*, **53** (1961).

U.S. Public Health Service, *A Manual of Practice Recommended by the Public Health Service*, U.S. Public Health Service Publ. 1820 (1969).

Water Sources

American Society of Civil Engineers, Ground Water Basin Management, *ASCE Manual of Engineering Practice*, **40** (1961).

Babbitt, H. E., et al., *Water Supply Engineering*, McGraw-Hill, New York (1962).

Linsley, R. K. and J. B. Franzini, *Water Resources Engineering*, McGraw-Hill, New York (1964).

Water Quantity

National Board of Fire Underwriters, *Standard Schedule For Grading Cities and Towns of The U.S. With Reference to Their Fire Defenses and Physical Conditions*, New York (1956).

Schmitt, R. C., Forecasting Population By The Ratio Method, *J. AWWA*, **46** (1954)

Wolff, J. B., Peak Demands in Residential Areas, *J. AWWA*, **53** (1961).

Wolff, J. B. and J. F. Loos, Analysis of Peak Water Demand, *Public Works* (1956).

Water Treatment (General)

American Water Works Association, *Water Treatment Plant Design Manual* (1969).

American Water Works Association, *Water Quality and Treatment*, 3rd ed., McGraw-Hill, New York (1971).

Great Lakes–Upper Mississippi River Board of State Sanitary Engineers, *Recom-*

mended Standards For Water Works, Health Education Service, Albany, New York (1968).

Hopkins, E. S. and E. L. Bean, *Water Purification Control*, Williams and Wilkins, Baltimore (1966).

Sawyer, C. N. and P. L. McCarty, *Chemistry For Sanitary Engineers*, 2nd ed., McGraw-Hill, New York (1967).

U.S. Public Health Service, *Manual of Recommended Water Sanitation Practice*, U.S. Pub. Health Service Publ. 296 (1946).

Water Treatment (Specific Processes)

Black, A. P., Theory of Coagulation, *Water and Sewage Works*, **108** (1961).

Camp, T. R., Sedimentation and The Design of Settling Tanks, *Trans. ASCE*, **111** (1946).

Committee Report, Capacity and Loadings of Suspended Solids Contact Units, *J. AWWA*, **43** (1951).

Culp, G. S., et al., High Rate Sedimentation in Water Treatment Works, *J. AWWA*, **60** (1968).

Ghosh, M. M., et al., Precipitation of Iron in Aerated Ground Water, *J. San. Eng. Div. ASCE*, **92**, **SA1** (1966).

Howson, L. R., Lagoon Disposal of Lime Sludge, *J. AWWA*, **53** (1961)

Laubusch, E. J., Water Disinfection Practices in the United States, *J. AWWA*, **52** (1960).

Mackrle, S., Mechanism of Coagulation in Water Treatment, *J. San. Eng. Div. ASCE*, **SA3** (1962).

Water Treatment, Specific Processes

Stumm, W., Chemistry of Natural Waters in Relation to Water Quality, U.S. Public Health Service Publ. 999-WP-15 (1964).

Stumm, W. and C. R. O'Melia, Stoichiometry of Coagulation, *J. AWWA*, **60** (1968).

Waring, F. H., Methods of Lime Softening Sludge Disposal, *J. AWWA*, **47** (1955).

White, G. C., Chlorination and Dechlorination: A Scientific and Practical Approach, *J. AWWA*, **60** (1968).

Water Distribution

Appleyard, V. A. and F. P. Linaweaver, Jr., The McIlroy Fluid Analyzer in Water Works Practice, *J. AWWA*, **49** (1957).

Clark, J. W., W. Viessman, Jr., and M. J. Hammer, *Water Supply and Pollution Control*, 2nd ed., International Textbook Co., Scranton, Pa. (1971).

Cross, H., Analysis of Flow in Networks of Conduits or Conductors, Univ. Illinois Bull. 286 (1936).

Kiker, J. E., Design Criteria For Water Distribution Storage, *Public Works* (1964).

Kincaid, R. G., Analyzing Your Distribution System, *Water Works Engineering*, **97** (1944).

McPherson, M. G., Generalized Distribution Network Head Loss Characteristics, J. Hydrol. Div. ASCE, **86 HY1** (1960).

McPherson, M. G. and J. V. Radziul, Water Distribution Design and The McIlroy Network Analyzer, *Proc. ASCE*, **84**, Paper 1588 (1958).

VC. SANITARY SEWAGE AND LIQUID WASTES

Sanitary sewage and other liquid wastes represent one of the major sources of surface water pollution. While agricultural runoff contributes significantly to both inorganic and organic pollution during periods of high flow, and mineral deposits will contribute inorganic pollutants (primarily during low flow), these sources are frequently not subject to economically feasible control methods.

Sewage, on the other hand, because of the collection systems in common use, usually represents a point source, or at most, a limited number of point sources in any community. The effects of such contamination are frequently painfully evident and directly traceable to the point of origin. Additionally, point sources are self-evidently subject to control—the simplest method (which is occasionally suggested) being the halting of the flow. Reasonable men, while recognizing that the production of wastes is a concomittant of any society, may still be gravely concerned with the effect of such wastes upon the environment. Uncontrolled introduction of pollutants into our natural waters can indeed produce the evils so vividly portrayed of late by both the press and television. Depletion of dissolved oxygen in streams with resultant odors and fishkills, eutrophication of lakes, and contamination of public water supplies have all occurred and will continue to occur until suitable control measures are instituted.

Fortunately the work of sanitary engineers during the last century has led to the development of waste treatment methods and systems which, while admittedly less than perfect, afford the means of controlling the major constituents of water pollution. It must be emphasized that the solution to the waste problems inherent in an industrial society can only lie in the hands of those with the required technical ability, not in those of the people, however well-intentioned, who so vociferously protest a situation which they evidently believe occurred within the last year and thus should be as quickly improved.

Ultimately the responsibility for waste treatment will devolve to the sanitary engineer. As he has done in the past, so he will continue in the future to provide whatever degree of treatment is required in the most efficient manner.

Government Regulations

Federal regulations, through both legislation and administrative and court interpretations, have become more and more stringent in recent years with respect to waters deemed subject to Federal control and water quality standards. The first attempts to control water pollution resulted in the River and Harbor Act of 1899, which prohibited discharge of wastes other than storm or sanitary sewage into navigable waters. Section 13 of this law, known as the Refuse Act, is the basis of current suits being brought against the organizations responsible for certain industrial discharges. The law has been somewhat strengthened by Executive Order No. 11574, December 25, 1970, 35 F.R. 19627, directing the Secretary of the Army to issue permits for future discharges. Disclosure of the nature of the waste will be required and the Secretary may *consider factors other than water quality* in granting or denying such permits.

Following the River and Harbor Act of 1899, additional legislation was enacted in 1912, 1924, 1948, and 1953 which led to the Federal Water Pollution Control Act of 1956 (Public Law 660, 84th Congress), the basic law governing federally controlled waters. Amendments since that time have included the Federal Water Pollution Control Act Amendments of 1961, (Public Law 87-88) which extended federal authority to include both navigable and interstate waters; the Water Quality Act of 1965 (Public Law 89-234), which created the FWQA and required the establishment of water quality standards for interstate waters; the Clean Water Restoration Act of 1966 (Public Law 89-753), which provided for Federal assistance in the construction of municipal waste water treatment plants; and the Water Quality Improvement Act of 1970 (Public Law 91-224). In addition, the President's Reorganization Plan No. 3 of 1970, dated July 9, 1970, established the Environmental Protection Agency in December of that year as the responsible agency under the above statutes.

The trend of recent legislation has been to increase the range of authority of the Federal government from interstate waters used as highways of commerce, to all navigable waters, then to all interstate waters, and to boundary waters. Current administrative proposals contemplate extension of the law to include tributaries of any of the above waters, ground water, territorial waters and, in some instances, the high seas.

In viewing the development of Federal law the engineer must anticipate that within a short time *all* discharges of waste waters will be regulated, whatever be their source or point of disposal. The form which such regulations might take may be judged from those currently in force (see the Water Quality Act of 1965, etc.). Reasonable projections would include:

1. No discharge of waste sludges, organic or inorganic, will be permitted.
2. No discharge of soluble wastes which adversely affect the quality of the receiving body of water will be permitted.
3. Water quality standards will be established for all waters, in no case at a level below that which currently exists.

Additionally, it is unlikely that toxic pollutants will be allowed to be discharged at any level which depends upon the dilution afforded by the receiving body of water in order to meet the established water quality standards.

Quantity of Sewage

The quantity of sanitary sewage which may be expected will clearly be related to the population and the water consumption of a community. The quantity of water pumped from the water treatment plant will therefore serve as a good first estimate of sewage flow; indeed, in many communities sewer fees are based upon water consumption. Such a scheme is, in fact, inequitable, as shall be seen, but that is not at issue here. All of the water pumped to consumers within a community does not find its way into the sanitary sewers. A portion may be used in industrial production, a portion may be lost in evaporative cooling, and a significant portion of domestic consumption for individual residences during warm weather will be used for watering lawns and gardens.

Additionally, since sewers are never truly watertight the flow may be either augmented by ground water intrusion or reduced by leakage. The amount of infiltration or exfiltration to be expected will depend on the condition of the sewer, the type of soil and the level of the ground water table. Roof drains may also be connected to sanitary sewers although the practice is illegal in most communities. Infiltration rates may be expected to vary from 10,000 to 50,000 gal/day/mile of sewer. Permissible rates of infiltration in new sewers are normally specified to be between 2,000 and 10,000 gal/day/mile of sewer but the designer should recognize that as the sewer ages the rate of infiltration will increase. It appears unreasonable to assume a rate of infiltration less than 10,000 gal/day/mile unless the sewer is exceptionally well constructed, the climate unusually dry, or a significant amount of local data is available. Some designers routinely assume an infiltration rate of as much as 30,000 gal/day/mile.

The rate of flow of sewage will exhibit fluctuations similar to those shown for water demand with low flows at night and a rapid increase in rate in the early morning. The larger the community is, the smaller the fluctuations will be. Similarly, branch sewer lines will exhibit larger variations in flow rate than those observed in trunk lines. The ratios of maximum and minimum flows to the average flow which may be expected are shown in Table C1.

Sanitary sewers are frequently designed on either of two bases:

1. A flow of 250 gpcd plus infiltration for laterals and submains and 150 gpcd plus infiltration for mains, trunks, and outfalls, or;

TABLE C1. VARIATION IN SEWAGE FLOW.

Flow	Ratio to average
Residential area	
maximum day	2.25:1
maximum hour	3.00:1
minimum day	0.75:1
minimum hour	0.50:1
Average community	
maximum day	1.50:1
maximum hour	2.25:1
minimum day	0.75:1
minimum hour	0.50:1

2. A flow of 400 gpcd for laterals and submains and 250 gpcd for mains, trunks, and outfalls with the infiltration assumed to be included in the higher per capita rates.

The first method has the sounder theoretical basis but the two will frequently yield similar results.

Major point sources of sanitary sewage must also be considered in the design of a sewage system. Industrial activities of all kinds produce large volumes of waste water which must be carried away. Approximate flows which might be expected from various industries are tabulated in Table C2. Actual flows should be determined by gauging or by analysis of plant processes. The flows tabulated do not include cooling waters which should not be discharged to sanitary sewers in any event. If such discharge of cooling water is permitted the flows in such industries as steel production will be greatly increased. The actual water use in steel production, for example, may exceed 60,000 gal/ton. Cooling waters are usually contaminated only in a thermal sense and will thus be unimproved by passage through a sewage treatment plant. Their introduction into a sanitary sewage system will result in hydraulic overload of both sewers and treatment facilities with resultant problems of surcharging and loss of treatment efficiency.

TABLE C2. INDUSTRIAL WASTEWATER FLOWS.

Industry	Flow
Dairies	800 gal/ton raw milk
Canneries	15,000 gal/ton of cannage
Wool processing	150,000 gal/ton of wool
Distilleries	2,500 gal/ton of grain
Paper processing	80,000 gal/ton of paper
Sugar-beet processing	3,000 gal/ton of beets
Tannery	40,000 gal/ton wet hide
Coffee processing	20,000 gal/ton of coffee
Laundries	8,000 gal/ton of clothes
Meat processing	
Cattle	2,500 gal/animal
Hogs	700 gal/animal
Poultry	5 gal/bird
Steel	1,000 gal/ton
Refineries	800 gal/bbl of oil

Sanitary Sewer Design

The design of sanitary sewers is normally based upon open channel flow as presented in the section dealing with hydraulics. Design flows are based upon tributary population and infiltration plus the flow from any major point sources.

Sewer grades are selected to provide a minimum velocity of 2 fps and a maximum velocity of 8 fps. The minimum velocity is selected to prevent deposition of solids in the lines while the maximum is intended to prevent excessive scour of the lines. Sewage flow is normally controlled by gravity alone; however, in special circumstances the lines may be pressurized over short distances. The determination of pipe sizes for sanitary sewers may be outlined as follows:

1. Draw a profile of the proposed sewer route with notation of any obstacles and existing sewer grades which must be met.
2. From the hydraulic grade at the downstream end of the sewer construct a tentative gradient roughly parallel to the ground surface. The grade may be other than parallel to the ground surface where this slope will be insufficient to maintain a self-cleansing velocity in the sewer or where obstructions block this path.
3. From the area served and the population density, determine the sanitary sewage flow plus infiltration, including any major point sources of waste water.
4. From the flow and the tentative grade select the pipe which most nearly will carry the flow. It is only by chance that a standard size will precisely carry the flow on the tentative grade; hence the usual practice is to select the next larger standard size or increase the slope so the next smaller size will be able to carry the flow. The choice of method will depend upon the relative cost of excavation and the overall desirability of a steeper or lesser grade with respect to the rest of the system.
5. The selection of size and grade for each length of pipe will yield the starting elevation for the next. Where sharp changes in grade, pipe size, or direction are encountered the following provisions for head loss should be made:
   a. At a change in line size either the crown elevations or 0.8 pt elevations of both lines should be equal.
   b. At a change in direction (this always implies the existence of a manhole) the drop in elevation across the change should be at least 0.04 ft.

Pumping of sewage may be necessary where the surface grade is relatively flat and sewer lines are long. Under such circumstances it may be necessary to install lines far below the ground surface if pumping is not used. Since excavation costs will increase rapidly with trench depth, lift stations, even though expensive, are frequently the most economical solution. Pumping stations installed in such circumstances are usually designed only to lift the sewage to near the ground surface, from which point it again flows under the action of gravity. The only section under pressure is the lift pipe itself, which is usually vertical.

Where large changes in grade must be traversed by sewer lines, pressure systems may be used. Pressurized sewer lines are normally kept to a minimum and must be designed to standards similar to those used in water transmission.

Pumping is also frequently required at the waste treatment plant to raise the flow to ground level for passage through the plant. Centrifugal pumps are best suited to all pumping applications on raw sewage. Their selection is detailed elsewhere.

Other sections of sewer lines which are under pressure will occur when it is necessary to construct the sewer below the hydraulic gradeline in order to bypass an obstruction such as a railway or highway cut or a stream. As in other pressurized lines the sewer in such locations must be designed to standards similar to those used in water transmission. Velocities of 3 fps or more should be maintained and it is therefore desirable to use several lines in parallel at staggered elevations in order to insure self-cleansing velocities at all flow rates.

Manholes are normally installed at intervals of 300–500 ft, at changes in pipe diameter, at changes in direction of flow, and at sharp changes in grade. Drop manholes are used where the difference in grade between two intersecting sewer lines exceeds two or three feet. Manhole covers and frames are readily available from foundries in standard sizes. Designs other than the standard circular cover are undesirable since other shapes may fall through the opening.

Sewer Materials

Sewer lines may be constructed of cast iron or steel under special circumstances such as those encountered in pressurized sections, areas of high ground water, or extraordinary loading conditions. The majority of sewers however are constructed of the cheapest material which will serve the purpose. Vitrified clay pipe and concrete pipe are normally the materials selected.

Clay sewer pipe is extremely resistant to corrosion caused by acid materials which may be discharged into sewers or be produced by the anaerobic decomposition of sewage within the lines. The material is brittle, however, and may be damaged during installation if carelessly handled. It is available in diameters of 4–36 in. in standard and extra strengths. The specified strength of these sections is as given in Table C3.

Clay fittings in standard bell and spigot sections are readily available in wyes, tees, double wyes, reducers, increasers, slants, $\frac{1}{16}$ bends, $\frac{1}{8}$ bends, $\frac{1}{4}$ bends, and other shapes.

Concrete sewer pipe is available in sizes from 12 to 108 in. in diameter and should always be reinforced. Reinforced concrete pipe is made in 4-ft lengths for diameters up to 72 in. and in 5-ft lengths for diameters from 78 to 108 in. Concrete pipe may be susceptible to corrosion if the sewer is unlined and subjected to acid wastes or the products of anaerobic decomposition. Fittings are usually not required with connections being made by cutting the pipe and mortaring the junction. The strength of reinforced concrete pipe is usually specified as shown in Table C4.

TABLE C3. CRUSHING STRENGTH OF CLAY PIPE.

Diameter (in.)	Average strength (plf)			
	Standard strength[1]		Extra strength[2]	
	3-Edge bearing	Sand bearing	3-Edge bearing	Sand bearing
4	1200	1800	2000	3000
6	1200	1800	2000	3000
8	1400	2100	2200	3300
10	1600	2400	2400	3600
12	1800	2700	2600	3900
15	2000	3000	2900	4350
18	2200	3300	3300	4950
21	2400	3600	3850	5775
24	2600	3900	4400	6600
27	2800	4200	4700	7050
30	3300	4500	5000	7500
33	3600	5400	5500	8250
36	4000	6000	6000	9000

[1] "Tentative specifications for Standard Strength Clay Sewer Pipe" (ASTM: C13-65T)
[2] "Tentative specifications for Extra Strength Clay Pipe" (ASTM: C200-65T)

TABLE C4. ULTIMATE STRENGTH OF REINFORCED CONCRETE PIPE.

Diameter (in.)	Test strength (plf)	
	3-Edge bearing	Sand bearing
12	2700	4050
15	3000	4500
18	3300	4950
21	3600	5400
24	3600	5400
27	3800	5700
30	4050	6100
33	4300	6400
36	4500	6750
42	4800	7200
48	5100	7650
54	5550	8300
60	6000	9000
66	6350	9550
72	6750	10100

Asbestos cement sewer pipe has the advantages of a lower coefficient of friction and tighter connections using rubber rings and slip joints. Additionally it is made in longer lengths (10 or 13 ft) than other materials, thus reducing the number of joints. Unlined asbestos cement sewers may be expected to be subject to corrosion in much the same fashion as reinforced concrete. Such pipe is

TABLE C5. CRUSHING STRENGTH OF ASBESTOS CEMENT PIPE.

Diameter (in.)	Crushing strength (plf) 3-Edge bearing Class 1	Class 2
6	2600	
8	2500	
10	2400	3000
12	2300	3000
12	2200	3000
16	2200	3700
18	2200	4000
20	2200	4000
24	2400	4500
30	3200	5200
36	4000	6000

available in diameters of 4–36 in. and has a crushing strength equivalent to that of clay and concrete lines as shown in Table C5.

Cast iron pipe is used in those sewer lines which are pressurized. Under such conditions the lines flow full, hence the mechanism of corrosion common to normal sewers cannot occur. Acid wastes may, however, cause corrosion of uncoated lines.

Sewers larger than those available in precast shapes are normally built in place. Brick and masonry have been used but modern sewers are usually constructed of reinforced concrete. The interiors of such sewers may be coated in place for corrosion protection. Various cross sections may be used.

Corrosion of sewers is primarily a result of the anaerobic decomposition of fatty materials in the sewage. At low flow rates the grease will tend to coat the walls of the sewer as shown in Fig. C1. Under anaerobic conditions, which it is

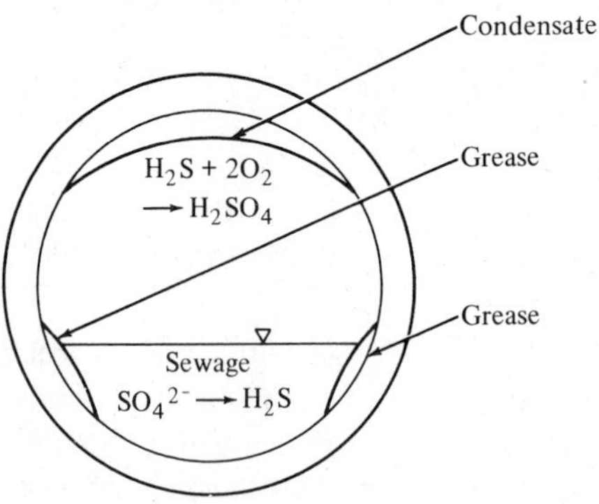

Fig. C1. Mechanism of sewer corrosion.

virtually impossible to avoid, the sulfates present will be reduced to hydrogen sulfide by *Desulfovibrio* bacteria.

The hydrogen sulfide, upon release from the sewage, will dissolve in the condensation water on the crown of the sewer where it may be oxidized by *Thiobacillus* to sulfuric acid. Sulfuric acid will react with the lime in the cement to form calcium sulfate which is relatively friable when dry and quite soluble in water as well. The net result of this process is the destruction of the sewer crown as shown in Fig. C1.

Corrosion of sewers may be prevented by the following techniques:

1. Ventilation to prevent excessive crown condensation and to carry off hydrogen sulfide from the sewer. Blowers may be used to insure a positive draft over particularly troublesome sections.
2. Prevention of biological action by chlorination. The dosages required are large and must be repeated at regular intervals in long lines.
3. Design of sewers to maintain higher, self-cleansing velocities and to reduce residence time of sewage in the system.
4. Protective coatings such as asphaltic or bituminous materials.

Vitrified clay pipe, of course, is not affected by acid corrosion and hence requires no protective technique.

Sewage Characteristics

Ordinary domestic sewage is actually a quite dilute waste: The concentration of degradable organic material is normally less than $\frac{1}{4}\%$ of the total flow. The material present, however, may have extremely deleterious effects upon a receiving stream and hence should be removed before the flow is discharged. On average, the per capita contribution of degradable organic material is about 0.2 lb per day, which is associated with a like quantity of suspended solids. While this per capita contribution will be fairly constant nationwide, the concentration will vary widely with the flow. In small communities or subdivisions of a larger community the concentration of pollutants will be higher than the average values presented below, while in communities with high infiltration rates and little industry the concentration may be far lower than shown. Fluctuations in concentration about the average will also occur just as fluctuations in flow rate will be observed as mentioned above. The values presented in Table C6 are averages for many communities over a period of years and instantaneous values at any individual treatment plant may be expected to vary widely from those shown. Concentrations may be expressed as parts per million or milligrams per liter. For practical purposes these terms are interchangeable in dilute solution.

The biochemical oxygen demand test is the most commonly used technique for determining the pollutional strength of wastes. Essentially, the test measures

TABLE C6. SEWAGE CONSTITUENTS.

Parameter	Concentration (mg/l)	
	Average	Range
BOD	200	100–500
Suspended solids	200	100–400
Volatile suspended solids	140	75–220
Total solids	750	400–1300
Total volatile solids	400	250–800
Ammonia nitrogen	10	5–20
Organic nitrogen	20	10–40
Fats and grease	50	20–100
Phosphate	20	10–50

the amount of oxygen consumed in the metabolism of the waste by microscopic organisms in five days at a temperature of 20°C. The test results are highly variable and even with the best technique have an accuracy of no better than ±10%. Additionally, the five-day period is usually insufficient for the complete metabolism of the waste and thus does not truly measure its strength. The test is widely used and will doubtless continue to be used in the future. The engineer, however, should consider BOD results to be only approximate at best unless he has personal knowledge of the ability of the analyst and of the technique used.

The chemical oxygen demand test is sometimes used as a substitute for or a supplement to the BOD determination. The analysis involves an acid oxidation of the organic material with potassium dichromate. The procedure is straightforward, rapid, and readily reproducible; however, it has not met with universal acceptance. The major difficulty appears to be that since the COD procedure will oxidize materials which are not oxidized by the BOD test, the COD result will nearly always be higher than the BOD. This, and the fact that design loadings and efficiencies of treatment have historically been expressed in terms of BOD have led to the lack of acceptance of an apparently superior test.

The total organic carbon determination, while only in use for a few years, shows promise as an alternative procedure for determining waste strength. The apparatus required is expensive; thus it is unlikely that the TOC determination will supplant either the BOD or COD.

Solids determinations are additional indications of the strength of a waste stream. The volatile solids determinations, both total and suspended, serve as a rough measure of the concentration of organic material. The quantity actually measured is the solids ignitable at 550°C, with the residue consisting of ash and inert materials. Suspended solids are determined either by filtration through an asbestos mat in a Gooch crucible or through a membrane filter. The membrane filter has advantages in ease of preparation and speed of drying and yields more reproducible results.

Nitrogen determinations have been used historically to indicate the degree of stabilization of wastes with a decrease in ammonia nitrogen and an increase in nitrate nitrogen indicating an increase in stability. The measurements are seldom or never made for that purpose in modern treatment, but rather first, to determine if sufficient nitrogen is present for the operation of biological treatment systems, and second, to measure the quantity of nitrogen being discharged from the treatment plant, since nitrogen itself may contribute to water pollution.

Phosphate, like nitrogen, is essential to the operation of biological waste treatment systems; however, its major significance lies in its possible pollutional effects if it is discharged to surface waters. The average phosphate concentration in domestic sewage has increased with use of phosphate-based detergents. Some cities have now prohibited the use of such products, and other detergents with less (or in some cases no) phosphate have been developed. The phosphate concentration in sewage may thus decrease somewhat in the future. Curiously, it now appears that the indictment of phosphates as the causative agent in the eutrophication of lakes was somewhat premature.

Other analyses of sewage and treatment plant effluents may occasionally be made, particularly in those cases in which toxic industrial wastes are treated with domestic sewage. The analyses cited are those generally used to evaluate the strength of a waste stream and the efficiency of a treatment process. Additional analyses may be required in the control of treatment processes. These are discussed where applicable to individual treatment systems.

Treatment Systems

The treatment of waste waters is customarily divided into primary, secondary, and tertiary systems. Efficiency of treatment is normally expressed as percentage removal, which is an unfortunate practice since it is not the percentage removal which is important, but rather the absolute quality of the effluent and the total quantity of pollutant which is discharged to the receiving stream. Efficiencies of systems will be expressed here in terms of percentage removal for purposes of comparison, but the designer should be aware that it is not efficiency of treatment but rather protection of water quality which should be his primary goal.

Primary Treatment

Primary treatment consists of systems and techniques directed toward reducing the concentration of suspended solids in the waste stream. The systems used include racks, screens, comminutors, skimming tanks, grit chambers, sedimentation tanks, and chemical precipitation tanks.

Screening devices include racks, screens, and comminutors (Figs. C2–C6). Such

Fig. C2. Mechanically cleaned bar screen (courtesy Rex Chainbelt, Inc., Process Equipment Division).

units are intended to remove floating material and large suspended solids from the flow prior to further treatment. Racks are constructed of vertical bars with openings of $\frac{1}{2}$-2 in. with a 1-in. opening being most common. In general it is desirable to prevent velocities between the bars in excess of 3 fps since higher velocities may force the screenings through the rack. The racks may be cleaned either mechanically or by hand, although hand-cleaned racks are now rare in all but the smallest plants.

Screens are normally installed on drums or on movable belts so that the screening surface may be replaced as required. The screenings are carried out of the flow by the motion of the screen, where they are then deposited in a hopper. The openings in the screens may range from $\frac{1}{16}$ to 1 in., with the finer screens seldom used.

Comminutors, rather than removing the coarse suspended solids, reduce them

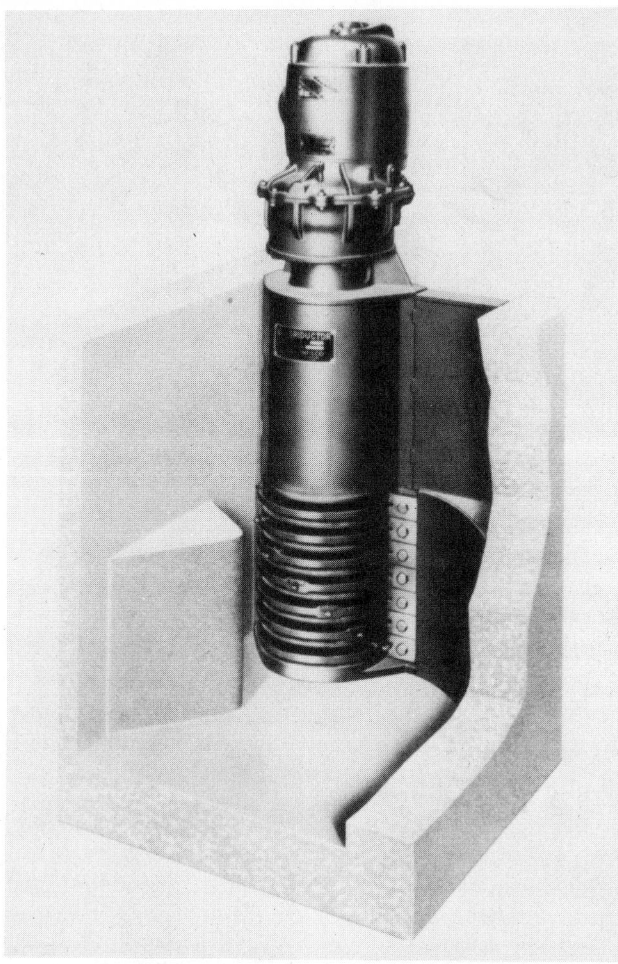

Fig. C3. "Griductor" comminutor (courtesy Westinghouse Electric Corp., Infilco Division).

to a smaller size by shredding them through a fine screen with openings of about $\frac{1}{4}$ in. Although such devices do not remove the large materials, they serve the same purpose as racks and screens; that is, they protect pumps from materials which might jam them.

The materials removed by racks and screens are quite offensive in nature and may contain particles of fecal material as well as rags, garbage, and wood. This material should be regularly removed before it can undergo appreciable decay. The most common disposal technique is burial either in trenches or in a sanitary

Fig. C4. Comminutor installation (courtesy Westinghouse Electric Corp., Infilco Division).

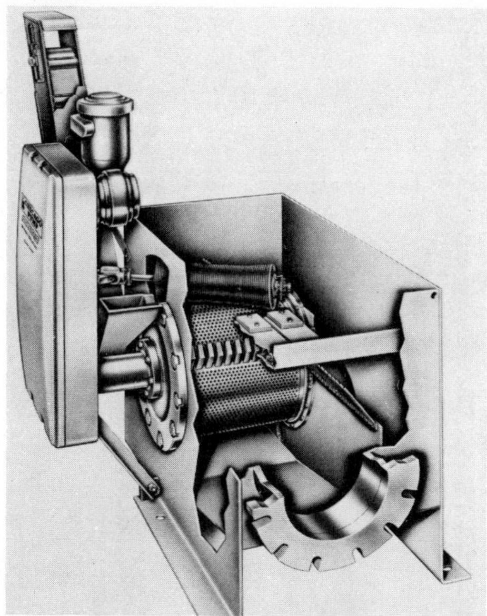

Fig. C5. Selectostrainer (courtesy FMC Corp., Environmental Equipment Division).

landfill. Other techniques include introduction into an anaerobic digester or incineration. These techniques are discussed below.

Skimming tanks are intended to remove floating material including particulates and grease. Air may be utilized to aid in coagulation of the grease and, by entrapment of air bubbles, to increase the rate of rise to the surface. The floating material is then skimmed from the surface and held for further treatment. Retention times of 15 min. and aeration rates of about 0.1 ft^3 of air per gallon of flow are commonly used.

Sedimentation. The theory of sedimentation discussed in the subsection on water treatment is also applicable to the sedimentation of discrete particles in sewage. Denser suspensions in which flocculation of the particles may occur during sedimentation cannot be mathematically described with any precision, nor can the interaction of the particles when the suspension reaches a density such that they strike one another with significant frequency. Design values for such sedimentation processes are based upon empirical results of settling column analyses in which a suspension is allowed to settle quiescently with samples withdrawn at various levels at predetermined times. In ordinary practice grit chambers are used for the removal of sand and other inert materials with a specific gravity greater than 2.5. Particles having a diameter greater than about 0.2 mm are removed. The concept of the ideal sedimentation basin may be applied in the design of grit removal facilities. Sedimentation tanks designed for removal

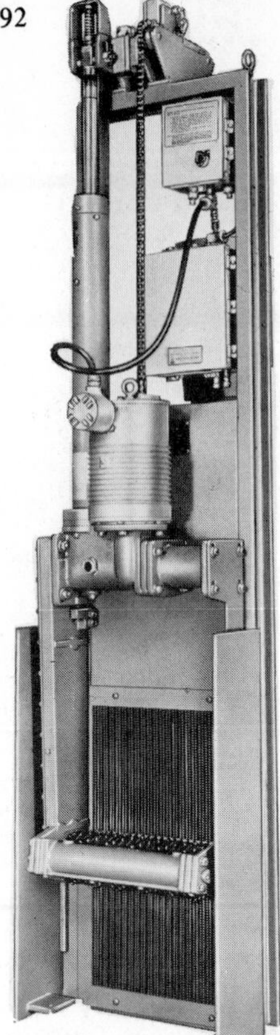

Fig. C6. Barminutor (courtesy FMC Corp., Environmental Equipment Division).

of organic suspended solids with a specific gravity of less than 1.2 may operate either as ideal sedimentation basins or flocculent settling tanks depending upon the density of the suspension and the character of the particles. In the lower regions of such tanks hindered settling may be expected.

Grit chambers are designed to remove inorganic materials with a minimum inclusion of organic particles (Figs. C7–C9). Inorganics will not benefit from secondary treatment processes and their inclusion in such systems may lead to heavy accumulations of inert solids which can interfere with their proper operation. Those particles having a settling velocity of about 0.75 in./sec or more will, in sewage, embrace the bulk of the inert materials and few of the organic solids. Grit chambers are therefore specifically designed for the removal of such particles. Consideration of the equations describing the ideal sedimentation basin leads to the conclusion that for a horizontal velocity of 0.75–1.0 fps (the scouring velocity of the particles removed) such a basin should have a length of about 12 times its depth or 1 ft of length for each inch of depth. It is important to

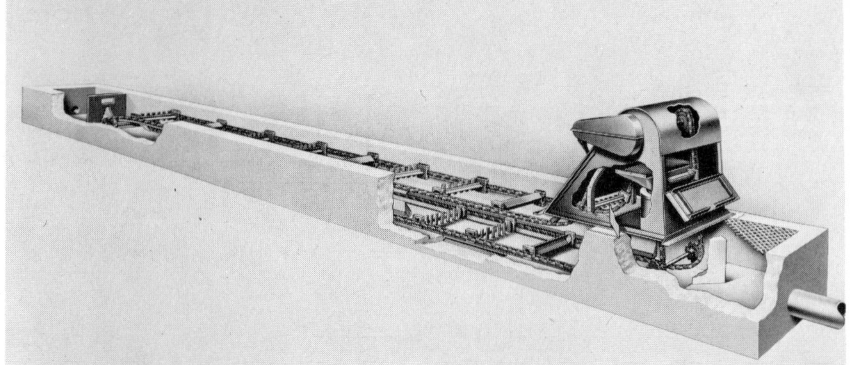

Fig. C7. Channel grit chamber (courtesy Rex Chainbelt, Inc., Process Equipment Division).

maintain a constant horizontal velocity in grit chambers despite variations in flow depth, hence proportional flow weirs are frequently used. The amount of grit removed will average about 6 ft^3/1,000,000 gal but may vary widely about this value particularly if significant quantities of storm water can enter the sewers. The accumulated grit may be stored for short periods in the bottom of the chamber. Removal of the grit may be effected by hand in small plants but automatic machinery is normally used for this purpose, since this removes the necessity of draining the basin for cleaning. If the motion of the grit in removal is counter to that of the water the slight agitation will aid in the removal of such organic material as may be entrapped during sedimentation.

Aerated grit chambers utilize air to maintain the lighter organic materials in suspension while permitting the heavier inorganics to settle. Such units have the additional advantage of coagulating grease in the flow which will aid in its removal in later treatment and, by removing the grease from other particles, increase their rate of sedimentation. The grit is usually removed from such units by air lift, although auger arrangements are also employed.

The material removed from a properly designed and operated grit chamber will be low in organic content and moderately inoffensive. It may be landfilled without difficulty but has been used for other purposes such as covering walks and access roads within the plant grounds.

Sedimentation tanks are normally classified according to their intended use as (1) primary, for the settling of raw sewage, (2) secondary, for the clarification of mixed liquor from secondary treatment systems, and (3) combined, for the clarification of primary sewage and waste solids from secondary treatment.

Sedimentation tanks are also sometimes placed between the individual units in a multistage trickling filter plant. Typical retention times and surface loading rates are as shown in Table C7. Weir loadings are usually about 10,000 gal/ft/day and should not exceed 15,000 gal/ft/day.

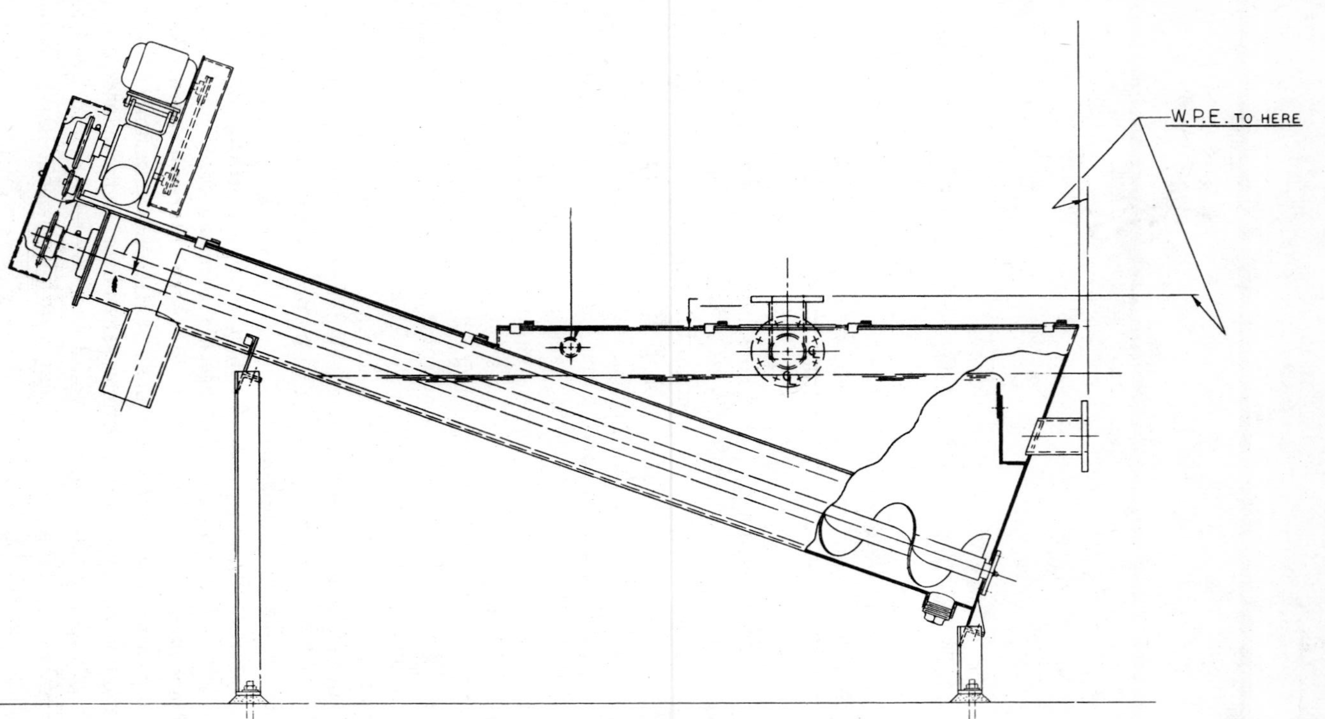

Fig. C8. Counter-flow grit washer (courtesy Walker Process Equipment).

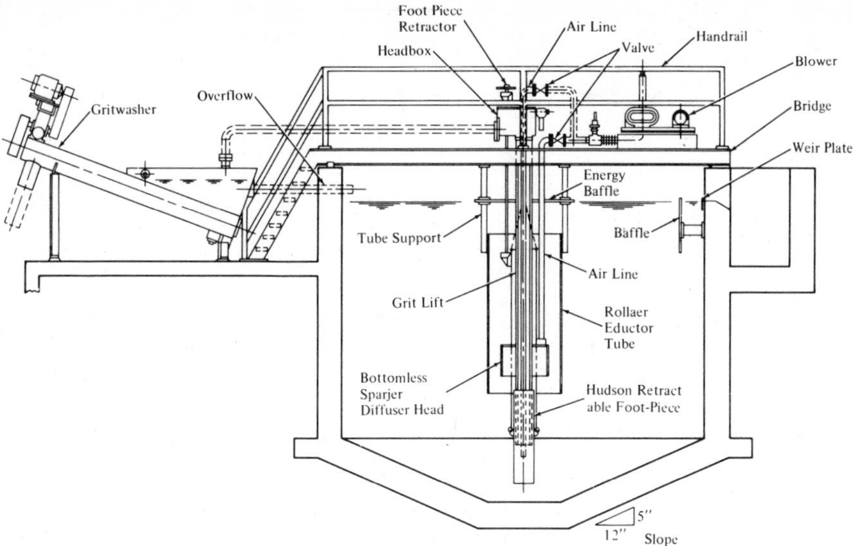

Fig. C9. Aerated grit basin (courtesy Walker Process Equipment).

TABLE C7. SEDIMENTATION BASIN DESIGN PARAMETERS.

Type of basin	Retention time (hs)	Surface overflow rate (gal/ft^2/day)
Primary before trickling filters	2.0	900
Primary before activated sludge	1.0	1500
Intermediate between trickling filters	2.0	1000
Final after activated sludge		
conventional processes	2.5	700
contact stabilization	3.0	600
extended aeration	3.5	300
Final after trickling filters		
standard trickling filters	2.0	1000
high-rate trickling filters	2.0	800

Sedimentation basins may be expected to give BOD removals of about 25-35% on raw sewage. The removal obtained will be related primarily to the surface overflow rate, as shown in Fig. C10. The values presented are for average domestic sewage and will not be representative for secondary settling basins nor for wastes which depart significantly from the composition of ordinary sewage.

The actual tank dimensions are unspecified here and will be generally dictated by the plant layout and the type of sludge removal equipment used. The tanks may be square, rectangular, or circular, but circular tanks appear to have certain advantages and are most frequently used in modern practice. To cite but one

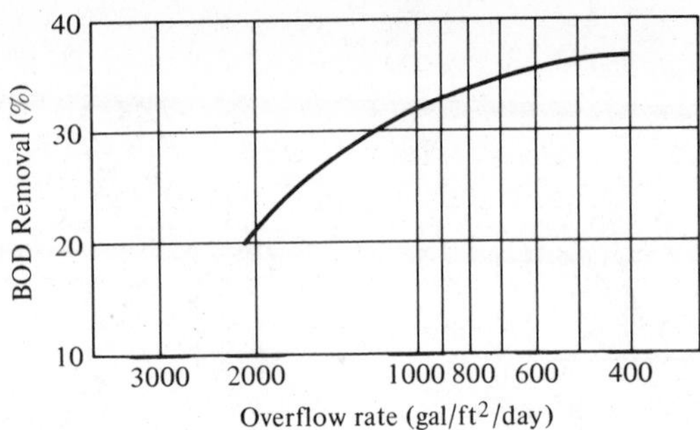

Fig. C10. BOD removal in primary settling tanks (Great Lakes–Upper Mississippi Board of State Sanitary Engineers).

advantage, for a given area (which is the most important design parameter for sedimentation tanks) a circular basin will have the minimum perimeter and hence will have the shortest length of wall. Additionally the sludge removal mechanisms used in circular tanks are less complex than those required for other shapes.

The sludge is normally carried to a hopper from which it may be withdrawn by gravity alone if sufficient head is available, or, more usually, may be pumped. The rate of sludge removal is low and pumping is frequently intermittent in order to permit further consolidation of the settled material.

As mentioned above, the per capita contribution of solids to the waste flow will be about 0.20 lb per day. Industrial wastes and widespread use of garbage grinders may increase this by as much as 50%. The solids removed in sedimentation will usually be from 60% to 70% of the total; thus the sludge removed from the tank may be expected to have from 0.12 to 0.21 lb per person per day of discrete solids. Since the sludge will normally be only about 5-6% solids this represents a volume of 0.06-0.07/ft^3/person/day.

The sludge produced in secondary treatment processes will vary in quantity and quality depending on the process used. Activated sludge processes will produce sludges ranging from about 1% to 5% solids. Trickling filter sludges are normally 3-4% solids. The actual dry weight of solids produced will be highest in the high-rate processes, and may approach the quantity removed in primary sedimentation.

Scum consists of solids lighter than water and the grease and fats normally present in sewage. It will float on the top of sedimentation tanks and will pass out with the effluent unless baffles are placed before the weirs. Scum may be skimmed by hand or by mechanical devices. It may be landfilled directly or may be combined with other waste solids for further treatment.

The sludge obtained from sedimentation basins will contain from 40%-70% volatile matter and hence will decay rapidly with resulting production of offensive odors. Primary sludge may be 95% water while secondary sludges may be 99% water. These sludges do not dewater readily and are too bulky for handling in their raw state, hence they require further treatment before final disposal. Their handling is detailed in the discussion of *sludge treatment* (below).

Combination units intended to provide both sedimentation and sludge digestion have been widely used in the past and are still in use in some cities. Septic tanks and Imhoff tanks will produce improvements in sewage quality but the improvement is minor when compared to the cost of construction and difficulty of operation. These systems are not recommended for any new treatment facility, and many old Imhoff tanks are now being converted to activated sludge units.

Sedimentation rates and efficiency of treatment may be improved by the addition of chemical coagulants, particularly ferric chloride and lime. BOD reductions of up to 75% have been reported for such treatment. The coagulation reactions are identical to those presented in the discussion of *water treatment*. Coagulation may be useful in improving the efficiency of a plant which incorporates only primary treatment but is probably not justified as a component of a secondary treatment plant since a satisfactory effluent will normally be attainable without resorting to this expedient.

Secondary Treatment

Secondary treatment of sewage implies the use of a biological system in which bacteria and higher microscopic forms metabolize the soluble organic material to form cell mass, thus removing it from the waste stream. Proper design and operation of such systems is dependent upon an understanding of the environmental requirements of the microbial mass and of the interrelationships among the various species.

Intermittent sand filters were developed in the last century for the treatment of sewage and may still be found in use. Dosages range from 75,000 to about 200,000 gal/acre/day with BOD removals of over 90%. The area required for such units is very large (a city of 100,000 would require a total filter area of about 150 acres) and they are not a realistic alternative to modern sewage treatment techniques. *Contact beds* are also of interest only in those areas in which they might still be in operation. Dosages range from 100,000 to 800,000 gal/acre/day with BOD removals of less than 50%.

Trickling filters consist of beds of crushed stone, rock, or occasionally other material which acts as a support for the biological growth which actually removes the organic material from the waste. The mechanism of treatment may be best understood by consideration of Fig. C11, which represents the surface of a single rock. The microbial growth on the surface is nourished by the nutrients held in the fixed water layer, which are intermittently restored by the moving

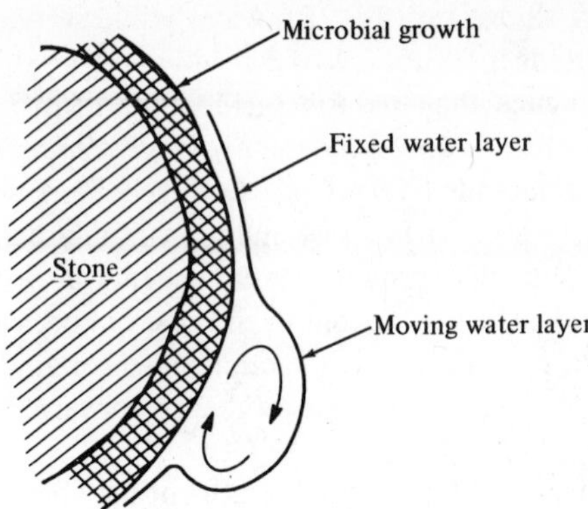

Fig. C11. Schematic of trickling filter media.

layer as the filter is dosed. The growth will gradually thicken until the layers next to the rock surface reach a condition in which they are limited with respect to both food and oxygen, both of which must be transferred from the free surface. When this occurs the lower layers will begin anaerobic endogenous metabolism of their own cell structure until they finally die and lyse with release of further nutrients. When a large percentage of the layer at the rock surface has lysed the entire biological layer will slough off and be carried out of the filter by the moving water layer. The suspended solids found in the effluent of trickling filters will consist primarily of such sloughed microbial solids. The rocks used in trickling filters are normally 2-5 in. in major dimension and are normally 4-6 ft in depth. Although a smaller rock size would afford a larger area for microbial growth, the transfer of air through the bed is usually the limiting factor in operation of trickling filters and the larger voids associated with large rocks aid in establishing the proper circulation pattern.

Plastic filter media (Fig. C12) are now being used in some trickling filters. These media have a high ratio of surface area to volume and very large void spaces, hence a desirable air circulation pattern is readily obtained. Filters using plastic media commonly employ a very high recirculation rate (see below). This high hydraulic load coupled with the smooth surface of the media leads to a mode of operation which might be considered more typical of a mechanically aerated activated sludge system. The distinction is perhaps unimportant if a satisfactory effluent is produced, but it should be noted that the sedimentation tank design is somewhat different for trickling filters than for activated sludge. A plastic media trickling filter might well require a sedimentation tank designed for a lower surface overflow rate than that common to trickling filters.

Pretreatment of sewage before application to trickling filters should always be provided in order to reduce the load on the filter and prevent plugging which might be caused by large suspended solids. From Table C7 a retention time of

Fig. C12. Plastic trickling filter medium (courtesy Koch Engineering Company, Inc.).

2 h and surface overflow rate of 800–1000 gal/ft^2/day will be adequate, however, where recirculation is incorporated in the design the return flow must be added to the influent flow or the basin will be hydraulically overloaded.

Trickling filter systems may be divided into low-rate, high-rate, and two-stage designs. Low-rate filters usually consist of a presedimentation basin, the filter, and a final clarifier, the solids from which are returned to the head of the plant. Figure C13 illustrates such a system.

High-rate filters are characterized by recirculation of the sewage, usually in a continuous fashion, although in some designs recirculation is used only to augment low flows. A schematic of such a plant is shown in Fig. C14. Various recirculation schemes (shown dashed) may be utilized.

The total flow through a high-rate trickling filter (raw sewage plus recirculation) will be between 5 and 15 times that through a low-rate filter. Two-stage filters may utilize various schemes of recirculation, as shown in Fig. C15.

Recirculation is desirable in trickling filter plants for the following reasons:

1. It permits continuous dosage of the beds and helps to equalize and reduce loading.

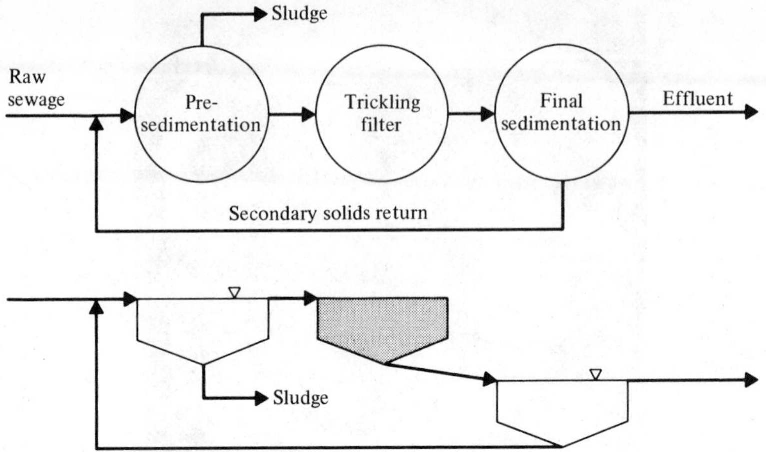

Fig. C13. Low-rate trickling filter plant.

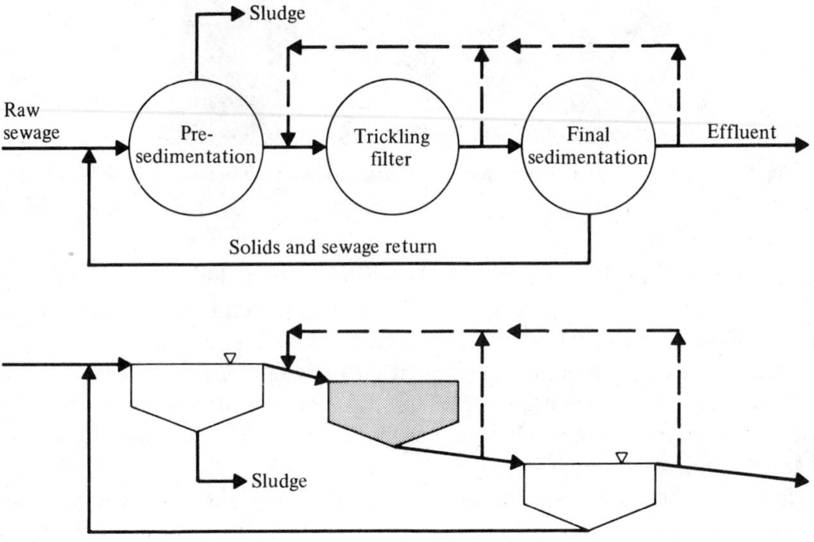

Fig. C14. High-rate trickling filter plant.

2. It dilutes the influent with a relatively pure flow containing dissolved oxygen and thus reduces odors.
3. The higher hydraulic rate tends to reduce the thickness of the biological film and minimize sudden widespread sloughing.
4. The return flow seeds the filter with microorganisms suited to metabolism of the waste.

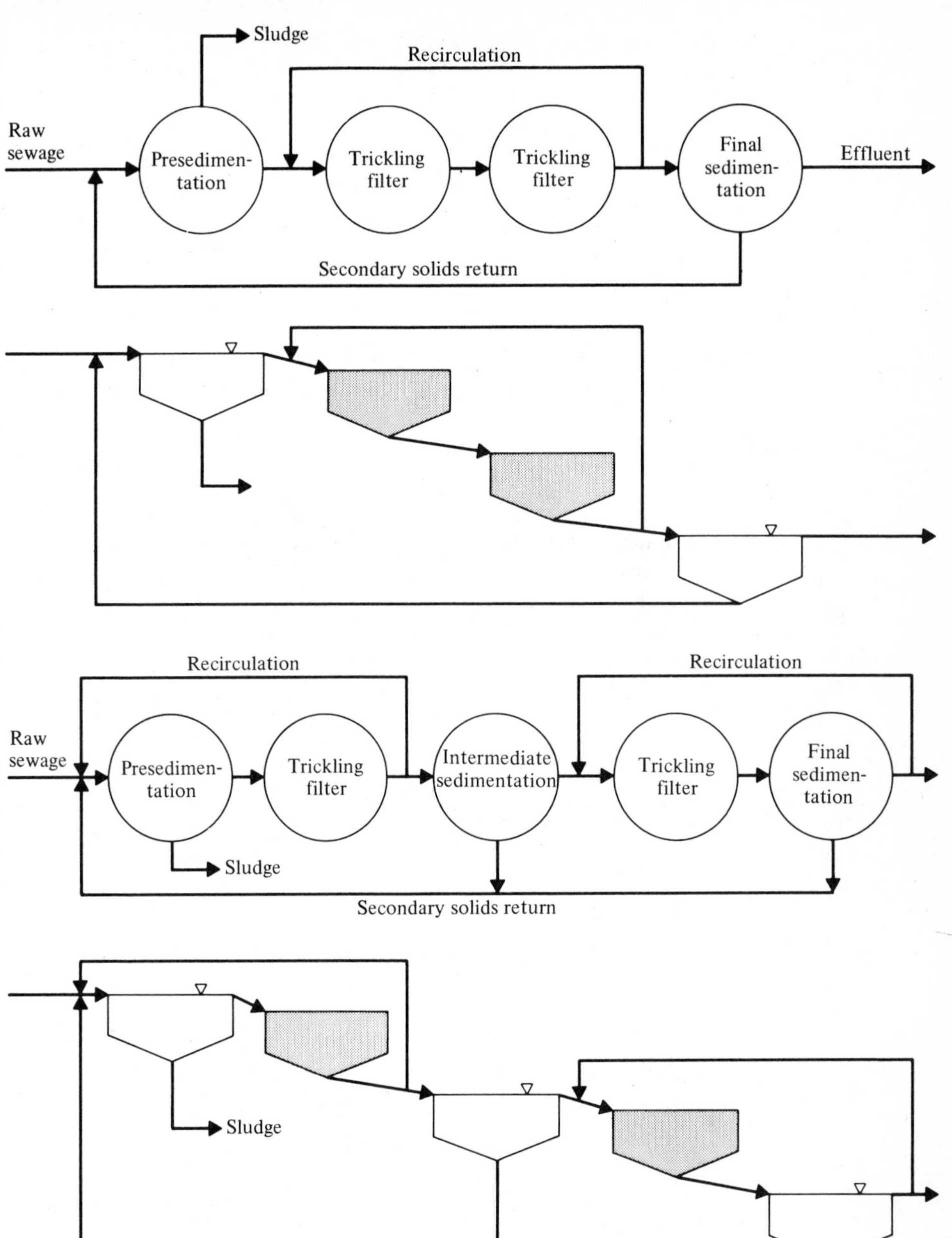

Fig. C15. Two-stage filtration designs.

Filter loadings which are commonly used are expressed in pounds of BOD per day per 1000 ft^3 of filter media or in pounds of BOD per acre-foot per day. The two units are directly convertible and hence should not lead to confusion.

TABLE C8. TRICKLING FILTER LOADING RATES.

Parameter	Low Rate	High rate	Two-stage
BOD loading			
lb/1000ft^3/day	5-25	30-90	45-70
lb/acre-ft/day	200-1100	1300-3900	2000-3000
Hydraulic loading			
mgad	2-5	10-30	10-30
gal/ft^2/day	25-100	200-1000	200-1000
Recirculation ratio	0	$\frac{1}{2}$-3	$\frac{1}{2}$-4

Hydraulic loading is normally expressed in millions of gallons per acre per day (mgad) or gallons per square foot per day. Typical loadings are presented in Table C8.

BOD loadings are based upon the raw flow neglecting the BOD of the recirculated flow. Hydraulic loadings on the other hand do include the recirculated flow. Plastic media filters have been loaded successfully at 50-100 lb of BOD per 1000 ft^3 per day and at hydraulic rates of 100 mgad.

The efficiency of trickling filters may be expected to vary as shown in Fig. C16. Efficiencies of 80-90% BOD removal are usually readily obtained.

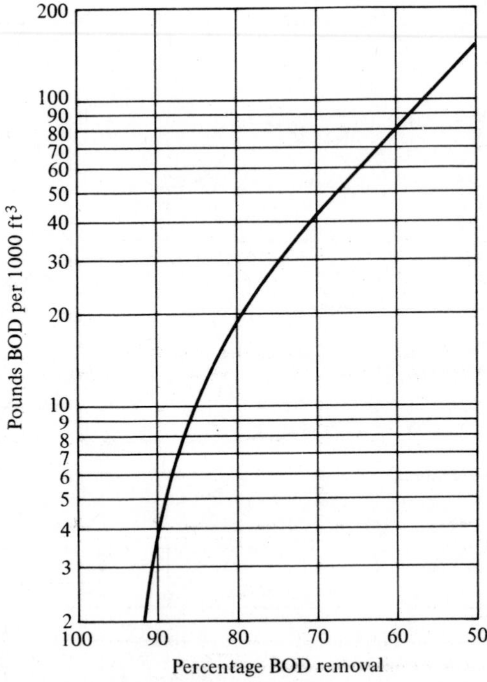

Fig. C16. Trickling filter efficiency.

Trickling filter designs have sometimes been based upon the results of detailed studies of such systems at military bases during World War II. The formula which was developed was:

$$E = \frac{100}{1 + (0.0085w/VF)} \tag{1}$$

in which E is the efficiency of BOD removal, w is the BOD applied to the filter in pounds per day, V is the volume of the filter in acre-feet, and F is the number of effective cycles of the sewage through the filter, which may be calculated from the following relation:

$$F = \frac{1 + R}{(1 + 0.1R)^2} \tag{2}$$

in which R is the Recirculation ratio.

It must be remembered that this equation is based upon the waste of atypical communities, the sewage of which may not be representative of others in either content or flow pattern.

The operation of trickling filter plants is normally trouble free with the following exceptions:

1. Heavy biological growth near the surface (usually algae) may block portions of the filter and cause ponding. Such ponding effectively reduces the area and volume of the filter and may lead to short circuiting. Dosing the filter with chlorine or copper sulfate will kill the algae and cause the accumulated material to slough. Heavy doses of chlorine however should be avoided if possible since this will also kill the bacteria essential for waste treatment.
2. Since a trickling filter is not a completely aerobic treatment process, odors may be encountered particularly in warm weather and during periods of high organic loading. Such problems may be reduced by increasing the recirculation rate or by using blowers to force air upward through the bed. Increased recirculation is probably the more effective technique.
3. *Psychoda alternata* (commonly called filter flies) may breed in trickling filters and by swarming in the area of the plant produce a public nuisance. Control techniques have included flooding the filter for periods of up to a day in order to drown the larvae, and dosing with DDT. Flooding the filter essentially takes it out of operation and thus reduces the capacity of the plant. Dosing with DDT in the quantities required (up to 5 lb/acre), if not an affront to the environment, would certainly be an affront to the environmentalists. High-rate trickling filters are little troubled by *Psychoda*, particularly if care is taken that the flow is spread over the entire surface of the filter so that no areas are permitted to dry. The higher hydraulic rate keeps the biological layer to a minimum and washes such eggs as are laid between dosings out of the filter.

4. In northern climates trickling filters may freeze during the winter months. Icing of the surface, like heavy biological growths, may cause short circuiting with resultant loss of efficiency. Since the filters may be quite large (over 100 ft in diameter) construction of covers has proven to be impractical until quite recently. Within the past few years fiberglass covers have been tested and may provide a solution to this problem. Such covers may also help to contain odors produced during the summer months.

Activated Sludge

The activated sludge process is at the same time one of the simplest and one of the most complex waste treatment processes. It may be used to produce an effluent with virtually any desired residual BOD and is thus the most versatile and most valuable system available to the sanitary engineer. Although developed in the early twentieth century, it is only in recent years that the activated sludge process has been truly understood and its widespread applicability to both industrial and domestic wastes has been fully appreciated.

From a physical standpoint activated sludge may be described as the aeration of a waste for a sufficient period of time to reduce the BOD. This description, while accurate, is far from adequate and can lead to a misunderstanding of the process, resulting in poor design and operation. Aeration of a sterile waste will lead to no improvement in its quality however long the aeration is continued. It is not then the aeration itself, but rather the utilization of the air supplied, by the microorganisms present, which leads to the stabilization of the waste. It is these microorganisms which are the key to the process, and its control therefore consists primarily in maintaining an environment favorable to their growth.

The microorganisms present in activated sludge consist primarily of bacteria, fungi, protozoa, rotifers, and sometimes worms and higher forms. It is the bacteria which are most important since it is they which metabolize the organic material and form the flocculent solids which are later removed. Various species of bacteria may predominate in wastes of different character since those best suited to the metabolism of the waste will have a distinct advantage over others initially present. In general, precise identification of the predominant species is unnecessary. *Zoogloea ramigera,* although it may be found in activated sludge systems, is not, as has been suggested, essential to the process.

Heavy growths of fungi are undesirable in activated sludge since these filamentous organisms cause "bulking" of the sludge, resulting in poor settling characteristics. In general the growth of fungi will be stimulated by a waste with an unusually high carbohydrate concentration, by acid wastes, which produce a low pH, and by nutritional deficiencies. The predominance of fungi over bacteria under these circumstances is produced by distinct differences in their cellular structure and can be avoided only by changing the environment, i.e., raising the pH and adding trace nutrients. The optimum pH range and the trace nutrients required to avoid fungal predominance are discussed below.

The protozoa feed primarily not on the raw waste itself, but rather on the bacteria which multiply while feeding on the waste. The free-swimming ciliated protozoa will tend to predominate in the initial stages of waste treatment but will decrease in numbers as the population of free-swimming bacteria decreases. Stalked ciliated protozoa will tend to increase in numbers as the free ciliates die off. When the waste reaches a high degree of stability, rotifers will be found feeding on the solid residue. The presence of these higher animals, which are easily seen and readily identifiable under microscopic examination, can be an excellent indication of the efficiency of a waste treatment system. Microscopic examinations can thus constitute a valuable addition to the operational tests normally performed in the control of activated sludge systems. While such examinations are not quantitative they provide qualitative evidence which is at least as convincing as the proverbial trout in the milk.

Theory of Operation Activated sludge may be produced for any waste whether it contains solids like domestic sewage or is completely soluble as are some industrial wastes. The sole requirement is that the waste be nutritionally balanced. Most municipal sewages, even with significant industrial components, will contain sufficient trace nutrients. Certain industrial wastes may be deficient in one or more elements. The approximate nutritional requirements are as listed in Table C9. The values listed are not absolute minima, but rather those which are known to be at least adequate.

The bacterial innoculum necessary to produce activated sludge will always be present in domestic sewage. Industrial wastes which are uncontaminated may require seeding with either domestic sewage or soil. Either source will furnish

TABLE C9. APPROXIMATE NUTRITIONAL REQUIREMENTS IN ACTIVATED SLUDGE (MG/L PER MG/L OF BOD)

Element	Concentration
Nitrogen	0.050
Phosphorus	0.016
Sulfur	0.004
Sodium	0.004
Potassium	0.003
Calcium	0.004
Magnesium	0.003
Iron	0.001
Molybdenum	Trace
Cobalt	Trace
Zinc	Trace
Copper	Trace

the bacteria required. Searching for specialized cultures or addition of "enzymes" is a futile endeavor. The waste will always select the bacterial species best able to metabolize it, since that species will always predominate.

At the beginning of aeration of a waste containing a suitable innoculum and the required nutrients, bacterial growth will be extremely rapid. As long as the food available is in excess of that required for cell maintenance the bacterial mass will increase at a logarithmic rate. While growth is in this log phase the oxygen requirements will be extremely high and the cells will be dispersed.

As the available food is decreased by the production of new cells the concentration of organic material begins to limit the rate of increase. As the concentration of food slowly decreases and the cell mass slowly increases, the point eventually is reached at which the cells die at the same rate at which they are produced. In this phase of growth the amount of oxygen required is greatly decreased and the cells, being at a lower energy level, will tend to agglomerate into flocculent masses readily visible to the naked eye.

If the aeration were to be continued beyond this point, eventually the individual cells would greatly reduce their mass through endogeneous metabolism, as would the protozoa and higher forms which feed on them, until finally only the inert portions of the bacterial cells would remain. Activated sludge systems are never carried to this point, nor are they run in such a fashion that each unit of waste entering the plant passes through the entire cycle. Rather, the cell mass generated is returned to the incoming waste stream in order to maintain a high ratio of microorganisms to organic matter and thus avoid the log growth phase where oxygen demand is high and settling of the bacterial cells is difficult.

Environmental Control The major environmental factors which will govern the operation of activated sludge systems are the pH, the temperature, and the dissolved oxygen concentration. The latter two factors are interrelated as shown below.

In any biological system it is safe to assume good operation at a pH close to neutrality. In general, activated sludge systems will operate satisfactorily in a pH range of about 6 to 9. Below pH 6 fungi will begin to compete with bacteria and may predominate if the pH drops below 5. As mentioned above, fungi will cause sludge bulking and thus are undesirable. Below pH 4.5 and above pH 9 the presence of free hydrogen or hydroxyl ions will significantly reduce the metabolic rate of nearly all microorganisms.

Most natural waters contain significant quantities of carbonate-bicarbonate alkalinity, and the buffer system provided by these salts will usually be sufficient to maintain the pH in the desired range. If problems are encountered, the buffer capacity may be increased by the addition of soda ash. The relation between the pH and the concentrations of carbonate and bicarbonate ions is given by the following equations.

$$\mathrm{pH} = 10.33 - \log \frac{[\mathrm{HCO_3^-}]}{[\mathrm{CO_3^{2-}}]} \tag{3}$$

$$\mathrm{pH} = 6.35 - \log \frac{[\mathrm{CO_2^{2-}}]}{[\mathrm{HCO_3^-}]} \tag{4}$$

All biological reactions are affected strongly by temperature. Over a considerable range such reactions may be expected to double their rate with each increase in temperature of 10°C. This increase in rate of activity will produce a similar increase in oxygen demand–a demand which may be difficult to supply. The oxygen available for biological activity is that which is dissolved in the water and the solubility of oxygen is strongly influenced by temperature, as shown in Table C10.

TABLE C10. SOLUBILITY OF OXYGEN IN WATER.

Temperature (°C)	Concentration (mg/l)	Temperature (°C)	Concentration (mg/l)
0	14.6	16	10.0
2	13.8	18	9.5
4	13.1	20	9.2
6	12.5	22	8.8
8	11.9	24	8.5
10	11.3	26	8.2
12	10.8	28	7.9
14	10.4	30	7.6

The rate at which oxygen may be transferred into solution is a direct function of the oxygen deficit, i.e.,

$$\frac{d[\mathrm{O_2}]}{dt} = K([\mathrm{O_2}]_{\mathrm{saturation}} - [\mathrm{O_2}]) \tag{5}$$

As the temperature increases, $[\mathrm{O_2}]_{\mathrm{saturation}}$ decreases; thus the rate of transfer will decrease. At the same time the oxygen demand of the biological system will increase as noted above. For example, if the temperature increases from 20°C to 30°C the rate of oxygen demand will double, but the rate of oxygen transfer will decrease by a factor of 1.2. This combination of factors may occasionally lead to operational difficulties if it is not taken into account in the design of the system. Failure to maintain sufficient dissolved oxygen (0.5-1.0 mg/l) in the aeration unit will lead to production of odors and loss of efficiency.

Oxygen may be supplied to activated sludge systems either by mechanical aeration or by bubbling air through the mixed liquor (Fig. C17). The oxygen in the air is transferred to the liquid through the surface layer between the

Fig. C17. Swingfuser aerator (courtesy FMC Corp., Environmental Equipment Division).

bubble and the liquid. The rate of transfer will be dependent not only on the temperature and oxygen deficit which are discussed above, but also on the total area over which transfer can take place. Thus, up to a point, the rate of oxygen transfer will increase with decreasing bubble size. Additionally, smaller bubbles will rise through the liquid less rapidly and thus will be in contact with the sewage for a longer period of time. This increase in contact time will also improve the efficiency of oxygen transfer.

Air diffusion systems have been developed which attempt to take advantage of the improved oxygen transfer theoretically attainable through the use of smaller bubbles. In practice, such systems may have disadvantages which outweigh their benefits. Biological growth may plug the diffusers, necessitating frequent cleaning or replacement, and the head loss through fine diffusers is quite large. Most modern plants use a rather large bubble size which may or may not be further broken down by mechanical stirring. The efficiency of oxygen transfer actually obtainable in practice will seldom exceed 6%. It must be remembered in selecting aeration equipment that such equipment is usually rated for *water* at 20°C and at a dissolved oxygen concentration of zero. The transfer rates in sewage will be less than those in water in any event, and no activated sludge system is designed to operate at zero dissolved oxygen. The claimed efficiencies of oxygen transfer will thus be difficult or impossible to attain in practice. It is advisable in this regard to specify aeration equipment not on the basis of horsepower, cubic feet of air per minute, etc., but rather on the basis of demonstrated oxygen transfer in the activated sludge unit at a specified temperature and suspended solids concentration. The furnished equipment may then be accepted or rejected on the basis of its performance in the unit for which it is intended.

Process Details The standard activated sludge process consists of a long relatively narrow tank in which the sewage enters at one end after being mixed with returned solids and travels through the aeration chamber more or less in a plug flow pattern. The mixed liquor then passes to a sedimentation basin in which the solids are removed for return or wasting. The standard process is diagrammed in Fig. C18. In units following this design the oxygen requirements

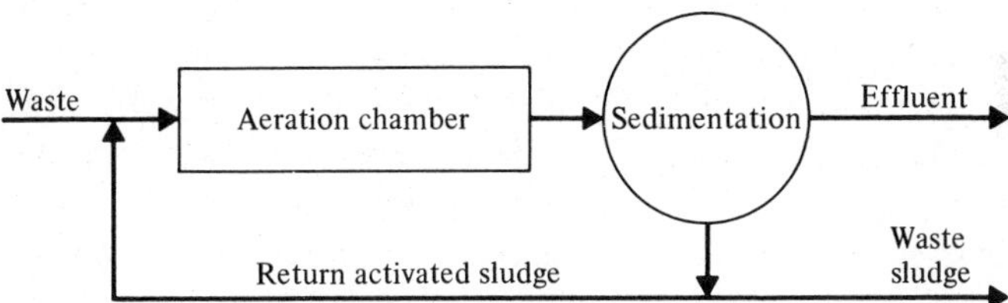

Fig. C18. Standard activated sludge.

vary widely along the length of the tank and may, in fact, be impossible to satisfy at the head of the basin. Additionally, such units respond poorly to the normal fluctuations observed in the strength and flow of domestic sewage. The difficulties encountered with this process have led to a number of modifications which are outlined below.

Tapered aeration is an attempt to provide for the variation in oxygen demand along the length of the aeration chamber by delivering more air (usually by installation of more diffusers) at the beginning of the tank. Most standard activated sludge plants have been modified in this fashion. The modification is partially effective in that air which is not necessary at the far end of the chamber is conserved. It still may be difficult, however, to meet the oxygen demand at the head of the basin.

Step aeration also is the result of attempts to equalize the oxygen demand throughout the length of the aeration chamber. As shown in Fig. C19, the influent sewage flow is split into four or more separate flows which are added at different locations through the basin. The net effect is to distribute the organic load, and thus the oxygen demand, more equally throughout the unit. Existing standard activated sludge plants may be modified to operate as step aeration units with significant improvement in their operation.

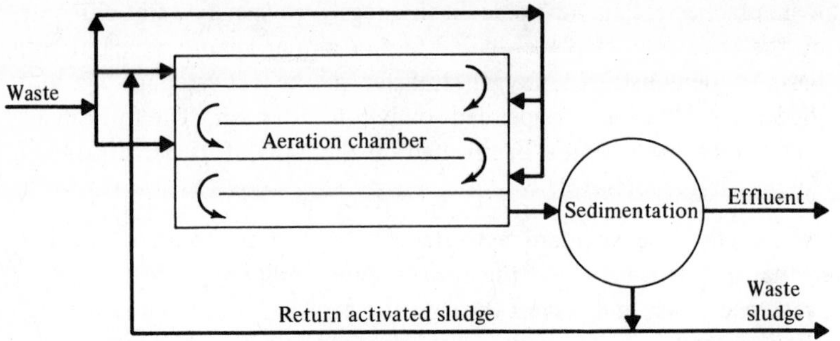

Fig. C19. Step aeration.

High-rate activated sludge is a basically inefficient process which has been used to partially treat sewage in some cities where more complete treatment was found unnecessary. The system is basically a short-aeration, high-synthesis design with low return of solids. Problems inherent in the process are difficulty of attaining the required rate of oxygen transfer and poor settling of the biological floc.

Contact adsorption or *biosorption* is a process which takes advantage of the experimental observation that the soluble BOD drops very rapidly when a waste stream is mixed with a biological floc, and later increases to near its original level before being removed by normal metabolism. The variation in soluble BOD which may be expected is as shown in Fig. C20. This variation is attributed to the adsorption of the soluble organic material on the surface of the cells and its subsequent release during metabolism. In the contact adsorption process the raw waste is mixed with the return activated sludge for 15–30 min and then permitted to settle. The solids are then aerated to permit the metabolism of the

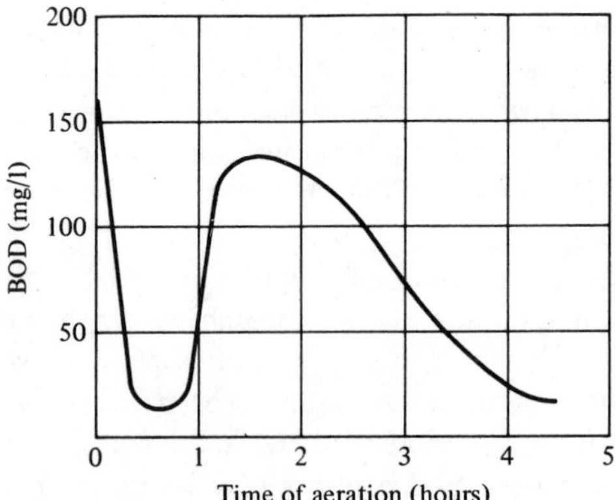

Fig. C20. Variation in soluble BOD.

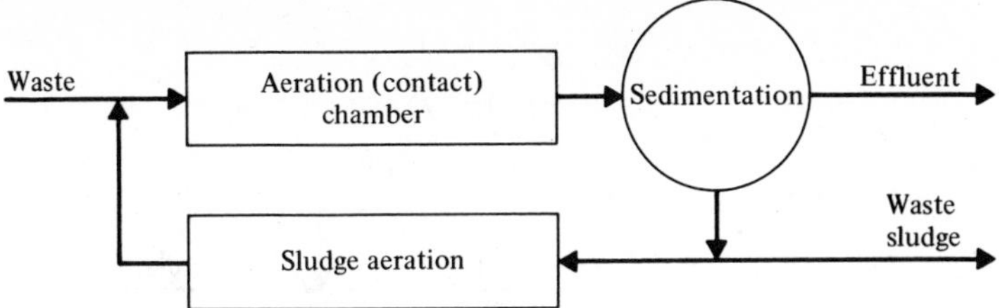

Fig. C21. Contact adsorption.

adsorbed material. The process is diagrammed in Fig. 21. The advantages of the contact adsorption system are significant. First, the retention time of the liquid waste is far less than in the standard activated sludge process, hence the required size of the treatment units is far smaller. The volume of the sludge (which must be aerated for a time equivalent to that of activated sludge) will be at most a tenth that of the raw flow. Contact adsorption thus, as well as being an economical process for initial design, is also a valuable modification for extending the capacity of existing plants which are reaching or have reached an overloaded condition.

The *Kraus process* has been used in the treatment of wastes deficient in nitrogen. The process utilizes liquid taken from an anaerobic digester to supply the nutritional deficiency. In anaerobic digestion (see the discussion of sludge treatment below) large quantities of ammonia are released as a result of the degradation of protein. The concentration of ammonia nitrogen in such units may reach several thousand milligrams per liter, hence they are a significant source of nitrogen. The system is presented in Fig. C22. The addition of nitrogen is not required for the treatment of ordinary sewage and anaerobic digesters will not always be used in modern waste treatment plants. However,

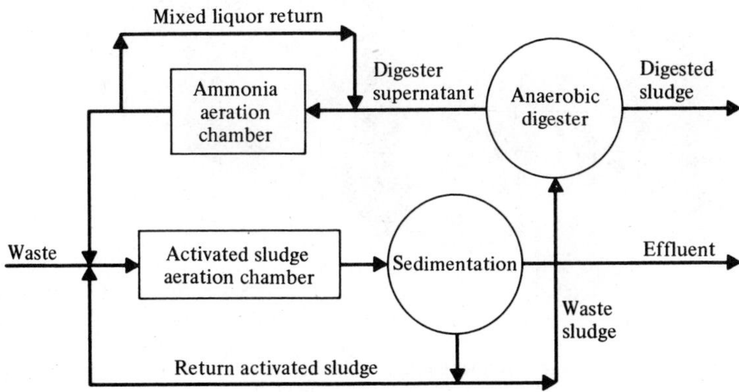

Fig. C22. Kraus process.

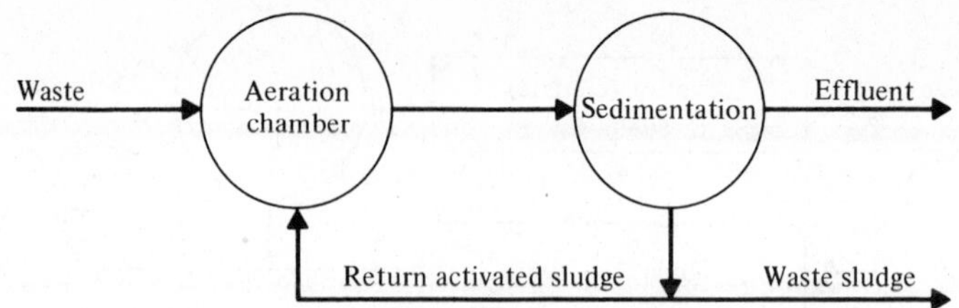

Fig. C23. Completely mixed activated sludge.

the process may have considerable value in treating wastes which have an unusually high carbohydrate concentration.

Completely mixed activated sludge was developed, perhaps inadvertently, in the design of small activated sludge plants such as those used in subdevelopments and trailer courts. In completely mixed systems the incoming waste is dispersed uniformly throughout the aeration chamber, and thus the oxygen demand is also uniform. Such units have the ability to absorb shock organic and hydraulic loads and function well under widely varying conditions. The completely mixed activated sludge process can be described mathematically, which may be a significant advantage in design. A schematic diagram of this system is presented in Fig. C23. Some small plant details are shown in Figs. C24–C26.

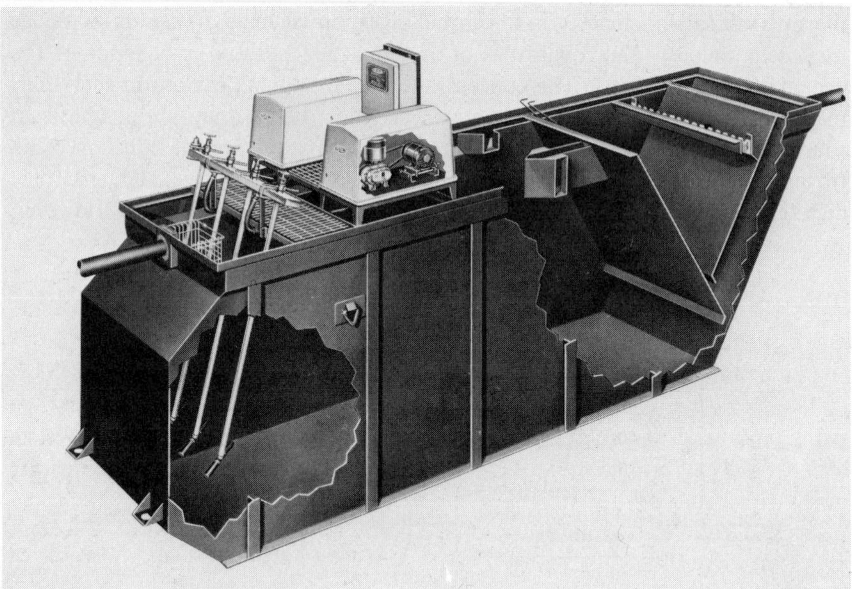

Fig. C24. "Oxigest" sewage treatment plant (courtesy Ecodyne Corp., Smith and Loveless Division).

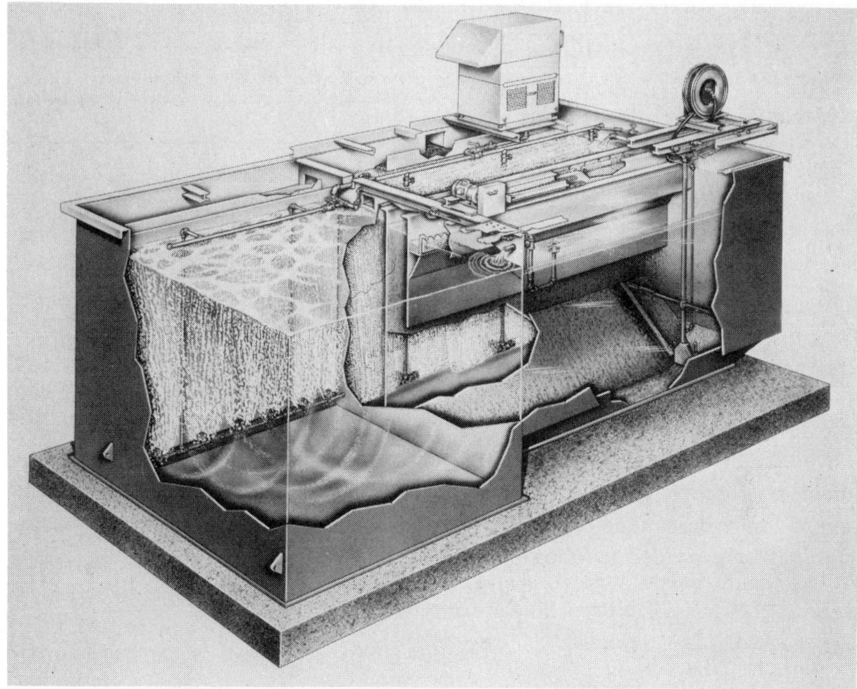

Fig. C25. "Completaire" sewage treatment plant (courtesy FMC Corp., Environmental Equipment Division).

Process Design The design of activated sludge systems is usually based upon both hydraulic and organic loading. Liquid retention times and BOD loadings are based upon the average raw flow and do not include the recycled flow. The actual hydraulic loadings are thus greater than the nominal value. The actual control of the process does not lie in maintaining a given liquid retention time, but rather in maintaining a large population of microorganisms in an aerobic environment. Thus loadings may also be expressed in organic matter per unit of biological mass. The units normally used for this parameter are pounds of BOD per pound of mixed liquor suspended solids (lb BOD/lb MLSS). Typical design values are presented in Table C11. The range given is that used in plants now in operation.

The mixed liquor suspended solids (MLSS) is an important operational parameter in activated sludge systems; however, it is not only the total quantity, but also the quality of the solids which is important. The individual cells in the biological floc will gradually age and lyse, producing a slow accumulation of relatively inert products of cellular decomposition. These inerts in no way aid

TABLE C11. ACTIVATED SLUDGE DESIGN PARAMETER.

Process	Organic loading $\left(\frac{\text{lb BOD}}{1000\ \text{ft}^3}\right)$	Organic loading $\left(\frac{\text{lb BOD}}{\text{lb MLSS}}\right)$	MLSS (mg/l)	Aeration period, (h)	Return sludge (%)	Efficiency (%)
Standard, incl. tapered aeration	30–40	0.2–0.5	1500–3000	5–6	30–50	90–95
Step aeration	30–50	0.2–0.5	1000–2000	4–6	50–100	90–95
High rate	100–150	0.5–2.0	500–750	2–4	25–50	60–75
Contact adsorption						
contact tank	300–500	1.0–2.0	4000–5000	$\frac{1}{4}$–$\frac{1}{2}$	100	90–95
sludge aeration	N/A	N/A	40,000–50,000	5–6	N/A	N/A
Completely mixed	30–50	0.2–0.5	1500–3000	3–4	100	≥95

the process and the accumulation may overburden sedimentation basins and pumps. If sludge is not wasted regularly the system may reach a point at which the returned sludge is relatively useless. The quality of the sludge is generally obviously demonstrated in the efficiency of treatment and is readily observed under microscopic examination. *Sludge age* is usually calculated by relating the mixed liquor suspended solids to the influent suspended solids. Mathematically this may be expressed as follows:

$$\text{Sludge age} = \frac{\text{MLSS} \times V}{\text{SS} \times Q} \tag{6}$$

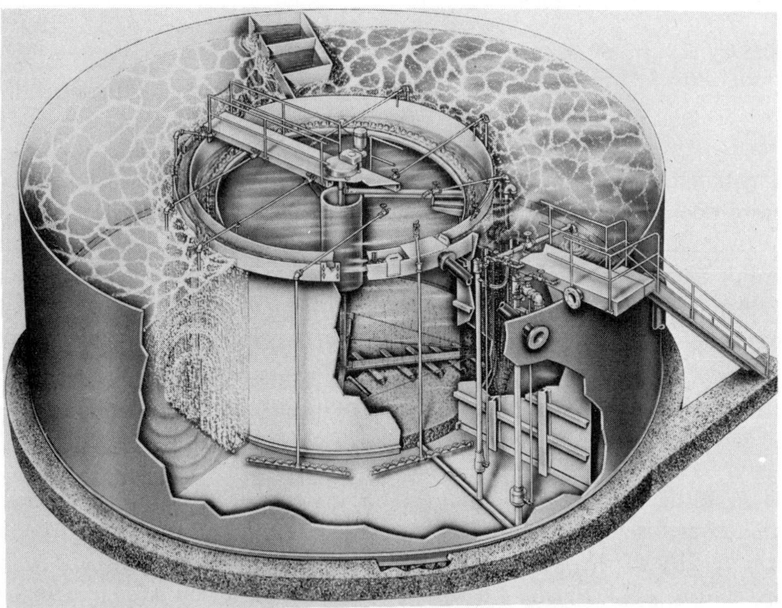

Fig. C26. "Stepaire" sewage treatment plant (courtesy FMC Corp., Environmental Equipment Division).

in which MLSS is the mixed liquor suspended solids in mg/l, V is the aeration chamber volume in gallons, SS is the influent suspended solids in mg/l, and Q is the raw waste flow in gallons per day. It should be recognized that this theoretical sludge age is not the actual average age of the biological mass, since this material does not enter with the waste but is generated in the aeration chamber. A more reasonable formulation would relate the sludge age to the mixed liquor suspended solids and the solids lost from the system in both the effluent and in waste sludge. On this basis the sludge age is

$$\text{Sludge age} = \frac{\text{MLSS} \times V}{(\text{SS}_e + \text{SS}_s r)Q} \tag{7}$$

in which SS_e is the suspended solids in the effluent in mg/l, SS_s is the suspended solids concentration of the waste sludge in mg/l, r is the ratio of waste sludge flow to raw waste flow, and the other quantities are as expressed in Eq. (6).

The aeration requirements in activated sludge systems will be dependent upon the efficiency of oxygen transfer of the particular system used. As noted above, the efficiency of a diffused air system may be less than 5-6%. Mechanical sparging may increase this significantly, but only at an added power cost. The excess air required should not be considered as a total loss since it aids in the mixing of the basin, which is also essential to the proper operation of the process. Typical aeration requirements are listed in Table C12. With highly efficient aeration equipment these requirements may be significantly reduced. The higher values listed are those frequently specified by regulatory agencies. Most operational problems associated with aeration stem from too little rather than too much air. Conditions attributed to "over aeration" in the past have doubtless resulted from the use of fixed design standards which did not take into consideration the variations in strength observed among wastes from different communities. Activated sludge plants should be designed so that the operator can vary the aeration rate. This provides additional flexibility in operation and permits savings in aeration cost.

Aeration designs for activated sludge are always high-volume, low-pressure systems. The compressors which should be used then are dictated by the nature

TABLE C12. ACTIVATED SLUDGE AERATION REQUIREMENTS.

Process	Air required (ft^3/lb BOD)
Standard, including tapered aeration	1000–1500
Step aeration	1000–1500
High rate	500–1000
Contact adsorption	1000–1500
Completely mixed	1500–2000

of the design. Centrifugal blowers should always be used rather than positive displacement compressors. The maximum pressure required at the diffusers will rarely exceed 10 psi and may often be less depending on the depth of the basin, the location of the diffusers, and the type of diffuser used. Delivery lines should be large in order to avoid excessive head losses. Velocities of less than 5 fps are commonly used.

Compressors should be selected with consideration given to flexibility of operation. Centrifugal blowers may be operated at other than their design speed at a loss in efficiency, but a good design will usually employ a series of blowers which may be cycled on and off as required. The compressors should be able to meet the maximum air demand with the largest unit out of service. This largest unit may be chosen to meet the minimum anticipated requirement.

Process Operation and Control Operational problems in activated sludge systems are normally associated with either inability to transfer the required oxygen or difficulty in maintaining the desired suspended solids concentration. Flexibility in aeration control and solids return are of prime importance in design.

Difficulty in maintaining the required solids level is frequently caused by "bulking" sludge, i.e., a light filamentous sludge which does not settle well. The sludge volume index and sludge density index may be used interchangeably to describe the settling characteristics of a sludge. The two quantities are related by the following equation:

$$SVI = 100/SDI \tag{8}$$

The sludge volume index is the volume in milliliters occupied by one gram dry weight of solids after settling for 30 minutes. The measurement is commonly made by allowing one liter of mixed liquor to settle in a graduated cylinder for 30 minutes, at which time the volume occupied by the sludge is recorded. The suspended solids concentration in the mixed liquor is then determined as described above and the sludge volume index is calculated as follows:

$$SVI = \frac{SV}{MLSS \times 1000} \tag{9}$$

in which SV is the sludge volume in milliliters per liter and MLSS is the mixed liquor suspended solids in milligrams per liter. The normal range of the sludge volume index for activated sludge plants is 50-200. If it increases beyond this range bulking may be expected. It is particularly important to monitor the SVI continually in order that trends toward bulking may be observed before the problem becomes critical.

The test schedule given in Table C13 is suggested for the control of activated sludge plants. In small plants with limited personnel and equipment the tests marked with an asterisk may be omitted. This should not be taken to imply

TABLE C13. ACTIVATED SLUDGE PROCESS CONTROL TESTS.

Test	Sample source			
	Raw waste	Mixed liquor	Settled solids	Effluent
BOD	X	–	–	X
Suspended solids	X	X	X	X
Settleable solids	X	X	X	X
Dissolved oxygen	–	X*	–	X*
SVI	–	X	X	–
pH	X	X	–	X
Alkalinity	X*	X*	–	X*

*May be omitted in small plants with limited personnel and equipment.

that they are unnecessary. Where possible, all determinations should be made routinely.

Oxidation Ponds

The simplest biological waste treatment system from a standpoint of operation is unquestionably the oxidation pond. The simplicity is only apparent, however, since the same biological reactions which occur in activated sludge systems also occur in oxidation ponds. In fact, the biological system may be even more complex, involving anaerobic digestion, aerobic stabilization, and a complex symbiosis between bacteria and algae.

Physically an oxidation pond consists of a large shallow basin to which the wastes are added, and from which the effluent is withdrawn, usually at two single points. The pond depth may range from 2 to 5 ft and may be varied during the year. Since waste stabilization occurs slowly during periods of low temperature it may be advisable to permit the pond to fill slowly during the winter and to decrease the depth of water during the summer months. The minimum depth will be that required to prevent weed growth, normally about 2 ft.

The biological reactions which occur in an oxidation pond are presented schematically in Fig. C27. The raw waste entering the pond will separate by sedimentation, with the settleable solids falling to the bottom and forming a relatively concentrated layer of organic material. Since it is impossible to obtain the necessary transfer of oxygen into this layer it will undergo anaerobic decomposition. This anaerobic decomposition will release carbon dioxide, methane, ammonia, and hydrogen sulfide as well as some soluble organic materials, which are intermediates in the anaerobic process, to the liquid layer above. Under the aerobic conditions generally maintained in the liquid layer the soluble organic material in the raw waste and the intermediate degradation products released from the lower layer will be oxidized to carbon dioxide,

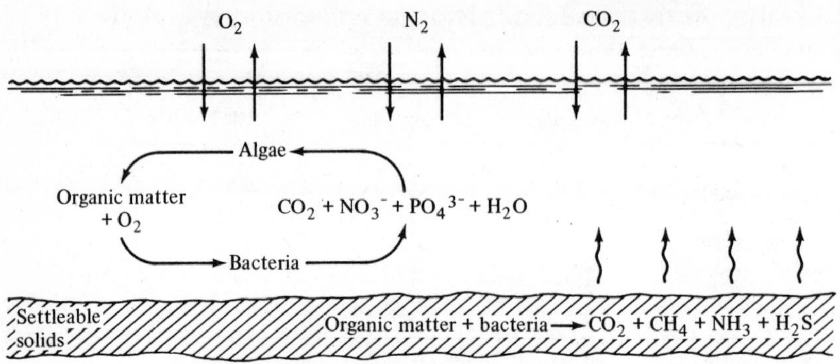

Fig. C27. Oxidation pond schematic.

nitrate, orthophosphate, and water. The ammonia released from the anaerobic zone may also be oxidized to nitrate and the hydrogen sulfide to sulfate. These reactions are of course dependent on the presence of *Nitrosobacter* and *Nitrobacter* for the ammonia and *Thiobacillus* for the sulfide.

Thus far the reactions are equivalent to those observed in other waste treatment systems. The shallow depth leads, however, to an additional complication. The penetration of sunlight provides the only element otherwise lacking for the growth of algae. The algae are microscopic photosynthetic organisms which in the presence of light are able to synthesize organic matter from carbon dioxide, nitrogen, and phosphate. All the inorganic materials are naturally present in sewage and are released in even greater quantities by the bacterial metabolism. Nearly all natural watérs contain significant carbonate-bicarbonate alkalinity, which also serves as a source of carbon to the algae, and some species of algae are able to fix nitrogen from the atmosphere. Thus algae will grow rapidly in shallow oxidation ponds.

The algae do not actually stabilize any organic material and may in fact produce an actual increase in the total organic material in the pond. They do however serve a useful purpose since one of the end products of photosynthesis is oxygen, which is thus made available to the bacteria. The amount of oxygen produced is sufficient at times to actually produce a supersaturated condition in the pond, with dissolved oxygen levels of as much as 15 mg/l.

During the absence of sunlight the algae cease photosynthesis and actually consume oxygen. At night therefore the dissolved oxygen level in the pond may approach zero with all oxygen transfer being from the liquid-air interface. If the dissolved oxygen level approaches zero the nitrates produced by the daylight aerobic bacterial metabolism will be reduced to nitrogen gas which may then be released from solution.

Stabilization of wastes in an oxidation pond is necessarily a long-term process because of the symbiosis between algae and bacteria, which leads to the recurent breakdown and resynthesis of organic material. In ponds of the required size the bacterial and algal cells eventually settle out before reaching the outlet; thus the organic material is retained in the treatment unit. Oxidation ponds will eventually fill with the end products of the algal and bacterial metabolism and must be dredged or rebuilt periodically.

Oxidation ponds have certain advantages and disadvantages which should be considered in selecting a waste treatment system for a community. The major advantages include the low operating cost and the minimum of operator skill which is required. The first cost may be lower than that of other systems, depending on land values and soil conditions. The more significant disadvantages include the large land area required, occasional odor production due to dissolved oxygen deficits, and poor treatment of some industrial wastes. The odors produced usually dictate a remote location which may lead to high costs for sewer lines and lift stations. Growth of the community may lead to encroachment of residential areas on the pond site with resultant public dissatisfaction.

The capacity of oxidation ponds may be increased by use of mechanical or diffused aeration systems. Plastic tubes laid on the bottom to bubble air gently through the liquid have been used with mixed success. Biological growth and other solids may cover or otherwise plug the tubing and cleaning is difficult and appears to be only temporarily effective. Floating mechanical aerators appear to be the most efficient method currently available. It is only necessary to provide sufficient dissolved oxygen to prevent odors (about 0.5-1.0 mg/l); thus the quantities of air required will be far less than in activated sludge. Aeration should normally not be required during the daylight hours.

Design Criteria Oxidation ponds are normally designed on the basis of BOD loading per acre per day or persons per acre. The population value should only be used for communities which are purely residential in character since industrial wastes will considerably increase the average strength and flow of the waste. Currently recommended design loadings are approximately 20 lb BOD/acre/day or 100-120 persons/acre for an average domestic waste (0.17-0.20 lb BOD/person/day). Oxidation ponds have been operated at loadings of up to 50 lb BOD/acre/day in the southwestern plains of the United States where the long day and continuous winds aid in maintaining the required dissolved oxygen concentration.

Oxidation ponds should be located insofar as possible in open terrain. There is a natural tendency to conceal waste treatment facilities, but building oxidation ponds in ravines or shielding them from view with trees will interfere with proper operation. It is safe to say that oxidation ponds which cannot be seen will frequently be smelled. The gentle mixing generated by wind currents is essential to the surface transfer of oxygen.

Because of the long retention times involved, the inlet and outlet structures of oxidation ponds may be very simple in nature. Corrugated metal pipe or

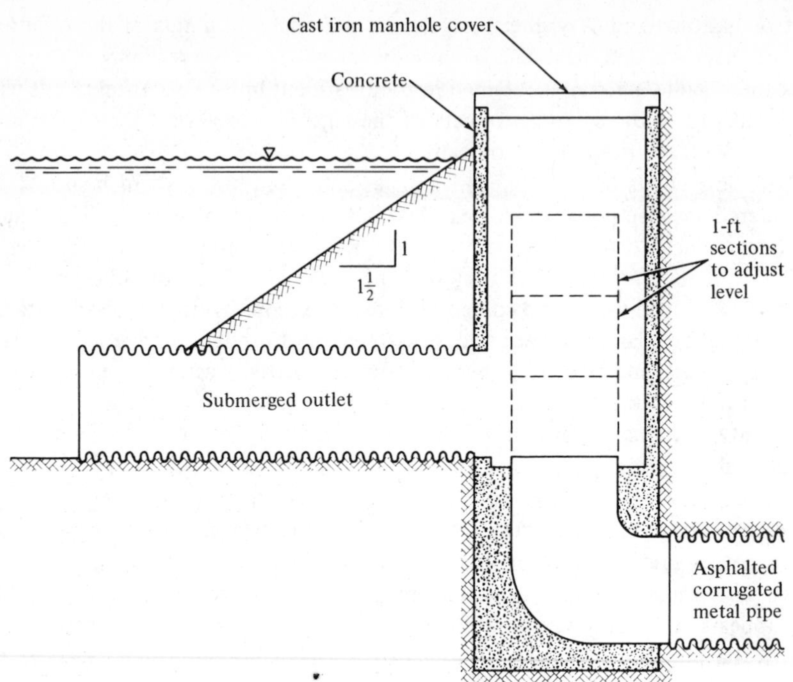

Fig. C28. Oxidation pond outlet.

similar inexpensive materials are usually suitable. An oxidation pond loaded at a rate of 20 lb BOD/acre/day will, for ordinary domestic sewage, have a retention time of approximately 16 days per foot of depth. A 4-ft pond would thus have a retention time of over 60 days, hence short circuiting from inlet to outlet will not occur. The flow pattern in the pond will be governed by wind currents, not hydraulic loading. The inlet is normally located in the center of the pond, the pond being either circular or square in plan view. A simple outlet structure which permits varying the liquid depth and draining when necessary is shown in Fig. C28. The submerged outlet helps to prevent icing in northern climates.

Where soils are permeable it will be necessary to line oxidation ponds to prevent contamination of the ground water. Linings which have been used include alternate layers of tar and burlap and spraying with mortar–asbestos cement. The most popular method at present is lining with sheet plastic. Such materials are readily available from a number of manufacturers. The sealant chosen should be relatively impermeable, easily installed, and moderate in cost. Lining of oxidation ponds may constitute a significant fraction of their cost.

Sludge Treatment

The solids naturally present in sewage and those generated in the secondary treatment of wastes constitute one of the major handling problems encountered in waste treatment. Primary sewage solids will contain more than 90% water while secondary solids may contain more than 95% water. The solids present will be highly putrescible and are extremely difficult to dewater. The major treatment methods currently in use include aerobic digestion, anaerobic digestion, vacuum filtration (which may be used either on the raw solids or the solid residue after aerobic or anaerobic digestion), wet oxidation, and incineration.

Aerobic Digestion Aerobic digestion is simply the long-term aeration of biological solids in a food-limiting condition. Under these circumstances the cells will undergo endogenous metabolism with a resulting reduction in mass. As individual cells die and lyse they will release small quantities of nutrients for the remaining cells which will help to maintain their existence. Thus the digestion period will be lengthy. Since there will be little metabolism of extracellular food the oxygen demand will be far less than that required for activated sludge. Design criteria will vary with the final disposal method intended for the solid residue, the shorter retention times and higher loading rates being used in conjunction with vacuum filtration and the longer times and lower loading rates in conjunction with drying on sand beds. Typical loading and aeration rates are presented in Table C14. At higher temperatures the rate of digestion will be in-

TABLE C14. AEROBIC DIGESTION DESIGN CRITERIA.

Volatile solids loading	0.05–0.10 $lb/ft^3/day$
Retention time	10–15 days
Aeration required	15–30 $ft^3/min/1000\ ft^3$

creased, hence retention times may be decreased. It should be noted, however, that if the rate of digestion is increased, the oxygen demand will be similarly increased. Aerobic digestion will normally yield reductions in BOD and volatile suspended solids of 40–50%. The digested sludge dewaters readily and is relatively odor free. It may be landfilled or incinerated but has little or no fertilizer value. Sludge drying beds for aerobically digested sludge may be constructed to the specifications listed under anaerobic digestion.

Anaerobic Digestion Despite the operational difficulties frequently encountered in its use, anaerobic digestion remains a popular method of treating sewage sludges and other concentrated wastes. The anaerobic process is the only waste treatment system which can be said to pay its own way, in that it produces more useable energy than is necessary for its maintenance. While the process has been used for many years, the design and operation of modern anaerobic systems requires an understanding of the basic biological phenomena which lead to waste stabilization in the absence of free oxygen.

Anaerobic systems treating complex organic wastes are dependent upon the interrelationship between two types of bacteria: the facultatively anaerobic microorganisms, which break down the complex organic materials to simpler, short-chain compounds such as the volatile fatty acids; and the obligate anaerobes, including *Methanobacterium*, *Methanosarcina*, and *Methanococcus*, which reduce the short-chain materials to methane, carbon dioxide, and ammonia. The acid-forming bacteria are essential to the process, but it is the methane formers which appear to be most important since it is their failure which generally leads to digester upsets. The breakdown of the complex organic material is usually rapid while the methane fermentation is slower and evidently more influenced by environmental factors. This difference in the rate of metabolism of the two species can lead to an accumulation of intermediate degradation products, notably the volatile fatty acids. These in turn can cause a sudden drop in pH as the buffer capacity of the system is exhausted. This decrease in pH can by itself produce a cessation of biological activity, but the volatile acids themselves have been thought to be toxic if present in sufficient concentration. It appears in the light of recent research that it is not the acids themselves but rather the associated cations (normally ammonium) which are toxic to the methane formers. Nonetheless, the volatile acid concentration is an important indicator of proper digester function.

The methane fermentation has been demonstrated to be particularly sensitive to the effects of temperature and pH. The optimum temperature for the process appears to be about 35°C. Higher rates of metabolism may be obtained at about 50°C but the energy requirements for heating are significantly higher and the process is reportedly more subject to upsets caused by slight fluctuations in temperature. The methane fermentation will continue at temperatures as low as the freezing point of water, but at a greatly reduced rate. From 30°C to 15°C the rate of metabolism decreases by approximately one-half for each 10°C drop in temperature. Below 15°C the rate of metabolism decreases still more rapidly and design data are lacking for this temperature range. Since digester heating is readily effected and since the process works best at 35°C it is common practice to heat digesters to that temperature.

The anaerobic process operates best at a pH close to neutrality, but functions well in a pH range of about 6.5–8.5. In any system the pH will tend to rise and drop slightly with variations in the volatile acid concentration and this is not, in itself, cause for concern.

A major end product of anaerobic digestion is methane. The gases produced will normally amount to about 15–20 ft^3 per pound of volatile solids destroyed.

Methane will constitute about 70% of the gas produced with the remainder being largely carbon dioxide with traces of hydrogen sulfide and nitrogen. This gas has a fuel value of about 700 BTU/ft^3 and hence is a useful energy source. Sludge gases are commonly burned to heat the digesters but have other applications as well, particularly since more gas than is required for temperature control is usually produced. Some treatment plants burn the digester gas in modified gasoline engines which drive the centrifugal blowers which produce the air for activated sludge units. The cooling water from these engines is then used to heat the digesters. Such a design can lead to significant economies in plant operation, but must include provision for the use of natural gas in the event of insufficient methane production. Gas storage at the plant must also be provided.

Design criteria commonly used for anaerobic digesters are presented in Table C15.

TABLE C15. ANAEROBIC DIGESTION DESIGN CRITERIA.

Temperature	30–35°C
pH	6.5–8.5
Loading (lb VS/ft^3/day)	0.05–0.20
Retention time (Days)	10–30

Mixing is one of the most important considerations in the design of anaerobic digesters. Unmixed digesters require retention times of 30-90 days to obtain the same degree of treatment obtainable in 10 days in a well mixed unit. Mixing brings the microbial mass into intimate contact with the material to be degraded and assures uniform distribution of the bacteria and the waste. Mixing may be effected by mounting motor-driven stirrers in the digester cover, by recirculating gas to the bottom of the unit, and by recirculating mixed liquor in the digester itself. The last method is most frequently used since it is advisable to pump the contents from the digester prior to heating in any event, but it may not be sufficient. The mixing should be as complete as possible. The digester should not be used as a sedimentation basin, but should be considered as a biological reactor like an activated sludge tank.

Heating is critical to proper digester function, with a rather narrow temperature range defining the optimum. Internal heating units have proven to be unsatisfactory since the sludge rapidly coats the coils with great reductions in heat transfer efficiency. Cleaning such coils requires emptying the digester, which is not a chore to be lightly undertaken. External heat exchangers, in which the sludge is pumped from the digesters, heated, and returned, have proven far more satisfactory. The sludge still will dry on the heated surfaces but the accumulation will tend to be less because of the relatively high velocity in the pipes and the lines may be readily cleaned when necessary. Such sludge recirculation will aid in mixing the digester but in most instances is insufficient by itself.

Uniformity of feeding is particularly significant in view of the disparate metabolic rates of the acid-forming and methane-forming bacteria. Slug loads of sludge will be rapidly broken down to volatile acids and the rate of acid production may readily exceed the rate at which the acids can be broken down to methane and carbon dioxide. The peak volatile acid concentration will occur within two hours after a batch feeding, which indicates the rapidity of the first stage of digestion. Provision for continuous or nearly continuous feeding should be incorporated in any digester design.

Under conditions of uniform feeding, complete mixing, and good thermal equilibrium, anaerobic digesters will function like completely mixed activated sludge units. That is, each addition of sludge will displace a like volume of mixed liquor. Attempts have been made to recirculate solids as in activated sludge systems in order to shorten the required retention time. Such recirculation is unfortunately futile since the solids in an anaerobic digester treating sewage solids are largely inert and the pumps recirculate grit while the dispersed bacteria pass out in the effluent. The required microbial population can only be obtained (in the current state of the art) by maintaining a retention time in excess of 5 days. This may not be the case for soluble industrial wastes.

Following passage through the digester the digested solids are allowed to settle, and then pumped either to drying beds, filters, or an incinerator. The supernatant liquid may be recycled through the treatment plant. The BOD of digester supernatant is primarily nitrogenous, i.e., due to the high ammonia concentration. Volatile solids reduction may be expected to be approximately 50%. The digested sludge dewaters readily and is not particularly offensive in odor.

Process control in anaerobic digestion requires continuous monitoring of temperature, volatile solids, pH, alkalinity, volatile acids, and gas production and composition. Temperature is subject to operator control, while the other determinations measure the efficiency of digestion. The volatile acid and alkalinity tests are particularly important since the pH may remain relatively constant over a wide range of volatile acid concentration. Continuous monitoring of volatile acid concentration will indicate the beginning of hindered digestion and corrective measures can then be taken. In normal operation the volatile acid concentration should not exceed 200–500 mg/l as acetic acid. Frequently the level may be far less. Occasional minor upsets may produce concentrations of up to 2000 mg/l. These should not be harmful, nor should even higher concentrations, provided the digester is adequately buffered.

The major buffer system in anaerobic digesters is naturally produced within the unit and is normally adequate to insure pH control. The anaerobic degradation of protein results in the production of ammonia which will combine with carbon dioxide and water to produce ammonium bicarbonate:

$$NH_3 + CO_2 + H_2O \longrightarrow NH_4(HCO_3) \tag{10}$$

The resultant alkalinity will normally be in the range of several thousand milligrams per liter as $CaCO_3$ and is hence sufficient to buffer a considerable increase in volatile acid concentration. Care must be taken in measuring the alkalinity when the volatile acid concentration is known to be high since the volatile acid salts will also be titrated as alkalinity and will produce erroneously high results. If the volatile acid concentration approaches the alkalinity in magnitude it may be necessary to lime the digester to maintain the proper pH. This has been done successfully, but a better procedure appears to be to lower the loading rate slightly when an increasing trend is first observed in the volatile acid concentration.

Sludge drying beds normally consist of a layer of gravel $\frac{1}{4}$-1 in. in size approximately 12 in. deep covered with a layer of sand approximately 12 in. deep. The bed is underlain with tile or perforated pipe underdrains which may drain to the treatment plant wet-well. Sludge is discharged to the beds normally to a depth of 12 in. and permitted to dry. When the depth has been reduced to about 5 in. the sludge (which will be cracked in appearance) will be sufficiently dry for solid handling techniques. The dried sludge may be removed either mechanically or by hand. A small amount of sand is lost when each bed is cleaned, hence the sand must be replenished occasionally. The drying rate and hence the area required for sludge dewatering will be dependent on the local climate. Drying may be facilitated by covering the bed with glass to prevent rewetting by rain and snow. Bed areas are normally specified in square feet per capita. Typical areas are listed in Table C16. In dry climates such as that in the southwestern United

TABLE C16. SLUDGE DRYING BED AREA.

	Area (Ft^2 per capita)
Primary only	1.00
Standard trickling filter	1.25
High-rate trickling filter	1.50
Activated sludge	1.75

States the table values may be reduced by a factor of two. The dried sludge may be landfilled or incinerated. Dried sewage sludge has been sold locally as a fertilizer or soil conditioner but the market for this material is limited. Putting aside the esthetic objections of many potential users, it is really not a very good fertilizer and cannot compete with nonorganic additives.

Vacuum filtration, usually with the aid of chemical coagulants, has proven successful in the rapid dewatering of sewage sludges. Vacuum filters employ a porous screen on which the solids are retained while the liquid is drawn through by the differential pressure. Screening materials include cloth, metal mesh, and closely spaced steel coils. The filters are normally cylindrical and rotate about a horizontal axis with about one-quarter of the surface immersed in the wet sludge. As the drum rotates the dewatered sludge is scraped from the filter surface before it reenters the wet sludge. Filter yields are normally expressed pounds per square foot per hour and range from 3–8 lb/ft^2/h with a moisture content of about 75%, depending upon the type of sludge and the chemical coagulant used. Common coagulants include lime and ferric chloride and polyelectrolytes.

Filtered sludge will be similar to that obtained from drying beds following aerobic or anaerobic digestion. Filtered undigested solids will be odorous, particularly if rewetted. While dry, however, they may be readily landfilled or incinerated.

Wet oxidation is a low-temperature combustion process which has been used successfully to destroy organic sludges. The process is dependent upon the introduction of air to oxidize the organic material, maintenance of a temperature above 100°C, and preferably between 200°C and 374°C, containment of the reaction in a pressure vessel, and pumps capable of delivering the sludge against the internal pressure of the system.

A schematic diagram of the patented *Zimmerman process* of wet oxidation is presented in Fig. C29. The heat exchangers shown recover a portion of the heat generated in the combustion process and thus the reaction can be sustained without the addition of outside heat once the system is started. The water in the sludge is essential to the process and drying is thus unnecessary prior to such treatment.

In order for the reaction to be selfsustaining the sludge must have a minimum volatile solids content. Weaker sludges will require the addition of external heat.

The effluent of the wet oxidation process will be a moderately clear liquid which is usually recirculated through the treatment plant. Any unoxidized organic material should be readily removed by aerobic biological treatment. A typical wet oxidation unit is illustrated in Fig. C30.

Incineration of primary sewage sludges is normally a self-sustaining process with no requirement for additional heat once combustion is begun (Fig. C31). The heat produced is frequently used to dry the sludge prior to its introduction into the incinerator. Digested sludges may not burn without the addition of further heat. The more thorough the digestion process has been, the less likely it is that the sludge will sustain its own combustion.

The sludge is delivered to the incinerator after dewatering either by vacuum filtration or by drying on a sand bed. The sludge will thus have a moisture content of 60–85% and must be dried before it can be ignited. Specialized incinerators have been designed for sewage solids, but it is not unusual to consider the sludge as simply one portion of a community's solid waste problem and incinerate sewage solids and garbage together. The ash which results from the incineration of sewage solids is completely odorless and may be landfilled without difficulty.

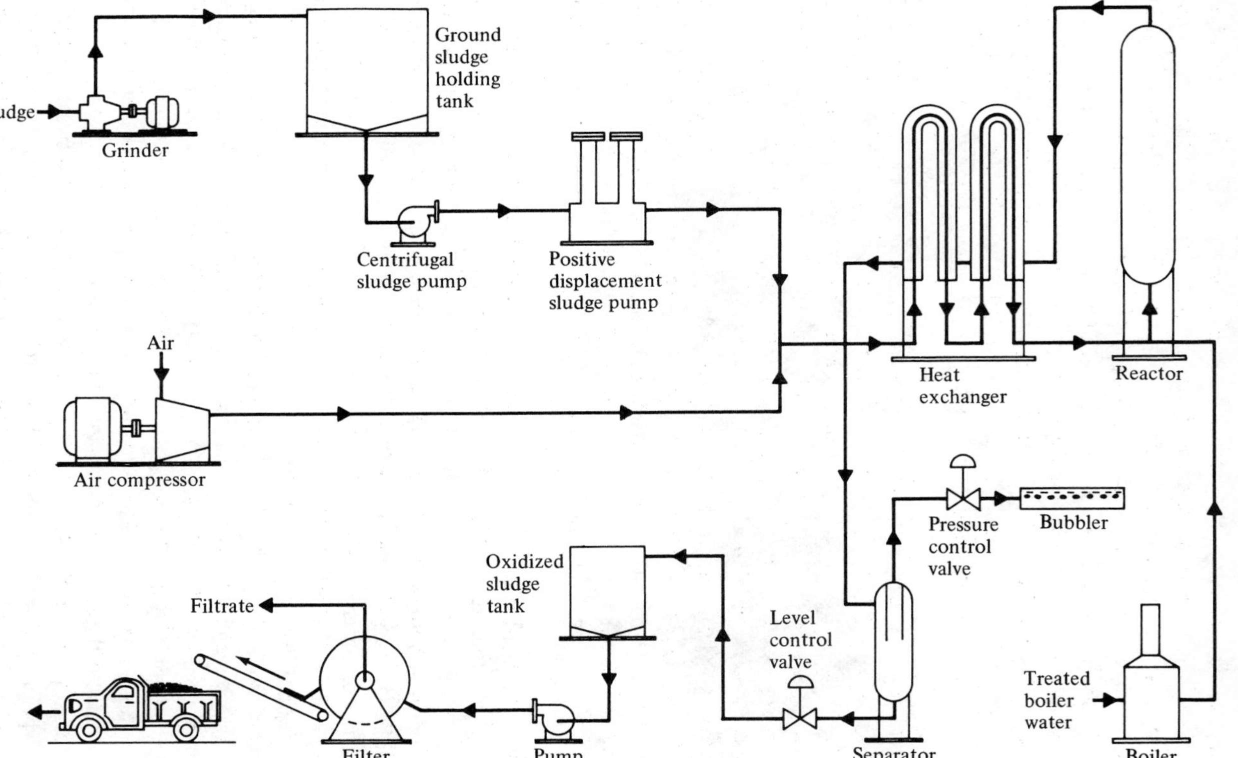

Fig. C29. Typical Zimpro sludge-conditioning unit (courtesy Zimpro Inc., Subsidiary of Sterling Drug Inc.).

Fig. C30. Wet oxidation plant (courtesy Zimpro Inc., Subsidiary of Sterling Drug, Inc.).

Tertiary Treatment

Tertiary treatment properly includes any treatment which follows the secondary biological processes discussed above. The techniques used all have the goal of further reducing the pollutional load upon the receiving body of water either through removal of suspended solids, organic materials both soluble and colloidal, inorganic salts, inorganic nutrients, or a combination of these (Figs. C32–C37). Chlorination or other disinfection is sometimes referred to as tertiary treatment, but by the above definition it is not, since such treatment in no way reduces the concentration of the materials cited.

Suspended solids not removed by sedimentation may be reduced by passage through fine screens; filtration, usually through diatomaceous earth filters similar to those used in water treatment; or coagulation, either with the usual chemical coagulants or polyelectrolytes. The techniques used are usually quite similar to those applied in water treatment since there is little difference between the raw material and the desired product in the two cases.

Trace organic materials may be removed by adsorption on activated carbon. The adsorption follows the same pattern as that in water treatment, with the required dosages being determined from the Freundlich isotherm.

Inorganic salts are removed when necessary by application of either distillation, dialysis, ion exchange, or reverse osmosis. All the techniques presented thus far

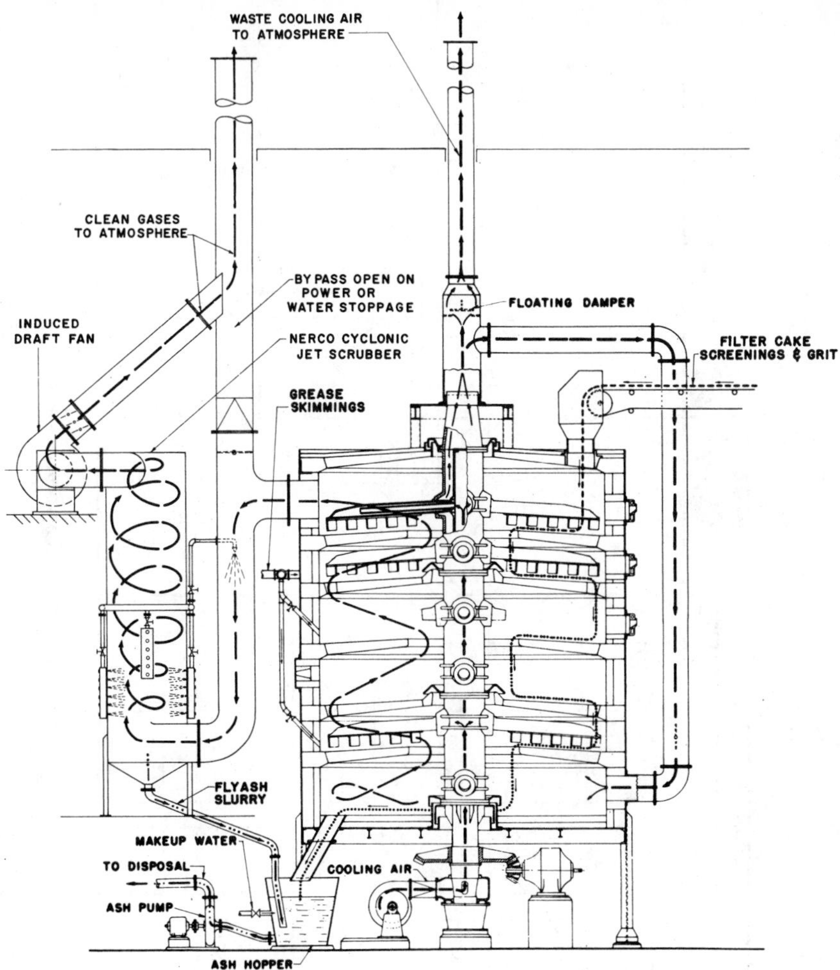

Fig. C31. Sludge furnace (courtesy Nichols Engineering and Research Corp.).

are identical in purpose and method with those discussed in the subsection on water treatment. The removal of nitrogen and phosphorus is, however, peculiar to the treatment of wastes.

Phosphorus and nitrogen have been implicated as being among the major causes of lake eutrophication. As mentioned elsewhere, the placing of total blame upon these elements may have been premature, but it is certain that they are critical nutrients for all plant life.

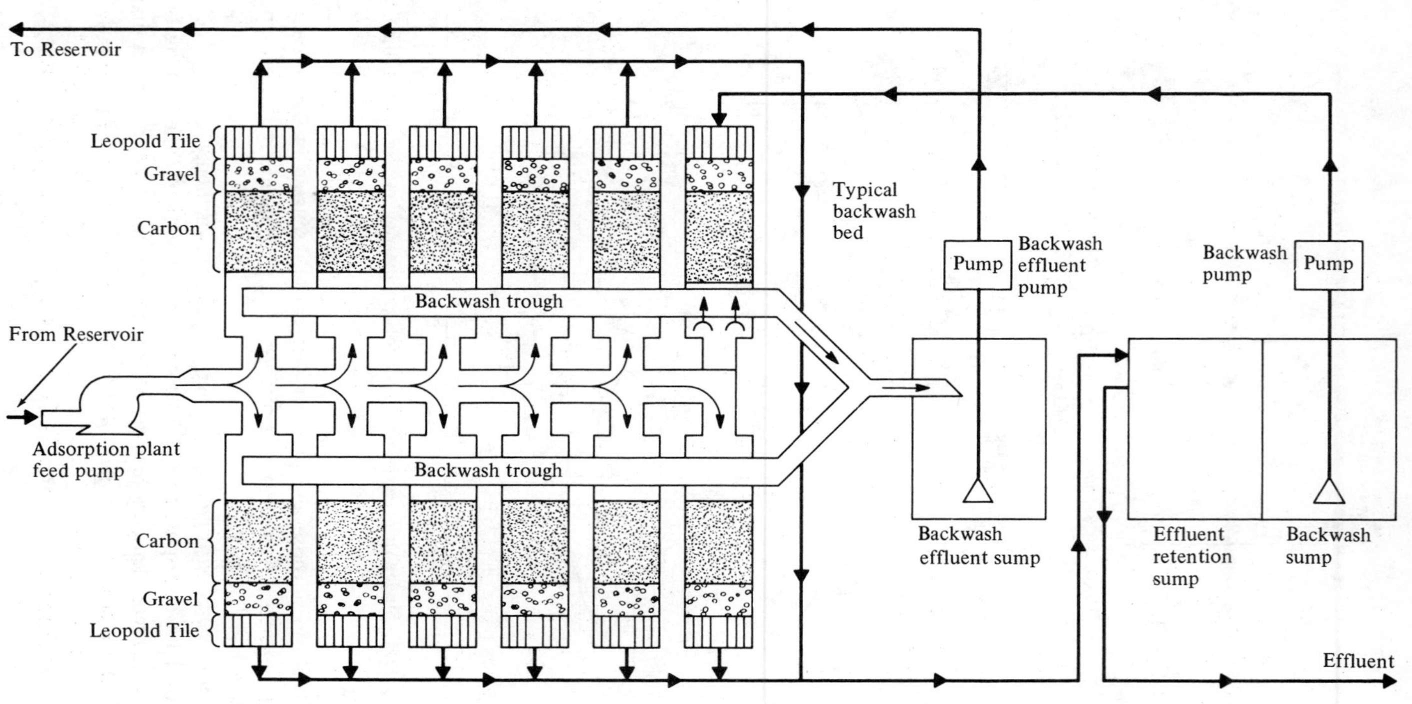

Fig. C32. Waste water activated carbon treatment plant (courtesy Calgon Corp. and Atlantic Richfield Company).

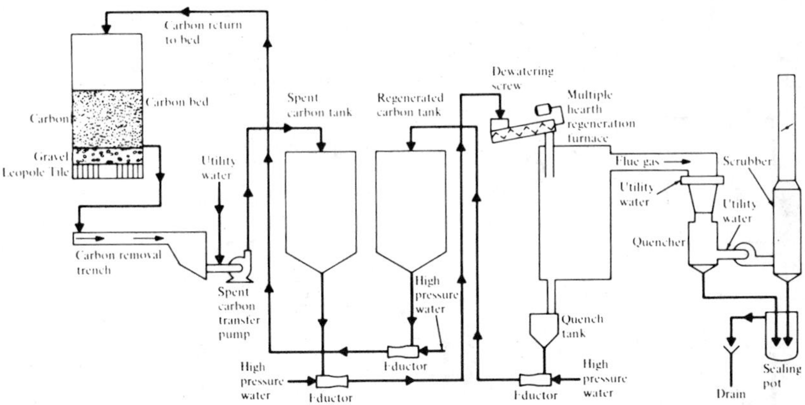

Fig. C33. Waste water activated carbon treatment plant, carbon transfer and regeneration system (courtesy Calgon Corp. and Atlantic Richfield Company).

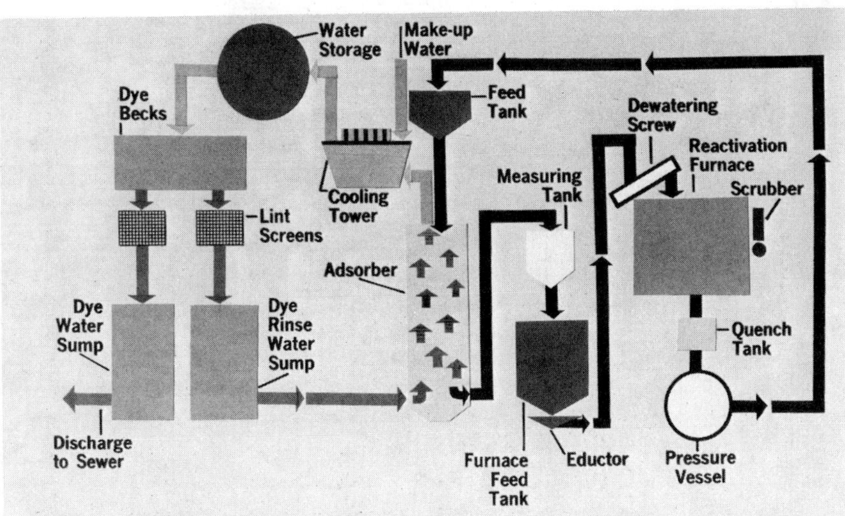

Fig. C34. Schematic of granular carbon waste water reclamation system (courtesy Calgon Corp.).

The sources of nitrogen and phosphorus entering surface waters are not limited to waste-water outfalls, but these are the only source which may be readily controlled. More attention has been paid to the removal of phosphorus than to the removal of nitrogen from waste waters. The reason for this emphasis is twofold. First, the phosphate is far more easily removed by straightforward chemical tech-

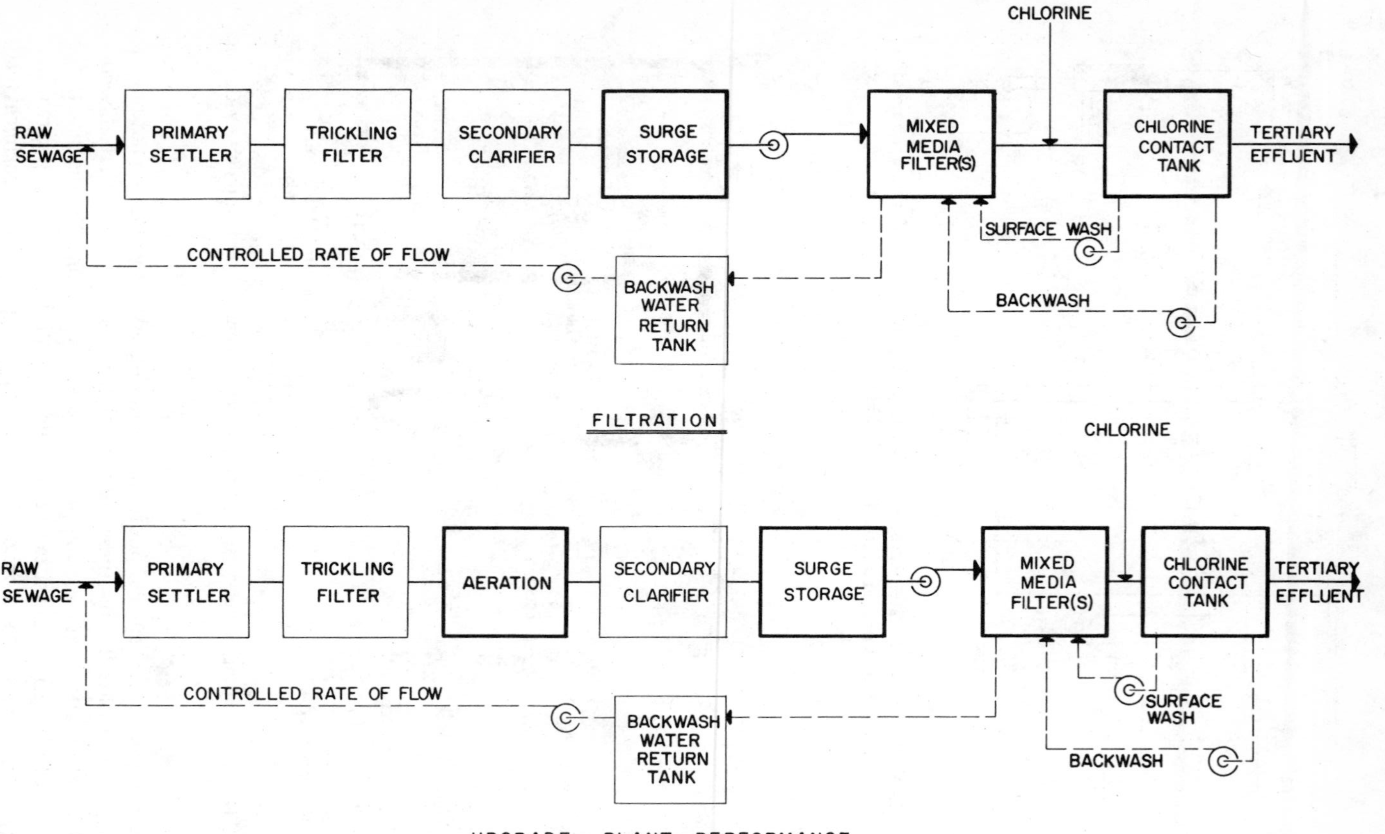

Fig. C35. Schematic of advanced waste treatment for trickling filter plants (courtesy Neptune Microfloc, Inc.).

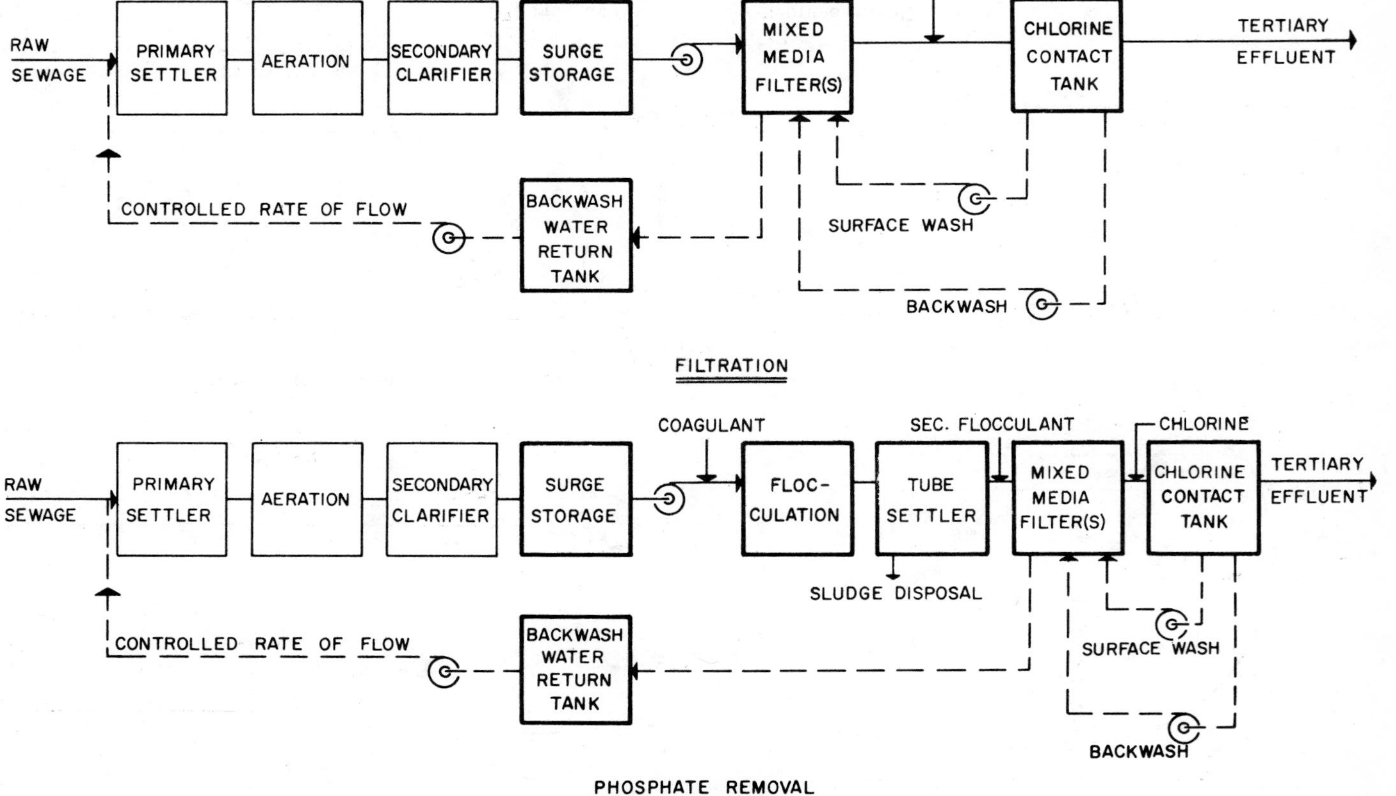

Fig. C36. Schematic of advanced waste treatment for activated sludge plants (courtesy Neptune Microfloc, Inc.).

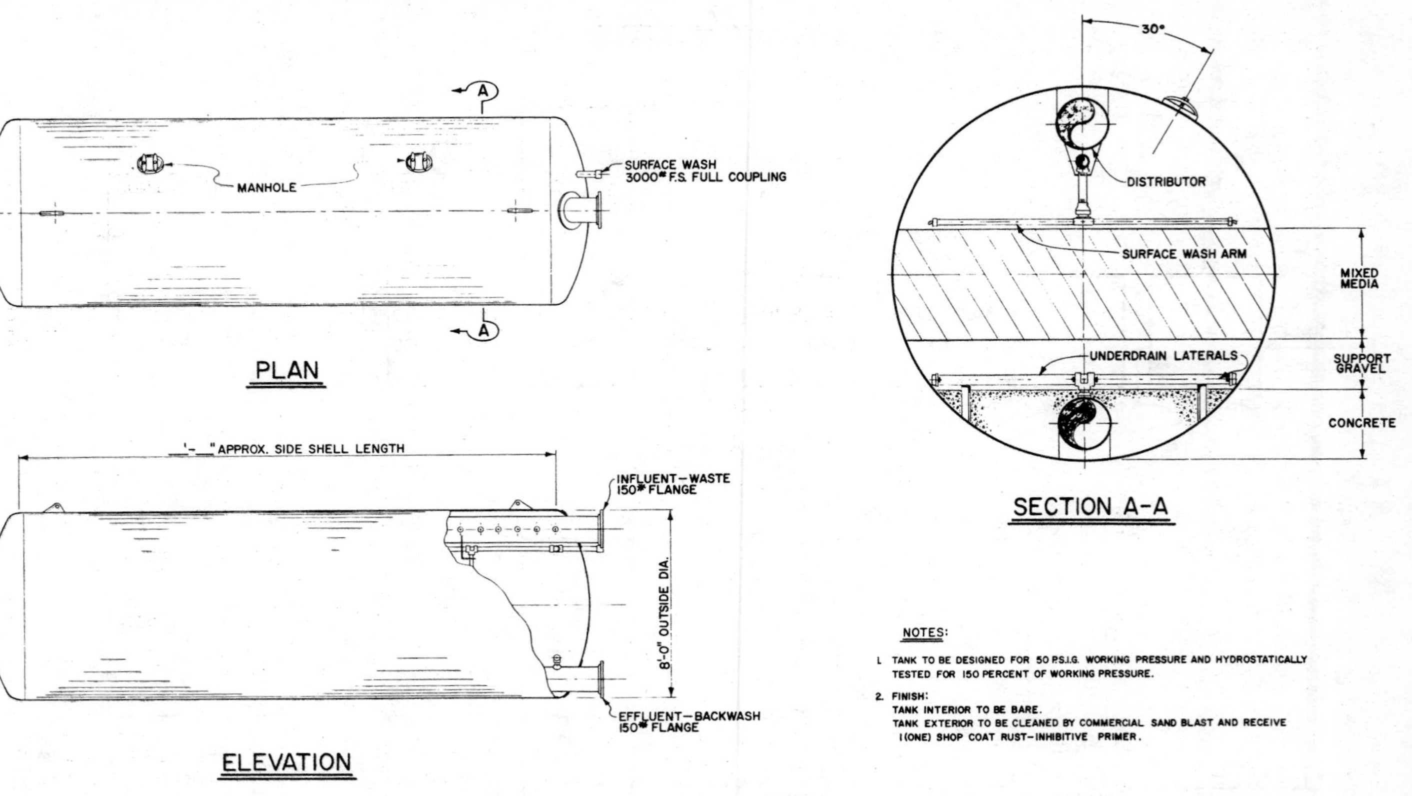

Fig. C37. Pressure filter (courtesy Neptune Microfloc, Inc.).

niques, and, second, total removal of phosphate will prevent plant growth while total removal of nitrogen will not, since some algae are able to fix nitrogen from the atmosphere. It now appears that the nitrogen may in fact be at least as important as the phosphate and it is likely that further attention will be paid to its removal. It should be noted that in the final analysis any material may be removed from water if it is sufficiently important to do so. The techniques mentioned here are methods which attempt to strike a balance between cost and efficiency.

Phosphorus may be removed by both biological and chemical means. Activated sludge and other aerobic waste treatment systems will complex phosphorus within the cell mass, primarily in the form of adenosine triphosphate. Since normal domestic sewage contains a surplus of phosphorus the removal by this means will usually not be sufficient. Phosphate removals will range from 1 to 2 mg/l per 100 mg/l of BOD removed. It must be noted that if the sludge is later digested or subjected to wet oxidation the phosphorus will be released and no net removal will be effected. To achieve phosphorus removal the sludge must be filtered and land-filled or incinerated.

Chemical precipitation of phosphorus is moderately effective, but near total removals require very heavy chemical dosages. Alum and ferric salts will precipitate phosphate as will lime. Typical dosages required are presented in Table C17. The required dosages may be expected to vary for different wastes.

TABLE C17. CHEMICAL DOSAGES FOR PHOSPHATE REMOVAL.

Chemical	Dosage for 80% PO_4 removal (mg/l)
$Al_2(SO_4)_3 \cdot 18H_2O$	300
$FeCl_3$	100
$Ca(OH)_2$	150

A combined chemical-biological process has been used for phosphate removal. The first stage includes dosage with lime to reduce the phosphate concentration to a level which can be nearly totally removed by biological uptake in the second biological treatment stage. Studies are also being conducted on dosage of alum in activated sludge tanks. This dosage not only reduces the phosphate concentration but aids in sedimentation of the biological floc. It is to be expected that further development along these and other lines will lead to improved phosphate removal techniques.

Nitrogen removal is possible by at least three separate methods. In any aerobic system nitrogen will be incorporated in the biological cell mass in a ratio of approximately 5 mg/l nitrogen per 100 mg/l BOD removed. As with phosphate, the nitrogen will be released if the sludge is digested or subjected to wet oxidation.

Algae have been cultured in secondary treatment plant effluents with resultant removal of both nitrogen and phosphorus. The process requires energy inputs both for mixing and light during the night. The system is similar to activated sludge in overall appearance, with a mixed illuminated chamber, a sedimentation basin for solids removal, and a solids recycling system. The solids produced, like those in the processes above, cannot be digested without release of the nutrients, and therefore must be vacuum filtered and landfilled or incinerated.

When sewage is subjected to long-term aeration at high solids retention times as in aerobic digestion or extended aeration, significant populations of *Nitrosobacter* and *Nitrobacter* will be developed. These microorganisms oxidize ammonia to nitrite and nitrite to nitrate, respectively. Thus all the nitrogen present which is not incorporated in cell mass will be oxidized to nitrate. If the system is then permitted to become anaerobic the nitrate will be reduced to nitrogen gas which will then be released to the atmosphere, thus being removed from the system. Such biological denitrification has been incorporated in some waste treatment systems. The biological sludges may be handled in any fashion since any nitrogen released in digestion will be removed by the nitrification–dentrification process.

A number of levels of advanced treatment for domestic wastes are outlined in Table C18.

TABLE C18. AVAILABLE LEVELS OF ADVANCED TREATMENT FOR DOMESTIC WASTES (COURTESY NEPTUNE MICROFLOC, INC.).

SECONDARY TREATMENT	ADDITIONAL TREATMENT	TYPICAL EFFLUENT QUALITY						
		S.S. MG/L	BOD MG/L	COD MG/L	TURBIDITY MG/L	COLOR UNITS	PO_4 MG/L	TOT MG
ACTIVATED SLUDGE	MIXED-MEDIA FILTRATION	< 5-10	<5-10	40–70	0.3–5	15–60	20–40	20-
	MIXED-MEDIA FILTRATION, CARBON COLUMN	< 3	< 1	5–15	0.3–3	5	20–40	15-
	COAGULATION PLUS SETTLING	< 5	<5-10	40–70	< 10	10–30	1–2	15-
	COAGULATION PLUS SETTLING AND MIXED-MEDIA FILTRATION	< 1	< 5	30–60	0.1–1.0	10–30	0.1–1.0*	15-
	COAGULATION, SETTLING, FILTRATION, AMMONIA STRIPPING	< 1	< 5	30–60	0.1–1.0	10–30	0.1–1.0*	1-
	COAGULATION, SETTLING, FILTRATION, AMMONIA STRIPPING, CARBON COLUMNS	< 1	< 1	1–15	0.1–1.0	< 5	0.1–1.0*	1
TRICKLING FILTER	MIXED-MEDIA FILTRATION	15–25	15–25	30–60	< 10	15–60	20–40	20-
	AERATION, SETTLING, MIXED-MEDIA FILTRATION	< 5-10	<5-10	30–60	0.5–5	15–60	20–40	20-

* REDUCTION OF PO_4 TO THIS LEVEL REQUIREDS 200 PPM OF ALUM OR 400 PPM OF LIME; IF GREATER PO_4 CONCENTRATIONS CAN BE TOLERATED, COAGULANT DOSAGE IS DECREASED.

** REQUIRES ELEVATING pH TO OVER 10.5 TO CONVERT NITROGEN TO AMMONIA.

Miscellaneous Waste Treatment Techniques

A number of waste treatment methods are occasionally used. These include chlorination of sewage and sewage plant effluents and relatively inefficient biological treatment techniques used primarily in rural areas where the first cost and operating costs of modern treatment are prohibitive and the environment is able to absorb the poorly treated wastes without ill effect.

Chlorination may be employed for disinfection, to prevent odor production, and to assist in BOD reduction. Disinfection of treated sewage is practiced where there is danger of contamination of drinking waters, bathing areas, or shellfish breeding areas. The chlorine dosage required will be dependent on the concentration of organic material, hence well treated sewage will require lower dosages than poorly treated effluents. Additionally, well treated sewage will be relatively low in pathogenic microorganisms prior to disinfection. Typical chlorine dosages which have been used are presented in Table C19.

TABLE C19. CHLORINATION OF SEWAGE EFFLUENTS.

Treatment	Chlorine dosage (mg/l)
Primary only	20
Trickling filters	15
Activated sludge	8

Since chlorine is a strong oxidizing agent, its introduction into polluted water will oxidize some of the organic material present with a resultant decrease in BOD. The BOD should be reduced by approximately 1 mg/l for each 2 mg/l of chlorine which reacts with the organic material. Such dosages should not be considered as an acceptable substitute for proper biological treatment but may aid in operation when a plant is temporarily overloaded.

Chlorination is also useful in prevention of odors and corrosion in long sewer lines in which the sewage is retained for a sufficient length of time for significant anaerobic degradation to occur. Chlorination will reduce the population of the normal sewage bacteria and thus prevent the decay of the organic material. Repeated dosages may be required, particularly in warm weather.

Rural waste treatment techniques have in the past included privies of various designs which are really retention rather than treatment devices. It is unlikely that such primitive conveniences will survive much longer except on private property. Most rural homes in the United States now have running water and flush toilets, so the wastes produced in such homes will be similar in quantity and quality to those found in larger communities. The BOD and suspended solids output per capita will be more or less invariant but the quantity of carriage water may be expected to vary. The actual sewage flow may be 50 gal per capita per day or even less. The higher flows in larger communities are due to industrial contributions and infiltration into the lengthy sewer lines. The wastes from individual dwellings or small communities will thus be somewhat less in volume but greater in strength than those usually encountered. The fluctuations in sewage flow will also be more strongly pronounced.

The most popular method of rural waste disposal couples the septic tank and subsurface disposal field, although this system is not applicable in all soil conditions. The septic tank is perhaps the most primitive biological system in current use and may in fact provide little or no treatment. Essentially a septic tank attempts to combine sedimentation and grease removal with anaerobic digestion of the separated solids. Multi-function waste treatment systems are seldom successful and the septic tank is no exception to this rule. A schematic diagram of a septic tank is shown in Fig. C38. The liquid retention time in the tank may be more than 24 hours. The settled sludge will undergo anaerobic digestion to an extent, but the extent is unpredictable. The sludge is neither heated nor mixed nor is any control of pH etc. possible, hence the progress of the digestion is uncertain. Additives which are advertised as beneficial to septic tank digestion are currently being marketed. These additives are purportedly enzymes which will speed liquification of solids. No lasting benefit has been observed in some laboratory studies of these materials and their usefulness is open to question. Even if they worked as advertised the gain would be marginal since no major sludge volume reduction occurs even in well operated septic tanks. Such additives will clearly be useless in septic tanks which have already soured. It appears more reasonable to consider septic tanks as simple sedimentation tanks with a large storage capacity and not to rely upon any significant volume reduction through biological action. Such biological activity as does occur is actually detrimental to the operation of the tank since the gas produced will tend to buoy up and resuspend those solids which have settled. These resuspended solids may then be carried out in the effluent and hinder further treatment.

Septic tanks are commonly precast in sections and then assembled by mortar-

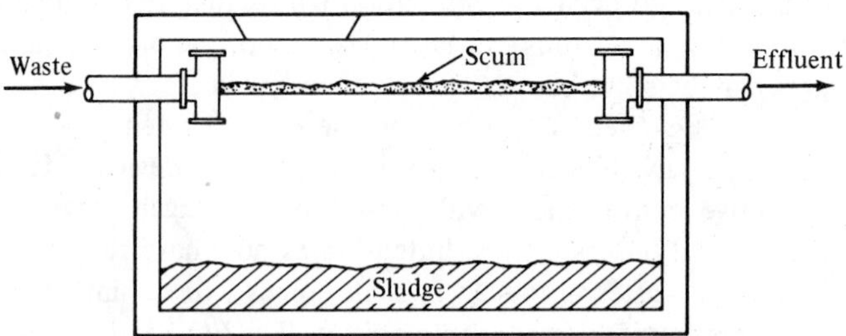

Fig. C38. Septic tank schematic.

ing the joints at the construction site. These tanks may include baffles at the entrance and exit, but at the low hydraulic loadings used tee sections with the lower ends submerged to contain the floating materials are adequate. It is important that the gases which may be produced be vented. The tee section also serves this purpose since the gases may pass through the upper end, then through the sewer, and to the house sewer vent.

Septic tanks should be readily accessible since the accumulated sludge must be pumped out at regular intervals. Intervals between pumping commonly range from 5 to 10 years with the shorter period being advisable. If the tank fills with sludge large quantities of solids will be carried out into the disposal field, plugging the tile and making it necessary to excavate the disposal field. The sludge which is removed may be buried in trenches or discharged at a sewage treatment plant where it can be properly handled. Typical design requirements for septic tanks are presented in Tables C20 and C21. The design flows from Table C21 may be used to determine the required tank volume based upon the following formulas:

$$V = 1.5Q \tag{11}$$

or

$$V = 1125 + 0.75Q \tag{12}$$

TABLE C20. SEPTIC TANK VOLUME FOR HOUSEHOLDS (INCLUDING MAJOR HOUSEHOLD APPLIANCES).

Number of bedrooms	Minimum tank volume (Gal)
2 or less	750
3	900
4 or more	250 per bedroom

TABLE C21. DESIGN FLOW FOR SEPTIC TANKS.

Source of waste	Gallons per capita per day
Single family dwelling	75
Single family dwelling (seasonally occupied)	50
Apartments	60
Hotels	60
Mobile home parks	50
Hospitals	250
Factories	15
Luxury residences	100
Motels	40

in which V is the required tank volume in gallons and Q is the daily design flow in gallons per day. Equation (11) is to be used for flows less than 1500 gal/day, and (12) for flows over 1500 gal/day. In no case should the final volume be less than 750 gal. Tables C20 and C21 yield equivalent results in most cases. For example, a 4-bedroom residence from Table C20 would require a septic tank volume of 1000 gal. From Table C21, assuming 8 persons in the household, the design flow would be 600 gal/day, and the tank volume, from Eq. (11), would be 900 gal.

The effectiveness of subsurface disposal fields is dependent upon the characteristics of soil, particularly its permeability to water. Percolation tests are always required prior to the design of a drainage field. The test procedure may be outlined as follows:

1. A hole 4 in. or more in diameter is dug or bored to the depth of the disposal field.
2. The sides of the hole are scratched and all loose earth is removed.
3. About 2 in. of fine gravel or coarse sand is placed in the bottom of the hole which is then filled with water to a depth of 12 in. above the gravel.
4. The hole is refilled as necessary to keep water in the hole for at least 4 h and preferably overnight.
5. If the water remains in the hole overnight the level is adjusted to 6 in. over the sand and the fall in water level in 30 min is measured.
6. If the hole is empty it is refilled to a depth of 6 in. over the sand and the fall in water level is recorded at 30-min intervals for 4 h, with the hole being refilled as necessary. The fall during the final 30-min period is used to calculate the percolation rate.
7. In sandy soils in which the hole drains completely in less than 30 min after the overnight period, the hole is filled to a depth of 6 in. over the sand and the fall in water level is recorded at 10-min intervals for 1 h, with the hole being refilled as necessary. The fall during the final 10-min period is used to calculate the percolation rate.

From the percolation rate (the time in minutes required for the water to fall 1 in.) the required adsorption area may be determined from Table C22. In no case should a field be designed for less than two bedrooms. The areas given are trench areas, not total field area. Trenches are usually 18–24 in. in width and have a minimum depth of 18 in. Lines should be separated by at least 6 ft. The trenches are filled to a depth of 6 in. with gravel upon which the distribution lines are laid at a grade of 2–4 in. per 100 ft. The lines are then covered with at least 2 in. of gravel after which the trenches are backfilled. The distribution lines are constructed of agricultural drain tile, unjointed vitrified clay sewer pipe, or perforated fiber or plastic pipe. Where several lines are to be used a distribution box will be necessary to insure equal distribution of flow. Such boxes are

TABLE C22. ADSORPTION FIELD AREA.

Percolation Rate (min/in.)	Maximum rate of application (gal/ft^2/day)	Required trench area (ft^2/bedroom)
≤1	5.0	70
2	3.5	85
3	2.9	100
4	2.5	115
5	2.2	125
10	1.6	165
15	1.3	190
30	0.9	250
45	0.8	300
60	0.6	330

available in precast concrete with "knock-out" openings for as many lines as may be required. These openings are somewhat larger than the 4-in. pipe line normally used and equal invert elevations may be obtained by mortaring the junctions, letting the mortar set up briefly, then partially filling the box with water and gently striking the outlet pipes to the same elevation.

In those areas in which the percolation rate exceeds 60 min/in. disposal fields are not practical. Consideration in such circumstances should be given to installation of an underground sand filter. Such filters contain, in addition to the distribution tile and gravel listed above, a 30-in. layer of coarse filter sand on an underdrain system of drain tile or perforated pipe in 8 in. of gravel. Distribution tile and underdrains are usually placed 6 ft on centers and the entire filter area is considered to be effective. Dosing rates for such filters are about 1 gal/day/ft^2. The effluent collected in the underdrains must be discharged either to a ditch leading to a stream or into a natural watercourse.

Both subsurface disposal fields and underground filters depend upon bacterial degradation for their effectiveness. As the liquid percolates through the soil or sand the soil bacteria will oxidize the pollutants in much the same fashion as in trickling filters. The low dosage rate prevents any appreciable build-up of bacterial solids which might plug the soil. The process is dependent on the slow transfer of air through the upper layers of soil, hence care must be taken so that this transfer may not be impeded. The area required for the disposal field should be devoted to that purpose alone. A grass cover will help to prevent build-up of inorganic nutrients but any use which might compact the upper level of the soil should be avoided.

Storm Sewage

Storm runoff from urban areas may constitute a large volume of flow and this flow is usually concentrated during a relatively brief period. The strength of such wastes is highly variable, ranging from more than that of domestic sewage to little more than that of normal surface waters. Normally the first runoff collected during a storm will be far higher in organic strength than later flows.

The rational method of estimating storm runoff is far more widely used than the empirical formulas which were developed in the past. The rational method may be expressed by the equation

$$Q = ciA \tag{13}$$

in which Q is the runoff in cubic feet per second, A is the area in acres, c is the runoff coefficient, and i is the rainfall intensity in inches per hour. The units are not commensurate; however, inches per hour is very nearly equal to cubic feet per second per acre (1 in./h = 1.008 ft^3/sec/acre), hence no constant is used. All the factors in question may be readily determined, thus the method is simple to apply.

The runoff coefficient C accounts for water held in soil or other porous materials, water lost due to evaporation, and water held in depressions. Commonly used coefficients of runoff are shown in Table C23.

TABLE C23. TYPICAL RUNOFF COEFFICIENTS.

Type of surface	Runoff coefficient, C
Asphalt streets	0.85–0.95
Concrete streets	0.85–0.95
Paved driveways, walks	0.75–0.85
Gravel driveways, walks	0.15–0.30
Watertight roofs	0.70–0.95
Lawns	
flat	0.05–0.17
average	0.10–0.22
steep	0.15–0.35

The runoff coefficient for a given area may then be determined by estimating the proportion of different surfaces present. Average figures sometimes used for areas of fixed character are presented in Table C24. Runoff coefficients will tend to vary during a storm event because of saturation of the soil and filling of depressions. Adjustments may be made for such increase in runoff with time in areas where sufficient data has been accumulated.

In selecting the rainfall intensity for runoff calculations a number of factors must be considered. First, the shorter the duration of a rainfall the higher the intensity will be. The critical duration for any area will be that which produces the maximum runoff. This will be the shortest time in which rainfall from the entire area contributes to the runoff, since longer periods will have a lower average intensity. This shortest period is called the time of concentration and is

TABLE C24. AVERAGED RUNOFF COEFFICIENTS.

Character of area	Runoff coefficient, C
High density	0.70–0.90
Medium density	0.50–0.70
Residential	0.25–0.50
Surburban	0.10–0.25

equal to the time required for a drop of water to travel from the most distant point in the watershed to the point for which the runoff is being calculated. For storm sewer design the time of concentration for the most remote sewer inlet is normally taken to be 5–10 min. As sewer lines increase in length this time becomes less important and the time of flow in the sewer governs the time of concentration.

Intensity curves which relate the intensity to be expected to the storm duration and to the frequency with which such intensities will occur have been constructed from rainfall data for many areas. A typical curve of this type is shown in Fig. C39. The equations of such curves may also be written for use in direct

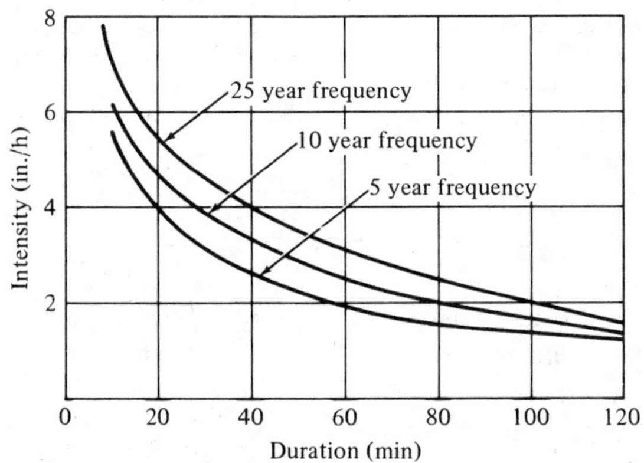

Fig. C39. Intensity–duration curves.

calculation of intensity. The equations usually have the general form

$$i = \frac{A}{t + B} \tag{14}$$

in which i is the maximum rate of precipitation in inches per hour during a time of t minutes, and A and B are constants. Specific formulas may be available for individual areas (Table C25) or may be derived from local data. General equa-

TABLE C25. RAINFALL INTENSITY FORMULAE.

Area	25 Years	10 Years	5 Years
Gulf Coast–Florida; Southern Alabama, Southern Mississippi, Louisiana, Southeast Texas.	$\frac{325}{t+35}$	$\frac{275}{t+32}$	$\frac{250}{t+30}$
Atlantic Coast, Rhode Island to Georgia; Northern Alabama; Mississippi; Arkansas; Missouri; Oklahoma; Southeastern Nebraska; Eastern Kansas.	$\frac{250}{t+28}$	$\frac{215}{t+26}$	$\frac{190}{t+25}$
Central New England, West to Eastern Nebraska and Eastern North and South Dakota.	$\frac{180}{t+20}$	$\frac{150}{t+19}$	$\frac{130}{t+19}$
Northern New England and New York and Areas Adjacent to the Great Lakes	$\frac{160}{t+20}$	$\frac{130}{t+19}$	$\frac{108}{t+18}$
Eastern Montana and Wyoming, Western Kansas, Nebraska, and North and South Dakota, Eastern Colorado.	$\frac{125}{t+14}$	$\frac{105}{t+13}$	$\frac{90}{t+13}$
Far Western States.	$\frac{155}{t+25}$	$\frac{125}{t+25}$	$\frac{80}{t+10}$

tions which have been used are

$$i = \frac{105}{t + 15} \tag{15}$$

for average storms, and

$$i = \frac{360}{t + 30} \tag{16}$$

for extreme storms. These equations are not applicable to all areas of the United States but may be useful in those cases where better information is not available.

In designing storms sewers the time of concentration for each line other than the first is dependent on the flow time in the preceding lines. The design is therefore sequential with the time of concentration for a given line being equal to the longest flow time in the lines which are its tributaries. This flow time cannot be determined until the preceding lines have been designed and their size and slope established.

Various general equations for rainfall intensity have been suggested for different areas of the United States. These equations should be used with caution since many variations may be expected within such large sections of the country. Local rainfall data should always be used when available.

The design of storm sewer systems may be outlined as follows:

1. Prepare a contour map of the drainage area showing streets, drainage limits, and direction of surface flow.

2. Divide the area into the subareas tributary to the proposed sewer inlets. These inlets should be located
   a. at reversals of road grade from negative to positive,
   b. at street intersections, and
   c. at intermediate points along long streets sufficient to prevent overflow of gutters during the design storm.
3. Determine the acreage and imperviousness of each area.
4. Calculate the required capacity of each inlet using the appropriate time of concentration, the tributary area, and the rational method.
5. Beginning at the highest elevation compute the flow to be carried by each line. The time of concentration for each line other than the first in a series is the sum of the time of concentration to the inlet next upstream and the flow time in the connecting pipe. Where more than two lines meet, the time of concentration to be used for the succeeding line is the longest time in the lines meeting. Each line will thus require calculation of time of concentration, tributary area (all upstream areas), and flow.
6. Select tentative pipe sizes and grades using the nomograph presented previously. Each line must be selected in order since the time of concentration for subsequent lines will be dependent upon the time of flow in all upstream lines. In general storm sewers are held as close to surface grade as possible with a minimum cover of about two feet.

Treatment and Disposal

Storm sewage, although it consists primarily of rainwater, will also contain significant quantities of suspended solids and organic material washed from yards and streets. The strength of storm sewage, particularly at the beginning of runoff, may exceed that of ordinary domestic sewage. It should be observed that similar material will be carried into streams from rural runoff and it would appear unfair to require communities to treat storm runoff simply because they are in a position to do so after having channeled the flow into sewers.

It is also significant that at the time of such runoff the stream flow will frequently, although not always, be high. Dilution of municipal runoff may thus be afforded by the stream, although the quality of the stream will usually be adversely affected by uncontrolled rural runoff. The situation may thus be one of relying upon an otherwise polluted stream to dilute a moderately strong waste, with neither having any significant effect upon the other.

Normally storm sewage is discharged directly into the receiving stream; however, where it is deemed necessary to protect water quality, storm sewage may be treated by the same methods used to treat domestic wastes. Since the flow will be highly erratic and instantaneous values very high, some sort of flow equalization will be necessary if treatment is intended. Also, in designing a storm sewage treatment facility it should be recognized that the initial flow will be far higher in suspended solids and BOD than later flows and hence might be handled differently. It appears reasonable therefore to design such a facility to retain and treat the first flow, perhaps the first $\frac{1}{4}$ in. of runoff from the sewered area, and to bypass the remainder of the flow. Even this would require a considerable storage volume (about half a million cubic feet per square mile of sewered area) and a major expense to the community.

The only treatment which can be reasonably applied to storm sewage is primary in nature. Grit chambers and primary sedimentation basins can be operated successfully on such intermittent flows but biological systems cannot. Some communities which have large natural storage capacity available have stored storm sewage and then pumped it to the sewage treatment plant during periods of low domestic flow. While this may be desirable, it appears unrealistic to require it of all communities.

Industrial Wastes

The wastes discharged by industrial operations may range from relatively dilute flows similar to domestic sewage to highly polluted waters which are amenable to normal treatment techniques and chemical discharges which not only fail to respond to usual techniques but may interfere with their proper operation. The flow from industries is in general dependent on the process used and may frequently be reduced by slight procedural variations and by establishment of better housekeeping techniques.

Waste Characteristics The constituents of industrial wastes will vary widely and should be determined individually for each treatment system. Approximate values which have been obtained in the past for particular wastes are presented in Table C26. These may be helpful as a basis for initial investigations but should not be considered as design criteria since wide variations may occur. Table C27 presents general characteristics for specific industrial wastes and treatment processes which have been used in their treatment. The list of treatment techniques is not exclusive, and other processes may be useful.

Physical Treatment The physical treatment processes used in the treatment of domestic sewage may also be applied to industrial wastes. In certain cases screens with a greater selectivity than those used for sewage may be used for separation of different classes of suspended solids. For example, foundries may recover casting sand from the mixed clay and sand in the waste stream.

Chemical Treatment The techniques of precipitation, adsorption, and coagulation used in sewage treatment are also applicable to industrial wastes. Many other chemical techniques may also be required for treatment of industrial wastes which are either toxic to, or not benefited by, biological treatment.

TABLE C26. APPROXIMATE STRENGTHS OF INDUSTRIAL WASTES.

Source	BOD
Food Processing	
breweries	400–2000
canneries	100–3000
dairies	1000–2000
slaughterhouse	
cattle	1000–2000
poultry	3500–4000
soft drinks	500–700
Textiles	500–8000
Paper	200–1000

TABLE C27. INDUSTRIAL WASTES AND TREATMENT PROCESSES.

Source	Characteristics	Treatment
Food industries		
breweries	high nitrogen and carbohydrates	biological treatment, recovery, animal feed
canneries	high solids and BOD	screens, lagoons, irrigation
dairies	high fat, protein, and carbohydrate	biological treatment
fish	high BOD, odor	evaporation, burial, animal feed
meat processing	high protein and fat (may be warm)	screens, sedimentation, flotation, biological treatment
pickles	high BOD, high or low pH, high solids	screening, flow equalization, biological treatment
soft drinks	high BOD and solids	screening, biological treatment
sugar	high carbohydrate	lagoons, biological treatment
Chemical industries		
acids	low pH, low organics	neutralization
detergents	high BOD and phosphate	floatation, precipitation
explosives	high organics and nitrogen, trinitrotoluene	chlorination or TNT precipitation
insecticides	high organics, toxic to biological systems	dilution, adsorption, chlorination at high pH.
Materials industries		
foundries	high solids (sand, clay, and coal)	screens, drying
oil	dissolved solids, high BOD, odor, phenols	injection, recovery of oils
paper	variable pH, high solids	sedimentation, biological treatment
plating	acid, heavy metals, cyanide toxic to biological systems	oxidation, reduction, precipitation, neutralization
rubber	high BOD and solids, odor	biological treatment
steel	low pH, phenols, high suspended solids	neutralization, coagulation
textiles	high pH, and BOD, high solids	neutralization, precipitation, biological treatment

Oxidation-reduction reactions are particularly useful in the treatment of industrial wastes containing toxic materials. Typical reactions include:

1. Hexavalent chromium reduction and precipitation:

$$H_2Cr_2O_7 + 6FeSO_4 + 6H_2SO_4 \longrightarrow Cr_2(SO_4)_3 + 3Fe_2(SO_4)_3 + 7H_2O \quad (17)$$

$$Cr_2(SO_4)_3 + 3Ca(OH)_2 \longrightarrow 2Cr(OH)_3 + 3CaSO_4 \quad (18)$$

2. Cyanide oxidation:

$$2NaCN + 5Cl_2 + 12NaOH \longrightarrow N_2 + 2Na_2CO_3 + 10NaCl + 6H_2O \quad (19)$$

In these reactions addition of the chemical quantities indicated by the stoichiometric equations may be insufficient for production of the desired result. For example, the equation

$$H_2Cr_2O_7 + 6FeSO_4 + 6H_2SO_4 \longrightarrow Cr_2(SO_4)_3 + 3Fe_2(SO_4)_3 + 7H_2O$$

indicates that 1 mg/l of chromium will require about 8.8 mg/l ferrous sulfate plus 5.6 mg/l sulfuric acid. The actual dosages required will usually be 15-16 mg/l ferrous sulfate and 6 mg/l sulfuric acid per mg/l of chromium reduced. Similarly, the theoretical lime dosage is about 2 mg/l per mg/l of tetravalent chromium while the practical dosage is 9-10 mg/l.

For the cyanide oxidation, practical dosages will usually be 6 mg/l of chlorine and 5-6 mg/l sodium hydroxide per mg/l of cyanide oxidized. The time required for this reaction to go to completion may approach 24 h.

Biological Treatment

The biological systems used in the treatment of ordinary sewage are applicable to many industrial wastes–directly in some cases. For other wastes, pretreatment or modification of the system may be necessary.

Unusually strong wastes cannot be treated at the hydraulic loading rates given for domestic sewage. It is not practically possible either to maintain the required mixed liquor suspended solids or to transfer the necessary oxygen. Design rates in such cases should be based upon organic loadings (pounds BOD per unit volume) equivalent to those used in the standard processes.

Wastes which are discharged at high temperatures are difficult to treat aerobically because of the reduced solubility of oxygen, but may be ideally suited to anaerobic treatment. As noted above, the anaerobic process functions best at a temperature of about 35°C and hence should always be considered in selecting a system for treating warm wastes. Slaughterhouse and other food processing wastes have been treated successfully by such processes.

Many industrial wastes may be deficient in trace nutrients required for proper biological treatment. Nitrogen deficiencies are a particularly common source of difficulty with predominance of filamentous organisms and a bulking sludge being the normal result. The major trace nutrients known to be required are tabulated in the section describing the activated sludge process. Before designing a treatment system for a waste of unknown composition a small unit should be operated to ensure the presence of the required nutrients. Operation of such a system is frequently simpler than testing directly for the various ions. If treatment is ineffective small amounts of different salts may be added and the response of the system noted. Normally nitrogen and phosphorus are tested for, since these are of prime importance and the determinations are relatively simple.

Some organic wastes may be difficult to treat by biological means. Where such wastes are encountered it may be necessary to produce an acclimated microbial seed before the process will be effective. Since microorganisms reproduce rapidly, any bacterium which can metabolize the waste or any form which undergoes a favorable mutation will tend to predominate.

A method which has been used to generate an acclimated biological mass may be outlined as follows:

1. The unit is started on an easily metabolized waste such as domestic sewage and operated until the suspended solids concentration reaches the normal operating range (2000–2500 mg/1).
2. The feed is then gradually switched from the easily degradable waste to the difficult material, usually at a rate of about 10%/day.
3. The system is monitored daily with observation of treatment efficiency. If necessary the process is repeated at a slower changeover rate.

Flow segregation is of importance in both the treatment and disposal of industrial wastes. Cooling waters do not require biological treatment and may produce hydraulic overloads in such units leading to either poor treatment or high costs. Acid and alkaline wastes may be diluted sufficiently by other flows to permit their discharge to the sewers but it is frequently advisable to separate them since normal sewage treatment processes will provide little or no improvement.

In general, concentrated wastes of any kind may be more efficiently treated prior to dilution, and many communities prohibit the discharge of toxic or dangerous materials to the sewage system. Segregation of different flows is thus desirable.

Material recovery will not only reduce the strength of a waste stream, and thus the cost of its treatment, but may also produce savings in plant operational cost. Metals may frequently be recovered from plating operations, fibers from paper making, salts from pickling plants, etc. Such material recovery and process changes in order to minimize waste constitute an important part of the design of an industrial waste treatment system.

Joint Treatment

Many industrial wastes are amenable to the ordinary treatment techniques applied to domestic sewage and may thus be discharged to the sewers for treatment by the community. Whether or not such joint treatment is desirable will depend upon the industries in question, the capacity of the municipal treatment plant, and the assessment of costs to the community and the industry. Some cities may offer free waste treatment for a period of time in order to attract new industries. Such treatment is of course not free, but rather constitutes a subsidy to the new plant.

In the normal course of events industrial sewer charges will be based upon the concept of the *population equivalent*, i.e., the number of persons who might be expected to contribute a similar waste. The calculation of the population equivalent may be based either upon waste strength (BOD or suspended solids) or flow, depending upon the waste treatment process in question. In general, industrial wastes will have BODs in excess of that of ordinary sewage; thus charges for biological treatment will be based on BOD. If the waste is highly dilute the hydraulic load may become limiting, in which case charges will be based upon flow. The usual bases for calculation of population equivalents for the different components of a sewage system are presented in Table C28.

TABLE C28. POPULATION EQUIVALENT BASES.

System component	Equivalent basis
Sewers and appurtenances	Flow
Primary sedimentation	Flow
Secondary treatment	BOD
Solids digestion and handling	
primary treatment only	suspended solids
primary and secondary	suspended solids and BOD

The average per capita contributions of BOD and suspended solids are about 0.17 and 0.20 lb/day, respectively. The average flow (including infiltration but excluding all industrial flows) will be about 100 gal per capita per day. Population equivalents of industrial flows are frequently based upon these values.

EXAMPLE

A cheese factory in a community of 3000 persons produces a waste with a flow of 100,000 gal/day, suspended solids of 500 lb/day, and a BOD of 2000 lb/day. Determine the population equivalent of the waste and the portion of waste treatment costs which might be equitably assessed against the cheese factory.

The population equivalent will be:
Based on flow

$$100{,}000/100 = 1000 \text{ persons}$$

Based on BOD

$$2000/0.17 = 11{,}760 \text{ persons}$$

Based on SS

$$500/0.20 = 2500 \text{ persons}$$

The percentages of cost assessable to the factory are:
Primary sedimentation

$$1000/4000 = 25.0\%$$

Biological treatment

$$11{,}760/14{,}760 = 79.6\%$$

It may frequently be desirable or even necessary to pretreat industrial wastes prior to discharge into the municipal sewers. Significant reductions in BOD and suspended solids may be more readily effected when such wastes are concentrated than after they have been diluted with domestic sewage. Each industry must establish the most economical method of handling its own wastes.

There are certain industrial wastes which should not be discharged to municipal sewers in any event. Gasoline, kerosene, and solvents such as methyl ethyl ketone pose a definite fire and explosion hazard and should be excluded from the sewers. Wastes with a high concentration of suspended solids or solids with a high density or particle size will settle in the sewers and decrease their capacity. Toxic materials such as heavy metals, phenols, etc. may interfere with the proper operation of biological waste treatment systems and will not be removed by ordinary waste treatment techniques. Such wastes, even if they do not hinder treatment, should not be discharged to sewers since this is tantamount to discharge into the receiving body of water.

REFERENCES

Quantity of Sewage

American Society of Civil Engineers, Design and Construction of Sanitary and Storm Sewers, *ASCE Manual of Engineering Practice*, **37** (1960).

Clark, J. W., W. Viessman, Jr., and M. J. Hammer, *Water Supply and Pollution Control*, 2nd ed., International Textbook Co., Scranton, Pa. (1971).

Geyer, J. C. and J. J. Lentz, An Evaluation of The Problems of Sanitary Sewer System Design, *J. WPCF*, **38** (1966).

Sewer Design

American Society of Civil Engineers, Design and Construction of Sanitary and Storm Sewers, *ASCE Manual of Engineering Practice*, **37** (1960).

Camp, T. R., Design of Sewers to Facilitate Flow, *Sewage Works J.*, **18** (1946).

Clark, J. W., W. Viessman Jr., and M. J. Hammer, *Water Supply and Pollution Control*, 2nd ed., International Textbook Co., Scranton, Pa. (1971).

Woodward, S. M. and C. J. Posey, *Hydraulics of Steady Flow in Open Channels*, John Wiley & Sons Inc., New York (1955).

Sewage Treatment (General)

American Society of Civil Engineers, Sewage Treatment Plant Design, *ASCE Manual of Engineering Practice*, **36** (1959).

Great Lakes–Upper Mississippi River Board of State Sanitary Engineers, *Recommended Standards For Sewage Works*, Health Education Service, Albany, N.Y. (1968).

McKinney, Ross E., *Microbiology For Sanitary Engineers*, McGraw-Hill, New York (1962).

Rich, Lig., *Unit Operations of Sanitary Engineering*, John Wiley & Sons, Inc., New York (1961).

Sewage Treatment (Specific Processes)

Balakrishnan, S., et al., Organics Removal by a Selected Trickling Filter Media, *Water and Wastes Eng.*, **6** (1969).

Cotteral, J. A. and D. P. Norris, Septic Tank Systems, *J. San. Eng. Div. ASCE*, **95 SA4** (1969).

Culp, R. L., Water Reclamation at South Tahoe, *Water and Wastes Eng.*, **6** (1969).

Eye, J. D., et al., Field Evaluation of the Performance of Extended Aeration Plants, *J. WPCF*, **41** (1969).

Hansen, S. P., et al., Practical Application of Idealized Sedimentation Theory in Wastewater Treatment, *J. WPCF*, **41** (1969).

Thirumurthi, D., Design Principles of Waste Stabilization Ponds, *J. San. Eng. Div. ASCE*, **95 SA2** (1969).

Storm Sewage

American Society of Civil Engineers, Design and Construction of Sanitary and Storm Sewers, *ASCE Manual of Engineering Practice*, **37** (1960).

Clark, J. W., W. Viessman Jr., and M. J. Hammer, *Water Supply and Pollution Control*, 2nd ed., International Textbook Co., Scranton, Pa. (1971).

U.S. Dept. of the Interior, Water Pollution Aspects of Urban Runoff, FWPCA Publ WP-20-5 (1969).

Viessman, W., Assessing the Quality of Urban Drainage, *Public Works*, **100** (1969).

Industrial Wastes (General)

Eckenfelder, W. W., Jr., *Industrial Water Pollution Control*, McGraw-Hill, New York (1966).

Nemerow, N. L., *Theories and Practices of Industrial Waste Treatment*, Addison-Wesley, Reading, Mass. (1963).
Water Pollution Control Federation, Regulation of Sewer Use, *WPCF Manual of Practice*, **3** (1963).

Industrial Wastes (Specific Processes)

Burkhead, C. E., et al., Pollution Abatement of a Distillery Waste, *Water and Waste Eng.*, **6** (1969).
Eliassen, R. and D. F. Coburn, Versatility and Expandability of Pretreatment, *J. San. Eng. Div. ASCE*, **95 SA2** (1969).
Ludzack, F. J. and M. B. Ettinger, Chemical Structures Resistant to Aerobic Biochemical Stabilization, *J. WPCF*, **32** (1960).
Meriwether, G. B., Treatment of Mixed Industrial Wastes at Bayport's Industrial Complex, *J. WPCF*, **41** (1969).
Moore, A. W., et al., Effects of Chromium on The Activated Sludge Process, *J. WPCF*, **33** (1961).
Nemerow, N. L. and R. Armstrong, Combined Treatment of Tannery and Municipal Wastes, *Water and Wastes Eng.*, **6** (1969).

VD. SOLID WASTES

Solid wastes constitute an enormous quantity of material which must be effectively handled if the quality of the environment is to be protected. These wastes may be variously classified with respect to their nature as shown in Table D1.

It would be desirable if solid wastes were as easily segregated in practice as they are on paper since, from the table, it is evident that their characteristics vary widely. However, in most cases, these wastes, particularly garbage and rubbish, are handled in the same fashion with no attempt to separate the various constituents before collection. The other materials listed in the table are normally collected separately but may frequently be handled in the same fashion as normal domestic wastes.

Quantity of Solid Wastes

The quantity of solid waste to be expected will vary widely among communities and can in no way be accurately predicted without local data. The values presented in Table D2 cannot therefore be used with confidence in estimating waste quantities for any particular area. Actual quantities may differ by 100% or more from those given. Where possible, waste volume may be best determined by solid waste surveys which may be as complete or cursory as finances will permit. Approximate volumes may be obtained from the records

TABLE D1. CLASSIFICATION OF SOLID WASTES.

Waste	Nature	Sources
Garbage	food wastes	households, stores, hotels, restaurants
Combustible rubbish	paper, cardboard, wood, plastics, cloth, leather, rubber, grass, leaves	
Noncombustible rubbish	metals, dirt, stone, crockery, glass	
Ashes	inerts	
Bulky materials	automobiles, household appliances, furniture, trees	streets, alleys, vacant lots
Street debris	sweepings, leaves, catch basin holdings, litter	
Dead animals		
Construction and demolition wastes	lumber, broken concrete, brick and plaster, pipe, wire, etc.	construction sites
Industrial solid wastes	process wastes dependent on industry, cinders, sweepings, typical combustible rubbish	factories
Agricultural wastes	manure, silage	farms, feeding operations
Sewage solids	screenings, grit, sludge, ashes	sewage treatment plants
Water treatment solids	screenings, sediments, chemical precipitates	water treatment plants
Hazardous wastes	explosives, radioactives, toxic materials	various

TABLE D2. WASTE QUANTITIES.

Waste	Quantity
Garbage	250 lb/capita/year
Rubbish	100 lb/capita/year
Animals	15–50/1000 population/year
Sweepings	3 ft^3/capita/year

of the waste collectors; more precise information from studies of volume, composition, weight, etc. in small representative areas.

The constituents of domestic solid waste have changed considerably with the great increase in convenience packaging of food products. Differences in waste composition and quantity have also been found between areas inhabited by segments of the population at different economic levels. It is to be expected that waste characteristics will change in the future as well. Convenience packaging may increase the percentage and total volume of rubbish, but the current popular interest in pollution and environmental protection may actually cause a decrease in solid waste quantities. Such a development is clearly desirable as would be any diminution of the problem.

Storage

Solid wastes are normally stored for short periods of time at or near the point at which they are produced. This storage may be in metal or plastic cans, paper or plastic sacks, or mobile containers. In the selection of containers a number of factors must be considered. Some of the advantages and disadvantages of different systems are presented in Table D3.

TABLE D3. COMPARISON OF STORAGE SYSTEMS.

	System		
Factor	Paper or plastic bags	Metal or plastic cans	Mobile containers
Spillage	likely	possible	unlikely
Collection	no return necessary	return necessary	return necessary
Cleaning	not required	necessary	necessary
Adjustibility to variations in volume	excellent	poor	poor
Handling	simple	may be heavy	machinery required
Noise of collection	low	low–moderate	low–moderate
Animal protection	poor	fair–large animals good–small animals	excellent
Cost	high	low	low
Increases waste	yes	no	no

Home compactor units have recently been placed on the market and may, in time, become more widely used. These units appear to have primarily a curiosity value aside from their small value to the individual homeowner in reduction of required storage capacity. They do not reduce in any way the required capacity of refuse collection and disposal systems where compactor trucks are used.

Collection

Collection may be at curb side, in alleys, from the building with or without return of storage containers, or carriage from the building in separate containers. In general, those methods requiring the least labor by the collection crew will be least expensive. However, in communities with existing collection systems, changes to methods of collection which are less convenient to the public will not meet with ready acceptance. At either end of the range of possible services there will be curb-side pickup with minimum cost and minimum convenience to the public and carriage from the building either with return of containers or in separate containers with both greater cost and greater convenience.

Collection may also be effected by either pneumatic or water carriage systems. Such devices as the household garbage grinder are not disposal techniques. The waste material must still be separated from the carriage system and will require further treatment or handling.

Collection Equipment

Collection systems may include hand-loaded compactor trucks, self-loading fork-lift trucks for use with mobile containers, scooter systems which carry small loads to a compactor, or truck trains serviced by a compactor. Open trucks are not to be recommended for routine collection of garbage and rubbish; they are of use primarily in the collection of bulky items which will not fit into ordinary compactor trucks. Transfer stations are of value when the haul distance to the disposal site becomes excessive. At such stations the waste is transferred to trailers, being compacted either in the trailer or at the transfer site.

Solid wastes are also occasionally transported by rail or by barge. Normally, rail transport is used primarily in those locations where haul distances are unusually long. The current practice in some coastal cities of hauling garbage and rubbish in barges to off-shore sites has produced public nuisances in some areas and is, at the very least, esthetically displeasing. It is likely that such offshore dumping will be prohibited in the near future, hence no long-term solid waste disposal plan should rely upon the availability of such sites.

Volume Reduction and Disposal

Some decrease in waste volume is achieved by the use of compactor devices on the collection vehicles but far greater reductions may be attained by salvage operations, incineration, composting, or combinations of the three. All of these will yield a residue and hence cannot be considered as disposal techniques. Disposal will normally consist of return of the waste materials to the soil, most commonly via a landfill operation.

Salvage operations are intended both to reduce the volume of the waste and to yield a saleable product. In assessing the feasibility of a salvage system the engineer should consider not only the value of the salvaged material, but also the savings in handling costs occasioned by the reduction in volume. It may also be proper in some cases to consider the secondary benefits of decreased use of land and resources.

TABLE D4. SALVAGE TECHNIQUES.

Material	Salvage technique	Salvage use
Iron and steel*	magnetic separation	iron and steel manufacture
Aluminum	hand separation	aluminum manufacture
Glass*	hand separation	glass manufacture, road surfacing, container reuse

*Market may be limited.

Various salvage schemes have been and are being evaluated. The simplest for the collector requires the householder to separate the garbage and rubbish prior to collection but this seldom meets with public acceptance. Salvage techniques applicable to ordinary solid wastes are listed in Table D4.

Composting is a biological degradation technique applicable to garbage and paper wastes. The process requires rather careful control of moisture, pH, and temperature and intermittent mixing (Table D5). The end product is a humus-like material useful as a soil conditioner or gardening aid. The market for this material is rather limited, greenhouses and mushroom growers being the largest users. The advantages of composting are a small reduction in volume and the production of a product which may be saleable.

The disadvantages include the cost of the treatment, the large area of land required, the occasional production of unpleasant odors by biological action, the necessity of separating inorganic material from the waste prior to treatment, the impracticability of the process in cold climates, and the necessity for disposal of any unsaleable compost produced. In some cases a combination of salvage and composting may prove practical where a market for the end products exists.

TABLE D5. COMPOSTING VARIABLES.

Variable	Max.	Min.	Optimum
pH	9.5	4.5	6.5
Moisture	75%	40%	60%
Temperature	–	–	65°C

Incineration will provide a considerable reduction in solid waste volume—approaching 100% for some wastes such as sewage sludge. In general the residue upon incineration will consist of the inorganic material present, which may be partially fused, and ash. This material must be further handled. For best efficiency operation should be continuous, with a storage pit serving to equalize the intermittent delivery of wastes. The stack gases must be scrubbed or otherwise purified in order to avoid distributing the ash over the surrounding area.

The plume associated with a properly operating incinerator is primarily water vapor.

Several specialized incinerators have been adapted to solid waste handling. Normally these are more applicable to such wastes as sewage sludge than to heterogeneous materials such as domestic refuse. The heat produced by the combustion of the waste is normally sufficient to maintain the temperature without the use of auxiliary fuel.

The advantages of incineration include the small land area required, the possibility of central location, the independence of weather or climate, the reduction in waste volume, and the production of a stable residue. The major disadvantages are the high first cost and operating cost, the need for continuous maintenance and skilled operation, and the need for further handling of the residue.

The preferred method for solid waste disposal remains the sanitary landfill. Materials which may be landfilled successfully include the wastes as collected or the residue of any of the processes above. A sanitary landfill must incorporate the following features:

1. The landfill must not intrude upon ground water.
2. The fill must not interfere with proper drainage on adjacent land.
3. The fill area must be fenced to prevent blowing of paper and other light material.
4. No burning may be permitted. Care must be taken in filling incinerator ashes with unburned material.
5. All material delivered to the site must be covered on the same day it is received.

In addition, it is desirable that the filled material and cover material be compacted as thoroughly as possible. In this regard it is frequently necessary to remind landfill operators that a tracked vehicle is specifically designed to exert a low soil pressure and is thus one of the poorest compaction devices available. A sheepsfoot roller is a valuable addition to a landfill operation.

REFERENCES

Solid Wastes

Anderson, R. J., The Public Health Aspects of Solid Waste Disposal, *Public Health Reports*, **79** (1964).

Black, R. J. and P. L. Davis, *Refuse Collection and Disposal: An Annotated Bibliography,* U.S. Public Health Service Publ. 91 (Rev. 1966).

Clark, R. M. and R. O. Toftner, Financing Municipal Solid Waste Management Systems, *J. San. Eng. Div. ASCE*, **96 SA4** (1970).

Gunnerson, C. G., et al., Marine Disposal of Solid Wastes, *J. San. Eng. Div. ASCE*, **96 SA6** (1970).

Sorg, T. A. and H. L. Hickman, Jr., *Sanitary Landfill Facts*, 2nd ed., U.S. Public Health Service Publ. 1972 (1970).

Steiner, R. L. and R. Kantz, *Sanitary Landfill: A Bibliography*, U.S. Public Health Service Publ. 1819 (1968).

Wiley, J. S. and O. W. Kochtitzky, Composting Developments in The United States, *Compost Science*, **6** (1965).

SECTION VI

Transportation

Herman A. J. Kuhn Ph.D., *P.E., M. ASCE*
Associate Professor of Civil and Environmental Engineering
The University of Wisconsin
Madison, Wisconsin

Historically, transportation has been a shaper of man's destiny. It has been the catalyst which has served to develop cities, open and exploit virgin territories. Militarily, it has played an important role in man's conquest over man.

The role of transportation today is no less significant. Without it man would not have set foot on the moon, nor would he be able to undertake many of his normal day-to-day activities. Transportation has an important impact on every facet of man's life. It affects his economic, biological, social and psychological well being.

THE TECHNOLOGY–MAN–SOCIETY INTERFACE

Thomas and Schofer (1) in depicting a regional transportation system model (Fig. 1) note some of the activities which occur in transportation and the interactions between the transportation system and the environment in which the system operates. The transportation processer shown in Fig. 1 combines the human and physical entities and activities to produce movement. In it the physical system includes such things as vehicles, operating networks (streets, and other rights-of-way), structures, terminals, and control elements. The activity subsystem is related to the movement function itself and includes such things as driving or riding, loading and unloading vehicles, and those activities necessary to maintain and operate the system. The human subsystem includes drivers and their passengers, operators of vehicles, persons involved in controlling the movement activity, and the persons (not the activity) involved in constructing and maintaining the physical subsystem. As Fig. 1 notes, outputs of the transportation system are of two kinds: performance and concomitant. Performance outputs are those which relate directly to the purpose for which the system is designed and operated. It includes such things as movement or the potential for movement (accessibility). Concomitant outputs are in essence the by-products of the transportation system. They are necessary but unintentional results of the performance of the system and include noise, air pollution, visual discord

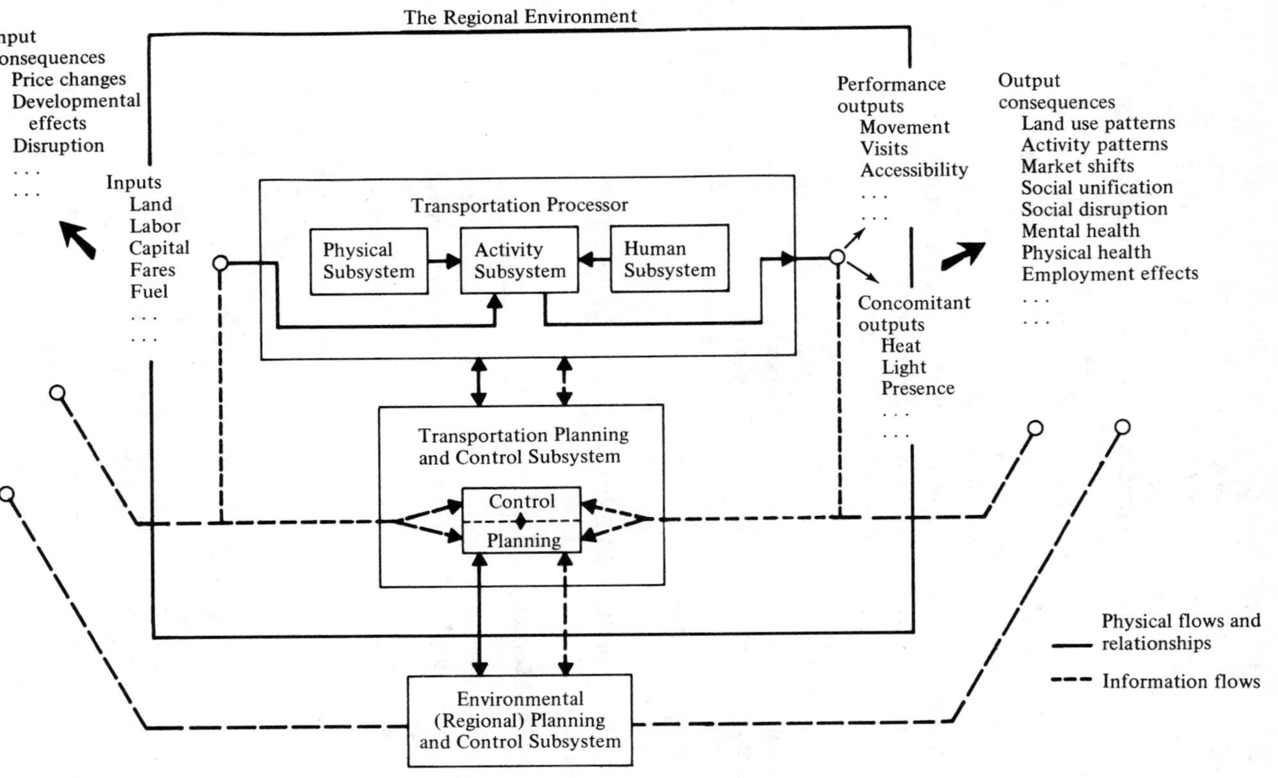

Fig. 1. Transportation system model.

(both of the system and from the system), safety by-products (deaths, injuries, and property damage) and the use of space. Both the performance and concomitant outputs have some impact on all facets of life and living. However, it is largely the concomitant outputs, e.g., noise and air pollution, social disruption, etc., which result in the current negative attitudes and disruptive actions directed toward proposed new transportation facilities.

In order for the transportation system to function and the transportation activity to be carried on, there must be an expenditure of resources, energies and efforts which occur in time and space. These are the inputs to the system, the use or depletion of which result in some of the input consequences shown.

The planning and control mechanism noted in the transportation system model depicts the activities and decisions which result in short-term changes to the system (controls) and those which result in long-term changes to the system (planning).

In order for the system to function properly there is need for continual movement through the system of information of all types. Information on its performance outputs, such as traffic volumes and speeds and fare collection, and on its concomitant outputs, such as noise and air pollution, are needed to assure proper planning and control.

SOCIAL, ECONOMIC, AND ENVIRONMENTAL EFFECTS OF TRANSPORTATION SYSTEMS

Environmental Capacity (2), (3)

Figure 2 denotes the concept of environmental capacity for streets and highways. Simply stated, the concept means that a given facility, operating at a

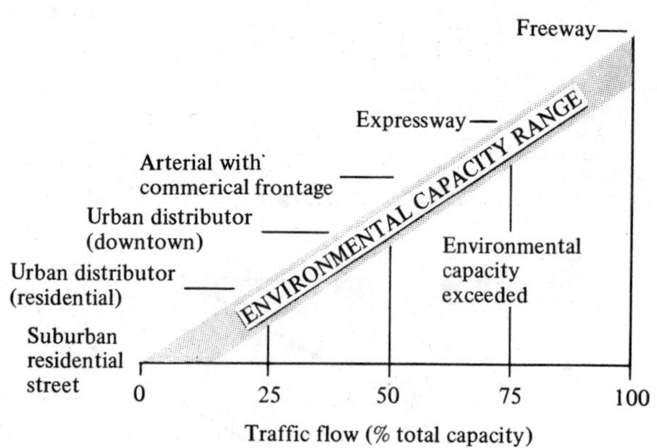

Fig. 2. Concept of environmental capacity[2,3]

particular volume, produces a level of outputs which may be in discord with the ongoing human activities adjacent to the facility, even though those volumes are within its traffic capacity. For example, a highway facility carrying a relatively high volume of traffic through a commercial area would have a less detrimental effect on the quality of living or activity in that area than would the comparable operating conditions through a residential area. It can be said, therefore, that the commercial area has a higher environmental capacity than the residential area.

Community and Human Needs Related to Transportation

Table 1 lists some of the basic social, environmental, access, and economic needs of people as they relate to transportation.

Environmental Policy Act of 1969

The recent past has shown a great upsurge in man's concern for his environment, a concern which has resulted in a broad frontal attack by Government aimed at preserving, protecting, and enhancing the environment. The six goals stated below serve to emphasize the national concern for the environment (5).

The act states that "It is the continuing responsibility of the Federal Government to use all practical means, consistent with other essential considerations of National policy, to improve and coordinate Federal plans, functions, programs and resources to the end that the Nation may–

1. Fulfill the responsibilities of each generation as trustee of the environment for succeeding generations;
2. Assure for all Americans safe, helpful, productive and aesthetically and culturally pleasing surroundings;
3. Obtain the widest range of beneficial uses of the environment without degradation, risk to health or safety, or other undesirable and unattended consequences;
4. Preserve important historic, cultural and natural aspects of our natural heritage, and maintain, wherever possible, an environment which supports diversity and variety of individual choice;
5. Achieve a balance between population and resource use which will permit high standards of living and wide sharing of life's amenities;
6. Enhance the quality of renewable resources and approach the maximum obtainable recycling of depletable resources."

The foregoing statement of environmental concern notes both areas of concern and areas in which transportation and the environment interact–interaction which may be detrimental or beneficial.

TABLE 1. BASIC SOCIAL, ENVIRONMENTAL, ACCESS, AND ECONOMIC NEEDS RELATED TO TRANSPORTATION (4).

A. Basic Social Needs
   1. Personal identity and recognition
   2. Control over own destinies–a voice in decision-making; involvement and participation
   3. A sense of community or belonging (at the local level)
   4. Territoriality–identification with a bounded "turf" or neighborhood
   5. A sense of being part of a united society at the metropolitan level
   6. Compatible neighbors
   7. Compatible playmates for children
   8. Stability and security; lack of anxiety

B. Basic Environmental Needs
   1. Clean air, unpolluted water, trash-free land
   2. Low levels of noise and vibration
   3. Conveniently situated local services: parks, schools, shops, churches
   4. Compatible mixtures of land uses
   5. Adequate shelter
   6. Privacy
   7. Uncongested transportation systems (in the locality)
   8. Preservation of buildings and sites of unusual beauty or historical and architectural interest
   9. Preservation of established neighborhoods
   10. Environment allowing social contact within the neighborhood
   11. Safety and security, especially for children
   12. Avoidance of commotion, such as during major construction

C. Basic Access Needs
   1. Access to employment, wehther one has an automobile or not
   2. Access to the facilities and services of an entire city, whether one has an automobile or not; mobility, opportunity, and variety
   3. Low travel times
   4. Low travel costs
   5. Safety while traveling
   6. Reliable means of travel
   7. Comfort and convenience in travel
   8. Choice of mode of travel
   9. A transportation system that is comprehensible because it is orderly; one can find one's way around easily

D. Basic Economic Needs
   1. Avoidance of financial losses occasioned by the construction of transportation facilities
   2. Preservation of community tax base (municipal or county)
   3. Maintenance of economic stability of a community
   4. Low transportation costs, both capital and operating
   5. Encouragement of economic growth, especially for the lower income and minority groups

Environmental Effects of Transportation Systems

Figure 3 depicts one way of showing the environmental system, in this case made up of a human subsystem and a natural subsystem with components below each. Virtually everything that man does has some effect, good or bad, on the environment. Some of the more common, related to transportation, are noise,

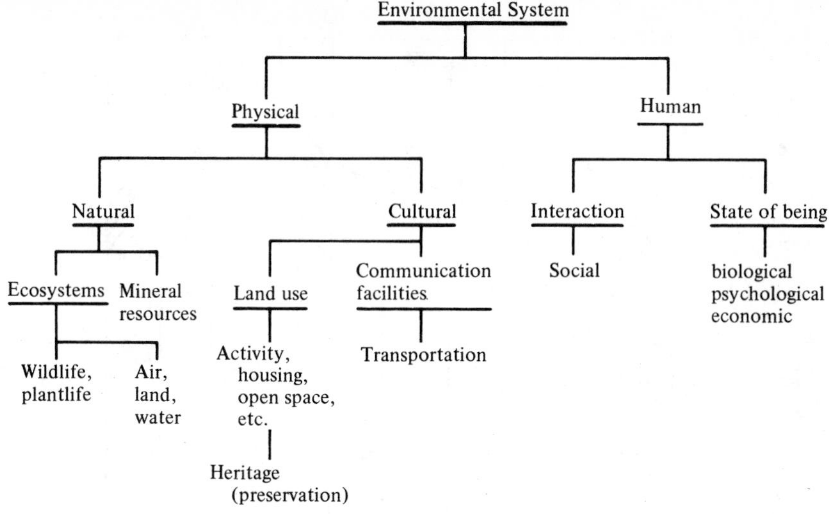

Fig. 3. Environmental system.

air, and visual pollution, and the disruptive effects, social and otherwise, of transportation facilities.

Noise Noise can be defined as unwanted sound. Noise, or more accurately sound, is a measure of the pressure intensity of the noise on the ear in dynes per square centimeter. It is expressed as a sound pressure in decibels (dB), where (6)

$$\text{dB} = 20 \log_{10}\left(\frac{P}{P_0}\right)$$

where

P = sound pressure at ear in dynes per square centimeter

P_0 = a reference sound pressure level (at the threshold of hearing) in dynes per square centimeter; $P_0 = 2 \times 10^{-4}$ dynes/cm^2

Note that the decibel is a logarithmic measure of noise and the increase in intensity therefore is not linear. What this means is that for each increase of approximately 6 dB the noise level (intensity) doubles. Since human beings react differently to different noise frequencies, that is to say, they are more annoyed by high-frequency noise than by low-frequency noise, noise is measured by sound level meters which weight the different noise frequencies emanating from a sound source so that the sound level measurements correlate well with human judgment as to that noise level. Figure 4 depicts the International Standard A, B, and C weighting curves for sound level meters and also a proposed D weighting curve for monitoring jet aircraft noise.

Highway noise (that generated by motor vehicles) is currently measured on the A scale with the readings given in dBA (decibels on the A scale).

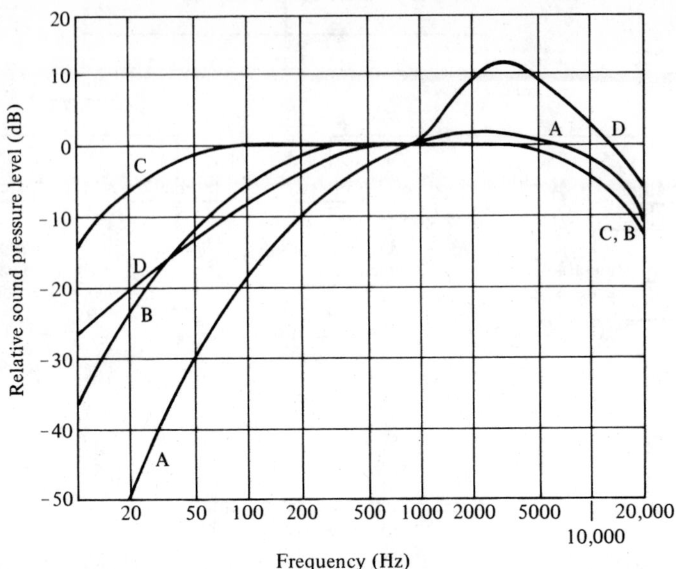

Fig. 4. Noise level–frequency weighting curves[6].

TABLE 2. SAMPLE NOISE LEVELS (7).

Sound	Noise level (dBA)
Rustling leaves	20
Window air conditioner	55
Conversational speech	60
Beginning of hearing damage (if prolonged exposure)	85
Heavy city traffic	90
New York City subway platform (average)	102
Jet airliner, 500 ft overhead	115
Threshold of human pain	120
Loud rock music	130
Laboratory rats died after prolonged exposure	150
Saturn V moon rocket (at launch pad)	180

Sample Noise Levels (7) Table 2 shows typical noise levels measured on the A scale for common noise sources.

Criteria for Measuring Highway Noise (8), (16) Human response to noise, i.e., the level of annoyance that is felt, or perceived, is subjective, and depends on:

1. The relation of the noise to the background or existing continuous noise level. Figure 5 depicts typical continuous background (ambient) noise levels for different activity areas in the metropolitan area.

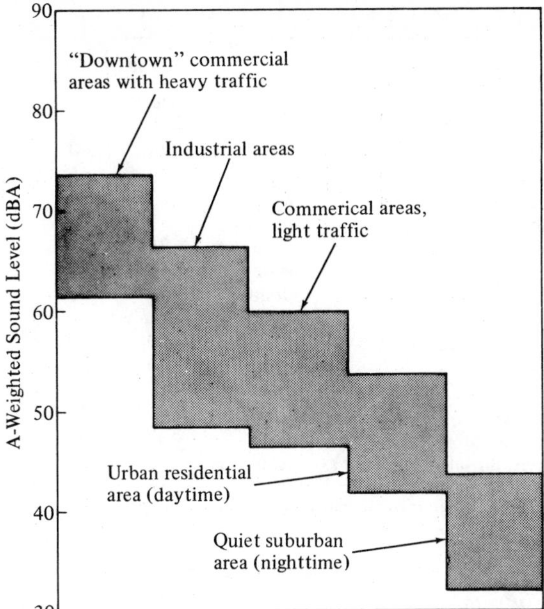

Fig. 5. Typical ambient (background) noise levels[8].

2. The task being interfered with, e.g., sleep, speech, learning, etc. For example, one can tolerate loud traffic noise while at work in a production activity but could not tolerate the same noise level while trying to sleep.
3. The general individual (personal) noise tolerance levels. The subjective satisfaction or dissatisfaction with the environment varies from individual to individual and from group to group. A recent study (8) reports that property owners with higher social-economic status have a greater sensitivity than do apartment dwellers with low social-economic status to the same level of noise.

Recommended Noise Level Design Criteria Table 3 lists recommended levels of noise for design purposes for noises generated by traffic sources. The levels are given in decibels on the A scale exceeded 50% of the time (L_{50}) and in decibels on the A scale exceeded 10% of the time (L_{10}).

Factors Affecting Highway Generated Noise Levels Heard by an Observer The primary factors which affect the amount of noise heard by an observer are listed in Table 4.

Predicting Highway Noise The highway noise prediction methodology currently available evaluates the noise generated by automobiles and trucks separately at a distance of 100 ft. Figures 6 and 7 show the L_{50} reference noise levels at 100 ft

TABLE 3. RECOMMENDED NOISE LEVEL DESIGN CRITERIA (8).

Observer Category	Structure		L_{50} (dBA) Day	L_{50} (dBA) Night	L_{10} (dBA) Day	L_{10} (dBA) Night
1	Residences	inside*	45	40	51	46
2	Residences	outside*	50	45	56	51
3	Schools	inside*	40	40	46	46
4	Schools	outside*	55	–	61	–
5	Churches	inside	35	35	41	41
6	Hospitals,	inside	40	35	46	41
7	convalescent homes	outside	50	45	56	51
8	Offices:					
	stenographic	inside	50	50	56	56
	private	inside	40	40	46	46
9	Theaters:					
	movies	inside	40	40	46	46
	legitimate	inside	30	30	36	36
10	Hotels, motels	inside	50	45	56	51

*Either inside or outside design criteria can be used, depending on the utility being evaluated.

TABLE 4. FACTORS AFFECTING HIGHWAY NOISE HEARD BY AN OBSERVER (8).

Traffic Characteristics:
- Volume (vehicles per hour)
- Mix (proportion of trucks)
- Average speed

Roadway Characteristics:
- Pavement width (distance across all lanes)
- Vertical configuration (elevated, depressed, at grade)
- Flow characteristics (flow interruption imposed by roadway design)
- Grade (%, if greater than 2)
- Surface condition (smooth, normal, rough)

Observer Characteristics:
- Distance from road (measured perpendicular to road)
- Element size (angle of exposure subtended at observer by roadway sound sources
- Shielding (acoustical shielding, buildings, landscaping)
- Observer height (vertical position with respect to road)

generated by a stream of automobiles and a stream of trucks, respectively. Adjustments are made to these levels to account for the prevailing roadway and observer characteristics noted in Table 4. The result is a predicted median noise level at the observer, L_{50}. The L_{50} level is then converted to the L_{10} level and the respective noise levels for autos and trucks added (not arithmetically) to give a composite predicted noise level at the L_{50} and L_{10} levels. These levels can then be compared with the ambient noise level in decibels and the criteria level in decibels to give some indication of the response or impact that the noise levels will result in socially. Table 5 gives impact evaluations based on these criteria.

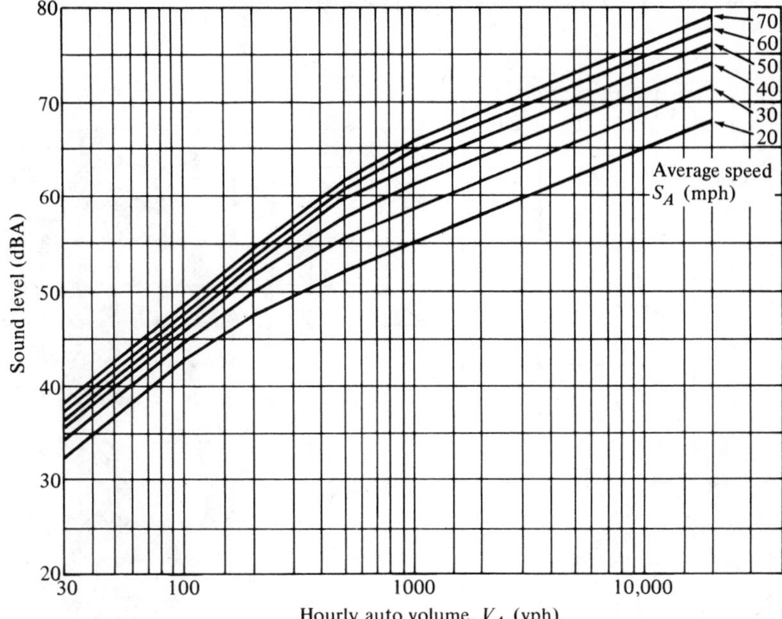

Fig. 6. Noise generated by automobiles at 100 ft, exceeded 50% of the time $(L_{50})^8$.

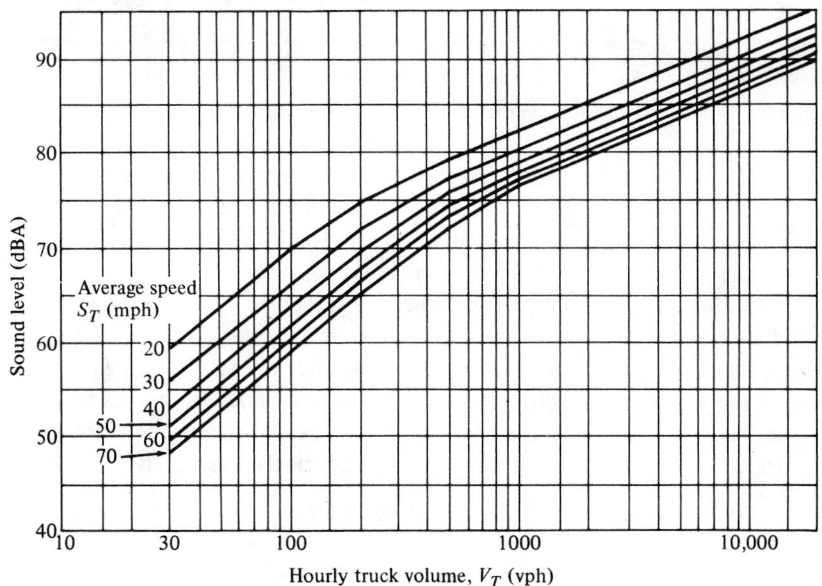

Fig. 7. Noise generated by trucks at 100 ft, exceeded 50% of the time $(L_{50})^8$.

TABLE 5. NOISE IMPACT, RELATING PREDICTED CRITERIA AND AMBIENT LEVELS[8].

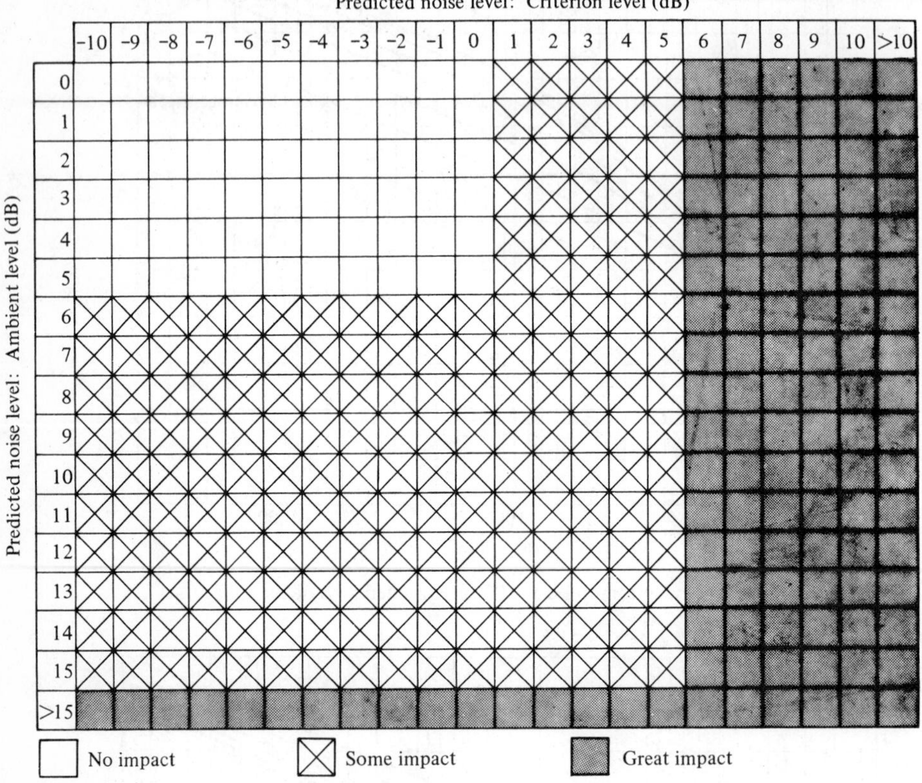

Aircraft Noise Aircraft noise in one noise measuring procedure is measured in perceived noise level (PNdB), a noise weighting scale which gives higher weight to higher frequencies because of the higher-pitched sounds generated by jet aircraft.

Table 6a illustrates the relatively greater subjective noise impact which results from the higher-frequency sounds generated by jets. It indicates a greater subjective annoyance from jet aircraft because of the greater effect of the higher frequencies on the ear; the three aircraft sounds are at the same 100 dB sound pressure level. As a rule of thumb, the perceived noise level for jets can be taken as approximately the sound pressure level generated in decibels plus 13 dB.

Airport Noise Level Sources (10) Noise generated at airports is generally from three sources:

1. Takeoffs and landings
2. Engine runups
3. Ground power units and other support activity noises

TABLE 6A. PERCEIVED NOISE LEVEL (PNdB) FOR SAME SOUND PRESSURE INTENSITY.

Noise source	Sound pressure level (dB)	Perceived noise level (PNdB)
Propeller aircraft, takeoff (low-frequency sounds)	100	102
Civil jet, takeoff (richer in high frequencies)	100	111
Civil jet, landing (higher frequency, compressor whine)	100	114

Factors Influencing Noise Level at Observer (10) The factors affecting the perceived noise level at an observer include:

1. Aircraft type
   a. Turbojets
   b. Turbofans
   c. 4-Engine piston
   d. 4-Engine Turbo prop
   e. Helicopters
2. Type of operation
   a. Takeoff
   b. Landing
   c. Engine runup
3. Aircraft trip range, divided into two trip-length categories: trips under 2000 miles and over 2000 miles. Trip length, because of greater fuel requirements, determines the overall takeoff weight and the power necessary for takeoff. This, in turn, affects the amount of noise generated. Both turbojet and turbofan operations can be broadly categorized for noise analysis into these two trip-length categories.
4. Distance from noise source: perceived noise level contours. Figures 8, 9, and 10 show typical perceived noise level contours for takeoffs, landings and engine runups of civil jet transports. For the takeoff operations two contours are given for turbojet type aircraft—one for trips under 2000 miles in length and another for trips over 2000 miles in length. For turbofans the takeoff chart values are increased by 5 dB. Such contours depict average noise levels for typical landing and takeoff configurations measured from the beginning of the takeoff roll (for takeoffs) and from the runway threshold (for landings).

For an example of noise determination, assume a concern with the noise generated in a residential area by turbojets taking off from a nearby airport. The

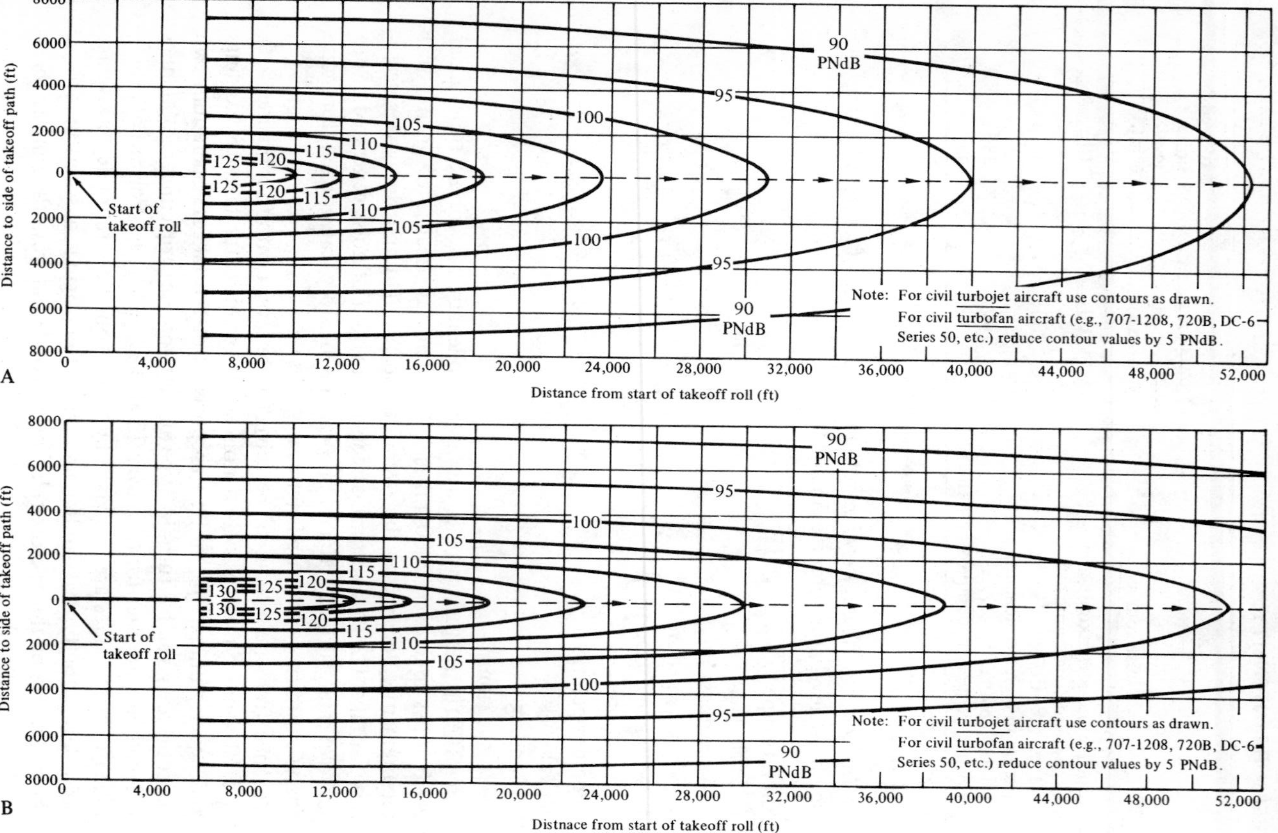

Fig. 8. Perceived noise level contours for civil jet takeoffs[10]. (A) Trip length less than 2000 miles. (B) Trip length more than 2000 miles.

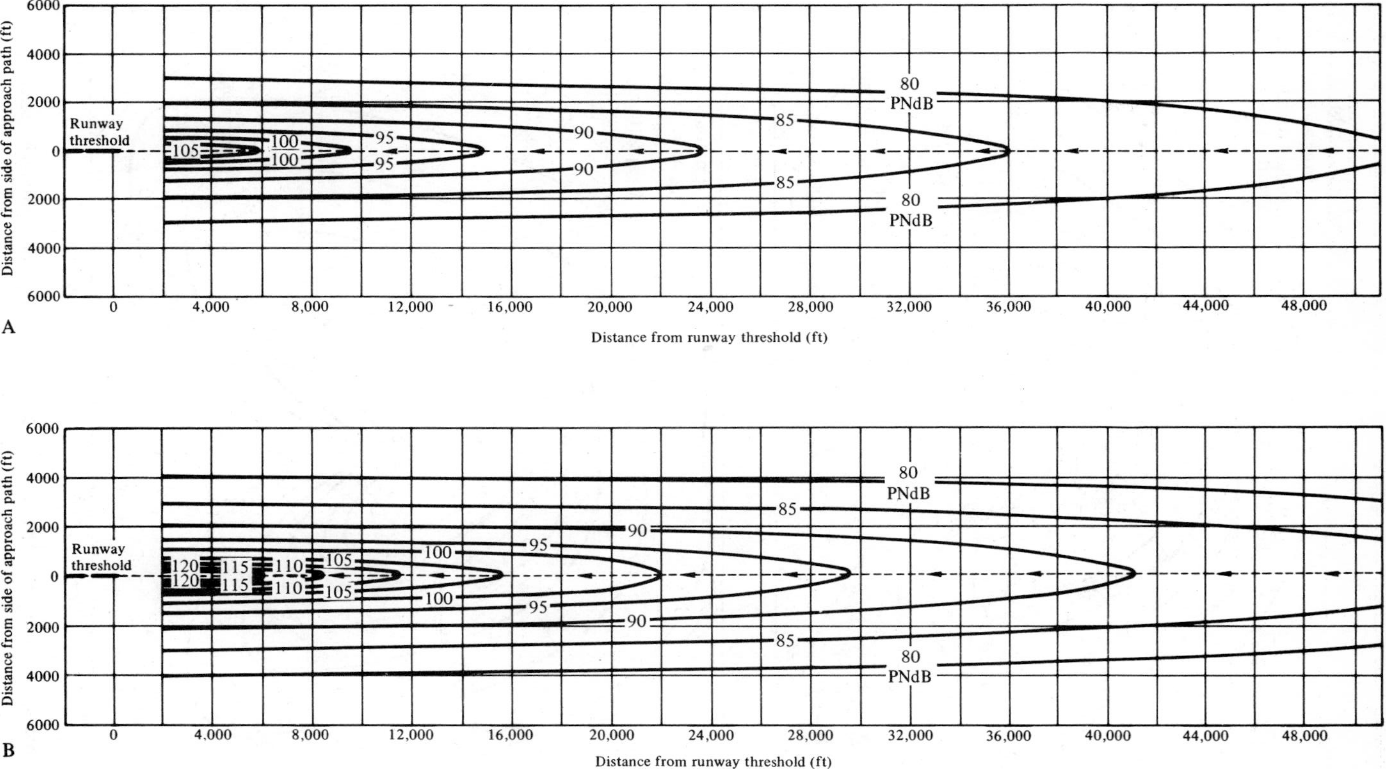

Fig. 9. Perceived noise level contours for civil and military landings[10]. (A) Four-engine piston and turboprop aircraft. (B) Turbojet and turbofan aircraft.

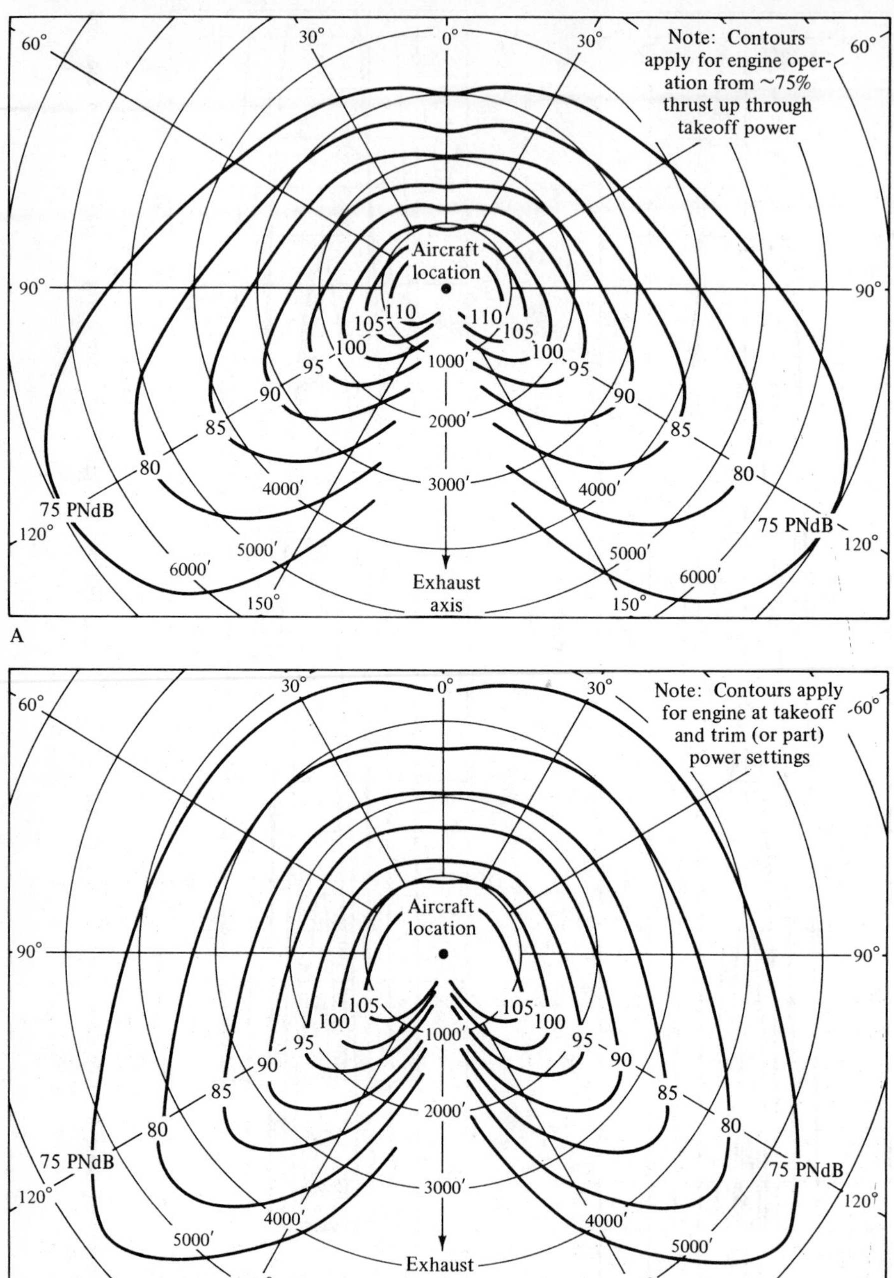

Fig. 10. Perceived noise levels for one-engine runup[10]. (A) Civil and military jet aircraft with turbofan engines. (B) Civil jet aircraft, non-turbofan engines.

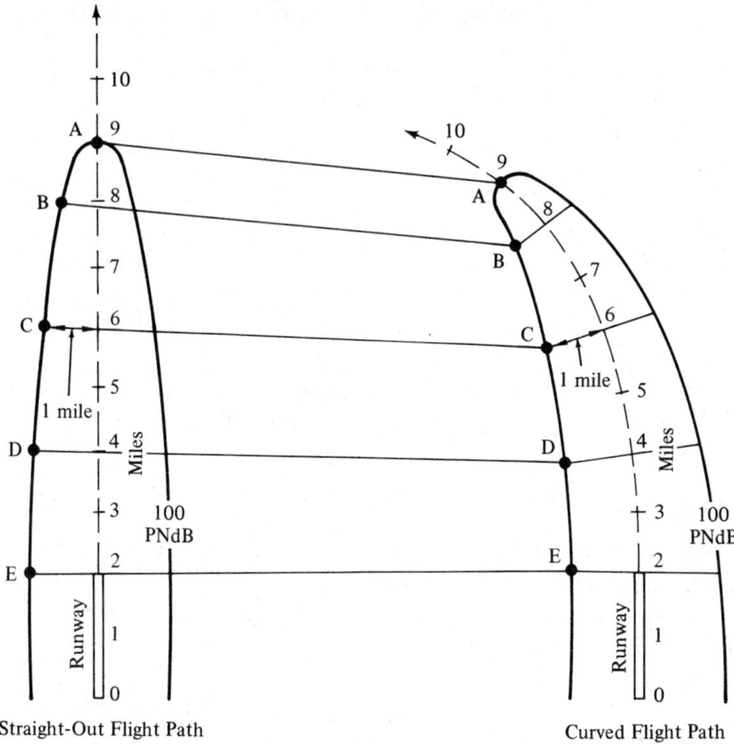

Fig. 10a. Method of adapting PNdB charts for curved flight paths[10].

point under consideration is 20,000 ft away from the start of the takeoff roll and 2,500 ft to one side. Using the perceived noise level contour (Fig. 8) the perceived noise level is approximately 103 PNdB. Figure 10a illustrates the method of adapting the perceived noise level contours to curved flight-path operations.

Composite Noise Ratings (10) Since aircraft noise levels are not time-constant, the composite noise rating (CNR) was developed to estimate the level of exposure of a given area to aircraft generated noise. The composite noise rating considers:

1. number of aircraft operations,
2. runway utilization (percentage of time that a runway is used), and
3. time of day, since time of day influences degree of annoyance with noise.

Community Response to Noise Using case histories of noise problems at various airports, both civil and military, an empirical relationship relating composite noise rating to expected community response has been developed. Table 6b notes the general type of response that can be expected for different composite noise ratings.

TABLE 6B. CHART FOR ESTIMATING RESPONSE OF RESIDENTIAL COMMUNITIES TO COMPOSITE NOISE RATING (10).

Composite noise rating			
Takeoffs and landings	Runups	Zone	Description of expected response
Less than 100	Less than 80	1	Essentially no complaints would be expected. The noise may, however, interfere occasionally with certain activities of the residents.
100 to 115	80 to 95	2	Individuals may complain, perhaps vigorously. Concerted group action is possible.
Greater than 115	Greater than 95	3	Individual reactions would likely include repeated, vigorous complaints. Concerted group action might be expected.

Air Pollution Effects of Automobile Operations

The automobile is a major contributor of air pollutants. Table 7 depicts the estimated nationwide emissions of five major pollutants in 1968 and the proportion of these generated by the motor vehicle. The automobile is the major contributor of carbon monoxide and hydrocarbon pollutants and makes a very significant contribution to nitrogen oxide pollution.

It is readily apparent that a significant improvement in air quality would result in many national benefits. A partial listing of some of these benefits is included in Table 8.

Pollution Generation Related to Vehicle Operation Figures 11, 12, and 13 show the relationships between automobile speed and the three major auto caused emissions: carbon monoxide, hydrocarbons, and nitrogen oxides. Note that the emission of carbon monoxide and hydrocarbon emissions decreases as speed increases, whereas the nitrogen oxide emission increases as speed decreases. These

TABLE 7. ESTIMATED NATIONWIDE AIR POLLUTION BY CONTRIBUTOR, 1968 (11).

Pollutant	Transportation	Fuel combustions in stationary sources	Industrial Processes	Solid waste disposal	Other	Total	Percentage of total by motor vehicle
Carbon monoxide	63.8	1.9	9.7	7.8	16.9	100.1	63
Hydrocarbons	16.6	0.7	4.6	1.6	8.5	32.0	52
Nitrogen oxides	8.1	10.0	0.2	0.6	1.7	20.6	39
Particulates	1.2	8.9	7.5	1.1	9.6	28.3	4
Sulfur oxides	0.8	24.4	7.3	0.1	0.6	33.2	2

Note: Amounts are in millions of tons.

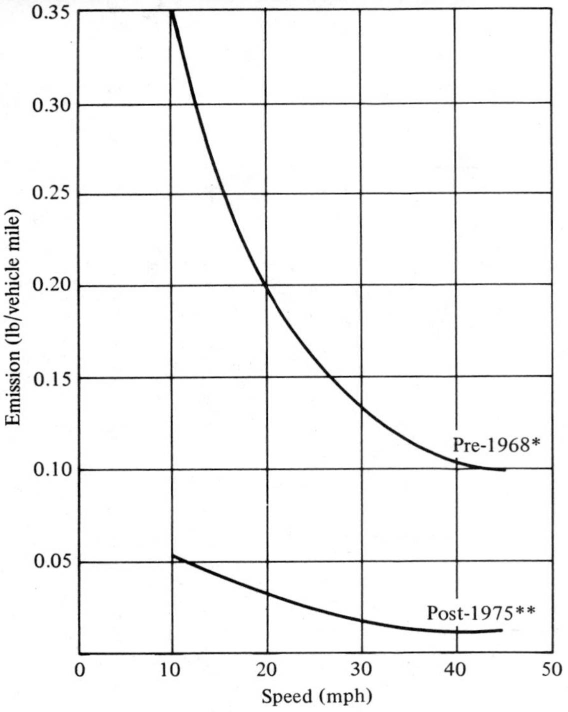

*Source: 2nd Report Secretary of H.E.W. to U.S. Congress pursuant to P.L. 88-206 Clean Air Act, 2/19/65, Table 1.

**Based on 1975 emission standards set by U.S. Congress.

Fig. 11. Relationship between auto speed and carbon monoxide emission[11].

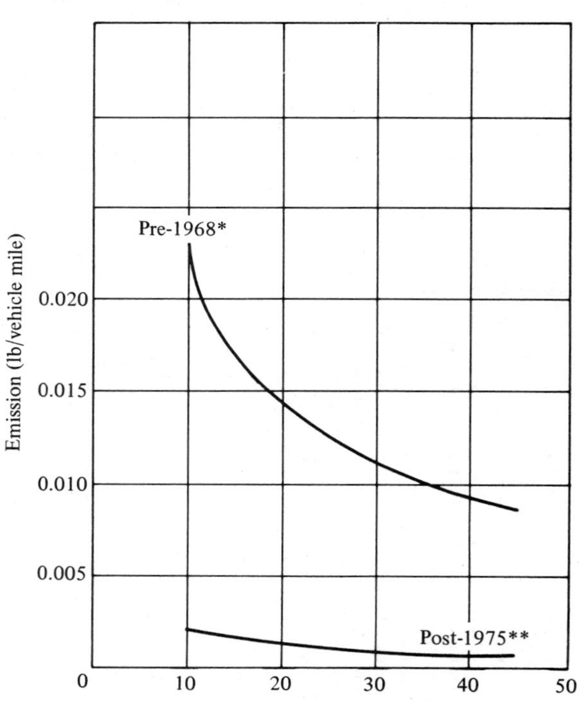

*Source: 2nd Report Secretary of H.E.W. to U.S. Congress pursuant to P.O. 88-206 Clean Air Act, 2/19/65, Table 1.

**Based on 1975 emission standards set by U.S. Congress.

Fig. 12. Relationship between auto speed and hydrocarbon emission[11].

TABLE 8. PARTIAL LIST OF BENEFITS RESULTING FROM AIR POLLUTION REDUCTION (11).

Health Benefits
- a. Reduction in colds, throat irritations, bronchial and other respiratory diseases.
- b. Reduction in the incidence of headaches and physical stress on healthy persons.
- c. Reduction in deaths from cardiovascular and pulmonary causes.
- d. Improved reflexes and greater clarity of thought and sight.
- e. Generalized improvement in body functioning.

Other Environmental Benefits
- a. Reduced damage to vegetation.
- b. Reduced smog and other air-pollution-caused restraints on visibility.
- c. Elimination of blackening to certain types of paints.
- d. Cleaner environment.

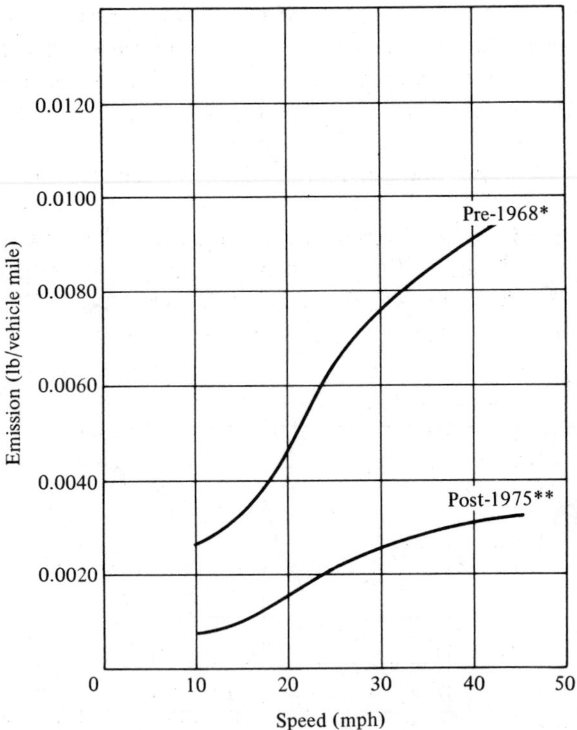

*Source: 2nd Report Secretary of H.E.W. to U.S. Congress pursuant to P.L. 88-206 Clean Air Act, 2/19/65, Table 1.

**Based on 1975 emission standards set by U.S. Congress.

Fig. 13. Relationship between auto speed and nitrogen oxide emission[11].

relationships are for constant-speed operation. During idling, accelerating, and decelerating, emissions are considerably higher than those produced during constant-speed operation. Studies have shown that, during idling, emissions are one and one-half times those during normal cruising; during deceleration, emissions are approximately nine times those obtained for constant-speed operation (11).

Methods of Reducing Automotive Air Pollution Apart from measures and mechanical devices which will result in the production of few emissions by the automobile (e.g., use of "cleaner" engines and exhaust systems), certain things can be done in the design of transportation facilities and the location of adjoining land uses which will result in reduced levels of automobile induced air pollution.

For example, a highway transportation system design which permits more uniform (constant-speed) operation will result in an overall reduction of air pollution. Since the pollution level also varies with distance from the source, placement of activities some distance away from the edge of the roadway will result in a considerable reduction of the level of pollution. Figures 14 and 15 note the reduction in carbon monoxide concentration related to the distance from the edge of the roadway or the height above the roadway. Note that at a distance of 100 ft from the edge of the roadway, the carbon monoxide concentration is approximately 20% of that generated at the roadway itself. However, atmospheric conditions have considerable influence on such concentrations.

The effect of uniformity of operation on pollution reduction can be shown rather dramatically by comparing the relative emission of carbon monoxide for a steady speed flow on a freeway facility compared to flow on a city street under

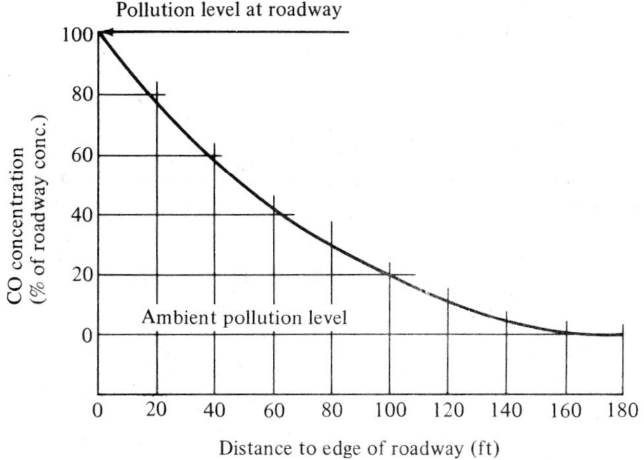

Fig. 14. Carbon monoxide concentration related to distance from roadway edge[11].

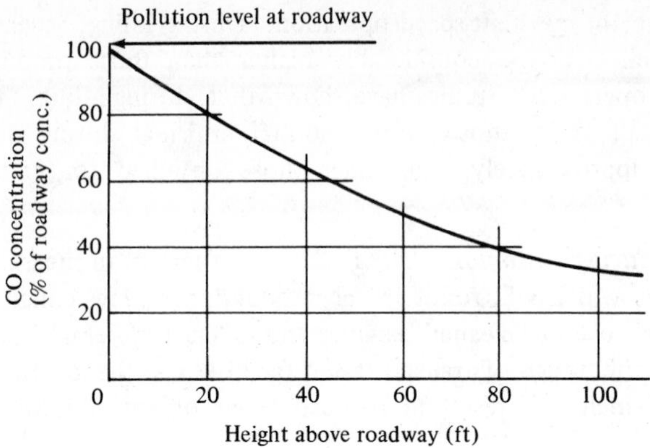

Fig. 15. Carbon monoxide concentration above a highway[11].

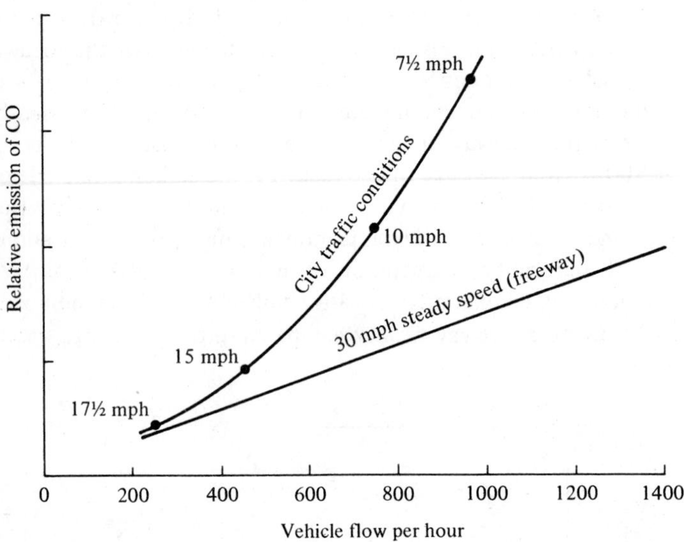

Fig. 16. Carbon monoxide emissions for uniform-flow (freeways) and interrupted-flow (city streets) facilities[11].

traffic conditions. As Fig. 16 shows, an appreciable reduction in carbon monoxide pollution accrues when traffic is moved on freeways rather than city streets.

Water Quality

The major detrimental effect on water quality results from soil erosion occurring during and following construction of transportation facilities and the draining of pollution-carrying storm waters from the transportation facility. Construction practices which will help maintain water quality and reduce possible soil erosion include:

1. Limiting the area of erodible earth exposed at any time.
2. Shaping the top of any earthwork to permit rainwater runoff with a minimum of erosion prior to any time that construction activities are suspended for any length of time.
3. Construction of temporary erosion and sediment control devices such as dykes and berms.
4. Where necessary constructing silt basins to settle out waterborne silt.
5. Minimizing the frequent fording of streams with construction equipment and using, where necessary, temporary structures for crossing streams.
6. Planting erosion control grasses and plants at a time when they will be useful for erosion control.
7. Reasonable restrictions on bridge and culvert construction activities.

In addition to erosion control during the construction activity there is the need to provide a facility which, during its operational lifetime, results in a minimum of erosion. Through proper geometric design and good drainage and landscape development, post-construction erosion can be controlled to a great extent. The use of flat side slopes which are rounded and blend with the existing terrain, provision of adequately designed drainage channels, and the use of protective treatments and devices along with protective ground covers and plantings will all result in a reduction in erosion.

Storm water drainage, in addition to carrying waterborne silts, can pollute water courses with chemicals, e.g., salts used for deicing, raw sewage, and other deleterious materials. One effective and efficient means of controlling water pollution which results from storm-water runoff is to construct separate storm and sanitary sewer systems.

The use of salt for the deicing of transportation facilities, e.g., highways and airport runways, results in detrimental effects to the natural animal and vegetative environment when the salts are washed away with runoff. The severity of this problem is pointed up by the fact that, in 1970, U.S. production of salt totaled 19.6 million tons, of which 9.5 million tons were used for deicing transportation facilities. Only 3% of the total salt production goes into salt shakers. Projections by the Salt Institute indicate that by the year 2000, 50 million tons of salt per year will be used for deicing (13).

Displacement of Families and Businesses

One of the major problems created by the development or modification of transportation facilities is the necessary displacement of some families and businesses. Even though the transportation facility planning proceeds on the philosophy that displacement and disruption are to be held to a minimum, some displacement is inevitable. To enable displaced owners and renters to buy or rent dwellings that are at least comparable to those from which they were displaced, the Uniform Relocation Assistance and Real Property Acquisition Policies Act of 1970 was enacted (14). The Act provides uniform relocation assistance treatment for all persons displaced by Federal or Federally assisted land acquisition programs, including:

a. Replacement housing for home owners: An additional payment not exceeding $15,000 may be made to a person who is displaced from a dwelling unit actually owned and occupied by himself. This additional payment is to permit the purchase of comparable replacement housing, since such housing often cannot be obtained at the so-called fair market value. Additional costs for the acquisition of replacement housing, including increased interest costs and other expenses such as title insurance, recording fees, and other closing costs may also be included.
b. Replacement rental housing for renters: In order to provide comparable replacement rental housing for tenants who have been displaced by Federally assisted projects an amount not to exceed $4,000 can be paid to enable a displaced person to lease or rent for a period not to exceed four years.
c. Other provisions: The act also provides for relocation assistance advisory services, housing replacement by the Federal Agency as a last resort, planning assistance for the development or rehabilitation of housing to meet the needs of displaced persons, and Federal cost sharing to State agencies.

Miscellaneous Environmental-Social Effects

Many other areas, among them economics, land use, health and safety, social and psychological well being, community political structure, and the natural and wildlife ecology are in some way or other affected by transportation facilities and their operation.

REFERENCES

1. Thomas, Edwin N. and Schofer, Joseph L., Strategies for the Evaluation of Alternative Transportation Plans, NCHRP Report No. 96 (1970).
2. Simpson and Curtin, Transportation for Madison's Future, Madison Area Transportation Study (MATS) Special Report No. 17, Prepared for MATS Technical Coordinating Committee (Philadelphia) (May 1970).
3. Buchanan, Colin, *Traffic in Towns*, Her Majesty's Stationery Office (London) (1963).
4. Transportation and Community Values, Report of Conference at Warrington, Va., Highway Research Board Special Report No. 105, (Washington) (1969).

5. The National Environmental Policy Act of 1969, Pub. L. 91–190, 83 Stat. 852, 42 U.S.C. § 4331, U.S. Congress (Washington) (1969).
6. Hewlett-Packard Company, *Acoustics Handbook*, Hewlett-Packard Company (Palo Alto, California) (1968).
7. Mecklin, John M., It's Time to Turn Down All That Noise, in *The Environment* (ed. by Fortune Magazine), Harper and Row, New York (1970).
8. Gordon, Colin, et al., Highway Noise: A Design Guide for Highway Engineers, NCHRP Report No. 117 (1971).
9. Arde, Inc. and Town and City, Inc., A Study of the Optimum Use of Land Exposed to Aircraft Landing and Takeoff Noise, NASA Report CR-410 (1966).
10. Bolt, Beranek and Newman, Inc., Land Use Planning Related to Aircraft Noise, Technical Report For Federal Aviation Administration, U.S. Government Printing Office, Washington, D.C. (1964).
11. Bellomo, Salvatore J. and Edgerley, Edward, Jr., Ways to Reduce Air Pollution through Planning, Design and Operations, Highway Research Record 356 Washington, D.C. (1971).
12. Steering Group and Working Group appointed by the Minister of Transport, *Cars For Cities*, Her Majesty's Stationery Office (London) (1967).
13. *Nation's Business*, December 1971, p. 79.
14. Uniform Relocation Assistance and Real Property Acquisition Policies Act of 1970, U.S. Congress Washington, D.C. (1970).
15. Sturman, G. M., The Effects of Highways on the Environment, May 1970.
16. Kugler, Andrew B. and Piersol, Allan G., "Highway Noise: A Field Evaluation of Traffic Noise Reduction Measures," NCHRP Report No. 144 (1973).
17. Galloway, William J. and Bishop, Dwight E., Bolt, Beranek and Newman, Inc., "Noise Exposure Forecasts: Evolution, Evaluation, Extensions, and Land Use Interpretations," Report No. FAA-NO-70-9, FAA, 1970.
18. Cruz, J. E., "Aircraft Sound Description System: Background and Applications," FAA (1973).

AUTO-TRUCK TRANSPORTATION

Movement Systems (Networks)

Movement systems can be classified into several major network categories, among them geometric, functional, jurisdictional, and financial.

Geometric Classification The geometric configuration denotes the physical pattern that the major street network assumes in terms of its geometry. Some of the major configurations are described below.

Grid or Rectangular Figure 17 illustrates a typical grid system. Although the grid system has right-angled intersections with rigid lot and street relationships, it does not respond well to topography. Because of this rigid pattern it does not serve varying land-use densities as well as other patterns. Although it overcomes the problem of central focus, or convergence on a central area, diagonal

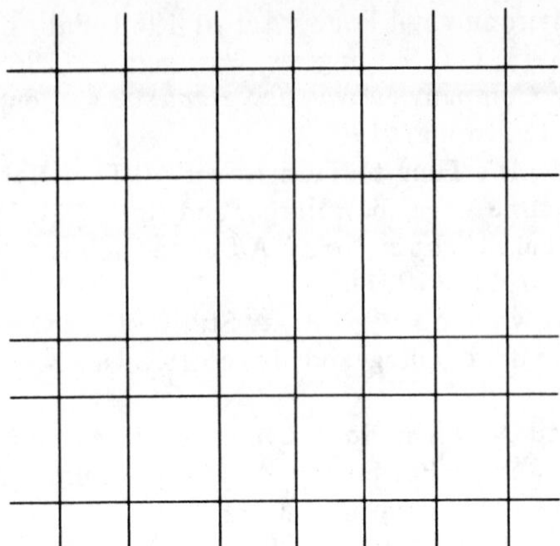

Fig. 17. Grid or rectangular street system.

movement over the network is difficult because of the zigzag movement pattern required.

Radial Systems Figure 18 illustrates a typical radial configuration. Radial street systems, although they have the advantage of permitting direct movement between major activity centers, create problems in terms of odd-shaped and -sized lots, and difficulty in operating through varying-angle intersections. Convergence of movement is on the center with the attendant problems of increased loadings on networks.

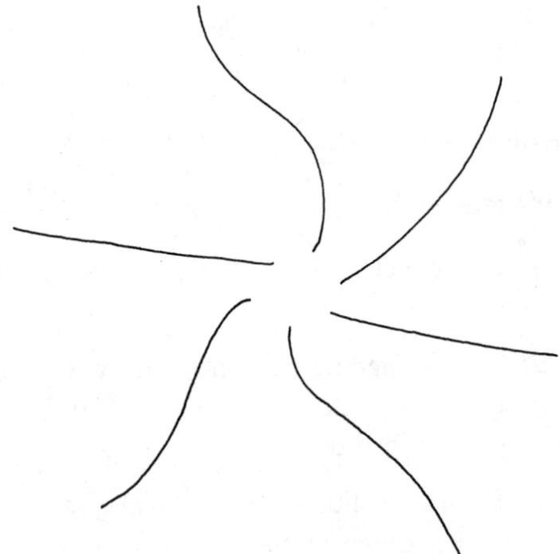

Fig. 18. Radial street system.

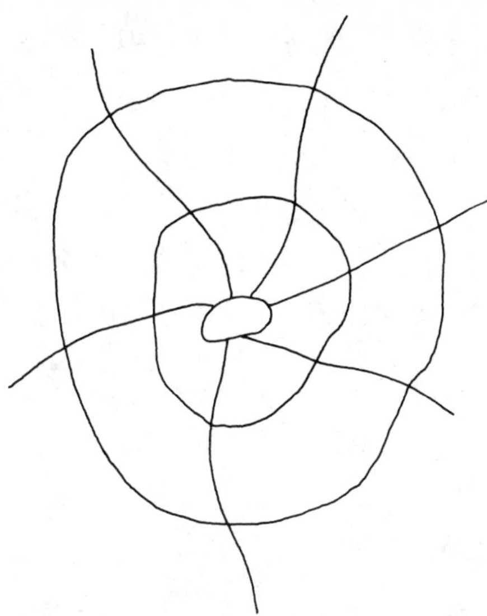

Fig. 19. Circumferential or ring and radial street system.

Radial/Circumferential or Ring and Radial Systems Figure 19 shows a ring and radial street pattern. The ring and radial system not only provides a focus on a center and as a result movement toward the center, but also facilitates necessary crosstown movements. It results in difficult intersections and may, because it focuses traffic on the center, create capacity and parking problems in the central sections of the metropolitan area. It does, however, allow for an even distribution of capacity relative to demand. As the system approaches the center, the lines converge and the major arterials are thus more closely spaced where the traffic demand is highest.

Curvilinear–Irregular or Organic Figure 20 illustrates a typical curvilinear street system. Curvilinear streets make good use of topography and, by following the contours of the land, create interesting development patterns. The irregular nature of the pattern discourages through traffic, an advantage in residential areas. It is not without its problems, however. The irregular pattern makes for irregular lot lines and lot shapes, although in residential areas this can be an advantage. It results in difficulty in intersection treatment and requires, at some times, large amounts of indirection in going from one point to another.

Combination Patterns Although the major arterial street system may assume a pattern such as one of those noted, various patterns are often superimposed or intermixed; within a major dominant pattern, one or several of the others may in fact be prevalent.

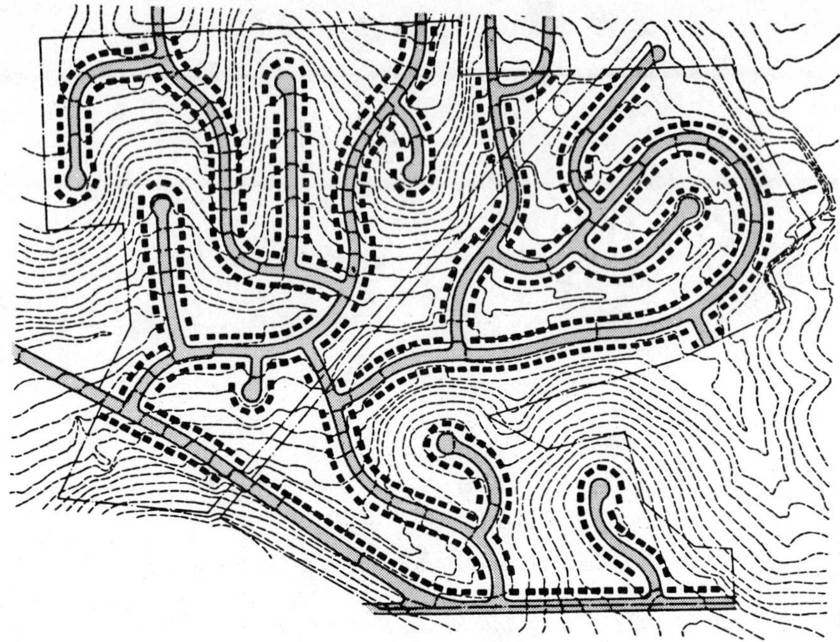

Fig. 20. Curvilinear street system[1].

Functional Classification Functional classifications categorize street systems by function, relating relative mobility and land access. Figure 21 shows in schematic form the relationship between mobility and land access for the three major types of functional street classifications: arterials, collectors, and locals. Each of these categories may be further refined, depending on the precise classification needs of the using agency. Table 9 illustrates the evolutionary phases of a

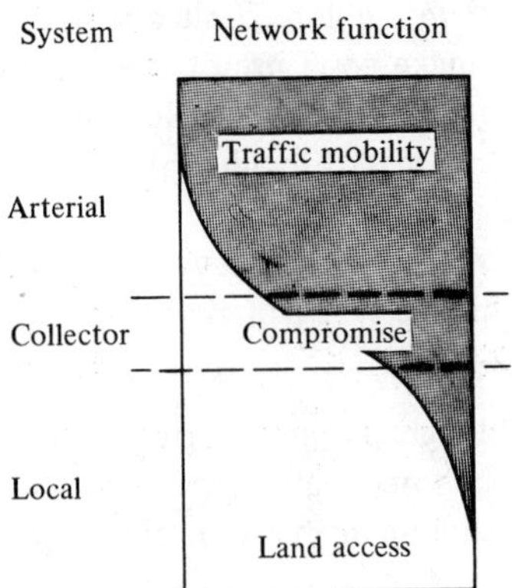

Fig. 21. Relationship between functional system and land access traffic mobility[2,4].

TABLE 9. DEVELOPMENT PHASES OF FUNCTIONAL SYSTEM TERMINOLOGY (2) (6).

Major functional type	State of Wisconsin (2)						U.S. needs study (6)	
	(1962) 1	2	(1965) 3	4	5	(1966) 6	Rural	Urban
Arterial	freeway	primary arterial	primary (closed) arterial	principal arterial	principal arterial	principal arterial	principal arterial interstate	principal arterial
								interstate
	expressway		primary arterial	primary arterial	primary arterial	primary arterial	principal arterial other	principal arterial
								other freeway
	thruway	standard arterial	standard arterial	standard arterial	standard arterial	standard arterial	minor arterial	principal arterial
								other
Collector	connector	connector	connector	connector	high connector	minor arterial	major collector	collector
					low connector	high collector	minor collector	
	collector-feeder	collector	collector	distributor	distributor	low collector	local	local
Local	local	local	local	local	local	local		

functional system terminology developed by one agency (2) and that used for a recent national system needs study (6).

Arterial systems range from freeway types of facilities, with complete control of access, to major urban streets. Their primary purposes are to carry high volumes of traffic in a safe manner with a minimum of delay. With the smallest total system mileage, arterials accommodate a high percentage of the total vehicle miles of travel.

Collectors act as links between major arterials and freeways and local streets. Although they provide some access to abutting properties, their function is a balance or compromise between the need for movement and the need for providing land access.

Local systems, which provide for 65-80% of local street mileage, primarily provide access to abutting properties.

Figure 22a compares some of the characteristics of each of the major functional systems (2).

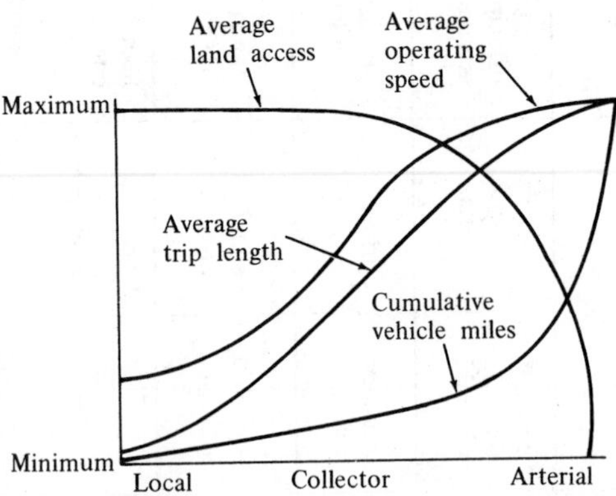

Fig. 22a. Functional highway system: comparison of characteristics[2].

Jurisdictional Classification Jurisdictional systems define the institution responsible for either the construction or maintenance or both of the system.

Financial Aid Classification Financial aid systems denote the availability of various financial aids for different portions of the movement system. These include the Federal aid systems, including the ABC system, and the interstate system, and several other types of systems including the former Type 2 TOPICS (Traffic Operation Procedures to Increase Capacity and Safety) systems.

Flow Characteristics

Volume Information on the volume of traffic using various types of facilities is important for a number of purposes, among them: (a) estimating annual travel

(b) estimating expected highway user revenues; (c) computing accident rates (e.g., accidents per 100 million vehicle miles); (d) indicating trends in volume as an aid in predicting future highway use; (e) measuring demand on system components; (f) evaluating current flow and deficiencies for various system components; (g) as an aid in programming capital improvements; (h) as an indicator or warrant for establishing different types of traffic controls (e.g., signs, signals, parking prohibition); and (i) as the basis for the geometric design of highway facilities.

Types of Volumes–Definitions.

AADT (Annual Average Daily Traffic): This is the total amount of traffic passing a point on a roadway during the full year averaged on a daily basis.

ADT (Average Daily Traffic): This is the volume of traffic expected for an average day of the year. It is essentially identical to AADT except that it is obtained by counting over a shorter time period. ADT is a volume figure commonly used for highway design and other traffic and transportation analysis.

30HV (30th Highest Hour): This is the 30th highest hourly volume of traffic recorded at a location during the year.

DHV (Design Hourly Volume): This is the hourly volume of traffic typically used for the design of a street or highway facility. Normally an estimate is made of the future DHV for design purposes based on a knowledge of the relationships between daily and hourly volumes.

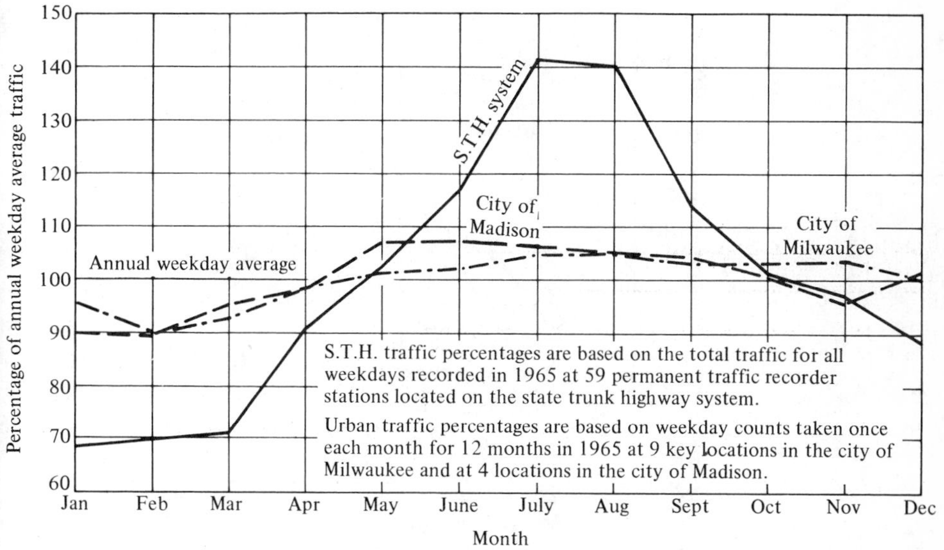

Fig. 22b. Average weekday traffic by months (1965) expressed as percentages of the annual weekday average[8].

Characteristics of Volume Traffic volumes fluctuate over time and by type of route. The composition (vehicle type mix) varies, the directional split varies and for multi-lane facilities, the volume and composition distribution varies by lane.

Time Fluctuations–Seasonal Figure 22b shows the manner in which annual average weekday traffic varies by season of the year and by type of road. In urban areas, daily traffic flows are fairly uniform throughout the year and do not vary a great deal around the annual weekday average denoted as 100% on the chart. Rural traffic volumes, on the other hand, vary considerably, with the low point occurring in the winter months, and the high volumes in the summer months. There is also a considerable variation in rural areas depending on the nature or type of roadway considered. Figure 23 shows typical curves for two rural highway systems of varying character and depicts the very large difference in traffic variation based on the function and the character of the traffic traveling on the facility. One important characteristic of this variation is that over the years (see Fig. 24) it is relatively time stable.

Daily Traffic Volume Fluctuations Figure 25 tabulates the variation in daily traffic as a percentage of the average weekday traffic for different days of the

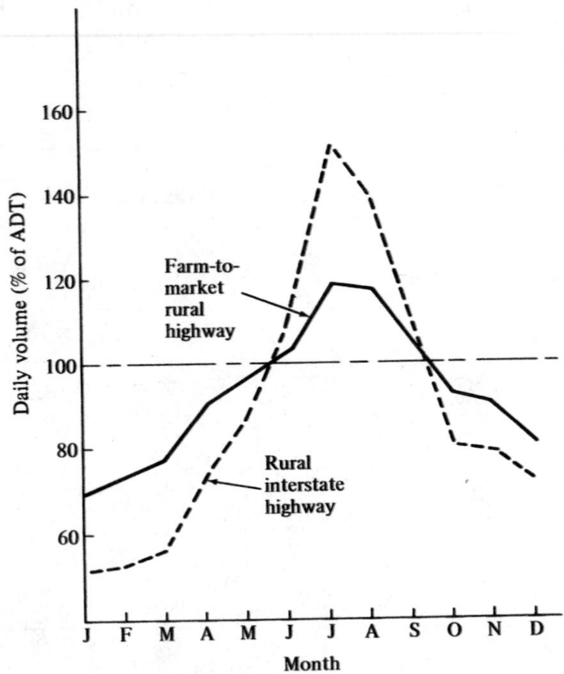

Fig. 23. Average daily traffic, seasonal variation by type of road[9].

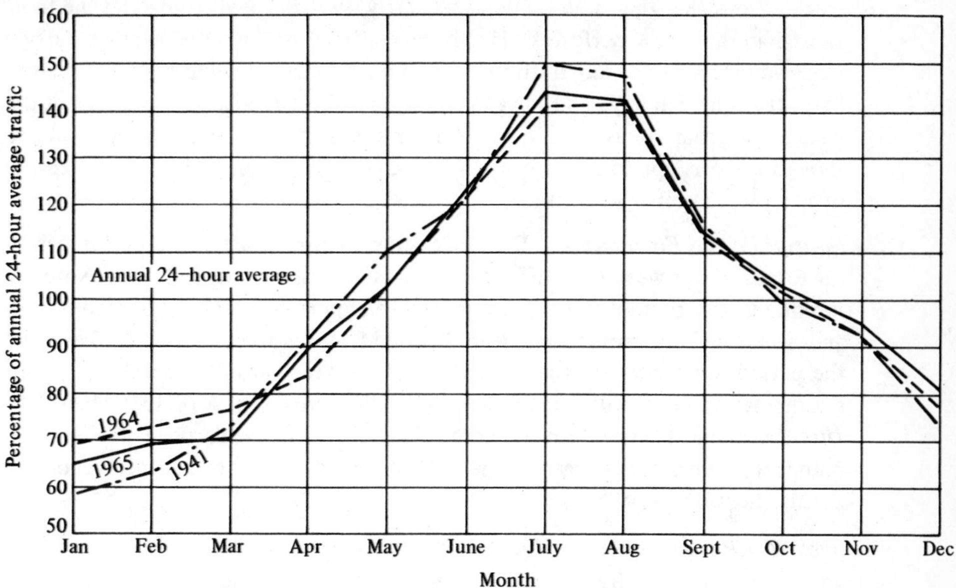

Fig. 24. Relative time stability of daily traffic related to ADT[8].

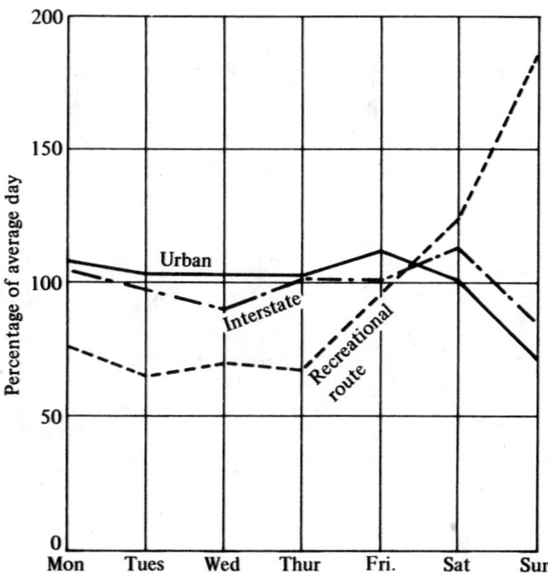

Fig. 25. Typical patterns of traffic-volume variation by day of week[3].

week. Note that the variation depends partly on the type of route and the function that the route performs. Highly recreational routes, for example, carry a large volume of weekend traffic in the summer time, showing high volumes on Saturdays and Sundays and lower volumes during the remaining portions of the week. In urban areas, traffic flows are more stable throughout the week, although there is apt to be an overall decrease on Saturdays and Sundays because of the lower number of work trips.

Hourly Traffic Fluctuations Figure 26 shows hourly traffic as a percentage of the total 24-hour weekday traffic for typical urban and rural roadways. Note the peaking which occurs in the morning between approximately 7 A.M. and 9 A.M. and again in the afternoon between 3:30 P.M. and 5:30 or 6:00 P.M. These are the periods during which the so-called transportation problem occurs in the urban areas. There is also apt to be, depending on the size of the urban area, minor peaking during the noon hour. This occurs particularly in smaller communities where a fairly large volume of lunch and shopping or other travel occurs during the noon hour.

Short-Term (Less than One Hour) Volume Fluctuations Figure 27 shows traffic volume fluctuations by 5-minute increments for two typical urban intersection approaches. Because traffic bottlenecks and other movement problems often oc-

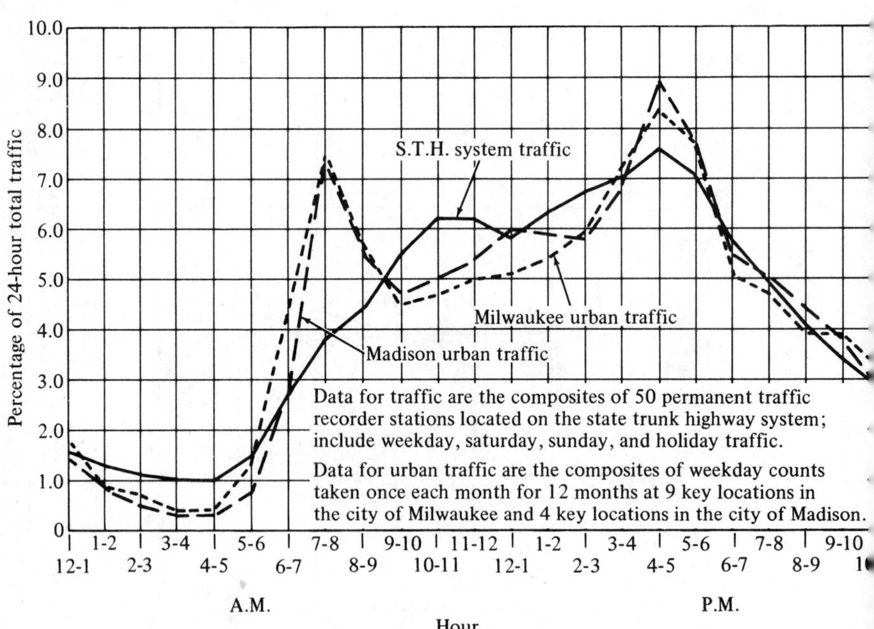

Fig. 26. Percentage of 24-hour traffic by hours[8].

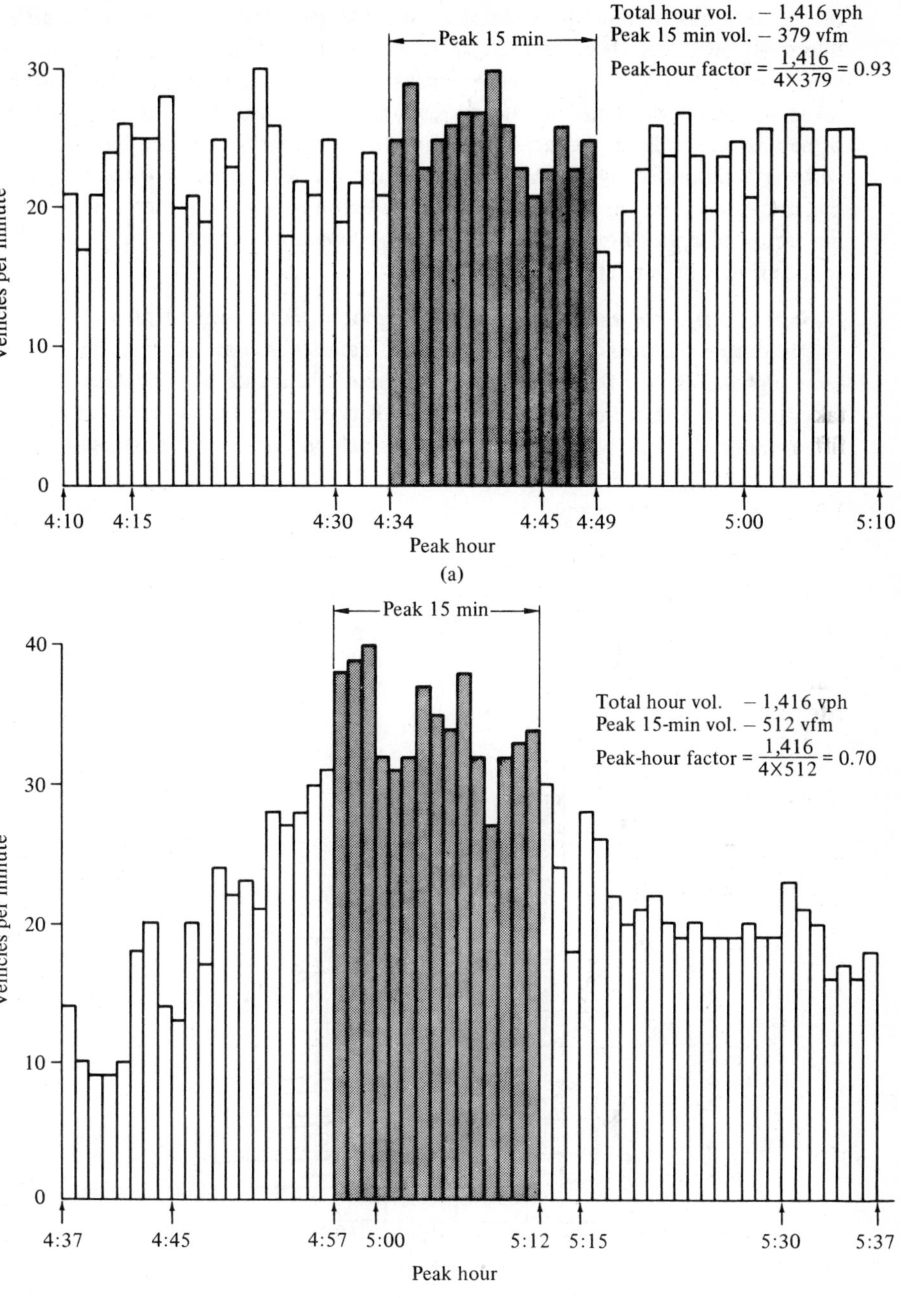

Fig. 27. Traffic flow at intersection approach with (a) high peak-hour factor and (b) low peak-hour factor[5].

cur in response to short-time fluctuations, it is necessary to know how traffic fluctuates over very short periods. The degree of this fluctuation related to an hourly volume is often expressed as a peak hour factor. For intersections this is taken as the peak hour volume divided by four times the peak consecutive fifteen-minute volume during the peak hour. On freeways and other arterials peaking is described by the peak hour volume divided by twelve times the peak five-minute volume. As can be seen from the figure, a peak hour factor approaching unity describes a rather uniform flow, whereas a low peak hour factor describes flow with a relatively intense peak.

Highest Hourly Volumes, Consecutive A plot of consecutive highest hours of traffic during the year for various kinds of facilities is shown in Fig. 28. It relates highest hour of the year as a percentage of the annual average daily traffic. In practice, the relationship depicted by this type of figure provides a guide for determining the hourly volume of traffic which will be used for transportation

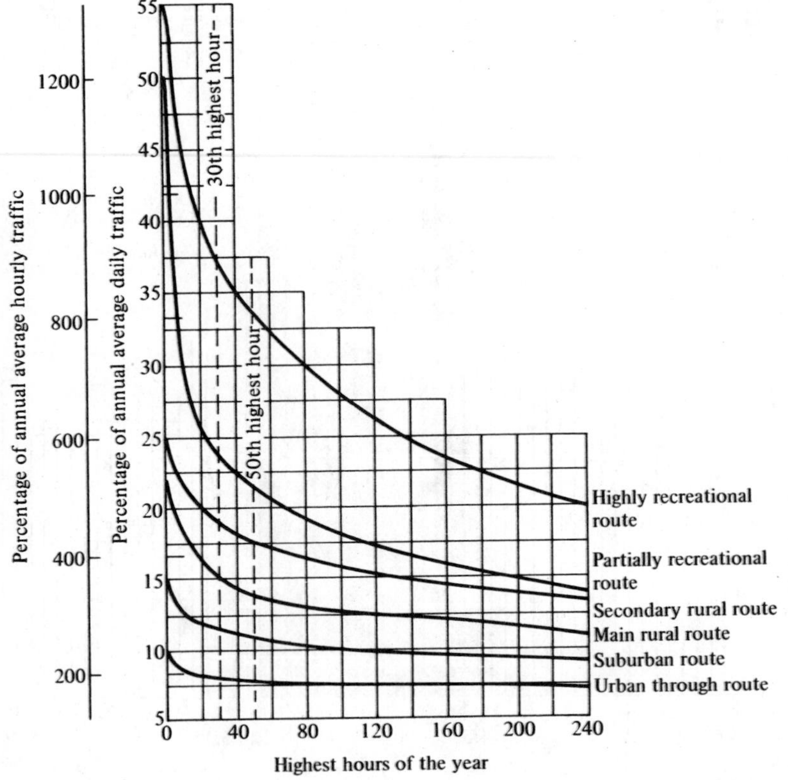

Fig. 28. Hourly volume expressed as percentages of average daily traffic, for typical types of highways[12].

facility design. For many facilities, the curve has a knee somewhere in the vicinity of the 30th highest hour. This indicates that the 30th highest hour (or some other) is, in effect, a point of diminishing returns where only a few additional hours of traffic are accommodated by a considerable increase in roadway capacity. Since most rural and some urban counting programs obtain traffic volumes on a daily basis, a knowledge of the relationship between the 30th (or some other) highest hour and the daily volume provides a basis for obtaining volumes used for design. Average values for the relationship between the 30th HV and the annual average daily traffic, commonly denoted as K, or the design hour factor, are 15% for rural roads, 7–8% for urban roadways, and 11–12% for suburban roadways. From Fig. 26, showing the hourly vs. daily volume relationship for urban areas, it is evident that the typical K factor for urban areas closely approximates the peak hour to average daily traffic volume ratio.

Directional Distribution of Traffic On two-lane rural highways the design hour volume (DHV) is typically that for both directions of travel. Where more than two lanes are required, or in special instances on two-lane highways with important intersections or other access problems, the volume used for design (DHV) must be considered separately for each direction. Since basic traffic volume data are typically two-way, knowledge of the directional distribution of traffic is needed, in addition to the knowledge of the relationship between the design or peak hour volume and the total day's traffic. Also, the current measured directional distribution is reasonably time stable (except in areas of intensive growth) and as a result can be applied to future design year traffic. Typical directional splits vary from 50% to 80% during the design hour with an average for rural highways of approximately 67%. Directional split data for selected highways are noted in Tables 10a and 10b (5).

Composition of Traffic Since vehicles of different sizes and weights have different operating characteristics, the relative proportion of trucks and passenger cars in the traffic stream must often be considered in design. For example, because trucks are slower and occupy more space on the roadway, they affect a reduction in the ability of the roadway to carry traffic. Depending on the situation, one truck acts as the equivalent of several passenger vehicles. The more important characteristics which influence the restrictive effect that trucks have on the traffic stream are the steepness and length of highway grade, available passing sight distance on two-lane roadways, and the proportion of trucks in the traffic stream. In addition to their effect on capacity, the proportion of trucks and their type may dictate the required design treatment on turning roadways, in channelized intersections, and in other special situations.

Vehicle types, as they affect traffic operations, can be grouped into two broad classes: (a) passenger cars, including passenger cars plus light single-unit trucks such as pick-up trucks and panel trucks, since the latter have size and operating

TABLE 10A. HIGHEST REPORTED HOURLY VOLUMES ON FOUR-LANE, TWO-WAY HIGHWAYS IN THE UNITED STATES, 1961. (5)

Route and location	Average lane width (ft)	Avg. volume (vph/lane) Light direction	Avg. volume (vph/lane) Heavy direction	ADT for both directions
Urban Freeways				
U.S. 40 (Trk.), Red Feather Expressway, St. Louis, Mo.	12.0	862	2,030	–
No. Sacramento Freeway, Sacramento, Calif.	12.0	860	1,900	64,000
Eastshore Freeway, Oakland, Calif.	12.0	1,315	1,850	66,000
Atlanta Expressway (N.E. Section), Atlanta, Ga.	12.0	950	1,800	50,300
Conn. 15, E. of Silver Lane, Hartford, Conn.	12.0	–	1,794	36,000
Rural Freeways				
Shirley Highway, Arlington, Va.	12.0	789	1,684	60,400
I-96, Grand River, Livingston Co., Mich.	12.0	214	1,518	15,200
Rt. 128, Circumferential Highway, Newton, Mass.	12.0	1,070	1,435	38,259
New Hampshire Turnpike, Hampton, N.H.	12.0	224	1,144	12,706
I-94, 6 mi. W. of U.S. 24, Wayne Co., Mich.	12.0	258	1,112	24,263
Urban Expressways at Grade				
Lake Shore Drive, S. of 57th Drive, Chicago, Ill.	12.0	445	2,236	75,000
N.J. 4, Paramus, N.J.	10.0	1,438	1,498	62,480
Olentangy River Road, Columbus, Ohio	14.4	851	1,345	42,259
U.S. 6, West 6th Ave., Denver, Colo.	12.0	587	1,177	30,000
U.S. 6, N. of Denver, Colo.	12.0	432	1,107	26,300
Rural Highways				
N.J. 3, Clifton, Passaic Co., N.J.	12.0	–	1,774	40,800
Rt. 90, Pearl City to Aiea, Haw.	11.0	739	1,289	37,728
U.S. 46, Ledgewood, N.J.	12.0	–	1,220	25,932
U.S. 75, W. of Galveston, Tex.	11.0	590	1,193	20,170
Major City Streets				
Sepulveda Blvd., S. of Mulholland Dr., Los Angeles, Calif.	12.5	737	1,742	45,000
U.S. 12, Wayzata Blvd., Minneapolis, Minn.	14.0	420	1,431	32,145
Fla. 9, 27th Ave., N.W., Miami, Fla.	11.5	785	1,195	43,851
Charles St., Baltimore, Md.	9.8	379	1,174	–
Aurora Ave., Seattle, Wash.	11.0	294	1,152	35,758

characteristics closely approaching those of passenger cars; (b) trucks, including all busses and single unit trucks with six or more tires (dual tires on a rear axle) and truck-combinations. For capacity analysis purposes they are all grouped together as trucks. However, it may sometimes be desirable to know the detailed breakdown of trucks in the traffic stream because a large proportion of truck combinations, for example, may require special geometric design consider-

TABLE 10B. HIGHEST REPORTED HOURLY VOLUMES ON TWO-LANE, TWO-WAY HIGHWAYS IN THE UNITED STATES, 1961. (5)

Route and location	Average lane width (ft)	Volume (vph) Heavy direction	Volume (vph) Both directions	ADT both directions
Urban Expressways at Grade				
N.J. 208, Fairlawn, Bergen Co., N.J.	12.0	1,090	2,056	16,028
P.R. 21, San Juan, Puerto Rico	9.1	–	1,482	19,201
Rural Highways				
Md. 5, Woods Corner, Prince Georges Co., Md.	12.0	1,099	1,871	18,825
Md. 26 (Liberty Road), Baltimore, Md.	10.0	1,224	1,777	21,500
U.S. 40, West of Denver, Colo.	12.0	–	1,760	5,950
Md. 3, Glen Burnie, Anne Arundel Co., Md.	12.0	855	1,680	22,275
Del. 141, New Bridge Rd., New Castle Co., Del.	12.0	963	1,605	15,935
Major City Streets				
U.S. 95, Bonanza Road, Las Vegas, Nev.	12.0	–	2,297	20,064
U.S. 60, Washington St., Charleston, W. Va.	15.0	1,125	2,062	19,850
U.S. 27, Clinton St., Ft. Wayne. Ind.	10.0	1,063	2,024	20,041
Coldwater Canyon Dr., Los Angeles, Calif.	15.0	1,586	1,985	15,000
Rt. TT, Brown Road, St. Louis Co., Mo.	11.0	1,223	1,970	–
Bridges and Tunnels				
Posey Tube, Oakland, Calif.	11.0	1,303	2,595	27,163
Lake St. Bridge, Minneapolis-St. Paul, Minn.	14.0	1,515	2,570	25,024
Broadway Ave. Bridge, Minneapolis-St. Paul, Minn.	14.0	1,498	2,373	17,956
C & O Bridge, Cincinnati, Ohio	13.0	1,397	2,281	31,088
Plymouth Ave. Bridge, Minneaplis-St. Paul, Minn.	15.0	1,182	2,262	14,062

ations. Typical percentage distributions of vehicle miles of truck travel on main and rural highways are given in Table 11.

Typical hourly traffic volumes related to 24-hour average weekday traffic for different vehicle types in rural and urban areas are given in Fig. 29 and 30.

Lane Distribution, Multi-Lane Highways Data on the volume and composition distribution of traffic by lane may be important in both the design and operation of divided roadways. The lane distribution affects not only the overall carrying capability but also the entering and exiting problems at interchange ramps. Figure 31 relates hourly traffic volume on six-lane divided facilities to the percentage of vehicles in each lane.

Although not now considered in design and capacity analysis, the great increase in the proportion of cars pulling boat trailers and campers and pick-up trucks carrying camper bodies, may require special consideration in design. This is particularly true on highways carrying large volumes of summer recreation travel.

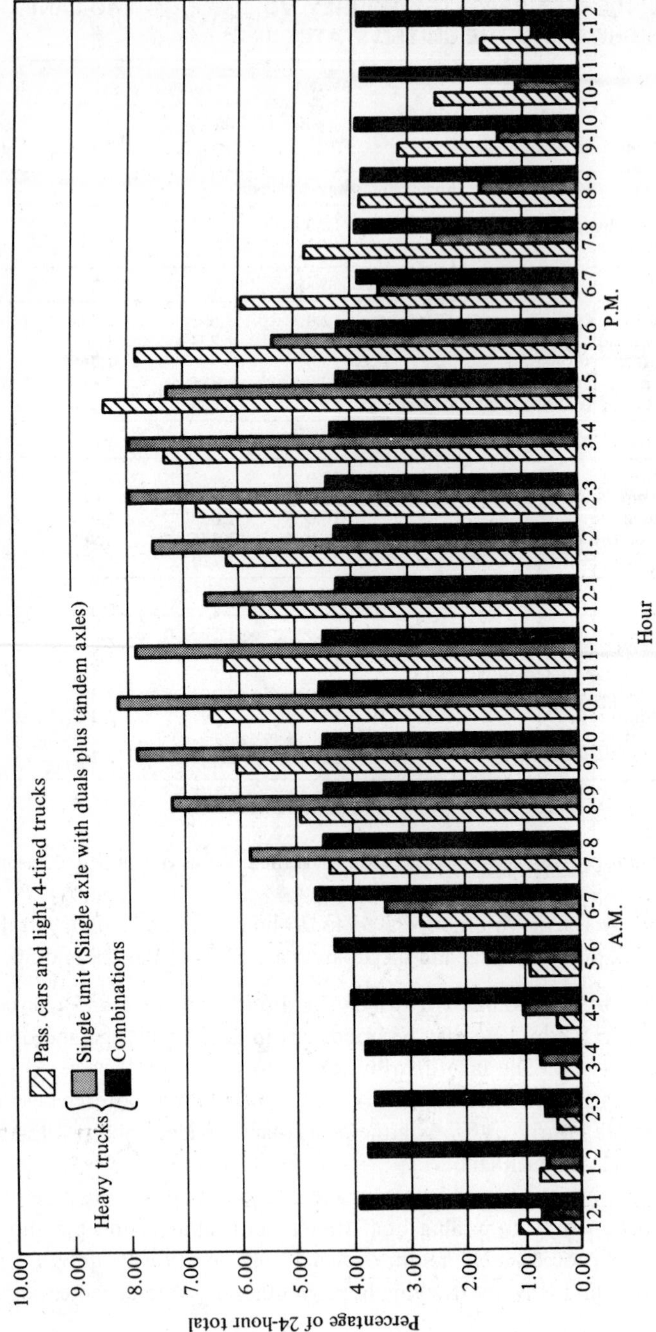

Fig. 29. Percentages of average 24-hour weekday traffic by hours, rural areas, 1965.

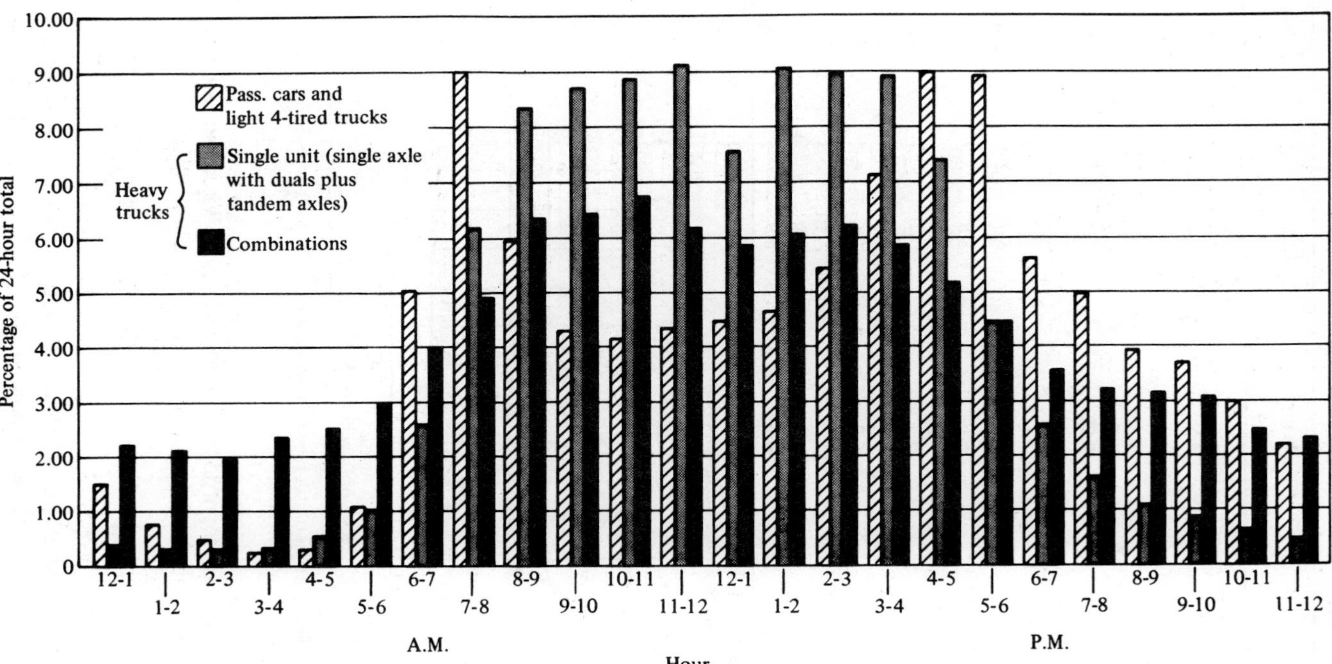

Fig. 30. Percentages of average 24-hour weekday traffic by hours, urban areas, 1965. Data for 5 permanent traffic recorder and 7 loadometer stations located on urban interstate and state trunk highway systems–winter, spring, summer, and fall seasons combined.

TABLE 11. TRUCK TRAVEL ON MAIN RURAL HIGHWAYS (7).

Year	Buses	Single unit trucks	Truck combinations	Total truck class
1947	1.1	9.2	5.7	16.0
1948	1.0	9.5	5.8	16.0
1949	1.1	8.5	6.0	15.6
1950	1.0	8.5	7.2	16.7
1951	0.9	7.7	6.7	15.3
1952	0.9	7.9	6.5	15.3
1953	0.8	7.4	6.9	15.1
1954	0.7	7.1	6.7	14.5
1955	0.7	6.7	6.7	14.1
1956	0.7	6.6	6.8	14.1
1957	0.6	6.5	6.4	13.5
1958	0.6	6.5	6.2	13.3
1959	0.5	6.5	6.6	13.6
1960	0.5	6.3	6.6	13.4
1961	0.5	6.1	6.8	13.4
1962	0.5	6.1	6.8	13.4
1963	0.5	6.0	6.9	13.4

Note: Panels and pickups not included; other 4-tired trucks included. Data for Alaska and Hawaii included beginning 1960.

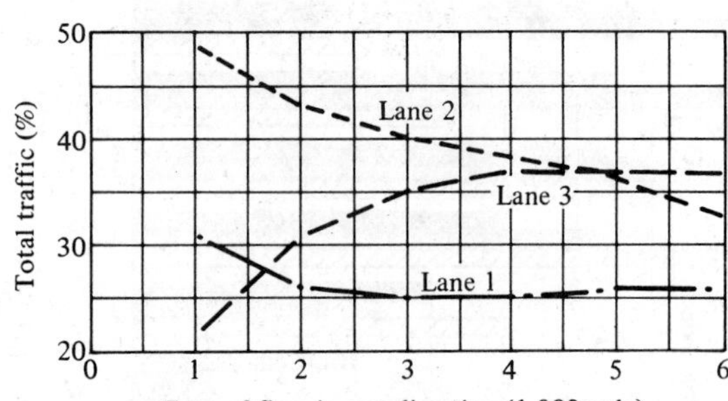

Fig. 31. Lane distribution on six-lane divided facilities[5]. Lane 1, shoulder lane; Lane 2, center lane; Lane 3, median lane.

Table 12 gives typical proportions of these vehicles on Wisconsin Interstate highways during the summer months (10).

Speed

Data on speed distributions of traffic are useful in helping to determine their effect on the need for traffic devices such as warning signs, posted advisory

TABLE 12. IMPACT OF RECREATIONAL VEHICLES ON TRAVEL MANUAL CLASSIFICATION COUNT (10).

11 A.M.-12 A.M., SATURDAY, 7/31/1971–WESTBOUND I-94, JACKSON COUNTY, WISCONSIN (TRAFFIC BYPASSING REST AREA).

Vehicle type	Number of vehicles		Percentage of total	
Passenger cars				
without trailer		762		79.9
with boat trailer	55		5.8	
with camper trailer	29		3.1	
with house trailer	24		2.5	
with other trailer	23		2.4	
	131		13.8	
Pickup trucks				
without camper or trailer		7		0.7
with camper	24		2.5	
Other single-unit trucks				
without camper or trailer		2		0.2
with camper		4		0.4
Large truck combinations		24		2.5
Totals	155	799	16.3	83.7
	954		100	

speeds, and speed zoning. Studies of vehicle speed operating relationships can also be related to such things as design features, accidents, and other operational characteristics of the roadway. In addition, speed data are useful in providing travel time and delay information as an indicator of problem areas and as a measure of the efficiency and quality of flow on a transportation system.

Types of Speed:

Spot Speed. This is the instantaneous speed of vehicles passing a point on a roadway. Spot speed data are normally obtained by sampling vehicles passing selected points on a roadway during a peak hour, an off-peak hour, or periods which would give data representative of the average day. Spot speed data are typically presented in two forms. Table 13 gives typical spot speed data for Wisconsin motorists driving on interstate highways (11). Figures 32 and 33 illustrate two ways of showing such data—as an individual frequency distribution and as a plot of a cumulative distribution curve.

From the speed distribution curves and associated data, certain speed characteristics having significance for design and traffic control purposes can be obtained. These include:

Mean speed: average speed (from Table 13).

Modal speed: the speed at which most vehicles are traveling on the roadway (from Fig. 32).

TABLE 13. SPOT SPEED SUMMARY, WISCONSIN PASSENGER CARS, 1969 (11).

Speed range	Midpoint	No. of observations	Percentage	Cumulative percentage
30–34.9	32.5	1	0	0
35–39.9	37.5	1	0	0
40–44.9	42.5	16	0.4	0.4
45–49.9	47.5	44	1.0	1.4
50–54.9	52.5	173	3.8	5.2
55–59.9	57.5	413	9.1	14.3
60–64.9	62.5	1083	23.7	38.0
65–69.9	67.5	1493	32.7	70.7
70–74.9	72.5	1050	23.0	93.7
75–79.9	77.5	242	5.3	99.0
80–84.9	82.5	35	0.8	99.8
85–89.9	87.5	5	0.1	99.9
90–94.9	92.5	5	0.1	100.0
		4563		

Mean spot speed: 66.4 mph.
Posted speed: 70 mph.
Daytime interstate composite.

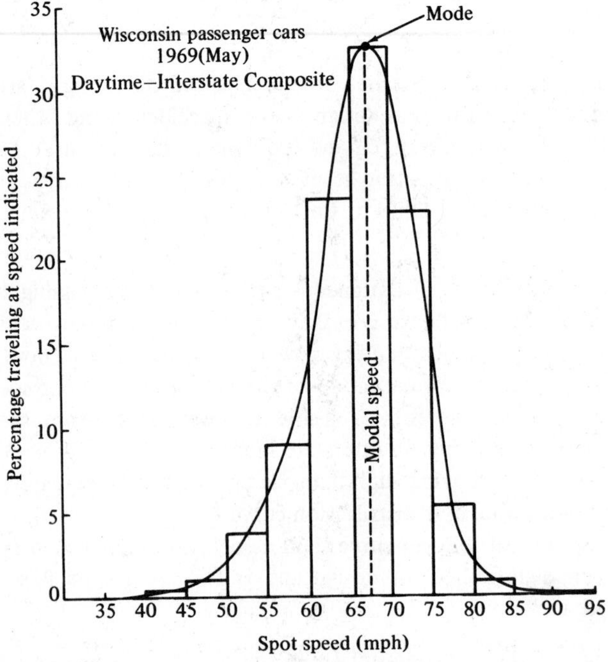

Fig. 32. Frequency–speed distribution curve (Wisconsin passenger cars, May 1969, daytime-interstate composite)[11].

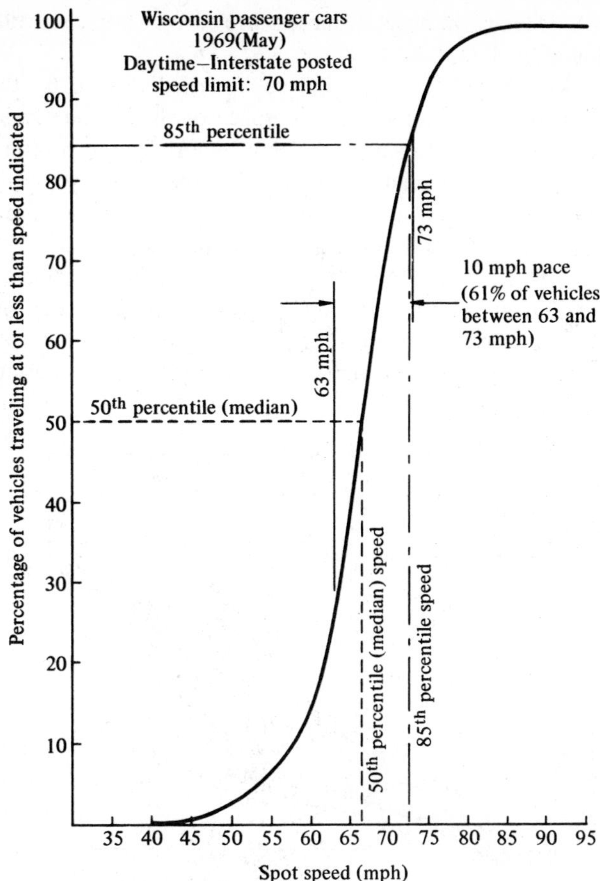

Fig. 33. Cumulative speed distribution curve (Wisconsin passenger cars, May 1969, daytime-interstate, posted speed limit: 70 mph)[11].

10-mph pace: the 10-mph speed increment covering the greatest number of observed vehicles (from Fig. 33).

50th-percentile speed: the speed above which 50% of vehicles are going and below which 50% are going (from Fig. 33).

85th-percentile speed: the speed below which 85% of the vehicles are operating, or conversely, the speed which 15% of the vehicles are exceeding–often used as a guide in helping to establish speed limits.

95th-percentile speed: the speed below which 95% of the drivers are driving, or conversely, the speed exceeded by 5% of the drivers. This closely represents the design speed used in speed–curvature relationships (7).

Overall Speed. This is the total distance traveled by a vehicle divided by the total time required, including traffic delays.

Running Speed. This is the total distance traveled divided by the running time, exclusive of delays. Average running speed or the en route vehicle speed on a highway is one of the factors upon which geometric design criteria are based; it is also used for evaluating road user costs and benefits and for evaluating the effectiveness of operation on a roadway. Running speed for various volume conditions can be related to the design speed of a highway.

Design Speed. This is the speed selected for design which correlates to curvature, superelevation, and passing and stopping sight distance–design speed is the speed selected for safe operation of a facility. The relationship between design speed and running speed for main rural highways for the light and intermediate traffic is noted in Fig. 34 and Table 14.

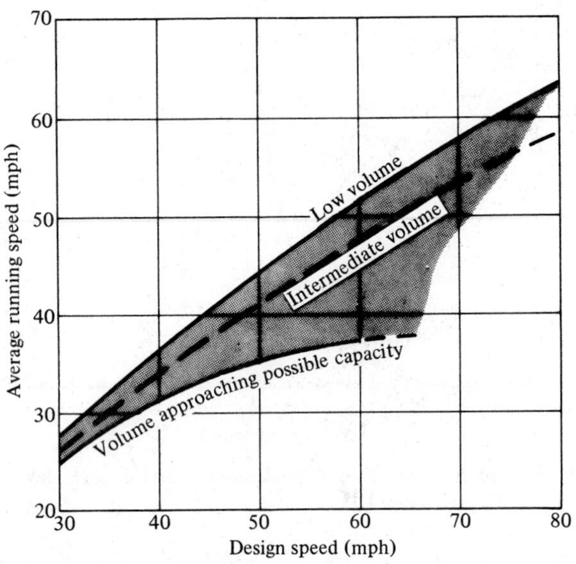

Fig. 34. Relation of average running speed and volume conditions[7]. *Running speed* is the speed of an individual vehicle over a specified section of highway, being the distance divided by running time. *Average running speed* is the average for all traffic or a component of traffic, being the summation of distances divided by the summation of running times. It is approximately equal to the average of the running speeds of all vehicles being considered.

Operating Speed. This is the highest overall speed at which a driver can travel on a given highway under favorable weather conditions and under prevailing traffic conditions without exceeding the safe speed as determined by the design speed.

TABLE 14. RELATIONSHIP BETWEEN DESIGN SPEED AND RUNNING SPEED (7).

Design speed (mph)	Average running speed (mph)		
	Low volume	Intermediate volume	Approaching capacity
30	28	26	25
40	36	34	31
50	44	41	35
60	52	47	37
65	55	50	
70	58	54	
75	61	56	
80	64	59	

Speed Variations Speeds vary on a highway by time of day, by weather, in relation to the existing geometric conditions, and in relation to existing volume on the roadway. Figure 35 shows typical cumulative spot speed relationships for different volume conditions. It is interesting to note that at the higher levels of service (higher quality of flow), speeds are generally higher and there is a wider range of speeds at which different vehicles are able to operate. At the lower

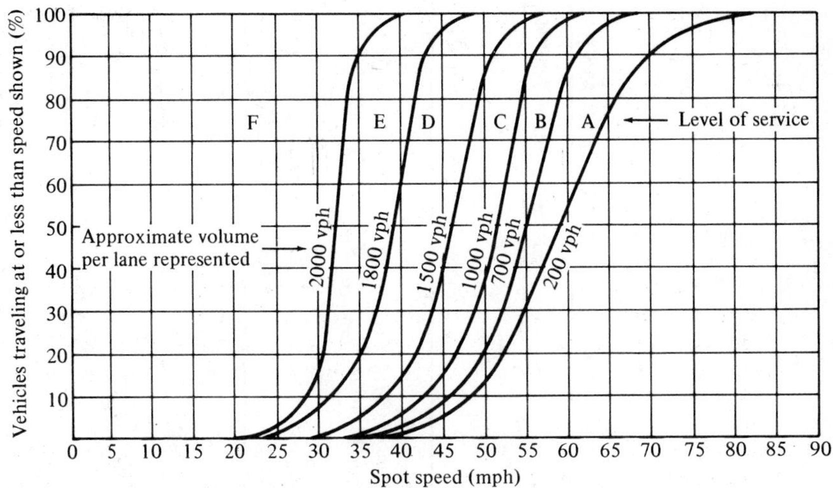

Fig. 35. Typical distribution of passenger car speeds in one direction of travel under ideal uninterrupted flow conditions on freeways and expressways[5].

TABLE 15. 1968 VEHICLE CLASSIFICATION—AVERAGE 24 HOUR WEEKDAY TRAFFIC (AWT).

	Rural interstate: I-90/94 (Columbia Co.)					Urban interstate: I-94 Milwaukee				
	Winter	Spring	Summer	Fall	All-season annual AWT	Winter	Spring	Summer	Fall	All-season annual AWT
Local passenger (%)	50	45	40	45	44	90	90	87	88	89
Foreign passenger (%)	16	32	43	17	30	1	—	3	2	1
Light single-unit trucks (%)	3	5	4	4	4	4	4	4	4	4
Buses (%)	—	—	—	—	—	—	—	—	—	—
Heavy S.U. (%)	3	2	2	4	3	3	3	3	3	3
Combinations (%)	28	16	11	30	19	3	3	3	3	3
AWT (vehicles)	8,714	14,061	19,703	12,367	13,711	84,874	89,286	87,697	85,219	86,769

levels of service, average and median speeds are lower and speed distributions are constrained to a very narrow band.

Composition

The preponderance of vehicles of different types also varies by season, time of day, day of week, type of highway, etc. Table 15 shows vehicle distributions by percentage of the total stream for a rural and an urban interstate, separately by season and as a composite average weekday.

REFERENCES

1. Tunnard, Christopher and Pushkarev, Boris. *Man Made America: Chaos or Control*, Yale Univ. Press, New Haven (1963).
2. *Highways I: The Basis For Planning*, Wisconsin Dept. of Transportation, Madison (1967).
3. *Traffic Engineering Handbook*, Institute of Traffic Engineers, Washington, D.C. (1965).
4. Marks, Harold, Protection of Highway Utility, NCHRP Report No. 121 (1971).
5. *Highway Capacity Manual–1965*, HRB Special Report No. 87, Highway Research Board, Washington, D.C. (1965).
6. *National Highway Functional Classification and Needs Study Manual:* 1970–1990, U.S. Dept. of Transportation, Washington, D.C. (1970).
7. *A Policy on Geometric Design of Rural Highways*, American Association of State Highway Officials, Washington, D.C. (1965).
8. *Wisconsin Highway Traffic–1965–Volume I*, Wisconsin Department of Transportation, Madison (1965).
9. *Permanent Traffic Counter (PTR) Group Adjustment Factors* (internal data), Wisconsin Department of Transportation, Madison (1967).
10. *Manual Traffic Classification Count Field Sheets 1971*, Wisconsin Department of Transportation, Madison (1971).
11. *Biennial Traffic Speed Study–State Trunk Highway System–1969*, Traffic Services Section, Division of Highways, Wisconsin Department of Transportation, Madison (1969).
12. Matson, T. M., W. S. Smith, and F. W. Hurd, *Traffic Engineering*, McGraw-Hill Book Company, Inc., New York (1955).
13. "A Policy on Design of Urban Highways and Arterial Streets," American Association of State Highway and Transportation Officials, Washington, D.C. (1973).

SYSTEM PLANNING–URBAN

The Planning Process

The urban transportation planning process, as currently used, historically results from new concepts reasoned out in the abstract and others developed in

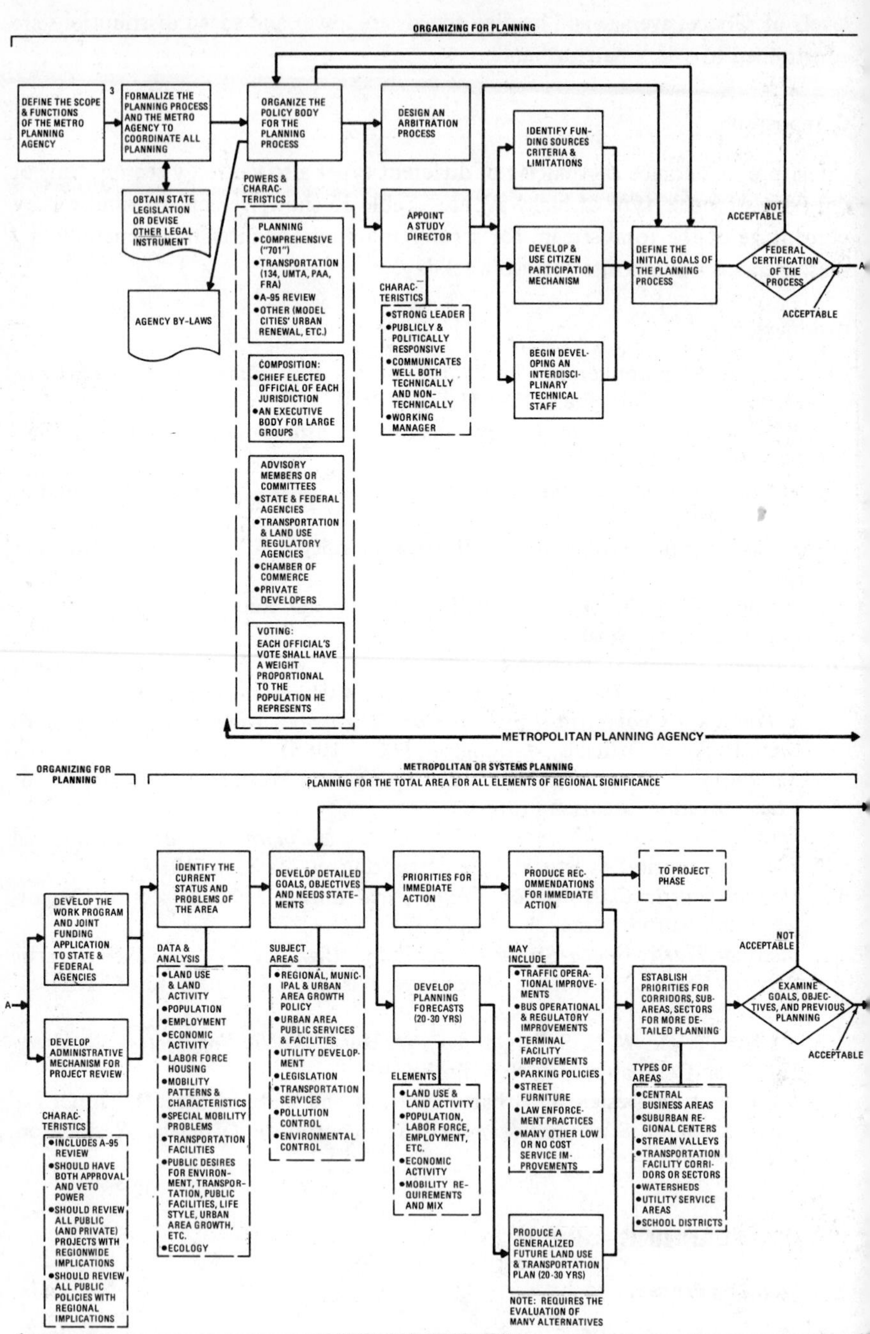

Fig. 36. Ideal urban planning process[1].

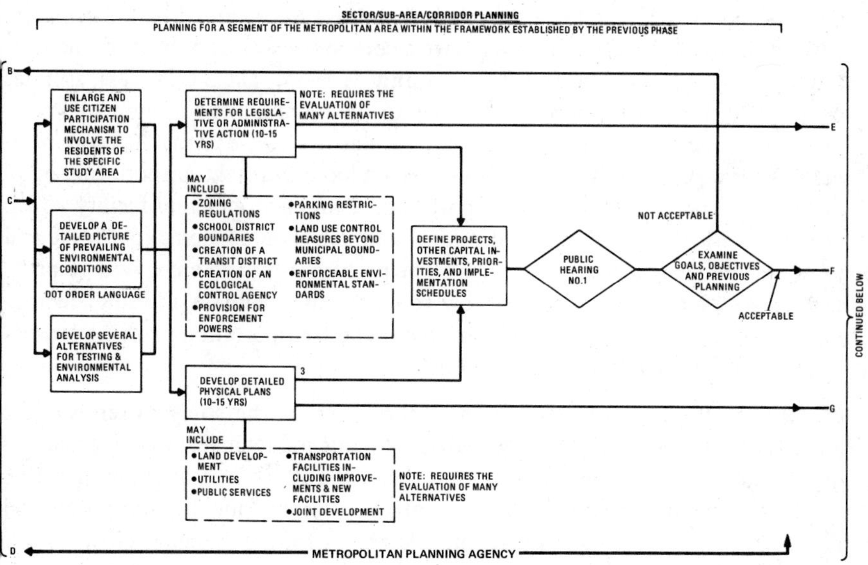

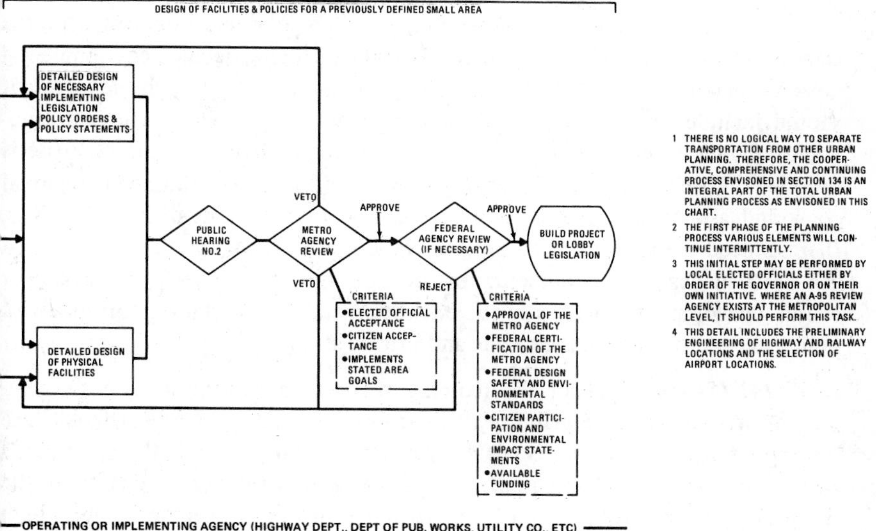

1 THERE IS NO LOGICAL WAY TO SEPARATE TRANSPORTATION FROM OTHER URBAN PLANNING. THEREFORE, THE COOPERATIVE, COMPREHENSIVE AND CONTINUING PROCESS ENVISONED IN SECTION 134 IS AN INTEGRAL PART OF THE TOTAL URBAN PLANNING PROCESS AS ENVISONED IN THIS CHART.

2 THE FIRST PHASE OF THE PLANNING PROCESS VARIOUS ELEMENTS WILL CONTINUE INTERMITTENTLY.

3 THIS INITIAL STEP MAY BE PERFORMED BY LOCAL ELECTED OFFICIALS EITHER BY ORDER OF THE GOVERNOR OR ON THEIR OWN INITIATIVE. WHERE AN A-95 REVIEW AGENCY EXISTS AT THE METROPOLITAN LEVEL, IT SHOULD PERFORM THIS TASK.

4 THIS DETAIL INCLUDES THE PRELIMINARY ENGINEERING OF HIGHWAY AND RAILWAY LOCATIONS AND THE SELECTION OF AIRPORT LOCATIONS.

the course of practical planning and transportation studies. The major impetus for the development of the process, as now used, came as a result of the enactment of the 1962 Federal Highway Act. The 1962 Act required that urban areas with more than 50,000 population have a comprehensive, coordinated, and continuing planning program, in process by July 1, 1965. The 3-C Process, as it has been called, has been a primary stimulus to the development of improved planning techniques. Although geared from the start for conventional highway planning, the process is evolving toward one which looks at total transportation of all modes and the need to view transportation as an environmental facility which interacts with the urban area and can be used as a tool in helping to achieve and promote the goals of the urban area. The problem is not so much how to make man move more efficiently, but how to create and maintain a livable environment. Figures 36 and 37 conceptualize the urban transportation planning process and some of the major activities with it.

Planning Study Organization Defining the scope of the study in terms of its area, function, and the generalized advisory body and technical working staff requirements is the starting point. The study should be metropolitan in scope, with a strong metropolitan institutional base for making transportation and planning decisions at the metropolitan level. Also, the planning process itself must fully integrate all community activities; it must recognize that decisions in every area affect some other area of man's environment. Thus, the process has to be the result of interaction by many agencies. Because transportation improvements have many social, economic, environmental and other impacts, the staff responsible for conducting the study and developing detailed environmental corridor designs must be multidisciplinary in nature. Among the members it should include are transportation planners, land use planners, architects, landscape architects, sociologists, traffic engineers, and whatever other expertise is required. To be certain that plans agree with and help fulfill local and regional goals and aspirations, and are truly responsive to the needs of human beings, there is need for sound citizen input at various stages of policy formulation and the detailed design development. In addition, there is the need to provide for sound development practices in the environmental transportation corridor with inputs needed from both the public and private sector.

Goals (4) (5) (6) A major shortcoming of most transportation-land use planning efforts today is the inability of metropolitan areas to clearly articulate the metropolitan area's goals. Goals are generalized statements of the basic ideals and value structure of the community. Because they only broadly relate to the physical environment, they must be translated into more specific criteria which can be used in evaluating the degree by which plans fulfill the metropolitan area's goals. Generally, objectives are developed based on the stated community goals and then criteria and standards are defined which provide a means of testing the attainment of the stated objectives.

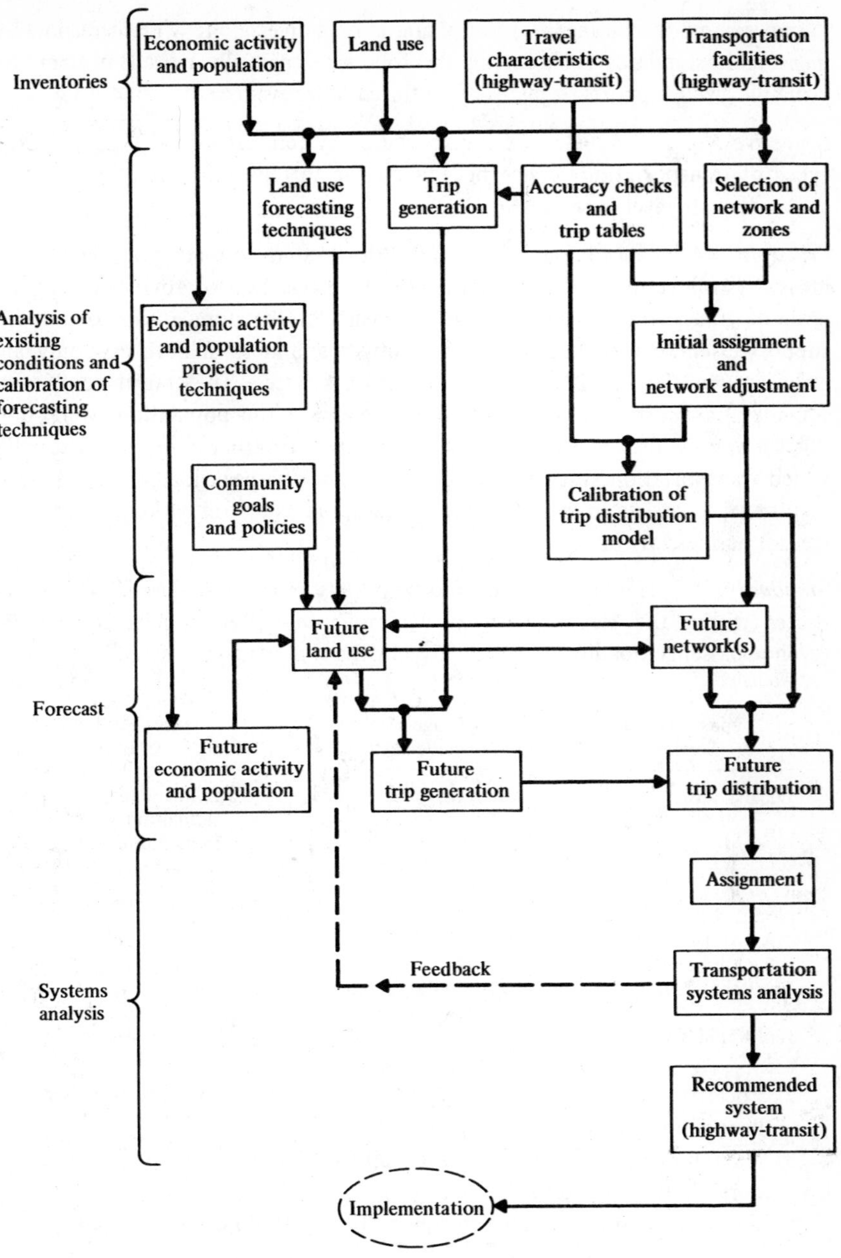

Fig. 37. Urban transportation planning process[2].

Two examples of transportation planning objectives along with corresponding measurable standards are given in the following (partially edited) extract (the principle given supports or justifies the stated objective) (4)

Objective No. 2 "A balanced transportation system providing the appropriate types of transportation service needed by the various subareas of the Region at an adequate level of service.

Principle "A balanced regional transportation system consisting of highway and transit transportation and terminal facilities is necessary to provide an adequate level of transportation service to all segments of the population, to properly support essential economic and social activities, and to achieve economy and efficiency in the provision of transportation service. The transit component provides transportation service to that segment of the population which does not for various reasons own and operate an auto. Furthermore, transit supplies added transportation system capacity to alleviate the peak loadings on highway facilities and assists in reducing the land use demand for parking facilities in central business districts.

Standards "1. Transit service of an appropriate type should be provided for all routes within the Region wherein the minimum potential average weekday revenue passenger loading equals or exceeds the following values:

Type of Transit Service	Minimum Potential Average Weekday Revenue Passengers	Transit Service Area Radius (miles)
Local Transit[b]	600/day/bus[c]	$\frac{1}{4}$ in high-density residential areas[d] $\frac{1}{2}$ in medium- and low-density residential areas[d]
Modified Rapid Transit[e]		
A. All Day[f]	600/day/bus[c]	3
B. Limited	300/4-hrs./bus[c]	3
Bus Rapid Transit[g]	21,000/day/preempted freeway lane	3
Rail Rapid Transit[h]		3

"2. Local transit routes should be provided at intervals of no more than one-half mile in all high-density residential areas.

"3. Maximum operating headways for all transit service throughout the daylight hours[i] should not exceed 1 hour.

"4. The average distance between transit stops should not be less than:

Type of Transit Service	Average Distance Between Stops
Local Transit	660 feet
Modified Rapid Transit	No stops between terminal areas
Bus Rapid Transit	2 miles (for line haul sections)
Rail Rapid Transit	2 miles (for line haul sections)

"5. Loading factors should not exceed:

Type of Transit Service	Maximum Loading Factor For Periods Exceeding 10 Minutes[j] (percent)
Local Transit	
A. 10 minute headway on route	100
B. 5–10 minute headway on route	125
C. Less than 5 minute headway on route	140
Modified Rapid Transit	100
Bus Rapid Transit	100
Rail Rapid Transit	100

"6. Transit routes should be direct in alignment, with a minimum number of turning movements, and arranged to minimize transfers and duplication of service.

"7. The proportion of transit ridership to the central business district of each urbanized area within the Region should be maintained at least at the present level and increased if possible.

"8. Modified rapid transit or rapid transit service should be provided as necessary to reduce peak loadings on arterial streets and highways in order to maintain a desirable level of transportation service between component parts of the Region.

"9. Parking should be provided at park-and-ride transit stations to accommodate the total parking demand generated by trips which change from auto to transit modes at each such station.

"10. Freeways or expressways should be provided for all routes within the Region where all of the following criteria are met:

a. The route provides intercommunity service;
b. The desired speeds or a volume to capacity ratio of 1.0 requires control of access and uninterrupted flow;
c. Alternate routes exist or will be provided to adequately serve local traffic; and
d. Potential average weekday traffic exceeds 25,000[k] vpd. in urban areas and 15,000[k] in rural areas.

"11. Arterial streets and highways should be provided at intervals of no more than one-half mile in each direction in high-density residential areas, at intervals of no more than one mile in each direction in medium-density residential areas, and at intervals of no more than two miles in each direction in all low-density residential areas.

"12. In the major central business districts of the Region, parking should be provided sufficiently near concentrations of demand so that 80 percent of the short-term parkers need walk no more than one block.[1]

"13. On a gross area basis, parking in the major central business districts of the Region should be provided at the following minimum levels:

Urbanized Area Population	Spaces Per 1,000 Auto CBD Destinations[m]
50,000	110
100,000	140
500,000	210
1,000,000	235
2,000,000	255

Objective No. 3 "The alleviation of traffic congestion and the reduction of travel time between component parts of the Region.

Principle "To support the everyday activities of business, shopping, and social intercourse, a transportation system which provides for reasonably fast, convenient travel is essential. Furthermore, congestion increases the cost of transportation, including the cost of the journey to work, which is necessarily reflected in higher production costs and thereby adversely affects the relative market advantages of businesses and industries within the Region.

Standards "1. The total vehicle-hours of travel within the Region should be minimized.

"2. Adequate capacity and a sufficiently high level of geometric design should be provided to achieve the following overall speeds based on potential 24-hour average weekday traffic volumes for arterial street and highway facilities:

	Overall Speed[n] in M.P.H. for Various Type Areas[o]			
Type of Facility	Downtown	Inter-mediate	Outlying	Rural
A. Arterials:				
1. Freeway	35–55	40–55	55–65	60–70
2. Expressway	25–40	30–45	40–50	50–65
3. Standard Arterials:				
a. Divided	15–25	25–35	35–45	45–60
b. Undivided	15–25	20–35	25–40	40–50
B. Collectors	10–20	15–30	20–35	40–50
C. Locals	5–15	10–20	15–25	30–40

[b]Local transit is defined as the transportation of persons by bus providing relatively frequent service to the general public on regular schedules over prescribed surface streets.

[c]A transit route may be serviced by a single bus if it can make a round trip in one hour or less. As the route length and/or the potential revenue passengers increase, additional buses may be required to service the route.

[d]High-density is defined as 39.5 persons and 12.0 dwelling units per net residential acre; medium-density is defined as 14.3 persons and 4.3 dwelling units per net residential acre; low-density is defined as 4.0 persons and 1.2 dwelling units per net residential acre.

[e]Modified rapid transit is defined as the transportation of persons by buses operating over freeways in mixed traffic lanes.

[f]Daylight hours are defined as the hours between 6:00 a.m. and 8:00 p.m.

"3. The proportion of total travel on freeway, expressway, and rapid and modified rapid transit facilities should be maximized.

Information Analysis

Information analysis for the continuing comprehensive transportation planning process involves collecting data and evaluating these data to assist in understanding the nature of the metropolitan area and its activity systems. The data are also used as a basis for forecasting future activity and for developing simulation models which explain activity in the area.

Some of the types of data normally collected include:

1. Travel data—demand
2. Transportation facilities data—supply
3. Land use characteristics
4. Economic characteristics
5. Population data
6. Miscellaneous
   a. Fiscal data
   b. Public policy
   c. Legal and statutory information
   d. Community goal and value structure
   e. Visual and aesthetic features
   f. Areas of historical or other significance

---

[g]Bus rapid transit is defined as the transportation of persons by buses operating over exclusive freeway lanes or private rights-of-way to provide high-speed service.

[h]Rail rapid transit is defined as the transportation of persons by single or dual track rail car trains operating over exclusive grade-separated rights-of-way to provide high-speed service.

[i]Ibid, footnote f.

[j]These maximum loading factors may be exceeded for periods of up to 10 minutes.

[k]T = 10 percent, K = 8 percent for freeways and 10 percent for all other streets and highways, D = 60/40 for non-CBD links and 50/50 for CBD links.

[l]In 1963 over 85 percent of the downtown parkers walked less than one block in the Milwaukee, Racine, and Kenosha CBD's.

[m]

CBD	Present Level in Population of Urbanized Area	Existing Space Per 1,000 Auto Destination
Milwaukee	1,200,000	350
Racine	108,000	200
Kenosha	82,000	270

[n]Overall speed is defined as average speed over the transportation system not including terminal time, is expressed in miles per hour based on 24-hour average weekday traffic and should not be confused with posted speed limits.

[o]Type of Area is defined in the *Highway Capacity Manual*, page 20, and discussed in "Capacity of Arterial Network Links," SEWRPC *Technical Record* Vol. 2 – No. 2."

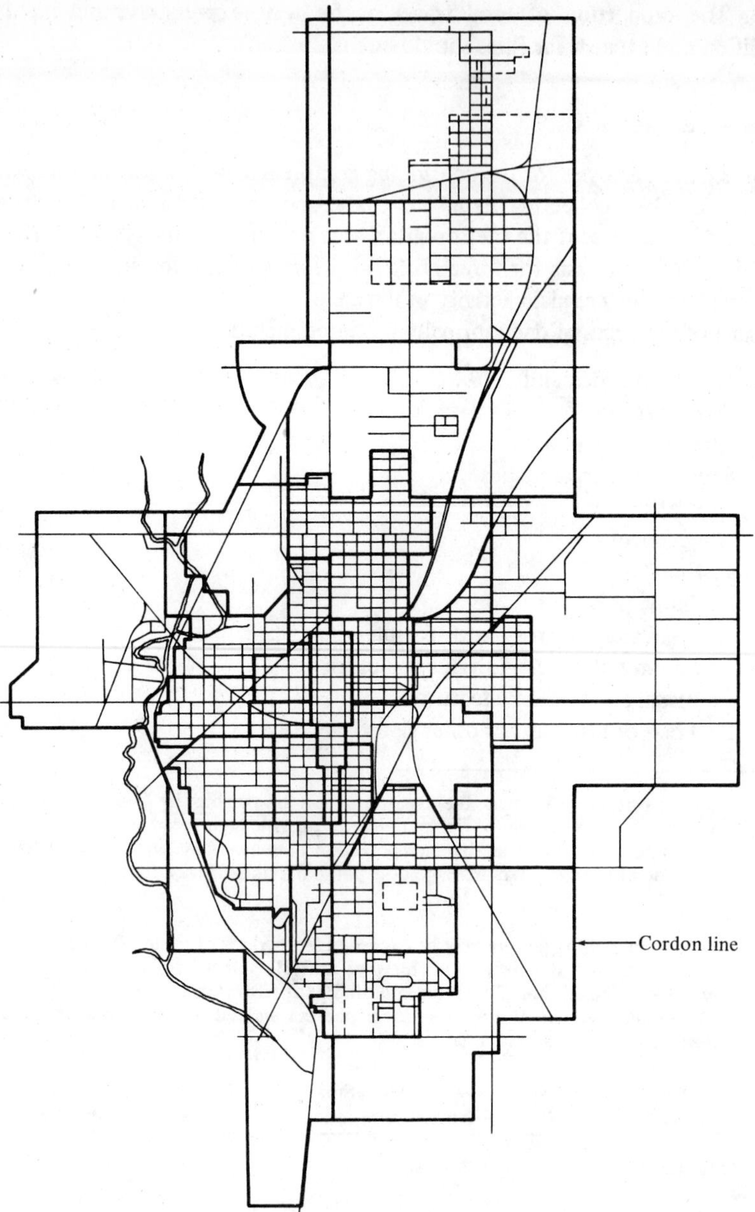

Fig. 38. Sample study area, zone map[8].

The data are collected for a study area normally defined by a cordon line which delimits the outward expansion of the urban area for the target planning year, normally a period 20–30 years in the future. Figure 38 shows a "typical" urban study area delimited by an outer cordon line; the smaller geographic zones within the cordon line are subareas for analysis purposes.

Travel and Transportation Facilities Inventory Information is obtained for travel performed in the area and the characteristics of the travelers (including person movement and goods movement) and on the physical and operating facilities over which travel is performed.

Travel Information Studies Information about the travel performed within the area of study is obtained by extensive interviewing of two basic types: internal home interviews and external roadside interviews. Most studies also include an inventory of taxi movements and truck movements.

Internal Home Interviews The home interview survey defines two types of trips: internal trips, those made solely within the cordon area (see Fig. 39); and external trips, those for which one end is inside and the other outside of the study area. Data of this type are normally obtained by taking a sampling of households within the cordon area. The sample size depends on the size of the area and the degree of sample accuracy desired.

Type of Information Obtained, Home Interview Survey Figures 40a and b show a standard home interview survey form. Two basic types of information are obtained, characteristics of the home and travel information. The characteristics of the home (Fig. 40a) include type of dwelling unit and for each dwelling unit: the number of residents (grouped by age and sex), the number of

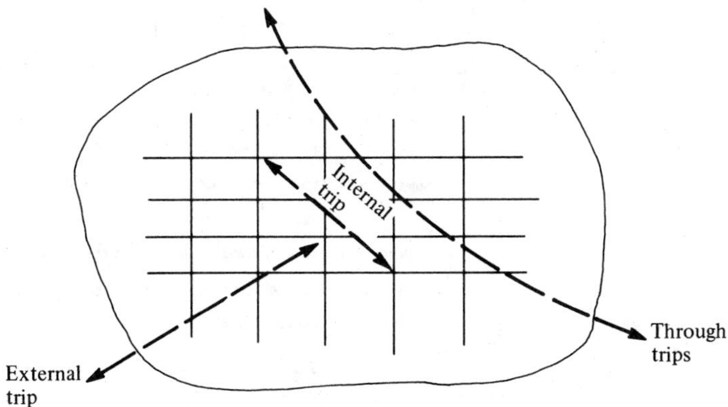

Fig. 39. Map of urban study area showing trip types.

P-E-520-69

TRANSPORTATION STUDY

HOME INTERVIEW ADDRESS SUMMARY

SECTION I

A. TRAVEL DATE ______ DISTRICT ______ BLOCK ______

B. INTERVIEW ADDRESS ______ (STREET) ______ (CITY)

C. STRUCTURE TYPE ______

1. SINGLE FAMILY
2. TWO FAMILY
3. 3-4 APARTMENTS
4. 5 OR MORE APARTMENTS
5. ROOMING HOUSE, DORMITORY
6. HOTEL

SECTION II

A. HOW MANY PASSENGER CARS ARE OWNED BY RESIDENTS AT THIS ADDRESS? (INCLUDE COMPANY-OWNED CARS) ______

B. HOW MANY PERSONS LIVE HERE? ______

C. HOW MANY PERSONS 5 YEARS OF AGE AND OVER LIVE HERE? ______

D. HOW MANY OUT-OF-AREA VISITORS 5 YEARS OF AGE AND OVER ARE STAYING HERE? ______

E. HOW LONG HAS HEAD OF HOUSEHOLD LIVED AT THIS ADDRESS? YEARS ______ MONTHS ______

F. HOUSEHOLD INFORMATION (COMPLETE ITEMS BELOW FOR EACH PERSON):

PERSON IDENTIFICATION (5 YEARS OR OVER)	✓ IF INTER-VIEWED	PERSON NUMBER	SEX AND RACE	AGE	DRIVE A CAR? 1. YES 2. NO	OCCUPATION	CODE	CODE
		1						
		2						
		3						
		4						
		5						
		6						
		7						
		8						
		9						
		0						

0

SECTION III (OFFICE USE ONLY)

A. NUMBER OF TRIPS REPORTED AT THIS ADDRESS: ALL TRIPS ______

B. NUMBER OF PERSONS 5 YEARS OF AGE AND OLDER MAKING TRIPS? ______

C. NUMBER OF PERSONS 5 YEARS OF AGE AND OLDER MAKING NO TRIPS? ______

D. NUMBER OF PERSONS 5 YEARS OF AGE AND OLDER WITH TRIPS UNKNOWN ______

E. NUMBER OF PERSONS 16 YEARS OF AGE AND OLDER? ______

F. NUMBER OF PERSONS 16 YEARS OF AGE AND OLDER THAT DRIVE A CAR? ______

G. COMPLETED OR NON-INTERVIEW CODE: ______

SEX AND RACE CODE:

1. M—W
2. F—W
3. M—N
4. F—N
5. M—O
6. F—O

Fig. 40a. Transportation study form[9].

STATE OF WISCONSIN/DEPARTMENT OF TRANSPORTATION

CITY AND YEAR

SAMPLE

SUBZONE

7. MOTEL
8. TRAILER
9. INSTITUTION
0. OTHER (SPECIFY)

INDUSTRY	TRIPS MADE 1-YES 2-NO 3-UNKNOWN	WORK AT HOME 1-YES 2-NO

AUTO DRIVER TRIPS

SECTION IV ADMINISTRATIVE RECORD

CONFIDENTIAL INTERVIEW DATA

SAMPLE PHONE NUMBER:

CALLS MADE		APPOINTMENTS	
DATE	TIME	DATE	TIME
1.			
2.			
3.			
4.			
5.			
6.			

NOTES AND COMMENTS

SUPERVISOR'S COMMENT:

INITIAL

INTERVIEWERS COMMENT

INTERVIEWER:

DATE COMPLETE NO.

INTERVIEWS CHECKED: (DATE) (INITIAL)

CODED BY: (DATE) (INITIAL)

CODING CHECKED BY: (DATE) (INITIAL)

Fig. 40a (*Continued*)

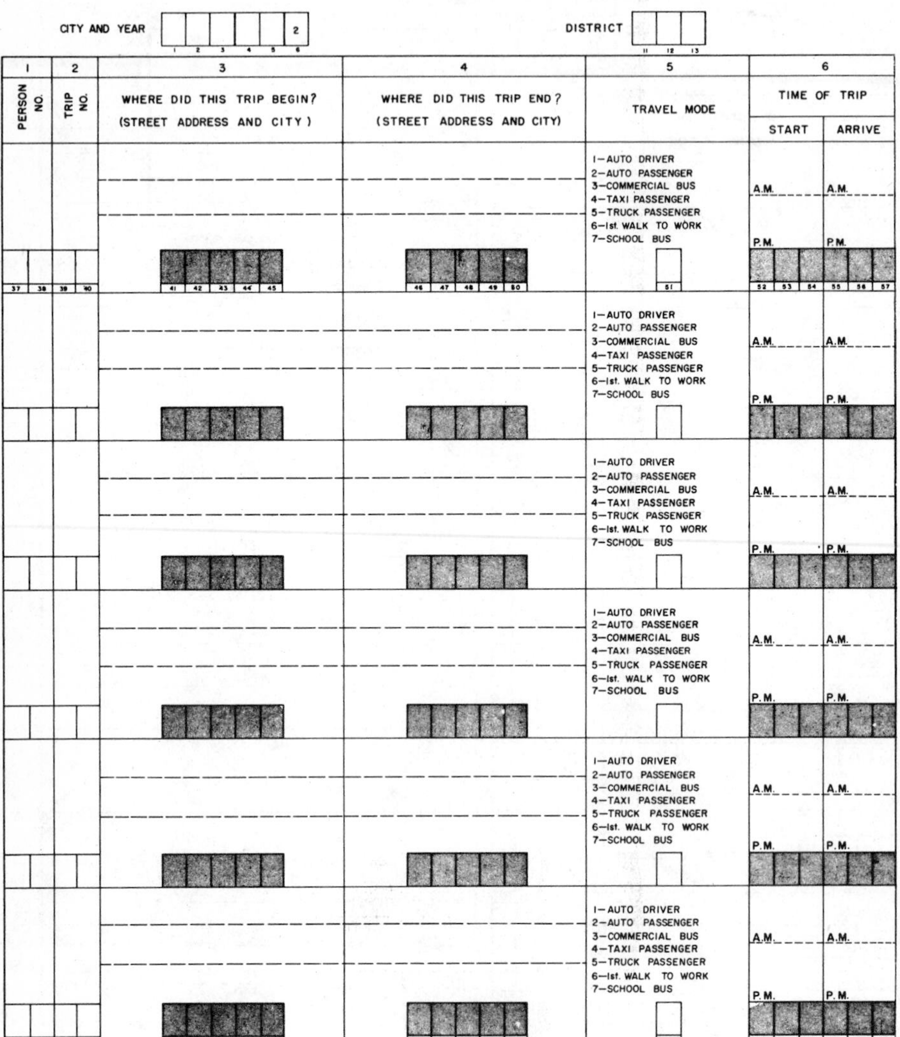

P-E-521-69

TRANSPORTATION STUDY

HOME INTERVIEW TRIP REPORT

CITY AND YEAR [| | | | | 2] (1 2 3 4 5 6)   DISTRICT [| |] (11 12 13)

1 PERSON NO.	2 TRIP NO.	3 WHERE DID THIS TRIP BEGIN? (STREET ADDRESS AND CITY)	4 WHERE DID THIS TRIP END? (STREET ADDRESS AND CITY)	5 TRAVEL MODE	6 TIME OF TRIP START	6 TIME OF TRIP ARRIVE
37 38	39 40	41 42 43 44 45	46 47 48 49 50	1—AUTO DRIVER 2—AUTO PASSENGER 3—COMMERCIAL BUS 4—TAXI PASSENGER 5—TRUCK PASSENGER 6—Ist. WALK TO WORK 7—SCHOOL BUS 51	A.M. P.M. 52 53 54	A.M. P.M. 55 56 57
				1—AUTO DRIVER 2—AUTO PASSENGER 3—COMMERCIAL BUS 4—TAXI PASSENGER 5—TRUCK PASSENGER 6—Ist. WALK TO WORK 7—SCHOOL BUS	A.M. P.M.	A.M. P.M.
				1—AUTO DRIVER 2—AUTO PASSENGER 3—COMMERCIAL BUS 4—TAXI PASSENGER 5—TRUCK PASSENGER 6—Ist. WALK TO WORK 7—SCHOOL BUS	A.M. P.M.	A.M. P.M.
				1—AUTO DRIVER 2—AUTO PASSENGER 3—COMMERCIAL BUS 4—TAXI PASSENGER 5—TRUCK PASSENGER 6—Ist. WALK TO WORK 7—SCHOOL BUS	A.M. P.M.	A.M. P.M.
				1—AUTO DRIVER 2—AUTO PASSENGER 3—COMMERCIAL BUS 4—TAXI PASSENGER 5—TRUCK PASSENGER 6—Ist. WALK TO WORK 7—SCHOOL BUS	A.M. P.M.	A.M. P.M.
37 38	39 40	41 42 43 44 45	46 47 48 49 50	1—AUTO DRIVER 2—AUTO PASSENGER 3—COMMERCIAL BUS 4—TAXI PASSENGER 5—TRUCK PASSENGER 6—Ist. WALK TO WORK 7—SCHOOL BUS 51	A.M. P.M. 52 53 54	A.M. P.M. 55 56 57

Fig. 40b. Transportation study form[9].

STATE OF WISCONSIN\DEPARTMENT OF TRANSPORTATION

PAGE ____ OF ____

SAMPLE [17] [18] [19] [20] [21]

7		8	9		10	11	12	13	14
TRIP PURPOSE		LAND USE AT	BLOCKS WALKED AT		AUTO DRIVERS ONLY				
					CAR POOL	TOTAL NO. IN CAR	PARKING AT DESTINATION		
FROM	TO		"O"	"D"			LOCATION (ADDRESS)	KIND	RATE
0-HOME 1-WORK 2-PERSONAL BUSINESS 3-MEDICAL — DENTAL 4-SCHOOL 5-SOCIAL - EAT MEAL 6-CHANGE MODE 7-SERVE PASS. 8-SHOPPING 9-RECREATION	0 1 2 3 4 5 6 7 8 9	ORIGIN____ DEST.____			YES 1 NO 2				
[58]	[59]	[60 61 62] [63 64 65]	[66]	[67]	[68]	[69]	[70 71 72 73 74]	[75]	[76]
0-HOME 1-WORK 2-PERSONAL BUSINESS 3-MEDICAL — DENTAL 4-SCHOOL 5-SOCIAL - EAT MEAL 6-CHANGE MODE 7-SERVE PASS. 8-SHOPPING 9-RECREATION	0 1 2 3 4 5 6 7 8 9	ORIGIN____ DEST.____			YES 1 NO 2				
0-HOME 1-WORK 2-PERSONAL BUSINESS 3-MEDICAL — DENTAL 4-SCHOOL 5-SOCIAL - EAT MEAL 6-CHANGE MODE 7-SERVE PASS. 8-SHOPPING 9-RECREATION	0 1 2 3 4 5 6 7 8 9	ORIGIN____ DEST.____			YES 1 NO 2				
0-HOME 1-WORK 2-PERSONAL BUSINESS 3-MEDICAL — DENTAL 4-SCHOOL 5-SOCIAL - EAT MEAL 6-CHANGE MODE 7-SERVE PASS. 8-SHOPPING 9-RECREATION	0 1 2 3 4 5 6 7 8 9	ORIGIN____ DEST.____			YES 1 NO 2				
0-HOME 1-WORK 2-PERSONAL BUSINESS 3-MEDICAL — DENTAL 4-SCHOOL 5-SOCIAL - EAT MEAL 6-CHANGE MODE 7-SERVE PASS. 8-SHOPPING 9-RECREATION	0 1 2 3 4 5 6 7 8 9	ORIGIN____ DEST.____			YES 1 NO 2				
0-HOME 1-WORK 2-PERSONAL BUSINESS 3-MEDICAL — DENTAL 4-SCHOOL 5-SOCIAL - EAT MEAL 6-CHANGE MODE 7-SERVE PASS. 8-SHOPPING 9-RECREATION	0 1 2 3 4 5 6 7 8 9	ORIGIN____ DEST.____			YES 1 NO 2				
[58]	[59]	[60 61 62] [63 64 65]	[66]	[67]	[68]	[69]	[70 71 72 73 74]	[75]	[76]

PARKING CODES

COLUMN 13-KIND		COLUMN 14-RATE
1. STREET	5. RES. PROPERTY	1. HOUR
2. LOT	6. CRUISED	2. DAY
3. PKNG. GARAGE	7. NOT PARKED	3. MONTH
4. SERVICE OR REPAIRS	8. RAMP	4. METER
		5. FREE
		0. NONE

Fig. 40b *(Continued)*

autos owned and the number of drivers, where possible the family income, and miscellaneous other data. The other basic type of information (Fig. 40b) concerns the trips made by persons residing in each sampled dwelling unit. This includes information for each person in the family on the origin and the destination for each trip made during the day prior to the interview; the purpose for which the trip was made; the time of day during which the trip was made; the mode of travel, i.e., auto or mass transit; and the number of persons making the trip. This information is indicative of the who, why, when, and where of travel.

Sampling Process and Size Home interview sampling in the internal study area is normally obtained by sampling selected households within the cordon area. Because the sampling process involves a major portion of both the dollar and time resource expenditure for the study, it is not possible to sample 100% of households. Figure 41 shows typical sampling rates based on the degree of accuracy required (expressed in terms of root mean square error).

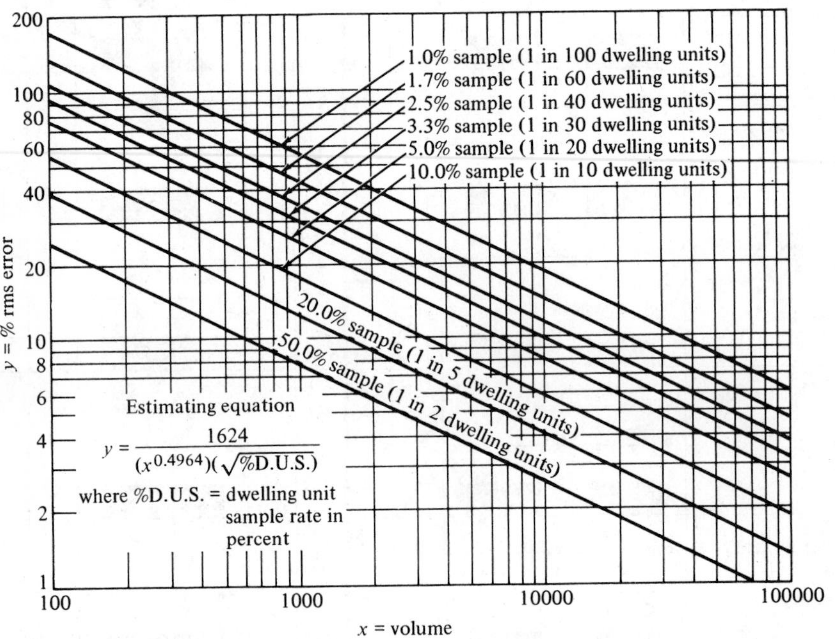

Fig. 41. Relation of percent root-mean-square error and volume for various dwelling-unit sample rates[2].

Data Aggregation by Geographic Area The area within the cordon is normally divided into a number of small geographic areas. For a unit stratification these are typically called districts, sectors, zones, and subzones. Figure 42 notes the

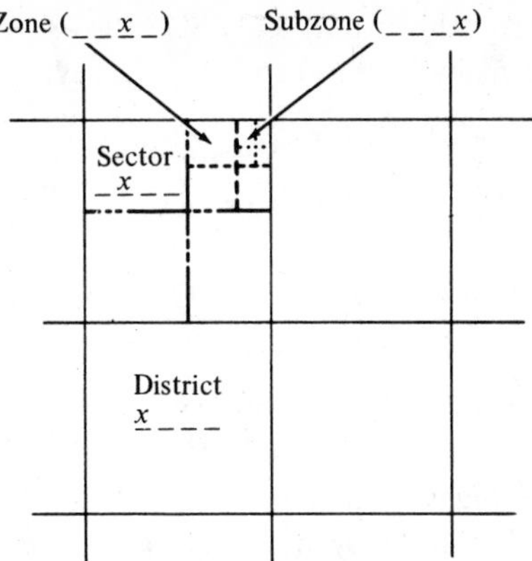

x defines subunit identification for four-digit number system

District: x _ _ _
Sector: _ x _ _
Zone: _ _ x _
Subzone: _ _ _ x

Fig. 42. System for subdividing study area into subunits.

interrelationship between these geographical subunits and how the numbering system is developed for computer analysis identification.

Criteria guiding the selection of boundaries of these small geographic areas to which travel data are aggregated are:

1. The location of natural barriers such as rivers, railroads, rough terrain, arterials, etc.
2. Homogeneity of land use.
3. Boundaries of existing political and census tracts.
4. The desirability of maintaining approximately square-shaped areas.
5. The fact that the zones should not straddle a screenline.*
6. Desirable maximum size: 1 square mile.
7. Desirable maximum population: 1000.

External Cordon Survey Information on travel, where one or both trip ends are outside of the study area, must also be obtained. This is done by stopping vehicles where they cross the external cordon line and asking each driver

*A screenline is an imaginary line, usually along a physical boundary, such as a river, which splits the study area into two (or more) parts. By comparing actual volumes of traffic against O-D volumes crossing the screenline the accuracy of the interview process can be checked.

pertinent questions about his travel. Figure 43 shows an external cordon survey form of the type normally used. From the survey, such information as trip origin and destination, land use at each trip end, the route of exit or entrance to the study area, the type of vehicle, the number of occupants in the vehicle, the purpose of the trip, and information about stops within the internal area are obtained.

Sampling rates vary up to a high of 100%, depending on the character and volume of traffic on the route upon which the sample is obtained.

Travel Facilities Inventory Besides information on the travel performed, detailed information must also be collected on the physical and operational characteristics of the highway and mass transit systems existing or proposed for the urban area. For highway networks, the network is explicitly defined by functional classification so as to define its basic land access and mobility characteristics. In addition, travel time information and capacity data are obtained for each link in the system. For mass transit, information is obtained about such things as: the transit network, route capacities, equipment characteristics, schedules, and use.

Where appropriate, information about other movement systems and their use may also be obtained.

Land-use Inventory It is also necessary to obtain detailed information on the characteristics of land use activity in the metropolitan area. For occupied land, this includes information on type of use (i.e., residential, business, industrial), intensity of use (i.e., area devoted to certain activities or numbers of employees per unit of area), and the location with respect to other uses. It is also necessary to define the characteristics of undeveloped vacant land to determine its potential for future development. Some of this information includes the availability or potential availability of utilities and other improvements, the accessibility of such lands, topographic characteristics, and drainage and soils problems. Such information, in identifying suitability for future uses, can be used as a guide in defining the future land use of undeveloped areas.

Oftentimes information is also obtained on the structural and environmental characteristics of existing uses. Such studies identify blighted areas and potential renewal areas.

Population and Economic Studies It is necessary to obtain detailed information on the characteristics of resident population and economic activity forecasts. The latter will permit an estimate of future land use needs by type and, as a result, the general travel activity for the area. Details of population and economic studies will not be dealt with here. For further information on these areas see references (11), (12), (13), and (14).

System Simulation–Modeling Travel The simplified flow chart identified as Fig. 44 depicts the general overall mathematical modeling procedure which is

INTERVIEWER ______ STATION NO. ______ DAY ______ HOUR PERIOD ______ A M INBOUND ______

STUDY HIGHWAY DATE P M OUTBOUND ______

1	2	3	4		6		7	8		9	10	11	12	13		14
			ORIGIN	5	DESTINATION			PURPOSE			← ASK OF THRU TRIPS ONLY →	INTERMEDIATE STOP				
SERIAL NUMBER	VEH. TYPE	NO. OF OCCUP.	ADDRESS	LAND USE	ADDRESS		LAND USE	O	D	WHERE IS THIS VEHICLE GARAGED?	ROUTE EXIT OR ENTRANCE	LOCATION	LAND USE	STOP PURPOSE AUTO	TRUCK	
	PC-W PC-O Ø			Ø			Ø			O D			Ø	1. WORK OR BUS. 2. SHOPPING 3. RECREATION 4. SERVE PASS. 5. EAT MEAL 6. GAS, OIL 7. OVERNIGHT	Ø	Ø
18 19 20	21 22	23	24 25 26 27 28 29 30 31 32 33	34 35 36	37 38 39 40 41 42 43 44 45 46		47 48 49	50	51	52	53 54	55 56 57 58 59	60 61 62		63 64	65
	PC-W PC-O Ø			Ø			Ø			O D			Ø	1. WORK OR BUS. 2. SHOPPING 3. RECREATION 4. SERVE PASS. 5. EAT MEAL 6. GAS, OIL 7. OVERNIGHT	Ø	Ø
	PC-W PC-O Ø			Ø			Ø			O D			Ø	1. WORK OR BUS. 2. SHOPPING 3. RECREATION 4. SERVE PASS. 5. EAT MEAL 6. GAS, OIL 7. OVERNIGHT	Ø	Ø
	PC-W PC-O Ø			Ø			Ø			O D			Ø	1. WORK OR BUS. 2. SHOPPING 3. RECREATION 4. SERVE PASS. 5. EAT MEAL 6. GAS, OIL 7. OVERNIGHT	Ø	Ø
	PC-W PC-O Ø			Ø			Ø			O D			Ø	1. WORK OR BUS. 2. SHOPPING 3. RECREATION 4. SERVE PASS. 5. EAT MEAL 6. GAS, OIL 7. OVERNIGHT	Ø	Ø
	PC-W PC-O Ø			Ø			Ø			O D			Ø	1. WORK OR BUS. 2. SHOPPING 3. RECREATION 4. SERVE PASS. 5. EAT MEAL 6. GAS, OIL 7. OVERNIGHT	Ø	Ø
18 19 20	21 22	23	24 25 26 27 28 29 30 31 32 33	34 35 36	37 38 39 40 41 42 43 44 45 46		47 48 49	50	51	52	53 54	55 56 57 58 59	60 61 62		63 64	65

COL. 2 VEHICLE TYPE

PC-W 1. PASSENGER CAR - WISCONSIN
PC-O 2. PASSENGER CAR - OTHER
3. PICK-UP, PANEL OR SINGLE UNIT - SINGLE REAR TIRE
4. SINGLE UNIT - DUAL REAR TIRE
5. SINGLE UNIT - THREE OR MORE AXLES
6. COMBINATIONS
7. BUS
8. TAXI

TRIP PURPOSES

PASSENGER CARS	TRUCKS & BUSES
0. HOME	A. PICK UP GOODS
1. WORK	B. DELIVER GOODS
2. PERSONAL BUSINESS	C. PICK UP & DELIVER GOODS
3. MEDICAL-DENTAL	D. OTHER WORK-CONNECTED BUSINESS
4. SCHOOL	E. TO BASE OF OPERATION
5. SOCIAL & EAT MEAL	F. PERSONAL BUSINESS
6. CHANGE TRAVEL MODE	G. SERVICE CALL
7. SERVE PASSENGER	H. SERVE PASSENGER
8. SHOPPING	
9. RECREATION	

TRUCKS & BUSES STOP PURPOSES

A. PICK UP GOODS
B. DELIVER GOODS
C. PICK UP & DELIVER GOODS
D. OTHER WORK-CONNECTED BUSINESS
E. TO BASE OF OPERATION
F. PERSONAL BUSINESS
G. SERVICE CALL
H. SERVE PASSENGER

Fig. 43. Transportation study, trip report, external survey[9]. Modal split is noramlly made at one of the two locations shown in the simulation process.

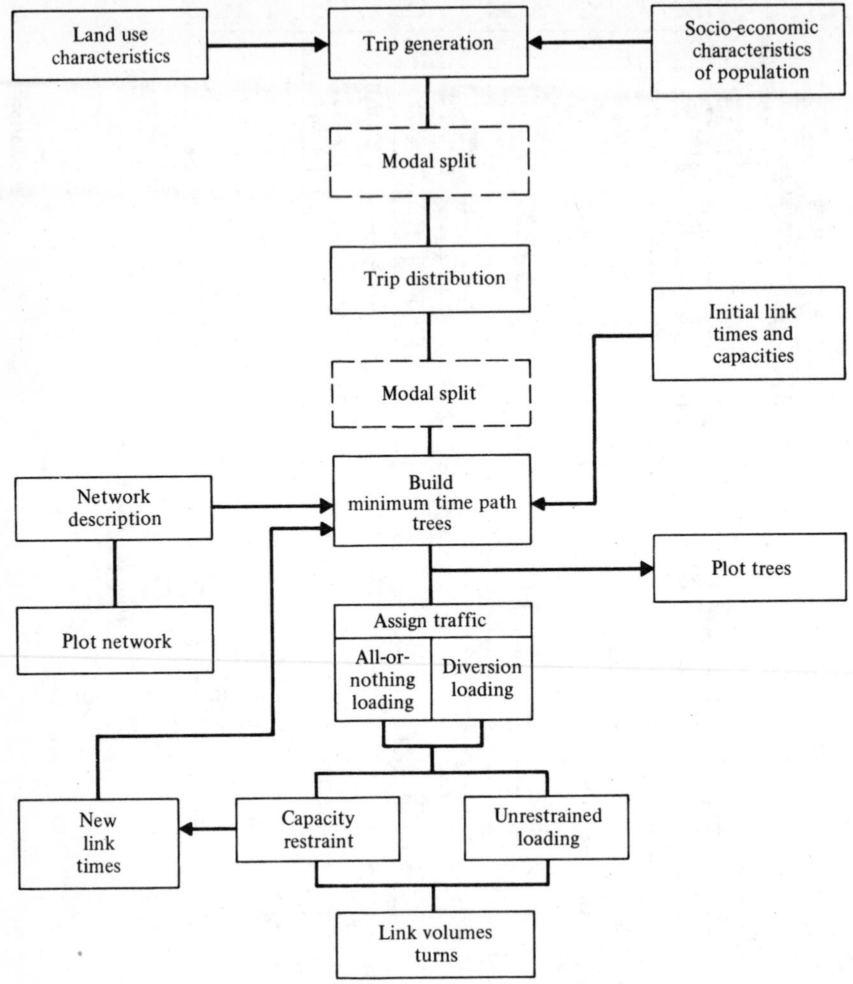

Fig. 44. Generalized model for network demand determination. Modal split is normally made at one of the two locations shown in the simulation process.

used for developing travel loadings on street/highway networks and mass transportation lines. The basic procedure includes three algorithms: trip generation, trip distribution, and traffic assignment. Figure 45 schematically shows the function of each of these algorithms.

Trip generation is the procedure which develops, for specific land use zones, the number of trips produced by the zone and attracted into the zone. This is normally done by trip purpose and often in terms of person trips.

Trip distribution is the procedure which connects the trip productions and trip attractions for each zone, irrespective of the highway or mass transit net-

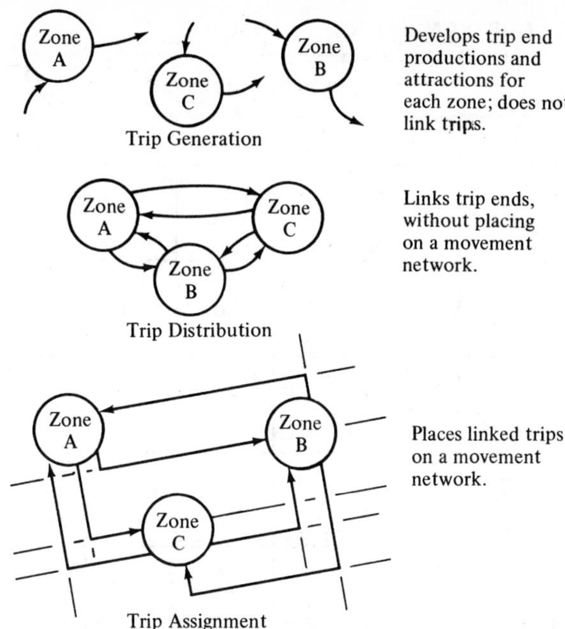
Zone A
Zone C
Zone B
Trip Generation
Develops trip end productions and attractions for each zone; does not link trips.
Zone A
Zone C
Zone B
Trip Distribution
Links trip ends, without placing on a movement network.
Zone A
Zone B
Zone C
Trip Assignment
Places linked trips on a movement network.

Fig. 45. Function of generation, distribution, and assignment models.

work over which these trips will travel. The result of the distribution process is a matrix of trip interchanges similar to that shown in Table 16.

Traffic assignment is the process by which the actual numbers of vehicles, auto or mass transit, traveling on each link in the transportation network is determined.

A fourth algorithm, modal split, describes, based on past travel performance, the number of person trips which will be made by auto and the number which will be made by some form of mass transit. Modal split is normally done at either the trip generation phase or the trip distribution phase.

Trip Generation Model (15)

The trip generation modeling process develops a functional relationship between trip end volumes (both productions and attractions) and land use and socio-economic characteristics at the land uses at both ends. It assumes that urban travel patterns now, and in the future, are and will continue to be a function of these characteristics and that the interrelationships developed will not materially change over time.

The nature of travel in the urban area is such that most travel, about 34% during the average day (Table 17 and Fig. 46), is work related; during the morning and afternoon peak travel period an even greater proportion is work

TABLE 16. TRIP DISTRIBUTION MATRIX.

Origins (by zone) \ Destinations (by zone)	1	2	3	4	5	6	Total
1	10	2	1	7	3	8	31
2	8	13	2	6	4	12	45
3	12	9	7	14	13	2	57
4	6	8	5	13	14	10	56
5	15	7	12	11	3	6	54
6	14	12	10	9	10	4	59
Total	65	51	37	60	47	42	302

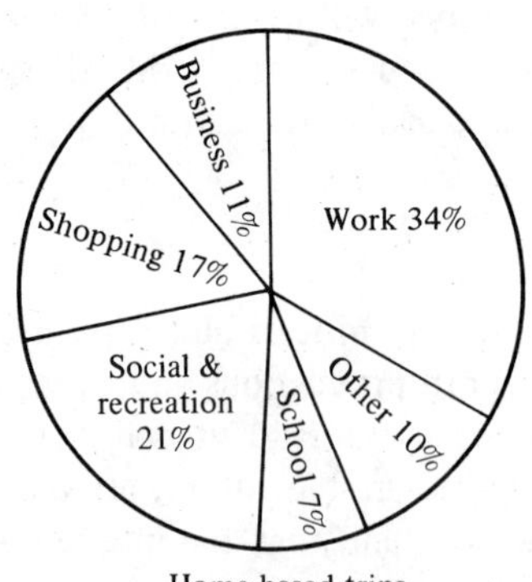

Fig. 46. Home-based trips, daily distribution[8].

TABLE 17. HOME-BASED TRIPS BY URBAN RESIDENTS IN STUDY AREAS ACCORDING TO PURPOSE[1] (19).

		Percentage of home-based trips to and from							
Urban area	Home-based trips (% of all linked trips)	Work	Business	Shopping	Social-recreational	School	Other	All Purposes	Total home-based trips per dwelling unit
Chicago[2]	86.8	37.5	9.7	18.9	22.8	4.0	7.1	100.0	5.17
Detroit	87.0	41.6	8.6	13.9	20.1	6.3	9.5	100.0	4.67
Washington	91.6	43.1	9.6	14.2	12.5	9.4	11.2	100.0	4.23
Pittsburgh[2]	87.0	37.7	21.6	14.9	13.8	12.0	–[3]	100.0	4.21
St. Louis	91.3	37.5	8.1	17.3	21.5	6.4	9.2	100.0	4.90
Houston	91.0	33.1	8.9	17.3	18.6	10.8	11.3	100.0	5.51
Kansas City	88.2	33.4	8.8	17.2	22.7	6.0	11.9	100.0	5.14
Phoenix	85.3	25.2	10.2	19.7	20.0	11.6	13.3	100.0	4.76
Nashville	85.5	30.3	8.5	16.9	23.9	7.4	13.0	100.0	5.48
Ft. Lauderdale	86.5	27.9	15.3	24.0	22.9	0.9	9.0	100.0	2.82
Charlotte	83.9	32.2	8.0	15.6	23.8	6.6	13.8	100.0	5.56
Reno	86.5	29.2	12.7	18.1	26.3	0.5	13.2	100.0	4.88
Avg. Per Cent	87.6	34.0	10.8	17.4	20.8	6.8	10.2	100.0	4.78[4]

[1] Source: Origin-destination studies in each area.
[2] Chicago and Pittsburgh data are for "linked" trips.
[3] All trips in Pittsburgh have been identified with one of the listed purposes.
[4] Unweighted average.

related (see Fig. 47). Since these are also the trips which are most stable and for which we have the most accurate information, they are most readily predicted for the future.

Model Parameters The parameters which best describe the number of trip ends produced and attracted by different types of zones are *intensity* of land use, *character* of the area and its inhabitants or employees, and *location.* Knowing these general attributes, the transportation planner can define specific relationships between trip production and the particular attribute being described.

For example, the relationship between the intensity of land use and trip making can be described as a function of the number of dwelling units per net residential acre. As Fig. 48 indicates, trip making increases as residential density decreases. This holds for auto trips and for total person trips. Similar relationships between intensity and trip making can also be developed for nonresidential uses.

Character of land use is also necessary to describe trip making in residential areas because residential land use intensity is not sufficient to form the entire basis of the residential trip generation rate structure. Character partly defines such things as the socio-economic identity of the households being considered. For example, there is a very distinct relationship between income and auto ownership and correspondingly between auto ownership and trip making. Figures 49 and 50 show the relationships between income and auto ownership

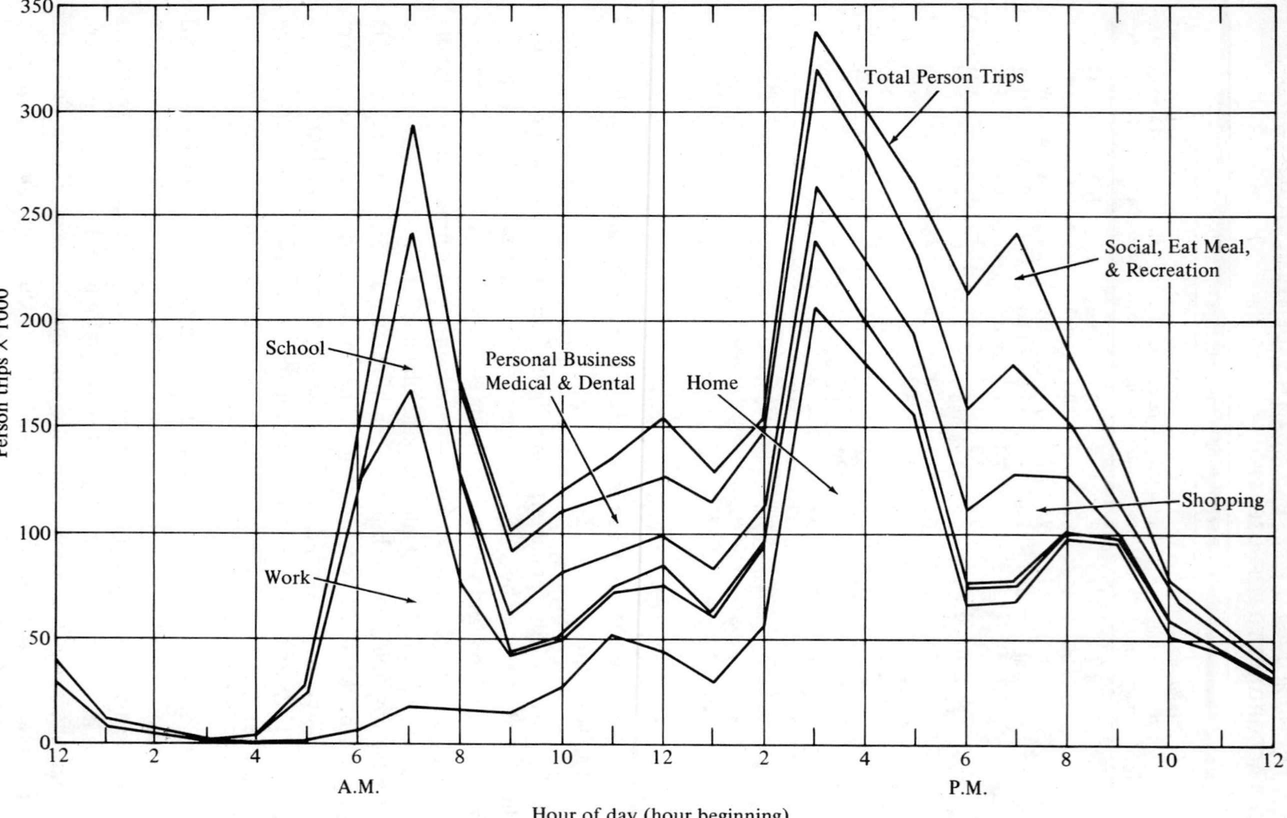

Fig. 47. Hourly variation of unlinked internal person trips by purpose at destination, 1963[7].

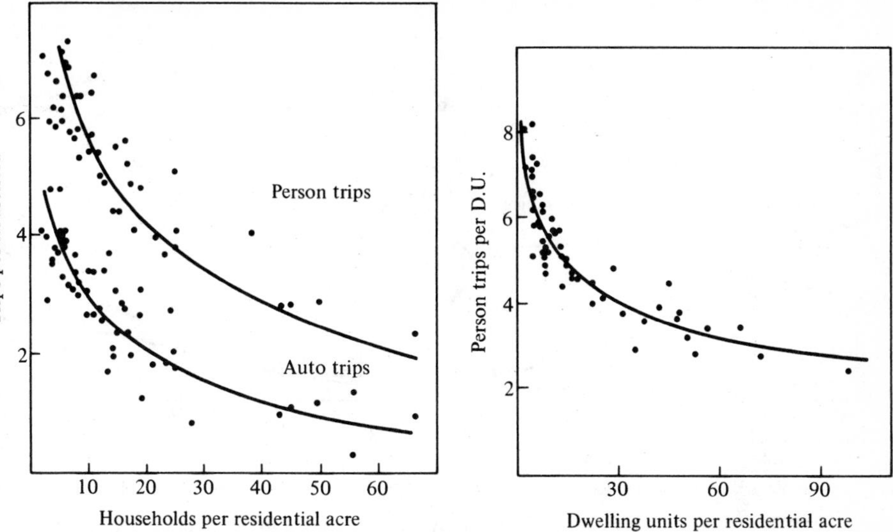

Fig. 48. Effect of residential density on trip production by districts: (a) Pittsburgh. (b) Washington, D.C.[15].

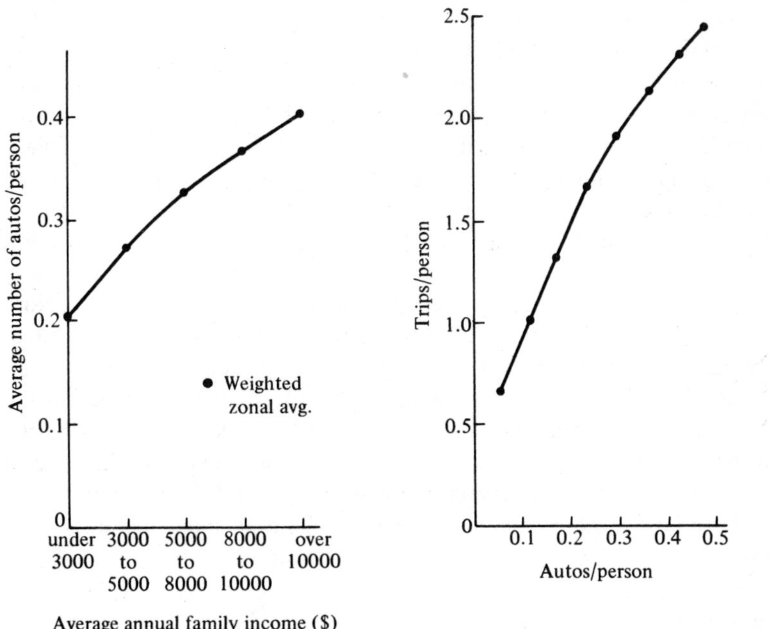

Fig. 49. Relationship between cars per person and average family income by zone and cars per person and trips per person by dwelling unit, Kansas City[5].

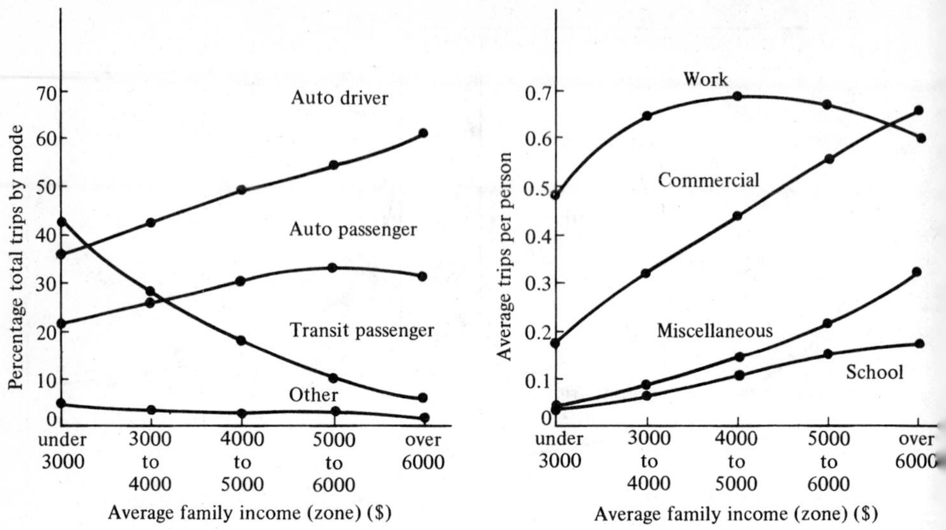

Fig. 50. Relationship between median family income and trip production by mode of travel and by trip purpose, St. Louis[5].

and trip production, and those between median family income and trip production by mode of travel and trip purpose.

The third characteristic, location of land use activity, often has a significant effect on trip generation. This is particularly true for residential areas where location is also likely to be related to family size, stage in the family cycle (e.g., young families with small children compared to middle aged families with grown children who are no longer residing at home), income, and a number of other characteristics.

For nonresidential uses, intensity, character, and location variables normally take on a somewhat different form. Table 18 identifies some of the typical variables used for describing residential and nonresidential travel. Note that these variables are for zonal and dwelling unit analysis procedures. These are the two methods currently used for estimating trip productions and attractions. The zonal procedure is most commonly used and uses average zonal socioeconomic data to obtain zonal trip making. The dwelling unit procedure develops trips by average household characteristics and expands this information into zonal trip making. Generation models are typically developed for different trip purposes; those commonly used for generation modeling are as noted in Table 19.

Trip Generation Modeling Procedures

Three basic modeling procedures exist for forecasting trip ends by areal units within the study area. These are land area trip rate analysis, cross classification analysis, and multiple linear regression analysis.

TABLE 18. SELECTED VARIABLES FOUND SIGNIFICANT IN URBAN TRANSPORTATION PLANNING (16).

Variables	Weight given for use in trip generation analysis*
A. Variables found significant in zonal trip generation analysis	
1. Demographic data	
a. Total population	1
b. Age, sex, race, etc.	3
c. No. of household units	1
d. School enrollment	2
e. Family life cycle	3
2. Economic data	
a. Total employment	1
b. Selected employment	1
c. Employment by industry	3
d. Employees by residence	1
e. Labor force	3
f. Labor force by occupation and industry	3
g. Median income	1
h. Income stratified	3
i. Automobile ownership	1
j. Dwellings without autos	2
k. Retail sales	2
l. Average home value	3
3. Land use data	
a. Specific activities	3
b. Selected categories	1
B. Variables found significant in dwelling unit trip generation analysis	
a. Car ownership	1
b. Family size	1
c. No. of persons 5 years old, and over in household	1
d. Length of residence	3
e. Family income	2
f. No. of persons 16 years old and over	2
g. No. of persons 16 years old and over who drive	1
h. Age of head of the household	2
i. Distance from the CBD	3
j. Stage in the family life cycle	1
k. Occupation of head of household	1
l. Structure type	1

*Key to weights: 1 = Essential data; 2 = Desirable data; 3 = Useful data.

Trip Rate Analysis Land area trip rate analysis develops average trip generation rates which reflect the character, location and intensity of land use for the zones considered. Table 20 gives an example of typical land area person trip generation rates for a particular study and Table 21 gives a generalized trip generation vocabulary and rates. One problem with land area trip generation rates is that they are gross estimates of trip making and often yield questionable results.

TABLE 19. TRIP PURPOSE CLASSIFICATIONS FOR MODELING (2)

Large Urban Areas (> 100,000)
1. Home-based work
2. Home-based shop
3. Home-based social-recreational
4. Home-based school
5. Home-based miscellaneous
6. Non-home-based
7. Truck trips
8. Taxi trips
Small Urban Areas (< 100,000)
1. Home-based work
2. Home-based nonwork
3. Non-home-based

TABLE 20. TYPICAL LAND AREA GENERATOR RATES FOR PERSON TRIPS, CHICAGO AREA, 1956 (16).

		Person trip destinations per acre					
Ring	Average distance from loop (miles)	Residential	Manufac-turing	Transpor-tation	Commercial	Public buildings	Public open space
0	0.0	2228.5	3544.7	273.1	2132.2	2013.8	98.5
1	1.5	224.2	243.2	36.9	188.7	255.5	28.8
2	3.5	127.3	80.0	15.9	122.1	123.5	26.5
3	5.5	106.2	86.9	10.8	143.3	100.7	27.8
4	8.5	68.3	50.9	12.8	212.4	77.7	13.5
5	12.5	43.0	26.8	5.8	178.7	58.1	6.1
6	16.0	31.2	15.7	2.6	132.5	46.6	2.5
7	24.0	21.1	18.2	6.4	131.9	14.4	1.5
Avg. for study area		48.5	49.4	8.6	181.4	52.8	4.2

Cross Classification Analysis Cross classification analysis is a technique which uses a matrix to describe how two or more variables effect trip making. Table 22 is a typical cross classification matrix relating total person trips per dwelling unit to auto ownership and the number of persons residing in each dwelling unit. One difficulty with cross classification analysis is the fact that there is no simple way of measuring the amount of variation in the dependent variable (that which is being predicted), which is "explained" by the independent variables. Another is the fact that the higher the degree of stratification, the larger the sample size must be. Cross classification does have the advantage that it helps determine which are the most significant variables in explaining travel.

A simple example will show how the cross classification data are used by zone. Assume a zone with 500 dwelling units, in which 50% of the dwelling units have 3 persons and one car. From Table 22, 250 dwelling units (50% times 500) would produce 1790 trips. This is obtained by multiplying the 7.16 total person trips per dwelling unit corresponding to 3-person households with one car by the 250 dwelling units. For the remainder of the dwelling units, other trip rates would be applied based on corresponding auto ownership and dwelling unit occupancy data.

Multiple Regression Analysis It is possible to develop, through multiple linear regression analysis, equations which explain trip making for various trip purposes related to those parameters which best explain the trip making. Multiple regression analysis is the technique most commonly used for trip generation analysis.

Criteria for the selection of variables used as "explainers" in the generation equations are listed below:

1. The variable selected should be highly correlated with trip making in a statistical sense.
2. A strong logical relationship should exist in a causal sense.
3. Variables should be used which are not difficult to obtain or forecast.
4. Variables should be stable over time.
5. The variables used should be those commonly used by operational studies in trip generation analysis.
6. The number of variables should be limited so as not to distort the analysis with too many variables which are interrelated. Typically, only two or three variables are used. Table 18 lists variables commonly used.
7. The variables which are used must be compatible with the techniques being tested.

Table 23a and 23b display typical multilinear regression trip generation equations and variable descriptions.

Trip Distribution

Trip distribution is the mathematical modeling process which allocates the trip ends determined from the trip generation process to specific zonal interchanges. As noted in Fig. 51, it links the trip ends produced and attracted in the generation process. This is done without regard to specific route.

Two basic types of model are used for distributing trips; the growth factor models and the so-called inter-area travel models. Growth factor models in today's modeling process are used to distribute through trips, whereas inter-area models are used to distribute internal trips and external trips. (See Fig. 51.)

TABLE 21. TRAFFIC GENERATION VOCABULARY AND TRIP GENERATION RATES (11).

Land use	Density		Traffic generation rate (veh. trip ends per day)				
			Number per acre		Number per unit		
	Unit	Number	Range	Typical	Unit	Range	Typical
Residential	dwelling units/acre				dwelling unit		
low density (single-family homes)		1–5	5–65	40		7–12	9
medium density (patio houses, duplexes, townhouses)		5–15	40–150	75		5–8	7
high density (apartments)		15–60	85–400	180		3–7	5
Commercial							
retail commercial	acres				1000 sq. ft F.A.		
neighborhood retail (supermarket)		10	800–1,400	1000		70–240	130
community retail (junior department store)		10–30	700–1,000	900		60–140	80
regional retail (regional shopping center)		30	400–700	600		30–50	40
central area retail	high density		600–1,300	900		10–50	40
highway-oriented commercial (motels, service stations)	low density		100–300	240		4–12	10
service commercial (office buildings)	FAR[a]				1000 sq. ft F.A.		
single-story bldg. with surface parking		0.5:1	120–1,200	300		6–60	14
two-story bldg. with surface parking		1:1	240–2,400	600		6–60	14
three- to four-story bldg. with deck parking		2:1	360–6,000	1200		6–60	14
three- to six-story bldg. with surface parking		5:1	1,200–12,000	2600		6–60	14
high-rise office bldg. (more than 10 stories) with structure parking		10:1	2,400–20,000			6–60	14

Industrial	employees/acre				1000 sq. ft F.A.		
highly automated industry; low employee density (refinery, warehouse)		5	2–8	4		0.2–1.0	0.6
light service industry, single-lot industry (lumber yard)		5–20	6–30	16		0.4–1.2	0.8
industrial tract (5 acres) (machinery factory)		20–100	30–160	70		0.6–4.0	2.0
office campus: research and development (research industry)		100	150–200	170		3–8	4
mixed central industry; small industrial plants		Varies	10–100			1–4	
Public and semi-public uses		Varies					
schools and colleges	no. of students		7–600 (colleges)	60	student	0.4–1.0	0.8
places of public assembly (theater, stadium, convention center)	no. attending				4 seats (stadia)		2
administration facilities (city hall, state offices, post offices)	FAR[a]		70–600	200	1,000 sq. ft F.A.	10–60	20
recreation facilities (park, zoo, beach, golf course)			1–10 (parks)	4	acre (golf course)	2–10	8
terminals (bus terminal, airport)			3–30	15	based aircraft[b]	6–12	8
hospitals	No. of beds		16–70	40	bed[c]	6–16	10

[a]Floor area ratio.
[b]Local airport.
[c]Person trip ends.

TABLE 22. RELATIONSHIP OF FAMILY SIZE AND AUTO OWNERSHIP TO AVERAGE TOTAL PERSON TRIPS PER DWELLING UNIT* (16).

	Average total person trips per D.U.				
	Number of autos owned per D.U.				
Number of persons per D.U.	0	1	2	3 and over	Weighted average
1	1.03	2.68	4.37**	–	1.72
2	1.52	5.13	7.04	2.00**	4.38
3	3.08	7.16	9.26	10.47	7.46
4	3.16	7.98	11.56	12.75	9.10
5	3.46**	8.54	12.36	17.73**	10.16
6–7	7.11**	9.82	12.62	16.77**	11.00
8 and over	7.00**	9.66	17.29	22.00**	12.24
Weighted average	1.60	6.62	10.53	13.68	6.58

*Based on 1962 O–D survey data supplied by the Madison Area Transportation Study, Madison, Wisconsin.
**Average based on fewer than 25 samples.

TABLE 23A. TRIP GENERATION EQUATIONS (20).*

Home-based work productions	= –0.26166 + 1.34883 AUTOS +0.15847 LABFOR
Home-based work attractions	= 79.59052 + 1.260368 TOTEMP
Home-based shop productions	= –8.17407 + 1.03587 AUTOS
Home-based shop attractions	
central business district	= 414.00524 COMLAN
shopping center	= 181.52416 COMLAN
other (neighborhood)	= 41.20011 COMLAN
Home-based social-recreation productions	= 18.08165 + 1.41352 AUTOS
Home-based social-recreation attractions	= 110.18328 + 0.24067 POP + 0.53633 EMPT + S
Home-based school productions	= 25.82347 + 0.33401 ENROLL
Home-based school attractions	= 0.57960 ATTEND
Home-based other productions	= 5.81557 + 1.58387 AUTOS
Home-based other attractions	= 114.67008 + 0.72150 DU + 0.32343 TOTEMP
Non-home-based productions	= 177.40781 + 0.75895 DU + 0.40813 TOTEMP
Non-home-based attractions	= 175.89957 + 0.88025 DU + 0.30663 TOTEMP

*Variables are described in Table 23b.

Growth Factor Models The growth factor models assume that travel patterns can be projected into the future on the basis of anticipated differential zonal growth rates and existing interzonal movements. One major drawback of the growth factor procedure is the fact that it requires comprehensive origin-

TABLE 23B. VARIABLE DESCRIPTION FOR TRIP GENERATION ANALYSIS (20).

Mnemonic	Variable	Variable type
POP	Population	independent
DU	Dwelling units	independent
AUTOS	Number of registered autos	independent
INCOME	Average annual family income	independent
LABFOR	Resident labor force	independent
RESLAN	Residential land use (acres)	independent
NETLAN	Net land use (acres)	independent
COMLAN	Commercial land use (acres)	independent
TOTEMP	Total employment	independent
EMPT + S	Employment in trades and services	independent
ENROLL	Student enrollment*	independent
ATTEND	Student attendance	independent
HBWP	Home-based work productions	dependent
HBWA	Home-based work attractions	dependent
HBSHP	Home-based shop productions	dependent
HBSHA	Home-based shop attractions	dependent
HBSRP	Home-based social recreation productions	dependent
HBSRA	Home-based social recreation attractions	dependent
HBSCP	Home-based school productions	dependent
HBSCA	Home-based school attractions	dependent
HBOP	Home-based other productions	dependent
HBOA	Home-based other attractions	dependent
NHBP	Non-home-based productions	dependent
NHBA	Non-home-based attractions	dependent

*For purposes of this analysis, enrollment is defined as the number of students residing in a zone, while attendance is based on the zonal school location.

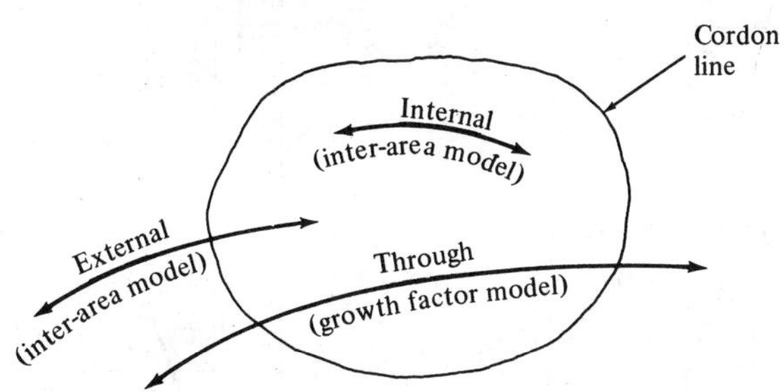

Fig. 51. Trip types and trip distribution procedure.

destination survey data. Because the procedure is based on expanding existing trips, travel to and from zones to which no trips presently exist but for which there will be major trip generation in the future, will show no trip interchange as a result of the distribution process.

The growth factor methods used are: the uniform factor, the average factor, the Detroit method and the Fratar (16) method.

Largely because of computational efficiency, the Fratar method is the one which finds most predominant use. The growth factor distribution process is an iterative one in which calculations are repeated until the entire process converges on a "true" value. As Fig. 52 shows, the Fratar method converges more rapidly than the other three.

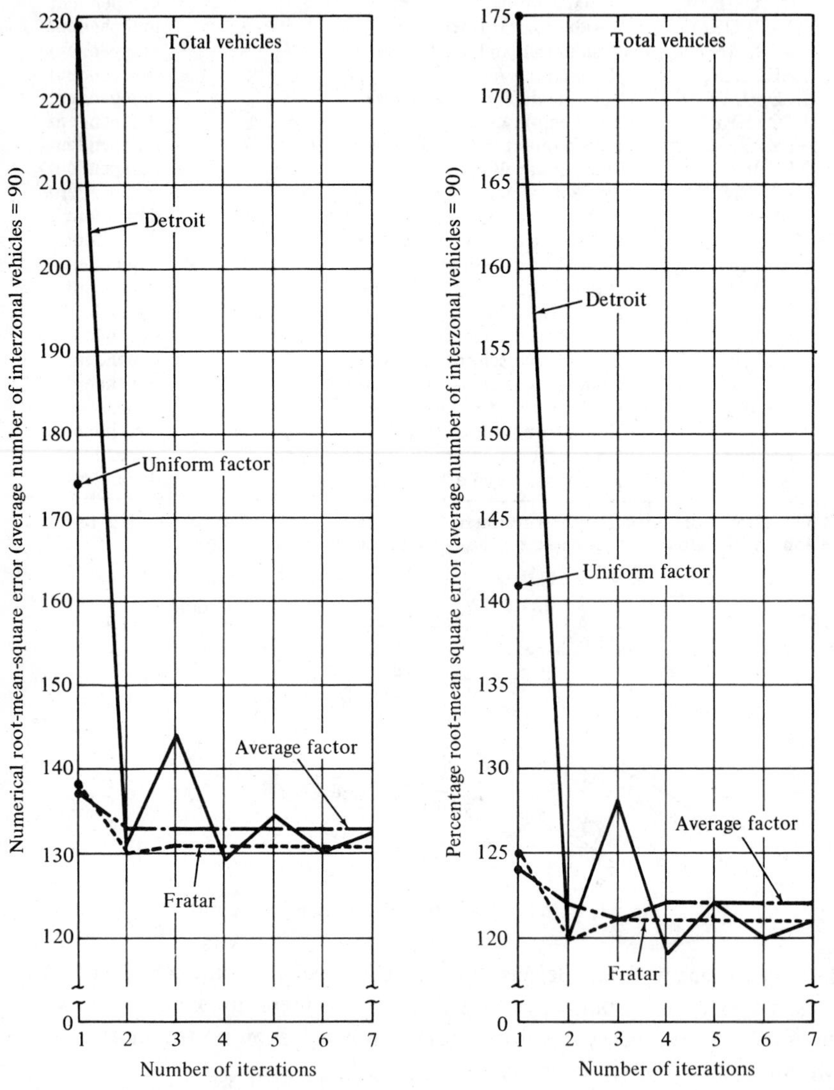

Fig. 52. Root-mean-square error vs. the number of iterations required to predict interzonal transfers, using various growth factor methods[7].

Fratar Method, Growth Factor (16) The Fratar method assumes that future trips can be estimated for any zone and distributed to movement involving that zone based on the existing trip interchange between that zone and every other zone and in proportion to the expected growth of every other zone. Doing this for each zone results in directional movements which are averaged for each pair of interzonal transfers. The entire process is then repeated.

Figure 53 illustrates a simplified three-zone area to which trips are distributed by the Fratar process. Following the theoretical formulation of the process is an example problem to demonstrate the theory of the Fratar method.

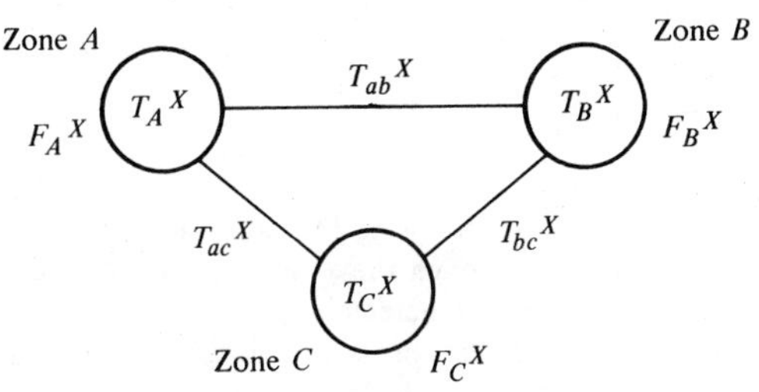

where

T_A^x, T_B^x, T_C^x = trip ends (productions plus attractions; zones A, B, and C at the end of iteration x)

$T_{ab}^x, T_{bc}^x, T_{ac}^x$ = two-way trips between zones at end of iteration x

T_{a-b}^x, T_{b-a}^x = one direction movement from A to B and B to A at end of iteration x

F_A^x, F_B^x, F_C^x = zonal growth factor at end of iteration x

Theory:

$x = 0$ will be used as initial values (those obtained in origin-destination study.

First Iteration ($x = 1$) Values:

$$T_{a-b}^1 = F_A^0 T_A^0 \left[\frac{T_{ab}^0 F_B^0}{T_{ab}^0 F_B^0 + T_{ac}^0 F_C^0}\right]$$

$$T_{b-a}^1 = F_B^0 T_B^0 \left[\frac{T_{ab}^0 F_A^0}{T_{ab}^0 F_A^0 + T_{bc}^0 F_C^0}\right]$$

$$T_{ab}^1 = \frac{T_{a-b}^1 + T_{b-a}^1}{2}$$

Similarly

$$T_{a-c}^1 = F_A^0 T_A^0 \left[\frac{T_{ac}^0 F_C^0}{T_{ac}^0 F_C^0 + T_{ab}^0 F_B^0}\right]$$

$$T_{c-a}^1 = F_C^0 T_C^0 \left[\frac{T_{ac}^0 F_A^0}{T_{ac}^0 F_A^0 + T_{bc}^0 F_B^0}\right]$$

$$T_{ac}^1 = \frac{T_{a-c}^1 + T_{c-a}^1}{2}$$

Fig. 53. Fratar process distribution model.

Similarly for $T^1_{b-c}, T^1_{c-b}, T^1_{bc}$

$$\left.\begin{aligned} T^1_A &= T^1_{ab} + T^1_{ac} \\ T^1_B &= T^1_{ab} + T^1_{bc} \\ T^1_C &= T^1_{ac} + T^1_{bc} \end{aligned}\right\} \text{First iteration, trip ends}$$

$$\left.\begin{aligned} F^1_A &= \frac{F^0_A T^0_A}{T^1_A} \\ F^1_B &= \frac{F^0_B T^0_B}{T^1_B} \\ F^1_C &= \frac{F^0_C T^0_C}{T^1_C} \end{aligned}\right\} \text{First iteration, adjusted growth factors}$$

Second Iteration ($x = 2$) *Values*

Second iteration values are obtained by using the calculated first iteration values as input to the calculating equations (just as though they were a new set of values). In effect, the iteration exponent (x) in the equations is incremented by one for each iteration.

To illustrate, the second iteration $a - b$ value is obtained from:

$$T^2_{a-b} = F^1_A T^1_A \left[\frac{T^1_{ab} F^1_B}{T^1_{ab} F^1_B + T^1_{ac} F^1_C}\right]$$

Note that

$$F^0_A T^0_A = F^1_A T^1_A = F^2_A T^2_A = F^x_A T^x_A$$

= (Future forecasted trip ends in zone A—similarly all other zones)

Sample Fratar Calculation:

A sample calculation for two iterations follows, with initial values (for iteration exponent $x = 0$) as noted:

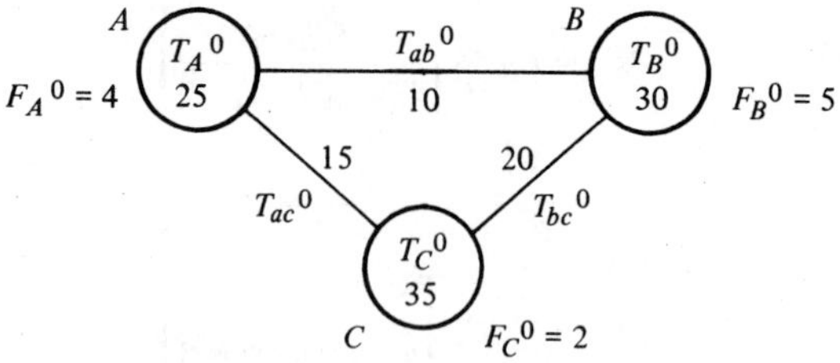

A, $T_A{}^0$ = 25, $F_A{}^0 = 4$; B, $T_B{}^0$ = 30, $F_B{}^0 = 5$; C, $T_C{}^0$ = 35, $F_C{}^0 = 2$; $T_{ab}{}^0$ = 10; $T_{ac}{}^0$ = 15; $T_{bc}{}^0$ = 20

$$T^1_{a-b} = 25(4)\left[\frac{10(5)}{10(5) + 15(2)}\right] = 62.50$$

$$T^1_{b-a} = 30(5)\left[\frac{10(4)}{10(4) + 20(2)}\right] = 75.00$$

68.75

Fig. 53 (*Continued*)

$$\left.\begin{array}{l} T^1_{a-c} = 25(4)\left[\dfrac{15(2)}{15(2) + 10(5)}\right] = 37.50 \\[2ex] T^1_{c-a} = 35(2)\left[\dfrac{15(4)}{15(4) + 20(5)}\right] = 26.25 \end{array}\right\rangle 31.88$$

$$\left.\begin{array}{l} T^1_{b-c} = 30(5)\left[\dfrac{20(2)}{20(2) + 10(4)}\right] = 75.00 \\[2ex] T^1_{c-b} = 35(2)\left[\dfrac{20(5)}{20(5) + 15(4)}\right] = 43.75 \end{array}\right\rangle 59.38$$

$T^1_{ab} = 68.75$

$T^1_{ac} = 31.88$

$T^1_{bc} = 59.38$

$$T^1_A = 100.63; \quad F^1_A = \frac{100}{100.63} = 0.994$$

$$T^1_B = 128.13; \quad F^1_B = \frac{150}{128.13} = 1.171$$

$$T^1_C = 91.26; \quad F^1_C = \frac{70}{91.26} = 0.767$$

Second Iteration:

In the second iteration $F^\circ_A T^\circ_A$ and $F^\circ_C T^\circ_C$ were used in place of $F'_A T'_A$ and $F'_C T'_C$ since they are equal.

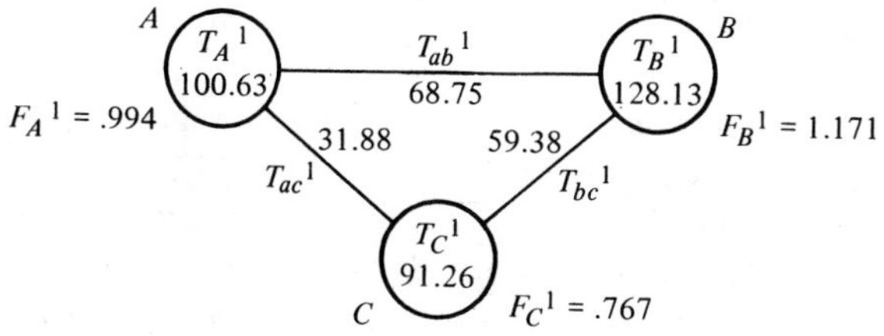

$$\left.\begin{array}{l} T^2_{a-b} = 25(4)\left[\dfrac{68.75(1.171)}{68.75(1.171) + 31.88(0.767)}\right] = 76.71 \\[2ex] T^2_{b-a} = 30(5)\left[\dfrac{68.75(0.994)}{68.75(0.994) + 59.38(0.767)}\right] = 90.02 \end{array}\right\rangle 83.37$$

Fig. 53 (*Continued*)

$$\left.\begin{array}{l} T_{a-c}^2 = 25(4)\left[\dfrac{31.88(0.767)}{31.88(0.767) + 68.75(1.171)}\right] = 23.29 \\[2ex] T_{c-a}^2 = 35(2)\left[\dfrac{31.88(0.994)}{31.88(0.994) + 59.38(1.171)}\right] = 21.92 \end{array}\right\rangle 22.61$$

$$\left.\begin{array}{l} T_{b-c}^2 = 30(5)\left[\dfrac{59.38(0.767)}{59.38(0.767) + 68.75(0.994)}\right] = 59.98 \\[2ex] T_{c-b}^2 = 35(2)\left[\dfrac{59.38(1.171)}{59.38(1.171) + 31.88(0.994)}\right] = 48.08 \end{array}\right\rangle 54.03$$

$$T_{ab}^2 = 83.37$$

$$T_{ac}^2 = 22.61$$

$$T_{bc}^2 = 54.03$$

$$T_A^2 = 22.61 + 83.37 = 105.98$$

$$T_B^2 = 83.37 + 54.03 = 137.40$$

$$T_C^2 = 22.61 + 54.03 = 76.64$$

$$F_A^2 = \frac{100}{105.98} = 0.944$$

$$F_B^2 = \frac{150}{137.40} = 1.092$$

$$F_C^2 = \frac{70}{76.64} = 0.913$$

Fig. 53 (*Continued*)

Table 24 summarizes the results of four Fratar iterations for the example problem. Note the convergence of the T_A, T_B, and T_C values on the ultimate expanded values of 100 trips (4×25), 150 trips (5×30), and 70 trips (2×35), respectively, and the convergence of successive iteration growth factors (F_A, F_B, F_C) on unity. This convergence indicates convergence on system balance. The iterations are continued until a satisfactory degree of convergence has been obtained.

Gravity Model, Interarea Distribution Formula The gravity model is used to distribute external and internal trips. Normally, different gravity models are used for external and internal trips by trip purpose.

The gravity model is based on the theory that trip interchange between zones is directly proportional to the relative attraction between zones and inversely proportional to some function of the spatial separation or travel friction between

TABLE 24. SUMMARY FOR FOUR ITERATIONS, FRATAR DISTRIBUTION.

	Iteration Number			
	1	2	3	4
T_{ab}	68.75	83.38	86.86	87.82
T_{ac}	31.88	22.61	18.54	16.20
T_{bc}	59.38	54.04	54.61	55.95
T_A	100.63	105.98	105.40	104.02
T_B	128.13	137.42	141.47	143.77
T_C	91.26	76.65	73.15	72.15
F_A	0.994	0.944	0.949	0.961
F_B	1.171	1.092	1.060	1.043
F_C	0.767	0.913	0.957	0.970

zones. The general formulation of the model is

$$T_{ij} = P_i \left[\frac{A_j/d_{ij}^{\,b}}{(A_1/d_{i1}^{\,b}) + (A_2/d_{i2}^{\,b}) + \cdots + (A_j/d_{ij}^{\,b}) + \cdots + (A_n/d_{in}^{\,b})} \right]$$

where

T_{ij} = trip interchange from zone i to j
P_i = trips produced by zone i
A_j = trips attracted to zone j
$d_{ij}^{\,b}$ = adjusted spatial separation between zones i and j
n = the total number of zones

The exponent b expresses an impedance or tolerance level for a certain type of trip. For example, a person will tolerate more travel impedance for a work trip than for a shopping trip; this is reflected in a lower b value for the work trip.

In applying the gravity model, the basic theoretical formula has been modified to the following form:

$$T_{ij} = \frac{P_i A_j F_{ij} K_{ij}}{\sum_{j=1}^{n} (A_j F_{ij} K_{ij})}$$

where

T_{ij} = trip interchange between zones i and j
P_i = trips produced in zone i
A_j = trips attracted to zone j
F_{ij} = interzonal trip impedance factor
K_{ij} = interzonal adjustment factor

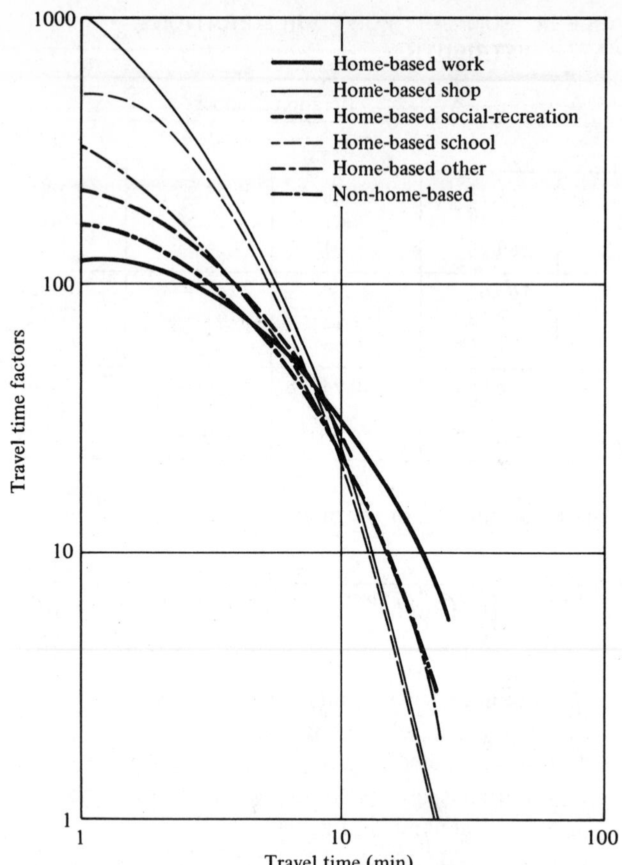

Fig. 54. Travel time factors vs. travel time.

F_{ij} approximates $1/(\text{time})^b$; see Fig. 54 for typical F_{ij} values from a recent study. The interzonal adjustment factor is used for calibrating the model to account for unidentified (in the model) peculiarities of zone pairs as they influence travel.

SAMPLE PROBLEM–GRAVITY MODEL

Figure 55 and Table 25 demonstrate the application of the gravity concept to the distribution of trips. It should be noted that the trip attractions and trip productions used are values obtained from the trip generation phase of the modeling process.

Zone I produces 1000 work trips per day which are distributed to zones I, II, III, and IV. Zones I, II, III, and IV attract 1000, 700, 6000, and 500 total work trips, respectively. Terminal times and interzonal travel times are as noted in Fig. 55.

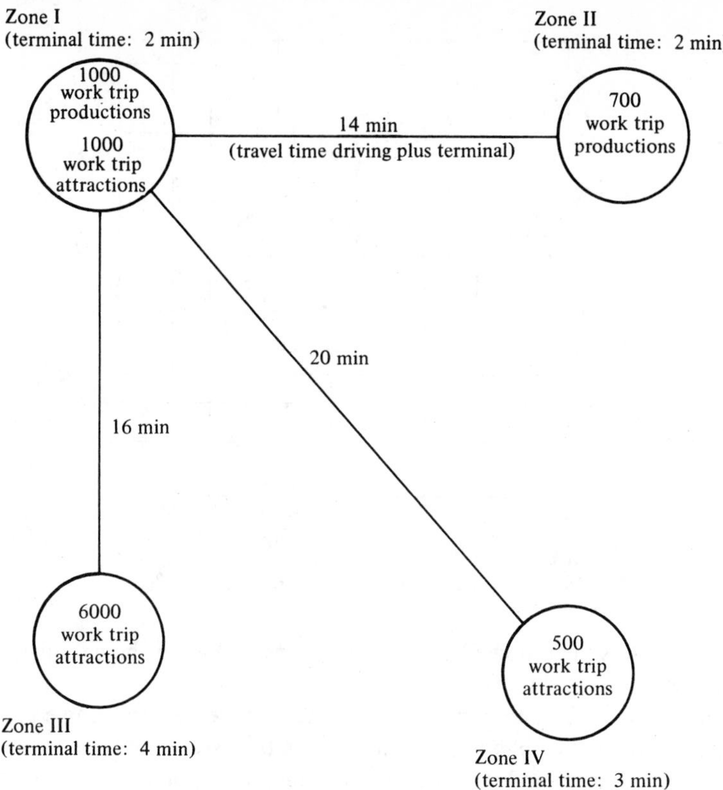

Fig. 55. Hypothetical four-zone problem, gravity model distribution[2].

Interzonal adjustment factors (K_{ij}) will not be used. Travel time factors (F_{ij}) used in the solution tabulation in Table 25 are obtained from Fig. 56.

Trip Purpose Classification For transportation studies in large urban areas (greater than 100,000 population), separate gravity models might be used for: (a) home-based work trips; (b) home-based shopping trips; (c) home-based social-

TABLE 25. PROBLEM SOLUTION

Zone	A_j	Terminal time	Driving time	Travel time*	F_{ij}	$A_{ij}F_{ij}/100$†	T_{ij}
I	1000	2	3	7	100	1000	186
II	700	2	10	14	68	476	88
III	6000	4	10	16	61	3660	680
IV	500	3	15	20	49	245	46
					Total	5381	1000

*The travel time consists of a driving time plus the terminal time at each trip end.
†Divide by 100 because values in Fig. 56 are multiplied by 100 for ease in computation.

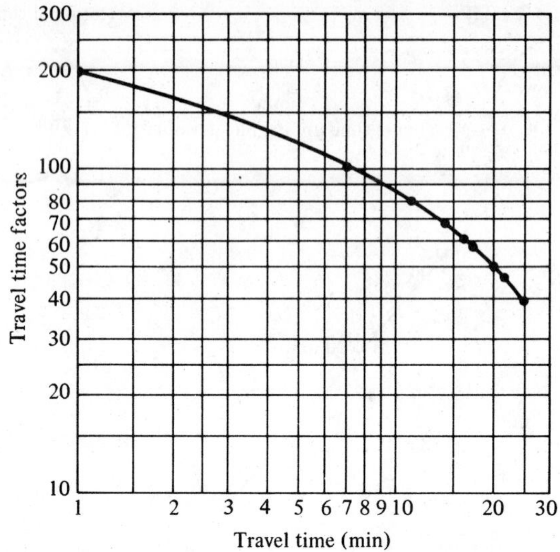

Fig. 56. Plot of travel time factors for sample problem[2].

recreational trips; (d) home-based school trips; (e) home-based other trips; (f) non-home-based trips; (g) truck trips; and (h) taxi trips.

In some areas, for example Washington, D.C., where there is a large predominance of government employment, it might even be desirable to stratify the work trip into a government and nongovernment employment sector.

For studies in small urban areas (less than 100,000 population) the same level of stratification is generally not necessary. Separate gravity models might be developed only for the following types of trips; (a) home-based work trips; (b) home-based non-work (other) trips; and (c) non-home-based trips.

Type of Model Used Some studies have modeled vehicle trips in the distribution phase whereas others have modeled person trips.

Use and Calibration of the Gravity Model Before the gravity model can be used for predicting interzonal travel, it must be "fine tuned" so that the output of the gravity model process replicates actual travel experience in the urban area being studied. Basically, this is done by developing the F_{ij} and the K_{ij} in a manner such that the trip table matrix produced by the gravity model closely approximates that obtained in the origin–destination survey.

The first adjustment factor to be developed is the travel time factor F_{ij}. This is a trial and error adjustment normally done separately for each trip purpose category. Typically, the gravity model is run using travel time factors from other comparable studies as initial values. A trip matrix is developed based on the gravity model trip distribution using the first run travel time factors. Then the proportion of total trips within each one-minute interzonal time increment category is compared with the proportion of total trips from the origin destination survey in comparable time increment categories. A first and fourth iteration comparison for one study is shown in Fig. 57. From the first-iteration F_{ij}'s new F_{ij}'s are developed based on the ratio

$$F_{ij}\,(\text{new}) = F_{ij}\,(\text{old}) \left(\frac{\%\ \text{Gravity model trips}}{\%\ \text{O–D trips}} \right)$$

This is done for each one-minute time increment. The process is repeated until the gravity model trip matrix replicates the origin–destination survey experience with reasonable accuracy. Typical travel time factors used in an urban transportation study are shown in Fig. 54.

Often there is sufficient variation in the socio-economic characteristics of individual zones that the gravity model results, even with properly applied travel

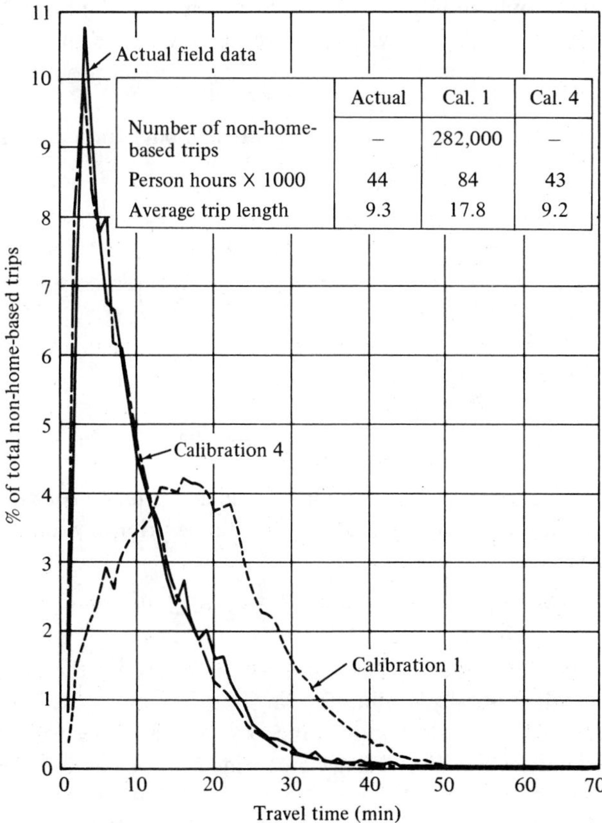

	Actual	Cal. 1	Cal. 4
Number of non-home-based trips	–	282,000	–
Person hours X 1000	44	84	43
Average trip length	9.3	17.8	9.2

Fig. 57. Trip length frequency for non-home-based trips, Washington, D.C., 1955[2].

time factors, will not adequately replicate the real life picture without some further adjustment. Consequently, individual zone-to-zone adjustment factors (K_{ij}) are developed for each zone to bring the gravity model results in line with those obtained from the origin–destination survey.

Once the gravity model is calibrated, i.e., the model accurately synthesizes current zone-to-zone travel interchange, it can then be used for synthesizing current and future travel interchange based on various transportation and land use alternatives.

Intervening Opportunities Model, Interarea Distribution Formula The intervening opportunities model was developed partly in response to the need for a travel interchange model which was more in line with human behavior as an explainer of travel interchange. The gravity model does not do this since it is basically an abstraction of a law of physics which only does the job of synthesizing travel interchange between zones.

The intervening opportunities model is based on the premise that a traveler seeking to satisfy a travel need will minimize his travel time subject to the condition that each destination point considered has a certain probability or likelihood of fulfilling his travel purpose.

The two forms of the intervening opportunities model formulation noted below are mathematically equivalent:

$$T_{ij} = O_i \, [e^{-LV_j} - e^{-LV_{j+1}}]$$

or

$$T_{ij} = O_i \, [(1 - L)^{V_j} - (1 - L)^{V_{j+1}}]$$

where

T_{ij} = trips from i to j
O_i = volume of trips originating in zone i
L = constant probability of a possible destination being accepted if it is being considered
V_j = the number of trip destinations considered prior to zone j
V_{j+1} = the number of trip destinations considered prior to zone j plus the zone j destinations.

In the formula the term in the brackets is the probability that a trip will end in zone j.

As with the gravity model, calibration is also necessary for the intervening opportunities model. Basically, calibration involves developing the L value for use in the model. This too is an iterative process with the model being run until it satisfactorily simulates real life. A single area-wide L value can be estimated from the equation

$$\bar{r} = K \sqrt{1/PL}$$

where

$\bar{r}$ = average trip length
K = a constant: approximately 2π
L = probability factor (probability that a destination being considered will be accepted)
P = density factor in trip ends per square mile

SAMPLE PROBLEM–INTERVENING OPPORTUNITIES MODEL (22)

Zone I produces 100 work trip origins per day which can be satisfied by work opportunities (destinations) in Zones II, III, and IV amounting to 50, 150, and 100 destinations, respectively. Zone I is strictly residential and has no work trip destinations. Figure 58 illustrates the problem situation along with travel times

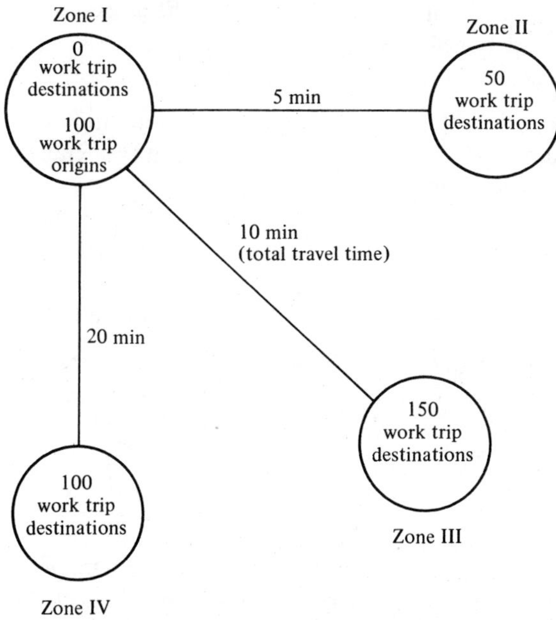

Fig. 58. Hypothetical four-zone problem, intervening-opportunities model[23].

between zones. The significance of the travel times is that they establish the order in which each zone is encountered as a possible destination for the trip production. Calibration of the model results in an L factor of 0.013.

Table 26 shows the development of the problem solution.

TABLE 26. PROBLEM SOLUTION, INTERVENING OPPORTUNITIES

Destination zone in Time Sequence from zone I	Destinations available in zone	V_j	V_{j+1}	e^{-LV_j} or $(1-L)^{V_j}$	$e^{-LV_{j+1}}$ or $(1-L)^{V_{j+1}}$	$e^{-LV_j} - e^{-LV_{j+1}}$	P_I	T_{I-j}
I	0	0	0	1.000	1.000	0.000	100	0
II	50	0	50	1.000	0.522	0.478	100	48
III	150	50	200	0.522	0.074	0.448	100	45
IV	100	200	300	0.074	0.020	0.054	100	6
$L = 0.013$							Total	100

Trip Assignment (8), (23)

The trip (traffic) assignment process is the procedure by which the interzonal trip interchanges are allocated to specific routes. The process basically quantifies the driver's or mass transit rider's route choice decision. Some of the purposes of traffic assignment are to:

1. Help determine deficiencies in the existing transportation system,
2. Assist in developing new systems by allowing an evaluation of the effect of various improvements,
3. Help in the decision process by which construction priorities are set,
4. Provide a systematic and reproducible test for evaluating alternate systems proposals, and
5. Provide the designer with specific volumes upon which to base his design.

Development of the traffic assignment process, as currently used, was initially blocked by the inability of the modeling process to select a minimum travel time path over which a trip could be routed. The need for such a network arises from the assumption that most drivers will select, given the opportunity, the path of least resistance in traveling from one location to another. Least resistance most often is measured in terms of least travel time.

Network Building (25) The development of the minimum-path technique (Moore's algorithm) allows the selection of a least-time path through a street system from each origin to every destination. A simplified network and network terminology are shown in Fig. 59.

Network definitions include the following:

Node: the intersection of two or more route sections (links) where there is the possibility of a turn or the beginning or terminating point for a trip.

Link: a one-way route section lying between two intersections or nodes.

Zone centroid: the node at which all trips are assumed to begin or end. There is normally one for each zone and it is not necessarily in the geographic

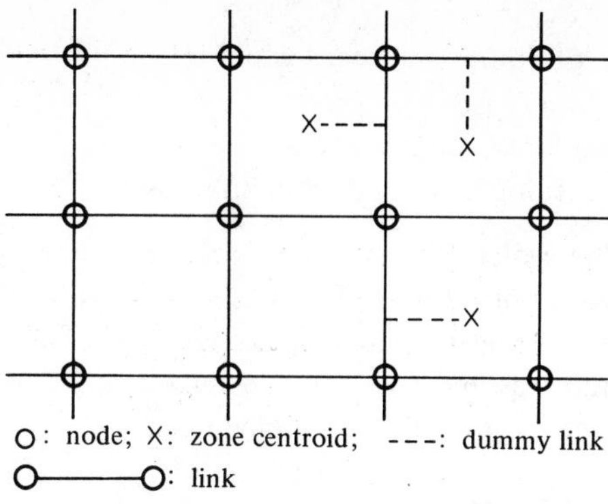

Fig. 59. Hypothetical network.

center of the zone. It is theoretically the center of the trip-generating activity.

Dummy link: a link of zero (usually) time used to connect the zone centroid and the remaining network.

Route or trace: a series of links which provide a minimum path between any two zones.

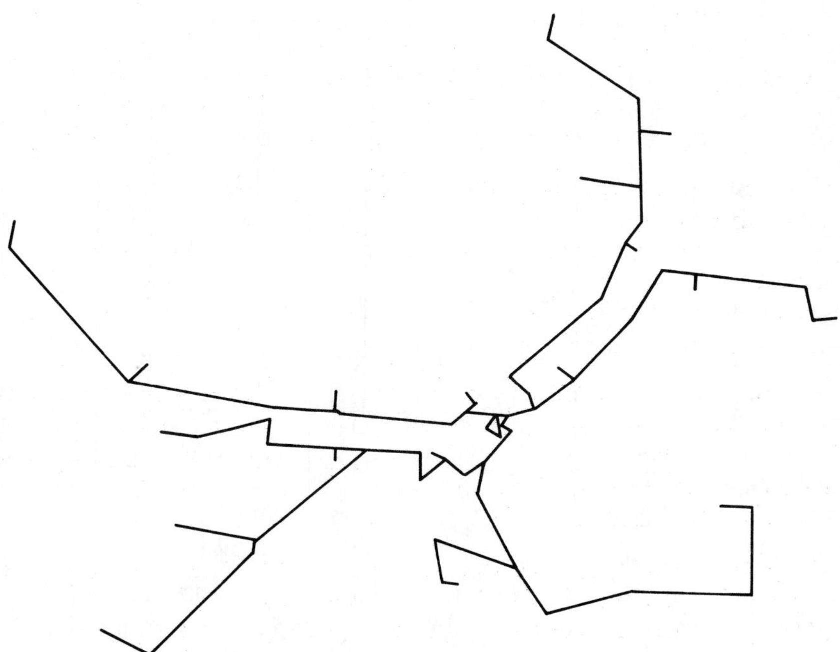

Fig. 60. Minimum-time-path tree for central business district home node[27].

Tree: a collection of routes or traces which make up the shortest-time path from one node to every other node.
Home node: a node for which a particular tree is built.
Back node: the preceding node in the minimum path.

Before trips can be assigned to the network, the minimum-time-path trees must be built for every home node. Initial trees are built using the original link travel times obtained during the travel analysis phase of the study. Figure 60 shows a typical minimum-time tree for one home node. A similar tree is developed by the computer for every other home node.

Assignment Methods (23) (24)

The traffic assignment procedure or network loading is normally done by one of two procedures. The most commonly used is an "all-or-nothing" procedure by which all trips between individual zones are loaded to the minimum-time path between those zones. A diversion program is also available which permits the assignment of a certain proportion of interzonal trips to the normal arterial system and the remainder to another system—possibly a freeway. The proportion of trips diverted is most often determined through the use of a time ratio diversion curve of the type shown on Fig. 61. The diversion data are based on

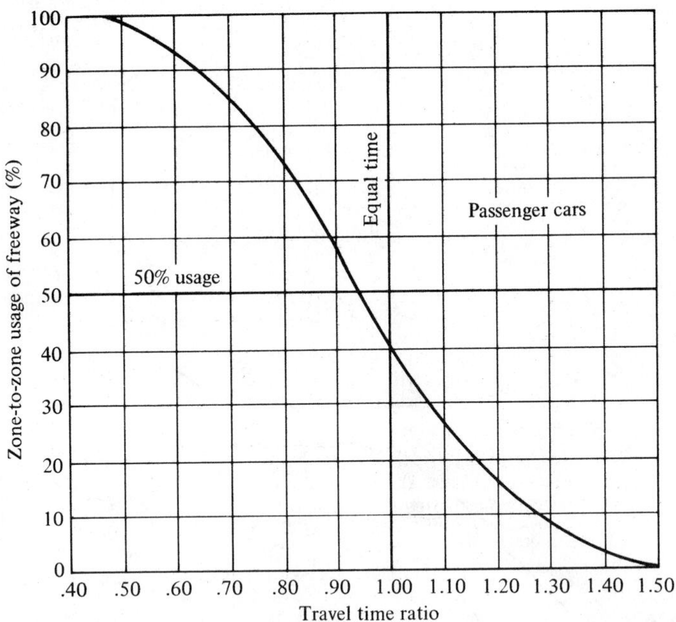

Fig. 61. Typical diversion curve (Federal Highway Administration)[8]. Travel time ratio = Time via freeway ÷ Time via quickest arterial route. Zone-to-zone usage = (Vehicles using freeway ÷ Total vehicles using all routes) × 100%.

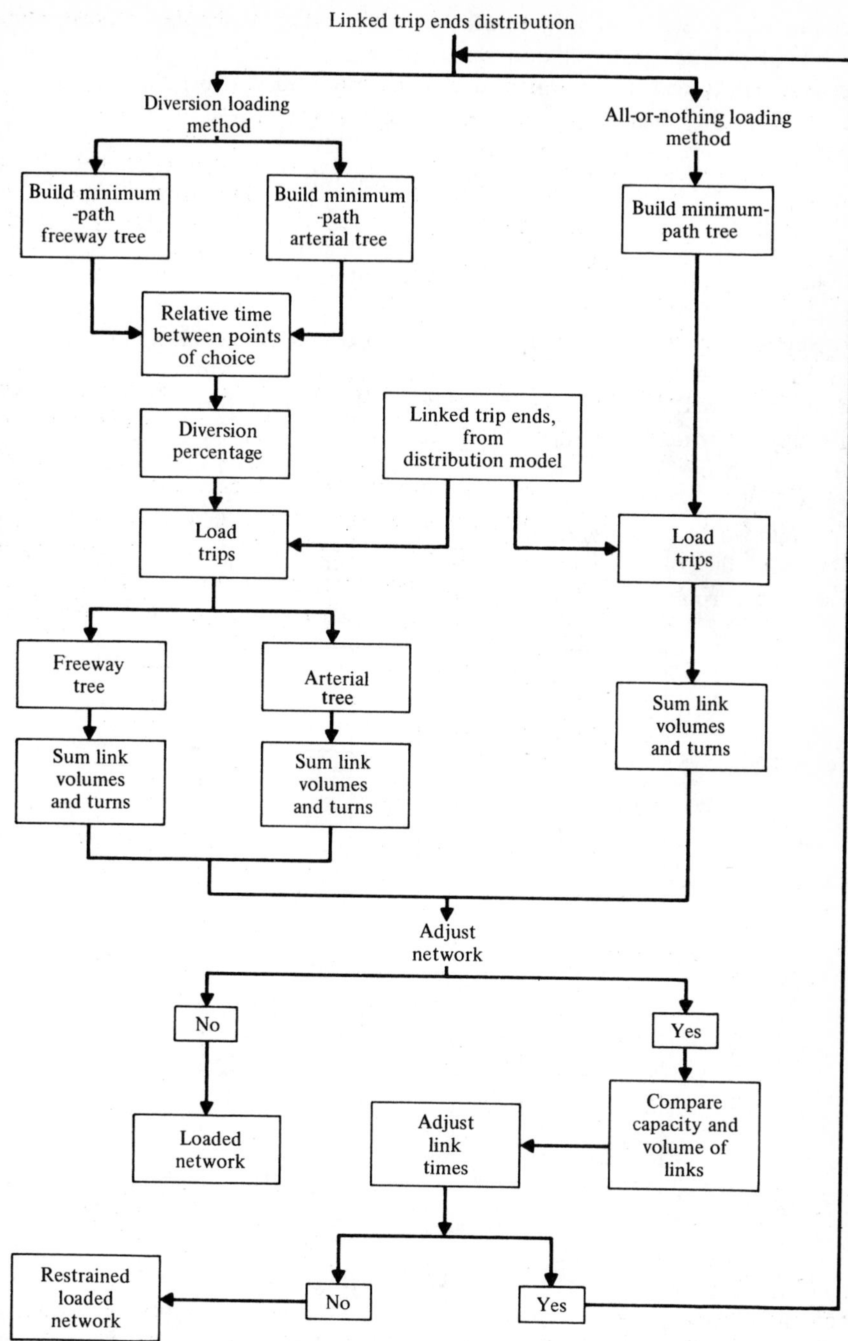

Fig. 62. Traffic assignment process.

empirical studies. Figure 62 demonstrates the assignment process for either a diversion program or the all-or-nothing program. All of the interzonal trip interchanges are loaded to the links and summed or averaged, depending on the process, with the final summation becoming the synthesized volume of travel anticipated for each link in a network.

Since the assignment process itself is not sensitive to the overloading of individual links, some procedure should be necessary to adjust the link volumes so that they are more in line with the carrying capability of each of the individual links. This is done through a procedure called capacity restraint. Capacity restraint permits the adjusting of individual link travel times to bring them more in line with what would actually occur on an operating network. For example, as traffic volumes increase on an arterial street, congestion slowly sets in, speeds drop, and travel times increase. This link travel time adjustment is made through a capacity restraint procedure which compares the assigned traffic volume on the link to the design capacity of the link as determined in the information analysis phase of the study.

A number of capacity restraint functions exist; among them are the Smock function, the BPR function, the Schneider function, and the TRC (Traffic Research Corporation) capacity restraint function (23). The Smock v/c function

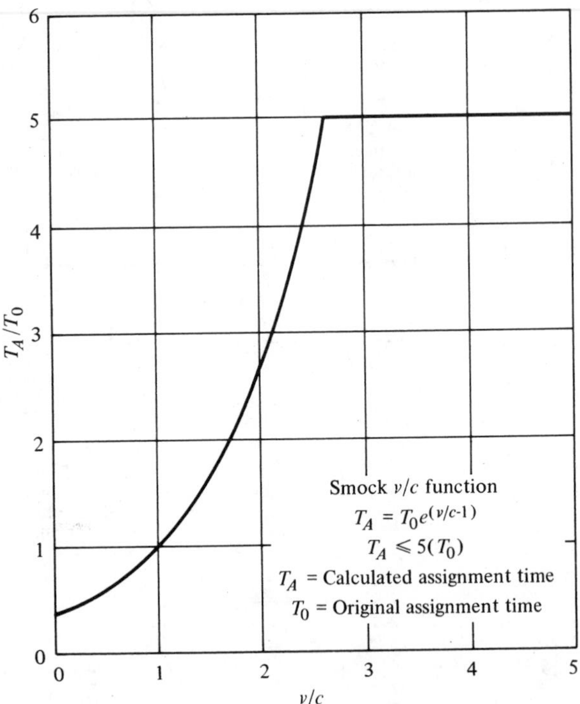

Fig. 63. Smock capacity-restraint function[24].

is shown in Fig. 63. It is used as follows: After the all-or-nothing traffic assignment has been made, link times are modified by obtaining a new travel time T_A for each link. New minimum-time-path trees based on the new T_A's are built and traffic volumes are again assigned on an all-or-nothing basis. This is done several times in an iterative fashion, with four iterations normally required for link assignment stability. The final link volume for the Smock procedure is the average of the individual link loadings outputted from each iteration.

Modal Split (27) (28)

The portion of the transportation modeling process described thus far makes no distinction between trips made by private vehicle and those made by some form of mass transportation. Modal split is the process which allocates a portion of total travel to automobile and mass transit. Typical models use data based on current mass transit and auto travel as explainers of the proportion of travel which will use mass transit. One of the problems with this approach is the fact that there is no assurance that today's travel choice experience will be the same as that in the future.

Types of Models Two basic concepts for the allocation of travel by mode prevail. One is the "trip end" modeling shown in Fig. 64 and the other is the "trip interchange" modeling process shown in Fig. 65.

The trip end philosophy allocates a portion of total person trip origins and destinations to alternative modes of transportation. Then, based on characteristics of the transit system and characteristics of the highway transportation system, trips are separately distributed by mode. The output is a trip interchange matrix, noting interzonal trip interchanges for each mode. The interzonal travel is then assigned to the highway network or the transit network as appropriate.

The trip interchange model allocates the interzonal trip movements which result from the trip distribution process to each mode of transportation using characteristics of the transit system and highway system as determinants of the portion of trips using each mode.

Variables Used as Explainers for Modal Split Models Table 27 denotes trip explainer variables used for trip end models and trip interchange models in some of the transportation planning studies done to date. These characteristics broadly group into three categories: trip characteristics, that is to say information about the travel being performed; trip maker characteristics, or information about the person or persons performing the travel; and characteristics of the transportation system. Table 28 shows a set of modal split regression models by purpose (19).

Advantages and Disadvantages of Today's Modeling Process Critics of today's modeling process suggest that there is a built-in bias favoring automobile usage because the models are based on existing auto use and the current nature of

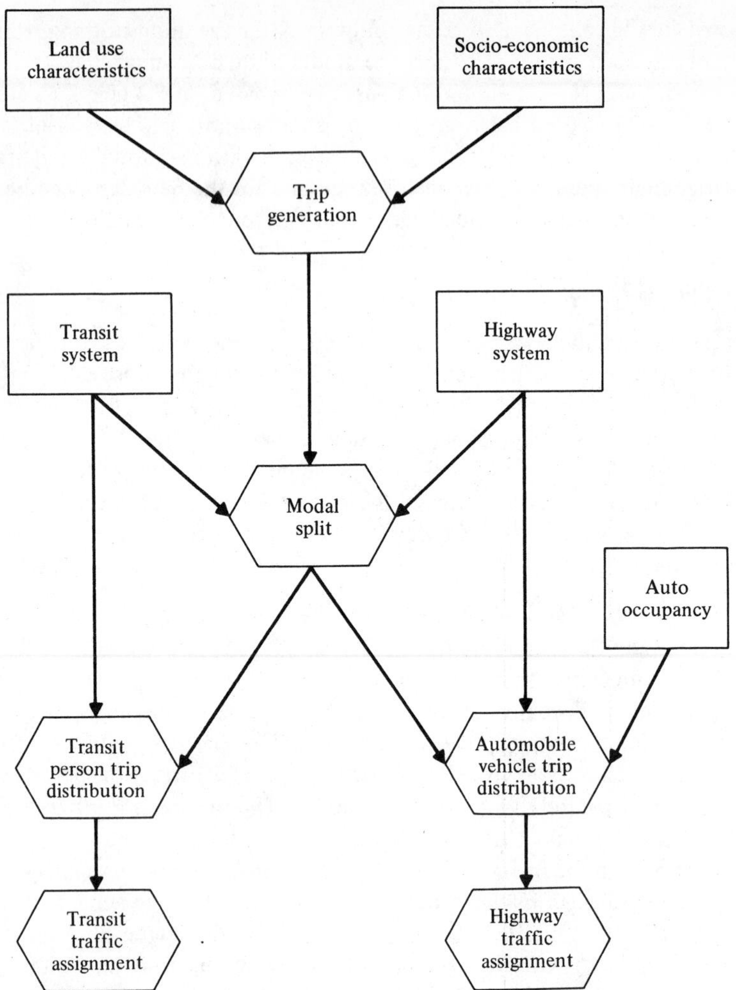

Fig. 64. Generalized diagram of trip-end modal split model[28].

transit systems. They also suggest that the modeling process is unresponsive to changes in the transit system and is unable to determine what effect such things as reduced fare structures, improved service, and other changes would have on transit use.

Supporters of the current modeling process, on the other hand, suggest that the facts of life are such that significant changes in travel habits are unlikely, and that man, unless drastic changes are made in his value structure, will continue to rely primarily on some form of private conveyance.

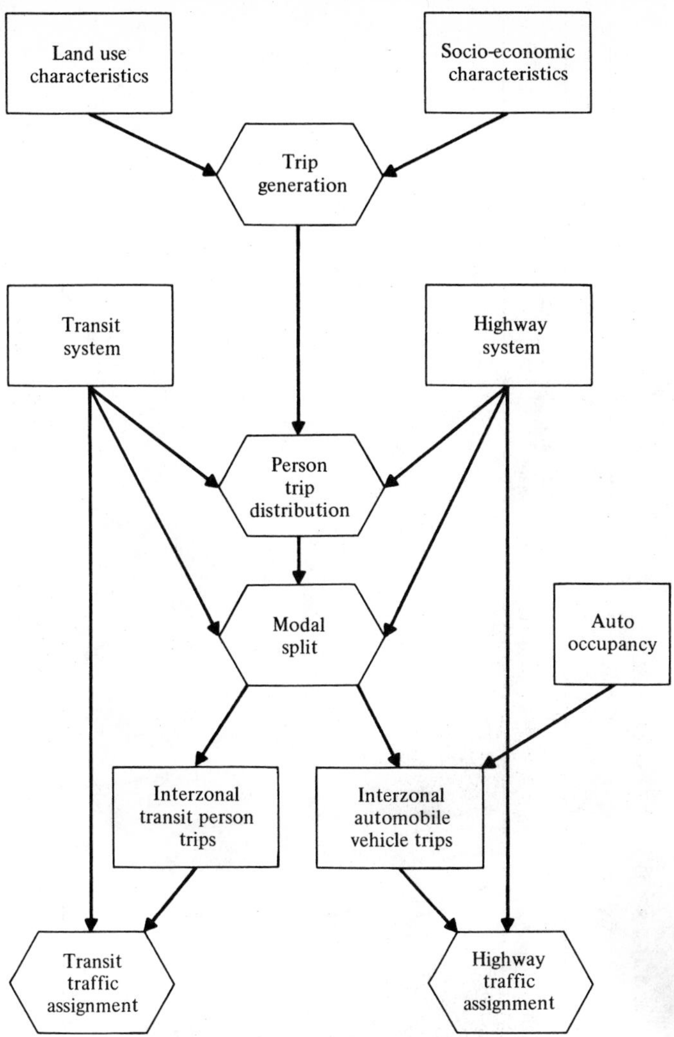

Fig. 65. Generalized diagram of trip-interchange modal split model[28].

Use of Simulation Models The modeling process previously described provides information on the anticipated loading of transportation networks, i.e., the numbers of automobiles on streets and the numbers of mass transit vehicles on mass transit networks. Such data is useful for estimating present usage of systems and the current efficiency of those systems and for estimating their future use. The future forecasts use the models developed on the basis of current experience,

TABLE 27. VARIABLES USED IN MODAL SPLIT MODELS (28).

Variables	Trip-end models					Trip-interchange models			
	Chicago	Pittsburgh	Erie	Puget Sound	Southeastern Wisconsin	Washington D.C.	Twin Cities	San Juan	Buffalo
Trip characteristics									
Number of trip purposes used	2	3	1	4	7	2	3	2	2
Length of trip									X
Time of day						X			X
Orientation to CBD	X	X							
Trip maker characteristics									
Auto ownership	X	X		X	X			X	X
Residential density		X		X			X		
Income				X		X	X		
Workers per household									X
Distance to CBD		X							
Employment density							X	X	X
Transportation system characteristics								X	
Travel time						X	X		X
Travel cost						X			
Parking cost						X	X		
Excess travel time[1]						X			
Accessibility[2]			X	X	X				

[1] Time spent outside the vehicle during a trip: walk, wait, and transfer times for transit trips, and parking delay time for auto trips.
[2] A measure of the level of travel service provided by the transit or highway system to trip ends in the study area.

TABLE 28. MODAL SPLIT MODELS; MADISON, WISCONSIN, 1970 (20).

Trip purpose	Modal split model
Work	Percentage = 0.908 + 6.898 PARCOS + 0.362 POP/RL – 0.094 (TDW + TT)
Shop	Percentage = –4.469 + 7.753 PARCOS + 12.288 AU/NR
Combined*	Percentage = 2.567 + 2.100 PARCOS + 7.163 AU/NR – 3.073 AUT/DU
University	Percentage = 11.862 + 59.788 AU/NR – 84.259 AUT/PO
Jr. and sr. high	Percentage = 47.293

PARCOS: Average Hourly Parking Cost
POP/RL: Population per residential land area (acres)
TDW + TT: Transit travel time plus Time to wait for bus plus Time to transfer Bus(es)–Auto travel time
AU/NR: Auto travel time/Time to get to bus stop plus Transfer waiting time(s)
AUT/DU: Registered autos/Dwelling units
AUT/PO: Registered autos/Population
*"Combined" includes home-based social-recreation, home-based other, and non-home-based.

and input into them based on land use, economic and population forecasts, information on land use activity, characteristics of the future economy and population.

System/Network Design (11) (29) (6) Although the design of new transportation networks or modification of existing ones is still very much an art, it is fast becoming a science. Some of the planning and design criteria guiding the development of such facilities include:

System Continuity. Safe and direct movement is most readily made over systems which are smooth, do not jog, and are continuous in type and marking.

System Efficiency. One criterion of efficiency is lowest total cost—construction, user, social, etc. Optimum spacing criteria related to some costs have been developed. The approach that follows relates minimum cost spacing to construction and right of way costs, time costs, trip density and travel speeds and facility use (demand) (11):

$$Z = 2.24\sqrt{\frac{C_z}{KDV_{yz}P_s}}$$

where

Z = optimum expressway spacing in miles
C_z = construction and right-of-way cost of expressways in dollars per mile
K = a constant capitalizing the value of time
V_{yz} = $(1/V_y - 1/V_z)$ where V_y and V_z are, respectively, the arterial and expressway speeds in miles per hour
P_s = proportion of trips whose length is greater than the length where trips start to use expressways for some portion of their journeys
D = number of trip destinations per square mile

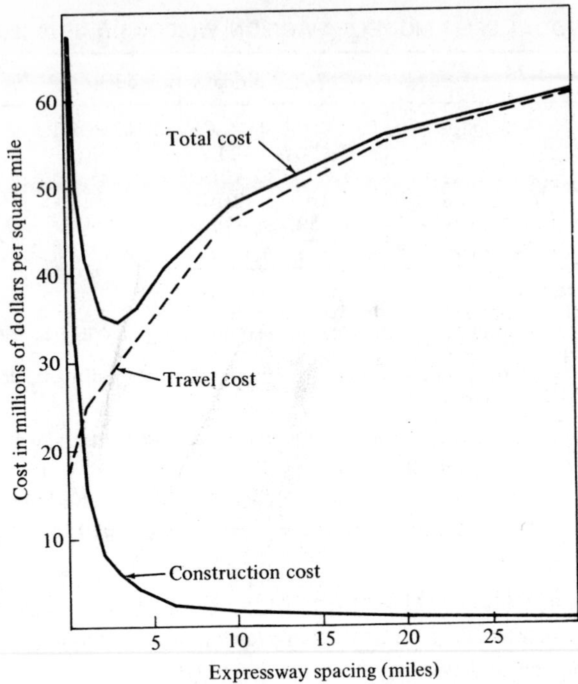

Fig. 66. Minimum-cost spacing in a high-density region[12].

Two graphical methods relating minimum cost spacing in high- and low-density regions are shown in Fig. 66 and 67. At the left side of the graphs, where expressways are closely spaced, construction costs are high; to the right, travel costs are high because long distances are traveled on local and arterial streets. At some point in between, the point of optimum freeway spacing overall, costs are a minimum.

Environmental Compatibility. Proper harmony with the total environment (the natural, cultural and social) requires that facilities be attractive, create as little disruption as possible, make minimum use of exploitable resources, and preserve and enhance the environment in fulfilling the goals of the community.

Multipurpose Capability. Facility location and design should aid and encourage *sound* new development and redevelopment through joint development and multiple use concepts. Joint development involves the cooperation of more than one agency or interest in conceiving and developing integrated and compatible uses; multiple development is the use of air rights and adjacent space for the integrated development of more than one activity. Figures 68 and 69 illustrate several recent notable multiple use developments.

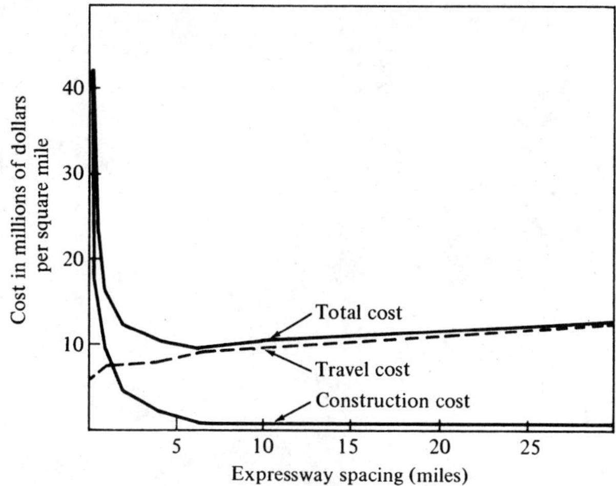

Fig. 67. Minimum-cost spacing in a low-density region[12].

System Integration. The system must make adequate connections with other components of the same system and must adequately interface with other modes and terminals.

System Multimodal Capability. The system (network) must be designed and located to optimize the use of autos and mass transit in the same corridor.

System Adaptability and Flexibility. The system by design and location must have the ability to change in response to changes in demand and changes in movement technology.

System Evaluation (6) (11) (30) (31) (32) To date, alternative transportation systems have been evaluated largely from an economic standpoint. Capital cost of the improvement, subsequent maintenance costs, and travel costs or savings have largely been the determinants used in system evaluation. Among the more important of the economic evaluation procedures are: (30) (31) (32)

1. Annual cost method.
2. Rate of return.
3. Present Worth.
4. Benefit–cost ratio.

Other procedures are being developed which, in evaluating alternative proposals, look not only at the tangible measurable costs and benefits but also at the intangibles–those costs and benefits which cannot be qualified. Four methods of looking at intangibles are: (a) use of some type of subjective scaling technique in the analysis; (b) where subjective scaling is not possible, the use of verbal de-

This restaurant over the highway has been proposed and approved for development in conjunction with Interstate 80 in Sacramento. It will provide a spectacular combination of multiple uses within the right-of-way as well as add to the local tax rolls and provide a visual landmark for drivers and an interesting focal point for local businesses and residents. (Sacramento, California)

This photo shows Interstate 95 crossing New York City, spanning Manhattan in a 12-lane depressed freeway linking the George Washington Bridge (top left corner) and the Alexander Hamilton Bridge (bottom center). In the lower center of the picture are the interchange ramps connecting Interstate 95 and the local streets, the Harlem River Drive and the Washington Bridge (bottom right corner). Use of airspace over Interstate 95 was made by the building of four 32-story apartments for 960 middle-income families (center left) and a bi-State bus terminal (above and left of buildings) which is also over a subway terminal below the highway's surface. (New York City)

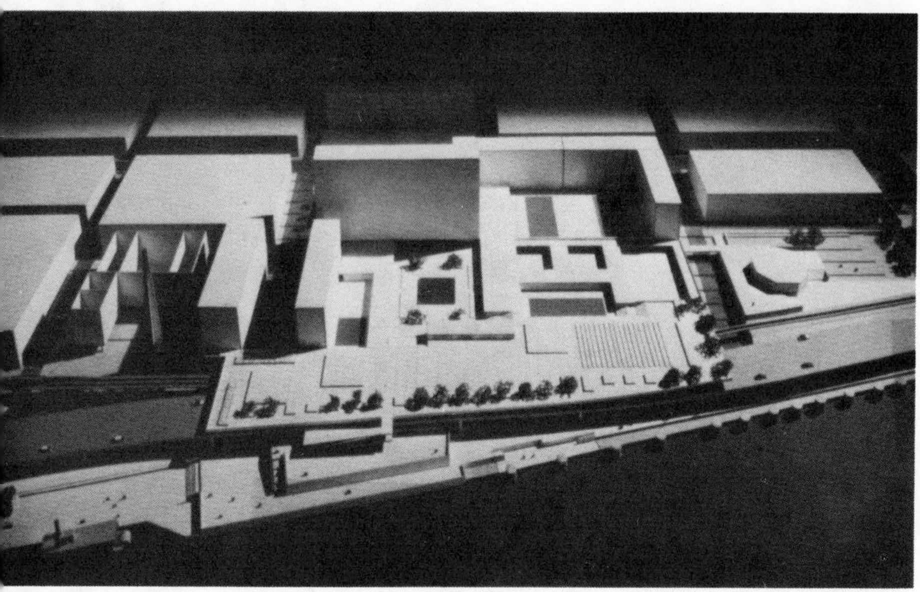

Although reduced somewhat in length from the illustration, this belvedere structure and wharf above and adjacent to the Riverside Parkway in Louisville has advanced to the final design stage. (Louisville, Kentucky)

Fig. 68. Examples of multiple-use concept[25].

scription of the impacts; (c) cost effectiveness; and (d) closing one's eyes to the intangibles (7). All too often the latter is the course of action taken and public dissatisfaction and controversy result.

Demand Reduction Most attempts to alleviate transporation problems have resulted in proposals for new physical improvements—of either major or minor proportions. There are a number of other methods by which traffic volumes can be reduced or traffic flow facilitated. These are noted as procedures of an operational nature or policy nature.

Operational Procedures

1. Surveillance and Control. A procedure used primarily on urban freeway systems requiring continual monitoring of traffic and the application of some form of flow control. Among the systems currently gaining favor are ones which control the flow of vehicles at on-ramps to the system. One of the earliest such systems was the ramp control system placed in operation on the Eisenhower Expressway in Chicago. Through surveillance of traffic and ramp metering, the efficiency of traffic flow has been improved, total volume carried increased, delay decreased, and accidents decreased.

A study of Interstate 95 between New York and Massachusetts Avenues has suggested this joint development of the highway to provide for the families to be displaced by the highway as well as to increase the total number of housing units available. (Washington, D.C.)

A northeasterly view looking across the Southwest Freeway (Interstate 95 at bottom) toward the proposed office-hotel structure at upper right center. A proposed office-hotel structure will be built in airspace over the 9th Street Expressway (right center). (Washington, D.C.)

Fig. 69. Examples of multiple-use concept[25].

2. Computerized Signal Networks. A number of major urban areas are installing computer-controlled signal systems in the urban area which will optimize traffic flow on urban arterials. Optimization will be measured by increasing overall speeds and reduction in accidents.

Policy Procedures for Reducing Demand

1. Staggered Work Hours. Since much of the movement problem occurs during morning and afternoon peaks, in areas where these peaks can be spread out, i.e., lengthened in total time of occurrence, the demand on the system at any particular time could be reduced. Some areas have greater potential for use of staggered work hours than others. Staggered work hours require that the work starting times and work ending times for various industries be arranged so as to spread the travel load on the transportation networks.

2. Improved Auto Ridership. If methods can be devised to improve the auto occupancy rates currently experienced, a great deal will be done toward lessening the vehicle demand on the street and highway system. Current automobile occupancies vary depending on area from between approximately 1.5 persons per vehicle to 1.8 or 1.9. If this figure could be doubled, there would be a considerable reduction in the number of automobiles using the streets. One way of improving auto occupancy would be through the use of incentives to drivers. This might be done, for example, by providing free parking for vehicles which had four or more riders and charging very high hourly rates for vehicles with two or less riders.

3. Mass Transit Incentives. Many studies indicate that increased mass transit ridership will not always greatly affect the volume of personal vehicles on the roads. There are certain travel corridors on which the increased use of mass transit could, however, provide some significant relief to already overloaded systems. This might be done through reduced fare structures or even free rides, providing passes to employees as one of their fringe benefits, or by providing vastly improved mass transit service.

4. Parking Policy. By controlling, as a matter of public policy, the supply, in terms of number, location, and rate structure, of parking space, more travellers might decide to use mass transit or obtain increased use of their own autos through car pooling, etc.

5. Operating Restrictions. During certain hours of the day, it might be wise to apply restrictions to the operation of vehicles in certain areas to help alleviate the movement problem.

6. Development Policy. Much work needs to be done in developing definitive development policies so that urban development is carried out in a manner which both minimizes the need for movement throughout the urban area and optimizes the manner in which transportation and other land uses interact with the scale of human living. Thought should also be given, where it is appropriate, to limiting the size of urban areas to that which is optimum in terms of livability.

REFERENCES

1. Cafferty, Michael, Urban Goals and Priorities: The Increasing Role of Transportation Planning, *Traffic Quarterly*, Volume XXV, No. 3 (July 1971).
2. U.S. Department of Transportation, *Calibrating and Testing a Gravity Model for Any Size Urban Area*, U.S. Government Printing Office, Washington, D.C. (1968).
3. Smith, Douglas C., *Urban Highway Design Teams,* Highway Users Federation for Safety and Mobility, Washington, D.C. (1970).
4. *Forecasts and Alternative Plans–1990*, Planning Report No. 7, Volume II, Southeastern Wisconsin Regional Planning Commission, Waukesha, Wisc. (1966).
5. *Transportation and Community Values,* Report of Conference at Warrington, Va., Highway Research Board Special Report No. 105, Washington, D.C. (1969).
6. Thomas, Edwin N. and Schofer, Joseph L., *Strategies for the Evaluation of Alternative Transportation Plans*, NCHRP Report No. 96 (1970).
7. *Inventory Findings–1963*, Planning Report No. 7, Volume I, Southeastern Wisconsin Regional Planning Commission, Waukesha, Wisc. (1965).
8. *Traffic Assignment Manual, U.S. Department of Commerce,* Bureau of Public Roads, Washington, D.C. (1964).
9. Wisconsin Department of Transportation, Transportation Study Survey Forms.
10. Marks, Harold, *Protection of Highway Utility,* NCHRP Report No. 121 (1971).
11. Creighton, Roger L., Urban Transportation Planning, Univ. Illinois Press, Urbana (1970).
12. U.S. Department of Commerce, *The Role of Economic Studies in Urban Transportation Planning,* U.S. Government Printing Office, Washington, D.C. (1965).
13. Chapin, Stuart, *Urban Land Use Planning,* Univ. Illinois Press, Urbana (1965).
14. Isard, Walter, *Methods of Regional Analysis: An Introduction to Regional Science*, MIT Press and John Wiley & Sons (1960).
15. U.S. Department of Transportation, *Guidelines for Trip Generation Analysis*, U.S. Government Printing Office, Washington, D.C. (1967).
16. Fratar, T. J., Vehicular Trip Distribution by Successive Approximations, *Traffic Quarterly*, Volume 8, N. 1(Jan. 1954).
17. Brokke, Glen E. and Mertz, William L., Evaluating Trip Forecasting Methods with an Electronic Computer, HRB Bulletin 203. Highway Research Board, Washington, D.C. (1958).
18. Smith, Wilbur and Associates, *Future Highways and Urban Growth*, Automobile Manufacturers Association, Detroit (1961).
19. Simpson and Curtin, *Transportation for Madison's Future*, Madison Area Transportation Study Special Report No. 17, Madison, Wisc. (1970).
20. Heanue, Kevin E. and Pyers, Clyde E., *A Comparative Evaluation of Trip Distribution Procedures.* Highway Research Record 114. Highway Research Board, Washington, D.C. (1966) pp. 20–37.

21. Jarema, Frank E., Pyers, Clyde E., and Reed, Harry A., *Evaluation of Trip Distribution and Calibration Procedures,* Highway Research Record 191, Highway Research Board, Washington, D.C. (1967), pp. 106–129.
22. U.S. Department of Commerce, *Calibrating and Testing a Gravity Model with a Small Computer,* U.S. Government Printing Office, Washington, D.C. (1963).
23. Huber, Matthew J., Boutwell, Harvey B., and Witheford, David K., *Comparative Analysis of Traffic Assignment Techniques with Actual Highway Use,* NCHRP Report No. 58 (1968).
24. U.S. Department of Transportation, *Highway Joint Development and Multiple Use,* U.S. Government Printing Office, Washington, D.C.
25. Moore, E. F., The Shortest Path Through a Maze, *International Symposium on the Theory of Switching, Proc.,* Harvard Univ., April 2–5, 1957, pp. 285–292.
26. Madison Area Transportation Study, *Semi Final Plan*, MATS Special Report No. 11.
27. U.S. Department of Transportation, *Modal Split*, U.S. Government Printing Office, Washington, D.C. (1970).
28. Weiner, Edward, Modal Split Revisited, *Traffic Quarterly,* Volume XXIII, No. 1, (Jan. 1969).
29. Institute of Traffic Engineers, "System Considerations for Urban Freeways," Washington, D.C. 1967.
30. Wohl, Martin and Martin, Brian V., *Evaluation of Mutually Exclusive Design Projects,* HRB Special Report No. 92, Highway Research Board, Washington, D.C. (1967).
31. Winfrey, Robley, *Economic Analysis for Highways,* International Text Book, Co., Scranton, Pa. (1969).
32. Wohl, Martin and Martin, Brian V., *Traffic System Analysis for Engineers and Planners,* McGraw-Hill Book Co., New York (1967).

GEOMETRIC DESIGN

Geometric design involves the design of the typical roadway features which provide for the safe and efficient operation of vehicular traffic. In addition to providing for the operational needs of traffic, the roadway, intersection, and interchange facilities must be designed to properly interact with the adjacent man-made and natural environment and provide a pleasing view both from the road and of the road. The location of a highway or street and its actual elements of design, therefore, are influenced to a large extent by the existing topography and development, the function of the roadway and traffic volumes, and the general carrying capability (capacity) of the roadways and intersections. These design controls are important in influencing the criteria and elements of the actual design.

Design Controls

Highway Function Different highway systems provide differing degrees of mobility and land access, two functions which are essentially conflicting. For example, a highway which provides a high degree of access to adjacent land cannot provide the degree of mobility required of high-type movement systems such as urban expressways and rural freeways. As one major design control, it is necessary to define the highway's function; this in turn is helpful in determining necessary design criteria and design elements. The development and adoption of a functional state highway plan such as that adopted by the State of Wisconsin, Department of Transportation (Fig. 70) provides a sound basis for the development of a continuous network which provides uniform and consistent treatment of all its segments. The functional classifications can be related specifically to the geometric design level, i.e., high- or low-type design, general

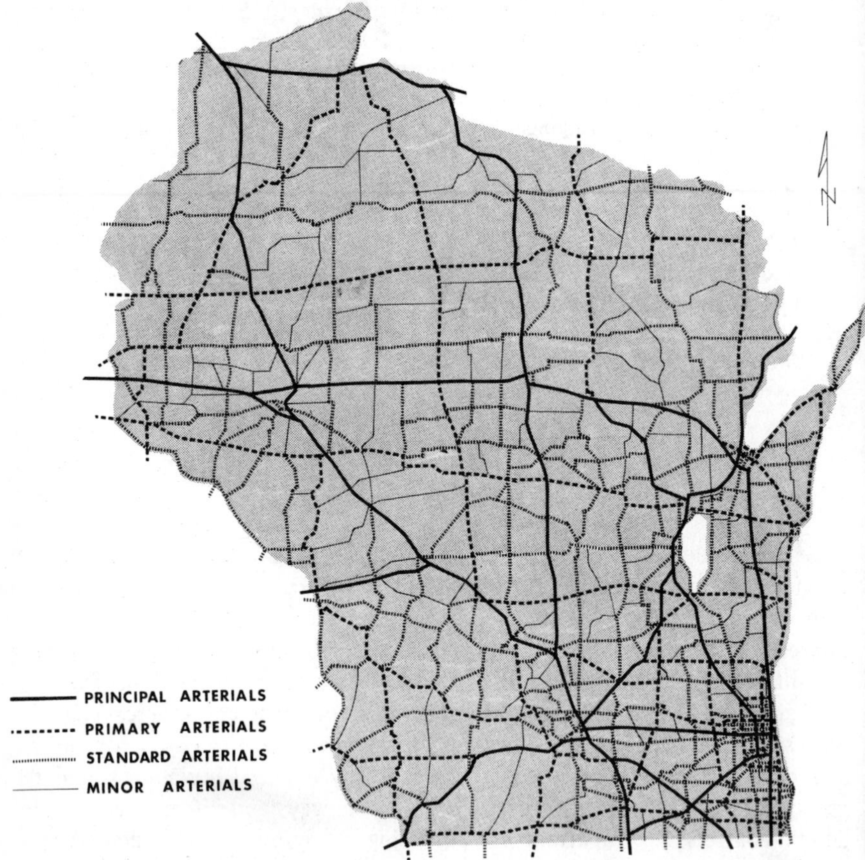

Fig. 70. Wisconsin Department of Transportation 1990 functional plan[1].

right-of-way requirements, access control requirements, and other requirements of design.

Traffic Volumes A street or highway is designed to safely accommodate actual traffic volumes anticipated in a design year, normally taken as 20 years after the date that the facility is opened to traffic. Since highway planning and design activities—beginning with the determination of the actual need for a facility, through the design, the acquisition of necessary right-of-way, and construction—normally involves a lead time of between 5 and 10 years, it is necessary to forecast traffic demand almost 30 years into the future. These volumes, normally stated in terms of a design hour volume (DHV), provide one basis for selecting the level of transportation service to be provided. In some cases, on lower-volume roads, current average daily traffic (ADT) is used as the traffic volume warrants.

Traffic data used for design (DHV) are usually obtained by relating future average daily traffic (ADT) to DHV through the application of factors relating to the peaking and directional distribution characteristics of traffic flow. These factors are normally expressed as a K factor and a D factor, respectively.

The K factor, or percentage relationship between DHV and ADT, for two-way roadways in rural areas has a normal range between 12% and 18% and for one-way roadways between 16% and 24%. This is based on a DHV which approximates the 30th highest hourly volume in the future design year. Figure 71 shows the relationship between highest hourly volumes and ADT for different roadway types. DHVs customarily used for design approximate the 30th highest hourly volume; however, it may be appropriate in some situations to design for a lower higher hour such as the 50th or the 100th. This is particularly true in recreational areas where travel demand has a higher summer peak and during the remaining nine months of the year carries relatively little traffic. Figure 71 also shows the relationship between DHV and ADT for urban areas. The 6–8% K factor relates very well to the peak hour to 24-hour average weekday relationship which occurs in many urban areas.

Directional distribution D expresses the one-way volume in the predominant direction of travel as a percentage of the two-way design hour volume. Normal values during the design hour range between 50% and 80% with an average somewhere in the vicinity of 60% and 70%.

Since the turning and operational characteristics of trucks must be considered in design, it is important to know the composition of traffic. For general traffic estimating purposes, traffic is split into two groupings: passenger vehicles (including light delivery trucks and pick-up trucks) and trucks. The truck category includes all vehicles with six or more tires.

Vehicle Operating Characteristics (3) (7) Since different vehicle types have different operating characteristics and different size characteristics it is necessary to select representative vehicles upon which to base design. This is particularly

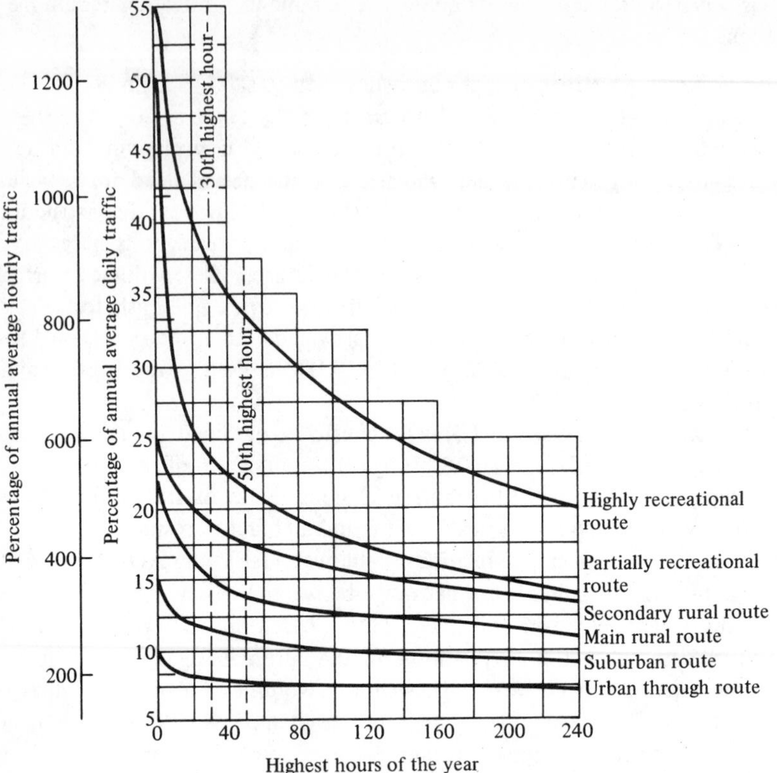

Fig. 71. Hourly volume expressed as a percentage of average daily traffic[2].

true in the design of turning roadways and intersections. Normally, the vehicle selected for design purposes should be one which has dimensions and minimum turning radius characteristics that are larger than those of the vehicles which must be accommodated. Five standard design vehicles have been developed for design use. They are: (a) passenger design vehicle, P; (b) single-unit design vehicle, SU; (c) medium to large truck-tractor, semi-trailer combinations, WB-40; (d) truck-tractor, semi-trailer combinations inclusive of almost all combinations currently in use, WB-50; and (e) very large semi-trailer combinations and "double bottoms," WB-60.* Figures 72–75a show the dimensions and crawl-speed (less than 10 mph) turning characteristics for each of the design vehicles (3).

Roadway and Intersection Service Volume and Level of Service Capacity, by definition, is the maximum number of vehicles that a roadway or intersection can accommodate during an hour of operation under existing roadway condi-

***A "Bus Design Vehicle" has been added for design use.**

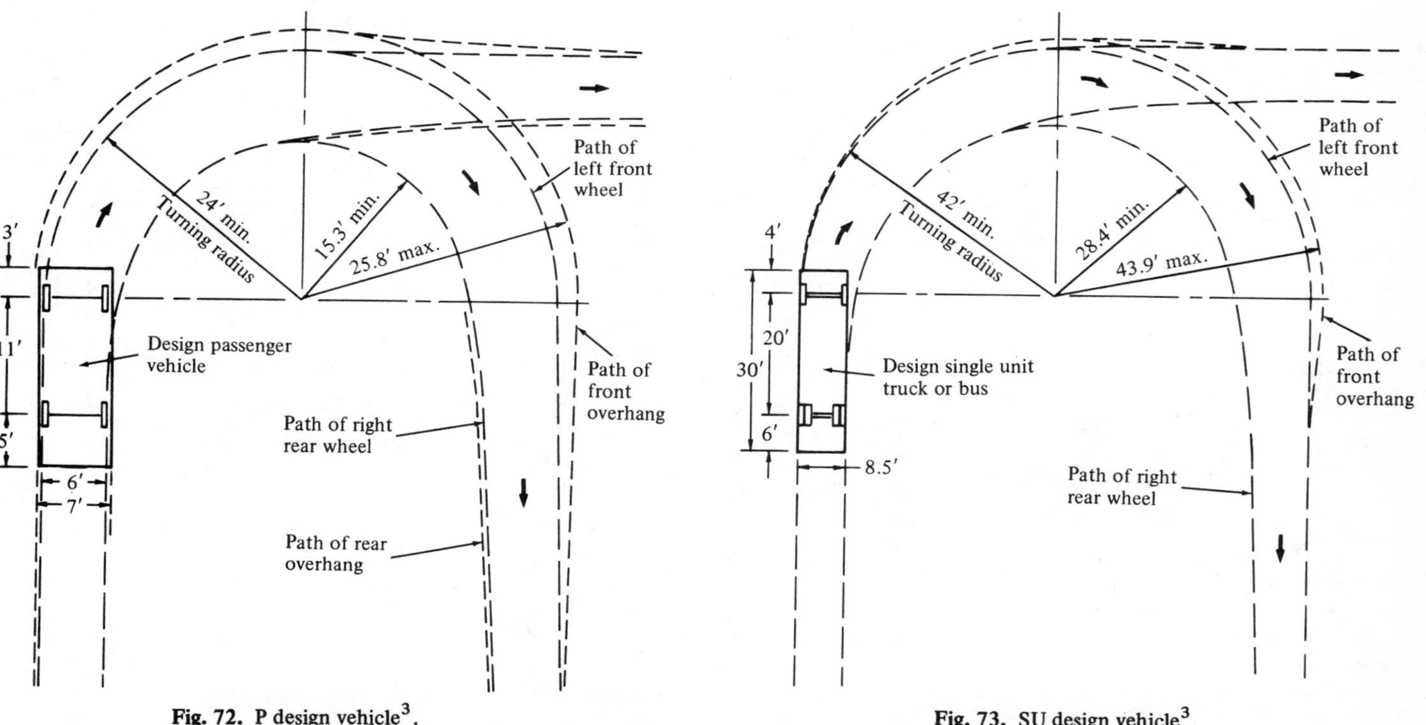

Fig. 72. P design vehicle[3].

Fig. 73. SU design vehicle[3].

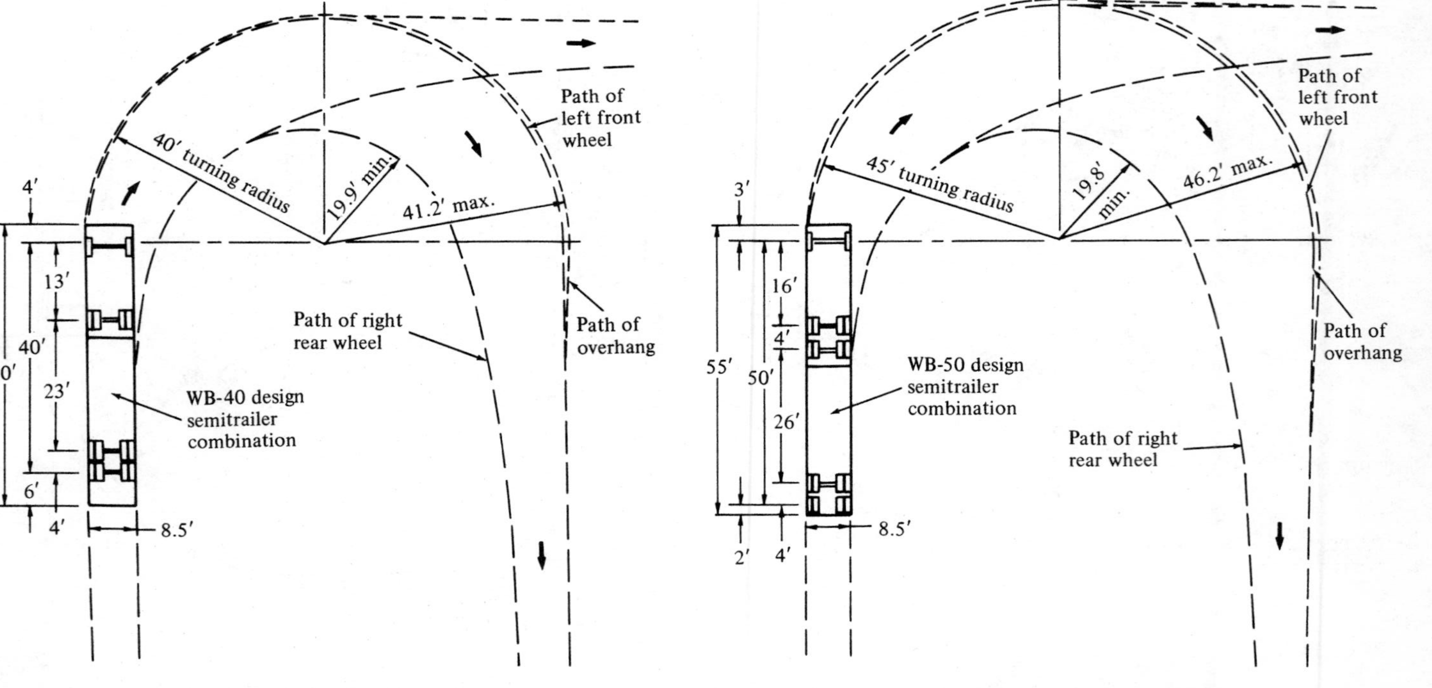

Fig. 74. WB-40 design vehicle[3].

Fig. 75. WB-50 design vehicle[3].

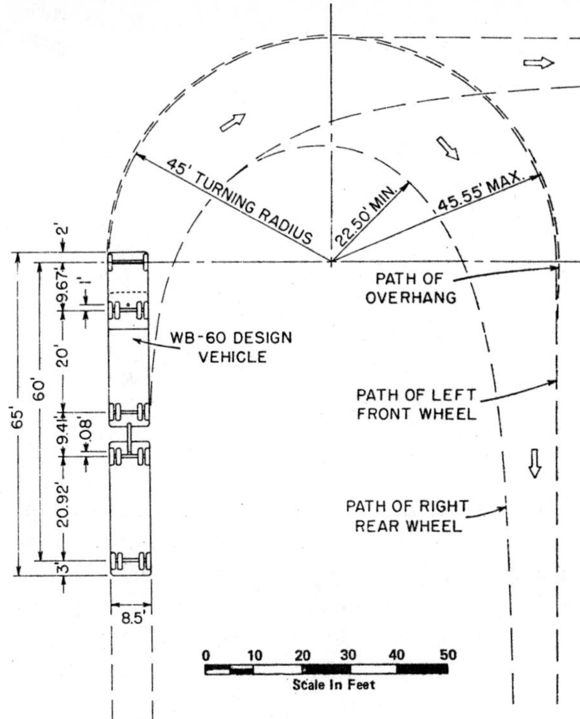

Fig. 75a. WB-60 design vehicle[3].

tions. Capacity itself, however, is an undesirable and unstable condition of flow–one in which individual vehicle operations are severely constrained by other vehicles in the stream; drivers are unable to select their most desirable operating speed; and the driving task is fraught with inconvenience, delay, and personal hazard. Therefore, it is desirable to design facilities on which the design hour volume is less than capacity. This lower-than-capacity roadway or intersection service volume results in a higher overall quality of flow.

Environmental Capacity Vs. Real Capacity (4) (5) Transportation and traffic engineers are becoming aware of the fact that roadways have an environmental capacity as well as a traffic capacity. The concept of environmental capacity, as Fig. 76 shows, states that streets and highways–depending on their function and location–have a maximum desirable traffic loading which may be considerably less than that which the facility can actually handle, even under so-called desirable levels of service. This is because of the manner in which the effects of traffic e.g., noise, air pollution, and hazard, among others, affect the quality of living in the area through which the facility passes.

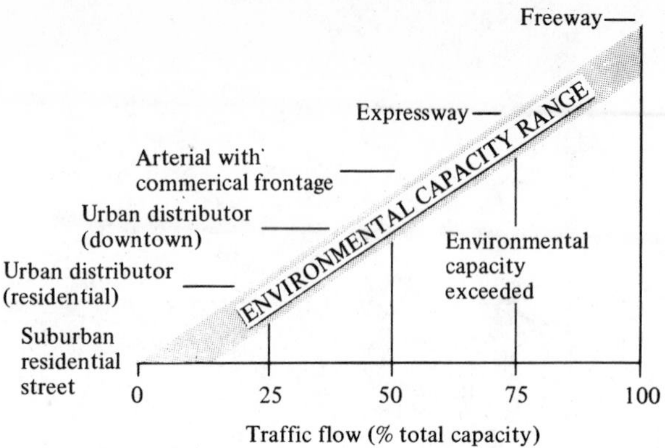

Fig. 76. Environmental capacity[4].

Level of Service Level of service is a measure of the general quality of service provided by a facility based on travel speeds, delay, safety, comfort and convenience, and a number of other subjective traits which can be ascribed to the traffic stream. Levels of service are commonly defined by the letter designations A through F, with A being the best and F the poorest flow quality. Figure 77 shows the general conceptual relationship between freeway level of service and quality of flow as described by operating speed and a volume-to-capacity ratio.

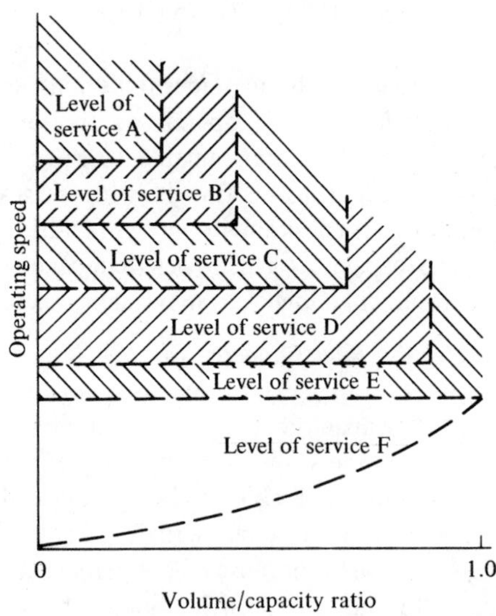

Fig. 77. General concept of relationship of levels of service to operating speed and volume/capacity ratio (not to scale)[6].

Figure 78 shows the six levels of service, A–F, for an urban freeway. The following description of the levels of service is from reference 6.

"Level of service A describes a condition of free flow, with low volumes and high speeds. Traffic density is low, with speeds controlled by driver desires, speed limits, and physical roadway conditions. There is little or no restriction in maneuverability due to the presence of other vehicles, and drivers can maintain their desired speeds with little or no delay.

"Level of service B is in the zone of stable flow, with operating speeds beginning to be restricted somewhat by traffic conditions. Drivers still have reasonable freedom to select their speed and lane of operation. Reductions in speed are not unreasonable, with a low probability of traffic flow being restricted. The lower limit (lowest speed, highest volume) of this level of service has been associated with service volumes used in the design of rural highways.

"Level of service C is still in the zone of stable flow, but speeds and maneuverability are more closely controlled by the higher volumes. Most of the drivers are restricted in their freedom to select their own speed, change lanes, or pass. A relatively satisfactory operating speed is still obtained, with service volumes perhaps suitable for urban design practice.

"Level of service D approaches unstable flow, with tolerable operating speeds being maintained though considerably affected by changes in operating conditions. Fluctuations in volume and temporary restrictions to flow may cause substantial drops in operating speeds. Drivers have little freedom to maneuver, and comfort and convenience are low, but conditions can be tolerated for short periods of time.

"Level of service E cannot be described by speed alone, but represents operations at even lower operating speeds than in level D, with volumes at or near the capacity of the highway. At capacity, speeds are typically, but not always, in the neighborhood of 30 mph. Flow is unstable, and there may be stoppages of momentary duration.

"Level of service F describes forced flow operation at low speeds, where volumes are below capacity. These conditions usually result from queues of vehicles backing up from a restriction downstream. The section under study will be serving as a storage area during parts or all of the peak hour. Speeds are reduced substantially and stoppages may occur for short or long periods of time because of the downstream congestion. In the extreme, both speed and volume can drop to zero."

Design Level of Service In rural areas, roadways and intersections are normally designed to accommodate traffic in the future design year at level of service B. In urban areas, the design level of service is normally set at level of service C. Capacity, considered by itself, is the operation at level of service E—a rate at which the quality of traffic service is very low.

Factors Affecting Intersection Capacity The factors listed in Table 29 are those which affect the carrying capability of signalized intersections. They involve characteristics which can be observed or measured for each situation.

Level of service A

Level of service B

Level of service C

Fig. 78. Levels of service, urban freeway looking upstream[6].

Level of service D

Level of service E

Level of service F

Fig. 78 (*Continued*)

TABLE 29. FACTORS AFFECTING INTERSECTION SERVICE VOLUME.

Physical and operating conditions

Approach width
- One-way streets, curb to curb
- Two-way undivided, curb to division line
- Two-way divided, curb to curb or curb to median edge

Type of operation (one-way or two-way)

Parking condition:
- if within 250 ft of intersection, Parking
- if beyond 250 ft of intersection, No Parking

Environmental Characteristics

Metropolitan area population

Location in metropolitan area
- Central business district (CBD)
- Fringe area–area adjacent to CBD (FRNG)
- Outlying business district (OBD)
- Residential (RES)

Intersection load factor

Load factor measures degree of utilization of a cycle

$$\text{Load factor} = \frac{\text{number of fully used greens in peak hour}}{\text{total number of green times in peak hour}}$$

Load factor is the single indicator used to define (metricize) level of service at an intersection.

Peak hour factor (PHF):

Peak hour factor measures consistency of demand.

$$\text{Peak Hour Factor} = \frac{\text{Peak hour volume}}{4 \times \text{Peak 15-min volume during peak hour}}$$

Traffic characteristics

Turns
- Percentage right turns
- Percentage left turns

Trucks (percentage)

Trucks include all vehicles with 6 or more tires (except local transit buses which stop at intersection).

Local transit

Number of buses per hour

Bus stop location
- Near side
- Far side

Traffic control measures

Signal operation

$$G/C \text{ ratio} = \frac{\text{Approach green time}}{\text{Cycle time}}$$

Degree of channelization and approach lane markings–particularly special turn bays.

Factors Affecting Unencumbered Sections of Freeway Capacity Analysis of traffic flow capacity for freeways is done separately for unencumbered sections of freeways and separately for ramps and weaving sections. An unencumbered freeway section is one which has no interruption from ramps, weaving sections, or other causes. This discussion will deal only with analysis of capacity on unencumbered sections of freeways. For analysis techniques applicable to ramps and weaving sections the reader is referred to reference (6).

The basic factors which influence the carrying capability of an unencumbered freeway section are listed below:

1. Lane width and adequacy of shoulder
2. Roadway lateral clearance restriction–obstructions within 6 ft of the right or left pavement edge
3. The proportion of trucks in the traffic stream
4. The influence of grade on the operation–length of grade and percent grade
5. Horizontal and vertical curvature alignment restrictions

Capacity Analysis of Signalized Intersections The basic procedure for estimating intersection capacity service volume, or level of service, involves making adjustments to empirically derived capacities and service volumes for average intersections. An average intersection has been defined as one in the central business district (CBD) of a metropolitan area of 250,000 population with a peak hour factor (PHF) of 0.85 (the average PHF for all of the observed data). Average traffic conditions include: 10% left turns, 10% right turns, 5% trucks and through buses, and no local transit buses. Average charts were developed separately for two-way streets with parking, two-way streets without parking, one-way streets without parking, and one-way streets with parking on one side and on both sides. The average condition charts shown in Figs. 79–83 relate approach width to the approach volume in vehicles per hour of green for various load factors or levels of service. Table 30 relates intersection level of service to the corresponding load

TABLE 30. INTERSECTION LEVEL OF SERVICE-LOAD FACTOR OR RELATIONSHIP.

Level of Service	Capacity descriptor	Flow description	Load factor
A		free	0.0
B	rural design capacity	stable	≤0.1
C	urban design capacity	stable	≤0.3
D		approaching unstable	≤0.7
E	possible capacity	unstable	≤1.0*
F		forced	—

*Sustained flow at a load factor of 1.00 is almost impossible to obtain. A lesser factor of 0.85 has been observed as indicative of the maximum loading which can be achieved and sustained over a period of one hour and is customarily used to represent possible capacity.

factor. Solution of a capacity-service volume problem requires use of the "average condition" charts and the application of adjustment factors which account for the existing conditions at the site being studied. Adjustments are made for metropolitan area population, peak hour factor, location within the metro-

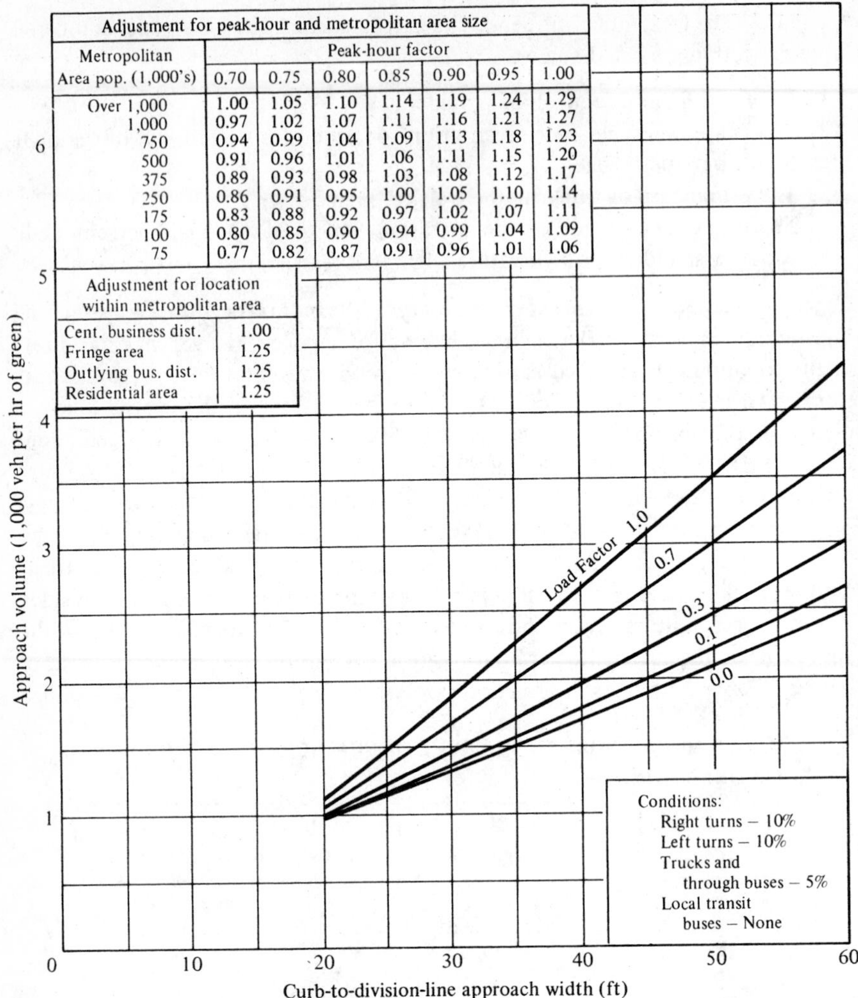

Adjustment for peak-hour and metropolitan area size

Metropolitan Area pop. (1,000's)	Peak-hour factor 0.70	0.75	0.80	0.85	0.90	0.95	1.00
Over 1,000	1.00	1.05	1.10	1.14	1.19	1.24	1.29
1,000	0.97	1.02	1.07	1.11	1.16	1.21	1.27
750	0.94	0.99	1.04	1.09	1.13	1.18	1.23
500	0.91	0.96	1.01	1.06	1.11	1.15	1.20
375	0.89	0.93	0.98	1.03	1.08	1.12	1.17
250	0.86	0.91	0.95	1.00	1.05	1.10	1.14
175	0.83	0.88	0.92	0.97	1.02	1.07	1.11
100	0.80	0.85	0.90	0.94	0.99	1.04	1.09
75	0.77	0.82	0.87	0.91	0.96	1.01	1.06

Adjustment for location within metropolitan area

Cent. business dist.	1.00
Fringe area	1.25
Outlying bus. dist.	1.25
Residential area	1.25

Fig. 79. Urban intersection approach service volume, in vehicles per hour of green signal time, for two-way streets with parking[6].

politan area, trucks (see Table 31), right and left turns (see Tables 32 and 33), local transit, and available green time.

Nomographs which permit rapid calculation of capacities, service volumes, and other capacity-related characteristics for signalized intersections have also been developed. Figures 84 and 85 show typical intersection capacity nomographs for a two-way street with parking in fringe, outlying business, and residential sections

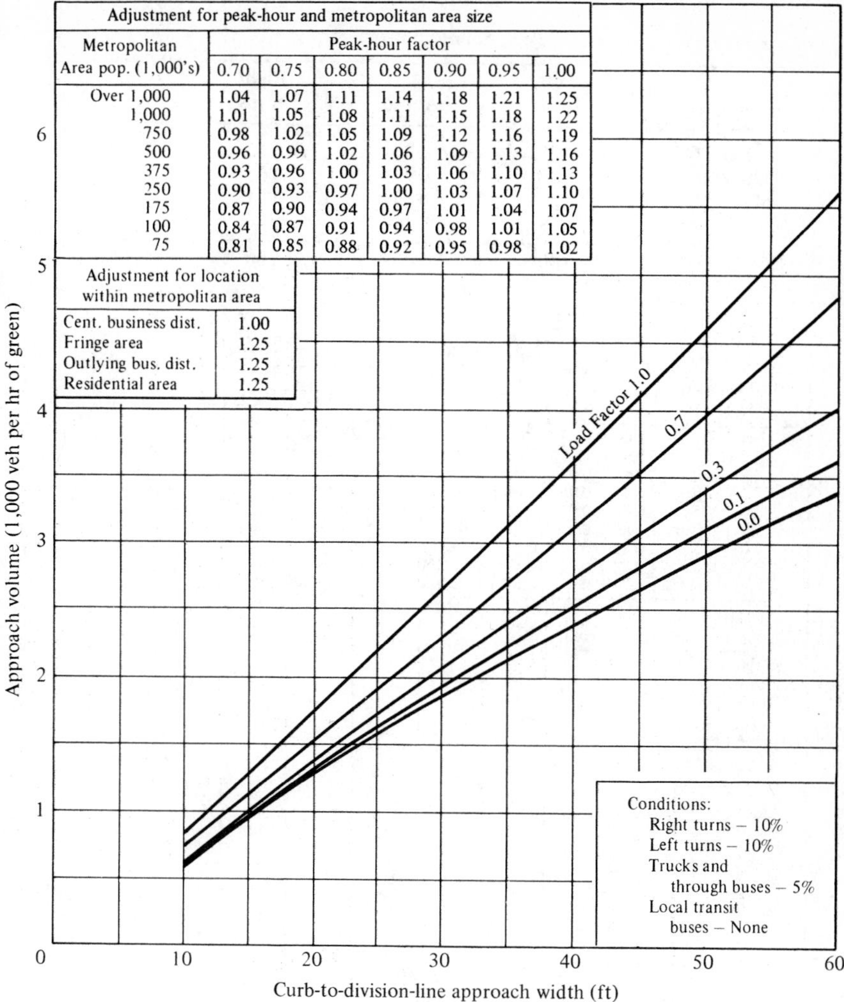

Adjustment for peak-hour and metropolitan area size

Metropolitan Area pop. (1,000's)	Peak-hour factor 0.70	0.75	0.80	0.85	0.90	0.95	1.00
Over 1,000	1.04	1.07	1.11	1.14	1.18	1.21	1.25
1,000	1.01	1.05	1.08	1.11	1.15	1.18	1.22
750	0.98	1.02	1.05	1.09	1.12	1.16	1.19
500	0.96	0.99	1.02	1.06	1.09	1.13	1.16
375	0.93	0.96	1.00	1.03	1.06	1.10	1.13
250	0.90	0.93	0.97	1.00	1.03	1.07	1.10
175	0.87	0.90	0.94	0.97	1.01	1.04	1.07
100	0.84	0.87	0.91	0.94	0.98	1.01	1.05
75	0.81	0.85	0.88	0.92	0.95	0.98	1.02

Adjustment for location within metropolitan area

Location	Factor
Cent. business dist.	1.00
Fringe area	1.25
Outlying bus. dist.	1.25
Residential area	1.25

Fig. 80. Urban intersection approach volume, in vehicles per hour of green signal time, for two-way streets with no parking[6].

of a city, and a one-way street without parking in a residential area. The nomographs permit rapid and easy solution of capacity problems.

The sample problem below demonstrates the general capacity analysis procedure using both the basic charts and the nomographs.

Signalized Intersection Capacity Problem The following peak hour traffic volume data were obtained for one approach to a signalized intersection in a *residential* area of a city of *500,000* population.

Peak hour volumes

Time period	Left turns	Ahead	Right turns	Total
7:15–7:30	25	183	45	253
7:30–7:45	30	199	65	294
7:45–8:00	26	182	54	262
8:00–8:15	19	136	36	191
Totals	100	700	200	1000

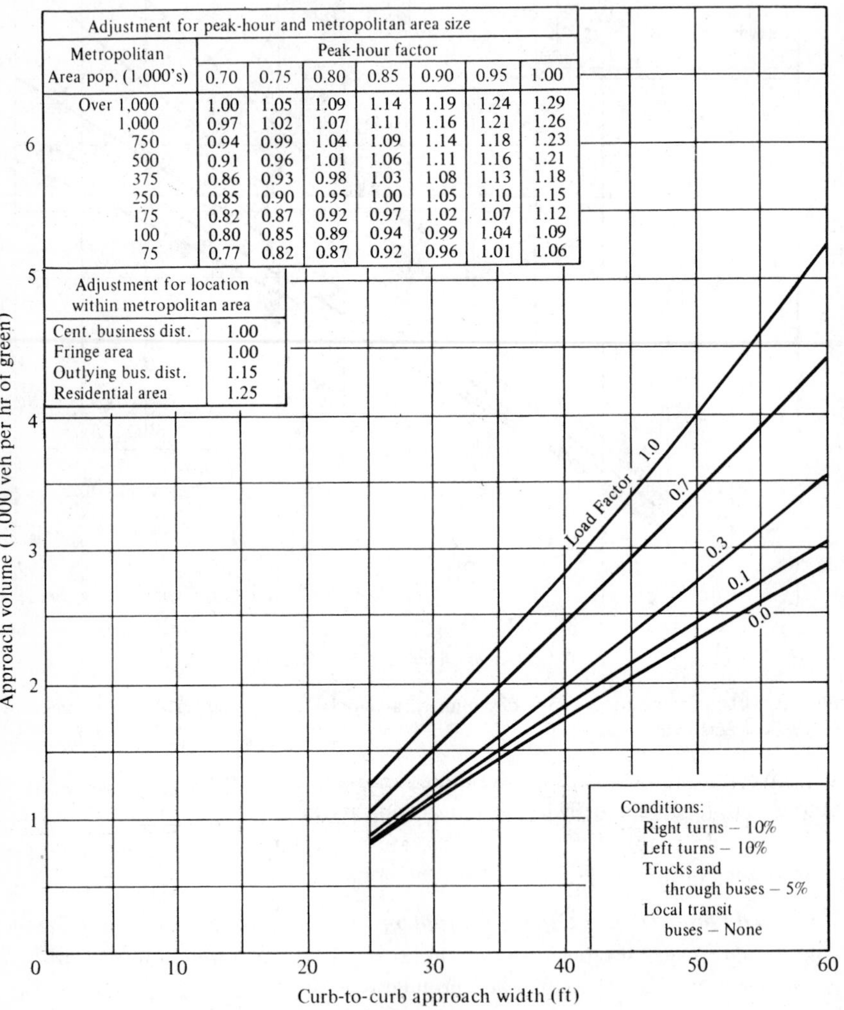

Fig. 81. Urban intersection approach service volume, in vehicles per hour of green signal time, for one-way streets with parking on both sides[6].

TABLE 31. TRUCK AND THROUGH BUS ADJUSTMENT FACTORS (6).

Trucks and through buses (%)	Correction factor	Trucks and through buses (%)	Correction factor	Trucks and through buses (%)	Correction factor
0	1.05	7	0.98	14	0.91
1	1.04	8	0.97	15	0.90
2	1.03	9	0.96	16	0.89
3	1.02	10	0.95	17	0.88
4	1.01	11	0.94	18	0.87
5	1.00	12	0.93	19	0.86
6	0.99	13	0.92	20	0.85

TABLE 32. ADJUSTMENT FACTORS FOR RIGHT TURNS ON TWO-WAY STREETS,[a] RIGHT TURNS ON ONE-WAY STREETS,[a] AND LEFT TURNS ON ONE-WAY STREETS[a] (6).

	Adjustment factor					
	With no parking[c]			With parking[d]		
Turns[b] (%)	Approach width ⩽15 ft	Approach width 16 to 24 ft	Approach width 25 to 34 ft	Approach width ⩽20 ft	Approach width 21 to 29 ft	Approach width 30 to 39 ft
0	1.20	1.050	1.025	1.20	1.050	1.025
1	1.18	1.045	1.020	1.18	1.045	1.020
2	1.16	1.040	1.020	1.16	1.040	1.020
3	1.14	1.035	1.015	1.14	1.035	1.015
4	1.12	1.030	1.015	1.12	1.030	1.015
5	1.10	1.025	1.010	1.10	1.025	1.010
6	1.08	1.020	1.010	1.08	1.020	1.010
7	1.06	1.015	1.005	1.06	1.015	1.005
8	1.04	1.010	1.005	1.04	1.010	1.005
9	1.02	1.005	1.000	1.02	1.005	1.000
10	1.00	1.000	1.000	1.00	1.000	1.000
11	0.99	0.995	1.000	0.99	0.995	1.000
12	0.98	0.990	0.995	0.98	0.990	0.995
13	0.97	0.985	0.995	0.97	0.985	0.995
14	0.96	0.980	0.990	0.96	0.980	0.990
15	0.95	0.975	0.990	0.95	0.975	0.990
16	0.94	0.970	0.985	0.94	0.970	0.985
17	0.93	0.965	0.985	0.93	0.965	0.985
18	0.92	0.960	0.980	0.92	0.960	0.980
19	0.91	0.955	0.980	0.91	0.955	0.980
20	0.90	0.950	0.975	0.90	0.950	0.975
22	0.89	0.940	0.980	0.89	0.940	0.980
24	0.88	0.930	0.985	0.88	0.930	0.985
26	0.87	0.920	0.990	0.87	0.920	0.990
28	0.86	0.910	0.995	0.86	0.910	0.995
30+	0.85	0.900	1.000	0.85	0.900	1.000

[a]No separate turning lanes or separate signal indications.
[b]Handle right turns and left turns separately in all computations; do not sum.
[c]No adjustment necessary for approach width of 35 ft or more; that is, use factor of 1.000.
[d]No adjustment necessary for approach width of 40 ft or more; that is, use factor of 1.000.

TABLE 33. ADJUSTMENT FACTORS FOR LEFT TURNS ON TWO-WAY STREETS[a] (6).

	Adjustment factor					
	With no parking			With parking		
Turns (%)	Approach width ⩽15 ft	Approach width 16 to 34 ft	Approach width ⩾35 ft	Approach width ⩽20 ft	Approach width 21 to 39 ft	Approach width ⩾40 ft
0	1.30	1.10	1.050	1.30	1.10	1.050
1	1.27	1.09	1.045	1.27	1.09	1.045
2	1.24	1.08	1.040	1.24	1.08	1.040
3	1.21	1.07	1.035	1.21	1.07	1.035
4	1.18	1.06	1.030	1.18	1.06	1.030
5	1.15	1.05	1.025	1.15	1.05	1.025
6	1.12	1.04	1.020	1.12	1.04	1.020
7	1.09	1.03	1.015	1.09	1.03	1.015
8	1.06	1.02	1.010	1.06	1.02	1.010
9	1.03	1.01	1.005	1.03	1.01	1.005
10	1.00	1.00	1.000	1.00	1.00	1.000
11	0.98	0.99	0.995	0.98	0.99	0.995
12	0.96	0.98	0.990	0.96	0.98	0.990
13	0.94	0.97	0.985	0.94	0.97	0.985
14	0.92	0.96	0.980	0.92	0.96	0.980
15	0.90	0.95	0.975	0.90	0.95	0.975
16	0.89	0.94	0.970	0.89	0.94	0.970
17	0.88	0.93	0.965	0.88	0.93	0.965
18	0.87	0.92	0.960	0.87	0.92	0.960
19	0.86	0.91	0.955	0.86	0.91	0.955
20	0.85	0.90	0.950	0.85	0.90	0.950
22	0.84	0.89	0.940	0.84	0.89	0.940
24	0.83	0.88	0.930	0.83	0.88	0.930
26	0.82	0.87	0.920	0.82	0.87	0.920
28	0.81	0.86	0.910	0.81	0.86	0.910
30+	0.80	0.85	0.900	0.80	0.85	0.900

[a]No separate turning lanes or separate signal indications.

Additional intersection approach data include the site diagram:

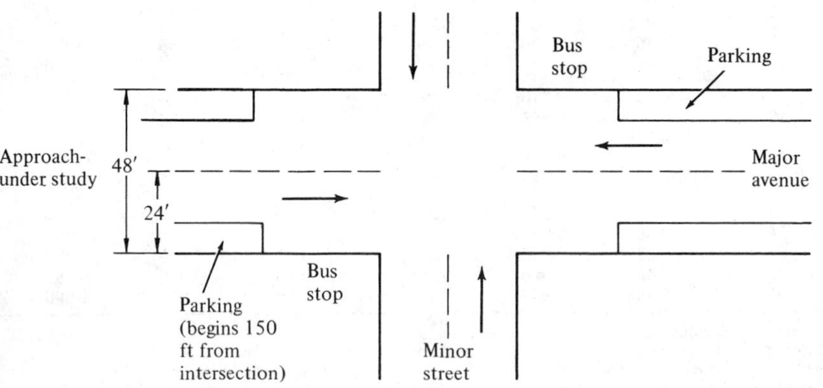

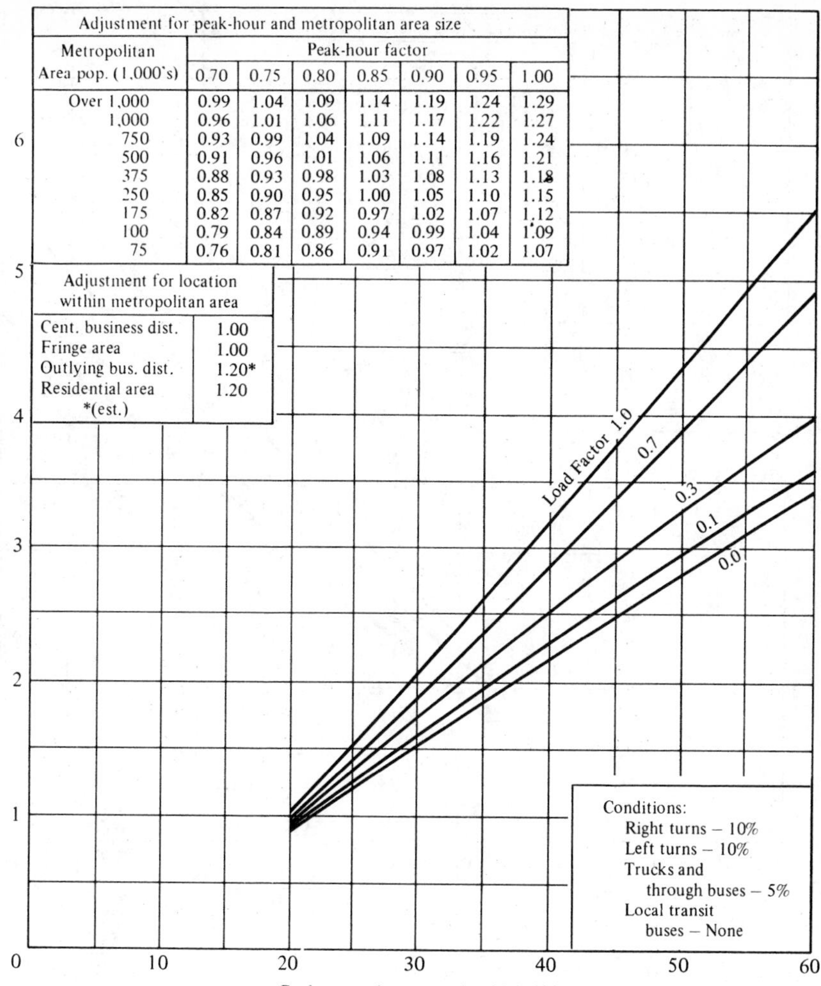

Adjustment for peak-hour and metropolitan area size							
Metropolitan	Peak-hour factor						
Area pop. (1,000's)	0.70	0.75	0.80	0.85	0.90	0.95	1.00
Over 1,000	0.99	1.04	1.09	1.14	1.19	1.24	1.29
1,000	0.96	1.01	1.06	1.11	1.17	1.22	1.27
750	0.93	0.99	1.04	1.09	1.14	1.19	1.24
500	0.91	0.96	1.01	1.06	1.11	1.16	1.21
375	0.88	0.93	0.98	1.03	1.08	1.13	1.18
250	0.85	0.90	0.95	1.00	1.05	1.10	1.15
175	0.82	0.87	0.92	0.97	1.02	1.07	1.12
100	0.79	0.84	0.89	0.94	0.99	1.04	1.09
75	0.76	0.81	0.86	0.91	0.97	1.02	1.07

Adjustment for location within metropolitan area	
Cent. business dist.	1.00
Fringe area	1.00
Outlying bus. dist.	1.20*
Residential area	1.20
*(est.)	

Fig. 82. **Urban intersection approach service volume, in vehicles per hour of green signal time, for one-way streets with parking on one side[6].**

Conditions:

Trucks: 5%
Transit buses: 50 per hour
Green time: 50 sec
Cycle time: 80 sec (required for major avenue progression)

Determine the design capacity and true capacity (possible capacity) of the approach.

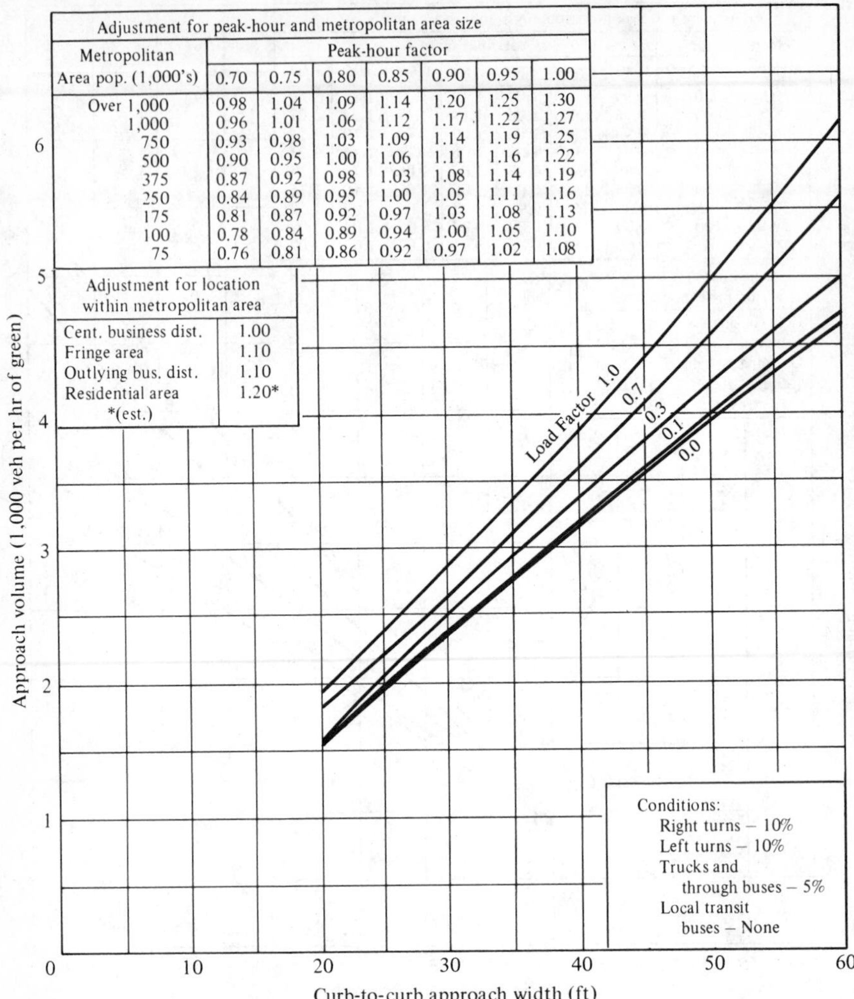

Fig. 83. Urban intersection approach service volume, in vehicles per hour of green signal time, for one-way streets with no parking[6].

Solution by Basic Charts:

Peak hour factor (PHF) (Determined from traffic count data):

$$\text{PHF} = \frac{\text{Peak Hour Volume}}{\text{4 Times 15-Minute Peak}} = \frac{1{,}000}{4(294)} = 0.85$$

Load factor (from Table 30):

at design capacity (level of service C) = 0.3

at possible capacity (level of service E) = 1.0

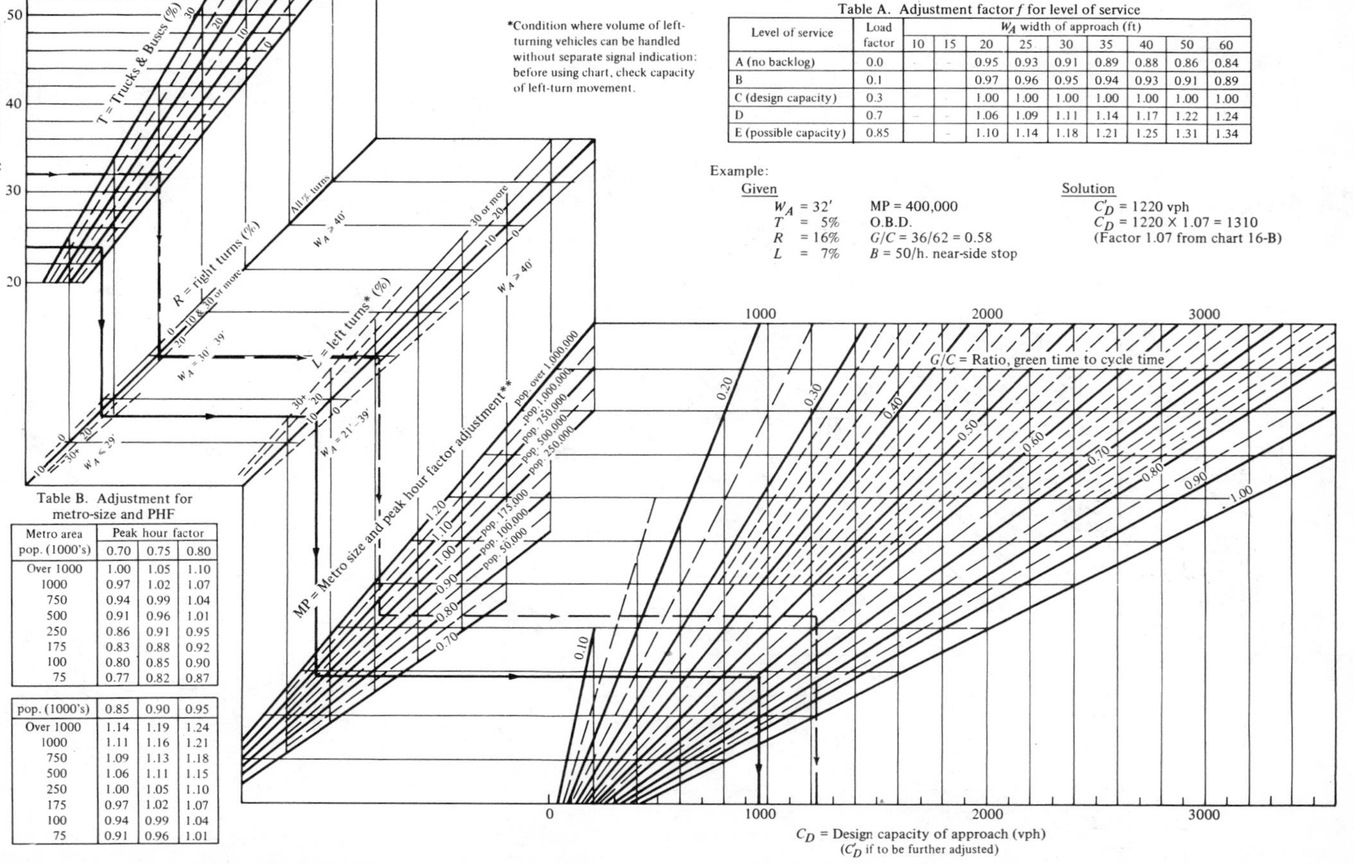

*Condition where volume of left-turning vehicles can be handled without separate signal indication: before using chart, check capacity of left-turn movement.

Table A. Adjustment factor f for level of service

Level of service	Load factor	W_A width of approach (ft) 10	15	20	25	30	35	40	50	60
A (no backlog)	0.0	-	-	0.95	0.93	0.91	0.89	0.88	0.86	0.84
B	0.1			0.97	0.96	0.95	0.94	0.93	0.91	0.89
C (design capacity)	0.3			1.00	1.00	1.00	1.00	1.00	1.00	1.00
D	0.7	-	-	1.06	1.09	1.11	1.14	1.17	1.22	1.24
E (possible capacity)	0.85		-	1.10	1.14	1.18	1.21	1.25	1.31	1.34

Example:

Given

W_A = 32' MP = 400,000
T = 5% O.B.D.
R = 16% G/C = 36/62 = 0.58
L = 7% B = 50/h. near-side stop

Solution

C'_D = 1220 vph
C_D = 1220 × 1.07 = 1310
(Factor 1.07 from chart 16-B)

Table B. Adjustment for metro-size and PHF

Metro area pop. (1000's)	Peak hour factor 0.70	0.75	0.80
Over 1000	1.00	1.05	1.10
1000	0.97	1.02	1.07
750	0.94	0.99	1.04
500	0.91	0.96	1.01
250	0.86	0.91	0.95
175	0.83	0.88	0.92
100	0.80	0.85	0.90
75	0.77	0.82	0.87

pop. (1000's)	0.85	0.90	0.95
Over 1000	1.14	1.19	1.24
1000	1.11	1.16	1.21
750	1.09	1.13	1.18
500	1.06	1.11	1.15
250	1.00	1.05	1.10
175	0.97	1.02	1.07
100	0.94	0.99	1.04
75	0.91	0.96	1.01

**Use Table B if PHF is known to find adjust. factor; otherwise use population directly.

Fig. 84. Signalized intersection capacity nomograph[7]. Two-way street, with parking; OBD, fringe, and residential area.

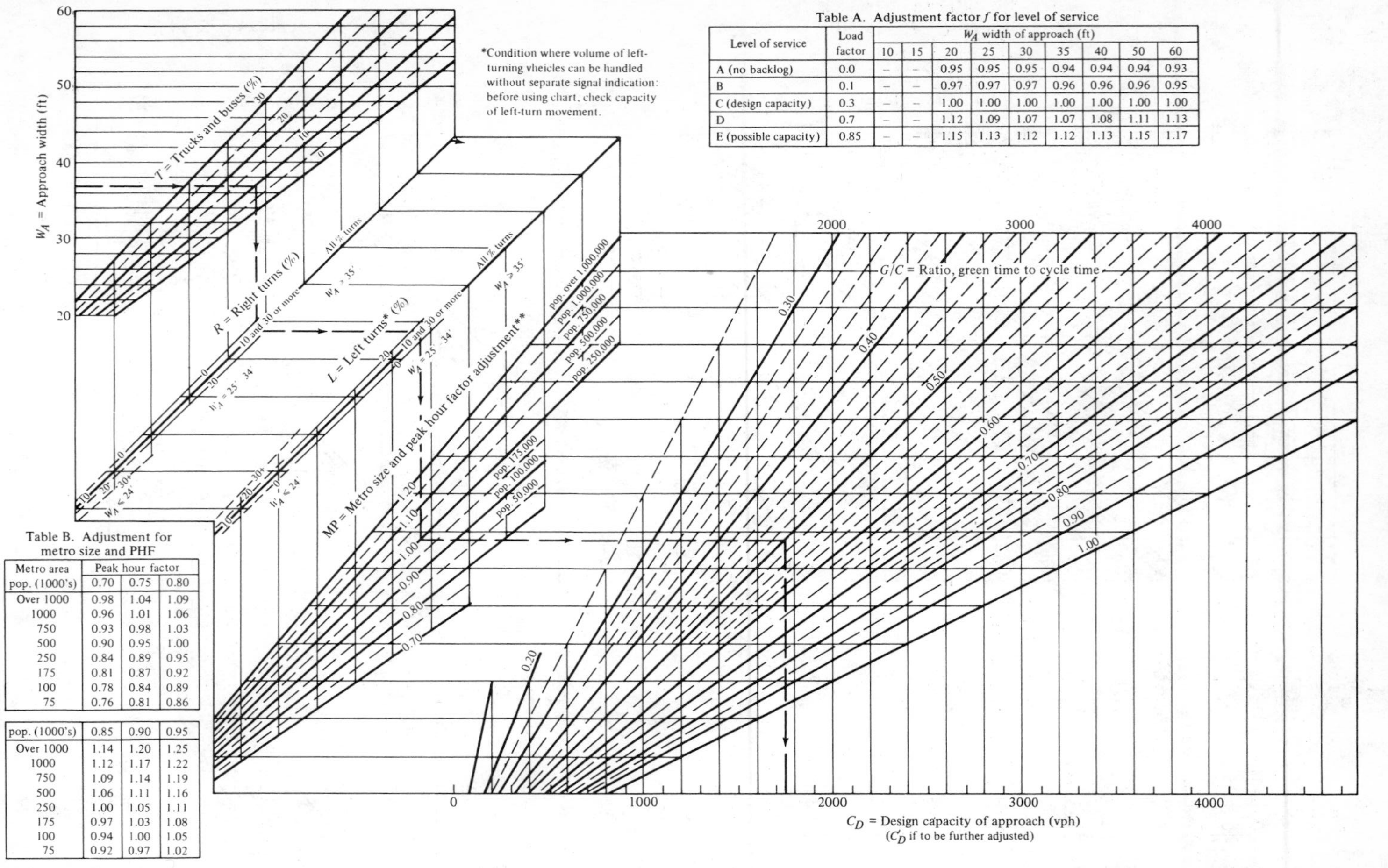

*Condition where volume of left-turning vheicles can be handled without separate signal indication: before using chart, check capacity of left-turn movement.

Table A. Adjustment factor f for level of service

Level of service	Load factor	W_A width of approach (ft) 10	15	20	25	30	35	40	50	60
A (no backlog)	0.0	–	–	0.95	0.95	0.95	0.94	0.94	0.94	0.93
B	0.1	–	–	0.97	0.97	0.97	0.96	0.96	0.96	0.95
C (design capacity)	0.3	–	–	1.00	1.00	1.00	1.00	1.00	1.00	1.00
D	0.7	–	–	1.12	1.09	1.07	1.07	1.08	1.11	1.13
E (possible capacity)	0.85	–	–	1.15	1.13	1.12	1.12	1.13	1.15	1.17

Table B. Adjustment for metro size and PHF

Metro area pop. (1000's)	Peak hour factor 0.70	0.75	0.80
Over 1000	0.98	1.04	1.09
1000	0.96	1.01	1.06
750	0.93	0.98	1.03
500	0.90	0.95	1.00
250	0.84	0.89	0.95
175	0.81	0.87	0.92
100	0.78	0.84	0.89
75	0.76	0.81	0.86

pop. (1000's)	0.85	0.90	0.95
Over 1000	1.14	1.20	1.25
1000	1.12	1.17	1.22
750	1.09	1.14	1.19
500	1.06	1.11	1.16
250	1.00	1.05	1.11
175	0.97	1.03	1.08
100	0.94	1.00	1.05
75	0.92	0.97	1.02

**Use Table B if PHF is known to find adjust. factor; otherwise use population directly.

Turns:
determined from traffic count data
right turns: (200/1,000): 20%
left turns: (100/1,000): 10%

G/C ratio:
50/80 = 0.625

1. Enter chart for two-way street with parking (Fig. 79) at approach width of 24 ft, reflect off 0.3 and 1.0 load factor lines to obtain chart volumes of 1200 and 1425 vehicles per hour of green for design and possible capacities, respectively.

Adjustment factors, individual	
Population and PHF (top upper left chart, Fig. 79)	1.060
Residential area (bottom upper left chart, Fig. 79)	1.250
Right turns (Table 32)	.950
Left turns (Table 33)	1.000
Trucks and through buses (Table 31)	1.000
G/C ratio	0.625
Composite factor (individual factors multiplied together):	0.787

2. Chart volume times composite factor gives service volume at respective level of service without adjustment for effect of local buses (without transit bus influence).

Design capacity = (1200) (0.787) = 944 veh/h
Possible capacity = (1425) (0.787) = 1121 veh/h

3. Bus Adjustment Factor. From Fig. 86, near-side bus stop with parking, enter Fig. 86B with 50 buses/h, reflect downward to approach width *W*, vane for 24 ft, reflect horizontally to set of vanes corresponding to 24-ft approach width to combined percentage of right and left turns of 30%, and reflect upward to bus adjustment factor = 1.09.

Comment: All bus adjustment charts, except the near-side bus stop with parking approach widths intermediate to those shown, can be used by interpolating the width on the charts. For intermediate approach widths on the near-side bus stop with parking, determine the bus adjustment factors for each nearest chart width and then interpolate.

The near-side bus stop with parking can help to increase service volumes (factors greater than 1.0) because where bus volumes are relatively low the bus stop area acts as a special right-turn lane part of the time and results in improved flow.

4. Service Volumes (with Bus Adjustment)

Design capacity = (944) (1.09) = 1029 veh/h
Possible capacity = (1121) (1.09) = 1222 veh/h

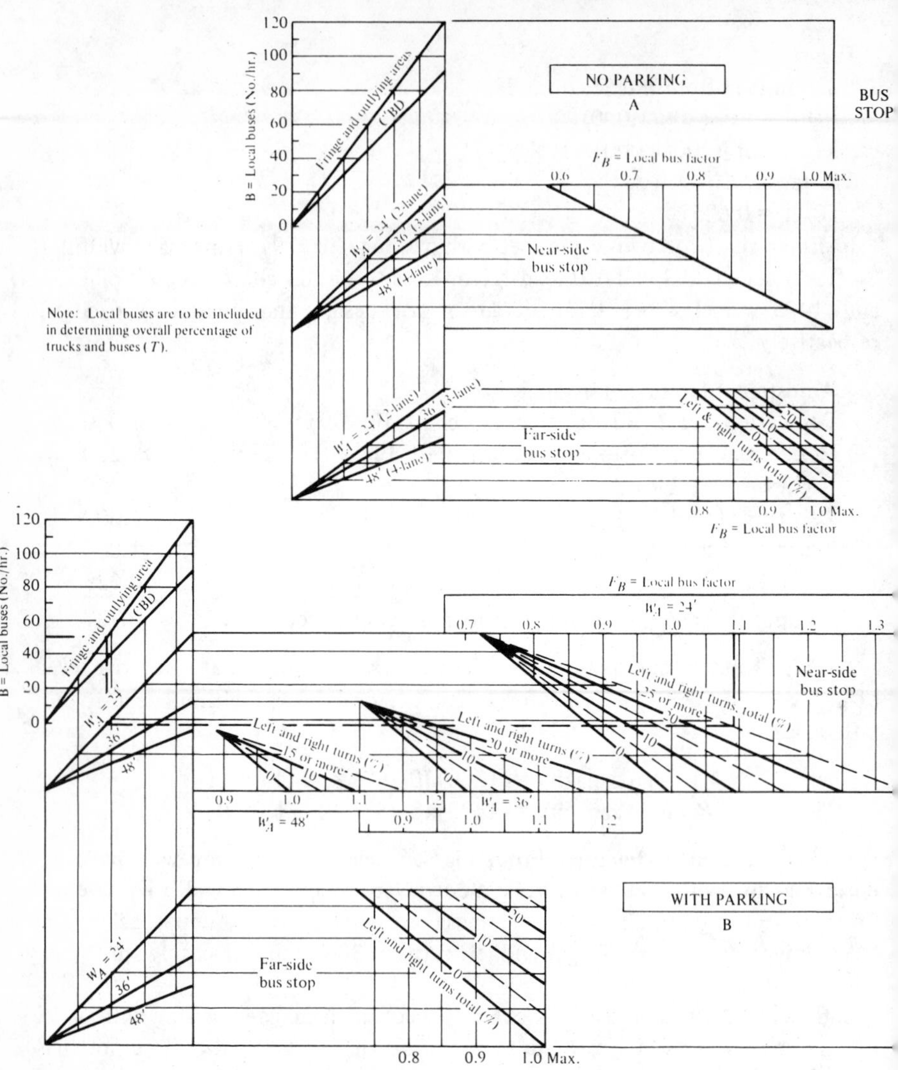

Fig. 86. Bus adjustment nomograph[7]. Design capacity of signalized intersections, Local Bus Factor.

Comment: The resulting service volumes are in mixed vehicles (autos, trucks, and buses) for one hour of operation (includes alternating greens, ambers, and reds).

Solution by Nomograph:

Enter two-way chart with parking, residential (Fig. 84), with approach width = 24 ft.

Turn at 5% truck vane.

Turn at 20% right turns (use set of right-turn vanes for proper W_A).

Turn at 10% left turns (use set of left-turn vanes for proper W_A).

Turn at MP factor = 1.06. MP factor obtained from Table B using population of 500,000 and PHF of 0.85.

Turn at $G/C = 0.625$.

Obtain chart volume of 960 vehicles per hour without bus adjustment.

Bus adjustment (same procedure as used previously) = 1.09.

Since nomographs are developed for design capacity, urban (level of service C), other service volumes are obtained by using the adjustment factor from Table A of the nomograph. Note that level of service E volumes are developed using a more realistic load factor of 0.85 in the nomographs as opposed to the 1.00 used in the basic charts. Level of service E volumes will, as a result, differ somewhat from those previously determined.

$$\text{Design capacity} = SV_C = \underset{\text{chart vol.}}{960} \times \underset{\text{bus factor}}{1.09} = 1050 \text{ veh/h}$$

$$\text{Possible capacity} = SV_E = SV_C \times \text{Service volume adjustment}$$

$$SV_E = 1050 \times 1.13 \text{ (from Table A)}$$
$$= 1190 \text{ veh/h}$$

Comment on Solution of Capacity Problem Information other than service volumes can be obtained using the capacity analysis procedure. Knowing demand volume (service volume) and green time and cycle time, required approach width can be obtained. Knowing demand and approach width, required G/C can be obtained, etc.

Capacity Analysis, Freeway (Unencumbered Sections) Table 34 shows levels of service and maximum service volumes for freeways and expressways under uninterrupted flow conditions. The volume data given on the right-hand portion of the table are for freeways operating under ideal conditions, i.e., all passenger cars, 12-ft lanes with adequate shoulders, no lateral restrictions within 6 ft of the pavement edge, and a horizontal and vertical alignment design permitting 70 mph operation. For conditions other than ideal, adjustments are made for the effect of shoulder width, lateral clearance, and truck operation.

The basic formula for determining service volumes on freeways is

$$SV_x = 2000\,(N)\,(v/c)\,(W)\,(T_L)\,(B_L)$$

where

SV_x = service volume at a specified level of service x

N = the number of lanes in one direction

TABLE 34. LEVELS OF SERVICE AND MAXIMUM SERVICE VOLUMES FOR

Level of service	Traffic flow conditions: Description	Traffic flow conditions: Operating speed[d] (mph)	Service volume/capacity (v/c) ratio[a]: Basic limiting value for average highway speed (AHS) of 70 mph, for: 4-lane freeway (2 lanes/ direction)	6-lane freeway (3 lanes/ direction)	8-lane freeway (4 lanes/ direction)	Approximate working value for any number of lanes for restricted average highway speed of 60 mph	50 mph
A	Free flow	≥60	≤0.35	≤0.40	≤0.43	—[b]	—[b]
B	Stable flow (upper speed range)	≥55	≤0.50	≤0.58	≤0.63	≤0.25	—[b]
	Peak-hour factor (PHF)[e]						
C	Stable flow	≥50	≤0.75 (PHF)	≤0.80 (PHF)	≤0.83 (PHF)	≤0.45 (PHF)	—[b]
D	Approaching unstable flow	≥40	≤0.90 (PHF)			≤0.80 (PHF)	≤0.45 (PHF)
E[f]	Unstable flow	30–35[e]	≤1.00				
F	Forced flow	<30[e]	← Not meaningful →				

[a]Operating speed and basic v/c ratio are independent measures of level of service; both limits must be satisfied in any determination of level.
[b]Operating speed required for this level is not attainable even at low volumes.
[c]Peak-hour factor for freeways is the ratio of the whole-hour volume to the highest rate of flow occurring during a 5-min interval within the peak hour.
[d]A peak-hour factor of 1.00 is seldom attained; the values listed here should be considered as maximum average flow rates likely to be obtained during the peak 5-min interval within the peak hour.
[e]Approximately.
[f]Capacity.

v/c = volume/capacity ratio
W = an adjustment factor for the effect of lane width and lateral clearance restriction
T_L = an adjustment factor for the influence of trucks in the traffic stream: this is based on the percentage of trucks and the (percentage) steepness and length of grade or general terrain condition, i.e., level, rolling or mountainous
B_L = an adjustment factor for the influence of large buses where their proportion is high in the traffic stream

Note on Table 34 that the service volume/capacity (V/C), ratio at levels of service C and D include the metropolitan area traffic peaking characteristic in the form of the peak hour factor. At levels of service C and D, traffic peaking has considerable influence on the total hourly service volume and consequently must be used in the calculation. Average peak hour factors related to metropolitan area population are noted in Table 35.

FREEWAYS AND EXPRESSWAYS UNDER UNINTERRUPTED FLOW CONDITIONS (6).

Maximum service volume under ideal conditions, including 70-mph average highway speed (total passenger cars per hour, one direction)

4-lane freeway (2 lanes one direction)				6-lane freeway (3 lanes one direction)				8-lane freeway (4 lanes one direction)				Each additional lane above four in one direction			
1400				2400				3400				1000			
2000				3500				5000				1500			
0.77	0.83	0.91	1.00[d]	0.77	0.83	0.91	1.00[d]	0.77	0.83	0.91	1.00[d]	0.77	0.83	0.91	1.00[d]
2300	2500	2750	3000	3700	4000	4350	4800	5100	5500	6000	6600	1400	1500	1650	1800
2800	3000	3300	3600	4150	4500	4900	5400	5600	6000	6600	7200	1400	1500	1650	1800
4000[e]				6000[e]				8000[e]				2000[e]			
← Widely variable (0 to capacity) →															

TABLE 35. AVERAGE PEAK HOUR FACTOR FOR DIFFERENT CITY SIZE.

Population	Peak hour factor
Greater than 1,000,000	0.91
500,000 to 1,000,000	0.83
Less than 500,000	0.77

Tables 36–40 are used for determining adjustment factors for the effect of lane width and lateral restriction and for the restrictive effect of grades on truck and bus operations. Note that truck operation tables are given for general terrain conditions, e.g., level, rolling and mountainous (Tables 37a and 37b), and specific grade conditions, e.g., length of grade and percentage grade (Tables 38, 39, and 40). The truck tables are used to convert trucks to equivalent passenger vehicles for the specific conditions at hand.

TABLE 36. COMBINED EFFECT OF LANE WIDTH AND RESTRICTED LATERAL CLEARANCE ON CAPACITY AND SERVICE VOLUMES OF DIVIDED FREEWAYS AND EXPRESSWAYS WITH UNINTERRUPTED FLOW (6).

Distance from traffic lane edge to obstruction (ft)	Adjustment factor,[a] W, for lane width and lateral clearance							
	Obstruction on one side of one-direction roadway				Obstructions on both sides of one-direction roadway			
	12-ft Lanes	11-ft Lanes	10-ft Lanes	9-ft Lanes	12-ft Lanes	11-ft Lanes	10-ft Lanes	9-ft Lanes
	(*a*) 4-lane divided freeway, one direction of travel							
6	1.00	0.97	0.91	0.81	1.00	0.97	0.91	0.81
4	0.99	0.96	0.90	0.80	0.98	0.95	0.89	0.79
2	0.97	0.94	0.88	0.79	0.94	0.91	0.86	0.76
0	0.90	0.87	0.82	0.73	0.81	0.79	0.74	0.66
	(*b*) 6- and 8-lane divided freeway, one direction of travel							
6	1.00	0.96	0.89	0.78	1.00	0.96	0.89	0.78
4	0.99	0.95	0.88	0.77	0.98	0.94	0.87	0.77
2	0.97	0.93	0.87	0.76	0.96	0.92	0.85	0.75
0	0.94	0.91	0.85	0.74	0.91	0.87	0.81	0.70

[a]Same adjustments for capacity and all levels of service.

TABLE 37a. AVERAGE GENERALIZED PASSENGER CAR EQUIVALENTS OF TRUCKS AND BUSES ON FREEWAYS AND EXPRESSWAYS, OVER EXTENDED SECTION LENGTHS (INCLUDING UPGRADES, DOWNGRADES, AND LEVEL SUBSECTIONS) (6).

Level of service	Equivalent, E, for: Level Terrain	Rolling Terrain	Mountainous Terrain
A	widely variable; one or more trucks have same total effect, causing other traffic to shift to other lanes. Use equivalent for remaining levels in problems.		
B through E			
E_T, for trucks	2	4	8
E_B, for buses[a]	1.6	3	5

[a]Separate consideration not warranted in most problems; use only where bus volumes are significant.

TABLE 37b. AVERAGE GENERALIZED ADJUSTMENT FACTORS FOR TRUCKS[b] ON FREEWAYS AND EXPRESSWAYS, OVER EXTENDED SECTION LENGTHS (6).

Percentage of trucks, P_T	Factor, T, for all levels of service		
	Level terrain	Rolling terrain	Mountainous terrain
1	0.99	0.97	0.93
2	0.98	0.94	0.88
3	0.97	0.92	0.83
4	0.96	0.89	0.78
5	0.95	0.87	0.74
6	0.94	0.85	0.70
7	0.93	0.83	0.67
8	0.93	0.81	0.64
9	0.92	0.79	0.61
10	0.91	0.77	0.59
12	0.89	0.74	0.54
14	0.88	0.70	0.51
16	0.86	0.68	0.47
18	0.85	0.65	0.44
20	0.83	0.63	0.42

[b]Not applicable to buses where they are given separate specific consideration; use instead Table 37a in conjunction with Table 40.

Freeway Capacity Problem Determine the service volume at level of service A and at level of service C on a section of four-lane urban freeway which has a 3% upgrade $1\frac{1}{2}$ miles long. A narrow center median requires the use of double-faced beam guard, the face of which is 4 ft from the left edge of the median traffic lane. The right side of the travelled way has a 10-ft shoulder. The percentage of trucks in the traffic stream is 5%, and the alignment is designed for 70 mph operation. Additional data include:

Metropolitan area size: 750,000
Lane widths: 12 ft
Buses: none

Solution: Service volume at level of service A, SV_A

$$SV_A = 2000(N)(V/C)(W)(T_L)(B_L)$$

$N = 2$

$V/C = 0.35$ (from Table 34)

$W = 0.99$ (from Table 36)

Equivalent passenger cars, $E_T = 9$ (Table 38)

$T_L = 0.71$ (using $E_T = 9$ and truck % = 5 in Table 40)

$B_L = 1.00$ (no buses)

Therefore

$$SV_A = 2000(2)(0.35)(0.99)(0.71)(1.00) = 985 \text{ veh/h}.$$

TABLE 38. PASSENGER CAR EQUIVALENTS OF TRUCKS ON FREEWAYS AND EXPRESSWAYS, ON SPECIFIC INDIVIDUAL SUBSECTIONS OR GRADES (6).

		Passenger car equivalent, E_r									
	Length of	Levels of service A through C for:					Levels of service D and E (capacity) for:				
Grade (%)	grade (mi)	3% Trucks	5% Trucks	10% Trucks	15% Trucks	20% Trucks	3% Trucks	5% Trucks	10% Trucks	15% Trucks	20% Trucks
0–1	All	2	2	2	2	2	2	2	2	2	2
2	¼–½	5	4	4	3	3	5	4	4	3	3
	¾–1	7	5	5	4	4	7	5	5	4	4
	1½–2	7	6	6	6	6	7	6	6	6	6
	3–4	7	7	8	8	8	7	7	8	8	8
3	¼	10	8	5	4	3	10	8	5	4	3
	½	10	8	5	4	4	10	8	5	4	4
	¾	10	8	6	5	5	10	8	5	4	5
	1	10	8	6	5	6	10	8	6	5	6
	1½	10	9	7	7	7	10	9	7	7	7
	2	10	9	8	8	8	10	9	8	8	8
	3	10	10	10	10	10	10	10	10	10	10
	4	10	10	11	11	11	10	10	11	11	11
4	¼	12	9	5	4	3	13	9	5	4	3
	½	12	9	5	5	5	13	9	5	5	5
	¾	12	9	7	7	7	13	9	7	7	7
	1	12	10	8	8	8	13	10	8	8	8
	1½	12	11	10	10	10	13	11	10	10	10
	2	12	11	11	11	11	13	12	11	11	11
	3	12	12	13	13	13	13	13	14	14	14
	4	12	13	15	15	14	13	14	16	16	15
5	¼	13	10	6	4	3	14	10	6	4	3
	½	13	11	7	7	7	14	11	7	7	7
	¾	13	11	9	8	8	14	11	9	8	8
	1	13	12	10	10	10	14	13	10	10	10
	1½	13	13	12	12	12	14	14	13	13	13
	2	13	14	14	14	14	14	15	15	15	15
	3	13	15	16	16	15	14	17	17	17	17
	4	15	17	19	19	17	16	19	22	21	19
6	¼	14	10	6	4	3	15	10	6	4	3
	½	14	11	8	8	8	15	11	8	8	8
	¾	14	12	10	10	10	15	12	10	10	10
	1	14	13	12	12	11	15	14	13	13	11
	1½	14	14	14	14	13	15	16	15	15	14
	2	14	15	16	16	15	15	18	18	18	16
	3	14	16	18	18	17	15	20	20	20	19
	4	19	19	20	20	20	20	23	23	23	23

Service volume at level of service C, SV_C:

$N = 2$

Peak hour factor = 0.83 (average for population group 500,000–1,000,000 from Table 35).

V/C = [0.75 PHF] (Table 34)

$= 0.75(0.83) = 0.62$

Equivalent passenger cars, $E_T = 9$ (Table 38)

$T_L = 0.71$ (using $E_T = 9$ and truck % = 5 in Table 40)

$B_L = 1.00$ (no buses)

TABLE 39. PASSENGER CAR EQUIVALENTS OF INTERCITY BUSES ON FREEWAYS AND EXPRESSWAYS, ON SPECIFIC INDIVIDUAL SUBSECTIONS OR GRADES (6).

	Passenger car equivalent,[b] E_B	
Grade[a] (%)	Levels of service A through C	Levels of service D and E (capacity)
0–4	1.6	1.6
5[c]	4	2
6[c]	7	4
7[c]	12	10

[a]All lengths.
[b]For all percentages of buses.
[c]Use generally restricted to grades over ½ mile long.

TABLE 40. ADJUSTMENT FACTORS[a] FOR TRUCKS AND BUSES ON INDIVIDUAL ROADWAY SUBSECTIONS OR GRADES ON FREEWAYS AND EXPRESSWAYS (INCORPORATING PASSENGER CAR EQUIVALENT AND PERCENTAGE OF TRUCKS OR BUSES)[b] (6).

Passenger car equivalent, E_T or $E_B{}^c$	Truck adjustment factor T_c or T_L (B_c or B_L for buses)[d] Percentage of trucks, P_T (or of buses, P_B) of:														
	1	2	3	4	5	6	7	8	9	10	12	14	16	18	20
2	0.99	0.98	0.97	0.96	0.95	0.94	0.93	0.93	0.92	0.91	0.89	0.88	0.86	0.85	0.83
3	0.98	0.96	0.94	0.93	0.91	0.89	0.88	0.86	0.85	0.83	0.81	0.78	0.76	0.74	0.71
4	0.97	0.94	0.92	0.89	0.87	0.85	0.83	0.81	0.79	0.77	0.74	0.70	0.68	0.65	0.63
5	0.96	0.93	0.89	0.86	0.83	0.81	0.78	0.76	0.74	0.71	0.68	0.64	0.61	0.58	0.56
6	0.95	0.91	0.87	0.83	0.80	0.77	0.74	0.71	0.69	0.67	0.63	0.59	0.56	0.53	0.50
7	0.94	0.89	0.85	0.81	0.77	0.74	0.70	0.68	0.65	0.63	0.58	0.54	0.51	0.48	0.45
8	0.93	0.88	0.83	0.78	0.74	0.70	0.67	0.64	0.61	0.59	0.54	0.51	0.47	0.44	0.42
9	0.93	0.86	0.81	0.76	0.71	0.68	0.64	0.61	0.58	0.56	0.51	0.47	0.44	0.41	0.38
10	0.92	0.85	0.79	0.74	0.69	0.65	0.61	0.58	0.55	0.53	0.48	0.44	0.41	0.38	0.36
11	0.91	0.83	0.77	0.71	0.67	0.63	0.59	0.56	0.53	0.50	0.45	0.42	0.38	0.36	0.33
12	0.90	0.82	0.75	0.69	0.65	0.60	0.57	0.53	0.50	0.48	0.43	0.39	0.36	0.34	0.31
13	0.89	0.81	0.74	0.68	0.63	0.58	0.54	0.51	0.48	0.45	0.41	0.37	0.34	0.32	0.29
14	0.88	0.79	0.72	0.66	0.61	0.56	0.52	0.49	0.46	0.43	0.39	0.35	0.32	0.30	0.28
15	0.88	0.78	0.70	0.64	0.59	0.54	0.51	0.47	0.44	0.42	0.37	0.34	0.31	0.28	0.26
16	0.87	0.77	0.69	0.63	0.57	0.53	0.49	0.45	0.43	0.40	0.36	0.32	0.29	0.27	0.25
17	0.86	0.76	0.68	0.61	0.56	0.51	0.47	0.44	0.41	0.38	0.34	0.31	0.28	0.26	0.24
18	0.85	0.75	0.66	0.60	0.54	0.49	0.46	0.42	0.40	0.37	0.33	0.30	0.27	0.25	0.23
19	0.85	0.74	0.65	0.58	0.53	0.48	0.44	0.41	0.38	0.36	0.32	0.28	0.26	0.24	0.22
20	0.84	0.72	0.64	0.57	0.51	0.47	0.42	0.40	0.37	0.34	0.30	0.27	0.25	0.23	0.21
21	0.83	0.71	0.63	0.56	0.50	0.45	0.41	0.38	0.36	0.33	0.29	0.26	0.24	0.22	0.20
22	0.83	0.70	0.61	0.54	0.49	0.44	0.40	0.37	0.35	0.32	0.28	0.25	0.23	0.21	0.19
23	0.82	0.69	0.60	0.53	0.48	0.43	0.39	0.36	0.34	0.31	0.27	0.25	0.22	0.20	0.19
24	0.81	0.68	0.59	0.52	0.47	0.42	0.38	0.35	0.33	0.30	0.27	0.24	0.21	0.19	0.18
25	0.80	0.67	0.58	0.51	0.46	0.41	0.37	0.34	0.32	0.29	0.26	0.23	0.20	0.18	0.17

[a]Computed by $100/(100 - P_T + E_T P_T)$, or $100/(100 - P_B + E_B P_B)$. Use this formula for larger percentages.
[b]Used to convert equivalent passenger car volumes to actual mixed traffic; use reciprocal of these values to convert mixed traffic to equivalent passenger cars.
[c]From Table 38 or Table 39.
[d]Trucks and buses should not be combined in entering this table where separate consideration of buses has been established as required, because passenger car equivalents differ.

Therefore

$$SV_C = 2000(2)[(0.75)(0.83)](0.99)(0.71)(1.00) = 1750 \text{ veh/h}.$$

Elements of Design

Design Speed Setting the functional classification of a roadway establishes a broad range of design elements suitable for the roadway design. Further specificity can be obtained by defining a design class which relates geometric design values to traffic volumes and design speeds. Since more difficult terrain (rolling or mountainous) often justifies the use of somewhat lower design speeds, the design speed within a particular design class is often related to the terrain type. Minimum design speeds for varying traffic volume conditions and terrain conditions are noted in Table 41 (8). The terrain condition pertains to the general terrain characteristic in the corridor under design. Level terrain is normally that

TABLE 41. MINIMUM DESIGN SPEEDS (8).

Type of terrain	Minimum design speeds (mph) for design volumes of					
	Current ADT* under 50	Current ADT 50–250	Current ADT 250–400	Current ADT 400–750 DHV 100–200	DHV* 200–400	DHV 400 and over
Level	40	40	50	50	50	50
Rolling	30	30	40	40	40	40
Mountainous	20	20	20	30	30	30

*Current ADT is the annual average daily traffic expected after completion. DHV is the design hourly volume for the future design year, normally the 30th highest hourly volume about 20 years after completion.

in which long vertical and horizontal sight distances can be provided at little expense or with little construction difficulty. Rolling terrain is a condition in which the general terrain consistently rises and falls above and below the highway grade line, and for which trucks would normally have to reduce their speeds somewhat below that of normal passenger car operation. Mountainous terrain is terrain where abrupt changes in longitudinal and transverse slope occur, where frequent benching and side-hill excavation are required, and where truck operations are often at crawl speed (3). It should be emphasized that these standards relate to minimum specifications. Where possible, it is desirable to provide a higher-type design than that noted by the minimums.

Maximum Grade Maximum grades for rural main highways and rural low-volume roads related to design speed and terrain condition are as noted in Tables 42 and 43. It should be noted that these are maximum grades for design and that

TABLE 42. RELATION OF MAXIMUM GRADES TO DESIGN SPEED, MAIN HIGHWAYS (2).

	Design speed (mph)							
Type of topography	30	40	50	60	65	70	75	80
Flat	6	5	4	3	3	3	3	3
Rolling	7	6	5	4	4	4	4	4
Mountainous	9	8	7	6	6	5	–	–

TABLE 43. MAXIMUM GRADES, LOW VOLUME RURAL HIGHWAYS (7).

	Design speed (mph)				
Type of terrain	20	30	40	50	60
Flat	7	7	7	6	5
Rolling	10	9	8	7	6
Mountainous	12	10	10	9	

Note: For highways with ADTs below 250, grades of relatively short lengths may be increased to 150% of the value shown.

they should be used infrequently rather than as a rule. In most cases where lesser grades are possible, lesser grades should be provided. Design practice for main rural highways normally permits, for short grades of less than 500 ft and for one-way downgrades, a maximum gradient about 1% steeper than those noted in the table. For low-volume roads in rural areas maximum grades may be 2% steeper.

Sight Distance Sight distance is the distance in front of the driver visible from the driver's seat. Stopping sight distance, measured from an eye height of 3.75 ft to an object 0.5 ft high lying on the roadway some distance ahead of the driver, will permit an average driver to stop on a wet pavement in sufficient time to avoid colliding with the object. The total stopping distance is the result of a perception brake reaction distance and a braking distance. The "design" perception-reaction time upon which the distance is based is 2.5 sec; of this the perception time is 1.5 sec. This is a value large enough to include the time required by nearly all drivers under most highway conditions. The approximate total stopping sight distance necessary may be determined by use of the formula:

$$SD_s = 1.47T_{p-r}V_i + \frac{V_i^2 - V_f^2}{30(f \pm g)}$$

where

SD_s = stopping sight distance in feet

T_{p-r} = design perception reaction time: assumed as 2.5 sec
V_i = the initial assumed running speed in miles per hour
V_f = the final running speed in miles per hour (equals zero for braking to a stop condition)
f = the wet pavement coefficient of friction, which varies with speed; typical values: 40 mph; 0.32; 80 mph; 0.27
g = the roadway gradient as a decimal

Tables 44 and 45 give typical minimum and desirable stopping sight distances used for design. The minimum values are based on an assumed initial speed somewhat less than the design because early research indicated that drivers drive more slowly in wet weather. Current research no longer supports that assumption. Hence, the new values labeled "desirable" were developed using actual design speed (12).

TABLE 44. MINIMUM STOPPING SIGHT DISTANCE (DESIGN CRITERION: WET PAVEMENTS) (3).

Design speed (mph)	Assumed speed for condition (mph)	Perception and brake reaction		Coefficient of friction, f	Braking distance on level (ft)	Stopping sight distance	
		Time (sec)	Distance (ft)			Computed (ft)	Rounded for design (ft)
30	28	2.5	103	0.36	73	176	200
40	36	2.5	132	0.33	131	263	275
50	44	2.5	161	0.31	208	369	350
60	52	2.5	191	0.30	300	491	475
65	55	2.5	202	0.30	336	538	550
70	58	2.5	213	0.29	387	600	600
75	61	2.5	224	0.28	443	667	675
80	64	2.5	235	0.27	506	741	750

TABLE 45. DESIRABLE STOPPING SIGHT DISTANCES (12).

Design speed (initial speed) (mph)	Perception and brake reaction		Coefficient of friction, wet pavements, f	Braking distance on level (ft)	Stopping sight distance	
	Time (sec)	Distance (ft)			Computed (ft)	Rounded for design (ft)
30	2.5	110	0.35	86	196	200
40	2.5	147	0.32	167	314	300
50	2.5	183	0.30	278	461	450
60	2.5	220	0.29	414	634	650
65	2.5	238	0.29	485	723	750
70	2.5	257	0.28	584	841	850
75	2.5	275	0.28	670	945	950
80	2.5	293	0.27	790	1083	1050

TABLE 46. MINIMUM PASSING SIGHT DISTANCE FOR DESIGN OF 2-LANE HIGHWAYS (3).

Design speed (mph)	Assumed speeds		Minimum passing sight distance (ft)	
	Passed vehicle (mph)	Passing vehicle (mph)	Computed	Rounded
30	26	36	1090	1100
40	34	44	1480	1500
50	41	51	1840	1800
60	47	57	2140	2100
65	50	60	2310	2500
70	54	64	2490	2500
75*	56	66	2600	2600
80*	59	69	2740	2700

*Design speeds of 75 and 80 mph are applicable only to highways with full control of access or where such control is planned in the future.

Passing sight distance, measured from an eye height of 3.75 ft to an object 4.5 ft in height (the height of an oncoming design vehicle) is the distance on a two-way roadway needed to permit a vehicle to safely pass another vehicle and return to its own roadway with some margin for safety before meeting the oncoming vehicle. Table 46 shows design passing sight distance for two-lane highways.

Both stopping sight distance and passing sight distance are measured across vertical and horizontal curves. Stopping sight distance is usually the basis for design of divided roadways whereas passing sight distance is the basis for design of counter-directional (two-way undivided) roadways.

For design purposes, stopping and passing sight distance requirements are related to minimum lengths of crest vertical curve through the use of a factor K which for a given speed relates curve length to the algebraic difference in tangent grades as follows:

$$L = KA$$

where

L = curve length in stations
K = horizontal distance in feet necessary to affect a 1% change in grade
A = algebraic difference between intersecting tangent grades

Tables 47, 48, and 49 give design K values for desirable stopping sight distance on crest and sag vertical curves (sag vertical curve values are based on headlight distance) and passing sight distance for crest vertical curves. Curve lengths for design are obtained by multiplying the algebraic difference in grades by the appropriate coefficient K. Use of K values facilitates selection of the correct curve length for design and allows easy checking of plans for adequacy of sight distance.

TABLE 47. DESIGN CONTROLS FOR CREST VERTICAL CURVES BASED ON DESIRABLE STOPPING SIGHT DISTANCES (12).

Design speed (mph)	Desirable stopping sight distance (ft)	K = Rate of vertical curvature, length in feet per percent of A	
		Calculated	Rounded
30	200	28.6	28
40	300	64.4	65
50	450	144.8	145
60	650	302.0	300
65	750	402.4	400
70	850	516.8	515
75	950	645.6	645
80	1050	788.6	780

TABLE 48. DESIGN CONTROLS FOR SAG VERTICAL CURVES BASED ON DESIRABLE STOPPING SIGHT DISTANCES (12).

Design speed (mph)	Desirable stopping sight distance (ft)	K = Rate of vertical curvature, length in feet per percent of A	
		Calculated	Rounded
30	200	36.4	35
40	300	62.1	60
50	450	102.5	100
60	650	157.9	155
70	850	214.0	215
75	950	242.3	240
80	1050	270.6	270

TABLE 49. DESIGN CONTROLS FOR CREST VERTICAL CURVES BASED ON PASSING SIGHT DISTANCE (3).

Design speed (mph)	Minimum passing distance (ft)	K = Rate of vertical curvature: length (ft) per percent of A, rounded
30	1100	365
40	1500	686
50	1800	985
60	2100	1340
65	2300	1605
70	2500	1895
75	2600	2050
80	2700	2210

Horizontal Alignment, Curvature The maximum curvature for a given design speed relates the safe and comfortable operation of a vehicle on a curve to the safe side friction and degree of superelevation on the curve. It is based on the formula

$$R = V^2/15(e + f)$$

where

R = maximum radius of curvature in feet
V = design speed in miles per hour
e = maximum superelevation rate, e.g., pavement cross slope in inches per inch or feet per foot
f = coefficient of safe side friction; varies with speed between limiting values of 0.11 at 80 mph and 0.16 at 30 mph

The maximum superelevation rate e used for design is controlled by several factors, including climatic conditions, type of terrain, type of area, and the number of very slow-moving vehicles using the roadway. For example, in areas in which ice and snow are prevalent during the winter months, maximum superelevation rates on the order 0.08 are used in order to minimize side slipping across the pavement for slow-moving vehicles on slippery winter roads. In urban areas, maximum superelevation rates are customarily 0.06 ft per foot. Table 50 notes maximum degree of curve related to design speed for different limiting values of superelevation.

Cross-Section Elements Figures 87 and 88 depict typical cross sections and major cross-section elements for two-way rural and urban roadways and multilane rural and urban roadways. The dimensions of the major elements of the cross section, i.e., surface widths and shoulder widths, are related to design speed and traffic volume (current ADT or DHV) as noted in Tables 51 and 52.

Types of Cross-Section Crown Figure 89 illustrates the basic types of cross-section crowns constructed today, while Table 53 lists typical cross-section cross slopes necessary to provide adequate runoff for various pavement types.

Safety Section The safety section shown on the typical cross section is a distance from the pavement edge within which, if properly designed, a vehicle which has accidentally left the roadway could return safely to the roadway without losing control or hitting an obstruction. A study of "ran-off-the-road" accidents indicates that 80% of all vehicles which run off the road recover control within 30 ft of the edge of the pavement if the side slope is sufficiently flat (no greater than 6 to 1) and if no obstacles are present. Therefore, it is desirable to provide an unencumbered recovery area or safety section as shown. Where signposts and other obstacles must be placed inside the safety section boundary,

TABLE 50. MAXIMUM DEGREE OF CURVE AND MINIMUM RADIUS DETERMINED FOR LIMITING VALUES OF *e* AND *f* (3).

	Design speed	Maximum *c*	Maximum *f*	Total $(e+f)$	Minimum radius	Max. degree of curve	Max. degree of curve, rounded
Urban maximum	30	0.06	0.16	0.22	273	21.0	21.0
	40	0.06	0.15	0.21	508	11.3	11.5
	50	0.06	0.14	0.20	833	6.9	7.0
	60	0.06	0.13	0.19	1263	4.5	4.5
	65	0.06	0.13	0.19	1483	3.9	4.0
	70	0.06	0.12	0.18	1815	3.2	3.0
	75	0.06	0.11	0.17	2206	2.6	2.5
	80	0.06	0.11	0.17	2510	2.3	2.5
Maximum for winter climates	30	0.08	0.16	0.24	250	22.9	23.0
	40	0.08	0.15	0.23	464	12.4	12.5
	50	0.08	0.14	0.22	758	7.6	7.5
	60	0.08	0.13	0.21	1143	5.0	5.0
	65	0.08	0.13	0.21	1341	4.3	4.5
	70	0.08	0.12	0.20	1633	3.5	3:5
	75	0.08	0.11	0.19	1974	2.9	3.0
	80	0.08	0.11	0.19	2246	2.5	2.5
Nationally representative maximum values	30	0.10	0.16	0.26	231	24.8	25.0
	40	0.10	0.15	0.25	427	13.4	13.5
	50	0.10	0.14	0.24	694	8.3	8.5
	60	0.10	0.13	0.23	1043	5.5	5.5
	65	0.10	0.13	0.23	1225	4.7	4.5
	70	0.10	0.12	0.22	1485	3.9	4.0
	75	0.10	0.11	0.21	1786	3.2	3.0
	80	0.10	0.11	0.21	2032	2.8	3.0
	30	0.12	0.16	0.28	214	26.7	26.5
	40	0.12	0.15	0.27	395	14.5	14.5
	50	0.12	0.14	0.26	641	8.9	9.0
	60	0.12	0.13	0.25	960	6.0	6.0
	65	0.12	0.13	0.25	1127	5.1	5.0
	70	0.12	0.12	0.24	1361	4.2	4.0
	75	0.12	0.11	0.23	1630	3.5	3.5
	80	0.12	0.11	0.23	1855	3.1	3.0

the signs, etc., should be on accepted break-away mounts and should be mounted on footings which are flush with the ground (9).

Superelevation Superelevation is the amount of cross slope on the pavement through a curve. It acts in conjunction with side friction to counter the centrifugal force of a vehicle rounding the curve and helps hold the vehicle on the curve without sliding to the side at various operating speeds. The maximum superele-

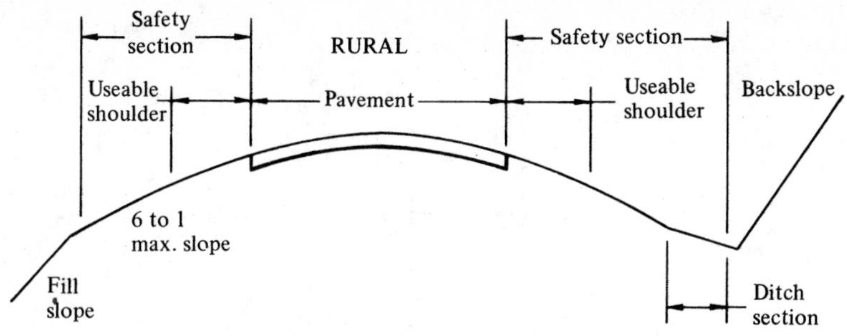

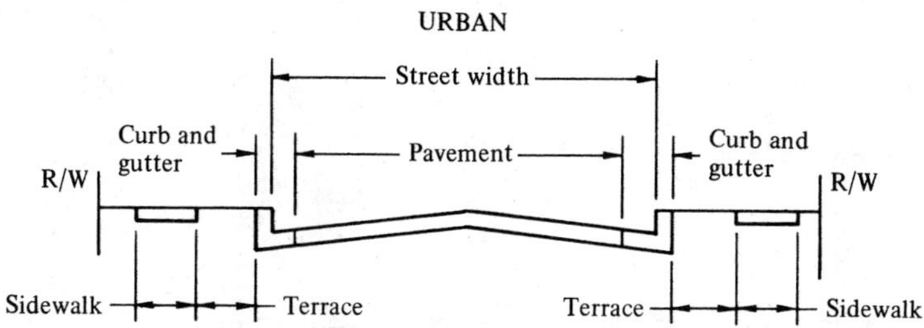

Fig. 87. Cross-section elements, single roadway.

TABLE 51. MINIMUM WIDTHS OF SURFACING FOR 2-LANE HIGHWAYS (3).

Design speed (mph)	Minimum widths of surfacing, in feet, for design volumes of:*				
	Current ADT 50–250	Current ADT 250–400	Current ADT 400–750, DHV 100–200	DHV 200–400	DHV 400 and over
30	20	20	20	22	24
40	20	20	22	22	24
50	20	20	22	24	24
60	20	22	22	24	24
65	20	22	24	24	24
70	20	22	24	24	24
75	24	24	24	24	24
80	24	24	24	24	24

*For design speeds of 30, 40, and 50 mph, surfacing widths that are 2 ft narrower may be used on minor roads with few trucks.

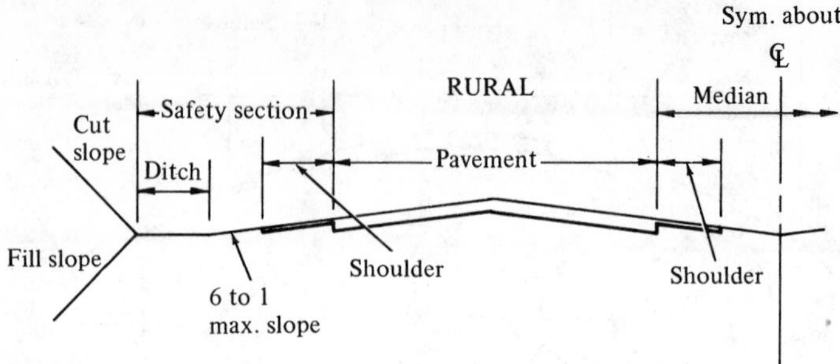

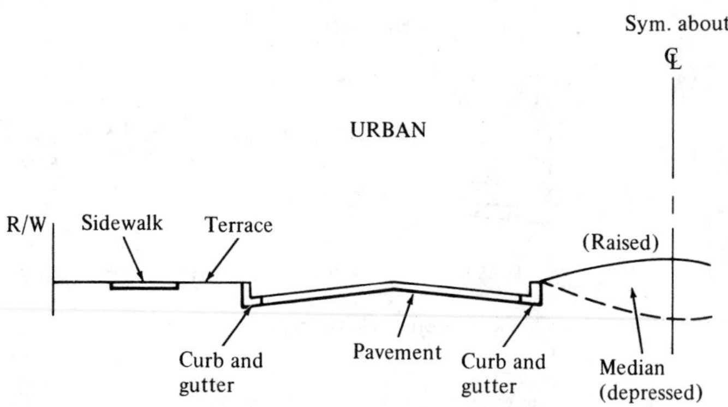

Fig. 88. Cross-section elements, multilane.

TABLE 52. SHOULDER WIDTHS FOR 2-LANE RURAL HIGHWAYS.

Design volume		Usable shoulder width (ft)*	
Current ADT	DHV	Minimum	Desirable
50–250	–	4	6
250–400	–	4	8
400–750	100–200	6	10
–	200–400	8	10
–	400 and over	10	12

*The "usable" width of shoulder is that which can be used when a driver makes an emergency stop.

vation rate in common use for open highways is 0.12. However, this maximum is only for those areas which experience no snow or icing conditions. In winter climates maximum superelevation rates for open highways are normally on the order of 0.08. Anything greater could result in slipping sideways across the high-

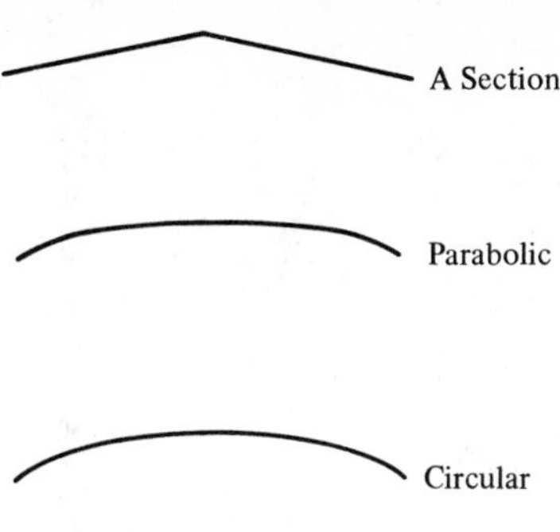

Fig. 89. Basic roadway crowns.

TABLE 53. NORMAL PAVEMENT CROWN SLOPES (3).

Type of surface	Texture	Pavement type	Range of slope (in./ft)	Range of slope (ft/ft)
High	smooth to fine	p.c. concrete, bit. concrete	$\frac{1}{8}$–$\frac{1}{4}$	0.01–0.02
Intermediate	fine to coarse	low-type hot mixes, road mixes	$\frac{3}{16}$–$\frac{3}{8}$	0.015–0.03
Low	very coarse	loose gravel, stone	$\frac{1}{4}$–$\frac{1}{2}$	0.02–0.04

way when vehicles are moving slowly or stopped because of winter driving conditions. In urban areas a practical maximum superelevation rate of 0.06 ft per foot is commonly used.

Superelevation runoff denotes a length of highway used for making a transition from the normal pavement crown section to the full superelevation section in a curve. As Fig. 90 shows, methods of attaining superelevation vary. The three most commonly used procedures for two-way undivided roadways are:

1. Revolving the pavement about the centerline.
2. Revolving the pavement around the inside edge.
3. Revolving the pavement around the outside edge.

Rotation about the centerline is the preferred procedure since the other two result in distorted views of the pavement ahead of the driver; for very sharp curves they may make the pavement appear as though it were becoming airborne or as though it were dropping into a hole. Tables 54–57, in addition to noting superelevation rates, note lengths of runoff commonly used. As a general guideline, for pavements rotated around the center line, the change in longitudinal grade for the profile of each pavement edge should be at a relative slope of approximately 1 in 200 to the center line, or flatter. Figure 91 illustrates typical

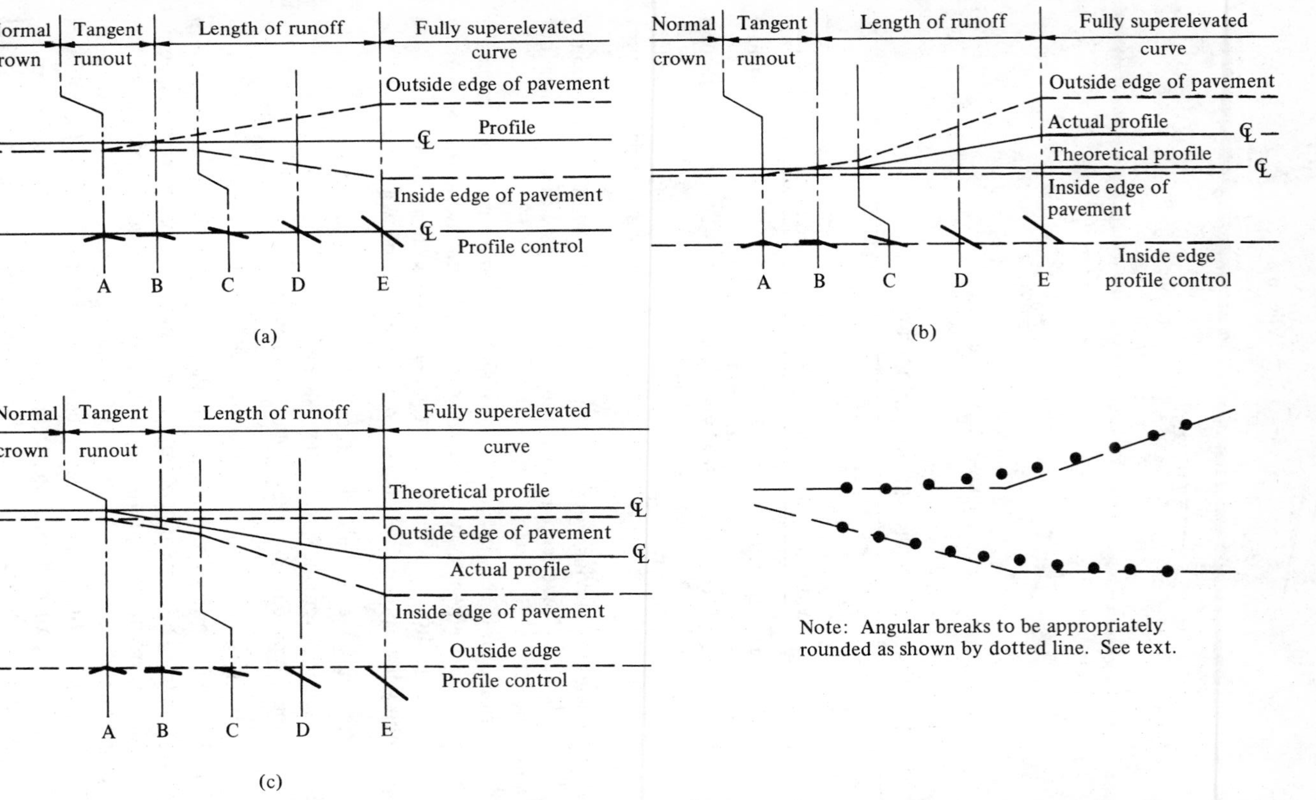

Fig. 90. Diagrammatic profiles showing methods of attaining superelevation[3]. (a) Pavement revolved about centerline. (b) Pavement revolved about inside edge. (c) Pavement revolved about outside edge. Conditions (b) and (c) may appear distorted.

D	R	V=30 mph			V=40 mph			V=50 mph			V=60 mph			V=65 mph			V=70 mph			V=75 mph			V=80 mph		
		e	L-Feet		e	L-Feet		e	L-Feet		e	L-Feet		e	L-Feet		e	L-Feet		e	L-Feet		e	L-Feet	
			2-lane	4-lane		2-lane	4-lane		2-lane	4-lane		2-lane	4-lane		2-lane	4-lane		2-lane	4-lane		2-lane	4-lane		2-lane	4-lane
0° 15′	22918′	NC	0	0	NC	0	0	NC	0	0	NC	0	0	NC	0	0	NC	0	0	NC	0	0	RC	240	240
0° 30′	11459′	NC	0	0	NC	0	0	NC	0	0	RC	175	175	RC	190	190	RC	200	200	.021	220	220	.023	240	240
0° 45′	7639′	NC	0	0	NC	0	0	RC	0	0	.021	175	175	.023	190	190	.026	200	200	.030	220	220	.033	240	240
1° 00	5730′	NC	0	0	RC	125	125	.020	150	150	.027	175	175	.029	190	190	.033	200	200	.037	220	220	.041	240	240
1° 30′	3820′	RC	100	100	.020	125	125	.028	150	150	.036	175	175	.040	190	190	.044	200	200	.050	220	240	.053	240	260
2° 00′	2865′	RC	100	100	.026	125	125	.035	150	150	.044	175	180	.048	190	210	.052	200	230	.057	220	270	.059	240	290
2° 30′	2292′	.020	100	100	.031	125	125	.040	150	150	.050	175	200	.053	190	230	.057	200	260	.060	220	290	.060	240	300
3° 00′	1910′	.023	100	100	.035	125	125	.044	150	160	.054	175	220	.057	190	250	.060	200	270	.060	220	290	.060	240	300
3° 30′	1637′	.026	100	100	.038	125	125	.048	150	170	.057	175	230	.059	190	250	.060	200	270	D max=2.5°			D max=2.5°		
4° 00′	1432′	.029	100	100	.041	125	130	.051	150	180	.059	175	240	.060	190	260	D max=3.0°								
5° 00′	1146′	.034	100	100	.046	125	140	.056	150	200	.060	175	240	.060	190	260									
6° 00′	955′	.038	100	100	.050	125	160	.059	150	210	D max=4.5°			D max=4.0°											
7° 00′	819′	.041	100	110	.054	125	170	.060	150	220															
8° 00′	716′	.043	100	120	.056	125	180	.060	150	220															
9° 00′	637′	.046	100	120	.058	125	180	D max=7.0°																	
10° 00′	573′	.048	100	130	.059	125	190																		
11° 00′	521′	.050	100	140	.060	130	190																		
12° 00′	477′	.052	100	140	.060	130	190																		
13° 00′	441′	.053	100	140	D max=11.5°																				
14° 00′	409′	.055	100	150																					
16° 00′	358′	.058	100	160																					
18° 00′	318′	.059	110	160																					
20° 00′	286′	.060	110	160																					
		.060	110	160																					
		D max=21.0°																							

$e_{max}=0.06$

D—Degree of curve
R—Radius of curve
V—Assumed design speed
e—Rate of superelevation
L—Minimum length of runoff of spiral curve
NC—Normal crown section
RC—Remove adverse crown, superelevate at normal crown slope
Spirals desirable but not as essential above heavy line.
Lengths rounded in multiples of 25 or 50 feet permit simpler calculations.

TABLE 55. VALUES FOR DESIGN ELEMENTS RELATED TO DESIGN SPEED AND HORIZONTAL CURVATURE[3].

D	R	V=30 mph			V=40 mph			V=50 mph			V=60 mph			V=65 mph			V=70 mph			V=75 mph			V=80 mph		
		e	L-Feet		e	L-Feet		e	L-Feet		e	L-Feet		e	L-Feet		e	L-Feet		e	L-Feet		e	L-Feet	
			2-lane	4-lane		2-lane	4-lane		2-lane	4-lane		2-lane	4-lane		2-lane	4-lane		2-lane	4-lane		2-lane	4-lane		2-lane	4-lane
0° 15′	22918′	NC	0	0	NC	0	0	NC	0	0	NC	0	0	NC	0	0	NC	0	0	NC	0	0	RC	240	240
0° 30′	11459′	NC	0	0	NC	0	0	NC	0	0	RC	175	175	RC	190	190	RC	200	200	.022	220	220	.024	240	240
0° 45′	7639′	NC	0	0	NC	0	0	RC	150	150	.022	175	175	.025	190	190	.029	200	200	.032	220	220	.036	240	240
1° 00′	5730′	NC	0	0	RC	125	125	.021	150	150	.029	175	175	.033	190	190	.038	200	200	.043	220	220	.047	240	240
1° 30′	3820′	RC	100	100	.021	125	125	.030	150	150	.040	175	175	.046	190	200	.053	200	240	.060	220	290	.065	240	320
2° 00′	2865′	RC	100	100	.027	125	125	.038	150	150	.051	175	210	.057	190	250	.065	200	290	.072	230	340	.076	250	380
2° 30′	2292′	.021	100	100	.033	125	125	.046	150	170	.060	175	240	.066	190	290	.073	220	330	.078	250	370	.080	260	400
3° 00′	1910′	.025	100	100	.038	125	125	.053	150	190	.067	180	270	.073	210	320	.078	230	350	.080	250	380	.080	260	400
3° 30′	1637′	.028	100	100	.043	125	140	.058	150	210	.073	200	300	.077	220	330	.080	240	360	.080	250	380	D max=2.5°		
4° 00′	1432′	.032	100	100	.047	125	150	.063	150	230	.077	210	310	.079	230	340	.080	240	360	D max=3.0°					
5° 00′	1146′	.038	100	100	.055	125	170	.071	170	260	.080	220	320	.080	230	350	D max=3.5°								
6° 00′	955′	.043	100	120	.061	130	190	.077	180	280	.080	220	320	D max=4.5°											
7° 00′	819′	.048	100	130	.067	140	210	.079	190	280	D max=5.0°														
8° 00′	716′	.052	100	140	.071	150	220	.080	190	290															
9° 00′	637′	.056	100	150	.075	160	240	D max=7.5°																	
10° 00′	573′	.059	110	160	.077	160	240																		
11° 00′	521′	.063	110	170	.079	170	250																		
12° 00′	477′	.066	120	180	.080	170	250																		
13° 00′	441′	.068	120	180	.080	170	250																		
14° 00′	409′	.070	130	190	D max=12.5°																				
16° 00′	358′	.074	130	200																					
18° 00′	318′	.077	140	210																					
20° 00′	286′	.079	140	210																					
22° 00′	260′	.080	140	220																					
		.080	140	220																					
		D max=23.0°																							

$e_{max}=0.08$

D—Degree of curve
R—Radius of curve
V—Assumed design speed
e—Rate of superelevation
L—Minimum length of runoff of spiral curve
NC—Normal crown section
RC—Remove adverse crown, superelevate at normal crown slope

Spirals desirable but not as essential above heavy line.
Lengths rounded in multiples of 25 or 50 feet permit simpler

D	R	V=30 mph			V=40 mph			V=50 mph			V=60 mph			V=65 mph			V=70 mph			V=75 mph			V=80 mph		
		e	L-Feet		e	L-Feet		e	L-Feet		e	L-Feet		e	L-Feet		e	L-Feet		e	L-Feet		e	L-Feet	
			2-lane	4-lane		2-lane	4-lane		2-lane	4-lane		2-lane	4-lane		2-lane	4-lane		2-lane	4-lane		2-lane	4-lane		2-lane	4-lane
0° 15′	22918′	NC	0	0	NC	0	0	NC	0	0	NC	0	0	NC	0	0	NC	0	0	NC	0	0	RC	240	240
0° 30′	11459′	NC	0	0	NC	0	0	NC	0	0	RC	175	175	RC	190	190	RC	200	200	.022	220	220	.024	240	240
0° 45′	7639′	NC	0	0	NC	0	0	RC	150	150	.024	175	175	.027	190	190	.029	200	200	.033	220	220	.036	240	240
1° 00′	5730′	NC	0	0	RC	125	125	.023	150	150	.032	175	175	.035	190	190	.039	200	200	.044	220	220	.048	240	240
1° 30′	3820′	RC	100	100	.021	125	125	.033	150	150	.046	175	190	.052	190	220	.058	200	260	.065	220	310	.071	240	350
2° 00′	2865′	RC	100	100	.028	125	125	.042	150	150	.058	175	230	.066	190	290	.074	220	330	.082	260	390	.089	290	440
2° 30′	2292′	.021	100	100	.034	125	125	.051	150	180	.069	190	280	.077	220	330	.086	260	390	.094	300	450	.099	330	490
3° 00′	1910′	.025	100	100	.040	125	125	.059	150	210	.079	210	320	.087	250	380	.094	280	420	.100	320	480	.100	330	500
3° 30′	1637′	.029	100	100	.046	125	140	.067	160	240	.087	230	350	.093	270	400	.099	300	450	.100	320	480	.100	330	500
4° 00′	1432′	.033	100	100	.051	125	160	.073	180	260	.093	250	380	.098	280	420	.100	300	450	D max=3.0°			D max=3.0°		
5° 00′	1146′	.040	100	110	.061	130	190	.084	200	300	.099	270	400	.100	290	430	.100	300	450						
6° 00′	955′	.046	100	120	.070	150	220	.092	220	330	.100	270	410	D max=4.5°			D max=4.0°								
7° 00′	819′	.053	100	140	.077	160	240	.098	240	350	D max=5.5°														
8° 00′	716′	.059	110	160	.084	180	260	.100	240	360															
9° 00′	637′	.064	120	170	.089	190	280	.100	240	360															
10° 00′	573′	.068	120	180	.093	200	290	D max=8.5°																	
11° 00′	521′	.073	130	200	.097	200	310																		
12° 00′	477′	.077	140	210	.099	210	310																		
13° 00′	441′	.080	140	220	.100	210	320																		
14° 00′	409′	.083	150	220	.100	210	320																		
16° 00′	358′	.089	160	240	D max=13.5°																				
18° 00′	318′	.093	170	250																					
20° 00′	286′	.097	170	260																					
22° 00′	260′	.099	180	270																					
24° 00′	239′	.100	180	270																					
		.100	180	270																					
		D max=25.0°																							

$e_{max}=0.10$

D—Degree of curve
R—Radius of curve
V—Assumed design speed
e—Rate of superelevation
L—Minimum length of runoff of spiral curve
NC—Normal crown section
RC—Remove adverse crown, superelevate at normal crown slope
Spirals desirable but not as essential above heavy line.
Lengths rounded in multiples of 25 or 50 feet permit simpler calculations.

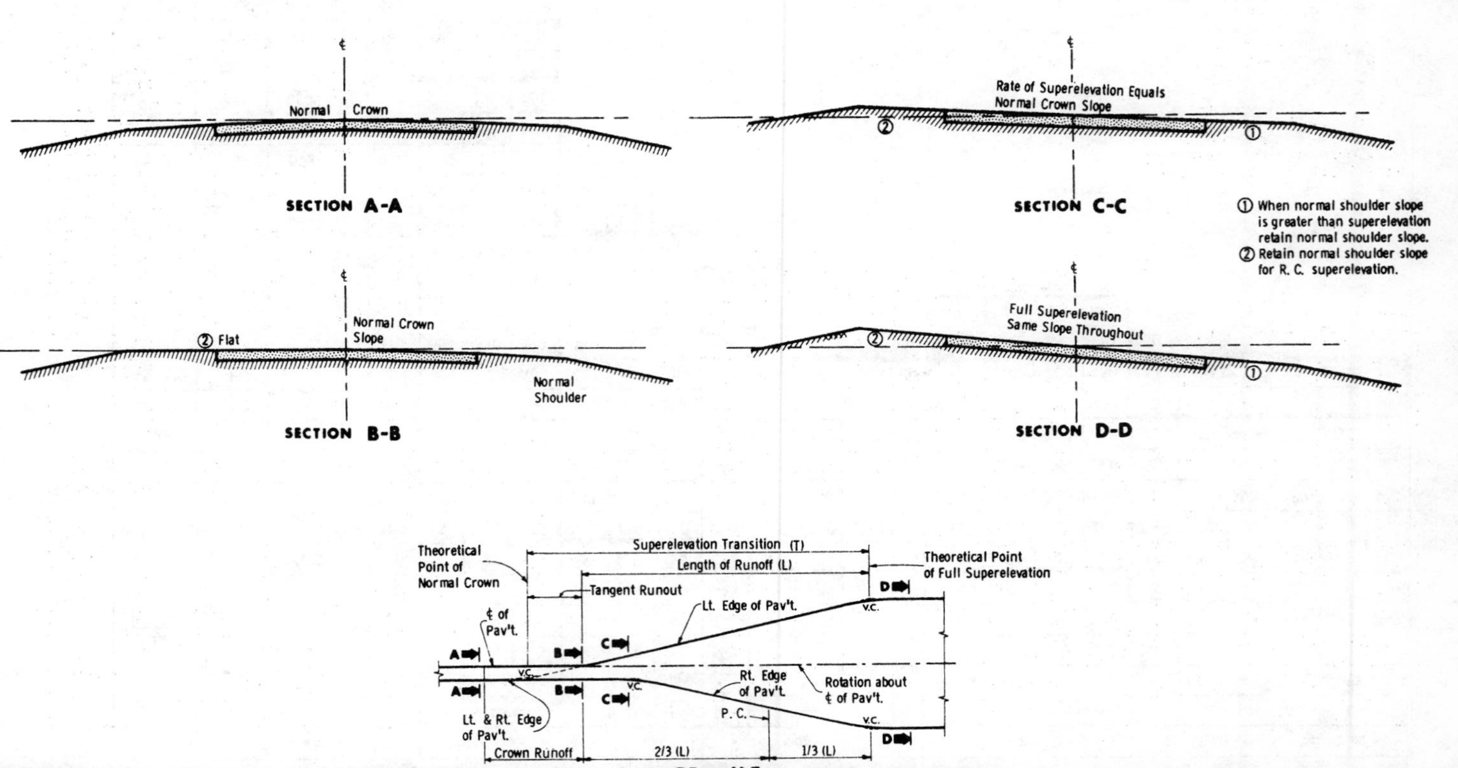
Normal Crown
SECTION A-A
Rate of Superelevation Equals
Normal Crown Slope
SECTION C-C
① When normal shoulder slope is greater than superelevation retain normal shoulder slope.
② Retain normal shoulder slope for R. C. superelevation.
② Flat
Normal Crown
Slope
Normal
Shoulder
SECTION B-B
Full Superelevation
Same Slope Throughout
SECTION D-D
Theoretical
Point of
Normal Crown
Superelevation Transition (T)
Length of Runoff (L)
Theoretical Point
of Full Superelevation
Tangent Runout
Lt. Edge of Pav't.
℄ of Pav't.
V.C.
Rt. Edge of Pav't.
Rotation about ℄ of Pav't.
P. C.
Lt. & Rt. Edge of Pav't.
Crown Runoff
2/3 (L)
1/3 (L)

SUPERELEVATION TABLE

Curve (D)	V = 40 mph				V = 50 mph				V = 60 mph				V = 65 mph				V = 70 mph			
	Superelevation				Superelevation				Superelevation				Superelevation				Superelevation			
	Rate (e) ft./ft.	(L) Length of Runoff (ft.)	Transition (T) Rigid Pavt. (ft.)	Transition (T) Flex. Pavt. (ft.)	Rate (e) ft./ft.	(L) Length of Runoff (ft.)	Transition (T) Rigid Pavt. (ft.)	Transition (T) Flex. Pavt. (ft.)	Rate (e) ft./ft.	(L) Length of Runoff (ft.)	Transition (T) Rigid Pavt. (ft.)	Transition (T) Flex. Pavt. (ft.)	Rate (e) ft./ft.	(L) Length of Runoff (ft.)	Transition (T) Rigid Pavt. (ft.)	Transition (T) Flex. Pavt. (ft.)	Rate (e) ft./ft.	(L) Length of Runoff (ft.)	Transition (T) Rigid Pavt. (ft.)	Transition (T) Flex. Pavt. (ft.)
0°-15'	NC	0	0	0	NC	0	0	0	NC	0	0	0	NC	0	0	0	NC	0	0	0
0°-30'	NC	0	0	0	NC	0	0	0	RC	175	350	350	RC	190	380	380	RC	200	400	400
0°-45'	NC	0	0	0	RC	150	300	300	.022	175	255	295	.025	190	265	305	.029	200	270	305
1°-00'	RC	125	250	250	.021	150	220	260	.029	175	235	265	.033	190	250	275	.038	200	255	280
1°-15'	.017	125	200	235	.026	150	210	235	.035	175	225	250	.040	190	240	260	.046	200	245	265
1°-30'	.021	125	185	215	.030	150	200	225	.040	175	220	240	.046	190	230	250	.053	200	240	255
1°-45'	.024	125	175	205	.034	150	195	215	.045	175	215	235	.052	190	225	245	.059	200	235	250
2°-00'	.027	125	170	195	.038	150	190	210	.051	175	210	225	.057	190	225	240	.065	200	230	245
2°-15'	.030	125	165	185	.042	150	185	205	.056	175	205	220	.062	190	220	235	.069	210	240	255
2°-30'	.033	125	165	180	.046	150	185	200	.060	175	205	220	.066	190	220	235	.073	220	250	265
2°-45'	.036	125	160	175	.050	150	180	195	.064	180	210	220	.070	200	230	245	.076	225	255	270
3°-00'	.038	125	160	175	.053	150	180	190	.067	180	205	220	.073	210	240	255	.078	230	260	275
3°-30'	.043	125	155	170	.058	150	175	190	.073	200	225	240	.077	220	250	265	.080	240	270	285
4°-00'	.047	125	150	165	.063	150	175	185	.077	210	235	250	.079	230	260	275	D Max. = 3°-30'			
4°-30'	.051	125	150	160	.067	160	185	195	.079	215	240	255	.080	230	260	275				
5°-00'	.055	125	150	160	.071	170	195	205	.080	220	250	260	D Max. = 4°-30'							
5°-30'	.058	125	145	155	.074	175	200	210	D Max. = 5°-00'											
6°-00'	.061	130	150	160	.077	180	205	215												
6°-30'	.064	135	155	165	.078	185	210	220												
7°-00'	.067	140	160	170	.079	190	215	225												
7°-30'	.069	145	165	175	.080	190	215	225												
8°-00'	.071	150	170	180	D Max. = 7°-30'															
9°-00'	.075	160	180	190																
10°-00'	.077	160	180	190																
11°-00'	.079	170	190	200																
12°-00'	.080	170	190	200																
12°-30'	.08	170	190	200																
	D Max. = 12°-30'																			

NOTE:

Superelevation rotation is about ¢ of pavement.
Curve shown is a horizontal curve to right

V = Design Speed
D = Degree of Horizontal Curve
N. C. = Normal Crown Slope, i. e.
Rigid - 0.010 ft./ft.
Flexible - 0.015 ft./ft.
R. C. = Remove Crown, Superelevate at Normal Crown Slope, Retain Slope on Both Shoulders
P. C. = Beginning of Horizontal Curve
V. C. = Vertical Curve

Fig. 91. Superelevation details for two-lane rural highways[13].

TABLE 57. VALUES FOR DESIGN ELEMENTS RELATED TO DESIGN SPEED AND HORIZONTAL CURVATURE[3].

D	R	V=30 mph			V=40 mph			V=50 mph			V=60 mph			V=65 mph			V=70 mph			V=75 mph			V=80 mph		
		e	L-Feet		e	L-Feet		e	L-Feet		e	L-Feet		e	L-Feet		e	L-Feet		e	L-Feet		e	L-Feet	
			2-lane	4-lane		2-lane	1-lane		2-lane	4-lane		2-lane	4-lane		2-lane	4-lane		2-lane	4-lane		2-lane	4-lane		2-lane	4-lane
0° 15′	22918′	NC	0	0	NC	0	0	NC	0	0	NC	0	0	NC	0	0	NC	0	0	NC	0	0	RC	240	240
0° 30′	11459′	NC	0	0	NC	0	0	NC	0	0	RC	175	175	RC	190	190	RC	200	200	.022	220	220	.024	240	240
0° 45′	7639′	NC	0	0	NC	0	0	RC	150	150	.024	175	175	.026	190	190	.029	200	200	.033	220	220	.036	240	240
1° 00′	5730′	NC	0	0	RC	125	125	.023	150	150	.031	175	175	.035	190	190	.039	200	200	.043	220	220	.048	240	240
1° 30′	3820′	RC	100	100	.022	125	125	.034	150	150	.047	175	190	.053	190	230	.059	200	270	.065	220	310	.072	240	360
2° 00′	2865′	RC	100	100	.030	125	125	.045	150	160	.062	175	250	.070	200	300	.078	230	350	.087	280	410	.095	310	470
2° 30′	2292′	.022	100	100	.037	125	125	.055	150	200	.076	210	310	.085	240	370	.095	290	430	.105	330	500	.113	370	560
3° 00′	1910′	.026	100	100	.044	125	140	.065	160	230	.088	240	360	.097	240	420	.108	320	490	.117	370	560	.120	400	600
3° 30′	1637′	.030	100	100	.050	125	160	.074	180	270	.098	260	400	.107	310	460	.116	350	520	.120	380	570	.120	400	600
4° 00′	1432′	.034	100	100	.057	125	180	.082	200	300	.106	290	430	.114	330	490	.120	360	540	.120	380	570	D max=3.0°		
5° 00′	1146′	.042	100	110	.068	140	210	.096	230	350	.117	320	470	.120	350	520	.120	360	540	D max=3.5°					
6° 00′	955′	.049	100	130	.079	170	250	.107	260	390	.120	320	490	.120	350	520	D max=4.0°								
7° 00′	819′	.055	100	150	.088	180	280	.114	270	410	.120	320	490	D max=5.0°											
8° 00′	716′	.062	110	170	.096	200	300	.119	290	430	D max=6.0°														
9° 00′	637′	.068	120	180	.103	220	320	.120	290	430															
10° 00′	573′	.074	130	200	.108	230	340	.120	290	430															
11° 00′	521′	.079	140	210	.113	240	360	D max=9.0°																	
12° 00′	477′	.084	150	230	.116	240	370																		
13° 00′	441′	.089	160	240	.119	250	370																		
14° 00′	409′	.093	170	250	.120	250	380																		
16° 00′	358′	.101	180	270	.120	250	380																		
18° 00′	318′	.108	190	290	D max=14.5°																				
20° 00′	286′	.113	200	310																					
22° 00′	260′	.117	210	320																					
26° 00′	220′	.120	220	320																					
		.120	220	320																					

e_{max}=0.12

D—Degree of curve
R—Radius of curve
V—Assumed design speed
e—Rate of superelevation
L—Minimum length of runoff of spiral curve
NC—Normal crown section
RC—Remove adverse crown, superelevate at normal crown slope

Spirals desirable but not as essential above heavy line.
Lengths rounded in multiples of 25 or 50 feet permit simpler

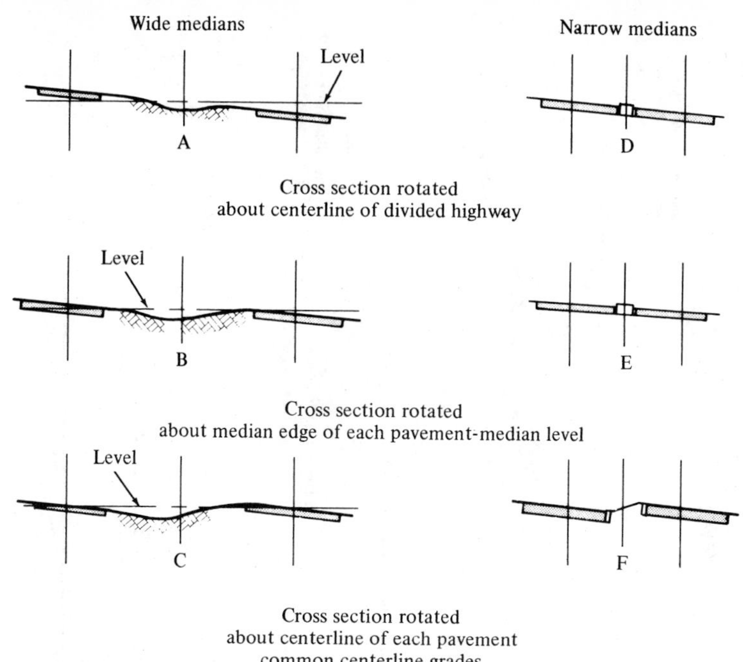

Fig. 92. Superelevated cross sections for divided highways[3].

practice for superelevation rate, length of runoff, and transition for rigid and flexible pavement for various design speed categories.

On divided highways (Fig. 92), superelevation is accomplished by:

1. Rotating the cross section about the centerline of the divided highway.
2. Rotating each pavement about its median edge.
3. Rotating each pavement individually about its own centerline.

In general, rotation about the median pavement edge provides the most pleasing appearance and best overall operation. Figure 93 shows a typical superelevation detail for rural divided pavements in which the pavement is rotated about the median pavement edge.

Roadway Design Strict adherence to vertical and horizontal alignment controls which are based solely on safe vehicle operation, i.e., stopping sight distance criteria and the speed–curvature relationship, will often result in the design of facilities which are functionally efficient but not aesthetically pleasing.

Alignment Types Three basic alignment types, (a) long tangent–short curve, (b) long curve–short tangent, and (c) continuous curvilinear alignment, are shown in Figure 94.

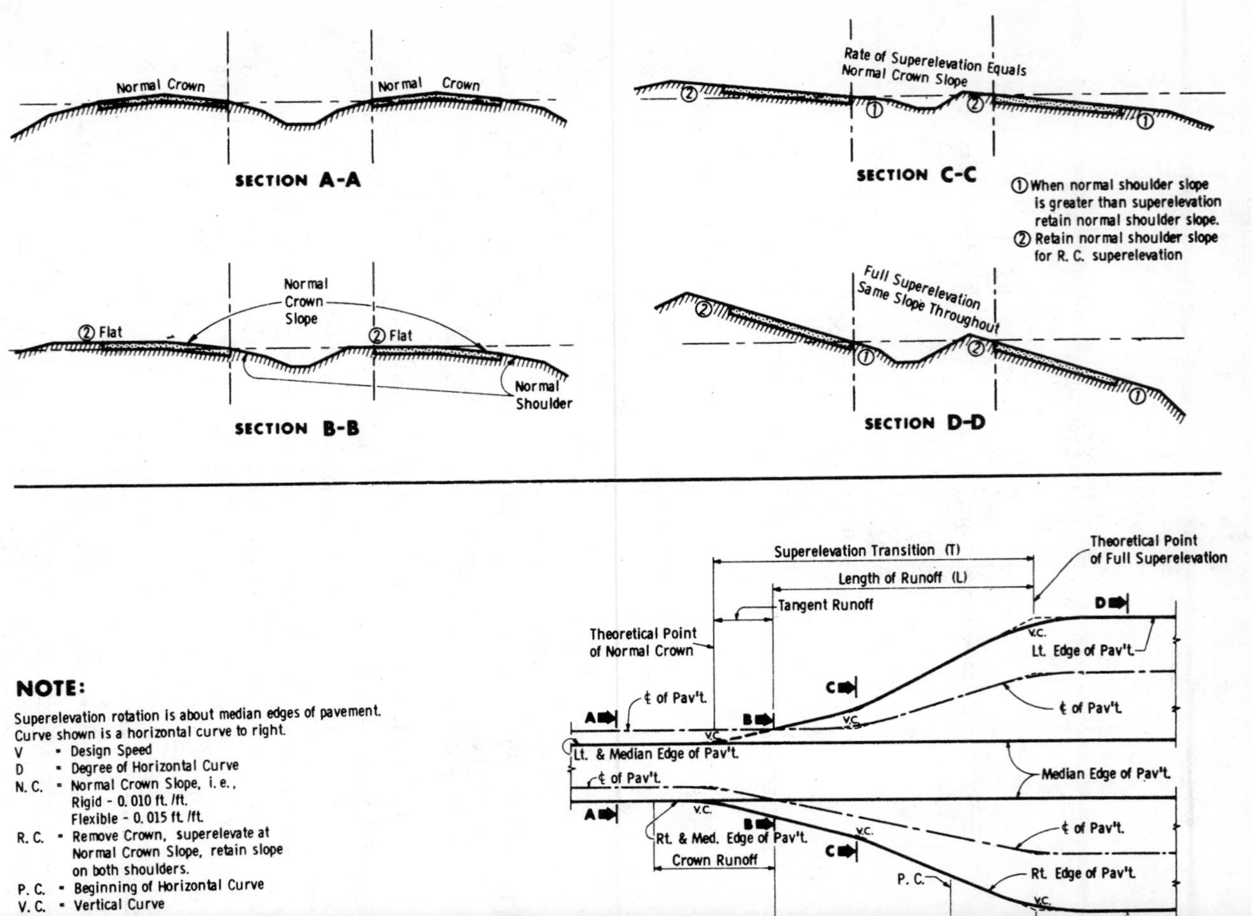
Normal Crown
Normal Crown
SECTION A-A
Rate of Superelevation Equals Normal Crown Slope
SECTION C-C
① When normal shoulder slope is greater than superelevation retain normal shoulder slope.
② Retain normal shoulder slope for R. C. superelevation
② Flat
Normal Crown Slope
② Flat
Normal Shoulder
SECTION B-B
Full Superelevation Same Slope Throughout
SECTION D-D
Superelevation Transition (T)
Length of Runoff (L)
Tangent Runoff
Theoretical Point of Full Superelevation
Theoretical Point of Normal Crown
Lt. Edge of Pav't.
℄ of Pav't.
℄ of Pav't.
Lt. & Median Edge of Pav't.
Median Edge of Pav't.
℄ of Pav't.
℄ of Pav't.
Rt. & Med. Edge of Pav't.
Crown Runoff
P. C.
Rt. Edge of Pav't.
V.C.
A
B
C
D
NOTE:
Superelevation rotation is about median edges of pavement.
Curve shown is a horizontal curve to right.
V - Design Speed
D - Degree of Horizontal Curve
N. C. - Normal Crown Slope, i. e.,
Rigid - 0.010 ft./ft.
Flexible - 0.015 ft./ft.
R. C. - Remove Crown, superelevate at Normal Crown Slope, retain slope on both shoulders.
P. C. - Beginning of Horizontal Curve
V. C. - Vertical Curve

SUPERELEVATION TABLE

Curve (D)	V = 65 mph				V = 70 mph				V = 75 mph				V = 80 mph			
	Superelevation				Superelevation				Superelevation				Superelevation			
	Rate (e) ft./ft.	(L) Length of Runoff (ft.)	Transition (T) Rigid Pavt. (ft.)	Transition (T) Flex. Pavt. (ft.)	Rate (e) ft./ft.	(L) Length of Runoff (ft.)	Transition (T) Rigid Pavt. (ft.)	Transition (T) Flex. Pavt. (ft.)	Rate (e) ft./ft.	(L) Length of Runoff (ft.)	Transition (T) Rigid Pavt. (ft.)	Transition (T) Flex. Pavt. (ft.)	Rate (e) ft./ft.	(L) Length of Runoff (ft.)	Transition (T) Rigid Pavt. (ft.)	Transition (T) Flex. Pavt. (ft.)
0°-15′	NC	0	0	0	NC	0	0	0	NC	0	0	0	RC	240	480	480
0°-30′	RC	190	380	380	RC	200	400	400	.022	220	320	370	.024	240	340	390
0°-45′	.025	190	265	305	.029	200	270	305	.032	220	290	325	.036	240	305	340
1°-00′	.033	190	250	275	.038	200	255	280	.043	220	270	295	.047	240	290	315
1°-15′	.040	200	250	275	.046	220	270	290	.052	260	310	335	.056	280	330	355
1°-30′	.046	200	245	265	.053	240	285	310	.060	290	340	365	.065	320	370	395
1°-45′	.052	230	275	295	.059	270	315	340	.066	320	370	390	.071	350	400	425
2°-00′	.057	250	295	315	.065	290	335	355	.072	340	385	410	.076	380	430	455
2°-15′	.062	270	315	335	.069	310	355	375	.075	355	400	425	.078	390	440	465
2°-30′	.066	290	335	355	.073	330	375	400	.078	370	415	440	.080	400	450	475
2°-45′	.070	310	355	375	.076	340	385	405	.079	375	420	445	D Max. = 2°-45′			
3°-00′	.073	320	365	385	.078	350	395	415	.080	380	430	450				
3°-30′	.077	330	375	395	.080	360	405	430	D Max. = 3°-00′							
4°-00′	.079	340	385	405	D Max. = 3°-30′											
4°-30′	.080	345	390	410												
	D Max. = 4°-30′															

Fig. 93. Superelevation details for rural divided highways[13].

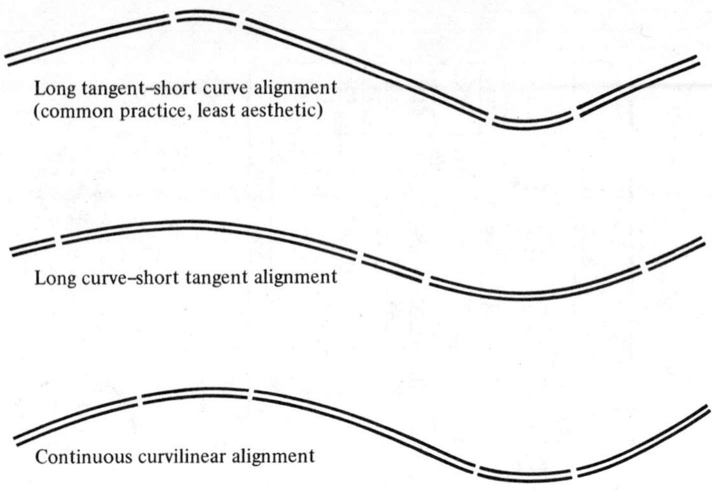

Fig. 94. Roadway alignment types[10].

For the long tangent–short curve alignment, approximately one-fifth of the total alignment is on curve and about four-fifths of the total alignment on tangent (Fig. 95). Tangents are from one to three miles in length, whereas curves are, on the average, 1,500 ft or so in length with radii anywhere from 2,000 to 12,000 feet (10). The curves and tangents are each seen separately and there is a definite visual discontinuity in the horizontal alignment.

For the long curve–short tangent alignment, two-thirds to four-fifths of the alignment is on curve and one-third to one-fifth of the alignment is on tangent. Tangents vary in length from 200 to 2,300 ft and curves vary in length up to 10,000 ft, with radii between 6,000 ft and 20,000 ft (10). Both the curves and tangents are still seen separately and a certain amount of discontinuity of alignment is apparent.

The continuous curvilinear alignment (Fig. 96) consists of long flat circular curves, both simple and compound, connected by fairly long spiral transitions. Approximately two-thirds of the alignment is on curvature and one-third on spiral (10). The continuous curvilinear alignment does not show any discontinuity of alignment, and the point of transition from one curve to another or between curve and spiral is not apparent to the eye.

Selecting the Alignment: Alignment Controls Selecting the precise alignment for a roadway between two terminal points within a broad corridor (see Fig. 97) is governed by a number of controls on the location. Among them are:

1. Land use patterns, existing and planned.

Fig. 95. Discontinuous, disjointed alignment on a long-tangent–short-curve alignment[10].

Fig. 96. Visually pleasing alignment on a continuous curvilinear alignment[10].

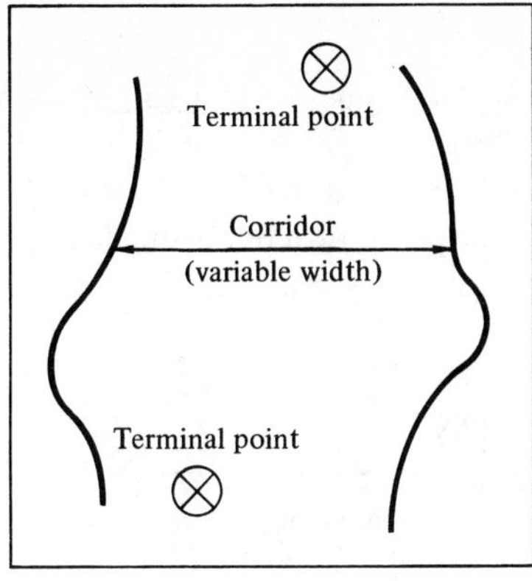

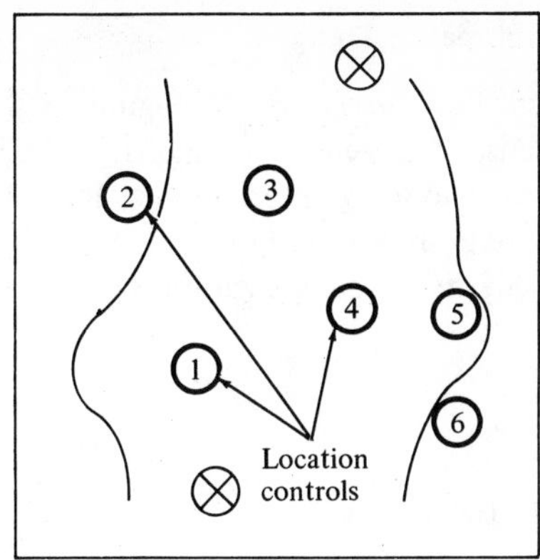

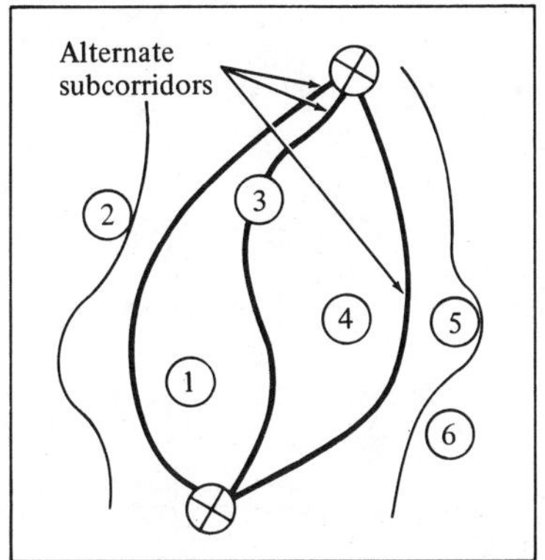

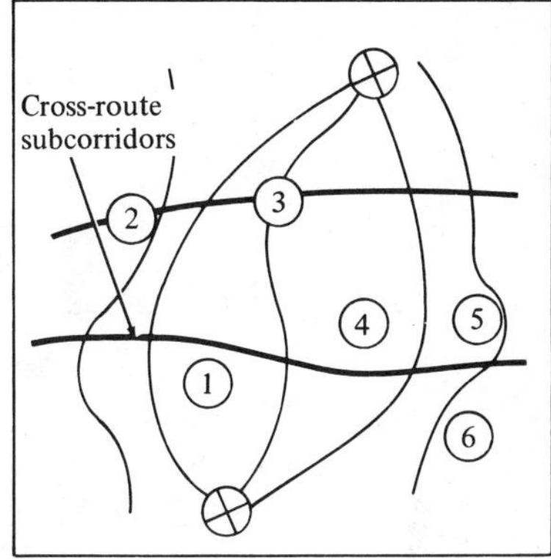

Fig. 97. Effect of controls on alignment selection[14].

2. Local unit needs: fire, utility, government, police.
3. Safety.
4. Aesthetics, emphasizing or screening views.
5. Conservation.
6. Community values.
7. Costs.
8. Multiple use, development opportunities, particularly in urban areas.
9. Level-of-service requirements.
10. Cross road requirements.

Intersection Design

Types of Intersections Figure 98 shows the general types of at-grade intersections. The general geometric form includes: three-leg, either T or Y, intersections; four-leg intersections; multi-leg intersections; and rotary intersections. Variations within these include flaring to provide special turn or passing lanes, or higher-type channelization for turning and through roadways.

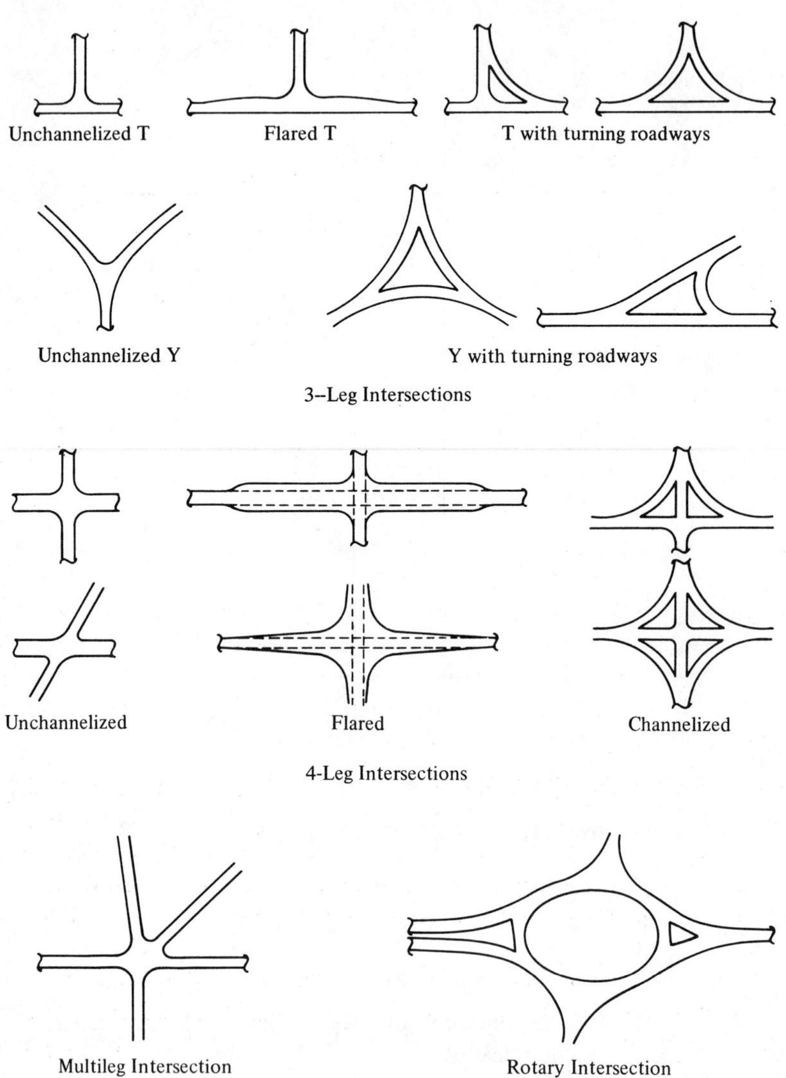

Fig. 98. General types of at-grade intersections[3].

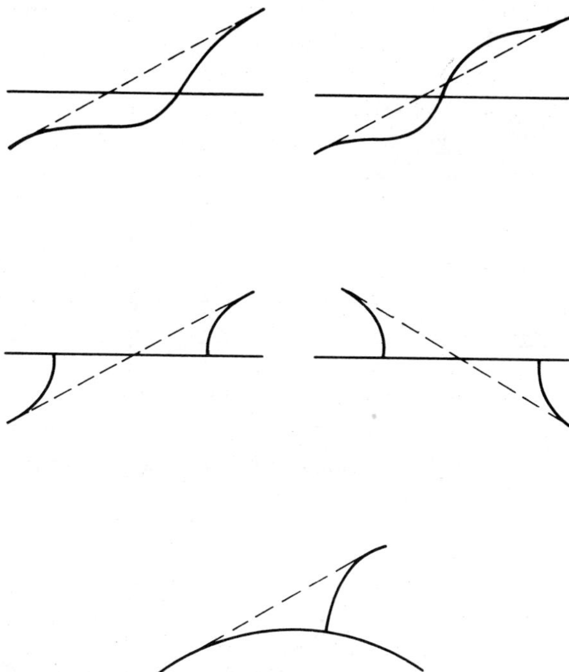

Fig. 99. Realignment variations at intersections, diagrammatic[3].

Although conditions at the site and the general alignment of both roadways dictate the location of the intersection, it may be desirable to compensate for a bad skew angle and modify the alignment to provide closer to a 90° intersection. Figure 99 shows several typical cases in which the intersection alignment might be modified to provide better intersection treatment.

Intersection Controls: Sight Distance For safety of operation it is necessary to provide the operator of a vehicle entering an at-grade intersection an unobstructed view of the intersecting roadways for sufficient length to enable him to avoid colliding with another vehicle. Three design cases (Fig. 100) provide a large enough minimum sight triangle at the intersection for the following purposes.

Case I: To enable vehicles to adjust their speed; the crossing is not controlled by yield signs, stop signs, or signal control. This case allows a driver approaching the intersection sufficient time to perceive the hazard and adjust his speed prior to reaching the intersection.

Case II: To enable vehicles to stop; the crossing is not controlled by yield signs, stop signs, or traffic signal controls. Sufficient sight distance must be pro-

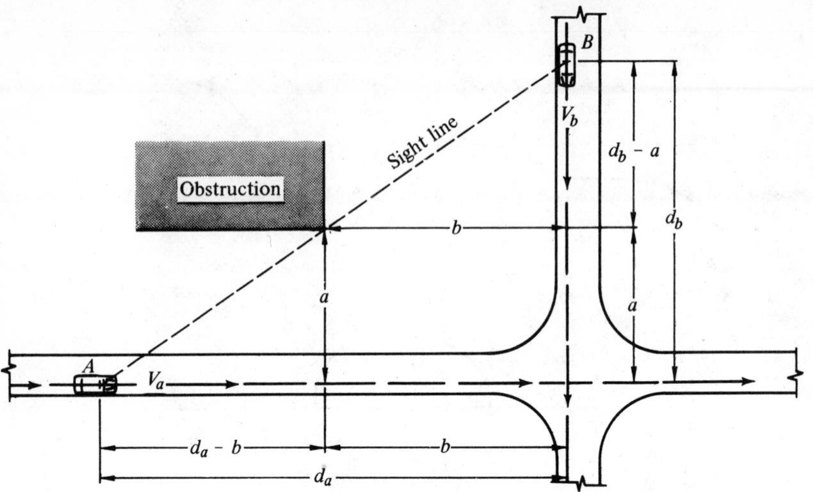

No stop or signal control at intersection Cases I and II

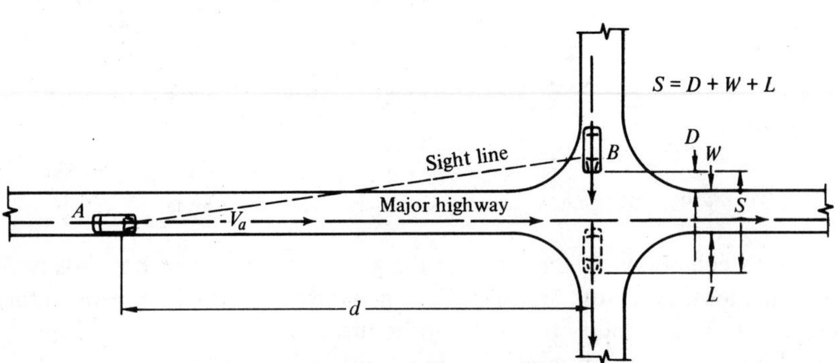

Stop control on minor road: Case III

Fig. 100. Sight distance at intersections, minimum sight triangle[3].

vided in this case to permit the driver approaching from either direction sufficient time to stop his vehicle before reaching the intersection.

Case III: To enable stopped vehicles to cross the major highway; this case applies to crossings at stop-controlled intersections. This condition permits the driver on the minor roadway sufficient vision along the major highway to enable him to cross it safely before the arrival of an oncoming vehicle on the crossroad. (3)

Table 58 and Figure 101 give representative design values for each of the three intersection vision requirements.

TABLE 58. INTERSECTION SIGHT DISTANCE, CASE I AND CASE II (3).

Speed (mph), V_a, V_b	Sight distance (ft) Case I, d_a, d_b	Case II,* d_a, d_b
20	90	
30	130	200
40	180	275
50	220	350
60	260	475
70	310	600

*These are minimum, not desirable, sight distances; refer to Tables 44 and 45.

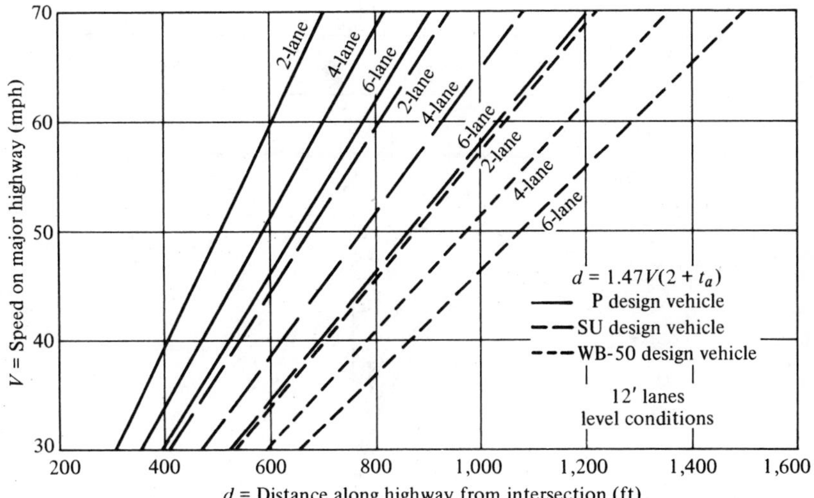

Fig. 101. Sight distance at intersections, Case III, required sight distance along major highway[3].

Turning Requirements Providing for turning vehicles at simple intersections or channelized intersections depends on:

1. The individual vehicle turning characteristics.
2. The type of operation permitted.
3. The speeds at which the turns are made.

Figures 72-75a[2] show the minimum-crawl-speed (less than 10 mph) turning paths for the five design vehicles currently used for design. Higher speeds will require longer transition curves, larger than the minimum turning radii, and possibly corner islands or more extensive channelization.

The type of turning operation permitted at an intersection also has considerable bearing on the radius required at the intersection. For example, traffic and other conditions may permit larger vehicles to swing wide, on the approach road, on the cross road, or on both, in making a turn. Figure 102 denotes four basic turning configurations possible at an intersection. Not numbered is the path when the vehicle approaches the turn within the proper lane and turns into the cross road on the proper lane.

Turns at Speeds Higher than Crawl Speed Generally it is desirable for vehicles turning at intersections to operate at speeds greater than crawl speed (10 mph or less) with the desirable turning speed for design normally taken as the average running speed (Table 59) of traffic on the highway approaching the turn. Table 60 denotes minimum radii for intersection curves for various design speeds.

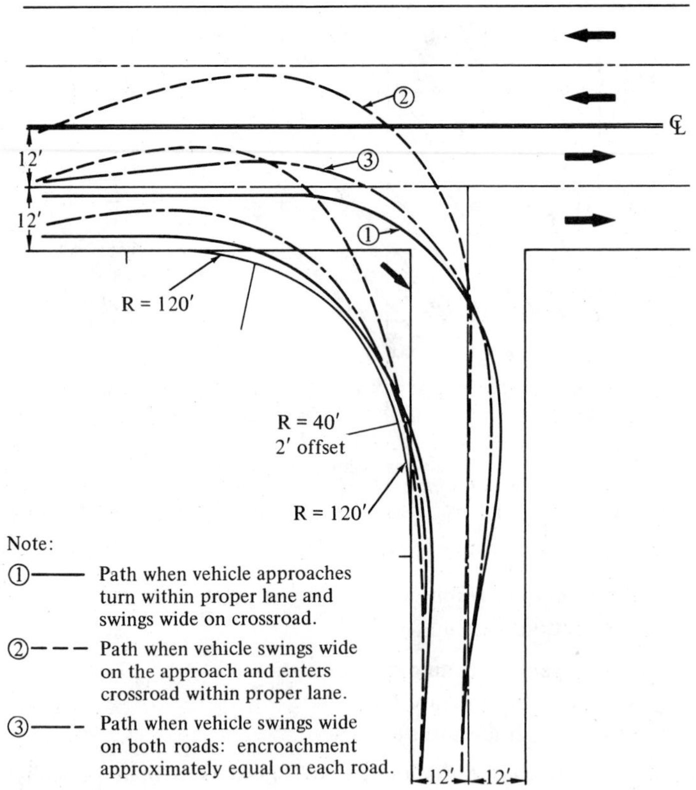

Fig. 102. Type of intersection turning movement required by WB-50 trucks[3].

TABLE 59. ASSUMED RELATIONSHIP BETWEEN DESIGN SPEED AND AVERAGE RUNNING SPEED–MAIN HIGHWAYS (3).

Design speed (mph)	Average Running Speed (mph)		
	Low volume	Intermediate volume	High volume
30	28	26	25
40	36	34	31
50	44	41	35
60	52	47	37
70	58	54	–*
80	64	59	–*

*At high volumes traffic congestion and interaction result in considerably lowered speeds.

TABLE 60. MINIMUM RADII FOR INTERSECTION CURVES WHEN C_{max} = 0.10 (3).

Design (turning) speed, V (mph)	15	20	25	30	35	40	50	60	70	80
Side friction factor, f	0.32	0.27	0.23	0.20	0.18	0.16	0.14	0.13	0.12	0.11
Assumed min. superelevation, e	.00	.02	.04	.06	.08	0.09	0.10	0.10	0.10	0.10
Total, $e + f$	.32	.29	.27	.26	.26	.25	0.24	0.23	0.22	0.21
Calculated min. radius, R (ft)	47	92	154	231	314	426	694	1043	1485	2032
Suggested curvature for design:										
Radius–minimum (ft)	50	90	150	230	310	430	700	1050	1500	2030
Degree of curve–maximum	–	64	38	25	18	13	8.5	5.5	4	3
Average running speed (mph)*	14	18	22	26	30	34	41	47	54	59

*Running speed for intermediate volumes.

Where turns are made at greater than crawl speed, turning radius requirements are such that excessively large intersections are apt to result and it may be desirable to provide minimum corner islands. Figure 103 shows minimum corner islands appropriate for passenger car operation at approximately 15 mph. (Crawl speed is 10 mph.) Large trucks, such as WB-40s, WB-50s, and WB-60s would have to operate at slightly lower speeds. The three cases, A, B, and C, refer to designs which provide for traffic mixes which are predominantly passenger car (Class A), single-unit vehicles (Class B), and large semi-trailers (Class C). Table 61 gives minimum designs for different angles of turn for each of the three traffic design classifications including the 90° case illustrated by Fig. 103.

At still higher speeds, larger turn radii are required (depending on the speed-curvature relationship). Table 62 provides design roadway widths for turning roadways, e.g., interchange ramps and channelized turns. Each of the design traffic conditions refers to an approximate traffic mix whereas each of the Cases (I, II, and III) refer to the manner of operation through the channel.

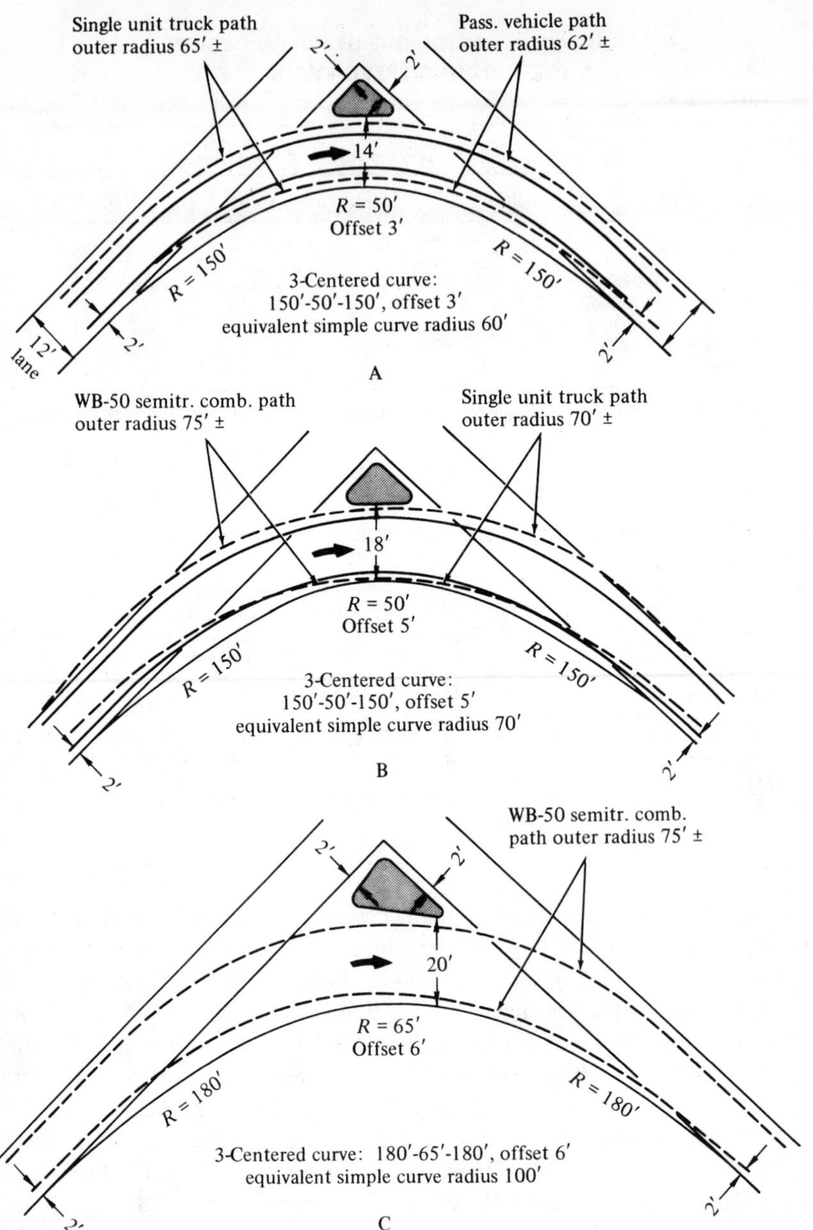

Fig. 103. Designs for turning roadways with minimum corner island, 90° right turn. Figures correspond to minimum design data given in Table 61[3].

TABLE 61. MINIMUM DESIGNS FOR TURNING ROADWAYS (3).

Angle of turn (degrees)	Design classification*	3-centered compound curve: Radii (ft)	3-centered compound curve: Offset (ft)	Width of lane (ft)	Approximate island size (ft^2)
	A	150-75-150	3.5	14	60
75	B	150-75-150	5.0	18	50
	C	180-90-180	3.5	20	50
	A	150-50-150	3.0	14	50
90	B	150-50-150	5.0	18	80
	C	180-65-180	6.0	20	125
	A	120-40-120	2.0	15	70
105	B	100-35-100	5.0	22	50
	C	180-45-180	8.0	30	60
	A	100-30-100	2.5	16	120
120	B	100-30-100	5.0	24	90
	C	180-40-180	8.5	34	220
	A	100-30-100	2.5	16	460
135	B	100-30-100	5.0	26	370
	C	160-35-160	9.0	35	640
	A	100-30-100	2.5	15	1400
150	B	100-30-100	6.0	30	1170
	C	160-35-160	7.1	38	1720

*A—Primarily passenger vehicles; permits occasional design single-unit truck to turn with restricted clearances.
B—Provides adequately for SU; permits occasional WB-50 to turn with slight encroachment on adjacent traffic lanes.
C—Provides fully for WB-50.

Asymmetric 3-centered compound curves and straight tapers with a simple curve can also be used without significantly altering the width of pavement or corner island size.

Designs are for speeds slightly greater than crawl speed. For passenger cars the designs are for approximately 15 mph, somewhat less for WB-40s and WB-50s.

Channelization (2) (3) (15) A channelized intersection is an at-grade intersection in which islands are used to define channels which control or guide the movement of the driver through the intersection. The islands can take any physical form: barrier and mountable curbs, buttons or raised jiggle bars, pavement areas marked by painting, cones, sand-bags, five-gallon buckets, etc.

Channelization accomplishes one or more of the following:

1. Controls the angle of conflict between vehicles.
2. Reduces the area of conflict.
3. Reduces the crossing time and hence exposure to opposing flows.

TABLE 62. DESIGN WIDTHS OF PAVEMENTS FOR TURNING ROADWAYS (3).

R, radius on inner edge of pavement (ft)	Pavement width (ft) for: Case I, one-lane, one-way operation—no provision for passing			Case II, one-lane, one-way operation—with provision for passing a stalled vehicle			Case III, two-lane operation —either one-way or two-way		
	Design traffic condition								
	A	B	C	A	B	C	A	B	C
50	18	18	23	23	25	29	31	35	42
75	16	17	19	21	23	27	29	33	37
100	15	16	18	20	22	25	28	31	35
150	14	16	17	19	21	24	27	30	33
200	13	16	16	19	21	23	27	29	31
300	13	15	16	18	20	22	26	28	30
400	13	15	16	18	20	22	26	28	29
500	12	15	15	18	20	22	26	28	29
Tangent	12	15	15	17	19	21	25	27	27
Width modification regarding edge of pavement treatment:									
No stabilized shoulder	none			none			none		
Mountable curb	none			none			none		
Barrier curb:									
one side	add 1 ft			none			add 1 ft		
two sides	add 2 ft			add 1 ft			add 2 ft		
Stabilized shoulder, one or both sides	none			deduct shoulder width; minimum pavement width as under Case I			deduct 2 ft where shoulder is 4 ft or wider		

Traffic Condition A: Predominantly P vehicles, but some consideration for SU trucks.
Traffic Condition B: Sufficient SU vehicles to govern design, but some consideration for semitrailer vehicles.
Traffic Condition C: Sufficient semitrailer, WB-40 or WB-50 vehicles to govern design.

4. Places the vehicle in a favorable position to judge the position and speed of approaching traffic.
5. Regulates the flow of traffic through the intersection.
6. Facilitates the predominant movement or maintains continuity of the movement system.
7. Provides a protected area for pedestrians.
8. Provides protection and storage for turning and crossing vehicles.
9. Provides a location for traffic control devices.

Channelization Design Criteria; Fundamentals of Channelization Listed below are basic fundamentals which should be followed in designing channelized intersections:

1. Crossings without merging or weaving should be made at angles approaching 90°. Where the skew angle is less than 60° the intersection configuration should be modified to provide a larger angle.
2. Streams should merge at very flat angles and vehicles generally should be pointed in the final direction which they are to assume.
3. Bending of the entering traffic stream can be used to control the speed of traffic.
4. Funneling, or a reduction in channel width, because of the psychological effect of the narrowing of the traffic lane, can be used to reduce the speed of traffic.
5. Areas which provide refuge for turning vehicles should be no less than 14–16 ft in width (11–12 ft for the turning lane and 4–5 ft for the channelizing island); for crossing vehicle refuge no less than 24–28 ft should be provided.
6. The vehicle conflict area should be such that drivers are exposed to only one conflict and one decision at a time.
7. Channelization can be used to block prohibited turns, e.g., turns into off-ramps.
8. In general, channelization should favor the predominant turning movement. There may be times, however, when highway system continuity requires favoring a lesser movement.
9. Multiple and compound diverging and merging maneuvers should be avoided.
10. Where the pavement has joints the pavement jointing should be continuous through the required movement channels.
11. Too many or too small islands should be avoided.
12. Do not channelize if it is not needed.
13. Avoid isolated channelization unless it is of major proportions.
14. Be sure that islands are visible to the approaching driver.
15. Because of visibility problems, do not introduce islands on crest vertical curves.
16. Channelization should be natural and in conformity with normal vehicle operations.
17. The approach end of channelizing islands should be offset from the lane and introduced by painting or jiggle bars.
18. The approach end should, if possible, be mountable.
19. Where signs are necessary in the islands, they should be on breakaway mounts.

Island Size and Arrangement The absolute minimum island size should be approximately 50 ft^2 with a desirable minimum of 75 ft^2. For a triangular island, this will result in a minimum side dimension of 8–12 ft, exclusive of rounding. Comparable dimensions for an elongated or divisional island are 12 ft to 20 ft in length, where the island width is 4 ft. In general, 4 ft should be the minimum island width for divisional islands; in some special circumstances, however, it may be necessary to go to an absolute minimum of 2 ft in width.

Interchange Design An interchange is a facility at the juncture or crossing of two or more roadways in which certain movements are grade separated. Separating certain movements (crossing and/or turning) provides a considerable increase in the service volume capability of the individual roadways, improves the overall level of service provided and enhances safety.

Types of Interchanges Figure 104 depicts the basic types of interchanges. Many variations exist within the basic types, particularly in the partial cloverleafs, diamonds, and directionals.

Principles of Interchange Design,
General:

1. Movements: all movements generally should be provided for no matter how small in volume.
2. Spacing: interchanges should not be spaced so closely that movements from one interchange interfere with those from another or that signing is made difficult.
3. Continuity of type: where a succession of interchanges occurs over a continuous system of high-type roadway, the same type of interchange should be used consistently where possible. In this way, drivers know what to expect as to type and become accustomed to operating on one kind of interchange. As a result, operation is generally more efficient and safe.
4. Lanedrops: traffic lanes should never be dropped as the roadways go through an interchange area, even though the roadway is not needed from a capacity-level of service standpoint. The lanes should be carried through the interchange and only then dropped. Figure 105 illustrates recommended practice for dropping and adding lanes.
5. Major forks: where high-volume flows depart to go through an interchange or join the roadway after the interchange, the flow can be best handled as a major fork rather than through the use of a standard on or off ramp treatment.

Details of Design

Speed-Change Lanes (9) A speed-change lane is that portion of an intersection or interchange where a turning roadway leaves or joins the through highway

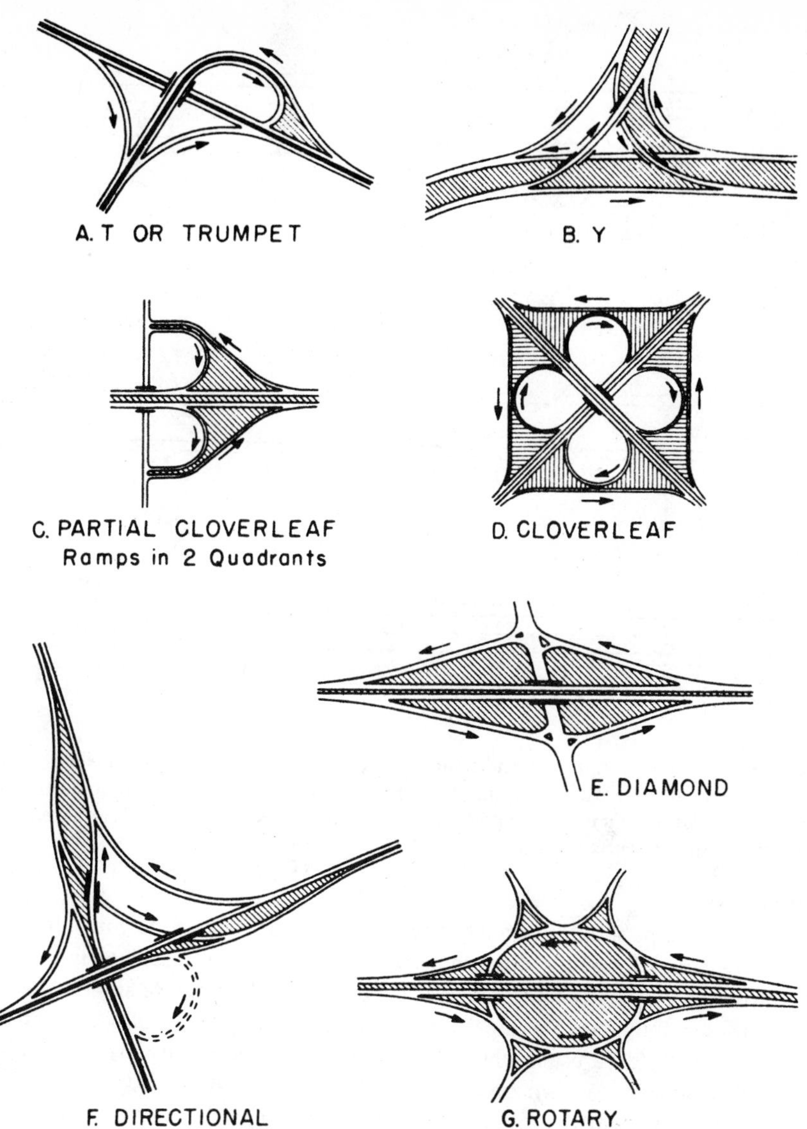

Fig. 104. General types of interchanges[3].

pavement. Normal operation usually requires that drivers leaving the mainline roadway reduce their speed before entering the turning roadway (or channelization) or, where entering the mainline roadway, accelerate from the lower speed of the turning roadway or ramp to the higher speed of the mainline traveled way. Design requirements of exit ramps, entrance ramps and slip ramps include (9):

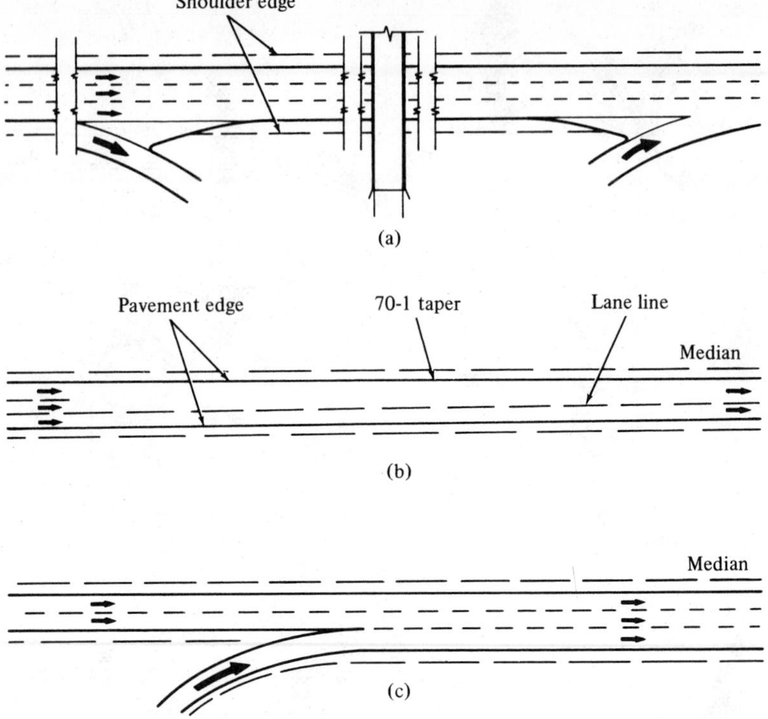

Fig. 105. Lane drops and lane additions, recommended practice[9]. (a) Retain all lanes through interchange area. (b) Lane drop from median side downstream from entrance ramp. Note: Lane drop more desirable on side adjoining and downstream of entrance ramp. (c) Lane addition at entrance ramp.

A. Exit Ramps

1. Exit ramps should normally have only a single lane at the gore, although they may be widened for increased capacity and storage shortly beyond the gore. The ramp terminal at the crossroad should have ample capacity to accommodate the stored vehicles. Exit ramps with two lanes at the gore should be used only where an additional outside lane is provided on the freeway, i.e., two existing lanes.
2. Exit ramps which leave the through roadway at an angle of 4°–5° have been found to operate substantially better than exits with parallel deceleration lanes, and hence are preferable.
3. Exit ramps should leave the mainline roadways at a point where there will be no vertical curvature to restrict visibility along the ramp to a value less than the standard nonpassing sight distance for the ramp design speed. Ramps that "drop out of sight" are a definite hazard and must be avoided.
4. Right-hand exits are markedly superior in their operational characteristics and safety to those that leave on the left.

5. Compound horizontal curvature on ramps, and sudden reversal of horizontal curvature without a sufficient intervening tangent section, should be avoided.
6. Exit ramps should always leave the mainline on a tangent section of the mainline, not at a point at which the ramp goes ahead on tangent while the mainline curves to the left. Ramps which leave where the through traffic alignment is on a right curve are likewise almost certain to be troublesome.
7. Sequential decision points on exit ramps, such as the exit from the mainline and the junction of the exit ramp with a collector–distributor roadway, should be separated by a minimum distance of 800 to 1000 ft to provide adequate distance for maneuvering and for signing.

B. Entrance Ramps

1. In general, entrance ramps should have a single lane at the entrance nose. Two-lane entrance ramps should be permitted only where an additional lane on the freeway is provided. This additional lane may be continuous, or under special circumstances may be terminated with the usual tapered section.
2. Undue variations in the width of entrance ramps can cause congestion and hazard. The provision of a two-lane throat at the intersection of the entrance ramp and the crossroad may be needed to accommodate turning maneuvers. However, a gradual reduction of this width to a single lane entrance at the junction with the through lanes is desirable.
3. The design of entrance ramps and merging areas should take into account the effect of ascending grade in cases where a substantial volume of heavy trucks is anticipated.
4. Entrance ramps from local streets should always join the main traffic stream from the right. Where two freeways join with direct connections, left-hand junctions are often a necessity. These latter situations should receive special study and treatment to assure the best possible correlation of design and traffic control. Drivers must always have full and adequately timed notice of any merging, lane changing, or other action required of them.
5. The metering of entrance ramp traffic onto the freeway or the complete closure of an entrance ramp, permanently or during certain hours of the day, has proved to be a valuable operational technique for preventing the build-up of extreme congestion due to the addition of very small volumes from critical ramps.

C. Slip Ramps

1. Braided intersections of two-way frontage roads and ramps (whether slip ramps or diamond-type ramps) are definitely discouraged because of operational problems. The slip ramp should completely separate existing traffic from the two-way frontage road.

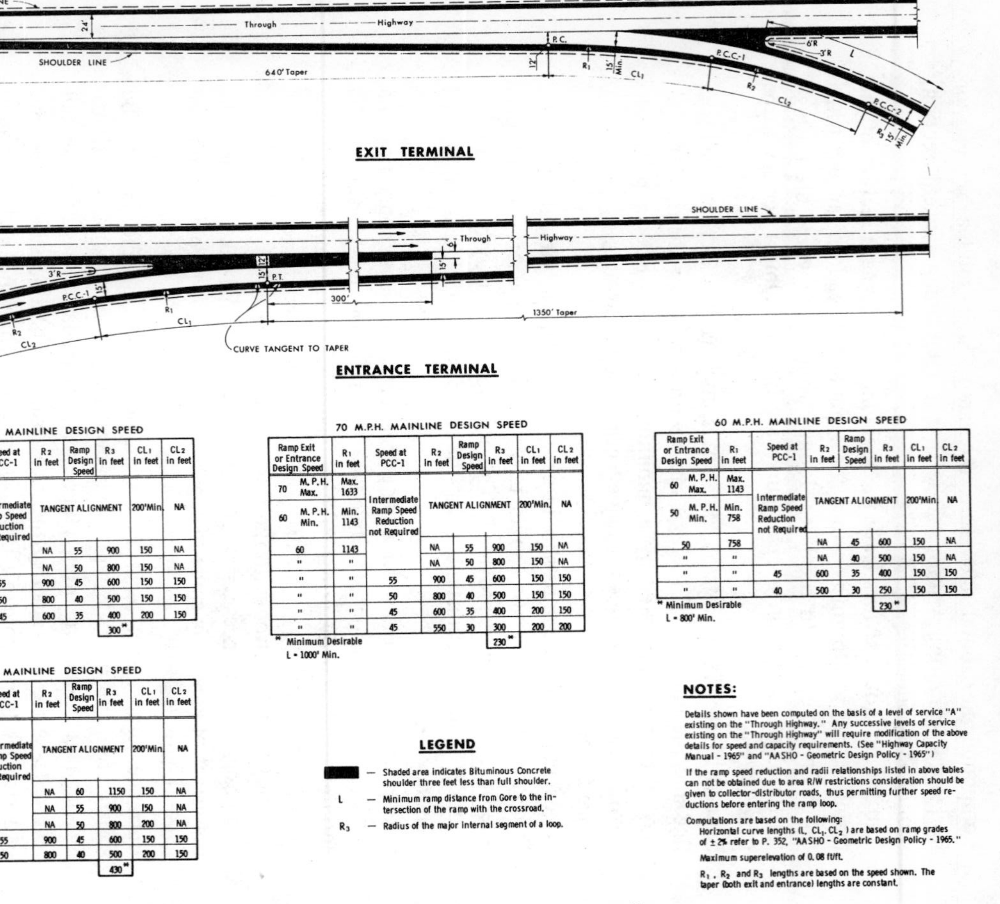

75 M.P.H. MAINLINE DESIGN SPEED

Ramp Exit or Entrance Design Speed	R_1 in feet	Speed at PCC-1	R_2 in feet	Ramp Design Speed	R_3 in feet	CL_1 in feet	CL_2 in feet
75 M.P.H. Max.	Max. 1974	Intermediate Ramp Speed Reduction not Required	TANGENT ALIGNMENT			200' Min.	NA
60 M.P.H. Min.	Min. 1143						
60	1143		NA	55	900	150	NA
"	"		NA	50	800	150	NA
"	"	55	900	45	600	150	150
"	"	50	800	40	500	150	150
"	"	45	600	35	400	200	150
					300**		

** Minimum Desirable
L = 1100' Min.

80 M.P.H. MAINLINE DESIGN SPEED

Ramp Exit or Entrance Design Speed	R_1 in feet	Speed at PCC-1	R_2 in feet	Ramp Design Speed	R_3 in feet	CL_1 in feet	CL_2 in feet
80 M.P.H. Max.	Max. 2246	Intermediate Ramp Speed Reduction not Required	TANGENT ALIGNMENT			200' Min.	NA
65 M.P.H. Min.	Min. 1341						
65	1341		NA	60	1150	150	NA
"	"		NA	55	900	150	NA
"	"		NA	50	800	200	NA
"	"	55	900	45	600	150	150
"	"	50	800	40	500	200	150
					430**		

** Minimum Desirable
L = 1200' Min.

70 M.P.H. MAINLINE DESIGN SPEED

Ramp Exit or Entrance Design Speed	R_1 in feet	Speed at PCC-1	R_2 in feet	Ramp Design Speed	R_3 in feet	CL_1 in feet	CL_2 in feet
70 M.P.H. Max.	Max. 1633	Intermediate Ramp Speed Reduction not Required	TANGENT ALIGNMENT			200' Min.	NA
60 M.P.H. Min.	Min. 1143						
60	1143		NA	55	900	150	NA
"	"		NA	50	800	150	NA
"	"	55	900	45	600	150	150
"	"	50	800	40	500	150	150
"	"	45	600	35	400	200	150
"	"	45	550	30	300	200	200
					230**		

** Minimum Desirable
L = 1000' Min.

60 M.P.H. MAINLINE DESIGN SPEED

Ramp Exit or Entrance Design Speed	R_1 in feet	Speed at PCC-1	R_2 in feet	Ramp Design Speed	R_3 in feet	CL_1 in feet	CL_2 in feet
60 M.P.H. Max.	Max. 1143	Intermediate Ramp Speed Reduction not Required	TANGENT ALIGNMENT			200' Min.	NA
50 M.P.H. Min.	Min. 758						
50	758		NA	45	600	150	NA
"	"		NA	40	500	150	NA
"	"	45	600	35	400	150	150
"	"	40	500	30	250	150	150
					230**		

** Minimum Desirable
L = 800' Min.

LEGEND

- [shaded] — Shaded area indicates Bituminous Concrete shoulder three feet less than full shoulder.
- L — Minimum ramp distance from Gore to the intersection of the ramp with the crossroad.
- R_3 — Radius of the major internal segment of a loop.

NOTES:

Details shown have been computed on the basis of a level of service "A" existing on the "Through Highway." Any successive levels of service existing on the "Through Highway" will require modification of the above details for speed and capacity requirements. (See "Highway Capacity Manual - 1965" and "AASHO - Geometric Design Policy - 1965")

If the ramp speed reduction and radii relationships listed in above tables can not be obtained due to area R/W restrictions consideration should be given to collector-distributor roads, thus permitting further speed reductions before entering the ramp loop.

Computations are based on the following:
Horizontal curve lengths (L, CL_1, CL_2) are based on ramp grades of ± 2% refer to P. 352, "AASHO - Geometric Design Policy - 1965."

Maximum superelevation of 0.08 ft/ft.

R_1, R_2 and R_3 lengths are based on the speed shown. The taper (both exit and entrance) lengths are constant.

Fig. 106. Layout details for entrance and exit terminals at interchanges[13].

2. Slip ramps on two-way frontage roads are generally unsatisfactory as they tend to induce wrong-way entry to the through lanes of the freeway.
3. Because slip ramps are relatively short, vehicles leaving the through lanes often enter the frontage road at excessive speeds, endangering both ramp traffic and frontage road traffic.
4. With one-way frontage roads, slip ramps should have a proper deceleration area, be long enough to enable traffic to slow down to the frontage road speed, and intersect with the frontage road sufficiently in advance of the crossroad intersection for an adequate weaving length for turning traffic.

Figure 106 details one good design configuration for speed change lanes.

Turning Roadways The turning roadway is that portion of a ramp between the speed change lanes and the crossroad terminal. Turning roadway requirements are the same as those noted in Table 60.

Terminals Where interchange ramps intersect the crossroad at grade, the ramp terminal should be designed essentially as the intersection of a one-way street. For off-ramps it is necessary to provide a design having sufficient sight distance along the crossroad and which discourages wrong-way entry into the off-ramp. Several typical terminal designs for the intersection of diamond ramp terminals with crossroads at 90° angles and at skew angles are shown in Fig. 107 and 108. Required crossroad sight distances at diamond ramp terminals are as shown in Fig. 109 and Table 63. A typical entrance and exit terminal detail used by one design organization is shown in Fig. 110.

TABLE 63. REQUIRED SIGHT DISTANCE ALONG THE CROSSROAD AT TERMINALS OF RAMPS AT DIAMOND INTERCHANGES, TWO-LANE, TWO-WAY HIGHWAYS (3).

Assumed design speed on crossroad through the interchange	Sight distance required to permit design vehicle to turn left from ramp to crossroad (ft)* — Design vehicle assumed at ramp terminal:			Sight distance available to entering vehicle when vertical curve on crossroad is designed for minimum stopping sight distance†	
	P	SU	WB-50	P	SU or WB-50
70	740	1060	1430	920	1040
60	630	910	1230	730	820
50	530	760	1030	540	600
40	420	610	820	420	480
30	320	460	620	310	350

*Sight distance measured from height of eye of 3.75 ft for P design vehicles and 6 ft for SU and WB-50 design vehicles to an object 4.5 ft high.
†Minimum available stopping sight distance based on the assumption that there is no horizontal sight obstruction and that $S < L$.

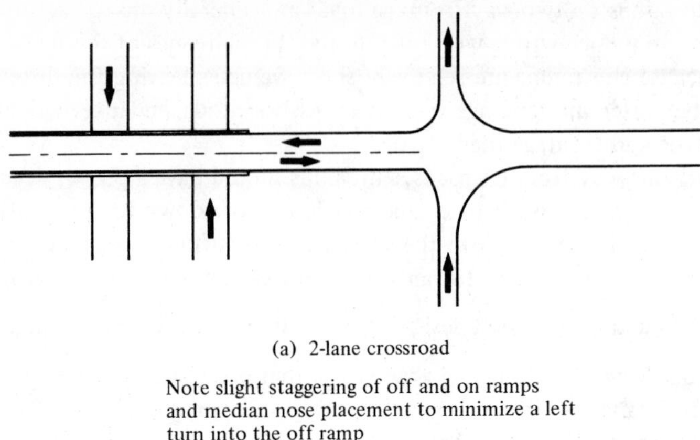

(a) 2-lane crossroad

Note slight staggering of off and on ramps and median nose placement to minimize a left turn into the off ramp

(b) 4-lane crossroad

Fig. 107. Intersections at diamond ramp terminals, 90° angles[3].

Dual Roadway–Single Roadway Transition It is often wise to provide a dual roadway through an interchange area, both to facilitate operation and provide a buffer against future increased demand. This is particularly true in or adjacent to urban areas, where dual roadways should be provided initially or designed into the facility for future staging without major reconstruction. Making the transition between the single roadway and dual roadway can best be done by a design similar to that shown in Fig. 111a. Note that at each end of the divided section the driver entering the dual roadway has a "direct shot" into the divided section; drivers leaving the divided section must, however, form a single lane before reaching the two-way roadway and make a distinct left-right jogging maneuver. With a treatment such as that shown in Fig. 111b the driver entering the dual roadways has more chance of accidentally entering the oncoming roadway. If such treatment is necessary, special, oversized, lighted signing may be required.

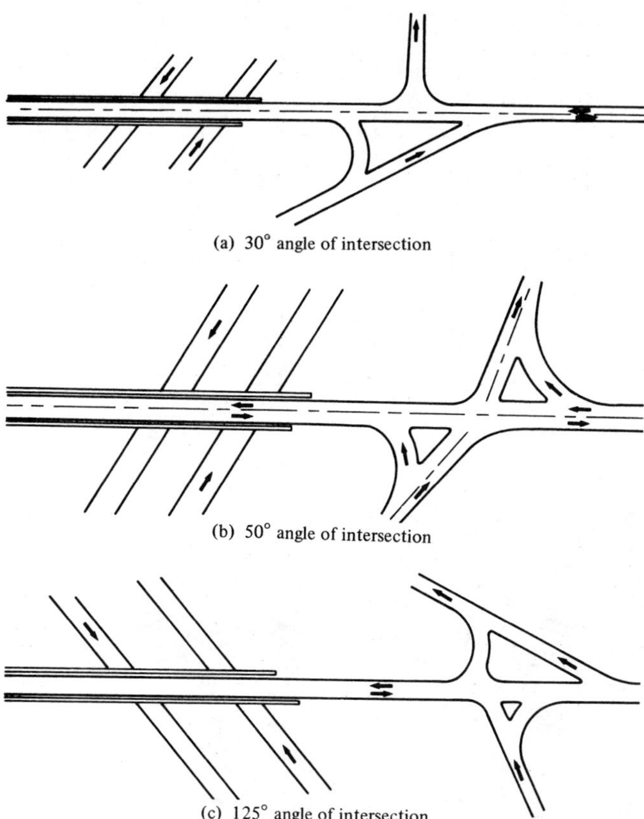

(a) 30° angle of intersection

(b) 50° angle of intersection

(c) 125° angle of intersection

Fig. 108. Intersections at diamond ramp terminals, skewed crossroads[1].

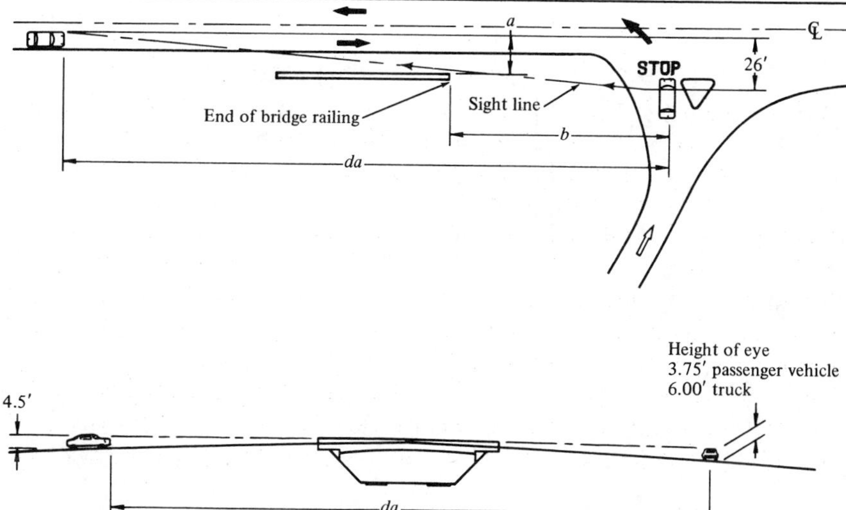

Fig. 109. Measurement of sight distance at diamond ramp terminals[3].

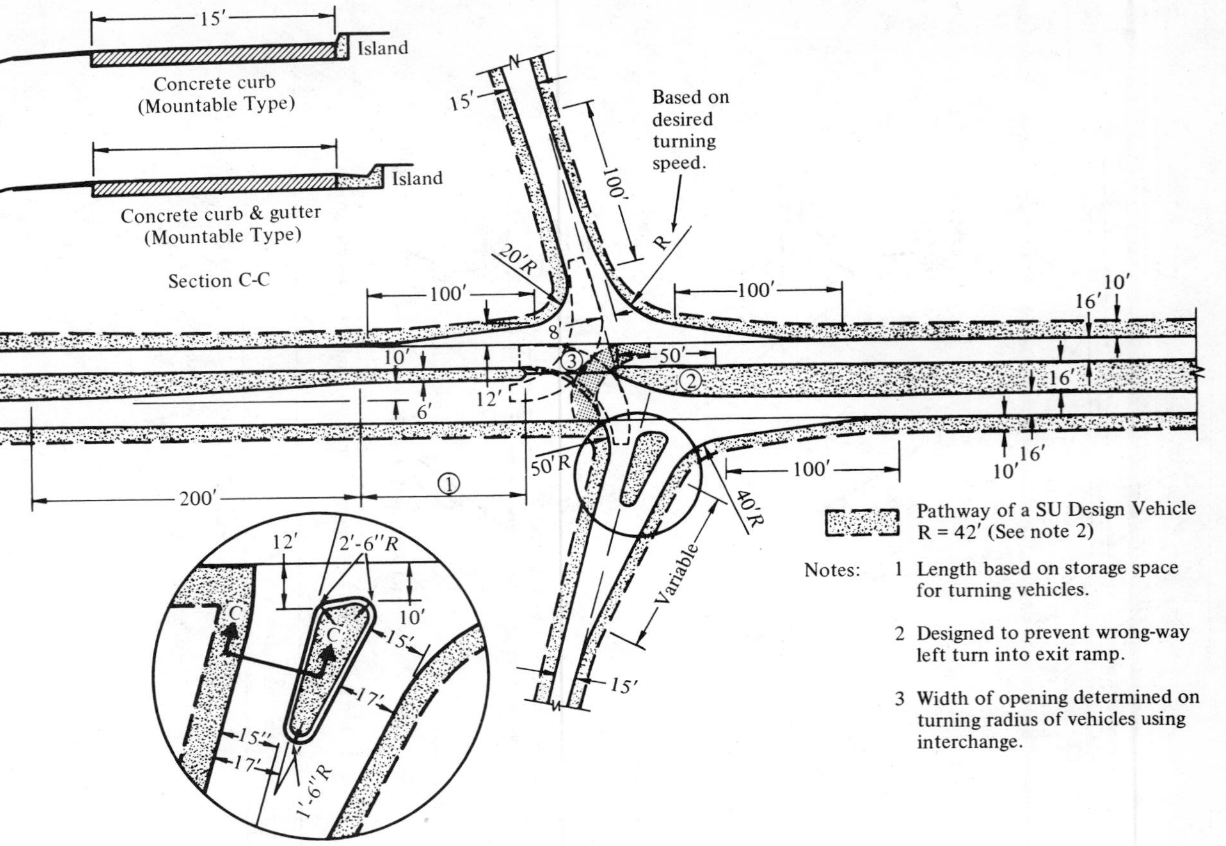

Fig. 110. Typical entrance and exit ramp terminal details[13].

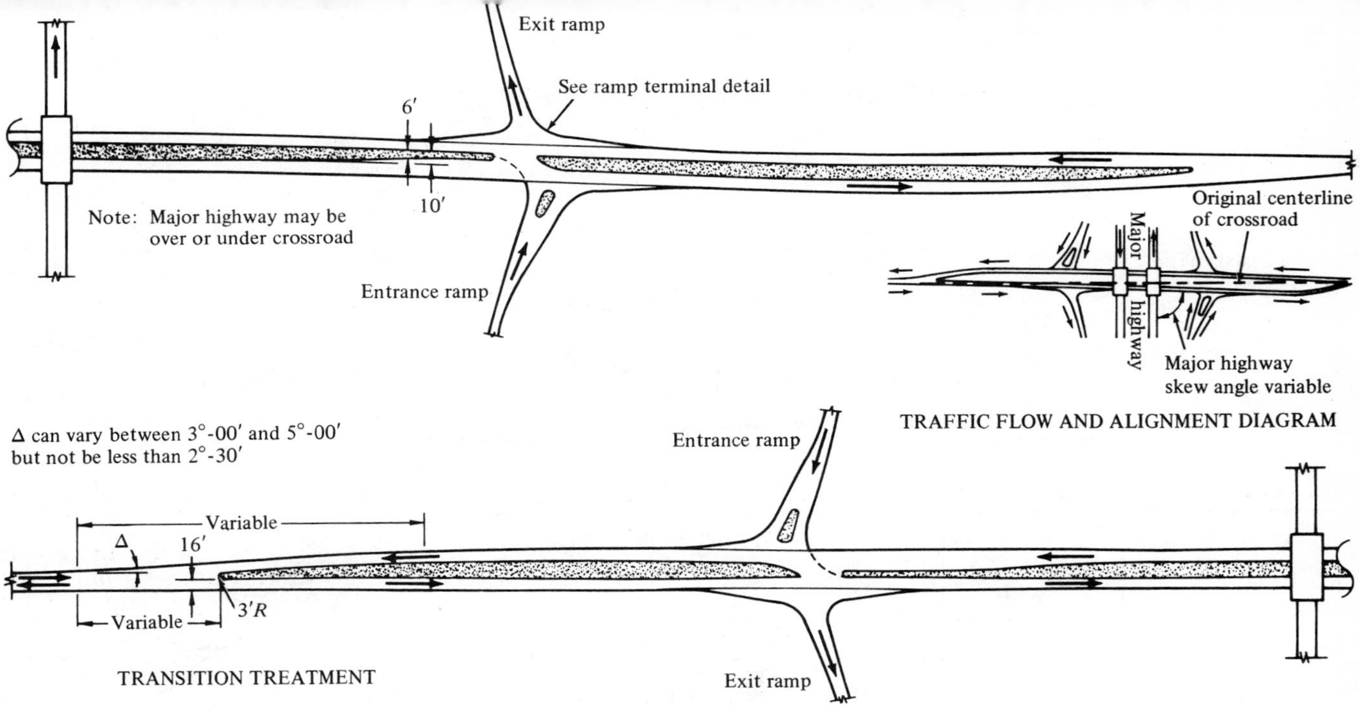

Fig. 111a. Recommended dual roadway–single roadway transition[13].

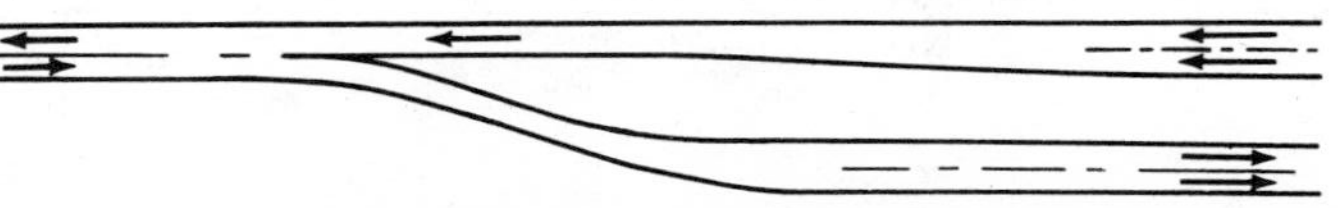

Fig. 111b. Not recommended dual roadway–single roadway transition.

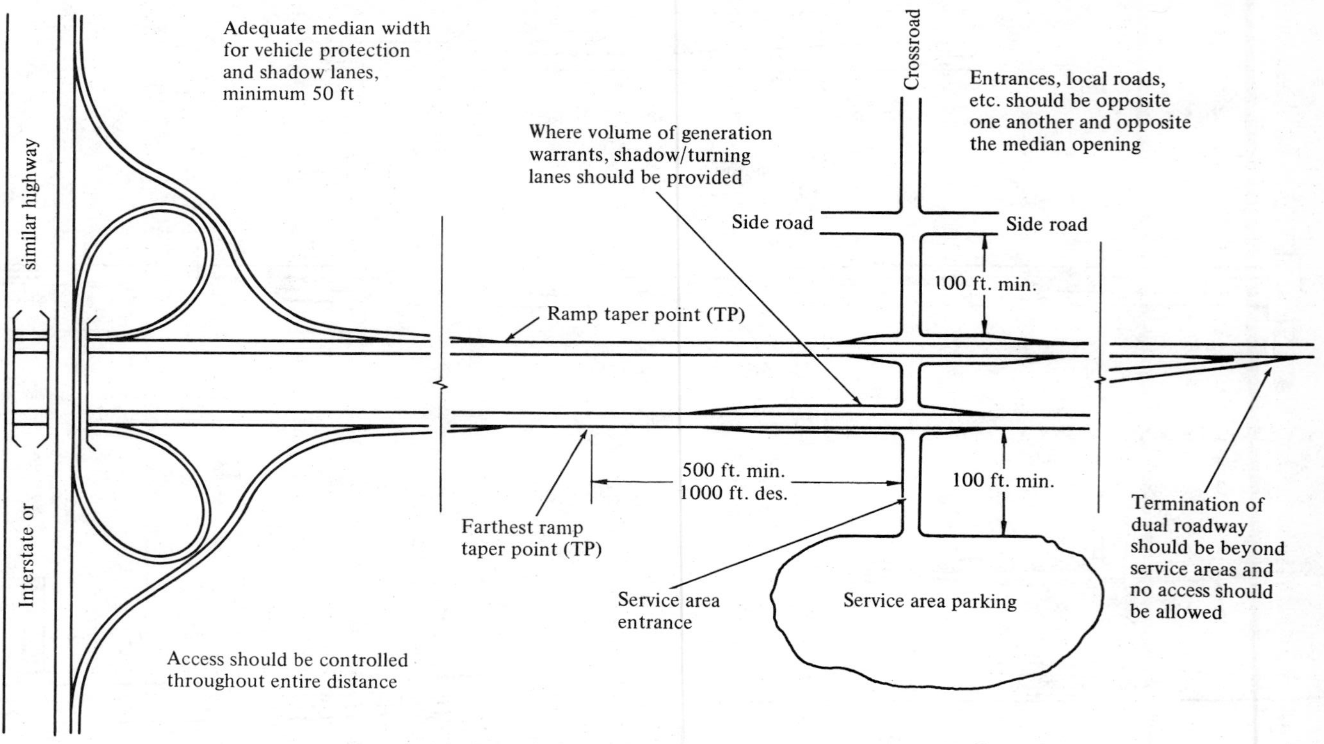

Fig. 112. General design and access control criteria for interchange area[11].

Associated Interchange Design-Planning Controls The highway interchange is often the site of intense traffic activity and extensive adjacent land use development. There is, therefore, a need to insure that the activities developed in the highway interchange area are compatible with one another, with adjacent land uses, and with the roadways themselves.

Preserving the operational integrity of the interchange itself should be a primary objective of the highway transportation designer. Although a number of land use controls are available (e.g., zoning, subdivision controls, driveway regulations, etc.), controlling access still remains the best tool available to the highway engineer. Figure 112 illustrates general control criteria for maintaining access and driveway control in the vicinity of interchanges.

REFERENCES

1. Wisconsin Department of Transportation. "State Highway 1990 Functional Plan." Wisconsin Department of Transportation (Madison) (1967).
2. Institute of Traffic Engineers. "Traffic Engineering Handbook." Institute of Traffic Engineers (ITE) (Washington) (1965).
3. American Association of State Highway Officials (AASHO). "A Policy on Geometric Design of Rural Highways 1965." AASHO (Washington) (1965).
4. Simpson and Curtin. "Transportation for Madison's Future." Madison Area Transportation Study (MATS) Special Report No. 17, Prepared for MATS Technical Coordinating Committee (Philadelphia) (May 1970).
5. Buchanan, Colin. "Traffic in Towns." Her Majesty's Stationery Office (London) (1963).
6. Highway Research Board. "Highway Capacity Manual–1965." HRB Special Report No. 87. HRB (Washington) (1965).
7. Leisch, Jack E. "Capacity Analysis Techniques for Design of Signalized Intersections." Reprinted from Aug. 1967 and Oct. 1967 issues of Public Roads. Government Printing Office, Washington, 1967.
8. American Association of State Highway Officials (AASHO). "Geometric Design Guide for Local Roads and Streets." AASHO (Washington) (1971).
9. American Association of State Highway Officials (AASHO). "Highway Design and Operational Practices Related to Highway Safety." AASHO (Washington) (1967).
10. Tunnard, Christopher and Pushkarev, Boris. "Man Made America: Chaos or Control." Yale University Press (New Haven) (1963).
11. Kuhn, Herman A. J. "Planning Implications of Urban and Rural Freeway Interchanges." Journal of the Urban Planning and Development Division. ASCE April 1969.
12. American Association of State Highway Officials (AASHO). "A Policy on Design Standards for Stopping Sight Distance." AASHO (Washington) (1971).

13. Wisconsin Department of Transportation. "Design Manual." Madison.
14. Wisconsin Department of Transportation.
15. Highway Research Board (HRB). "Channelization: The Design of Highway Intersections at Grade." HRB Special Report No. 74. HRB (Washington) (1962).
16. A Policy on Design of Urban Highways and Arterial Streets", American Association of State Highway and Transportation Officials, Washington, D.C. (1973).
[AASHO is now AASHTO, HRB is now TRB. The "T" represents "Transportation."]

SYSTEM OPERATION AND CONTROL

Highway movement systems must include mechanisms to provide operational control and guidance to motorists using the system. Among the most common are signals, where appropriate; signs; roadway markings; and the general operational regulations known as "rules of the road."

Signalization

A signal is a device, normally power operated, which warns or directs traffic to take some course of action. It does not include such things as construction and hazard flashers or signs and markings. Signals are not the panacea for all ills, and often should not be installed. On the plus side, signals accomplish the following:

1. Provide for the orderly movement of traffic.
2. Reduce certain types of accidents, in particular right angle and pedestrian accidents.
3. Facilitate, through signal coordination, continuous (or nearly continuous) traffic flow at preselected speeds.
4. Interrupt heavy through traffic to permit cross traffic and cross pedestrian movements.
5. Provide speed control through signal coordination.
6. Provide economy in traffic control when compared to manual methods.
7. Assign a specific right of way to a driver from a legal standpoint.

Signals also have some negative aspects. Among other things they:

1. Reduce total intersection capacity.
2. Increase total intersection delay.
3. Increase certain types of accidents, in particular rear-end collisions.

4. Encourage disrespect of travel control devices, causing unnecessary delay, irritation, and possibly increased accidents if installed where not warranted.

Signal Lenses and Arrangements The basic signal lens displays include the circular red, amber, and green balls in 8-in. and 12-in. diameter. The larger diameter should be used where:

1. The 85th percentile approach speeds exceed 40 mph.
2. Signalization is unexpected.
3. Background and adjacent signs and lighting conflict or compete.
4. Drivers view traffic control and lane-direction control simultaneously.
5. Arrow indications are used.

Figure 113 illustrates typical approved arrangements for signal lenses.

Location and Mounting of Signal Faces Signal faces should be placed for maximum driver visibility with a minimum of two signal faces continuously visible at least the distance $D_{\text{s.v.}}$ in advance of the stop line, where

$$D_{\text{s.v.}} = 15S_{85} - 200$$

where

$D_{\text{s.v.}}$ = signal visibility distance in feet
S_{85} = 85th percentile speed in miles per hour—minimum $S_{85} = 20$ mph

In general, at least one and preferably two signal faces shall be at least 40 ft but not more than 120 ft beyond the stop line. Where signals are post mounted there shall be two on the far side of the intersection, one on the right and one on the left; or, if practical, on a median island. Because of driver eye positioning the most desirable lateral positioning lies within a 40° arc subtending 20° equally on both sides of the extended approach centerline (Fig. 114).

Mounting heights to the bottom of the signal face housing shall be between 8 ft and 15 ft for post mounted installations, and between 15 ft and 19 ft for installations suspended over the roadway.

Signal Phases That part of the signal time cycle allocated to any traffic movement receiving the right of way is a signal phase:

Two-Phase. Two-phase signals are those normally encountered on four-leg intersections displaying only a green ball in each direction.

Three-Phase. Three-phase systems are often encountered at multi-throated intersections having five or more legs and those for which one of the legs requires a special green indication because of a heavy turning movement, an accident problem, or special pedestrian requirements.

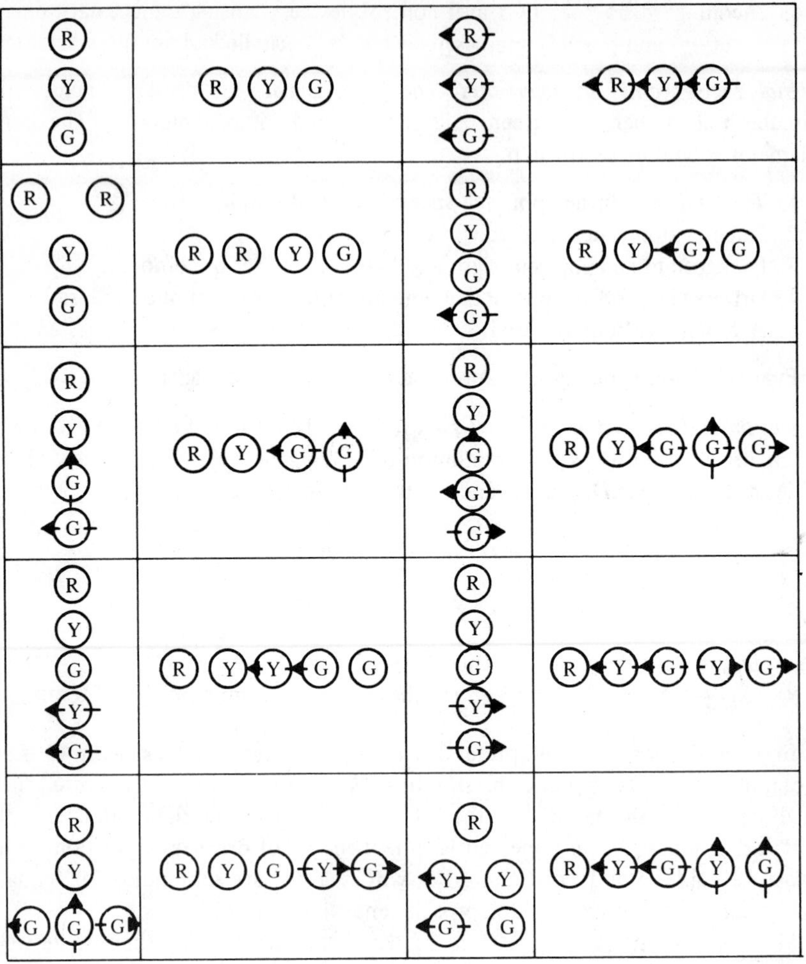

Fig. 113. Typical arrangements of lenses in signal faces[1].

Figure 115 schematically shows a typical two-phase and a typical three-phase operation for an intersection.

Where possible, alternate solutions to signalization may be practicable. These include (a) revising the intersection layout and/or operation, (b) using other existing streets for making turns, (c) prohibiting certain turns at intersections, and (d) providing special turning geometry for critical turning movements.

Signal Warrants, Pretimed (1) For pretimed signals, certain operations criteria have been established and are nationally accepted practice. Determining whether

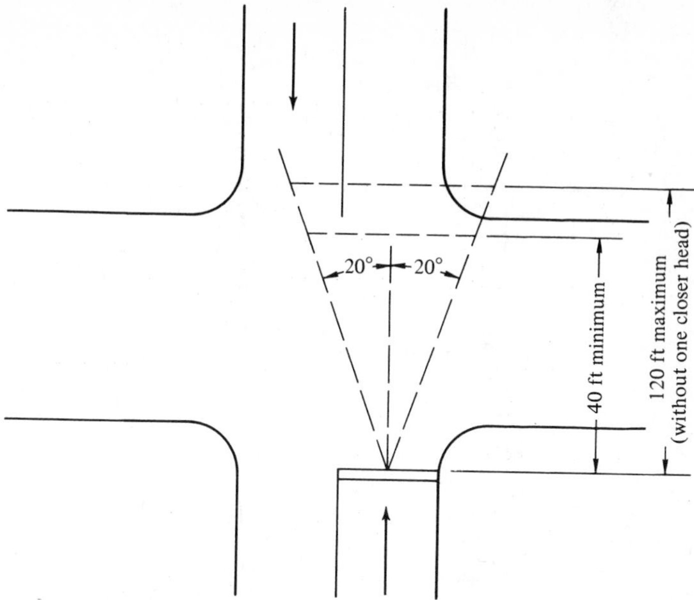

Fig. 114. Desirable location for signal faces[1].

or not a signal should be installed, is based on an evaluation of the extent each of the following warrants is met:

Warrant 1, Minimum Vehicular Volume The Minimum Vehicular Volume warrant is intended for application where the volume of intersecting traffic is the principal reason for consideration of signal installation. The warrant is satisfied when, for each of any 8 hours of an average day, the traffic volumes given in the table below exist on the major street and on the higher-volume minor-street approach to the intersection.

MINIMUM VEHICULAR VOLUMES FOR WARRANT 1

Number of lanes for moving traffic on each approach		Vehicles per hour on major street (total of both approaches)	Vehicles per hour on higher-volume minor-street approach (one direction only)
Major Street	Minor Street		
1	1	500	150
2 or more	1	600	150
2 or more	2 or more	600	200
1	2 or more	500	200

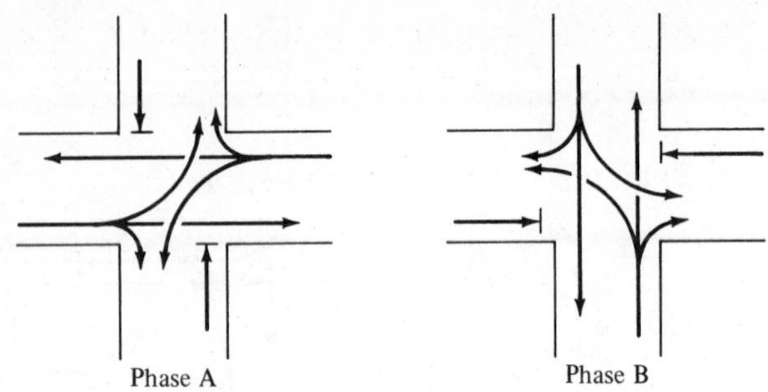

Two-Phase Operation

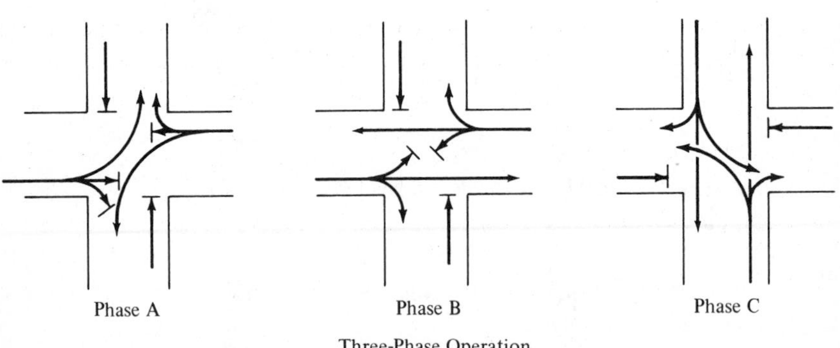

Three-Phase Operation

Fig. 115. Traffic movement during typical two-phase and three-phase signal operation.

These major-street and minor-street volumes are for the same 8 hours. During those 8 hours, the direction of higher volume on the minor street may be on one approach during some hours and on the opposite approach during other hours.

When the 85-percentile speed of major-street traffic exceeds 40 miles per hour, or when the intersection lies within the built-up area of an isolated community having a population of less than 10,000, the minimum vehicular volume warrant is 70 percent of the requirements above (in recognition of differences in the nature and operational characteristics of traffic in urban and rural environments and smaller municipalities).

Warrant 2, Interruption of Continuous Traffic The Interruption of Continuous Traffic warrant applies to operating conditions where the traffic volume on a major street is so heavy that traffic on a minor intersecting street suffers excessive delay or hazard in entering or crossing the major street. The warrant is satisfied when, for each of any 8 hours of an average day, the traffic volumes given in the table below exist on the major street and on the higher-volume minor-street approach to the intersection, and the signal installation will not seriously disrupt progressive traffic flow.

MINIMUM VEHICULAR VOLUMES FOR WARRANT 2

Number of lanes for moving traffic on each approach: Major Street	Minor Street	Vehicles per hour on major street (total of both approaches)	Vehicles per hour on higher-volume minor-street approach (one direction only)
1	1	750	75
2 or more	1	900	75
2 or more	2 or more	900	100
1	2 or more	750	100

These major-street and minor-street volumes are for the same 8 hours. During those 8 hours, the direction of higher volume on the minor street may be on one approach during some hours and on the opposite approach during other hours.

When the 85-percentile speed of major-street traffic exceeds 40 miles per hour, or when the intersection lies within the built-up area of an isolated community having a population of less than 10,000, the interruption of continuous traffic warrant is 70 percent of the requirements above (in recognition of differences in the nature and operational characteristics of traffic in urban and rural environments and smaller municipalities).

Warrant 3, Minimum Pedestrian Volume The Minimum Pedestrian Volume warrant is satisfied when, for each of any 8 hours of an average day, the following traffic volumes exist:

1. On the major street, 600 or more vehicles per hour enter the intersection (total of both approaches); or where there is a raised median island 4 feet or more in width, 1,000 or more vehicles per hour (total of both approaches) enter the intersection on the major street; and
2. During the same 8 hours as in paragraph (1) there are 150 or more pedestrians per hour on the highest volume crosswalk crossing the major street.

When the 85-percentile speed of major-street traffic exceeds 40 miles per hour, or when the intersection lies within the built-up area of an isolated community having a population of less than 10,000, the minimum pedestrian volume warrant is 70 percent of the requirements above (in recognition of differences in the nature and operational characteristics of traffic in urban and rural environments and smaller municipalities).

A signal installed under this warrant at an isolated intersection should be of the traffic-actuated type with push buttons for pedestrians crossing the main street. If such a signal is installed at an intersection within a signal system, it should be equipped and operated with control devices which provide proper coordination.

Signals installed according to this warrant shall be equipped with pedestrian indications conforming to requirements set forth in other sections of this Manual.

Signals may be installed at nonintersection locations (mid-block) provided the requirements of this warrant are met, and provided that the related crosswalk is not closer than 150′ to another established crosswalk. Curbside parking should be prohibited for 100′ in advance of and 20′ beyond the crosswalk. Phasing, coordination, and installation must conform to standards set forth in this Manual. Special attention should be given to the signal head placement and the signs and markings used at nonintersection locations to be sure drivers are aware of this special application.

Warrant 4, School Crossing A traffic control signal may be warranted at an established school crossing when a traffic engineering study of the frequency and adequacy of gaps in the vehicular traffic stream as related to the number and size of groups of school children at the school crossing shows that the number of adequate gaps in the traffic stream during the period when the children are using the crossing is less than the number of minutes in the same period (sec. 7A-3).

When traffic control signals are installed entirely under this warrant:

1. Pedestrian indications shall be provided at least for each crosswalk established as a school crossing.

2. At an intersection, the signal normally should be traffic-actuated. As a minimum, it should be semi-traffic-actuated, but full actuation with detectors on all approaches may be desirable. Intersection installations that can be fitted into progressive signal systems may have pretimed control.

3. At non-intersection crossings, the signal should be pedestrian-actuated, parking and other obstructions to view should be prohibited for at least 100 feet in advance of and 20 feet beyond the crosswalk, and the installation should include suitable standard signs and pavement markings. Special police supervision and/or enforcement should be provided for a new non-intersection installation.

Warrant 5, Progressive Movement Progressive movement control sometimes necessitates traffic signal installations at intersections where they would not otherwise be warranted, in order to maintain proper grouping of vehicles and effectively regulate group speed. The Progressive Movement warrant is satisfied when:

1. On a one-way street or a street which has predominantly unidirectional traffic, the adjacent signals are so far apart that they do not provide the necessary degree of vehicle platooning and speed control, or

2. On a two-way street, adjacent signals do not provide the necessary degree of platooning and speed control and the proposed and adjacent signals could constitute a progressive signal system.

The installation of a signal according to this warrant should be based on the 85-percentile speed unless an engineering study indicates that another speed is more desirable.

The installation of a signal according to this warrant should not be considered where the resultant signal spacing would be less than 1,000 feet.

Warrant 6, Accident Experience The Accident Experience warrant is satisfied when:

1. Adequate trial of less restrictive remedies with satisfactory observance and enforcement has failed to reduce the accident frequency; and
2. Five or more reported accidents, of types susceptible of correction by traffic signal control, have occurred within a 12-month period, each accident involving personal injury or property damage to an apparent extent of $100 or more; and
3. There exists a volume of vehicular and pedestrian traffic not less than 80 percent of the requirements specified either in the minimum vehicular volume warrant, the interruption of continuous traffic warrant, or the minimum pedestrian volume warrant; and
4. The signal installation will not seriously disrupt progressive traffic flow.

Any traffic signal installed solely on the Accident Experience warrant should be semi-traffic-actuated (with control devices which provide proper coordination if installed at an intersection within a coordinated system) and normally should be fully traffic-actuated if installed at an isolated intersection.

Warrant 7, Systems Warrant A traffic signal installation at some intersections may be warranted to encourage concentration and organization of traffic flow networks. The Systems warrant is applicable when the common intersection of two or more major routes has a total existing, or immediately projected, entering volume of at least 800 vehicles during the peak hour of a typical weekday, or each of any five hours of a Saturday and/or Sunday.

A major route as used in the above warrant has one or more of the following characteristics:

1. It is part of the street or highway system that serves as the principal network for through traffic flow;
2. It connects areas of principal traffic generation;
3. It includes rural or suburban highways outside of, entering or traversing a city;
4. It has surface street freeway or expressway ramp terminals;
5. It appears as a major route on an official plan such as a major street plan in an urban area traffic and transportation study.

Warrant 8, Combination of Warrants In exceptional cases, signals occasionally may be justified where no single warrant is satisfied but where two or more of Warrants 1, 2, and 3 are satisfied to the extent of 80 percent or more of the stated values.

Adequate trial of other remedial measures which cause less delay and inconvenience to traffic should precede installation of signals under this warrant.

Pretimed Controllers A pretimed controller is one for which each signal indication, e.g., green, amber, red, and/or special arrows, is preselected for a particular traffic flow condition and that set of indications used repetitively from one signal cycle to another. Advantages include:

1. More precise signal coordination than with actuated equipment.
2. Good speed control through coordination.
3. Operation which does not depend on traffic actuation and as a result is not adversely affected by abnormal flows, stopped vehicles, or construction.
4. Simple and easy maintenance.
5. Relatively low cost.
6. Capability of being programmed to handle most peak flows.

Disadvantages include:

1. Inability to respond to short-term variation in flow.
2. Serious delays during off-peak periods of flow.

Traffic-Actuated Control, Full or Semi- Traffic-actuated equipment relies on detectors buried in the pavement or suspended above it to respond to the presence of traffic and cause the signal indication to change to accommodate the intersection approach traffic. Advantages include the following:

1. Traffic-actuated controllers provide greater efficiency at intersections where:
   a. Traffic fluctuation cannot be anticipated or programmed in advance for pretimed control.
   b. The intersection is complex and certain movements are sporadic.
   c. Cross-street flows are minor and major-street interruptions are to be held to a minimum.
   d. A progressive pretimed system provides overall operational control and there is need to provide signalization at an unfavorably located intersection (in this case, interruptions to the major street flow will be held to a minimum through the use of traffic-actuated control).
2. Useful where signal control is only warranted for a few brief hours during the day.
3. Hazards associated with the arbitrary stopping of vehicles, which occurs for the pretimed type of control, are reduced for traffic-actuated control.
4. Increased major street capacity and reduced overall delay.
5. Applicable to multiphase applications.

Disadvantages include:

1. High cost.
2. Complicated equipment and complicated design procedures.

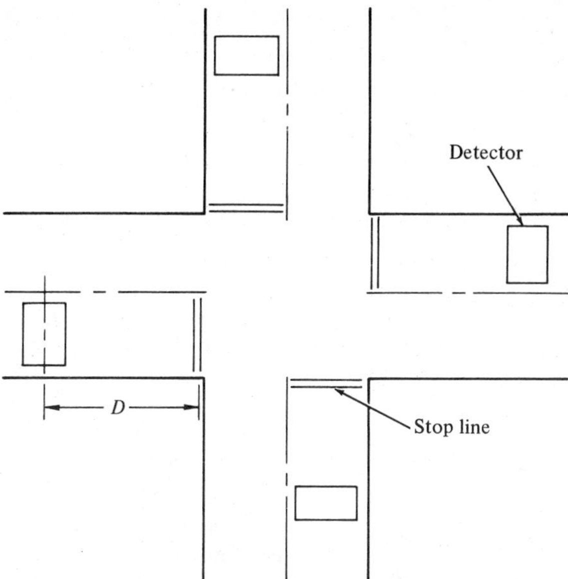

Fig. 116. Typical actuated signal site diagram[2]. D = detector-to-stop-line distance, in feet. Typical values: 100 ft at 25 mph operation; 180 ft at 40 mph operation. Detector placed to result in approximate 3-sec travel time (common amber time) from detector to stop line.

Actuated Signal Operation Figure 116 shows a four-leg intersection having fully actuated control, i.e., actuated control on each approach leg. Although considerable sophistication is possible, the basic actuated signal operation consists of a green period divided into two portions, an initial period, consisting of a fixed time length, and a vehicle period, which is an extendable time period that is renewed each time a vehicle crosses a detector within a certain preset time period or gap.

A maximum green time is also set for each approach so that when a vehicle is detected on the street having the red indication, the other street's green will end at the maximum time setting even if vehicles continue to arrive on it within the vehicle period. The maximum time begins to cycle when a vehicle is detected on the street having the red indication.

Traffic-Adjusted Control Traffic-adjusted control makes use of some of the best features of pretimed control and traffic-actuated control. In such a system, detectors are placed at representative locations to sample traffic and provide a master controller with regular information on traffic at each of these points. The controller then supervises the system of pretimed and/or actuated controllers by selecting a particular signal operational mode which best serves existing traffic.

Signal Timing Definitions

Cycle Time: time to complete one sequence of signal indications.

Phase: that part of a cycle allocated to a traffic movement or combination of movements which receive the designated right of way.

Interval: any one of several divisions of a time cycle during which a signal indication does not change.

Yellow or Warning (Clearance) Interval: the signal indication displayed immediately following the green interval and preceding the red interval.

All Red Interval: a time during which the red indication is displayed on all approaches following the clearance interval. The all red permits additional time for vehicles or pedestrians to clear large intersections and minimizes the need to use an excessively long yellow interval.

Types of Timing Systems

Isolated: a system in which each individual signal is timed independently of every other.

Progressive System: a system in which the time relationship between adjacent signals is coordinated to permit continuous operation of groups of vehicles at a planned rate of speed.

Timing Isolated Signals: Determining Cycle Length Several procedures are available for determining signal cycle timing requirements. Two of the more common ones are: (a) the determination of a green time–cycle time ratio for each approach using a capacity analysis procedure (4) and (b), a method recommended by the Institute of Traffic Engineers (ITE) (3).

ITE Signal Timing Method Determining the total cycle length requires a determination of the green time required for each approach and the necessary corresponding clearance or amber time for each approach. Minimum green times are based on pedestrian walking requirements and a practical vehicle operational minimum of approximately 15 sec. Beyond these minima, the necessary green times are determined by proportioning green times between both approaches so that the ratio between capacity and demand on each of the critical approaches is approximately the same and such that the green interval is of sufficient length to accommodate most of the traffic arriving during the peak period.

The ITE procedure for determining signal timing requirements is as follows:

1. Select a tentative phasing scheme for the intersection.
2. Compute the necessary clearance interval for each approach and select the longest required for each phase. The purpose of the clearance interval is: (a) to advise that the green is coming to an end and permit a vehicle to make a safe stop and (b) to allow a vehicle in the intersection to pass safely through. Table 64 lists clearance intervals necessary to satisfy requirements a and b for different approach speeds and different street widths.

TABLE 64. THEORETICAL MINIMUM CLEARANCE INTERVALS FOR DIFFERENT APPROACH SPEEDS AND CROSSING STREET WIDTHS* (3).

Approach speed (mph)	Minimum time to stop, y_1 (sec)	Minimum time to stop or clear intersection, y_2 (sec)				
		$w = 30$†	$w = 50$†	$w = 70$†	$w = 90$†	$w = 110$†
20	2.0	3.8	4.4	5.6	5.7	6.4
30	2.5	3.6	4.1	4.5	5.0	5.5
40	3.0	3.9	4.2	4.5	4.9	5.2
50	3.4	4.1	4.4	4.7	5.0	5.2
60	3.9	4.5	4.7	4.9	5.1	5.4

*Obtained from the formulas:

$y_1 = t + \frac{v}{2a}$ and $y_2 = t + \frac{v}{2a} + \frac{(w + l)}{v}$, when $t = 1$ sec, $a = 15$ ft per sec per sec, and $l = 20$ ft.

†Crossing street width in feet.

3. Compute the minimum green time required to accommodate pedestrians. Minimum green time for pedestrian crossing is normally based on a 5-sec walking speed. The minimum green time is therefore:

$$G_{\min} = 5 + \frac{d}{4} - y$$

where

G = the green time in seconds
d = the length of the longest crosswalk in feet
y = the associated clearance time in seconds.

Fifteen seconds is normally taken as a practical operational minimum; the minimum green time selected for the signal installation would be taken as the larger of the values calculated for the two approaches.

4. Calculate the capacity for each approach on the basis of a *full hour of green time.* This should be done using the service volume–capacity evaluation procedure noted in the sub-section on capacity analysis of signalized intersections.
5. Based on traffic counts for the intersection, obtain the average hourly volume for each approach for the period under consideration, normally the peak hour for a typical week day.
6. Calculate the ratio of the volume to the design capacity for each approach and determine the critical or highest ratio for each street.
7. For the phase having the lowest critical ratio, assign the green time determined in step 3.
8. The green times for the remaining phases are determined by multiplying the

green time determined in step 7 by the ratio of the second-phase critical ratio to the first-phase critical ratio.

Limitations of existing control equipment, particularly where mechanical equipment is used, may require adjusting the times obtained in steps 1-8.

9. Summing the clearance intervals and green times obtained in steps 2, and 7 and 8, respectively, will provide a total minimum cycle time.

The minimum cycle lengths obtained by using the previous procedure are just that and do not assure green times long enough to accommodate most traffic approaching the signal during a given cycle.

For random traffic flows, an assumption which is not completely valid, particularly at higher flow rates, is that a Poisson probability distribution of arrivals can be used to determine a level of performance for each approach. Use of the Poisson probability distribution of arrivals indicates the percentage likelihood that a certain number of vehicles will arrive during a given signal cycle time. By knowing how many vehicles are likely to arrive 90% of the time, for example, (90% of the cycles can handle the demand present) one can determine how long the green has to be to handle those vehicles. Figure 117 provides a procedure for determining the green time required per cycle in order to accommodate the approach demand *per lane* at a given probability of performance. Supplementing the prior determination of minimum green time, the procedure is as follows:

10. For the desired level of performance (e.g., 90% of the cycles are to serve the intersection demand) enter the upper portion of the figure with the critical *per lane* hourly volume for each phase. Utilizing a trial cycle time, reflect vertically to desired level of performance and then horizontally from the level of performance vane to the corresponding design capacity in vehicles per cycle (or directly to the required green time per cycle to accommodate the demand volume for the given cycle length) at the specified level of performance.

The required green time per cycle is composed of an initial starting delay for the first vehicle and a uniform average headway for successive vehicles. Values typically used are a 2.5-sec starting delay (3.7 sec was used in developing Fig. 117) and 2.1 sec for each vehicle headway. Hence if ten vehicles are at a signal they would require 23.6 sec of green [2.5 + 10(2.1)].

Sample Problem Determine the required signal timing necessary to handle the demand volumes noted in Fig. 118 at a 90% level of performance. Additional data are noted with the figure.

Solution Will be solved to nearest 10 sec total cycle length with individual interval times to the nearest second.

Step 1: Assume two-phase signal will be used

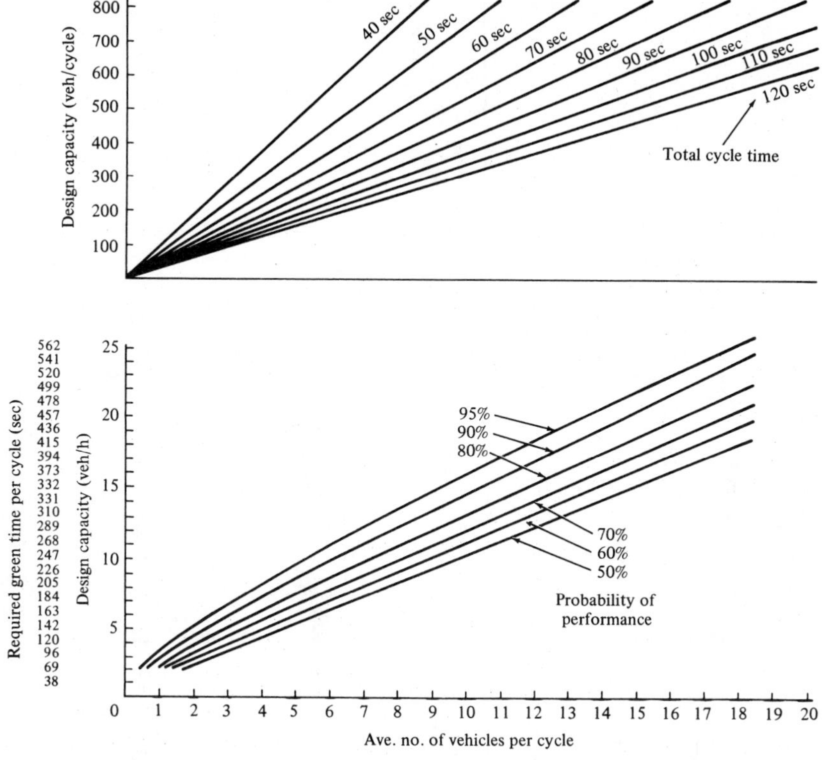

Fig. 117. Cycle capacity probability design curves for use in traffic signal timing[5].

Step 2: Clearance Interval (Yellow)

Aye Street: 3.98 sec from Table 64, using 25 mph and 40-ft width (Bee Street crossing); use 4.0 sec.

Bee Street: 4.19 sec from Table 64 using 25 mph and 48-ft width (Aye Street crossing); use 4.0 sec.

Step 3: Minimum Green Time (for pedestrians)

$$G_{\text{min}} = 5 + \frac{d}{4} - Y$$

Aye Street:

$$G_{\text{min}} = 5 + \frac{40}{4} - 4 = 11 \text{ sec}$$

Bee Street:

$$G_{\text{min}} = 5 + \frac{48}{4} - 4 = 13 \text{ sec}$$

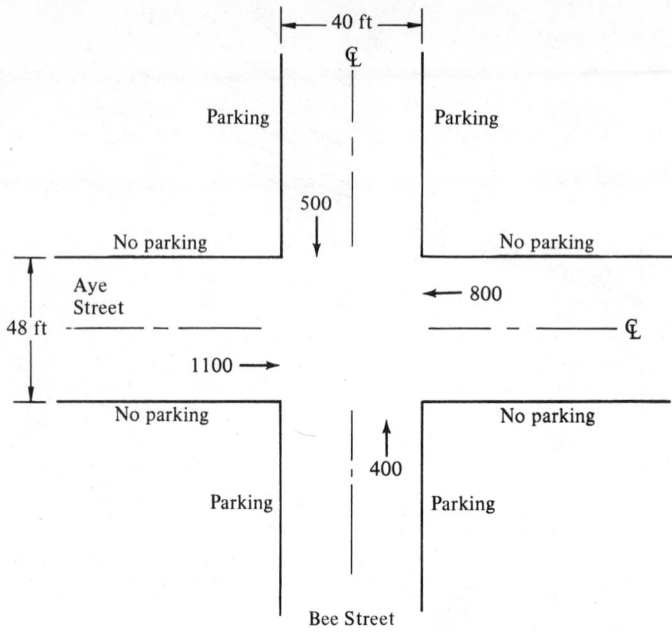

Fig. 118. Signalization problem. Volumes are total and include turns.

Area population: 250,000
Peak-hour factor (PHF): 0.85
10% left turns
10% right turns
5% trucks and through buses
No local transit buses
Speeds: 25 mph.

Assume minimum green on each approach as 15 sec (operational minimum). Note that width *d* used in calculation is crosswalk (cross-street) width.

Step 4: Capacity for full hour of green (at load factor = 1.0 for level of service E)* From capacity charts:

Aye Street Capacity: 2150 vph
Bee Street Capacity: 1150 vph

Step 5:

Aye Street Volume: 1100 vph
Bee Street Volume: 500 vph

Step 6:

[1] Aye Street Volume/Capacity = 1100/2150 = 0.512
[2] Bee Street Volume/Capacity = 500/1150 = 0.435

*Note that similar but not identical end results would be obtained using design capacities (at level of service C rather than actual).

Step 7: Assign 15-sec minimum green time to Bee Street since it has the lowest critical volume–capacity ratio.

Step 8: Assign to Aye Street a minimum green time in proportion to the ratio of the volume–capacity ratios, i.e., 1.18 × 15 = 17.7 sec. Use 18 sec.

Step 9: Total Minimum Cycle Time

	Aye Street	Bee Street
Green	18	15
Yellow	4	4
	22	19

Cycle: 41 sec

Aye St.	$G = 18$	$Y = 4$	Red (19 sec)	
Bee St.	Red (22 sec)		$G = 15$	$Y = 4$

Cycle = 41 sec

Step 10: Green times necessary for 90% level of performance:

Aye Street: Per lane demands without parking assumed as one-half of total demand or 1100/2 = 550 vph. Enter Fig. 116 with 550 vph; reflect downward off 40-sec cycle-length vane to 90% performance line; reflect horizontally to signal design capacity of 10 vehicles with a required green time of 25 sec. *Bee Street:* Demand on one moving lane, 500 vph. Enter Fig. 116 with 500 vph. Time required is 23 sec to handle 9 vehicles.

Summing the green times obtained in step 10 and the yellow clearance times previously obtained results in a total required cycle time of 56 sec. This is based on the time required to serve, 90% of the time, those vehicles which arrive in a random fashion (Poisson) during a 40-sec interval. Obviously the total service cycle time supplied must be the same as the time required or that during which the vehicles arrive, i.e., the 40-sec time and 56-sec time must agree. A series of trial solutions will draw these two times together.

A step 10 solution for varying cycle times is given in Table 65.

When the cycle time supplied equals the cycle time required, a solution has been obtained. This occurs at a 100-sec cycle time. Note that the green times on Aye Street and Bee Street, 48 sec and 44 sec, respectively, are in the ratio of

TABLE 65. GREEN TIME REQUIREMENTS FOR 90% LEVEL OF PERFORMANCE IN SAMPLE PROBLEM.

Cycle supplied	40	70	100	120
Required green times				
Aye Street	25	36	48	56
Bee Street	23	34	44	51
Yellow time	8	8	8	8
Cycle required	56	78	100	115

1.09, not quite that determined by the demand-capacity analysis in step 6. Nonetheless the solution is acceptable since it does meet the more important 90% performance requirement. Use of slightly longer total cycle length would permit use of the 1.18 ratio of step 6. For example, with a 120-sec cycle and two 4-sec yellow clearance intervals the respective greens for Aye Street and Bee Street which would provide the step 6 ratio are 60.6 sec and 51.4 sec. Since these values exceed those given in Table 65 for a 120-sec cycle such times actually provide a higher than 90% level of performance.

Progressive Systems A progressive signal system is one in which adjacent signals operate in a planned time sequence, permitting uninterrupted flow of groups of vehicles through a number of successive signals.

Definitions, Progressive System:

Progression: Time relationship between adjacent signals permitting continuous operation of groups of vehicles at a planned rate of speed

Offset: Number of seconds or percentage of time cycle that the green (or other) indication appears at a given signal in relation to a "zero" time base

Through Band: Time between first and last vehicle in a group at a designed speed of progression

Types of Progressive Systems

Simultaneous. A system in which all signals display their green at essentially the same time. At closely spaced intersections little or no progression is possible, stops are frequent and speeding is encouraged. Figure 119 shows a simultaneous progressive system for equally spaced blocks in which the split is 50-50 and a common cycle time has been used. The speed of progression is given as

$$V = D/1.47C$$

where

V = speed, in miles per hour
D = block spacing, in feet
C = cycle length, in seconds

Alternate (single or double). A system in which individual signals or groups (pairs) of signals give opposite indications at the same time. In a single alternate system, each signal alternates with those immediately adjacent. In the double alternate, pairs of signals alternate with adjacent pairs. Alternate systems may give excellent progressions depending on the block spacing. Figure 120 shows a single alternate progressive system for equally spaced blocks using a common signal time split of 50-50. The speed of progression is given as:

$$V = D/0.735C$$

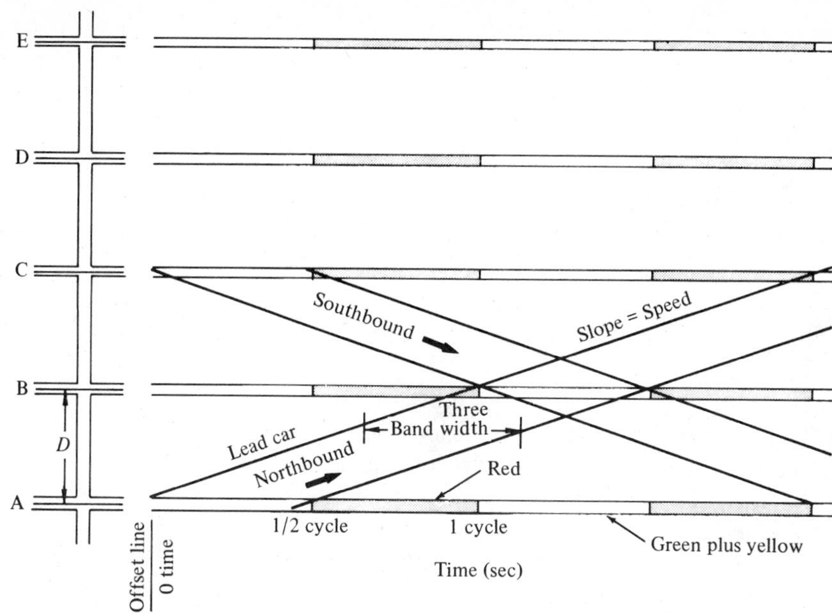

Fig. 119. Simultaneous signal system. Note: For the same block length, the speed of progression is half that of a single alternate system.

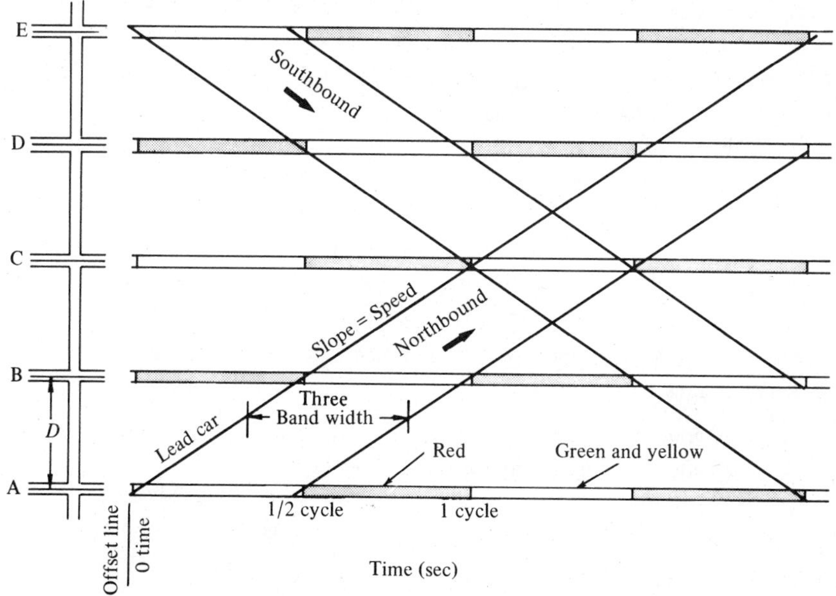

Fig. 120. Single alternate. Speed-slope of line is progression speed; normally the 85th percentile speed.

where

V = speed, in miles per hour
D = block spacing, in feet
C = cycle length, in seconds

Limited Progressive. Also known as a simple progression, the greens of the common cycle time are coordinated to permit continuous movement at a preplanned rate of speed. Different parts of the system may, however, have different progression speeds and block spacings are not uniform throughout the system. Figure 121 shows a series of three different limited progressive systems which make up a flexible progressive system when combined.

Flexible Progressive. This system provides a series of separate limited progressive systems each of which can be tailored to the demands of different periods. For example, Fig. 121 illustrates a flexible progressive system which by varying cycle time, offset, and split, or any combination thereof, accommodates a different flow need.

Signing and Marking

Traffic Signs and Markings (1) Signs and markings are used to regulate traffic, warn of potentially hazardous conditions or guide the motorist. To be effective, signs must be uniform in color, shape, message (word or symbol), and installation. The general sign colors and their meanings are as follows:

Red:	stop or prohibition
Black:	regulatory
White:	regulatory
Yellow:	warning
Orange:	maintenance and construction
Blue:	motorist services
Green:	permitted movements and guidance
Brown:	recreational, cultural, scenic

Sign shape plays an important role in conveying the sign's meaning. The standard sign shapes are as follows:

Octagon:	stop sign
Equilateral triangle, point down:	yield
Round:	railroad crossing, advance warning, civil defense evacuation route
Pennant, isosceles triangle on side:	no passing
Diamond:	hazard

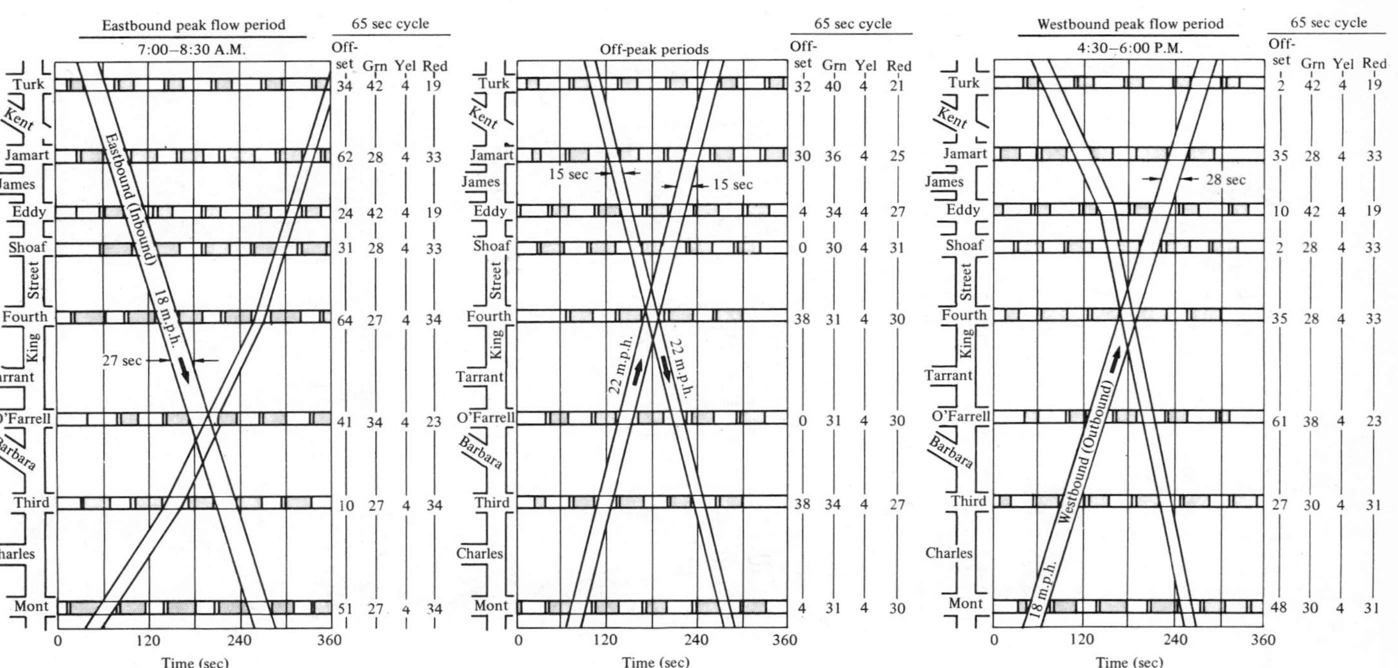

Fig. 121. Three time–space diagrams for a hypothetical street, illustrating the way a flexible progressive system may be timed to favor "inbound" traffic during the morning peak flow, "outbound" during the afternoon peak, and both directions equally during off-peak periods[3].

Fig. 122. Standard roadway signs. Colors in first line, official sign designation code on second line, size on third line[1].

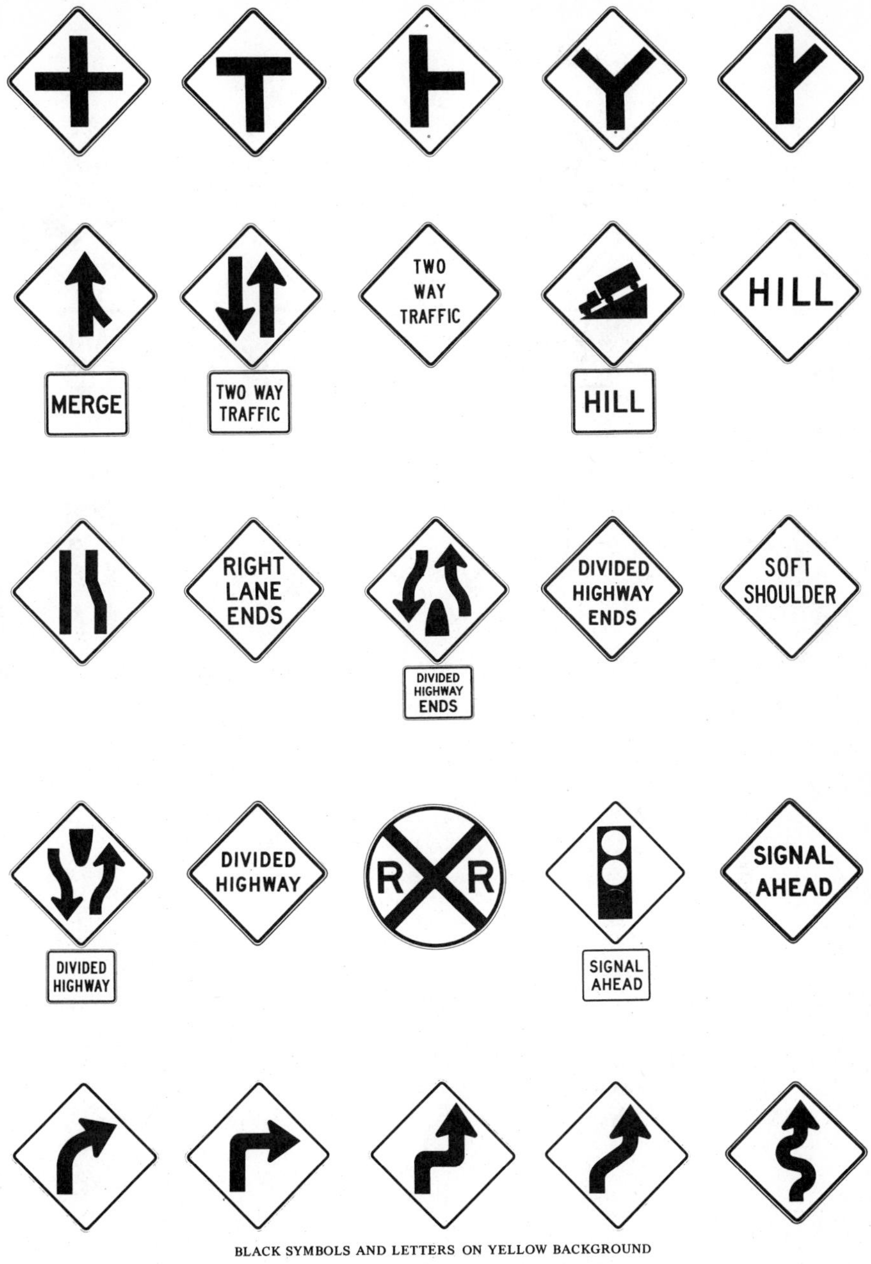

BLACK SYMBOLS AND LETTERS ON YELLOW BACKGROUND

Fig. 122 (*Continued*)

Fig. 122 (*Continued*)

Rectangle:	regulatory, guide
Pentagon-point up:	school crossings and advance warning
Trapezoid, short side down:	recreation area signs
Miscellaneous shapes:	other shapes are reserved for such things as route markers, railroad crossing cross-bucks, etc.

Apart from the message, general or specific, conveyed by a sign's color or shape, a word or symbol (or combination) is used to convey a specific meaning. Greater use of symbology is desirable because it permits more instant and universal understanding.

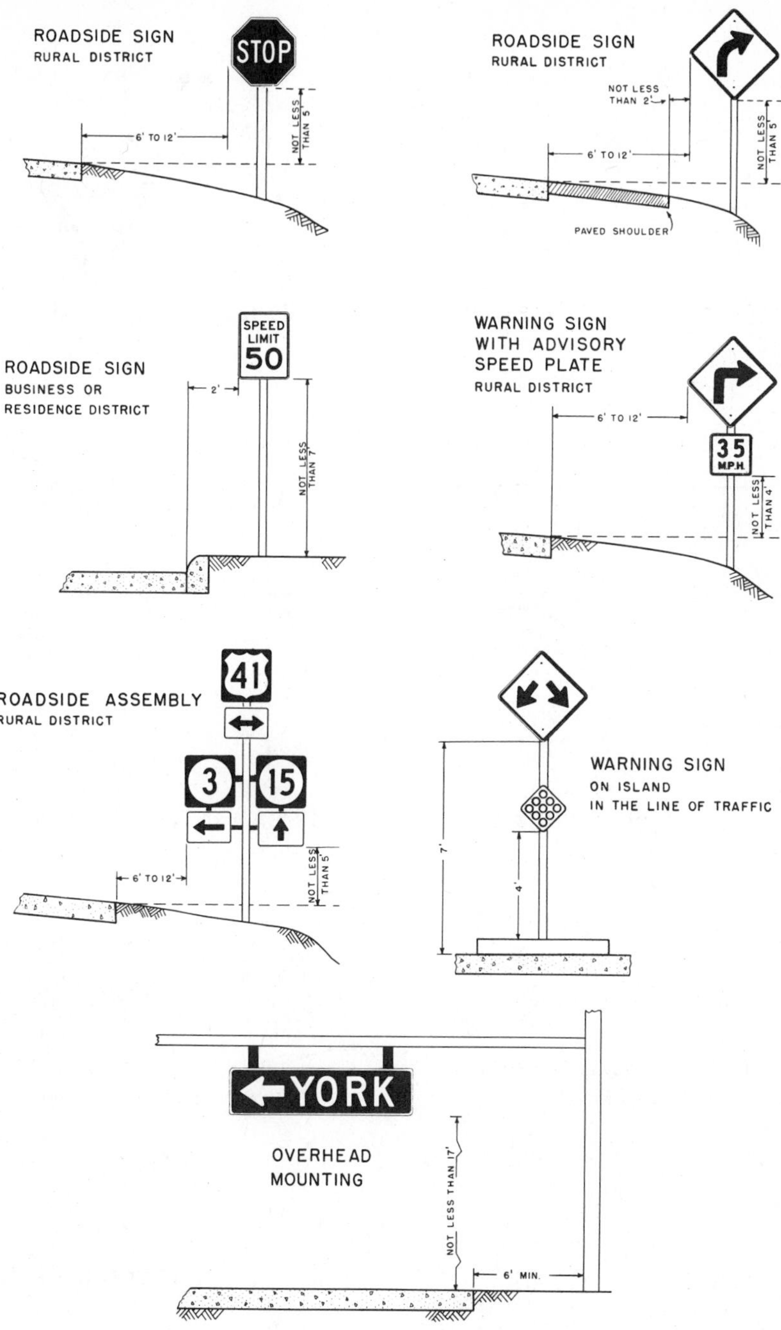

Fig. 123. Height and lateral location of signs, typical installations[1].

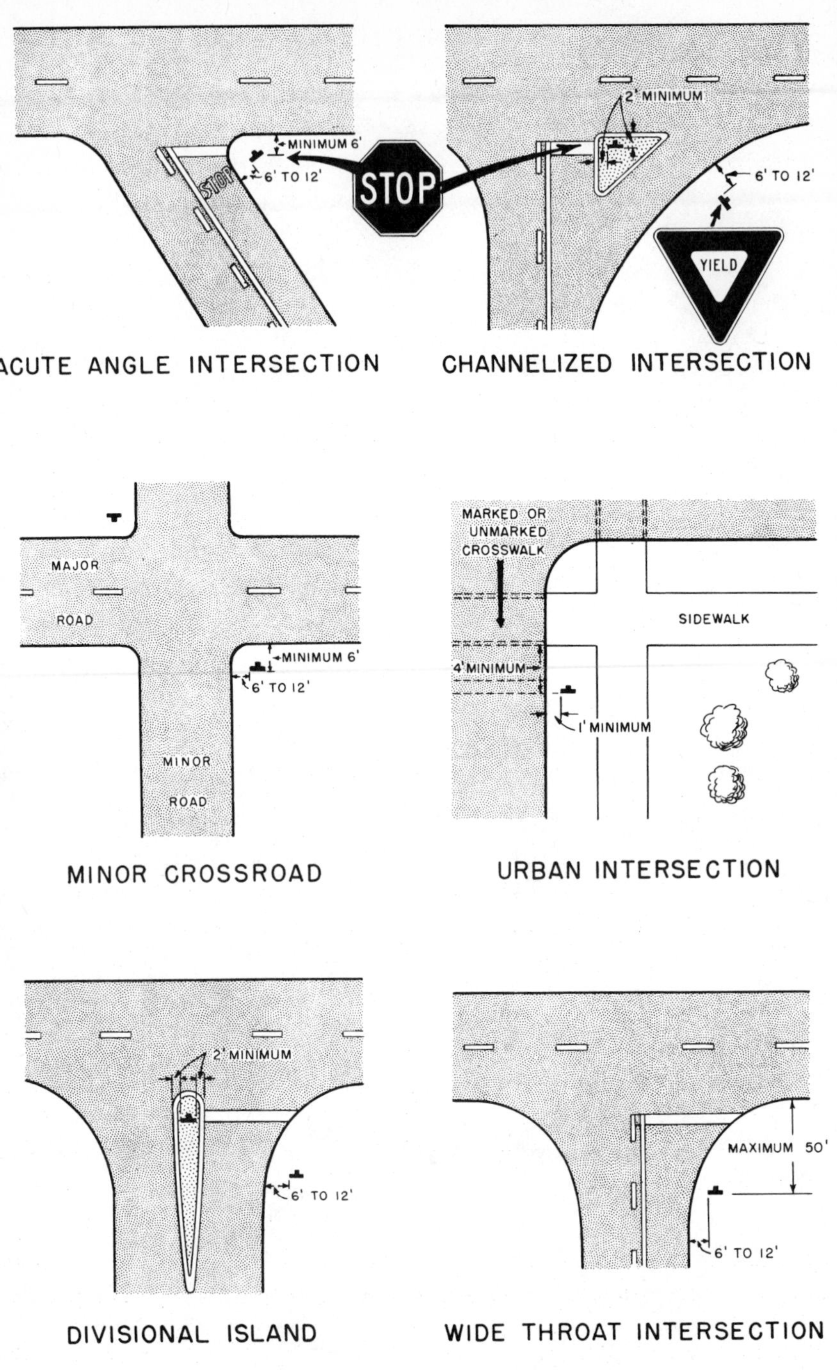

Fig. 124. Typical locations for stop signs and yield signs[1].

Typical Signs Figure 122 displays typical examples of newly approved road signs. Where symbology is being used for the first time the sign will have both a word and symbol message to permit public education during the sign transition period. Eventually, the word messages will be removed, leaving only the glance-legible symbol message. Figures 123 and 124 show typical height and lateral location criteria for sign placement and the typical location for stop and yield signs.

Marking Roadway markings supplement other controls or, in some instances, stand alone as means of regulating, warning, or guiding traffic. The most common marking materials are paints and plastics, with white, yellow, and red the normal marking colors. Black may be used in conjunction with either of these colors as a means of achieving good color contrast.

Most marking is of a longitudinal character; longitudinal markings should conform to the following basic concepts (1):

1. Yellow lines delineate the separation of traffic flows in opposing directions or mark the left boundary of the travel path at locations of particular hazard.
2. White lines delineate the separation of traffic flows in the same direction.
3. Red markings delineate roadways that shall not be entered or used by the viewer of those markings.
4. Broken lines are permissive in character.
5. Solid lines are restrictive in character.

(a) Typical two-way marking with a reversible center lane.

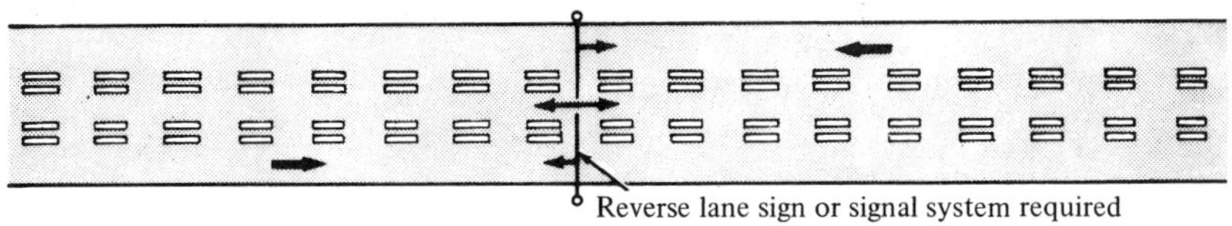

(b) Typical two-way marking where motorists in a single lane are permitted to pass.

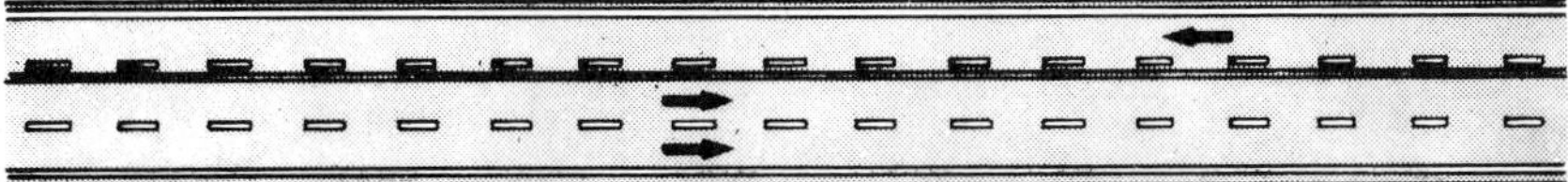

(c) Typical two-way marking where motorists in a single lane are not permitted to pass.

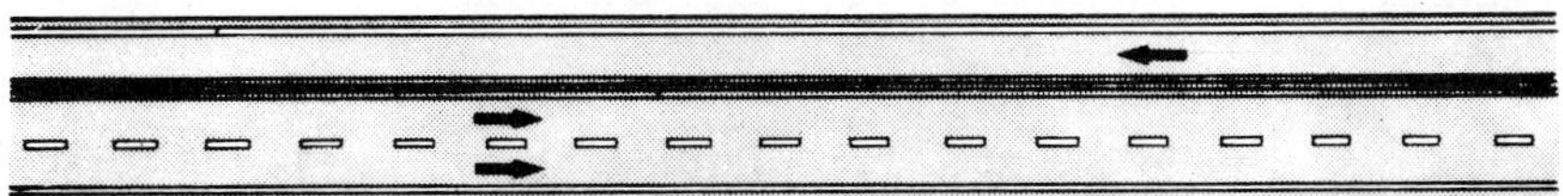

Fig. 125. Typical two-way marking applications[1].

6. Width of line indicates the degree of emphasis.

7. Double lines indicate maximum restrictions.

8. Markings which must be visible at night shall be reflectorized unless ambient illumination assures adequate visibility.

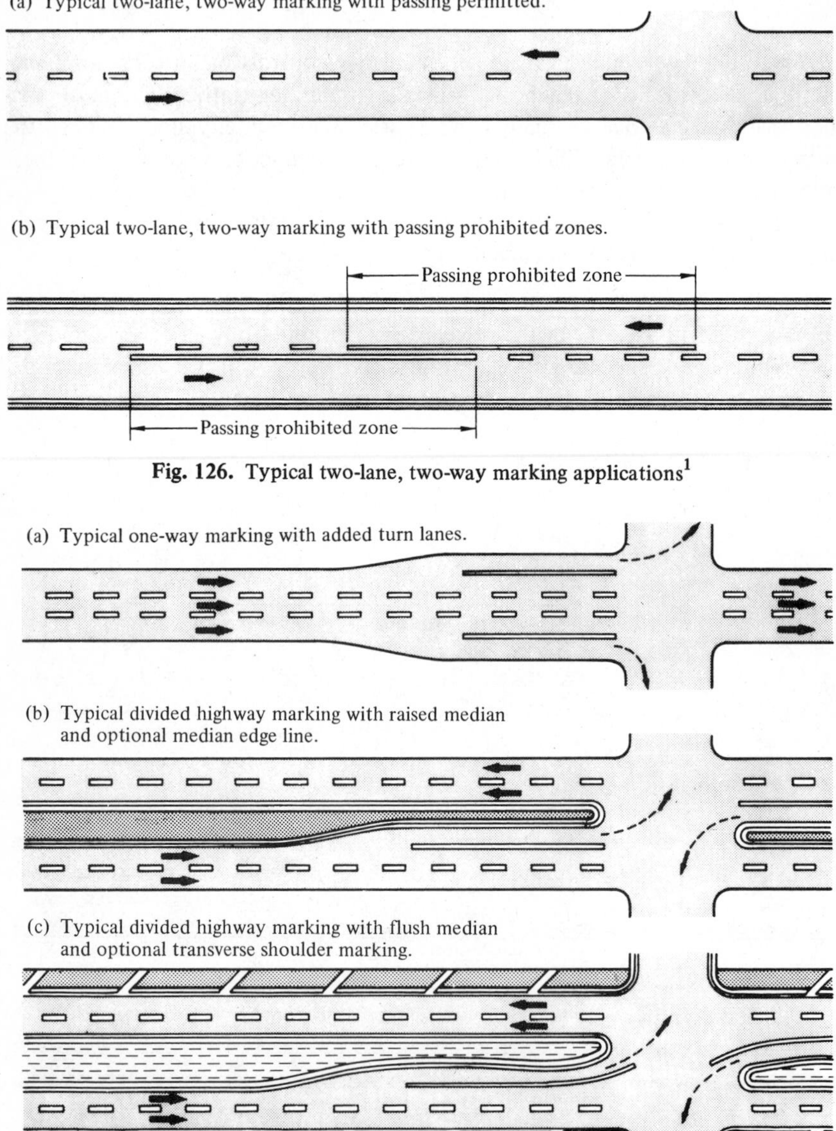

(a) Typical two-lane, two-way marking with passing permitted.
(b) Typical two-lane, two-way marking with passing prohibited zones.
Passing prohibited zone
Passing prohibited zone
(a) Typical one-way marking with added turn lanes.
(b) Typical divided highway marking with raised median and optional median edge line.
(c) Typical divided highway marking with flush median and optional transverse shoulder marking.

Fig. 126. Typical two-lane, two-way marking applications[1]

Fig. 127. Typical one-way and divided-highway marking applications[1].

Vertical Curve

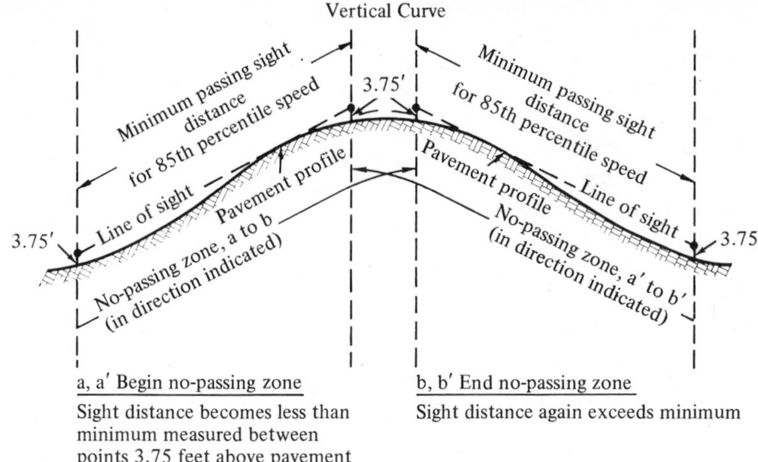

a, a′ Begin no-passing zone

Sight distance becomes less than minimum measured between points 3.75 feet above pavement

b, b′ End no-passing zone

Sight distance again exceeds minimum

Note: No-passing zones in opposite directions may or may not overlap depending on alignment.

Horizontal Curve

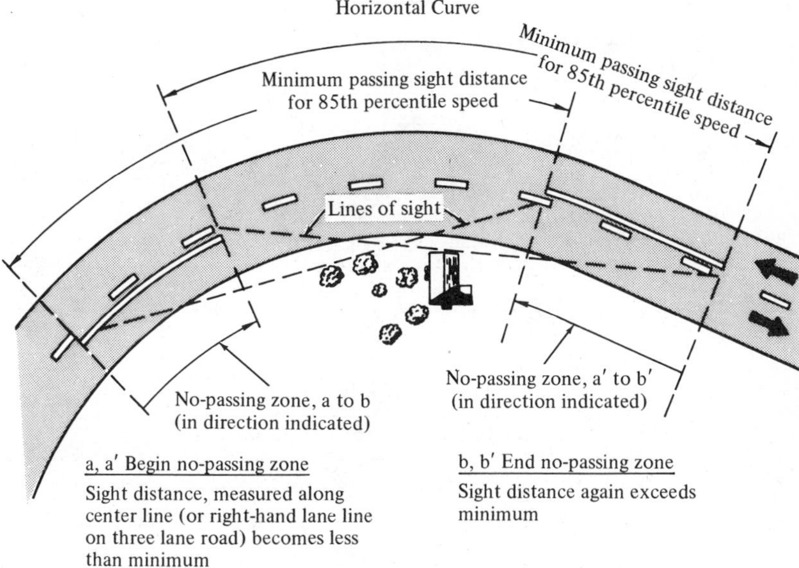

a, a′ Begin no-passing zone

Sight distance, measured along center line (or right-hand lane line on three lane road) becomes less than minimum

b, b′ End no-passing zone

Sight distance again exceeds minimum

Note: No-passing zones in opposite directions may or may not overlap, depending on alignment.

Fig. 128. Method of locating and determining the limits of no-passing zones at vertical and horizontal curves[1].

Common types of lines include:

Longitudinal:

1. Lane lines.
2. Center line.
3. Pavement edge lines.
4. No passing zones.
5. Lane reduction transitions.
6. Channelizing lines.
7. Median markings.

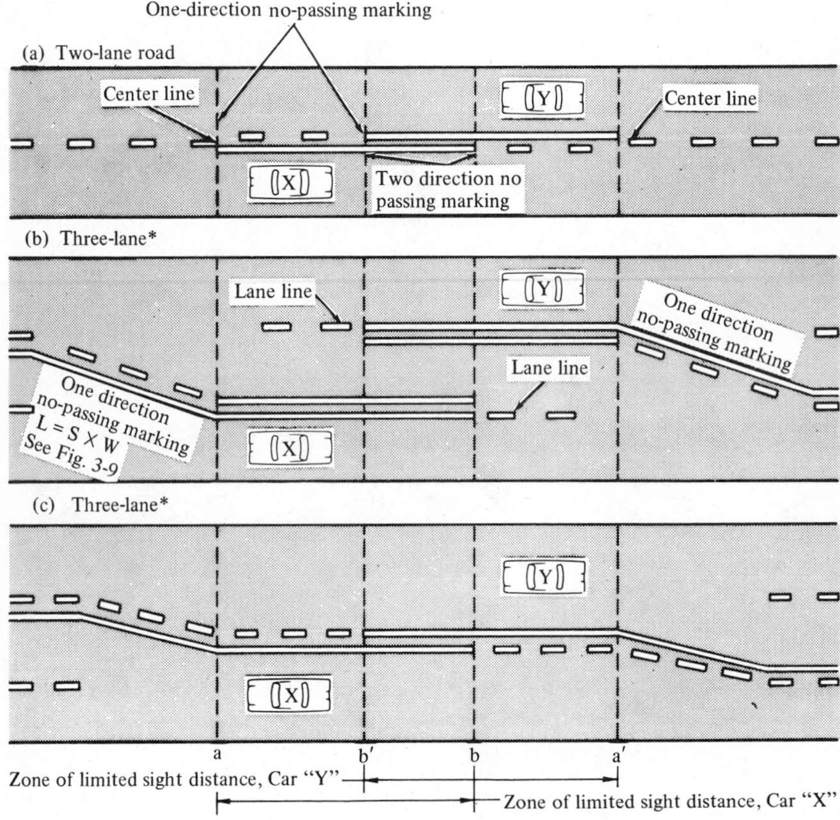

*This marking may also be used to alternate the preferred direction of 2 lanes on a 3 lane highway. In this case b to a′ & a to b′ = 300′, b′ to b = 50′.

Fig. 129. Standard pavement marking for no-passing zones[1].

8. Interchange ramp markings.
9. Approach to obstructions.
10. Curb marking.

Transverse:

1. Stop lines.
2. Crosswalks.
3. Railroad cross.
4. Parking space.

Figures 125 through 130 illustrate some typical marking practices.

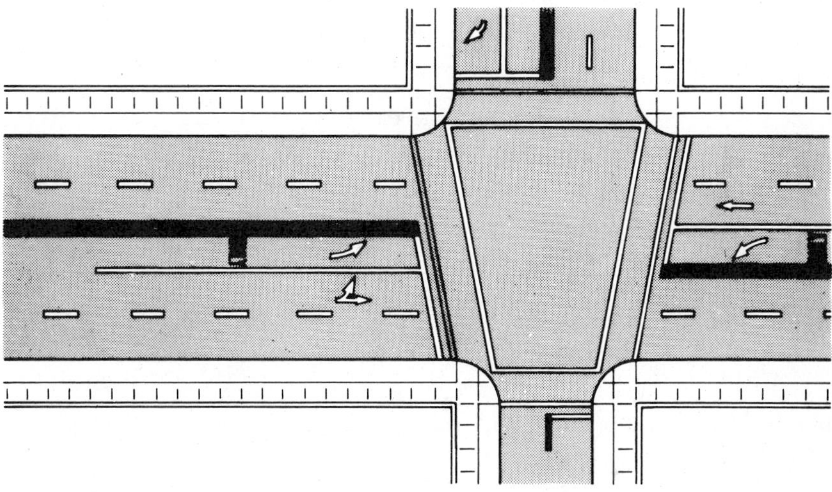

Fig. 130. Typical lane-use control word and symbol marking[1].

REFERENCES

1. *Manual on Uniform Traffic Devices for Streets and Highways,* U.S. Department of Transportation, Washington, D.C. (1971).
2. Rodgers, Lionel M. and Sands, Leo G., *Automobile Traffic Signal Control Systems,* Chilton Book Co., Philadelphia, Pa. (1969).
3. *Traffic Engineering Handbook*, Institute of Traffic Engineers, Washington, D.C. (1965).
4. *Highway Capacity Manual–1965,* HRB Special Report No. 87, Highway Research Board, Washington, D.C. (1965).
5. Davidson, Bruce M., Traffic Signal Timing Utilizing Probability Curves, *Traffic Engineering*, Volume 32, No. 2 (November 1961).
6. Pignataro, Louis J., "Traffic Engineering: Theory and Practice," Prentice-Hall, Inc., Englewood Cliffs, N.J., (1973).

MASS TRANSPORTATION—GROUND, INTRACITY

Over the last quarter century there has been a continued decline in mass transit usage. Figure 131 shows national urban transit patronage from 1935 through 1968. The decline in transit usage from a peak during World War II to the present has been a very notable one. This decline has been spurred by a number of developments, among them:

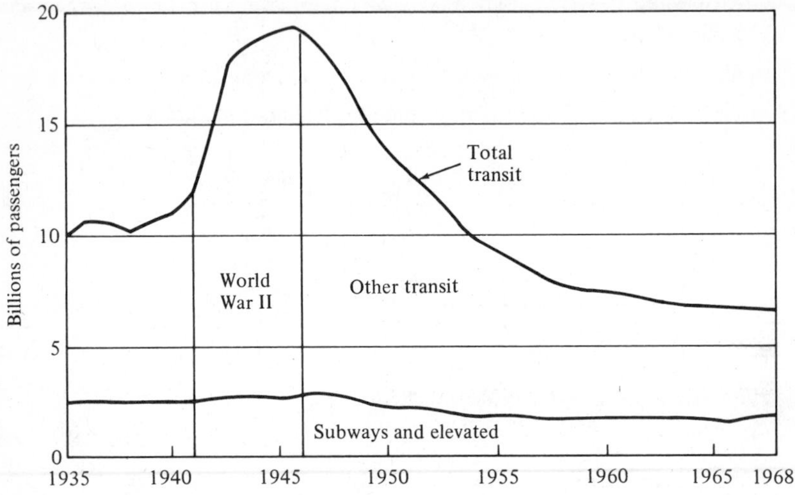

Fig. 131. Urban transit patronage (American Transit Association).

1. Increased availability and use of the private automobile.
2. Low density urban sprawl, which necessitated the use of the automobile and made service by mass transportation more difficult.
3. Financial difficulty of the transit companies resulting in reduced service and continued decline.
4. Changing work and recreational patterns.
5. High peak-period demand requiring large equipment and labor outlays coupled with low off-peak use and costly underutilization of equipment and labor.

Present Transit Use

Today, in cities over 50,000 population, 96% of all daily passenger trips are made by automobile, 3% by bus, and the remaining 1% by rail rapid transit (3). New York City, however, accounts for a large share of the total transit—

approximately one-third of the total public transit riding nationally and about 81% of all of the rail transit patronage in the nation (1).

Future Transit Use

Except for a few unique cities, current mass transit use and technology appears to favor the motor bus because of the following advantages:

1. Flexibility: ability to adapt to unforeseen changes in development, travel demand and patterns, and changes in technology.
2. Dual service capability: the same vehicle can be used for line-haul and feeder (local) service.
3. Lower capital costs.
4. Earlier implementation: rail systems often require 10 or more years of lead time between conception and implementation.
5. High volume capability: buses on exclusive rights-of-way can handle single-lane demands up to 30,000-40,000 passengers per hour (1) (4).
6. Demand routing potential (dial-a-bus): buses have the capability of demand-responsive routing—a type of service in which the rider "telephone orders" a bus, and the bus is dispatched by radio or some other means to the rider's door or nearest corner for pickup at a certain time.

Criteria for Rail Rapid Transit

Based on studies of existing and proposed rail rapid transit systems (1), the following are suggested as criteria for justifying the development of rail rapid transit:

1. High threshold population: metropolitan area population upwards of 1.5 million persons. Cleveland, Ohio, with a population of 1.8 million, is the smallest metropolitan area to have a rail rapid transit system.
2. High central city density: 14,000-20,000 persons per square mile.
3. Large downtown employment: 100,000 persons or more.
4. A central business district: one which acts as a regional center with a total central business district floor area of at least 50,000,000 square feet and central business district destinations of 300,000 or more persons per square mile.
5. High-density travel corridor: one which provides daily destinations of 70,000 persons or more to the central business district by all modes of travel.
6. Historical pattern of transit usage in dense high-volume corridors: one which will provide a built-in demand for rail rapid transit.
7. Linear urban development.

Predicting Demand

Current procedures for predicting mass transit usage rely generally upon an extrapolation of past historical patterns or the use of modal split models which (based on characteristics of the transit system, the network, the rider or potential user, and the trip) attempt to predict transit usage.

Improving Transit Usage

Although there has been a general decline historically in transit use, a number of things can be done to improve the use of mass transit and in some areas even reverse that decline. These include:

1. Improved amenities:
   a. Use of enclosed and heated bus shelters in areas of heavy demand (Fig. 132).
   b. Attractive bus stop signing (Fig. 133).
   c. Posting of schedules at bus stops.
2. Service and fare structure:
   a. Frequent service during peak travel periods and adequate service during off peaks, including 24-h service on certain lines.
   b. Fares structured to promote ridership: fare reductions through the use

Fig. 132. Closed and heated bus stop (Simpson & Curtin).

of subsidies to promote ridership during peak periods and make mass transit a more attractive alternative to the automobile. In addition, such things as reduced fares to certain groups, e.g., students and the elderly, reduced fares to all groups during off-peak periods to promote additional off-peak travel, and graduated fares based on length of ride with lower per mile costs for longer trips.

3. Other:
   a. Peripheral parking: facilities at no cost to the user at originating and terminating points of line-haul rapid-movement systems have proven successful in a number of areas. The City of Milwaukee, Wisconsin, has experienced considerable success with its Freeway Flyer Operation, where major shopping center parking lots are used as the peripheral

Fig. 133. Attractive bus stop signing (Simpson & Curtin).

parking pickup points for buses which use the freeway system for line-haul high-speed service into the CBD.

b. Parking restraints: in areas such as the central business district where land costs are high and land itself is at a premium, increased mass transit use can be encouraged by:
   i. Limiting parking supply.
   ii. Restricting parking basically to short-term parking.
   iii. Graduating the parking rates to discourage long-term parking usage.
c. Employee transit benefits:
   i. Bus passes given to employees as a fringe benefit.
   ii. Allocating existing parking spaces to car poolers. While not a transit incentive, increased car pooling would help alleviate traffic congestion.

Mass Transportation Technology, Non-Exclusive Guideway

Most mass transit systems operating on a scheduled or demand basis on existing streets depend on some form of bus technology. The *mini-bus* is a small, low-

Fig. 134. Mini bus[4]. Mini Bus Model MB-701 manufactured by Mini Bus, Inc., Huntington Park, California. *Dimensions:* Length, 20 ft; width 90 in.; height, 102 in. *Weight:* 6230 lb. *Accommodations:* Seated 15–25 (depending on seat arrangement). Maximum passengers including standees, approximately 30. *Performance:* Maximum speed 25–65 mph depending on engine, transmission, and gearing. (25–65 mph is the general speed range for small buses of this type).

speed, bus-type vehicle, used either singly or in trains to handle relatively low demand volumes in compact use areas. It has been used successfully in major metropolitan area downtown shopping loops at reduced fares, at airports, for service between activity centers and parking lots and for sight-seeing. Another application for such buses is in the so-called Dial-a-Bus service where a rider telephones for a bus which is then radio or radio-telephone dispatched for pickup at the rider's origin or a convenient pickup point. Figure 134 shows a typical mini-bus with appropriate physical and performance data. Figures 135 and 136 show exterior and interior views of another mini-type bus, the Ginkelvan, according to its manufacturers a "wickedly handsome" little bus.

A number of other companies also build mini- and conventional-type buses. In addition, other types of bus technology exist, including the articulated bus manufactured by the Flexible Company, Loudonville, Ohio, and the double-deck bus manufactured by Daimler Transport Vehicles, Coventry, England. Although not included here, current technology streetcars, on rubber tires or rails, are also in use, particularly in Europe. Figure 137 shows a *conventional motor bus* of the type which can be used for relatively high-demand service. It has the capability of performing in either a feeder or line-haul capacity or as a combination of the two.

One of the *new technology type bus concepts*, the Model RTX (Rapid Transit Experimental) under development by General Motors Corporation, is shown in

Fig. 135. Exterior view of Ginkelvan.

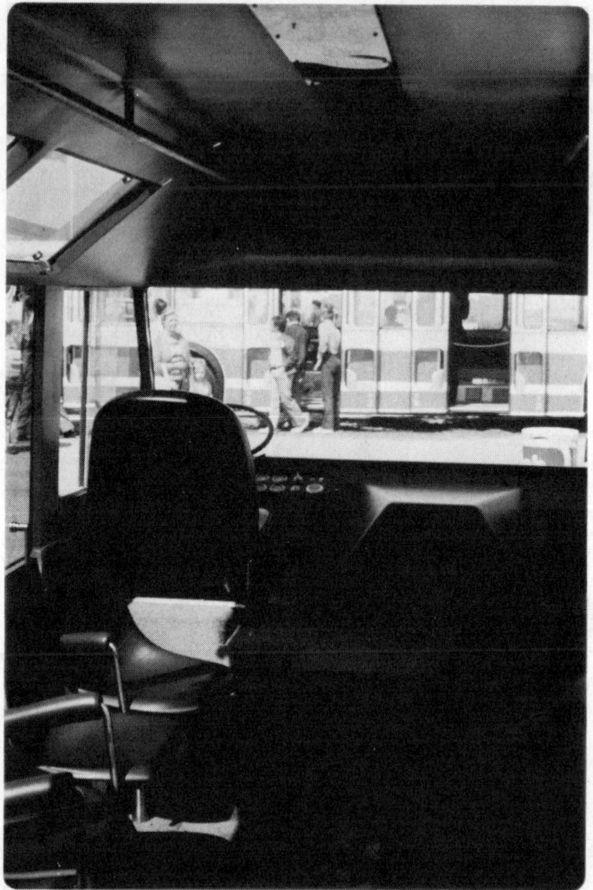

Fig. 136. Interior view of Ginkelvan.

Figs. 138–140. The RTX is quite advanced mechanically compared to current bus technology. Instead of the present popular semi-monocoque body structure, the RTX will use a stiff chassis frame with a light-weight body and very large doors and windows. This type of construction, along with the use of small-diameter tires and compact auxiliary equipment modules, will permit a low floor height and easy entering and exiting. Power will be supplied by a rear-mounted turbine engine driven through a variable-ratio transmission and torque converter. This will result in lower emission of pollutants and will provide a performance capability for traditional operation on city streets as well as high-speed service on freeways and exclusive bus ways.

Whether the GM-RTX will eventually see extended mass transit service is in

Fig. 137. Conventional bus[4]. *Dimensions* (for General Motors GM TDH-5303 model): Length, 40 ft; width 102 in.; height 120 in. *Weight:* 20,000 lb (empty). *Accommodations*: Typical seating configuration: 53 seats. *Performance:* Maximum speed, 45–60 mph, depending on engine and transmission.

Fig. 138. RTX bus, General Motors[4]. *Dimensions:* Length, 40 ft; width, 96 in.; height 106 in. *Accommodations:* Typical seating configuration: 45 seats. *Performance:* Maximum speed, 70 mph; acceleration 0–60 mph in 40 sec.

Fig. 139. Front- and rear-door sections of GM-RTX.

Fig. 140. Interior of GM-RTX.

doubt, because its cost is higher than that for conventional bus technology. This may make it difficult to accept in a market which has traditionally required a low bid.

Mass Transportation Technology, Exclusive Guideway (Right-of-Way) Systems

Figure 141 depicts the operation of high-speed buses (exclusive right-of-way–Metro Mode Concept) on exclusive rights of way. Through the use of special bus ways the nonstop (line-haul) portion of bus transit trips can be made rapidly at speeds and capacities approaching those of mass rail systems.

Through the use of retractable rail bogies (bus-rail, multi-modal operation) and various other modifications, the conventional highway bus can operate over rail networks or city streets (Fig. 142). The dual mode bus can therefore make use of the better operating features of buses and rail systems, including the ability to circulate through neighborhoods picking up and discharging passengers and high-speed line-haul travel over rails for certain portions of the trip. Where the dual mode vehicle uses its own tires (driving against the tops of the rails) for traction, winter snow and ice may cause serious disruptions in service.

Several *low-volume, low-speed guideway movement systems* appropriate for compact, high-use areas are available. Such systems involve relatively low capital

Fig. 141. Buses on exclusive rights-of-way[10].

Fig. 142. Multi-mode bus (conventional bus adapted to rail operation).

Fig. 143. Monorail[4]. *Dimensions:* Length (7-car train), 96 ft; width, 63 in.; height, 110 in. *Accommodations:* 12 passengers per car. *Performance:* Speed, 5–20 mph.

and operating costs, can be erected and dismantled rapidly, and have the ability to negotiate steep grades and sharp curves. As a result they are adaptable to many locations. Because of their light weight and relatively small size, they have light supporting structures and can be made to blend in very nicely, aesthetically, with the local environment. Figure 143 shows one such system, the Habegger Minirail. Other systems having similar low-speed to medium-speed, low-demand-volume capabilities, including the People Mover* have useful application in special dense areas requiring relatively short-distance movement.

A Westinghouse Corporation *medium-volume fixed-guideway system* development, the Transit Expressway (7), uses a rubber-tired vehicle running over a special guideway (Fig. 144), to provide medium-density fixed-guideway operation. The system has a number of advantages. They include lower capital and

*WED Enterprises, Anaheim, California.

Fig. 144. Transit expressway (skybus), Westinghouse Corp.[7]. *Dimensions:* Length (per car), 30.5 ft; width, 104 in.; height, 120 in. *Weight:* 19,500 lb. *Accommodations:* Seated, 28; standing, 42. *Performance:* Maximum speed, 50 mph.

Fig. 145. Exterior of Revenue Line Skybus-type mass transit car.

operating costs than rail; quieter, smoother operation; greater grade-climbing capability; and the ability to negotiate sharper curves. Its lighter structural framing requirements permit the development of more attractive guideway systems which better blend into the natural environment. Figures 145 and 146 show the exterior and interior decor and seating of a revenue car.

A new concept in inter- and intracity transportation, that of *personal rapid transit* (PRT), promises to become an attractive alternative to the automobile—particularly in traffic clogged sections of our cities such as the central business district and for special applications such as airports and universities. A PRT system works much the same as an automatic elevator—except that it carries passengers horizontally rather than vertically. After fare payment, which may involve the use of a prepurchased magnetically coded "credit" card, the station area is entered and a waiting car boarded. Once inside the car, a button indicating destination choice (floor on an elevator) is pushed and the car moves off. Seating configurations vary from 4 or 6 to more than 20 (depending on manufacturer) in comfortable, climate-controlled cars. Travel at speeds up to 70 mph

Fig. 146. Interior of Revenue Line Skybus-type mass transit car.

under control of fully automatic computerized command and control systems is possible.

Figures 147 and 148 show the Transportation Technology Incorporated (TTI) PRT. The vehicle is supported in its concrete guideway by a cushion of air and is propelled by a linear induction motor. Seating capacity is 4 or 5, with a stretch version for 16 persons.

Figure 149 shows another interesting PRT development, the six-passenger Monocab monorail (Monocab, Inc.). The Monocab rides suspended from rubber-tired wheels which ride inside an inverted U-channel guideway. Switching is by a unique mechanical system inside the guideway and does not require movement of massive sections of guideway. Guidance and control are affected by an optical scanning system inside the guideway.

The Alden StaRR car (Figs. 150 and 151) is typical of the rubber-tired vehicles which run in a concrete trough-like guideway. Figure 152 shows mechanical details for the vehicle including its ride-wheels, horizontal bogies (which position the vehicle laterally in the guideway and through computer control and left or right biasing select the vehicle's path when it is to be switched) and the electrical power pick-up system. Vehicle propulsion is by electric motor. A system similar to the Alden system has been constructed in Morgantown, West Virginia to serve

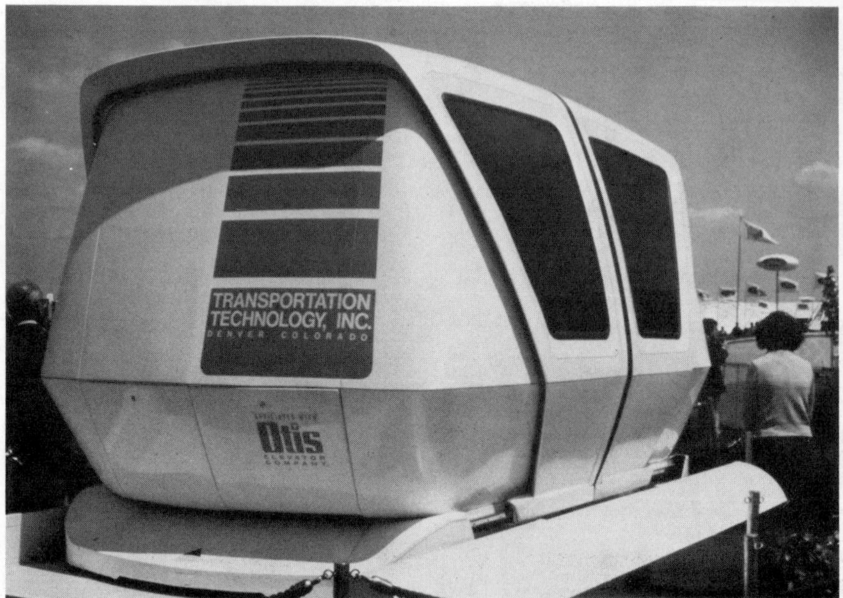

Fig. 147. Personal rapid transit (Transportation Technology, Inc.).

Fig. 148. Personal rapid transit (Transportation Technology, Inc.).

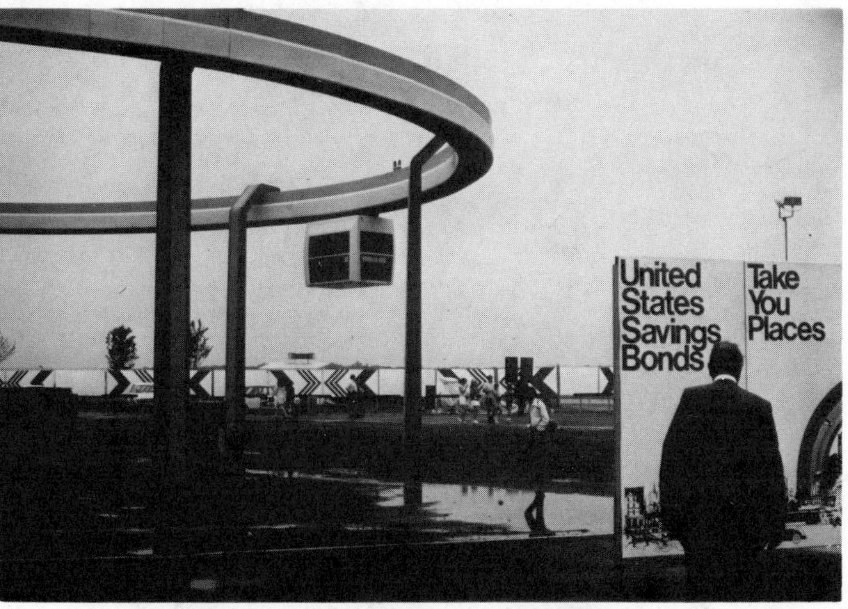

Fig. 149. Personal rapid transit (Monocab, Inc.).

Fig. 149 (*Continued*)

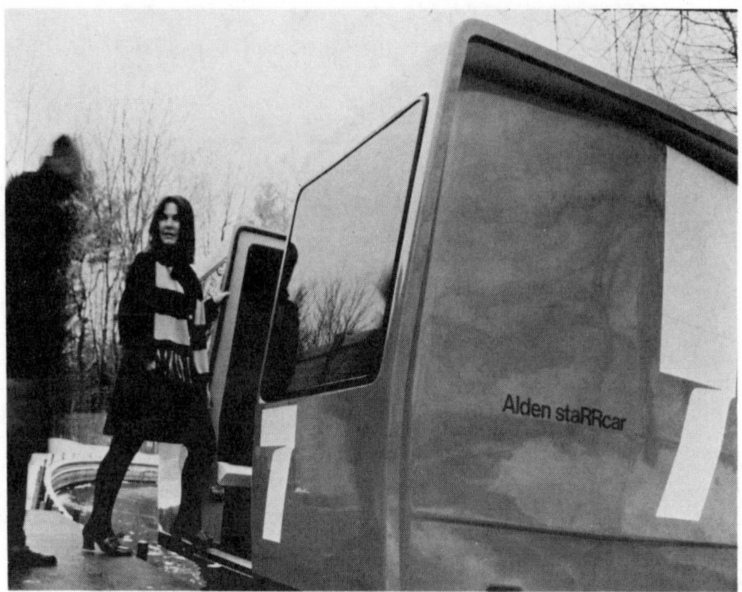

Fig. 150. Alden StaRRcar vehicle (Alden Self Transit Systems Corp.).

Fig. 151. Alden StaRRcar vehicle details (Alden Self Transit Systems Corp.).

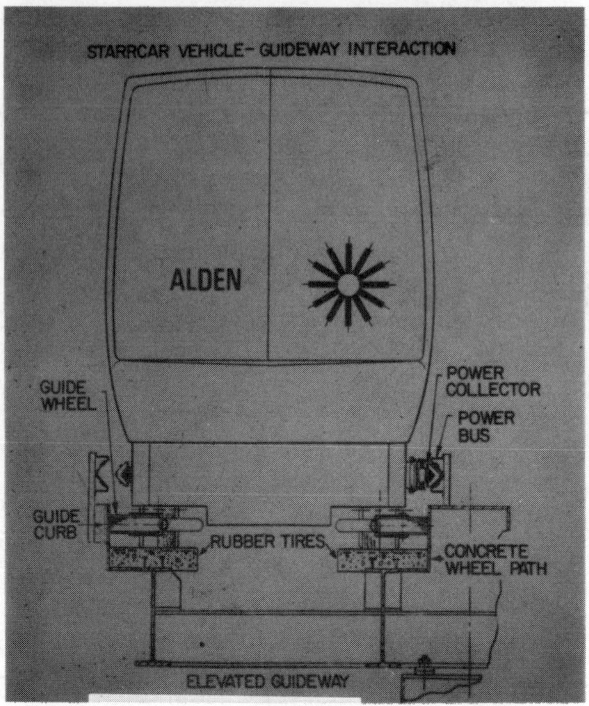

Fig. 152. Details of Alden StaRRcar vehicle showing suspension and drive details (Alden Self Transit Systems Corp.).

the downtown area and several campuses of the University of West Virginia. Figure 153 shows an artist's sketch of a terminal facility.

Conventional rail facilities, such as those in New York City, provide a high standard of service in terms of passenger movements per hour but at relatively low levels of comfort and with considerable sacrifice to other passenger amenities. Generally, except for BARTS (Bay Area Rapid Transit System), rail transit cars placed in service recently differ little from those placed in service 30 years ago. They are heavy, noisy, and relatively uncomfortable. One current type car presently in use is that shown in Fig. 154.

The *rail rapid transit system* for the San Francisco Bay area, BARTS, began operation in the early 70's as the most advanced system of its type in the world. BARTS involves the updating of 50 years of rail technology in the areas of propulsion, trucks, automatic train control, noise reduction, and passenger service and convenience. The BARTS trains (Fig. 155) are modern, comfortable, air-conditioned units of completely new design. The metal cars make extensive use

Fig. 153. Sketch of station for PRT System, Morgantown, West Virginia (Alden Self Transit Systems Corp.).

Fig. 154. Conventional rail rapid transit car, BMT–IND Lines, New York City[4]. *Dimensions:* Length (per car), 60 ft; width, 120 in.; height, 194 in. *Weight:* 80,000 lb (per car). *Accommodations:* Maximum passengers including standees, 300. *Performance:* Maximum speed, 50 mph.

Fig. 155. Bay Area Rapid Transit System (B car). See Table 66 for data[8].

TABLE 66. BAY AREA RAPID TRANSIT SYSTEM, "AS DESIGNED" SYSTEMS AND VEHICLE DATA (4) (8) (11).

Network Length:	75 miles
Subway:	25 miles
At-Grade:	25 miles
Elevated:	25 miles
Stations:	38
Capacity:	28,800 passengers/h/line at 90 sec headways
Speed:	maximum 80 mph; average including station stops; 45–50 mph
Car:	length 75 ft; width, 126 in.; weight (empty), 62,000 lb; passengers (seated), 72
Fare collection:	Automated, based on prepurchase (from vending machines) single or multiple magnetically coded ride ticket
Financing:	bond issue; Bay Bridge tolls; HUD

of plastics on both the interior and exterior, are lighted and air-conditioned, and have complete automation capability. Table 66 provides system and vehicle data for BARTS.

The two basic *monorail systems*—the suspended and the supported—are shown in Fig. 156. Monorails are faced with basic technical difficulties, notably sway

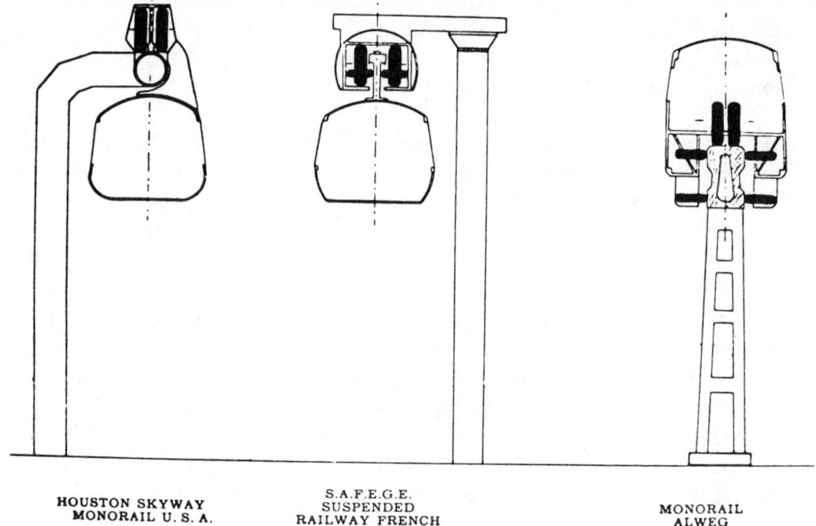

Fig. 156. Basic monorail types.

TABLE 67. CONSTRUCTION COSTS: MONORAIL VS. CONVENTIONAL RAIL (4).

Type of system	At-grade construction costs per mile	Elevated construction costs per mile
Rail, conventional	$2,100,000	$8,200,000
Monorail		
Alweg	3,500,000	4,000,000
SAFEGE	4,500,000	no change

and difficulty in switching. Therefore, they have not assumed anything more than specialty roles in the mass transportation picture. In addition, as the data in Table 67 indicate, monorail construction costs are normally higher than those for conventional rail type systems.

REFERENCES

1. Smith, Wilbur and Associates, *The Potential for Bus Rapid Transit*, The Automobile Manufacturers Association, Detroit (1970).
2. *Transit Fact Book* (Annual) American Transit Association, Washington, D.C.
3. Weiner, Edward, *Bus Transit and Rail Transit*, U.S. Dept. of Transportation (BPR), Washington, D.C. (1967).
4. Simpson and Curtin, *Transportation for Madison's Future*, Madison Area Transportation Study (MATS) Special Report No. 17, Prepared for MATS Technical Coordinating Committee, Madison, Wisc. (1970).

5. *Demand-Actuated Transportation Systems*, HRB Special Report No. 124, Highway Research Board, Washington, D.C. (1971).
6. *Tomorrow's Transportation: New Systems for the Urban Future.* U.S. Department of Housing and Urban Development (HUD), Washington, D.C. (1968).
7. MPC Corporation, *Report on Testing and Evaluation of the Transit Expressway*, for Port Authority of Allegheny County, Pittsburgh, Pa. (1967).
8. Cohen, Stanley, BART Makes Tracks to the Future, *Consulting Engineer* (January 1972).
9. Prytula, George, *Community Mobility Systems*, Special Report of Urban Land Institute, Washington, D.C. (1970).
10. Southeastern Wisconsin Regional Planning Commission, *Metro-Mode: A New Approach to Rapid Transit*, with General Motors Corp., Waukesha, Wis.
11. Hammond, David G., Overall Planning of the Bay Area Rapid Transit District Project, *Transportation Engineering J. ASCE*, Vol. 95, No. TE 2, New York (May 1969).
12. Levinson, Herbert S. et al, "Bus Use of Highways: State of the Art," NCHRP Report 143, Transportation Research Board, Washington, D.C. (1973).

AIR TRANSPORTATION

The planning for future airport facilities requires forecasts of the future airport activities and the establishment of airport requirements based on the relationships between the demand and capacity. Accurate forecasting of future activity is a difficult task requiring common sense, good judgment, keen analytical ability, and considerable experience. Some of the forecasting methods used are:

1. Mechanical extrapolation: extending past historical trends.
2. Surveys of expected consumer behavior.
3. Analytical forecasts: mathematical modeling which explains the interrelationships between air travel and the variables influencing air travel.
4. Forecasts based on informed judgment. Although a most difficult method to defend under close scrutiny, the professional judgment of personnel with long experience in the field of air transportation is useful in helping to make forecasts.

Factors Affecting Airport Capacities (5)

Some of the operational activities which must be considered in determining specific airport facility requirements are:

1. Departing passengers.
2. Tonnage of air cargo by type departing the airport.
3. Aircraft arrivals (landings) and departures (take-offs). Total aircraft movements are normally broken down into local operations, those flying for

business, pleasure, or practice within the local vicinity, and itinerant operations, including all arrivals and departures other than local flights. Air carrier operations are in the class of itinerant operations.

4. Peaking characteristics: busiest hours.
5. The total number of home-based aircraft, general aviation, and air carrier aircraft using the airport as a home base. The aircraft information should be categorized by both the number and type of engines, and the type of use—air carrier or general aviation.

Planning for the future operational requirements of an airport requires a knowledge of the type of aircraft which must be accommodated by the facility. Table 68 lists selected aircraft currently in operation and some of their size, weight, and operational characteristics.

An evaluation of the total air traffic capacity in the area of the airport will provide information about the capacity required for the airport being designed. The major components of the airport which require investigation, and for which information will be developed, include (5):

1. The Airfield:
   a. Number and configuration of runways and taxiways.
   b. Runway, taxiway, and apron dimensions.
   c. Lateral clearances between airport operational areas.
2. Terminal Area:
   a. Airline gate positions.
   b. Airline apron areas.
   c. Cargo apron areas.
   d. General aviation apron requirements.
   e. Airline passenger terminal requirements.
   f. General aviation terminals.
   g. Cargo buildings.
   h. Auto parking: loading and unloading.
   i. Aircraft maintenance facilities.
3. Air Space Capacity:
   a. Proximity to other area airports.
   b. Runway alignments and arrangements.
   c. Type of operation:
      i. IFR (instrument flight rules): IFR procedures are required of all flights when the ceiling is below 1000 ft, or visibility is below 3 miles. FAA (Federal Aviation Administration) rules require, however, that air passenger and air cargo operations always be under IFR flight rules.
      ii. VFR (visual flight rules): VFR procedures apply when the ceiling is 1000 ft or more, and visibility is more than 3 miles.

TABLE 68. TYPICAL DATA FOR SELECTED COMMERCIAL AIRCRAFT: 1936–PRESENT[a] (47) (53).

Aircraft number/ model	Common name	Manufacturer	Year Intro- duced into civil service	Engines Num- ber	Engines Type	Wing Span (ft)	Length (ft)	Maximum take-off weight (lb)	Maximum landing weight (lb)
DC 3	Dakota, Skytrain	Douglas	1936	2	P	95	64	25,200	24,400
DC 4	Skymaster	Douglas	1946	4	P	118	94	73,000	63,500
DC 6		Douglas	1947	4	P	118	101	97,200	80,000
377	Stratocruiser	Boeing	1948	4	P	141	110	145,800	129,000
L1049A	Super Constellation	Lockheed	1951	4	P	123	114	120,000	98,500
707-120	Stratoliner	Boeing	1958	4	J	131	145	247,000	175,000
707-420[f]		Boeing	1960	4	J	142	153	312,000	207,000
707-320B		Boeing	1962	4	J	146	153	328,000	207,000
DC 8-10		Douglas	1959	4	J	142	151	273,000	193,000
DC 8-63		McDonnell-Douglas	1967	4	J	148	157	350,000	245,000
727-100		Boeing	1963	3	J	108	133	160,000	137,500
727-200		Boeing	1969	3	J	108	153	172,000	150,000
DC 9-10		Douglas	1965	2	J	89	104	77,700	74,000
DC 9-40		McDonnell-Douglas	1968	2	J	93	125	114,400	102,000
737-200D		Boeing	1967	2	J	100	97	109,000	98,000
747C		Boeing	1969	4	J	196	231	775,000	564,000
DC 10-30		McDonnell-Douglas	1971	3	J	161	181	555,000	403,000
L1011-1	Tristar	Lockheed	1972	3	J	155	179	426,000	350,000
L-500[g]	Galaxy	Lockheed	1969	4	J	223	248	858,500	699,000
Concorde[h]		BAC-Aerospatiale	1969	4	J	84	204	385,000	240,000
TU-144[h]		Tupolev	1969	4	J	91	197	395,000	
2707-300[i]	SST	Boeing		4	J	280	142	635,000	

[a] Figures are approximate and may vary considerably depending on engines, weights, etc.
[b] Passenger capacity varies with proportion in various travel classes, seat widths, seat back angle, spacing, etc.
[c] Figures are for different types of operations, loadings etc.; hence are only approximate.
[d] "Typical" speed often that at economical cruise.
[e] Criteria for maximum range varies by manufacturer; some are for empty aircraft, others with "typical" loading.
[f] Boeing 707 manufactured for British Airlines.
[g] Civilian version of military C-5A; cargo only.
[h] British/French and Russian SSTs; not in civil service.
[i] U.S. prototype SST; not produced.

4. Airport Access Requirements: A major movement problem associated with aircraft is that of getting to and from the airport. As the statistics from an air travel study of O'Hare International Airport in Chicago show (Figs. 157 and 158) (52), most passengers arrive by private car. Their times of arrival, at least during weekdays, coincide with typical urban travel patterns (see Fig. 26), adding to the severity of already critical movement patterns. Figure 159 (51), with similar data for San Francisco International Airport, compares total vehicle arrivals (all purposes) with passenger arrivals and parking needs.

Passengers[b]		Approximate runway required for take off[c]	Cruising speed[d]		Ceiling, cruising	Range[e]		Cost	
Typical	Maximum	(ft)	Typical (mph)	Maximum (mph)	(ft)	With Maximum payload (miles)	Maximum (miles)	$ (where available)	Year (where available)
21	32	4,050	170		6,000	660	1,670	90,000	
	66	5,250	204		10,000	1,150	2,180	410,000	1946
	68	5,400	280	315	16,000	3,070	3,810	640,000	1947
81	100	7,200	285		20,000	2,750	4,300	1,470,000	1949
	92	6,230	255	320	25,000	3,100	4,250	1,260,000	1951
	179	10,200	526	585	38,000		3,750	4,200,000	1958
	143	10,700		600	37,000	4,720	6,955	5,860,000	1961
147	215	10,440	550	600	42,000	6,160	7,610		
132	179			542	42,000		4,300		
	189	11,500		600	42,000	4,500	7,700		
	135	4,980	570	605	37,400	1,900	2,760	4,500,000	1961
163	189	5,820	568	595	35,200	1,543	2,530		
60	90	5,300		561	30,000+		1,311		
	125	8,080		561	30,000+		1,685		
115	130	7,100	Mach 0.78	570	30,000+	1,854			
374	490	10,900		608	45,000	2,880	7,090	35,000,000	
270	345	11,050		570	32,700	4,272	6,909		
256	400	7,780		Mach 0.85	42,000	2,878	4,467		
cargo only		9,500	503	558	32,500	2,975	6,960		
144	128	9,850		1,450	65,000	4,020	4,400		
130	121	6,235		1,550	65,000		4,040		
280	321		1,800		60,000+				

Airport Facility Requirements

The construction and operation of a major airport will have a significant effect on the local environment. For example, at very large airports, site work will cover thousands of acres of land, and require major changes to flowing streams and other drainage courses, disruption of wildlife habitats, and other natural areas. In addition, providing the enormous quantities of native construction materials for constructing the facility, e.g., soils and aggregates, may require mines and quarries which result in further disruption of the land and a further

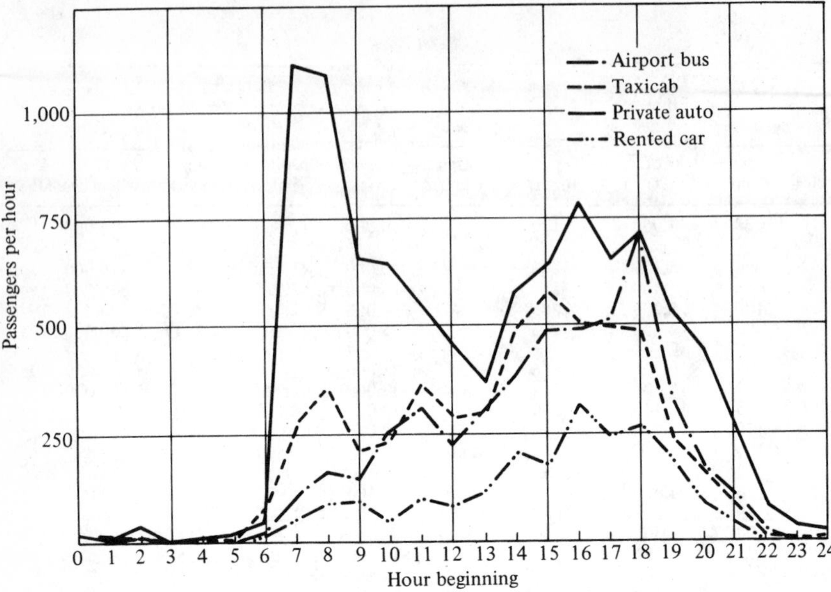

Fig. 157. Passengers by mode by hour of the day, Wednesday[52].

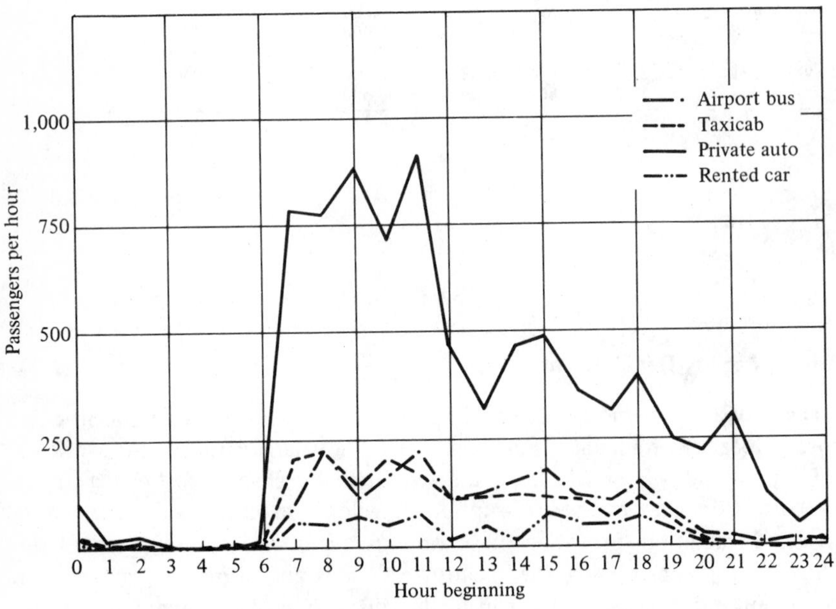

Fig. 158. Passengers by mode by hour of the day, Saturday[52].

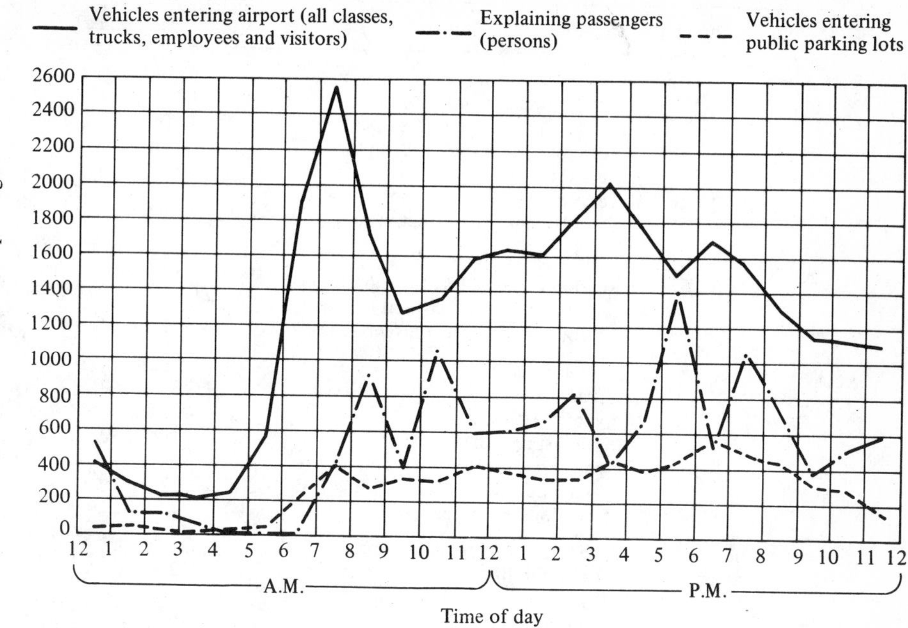

Fig. 159. Hourly variation of vehicle and person movements at San Franscisco International Airport, typical weekday, 1966[51].

depletion of local resources. Apart from the detrimental effects caused by the development of the airport site, the operation of the airport also has a number of environmental impacts. Table 69 lists some of the significant environmental impacts.

Table 70 lists the normal facilities required at an airport by major subject and the specific facilities. The table also notes basic reference sources, which provide the necessary information for the detailed development of the individual facility.

TABLE 69. ENVIRONMENTAL IMPACTS OF AIRPORT DEVELOPMENT AND OPERATION (5).

1. Increased ambient noise level.
2. Displacement of people and businesses.
3. Aesthetic impact.
4. Community disruption.
5. Destruction or other influence on recreational areas.
6. Alteration of behavior pattern for a widelife species.
7. Interference with wildlife breeding, nesting or feeding grounds.
8. Increased air pollution.
9. Increased water pollution.
10. Affect on surrounding area water table.

TABLE 70. REFERENCES FOR DETERMINATION OF FACILITY REQUIREMENTS.*

Subject	Items	Reference†
Runway	length	12, 17
	width, clearances	12, 20
	clear zones, approach slopes	1, 13
	orientation, crosswind runway	1, 21
	grades	16
	capacity, stage construction, delay and cost effectiveness	1, 2
Taxiways	width, clearances	12, 20
	exit design and location, grades ef-	22
	fect on runway capacity stage construction, cost effectiveness	1, 2
Terminal area	clearances	12, 20
	grades	23
	gate positions	40, 49, 50
	aircraft parking clearances	23, 49, 50
	space requirement in terminal and administration building for various activities	33
Service and hangar areas	service equipment buildings	31
	cargo facilities	32
	fire and rescue equipment buildings	10
Heliports	planing and design	34
	rooftop or elevated heliports	34
Obstructions	standards of approach, horizontal, and other control surfaces	12
	clear zones	46
Drainage	structures, layout	14
	grades	14
Paving	fillets	22
	jet blast protection	18
	pavement types and details	15
Lighting and marking	approach lighting	28
	runway lighting	27, 30
	taxiway lighting	25, 29
	runway and taxiway marking	24, 26
	helicopter landing area	34
	obstructions	35
Wind data	source of data	1, 21
Navaids	location, grading requirements	11
Airport types	utility airports	12
	general aviation airports	12
Environmental impact	noise, community consequences, zoning	5, 37, 38, 42, 43
General		44, 45
Aircraft	types, physical and operating characteristics	47, 48

*Adapted from source (5).
†Reference numbers refer to the publications listed in the bibliography.

TABLE 71. FACTORS AFFECTING SITE LOCATION (5).

Aeronautical demand and proximity to other airports
Aircraft performance characteristics
Access to ground transport
Meteorological conditions
Physical aspects of site including topography, elevation, and drainage
Nature of surrounding environment
Nuisance abatement requirements
Safety requirements
Economic factors

The *site selection* process involves selecting a site of adequate size, suitable to accommodate the needs of the local area. Each site is evaluated from an environmental, geographic, economic and engineering standpoint. Some of the factors which will influence the selection of the site and its location are listed in Table 71.

After the facility requirements have been specified and the site selected for the airport modification or construction, a scale layout of all proposed (and existing) airport facilities and land uses, their location within the airport itself, and information regarding clearance and other dimensional requirements, is developed. The airport plans will also establish the runway configuration, the placement and interrelationships of the taxiways and aprons, and the area on which the terminal facilities will be developed. In addition, approach zones for aircraft operation will be shown. The detailed steps required in developing the airport layout plan are specified in the Department of Transportation document "Airport Master Plans" (5).

Terminal Facilities

The terminal area is the location in which the ground and air travel modes interface. The terminal area includes terminal and cargo buildings, aircraft gates, hangars, shops and unloading facilities, motels and restaurants, and auxiliary facilities such as entrance and service routes. Factors which should be considered in the location and the development of the terminal area, and in determining the terminal type, are as follows (5):

1. Passengers:
   a. Adequate terminal area curb space for private and public transportation.
   b. Minimum walking distance, Automobile parking to ticket counter.
   c. Minimum walking distance, Ticket counter to passenger holding area.
   d. Minimum walking distance, Passenger holding area to aircraft.
   e. Passenger transportation, Where long distances must be traversed.
   f. Pedestrian walkways to aircraft, As backup to mechanical transportation systems for passengers.
   g. Efficiency of passenger interline connection.

h. Baggage handling, Enplaning.
i. Baggage handling, Deplaning.
j. Convenient hotel-motel accommodations.
k. Efficient handling of visitors and sightseers at the airport.

2. Passenger Vehicles:
   a. Public automobile flow separation from service and commercial traffic (a necessity at large airports).
   b. Public transportation to and from the airport.
   c. Public parking, long term (3 hours or more).
   d. Public parking, short term (less than 3 hours).
   e. Airport employee parking.
   f. Airline employee parking.
   g. Public auto service area.
   h. Rental car parking and service areas.
3. Airport Operations:
   a. Flexibility.
   b. Separation of apron vehicles from moving and parked aircraft.
   c. Passenger flow separation in the terminal building (departing and arriving).
   d. Passenger flow separation from apron activities.
   e. Concession availability and exposure to public.
   f. Airfield security and prevention of unauthorized access to apron and airfield.
   g. Air cargo and freight forwarder facilities.
   h. Airport maintenance shops and facilities.
   i. Airfield and apron drainage.
   j. Airfield and apron utility distribution.
   k. Utility plants, and heating and air conditioning systems.
   l. Fire and rescue facilities and equipment.
4. Aircraft:
   a. Efficient aircraft flow on aprons and between terminal aprons and taxiways.
   b. Easy and efficient maneuvering of aircraft parking at gate positions.
   c. Aircraft fueling.
   d. Heliport areas.
   e. General aviation areas.
   f. Noise, fume and blast control.
   g. Apron space for staging and maneuvering of aircraft service equipment.
5. Safety (for the following functions and areas):
   a. Enplaning and deplaning at aircraft.
   b. Elevators, escalators, stairs, and ramps as to location, speed, and methods of egress and ingress.

c. People-mover systems as to location, speed, and methods of egress and ingress.
d. Road crossings as to protection of pedestrians.
e. Provisions for disabled persons.

6. Expansion Capabilities to Accommodate:
a. Increase of passenger volume.
b. Increase in number of aircraft positions.
c. Increase in aircraft size.

7. Economics: To provide a proper balance between capital investment, aesthetics, operation and maintenance costs, and passengers and airport revenues.

Figure 160 depicts several *terminal concepts*. The following discussion of the general terminal concepts is taken from "Airport Master Plans" (5).

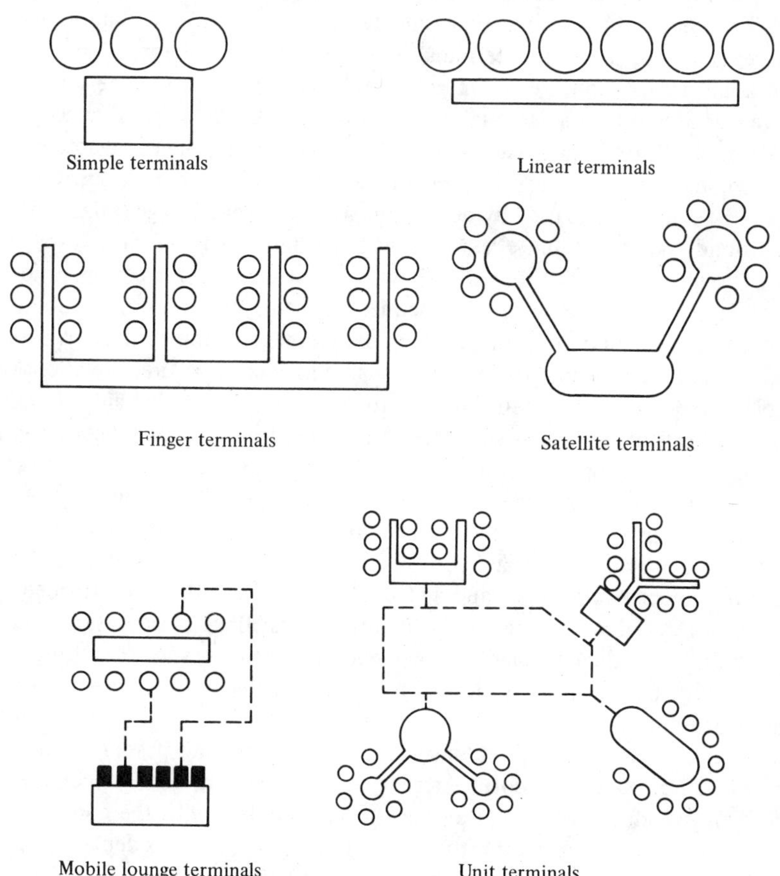

Fig. 160. Terminal area concepts[5].

"The *simple terminal* consists of a single common waiting and ticketing area with several exits onto a small aircraft parking apron. It is adaptable to airports with low airline activity and is also adaptable to general aviation operations, whether it is located as a separate entity on a large airline-served airport or is the operational center for an airport used exclusively by general aviation. Where the simple terminal serves airline operations, it will usually have an apron which provides close-in parking for three to six commercial transport aircraft. Where the simple terminal serves general aviation only, it should be within convenient walking distance of aircraft parking areas and should be adjacent to an aircraft service apron. The simple terminal will normally consist of a single level structure where access to aircraft is afforded by a walk across the aircraft parking apron. The layout of the simple terminal should take into account the possibility of linear extension for terminal expansion.

"The *linear terminal* concept is merely an extension of the simple terminal concept, that is, the simple terminal is repeated in a linear extension to provide additional apron frontage, more gates, and more room within the terminal for passenger processing. It is sometimes referred to as the gate arrival concept. The more sophisticated linear terminals often feature a two-level structure where enplaning passengers are processed direct from curb to aircraft on one level, while the other level is used by deplaning passengers for baggage claim and access to ground transportation. Passenger walking distance from curb through terminal to aircraft is short, usually 75 to 100 feet. The linear configuration also lends itself to the development of adequate, close-in public parking. Ample curb frontage for loading and unloading ground transportation vehicles is provided with each extension of the linear terminal, and there is a direct relationship of enplaning or deplaning curb frontage to departing or arriving aircraft. Linear terminals can be expanded with almost no interference to passenger processing or aircraft operations. Expansion may be accomplished by linear extension of existing structure or by developing two or more linear terminal units.

"The loading of aircraft may be accomplished by nose-in/push-out operations or by loading bridges. Aircraft can maneuver on apron areas with unobstructed flow if a maneuvering taxi lane and dual taxiway parallels are provided. The linear concept does not require long concourses, fingers, satellites, or service buildings, but it does not lend itself to common facilities such as waiting rooms, concessions, ticket counters, or hold rooms. These facilities are usually repeated with each linear extension. At large airports, the concept can also require an extensive system of directional signs since enplaning passengers must not only be directed to the correct airline area but also to the correct passenger processing module within that area. Another problem with the concept is that on a return flight to the airport, a passenger may find that his deplaning module is located a long distance from where he parked his car at his enplaning module. These factors must be taken into account in comparing the operating and construction costs of the linear terminal with other concepts. The configuration of the space occupied in the linear concept must also be compared with the space and configurations of other concepts in determining their compatibility with particular airport situations.

"The *finger or pier concept* evolved in the 1950's when gate concourses were added to simple central terminal buildings. Since then, very sophisticated forms of the concept have been developed with the addition of hold rooms at gates, jetways and aircraft loading bridges and vertical separation of the ticketing/check-in function from the baggage claim function. However, the basic concept has not changed in that the main central terminal building is used to process passengers and baggage while the finger or pier provides means of enclosed access from the central terminal to aircraft gate. Aircraft are parked at gates along the pier as opposed to the satellite concept where they are parked in a cluster at the end of a concourse.

"Walking distances through finger terminals are long, averaging 400 feet or more. Curb space must be carefully planned since it depends on the length of the central terminal and is not related to the total number of gates afforded by fingers. This is particularly true of deplaning curbs near centralized baggage claim facilities.

"Although the finger concept afforded one of the most economical means of of adding gate positions to existing terminals, its use for expansion should be limited. Existing fingers should not be extended at the expense of taxiway maneuverability nor should new fingers be added without providing adequate space for passenger processing in the main terminal. Most successful additions are effected by extending the main terminal and then increasing the number of fingers.

"Adequate space must be provided between fingers for the maneuvering of aircraft. Dual taxi capability between fingers is desirable and aircraft growth should be taken into account in planning separation. Since most aircraft maneuvering takes place between piers, outside taxiways are free of push-out operations.

"The primary feature of the *satellite concept* is the provision of a single central terminal (with all ticketing, baggage handling, and ancillary services) which is connected by concourses to one or more satellite structures. It is sometimes called the rotunda concept. The features of the satellite concept are very similar to those of the finger concept except that aircraft gates are located at the end of a long concourse. Satellite gates are usually served by a common hold room rather than individual hold rooms. Another feature is that the concourse can be located underground thereby providing space for aircraft taxi operations between the main terminal and the satellite.

"The distance from the main terminal to a satellite is usually well above the average distance to gates found with the finger concept. Therefore, people-mover systems are being provided between terminal and satellite at many installations to reduce walking distances.

"There is no direct relationship between the number of gates and curb space so that special care should be taken in planning enplaning and deplaning ramps for the central terminal to prevent curb overloads. One of the advantages of the satellite concept is that it lends itself to a compact central terminal with common areas for processing passengers. In some instances, where terminal area space is limited, structural parking is provided above the central terminal building.

"Aircraft maneuvering areas are required around satellites so that push-out tug operations do not cause aircraft to block active taxiways. Wedge shaped aircraft parking positions around the satellite also tend to crowd the operation of aircraft servicing equipment.

"Terminals developed under the satellite concept are difficult to expand without reducing ramp frontage or disrupting airport operations. Therefore, increases in terminal capacity are usually effected by the addition of terminal units rather than expansion of an existing unit.

"The *mobile lounge or passenger transporter concept* is in use at Dulles International Airport. It is sometimes called the remote aircraft parking concept. Aircraft parking aprons are remote from the terminal building. The mobile lounges transport passengers from the building to aircraft and can be used as hold rooms at terminal building gate positions. In this concept, the aircraft gate positions are placed in parallel rows at required spacings with mobile lounge and service vehicle roads running between the parallel rows of aircraft. Several sets of parallel aircraft parking rows can be provided for ultimate development of gate positions. Airline operations buildings must be provided adjacent to aircraft parking aprons.

"With the mobile lounge concept, walking distances are held to a minimum since the compact terminal building contains common passenger processing facilities and curb frontage can be located directly across the terminal building from mobile lounge gates. Building and curb length, which is established in part by the number of mobile lounge gates, must be carefully planned to provide adequate frontage for enplaning and deplaning passengers.

"The concept has good expansion capability in that capacity can be increased by the addition of mobile lounges and the main terminal and aprons can be expanded without the addition of extensive concourses, fingers or satellites. With the mobile lounge concept, additions can be made with little impedance to airport operations and aircraft movements.

"Aircraft maneuvering capability is excellent with this concept. Remote aircraft parking can reduce taxi time and distance to runways and avoid aircraft congestion next to terminal building facilities. This also removes the aircraft noise and jet blast problem from the building area. Mobile lounges must be capable of mating with various aircraft sill heights and terminal building floor heights.

"In comparing the mobile lounge concept with other concepts, the cost of independent terminal and service buildings and the purchase operation, and maintenance of mobile lounges must be considered. The time required to move passengers between terminal and aircraft by mobile lounge should also be taken into account.

"With the *unit terminal concept* described herein, the airlines build individual terminals around a system of interconnecting access and service roads. The terminals are spaced some distance apart and each terminal provides complete passenger processing and aircraft parking facilities. Airlines which provide limited service to an airport will sometimes combine their operations in a single unit terminal. The concept permits each airline to build a terminal to its own liking and provides for maximum airline identification. Kennedy International Airport is the best known example of the unit terminal concept.

"Walking is held to comfortable distances since unit terminals are much smaller than large joint-use terminals. For this same reason, adequate curb frontage can be easily designed into the unit terminal.

"Expansion capability with the unit terminal concept can be difficult because of the gross area required for each individual terminal. Unit terminals probably require the greatest acreage for development and expansion of any of the terminal area concepts. Construction costs are high because passenger processing facilities, aircraft parking aprons, and public parking must be repeated with each unit terminal.

"Aircraft maneuvering capability within the vicinity of unit terminals is usually good since the airlines have the opportunity to design aircraft aprons and passenger loading devices to suit their own operating requirements. With the unit terminal concept the relationship between adjacent terminals must be carefully planned if interference between aircraft flows is to be avoided and the capability of expanding individual terminals is to be preserved."

REFERENCES

Department of Transportation, Advisory Circulars (AC)

1. Airport Capacity Criteria Used in Preparing the National Airport Plan.
2. AC 150/5060-3A, Airport Capacity Criteria Used in Long Range Planning.
3. AC 150/5070-3, Planning the Airport Industrial Park.
4. AC 150/5070-5, Planning the Metropolitan Airport System; may be obtained from the Superintendent of Documents, U. S. Government Printing Office, Washington, D.C. 20402. Price $1.25–Catalog No. TD 4.108:M 56/2.
5. AC 150/5070-6, Airport Master Plans.
6. AC 150/5090-1, Regional Air Carrier Airport Planning.

7. AC 150/5100-3A, Federal-Aid Airport Program–Procedures Guide for Sponsors.
8. AC 150/5100-4, Airport Advance Planning and Engineering.
9. AC 150/5190-3, Model Airport Zoning Ordinance.
10. AC 150/5210-10, Airport Fire and Rescue Equipment Building Guide.
11. AC 150/5300-2A, Airport Design Standards–Site Requirements for Terminal Navigational Aids.
12. AC 150/5300-4A, Utility Airports; may be obtained from the Superintendent of Documents, U.S. Government Printing Office, Washington, D.C. 20402. Price $1.75–Catalog No. TD 4.8:Ai 7/968.
13. AC 150/5300-6, Airport Design Standards–General Aviation Airports–Basic and General Transport.
14. AC 150/5320-5B, Airport Drainage; may be obtained from the Superintendent of Documents, U.S. Government Printing Office, Washington, D.C. 20402. Price $1.00–Catalog No. TD 4.8:D 78/970.
15. AC 150/5320-6A, Airport Paving.
16. AC 150/5325-2B, Airport Design Standards–Air Carrier Airports–Surface Gradient and Line of Sight.
17. AC 150/5325-4, Runway Length Requirements for Airport Design.
18. AC 150/5325-6, Effects of Jet Blast.
19. AC 150/5325-7, Is Your Airport Ready for the Boeing 747?
20. AC 150/5330-2A, Runway/Taxiway Widths and Clearances for Airliner Airports.
21. AC 150/5330-3, Wind Effect on Runway Orientation.
22. AC 150/5335-1A, Airport Design Standards–Airports Served by Air Carriers–Taxiways.
23. AC 150/5335-2, Airport Aprons.
24. AC 150/5340-1C, Marking of Paved Areas on Airports.
25. AC 150/5340-4B, Installation Details for Runway Centerline and Touchdown Zone Lighting Systems.
26. AC 150/5340-5, Segmented Circle Airport Marker System.
27. AC 150/5340-13A, High Intensity Runway Lighting System.
28. AC 150/5340-14B, Economy Approach Lighting Aids.
29. AC 150/5340-15A, Taxiway Edge Lighting System.
30. AC 150/5340-16B, Medium Intensity Runway Lighting System and Visual Approach Slope Indicators for Utility Airports.
31. AC 150/5360-1, Airport Service Equipment Building.
32. AC 150/5360-2, Airport Cargo Facilities.
33. AC 150/5360-3, Federal Inspection Service Facilities at International Airports.
34. AC 150/5390-1A, Heliport Design Guide; may be obtained from the Superintendent of Documents, U.S. Government Printing Office, Washington, D.C. 20402. Price 75 cents–Catalog No. TD 4.108:H36.
35. AC 70/7460-1, Obstruction Marking and Lighting; may be obtained from the Superintendent of Documents, U.S. Government Printing Office, Washington, D.C. 20402. Price 60 cents–Catalog No. TD 4.8:06 7/968.

36. Planning the State Airport System; may be obtained from the Department of Transportation, Federal Aviation Administration, Environmental Planning Branch, AS-440, 800 Independence Ave., S.W., Washington, D.C. 20591.

Other Publications (These publications may be obtained as indicated.)

37. The Airport–Its Influence on the Community Economy; may be obtained from the U.S. Department of Transportation, Distribution Unit TAD-484.3, Washington, D.C. 20590. No charge.
38. Land Use Planning Relating to Aircraft Noise; and Appendix A; both may be obtained from the National Technical Information Service, Springfield, Va. 22151. Price $3.00 each. Order No. AD 615 015. Appendix A, order No. AD 617 954.
39. Alternative Approaches for Reducing Delays in Terminal Areas; may be obtained from the National Technical Information Service, Springfield, Va. 22151. Price $3.00. Order No. AD 663 089.
40. Aviation Demand and Airport Facility Requirements Forecasts for Medium Air Transportation Hubs Through 1980; may be obtained from the National Technical Information Service, Springfield, Va. 22151. Price $3.00. Order No. AD 688 826.
41. Aviation Demand and Airport Facility Requirement Forecast for Large Air Transportation Hubs Through 1980; may be obtained from the National Technical Information Service, Springfield, Va. 22151. Price $6.00. Order No. AD 684 811.
42. Noise Exposure Forecasts, Contours for 1967, 1970, 1975, Operations at Selected Airports, 1970; may be obtained from the National Technical Information Service, Springfield, Va. 22151. Price $3.00. Order No. AD 712 646.
43. Airport Environs: Land Use Controls; may be obtained from the Department of Housing and Urban Development. No charge.
44. Horonjeff, Robert, *The Planning and Design of Airports*, McGraw-Hill, New York (1962).
45. Paquette, Radnor J., Ashford, Norman, and Wright, Paul H., *Transportation Engineering–Planning and Design*, Ronald Press, New York (1972).
46. Federal Aviation Administration, *Federal Aviation Regulations*, Volume X, U.S. Government Printing Office, Washington, D.C. 20402.
47. *Jane's All The World's Aircraft* (Annual editions, 1945 through 1972), Jane's Yearbooks, London.
48. AC 150/5325-5A, Aircraft Data.
49. *Airport Terminals Reference Manual*, International Air Transport Association, Montreal (1970).
50. *Airport Terminal Facilities*, American Society of Civil Engineers and The Airport Operators Council International Specialty Conference, 1967, ASCE, New York.
51. Whitlock, E. M. and Cleary, E. F., *Planning Ground Transportation Facilities*

for Airports, Highway Research Record 274, Highway Research Board, Washington, D.C. (1969).

52. *O'Hare Passenger Survey–1969*, City of Chicago, Department of Public Works, Chicago, Ill. (1970).
53. Brooks, Peter W., *The World's Airliners*, Putnam, London (1962).

SECTION VII

Quantitative Engineering Management

Marvin M. Johnson, *Professor*
Industrial and Management Systems Engineering Department
College of Engineering and Technology
University of Nebraska–Lincoln

with

B. M. Radcliffe, *Professor, P.E., M. ASCE*
Construction Management Department
College of Engineering and Technology
University of Nebraska–Lincoln

and

R. G. Zilly, *Professor and Chairman, P.E., M. ASCE*
Construction Management Department
College of Engineering and Technology
University of Nebraska–Lincoln

INTRODUCTION

By mid-1971 there were probably close to 100,000 unemployed engineers in the United States. Based on generally accepted definitions, this represented about 10% of the total supply of over 1,000,000. As a group, civil engineers (nearly 70,000 are members of the American Society of Civil Engineers, which was founded in 1852) did not suffer proportionately with those in other engineering disciplines, but they were suddenly beset by criticism of what they had considered their finest contributions to American life—highways, large dams, power plants, and towering skyscrapers.

Public disfavor suddenly made the engineer acutely aware of something he had long suspected. He had too little chance to participate in the management decisions that affected the use of his knowledge and skills, and had often failed to move decisively when he had the opportunity. Too many engineers had simply become cogs in a huge labor force where, lulled by a long period of full employment, they had not concerned themselves with job security, loss of pensions with job changes, and inadequate pay raises. Meanwhile, pressed by the technology they themselves had created, they were finding that their skills were becoming obsolete at a rapidly increasing rate. A side effect of all this may well be the low enrollment in colleges of engineering. Though it was noticeably up in the 1969-1970 academic year over the previous year, the number of bachelor degrees awarded was still about 10,000 below the 1949-50 high of 52,732.

Within this framework, and under the bright spotlight of public scrutiny, the civil engineer must work out his destiny. He can fall back on unionism, although he probably will not in view of the fact that unionized engineers have represented only about 5% of the engineering population for many years; or he can seek to shore up the erosion that has worked to lower his professional status. There is strong evidence to support the latter direction, for most engineers are management oriented and fear the domination of technicians and other subprofessionals who control the engineering unions. It is reasonably safe to assume, then, that engineering management in the future will be highly professional in orientation.

In many ways, the engineer is a different breed of cat than the man for whom business school theory was developed. In many ways he is the same. Thus, though engineers have been accused of being weak in the human side of management practice, the emphasis here will be on quantitative management methods. This does not mean that other facets of management theory are not valuable to the manager of engineers performing engineering tasks. However, it does recognize the engineer's penchant for dealing with problems on a quantitative basis, and it assumes that he will be better able to relate himself to a management system that is as quantitative as it is humanly and humanely possible to make it.

Since it involves the broadest range in terms of both quantitative and qualitative measurement, the consulting engineering firm in private practice makes an excellent prototype in any general discussion of engineering management. An excellent statistical portrait of the private practice group has been painted in the first definitive study by the Federal Government, entitled "1967 Census of Business, Selected Services–Architectural and Engineering Firms; U.S. Department of Commerce, Bureau of the Census, BC 67-556, December 1970." Managers of engineering staffs in both the public and private sector of the economy would be well advised to study this report in its entirety.

According to this report, there were 235,165 employees in the 48,809 reporting firms in 1967. However, 217,812 of these employees were working for the 10,619 firms that claimed four or more employees. Similarly, \$3.84 billion of the reported \$4.51 billion income went to firms employing four or more. Thus, it is evident that the tendency for the big fish to eat more than the little fish is no different in professionally oriented business than in industry in general.

Looking at the legal form of business organization provides further insight into the structure of professional organizations. Of the 48,809 reporting firms,

37,253 are proprietorships, 5,073 partnerships, and 6,483 corporations or other forms. However, of the 10,619 firms with four or more employees, there are only 3,404 proprietorships, 2,450 partnerships, and 4,760 corporations or other. The engineer's affinity to the corporate form of business organization becomes even more evident when only consulting engineering firms are considered. Of the 3,970 firms who classed themselves as purely engineering, there were 1,002 proprietorships, 771 partnerships, and 2,197 corporate and other. By contrast, among architectural firms the partnership was the dominant form.

Though architectural firms tend to rate themselves poorly in the area of business acumen, they are superior to the more sophisticated engineer–assuming that use of the corporate structure is a measure of sophistication. Of all firms with four or more employees reporting in the census, expenses represented 79.4% of receipts. But, the architect emerges a clear-cut winner, as the following tabulation of expenses as a percentage of receipts shows:

Business form	Architectural firms	Engineering firms
All	74.6	81.7
Proprietorships	71.3	72.6
Partnerships	68.5	73.9
Corporations and other	82.4	84.2

There are many contributing factors behind these data, not the least of which is probably the fact that young engineers have tended to command a considerably higher starting salary than their counterparts in architecture.

Table A provides information on engineering firms by specialized types of engineering service and type and location of projects, while Table B does the same for class of client. For those firms classifying themselves as purely civil engineering, well over half the receipts come from projects involving water supply and sanitation facilities, roads, bridges, streets, and railroads. Since most of these types of projects are originated by government at some level, it follows–and Table B confirms–that over half of these firms' income should come from government. Thus, though the engineer in private practice frequently complains about competition from in-house government engineering departments, it is those same departments that generate much of his work. In fact, employees in private practice firms, government, and industry are often rated and remunerated on the basis of a common classification system.

In approaching the subject of engineering management from the quantitative point of view, one important precautionary note is necessary. Just as the rise of automation in industry failed to raise the level of skill of operators and free the better ones from close supervision, so the rise of the computer in civil engineering practice may also fail–if the human side of management is neglected. The engineer who must give up his slide rule for the key punch machine has hardly

TABLE A. ENGINEERING FIRMS BY SPECIALIZED TYPES OF ENGINEERING SERVICE AND TYPE AND LOCATION OF PROJECTS: 1967

(Based on data for engineering firms with 4 employees or more reporting specialized types of engineering service and type and location of projects. Data for all architectural, engineering, and land surveying firms are available in tables 1A, 1B, and 1C)

Specialized types of engineering service provided	Firms	Total receipts	Receipts for architectural, engineering, and land survey services[1]											
			Total	Single family dwelling	Multi-family dwelling	Nonresidential buildings	Water supply and sanitation facilities	Industrial plant, processes and systems	Roads, bridges, streets, and railroads	Civil airports	Power generating and transmission facilities	Flood control drainage, navagation, rivers and harbors	Mining and metallorgical	Other
	(number)	($1,000)	($1,000)	($1,000)	($1,000)	($1,000)	($1,000)	($1,000)	($1,000)	($1,000)	($1,000)	($1,000)	($1,000)	($1,000)
UNITED STATES, TOTAL	3 574	1 661 577	1 645 361	26 868	22 929	258 328	185 686	316 685	259 247	21 108	137 388	40 385	25 540	351 197
FIRMS PROVIDING 1 SPECIALTY														
TOTAL	1 539	411 214	405 076	16 725	8 576	66 299	59 991	62 385	62 406	1 929	11 881	13 521	4 545	96 818
TYPE OF SPECIALTY—														
CIVIL	723	170 472	168 975	14 905	4 524	8 270	49 900	2 890	55 111	1 216	2 792	11 771	178	17 418
SOIL AND FOUNDATION	41	15 661	15 430	1 540	835	4 048	1 141	3 335	2 261	316	682	825	92	355
MECHANICAL	180	37 733	37 192	53	561	10 493	397	15 501	120	48	443	-	600	8 976
ELECTRICAL	150	29 506	29 303	23	549	8 422	432	5 705	76	114	6 400	-	5	7 577
STRUCTURAL	205	40 968	40 520	126	1 960	29 179	401	4 905	1 197	136	255	70	-	2 291
CHEMICAL	13	12 553	12 553	-	-	26	53	12 244	-	-	-	-	-	230
INDUSTRIAL	58	22 469	22 150	5	5	657	400	11 518	63	14	27	22	27	9 412
MINING AND METALLURGICAL	15	4 806	4 707	-	-	-	-	-	-	-	11	-	3 486	1 210
OTHER[2]	154	77 046	74 246	73	142	5 204	7 267	6 287	3 578	85	1 271	833	157	49 349
FIRMS PROVIDING 2 SPECIALTIES														
TOTAL	866	266 179	263 297	7 538	7 627	92 872	33 306	30 237	44 252	2 558	5 470	5 548	2 371	31 518
SELECTED COMBINATIONS—														
CIVIL--SOIL AND FOUNDATIONS	52	12 472	12 471	1 649	352	1 755	1 720	364	4 440	416	75	808	12	880
CIVIL--STRUCTURAL	251	81 995	80 369	4 510	1 812	15 204	11 559	4 840	31 119	1 116	572	2 775	20	6 842
CIVIL--OTHER[2]	73	24 701	24 675	625	364	723	14 028	446	4 282	342	104	1 340	5	2 416
SOIL AND FOUNDATION--STRUCTURAL	30	6 822	6 822	18	179	4 397	34	687	1 405	1	36	7	-	58
MECHANICAL--ELECTRICAL	344	103 341	102 716	222	4 452	68 744	1 115	16 527	172	266	2 376	87	43	8 712
MECHANICAL--OTHER[2]	11	3 678	3 555	(N)	(N)	(N)	(N)	(N)	(N)	(N)	(N)	(N)	(N)	(N)

FIRMS PROVIDING 3 SPECIALTIES														
TOTAL.	477	179 265	178 092	1 885	3 339	28 181	21 851	26 291	49 949	2 094	11 140	5 344	1 336	26 682
SELECTED COMBINATIONS--														
CIVIL--SOIL AND FOUNDATION--STRUCTURAL	153	59 014	58 876	1 064	1 959	5 737	6 527	1 699	36 682	807	483	1 678	3	2 237
CIVIL--MECHANICAL--ELECTRICAL.	47	11 295	11 292	94	352	4 835	588	739	528	142	1 263	23	-	2 728
CIVIL--MECHANICAL--STRUCTURAL.	15	12 027	12 027	234	101	310	4 141	769	710	450	4 873	230	-	209
CIVIL--STRUCTURAL--OTHER[2]. . .	41	18 616	18 333	169	274	473	5 278	565	8 775	228	125	760	-	1 686
MECHANICAL--ELECTRICAL--STRUCTURAL	69	25 187	25 187	35	239	9 878	477	7 104	453	116	1 089	1 076	60	4 660
MECHANICAL--ELECTRICAL--INDUSTRIAL.	36	13 976	13 966	-	105	2 369	310	5 913	-	-	1 073	21	-	4 175
FIRMS PROVIDING 4 SPECIALTIES														
TOTAL.	274	195 723	192 852	412	1 071	18 019	17 685	47 658	21 252	4 818	31 339	2 539	13 060	34 999
SELECTED COMBINATIONS--														
CIVIL--SOIL AND FOUNDATION--STRUCTURAL--INDUSTRIAL . . .	26	6 375	6 375	28	103	527	965	874	3 098	94	141	133	-	412
CIVIL--SOIL AND FOUNDATION--STRUCTURAL--OTHER[2].	24	15 779	15 394	203	186	2 521	1 608	1 633	7 269	209	60	862	20	823
CIVIL--MECHANICAL--ELECTRICAL-STRUCTURAL	112	79 624	78 994	91	336	11 727	8 683	12 423	5 947	4 105	28 638	1 040	-	6 004
MECHANICAL--ELECTRICAL--STRUCTURAL--INDUSTRIAL . . .	18	18 659	18 654	-	-	402	33	15 587	12	-	41	-	-	2 579
FIRMS PROVIDING 5 SPECIALTIES														
TOTAL.	210	326 053	325 137	113	881	32 328	26 857	67 149	37 468	5 908	18 145	5 048	594	130 646
SELECTED COMBINATIONS---														
CIVIL--SOIL AND FOUNDATION--MECHANICAL--ELECTRICAL--STRUCTURAL	75	58 045	57 933	24	354	12 317	10 723	4 456	20 309	3 273	2 078	1 689	-	2 710
CIVIL--MECHANICAL--ELECTRICAL-STRUCTURAL--INDUSTRIAL. . . .	59	86 627	85 863	53	434	11 514	7 817	37 568	4 030	2 297	10 840	2 066	-	9 244
CIVIL--MECHANICAL--ELECTRICAL-STRUCTURAL--OTHER[2].	22	125 613	125 613	4	30	1 922	4 434	2 053	8 084	26	4 194	1 002	-	103 864
FIRMS PROVIDING 6 OR MORE SPECIALTIES														
TOTAL.	208	283 143	280 907	195	1 435	20 629	25 996	82 965	43 920	3 801	59 413	8 385	3 634	30 534

Standard Notes: - Represents zero. D Withheld to avoid disclosure. N Not shown since data may not be representative. NA Not available. X Not applicable.
[1]Includes receipts from all architectural, engineering, and land surveying services; not just engineering specialties.
[2]"Other specialty" is any specialty other than civil, soil and foundation, mechanical, electrical, structural, chemical, industrial, or mining and metalurgical.

TABLE B. ENGINEERING FIRMS BY SPECIALIZED TYPES OF ENGINEERING SERVICE AND CLASS OF CLIENT: 1967

(Based on data for engineering firms with 4 employees or more reporting specialized types of engineering service and receipts by class of client. Data for all architectural, engineering, and land surveying firms are available in tables 1A, 1B, and 1C)

Specialized types of engineering service provided	Number	Total receipts	Receipts from architectural, engineering, and land surveying services[1]								
			Total	From govern-ment	From individ-uals	From construc-tions firms	From archi-tects	From engi-neers	From business firms and farming	From indus-trial firms	Other
		($1,000)	($1,000)	($1,000)	($1,000)	($1,000)	($1,000)	($1,000)	($1,000)	($1,000)	($1,000)
UNITED STATES, TOTAL	3 427	1 384 308	1 371 774	522 020	46 271	57 027	154 146	45 825	70 128	390 745	85 612
FIRMS PROVIDING 1 SPECIALTY											
TOTAL	1 504	357 669	354 814	114 657	24 755	21 101	45 662	16 588	23 684	98 945	9 422
TYPES OF SPECIALTY--											
CIVIL	728	159 219	157 761	85 334	21 882	13 472	3 409	7 801	12 950	9 861	3 052
SOIL AND FOUNDATION	32	11 622	11 622	2 251	709	1 766	1 381	1 775	1 167	2 517	56
MECHANICAL	185	46 224	45 684	3 043	243	483	9 671	1 152	495	29 722	875
ELECTRICAL	146	26 843	26 640	3 707	250	583	7 220	2 344	2 486	9 460	590
STRUCTURAL	202	39 641	39 551	3 167	1 105	3 672	22 310	991	2 735	5 287	284
CHEMICAL	14	5 933	5 933	254	39	26	-	1 039	226	4 119	230
INDUSTRIAL	63	23 114	22 795	2 304	89	318	122	397	224	19 107	234
MINING AND METALLURGICAL	15	4 806	4 707	1 511	11	16	-	265	306	2 503	95
OTHER[2]	119	40 267	40 121	13 086	427	765	1 549	824	3 095	16 369	4 006
FIRMS PROVIDING 2 SPECIALTIES											
TOTAL	861	249 654	246 774	83 690	9 804	14 240	77 157	4 744	12 948	37 649	6 542
SELECTED COMBINATIONS--											
CIVIL--SOIL AND FOUNDATION	52	12 472	12 471	5 686	1 357	2 108	743	438	1 106	930	103
CIVIL--STRUCTURAL	239	70 228	68 602	36 086	4 159	5 411	10 356	1 778	3 542	5 675	1 595
CIVIL--OTHER[2]	70	21 912	21 886	14 474	1 206	793	292	758	1 845	870	1 648
SOIL AND FOUNDATION--STRUCTURAL	30	6 822	6 822	2 712	263	227	3 140	90	147	195	48
MECHANICAL--ELECTRICAL	338	102 108	101 485	12 667	1 704	3 275	61 250	1 294	3 541	16 658	1 096
MECHANICAL--OTHER[2]	24	4 397	4 274	1 241	46	68	70	21	438	2 337	53

FIRMS PROVIDING 3 SPECIALTIES											
TOTAL	456	156 253	155 085	73 654	5 674	3 180	18 821	2 728	9 185	38 137	3 706
SELECTED COMBINATIONS--											
CIVIL--SOIL AND FOUNDATION--STRUCTURAL	145	54 703	54 565	38 136	2 537	1 182	5 178	1 805	2 835	2 392	500
CIVIL--MECHANICAL--ELECTRICAL	45	10 753	10 750	3 204	380	211	3 923	72	1 537	1 144	279
CIVIL--MECHANICAL--STRUCTURAL	15	12 007	12 007	6 064	659	120	96	17	401	4 650	-
CIVIL--STRUCTURAL--OTHER[2]	38	12 848	12 565	8 511	1 212	595	477	89	913	730	38
MECHANICAL--ELECTRICAL--STRUCTURAL	68	22 318	22 318	5 483	121	704	3 527	101	1 634	8 097	2 651
MECHANICAL--ELECTRICAL--INDUSTRIAL	38	15 215	15 210	1 320	46	2	2 304	196	351	10 991	-
FIRMS PROVIDING 4 SPECIALTIES											
TOTAL	259	173 084	170 593	51 760	2 072	12 137	6 798	3 370	5 567	67 326	21 563
SELECTED COMBINATIONS--											
CIVIL--SOIL AND FOUNDATION--STRUCTURAL--INDUSTRIAL	24	4 195	4 195	1 887	152	49	556	215	144	1 183	9
CIVIL--SOIL AND FOUNDATION--STRUCTURAL--OTHER[2]	20	13 216	12 853	8 270	327	634	1 109	394	393	1 665	61
CIVIL--MECHANICAL--ELECTRICAL--STRUCTURAL	106	69 484	69 212	22 089	781	3 061	3 529	955	1 113	19 548	18 136
MECHANICAL--ELECTRICAL--STRUCTURAL--INDUSTRIAL	19	15 735	15 730	733	-	39	47	24	534	12 076	2 277
FIRMS PROVIDING 5 SPECIALTIES											
TOTAL	173	267 510	266 601	127 911	2 373	2 293	3 599	5 190	9 167	76 156	39 912
SELECTED COMBINATIONS--											
CIVIL--SOIL AND FOUNDATION--MECHANICAL--ELECTRICAL--STRUCTURAL	66	42 769	42 657	28 700	813	523	1 560	4 400	1 907	3 699	1 055
CIVIL--MECHANICAL--ELECTRICAL--STRUCTURAL--INDUSTRIAL	52	75 585	74 821	14 966	1 128	857	1 430	580	1 566	52 848	1 446
CIVIL--MECHANICAL--ELECTRICAL--STRUCTURAL--OTHER[2]	20	123 760	123 760	79 892	125	228	399	81	4 285	1 441	37 309
FIRMS PROVIDING 6 OR MORE SPECIALTIES											
TOTAL	174	180 138	177 907	70 348	1 593	4 076	2 109	13 205	9 577	72 532	4 467

Standard Notes: - Represents zero. D Withheld to avoid disclosure. NA Not available. X Not applicable.

[1]Includes receipts from all architectural, engineering, and land surveying services; not just engineering specialties.

[2]"Other specialty" is any specialty other than civil, soil and foundation, mechanical, electrical, structural, chemical, industrial, or mining and metalurgical.

achieved a new sense of personal worth. It must, then, be a credo of management that something be done to counter the impersonality of the computer via closer interpersonal relationships between management and employee–a more professional approach, if you will.

The general approach to business and industrial management involves:

1. Organization, both in the functional sense and in the legal sense of control via the proprietorship, partnership, and corporation.
2. Staffing, both in terms of selection of personnel and in terms of making it easy for members of the organization to "get along."
3. Directing, in the sense of setting broad overall goals.
4. Planning, procuring and assigning work within the limitations of staff capacity and capability.
5. Controlling, both in a qualitative and quantitative (money) sense.

Looked at another way, management can be said to involve two distinct functions: planning and executing. The former includes forecasting, planning, and organizing; the latter includes motivating, controlling, and coordinating. But, none of these functions can be handled effectively if there is poor communication–both up and down the hierarchy. Too often the design engineer is made to feel put upon by the superior who pushes for tight deadlines without giving adequate reason for the need. There is no ready panacea for this problem, but the professional employee who has a part in using modern tools of quantitative management is more likely to understand the need for haste. If he has, for example, been allowed to participate in the preparation of a critical path network for some major design project, he is more likely to understand and accept his own role and the limited time allotted for its completion.

Engineers have been born and bred on the "scientific method." At an early stage in their education they have been taught to:

1. Define the problem.
2. Gather pertinent data.
3. Develop a hypothesis.
4. Test the hypothesis.
5. Apply the hypothesis (if it proves workable) to solve the problem or control the process.

Yet engineers have been quite reluctant to apply the scientific method to "unscientific" areas and have often been scornful of the psychologists, sociologists, and others who have tried. This is a luxury they can no longer afford, for there is growing evidence that quantitative management methods can and will work for those who are willing to try. But they will only work if the engineer learns to gather data–to observe, if you will–with the same objectivity that he uses in scientific study when he is studying the way his professional employees work together within the organization. For it is a truism of management that managers manage people to gain objectives.

The growing interest of engineers in management is evidenced by the formation of the ASCE Committee of Engineering Management in 1971. By the end of its first year of existence, this committee had established the following definition of *engineering management*: "Civil engineering management is the use of technical and business management skill to govern the actions of professional, technical and support personnel working in the civil engineering field."

HISTORY OF THE QUANTITATIVE APPROACH TO MANAGEMENT

Planning and control have long been accepted as two of the primary functions of management. In the early stages of the development of man, these functions were comparatively simple. The Stone Age man of 6,000 to 27,000 years ago gradually evolved fields of specialization such as hunting and weapons production. He gradually added apprentices to perform many of the lesser tasks, but the scope of the activity was relatively small and the planning and control could be envisioned and carried out by the owner-manager.

The scope of managerial activity became more complex during the Chalcolithic Age, from about 4500 to 3000 B.C. During this period, copper was put into use, as was written communication in Babylonia. Of even greater importance, the first great buildings were erected during this period. Perhaps this was the real beginning of civil engineering. This was followed by other civil engineering achievements during the Bronze Age from about 3000 to 1200 B.C. Unfortunately, documentation is not available to provide us with a clear concept of the management planning and control approaches utilized to build the great structures of this era, but we can be comparatively certain that they provided the evolutionary basis for the management techniques now in use by large enterprises.

As evidence that management methods for large enterprises had been developing during the Bronze Age, a generalized organizational structure was included in the Bible in the book of Exodus (Exod. 18:21). About 1300 B.C., Moses told his father-in-law that he was acting as judge for his followers so that they would know the laws. He was admonished by his father-in-law, because the task was too great for only one man, in this fashion:

"Moreover thou shalt provide out of all the people able men, such as fear God, men of truth, hating covetousness; and place such over them, to be rulers of thousands, and rulers of hundreds, rulers of fifties, and rulers of tens."

In the following section (Exod. 18:22), Moses is told to use the basic concept of "management by exception."

"And let them judge the people at all seasons: and it shall be, that every great matter they shall bring unto thee, but every small matter they shall judge: so shall it be easier for thyself, and they shall bear the burden with thee."

The early civil engineering efforts for static types of construction involving surveying, planning irrigation canals, and designing structures had a "side effect." It is reported that surveying and building stimulated the development of geometry as recorded by Euclid about 300 B.C. Consequently, the quantitative approach and civil engineering have been synonymous throughout much of history.

About this same time (i.e., 300 B.C.), Hieron, King of Syracuse, asked Archimedes to establish some method to overcome the Roman naval siege of his city. Since this time, and probably much earlier, political and military leaders have sought the aid and counsel of scientists and engineers to solve war problems. This event is considered to be the beginning of operations research.

The industrial revolution provides the next milestone in the development of quantitative approaches to management. During this period, which varies by country, primary emphasis was on providing alternate power sources to replace beasts of burden and man, and to provide the basic machinery for utilization of this power. For industrial activities, this meant bringing together many workers under one roof for the first time. Thus, it created some of the managerial problems which had been previously encountered by the civil engineer.

Frequently, a discussion of the industrial revolution centers around the textile industry. For example, the invention of the cotton gin by Eli Whitney in 1793 is often heralded as its beginning in the United States. It would be expected that this would be subsequent to industrialization of the textile industry in Britain, since it was British policy to import raw materials from the colonies and export the finished products back to the colonies. In general, this appears to be valid. However, the most pertinent development occurred in France. In 1745, Joseph Jacquard developed a punched card to control the selection of threads in weaving designs. This was to be used subsequently for the input and output method for computational devices. Also, in an allied industry, Perronet made extensive time studies on the manufacture of No. 6 common pins during 1760. This appears to be the first recorded application of industrial engineering techniques.

As noted previously, one of the primary factors associated with the industrial revolution is the development of relatively efficient power sources. While Hero of Egypt devised the concept of the steam engine in 120 B.C., it was late in the 17th century before the concept was useful. The designer was not James Watt, as might be expected. Instead, it was Thomas Savery of England who designed such an engine to evacuate water from subsurface mines. Later, in 1776, Watt and Matthew Boulton designed and manufactured the steam engine to provide rotary power. However, Richard Trevithick is usually given credit for the design of the high-pressure steam engine for locomotion power.

While it is generally accepted that the application of steam-engine rotary power significantly changed the structure of industrial activity, the "spin-off" from the manufacture of these steam engines is frequently neglected. The Watt steam engine would not have been successful without the concurrent development of the boring mill by the Englishman John Wilkinson. Of perhaps greater importance to the application of quantitative techniques was the early utilization of scientific management practices at Watt's Soho Foundry. Fortunately, extensive records were maintained of this venture. First, however, let us consider Watt's background to learn where he acquired his basic concepts.

James Watt has been called a "mechanic" but he was much more. During the period 1769–1773, he was an engineer-surveyor on the Monkland canal. Not only did he probably encounter the application of the steam engine for drainage in this job; he also found himself concerned with management activities, since the contractors of this period shifted most of these burdens to the company's engineer. During the mid-18th century, the term "civil engineering" was first used and the exploits of this new profession became legendary not only because of what was achieved, but also because of how it was achieved. Many outstanding canals, tunnels, bridges, and drainage systems were built during this period. Further, canal-builder Thomas Telford from England formed the first large contracting firm which pioneered many management solutions such as the system of monthly payments and selection of specialized subcontractors.

The background Watt attained during this period was utilized to great advantage. Prior to erection in 1795, he and Boulton planned not only the entire Soho Foundry physical facility, but a complete method of operation for the production of the steam engine as well. In addition to accelerating industrial and machinery development, many quantitative management techniques were evolved and/or utilized. This included elaborate costing techniques, production planning, layout, component standardization, and labor skill subdivision. Thus, many of the quantitative approaches discussed in this section had their roots in civil engineering.

The next man on the scene was an exceedingly colorful Englishman named Charles Babbage. He was called a "scientific gadfly" because of his varied interests and endeavors. He was active in the development and application of statistics, mathematics, computers, manufacturing, industrial research and management. This listing is far from exhaustive because he also was concerned with river tunnel excavation, and invented such things as a lighthouse signaling system, submarine, and rocket apparatus. The list of his achievements is impressive, as was the list of his friends, which included such notables as Darwin, Laplace, Poisson, Fourier, and even Lord Byron's daughter. And, at the age of 36, he was appointed Lucasian Professor of Mathematics at Cambridge (1828–39).

Babbage, the man of many interests, has been called the father of modern computers, operations research, and scientific management. In addition, it is said that he was the anticipator of systems engineering, quality control, industrial statistical analysis, "pilot project" methods, and government subsidy of research.

It is uncertain what his personal contribution was to many of these activities, but he was a genius at utilizing existing knowledge to improve the "state-of-the art" in many diverse fields.

Babbage's contribution to quantitative approaches centers around the development and manufacture of the "analytical engine," which he first conceived about 1828 while he was laboriously validating logarithmic and trigonometric tables. Unfortunately, his mechanical computer was never completed. However, Scheutz, a Stockholm firm, did manufacture a simpler machine in 1855 based on Babbage's ideas.

A major problem which Babbage encountered was the lack of machinery and techniques for producing components to a precise tolerance. This probably motivated his many contributions to the design of machine tools, gauges, die casting, and other production techniques. Of subsequent importance to the discussion of Taylor's contributions in the United States, Babbage illustrated his analytical approach with the problem of two workmen removing earth with spades and wheelbarrows–an excavation task well known to all civil engineers. He noted that the efficiency of the men is a function of the weight of each shovelful, the man's strength, the size and design of the shovel, lift distance, and wheelbarrow design. This, and his work with metal cutting, has given Babbage a claim to the title, "father of scientific management."

At about the same time that Babbage was conceiving his mechanical computer, Eli Whitney was producing interchangeable parts for guns (which previously had been "tailor made" for each "piece"). He had previously contributed to the American industrial revolution with his invention of the cotton gin, but the concept of interchangeability was to have an even greater impact on industrial and quantitative activity throughout the world. Because Whitney encountered many of the same manufacturing problems faced by Babbage, he was forced to invent the milling machine in 1818.

At this point in time, technological and quantitative developments began to accelerate at an increasing rate. Unfortunately, a period of isolationism in the United States made achievements by workers in other countries almost unknown or completely disregarded. The lack of rapid communication was partially responsible for this failure, but the growing national pride of a relatively new political entity cannot be regarded lightly. Although this attitude did begin to dissipate rapidly after World War I, work in the United States proceeded with or without knowledge of the achievements abroad through the contributions of Taylor, Gantt, and the Gilbreths.

Frederick W. Taylor, a mechanical engineer, is often recognized as the father of scientific management and the initiator of the second industrial revolution. However, the term "scientific management" was not coined by him. It evolved in 1910 when Louis D. Brandeis was presenting the famous railroad rate case before the Interstate Commerce Commission. Brandeis contended that a rate increase would not be necessary if Taylor's techniques were applied to the railroad operations. It is not known who proposed the term to identify these techniques at a meeting which Gantt, but not Taylor, attended. But, the term became commonplace and was used by Taylor to identify his work measurement, incentive system, and basic approach to management. These techniques occupy an important position in the field of quantitative approaches to management.

Taylor started working at the Midvale Steel Company in 1878, and there developed his work measurement and incentive approach, performed extensive studies of metal-cutting, and devised the concept of the separation of planning and execution for organizational management. However, he is probably better known for his studies of shoveling and handling of pig iron performed later at the Bethlehem Steel Works, work which is reminiscent of that used by Babbage about fifty years earlier. However, in addition to his published works such as *Shop Management* (1903) and *Principles and Methods of Scientific Management* (1911), Taylor's greatest contribution was probably achieved by his spending the last 14 years of his life serving as an unpaid consultant and lecturer to explain and gain acceptance of his ideas.

It is difficult to separate the contributions of Taylor and Gantt to the quantitative approach. However, since they were fellow students at the Stevens Institute of Technology in Hoboken, New Jersey, this was to be expected. Also, Gantt went to work for Taylor at Midvale Steel during 1887. Later, in 1908, Gantt was retained as a consultant at Brighton Mills in Passaic, New Jersey, upon Taylor's recommendation, to apply scientific management concepts. It was here that Henry L. Gantt devised and tested his now famous Gantt chart for scheduling and controlling production activity. However, it was Wallace Clark who refined and internationally popularized the Gantt charting technique.

Though the work of Gantt and Taylor was similar, there were some notable differences. Gantt emphasized worker training, a "living wage" for all workers, man as the most important resource, and the concept of the enterprise as a system. Taylor emphasized the productive activity, but Gantt conceived the industrial activity as a dynamic system with many subsystems which must be considered as a whole. A statement which Gantt made in 1915 at an ASME meeting in Buffalo typified his basic philosophy. He noted: "It is better to do inefficiently that which should be done than to do efficiently that which should not be done." Indirectly, Gantt was reminding all who utilize quantitative approaches that the qualitative aspects must be considered in the desision process.

Most romanticists are fully aware of Frank and Lillian Gilbreth from the books and ensuing movies entitled *Cheaper by the Dozen* and *Belles on Their Toes*. But this couple did far more than raise six sons and six daughters. They were contemporaries of Taylor and Gantt, but the scope of their activity encompassed many more fields such as construction, sports, hospitals, rehabilitation, and kitchens.

Frank Gilbreth was a man of many achievements, including a number in civil engineering. He invented the gravity and rotary type concrete mixer, a fire-resistant concrete window, and an improved corrugated concrete pile. He started his illustrious career at the age of 17 as an apprentice bricklayer for the Thomas J. Whidden Company, contractors and builders in Boston. Renton Whidden had indicated that Frank might become a partner if he became more proficient in the trades than the workers. It was agreed that bricklaying was most important and Frank should start with that skill. Later, it was reported that he had worked as a journeyman in at least 50 trades.

Frank Gilbreth soon became aware that different bricklayers used different motions to perform the same task. Consequently, he noted that the time required to perform a task was a function of the method utilized. Thus, the field of "motion study" evolved as did his basic philosophy of finding the "one best way." By using a newly designed scaffold, hod-carriers, and improved methods (which reduced the number of motions from 18 to about 5), the number of bricks laid per man hour increased from 120 to 350.

The 1904 marriage of Lillian Moller, a psychologist, and Frank Gilbreth brought to the world far more than 12 children. Lillian injected a sense of compassion and a vitality for life into the partnership, which was reflected subsequently in their approach to and application of motion study. Together, they used motion-study techniques to assist handicapped soldiers and workers. After Frank's death in 1924, Lillian continued these efforts and expanded them to include kitchen activity for women with handicaps or organic ailments.

The Gilbreths developed many methods for motion recording and analysis. These included a classification of basic elements into therbligs (Gilbreth spelled more or less backward), micromotion study, and various recording techniques such as man and machine charts, and right- and left-hand charts. The interested reader is referred to *Frank and Lillian Gilbreth: Partners for Life,* written by E. Yost and published by the American Society of Mechanical Engineers.

After Euclid's efforts in the field of geometry, many mathematicians contributed to quantitative approaches. Thus, the classical methods for optimization had been established by the end of the 19th century. Perhaps Lagrange made the most notable contribution to optimization techniques when constraints are present. His technique establishes the necessary and sufficient conditions required for a global optimal solution to exist. Based on the required conditions of optimality as revealed by the Lagrangian function, many algorithms have been devised for various types of problems to assure attainment of optimality. English mathematician George Boole's book *The Laws of Thought*, published in 1854, is important to modern quantitative methods from another standpoint. His Boolean algebra provided the systematic basis for the computations performed by the modern digital computer.

As a continuing development of the application of statistics to industrial problems considered by Babbage, A. K. Erlang, who was working for the Copenhagen Telephone Company at the time, developed queueing (or waiting line) theory in 1908 primarily for telephone switching applications. This theory is widely used today for relatively small problems, but, because of solution difficulties, simulation is more frequently applied to the larger and more complex problems. This technique, initially called "stochastic sampling," was developed by some unknown mathematician to solve the problem of how many steps a drunkard would have to take on the average to traverse a given distance. While simulation may have been performed initially on an analog computer, the first simulation performed on a digital computer is credited to John von Neumann and Stanislaw Ulam. They used this technique, which they subsequently termed the Monte Carlo method, in 1942 to solve a specific problem for the Atomic Energy Commission.

While there were a number of earlier applications of quantitative statistics in fields such as physics, astronomy, biology and the social sciences, applications in engineering and industry did not occur before 1924. At this time, Walter A. Shewhart of the Bell Telephone Laboratories evolved a "control chart." The important aspect of this development was not the application of statistics to industrial problems so much as it was the simplification of statistical procedures so that shop personnel could understand and use them. As a result of Shewhart's developments and the contributions of many individuals who followed, today's engineer is aware of the random nature of many variables which were previously considered to be constants, and he has modified his analytical techniques accordingly.

It has been recognized that operations research sprang from military activities. In addition to the efforts of Archimedes, there were a number of attempts to utilize mathematics for the analysis of military operations prior to World War I. For example, by 1915 F. W. Lanchester of England had established a relationship between victory and numerical and firepower superiority. However, modern operations research evolved during World War II.

In 1939, the Royal Air Force of England assigned G. A. Roberts and E. C. Williams to study the potential integration of radar into the early aircraft warning system. Subsequently, in 1940, P. M. S. Blackett of the University of Manchester was selected to study the coordination of radar equipment at gun sites with some new device for determining aircraft altitude. To perform this task, he assembled an interdisciplinary group comprised of a surveyor, astrophysicist, general physicist, Army officer, two mathematicians, two mathematical physicists and three physiologists. For obvious reasons, this team acquired the title "Blackett's circus." The intent of using such an interdisciplinary team was to focus many different problem-solving techniques simultaneously on a single large problem. It was hoped that this would cause a cross-fertilization of ideas so that all aspects of the problem and all feasible solutions would be considered. The success of this approach is legendary and included, in addition to the radar studies, depth setting for depth charges for U-boats, ship and submarine searching, and the size of the merchant convoy. Later, this group, without Blackett, became the Army Operational Research Group.

In 1940, J. B. Conant, chairman of the National Defense Research Committee, visited England to investigate the new operational research approach. Shortly after the United States entered the war, both the Army Air Corps and the Navy were utilizing this approach. It was reported that by V-J Day there were 26 operations analysis groups at Air Force headquarters. For the Navy, a section of seven research men headed by P. M. Morse of MIT studied the effects of sea and air attacks against German U-boats in 1942. In addition, offensive mining studies were performed by the Naval Ordnance Laboratory Operational Research Group, headed by W. Michels. Both the RAND Corporation and the Operations Research Office of the Army, with E. A. Johnson as director, were outgrowths of the wartime efforts.

In general, operations research is concerned with problems which have a broad scope, utilize an interdisciplinary team effort and use mathematical models wherever practical. In this section, however, attention will be given to the quantitative approaches known as "management science." To provide a distinction between the two approaches, it is noted that management science is a philosophical approach to assist the manager in the timely formulation of better operational decisions through the application of mathematical models.

After World War II, operations research was readily accepted by business because of the existing shortages and the economic situation. Unfortunately, many of the initial efforts were too strongly research oriented for many managers. They were more concerned with solving current operational problems in a short period of time. Thus, the management science approach evolved to assist the decision maker. In addition, this early work in operations research gave rise to the concept of systems engineering.

Over the ensuing years, a number of quantitative techniques have been evolved so that models are available, many programmed on the computer, to assist the decision maker in a relatively short period of time. For example, von Neumann developed game theory, G. B. Dantzig (in collaboration with M. Wood, A. Orden, and others) developed the simplex method for the solution of linear optimization problems in 1947, and R. Bellman of the RAND Corporation is credited with developing dynamic programming during the 1950s. Some other useful techniques are PERT, devised to assist management control for the Polaris project; and CPM, developed almost concurrently in the late 1950s at duPont.

The foregoing list would be exceedingly long if an attempt were made to note all of the models, techniques, and individuals who have contributed to the field of management science. Such omissions, including the development of computers which are important to the utilization of some models, are for the sake of brevity and are not intended to diminish such contributions.

In summary, a review of the prior history of the quantitative approach to management reveals that many of the approaches, and the evolution of basic philosophies, began within civil engineering. However, other disciplines such as operations research, applied mathematics, and industrial engineering have rushed forward to utilize and further develop these techniques—many of which are finding a place in civil engineering. This is particularly true as the environmental aspects of civil engineering practice move into the public spotlight.

REFERENCES

1. Blair, R. N. and Whitson, C. W., *Elements of Industrial Systems Engineering,* Prentice-Hall, Englewood Cliffs, New Jersey, 1971.
2. McCloskey, J. F. and Trefethen, F. N. (Eds.), *Operations Research for Management,* The Johns Hopkins Press, Baltimore, Maryland, 1954.
3. Pollard, Sidney, *The Genesis of Modern Management, A Study of the Industrial Revolution in Great Britain,* Penguin Books, Baltimore, Maryland, 1965.
4. Davis, G. B., *An Introduction to Electronic Computers,* McGraw-Hill, New York, 1965.
5. Van Norman, R. W., Charles Babbage (1792–1871), *J. Industrial Engineering,* Vol. XVI, No. 1 (Jan–Feb. 1965), pp. 3–7.
6. Mullett, C. F., Charles Babbage: A Scientific Gadfly, *Scientific Monthly,* vol. 67, (November 1948), pp. 360–71.
7. Ritchey, J. A., Louis Dembitz Brandeis, *J. Industrial Engineering,* Vol. XVI, No. 3 (May–June 1965), pp. 159.
8. Scheel, H. V. R., Some Recollections of Henry Laurence Gantt, *J. Industrial Engineering,* Vol. XII, No. 3 (May–June 1961), pp. 220–21.
9. Sevenson, A. L., Henry Laurence Gantt: Industrial Conservationist, *J. Industrial Engineering,* Vol. XII, No. 3 (May–June 1961), pp. 222–23.
10. Provost, R. G., Sr., The Contributions of Henry Laurence Gantt to Scientific Management, *J. Industrial Engineering,* Vol. XII, No. 1 (Jan–Feb. 1961), pp. 62–65.
11. Neely, F. H., Henry Laurence Gantt, *J. Industrial Engineering,* Vol. XII, No. 3 (May–June 1961), pp. 219–21.
12. Yost, Edna, *Frank and Lillian Gilbreth: Partners for Life,* American Society of Mechanical Engineers, New York, 1949.
13. Hammond, R. W., *A Definitive Study of Your Future in Industrial Engineering,* Richard Rosen Press Inc., New York, 1965.
14. Davis, G. B., *An Introduction to Electronic Computers,* McGraw-Hill, New York, 1965.
15. Porter, A., IE in Retrospect and Prospect, *J. Industrial Engineering,* Vol. XIV, No. 5 (Sept–Oct. 1963), pp. 227–37.
16. Kahn, H., *Applications of Monte Carlo,* Project RAND, RM-1237-AEC, Santa Monica, California, April 19, 1954.
17. *Monte Carlo Method,* Applied Mathematics Seminar 12, U.S. Dept. of Commerce, National Bureau of Standards, June 11, 1951.
18. Churchman, C. W., Ackoff, R. L., and Arnoff, E. L., *Introduction to Operations Research,* John Wiley & Sons, New York, 1957.
19. Grant, E. L., *Statistical Quality Control,* McGraw-Hill, New York, 1946.

APPLICATION OF MODELING TO DECISION MAKING

Before discussing specific models and example applications, it is important to establish the basic philosophy, advantages, disadvantages, and limitations of modeling. Initially, let us reconsider the prior definition of management science. This definition was: "Management science is a philosophical approach to assist the manager in the timely formulation of better operational decisions through the application of mathematical models."

First, consider the phrase "philosophical approach." It is an approach rather than a specific discipline. Actually, it may be practiced directly by the manager, by an engineer, mathematician, physicist, or business administration specialist. Thus, the approach and many of the basic tools and techniques may be learned and applied by individuals of differing disciplines serving differing organizational functions. Further, it is a philosophical approach because it must be specifically oriented toward operational decision making. To visualize this philosophy, consider the following comparison of the attitudes of three different "professions" toward the utility of time, the desired decision confidence, and the desired model. For this discussion, when an engineer is serving in a managerial capacity rather than as an engineer, he should be considered as a manager. In actual practice, however, his attitudes probably would be a compromise between the two "professions," as shown in the following table:

"Profession"	Utility of time	Desired: Decision confidence	Desired: Model
Scientist	low	high	"best" possible
Engineer	deadline important	intermediate	"best" that works
Manager	high	relatively low	conceptual

The above table indicates that time is of relatively minor concern to the research scientist, but he wants to have a good model which can be verified by any colleague, and which accurately represents the real situation. The manager, on the other hand, must make his decisions in a relatively short period of time regardless of the information available. It is interesting to note that the operations research applications made within industry immediately after World War II were research oriented, and thus contained attitudes somewhere between the scientist and engineer. Considering the variations in attitudes, it is little wonder that the manager felt that operations research activities were not providing the decision making assistance which he desired. However, management science attitudes and philosophy are somewhere between the engineer and the manager. Consequently, the attitudes are relatively comparable to the engineer serving in a management capacity, and this is the situation encountered by many practicing civil engineers.

According to Dantzig, the USSR is currently attempting to replace middle management by computers and pre-programmed decisions. The Russians believe that this will enable them to overcome "inertia" to change. However, the approach in the United States is the opposite, for a number of reasons. First, not all variables are quantifiable and others are exceedingly difficult or expensive to quantify. For example, factors such as employee loyalty, customer "good will," and public relations may be difficult if not impossible to measure on a numerical basis. Nevertheless, such factors should be considered in the decision-making process. Second, some variables which might have a decided effect on the model have such a small probability of occurrence that inclusion in the model would not be economically desirable. Some examples are the probability of gaining energy self-sufficiency, the start of World War III, and changes in domestic and foreign policies, within the time duration of concern. Third, it is not computationally feasible to include all of the factors which influence the real situation because of the resulting model size. As a general observation, the inclusion of more variables makes data gathering and model construction more expensive, and more time is required to manipulate the model. Finally, a model which precisely conforms with historical data is not desired because this assumes that the past conditions will prevail in the future. Since random variables are actually present, such an approach would imply far more accuracy than really exists.

In consideration of the foregoing, the approach in the United States is to provide data on a management by exception basis, and to provide decision-maker guidance through the application of management science techniques. Since all variables are not included in the model, however, the manager must utilize "intuitive executive judgment" to assure consideration of the factors which have been omitted. Thus, the basic intent of the management science approach is to assist the manager in his decision-making process, and definitely not to replace him by a computer and pre-established decisions.

A review of the figure which portrays the attitudes of the "professions" supports the phrase "timely formulation" included in the definition of management science. If the manager is faced with deciding who should work overtime on a Saturday, information provided the following Monday to guide his decision is less than worthless. It was this lack of concern for the time element which made the manager view early operations research activities with reservation.

The definition goes on to state a concern with the formulation of better operational decisions. Note that it does not say the optimal, or best, decision. An optimal solution is very precise in mathematical terms. In the discussion of assistance to the decision-maker, it was noted that not all of the variables which have an influence on the real problem are included in the model. In addition, the data utilized to formulate, verify, and calibrate the model usually contain varying degrees of accuracy. These factors alone are enough to prevent determination of the optimal solution, but there are more reasons why only a better decision is expected. When problems are solved manually, computational errors are not uncommon, and when solved by computer, significant round-off errors may be involved. Further, a number of techniques (e.g., queueing, simulation, and engineering economics) only enable comparison of the conceived alternatives and are not intended to provide an optimal solution. These techniques simply aid in the selection of the best alternative of those studied.

An even greater problem with optimization exists than is indicated in the prior paragraph. This pertains to the objective, or goal, to be attained. It is difficult to obtain a consensus for the goals from the various functional activities of an enterprise unless a systems approach is being utilized, even though each activity is concerned with the welfare of the enterprise. To envision some of the problems encountered, consider the table presented below (the topics considered here are the number of different house designs and the inventory of building materials for a firm involved in a housing development):

Functional activity	Attitudes of Functional Activity	
	Different designs	Material inventory
Construction	few	large
Finance	moderate to few	small
Engineering	moderate	small

For such a firm, the task of construction would be simplified if a limited number of designs were involved and if the material inventory were sufficiently large so that the probability of work stoppage would be small. On the other hand, the people controlling the money would like to have as little inventory as possible because of the finance charges, but mixed feelings exist regarding the number of different designs. For the engineering attitude, the reader should evaluate the indicated responses to determine if he concurs. Under any circumstance, the point is that goals are viewed differently by the different functional activities. Unfortunately, the same optimal solution will not be attained for different objectives such as "maximum profit" and "minimum costs." Further, the attainment of a goal such as minimum costs in one function may actually result in greater cost to the total enterprise. This is called suboptimization. In summary, the intent of management science is to obtain a better operational decision as related to the overall objectives of the enterprise.

A primary concern at this point is whether obtaining only a better solution is adequate. Fortunately, most of the functions have a "dish-pan" appearance so that we can deviate a moderate amount around the optimal point and still not be far from the minimum or maximum. This is shown in Fig. 1 where quantity is plotted in relation to the average unit cost per quantity produced. Such a function is typical when both fixed costs and variable costs are encountered. Note that the function exhibits a gradual change around the minimum point, so that if the quantity is increased or decreased slightly, the average

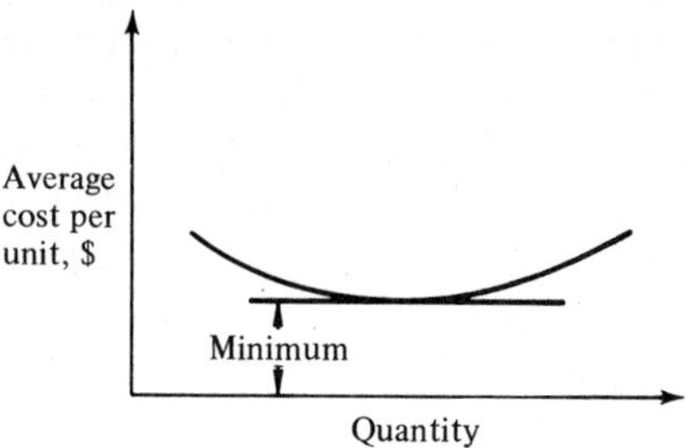

Fig. 1. A typical "dish-pan" plot.

cost per unit does not change appreciably. This means that even without considering all of the factors, and regardless of inherent errors in data, the solution may be comparatively close to the true optimal. Then, the "fine tuning" of the solution is performed by the decision maker to compensate for the other factors.

The last point of concern in the definition of management science is that all of this is to be achieved through the application of mathematical models. According to Churchman et al., there are three types of models: the iconic, or "looks like;" the analog, or "acts like;" and the symbolic, or "stands for." Operations research utilizes any of the three types of models as applicable to the specific problem but with heavy emphasis on the symbolic models. For management science, however, attention is focused on the mathematical models which are a subset of the symbolic category. Such models are relatively compact and can be manipulated readily and relatively economically to investigate the effect upon the system of changing various variables. This capability of manipulation is highly advantageous. It permits synthesis of the effect without disrupting and possibly causing irreparable damage to the enterprise. For example, an investigation could be made to determine how low available capital could become before a firm was forced into bankruptcy. To perform such an experiment on the actual firm, obviously would close its doors.

Before considering modeling and model classification, some of the modeling limitations associated with management science should be restated and additional ones mentioned as applicable. The economics of modeling must be given careful attention. In general, as the models become more complex and include more variables, more data and solution time are required. Obviously, it is impractical to spend more on data gathering, modeling, and solving the problem than can be saved by using the resulting solution. Consequently, not all quantifiable variables are included in the model and usually little attention is given to quantifying the qualifiable variables. In addition, because of the imposed time limitation, the available data which must be used may include inherent inaccuracies. In the most extreme case, historical data are not available for a new system. Thus, equivalent data from other firms, or mere conjectural data, must be used. Finally, even though the stated goal is consistent with the objectives of the total firm, the resulting mathematical solution may not be the best because restraints on the composite system may have been violated.

Regardless of the foregoing, the management science approach does assist in arriving at better decisions. However, the user must be aware of the limitations and specifically call to the attention of the decision-maker any which are especially applicable to the particular problem. Subsequently, the decision-maker can knowingly compensate for these aspects in the establishment of the final decision.

Now that the basic approach and philosophy has been established, specific methods of conforming with the management science definition will be considered. Initially, we must either select or build a mathematical model. While some standard models are available for specific purposes, usually values for the constants in the model must be established. Thus, in this case, the model building element in Fig. 2 requires only a minor effort. When existing models are not available, or do not apply directly, much more effort is required to build the model. In either case, however, the essential steps are as shown on the flow diagram.

The initial step involves the formulation of the model in an attempt to represent the real situation. This is the preliminary design activity. Unfortunately, some individuals will assume that this evolved model is correct and immediately utilize it to obtain a solution. Engineers, however, are aware that some testing must be performed on any designed product before it is released to "production." This is precisely what must be done with the model. Thus, the model should be validated at this point to provide assurance that the real system is adequately represented. This requires a comparison with the real system to determine if, at least within the applicable range, it will react in the intended manner.

When the model does not conform with reality, two courses of action are possible. First, if the variation is consistent between the model and the system, a calibration factor or "correction curve" may be utilized. However, if the error appears random, efforts should be expended to modify the model for achievement of the desired accuracy and repeatability. Then, when the model is considered to be sufficiently valid, manipulation may be performed to arrive at a solution.

The foregoing activity does not necessarily complete the task. In many instances, one investigation will create additional alternatives which, in turn, will be explored. Finally, the results from the model must be interpreted in terms of the real situation. This is where the intuitive executive judgment must be utilized.

Before considering any of the models, a method of classification should be established for a basis of reference and comparison. One possible classification method is by the intended application. As an alternative, the classification

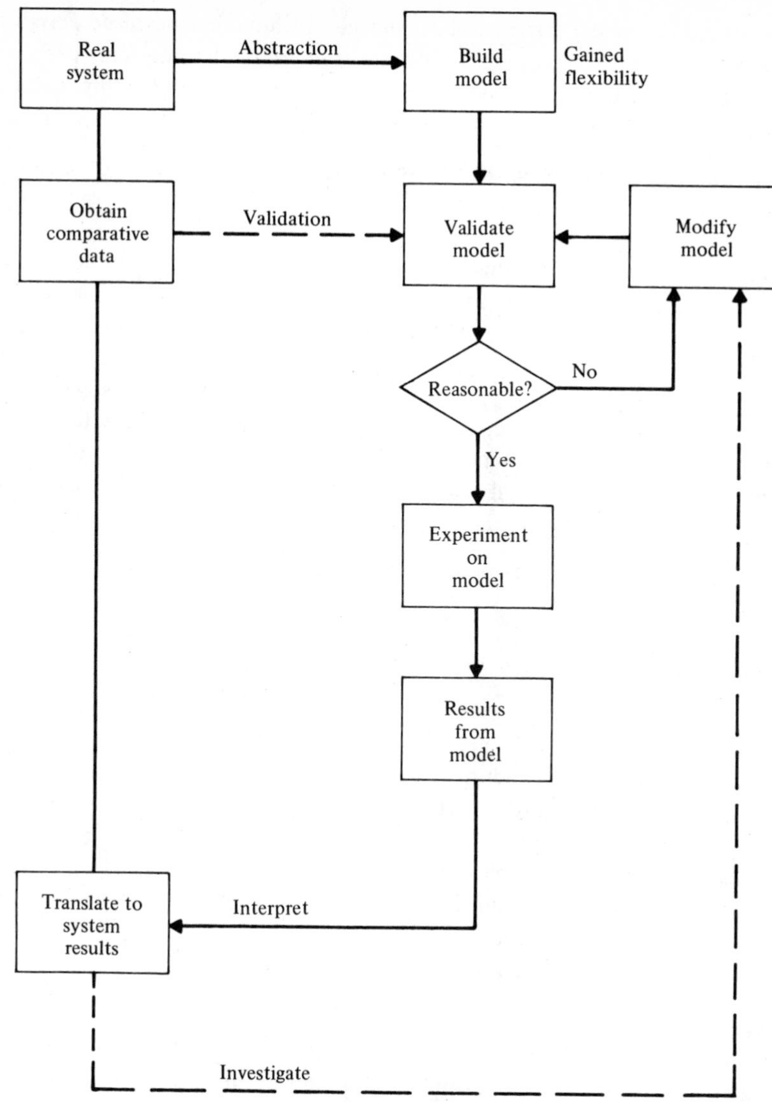

Fig. 2. "Modeling" flow diagram.

could indicate the functional activity where the models were to be utilized. For example, marketing, research and development, financing and production control might be used. An analysis of both the foregoing methods, however, reveals that a considerable amount of confusion would result, since many of the basic models and techniques would appear in more than one category. Consequently, some other method is desired.

The most logical model classification method which has been evolved pertains to the extent of the knowledge of the future state of the world. This pertains to our certainty regarding the events which will happen in the future. Thus, all of the decision models may be classified* into three distinct groupings:

1. *Decisions under certainty.* For these models, the future state of the world is known or assumed and only single-point, or "expected," values are utilized.
2. *Decisions under risk.* In this case, the exact future state is unknown but the probability of being in any specific future state is known or assumed. Thus, random variables and parametric statistical techniques are included in this grouping.
3. *Decisions under uncertainty.* For this classification, the future state of the world may be identifiable but the associated probabilities are unknown and not assumed. When random variables are involved, the probability distributions are unknown and not assumed. Therefore, this grouping includes nonparametric statistics.

To show how this system is used, some general models and techniques have been categorized as follows:

Model Classification

Certainty Models

1. *Calculus:* differential calculus, integral calculus, difference equations.
2. *Advanced calculus:* differential equations, Lagrange functions, topology.
3. *Network analysis:* Gantt charts, line of balance, critical path method, transhipment, dynamic programming, branch and bound.
4. *Linear optimization:* simplex, assignment, transportation.
5. *Nonlinear optimization:* quadratic programming, geometric programming, incremental and tangential approximations.
6. *Other:* engineering economics, break-even charts.

Risk Models

1. *Statistical decision theory:* parametric tests of hypothesis, correlation, regression, exponential smoothing, time series, curve fitting, analysis of variance, information theory, response surfaces, and competitive bidding.
2. *Stochastic Processes:* stochastic programming, queueing, Markovian processes, simulation.
3. *Network analysis:* PERT, GERT.

*Models for decision under risk are relatively scarce, but significant work is being done in the field. However, at the moment little is available in the way of techniques for optimization. Probability decision theory is rather well developed, but too sophisticated to be included in the limited space available.

Uncertainty Models

1. *Statistical decision theory:* Nonparametric and distribution-free tests of hypothesis, rank analysis of variance, rank correlation coefficient.
2. *Linear optimization:* sensitivity analysis, post-optimal analysis.
3. *Other:* game theory, adjustment techniques.

For illustrative purposes, consider the function

$$y = ax^2 + bx + c$$

Further, suppose that we want to obtain the minimum value of y. Using calculus, the first derivative would be set equal to zero and the second derivative would be tested to assure that it is equal to or greater than zero. However, when calculus was utilized, the coefficients frequently were assumed to be constants, e.g.,

$$y = 6x^2 - 18x + 15$$

Here, the minimum y value would be $\frac{3}{2}$ at $x = \frac{3}{2}$. Thus, when constant coefficients are specified, the "future state of the world" is assumed to be known, or a "decision under certainty" is involved.

Now assume that the coefficients a, b, and c are not constants but are random variables with explicit distributions. For example, assume that each variable has a continuous uniform distribution with a from 4 to 8, b from -10 to -24, and c from 5 to 25. The statement "continuous uniform distribution" implies that the variable may assume any fractional value within the indicated range, and it is equally likely that any value in the range will occur. From the foregoing, it is observed that the exact "future state" is unknown, but the probability associated with a given outcome is known or at least assumed. Thus, such a problem would be concerned with a "decision under risk."

The final classification pertains to "decisions under uncertainty." In the case of the previous function,

$$y = ax^2 + bx + c$$

the problem would be worked for all (or a large number) of the feasible values of a, b, and c. In the future, when it was believed that the variables would take on some specific values, the "decision" would be obtained from the previously solved general problem. Thus, when uncertainty models are involved, the solution procedure is usually more extensive.

The remainder of this section is devoted to an introductory presentation of a limited number of management science models which should be useful in the day-to-day practice of civil engineering.

REFERENCES

1. Dantzig, G. B., Management Science in the World of Today and Tomorrow, *Management Science*, Vol. 13, No. 6 (Feb. 1967), pp. C107–11.

2. Churchman, C. W., Ackoff, R. L., and Arnoff, E. L., *Introduction to Operations Research*, John Wiley & Sons, New York, 1957 [Ref. (18) on History].

LINEAR PROGRAMMING

The objective of linear programming is to mathematically optimize a linear equation subject to the restrictions of a set of linear inequalities and/or equalities. A number of iterative approaches to the solution of linear programming problems have been developed that are ideally suited to the talents of the computer. However, there remain some simple problems which are amenable to pencil and paper solution. Allocation and transportation problems are constantly being solved in the engineering organization, and numerous other problems arise for which optimal solutions can be obtained by use of the well known simplex technique.

Linear programming is, then, one of a variety of mathematical models developed for the purpose of optimizing functions. It deals with a linear equation called an objective function which has the general form

$$Z = C_1X_1 + C_2X_2 + \cdots + C_jX_j + \cdots + C_nX_n \tag{1}$$

and which is subject to limitations imposed by a set of linear inequalities of the general form

$$\begin{array}{l} A_{11}X_1 + A_{12}X_2 + \cdots + A_{1j}X_j + \cdots + A_{1n}X_n \leqslant B_1 \\ A_{21}X_1 + A_{22}X_2 + \cdots + A_{2j}X_j + \cdots + A_{2n}X_n \leqslant B_2 \\ \cdots\cdots\cdots\cdots\cdots\cdots\cdots\cdots\cdots\cdots \\ A_{i1}X_1 + \quad \cdots \quad + A_{ij}X_j + \cdots + A_{in}X_n \leqslant B_i \\ \cdots\cdots\cdots\cdots\cdots\cdots\cdots\cdots\cdots\cdots \\ A_{m1}X_1 + \quad \cdots \quad + A_{mj}X_j + \cdots + A_{mn}X_n \leqslant B_m \end{array} \tag{2}$$

Inherent in the simplex technique is the added constraint that

$$X_j \geqslant 0 \quad \text{for } j = 1, 2, \cdots, n \tag{3}$$

Thus, the primary requirement of the LP model is that all X_j decision variables be nonnegative. Further, the A_{ij} and B_i constants pertain to limited or scarce "resources." In the inequality set (2), A_{ij} represents the amount of the ith "resource" required to produce one unit of the jth variable, and B_i represents the limited total amount of ith resource available. The C_j constant may be considered as the cost or profit contribution to the objective function for one unit of the appropriate X_j decision variable.

There are many situations for which LP may be applied as an optimization model. The set of equations, inequalities, and limitations stated above apply to maximizing the objective function. The optimal goal could be one of minimization by simply reversing inequality set (2) to read $AX \geqslant B$.

The most common application of linear programming has been in the area of production management where the problem involves allocation of resources through several common processes in the manufacture of several alternative products. This has frequently involved total net profit as the objective function to be maximized. Such an example will serve here to explain the procedure of simplex, a mathematical iteration process solution devised by George B. Dantzig in 1948. Once the nature of linear programming is understood, the engineer may extend the method to more diverse and perhaps more complex problems of his own and of his clients.

Since inequalities are more difficult to work with than equations, the inequality set (2) can be converted to a set of simultaneous equations by adding a term called a slack variable to each expression. A slack variable may be defined as a number of imaginary "products" utilizing the remaining "resources" to completely satisfy the resource limit in each inequality. In the following examples, S_j will be used to symbolize the slack variables.

Most practical problems will be sufficiently complex (i.e., contain three or more variable activities or products) to make manual calculation impractical. Therefore, most real problems would be solved by computer. Simplex is perfectly structured for digital computer application and standard programs are available, including several for special cases and variations. However, simple problems of two and three dimensions will be used here to demonstrate the physical nature and algebraic iteration involved.

Graphic Solution of a Two-Dimensional Simplex Problem

Assume that a company has manufacturing capacity for two kinds of steel hangers for beams and girders, types A and B, for which profits of \$2.00 and \$1.60 have been well established. There appears to be no limit to demand as far as the company's operation is concerned (i.e., it can sell all the A- and/or B-type hangers it can make with its present production facilities). Both products require different amounts of time in the three common processes: x, y, and z. Average times per product per process have been established and the time capacity of each process is also known. Other notation will be as follows:

N = Number of products
P = Profit
t = Time (T for capacity)
Subscripts identify variables A, B, x, y, and z.

The general form of the objective function becomes

$$\text{Maximize} \quad P = N_A P_A + N_B P_B \tag{4}$$

The inequalities restricting solution are

$$N_A t_{Ax} + N_B t_{Bx} \leqslant T_x$$
$$N_A t_{Ay} + N_B t_{By} \leqslant T_y$$
$$N_A t_{Az} + N_B t_{Bz} \leqslant T_z \tag{5}$$

where

t_{Ax} = time required for product A in process x, etc.
T_x = time capacity for process x.

By introducing slack variables, the inequalities become algebraic equations:

$$N_A t_{Ax} + N_B t_{Bx} + S_x = T_x$$
$$N_A t_{Ay} + N_B t_{By} + S_y = T_y$$
$$N_A t_{Az} + N_B t_{Bz} + S_z = T_z \tag{6}$$

Finally, the nonnegative constraints are

$$N_A \geqslant 0$$
$$N_B \geqslant 0$$
$$S_j \geqslant 0 \quad \text{for } j = x, y, \text{ or } z. \tag{7}$$

The time coefficients and capacities for the specific example being presented are given in Table 1, and the objective function is

$$P_{max} = 2.00N_A + 1.60N_B$$

TABLE 1. TIME COEFFICIENT AND CAPACITIES.

Process	Product time requirement per process: A	Product time requirement per process: B	Process time capacity
x	3.1	6.1	2200
y	5.0	4.4	2000
z	6.0	3.2	2100

Now if we establish A and B as a set of cartesian coordinates (for the variables previously identified as N_A and N_B, respectively), it is a simple matter of dividing process time capacity by product time coefficients in each case to establish A and B intercepts for x, y, and z. These are given in Table 2.

Straight lines for x, y, and z limits may be drawn between appropriate intercept pairs, as shown in Fig. 3. These lines represent physical limits imposed by each capacity individually. For instance, any point falling within space

TABLE 2. INTERCEPTS.*

Product axis	Process intercept		
	x	y	z
A	710	400	350
B	361	455	656

*To facilitate presentation, values have been rounded to the nearest unit.

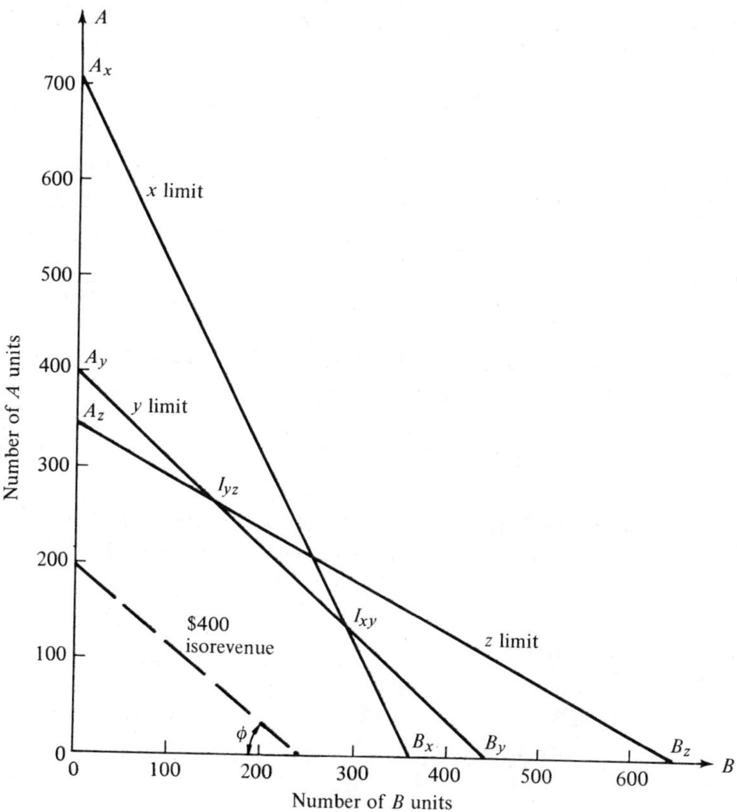

Fig. 3. Plot of data from Table 2.

OA_xB_xO would be possible for process x in manufacturing combinations of types A and B hanger. However, the same statement applies for all processes. Therefore, interaction of limitations of all three processes, indicated by segments of intersecting lines, describes the physical area of feasible production:

$$OA_zI_{yz}I_{xy}B_xO$$

In Fig. 3, based on Table 2, we assume no nonlinear functions exist. The objective function (4) in this case becomes:

$$P = 2.00N_A + 1.60N_B \tag{8}$$

Since the objective function is assumed to be linear, as it is in the preceding equation, an isorevenue line may be established by merely determining the intercepts on the A and B axis for an arbitrary assignment of the profit P. Then, connecting these two intercepts provides the isorevenue line. That is, the graph of the objective function determined in this manner identifies the combinations of A and B which yield the same revenue or profit. For example, if a profit of $400 is assumed,

$$400 = 2.00\,(200) + 1.60\,(0) \quad @\,A \text{ intercept}$$
$$400 = 2.00\,(0) + 1.60\,(250) \quad @\,B \text{ intercept}$$

This $400 isorevenue plot is shown in Fig. 3. From this, it is observed that the manufacture of only 200 type A, or 250 of type B, or a combination of 100 type A and 125 type B, or any combination of N_A and N_B specified by this line would yield a profit of $400.

Because the objective function is linear, all isorevenue lines are parallel. Further, for the optimal solution to be feasible, it must be within the previously identified physical manufacturing constraints. Consequently, the solution point usually can be identified graphically by using two triangles to move the isorevenue line in an increasing value direction until it is coincident with the most extreme point or line of the set of constraints. In this latter case, we would have a range of N_A, N_B possibilities instead of a single-point solution.

For the example of Fig. 3, a $600 profit would provide an increasing isorevenue in relation to the previously plotted $400 line. This would pass through the intercepts of 300 for A and 375 for B. It is observed that the extreme point B_x is between these two lines. Thus, the profit for manufacturing only 361 of type B is less than $600 (actually $577). Continuing this graphical search, it is observed that the isorevenue line which is furthest from the origin, but still has at least one point within the set of constraints, passes through point I_{yz}. This point is the intercept of the constraints for processes y and z. For the graphical solution at this point, $N_A = 273$ and $N_B = 144$. These values maximized the objective function:

$$P_{\max} = 2.00(273) + 1.60(144) = \$776$$

Before proceeding with the algebraic simplex example, some interesting observations may be developed from a study of Fig. 3. The constants or coefficients of individual resource or process determined the three line segments of the boundary, and the profits for A and B determine the slope of the isorevenue (and thus angle ϕ). With the same physical limit boundary but different unit profits, ϕ could be changed so that A_z, I_{yz}, I_{xy}, or B_x would yield P_{max}. It is also possible for different unit profits to have the P_{max} isorevenue fall along line A_zI_{yz}, line $I_{yz}I_{xy}$, or line $I_{xy}B_x$. In any of these three contingencies, the $N_A + N_B$ sum would be variable between line segment intercepts rather than having a fixed N_A and N_B, as in this case, to yield P_{max}.

These two factors concerning present fixed resources and profits suggest a multitude of possibilities if capacity and/or unit profits could be changed. For instance, if management decides that the status quo N_A, N_B, and P_{max} are acceptable under present operations, then either: (1) x capacity could be reduced so its limit line would pass through point I_{yz} and thus reduce overhead, or (2) the excess x could possibly be utilized in some other profitable way.

Subsequent analysis to determine the effect of changing process capacities and/or increasing profits by reducing costs or increasing revenue could be undertaken by utilizing a technique known as post-optimal analysis. In some cases, it may be possible to shift resources internally, and improve the function of a system without adding new resources. However, even with a simple two-dimensional problem, this kind of analysis is usually too complex for graphical solution techniques.

Manual Computation Process Involved in Simplex

Let us examine another simple two-dimensional case, similar to the previous manufacturing problem, but use analytical procedures rather than graphic. Again, assume the production of two products, A and B, each requiring three processes, x, y, and z. Table 3 summarizes time requirements and capacities as was done in Table 1 for the previous example.

TABLE 3. TIME REQUIREMENTS AND CAPACITIES.

Process	Product time requirement		Process time capacity
	A	*B*	
x	12.0	9.6	480
y	8.4	12.3	420
z	7.4	8.3	400

Product A has a unit profit of \$2.50; B, a unit profit of \$3.00. Therefore, if we select maximization of profit as the function to be optimized, the objective function becomes:

$$P = 2.50N_A + 3.00N_B \tag{9}$$

and the set of restriction equations, derived from inequalities to which slack variables have been added, may be written as

$$\begin{aligned} 12.0N_A + \ \ 9.6N_B + S_x &= 480 \\ 8.4N_A + 12.3N_B + S_y &= 420 \\ 7.4N_A + \ \ 8.3N_B + S_x &= 400 \end{aligned} \tag{10}$$

Again, it is obvious by inspection that even though slack variables have been introduced to create equations, the problem is not determinate by direct algebraic solution of simultaneous equations. This is because there are more unknown variables than equations. Consequently, iteration toward a best solution by means of the simplex method is in order. To simplify presentation of the solution, matrix tables will be used for each step of the simplex process. We begin by assuming all slack variables "in solution" as the first simplex step. (That is, for the solution of m equations, a maximum of m variables can be equal to other than zero, and these m variables are designated as being "in solution.") This amounts to assuming our first isorevenue passes through the origin in the previous geometric example. In each succeeding step, that available variable which indicates the greatest unit profit incremental improvement is brought into solution, replacing a prior "in solution" variable by means of a rational process of elimination, as will be demonstrated.

The second column in Table 4, labeled "in mix," indicates those variables in solution at the beginning of the first step. Since we begin at the origin, or $P = 0$, the variables "in solution" are all slack variables. The numbers to the left of each row indicate the unit profit corresponding to the "in solution" variable of that row. The next five row headings (starting with S_x) are variables which can be included in the solution. The numbers above each of the column headings represent the unit profit of the available variable as specified in the original objective function. Finally, column N_r contains the number of units of the "in mix" variable which are in the current solution. For the Table 4 example, $S_x = 480$, $S_y = 420$, and $S_z = 400$. Consequently, the profit for this iteration is $P = 0$ as previously indicated, and this is included on the top line above the N_r column.

TABLE 4. SIMPLEX COEFFICIENTS—STEP 1.

C_j		0	0	0	2.5	3.0	0
C_i	In mix	S_x	S_y	S_z	N_A	N_B	N_r
0	S_x	1	0	0	12.0	9.6	480
0	S_y	0	1	0	8.4	12.3	420
0	S_z	0	0	1	7.4	8.3	400

TABLE 5. SIMPLEX STEP 1.

C_j		0	0	0	2.5	3.0	0	
C_i	In mix	S_x	S_y	S_z	N_A	N_B	N_r	N_r'
0	S_x	1	0	0	12.0	9.6	480 ÷ 9.6 = 50.000	
0	S_y	0	1	0	8.4	12.3	420 ÷ 12.3 = 34.146	
0	S_z	0	0	1	7.4	8.3	400 ÷ 8.3 = 48.193	

	N_A	N_B
Bring in	(1)(2.5)	(1)(3.0)
Take out	(−12.0)(0) (−8.4)(0) (−7.4)(0)	(−9.6)(0) (−12.3)(0) (−8.3)(0)
Result	+2.5	+3.0

An explanation of Table 5 will demonstrate the selective iterative simplex applicable to the subsequent matrix tables. This simplex method has been designed to move progressively from one extreme point of the feasible set space to another by changing only one "in solution" variable at a time. Progressively, this requires a determination of whether the solution can be improved by the introduction of another variable, establishing which "in solution" variable should be removed, and the application of an "elimination" technique to perform the change. This results in the next matrix, which is treated in the same iterative manner.

Of these variables not currently "in solution," we would like to bring in the one which would make the greatest improvement in the value of the objective function. In an attempt to do so, an evaluation is performed to determine the unit profit change if the other variables (i.e., N_A and N_B for the example) are entered "in solution." This evaluation is shown at the bottom of Table 5. For example, to bring in one unit of A requires that 12.0 units of S_x, 8.4 units of S_y, and 7.4 units of S_z must "go out" of the solution to maintain the specified equalities. The net effect of such a change has been included in the row of Table 5 titled "Result." Thus, the introduction of N_A will increase the profit by \$2.50 per unit, and \$3.00 per unit for N_B. Therefore, it is obvious that we should bring "in solution" as many B's as possible.

The next step is concerned with the determination of the maximum number of B's which can enter the solution and still remain within the feasible set space (i.e., all variables must remain nonnegative). At the same time, this establishes which variable will be replaced "in solution" by N_B. To determine this maximum number, the N_r's are divided by the coefficient in the entering variable column for all positive coefficients. (When all coefficients in the entering variable column are nonpositive, the set space is unbounded. That is, the optimal solution for a maximization problem would be equal to plus infinity.) For the example in Table 5, the maximum N_B which may be introduced is 34.146, and this means that the slack variable S_y should be removed.

The last step associated with the simplex iteration involves an algebraic manipulation to bring "in solution" the new variable, set the variable to be removed equal to zero, and to redetermine the value of the objective function. While equations have been developed to permit this to be performed directly within the matrix, the algebraic equation solution technique will be presented here to better portray these calculations.

Initially, we need to solve for N_B in the prior S_y row so that the value of this new "in solution" variable is precisely stated. The prior equation was

$$S_y + 8.4N_A + 12.3N_B = 420$$

Solving for N_B;

$$N_B = (420 - S_y - 8.4N_A) \div 12.3,$$

or, in the original equation format,

$$0.081S_y + 0.683N_A + N_B = 34.146.$$

(Note: While only 3 decimal places are shown here, the calculations were performed to 5 places.) This result has been restated in the second row of Table 6, and it is noted that 34.146 corresponds with the prior limit of N_B calculated for Table 5.

Now that the amount of N_B to be included in the solution is known, we want to "eliminate" this variable from the other equations in a manner equivalent to that used for direct solution of simultaneous equations. Considering

TABLE 6. SIMPLEX STEP 2.

C_j		0	0	0	2.5	3.0	102.439	
C_i	In mix	S_x	S_y	S_z	N_A	N_B	N_r	N_r'
0	S_x	1	−0.780	0	+5.444	0	152.195 ÷ 5.444	= 27.957
3.0	N_B	0	+0.081	0	+0.683	1	34.146 ÷ 0.683	= 50.000
0	S_z	0	−0.675	1	+1.732	0	116.585 ÷ 1.732	= 67.322
	Bring in		(1)(0)		(1)(2.5)			
	Take out		−(−0.780)(0) −(0.081)(3.0) −(0.675)(0)		−(5.444)(0) −(0.683)(3.0) −(1.732)(0)			
	Result		−0.244		+0.451			

the S_x equation in the first row of Table 5, we had

$$S_x + 12.0N_A + 9.6N_B = 480$$

Substituting the prior solution for N_B,

$$S_x + 12.0N_A + 9.6(34.146 - 0.081S_y - 0.683N_A) = 480$$

and collecting common terms,

$$S_x - 0.780S_y + 5.444\,N_A = 152.195$$

A similar substitution for the prior S_z equation yields

$$-0.675S_y + S_z + 1.732N_A = 116.585$$

These results are recorded in rows 1 and 3, respectively, of Table 6.

To complete this iteration, the current value of the objective function for this solution (where $S_x = 152.195$, $N_B = 34.146$, and $S_z = 116.585$) is determined and recorded on the top line (Table 6) above N_r. To assist in these calculations, the unit profit of the variables "in mix," as specified in the original objective function, is recorded in the C_i column to the left of the matrix. Thus, the value of the objective function P at this iteration is the sum of the products of C_i and N_r. Here,

$$P = (0)(152.195) + (3.0)(34.146) + (0)(116.585) = 102.439$$

A subsequent analysis of Table 6 reveals that the value of the objective function may be increased by \$0.451 for each unit of N_A introduced "in solution," and S_x is the variable to be removed. The result of this iteration is presented in Table 7. Here, both evaluators for the variables not "in mix" are negative (i.e., -0.083 for S_x and -0.179 for S_y). Thus, if either of these variables were to be

TABLE 7. SIMPLEX STEP 3.

C_j		0	0	0	2.5	3.0	115.054
C_i	In mix	S_x	S_y	S_z	N_A	N_B	N_r
2.5	N_A	0.183	−0.143	0	1	0	27.957
3.0	N_B	−0.125	+0.179	0	0	1	15.054
0	S_z	−0.318	−0.427	1	0	0	68.171
	Bring in	+(1)(0)	+(1)(0)				
	Take out	−(0.183)(2.5) −(−0.125)(3.0) −(−0.318)(0)	−(−0.143)(2.5) −(0.179)(3.0) −(−0.427)(0)				
	Result	−0.083	−0.179				

brought "in solution" the value of the objective function would be reduced. Consequently, this is the maximum solution for the stated problem.

The optimal solution indicated in Table 7 would yield a profit of \$115.05. Can you tell from this matrix what the product mix would be to attain this profit? Well, we would manufacture about 28 units of product A and about 15 units of product B. Further, this optimal product selection would mean that about 68 units of the process z resource would be unused or could be assigned to some other product.

Admittedly, we have used a simple two-dimensional problem to demonstrate the nature of the simplex iteration process. No matter how many dimensions a problem has or how many resources are required, the simplex solution follows the same path of iteration steps to the optimal solution.

As problem dimensions and resources increase, the manual procedure demonstrated becomes long and costly compared to computer solution. Also, the sequence of iteration might not be so obvious as was the case in the simple demonstration problem.

Graphic Representation of the Three-Dimensional Case

As an extension of the graphic solution for a two-dimensional case, an isometric sketch showing the nature of the LP problem in three dimensions will aid in better understanding the n-dimensional general case. Again, for simplicity, assume a production problem involving three products (A, B, and C) requiring three processes (x, y, and z) in the amounts indicated in Table 8.

The objective function for $P_A = 2.00, P_B = 2.75, P_C = 3.00$ becomes a family of planes (feasible from zero to P_{max}):

$$P = 2.00N_A + 2.75N_B + 3.00N_C$$

The results are shown by isometric sketch in Fig. 4. The intercepts are tabulated in Table 9. The space of feasible or physically possible production is indicated by the quadrant bounds (i.e., A, B and $C \geqslant 0$), $B_zI_{yz}A_yI_{xy}C_xI_{xz}B_z$. In addition, the solution must be on an isorevenue plane which is parallel to

TABLE 8. PROCESS TIME REQUIREMENT BY PRODUCT.

Process	Process time requirement			Process capacity
	A	B	C	
x	6.2	5.0	11.3	1720
y	13.1	10.5	7.9	2400
z	5.5	12.5	11.0	2100

NOTE: A table of A, B, and C axis intercepts for planes x, y, and z could be found.

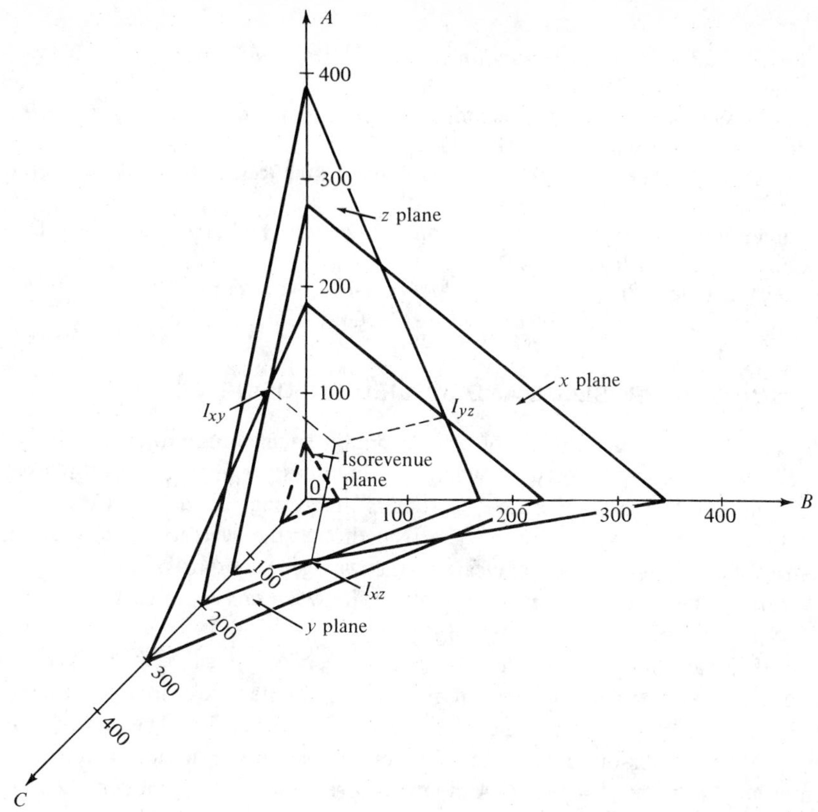

Fig. 4. Three-dimensional model.

TABLE 9. PROCESS INTERCEPTS ON PRODUCT AXIS.

Process	Process intercepts		
	A	*B*	*C*
x	277	344	152
y	183	229	304
z	382	168	191

(Values rounded to nearest digit.)

the plane shown. That is, the last isorevenue plane which is just within the feasible boundary space (by a point, a line or a plane section according to the orientation of things) would comprise the optimal case for maximizing profit.

REFERENCES

1. Brown and Nilsson, *Introduction to Linear Systems Analysis*, John Wiley & Sons, New York, 1962.
2. Charnes, Cooper, and Henderson, *An Introduction to Linear Programming*, John Wiley & Sons, New York, 1953.
3. Goddard, *Mathematical Techniques of Operations Research*, Addison-Wesley, Reading, Mass., 1964.
4. Hillier and Lieberman, *Introduction to Operations Research*, Holden-Day, San Francisco, 1967.
5. Pfeiffer, *Linear Systems Analysis*, McGraw-Hill, New York, 1961.
6. Swartz and Friedland, *Linear Systems*, McGraw-Hill, New York, 1965.

WAITING LINE PROBLEMS AND APPLIED QUEUEING THEORY

One of the most common problems arising in engineering projects is that of periodic situations of congestion where waiting lines (or queues) occur in a system from time to time because "customers" requiring "service" arrive at the facility at a period when it is busy. Even though the average capacity of the service facility is adequate to take care of the average rate of arrival of customers, the waiting line problem arises when intervals between successive customers tend to vary in random fashion.

Queueing terminology of "customer" and "service" must be broadly interpreted. A customer may be a person, machine, computer program input, item in production, or one of many other things too numerous to list. The service facility could be a check-out counter in a market, a repair or production facility, or a computer hardware complex. And, of course, congestion problems are not unique to engineering projects. Queueing theory has application in many fields including sociology, architecture and planning, economics, medicine, and business management. Due to his extensive mathematical training and quantitative approach to analyzing problems, the engineer is uniquely equipped to undertake the solution of complex queueing problems in many areas other than engineering. Thus, the engineer has become a key member of the operations research team in many areas.

The fundamental congestion problem, regardless of specific application, is always the same: how to optimize the balance of service requirements with the capacity of the service facility. Very often the optimization function involves minimizing total cost of operation. The solution procedure involves creating a mathematical model of the real situation and limiting the decision alternatives.

In some cases, application of queueing theory may provide direct quantitative solutions, but more commonly it provides a rational quantitative basis for making decisions rather than automatically providing the decision itself. Thus, queueing techniques fall within the realm of "management by exception" with the responsibility for selecting a "best solution" or optimum "trade-off" assigned to the systems engineer. It is also necessary to mention that many complex problems, particularly in the present sociological planning areas, have not as yet successfully applied queueing theory, although further research should produce far better approaches than many of the purely intuitive methods of the past. It must be the responsibility of practicing engineers to broaden their range of activity beyond traditional boundaries and lend their mathematical muscle and systematic quantitative methodology in team effort with professionals from other disciplines.

Due to limits of space, this section will provide an introduction to the fundamentals of queueing theory and rather simple problem models will be developed. The subject is extensive and many excellent books and papers are available for a more comprehensive study by the engineering practitioner. The bibliography includes references ranging in nature from introduction of fundamentals and simple application to complex mathematical treatment of the subject. Some problems lend themselves to the mathematics of statistical probability, but the engineer must be aware at all times of the necessity of empirical approximation based upon sound principles of observation and sampling theory.

Fundamental Elements of the Queueing Problem

The basic queueing system is depicted in the simplified diagram of Fig. 5. Although every waiting line problem may possess unique characteristics, all queueing systems have the basic elements depicted in Figure 5 which may be briefly defined as follows.

Arriving Population The population or universe from which customers arrive may be infinite or finite and arrivals may be singular or multiple. In actual practice, a population may be finite but large enough so that the distribution of arrivals rate is not influenced by the number of customers in the queue system, either waiting in line or being serviced. In such a case, the population can be assumed to be infinite. If, however, the population is finite, the arrival mechanism will be affected, and this must be taken into account and the problem is more complex. Only the case where the arrival population(s) can be assumed to be infinite will be discussed here.

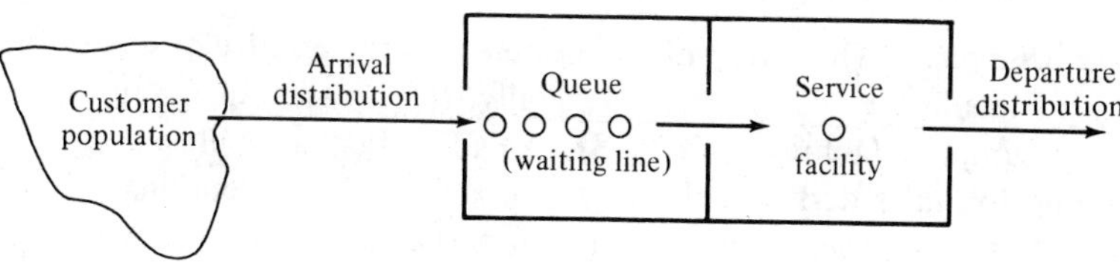

Fig. 5. Simple model of queue system.

Arrival Mechanism For most queueing situations, the average rate of arrivals per unit of time (λ) is assumed to be a constant, and the variable (n) which represents the number of arrivals is assumed to be randomly distributed in accordance with the Poisson distribution presented subsequently. In more complex situations (which are not discussed here), this arrival rate may be a function of both the number in the system and the time. When this arrival distribution exists, the time between arrivals (i.e., the inter-arrival time) conforms to an exponential distribution.

Although queueing theory was first applied on a large scale during World War II, and is generally assumed to be one of the modern tools of operations research, the elements of the theory were developed prior to World War I by a Danish engineer, A. K. Erlang, who was concerned with an economic balance between telephone trunk lines and switchboards. Most of the problems which will confront the engineer today can be approached successfully using the classic Erlang queueing model and resulting equations. Erlang's assumptions regarding random distribution of arrivals satisfy the requirements of the Poisson distribution of probability:

$$P_{n(t)} = \frac{e^{-\lambda t} (\lambda t)^n}{n!} \qquad \text{for } n = 0, 1, \ldots$$

where

$P_{n(t)}$ = probability of n arrivals per time t
e = 2.718. . . (base constant of the natural logarithms)
$n!$ = n factorial; or $1 \times 2 \times 3 \times \cdots \times n$
λ = average number (arithmetic mean) of arrivals per unit time
t = unit time

The derivation of Poisson's equation may be found in most standard probability texts.

It may be seen that λ is the reciprocal of the average time between arrivals. In an actual situation (for instance the problem of loading dump trucks in an excavation project), it would be advisable to verify the assumption of the Poisson distribution by actual time study during the steady-state operation. A sampling program would be recommended for a long-term situation, particularly if any "assignable cause" parameters may be introduced.

Queue Discipline The queue discipline refers to the priority of service for units or customers in the waiting line. Generally, this is based upon a first-come-first served assumption (abbreviated FCFS). A FCFS discipline will be assumed in the queueing formulas derived below. There are notable exceptions, for instance the processing of computer problems on a shared-time system where priority classifications are assigned. In any event, the engineer should carefully check any deviation of discipline from that of FCFS before using the equations developed here. Should another discipline apply, consult bibliography references for appropriate equations.

Probability Distribution of Customer Services Time It will be assumed that customer service times are independent random variables. The statistical distribution can be determined by observation in an actual on-going operation or estimated from available information in a planning problem. The symbol μ denotes service rate, which is the reciprocal of the average service time (assuming a constant service situation; i.e., there may be times when the service facility is empty, so that μ is not equal to the reciprocal of the average time between service completions).

Service Channel Configuration The formulas which will be derived below will apply to a single-channel queue system. This means a single-service-capacity system where customers enter the service facility one at a time on a FCFS basis and only one customer can be accommodated at any time in the facility. Other arrangements are common, such as (a) multi-channel-in-parallel systems, as would be the case when there are several check-out counters in a supermarket; (b) multi-channel-in-series systems where customers departing the service facility of one queueing system become the arrival mechanism for a subsequent queueing system; or (c) a combination of parallel and series channel systems such as is common for enrollment procedures for university courses. In the cases of multiple service channel systems, the mathematics becomes complex and numerical solutions are burdensome, although tables and monographs have been published where basic results are known. The use of computer solutions is certainly indicated for these more complex situations. Some programs have been developed and, as queueing theory becomes more commonly applied, more basic standard computer programs will be developed and made available.

Development of Single-Channel Queueing Formulas

Figure 6 illustrates a single-channel queue. (Assuming exponential arrivals and service resulting in average λ and μ values as above explained.)

Notation:

n = number of customers in the system (queue plus service facility)
$P_n(t)$ = probability of n customers in the system at time, t
Δt = small increment of time following time t
λ = average arrival rate (as previously defined)
μ = average service rate (as previously defined)

Summary of basic arrival and departure probabilities:

a. One arrival in $\Delta t = \lambda \Delta t$

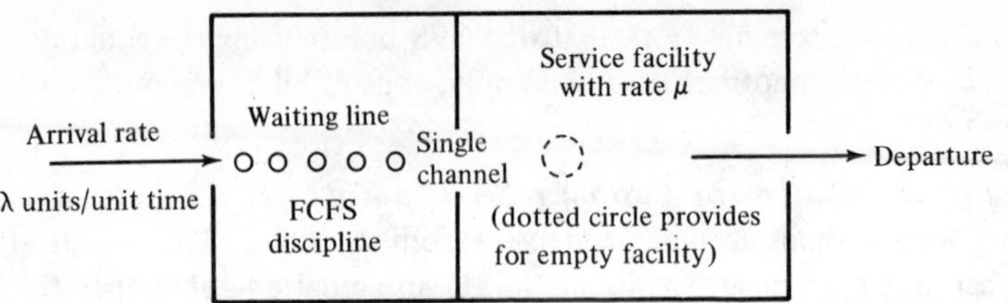

Fig. 6. Single-channel queue.

b. No arrivals in $\Delta t = (1 - \lambda \Delta t)$
c. One departure in $\Delta t = \mu \Delta t$
d. No departures in $\Delta t = (1 - \mu \Delta t)$

The various situations resulting in the probability that there would be n customers in the queue system, including waiting line and service facility, at time $(t + \Delta t)$ are limited to the following four conditions:

a. n in the system at time t with no arrivals or departures during Δt, or

$$P_n(t + \Delta t) = P_n(t)\,(1 - \lambda \Delta t)\,(1 - \mu \Delta t)$$

b. $(n - 1)$ in the system at t with one arrival and no departures during Δt, or

$$P_n(t + \Delta t) = P_{n-1}(t)\,(\lambda \Delta t)\,(1 - \mu \Delta t)$$

c. $(n + 1)$ in the system at t with no arrivals and one departure during Δt, or

$$P_n(t + \Delta t) = P_{n+1}(t)\,(1 - \lambda \Delta t)\,(\mu \Delta t)$$

d. n in the system at t with one arrival and one departure during Δt, or

$$P_n(t + \Delta t) = P_n(t)\,(\lambda \Delta t)\,(\mu \Delta t)$$

Since cases a–d include all possibilities for $P_n(t + \Delta t)$, the total probability can be obtained, in accord with probability theory, through summation of all components, or

$$P_n(t + \Delta t) = \sum_{i=d}^{i=a} [P_n(t + \Delta t)]_i \tag{1}$$

Performing this summation, collecting like terms and dropping infinitesimals of higher order, the results may be written in the following convenient form:

$$\frac{P_n(t + \Delta t) - P_n(t)}{\Delta t} = (\lambda + \mu)P_n(t) + \lambda P_{n-1}(t) + \mu P_{n+1}(t) \tag{2}$$

The left side of (2) represents a change of rate of probability. From calculus (allow $\Delta t \to 0$ in the limit), we can rewrite (2) as a differential equation:

$$\frac{dP_n(t)}{dt} = -(\lambda + \mu)P_n(t) + \lambda P_{n-1}(t) + \mu P_{n+1}(t) \tag{3}$$

Some simplification will be possible by changing our terminology so that $P_n(t) = P_n$, etc. for $n - 1$ and $n + 1$, since we are concerned only with the steady-state conditions (i.e., where $t \to \infty$). If transient conditions (i.e., "start up" and "shut down" process activities) are needed for the solution of a problem, more complex mathematical relationships than shown will be involved. In such a case, references included at the end of this subsection should be consulted.

The left-hand side of the equation, dP_n/dt, is physically interpreted as a rate of change of probability that there will be n customers in the system which is not congruent with the steady-state conditions of process. Therefore, it is obvious that $dP_n/dt = 0$, resulting in

$$(\lambda + \mu)P_n = \lambda P_{n-1} + \mu P_{n+1} \tag{4}$$

Now in the special case where $n = 0$,

$$(\lambda + \mu)P_0 = \lambda P_{-1} + \mu P_1 \tag{4a}$$

It is obvious that a probability for a negative number of people in the system, (P_{-1}), is impossible. Also when $n = 0$ the queue system is inoperative; thus there can be no departures and $\mu P_0 = 0$, reducing $(\lambda + \mu)P_0$ to λP_0.

Thus the relation between P_0 and P_1 becomes:

$$P_1 = \left(\frac{\lambda}{\mu}\right) P_0 \tag{5}$$

Now rewriting (4) in a form convenient for developing the series of probabilities for $n = 1, 2, 3, \ldots, k$,

$$P_{n+1} = \left[\frac{\lambda + \mu}{\mu}\right] P_n - \left[\frac{\lambda}{\mu}\right] P_{n-1} \tag{6}$$

Now let $n = 1$ in (6):

$$P_2 = \left[\frac{\lambda + \mu}{\mu}\right] P_1 - \left[\frac{\lambda}{\mu}\right] P_0$$

and, substituting (5),

$$P_2 = \left[\frac{\lambda}{\mu}\right]^2 P_0$$

In like manner, it is found that $P_3 = [\lambda/\mu]^3 P_0$, etc.

Finally, by induction, $P_k = [\lambda/\mu]^k P_0$ and from probability theory, $\Sigma P = 1.0$, we have:

$$P_0 + P_0 \left[\frac{\lambda}{\mu}\right] + P_0 \left[\frac{\lambda}{\mu}\right]^2 + \ldots + P_0 \left[\frac{\lambda}{\mu}\right]^k = 1$$

$$\therefore P_0 = \frac{1}{1 + [\lambda/\mu] + [\lambda/\mu]^2 + \cdots + [\lambda/\mu]^k}$$

It may be seen that the denominator of this equation is a geometric series of the form, $a + ar + ar^2 + ar^3 + \cdots + ar^k$, where $a = 1$ and $r = [\lambda/\mu]$. For the case of $r < 1$, such a series is convergent with a sum equal to $a/(1 - r)$; or in our terms:

$$\frac{1}{1 - [\lambda/\mu]}$$

and we find:

$$P_0 = \left[1 - \frac{\lambda}{\mu}\right] \tag{7}$$

From the expression for P_k above, let $k = n$, and utilizing (7) the mathematical model for the probability of n customers in the single-channel queue system finally becomes

$$P_n = \left[1 - \frac{\lambda}{\mu}\right] \left[\frac{\lambda}{\mu}\right]^n \tag{8}$$

subject to the important restriction that

$$0 < \frac{\lambda}{\mu} < 1 \tag{9}$$

This is the original contention that the facility capacity must be sufficient to accommodate the arrival rate. When $\lambda/\mu \geqslant 1$, we would have an infinite waiting line.)

Using (7) and (8), a set of simple algebraic equations may be derived for a single-channel queue system, as follows:

Number expected in the total queue system, N (waiting line and facility)

$$N = \sum_{i=1}^{n} i P_i = \sum_{i=1}^{n} i \left[1 - \frac{\lambda}{\mu}\right] \left[\frac{\lambda}{\mu}\right]^i$$

which represents a series of the form $ar + 2ar^2 + 3ar^3 + \cdots + nar^n$ that is convergent to a sum of $r/(1 - r^2)$ if $r < 1$. In this case $a = [1 - (\lambda/\mu)]$, $r = [\lambda/\mu]$ and $[\lambda/\mu] < 1$.

Therefore:

$$N = \left[1 - \frac{\lambda}{\mu}\right] \left[\frac{\lambda/\mu}{(1 - [\lambda/\mu])^2}\right] = \frac{\lambda}{\mu - \lambda} \tag{10}$$

Number expected in the waiting line, $N_w = N - N_F$, where N_F is the number in the facility. Assuming the facility can hold only one customer (single-channel restraint), then $N_F = 1$ times the probability that the facility is in use. This probability is most easily established as follows. If $n = 0$, the queue system (and therefore the facility) must be empty:

$$P_0 = \left[1 - \frac{\lambda}{\mu}\right]\left[\frac{\lambda}{\mu}\right]^0 = \left[1 - \frac{\lambda}{\mu}\right]$$

If there is a probability of n, then the probability that the facility will not be empty is $(1 - P_0)$ or $\{1 - [1 - (\lambda/\mu)]\} = \lambda/\mu$ and $N_F = 1 \cdot [\lambda/\mu]$, then

$$N_w = \left[\frac{\lambda}{\mu - \lambda}\right] - \left[\frac{\lambda}{\mu}\right] = \frac{\lambda^2}{\mu(\mu - \lambda)} \tag{11}$$

Utilization factor of facility efficiency, E_F, is a ratio developed as part of the derivation for N_w above:

$$E_F = 1 - P_0 = \frac{\lambda}{\mu} \tag{12}$$

In effect, (12) expresses the fractional (or percentage) utilization of the full capacity of the service facility.

Expected times in queue system or waiting line. Inasmuch as time is a direct function of cost in most engineering projects, the equations for time are easily determined from the equations relating to expected number, (10) and (11), as follows:

a. Expected time in the system, T. $N = T\lambda$; therefore, $T = N/\lambda$, or

$$T = \left[\frac{\lambda}{\mu - \lambda}\right]\Big/\lambda = \frac{1}{\mu - \lambda} \tag{13}$$

b. Expected time in waiting line, T_w. $T_w = T - T_F$, where $T_F = 1/\mu$ (average time in facility); therefore

$$T_w = \left[\frac{1}{\mu - \lambda}\right] - \left[\frac{1}{\mu}\right] = \frac{\lambda}{\mu(\mu - \lambda)} \tag{14}$$

Summary of single-channel queue formulas (with exponential arrival and service mechanisms and FCFS queue discipline):

1. Probability of n customers in queueing system (waiting line and service facility):

$$P_n = \left[1 - \frac{\lambda}{\mu}\right]\left[\frac{\lambda}{\mu}\right]^n \quad \text{for} \quad n = 0, 1, 2, \ldots$$

2. Expected number of customers in queueing system:

$$N = \left[\frac{\lambda}{\mu - \lambda}\right]$$

3. Expected number of customers in waiting line:

$$N_w = \left[\frac{\lambda^2}{\mu(\mu - \lambda)}\right]$$

4. Expected time per customer in queueing system:

$$T = \left[\frac{1}{\mu - \lambda}\right]$$

5. Expected waiting time per customer:

$$T_w = \left[\frac{\lambda}{\mu(\mu - \lambda)}\right]$$

6. Utilization percentage of service facility:

$$\%E_F = \left[\frac{\lambda}{\mu}\right] \times 100\%$$

In almost all practical applications of these formulas, the basic management problem is one of economics. Considering the operational cost of the service facility or facilities and the costs associated with the existence of the queue (or waiting line), models to select the alternative with minimum total cost are easily constructed using the equations given. Such optimization procedure, of course, would require the assumption of an "infinite" (including reasonably large finite) customer population(s).

SAMPLE PROBLEM OF A SINGLE-CHANNEL QUEUE SYSTEM

Assume a study has indicated that dump trucks will arrive in random fashion at a loading facility. The average rate of arrival is $\lambda = 16$ trucks per hour and and truck-driver unit cost is \$15/h. The average capacity of the loading facility is one truck in three minutes or $\mu = 20$ trucks/h. Facility cost is \$80/h. If we have a single-channel FCFS queueing situation, compare the idle costs for trucks and facility.

The efficiency of the loading facility is $(16/20) \times 100\% = 80\%$. Thus, 20% of the time the unit is idle. Therefore, idle cost for the facility is $0.20 \times \$80 =$ \$16/hour.

To find idle truck time, we first determine the number of trucks expected in the waiting line on the average:

$$N_w = \left[\frac{(16)^2}{20\,(20 - 16)}\right] = 3.2 \text{ trucks in line (average, or expected at any time)}$$

The idle truck cost, then, is $3.2 \times \$15 = \$48$/h.

REFERENCES

1. Benes, *General Stochastic Processes in the Theory of Queues*, Addison-Wesley, Reading, Mass., 1963.
2. Cox and Smith, *Queues*, Barnes & Noble, New York, 1961.
3. Cruon, *Queuing Theory*, American Elsevier, New York, 1967.
4. Goddard, *Mathematical Techniques of Operations Research*, Addison-Wesley, Reading, Mass., 1964.
5. Hillier and Lieberman, *Introduction to Operation Research*, Holden-Day, San Francisco, 1967.
6. Khinchin, *Introduction to the Theory of Queuing*, Hafner, New York, 1960.
7. Lee, *Applied Queuing Theory*, St. Martin's Press, New York, 1966.
8. Ruiz-Pala, et al., *Waiting Line Models*, Van Nostrand Reinhold, New York, 1967.
9. Saaty, T., *Elements of Queuing Theory*, McGraw-Hill, New York, 1959.
10. Saaty, T., *Elements of Queuing Theory with Applications,* McGraw-Hill, New York, 1961.

ENGINEERING ECONOMY

Engineering economy is probably the most frequently used quantitative technique to assist the decision-making process for capital equipment investments or in situations where expenditures and returns occur over a relatively long period of time. This contention is supported by a recent survey to ascertain the importance of subjects included in industrial engineering curriculums. Industrial firms indicated that engineering economy was the topic which should be given the greatest emphasis in the undergraduate curriculum. Most civil engineering schools have long been aware of the importance of this quantitative technique, as evidenced by inclusion of such a course as a requirement for the bachelor degree.

The importance attached to engineering economy arises from the simplicity, directness, and understandability of these models for decisions under certainty. If it is not possible to earn at least as much money on one's capital from a business enterprise as can be achieved from a savings account, bond, or other relatively secure investment, then the continuance of the enterprise should be seriously questioned. Further, when it is necessary to finance at least a portion of the activity by a bank loan or a bond issue, the return must be at least sufficient to repay the loan and applicable interest. In addition, some compensation is desired to counteract the effect of inflation.

To initiate this discussion, simple interest will be considered. In this case, interest is paid only on the amount of money owed. For example, assume that $10,000 has been borrowed at 8% interest to modernize a consulting engineering firm's offices. If this loan is repaid at the end of the first year, the payment would be

$$\$10{,}000\,(1.08) = \$10{,}800.$$

If, however, payment is at the end of the third year and simple interest is involved, 8% interest would be paid on the principal each year, or the total interest would be 24% (i.e., 3 × 8% = 24%). Thus, the resulting payment at the end of three years would be

$$\$10{,}000\,(1.24) = \$12{,}400$$

In general

$$S = P(1 + ni) \tag{1}$$

where S is the sum to be paid in the future, P represents the amount of money involved at time zero (i.e., present worth), n is the number of years that the loan remained unpaid, and i designates the yearly interest rate.

It would be nice for the borrower if simple interest were always involved. However, from the standpoint of the lender, earnings would be greater if full payment were demanded at the end of the first year and new loans issued at the going rate of interest. Obviously, such a procedure would be an inconvenience to both the borrower and the lender. To achieve the same end result, compound interest is used. For the prior example, the amount due at the end of the three years is shown in the following table of compound interest with total sum payment:

Year	Calculation	Amount due at year end
1	\$10,000 × 1.08	\$10,800.00
2	\$10,800 × 1.08	\$11,664.00
3	\$11,664 × 1.08	\$12,597.12

Thus, the single-payment, compound-interest amount would be \$12,597.12 for the use of the \$10,000 over a three-year period. Consequently, this interest payment would be \$197.12 more than that for simple interest to compensate the loan institution for the unavailability of the intermediate interest funds.

In general, this single-payment, compound-interest computation may be expressed as:

$$S = P(1 + i)^n \tag{2}$$

where the symbols have the same meanings as for (1). The expression $(1 + i)^n$ has been tabled in many books to further simplify calculations. However, while such tabled values are utilized for the computations in this section, reasonable accuracy for most desired comparisons may be performed on the slide-rule. Consequently, such tables are omitted from this presentation.

Now consider the reverse situation of that discussed in the preceding paragraph. Assume that someone is willing to make a "lump sum" payment of \$10,000 at the end of four years. If a lender wished to earn 7% interest, how much would he lend the borrower now? Using (2), we can solve for P:

$$P = S(1 + i)^{-n} \tag{3}$$

where P represents the present worth of a future single-payment:

$$P = \$10{,}000\,(1.07)^{-4} = \$7{,}628.95$$

Consequently, if \$7,628.95 were loaned now for a payment of \$10,000 at the end of four years, 7% interest compounded annually would be earned.

Frequently, the loan is to be paid by a series of equal payments as in the case of home mortgages. Consequently, the borrower does not have all of the funds at his disposal for the entire duration of the loan so it would be unjust to calculate the interest according to (2). To develop the applicable relationship for this case, let E represent an equal-payment series when compound interest is involved. This is portrayed in the following figure:

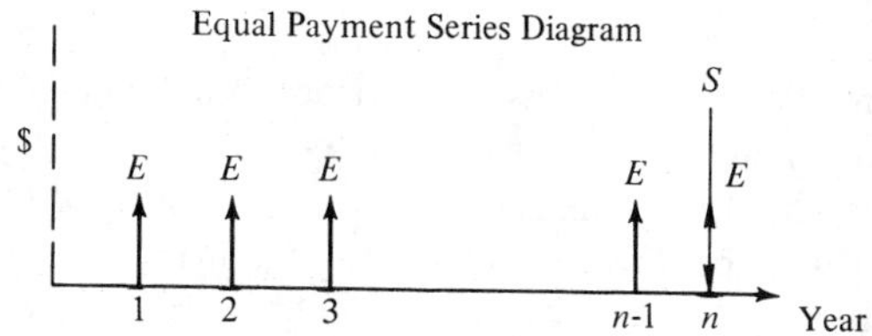

For relative simplicity in establishing the applicable function, the diagram presents the future sum S, which may be equated to n of the E end-of-period payments. Solving for S, it is noted that E, which is paid at the end of the nth year, does not involve interest. The payment at the end of the $(n - 1)$th year, however, accumulates interest for one period and $(n - 2)$th encounters two years of interest, etc. The resulting relationship is

$$S = E + E(1 + i) + E(1 + i)^2 + \cdots + E(1 + i)^{n-1}$$

However, this is an unwieldy relationship for calculations, so some method of summarization is desired. First, multiply both sides of the equation by $(1 + i)$. Then

$$S(1 + i) = E(1 + i) + E(1 + i)^2 + \cdots + E(1 + i)^n$$

Now subtract the second function from the first. This yields

$$S(1 + i) - S = E(1 + i) + \cdots + E(1 + i)^n - E - \cdots - E(1 + i)^{n-1}$$

Simplifying:

$$Si = E(1 + i)^n - E$$

or

$$S = E\left[\frac{(1+i)^n - 1}{i}\right] \tag{4}$$

This is termed the compound-interest future sum value of the equal-payment series E.

For an example of this, assume that you intend to deposit \$500 in a savings account at the end of each year during your 40-year working career to provide a retirement fund. Further, assume that the deposit is to be made at the end of the year and the bank pays 5% interest. The question is, of course, how much money would you have when the "golden years" arrive. Here

$$S = \$500\left[\frac{(1+0.05)^{40} - 1}{0.05}\right] = \$60{,}400$$

so that by saving \$500 a year during the usually productive life span and investing it in a 5% interest bearing savings account, you would have a "nest egg" of about \$60,000.

The remaining relationships for discrete-period compound interest are variations derived from (2) and (4). For example, to determine what equal-payment series is required so that the "bank account" will be a specified size after a given number of years, (4) may be inverted to yield the following equal-payment series, sinking-fund equation:

$$E = S\left[\frac{i}{(1+i)^n - 1}\right] \tag{5}$$

As an example, assume a husband wishes to attend a civil engineering conference in Paris three years from now, with his wife included in the tour. The air fare will be about \$600 per person, and housing, food, registration fees, and entertainment will average about \$75 per day for a two-week trip. However, it is obvious that a contingency fund needs to be established to cover such "minor" items as a designer dress, and various souvenirs for the children and relatives. About \$1,800 would cover these items, and an additional \$400 gift should appease grandma, who has agreed to stay with the children. This sums up as follows:

$$2 \times \$600 + 14 \times \$75 + \$1{,}800 + \$400 = \$4{,}450.$$

If savings can be banked at 5% interest, how much must be set aside on a year-end basis so that the money will be available at the end of three years for the trip?

$$E = \$4{,}450\left[\frac{0.05}{(1.05)^3 - 1}\right] = \$1{,}411.58$$

The foregoing solution is not overly realistic since many banks currently compound the interest more frequently than once a year and deposits would probably be made on a monthly basis. To envision the effect on the interest rate of more frequent compounding, let i^* represent the nominal annual interest rate (i.e., the interest rate without intermediate-period compounding) and let i represent the effective annual interest rate which is really involved. By definition, the effective interest is

$$i = \left[1 + \frac{i^*}{t}\right]^t - 1 \tag{6}$$

where t represents the number of times interest is compounded during the year. If quarterly compounding were involved for the previous example:

$$i = \left[1 + \frac{0.05}{4}\right]^4 - 1 = 0.050945$$

This is not a very dramatic change for just one year, but it may be if many years are involved. For the desired future sum, the deposit per year would be

$$E = \$4{,}450\left[\frac{0.050945}{(1.050945)^3 - 1}\right] = \$1{,}410.27$$

If it is assumed that interest is paid on the amount in the account at the beginning of the quarter and end of the period quarterly deposits are made, a deposit of \$346.02 per quarter for the three-year period would provide the funds for the trip.

As an extension of the foregoing, if the number of compounding periods increases to the limit, the resulting effective interest for this continuous compounding is:

$$i = \lim_{t\to\infty}\left[1 + \frac{i^*}{t}\right]^t - 1 = e^{i^*} - 1 \tag{7}$$

Considering other functions of pertinence, replace S in (3) with (4) and a relationship is yielded to convert an equal-payment series into an equivalent present worth:

$$P = S(1+i)^{-n} = E\left[\frac{(1+i)^n - 1}{i}\right](1+i)^{-n}$$

or

$$P = E\left[\frac{(1+i)^n - 1}{i(1+i)^n}\right] \tag{8}$$

Finally, the preceding relationship may be inverted to determine the equal-payment series required to repay a loan. For the person making the loan or investment, this is considered capital recovery and is expressed as

$$E = P\left[\frac{i(1+i)^n}{(1+i)^n - 1}\right] \tag{9}$$

To assist subsequent engineering economy calculations, the foregoing relationships are summarized in the table of conversion equations

Conversion Equations

Calculate	From	Multiplier	Relationship
For Simple interest			
S	P	$(1 + ni)$	future sum of present worth
For compound interest			
S	P	$(1 + i)^n$	compound interest, single payment
P	S	$(1 + i)^{-n}$	present worth of future sum
S	E	$\left[\frac{(1+i)^n - 1}{i}\right]$	future sum of equal series
E	S	$\left[\frac{i}{(1+i)^n - 1}\right]$	equal series of future sum
P	E	$\left[\frac{(1+i)^n - 1}{i(1+i)^n}\right]$	present worth of equal series
E	P	$\left[\frac{i(1+i)^n}{(1+i)^n - 1}\right]$	equal series of present worth
Other			
i		$\left[1 + \frac{i^*}{t}\right]^t - 1$	effective interest
i		$e^{i^*} - 1$	continuous compounding

Engineering economy is the technique which is frequently used to evaluate capital equipment investment alternatives. If the expenses associated with the acquisition of equipment such as a crane or bulldozer had to be paid from its use on only one job or for only one year, few labor saving devices would be purchased by any enterprise. Further, since the equipment or buildings will be used for a number of years and for different jobs, business and the Internal Revenue Service have adopted the concept of depreciation. This enables the enterprise to charge a portion of the costs each year as operating expenses for the "wear and tear" encountered. In actuality, many different depreciation models have been approved by IRS when the method is applied consistently by the enterprise. The most frequently used approach, and the only one considered here, is the straight-line depreciation model. Other approaches are known as the "sum of the years digits" and "declining balance."

Straight-line depreciation, as the name implies, assumes that the remaining value of any capital asset is represented by a linear function with a negative slope. The difference between the asset value at the beginning and end of the same year is considered as the cost of depreciation for that year. This value is identified as the capital recovery for the year and for the selected model it is constant for the usable life of the asset. However, an additional factor must be considered and this is the salvage value. Even though the asset may be of no further use to the original firm, it may be sold as scrap or to another firm for use on its projects. Obviously, this salvage value must be an estimate since the actual amount received will not be known when various alternatives are being considered initially. In equation form,

$$D = \frac{(P - V)}{n} \tag{10}$$

where D is the depreciation or capital recovered each year by the straight-line model, P is the initial cost (i.e., present worth) of the asset, V denotes the estimated salvage value to be received at the end of the useful life period, and n represents the estimated useful life of the equipment in years.

Now that the basic relationships have been indicated, the primary intent of engineering economy, which is to evaluate alternative investment decisions, will be explored. The different situations which may be encountered are the evaluation of single and multiple alternatives. In reality, however, the single-alternative problem does involve at least two choices, where the second is to not purchase the equipment. In addition, regardless of the alternative selected by engineering economy, usually the total amount of funds available for capital equipment expenditures is limited to something less than the total desired investment.

The only difference encountered for multiple rather than single alternatives is that the selected "measure of merit" is established for each alternative and the one selected exhibits the "best" value. Consequently, the subsequent discussion is limited to consideration of the single investment alternative. Several different comparative measures may be utilized, such as present worth, equivalent equal series, future sum, rate of return on investment, or service life.

To portray the various methods, assume that the procurement of a \$40,000 bulldozer which has an expected usable life of eight years is being considered. Assume also that the expected salvage value of this equipment after eight years of usage is \$8,000. Then,

$$D = \frac{(P - V)}{n} = \frac{\$40{,}000 - \$8{,}000}{8} = \$4{,}000 \text{ per year}$$

or the capital recovery (i.e., depreciation) for each of the eight years would be \$4,000. Now assume that an additional cost of \$5,000 for maintenance will be necessary during the fifth year and that a study reveals a potential reduction in operational costs of \$7,800 per year during the useful life of the bulldozer. These expenditures and returns are presented in the following table, where the salvage value has been combined with the returns for the eighth year:

Capital Investment Expenditures and Returns

Year	Expenditures (\$)	Returns (\$)
0	40,000	
1		7,800
2		7,800
3		7,800
4		7,800
5	5,000	7,800
6		7,800
7		7,800
8		15,800

The present-worth, equivalent annual series amount and the future sum are interrelated as shown in the conversion equations. Consequently, only the present-worth comparison is considered. Regardless of the method of comparison, the basic intent is to state the expenses and revenue in common terms by compensating for the "time value of money." Recognizing that an expenditure is a negative return and using an 8% interest rate,

$$P_e = -\ \$40{,}000 - \$5{,}000(1 + 0.08)^{-5} = -\ \$43{,}402.90$$

Thus, since the maintanence cost is a future sum expenditure which occurs during the fifth year, the present worth of these costs (i.e., P_e) is \$43,402.90. For the reduced operating costs and salvage value,

$$P_r = \$7{,}800\left[\frac{(1.08)^8 - 1}{0.08(1.08)^8}\right] + \$8{,}000(1.08)^{-8} = \$49{,}145.95$$

or the present worth of the return is \$49,145.95. To determine whether this is a desirable investment, the combined present worth is evaluated:

$$P = P_r + P_e = \$49{,}145.95 - \$43{,}402.90 = \$5{,}743.05$$

Since the present worth is positive, the recommended decision would be to purchase the bulldozer. If it were negative, however, the desired minimum return of 8% would not be realized and the decision would be to forgo procurement.

The rate-of-return comparison is somewhat different and is frequently utilized to enable ranking of various, often unrelated, investment opportunities. For this comparison, the interest rate required to equate the presentworth for income and expenditures is determined. When a single alternative is being considered, this interest rate is compared with some arbitrary value established by management as the minimal acceptable return. If the investment opportunity exceeds this standard, procurement of the equipment is recommended. Using the prior example, the equated present-worth relationships,

$$\$40{,}000 + \$5{,}000(1+i)^{-5} = \$7{,}800\left[\frac{(1+i)^8 - 1}{i(1+i)^8}\right] + \$8{,}000(1+i)^8$$

are solved for i, the interest rate. It is observed that this may have one or several solutions when such a polynomial is involved. Consequently, the use of this approach has been questioned by many. However, it is frequently argued that the approximate answer is known and this approach merely assists in locating a "logical" interest rate. Using this latter approach, it is found that $i = 11.5\%$ for this investment. Thus, the decision would be to purchase the equipment if the minimum desired return were 8%.

Another method used frequently in industry is to perform a service life comparison. In this case, a zero interest rate is usually used and the time period for the asset cost to be recovered from the return is determined. For the prior example,

$$k(\$7{,}800) + \$8{,}000 - \$5{,}000 = \$40{,}000$$

or

$$k = 4.7 \text{ years}$$

where the \$8,000 salvage value and the \$5,000 maintenance cost is assumed to be encountered within the applicable time span. Thus, almost 5 years would be required for the payout of this investment. While the foregoing analysis yields a measure for comparison with a standard, it violates the basic philosophy of engineering economy since it does not consider the "time value of money." Inclusion of this factor will tend to increase the payout period. Using 8% as before, the number of years required to recover the associated costs may be calculated from the following equation:

$$\$40{,}000 + \$5{,}000(1.08)^{-5} = \$7{,}800\left[\frac{(1.08)^n - 1}{0.08(1.08)^n}\right] + \$8{,}000(1.08)^{-n}$$

where it is assumed that at least a 5-year service life is required so that the maintenance cost will be encountered, and where it is assumed that the \$8,000 salvage value will be attained at the end of n years. Obviously, this salvage value would be a function of the age and condition of the bulldozer, but the foregoing assumption is utilized to simplify calculations. Using logarithms to solve this relationship, the resulting answer is 6.5 years. This is greater than the prior 4.7 years, as anticipated. Thus, if management had established a standard maximum payout period less than this, say 5 years, the recommendation would be to not consider this investment further.

REFERENCES

1. Morris, W. T., *The Analysis of Management Decisions* (revised ed.), Richard D. Irwin, Inc., Homewood, Illinois, 1964.
2. Fabrycky, W. J., and Torgersen, P. E., *Operations Economy, Industrial Applications of Operations Research*, Prentice-Hall, Inc., Englewood Cliffs, New Jersey, 1966.
3. Barish, N. N., *Economic Analysis for Engineering and Managerial Decision-Making*, McGraw-Hill, New York, 1962.
4. Kooy, E. D., Editor, "Industrial Engineering Subject Survey," *AIIE News of Facilities Planning and Design Division*, Vol. VI (Dec. 1971)

ECONOMIC PERFORMANCE CHARTING

The desirability of using an overall optimal approach is axiomatic. However, it is difficult to obtain an objective view of the "big picture" because such a large number of factors are involved, pertaining to both the external and internal environment of the firm. Consequently, it is relatively easy to become lost in the details associated with the complex, dynamic enterprise. Thus, an approach is needed to assist in the attainment of the proper perspective.

To perform such a comprehensive analysis of the system requires an initial establishment of the amount of detail to be considered, and the level of sophistication to be incorporated in the quantitative approach. Obviously, the greater the detail and level of sophistication, the more expensive will be the data accumulation, model building, and attainment of a solution. Further, the inclusion of more detail usually means that the model becomes obsolete faster. Thus, not only is more expense encountered in model building and manipulation when more detail is involved, but subsequent efforts required to maintain the desired level of accuracy are also greater.

In consideration of the foregoing, the approach which follows looks at the performance of the enterprise in a gross, generalized manner, and may involve only simple arithmetic, algebra, or differential calculus as required. This approach is called economic performance charting (alternately, the profit graph or the break-even chart), which is a model for decision under certainty.

For the most simple application of economic performance charting, only the cost of sales before income taxes for a given time period (usually a year) is considered. These costs are plotted as the Y variable and the sales revenue as the X variable. For example, assume that the following economic performance has been achieved by a medium-sized construction firm (cost and revenue are in millions of dollars):

Year	Sales revenue	Cost of Sales
1967	10.1	13.2
1968	34.3	28.4
1969	29.4	23.3
1970	17.5	15.7
1971	24.2	22.4
Total	115.5	103.0
Average/year	23.1	20.6

With such a limited number of datum points (5 for this example), it is considered undesirable to use other than a linear function to approximate the relationship over the applicable sales revenue range. Even when more data are available and a nonlinear function appears to be present, linear functions are usually used for the sake of simplicity. In such a case, however, the sales range may be subdivided so that the relationship is adequately approximated by linear functions within the subrange. For the example, assuming a sufficient number of datum points in each subrange, one function might be fitted for sales equal to or less than 21 million dollars and another function fitted for sales greater than this amount. Obviously, when this approach of splitting the sales range is utilized, care must be exercised to assure that the applicable function is used for the pertinent range.

The simplest approach is to "eyeball" a line through the points on the "scatter chart," Fig. 7. To do this, some basic rules should be followed so that the linear function will be reasonably accurate. The first rule is that the line must pass through the point represented by the average of the observed X and Y variables. These averages have been indicated for the example in the economic performance table foregoing (i.e., X (average) = 23.1 and Y (average) = 20.6). The second rule is that the sum of the distances from the line to the points above the line should be about equal to the equivalent sum of distances for the points below the line. To achieve this, a transparent plastic rule or triangle may be used. The straightedge is then pivoted about the average point until the variation above and below the line are about equal. Such a result for the example is presented in Fig. 8. The average point (or centroid) is shown as +, and the linear function is dotted beyond the sales revenue range for the observed data. These line segments are dotted to indicate that any results in these ranges should be treated cautiously because equivalent actual performance has not been observed.

If desired, more sophisticated techniques may be used to establish the functional relationship. For a simple linear regression, the line $y = a + bx$, where y is the independent variable to be predicted, may be established by the least-squares

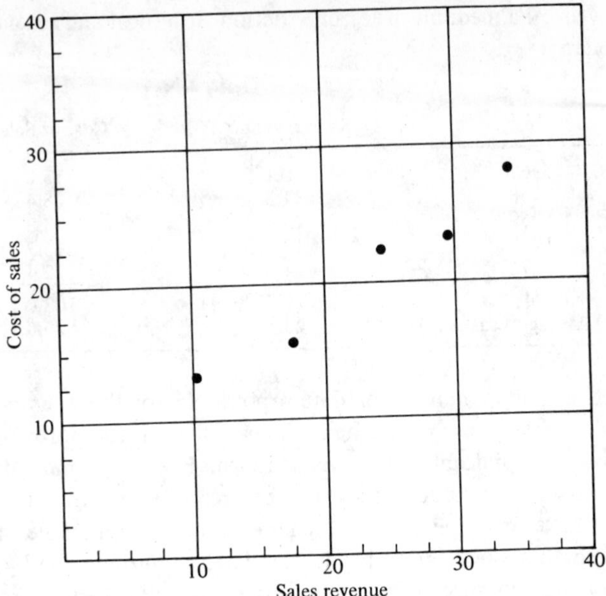

Fig. 7. Economic performance scatter chart (in million dollar units).

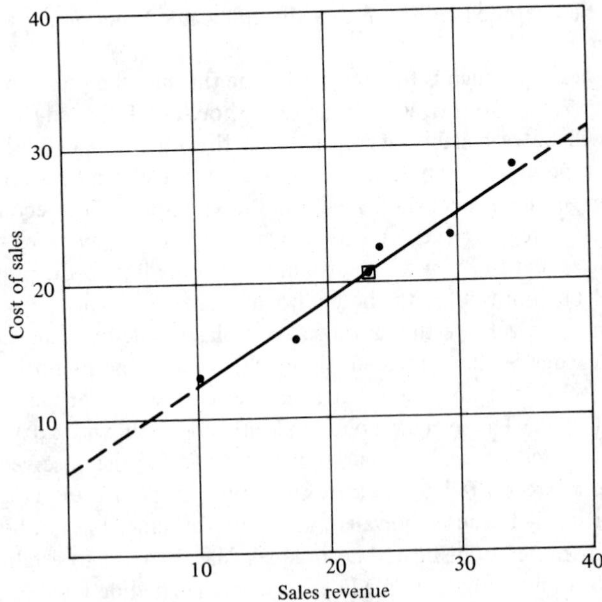

Fig. 8. Economic performance chart, "eyeball" linear function (in million dollar units).

technique. Here,

$$b = \frac{n\sum x_j y_j - \sum x_j \sum y_j}{n\sum x_j^2 - \left(\sum x_j\right)^2}$$

and

$$a = \bar{y} - b\bar{x}$$

where $\sum$ represents a summation over the n sets of paired data, and $\bar{y}$ and $\bar{x}$ represent the average values. Further statistical tests may be performed on a and b, but discussion of this is omitted.

Using the prior example, the basic and final calculations are provided for the least-squares linear function:

	X	Y	X^2	Y^2	XY
	10.1	13.2	102.01	174.24	133.32
	34.3	28.4	1,176.49	806.56	974.12
	29.4	23.3	864.36	542.89	685.02
	17.5	15.7	306.25	246.49	274.75
	24.2	22.4	585.64	501.76	542.08
Total	115.5	103.0	3,034.75	2,271.94	2,609.29

Noting that the sample size $n = 5$, the final calculations are

$$b = \frac{5(2{,}609.29) - 115.5(103.0)}{5(3{,}034.75) - (115.5)^2} = 0.62719$$

and

$$a = 20.6 - 0.62719(23.1) = 6.11195$$

Thus, $y = 6.11195 + 0.62719x$.

To verify that the "eyeball" fitting of the linear function is sufficiently accurate for the intended purposes, compare the two functions. Figure 8 shows an approximate intercept $a = 6.0$ and slope $b = 0.625$, or a linear equation of $y = 6.0 + 0.625x$. This compares favorably with the least-squares fit of $y = 6.11195 + 0.62719x$. Consequently, for the intended purposes of the economic performance charting, the "eyeball" fitting technique is usually sufficiently accurate.

Now consider the example linear function in greater detail as presented in Fig. 9. A line with a slope of +1 which passes through the origin has been added. This represents the sales revenue regardless of the associated costs. Thus, the difference between this revenue and the cost line represents the profit or loss at that sales volume. If the cost line appears above the revenue line, the difference

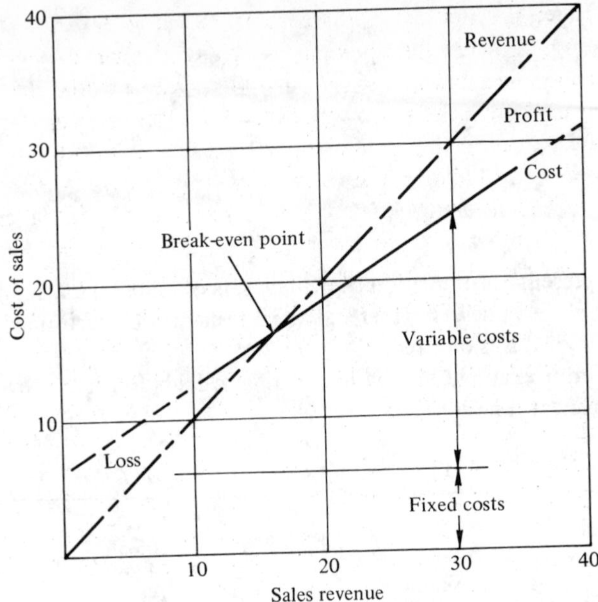

Fig. 9. Economic performance chart, example (in million dollar units).

represents a loss, as indicated on the left side of the figure. However, if the revenue is greater than the cost, as on the right side of the figure, the differential represents profit.

When the cost and revenue are equal, this is termed the "break-even point" because neither a profit or a loss is involved. This is at a sales revenue of 16 million dollars for our example. Thus, to have a profitable enterprise with the specified cost history, the sales volume must exceed this amount.

It is apparent that both the intercept and the slope of the cost function are important. These costs can be considered as being composed of two elements. The first is the fixed costs (i.e., the intercept), which may be envisioned as the stand-by costs encountered even though the enterprise is not functioning. For example, depreciation of owned equipment and buildings continues regardless of whether they are utilized. Obviously, while these costs are termed "fixed" for the appropriate range of the sales volume, this element may be changed by management decision. For example, equipment may be either sold or purchased. Such action would modify the indicated "fixed costs."

The second cost element is termed "variable" (i.e., the slope of the cost function) and it is considered directly proportional to the sales activity. For the example, the indicated variable cost was $0.625 for each dollar of sales revenue. This includes factors such as labor and material used in the construction projects. In other words, it is assumed that if the sales are doubled, an associated doubling of the labor and material requirements will be encountered. Thus, it should be recognized that the economic performance charting technique based on sales revenue and a linear cost function inherently assumes that the pricing (i.e., bidding) method is the same regardless of the type of project involved. As an alternative, if the pricing method is different (as it might be for example when both heavy and building construction are involved), it is assumed that the relative mix for the different types of projects remains constant regardless of the sales volume. If neither of these assumptions are acceptable, separate economic performance charts should be prepared for each type of project.

The importance of specifying the applicable sales range for the linear function cannot be overemphasized. To see why this is so, consider the fixed costs once again. It has been noted that equipment depreciation is included in these costs. Consequently, when the work requirement becomes sufficiently large, because of an increasing sales volume, an additional piece of equipment (or one with a greater productive capacity) must be acquired. Thus, a higher fixed cost is encountered beyond some specific level of activity. This same condition prevails for some of the other fixed costs such as supervision. As a result, these costs are actually "stair-step" over the entire range, as shown in Fig. 10.

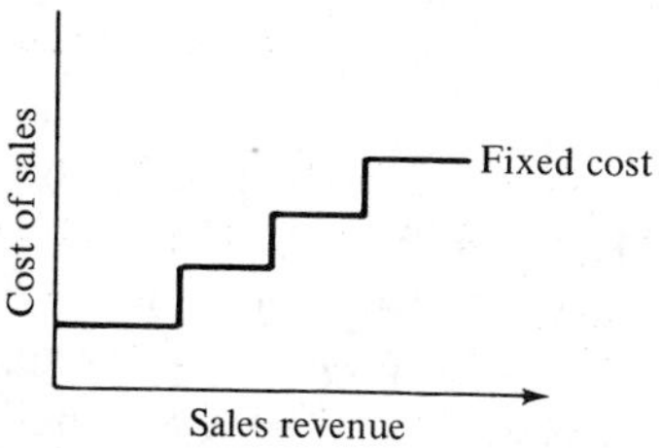

Fig. 10. Fixed cost by sales revenue range.

Variable costs also contribute to the need for specifying the applicable sales range. For example, most firms have some relatively inefficient and obsolete equipment available that is utilized only when the sales volume is such that the capacity of the more efficient equipment is exceeded. If the increased volume is expected to exist for only a short time period, or if the increased volume is not adequate to justify the acquisition of new equipment, this marginal older equipment may be used. Unfortunately, the incremental costs are greater under these circumstances (i.e., the slope of the cost line is greater), so that the marginal profit is less for these additional sales. Equivalent conditions prevail when additional sales require that less qualified personnel be utilized. Assuming a constant fixed cost over the pertinent sales revenue range, use of such marginal facilities is portrayed in Fig. 11.

The dashed line on Fig. 11 indicates what the costs would have been if the

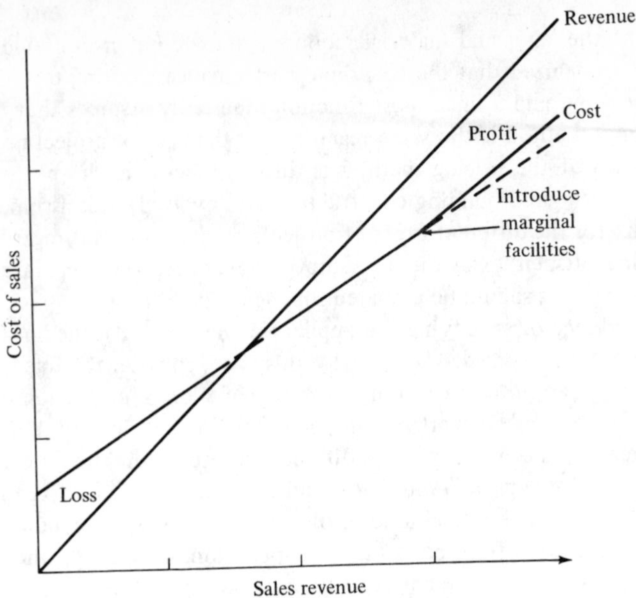

Fig. 11. Use of marginal facilities.

same variable costs were applicable to this greater sales revenue range. Thus, it appears that the potential profit has been reduced from that attainable with the more efficient facilities. However, it must be remembered that this sales volume was not attainable unless additional equipment was utilized. Consequently, the applicable alternatives are the acquisition of equivalent additional equipment or replacement of existing items with more efficient equipment which can handle the increased volume.

Initially, consider a comparison of the use of marginal equipment with the alternative which involves acquisition of an equivalent piece of equipment. If it is equivalent, the variable costs will remain the same. However, the fixed costs will increase, if for no other reason than the additional depreciation. Using the prior example, it is assumed that the variable cost for the marginal equipment, which is required to meet the sales demand in excess of 30 million dollars, is \$0.80 per sales dollar rather than the previously specified \$0.625. For the added equivalent equipment, it is assumed that the fixed costs are increased from the prior 6 to 8 million dollars per year, but the variable costs remain unchanged. While such an increase in the fixed cost is not overly realistic for most pieces of equipment, this cost is utilized so that the impact of selecting such an alternative can be readily observed.

A review of marginal vs. added equivalent facilities (Fig. 12) reveals that increasing the fixed cost while retaining the same variable cost has the effect of increasing the break-even point. Initially, the break-even point was 16 million

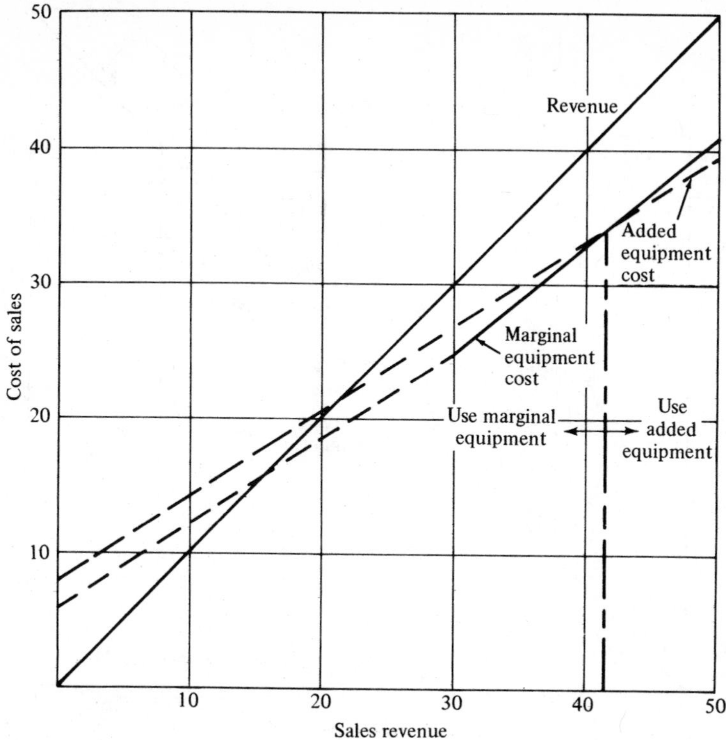

Fig. 12. Marginal vs. added equivalent facilities (in million dollar units).

dollars but the new fixed cost has increased it to 21.5. This would be a major cause for concern if sales revenue were expected to be between these two values for some of the years. Further, unless the revenue for at least most of the future years is expected to be greater than 41.5 million dollars, using the marginal equipment would be expected to provide the greatest profit.

Now consider the alternative of replacing the existing items with more efficient equipment which can handle the increased volume. Such an alternative usually decreases the variable costs at the expense of an increase in the fixed costs. Consequently, assume that the fixed costs for the new equipment will be 10 million dollars, but the variable costs will be only $0.55 per sales dollar. This is presented in Fig. 13, which indicates that the alternative increases the break-even point to 22 million dollars. (However, as opposed to the situation where the fixed cost is increased while the variable cost remains the same, the break-even point may actually decrease through the acquisition of more efficient facilities.) Further, the more efficient equipment is expected to yield a greater profit than the marginal equipment if the sales revenue exceeds 36 million dollars.

In the foregoing, two alternatives were compared separately with the marginal

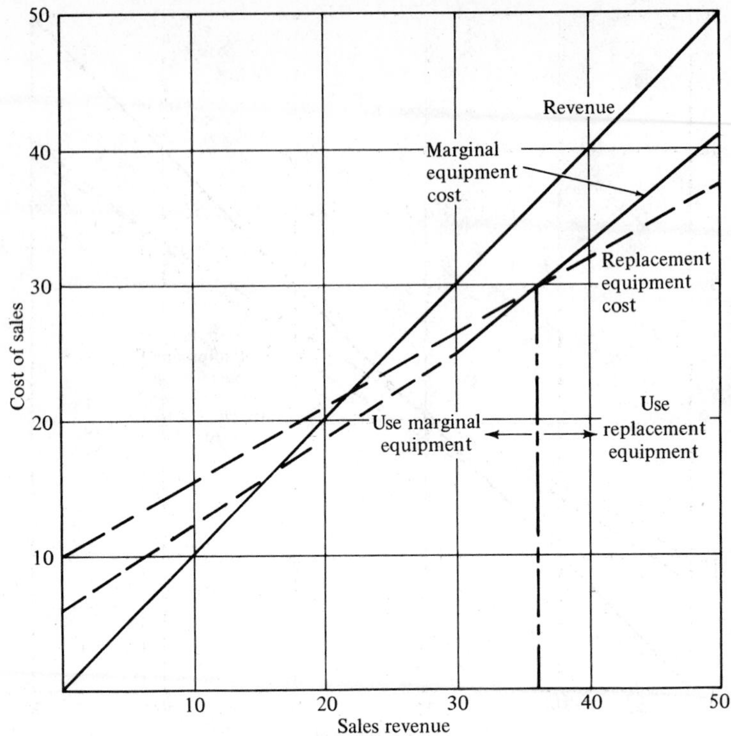

Fig. 13. **Marginal vs. replacement facilities (in million dollar units).**

equipment when the sales exceeded 30 million dollars. To arrive at a final decision, however, it is frequently desirable to make a direct comparison of the various alternatives, as shown in Fig. 14. This reveals that if the sales revenue is greater than 22 million dollars, the best alternative is the acquisition of more efficient equipment. Thus, the problem reverts to a choice between the marginal and the more efficient equipment. Considering the prior applicable comparison, the marginal equipment provides a greater profit if the majority of the annual sales were less than 36 million dollars. A review of the basic data used for this example reveals that the maximum sales revenue was 34.4 million dollars during 1968 and the most current sales were 24.4 for 1971. Consequently, unless a significant increase in sales volume is anticipated for future years, the most logical recommendation is to utilize the marginal equipment when required.

The economic performance charting technique is an exceedingly versatile tool, as may be seen from the prior example. (As a potential matter of interest, for manufacturing activities which involve common units, the production quantity replaces the sales revenue scale. For such an economic production chart, the primary difference is that the slope of the revenue line may be other than +1.)

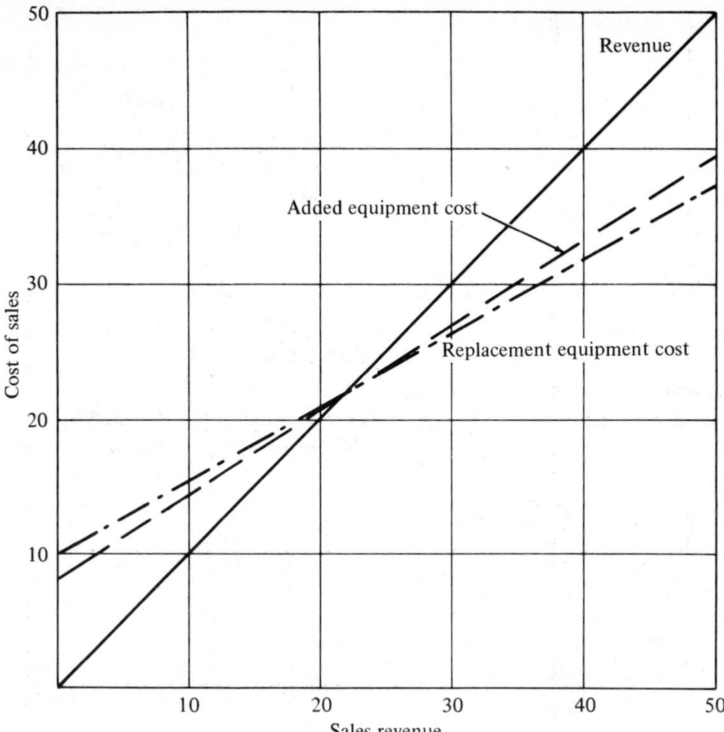

Fig. 14. Equivalent vs. replacement facilities (in million dollar units).

Not only can the technique assist in the formulation of major capital investment decisions, it can assist management in the pre-establishment of organizational, staffing, and resource allocation plans as a function of the sales revenue range. Such advance planning is inherent in the efficient operation of any enterprise. Otherwise, rapidly established and potentially ill-conceived plans, may be implemented when a new sales revenue level is encountered. These are often not in the long-range interest of the firm.

REFERENCES

1. Gardner, F. V., *Profit Management and Control*, McGraw-Hill Book Company, 1955.
2. Fabrycky, W. J., and Torgersen, P. E., *Operations Economy–Industrial Applications of Operations Research*, Prentice-Hall, Inc., Englewood Cliffs, New Jersey, 1966, pp. 71-94.
3. Johnson, N. L., and Leone, F. C., *Statistics and Experimental Design*, Volume 1, John Wiley & Sons, New York, 1964.
4. Johnson, M. M., Simple Linear Regression, *J. Quality Technology,* Vol. 3, No. 3 (July 1971), pp. 138-143.

DYNAMIC PROGRAMMING

While engineering economy will select the alternative for an investment opportunity, frequently many different and unrelated opportunities are involved. If the resources are unlimited, this does not present a problem because the best alternative for each investment may be selected. Unfortunately, however, only rarely are the funds sufficiently adequate to permit such action. Consequently, another approach is needed to assist in the formulation of the final set of decisions.

An obvious approach to the final selection of the set of alternatives is to consider all possible decision combinations. This would involve a complete enumeration of the decision sets and the evaluation of the measure of merit for each. Then, the set which exhibits the best measure for the available capital would yield the recommended decisions. Further, when more than one set has the same optimal value, other factors not included in the basic mathematical formulation of the problem may be introduced by the decision-maker to establish the final selection. However, if many investment opportunities are involved and each has several alternatives, the number of combinations to be considered can be beyond the solution capabilities of most computers. This quantity is a product of the number of alternatives. For example, if five such opportunities each with four alternatives were involved, there would be $4^5 = 1{,}024$ combinations, and there would be 1,048,576 combinations for ten such opportunities. Consequently a method is needed which considers all combinations but without explicitly evaluating each one. Dynamic programming is such a technique.

While dynamic programming has a firm mathematical basis, it appeals to one's intuitive logic. The approach involves making a series of decisions, each of which makes the best selection in consideration of the previously considered investments. An example will best illustrate this technique. Assume that three highway locations are being considered for improvement to reduce the accident rates. Obviously, in an actual case, many more locations would be involved within any state highway system. However, this example is adequate to portray the concept. Further, assume that only one alternative other than leaving the highway in its present condition is applicable for each location. The appropriate accident cost and the associated capital expenditures for the improvement are included in the following table, where the alternative indicated as 0 represents retaining the present road conditions:

Highway Project Data

Location	Alternative	Accident cost ($)	Improvement costs ($)
1	0	12,700	0
	1	3,900	25,000
2	0	34,800	0
	1	7,700	16,000
3	0	138,300	0
	1	46,600	51,200

Since only a limited number of combinations are possible (i.e., $2^3 = 8$), it is relatively easy to determine the optimal decision set by considering all possible combinations. This approach may be used to verify the dynamic programming solution which follows.

Assume initially that the amount of funds which will be made available for these improvements is unknown. Consequently, the problem is to be solved over the entire range of funds up to the full amount necessary for all of the improvements. Thus, the decisions will be investigated for the conditions when the funds range from nothing to $92,200. In solving the problem in this manner, we are considering it as one involving decisions under uncertainty.

To initiate the analysis, consider location 1 first. It is obvious that the decision here must be a function of the funds which are available at this stage. Consequently, the input and output of the scarce resource determines which alternative is feasible and provides the best result within the allotted funds. For this example, the intent is to minimize the accident costs within the constraint of funds. This is presented in the following table of optimal one-stage returns:

S_1, Stage 1 input ($)	D_1, decision 1	$f_1(S_1)$, stage 1 return ($)	S_0, stage 1 output ($)
0–24,999	0	12,700	0–24,999
25,000–92,200	1	3,900	0–67,200

It will be observed that if the available funds for location 1 are less than that required for the improvement (i.e., less than $25,000), the only possible choice is to do nothing. However, if the available funds are equal to or greater than this amount, the improvement alternative $D_1 = 1$ would be selected since the intent is to minimize the accident cost, which is shown as the stage 1 return. Further, the remaining funds after the applicable expenditure for the selected alternative is indicated as S_0, stage 1 output, which represents unspent funds.

This optimal one-stage return is easy to obtain since it is only a restatement of the original data. The optimal two-stage return is somewhat more complex, however, because the decision selection is made in a manner to optimize both the stage 2 and stage 1 returns within the limits of the resource entering stage 2 (i.e., S_2). As a result, a two-stage return analysis is usually prepared to assist in the calculations.

Two-Stage Return Analysis

S_2, stage 2 input ($)	D_2, decision 2	Two-stage return ($)	S_1, stage 2 output ($)
0–15,999	0[b]	34,800 + 12,700 = 47,500	0–15,999
16,000–92,200	0		(16,000–92,200)[a]
16,000–24,999	0	34,800 + 12,700 = 47,500	16,000–24,999
25,000–92,200	0	34,800 + 3,900 = 38,700	25,000–92,200
16,000–92,200	1		(0–76,200)[a]
16,000–40,999	1[b]	7,700 + 12,700 = 20,400	0–24,999
41,000–92,200	1[b]	7,700 + 3,900 = 11,600	25,000–76,200

[a] Stage 2 output range matches with more than one stage 1 input, so the stage 2 input must be split accordingly.
[b] Optimal decision for the indicated range of S_2 input.

For the two-stage return analysis, it is necessary to consider all possible stage 2 decisions in conjunction with the optimal one-stage return and then select the D_2 decision which has the minimum accident costs for stages 2 and 1 combined. For example, if less than \$16,000 is still available when stage 2 is reached (i.e., S_2 = 0 to \$15,999), it is not possible to make the improvement at the second highway location since the cost to do so is \$16,000. Consequently, the decision D_2 must be 0 (i.e., to do nothing and continue the current accident cost of \$34,800 per year). Sequentially, this means that nothing to \$15,999 is available for the improvement at the first highway location. Using the optimal one-stage return for this input range, the table indicates $D_1 = 0$ and $f_1(S_1)$, the stage 1 return, is \$12,700. Thus, the total return for this S_2 input range of 0 to \$15,999 is \$34,800 plus the \$12,700, or \$47,500, as shown on the first line of the two-stage return analysis table.

Now consider what happens if the input range S_2 is \$16,000 to \$92,200. For this range, the decision can be either 0 or 1, so both must be investigated. However, for both decisions, the output from stage 2 does not match with the input for stage 1 as shown on the optimal one-stage return table. That is, if D_2 is 0, then $S_1 = S_2$ but S_1 splits the total range at \$24,999. As a result, the stage 2 output must be split likewise. When decision D_2 is 1, which means that an expenditure of \$16,000 is to be made at stage 2, the input must be adjusted accordingly (i.e., the split must be \$24,999 + \$16,000 = \$40,999). These appropriate stage 2 input values have been included in the two-stage return analysis table.

After the computations have been completed for each feasible alternative within the specified input range, the results are reviewed to determine the decision which yields the optimal return (i.e., minimum cost for this example). These decisions have been identified by "b" in the two-stage return table. Since only D_2 is feasible for the 0 to \$15,999 range of S_2, this must be the optimal decision for this range. However, the determination is more involved for the rest of the stage 2 input funds because the input ranges are not the same for the different alternatives. Therefore, the ranges should be split at \$24,999 and at \$40,999 to enable a complete comparison. This is presented in the table of comparison of D_2 returns to substantiate the indicated decisions which yield the minimum costs. Thus, the optimal decision D_2 is 1 when the funds exceed \$15,999:

Comparison of D_2 Returns

S_2, stage 2 input (\$)	Two-stage return, \$	
	$D_2 = 0$	$D_2 = 1$
16,000–24,999	47,500	20,400[b]
25,000–40,999	38,700	20,400[b]
41,000–92,200	38,700	16,000[b]

[b]Optimal decision for the indicated range of S_2 input.

At this point, dynamic programming has not reduced the number of calculations required for the establishment of the optimal decisions because all feasible combinations for the two stages have been investigated. However, the calculations will be reduced for all other stages since the combinations not identified as optimal for the preceding stages need not be considered further. To simplify subsequent use of this information, the results are usually summarized as presented in the following table:

Optimal Two-Stage Return

S_2, stage 2 input ($)	D_2, decision 2	$f_2(S_2)$, two-stage return ($)	S_1, stage 2 output ($)
0–15,999	0	47,500	0–15,999
16,000–40,999	1	20,400	0–24,999
41,000–92,200	1	11,600	25,000–76,200

If it were known that exactly $92,200 were to be made available for the three stages, only this amount of input to stage 3 would have to be considered. This situation would involve decision under certainty. However, since the funds which will be appropriated are unknown, a decision under uncertainty is achieved by considering the entire range of potential appropriations from nothing to the maximum amount. Then, when the extent of the appropriation is known later, the decision set is selected from the generalized solution. Thus, it will be assumed that the input range S_3 is from nothing to $92,200 as noted previously. These results are included in the two tables of three-stage return.

Three Stage Return Analysis

S_3, stage 3 input ($)	D_3, decision 3	Three-stage return ($)	S_2, stage 3 output ($)
0–51,199	0		(0–51,199)[c]
0–15,999	0[d]	138,300 + 47,500 = 185,800	0–15,999
16,000–40,999	0[d]	138,300 + 20,400 = 158,700	16,000–40,999
41,000–51,199	0[d]	138,300 + 11,600 = 149,900	41,000–51,199
51,200–92,200	0	138,300 + 11,600 = 149,900	51,200–92,200
51,200–92,200	1		(0–41,000)[c]
51,200–67,199	1[d]	46,600 + 47,500 = 94,100	0–15,999
67,200–92,199	1[d]	46,600 + 20,400 = 67,000	16,000–40,999
92,200 and up	1[d]	46,600 + 11,600 = 58,200	41,000 and up

[c]Stage 3 output range matches with more than one stage 2 input, so the stage 3 input must be split accordingly.
[d]Optimal decision for the indicated range of S_3 input.

Optimal Three Stage Return

S_3, stage 3 input ($)	D_3, decision 3	$f_3(S_3)$, three-stage return ($)	S_2, stage 3 output ($)
0–15,999	0	185,800	0–15,999
16,000–40,999	0	158,700	16,000–40,999
41,000–51,199	0	149,900	41,000–51,199
51,200–67,199	1	94,100	0–15,999
67,200–92,199	1	67,000	16,000–40,999
92,200 and up	1	58,200	41,000 and up

As was the case for the two-stage analysis, it was necessary to subdivide the input range of S_3 so that the output would correspond with the S_2 input ranges. Of greater importance, however, is the method utilized to establish the three-stage return. Observe that this return is the sum of the stage 3 return associated with the indicated decision and the applicable optimal two-stage return for the compatible output. For example, when the input range is \$51,200 to \$67,199 and the selected decision $D_3 = 1$, the return for stage 3 is \$46,000. This decision yields an output of 0 to \$15,999 for the remaining funds. A review of the optimal two-stage return table indicates that the return for this input S_2 range is \$47,500. Thus, the three-stage return for this case is \$46,600 plus \$47,500, or \$94,100. This approach effectively reduces the total combinations as noted previously by considering only the optimal decisions selected for all of the previously analyzed stages. Observe that this is achieved by considering only the stage being analyzed and the immediately preceding composite optimal stage returns.

Now assume that \$75,000 is appropriated for the highway improvement projects; it is obvious that not all three improvements can be made since it would cost \$92,200 to do so. To determine the decision set, the optimal return tables are considered in reverse. The optimal three-stage return table indicates that the optimal decision D_3 is 1 for an S_3 input of \$75,000, which yields an output of \$16,000 to \$40,999. While the actual output is really \$23,800, it is not necessary to determine this precisely. Here, the resulting annual accident cost would be \$67,000 as opposed to the current \$185,800 annual cost. Further, considering the optimal two-stage return, the optimal decision D_2 is 1 for an S_2 input of \$16,000 to \$40,999, which, in turn, yields an output of 0 to \$24,999. Using this input, the optimal one-stage return decision D_1 is 0. Thus, the optimal decision set is to make the improvements on locations 2 and 3 while leaving location 1 in the current condition. The total expenditures for this would be \$67,200, leaving \$7,800 of the allocated funds unspent. Observe here that the optimal decisions have been established by considering the optimal return figures in the reverse sequence used for the analysis.

The foregoing approach for decision under uncertainty provides the solution regardless of the allocated funds. It is only necessary to interpret the decision set when the quantity of funds allocated are known. For example, if no funds are allocated, the decision set necessarily must be to do nothing at any of the locations. Further, if relatively unlimited funds are made available, the improvement which minimizes the accident cost at each location is selected. Thus, it is important that every alternative considered at each location yield an acceptable measure of merit by some preliminary technique such as engineering economy, otherwise an unacceptable alternative may be selected by dynamic programming.

For decisions under certainty, the funds to be allocated are known prior to solving the problem. Using the prior example, where S_3 = \$75,000 it is only necessary to perform the three-stage return analysis for the specific input amount. Thus, each feasible decision D_3 would be considered only for this specific S_3 input and the optimal decision would be selected from these results. Such a three-stage return analysis given S_3 follows. Obviously, the computational effort is greatly reduced for this last stage when the exact input is shown. As anticipated, however, the resulting decision set in this case is identical to that established previously for decision under uncertainty when S_3 = \$75,000.

Three Stage Return Analysis Given S_3

S_3, stage 3 input (\$)	D_3, decision 3	Three-stage return (\$)	S_2, stage 3 output (\$)
75,000	0	138,300 + 11,600 = 149,900	75,000
	1[d]	46,600 + 20,400 = 67,000	23,800

[d]Optimal decision for the indicated range of S_3 input.

The pertinent aspects of dynamic programming are the scarce resource S_j, the return $R_j(D_j, S_j)$, the decision D_j and the transformation $T_j(D_j, S_j)$ where the j subscript identifies the stage. These variables are presented in Fig. 15. In words, the decisions, transformations and returns at each stage must be independent of those for every other stage. Then, at each stage, a decision D_j is to be selected from the feasible set P_j in consideration of the input S_j. This is transformed into the S_{j-1} output by some function $T_j(D_j, S_j)$ which is a

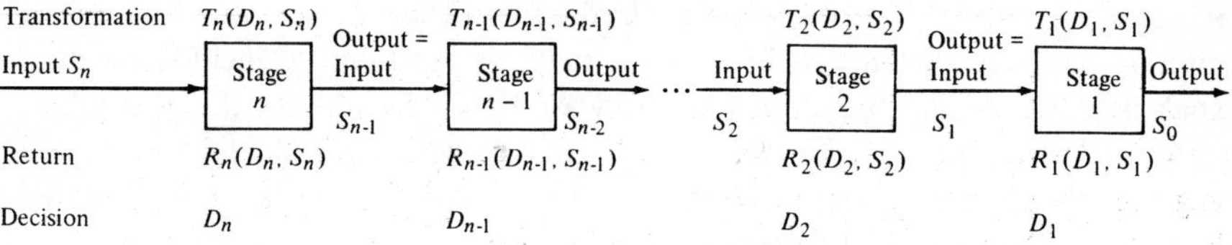

Fig. 15. *N*-stage decision problem.

function of both the stage input and decision. Further, this decision coupled with the input yields a given return $R_j(D_j, S_j)$.

The resulting problem may be restated mathematically in accordance with the following:

$$\text{Optimize } Z = \sum_{j=1}^{n} R_j(D_j, S_j)$$

For the type of problem used as an example, the solution is subject to a limitation on the availability of the scarce resource, say k:

$$\sum_{j=1}^{n} C_j(D_j) \leqslant k$$

where $C_j(D_j)$ is the cost associated with the selected decision D_j, and it is subject to

$$D_j \in P_j$$

or the decision D_j must be an element of the feasible decision set P_j for the jth stage. These constraints imply that $S_0 \geqslant 0$, or the output from stage 1 can not be negative.

As a variation of the foregoing, the input to the last stage, say S_n, may be considered as a variable and the output of stage 1 may be fixed. In this case, the output S_0 is usually set equal to zero, but it may be any specific value which is nonnegative. Here, the restraints are $S_0 = k$ and $D_j \in P_j$.

Finally, the basic solution approach of dynamic programming involves the establishment of the decisions on a sequential basis in consideration of the optimal decisions established for the previously analyzed stages, or

$$f_j(S_j) = \underset{D_j \in P_j}{\text{Optimum}} \{R_j(D_j, S_j) + f_{j-1}(S_{j-1})\}$$

where $f_{j-1}(S_{j-1})$ is the optimal cumulative return for all of the previously considered stages and $f_0(S_0) = 0$.

In summary, dynamic programming is an exceedingly versatile technique which assures that the overall optimal decision set is selected within the basic set of assumptions. Not only can this technique be used for both maximum and minimum problems, but it can be utilized when either, or both, continuous or discrete variables are involved. Also, it may be expanded to include network analysis with parallel branches and backward flow. Because of this versatility, only a relatively simple example could be illustrated here. Thus, the interested engineer should become acquainted with the more extensive material presented by Bellman, Nemhauser, or Kaufmann and Cruon.

REFERENCES

1. Nemhauser, G. L., *Introduction to Dynamic Programming*, John Wiley & Sons, New York, 1967.
2. Johnson, M. M., Dare, C. E., and Skinner, H. B., Dynamic Programming of Highway Safety Projects, *Transportation Engineering J.*, Vol. TE4, (November 1971).
3. Bellman, *Dynamic Programming*, Princeton Univ. Press, Princeton, New Jersey, 1957.
4. Kaufmann, A., and Cruon, R., *Dynamic Programming Sequential Scientific Management*, Academic Press, New York, 1967.

CRITICAL PATH METHOD

The critical path method (CPM) and program evaluation review technique (PERT) have been all the vogue in construction industry bull sessions. Yet, both have suffered from the "more said, less done" syndrome more than any other tools that have been developed for the design and construction process. Though not as many contractors and subcontractors are using either technique with the frequency they probably should, they are at least aware of them. However, with the exception of a few architects and engineers, this is not true of the design professions.

Though it may be off to a slow start, there is a growing awareness that the design process can benefit from both CPM and PERT. Network planning appears to be superior to reliance on the integrity and communications capability of a group of individuals. It pays off economically, and at the same time is conducive to improved morale among engineering personnel.

For example, the traditional check list reminds the designer of what remains to be done on a design project, but it does not help him at all to determine the time factors or the priorities involved. This results from the fact that people tend to ignore each other's needs until it becomes convenient for them to be concerned. Each of us tends to look at a system from our personal point of view rather than from a point of view that will benefit the total system. This, in spite of the fact that we all recognize that there is a critical path for activities that must be heeded to insure effective continuity of the work at hand.

From the personnel point of view, there is an unfavorable reaction within the informal structure of the design organization to the setting of deadlines without some guidelines for their reasonableness. This is particularly true among the nonprofessional staff. Thus, it would seem evident that the confidence and morale of the entire organization could be improved if a better method of project control were developed—as it could be with the proper use of CPM and PERT.

But, the benefits of network planning are not confined to improving morale. Increased efficiency is a natural by-product, based on the collection of important project data such as the ratio of time estimated for each phase of the project and job cost ratio, i.e., estimated cost to actual cost. Not only does network planning make it possible to do a better job of allocating manpower initially, it provides data for reallocation as the job progresses should difficulties arise.

To test this hypothesis, Paul Martin and Kenneth Livingston, at the University of Nebraska–Lincoln, College of Engineering and Architecture, ran a CPM model of the mechanical engineering design process through the computer. They found little difficulty in using time and cost programs designed for construction in the solution of problems related to design. Samples of cost control printouts along with an analog bar chart and the usual time-oriented CPM are contained in Martin's unpublished thesis.

As a result of this work, it is evident that the CPM approach is no more time consuming or difficult than other management control approaches. In addition, it presents conditions realistically and in a manner amenable to change. The following facts are evident.

1. Engineering aspects of building design–as well as heavy and highway–analyzed as a subsystem of the total project and the problems of planning, scheduling, and time-resource-cost control of engineering functions are susceptible to operations research methods such as CPM and PERT.

2. The techniques and procedures of CPM/PERT are as applicable to analysis of engineering design projects as they are to construction and production projects because engineering design involves definable, interrelated individual tasks requiring resource allocation and time assignment to develop a workable schedule for execution.

3. The CPM/PERT model of the engineering design subsystem provides a dynamic and responsive management tool from the standpoint of similitude, simulation, and optimization during the planning, scheduling, and control phases of project organization and operation.

4. The CPM/PERT system must be recognized as nothing more than a management tool. Each engineering firm must adapt the techniques and methods in a manner best fitted to the structure and operational philosophies it has established for itself.

The critical path method (CPM) is, then, the simplest and most effective technique yet developed for the planning, scheduling, and control of projects. Defined broadly, a project is any total job which is composed of a group of interrelated tasks or activities. CPM is an operations research tool for the allocation and control of personnel, resources, and money in total management of projects of virtually any description. Within the limits of space and intent of this discussion, only the graphic and mathematical nature of CPM will be presented. For this purpose it is advantageous to use a simplified abstract model as the project in which component tasks are defined by capital letters rather than actual descriptions of work or activity as in the case of an actual project. Although this approach has definite advantages in explaining the mechanics of the technique, the very important discussions of effective management decision processes must be left to more comprehensive treatises which are readily available in the literature. It should also be noted that expertise in CPM can be fully developed only through extensive experience in actual application.

The basic graphic model for CPM is a diagram which depicts the project by similitude, using an arrow network or a block flow diagram interconnected by arrows. The latter more recently developed model is commonly called a precedence diagram or network plan and will be the one used here since it has many practical advantages over the original arrow diagram.

Assembling the network diagram as a graphic plan of sequential action for the project is the first step of CPM. The project is broken down into convenient components of activity, each of which can be defined as a specific task with clear points of beginning and end, containing a definite amount of work to be accomplished, and having known or assigned sequential interrelationships with other such tasks. The graphic symbol for an activity is a rectangular block containing a brief but complete and accurate description of the task (simplified here by an arbitrary capital letter). At either end of the task block are so-called nodes containing integers which identify the task numerically. A typical task would appear as shown in Fig. 16. The node numbers, shown as i and j, may be

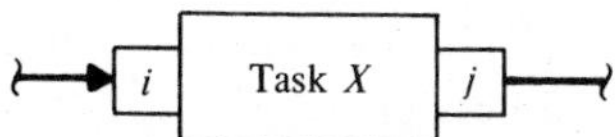

Fig. 16. Graphic symbol for an activity.

any sequential integers (such as 5 and 6 or 15 and 21) and the task may be identified as the i-j task as well as by an actual description X contained within the activity block.

The interrelationships among tasks in the real project are denoted in the graphic model by interconnecting arrows whose sense indicates sequence of action, as shown in Fig. 17. This diagram obviously indicates that tasks Y and Z cannot begin until the initial task X is complete and that beginning the last task W depends upon the completion of tasks Y and Z. Note that the sequential numbering of tasks (or rule that $j > i$) applies to the sequential flow of arrow restraints as well as nodes. No two node numbers may be the same and all must be integers increasing sequentially with the "flow" of the network. Though sequential, node numbering need not be continuous; i.e., numbers may be skipped in the sequence purposely to provide for additional tasks which may have to be added as the network plan is developed.

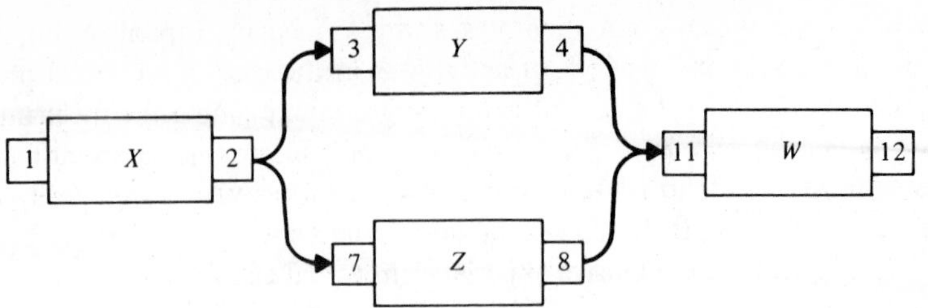

Fig. 17. Graphic model of a four-activity project.

Ignoring the numbers (except node numbers) for the moment, the sequential plan of action for the actual project is obvious from the graphical logic of the network model. All tasks must be completed within the total span of the project. Thus our network must be a "closed" system. In the event that several initial and/or several final tasks can actually be concurrent for the real project, initial and/or final dummy blocks should be employed in the diagram to indicate definite points of beginning and completion (and so that manual or computer computations can be made by the CPM technique explained later).

The dependence of one task on another, as shown by the interconnecting arrow, may be a physical or procedural restraint in the real situation or may be a management decision establishing the best sequence of activities. Putting together the sequence logic of the network plan is one of the primary responsibilities of the planning team, for once the network is complete, it represents the plan of action for the project. The cardinal logic rule is to ask of each task, "What tasks must or should be completed before this one begins?" Obviously, two or more tasks may be concurrent.

Each activity represents an actual accomplishment requiring time. The amount of time required depends on the nature of the task, the methods to be employed and the allocation of necessary resources. It is therefore possible to make a valid estimate of the time duration of each activity. Some convenient unit of time, such as hours or working days, will be necessary to apply to the project. In our illustration, the working days will be used as a convenient unit of time. The estimated duration time is shown directly beneath the block for each task, as shown in Fig. 18. The symbol d is used to denote task duration.

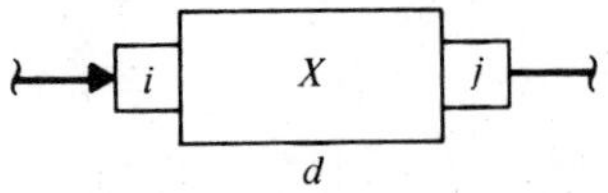

Fig. 18. Location of symbol d, indicating duration of task.

Now consider the X, Y, Z, W model with arbitrary time requirements assigned to each task, as shown in Fig. 19. The numbers just above the i nodes represent the earliest possible starting time for each activity. Day 1 is assigned for the start

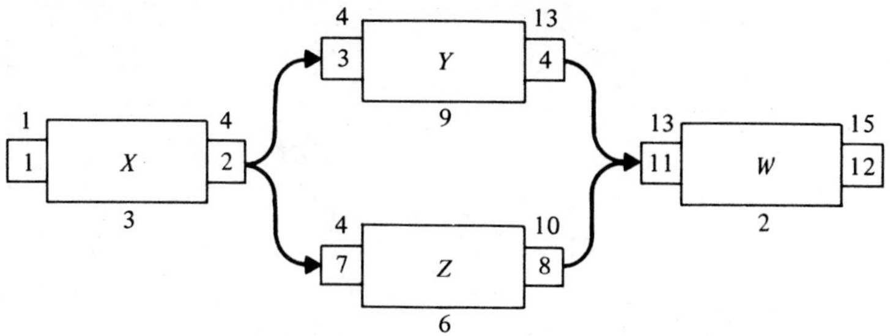

Fig. 19. Calculation of early start and early finish times.

of the project and thus appears above the *i* node of the first task *X*. The abbreviation ES is used to represent earliest possible starting time. The numbers above the *j* nodes are the earliest possible finish time for each task; abbreviated as EF in general terminology. For any task, it is obvious that EF = ES + *d*. So the beginning of working day 4 is the EF for task *X*. The beginning of working day 4 is therefore the ES time for both tasks *Y* and *Z* since they each may begin as soon as task *X* is complete. For task *Y*, EF = 4 + 9 = 13 and for task *Z*, EF = 4 + 6 = 10.

Now task *W* cannot begin until both tasks *Y* and *Z* are complete, hence the ES for task *W* can be no sooner than day 13—the greatest-precedence EF. The EF for activity *W* is day 15; and this also represents the end of the four-task project.

Thus, the rule of sequential flow used in assembling the logic network (or plan of action) simply extends to the ES and EF arithmetic of the CPM algorithm: no task may begin until all preceding tasks upon which it directly depends have been completed.

The EF of task *W* (beginning of day 15—or actually also the end of day 14 since these have the same real meaning) is a calculated completion date for the project and, by definition, is the latest finish time for the last task. The term LF is used to signify latest finish time. The LF of any task in the network is the time by which that task must be completed if the project is to be done by the calculated time of completion (EF = LF of final activity). The latest possible starting time for any task, abbreviated as LS, must therefore be LS = LF − *d*.

Computing LF and LS times is a reverse process of that for ES and EF times. We commence at the LF of the last task and calculate backward through the diagram. The LF of a task preceding other tasks as indicated by arrows is the least value of ES of these tasks. The simple four-task example will serve to demonstrate the process in Fig. 20.

The LF and LS values are shown directly below the *j* and *i* nodes for each task. For task *W*, LF equals 15 since this is the termination of the project and thus must have the same value of the EF of the last activity. The LS for task *W* is obviously 15 − 2 = 13. Since tasks *Y* and *Z* directly precede task *W*, both must be complete by no later than day 13 because this is the latest possible time task *W* may begin if the project is to end on time.

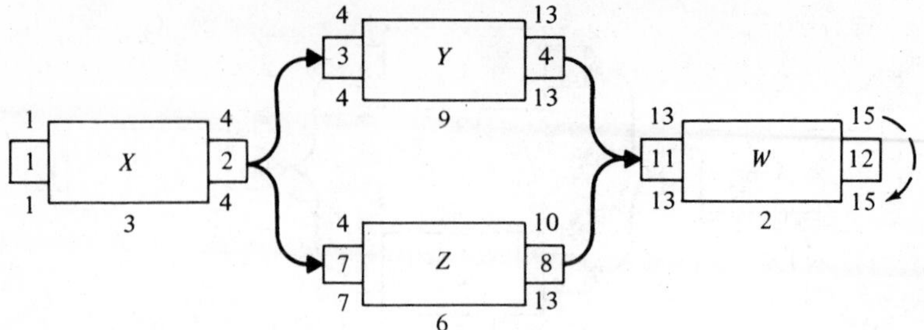

Fig. 20. Calculation of late finish and late start times.

The LS value for task Y is thus $13 - 9 = 4$ and for task Z is $13 - 6 = 7$. Task X must be complete in time for either task Y or Z to begin on their respective LS times. Thus LF for activity X must be day 4. The LS for task X is $4 - 3 = 1$. Note that this checks the CPM arithmetic since, by obvious definition, ES and LS for the beginning of the first task in the project have the same meaning.

One more time calculation must be made to determine the float, F, or time leeway for each task. Float is the difference between the time available during which a task must be done and the time required to do that task. Hence, for each task, $F = [(\text{LF} - \text{ES}) - d]$. It can be shown by algebraic substitutions that this may also be expressed as either $F = \text{LF} - \text{LS}$ or $F = \text{EF} - \text{ES}$. The float time is shown directly beneath duration time for each task in Fig. 21.

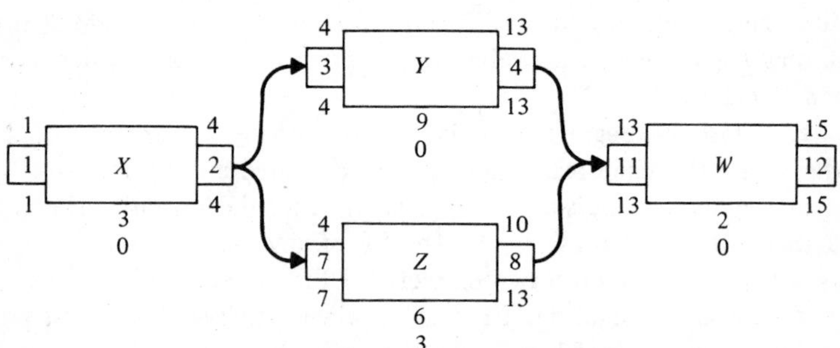

Fig. 21. Calculation of float, or time leeway for each task.

The term "critical" refers to a state of zero float time. Thus tasks X, Y, and W are "critical tasks" which must be started by their ES time and completed within the estimated duration times if the project is to terminate by the calculated completion time. Critical tasks lie along continuous paths through the project and these are called "critical paths." Such a chain is XYW. In a more complex network more than one critical path may exist.

Task Z has a float time for 3 days indicating that it may be started at any time between day 4 and 7 without affecting the completion day of the project. Float time may be used for convenience in scheduling or may be used during the project execution as contingency time leeway.

Float time actually applies to chains of noncritical tasks in a more general case. This may be demonstrated by examining the small project shown in Fig. 22. The critical path is *PQT*. Tasks *R* and *S* have 3 days of float time each, but this is shared float; in other words, the *RS* chain of tasks has a total of 3 days leeway. The actual starting time for task *S* could be changed from day 7 up to day 10 without affecting task *R* (assuming *R* is scheduled for its ES time). However, if the start time of task *R* is moved ahead, then the start time of task *S* must also be moved ahead by the same amount. If one relates to train cars on a siding or beads on a pool score wire, float is quite easily visualized.

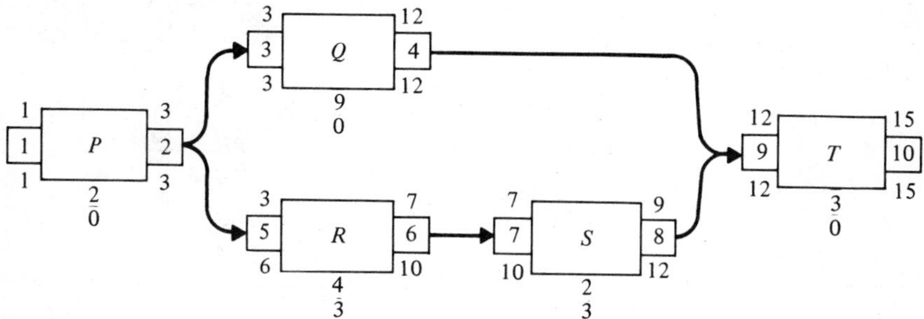

Fig. 22. Float in chains of noncritical tasks.

That is all there is to the mathematics of CPM time calculations. Consider the example project of Fig. 23. All of the CPM times are shown on the diagram, which now represents a quantitative model for time scheduling as well as a logic model for the sequential plan of action.

A convenient way of showing the nature of the CPM network solution is by means of an "analog bar chart," as in Fig. 24. The solid-line bars for each task are in the ES-EF position. For noncritical tasks, the float is represented by a dotted-line extension to the LF day. In order to indicate restraints (or sequential dependencies), lighter-line arrows are used to interconnect activity bars with the same logic that exists in the network diagram. Although the analog bar chart has the obvious advantage of being related to a time scale, there is a danger of succumbing to "bar chart" philosophy—that is, planners may revert to the bar chart as a planning tool, for which purpose it is very ill-suited. The most practical use of an analog bar chart, aside from being a valuable visual aid in understanding the nature of various CPM time definitions, would be as a scheduling and work monitoring document. It is also a useful device for scheduling and monitoring cash flow and other resources. Suffice it to say here that the analog bar chart may supplement the network diagram but in no way replaces it.

Scheduling in its simplest form involves relating the continuous arithmetic of

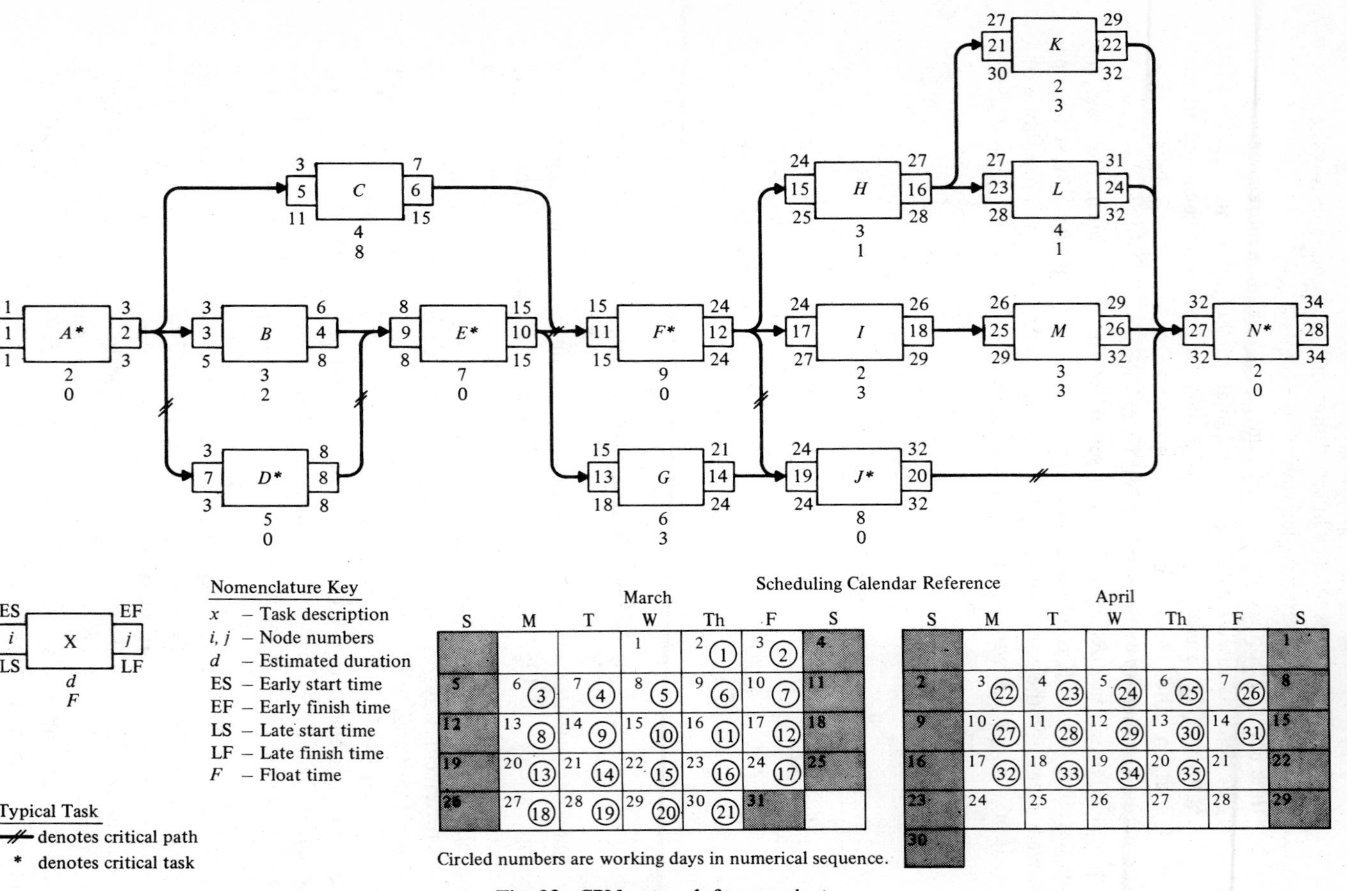

Fig. 23. CPM network for a project.

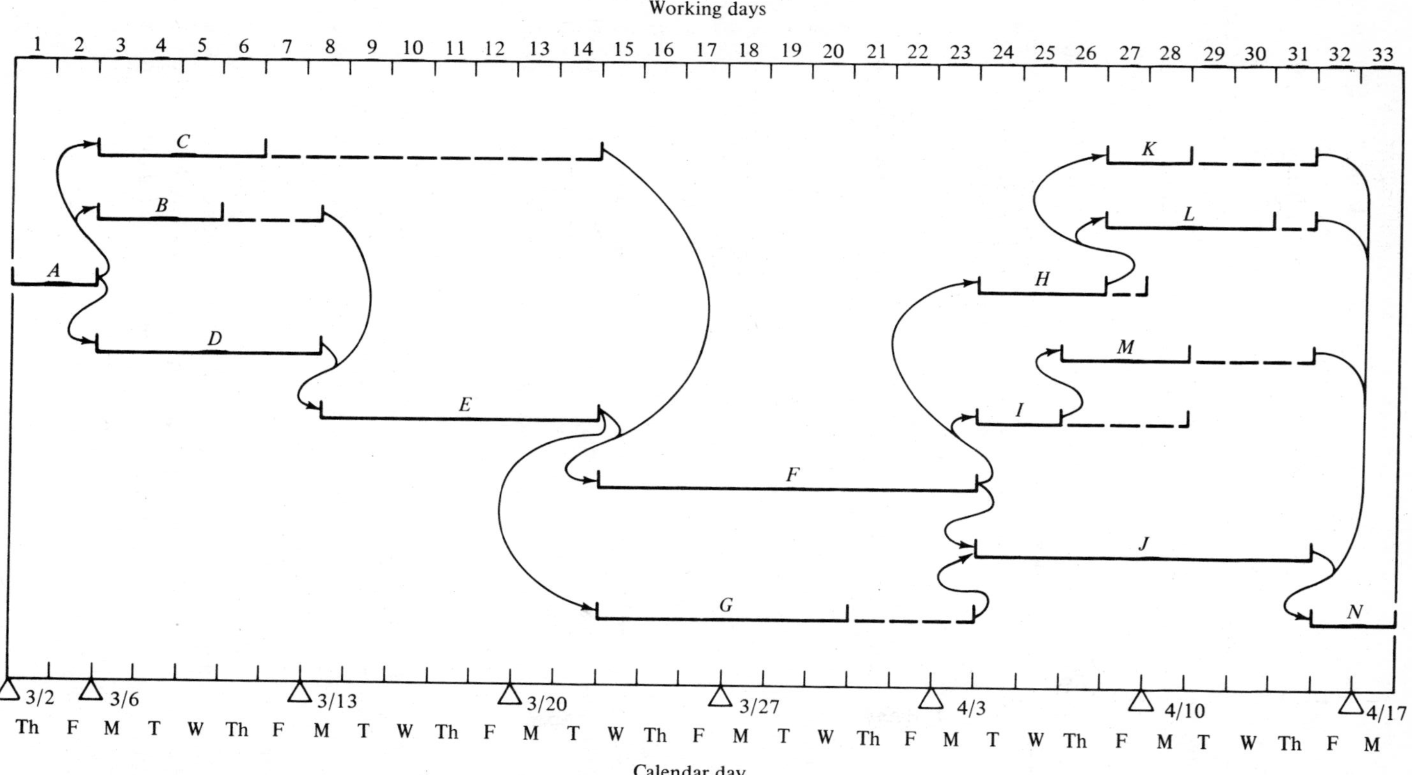

Fig. 24. Analog bar chart from CPM network solution.

working-day CPM times to actual calendar dates, which is a discontinuous system because of non-working periods such as weekends and holidays. (Different appropriate transformations apply if time units other than working days apply, of course.) In reality, scheduling is more complex than this because of problems involving availability and balancing of resources, both within and out of the planner's direct control. Float may be insufficient to compensate for such problems and more drastic action must be taken to alter the computer CPM network so that scheduled times accommodate these problems. Scheduling is thus an art beyond the limits of this discussion, but it is greatly facilitated by the analytical basis CPM provides.

As an example of the simplest kind of scheduling, a working day vs. calendar date scheme is shown below the network of Fig. 23. On actual calendar pages, working days are shown as circled numbers within calendar day blocks. On the analog bar chart, which has a continuous working day scale at the top, the corresponding calendar day scale can be shown at the bottom, as in Fig. 24.

The most important phase of CPM is the monitor–control effort once the project has begun. The network plan is the model through which the project is managed. As work is completed in the field, the progress is shown on the network by shading tasks in progress and complete. A regular monitor schedule should be followed which is sensitive enough in frequency to give management a clear picture of the current status—particularly in those areas where trouble may be indicated. Various color codes have been successfully employed which, by various color shading of work being done, indicate whether the job is on time, falling behind or ahead, or actually in trouble as would be the case when the LF time is exceeded. One of the powerful aspects of CPM monitoring is that trouble is pinpointed before it happens so that remedial action may be taken to bring the project back on schedule.

Every effort should be made to make the project progress follow the plan. Of course, some unforeseen things can happen on any job which cannot be corrected within reasonable cost limits. This calls for management decisions resulting in network surgery. A change in logic or in duration estimates (or both) may be required to reschedule the work. The CPM network lends itself easily to changes such as those involving recalculation of times and updating of the schedule. The computer has an advantage in speed and, usually, cost for such revisions. A wide variety of standard CPM programs are readily available for computer applications.

The fact that CPM networks are so easily revised can be dangerous as well as beneficial. Unless management is careful, the CPM network can wind up following the work rather than controlling it. Therefore, any change must be undertaken only as a last resort. And, when possible, changes should be made to eventually bring the work back on schedule rather than to simply extend the project completion date.

Unfortunately, it is not possible to discuss the monitor–control phase of CPM adequately in the space allotted here. In fact, skill in applying the network as a management tool for work control is gained by experience rather than dissertation.

The total use of CPM includes cash flow prediction and control as well as time control of work. The calculated network plan is an excellent framework for cost analysis. Each task has associated costs which may be estimated from material, manpower, equipment, and other resource needs. The total cost for an activity may be treated as such or broken down into conventional categories. Since the CPM times indicate the periods during which these costs will be incurred, it is a simple matter to establish a cost vs. time schedule from the network. After the work has begun, the actual vs. estimated costs can be monitored along with progress vs. time. There are thus two criteria for management control–time and cost. A project can be on schedule and running into cost problems or the costs may be under control but the progress out of schedule–or both cost and progress often may be causing trouble simultaneously. Management decisions for appropriate action must consider both parameters in a time-cost trade-off analysis for optimum solution to the problem.

From the standpoint of work–time analysis, all of the fundamentals have been covered above. Integration of cost and other resource allocation and control has been only touched upon but the principles involved are as simple as the time aspect of CPM. As is the case with many other operations research techniques, CPM is quite easy to understand but rather difficult to apply successfully. For a discussion of PERT, check the following references.

REFERENCES

1. Antill, J. M., and Woodhead, R. W., *Critical Path Methods in Construction Practice*, John Wiley & Sons, Inc., New York, 1965.
2. *CPM in Construction, A Manual for General Contractors*, Associated General Contractors of America, 1965.
3. Fondahl, J. W., *A Non-Computer Approach to the Critical Path Method for the Construction Industry* (2nd ed.), Dept. Civil Engineering, Stanford Univ., Stanford, Calif., 1962.
4. *Fundamentals of Network Planning and Analysis*, Remington Rand Univac, St. Paul, Minn., 1962.
5. *General Information Manual: PERT . . . A Dynamic Project Planning and Control Method*, Publication No. E 20-8067-1, IBM Corp., White Plains, N.Y., n.d.
6. Horowitz, J., *Critical Path Scheduling*, Ronald Press, New York, 1967.
7. Lockyer, K. G., *An Introduction to Critical Path Analysis*, Pitman Publishing Corp., New York, 1964.
8. O'Brien, J. J., *CPM in Construction Management*, McGraw-Hill, New York, 1965.

9. O'Brien, J. J., and Zilly, R. G., *Contractor's Management Handbook*, McGraw-Hill, New York, 1971.
10. *PERT Instruction Manual and System and Procedures for the Program Evaluation System*, Special Projects Office, Department of the Navy, Washington, D.C., 1960.
11. Radcliffe, B. M., Kawal, D. E., and Stephenson, R. J., *Critical Path Method*, Cahners, Chicago, 1967.
12. Shaffer, L. R., Ritter, J. B., and Meyer, W. L., *Introduction to the Critical Path Method*, Univ. of Ill., Urbana, 1963.
13. *The ABC's of the Critical Path Method*, Office of the Regional Director GSA, New York, n.d.
14. Waldron, A. J., *Fundamentals of Project Planning and Control*, A. James Waldron, Haddonfield, N.J., 1963.

GAMES AGAINST NATURE

In the economic performance charting subsection, it became obvious that there are a number of basic components to a decision problem. Among the more obvious, a problem which must be resolved by a decision-maker must arise. For a problem to exist, alternative courses of action must be possible from which the decision-maker can choose in accordance with some strategy. This implies that the decision-maker has one or more objectives, or goals, which may be defined as the desired outcome. Further, since mathematical techniques are to be utilized, some measure of effectiveness must be associated explicitly with each alternative.

Games against nature, alternatively titled decision theory, is less encompassing than the alternative name implies. A review of the model classification included in the previous discussion of the application of modeling to decision making clearly makes this point. These models may involve decisions under either uncertainty or risk. However, most of the approaches discussed here will pertain to decisions under uncertainty.

It should be emphasized that this topic does not include what is frequently termed "theory of games." The primary difference is that of the opponent encountered. For the approaches to be discussed here, the opponent is nature, but the opponent in the theory of games is a reasoning, logical decision-maker who is trying to win. Since most of such game theory applications have been to military activities, this topic is omitted.

In classifying the various models, the primary basis for categorization was the knowledge of future states of the world. It was stated previously for uncertainty models that the future states may be identifiable but the associated probabilities of occurrence are unknown and not assumed. This provides the basis for the establishment of a specific example, the problem presented in the discussion of economic performance charting.

The variable which represents the future state of the world for this problem is the sales revenue which might be realized. Obviously, the future sales may be any volume over the entire applicable range. However, to simplify the problem, assume that only three future states (i.e., 30, 40, and 50 million dollars) may exist. These indicated volumes are representative of the ranges previously found to be pertinent to the solution of the problem.

As before, the applicable measure of effectiveness is the profit associated with the specific alternative for the future state of the world. The alternatives, it will be remembered, were to use the marginal equipment when sales exceed 30 million dollars, to obtain equivalent additional equipment for the excess sales demand, or to acquire new and more efficient equipment which had sufficient capacity for the entire sales volume. The appropriate cost data for these alternatives and sales revenues as obtained from the figures included in the economic performance charting section are included in the following tabulation, where revenue is in millions of dollars:

Operational Costs

Equipment alternatives		Future states of the world (sales revenue)		
		30	40	50
Marginal	(a_1)	25	33	41
Added	(a_2)	27	33.3	39.5
New	(a_3)	26.5	32	37.5

This table presents the appropriate costs, but the measure of effectiveness was to be the profit. This variable is obtained by subtracting the cost from the appropriate sales revenue (again in millions of dollars):

Operational Profit

Equipment alternatives		Future states of the world (sales revenue)		
		30	40	50
Marginal	(a_1)	5	7	9
Added	(a_2)	3	6.7	10.5
New	(a_3)	3.5	8	12.5

If the future state of the world were known (i.e., if a decision under certainty were involved), solution of the example would be relatively simple. The decision would be to use the marginal equipment if sales were 30 million dollars, or to obtain the new and more efficient equipment for sales of either 40 or 50 million dollars since profit is to be maximized. However, because the future is uncertain, the choice of the alternative is more difficult. Nevertheless, it is observed that some other alternative is better than adding equivalent equipment for every future state. Thus, it is said that this alternative is dominated and may be removed from further consideration. This is the same conclusion as was reached in the prior analysis. Therefore, with minor exceptions, it is necessary to consider only the two remaining alternatives.

Now some decision criterion must be utilized to select an alternative. Initially, only approaches for decision under certainty will be considered. However, this is far more extensive than might be imagined because many different decision criteria may be utilized. In this section, the discussion will be limited to only five approaches. First, these criteria will be applied to the profit problem. Then, the operational costs will be considered to portray the changes required for minimization.

The expression "go for broke" is illustrative of the first approach, which is called the plunger criterion. As anticipated, this robust decision-maker has an eye toward the "pot of gold at the end of the rainbow" and formulates his decisions accordingly. Thus, he selects the alternative which yields the largest profit regardless of other consequences. Using i to represent the row alternative and j for the column future state of the world, the decision criterion is to select alternative i from $\max_i \max_j e_{ij}$, where max represents the maximum over the j columns and the i rows, and e_{ij} is the indicated profit element in the matrix. It has already been established that the alternative of obtaining equivalent additional equipment is dominated and will not be selected. Consequently, only the other two are considered:

Alternative	Plunger profit criterion, future states 30	40	50	$\max_j$	
a_1 (Marginal)	5	7	9	9	
a_3 (New)	3.5	8	12.5	12.5	$\max_i$

To utilize the plunger criterion for a profit problem, the maximum value in each row is selected first (i.e., $\max_j$). From this set, the decision is selected as the maximum value (i.e., $\max_i$), which is 12.5 for the example. Alternatively, the largest profit value in the matrix is 12.5 for alternative a_3 (i.e., for the alternative of purchasing new and more efficient equipment). This profit, however, would be realized only if the sales revenue were 50 million dollars for the year. Even though a lesser profit would be realized by this alternative if the sales were only 30 million dollars, a_3 is the decision selected when the plunger criterion is used.

The next approach is termed the maximin (or minimax for a cost problem) criterion or Wald's decision criterion. This is the most conservative approach available and, for this reason, is frequently identified as the banker's strategy.

Wald's strategy is to select the alternative which has the best of the worst possible outcomes. This section assures that the actual results will never be worse than the indicated profit and may be much better. Thus, the decision criterion is to obtain $\max_i \min_j e_{ij}$ and select the ith alternative. This is presented in the following table:

Wald Profit Criterion

Alternative	Future states 30	40	50	$\min_j$	
a_1	5	7	9	5	$\max_i$
a_2	3.5	8	12.5	3.5	

As indicated in the foregoing table, the worst results were obtained initially for each row (i.e., $\min_j$), and then the best of these was selected (i.e., $\max_i$). Thus, with the selection of alternative a_1 (i.e., use of marginal equipment), the firm is assured of at least 5 million dollars of profit and will never attain the lower value of 3.5. This selection, however, will prevent the realization of the maximum profit when the sales are greater than 30 million dollars. Because this is such a conservative approach when the adversary is nature, it is rarely used by industry or most business enterprises. While the approach may be a good one when the opponent is "playing to win the game," nature is usually somewhat more kind.

The Laplace decision criterion assumes that some probability is associated with each future state of the world, but this probability is unknown. Consequently, because of a lack of more specific information, it is assumed that each future state has an equal probability of occurrence. Using this approach, the alternative selected will have the best expected profit (i.e., maximum expectation). That is, the approach considers that the same or equivalent decision is to be made many times. Therefore, while the indicated expected result may not be attained (or even attainable) for the specific problem, the long-term average profit will conform with this "expected profit" if the assigned probabilities for the future states are valid.

The selection criterion for the Laplace approach may be restated as $\max_i E(a_i)$, where $E(a_i)$ represents the expected profit for the ith alternative, and $E(a_i) = \sum_j e_{ij} p_j$, where e_{ij} is the profit for the ith alternative and the jth future state, and p_j is the probability that the jth future state will occur. However, since equal probability is assumed, if there are n future states, $p_j = 1/n$ for $j = 1, 2, \ldots, n$. Using the prior profit values,

$$E(a_1) = 5(1/3) + 7(1/3) + 9(1/3) = 7.0$$

$$E(a_3) = 3.5(1/3) + 8(1/3) + 12.5(1/3) = 8.0$$

since there were three future states of the world (i.e., $n = 3$). Thus, using the Laplace decision criterion, the selection would be alternative a_3 since this expected profit (i.e., 8.0) is greater than that for a_1.

From the foregoing, it appears that the decision may be related to some other variable. In consideration of this, the Hurwicz decision criterion was formulated as a function of the relative optimism of the decision-maker. This variable is represented by u where $0 \leqslant u \leqslant 1$ and 1 represents the greatest degree of optimism. The resulting function to be considered is

$$H(a_i) = u\{\max e_{ij}\} + (1 - u)\{\min e_{ij}\}$$

for a profit problem. In general,

$$H(a_i) = u\{\text{Best outcome for } a_i\} + (1 - u)\{\text{Worst outcome for } a_i\}$$

When profit is involved, the maximum $H(a_i)$ alternative is selected.

Using the foregoing $H(a_i)$ equation and the prior profit values for the two alternatives,

$$H(a_1) = u\{9\} + (1 - u)\{5\} = 5 + 4u$$

$$H(a_3) = u\{12.5\} + (1 - u)\{3.5\} = 3.5 + 9u$$

The decision-maker can choose a specific degree of optimism dependent upon his "frame of mind" at the moment, solve for each $H(a_i)$ and select his decision accordingly. However, it is more common to determine which decision applies to each applicable range of optimism.

To permit better visualization of this situation, the two functions are plotted

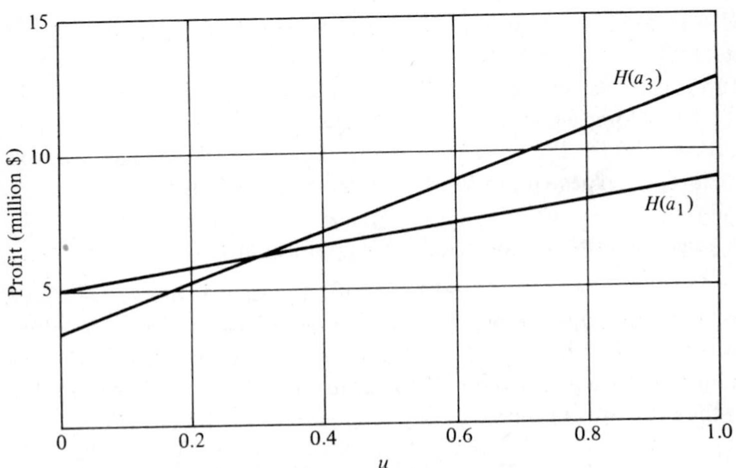

Fig. 25. Hurwicz profit analysis.

in Fig. 25. While only two alternatives are considered for this example, any number may be involved. Obviously, since the intent is to maximize profit, the desired choice is the alternative which has the greatest profit over each range of optimism. Thus, the ranges are determined by establishing the applicable points of intersection for the functions. For this example:

$$5 + 4u = 3.5 + 9u$$

and

$$1.5 = 5u$$

or

$$u = 1.5/5 = 0.3$$

Consequently, if the degree of optimism is equal to or less than 0.3, alternative a_1 is selected. For a greater degree of optimism, however, alternative a_3 would be the choice.

A review of the plunger criterion reveals that the decision-maker must have been very optimistic, say $u = 1$. Note that the same alternative (i.e., a_3) is selected by the Hurwicz criterion when $u = 1$. This is not fortuitous. Further, it was noted that the Wald criterion was the most conservative. The resulting selection was alternative a_1. Using $u = 0$, the Hurwicz criterion also chooses a_1. Thus, the Wald and plunger criteria are merely special cases of the Hurwicz approach.

Have you ever taken a given course of action and afterwards wished that you had made a different decision? Savage undoubtedly did so! Consequently, he formulated the Savage minimax regret criterion. The basic philosophy here is to formulate the decision in such a manner that the potential disappointment is minimal. To achieve this, the entire matrix must be reformulated in accordance with the relationship

$$r_{ij} = \max_k e_{kj} - e_{ij}$$

for a profit problem. That is, the regret element r_{ij} is established by subtracting each element in any column from the maximum element in that column. Because a new matrix is being formulated, it is desirable to consider all of the original alternatives. To visualize these computations, consider alternative a_2 and a sales revenue of 40 million dollars. The indicated profit for this element, as restated on the original profit matrix, is 6.7. Further, the maximum profit attainable when the sales are 40 million dollars is 8 million when alternative a_3 is selected. Thus, the regret here would be 1.3 million dollars (i.e., 8 - 6.7) if alternative a_2 had been selected but the sales were 40 million dollars. These regret elements are summarized in the profit regret matrix.

Original Profit Matrix

Alternative	Future state 30	40	50
a_1	5	7	9
a_2	3	6.7	10.5
a_3	3.5	8	12.5
$\max_k e_{kj}$	5	8	12.5

Profit Regret Matrix

Alternative	Future state 30	40	50	$\max_j$	
a_1	0	1	3.5	3.5	
a_2	2	1.3	2	2	
a_3	1.5	0	0	1.5	$\min_i$

As opposed to the Wald criterion for a profit maximization problem, the selection for the Savage technique involves choosing the alternative which minimized the maximum regret. This is to be expected, since the regret represents lost potential profit, which may be considered as a cost. Consequently, the maximum value for each row (alternative) is determined initially as shown on the profit regret matrix. Then the alternative is selected which yields the minimum of these maximum regrets. For the example, this is alternative a_3 with a minimum maximum regret of 1.5 million dollars.

Before discussing a modification which pertains to decision under risk, consider the changes required for solving a cost problem. A review of the original operational costs indicates that the alternative to add equivalent equipment (i.e., a_2) is dominated once again. The remaining pertinent costs are presented in the following cost matrix:

Cost Matrix

Alternative	Future state 30	40	50
a_1	25	33	41
a_3	26.5	32	37.5

When cost rather than profit is involved, the plunger criterion selects the alternative with the lowest cost (i.e., $\min_i \min_j e_{ij}$). Since this lowest cost is 25 million dollars for a sales revenue of 30 million, the applicable alternative selected is a_1.

The Wald criterion for a cost problem was introduced as the selection rule for the Savage regret technique. This involves the selection of the minimum from the maximum cost for each alternative (i.e., $\min_i \max_j e_{ij}$). For clarity, this is presented in the Wald cost criterion table. Here, the maximum costs were 41 and 37.5 for alternatives a_1 and a_3, respectively. Thus, since the minimum in this set is 37.5, alternative a_3 is selected.

Wald Cost Criterion

Alternative	Future state 30	40	50	$\max_j$	
a_1	25	33	41	41	
a_3	26.5	32	37.5	37.5	$\min_i$

When the Laplace criterion is applied to a cost problem, the expectations are calculated as before (i.e., $E(a_i) = \sum_j e_{ij}p_j$), where it is assumed that the probability of each future state is equal. Here,

$$E(a_i) = 25(1/3) + 33(1/3) + 41(1/3) = 33.0$$

and

$$E(a_3) = 26.5(1/3) + 32(1/3) + 37.5(1/3) = 32.0.$$

Thus, $\min_i E(a_i) = 32$ and the selected alternative is a_3.

The basic Hurwicz criterion was stated previously in general terms. This was $H(a_i) = u\,\{\text{Best outcome for } a_i\} + (1 - u)\,\{\text{Worst outcome for } a_i\}$. For a cost problem, this reduces to

$$H(a_i) = u\,\{\min_j e_{ij}\} + (1 - u)\,\{\max_j e_{ij}\}$$

and the ith alternative is selected in accordance with the scheme $\min_i H(a_i)$. For the example,

$$H(a_1) = u\,\{25\} + (1 - u)\,\{41\} = 41 - 16u$$

and

$$H(a_3) = u\,\{26.5\} + (1 - u)\,\{37.5\} = 37.5 - 11u.$$

These functions are presented in Fig. 26. For the cost problem, however, a minimum cost is desired. Consequently, the function which has the lowest cost over the specified range of the coefficient of optimism u yields the desired alternative. Equating the two functions to establish the intercept,

$$41 - 16u = 37.5 - 11u$$

so that $u = 0.7$. This means that if the coefficient of optimism is equal to or less than 0.7, the decision-maker should select alternative a_3. Otherwise, alternative a_1 should be chosen. Once again, note that $u = 0$ is equivalent to the Wald criterion and $u = 1$ is the plunger criterion.

For the Savage minimax regret criterion when a cost problem is involved, the regret computations differ. Here, $r_{ij} = e_{ij} - \min_k e_{kj}$, where the regret value is the

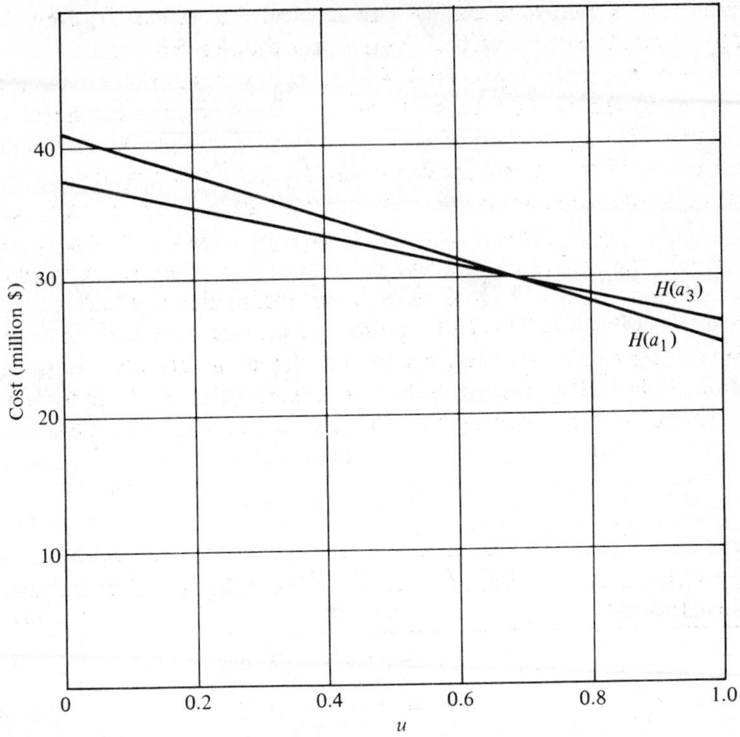

Fig. 26. Hurwicz cost analysis.

original element reduced by the minimum cost in the applicable column. As before, the entire original cost matrix and the resulting regret matrix is included:

Original Cost Matrix

	Future state		
Alternative	30	40	50
a_1	25	33	41
a_2	27	33.3	39.5
a_3	26.5	32	37.5
$\min_k e_{kj}$	25	32	37.5

Cost Regret Matrix

	Future state				
Alternative	30	40	50	$\max_j$	
a_1	0	1	3.5	3.5	
a_2	2	1.3	2	2	
a_3	1.5	0	0	1.5	$\min_i$

It is observed that the resulting cost regret matrix is identical with the prior regret matrix. This condition exists for the example because both the profit determination and the regret elements are functions of the future state of the world. Further, since the regret is a lost opportunity cost, the same minimax selection rule is used regardless of whether profit or cost is involved. Consequently, alternative a_3 is selected by the Savage decision criterion.

Because of the preceding method of presentation, it is difficult to visualize the differences in the various decision criteria. Therefore, the decision criterion techniques and a comparison of selected alternatives are presented in Table 10.

An analysis of the foregoing reveals that the process of arriving at a decision when the future is unknown does not merely involve a mathematical solution to the problem. The decision is dependent upon the criterion to be satisfied, and this requires the application of value judgment by the manager. Consequently, these techniques can only assist the manager in formulating a course of action.

When the probability of each future state of the world is known or assumed, a decision under risk is involved. A relatively simple modification to the Laplace criterion yields the Bayes criterion. For the Laplace criterion, it was assumed that every future state was equally probable because actual probabilities were un-

TABLE 10a. DECISION CRITERION TECHNIQUES.

	Profit		Cost	
Technique	Variable	Rule	Variable	Rule
Plunger	e_{ij}	$\max_i \max_j e_{ij}$	e_{ij}	$\min_i \min_j e_{ij}$
Wald	e_{ij}	$\max_i \min_j e_{ij}$	e_{ij}	$\min_i \max_j e_{ij}$
Laplace	$E(a_i) = \sum_{j=1}^{n} e_{ij}/n$	$\max_i E(a_i)$	$E(a_i) = \sum_{j=1}^{n} e_{ij}/n$	$\min_i E(a_i)$
Hurwicz	$H(a_i) = u \{\max_j e_{ij}\} + (1 - u) \{\min_j e_{ij}\}$	$\max_i H(a_i)$	$H(a_i) = u \{\min_j e_{ij}\} + (1 - u) \{\max_j e_{ij}\}$	$\min_i H(a_i)$
Savage	$r_{ij} = \max_k e_{kj} - e_{ij}$	$\min_i \max_j r_{ij}$	$r_{ij} = e_{ij} - \min_k e_{kj}$	$\min_i \max_j r_{ij}$

TABLE 10b. COMPARISON OF SELECTED ALTERNATIVES.

	Alternative selected	
Decision criterion	Profit	Cost
Plunger	3	1
Wald	1	3
Laplace	3	3
Hurwicz		
$0 \leq u \leq 0.3$	1	3
$0.3 \leq u \leq 0.7$	3	3
$0.7 \leq u \leq 1$	3	1
Savage	3	3

known. For the Bayes criterion, historical data are utilized to provide a prediction of the future state. Using the prior example for the profit problem, assume that a study indicated that the sales revenue will be 30, 40, and 50 million dollars for 4/7, 2/7, and 1/7 of the time, respectively. As for the Laplace criterion, the alternative which yields the maximum expected profit is to be selected (i.e., $\max_i E(a_i)$). Thus, the only change is the probability p_j for calculating $E(a_i) = \sum_j e_{ij}p_j$. Using the prior operational profit data,

$$E(a_1) = 5(4/7) + 7(2/7) + 9(1/7) = 6.14$$

$$E(a_2) = 3(4/7) + 6.7(2/7) + 10.5(1/7) = 5.13$$

$$E(a_3) = 3.5(4/7) + 8(2/7) + 12.5(1/7) = 6.07$$

Consequently, alternative a_1 which utilizes marginal equipment when the sales exceed 30 million dollars would be selected because $E(a_1)$ has the greatest expected profit. When a cost problem is involved, the alternative which has the minimum expected value is selected.

The prior application of the Bayes criterion for decision under risk assumes that the future will conform with the past. If, however, some other variable which is a relatively good predictor of the future could be observed immediately in advance of making the decision, the actual decision-making should be improved. This approach, in reality, is the one utilized by most managers to consider the many factors not included in the mathematical models. A Bayes procedure has been evolved to consider one such factor. For example, the sales revenue for next year may be highly correlated with the average sales for the preceding two years. Unfortunately, however, this procedure appreciably increases the complexity of the problem. Thus, it is not illustrated here.

REFERENCES

1. Gue, R. L., and Thomas, M. E., *Mathematical Methods in Operations Research*, Macmillan, New York, 1968, pp. 276–290.
2. Spivey, W. A., and Thrall, R. M., *Linear Optimization*, Holt, Rinehart and Winston, Inc., New York, 1970, pp. 344–347.
3. Hillier, F. S., and Lieberman, G. J., *Introduction to Operations Research*, Holden-Day, Inc., San Francisco, 1967, pp. 77–83.

BRANCH AND BOUND

The application of implicit enumeration for combinatorial problems was introduced in the dynamic programming section. The intent of such an approach is to reduce the number of combinations which must be explored to locate the optimal solution. In general, the desire is to reduce the set space in a sequential manner until the optimal decision set(s) have been located. While the technique to be explored here, branch and bound, has been evolved from mathematical set theory, the method appeals to one's intuitive logic.

This solution technique for combinatorial problems pertains to decision under certainty and is not as versatile as dynamic programming. However, when the structure of the problem conforms with that required for branch and bound, convergence to the optimal solution may be more rapid if the bounds are properly selected. Unfortunately, the constraints imposed on the problem are not an integral part of the basic solution and must be imposed as "side" considerations.

Since the branch and bound technique is such a logical approach to the solution of a problem, it is difficult if not impossible to identify all of the contributors. The technique may be classified as a network model, a combinatorial analysis model, a sequential solution model, etc. Further, the technique has been variously termed "implicit enumeration," "partial enumeration," and "decision tree theory." The reason for the latter name will become evident from the diagrams for the subsequent example. Regardless of the classification and name, the technique has gained wide acceptance because it is not necessary to follow some mystical algorithm and because of its ability to converge comparatively rapidly to the optimal solution.

Wagner presents applications of branch and bound to the solution of linear integer programming problems, the traveling salesman problem, and capital budgeting. For the sake of variety in this presentation, however, the example considered here is an assignment problem. This type of problem enables easier visualization of the structural aspects which are necessary for the branch and bound solution technique.

For the example, assume that a consulting firm has four engineers available and four contracts to be completed. Also, assume that each contract requires the effort of only one professional man. Because the tasks to be performed differ in technical content, and because the experience and capabilities of the engineers differ, the total profit which may be realized from these contracts is highly variable. Further, since some of the tasks require special knowledge and experience, some of the engineers cannot be assigned to specific contracts. Consequently, it is desirable to assign the men to the tasks in a manner which will maximize profit.

As a starting point, the contractual requirements and the capabilities of the staff are reviewed. Using this information, an estimate of the profit which should be attained by all applicable engineer–contract combinations can be made. When it is not considered desirable to assign a given engineer to a contract, an X has been placed on the engineer–contract assignment profit chart.

Engineer–Contract Assignment
(profit in \$100 units)

	Contract			
Engineer	1	2	3	4
A	43	69[a]	*X*	32
B	26[b]	*X*	34[b]	25
C	63[a]	15[b]	47[a]	34[a]
D	47	65	39	19[b]

[a]Maximum column value.
[b]Minimum column value.

The foregoing problem may be stated mathematically as follows:

$$\text{Maximize } Z = \sum_i \sum_j p_{ij} x_{ij}$$

subject to

$$\sum_i x_{ij} = 1,$$

$$\sum_j x_{ij} = 1,$$

and

$$x_{ij} = \begin{cases} 0 \text{ for no assignment} \\ 1 \text{ for an assignment} \end{cases}$$

where p_{ij} and x_{ij} are the profit and assignment, respectively, for the ith engineer and the jth contract.

To initiate the solution procedure, consider the structural aspects of the problem. It is observed that the set space for the solution can be subdivided by row or column. Thus, the upper and lower bounds may be calculated initially for the entire set space. Then, using the inherent structure of the problem, the bounds may be calculated for each subset, say for subdivision by column. Whenever possible, some subset space should be removed from further consideration when the applicable upper bound indicates that the subset space cannot contain the optimal solution. This is portrayed in Fig. 27 by the "*X*ed out" sections, which provide an indication of when the subset was removed from further consideration.

Now the point of concern is the method to be used to eliminate the subsets from further consideration to achieve the desired convergence to the optimal solution. Remembering that the constraints must be considered as side conditions which must be met, the bounds are calculated without regard to feasibility. Then, the solution which yields the bounds is investigated to determine whether it is feasible (i.e., if it is in conformance with the specified constraints). When feasibility does not exist, this is denoted by nf (i.e., nonfeasible) for the particular bound. Obviously, the optimal solution must be feasible. Further, when the

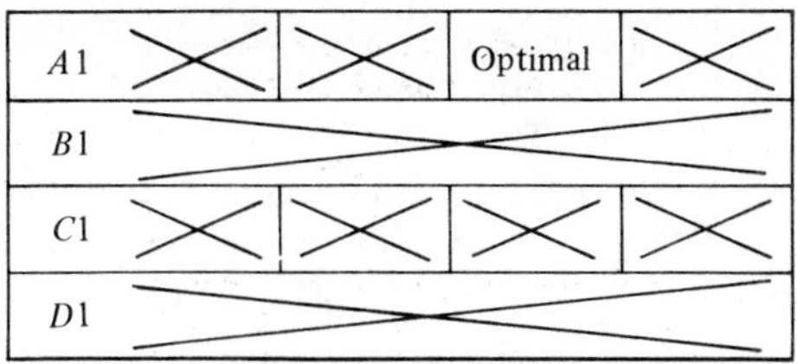

Fig. 27. Set space subdivision.

upper bound of any subset for a maximization problem is feasible, it is known that the optimal solution will be *at least* this large, if not larger. Consequently, as shown in Fig. 28, this feasible upper bound becomes the new lower bound. When more than one such feasible upper bound exists, the maximum value is the new lower bound. Then, when the upper bound of any subset is less than this lower bound, that subset need not be considered further since it cannot contain the maximum solution.

A further study of Fig. 28 shows that the upper bound for the problem decreases progressively as the sequential analysis continues. At any point in the analysis, this upper bound on the remaining set space is the maximum upper bound for the subsets which have not been removed from further consideration. While it portrays the actual change to the upper bound, in the solution of a maximization problem, it really is not necessary to worry about this total problem upper bound at each sequential analysis point, as will be shown in the example. Finally, when the upper bound is equal to the lower bound, the optimal solution has been located. This point of convergence is shown on the right side of Fig. 28.

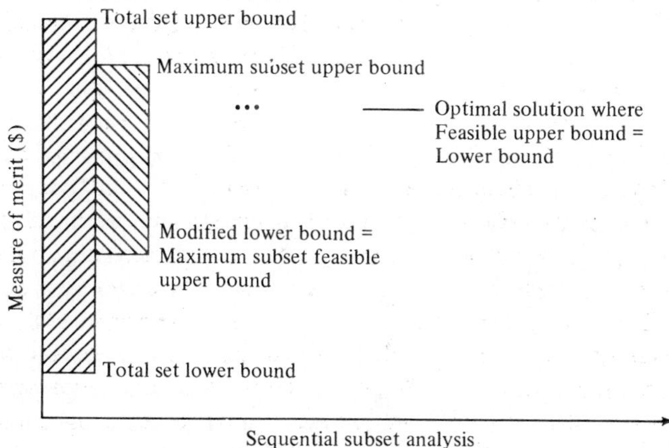

Fig. 28. General branch and bound maximum solution procedure.

Returning to the example, the upper bound on the total set space when determined by selecting the maximum value for each column is 63 + 69 + 47 + 34 = 213 nf (i.e., nonfeasible). This bound is nonfeasible because engineer C would be assigned to three contracts, while engineers B and D would not be assigned (refer to the "a" identifications on the original engineer–contract assignment table). Remember that for a feasible solution, each engineer must be assigned to one contract, and each contract must have one engineer assigned. (Observe that any undesirable assignment indicated by X is not considered for the establishment of the bounds. When such a problem is solved by computer, this X element is usually assigned a large negative value for a profit problem, or a large positive value for a cost problem. This method forces some other assignment to be made. However, when the solution is manual, such an assignment can be considered impossible.) Also, the lower bound for the total set space (determined by the minimum "b" value in each column) is 26 + 15 + 34 + 19 = 94 nf. This bound is nonfeasible because engineer B would be assigned two contracts while engineer A would be unoccupied. This initial condition is presented graphically in Fig. 29 as the beginning of a decision tree, where UB and LB represent the upper and lower bounds, respectively.

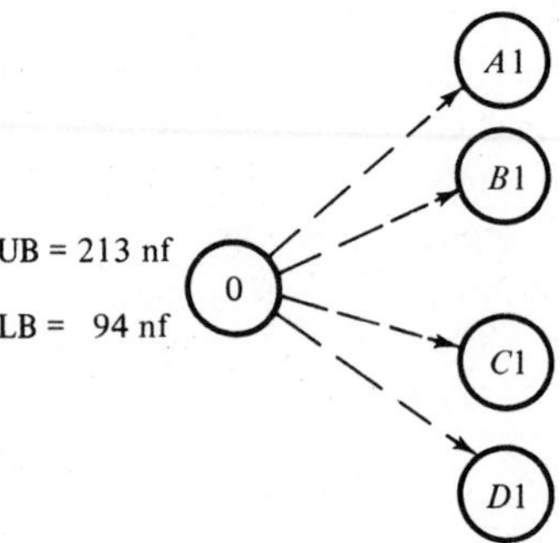

Fig. 29. Beginning of decision tree for example problem.

It is now known that the total set space includes feasible and nonfeasible solutions ranging from 92 to 213 (in $100 units). Now, the total set space is to be subdivided on the basis of some structural variable. Performing this subdivision on a columnar basis, all possible assignments of contract 1 are considered as shown in Fig. 29. Branches will be added to this decision tree diagram as the analysis progresses.

At this point, the upper bounds should be established for each branch (i.e., for each subset). Since the intent is to establish the maximum solution, it is not necessary to calculate the lower bound for each subset because the lower bound of concern is the maximum feasible upper bound if one exists. Although it is not necessary to do so, the subproblems will be restated as the solution progresses to provide clarity.

The first subproblem is for the *A*1 branch. That is, given that engineer *A* is assigned to contract 1, what is the appropriate upper bound? Since *A*1 has already been assigned, the profit for this element (i.e., 43) must be included in the bound. Further, it is desirable to assure that feasibility is attained when the end of any branch has been reached if it is explored to its conclusion. Consequently, since an engineer can be assigned to only one contract and any contract can be assigned to only one engineer, this choice (i.e., *A*1) removes the first row (i.e., engineer *A*) and the first column (i.e., contract 1) from the subproblem. Thus, the applicable subset is rewritten below. The resulting upper bound is (43) + 65 + 47 + 34 = 189 nf. The nonfeasibility results because engineer *C* has been assigned two contracts.

Profit Matrix Given *A*1

Engineer	Contract		
	2	3	4
B	*X*	34	25
C	15	47[a]	34[a]
D	65[a]	39	19

[a]Maximum column value.

The next subproblem assumes the initial assignment of contract 1 to engineer *B* (i.e., *B*1). This subproblem is rewritten below. For this subset, since the indicated profit for *B*1 was 26, the upper bound is (26) + 69 + 47 + 34 = 176 nf. Once again, engineer *C* has been assigned two contracts so that the bound is nonfeasible.

Profit Matrix Given *B*1

Engineer	Contract		
	2	3	4
A	69[a]	*X*	32
C	15	47[a]	34[a]
D	65	39	19

[a]Maximum column value

Considering both the *C*1 and *D*1 branches, the following rewritten matrices apply. The resulting upper bounds are (63) + 69 + 39 + 32 = 203 nf for *C*1, and (47) + 69 + 47 + 34 = 197 nf for *D*1. At this point, it should be obvious why these bounds are nonfeasible.

Profit Matrix Given C1

Engineer	Contract 2	3	4
A	69[a]	*X*	32[a]
B	*X*	34	25
D	65	39[a]	19

Profit Matrix Given D1

Engineer	Contract 2	3	4
A	69[a]	*X*	32
B	*X*	34	25
C	15	47[a]	34[a]

[a]Maximum column value.

To ascertain the next step in the solution procedure, it is helpful to "update" the decision tree, as shown in Fig. 30. This diagram reveals that the *C*1 branch has the largest upper bound of the subsets considered. For this decision tree, the solid lines represent the branches already explored, and the dashed lines indicate the next branches to be considered.

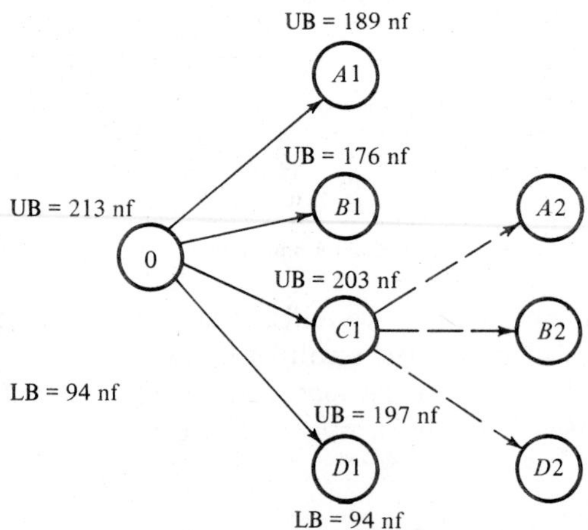

Fig. 30. First update of decision tree.

While it does not necessarily follow that the *C*1 branch must contain the maximum solution, it is logical to explore this branch next. If it does contain a feasible assignment in the region from 197 (i.e., the next largest upper limit was 197 for branch *D*1) to 203, the least amount of analysis would be involved in locating the maximum solution. Thus, Fig. 30 indicates the establishment of the next subsets by arbitrarily assigning contract 2 to engineer *A*, *B*, or *D*. Obviously, engineer *C* cannot be considered in this branching since he was previously assigned to contract 1. Further, a review of the profit matrix given *C*1 reveals that an *X* has been indicated for the profit of element *B*2, indicating an undesirable assignment. Therefore, this *C*1*B*2 branch will not be analyzed. The applicable resulting partial matrices follow:

Profit Matrix Given $C1A2$

Engineer	Contract 3	Contract 4
B	34	25[a]
D	39[a]	19

Profit Matrix Given $C1D2$

Engineer	Contract 3	Contract 4
A	X	32[a]
B	34[a]	25

[a]Maximum column value.

Using the foregoing profit matrices, the upper bound for the $C1A2$ branch is $(63 + 69) + 39 + 25 = 196$, and that for the $C1D2$ branch is $(63 + 65) + 34 + 32 = 194$. Observe that both of these solutions are feasible since only one engineer is assigned to each contract. This means that the lower bound can be reestablished now. Further, since the $C1A2$ branch has a greater upper bound than the $C1D2$ limb, this latter limb may be removed from further consideration. Thus, the lower bound is now 196. How many other branches can be "pruned" from the tree at this point? To investigate this, refer to Fig. 31, which reveals that the $A1$, $B1$, and $C1D2$ branches may be pruned since the smallest possible maximum solution is 196 and these subsets cannot include a solution with that great a profit. This is indicated by an X after the branch. If another branch did have an equivalent upper bound (i.e., 196 in this case), however, it would be necessary to explore that branch further since an alternative and equally optimal solution might exist in another subset. For this example, it is observed that only

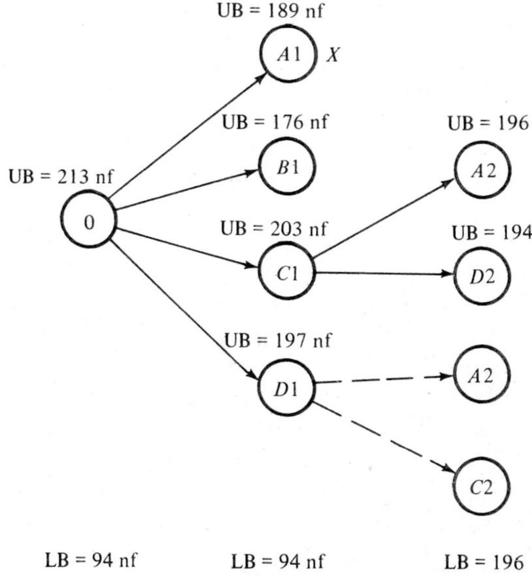

Fig. 31. Pruning branches in the decision tree.

the *D*1 branch must be considered further. The *D*1*B*2 branch need not be considered, however, since the assignment of engineer *B* to contract 2 was indicated in the engineer–contract assignment table as unacceptable (i.e., *X* profit). These two partial matrices are restated as follows:

Profit Matrix Given *D*1*A*2

	Contract	
Engineer	3	4
B	34	25
C	47[a]	34[a]

Profit Matrix Given *D*1*C*2

	Contract	
Engineer	3	4
A	*X*	32[a]
B	34[a]	25

[a]Maximum column value.

From the foregoing, the upper bounds are (47 + 69) + 47 + 34 = 197 nf and (47 + 14) + 34 + 32 = 128, for branches *D*1*A*2 and *D*1*C*2, respectively. Since the prior lower bound was 196, the *D*1*C*2 limb may be pruned. However, the upper bound for *D*1*A*2 exceeds 196, so this branch should be explored by further subdivision. At this point, the final solution is obvious, but, to show that feasibility is forced on the solution when the end of any branch has been reached, the following profit matrices are presented:

Profit Matrix Given *D*1-*A*2-*B*3

	Contract
Engineer	4
C	34[a]

Profit Matrix Given *D*1-*A*2-*C*3

	Contract
Engineer	4
B	25[a]

[a]Maximum column value.

The resulting upper bounds for these subsets are (47 + 69 + 34) + 34 = 184 for the first subset and (47 + 69 + 47) + 25 = 188 for the second. Both of these solutions to the maximization problem are feasible since only one contract is assigned to each engineer. However, both of these subsets have upper bounds which are less than the prior 196 lower bound. As a result, both subsets are dominated and must be disregarded. Consequently, the remaining branch has the same value for its upper bound as the lower bound of the problem. Thus, the optimal maximum solution for the assignment has been located and it is *C*1*A*2*D*3*B*4. So engineers *A*, *B*, *C*, and *D* would be assigned to contracts 2, 4, 1, and 3, respectively. The resulting expected profit from this manpower allocation would be $19,600, and a greater profit cannot be realized (assuming that the original estimates were correct) for this particular problem. The decision tree involved in this implicit enumeration technique is presented in Fig. 32.

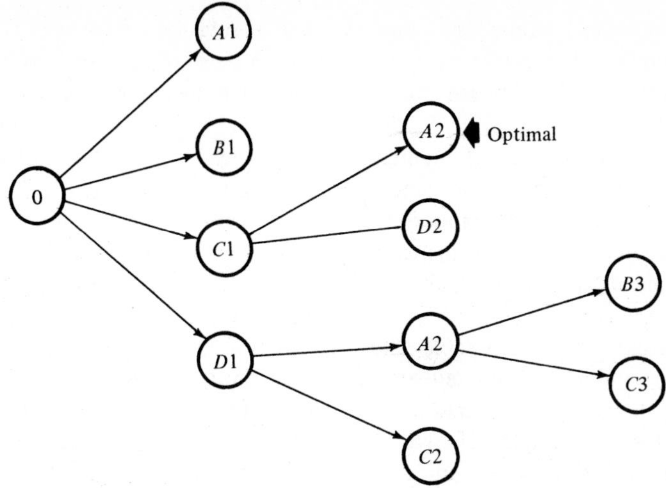

Fig. 32. Decision tree involved in implicit enumeration technique.

The foregoing example reveals the basic concepts and logic employed by the branch and bound technique to converge to a maximum solution. The only modification required when a minimization (i.e., cost) problem is involved is to force the return downward rather than upward. Thus, the upper and lower bounds change places and the lower bound is calculated for each subset for comparison with the established total set upper bound.

REFERENCES

1. Gue, R. L., and Thomas, M. E., *Mathematical Methods in Operations Research*, The Macmillan Company, New York, 1968.
2. Hillier, F. S., and Lieberman, G. J., *Introduction to Operations Research*, Holden-Day, Inc., San Francisco, 1967.
3. Wagner, H. M., *Principles of Operations Research*, Prentice-Hall, Englewood Cliffs, New Jersey, 1969.

STRATEGY OF COMPETITION

An interesting application of quantitative management analysis is that of studying uncertainty in a competition situation through the application of statistical probability theory. Competitive bidding in construction is an excellent example to use to explain this application. Consulting engineer Marvin Gates developed the following procedure of competitive bidding strategy which was published in the *ASCE Transactions*. It is not suggested that

Gates' probability model is a panacea for total solution of the highly complex problem of bidding, but it has merit in that the range of the variables to be considered can be materially reduced, thus minimizing the management decisions to be made.

First assume the case of bidding against a single known competitor against whom a firm has competed many times in previous bidding contests. Records would be available with competitor bid prices on many jobs for which the company had prepared cost estimates. Regardless of how many other firms were involved, and regardless of which of many contractors actually won each bid, the matched pairs of the competitor's bid amount B and the company's cost estimate for the same job E are reliable data for statistical analysis. For each contest a B/E ratio could be calculated which, by the simplest definition, would be an index of the competitor's markup on each job for which data is available. If a sufficiently large sample of B/E ratios (usually assumed to be 30 or more) is available, data can be conveniently grouped into classes or groups of uniform intervals of B/E values and a frequency distribution derived. The size of class interval, number of groups, and number of data in the various groups should be determined using accepted statistical techniques.

As an example, suppose a company has competed against Able Associates in fifty contests in the past two years for projects of reasonably similar size and nature. Calculated B/E ratios range from 0.91 to 1.24. If B/E class intervals of 0.05 increments are selected there will be seven groupings of data with reasonable numbers of data distributed among groups to yield an acceptable situation. The first two columns of Table 11 are thus established.

The third column in Table 11 is the percentage cumulative frequency computed to the lower limit in each class. Thus, for example, at a B/E ratio of 1.00, 94% of Able Associates' B/E markups have been higher than 1.00. Percentage cumulative frequency is actually a percentage probability that Able Associates' markup will be greater than the B/E ratio in any row. Profit equals (B/E –

TABLE 11. STATISTICAL ANALYSIS OF B/E DATA FOR A SINGLE COMPETITOR.

B/E class	Number of cases	Cumulative frequency	Percentage probability	Profit at class LL	Math. Exp. profit
0.90–0.949	1	50	100	–10	–10.0
0.95–0.999	2	49	98	–5	–4.9
1.00–1.049	7	47	94	0	0
1.05–1.099	9	40	80	5	+4.0
1.10–1.149	13	31	62	10	+6.2
1.15–1.199	10	18	36	15	+5.4
1.20–1.249	8	8	16	20	+3.2
1.25–1.299	0	0	0	25	0

$N = 50$

1.00) times 100%, and these values are shown in column 5 relative to the lower limit in each class row. The last column contains "mathematical expectancy" values of profit; or profit times probability (expressed as a decimal part of unity).

The results of Table 11 are shown graphically in Fig. 33. The curves show clearly the probability and mathematical expectancy parameters in bidding against a single known competitor. Point *C* in Fig. 33a represents the mean probability, or 50–50 chance, of beating Able Associates by adding a 12.5% profit to estimated costs (or *B*/*E* ratio of 1.125 as a markup) in calculating the bid. The situation at any other point, for instance point *A*, may be looked at from two points of view. A minimum markup, say 1.075 (a 7.5% profit), may be set before determining that the probability of winning is 70%. Or, a desired probability of winning, say 70%, may be established before finding a required markup of 1.075.

Mathematical expectancy (denoted as E_p) is a more realistic parameter for long-range profit planning and/or optimization of a single contest situation. The "expectation" calculated here, however, is a conservative value, since the lower limit for the cell rather than the midpoint is utilized. If a specific profit policy, say 10%, is established, Able Associates will lose 62% of the time and win 38. In the winning situation, the company would realize a dollar profit of 10% of the project cost. In a losing situation, it would make nothing. Therefore, in the long run, the real profit will be (0.62 × 10% + 0.38 × 0%) = 6.2% of the total amount of jobs bid. Since cash profit, not percentage, is the only real measure of gain, the mathematical expectancy of profit times probability of winning vs. *B*/*E* is a valuable management tool. These quantities, calculated in column 6 of Table 11, are plotted in the graph of Fig. 33b.

Two important facts are displayed in this graph. First, there will always be one maximum expected profit (based upon total bid rather than cost) at a mathematically determined *B*/*E* ratio. In this case 6.2% at a *B*/*E* of 1.12. This is shown as point *M*. Secondly, a specific long-run profit gain in dollars can be realized by few bids at low markup (point *P*, 3.5% at a *B*/*E* of 1.04) or more bids at higher markup (point *Q*, 3.5% at a *B*/*E* of 1.19). Thus, in light of this second observation, a company may be able to quantitatively arrive at a volume–profit policy to reach a desired gain by either of two means.

Engineer Gates defines a "typical competitor" for cases of bidding against unknown opponents as a composite probability situation. Rather than analyze the *B*/*E* data for one competitor, he uses all of the *B*/*E* data of all reliable competitors as a composite group representing a fictitious average "typical competitor." The analysis procedure is identical to that for a single known competitor.

The application of statistical probability in situations involving several competitors depends upon the definitions of degree of mutual exclusion and independence of events. Gates has proposed that the probability of winning against *n* known competitors could be determined from the following:

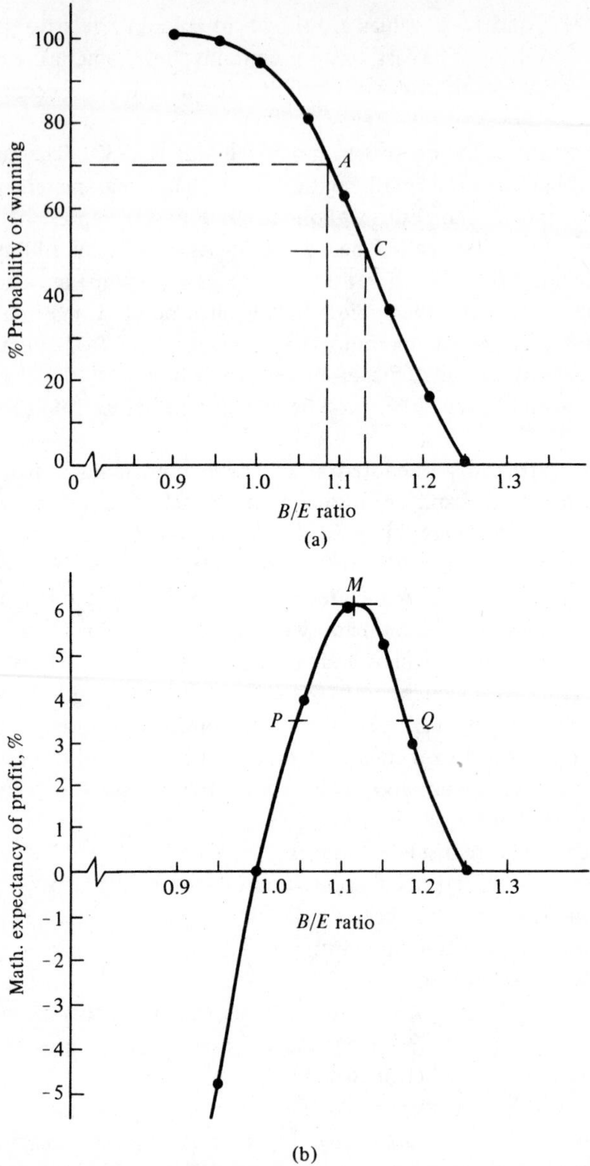

Fig. 33. Analysis of bidding against single known competitor. (a) *B/E* vs. probability. (b) *B/E* vs. mathematical expectancy profit.

$$p = \frac{1}{\dfrac{1 - p_A}{p_A} + \dfrac{1 - p_B}{p_B} + \dfrac{1 - p_C}{p_C} + \cdots + \dfrac{1 - p_n}{p_n} + 1}$$

where p = probability of winning; p_A, p_B, p_C, p_n = probability of winning over contractor $A, B, C, \ldots n$.

According to Gates, this formula yields reasonably valid predictions based upon actual applications. Of course, the mathematical expectancy calculations would apply here as with the case of a single competitor.

Some concluding remarks concerning application of a singular theoretical model such as presented above are in order. No mention was made of the probable variability of cost estimate; E was assumed to be a reliable constant. Before making a B/E vs. E_p analysis it would be suggested that a comprehensive statistical analysis be made of the company's actual vs. estimated costs. Some measure of confidence limits should be established from such an analysis and applied in forecasting probability of success against competitors.

It is also recognized that many factors such as the local economy, relative activity of the competitors, and desirability of particular projects must be taken into account in final decisions regarding markup. In each bidding situation it is advisable to weigh the relative significance of each of the factors applicable at that specific time, making the final bid price a management decision that is aided, but not made, by statistical analysis. A statistical probability analysis such as the model by Gates should at least narrow the consideration of alternatives for management decision because the expected action of competitors is best determined by past behavior. The economics of the moment would be more apt to simply shift the mean of the probability distribution rather than change its nature of variance.

REFERENCES

1. Casey, B. J., and Shaffer, L. R., *An Evaluation of Some Competitive Bid Strategy Models for Contractors*, Report No. 4, Dept. Civil Engineering, Univ. Illinois, Urbana, June 1964.
2. Clough, R. H., *Construction Contracting*, Wiley-Interscience, New York, 1969.
3. Freidman, L., A Competitive Bidding Strategy, *Operations Research*, Vol. 4 (February 1956).
4. Gates, M., Statistical and Economic Analysis of a Bidding Trend, *J. Construction Division ASCE*, Paper 2651 (November 1960).
5. Gates, M., Bidding Strategies and Probabilities, *J. Construction Division ASCE*, Paper 5159 (March 1967).
6. O'Brien, J. J., and Zilly, R. G., *Contractor's Management Handbook*, McGraw-Hill, New York, 1971.

7. Park, W. R., *The Strategy of Contracting for Profit*, Prentice-Hall, Englewood Cliffs, New Jersey, 1966.
8. Park, W. R., How Low to Bid to Get Both Job and Profit, *Engineering News-Record*, April 19 (1962).
9. Park, W. R., Less Bidding for Bigger Profits, *Engineering News-Record*, February 14 (1963).
10. Park, W. R., Profit Optimization Through Strategic Bidding, *Bulletin of the AACE*, December (1964).
11. Park, W. R., Bidders and Job Size Determine Your Optimum Markup, *Engineering News-Record*, June 20 (1968).
12. Stark, R. M., *Unbalanced Bidding Models*, Dept. Civil Engineering, Univ. Delaware, Newark, Delaware, August 1966.
13. Stark, R. M., and Mayer, R. H., Jr., *Static Models and Other Aspects of Closed Competitive Bidding,* Dept. Civil Engineering, Univ. Delaware, Newark, Delaware, January 1969.

Index

Index

AASHO (AASHTO)
  code provisions, 505
  live loads, 504
Aerobic digestion, sludge, 621
Air pollution
  automobile, 676
  methods of reducing, 679
Air transportation, 898
  airport capacity, 898
  airport facility requirements, 901
  terminal facilities, 905
Alignment types, roadway, 817
Anaerobic digestion, sludge, 621
Arches
  concrete, 254
  steel, 471
Architecton, 1
Atterberg limits, 90
Azimuth, 41

Baltimore & Ohio Railroad, 4
Beams, steel
  design of laterally unsupported, 381
  types of, 378
Bending moments, 287
  diagrams, 289
  framed beams, 291
Black body, 64
Bolts, 244, 400
Boring layout, soil testing, 141
Bow-string trusses, 447
Branch and bound, 1008
Bridges
  esthetics, 241
  single-span, steel beam, design of, 510
  types of, 499
Bridge structures, 256
  arch, 256
  box girder, 257
  cable-supported, 256
  deck, 257
  girder, 256
  through, 257
Bulkheads, anchored, 210

Caissons, 187
Channelization, traffic, 831
Chesapeake & Ohio Canal, 4
Chezy-Manning Equation, 520
Chlorination
  waste, 637
  water, 552
Civil engineer and environment, 1
Christensen, Richard W., 87
Clapp, James L., 11
Column design
  concrete, 424
    tables for, 430
  steel, 386
Columns, 372
  buckling of, 372
  Euler buckling load, 374
  with lateral load, 373
Compressibility, soil, 102
Compression, secondary, soil, 114
Concrete
  precast, 246
  reinforced, 245
    shear stresses, 408
    stress analysis, 406
    ultimate-strength design, 409
  strength of, 403
Conjugate beam, 319, 323
Connections, steel structural, 399
Continuous framing
  articulated, 447
  plastic or elastic, 448
Continuous structures, 292
Convergence of meridians, 49
Coordinate system, state plane, 57
Corrosion, pipe, 573
Critical path method, 987

Dams, earth, 227
Deflection
  beams, 317
  cantilever beam, 321
  rigid frames, 327
  simple beam, 322
    with uniform load, 324

Deflection (*Continued*)
solution by geometry, 318
trusses
rotation diagram, 341
virtual-work analogy, 333
Williot diagram, 336
Desalinization, water
dialysis, 558
reverse osmosis, 558
Design of structural steel elements, 375
Design of structures
concrete
retaining walls, 493
small, 477
tall, 482
steel, 446
arch buildings, 471
continuous articulated roof system, 457
office and apartment buildings, 460
single-span truss, 452
truss connections, 456
Design speed, geometric design, 800
Determinate structures, 268
Deviation
probable, 16
standard, 16
Direct stresses, beams, 311
Distribution, water, 563
equivalent pipe, 564
Hardy Cross, 567
Dynamic programming, 980

Economic performance charting, 970
Elastic constants, soil, 128
Elastic limit, 304
Elastic-weight analogy, 319
Empire State Building, 5, 9
Engineering, private practice, 916
Engineering economy
depreciation, 966
interest, 961
Engineering management
branch and bound, 1008
critical path method, 987
dynamic programming, 980
economic performance charting, 970
engineering economy, 961
games against nature, 998
history of, 923
linear programming, 940
modeling, 932
modeling classification, 938
quantitative, 915
queueing theory, 952
strategy of competition, 1017
Environmental concern
impact on civil engineer, 7
multidisciplinary approach, 8
the ethical question, 9
Environmental Policy, ASCE, 5
Equations
condition, 19
observation, 19
Equilibrium, structural, equations of, 262
Equivalent pipe, water distribution, 564
Erie Canal, 3

Filters
sewage
intermittent sand, 597
plastic media, 598
recirculating, 599
trickling, 597
water
diatomaceous earth, 563
mixed media, 561
pressure, 562
rapid sand, 559
Flexure formula, 313
Flow measurement, 524
Flumes, 524
Fluoridation, water (defluoridation), 553
Folded plate, concrete, 253
Footing design, concrete, 442
Force components, 260
Force triangle, 271
Forces
resultant of, 278
structural, 259
substitute, 273
Foundations, 147
deep
drilled piers or caissons, 187
piles, 187
types of, 186
shallow
consolidation settlements, 180
immediate settlement, 180
secondary compression, 185
settlement, 164
settlement computation in cohesive soils, 179
stress distribution beneath, 159
Foundation selection, 150

Fratar Method, trip distribution, 739
Free-body diagram, 262
Freeway capacity, 780
Functional system terminology, roadway, 687

Games against nature, 998
Geodimeter, 26
Geometric design of roadways, 769
  design controls, 770
  elements of design, 800
  signalized intersection, 783
Goodrich formula, 535
Gravity, center of, 279
Gravity model
  trip distribution, 744
  use and calibration, 748
Grid azimuth, 60
Ground swing, 33

Hardy Cross Method, pipe network analysis, 567
Hazen-Williams Formula, 516
Head loss, 519
Hydraulic cross sections, 521
Hydraulics, 516

Indeterminate structures, 268, 342
  moment-distribution, 353
  rigid frames, 344
  three span beam, 352
  with two or more unknowns, 348
Industrial wastes, 646
  biological treatment, 648
  joint treatment, 650
Inertia, moment of, 281
Influence values, vertical stress, soils, 160
Ingeniator, 1
Interchange design, roadway, 834
Intervening opportunities, model, trip distribution, 750
Ion exchange, 556

Johnson, Marvin M., 915

Ketchum, Milo S., 236
Klystron, 29
Krauss Process, 611
Kuhn, Herman A. J., 660

Lambert projection, 57
Least squares, method of, 18
Leveling
  cross section, 47
  profile, 46
Linear programming, 940
  graphic solutions, 941, 950
  manual solution (Simplex), 945
Liquid wastes, 577
Long-span joists, 450
Long-span truss systems, 450

McGhee, Terence J., 516
Mass transportation, ground
  exclusive guideway, 885
  intra-city, 876
  non-exclusive guideway, 880
  rail rapid transit, 877
  transit use, 876
Master builder, 1
Materials, structural, 242
Maxwell's Theorem, 350
Measurement
  accuracy of, 12
  adjustment of, 18
  angles, 40
    vertical, 42
    zenith, 42
  definition of terms, 12
  distance
    direct, 20
    electromagnetic, 26
    indirect, 23
    slope taping, 21
    stadia method, 24
    subtense method, 25
    taping, 20
    units of length, 20
  elevation, 34
    barometric leveling, 39
    curvature and refraction, 35
    differential leveling, 36
    reciprocal leveling, 38
    trigonometric leveling, 39
  engineering, 11
  errors, 13
    absolute, 13
    apparent absolute, 13
    discrepancy, 13
    discrimination, 13
    distribution of accidental, 14
    mathematical, 13
    percentage, 13

Measurement (*Continued*)
physical, 14
propagation of accidental, 16
relative, 13
systematic and random, 14
multiple, 12
precision, 15
reliability of, 12
sensitivity of, 12
single, 12
theory of, 12
Mercator projection, 57
Meridian, 40
central, 58
Modal split, travel allocation, 757
Modeling
application to decision making, 932
certainty, 938
classification, 938
flow diagram, 937
risk, 938
uncertainty, 939
Modulus of elasticity, 304
Mohr-Coulomb failure criterion, soil, 131
Mohr's Circle
soil stress, 117
structural applications, 307
Moment-area method, 320
Moment distribution, 353
beam continuous with columns, 363
constants, 355
derivation of, 356
continuous beam, 363
short cuts and simplifications, 366
Moments
statically determinate beams and frames, 282
structural, 259
Monadnock Building, 4
Monitors, atmospheric gas, 73
MPN determination, 551

Noise, transportation, 665
airport, 670
automobile, 669
composite ratings, 675
perceived level, 671
trucks, 669

Oedometer, 103
Open channel flow, 520
Orientation, relative and absolute, surveying, 84
Oxidation ponds, sewage, 617
Oxygen solubility, 607

Parallax, 80
Permeability, soil, 96
Photogrammetry
aerial cameras, 74
geometric, 74
mosaics, 80
photo interpretation, 74
radial triangulation, 79
vertical aerial photos, 75
Photography, 67
Piers
drilled, 187
driven, 186
Pile groups, 195
Pile load tests, 195
Pipe
cleaning and disinfection, 574
leakage, 574
water, materials and fittings, 571
asbestos cement, 573
cast iron, 572
concrete, 573
plastic, 573
steel, 572
Plate girders, steel, design of, 385
Poisson's Ratio, 311
Pollution generation, vehicle, 676
Population estimation
arithmetic increase, 530
geometric increase, 531
graphic projection, 530
Pore pressure, soil, 119
Pore pressure parameters, soil, 119
Post-tensioning, concrete, 247
Prestressed concrete, 246
Pretensioning, concrete, 247
Principal stresses, 309
Prismoidal Formula, 56
Prolonging straight line, 44
Psychoda Alternata (filter flies), 603
Pumps, 526

Quantitative Engineering Management, 915
history of, 923
Queuing theory, 952
elements of, 953
single-channel formulas, 955

Radar systems, 71
Radcliffe, B. M., 915
Radiometer, 68
Reactions, beam, 276
statically determinate, 282
Reinforcing steel, 401
Relative accuracy, measurement, 37
Relief displacement, 78
Remote sensing, 62
sensing sequence, 63
Retaining structures, earth, 200
anchored bulkheads, 210
braced sheet-pile walls, 214
effects of wall restraint and construction procedure, 207
flexible walls, 210
rigid structures, 200
translating or rotating walls, 209
Retaining walls, concrete, 493
Rigid frame, 291
buildings, 449
Rivets, 399
Rotation diagram, 341

Sanitary landfill, 658
Scanner, 69
multispectral, 70
Scientific method, 922
Screening, sewage, 587
Sedimentation, sewage, 593
Septic tank, waste disposal, 638
Settlement
consolidation, 180
foundation, 164
immediate, 180
secondary compression, 185
Sewage, 577
biological oxygen demand test, 585
characteristics of, 585
chemical oxygen demand test, 586
government regulations, 577
nitrogen determination, 587
phosphate determination, 587
quantity, 578
solids determination, 586
total organic carbon determination, 586
Sewage treatment, 587
activated sludge, 604
aeration, sludge, 609
completely mixed, sludge, 612
industrial wastes, 646
miscellaneous, 637
nitrogen removal, 635
oxidation ponds, 617
phosphorus removal, 635
primary, 587
process design, activated sludge, 613
secondary, 597
sludge treatment, 621
storm, 641
tertiary treatment, 628
Sewer design, sanitary, 580
Sewer lines, materials, 582
asbestos cement, 583
cast iron, 584
clay, 582
concrete, 582
corrosion of, 584
Shear, statically determinate beams and frames, 282
Shear diagrams, 287
Shear strength, soil, 130
Shear stress and strain, 305
Shear Stresses, 286
beams, 315
Sheet-pile walls, braced, 214
Shell, concrete, 252
Sight distance, geometric design, roadways, 801
Sign convention, trusses, 295
Signal warrants, roadway, 849
Signalization, 846
Signing and marking, roadway, 864
Single-span trusses, 499
Site investigation, soil, 140
Slabs, concrete, 251
flat, 251
flat plate, 252
joist, 252
one-way, 251
two-way, 251
waffle, 252
Slopes and embankments
analysis of forces, 219
chart solutions, 223
earth dams, 227
excavations in clay, 226
factor of safety, 218
over soft clay foundation, 225
stability analysis, 218
stability problems, 224
Sludge
activated, 604
aeration, 609

Sludge (*Continued*)
drying, 625
incineration, 626
treatment, 621
vacuum filtration, 625
wet oxidation, 626
Smock function, capacity restraint, 756
Soil classification, 91
Soil components, 87
gradation, 88
grain shape, 89
grain size, 88
weight-volume relationships, 89
Soil index properties, 90
Atterberg limits, 90
consistency, 90
Soil properties, 96
compressibility
one dimensional granular, 103
one dimensional saturated cohesive, 106
secondary, 114
time rate in saturated clays, 110
measurement of permeability, 97
permeability, 96
ranges of permeability, 99
shear, 117
stress at a point, 117
Solid wastes
collection, 655
disposal, 656
quantity, 653
Spectral signature, 65
Speed, traffic, 700
Stanley, C. Maxwell, 6
Statics, 259
Steel, structural, 242
Steel-Mill Buildings, 451
Steel sections, properties of, 377
Stefan-Boltzmann Law, 64
Stereoscope, 81
plotting instrument, 83
Stokes' Law, 538
Storm Sewage, 641
estimating runoff, 642
treatment and disposal, 645
Strain, 304
Strength of figure, 52
Strength of materials, 303
Stress, 286
Stress coefficients, parallel-chord trusses, 300
Stress distribution, foundation, 159
Stress-strain diagram, 304
Stress systems, soil, 120
anisotropic consolidation, 126
drainage conditions, 121
elastic constants, 128
granular soils, 132
Mohr-Coulomb failure criterion, 131
partially saturated cohesive soils, 138
saturated cohesive soils, 133
shear strength, 130
stiff fissured clays, 137
stress-strain in triaxial compression, 124
stress-strain properties, 123
Stresses, allowable, steel, 375
Stresses at a point, 306
Stress-path method, settlement, 165
Stress-strain properties, soil, 123
Structural design process, 237
Structural elements, design of reinforced concrete, 401
Structural engineer and the environment, 239
Structural Members, 248
arches, 249
beams, 248
cables, 249
columns, 248
footings, 250
girders, 248
girts, 250
plate girders, 248
purlins, 250
trusses, 248
Subsurface disposal, waste, 638
Subsurface exploration, soil, 142
Superelevation, roadway, 806
System operation and control, roadway, 846
sample signalization problem, 858
signalization, 846
signing and marking, 864

Tacoma Building, 5
Tanks, esthetics, 240
Tape, 20
Tellurometer, 29
Theodolite, 45
direction, 46
repeating, 46
Timber structures, 254

Torsion, 309
Towers, esthetics, 240
Traffic fluctuation, 642
Traffic volume definitions, 689
Transit, 43
Transportation
  air pollution, 676
  auto-truck
    flow volume characteristics, 688
    functional classification, 686
    geometric classification, 683
    growth factor models, 738
    hourly volume of traffic, 696
    information analysis, 714
    modal split, 757
    modeling procedures, 732
    movement systems, 683
    planning objectives, 712
    simulation models, 759
    speed, 700
    system evaluation, 763
    system/network design, 761
    system planning, urban, 707
    traffic assignment methods, 754
    trip distribution, 735
    trip generation model, 727
  community and human needs, 663
  displacement of businesses and families, 681
  effects of systems, 662
  environmental effects, 664
  general and environmental aspects, 660
  noise, 665
  policy act of 1969, 663
  technology-man-society interface, 660
  water quality effects, 680
Transportation study forms, 718
Transportation system model, 661
Traverse, 47
  area enclosed by, 55
  closed, 47
  computations, 53
  forward and inverse problems, 53
  open, 47
Triangulation, 49
  specifications for, 51
Triaxial soil compression, 124
Trilateration, 53
Trip generation rates, 736
Truss analysis, 294
Truss parts, 294
Turning requirements, roadway, 827

Ultimate-strength design, concrete, 409
  basic assumptions, 410
  capacity-reduction factors, 410
  load factors, 409
Unified Soil Classification, 92

Vehicles, design, 772
Virtual-work analogy, 333

Walls, concrete, 252
Wastes
  industrial, 646
    biological treatment, 648
    joint treatment, 650
  solid, 653
    collection, 655
    disposal, 656
    quantity, 653
Water, 526
  demand estimates, 528
  design volumes for demand, 535
  fire flow, 535
  sources, 528
  standards, 526
Water treatment, 536
  coagulation, 543
  desalting, 558
  disinfection, 551
  filtration, 559
  flocculation, 543
  fluoridation, 553
  ion exchange, 556
  iron and manganese removal, 550
  recarbonation, 547
  screening, 537
  sedimentation, 538
  softening, 545
  taste and odor control, 554
Wein Displacement Law, 65
Weirs, 524
Welds, 400
  structural, 244
Williot Diagram, 336

Yield stress, 304

Zilly, Robert G., 1,915
Zoogloea Ramigera, 604